Indole Ring Synthesis

Indole Ring Synthesis

From Natural Products to Drug Discovery

GORDON W. GRIBBLE

Department of Chemistry, Dartmouth College, USA

WILEY

This edition first published 2016

Registered Office
John Wiley & Sons, Ltd, The Atrium, Southern Gate, Chichester, West Sussex, PO19 8SQ, United Kingdom

For details of our global editorial offices, for customer services and for information about how to apply for permission to reuse the copyright material in this book please see our website at www.wiley.com.

Library of Congress Cataloging-in-Publication Data

Names: Gribble, Gordon W.
Title: Indole ring synthesis : from natural products to drug discovery / Gordon W. Gribble.
Description: Chichester, West Sussex : John Wiley & Sons, Inc., 2016. | Includes bibliographical references and index.
Identifiers: LCCN 2016007851 (print) | LCCN 2016010584 (ebook) | ISBN 9780470512180 (cloth) | ISBN 9781118683057 (pdf) | ISBN 9781118683859 (epub)
Subjects: LCSH: Organic compounds–Synthesis. | Indole. | Ring formation (Chemistry) | Transition metal catalysts.
Classification: LCC QD400.5.S95 G75 2016 (print) | LCC QD400.5.S95 (ebook) | DDC 661/.816–dc23
LC record available at http://lccn.loc.gov/2016007851

A catalogue record for this book is available from the British Library.

Set in 10/12pt Times by SPi Global, Pondicherry, India

1 2016

Dedicated to the memory of my beloved brother, Alan Paul Gribble, 1944–2014.

About the Author

Gordon W. Gribble is a native of San Francisco, California, and completed his undergraduate education from the University of California at Berkeley in 1963. He earned his PhD in organic chemistry from the University of Oregon in 1967, working with Lloyd Dolby on indole chemistry as a National Institutes of Health predoctoral fellow. After his National Cancer Institute postdoctoral fellowship in 1967–1968 with Frank Anet at the University of California, Los Angeles, he joined the faculty of Dartmouth College in 1968 where he has been full professor of chemistry since 1980. His awards include the National Science Foundation Professional Development Award (1977), the American Cyanamid Academic Award (1988), the Dartmouth College Distinguished Teaching Award (1997), the University of Oregon Chemistry Alumni Achievement Award (1998), Abraham Lincoln High School "Wall of Fame" (2004), and the Dartmouth College Arts and Sciences Graduate Faculty Mentoring Award (2006). He served as department chair from 1988 to 1991. In 2005, he was named to the inaugural endowed chair as "The Dartmouth Professor of Chemistry." He has been a visiting scholar/professor at Caltech; the University of Hawaii; the University of California, Santa Cruz; and Gettysburg College. He is a scientific adviser to the American Council on Science and Health and a member of the editorial boards of *Arkivoc* and *Current Organic Synthesis*. Dr. Gribble has published 370 papers on natural product synthesis, synthetic methodology, heterocyclic chemistry, polycyclic aromatic hydrocarbons, natural organohalogen compounds, organic chemical toxicity, and synthetic triterpenoids, one of which entered phase 3 clinical trials for the treatment of chronic kidney disease. He holds 34 patents. Since 1995 he has coedited the annual series *Progress in Heterocyclic Chemistry* and coauthored the second edition of *Palladium in Heterocyclic Chemistry*, published in 2007, along with Jack Li. He has written two monographs documenting more than 5000 naturally occurring organohalogen compounds. As a nationally ranked home winemaker for the past 38 years, he has a strong interest in the chemistry of wine and winemaking. He is a rated tournament chess player, enjoys scuba diving, and has a strong personal interest in the battles of Gettysburg and Iwo Jima, to which he has written about. He lives in Lebanon, New Hampshire, with his wife, and has two children, two grandsons, and two stepgrandsons.

Preface

Given the enormous resurgence in indole ring synthesis over the past decade—highlighted by the power of transition-metal catalysis—there is a need for a comprehensive presentation of the myriad methods for constructing the indole ring: from the ancient to the modern and from the well-known to the obscure. The organization that I have adopted follows that in my two earlier reviews on indole ring synthesis,[1,2] beginning with an Introduction on the importance of indoles and their role in society. Given space limitations, with a few exceptions I do not explicitly cover the synthesis of indolines (2,3-dihydroindoles), oxindoles (indolin-2-ones), indoxyls (indole-3-ols), isatins (indoline-2,3-diones), and azaindoles (pyrrolo[2,3-x]pyridines). However, carbazoles, carbolines, and their fused ring derivatives are covered.

---

[1] G.W. Gribble, *Contem. Org. Syn.*, 1994, 145–172.
[2] G.W. Gribble, *J. Chem. Soc., Perkin Trans. 1*, 2000, 1045–1075.

Contents

1

Introduction

1.1 Preview

From its early isolation by Baeyer from the reaction of indigo with a mixture of sulfuric acid and sulfuric anhydride [1], indole—*in*digo + *ol*eum—has a remarkable history and has made a huge impact on society, as we will see in this chapter. The reader is referred to several general reviews on the chemistry and synthesis of indoles [2–11] and their role in society [12]. Reviews devoted solely to indole ring synthesis are tabulated in Section 7 in this chapter.

1.2 Indole-Containing Natural Products

Indole (**1**) itself has several interesting natural sources, the most familiar of which is mammalian feces [13, 14], although its toxicity is low (LD_{50} = 1,100 kg/mg in rats) [15]. Indole has also been identified in significant amounts in flowers (jasmine, narcissus, lilac, Easter lily, lemon flower, tuberose, and honeysuckle) and in trace amounts in other flowers and foods (clove, orchid, gardenia, coffee flower, *Daphne odora*, tomato, molasses, sesame seed, rye bread, cheese, aged casein, and aging fish) [15]. Despite its objectionable and pervasive odor at high concentration, at low levels indole as been used by perfumers to augment fragrances. The odor threshold of indole is 140 parts per billion, significantly higher than, for example, methyl mercaptan (0.02 ppb) and dimethyl sulfide (0.30–1.00 ppb) [15]. Indole is also a component of human sweat [16] and breath [17]. Indeed, almost 30% of the volatile head space of sweat is due to indole [16]. Along with several other odorants, indole is attractive to mosquitos (*Anopheles gambiae*) [18].

Other well-known indoles that have various natural sources are skatole (3-methylindole) (**2**), serotonin (**3**), L-tryptophan (**4**), tryptamine (**5**), the plant growth hormones 3-indoleacetic acid (**6**) and 4-chloro-3-indoleacetic acid (**7**) [19], the mushroom hallucinogen psilocin (**8**), and the indole-derived ancient dyes indigo (**9**) [20] and Tyrian Purple (**10**) [19] (Scheme 1).

The vast marine environment, which covers 70% of Earth's surface, provides a wealth of naturally occurring indoles, and several reviews are available [21–24]. According to Hamann, 95% of the marine tropical biosphere accounts for 34 of the 36 phyla of life on Earth [24]. Some recently discovered marine indoles are depicted in Scheme 2. Several eusyntyelamides (e.g., D (**11**)) were isolated from the Arctic bryozoan *Tegella* cf. *spitzbergensis* [25], and the indole **12** was discovered in the marine fungus *Aspergillus sydowii* [26]. A New Zealand ascidian *Didemnum* sp. has furnished the β-carboline alkaloid didemnidine B (**13**) [27], and the toxin, bunodosine 391 (**14**) is part of the venom of the sea anemone *Bunodosoma cangicum* [28]. The Arctic hydrozoan *Thuiaria breitfussi* has yielded the novel breitfussin B (**15**) [29]. Tribromoindole (**16**) was found in the red alga *Laurencia similis* collected from Hainan Island, China, along with two other tribromoindoles [30].

Our terrestrial environment also contains a wealth of naturally produced indoles, and some recent examples are shown in Scheme 3 [31–38]. The novel thiazolylindole barakacin (**17**) was found in the ruminal bacterium *Pseudomonas aeruginosa* strain Z10 [31]. Spirobacillene A (**18**) was isolated from a culture of *Lysinibacillus fusiformis* KMC003 derived from coal mine acidic drainage [32]. The Chinese plant *Alocasia*

Indole Ring Synthesis: From Natural Products to Drug Discovery, First Edition. Gordon W. Gribble.

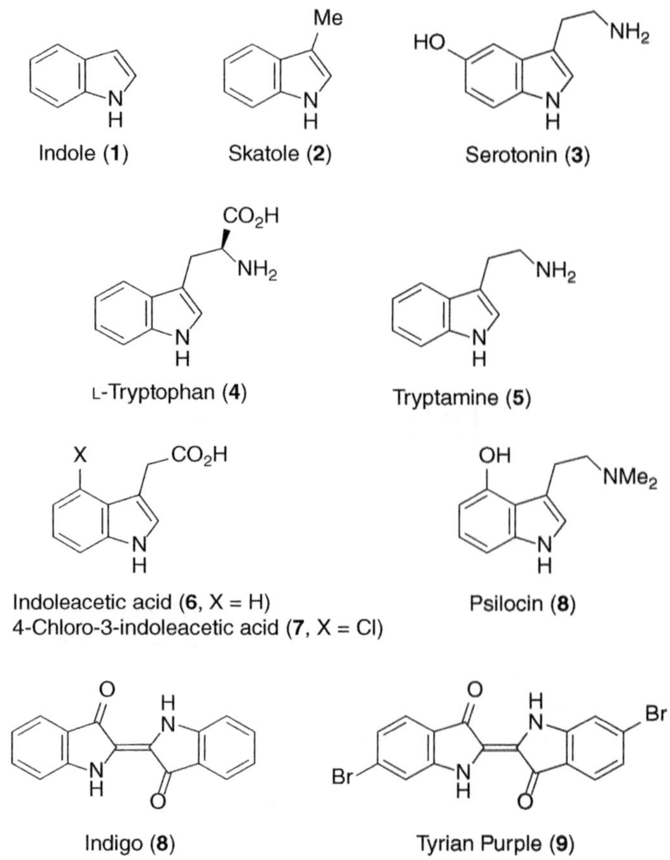

Scheme 1 *Well-Known Common Natural Indoles*

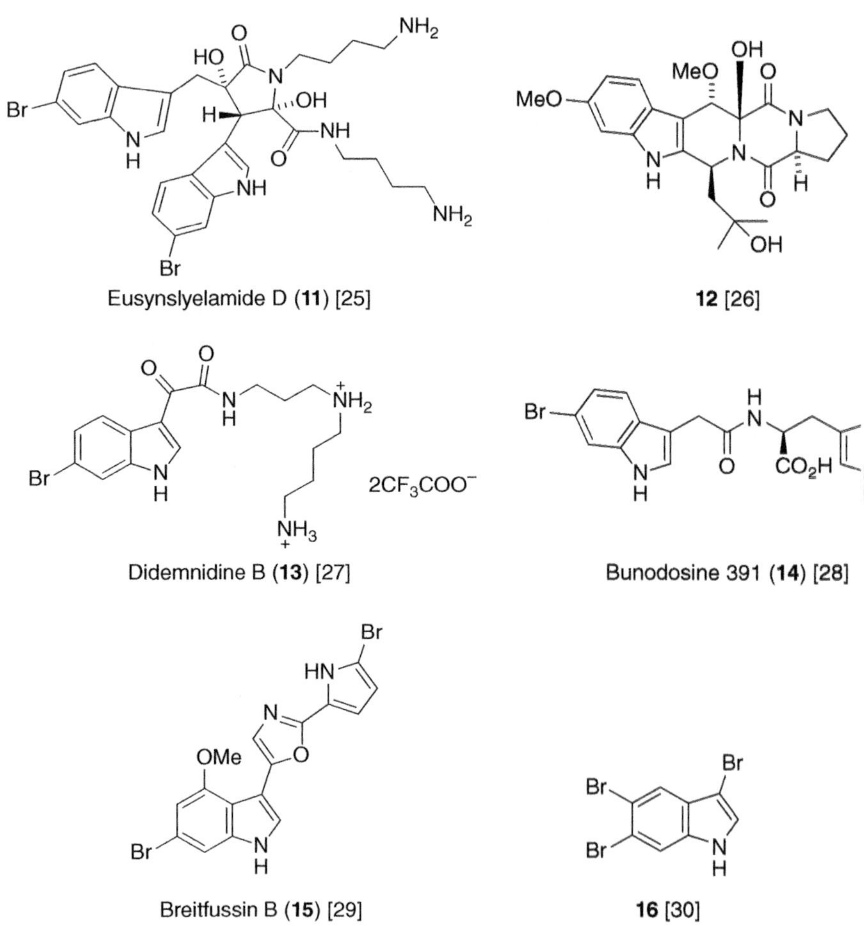

Scheme 2 *Representative Newly Discovered Marine Indoles*

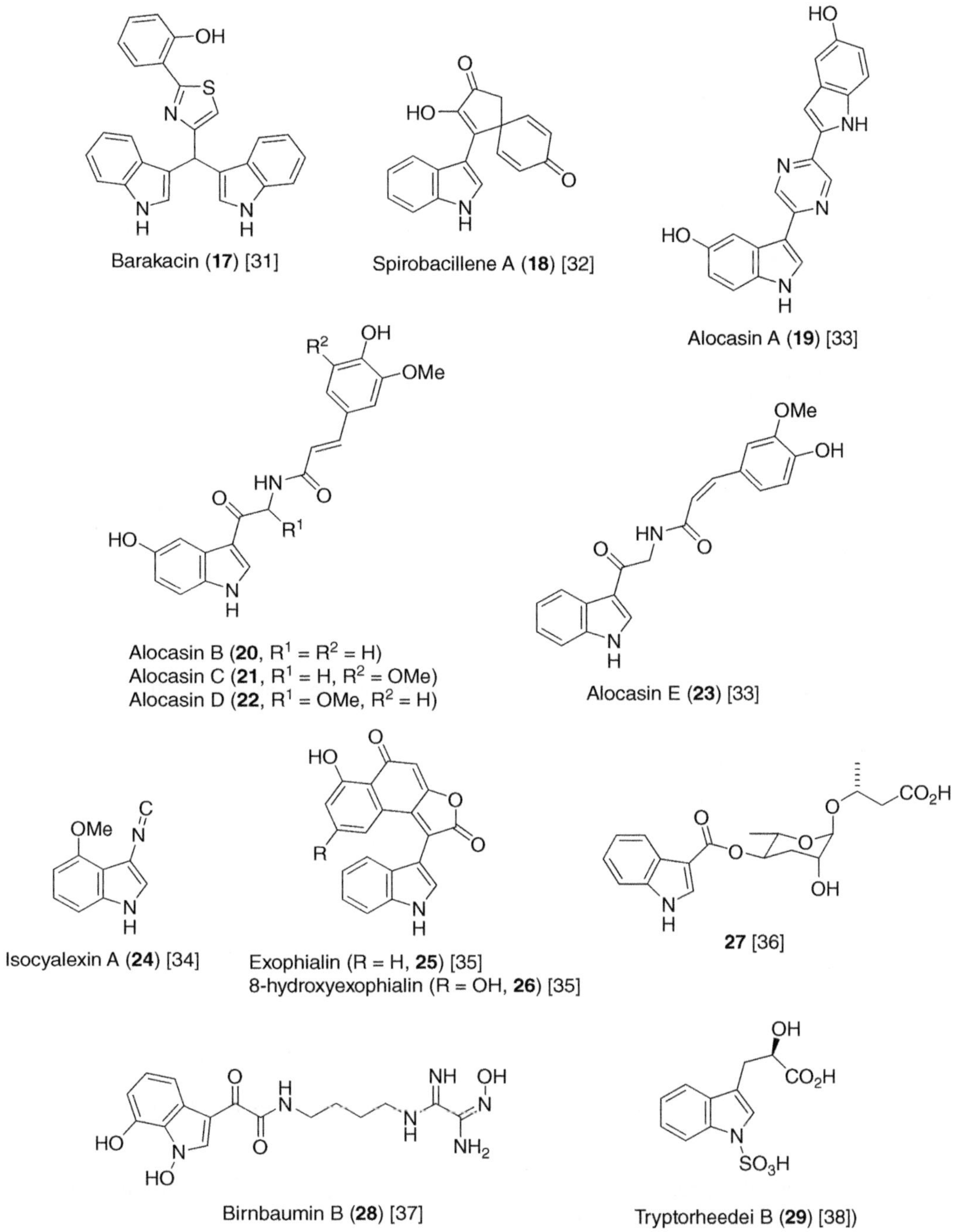

Scheme 3 *Representative Recently Discovered Terrestrial Indoles*

macrorrhiza has yielded the five new indole alkaloids alocasins A–E (**19–23**) [33]. Isocyalexin A (**24**) is the first plant-derived isocyanide to be discovered, isolated from rutabaga roots (*Braesica napobrassica*) [34]. The human pathogenic fungus *Exophiala dermatitidis* generates exophialin (**25**), and 8-hydroxyexophialin (**26**) is found in cultures of the mutant strain Me1-1 of *Exophiala dermatitidis* [35]. A component of the dauer larval stage pheromone of the nematode *Caenorhabditis elegans* is indole **27** [36]. The novel tryptorheedei B (**29**) is found in the seeds of *Entada rheedei*, a large woody liana growing in tropical Africa and Southeast Asia [38]. The corresponding *N*-sulfonyl-L-tryptophan (tryptorheedei A) accompanies **29**.

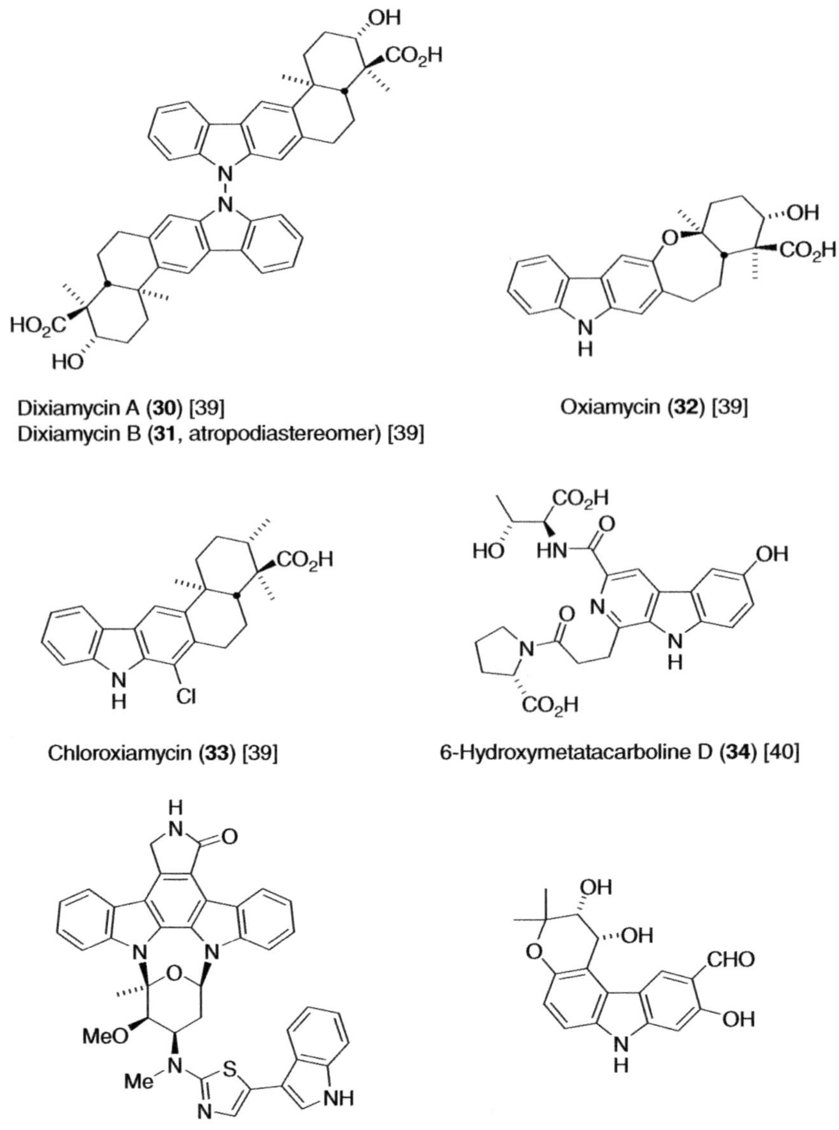

Dixiamycin A (**30**) [39]
Dixiamycin B (**31**, atropodiastereomer) [39]

Oxiamycin (**32**) [39]

Chloroxiamycin (**33**) [39]

6-Hydroxymetatacarboline D (**34**) [40]

Fradcarbazole A (**35**) [41]

Clausenawalline K (**36**) [42]

Scheme 4 *Representative Recently Discovered Carbazoles, Carbolines, and Indolocarbazoles*

Carbazoles and the related indolocarbazoles represent a huge collection of natural products, and some recently discovered examples are shown in Scheme 4. A marine *Streptomyces* sp. SCSIO02999 has yielded four new carbazolo-sesquiterpenes, dixiamycins A (**30**), B (**31**), oxiamycin (**32**), and chloroxiamycin (**33**) [39]. The novel β-carboline **34** is found in the mushroom *Mycena metata* [40], and the extraordinary fradcarbazole A (**35**) is one of three related indolocarbazoles produced by the marine *Streptomyces fradiae* [41]. A series of new carbazole alkaloids, clausenawallines G–K (e.g., **36**), was isolated from twigs of *Clausena wallichii*, a folk medicine plant distributed throughout Southeast Asia [42].

1.3 Biological Activity of Indoles

All indoles probably have some biological activity. Kumar and colleagues have briefly tabulated the range of activities that indoles possess [43]. More generally, Rosén and colleagues compare the chemical space that is occupied by natural products and bioactive compounds as a strategic starting point for drug discovery [44]. Section 3 presents biological activities of indoles, and Section 4 covers those bona fide indole-containing pharmaceuticals.

A growing worldwide problem is drug resistance to disease-inflicting bacteria, such as MRSA (methicillin-resistant *Staphylococcus aureus*) [45, 46]. Several indoles show promise in treating these bacterial infections, such as aryloxyindole **37** [47], 2-aryl-5-nitroindole **38** [48], cationic peptide **39** [49], and pacidamycin D (**40**) [50]. Biofilm infections cause 17 million new cases and up to 550,000 fatalities per year in the United States. Menthyl indole **41** is very active against biofilm formation induced by several strains of *S. aureus* [51] (Scheme 5).

Marine biofouling is a major problem to the shipping industry, but not to sponges, many of which produce antifouling compounds that inhibit settlement and smothering by barnacle larvae (*Balanus improvisus*). Some of these indole compounds are shown in Scheme 6. The novel cyclopeptide bromobenzisoxalone barettin **42** was isolated from the marine sponge *Geodia barretti* [52], and the marine ascidian *Stomoza murrayi* contains several brominated indole-3-carbaldehydes such as tribromoindole **43**, both of which prevent larval settlement or overgrowth by other marine species [53]. The physostigmine-like alkaloid urochordamine A (**44**) from the tunicate *Ciona savignyi* has potent larval settlement and metamorphosis-promoting activity at 2 μg/mL [54]. The Mediterranean gorgonian *Paramuricea clavata* contains several antifouling indoles, such as 2-bromo-*N*-methyltryptamine (**45**) [55].

Antifungal activity is seen with indole RWJ-61907 (**46**), which inhibits the growth of *Saccharomyces cerevisiae* and *Candida albicans* [56]. The *N*-methylcryptolepine salt **47** shows activity against *Cryptococcus neoformans* and *C. albicans*, two fungi associated with human immunodeficiency virus (HIV) and acquired immunodeficiency syndrome (AIDS), and *Aspergillus flavus* [57]. Antiparasitic activity is observed for several indole diamidines, such as **48**, which is active against *Trypanosoma brucei rhodesiense* and *Plasmodium falciparum* [58]. The glycosyl-isoindigo derivative **49** is active *in vitro* against *Trypanosoma brucei rhodesiense*, *Trypanosoma cruzi Tulahuen* (Chagas disease), *Plasmodium falciparum* (malaria), and *Leishmania donovani* (leishmaniasis [59]) (Scheme 7).

The final stage of HIV disease is AIDS. At the end of 2011 some 34 million people were living with HIV worldwide, and 1.7 million AIDS-related deaths were reported in 2011 [60]. Although these figures are lower than they were ten years ago, HIV drugs are still in great demand. Several indole derivatives show promise in this area (Scheme 8). Notably, indolyl aryl sulfones (e.g., **50** [61], **51** [62], **52** [63]), indole-3-sulfonamides (e.g., **53** [64]),

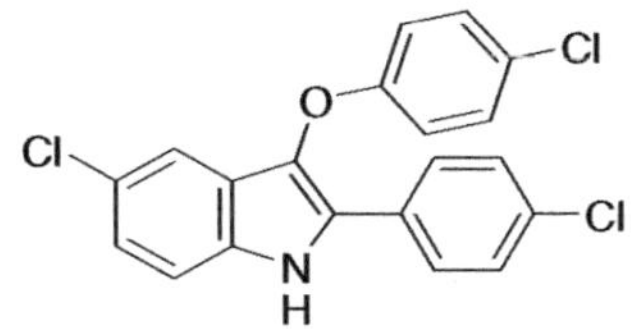

37 [47]

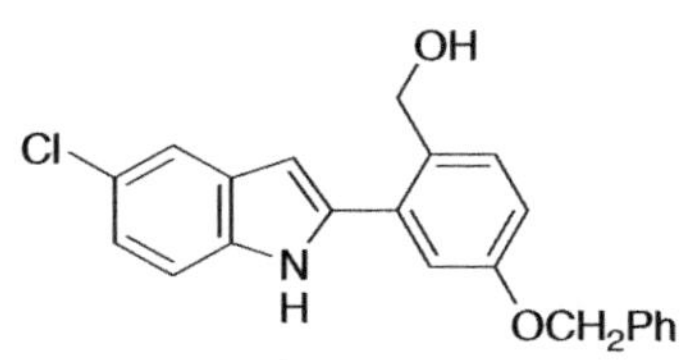

38 [48]

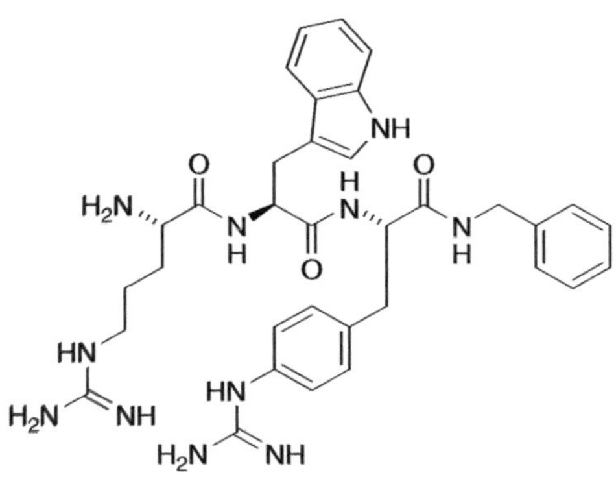

39 [49]

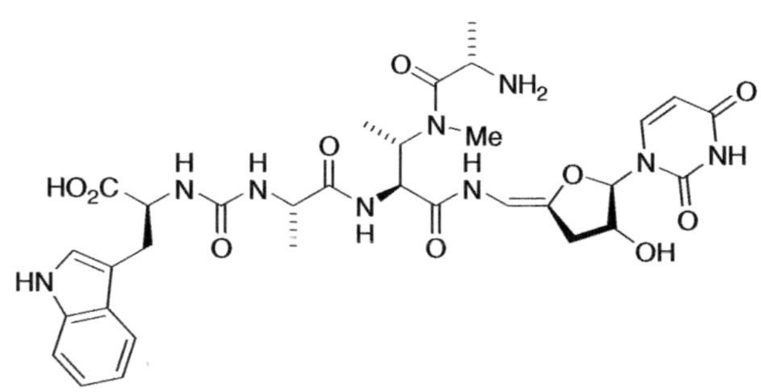

Pacidamycin D (**40**) [50]

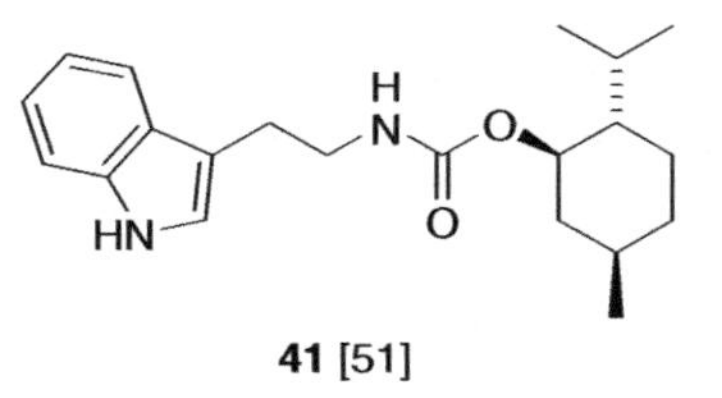

41 [51]

Scheme 5 *Representative Antibacterial Indoles*

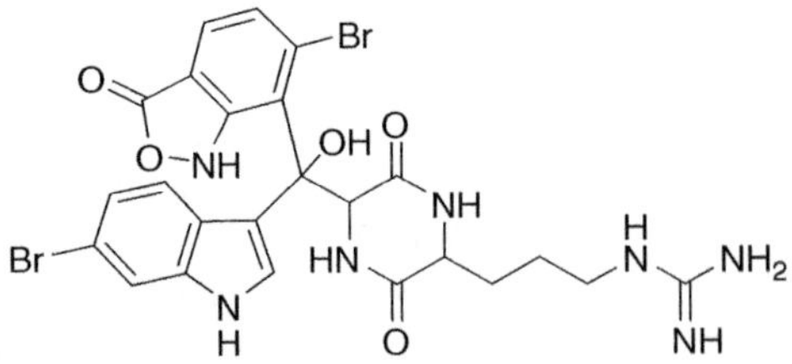

Bromobenzisoxazole barettin (**42**) [52]

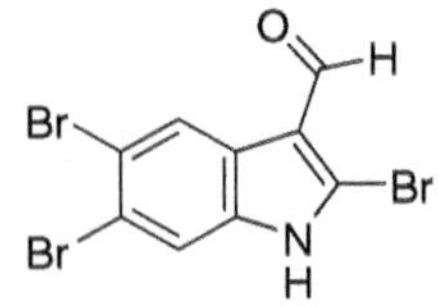

43 [53]

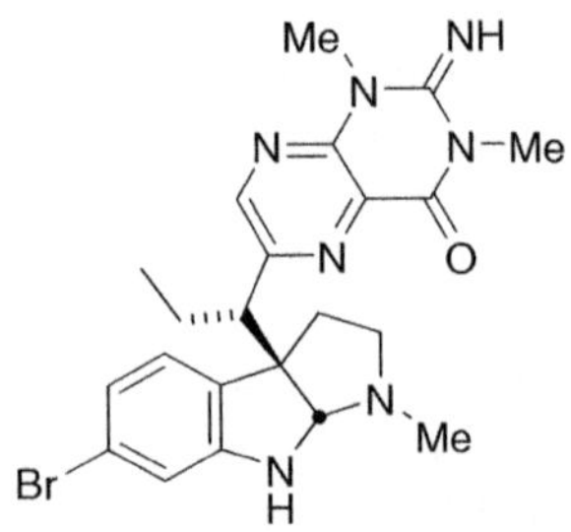

Urochordamine A (**44**) [54]

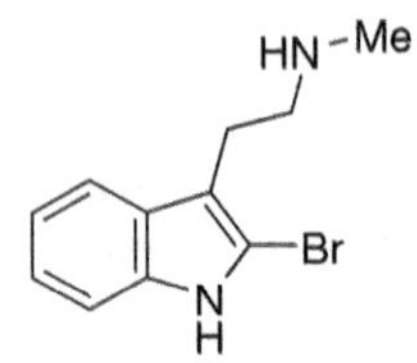

2-Bromo-*N*-methyltryptamine (**45**) [55]

Scheme 6 *Representative Antifouling Indoles*

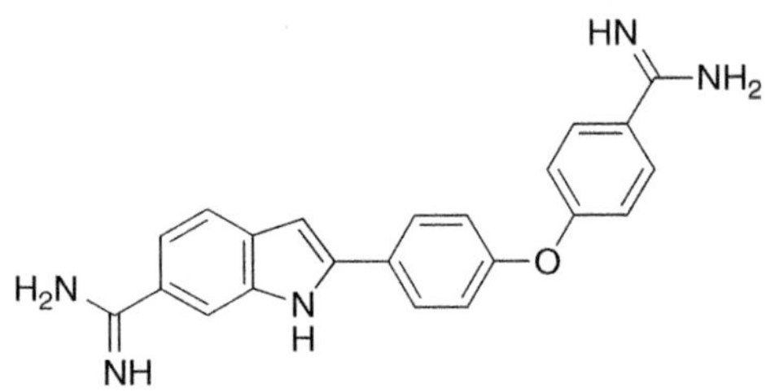

RWJ–61907 (**46**) [56]

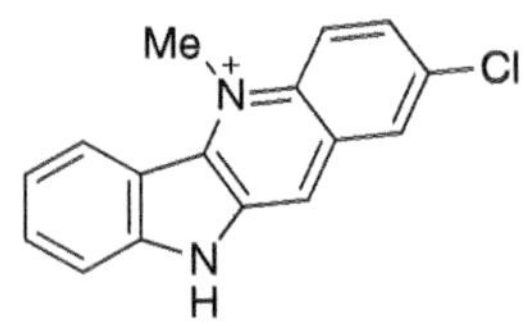

47 [57]

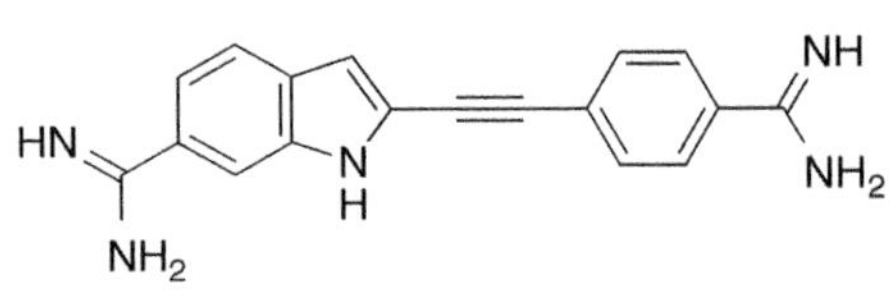

48 [58]

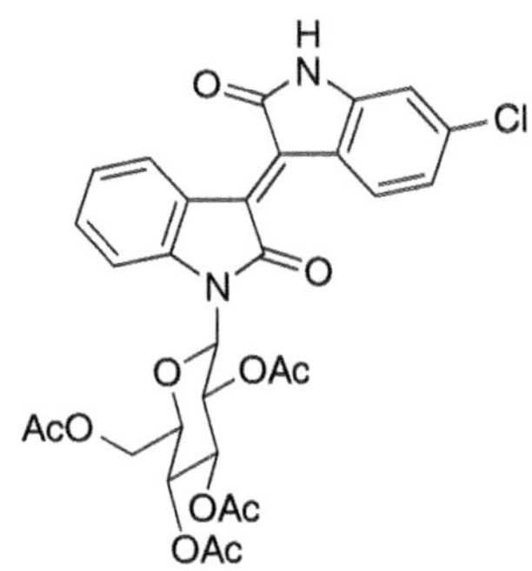

49 [59]

Scheme 7 *Representative Antifungal and Antiparasitic Indoles*

and pyrano[3,4-*b*]indoles (e.g., **54** [65]), are active as potent non-nucleoside reverse transcriptase inhibitors (**50–53**), and **54** is a selective hepatitis C virus (HCV) RNA polymerase inhibitor. The development of predictive quantitative structure–activity relationship (QSAR) models for anti-HIV indolyl aryl sulfones has been described [66].

A number of indoles and carbazoles possess antiinflammatory activity (Scheme 9). Thus, indoles **55–57** are three of several cyclooxygenase (COX) inhibitors based on the structure of thalidomide [67]. Whereas **55** shows no COX-1 activity and only weak COX-2 activity, indole **56** displays potent COX-1 activity and modest COX-2 activity. Indole **57** shows strong inhibition of both enzymes. Several

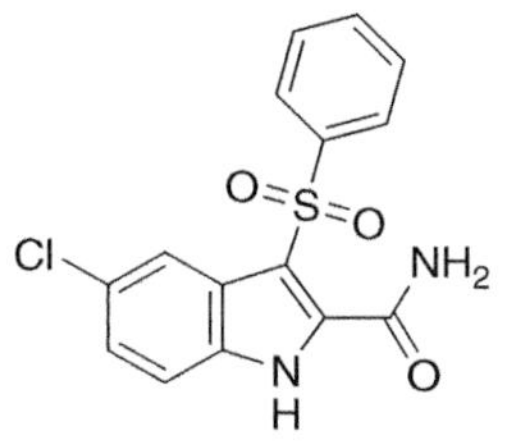

50 [61]

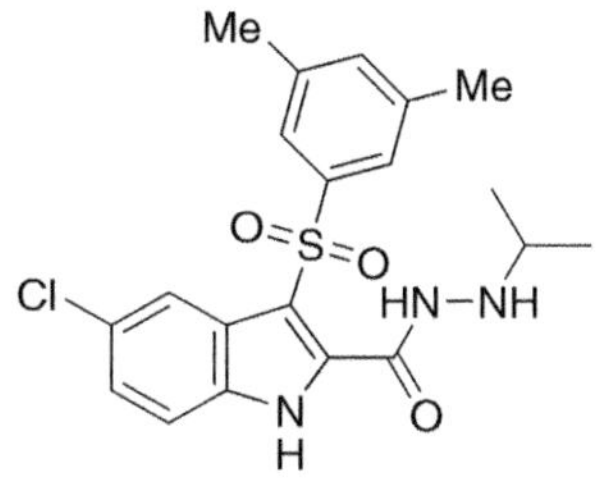

51 [62]

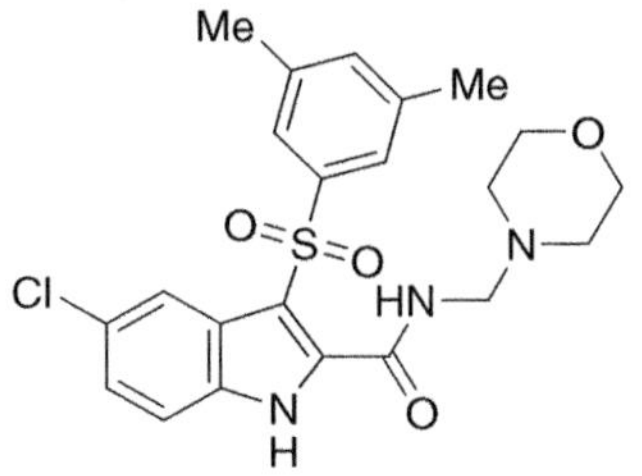

52 [63]

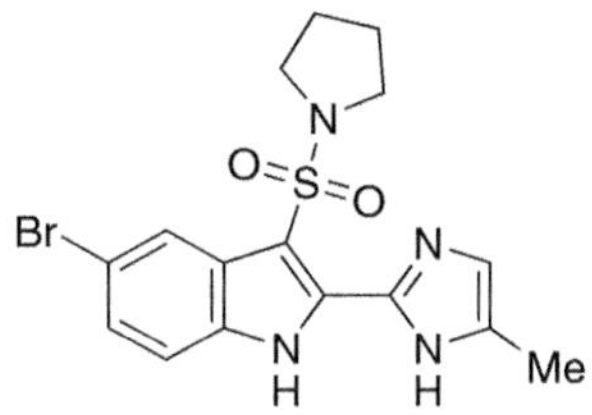

53 [64]

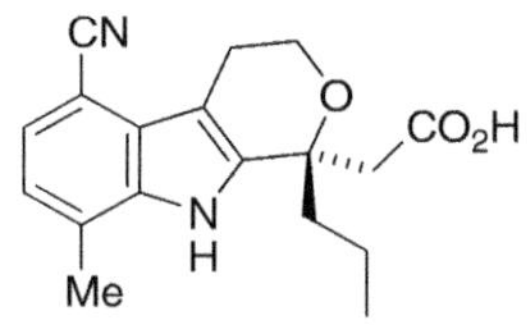

54 [65]

Scheme 8 *Representative HIV Active Indoles*

2-phenyl-3-(sulfonylphenyl)indoles (e.g., **58**) are potent and selective COX-2 inhibitors and possess higher activity than celecoxib [68]. Likewise, indole Schiff base **59** is a highly selective COX-2 inhibitor ($IC_{50}=0.32\,\mu M$; COX-1, $IC_{50}>100\,\mu M$) [69]. Furo[3,2-*b*]indole FI–302 (**60**) is a nonulcerogenic antiinflammatory compound with potency superior to that of the nonsteroidal antiinflammatory drugs (NSAIDs) mepirizole and tiaramide [70]. The carbazole carprofen (**61**) is a multitarget-directed ligand that inhibits COX-1, COX-2, and a fatty acid amide hydrolase (FAAH), and it is the starting point for the synthesis of many analogues [71]. Several indomethacin derivatives have been designed and synthesized to evaluate their inhibitory effects on COX, P-glycoprotein, and multidrug resistance [72]. Indole **62** is a potent inhibitor of matrix metalloproteinase–13 (MMP-13), a protein that functions in cartilage homeostasis [73]. The *Streptomyces* sp. HKI0231 indoles 0231A (**63**) and 0231B (**64**) are inhibitors of 3α-hydroxysteroid dehydrogenase, an enzyme involved in inflammatory processes [74, 75], and thus may be excellent lead structures as new antiinflammatory agents. The novel prostaglandin D_2 receptor antagonist **65** was developed for the treatment of allergic rhinitis, an inflammatory disease [76].

Cancer and cardiovascular disease not withstanding, obesity and diabetes are major global health problems. Several indoles have potential activity in this area (Scheme 10). Indoles **66** and **67** show significant antidyslipidemic activity and weight loss in hyperlipidemic rats, and these compounds represent a new class of hypolipidemic and antiobesity agents [77]. Tetracyclic indole **68** is a melanin-concentrating hormone receptor 1 (MCHR1) antagonist and is effective in reducing food intake in rats and monkeys [78]. *N*-Benzoylindole **69** is a potent liver X receptor (LXRβ) agonist and may exhibit antidiabetic activity of type 2 diabetes by reversing cholesterol accumulation and raising plasma high-density lipoprotein cholesterol (HDL) levels [79]. As a peptidomimetic agonist for the human orphan receptor BRS–3, indole **70** may find use in the treatment of obesity [80].

Serotonin (5-hydroxytryptamine [5-HT], **3**) receptors play an essential role in mediating neurotransmission and in so doing they influence memory, learning, sleep, aggression, anxiety, appetite, mood, and other neurological functions [81, 82]. These dozen or so receptors are targets for drugs to treat depression, pain, psychosis, sleep, learning disorders, insulin secretion, epilepsy, schizophrenia, and

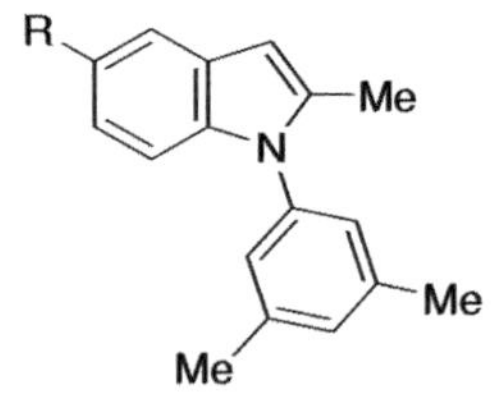

55 [67], R = NO_2
56 [67], R = NH_2

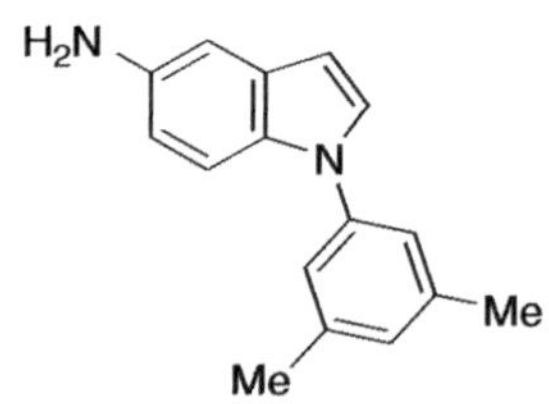

57 [67]

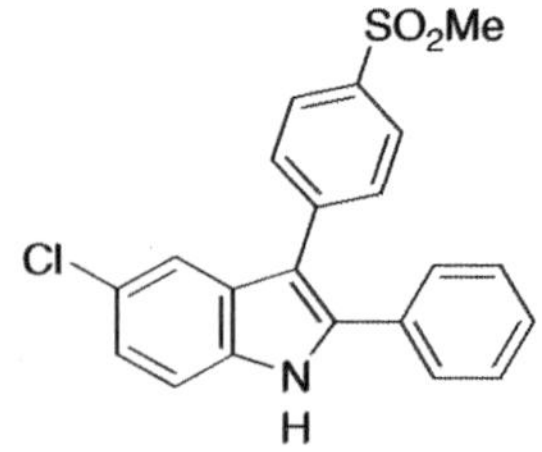

58 [68]

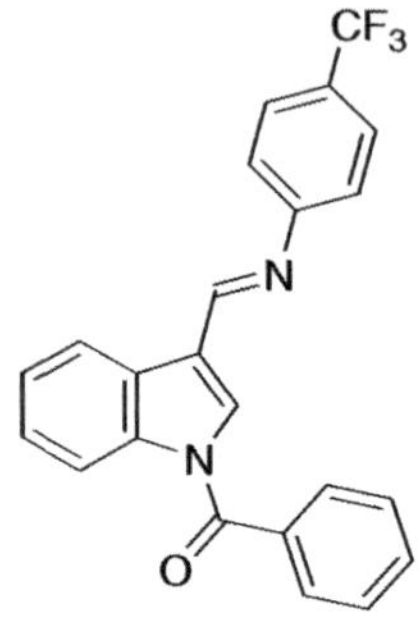

59 [69]

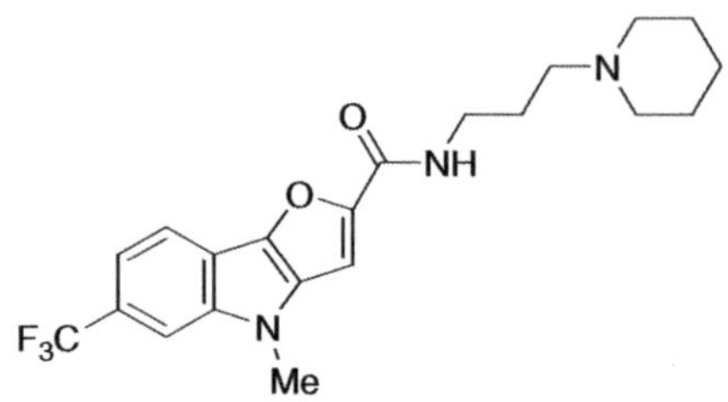

FI–302 (**60**) [70]

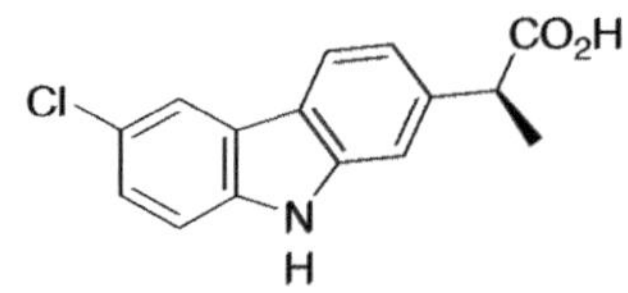

Carprofen (**61**) [71]

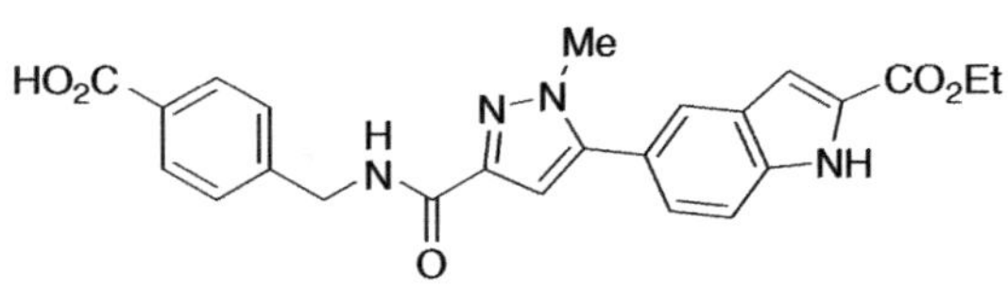

62 [73]

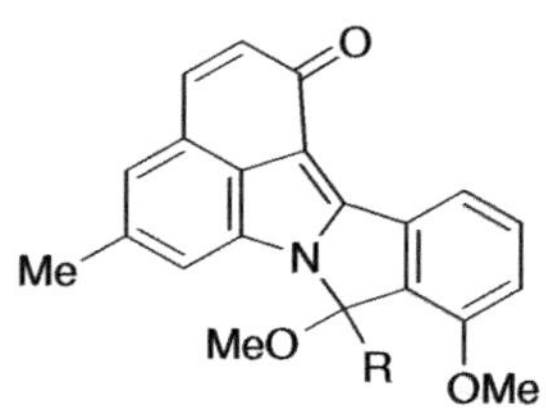

0231A, R = OMe (**63**) [74], [75]
0231B, R = H (**64**) [74], [75]

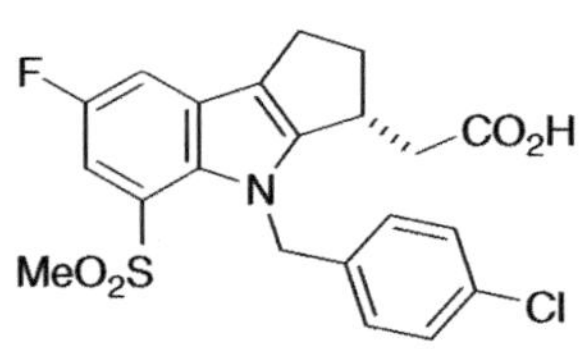

65 [75]

Scheme 9 *Representative Antiinflammatory Indoles*

other biological dysfunction. Several indoles bind to various 5-HT receptors (Scheme 11), including the drug sumatriptan (Section 4). A review of the 5-HT receptor subtype 5-HT_6 has appeared [83]. Indole **71** is remarkably selective as an agonist toward 5-HT_{1D} versus 5-HT_{1A} [84], and **72** is a potent and selective 5-HT_6 receptor antagonist having subnanomolar inhibition of the production of adenylate cyclase [85]. Indole **73** also has very high and selective affinity for 5-HT_6 as an agonist [86], and **74** is a selective antagonist for 5-HT_6 [87].

In addition to the anticancer indole alkaloids vinblastine and vincristine, discussed in the next section, many indoles display antitumor activity. Space does not allow complete coverage of these studies. Indole-3-carbinol (Scheme 12, **75**), found in vegetables of genus *Brassica* (kale, cauliflower, broccoli, turnip, collard, and others), its acid- and/or enzymatic-induced dimer, 3,3′-diindolylmethane (**76**), and its trimer, 2-(indol-3-ylmethyl)-3,3′-diindolylmethane (**77**), inhibit cancer cell proliferation and induce apoptosis in several cell lines [88–91]. However, one study reports that

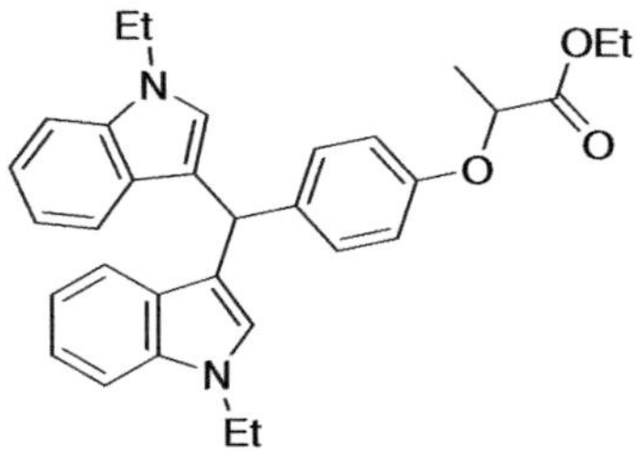

66 [77]

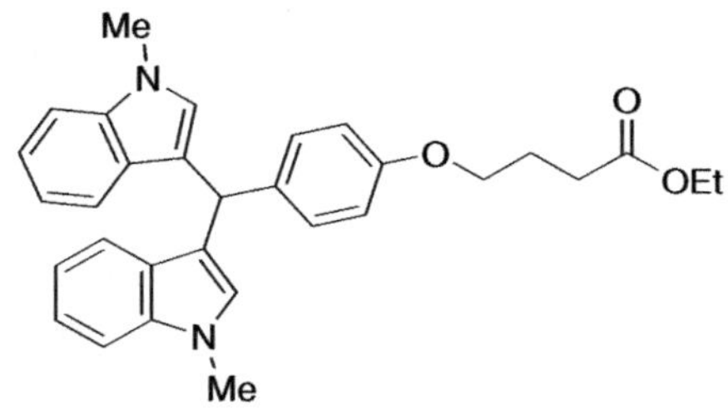

67 [77]

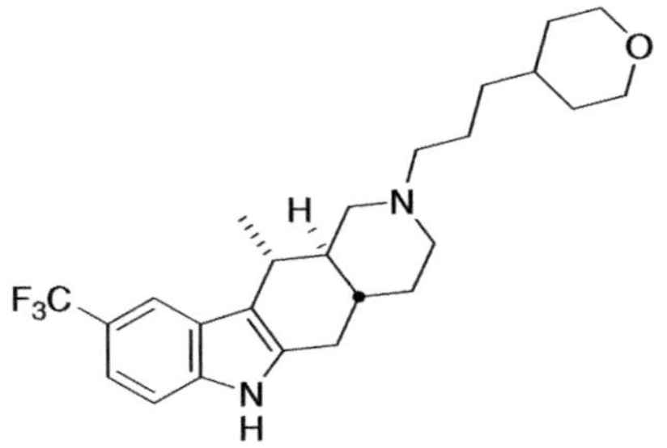

68 [78]

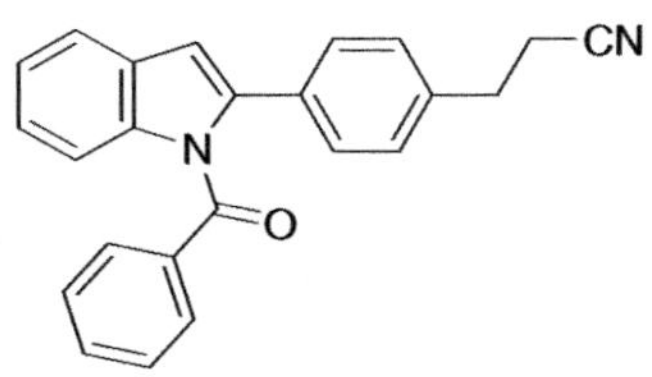

69 [79]

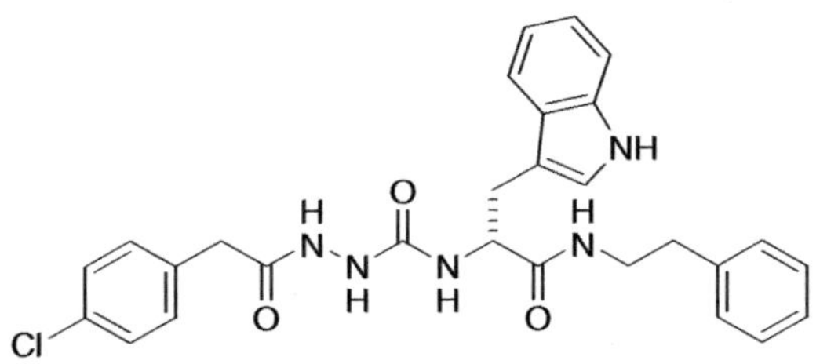

70 [80]

Scheme 10 *Representative Antiobesity Indoles*

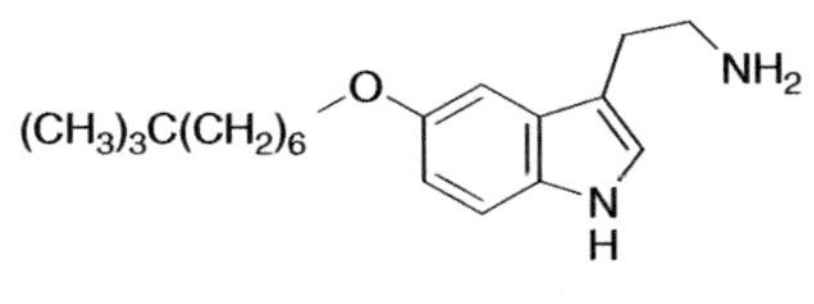

71 [84]

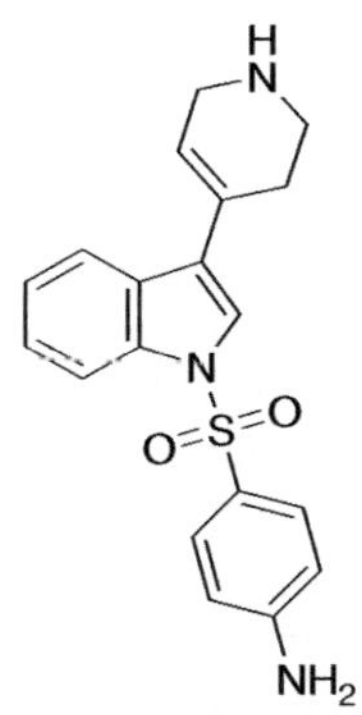

72 [85]

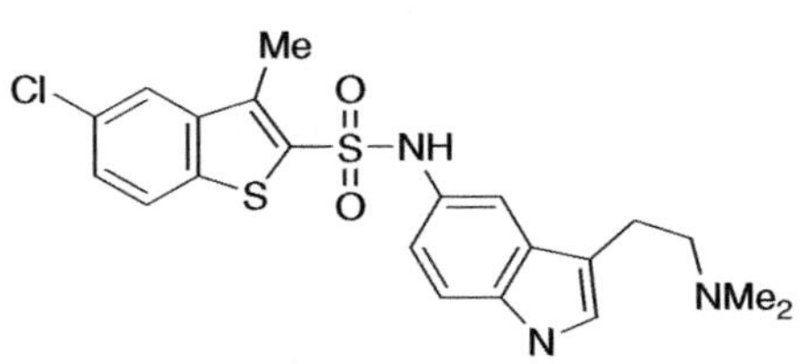

73 [86]

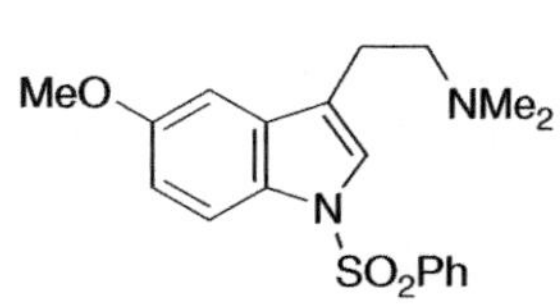

74 [87]

Scheme 11 *Representative Indoles that Bind to 5-HT Receptors with High Affinity*

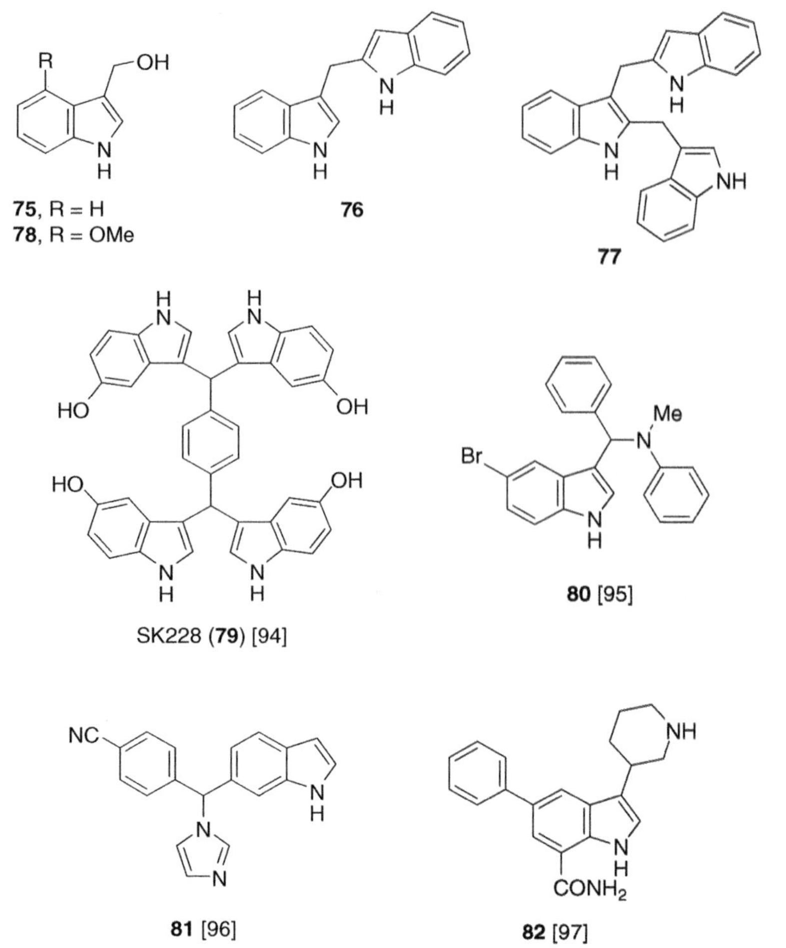

Scheme 12 *Representative Indoles with Antitumor Activity*

3,3′-diindolylmethane (**76**) is a liver carcinogen in trout by an estrogenic pathway [92]. A more potent inhibitor of human colon cancer cell proliferation than **75** is 4-methoxyindole-3-carbinol (**78**), which is a metabolite of 4-methoxyglucobrassicin formed during ingestion [93]. The novel 5-hydroxy tetraindole **79** (SK228) induces G_2 arrest and apoptosis in human breast cancer cells [94], and several indoles of type **80** inhibit cell proliferation of human colon cells (HT–29), human ovarian cells (SK–OV–5), and c-src kinase activity [95]. Indolyl imidazole **81** is a potent inhibitor of aromatase (CYP19) (IC_{50} = 11.5 nM), suggesting activity against breast cancer [96]. Indole-7-carboxamide **82** is a potent inhibitor of the serine–threonine kinase (IKK-β), which regulates an important signaling pathway [97].

An important strategy for the treatment of cancer is the modulation of microtubule assembly either by preventing its disassembly or by blocking tubulin polymerization, and an excellent review is available that discusses several indole leads [98]. For example, Silvestri and colleagues report that arylthioindoles are potent inhibitors of tubulin polymerization [99–101]. For example, **84** inhibits the growth of MCF-7 cells at IC_{50} = 13 nM [99], and **85** is the most potent antitubulin agent discovered thus far [101] (Scheme 13). A number of indole-3-carbaldehydes and their corresponding imines inhibit tubulin polymerization and inhibit the growth of breast cancer cells; for example, imine **86** (MCF-7, IC_{50} = 27 nM; MDA-MB 231, IC_{50} = 6 nM) [102]. 5′-Methoxyindirubin (**87**) induces cell death in human neuroblastoma cells (IMR-32, SK-N-SH, NB-39) without affecting normal cells (NHDF and HUVEC) [103].

A large number of naturally occurring indoles display antitumor activity, but only a limited number can be illustrated here (Scheme 14). Cultures of *Aspergillus ochraceus* WC76466 produce stephacidins A (**88**) and B (not shown), both of which are selective inhibitors of prostate LNCaP cells, and they also show activity against a panel of other tumor cell lines [104]. The Panamanian soil microbe *Nocardia aerocolonigenes* (now reclassified as *Saccharothrix aerocolonigenes)* produces rebeccamycin (**89**) and 4′-deschlororebeccamycin (**90**), which have potent anticancer activity [105, 106] and an analogue is in human cancer trials (Section 4). A deepwater Palauan

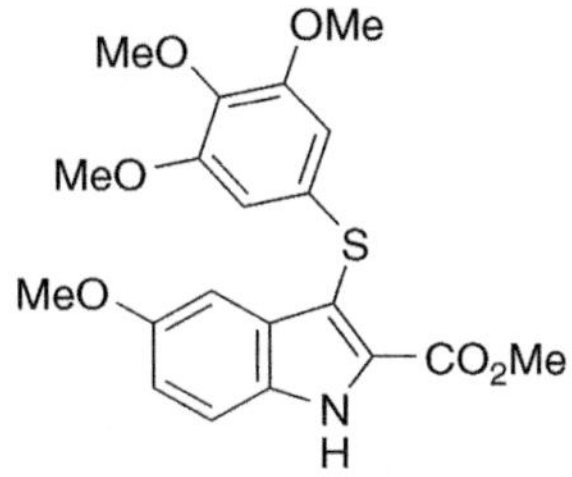

84 [99]

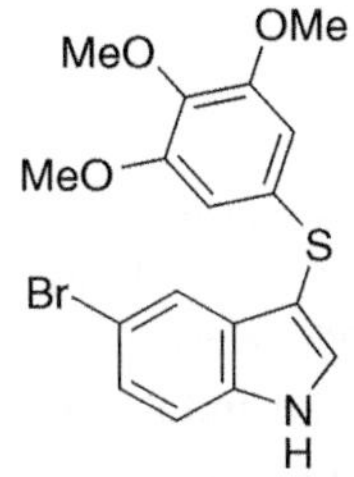

85 [101]

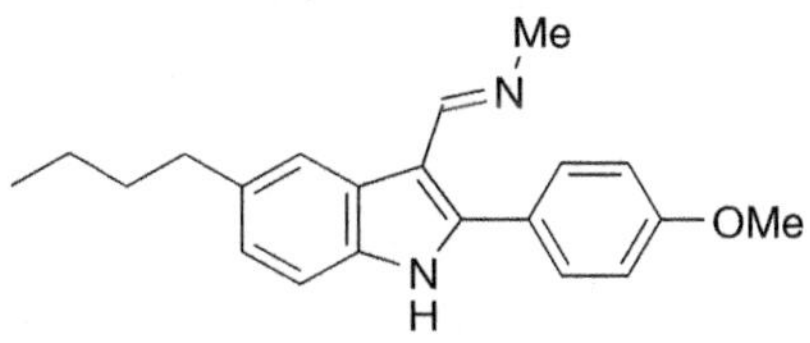

86 [102]

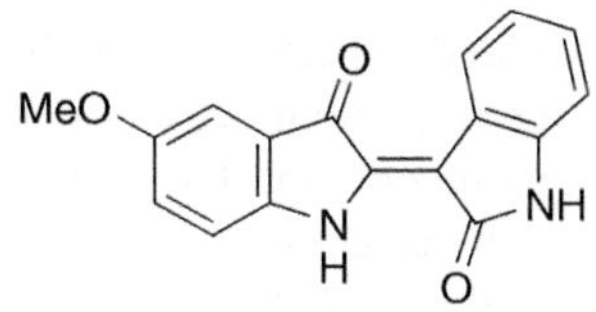

87 [103]

Scheme 13 *Representative Indoles with Tubulin Inhibitory Activity*

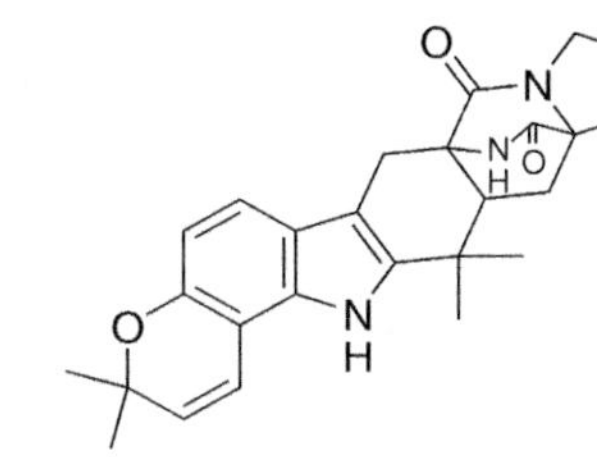

Stephacidin A (**88**) [104]

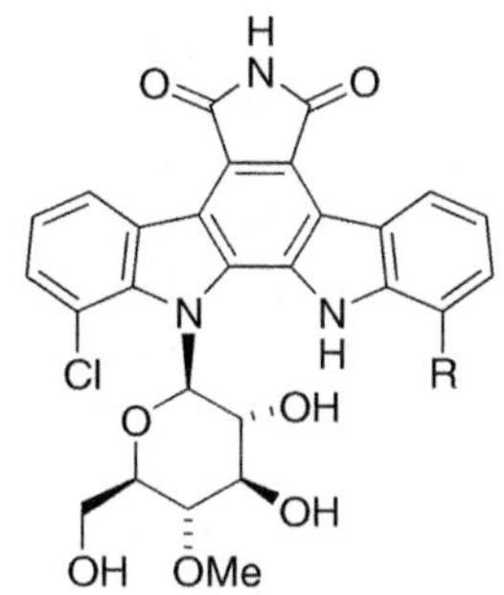

Rebeccamycin (**89**, R = Cl) [105]
Rebeccamycin (**90**, R = H) [106]

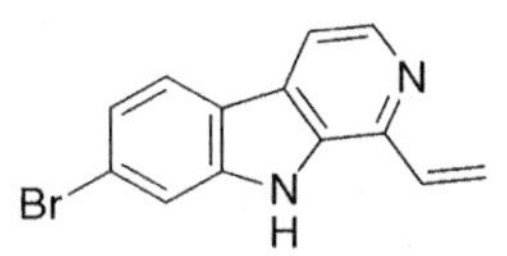

Plakortamine B (**91**) [107]

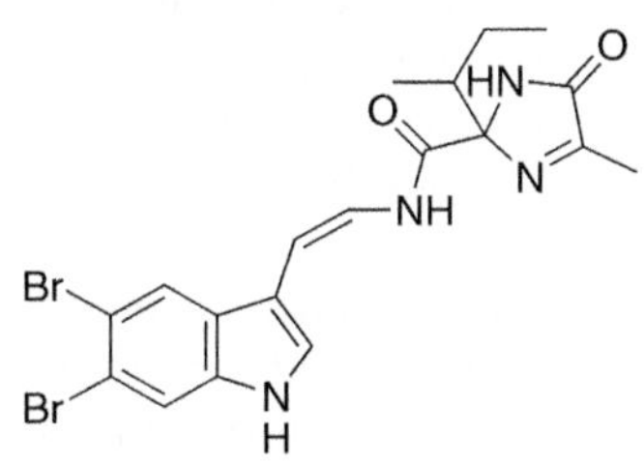

Kottamide D (**92**) [108]

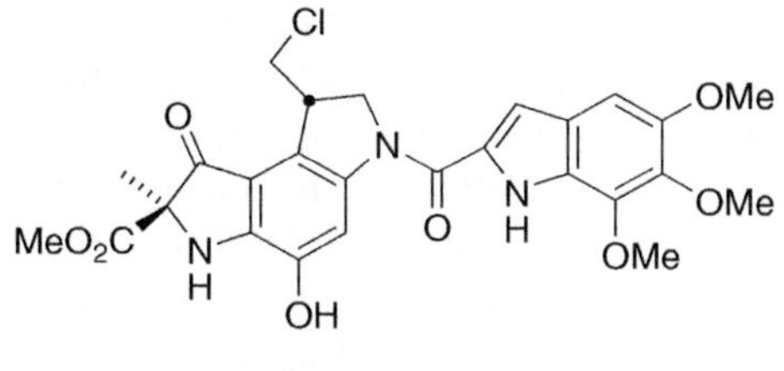

Pyrindamycin A (**93**) [110]

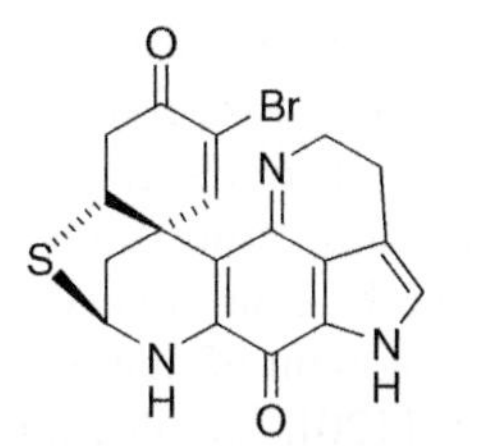

Discorhabdin A (**94**) [112]

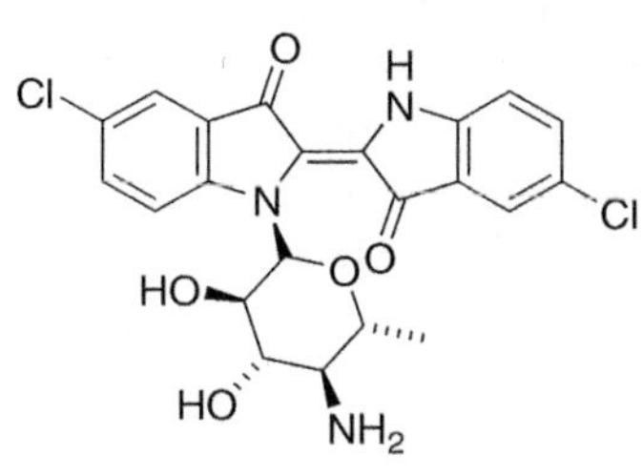

Akashin A (**95**) [113]

Scheme 14 *Representative Natural Indoles with Antitumor Activity*

sponge, *Plakortis nigra*, produces several plakortamines, the most active of which against HCT-116 human cancer cells is plakortamine B (**91**) [107]. The New Zealand ascidian *Pycnoclavella kottae* contains the indoles kottamides A–E [108, 109], one of which, kottamide D (**92**), inhibits the proliferation of HL60 cancer cells [108]. Pyrindamycin A (= duocarmycin C_2) (**93**) is a potent antitumor metabolite from *Streptomyces* SF2582 [110, 111]. The New Zealand sponge *Latruncula* sp. has yielded several discorhabdins, such as discorhabdin A (**94**), having potent cytotoxic activity [112]. A terrestrial *Streptomyces* sp. has furnished akashins A–C (e.g., **95**), which have antitumor activity against several human cancer cell lines [113].

In addition to the major diseases we have discussed, numerous other disease conditions and biological syndromes are affected by indoles. For example, several β-carbolines show acetylcholinesterase activity [114], and the natural nostocarboline (**96**), from the freshwater cyanobacterium *Nostoc* 78-12A, has butyrylcholinesterase inhibitory activity comparable to that of galanthamine, a drug approved for the treatment of Alzheimer's disease (Scheme 15) [115]. Several synthetic indirubins are inhibitors of glycogen synthase kinase-3 (GSK-3), a kinase involved in abnormal hyperphosphorylation of proteins and the production of β-amyloid peptides and neurofibrillary tangles, a cascade of events thought to develop into Alzheimer's disease. One such active GSK-3 inhibitor is indirubin **97**. These indirubins also inhibit cyclin-dependent kinases (CDKI/cyclin B and CDK5/p25) [116]. Dibromocarbazole P7C3 (**98**) is a neuroprotective synthetic compound that could find utility in the protection of the hippocampus, the degeneration of which is associated with Alzheimer's disease [117, 118]. Thus, P7C3 and analogues protect newly born neurons from apoptosis, and thus they may represent a new therapy for Alzheimer's patients.

A new set of dihydroindoles, related structurally to the neuroprotective stobadine, has been developed that diminishes the toxicity of stobadine. For example, hexahydro-1*H*-pyrido[4,3-*b*]indole **99** displays improved neurological efficacy over that of stobadine [119]. A new human neurokinin-1 (hNK_1) receptor antagonist, 2-arylindole **100,** is one of several simple compounds that exhibit both good receptor-binding affinity and brain penetration. The hNK_1 receptor in the central nervous system is a potential target for the treatment of depression, anxiety, and drug-induced emesis [120]. A collection of tetracyclic indoles, such as **101**, possesses anticonvulsant activity [121], and 4,6-dichloroindole **102** inhibits convulsions induced by *N*-methyl-D-aspartate (NMDA) in mice. This potent *in vivo* antagonist acts at the strychnine-insensitive glycine-binding site [122]. A similar indole with excellent affinity for the glycine site of the NMDA receptor is 2-indolecarboxylic acid **103** [123].

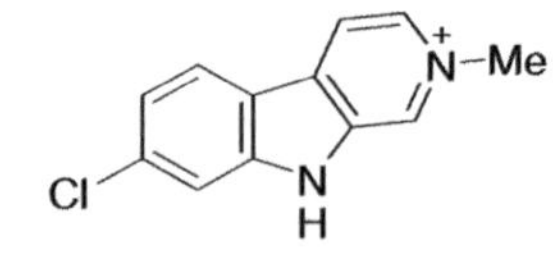

Nostocarboline (**96**) [115]

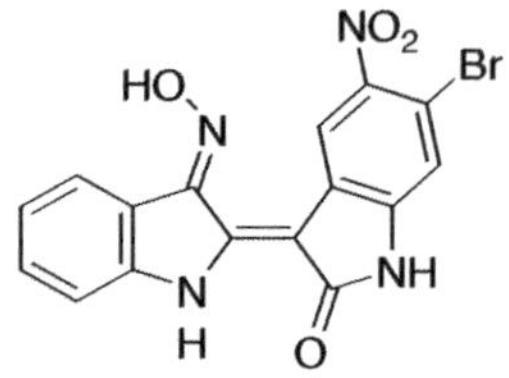

97 [116]

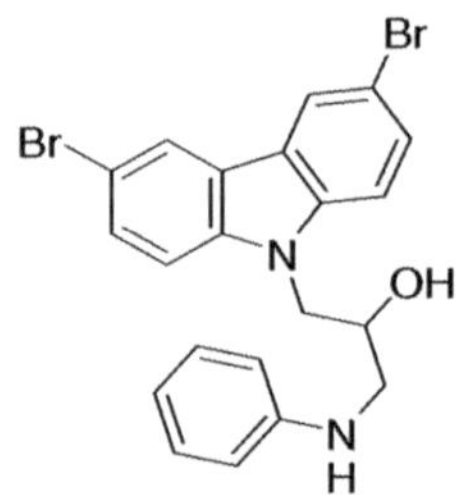

P7C3 (**98**) [117]

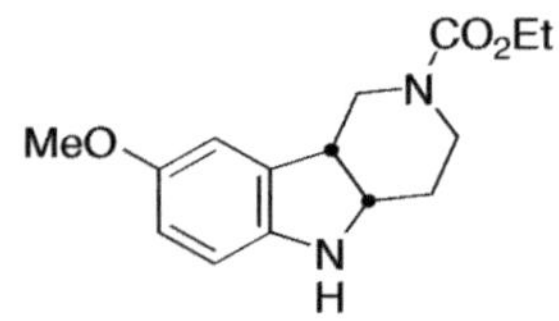

99 [119]

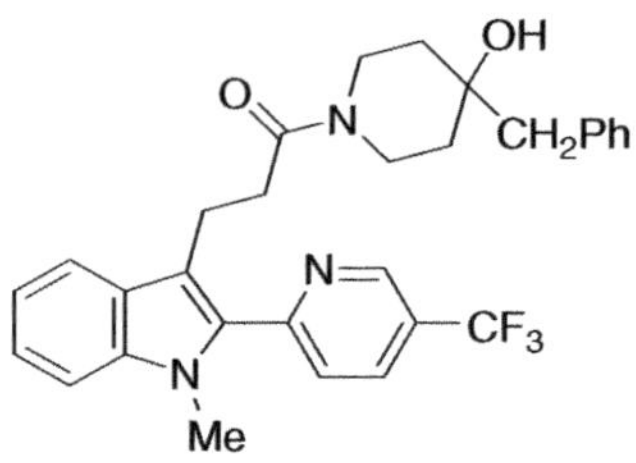

100 [120]

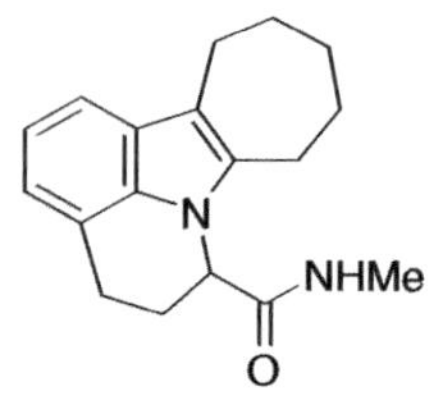

101 [121]

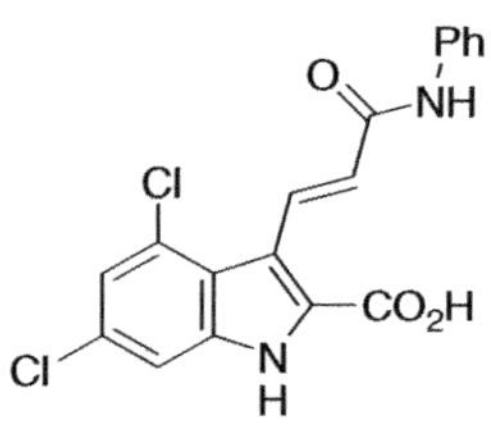

102 [122]

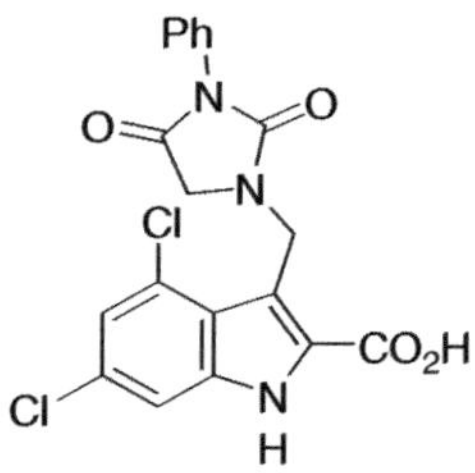

103 [123]

Scheme 15 *Representative Biologically Active Indoles – 1*

The new cannabimimetic phenylacetylindole cannabipiperidiethanone (Scheme 16, **104**) is an adulterant found along with two other previously known synthetic cannabinoids, JWH-122 (**105**) and JWH-081 (**106**), in a Japanese herbal product [124]. These illegal designer drugs have potent affinity for the cannabinoid CB_1 and CB_2 receptors. Huffman and colleagues have developed structure–activity relationships at both of these receptors for 3-(1-naphthoyl) indoles [125]. For example, one compound, JWH-416 (**107**), has the desirable combination of very good CB_2 affinity but low CB_1 affinity, although many others are also selective for the former receptor [125]. A review of this area is available [126]. A study of the melatoninergic binding site MT_3 has found that 4-nitroindole melatonin

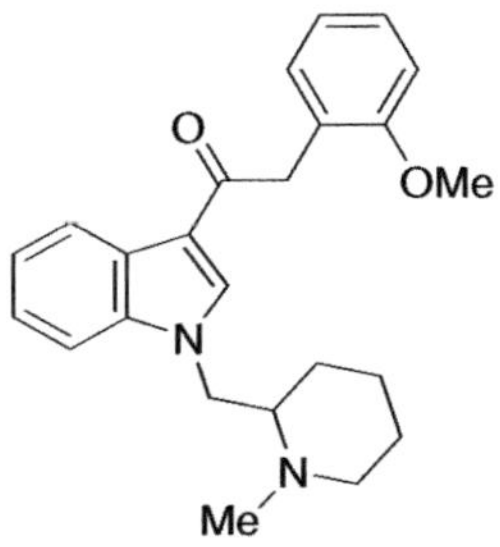

104 [124]

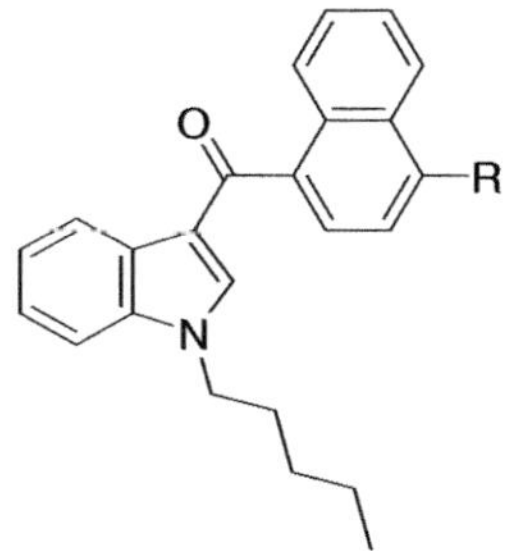

JWH-122, R = Me (**105**) [124]
JWH-081, R = OMe (**106**) [124]

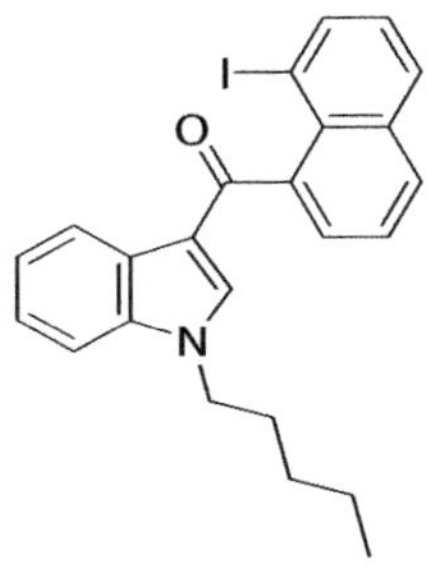

JWH-416 (**107**) [125]

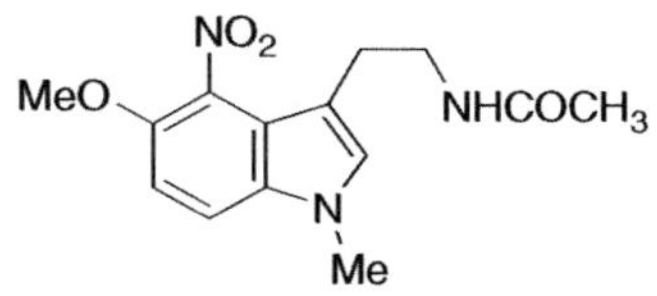

108 (R = H) [127]
109 (R = I) [127]

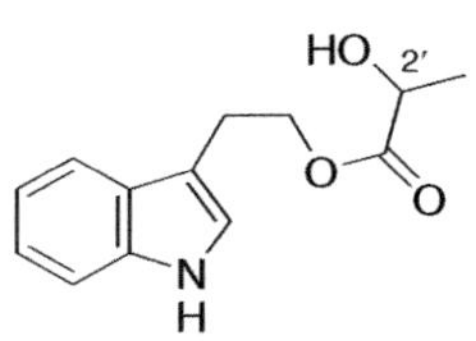

110 (2′ 5:8 *S*/*R*)

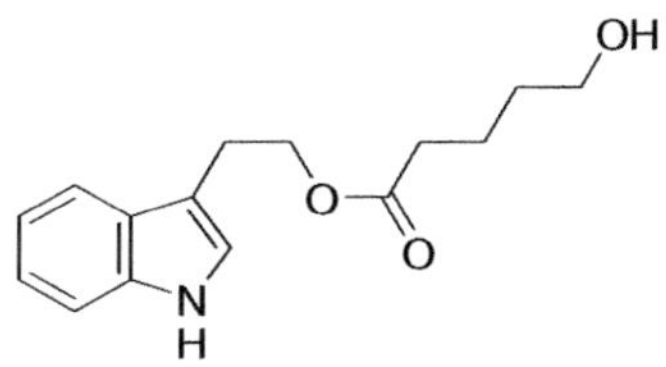

111 [128]

Scheme 16 *Representative Biologically Active Indoles—2*

derivative **108** is both potent and selective for the MT_3 receptor. The 2-iodo analogue **109** is also selective for MT_3, and the radiolabeled iodine-125 version might be used to characterize the MT_3 binding sites [127]. Indole esters **110** and **111** were isolated from the yeast *Pichia membranifaciens* living on the marine sponge *Halichondria okadai*. Both novel compounds have DPPH (2,2-diphenyl-1-picrylhydrazyl) radical scavenging activity [128].

Several tricyclic indoles have been discovered to be potent and selective $S1P_1$ agonists. *In vivo* these compounds significantly reduce circulating lymphocytes, which in turn depresses the circulation of autoreactive immune cells. The net effect is to slow progression of multiple sclerosis. Indole **112** (Scheme 17) is the most potent compound tested that lowered the flowing lymphocytes in mice with autoimmune disease [129]. Such compounds may also act on astrocytes to reduce inflammation in the central nervous system. The antiinflammatory leukotriene modifier MK-0591 (**113**) is a FLAP (5-LO-activating protein) inhibitor that was efficacious in initial human asthma trials but was later withdrawn [130–132]. FLAP is a protein that is essential for the biosynthesis of leukotriene [133]. Zafirlukast (**114**) is a leukotriene receptor antagonist that has been marketed for the treatment of asthma since 1996 in the United States and is now available in at least 60 countries [132]. This drug reduces mucus buildup and constriction of the airways and lungs. Some *Nephila* spider polyamine toxins have an indole anchor, such as NPTX-8 (**115**) and NPTX-1 (**116**) secreted by *Nephila clavata*. These toxins are the most potent polyamine kainate receptor antagonists reported to date. Moreover, NPTX-1 (**116**) shows an extraordinary selectivity for GluK1 receptors over GluN1/2A type NMDA (>40,000 fold) and GluA1 type AMPA (>150 fold) receptors [134]. The indole headgroup is important for maximum inhibitory activity because other spider polyamines with a phenolic (resorcinol) headgroup are much less active. Inhibition of these ionotropic glutamate (iGlu) receptors is a promising tactic for the treatment of pain, stroke, and Alzheimer's disease [135–137]. Synthetic analogues of the indole alkaloid

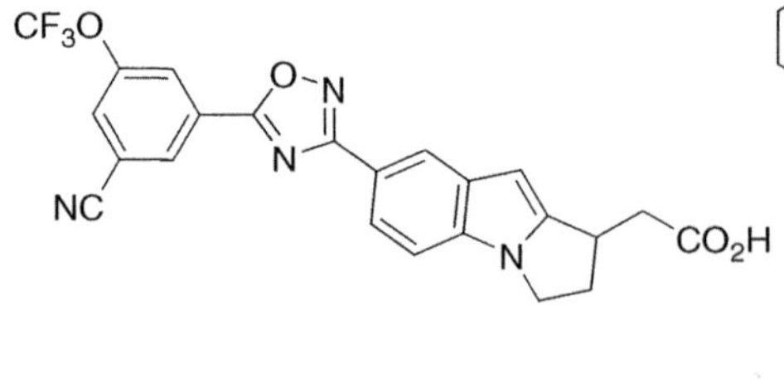

112 [129]

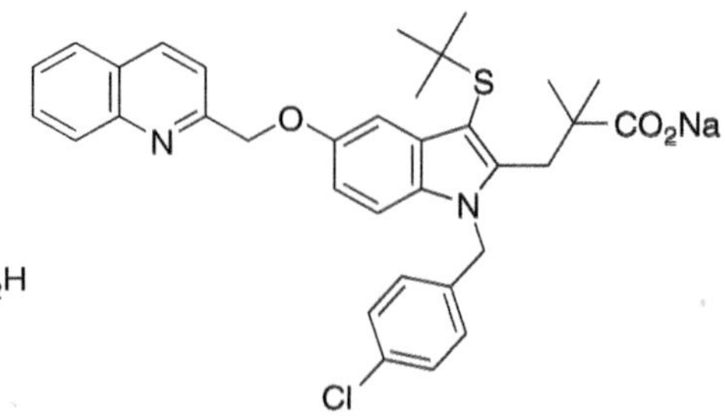

MK-0591 (**113**) [130], [131]

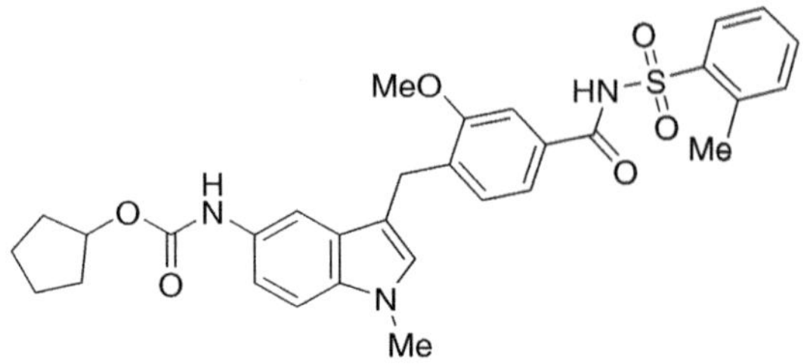

Zafirlukast (**114**) [133]

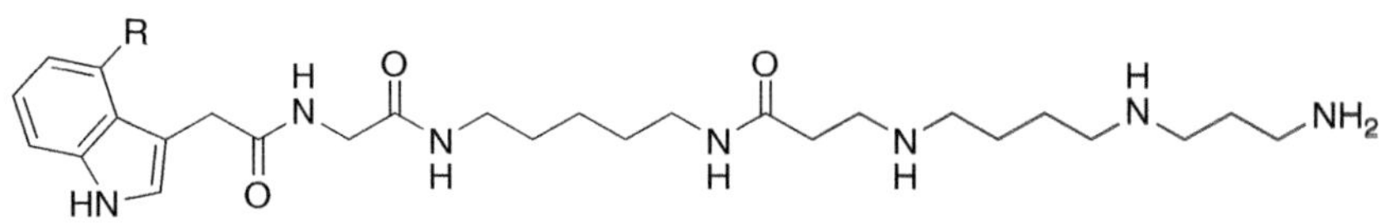

NPTX-8, R = H (**115**) [134]
NPTX-1, R = OH (**116**) [134]

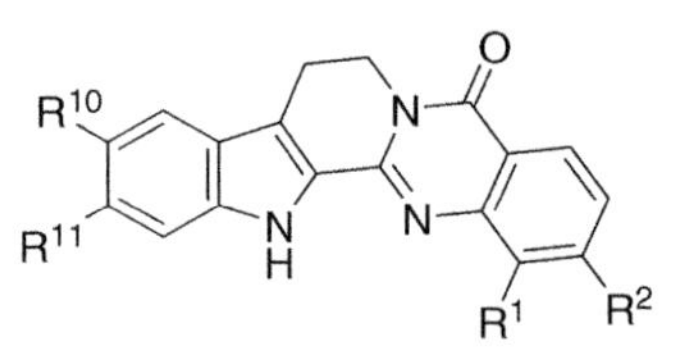

117, R^1 = OMe, R^2 = R^{10} = R^{11} = H
118, R^1 = R^2 = OMe, R^{10} = R^{11} = H
119, R^{10} = OMe, R^1 = R^2 = R^{11} = H
120, R^{11} = OMe, R^1 = R^2 = R^{10} = H

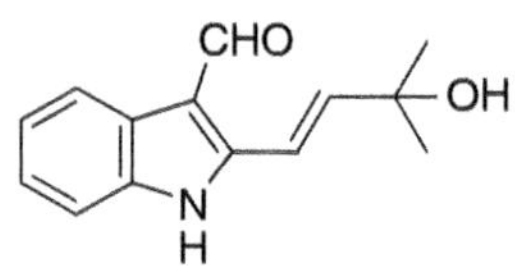

121

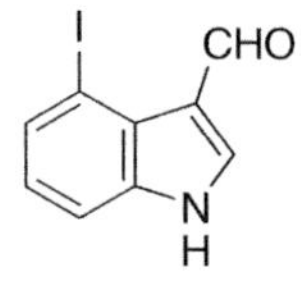

122

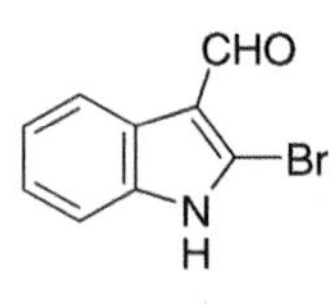

123

Scheme 17 *Representative Biologically Active Indoles—3*

rutaecarpine are found to be potent and selective inhibitors of members of the human cytochrome P450-1 (CYP1) family of enzymes [138]. Thus, 1-methoxyrutaecarpine (**117**) and 1,2-dimethoxyrutaecarpine (**118**) are highly selective for CYP1A2, whereas 10- (**119**) and 11-methoxyrutaecarpine (**120**) are most selective for CYPB1 enzyme. These P450 enzymes play a critical role in the detoxification of foreign chemicals such as polycyclic aromatic hydrocarbons. The plant growth regulator indole-3-acetic acid (**6**) and synthetic derivatives have been employed by Somei to convert desert areas into green tracts of plants and grass as a means to combat global warming and prevent further deforestation (desertification). Three of these SOMRE (*Som*ei, *r*oot, and *e*longation) indoles are shown (**121–123**) that are intended for the greening of, for example, the Gobi desert in China. Applications of SOMRE indoles to sword bean (*Cryptomeria japonica*) and Chinese medicine plants (*Forsythia viridissima* and *Glycyrrhiza*) are currently under way, and future target plants include sugar cane, rubber trees, and grasses [139].

1.4 Indole-Containing Pharmaceuticals

A number of indoles are in current use as pharmaceuticals, and the number is certain to grow in the future, as indole remains a fundamental privileged scaffold [140].

Several anticancer drugs are indoles, the most famous of which are the *Catharanthus roseus* plant alkaloids vinblastine (**124**) and vincristine (**125**) (Scheme 18), both of which continue to be important drugs for the treatment of Hodgkin's disease, childhood leukemia, and other cancers [141–143]. The rebeccamycin analogue NSC-655649 (**126**) is in phase II trials for the treatment of metastatic renal cell cancer [144]. An analogue of the antitumor CC-1065, bizelesin (**127**), is in human clinical trials for the treatment of solid tumors (kidney, stomach, colon, ovary, uterus, and others) [145]. One patient with advanced ovarian cancer

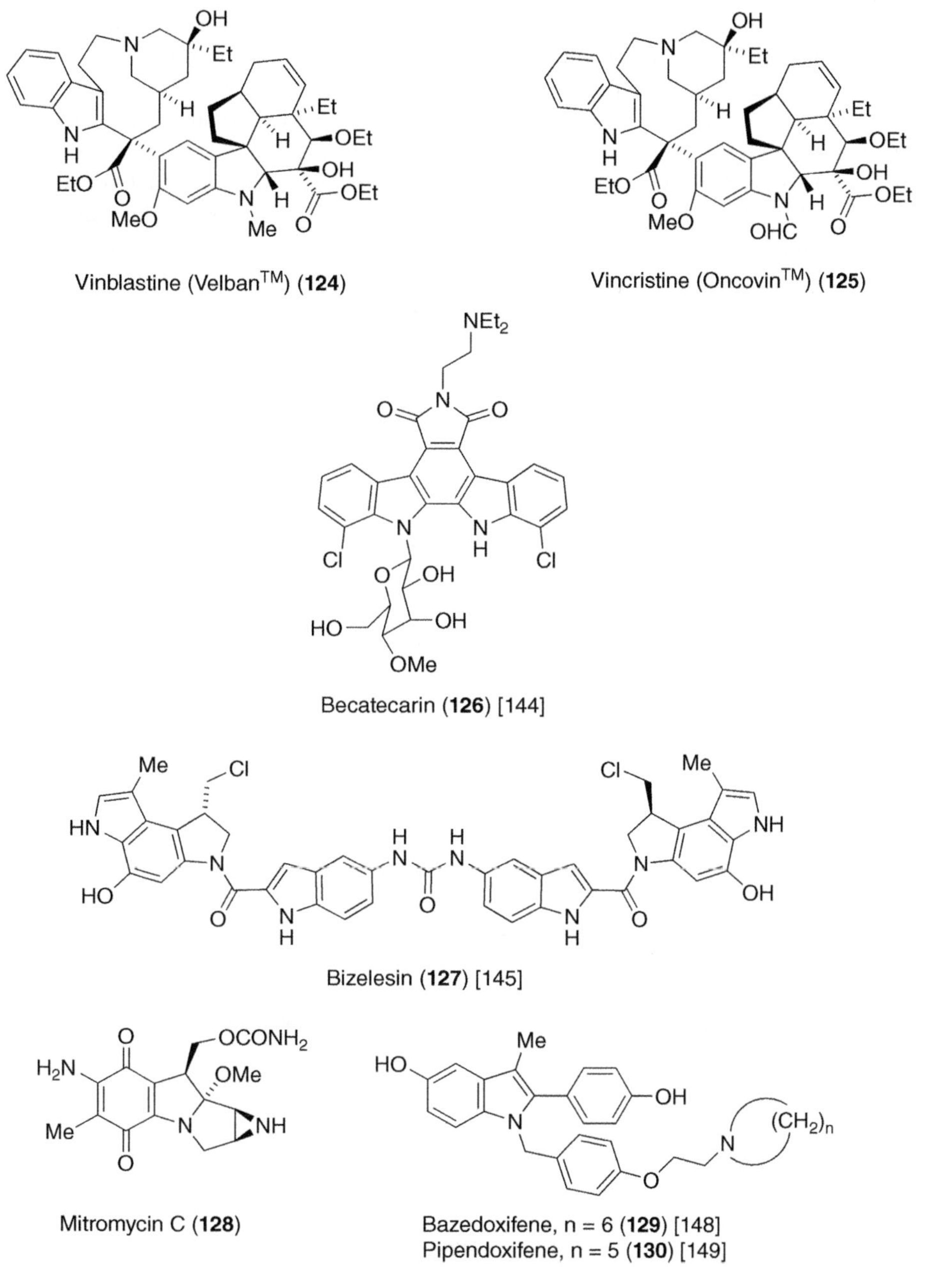

Scheme 18 *Anticancer Indole Drugs*

experienced a 40% reduction in her metastatic disease. Mitomycin C (**128**) is a widely used anticancer agent isolated from *Streptomyces caespitosus* and *S. lavendulae* [146, 147]. Bazedoxifene (**129**) [148] and pipendoxifene (**130**) [149] are related selective estrogen-receptor modulators (SERMs) and are being developed for osteoporosis and breast cancer, respectively. Pipendoxifene is in phase II trials for metastatic breast cancer [149, 150], and bazedoxifene is in final review by the U.S. Food and Drug Administration for the prevention and treatment of osteoporosis [150].

Panobinostat (Scheme 19, **131**) inhibits histone deacetylases (HDAC) and the proliferation of several tumors. It is in phase I/II trials for hematological cancers [150,151]. Brivanib (**132**) is a prodrug for BMS-540215 (**133**), and it inhibits the angiokinases VEGFR-2 and FGFR-1 [150,152]. The 4-fluoro-2-methylindole–related cediranib (**134**) is a tyrosine kinase inhibitor of all three VEGF receptors and is currently in phase II/III trials for advanced colorectal and non–small cell lung cancer [150, 153]. Poly(ADP-ribose) polymerase (PARP) is a protein involved in DNA repair and is elevated in some cancer patients. Indole AG-14699 (**135**) is an inhibitor of PARP and is currently in phase II studies for the treatment of ovarian and breast cancer and melanoma [150, 154]. The family of B-cell lymphoma-2 (Bcl-2) proteins is overexpressed in certain cancers such as non-Hodgkin's lymphoma. Obatoclax (**136**) inhibits Bcl-2 and is in phase II studies for the treatment of Hodgkin's lymphoma and related conditions [150, 155].

A major class of anticancer indoles is the indolocarbazoles, epitomized by the naturally occurring staurosporine (Scheme 20, **137**) and the earlier presented rebeccamycin (**89**) [156]. Indolocarbazoles typically inhibit members of the large family of protein kinase C (PKC) enzymes [157]. These serine–threonine kinases are involved in proliferation, differentiation, transcription, tumorigenesis, and angiogenesis, and overexpression of PKCs is implicated in numerous cancers. Staurosporine itself is too nonselective in its inhibition of PKCs, but the naturally occurring UCN-01 (7-hydroxystaurosporine) (**138**) is currently in phase II trials for the treatment of chronic lymphocytic leukemia (CLL). The related midostaurin (**139**) selectively inhibits the FLT-3 receptor tyrosine kinase and is currently in phase III trials for acute myelogenous leukemia (AML). Midostaurin is also broadly antiproliferative in other cancer cell lines [150]. A reduction product of the natural K-252a (not shown) is lestaurtinib (**140**), which is a

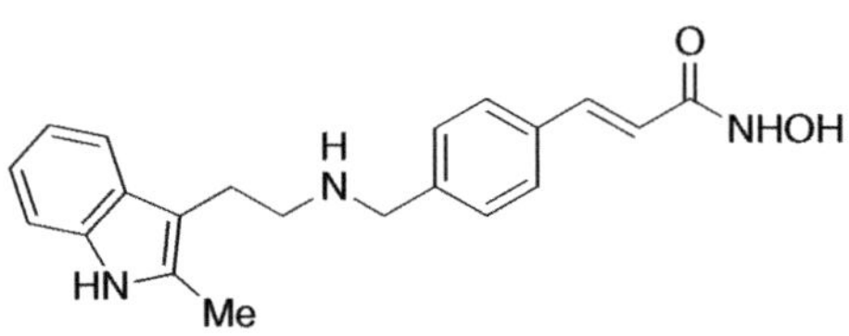

Panobinostat (**131**) [151]

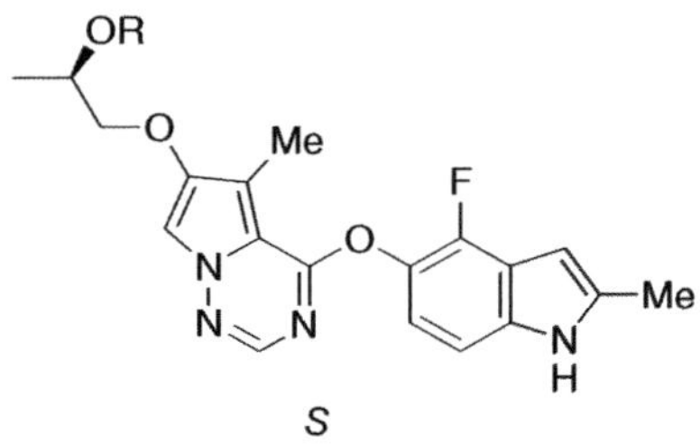

Brivanib, R = COCH(Me)NH_2 (**132**) [152]
BMS-540215, R = H (**133**) [152]

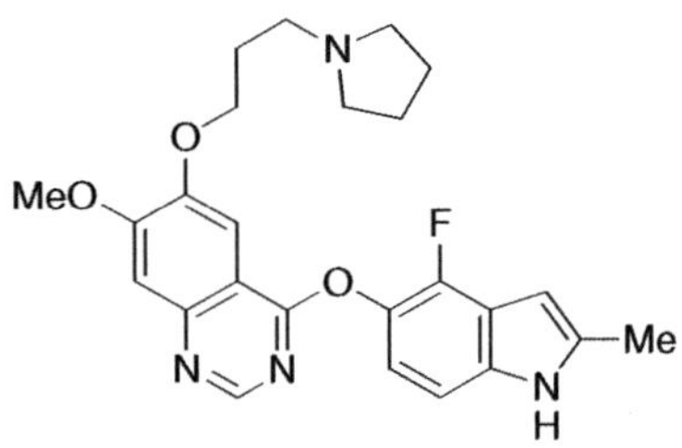

Cediranib (**134**) [153]

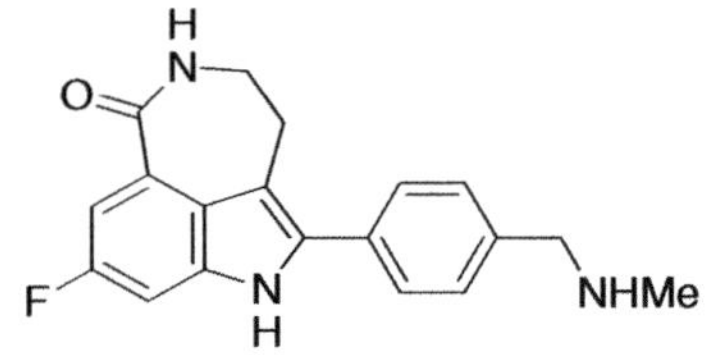

AG-14699 (**135**) [154]

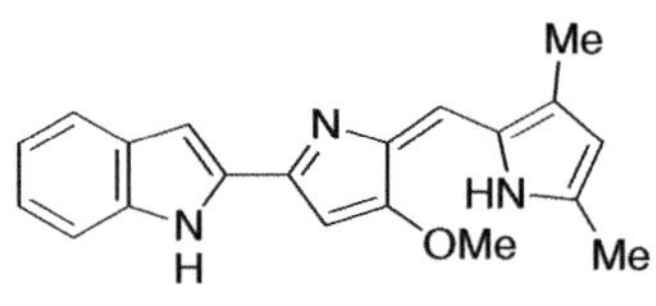

Obatoclax (**136**) [155]

Scheme 19 *Anticancer Indole Drugs in Clinical Trials*

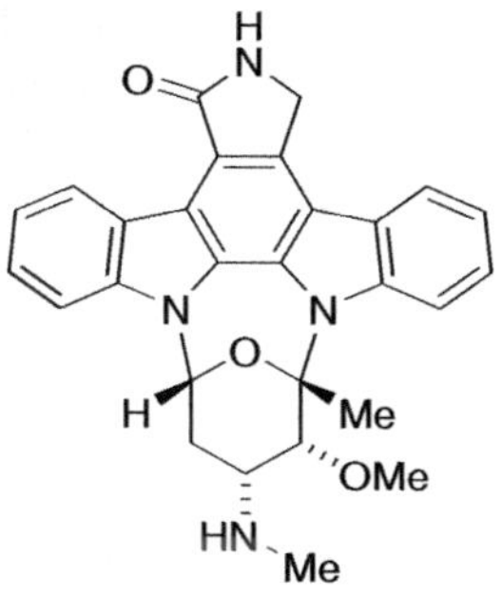

Staurosporine (**137**)

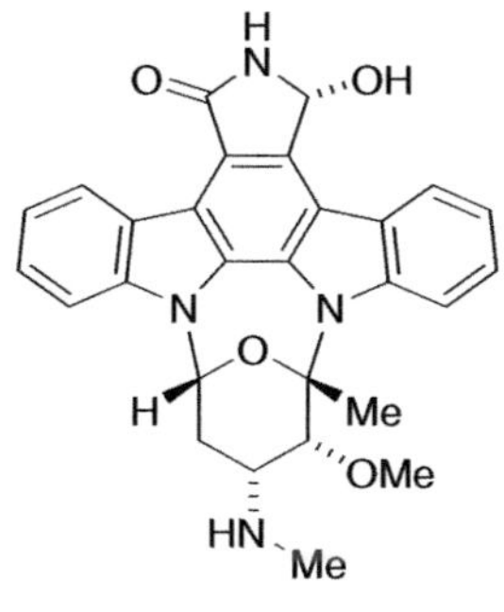

UCN-01 (**138**)

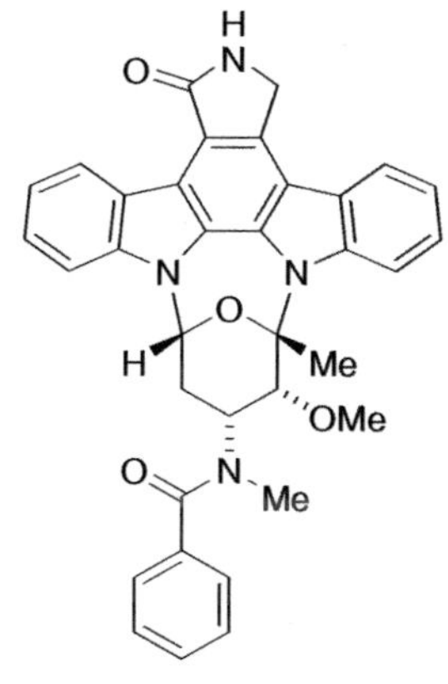

Midostaurin (**139**)

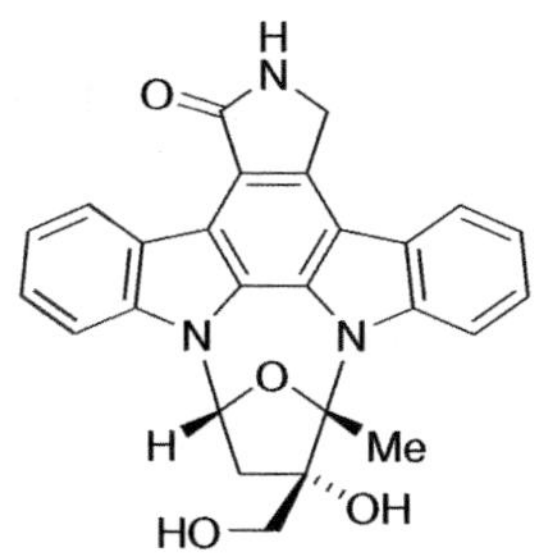

Lestaurtinib (**140**) [158]

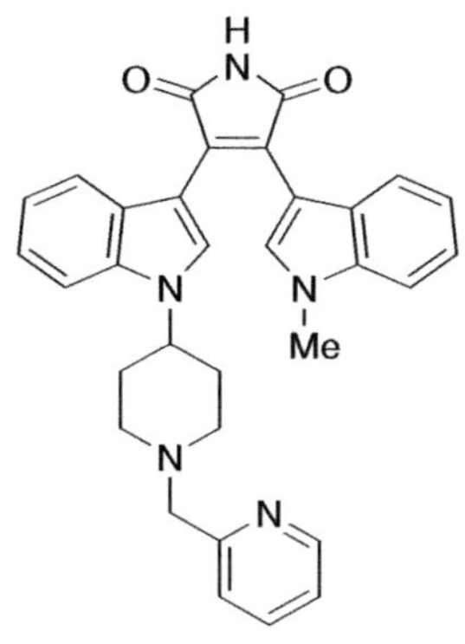

Enzastaurin (**141**) [159]

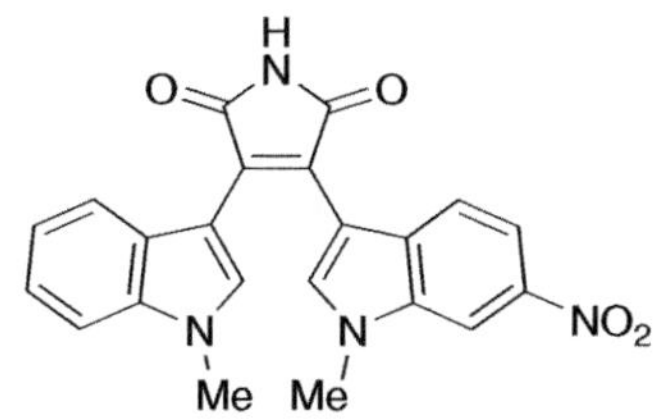

MKC-1 (**142**) [160]

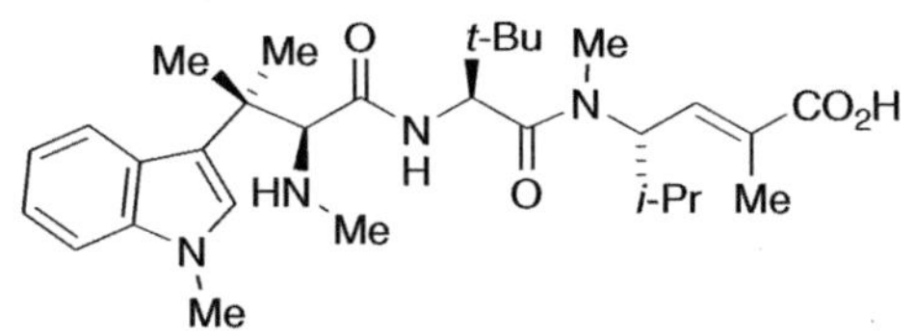

Hemiasterlin (**143**) [161]

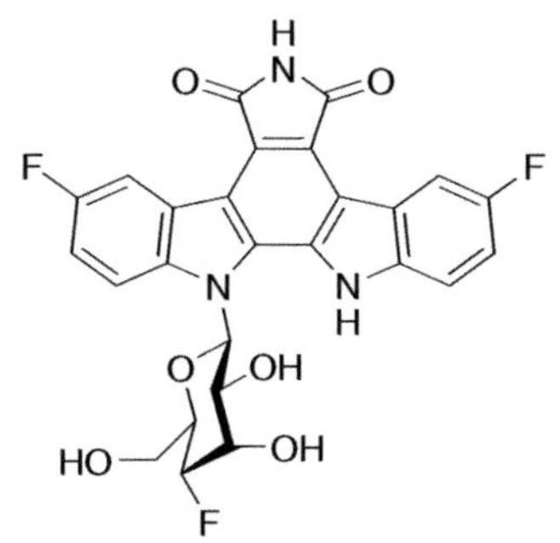

BMS-250749 (**144**) [163]

Scheme 20 *Anticancer Indolocarbazole and Related Drugs in Clinical Trials*

multiple tyrosine kinase inhibitor and is in phase III trials for several cancers [150, 158]. Enzastaurin (**141**) is an acyclic bisindolylmaleimide that is highly potent and selective toward protein kinase C β isoform (PKC β) and is currently in phase III trials for large B-cell lymphoma and in phase II trials for glioblastoma [150, 159].

Another bisindolylmaleimide, MKC-1 (**142**), induces cancer cell death (apoptosis) by targeting tubulin. This agent demonstrated activity in phase II trials against breast, ovarian, and non–small cell lung cancer and leukemia, but it was subsequently withdrawn from further study [150, 160]. Another tubulin-binding indole is the tripeptide hemiasterlin (**143**), which was isolated from the South African sponge *Hemiasterella minor* [161]. Although toxic, **143** has served as a lead compound for nonindole analogues in clinical trials. By binding to tubulin, hemiasterlin and its derivatives prevent tubulin polymerization and cause cell death [162]. The difluoro indolocarbazole BMS-250749 (**144**) is a selective topoisomerase I (topo I) inhibitor and has entered phase I trials as a broad-spectrum anticancer agent, including showing curative antitumor activity against Lewis lung carcinoma [163].

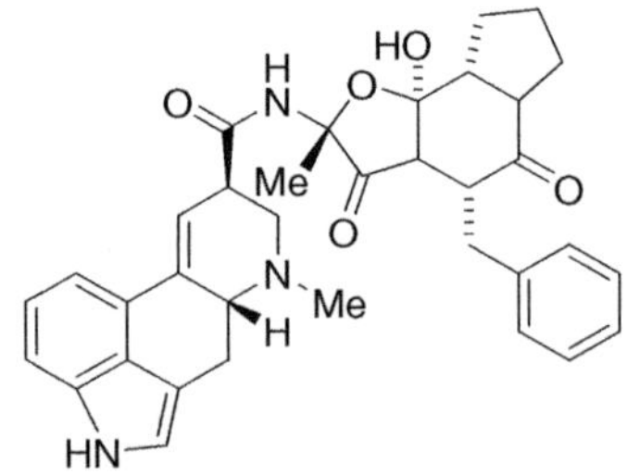

Ergotamine (Ergostat™) (**145**)

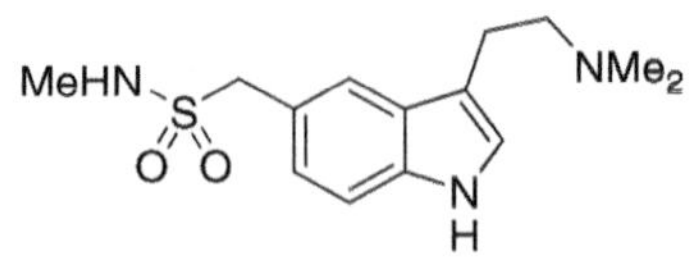

Sumatriptan (Imitrex™) (**146**)

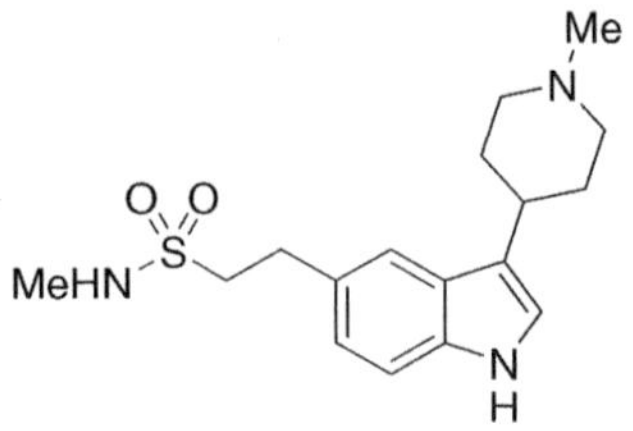

Naratriptan (Amerge™) (**147**)

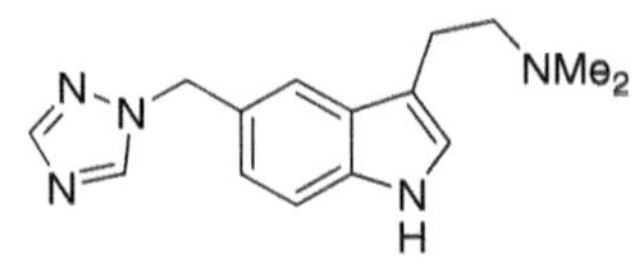

Rizatriptan (Maxalt™) (**148**)

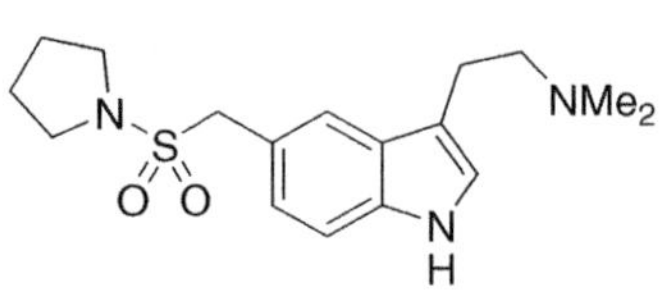

Almotriptan (Axert™) (**149**)

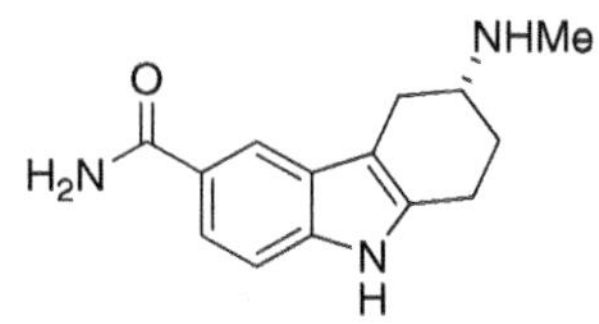

Frovatriptan (Frova™) (**150**)

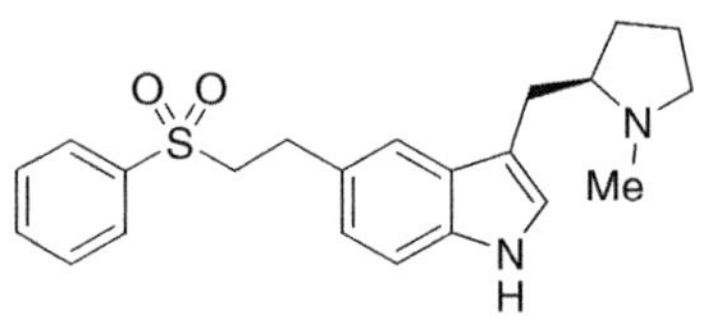

Eletriptan (Relpax™) (**151**)

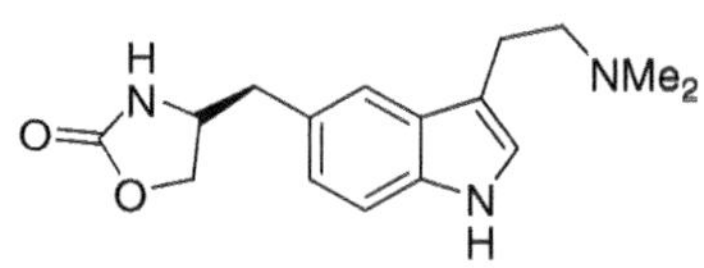

Zolmitriptan (Zomig™) (**152**)

Scheme 21 *Indole Triptans*

Several other diseases are prevented or managed using indole-based drugs. For example, the triptans are powerful antimigraine medicines, and several are known (Scheme 21) [150, 164]. These compounds are selective 5-$HT_{1B/10}$ agonists, and the ergot alkaloid ergotamine (**145**) provided a lead for the subsequent development of sumatriptan (**146**), naratriptan (**147**), rizatriptan (**148**), almotriptan (**149**), frovatriptan (**150**), eletriptan (**151**), and zolmitriptan (**152**). Sumatriptan was the first triptan to be approved in the United States (1991) as a drug to treat migraine.

Drugs that lower blood cholesterol by retarding its synthesis in the liver are called statins [165] and are epitomized by atorvastatin (Lipitor; a pyrrole, not an indole). These drugs function by inhibiting the enzyme hydroxymethylglutaryl-coenzyme A (HMG-CoA), crucial for a step in the cholesterol biosynthesis pathway. The one indole statin is fluvastatin (**153**), a minor player in this arena [165] (Scheme 22). Tadalafil (Cialis) (**154**) is a drug for the treatment of erectile dysfunction by inhibiting type 5 phosphodiesterase (PDE5) and relaxing smooth muscle to increase blood flow [150]. The 5-HT_4 agonist tegaserod (**155**) is a drug for irritable bowel syndrome and constipation [166]. The relatively new antipsychotic sertindole (**156**) acts against the dopamine D_2, serotonin 5-HT_2, and α_1-adrenergic receptors and is a drug for the treatment of schizophrenia [150]. The PKC inhibitor sotrastaurin (**157**) is under consideration as a transplant rejection preventive [167]. This indole acts synergistically with cyclosporine, a well-established immunosuppressive drug.

As mentioned earlier in Section 3, serotonin (**3**) activates a series of well-defined receptors in the central and peripheral nervous systems [81, 82]. The 5-HT_3 receptor comprises a

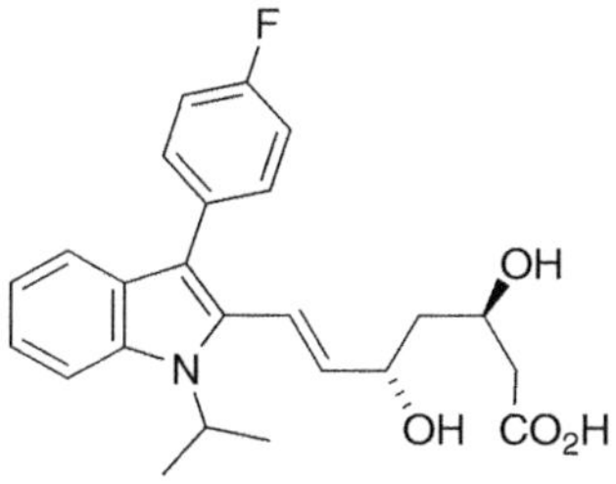

Fluvastatin (**153**) [165]

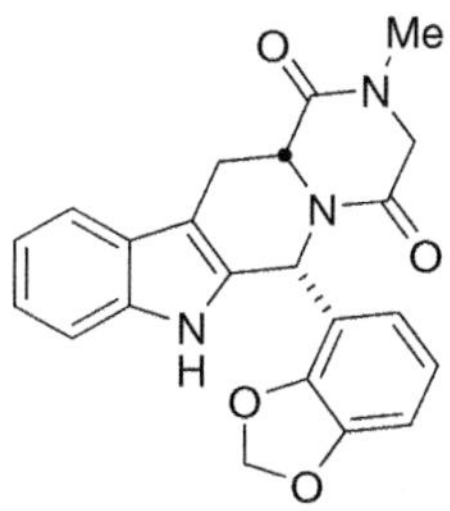

Tadalafil (**154**) [150]

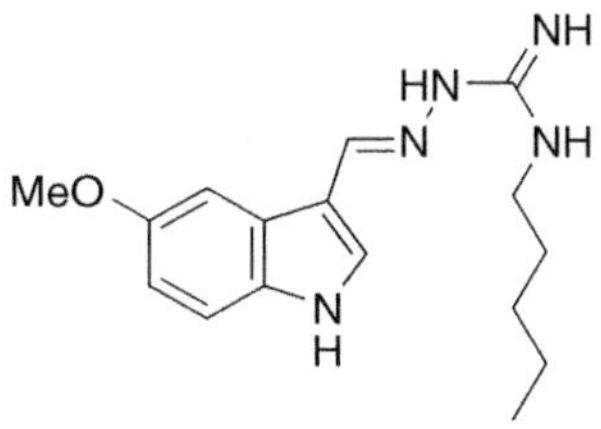

Tegaserod (**155**) [166]

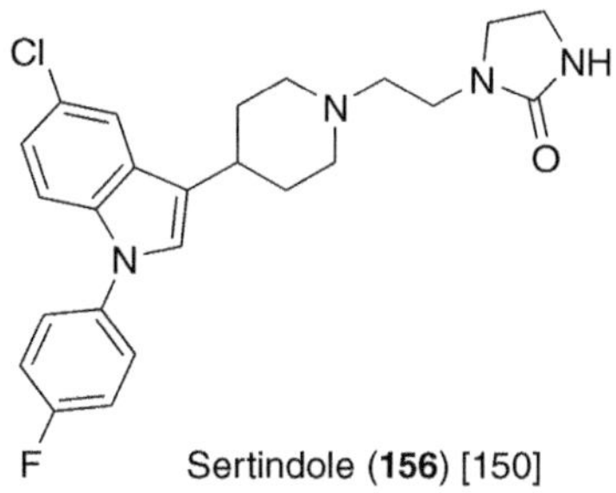

Sertindole (**156**) [150]

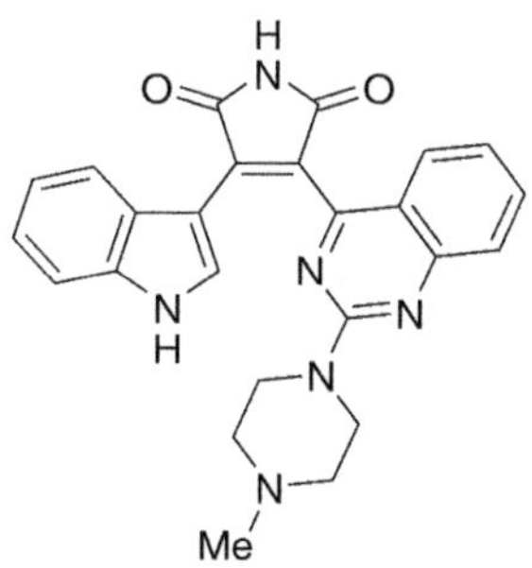

Sotrastaurin (**157**) [167]

Scheme 22 *Miscellaneous Indole Pharmaceuticals*

superfamily of ion channels that regulate neuronal depolarization by increasing sodium, potassium, and calcium ion flux [168]. Several indole pharmaceuticals have been developed as antiemetics to prevent nausea and vomiting following chemotherapy (Scheme 23)[150]. They function by diminishing the role of the vagus nerve to trigger vomiting in the medulla oblongata, and they also obstruct serotonin receptors in the chemoreceptor activation region. One of these 5-HT_3 antagonists, alosetron (**162**), is used to treat severe diarrhea in women with irritable bowel syndrome (IBS).

A few indole-based compounds are active against viruses (Scheme 24). The non-nucleoside reverse transcriptase inhibitor delaviridine (**163**) is used in antiretroviral therapy [150]. The blue-green alga *Dichothrix baueriana* has yielded bauerine B (**164**), a dichloro β-carboline active against herpes simplex virus type 1 [169]. Seeds of the plant *Solanum indicum* have yielded the new coumarinolignoid indicumine B (**165**), which possesses anti–hepatitis B activity [170]. Eudistomin K sulfoxide (**166**) displays activity against both herpes and polio viruses. This β-carboline was isolated from the New Zealand ascidian *Ritterella sigillinoides* [171]. Other marine indoles such as dragmacidin F, manzamine A, and microspinosamide have antiviral activity (HIV, herpes). The marine environment is a very promising source for new antiviral lead compounds [172].

Several natural and synthetic indole-containing cyclic peptides have pharmacological properties (Scheme 25). Daptomycin (**167**) was isolated from the soil microbe *Streptomyces roseosporus* and is used to treat gram-positive bacterial infections [150, 173]. The synthetic cyclic peptide eptifibatide (**168**) is a glycoprotein IIb/IIIa antagonist and platelet aggregation inhibitor. This drug is used to treat patients who have acute coronary syndrome (ACS) and are especially susceptible to blood clots [150,174].

Several other human diseases are favorably affected by indole-based drugs (Scheme 26). Vilazodone (**169**) is a

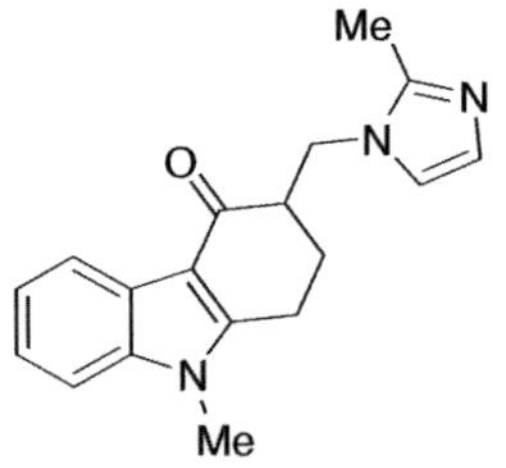

Ondansetron (Zofran™) (**158**)

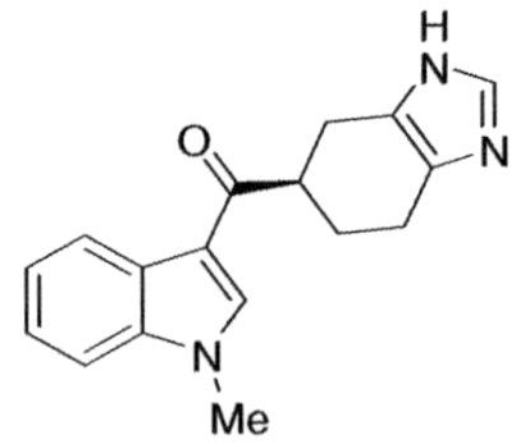

Romosetron (Nasea™) (**159**)

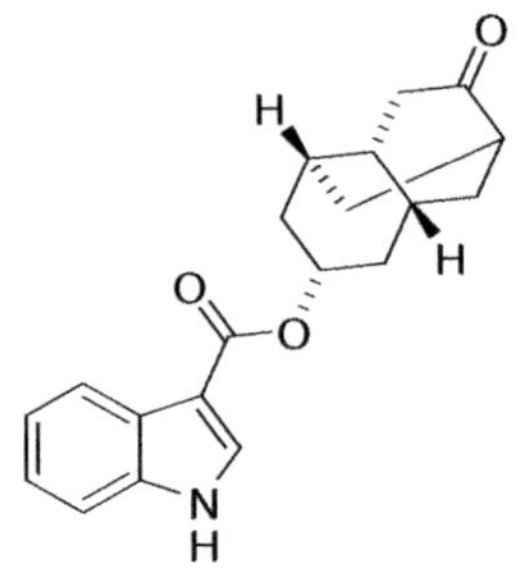

Dolasetron (Anzemet™) (**160**)

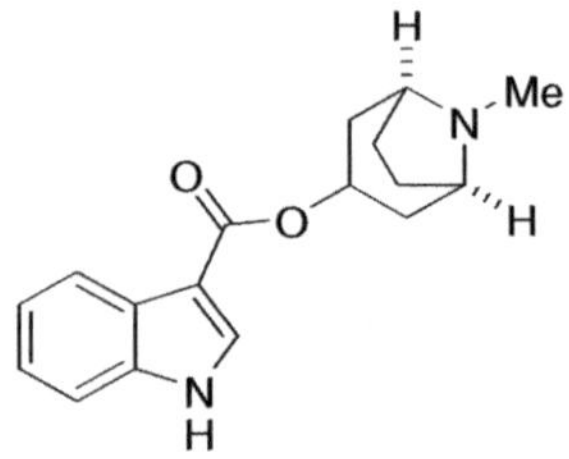

Tropisetron (Navoban™) (**161**)

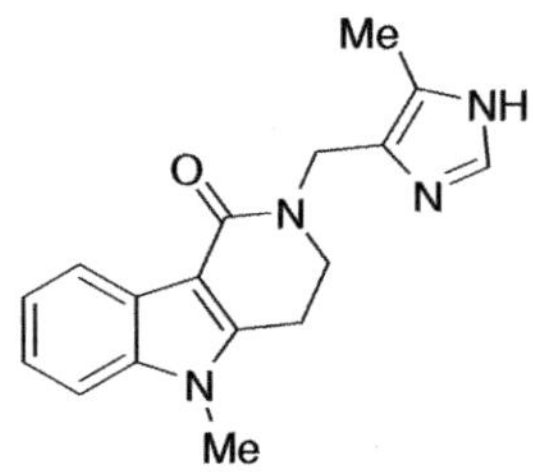

Alosetron (Lotronex™) (**162**)

Scheme 23 *Indole 5-HT$_3$ Receptor Antagonist Pharmaceuticals*

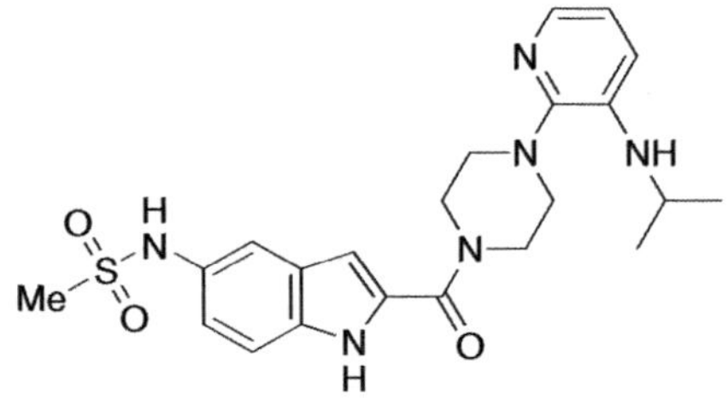

Delaviridine (**163**) [150]

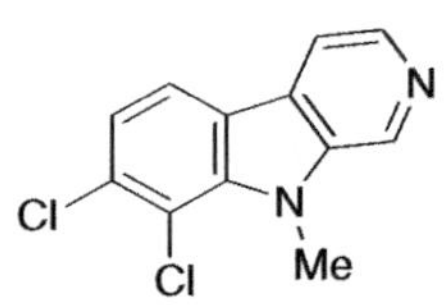

Bauerine B (**164**) [169]

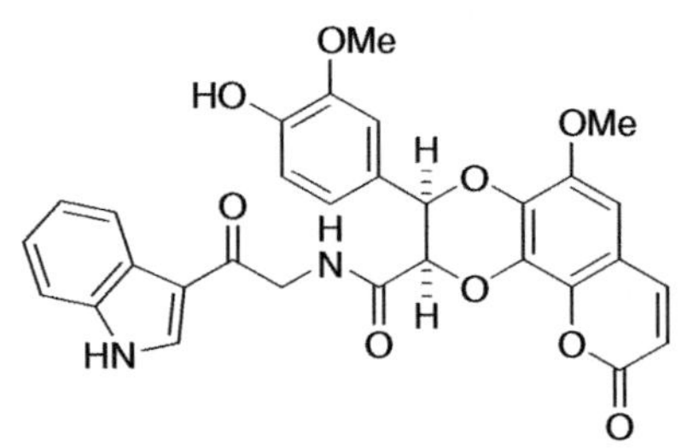

Indicumine B (**165**) [170]

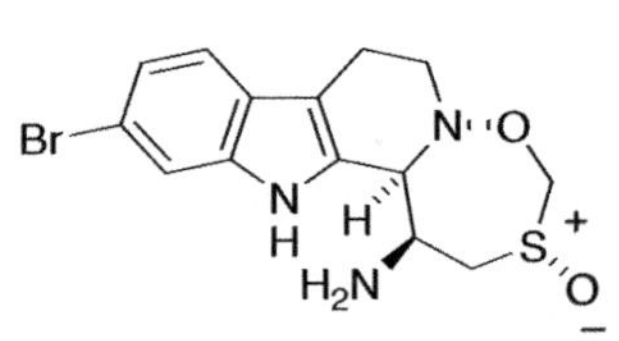

Eudistomin K Sulfoxide (**166**) [171]

Scheme 24 *Representative Antiviral Indoles*

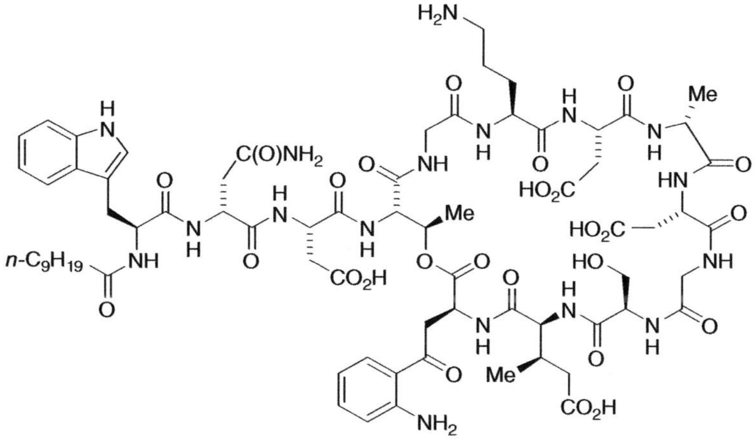

Daptomycin (Cubicin™) (**167**) [173]

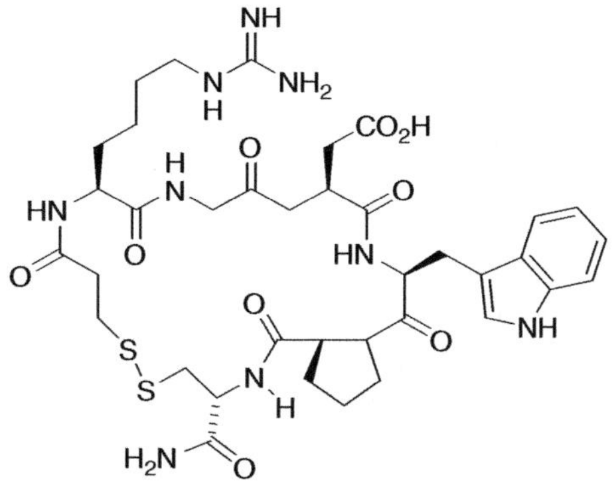

Eptifibatide (Integrilin™) (**168**) [174]

Scheme 25 *Indole-Based Cyclopeptides*

long-acting 5-HT_{1A} partial agonist and selective serotonin reuptake inhibitor (SSRI) that is in phase III trials for the treatment of depression [175]. The highly specific 5-HT_2 receptor antagonist pruvanserin (**170**) is in phase II clinical trials for the treatment of insomnia [150]. Tiplaxtinin (**171**) is a potent and selective inhibitor of plasminogen activator inhibitor 1 (PAI-1) and has efficacy in preventing acute arterial thrombosis [176]. The cyclic bisindolylmaleimide ruboxistaurin (**172**) is an inhibitor of PKCβ and is under consideration for the treatment of diabetic retinopathy [177]. An inhibitor of human nonpancreatic secretory phospholipase A2 (hnpoPLA2), LY-311727 (**173**), is a potential treatment for sepsis [178], and the M_4 muscarinic antagonist **174** may find utility against Parkinson's disease [179]. The indole thiophene DG-041 (**175**) is an antagonist for the EP3 receptor for prostaglandin E_2 and is in clinical studies for the treatment of peripheral artery disease (PAD) [180]. Lastly, one of the most familiar indole drugs is indomethacin (**176**), which is a classic powerful nonsteroidal antiinflammatory drug. It has also found use in controlling fever in patients with liver metastases of solid tumors [181].

This brief survey of indole pharmaceuticals and indole candidate drugs is necessarily incomplete. Many more indoles are known that have excellent biological activity, but space does not permit coverage. For example, myriad indole alkaloids have a range of biological activities, and some, like the antihypertensive reserpine, are clinically useful. In addition, the many indoline- and oxindole-based pharmaceuticals could not be included in this chapter.

1.5 Indole-Containing Materials

The enormous resurgence in recent years of the chemistry of materials has had a significant indole component, and several applications of indole-based materials are presented. Section 6 covers indole-designed ligands.

A major thrust in the area of indole materials has been anion complexation and sensing by indole-, carbazole-, biindole-, and indolocarbazole-designed receptors. An excellent feature article by Gale [182] precisely covers this literature to 2008, so the treatment here begins with 2009. d'Ischia

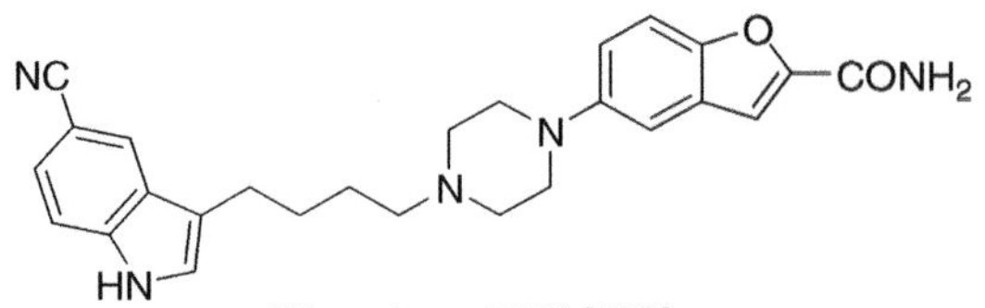

Vilazodone (**169**) [175]

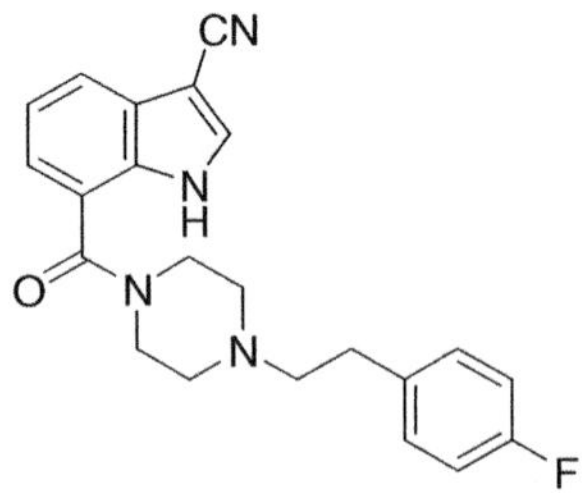

Pruvanserin (**170**) [150]

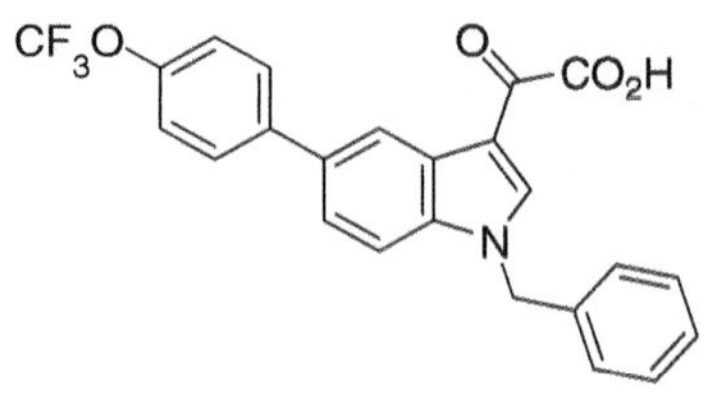

Tiplaxtinin (**171**) [176]

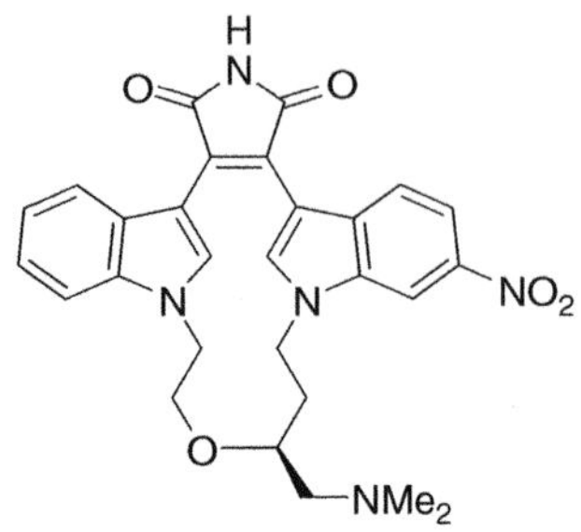

Ruboxistaurin (Arxxant^{TM}) (**172**) [177]

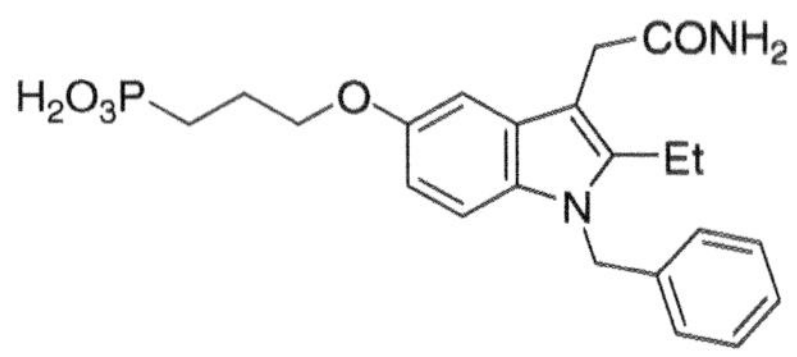

LY-311727 (**173**) [178]

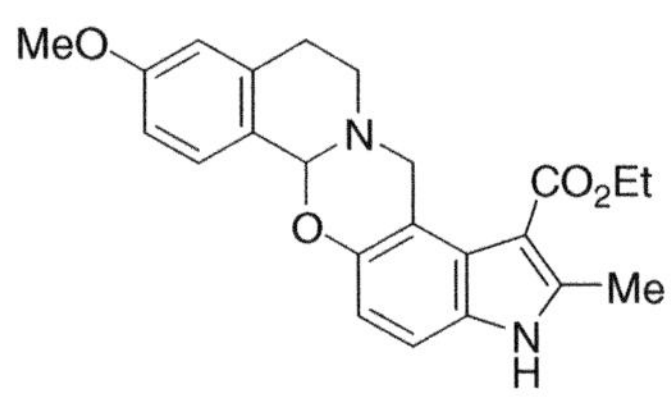

PD-102,807 (**174**) [179]

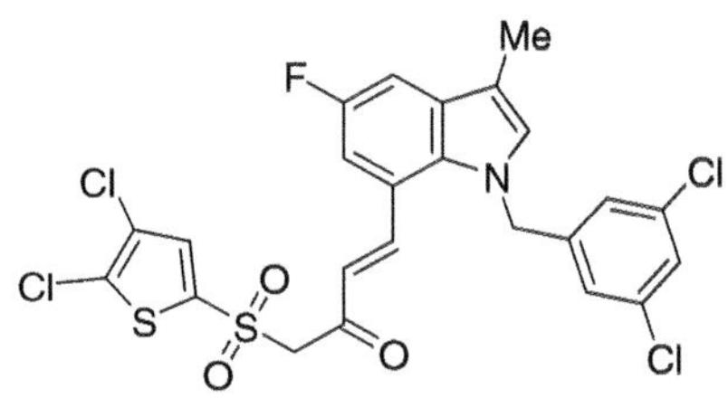

DG-041 (**175**) [180]

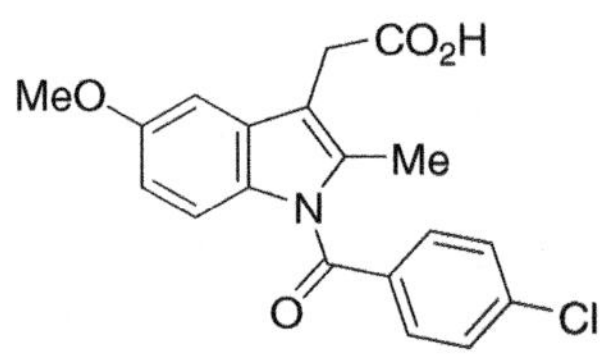

Indomethacin (**176**) [181]

Scheme 26 *Other Indoles with Pharmacological Activity*

and colleagues have synthesized the novel acetyl trimer **177** by acid-promoted trimerization of 5,6-dihydroxyindole followed by acetylation [183]. Compound **177** is a selective fluoride-sensing compound, as illustrated in **177–178** (Scheme 27). This fluoride-sensing scaffold displays marked fluorescence enhancement at 489 nm that is selective for fluoride. Other anions, including chloride, bromide, iodide, acetate, nitrite, and bisulfate, show no significant changes in the fluorescence spectra. A different fluoride-sensing system was described by Shiraishi involving indole-azadiene **179–180**, in a H-bonding 1:2 stoichiometry [184]. This system is highly sensitive for fluoride in both colorimetric and fluorometric analyses, but not for chloride, bromide, iodide, acetate, perchlorate, bisulfate, biphosphate, nitrate, and thiocyanate. Indole **179** was readily prepared by condensing indole-3-carbaldehyde with hydrazine.

Jeong and colleagues have synthesized the biindole–diazo conjugates **181** and **182** as new anion sensors (Scheme 28) [185].

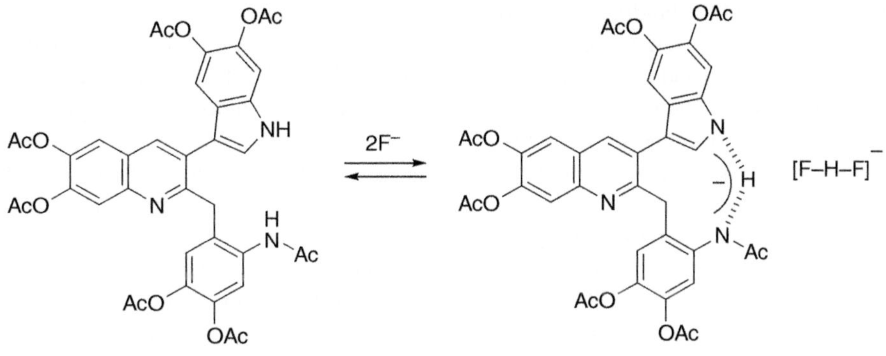

177 **178**

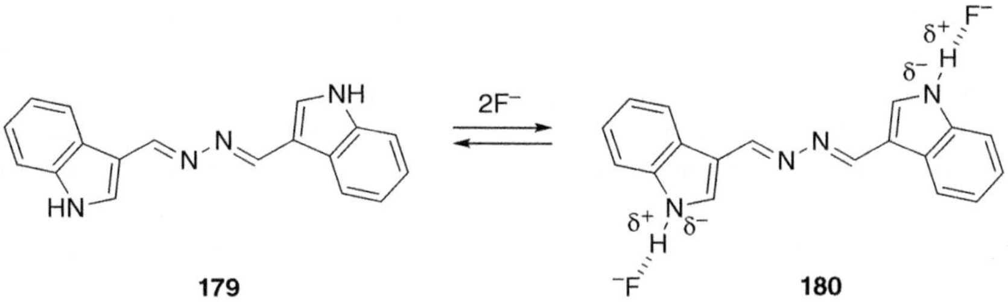

179 **180**

Scheme 27 *Indole Fluoride Sensors*

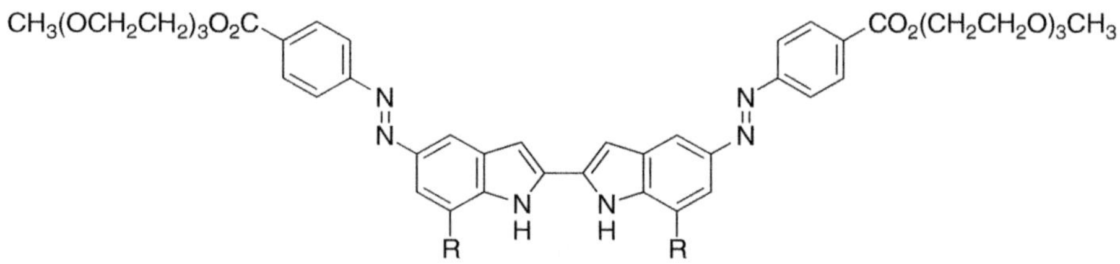

181, R = I

182, R = ≡— C(OH)$(CH_3)_2$

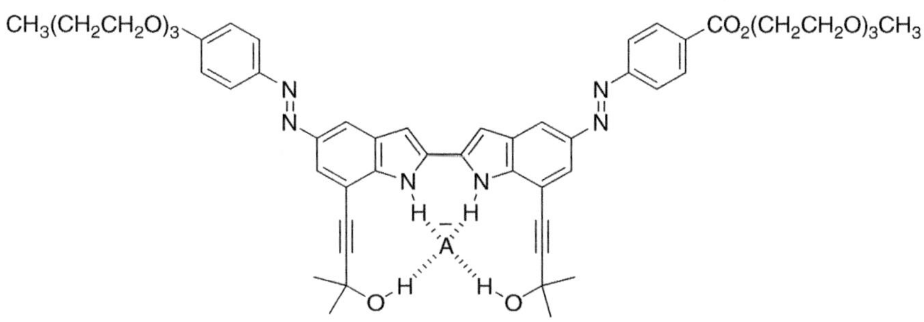

183

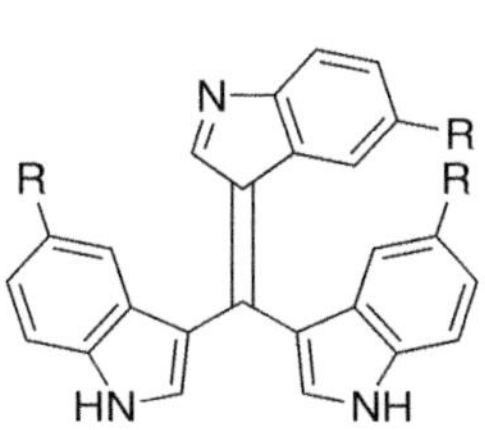

184, R = H
185, R = OMe
186, R = NO_2

Scheme 28 *Indole Anion Sensors*

Both compounds bind basic anions, but the binding of anions with **182** is up to 100 times stronger given the extensive hydrogen bonding as shown in **183**. Shao and colleagues prepared tris(indolyl)methanes **184–186** as anion receptors and colorimetric chemosensors as a function of R [186]. Thus, whereas all three systems show strong binding of fluoride, **184** and **185** show much higher binding affinity for biphosphate than for acetate, and **186** shows the strongest binding for fluoride. Other anions such as chloride, bromide, iodide, perchlorate, and bisulfate are much less bound to **186**.

The novel ion-pair indole-pyridine-amine–based receptor **187** binds both anions and cations [187]. For example, both lithium and chloride bind simultaneously to this molecule as shown in **188** and **189** (Scheme 29). The authors propose for chloride binding a weak H-bond to the C-3H on pyridine in addition to two strong H-bonds to the two indole nitrogen hydrogens. Hof and Whiting have developed a series of indole-derived hosts that bind various ammonium cations (e.g., **190**, **191**). Binding results from a combination of electrostatic attraction, cation–π interaction, and hydrophobicity [188]. Host **190** was designed as a mimic of tryptophan-endowed protein-binding cavities.

Several novel indole-formulated fluorescence devices have been described (Scheme 30). Müller has developed metal-selective luminescence sensors **192** that exhibit selective halochromic fluorescence of zinc and magnesium over calcium [189]. Whereas these 2,4-diarylpyrano[2,3-*b*]indoles display no fluorescence as the free base, halochromic green fluorescence results upon protonation, methylation, or selective metal cation complexation [189]. Ahn and colleagues synthesized LipidGreen (**193**), a fluorescent probe for *in vivo* lipid imaging in zebrafish. LipidGreen stains lipid droplets in 3T3L1 preadipocyte cell lines and zebrafish fat deposits. It also can be employed as a lipid marker in drug screening [190]. Manderville and colleagues show that indole-deoxyguanosine **194** is a fluorescent reporter of hydrogen-bonding specificity. Thus, the fluorescence in **194** is quenched on Watson–Crick hydrogen bonding to deoxycytosine, but the fluorescence is amplified upon hydrogen bonding to guanosine in Hoogsteen base pairing. The fluorescence of indole **194** is about ten times brighter than that of the corresponding pyrrole–deoxyguanosine [191]. Yao has prepared several novel pyrrolocoumarin fluorescent dyes

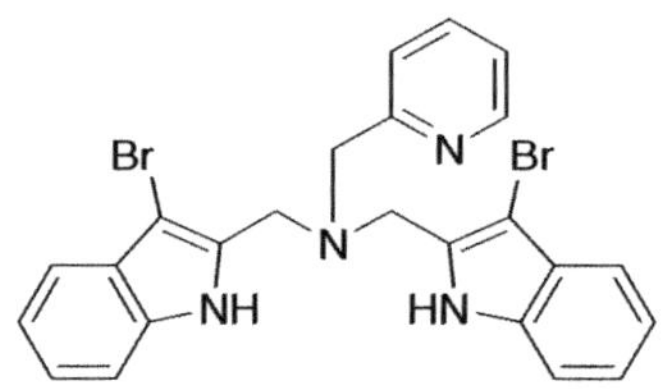

187 [187]

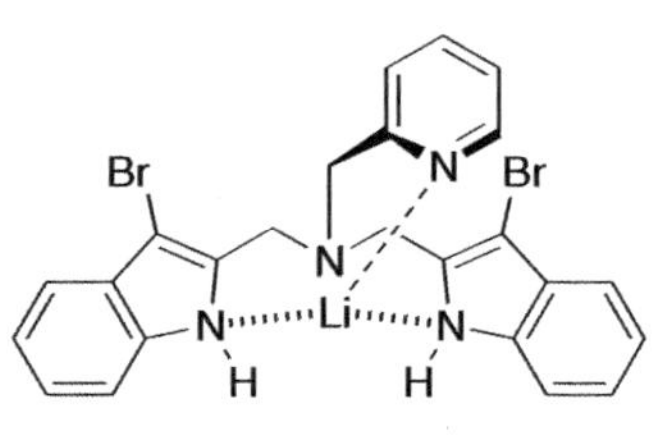

188 [187]

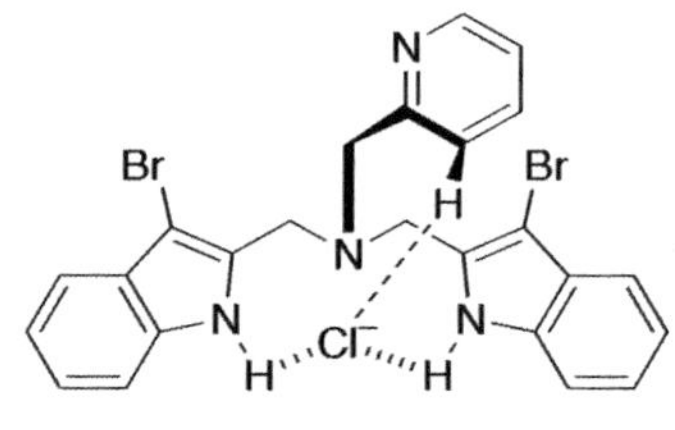

189 [187]

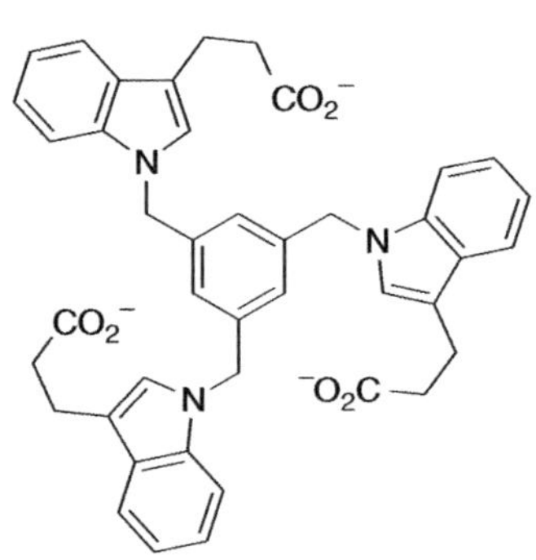

190 [188]

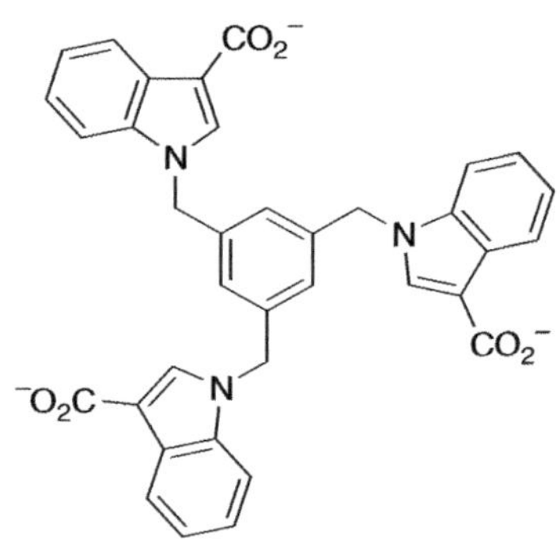

191 [188]

Scheme 29 *Indole-Based Receptors*

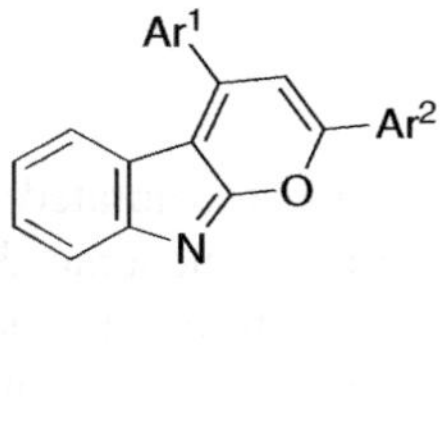

192 [189]

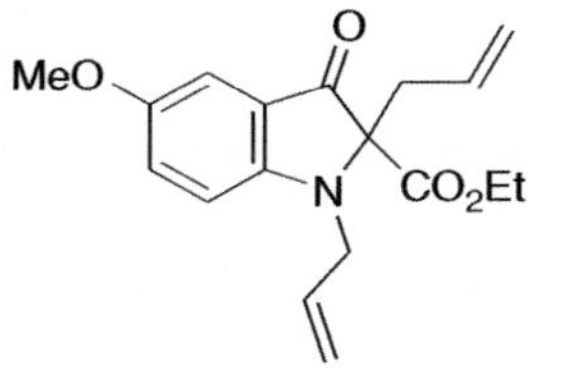

LipidGreen (**193**) [190]

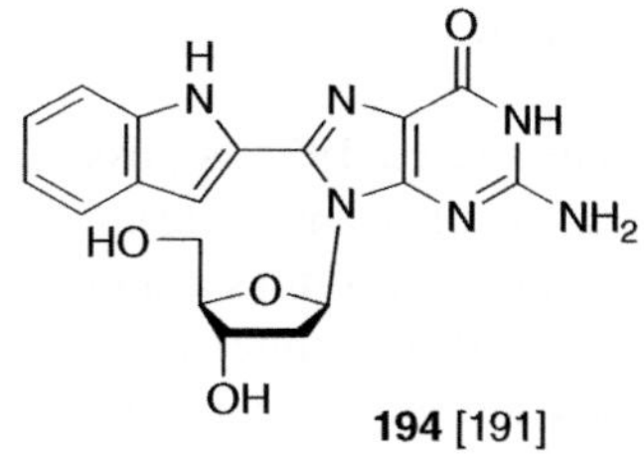

194 [191]

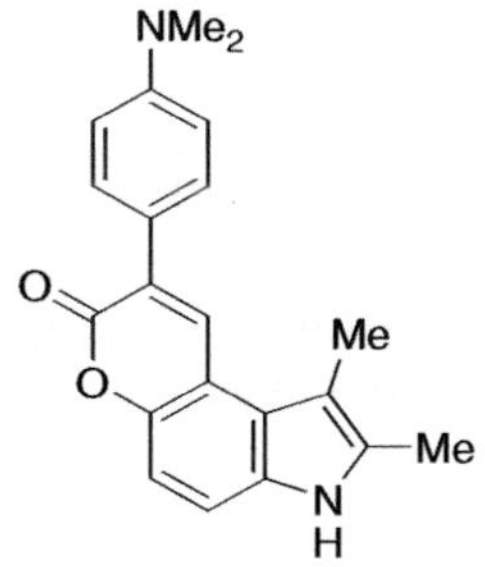

195 [192]

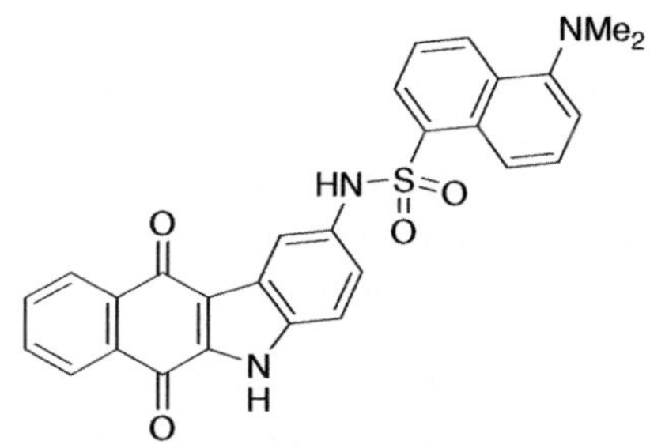

196 [193]

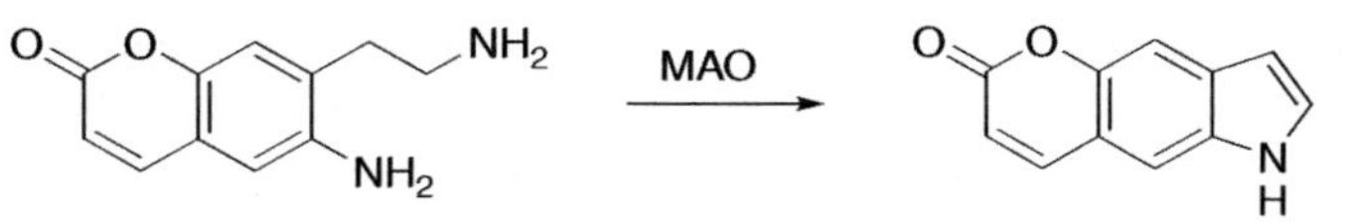

197 [194] **198**

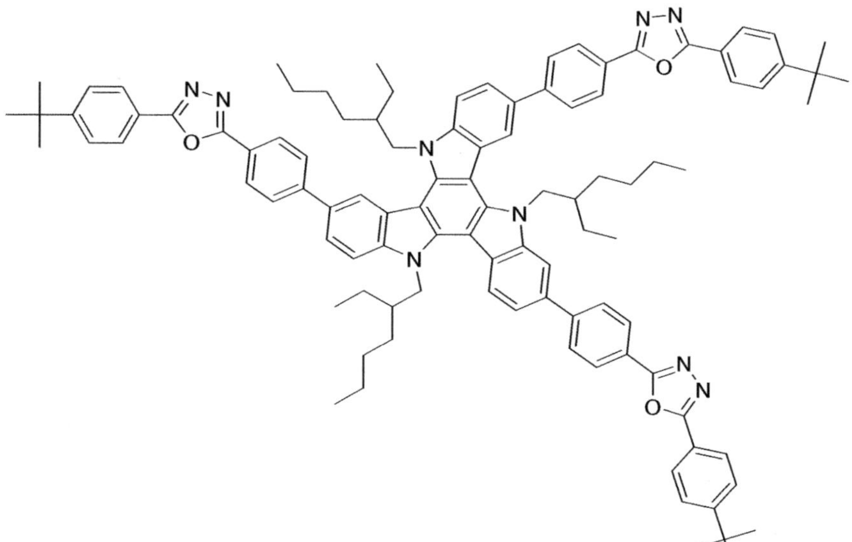

199 [195]

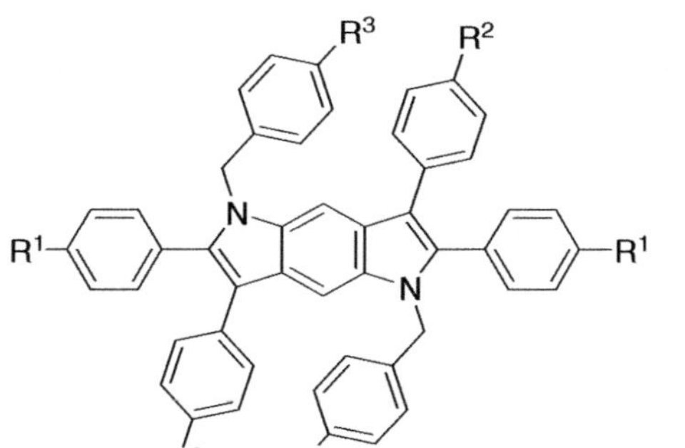

200 [196], R^1 = H, NPh_2; R^2 = H, NPh_2, CN;
R^3 = *n*-hexyl

Scheme 30 *Indole-Based Fluorescence Probes*

with large Stokes shifts (e.g., **195**) as possible biological imaging agents [192]. Bittner and colleagues have developed benzocarbazole quinone **196** as a fluorescent molecular switch (quinone–hydroquinone) for possible use in molecular electronics and recognition. The switching can be induced both chemically and electrochemically [193]. Sames designed a fluorescent reporter for monoamine oxidase (MAO) enzymes. Thus, nonfluorescent aminocoumarin **197** is switched (oxidized) to fluorescent indolocoumarin **198** upon exposure to MAO A and B [194]. Hiyoshi and colleagues prepared the donor-π-acceptor 1,3,4-oxadiazole triindole **199**. This material has good film-forming capability and shows promise as an electroluminescence device in a single-layer OLED (organic light-emitting diode) [195]. Another candidate molecule for an OLED is benzodipyrrole **200**, which was synthesized by Nakamura and represents a new class of organic electronic materials [196].

Indole-containing materials are used as organic field-effect transistors (OFET) (Scheme 31). Wudl and colleagues synthesized dinaphthocarbazoles (azaheptacenes) **201** and used them in solution-processed OFETs. These azaheptacenes are more robust than the standard heptacene and other polyacene derivatives [197]. Pei and colleagues find that the fluoro-substituted isoindigo polymer **202** has both improved stability and high performance over the nonfluorinated analogue. Fluorination lowers the LUMO level of the polymer and greatly amplifies electron mobility in this novel isoindigo-crafted donor-acceptor conjugated polymer [198]. The venerable natural dye indigo (**203**) has found use as an OFET by Irimia-Vladu and colleagues. By virtue of its planar structure, cross-conjugated π-system, and strong intermolecular interactions, indigo possesses excellent charge-transporting properties. Furthermore, the property of this compound to be oxidized and reduced reversibly provides transport ambipolarity [199]. Yao found that indigo carmine (**204**) is a positive-electrode material for rechargeable lithium batteries [200].

Wang, Andersson, and colleagues designed and synthesized the novel isoindigo-terthiophene copolymer **205** for use as a donor-acceptor polymer in a high-performance solar cell (Scheme 32). The 6.3% power conversion efficiency observed for **205** is the highest yet reported for an isoindigo-based polymer solar cell [201]. Fulgimide **206** undergoes photochromic fluorescence switching

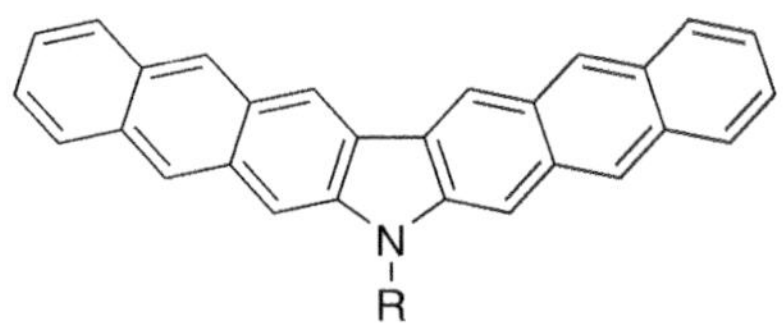

201 [197], R = benzyl, *n*-dodecyl

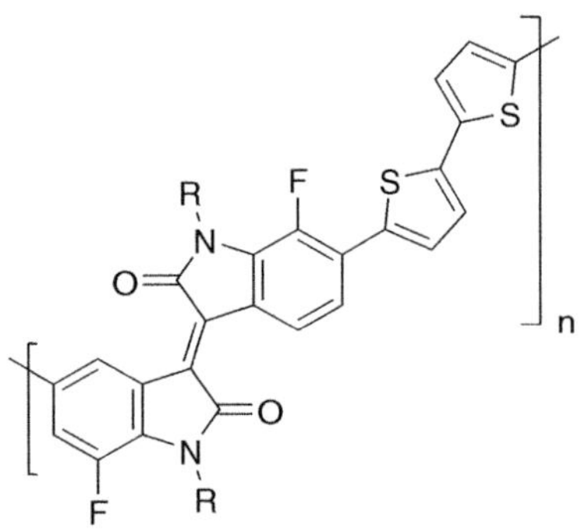

202 [198], R = $(CH_2)_3CH(C_{14}H_{29})_2$

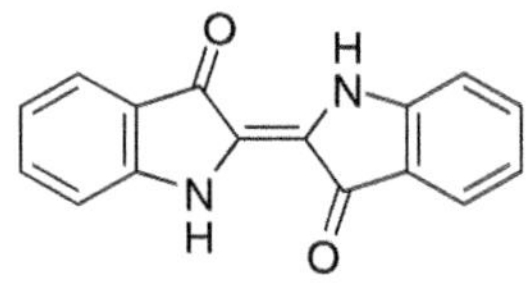

Indigo (**203**) [199]

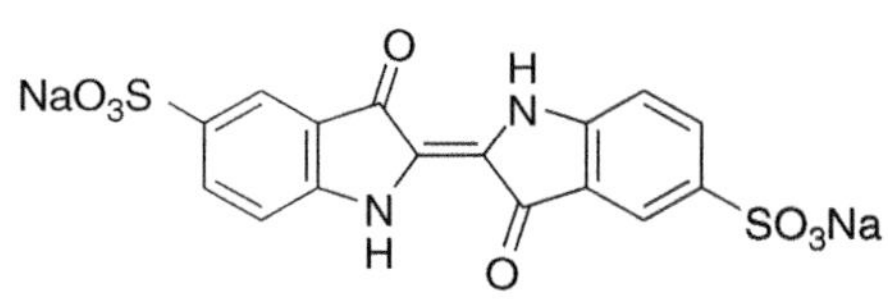

204 [200]

Scheme 31 *Indole-Based Transistors*

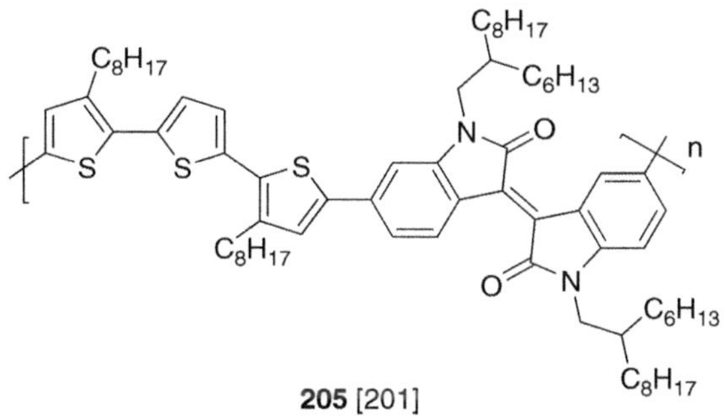

205 [201]

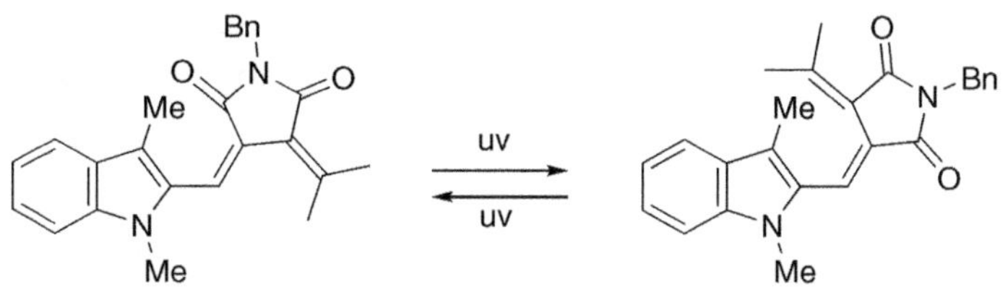

206 [202] **207**

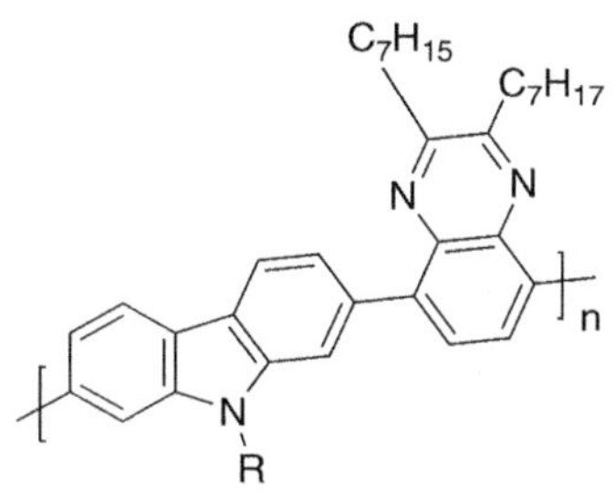

PCQ (**209**) [204], R = 2-ethylhexyl

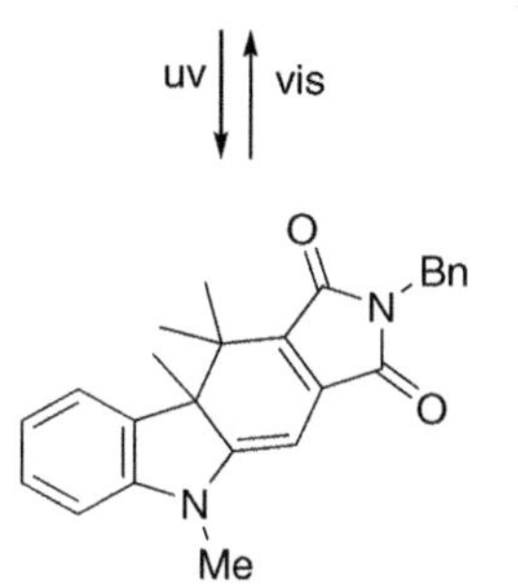

208

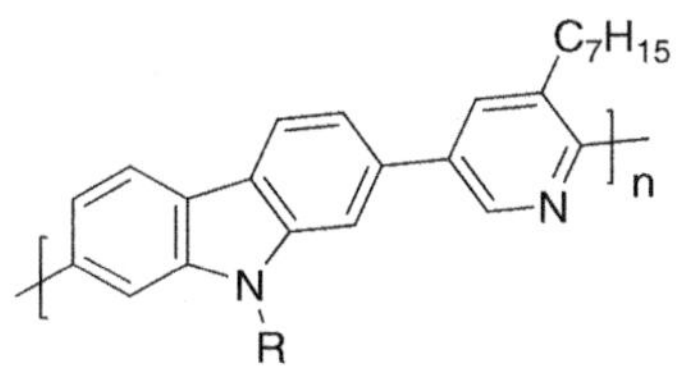

PCPy (**210**) [204], R = 2-ethylhexyl

Scheme 32 *Indole Based Electronic Devices*

(**206–207–208**) as shown by Andréasson and Pischel [202]. Leclere and colleagues have devised new blue, green, and red light-emitting conjugated poly(*N*-substituted-2,7-carbazole) derivatives. For example, PCQ (**209**) emits green light and PCPy (**210**) emits blue light. These materials are thermally stable, show good optical properties, and will soon be tested in light-emitting devices [203, 204].

Emrick and Ryu have used the thermolysis of diphenyl-1,2,3-triazoles to give phenylindoles as novel monomers leading to polymers **211** → **212** that possess low flammability (Scheme 33) [205]. Franceschin and colleagues prepared azatrux (**213**) as a selective binder to G-quadruplex DNA [206]. This family of nucleic acid secondary structures is stabilized by coplanar quartets of guanines, called G-quadruplexes, held together by hydrogen bonding (Hoogsteen bonding). Binding to G-quadruplex DNA may offer a novel anticancer strategy by forming telomeric G-quadruplexes, thus blocking telomerase, an enzyme that is overexpressed in most cancer cells and whose substrate is telomeric single-strand DNA. Indole has been incorporated into DNA as an artificial DNA base, as illustrated for the indole nucleoside of β-2′-deoxyribofuranoside **214** as well as the tetered analogues **215** and **216**. All three modified

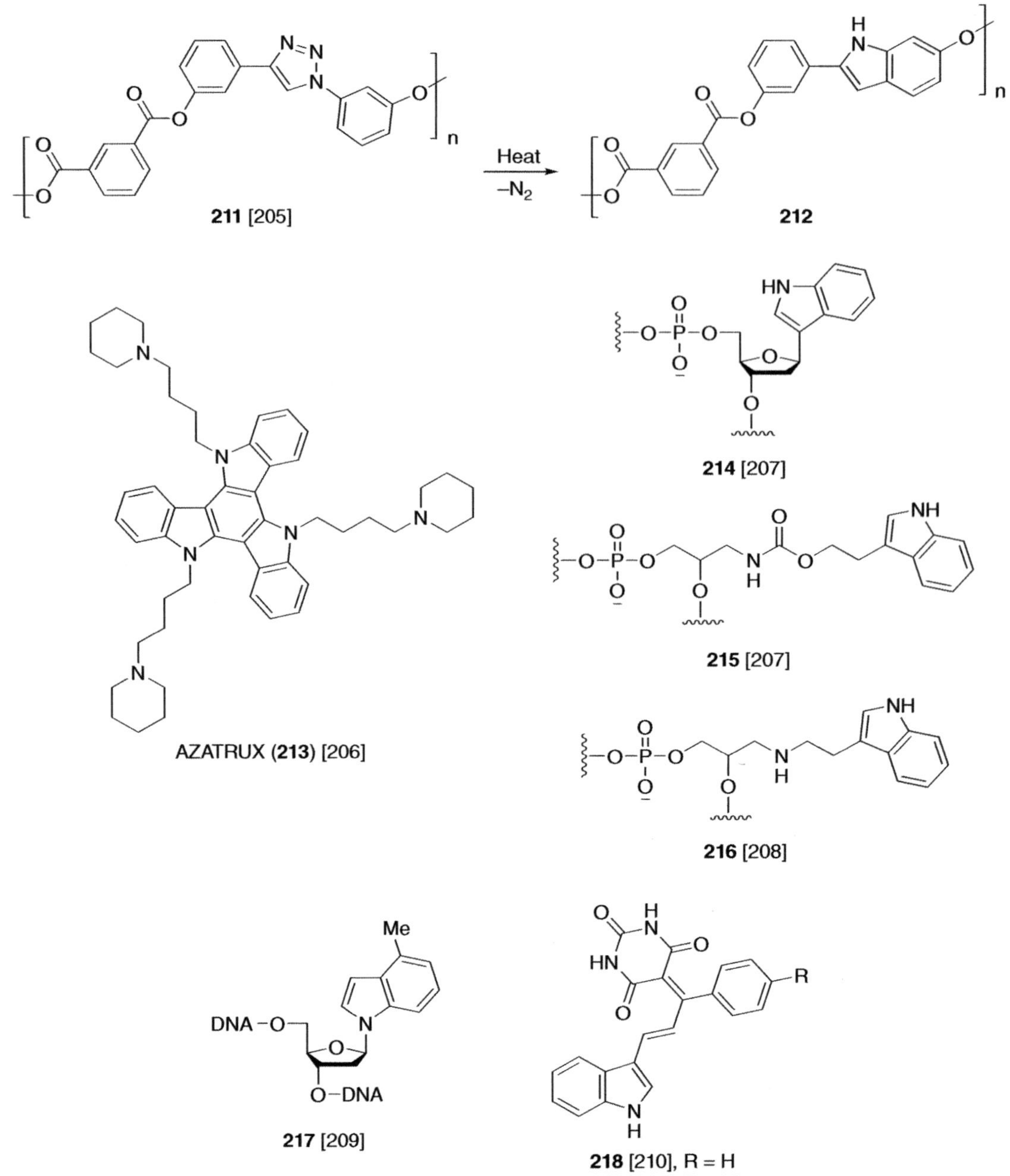

218 [210], R = H
219 [210], R = Me
220 [210], R = Cl

Scheme 33 *Miscellaneous Indole-Based Materials*

DNA duplexes are less stable than the natural DNA duplex [207, 208]. David and colleagues find that the DNA repair system adenine glucosylase MutY easily recognizes 4-methylindole β-deoxynucleoside (**217**) when it is opposite 7,8-dihydro-8-oxo-2′-deoxyguanosine and guanosine in DNA [209]. Biradar and colleagues designed the indole–barbitone compounds **218–220** and find that they exhibit antioxidant and DNA cleavage activity [210].

1.6 Indole-Containing Ligands

With the incredible increase in research devoted to metal-catalyzed coupling reactions, which we will encounter in subsequent chapters, and the necessity of employing specialized ligands for this chemistry, it is no surprise that numerous indole-based ligands have been invented. In this regard Bandini and Eichholzer published an excellent review, "Catalytic Functionalization of Indoles in a New Dimension," which emphasizes the advancements and triumphs in the selective catalytic carbon–carbon bond-forming reactions of indoles from 2005 to 2008 [211].

A very large number of indole-based ligands have been developed to facilitate palladium- and rhodium-catalyzed cross-coupling reactions. The research groups of Sannicolo, Benincori, Beller, Reek, Koskinen, Heo, Mino, Sarkar, Franzén, and Kwong have made enormous contributions in this area.

Sannicolo, Benincori, and colleagues have designed biindole-phosphine ligands BISCAP (**221**) [212] and 2-BINPO (**222**) (Scheme 34) [213]. Both are chiral, and **222** has been resolved into enantiomers. Reek and colleagues designed the phosphine–phosphoramidite ligand IndolPHOS (**223**), which is an excellent ligand for rhodium-catalyzed asymmetric hydrogenation and rhodium-catalyzed hydroformylation [214, 215]. This research team later designed INDOLPhospholes (**224**) as a novel catalyst for allylic alkylation [216]. Beller and colleagues synthesized

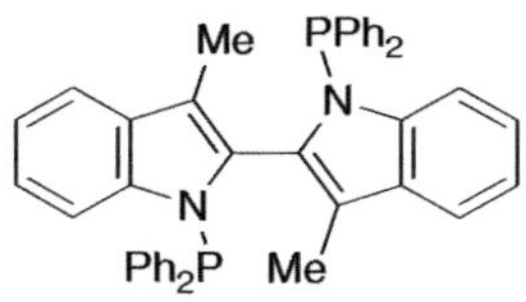

BISCAP (**221**) [212]

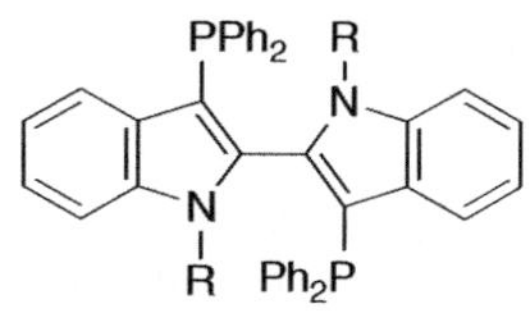

2-BINPO (**222**), R = Me, CH_2OMe [213]

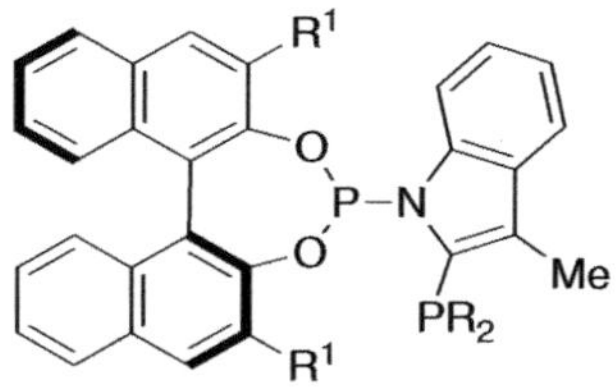

INDOLPHOS (**223**) [214], [215]

R	R^1
Ph	H
i-Pr	H
Cy	H
o-Tol	H
Ph	$SiMe_3$
i-Pr	Me

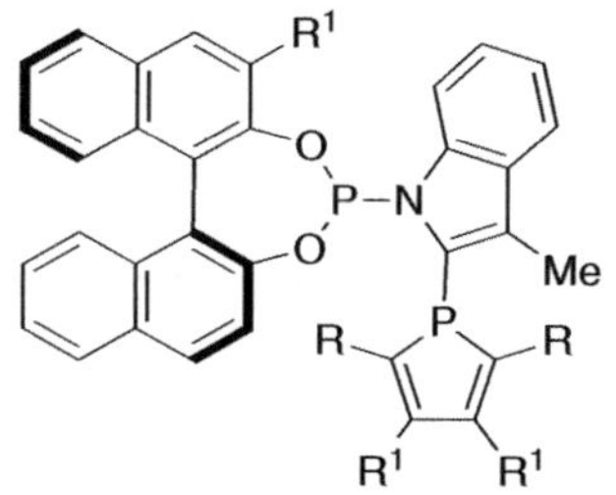

INDOLPhosphole (**224**) [216]

R	R^1
Ph	H
H	Me

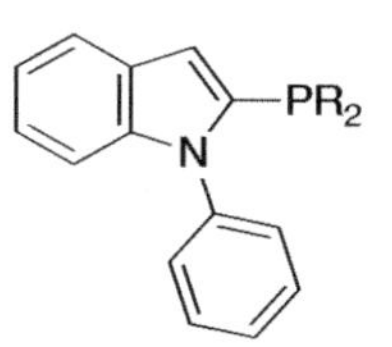

225 R = *t*-Bu, adamantyl [217]

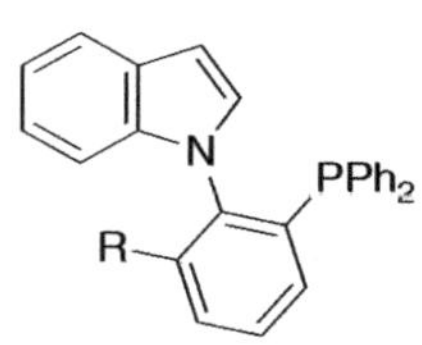

226 R = OMe, Me, CF_3 [219]

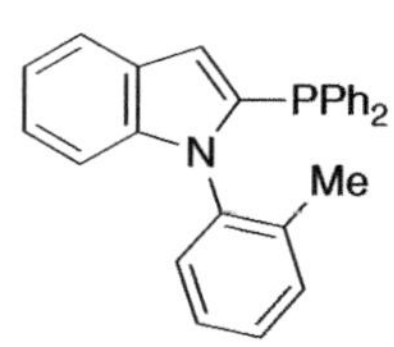

227 [219]

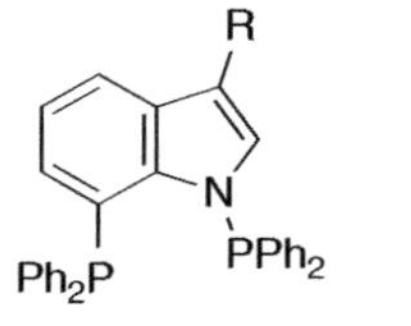

228 R = H [220], [221]
229 R = Me [222]

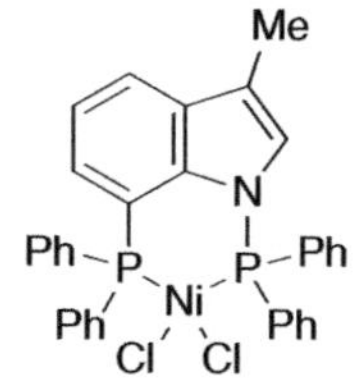

230 [222]

Scheme 34 *Indole-Phosphine Ligands*

indole–phosphine ligands **225** for the efficient palladium-catalyzed amination of aryl chlorides [217]. Heo used this ligand in Suzuki cross-coupling reactions of arylboronic acids and aryl halides [218]. Mino and colleagues synthesized the new indole–phosphine ligands **226** (chiral) and **227** for palladium-catalyzed allylic allylation [219]. Sarkar and colleagues prepared indole–phosphine ligands **228** and **229**, both of which are air stable. Ligand **228** was used to effect Suzuki cross-coupling of arylboronic acids and aryl/allyl chlorides [220] and the amination of allylic alcohols [221], and ligand **229** is a ligand for nickel-catalyzed Kumada coupling of aryl Grignards with aryl chlorides [222]. The nickel–phosphine complex **230** was isolated and characterized by x-ray crystallography.

Franzén and coworkers described a series of indole-phosphine-oxazoline (IndPHOX) ligands, **231–238** (Scheme 35). The power of these ligands is demonstrated by the palladium-catalyzed asymmetric allylic alkylation of 1,3-diphenyl-2-propenyl acetate with dimethyl malonate with 98% enantioselectivity [223]. These IndPHOX ligands have also been used in palladium-catalyzed asymmetric allylic amination with enantioselectivities of as high as 99% [224]. Subsequent work by this group led to the new ligands **239–245** [225]. The *N*-MOM and *N*-THP groups in these ligands afforded improved enantioselectivity of the palladium-catalyzed allylic amination reaction, with **243** giving higher percentage of enantiomeric excess than **244** (97% vs. 80%).

Another leader in this area of indole-based ligands is Kwong (Scheme 36). He and his colleagues have prepared the new ligands **246–251**, which are very efficient in Suzuki couplings [226]. Subsequent research by this group led to the preparation and evaluation of indole–phosphine ligands **251** (CM-Phos) [227], **253** [228], **254** [229], **255** [229], and **256–263** [230]. Of this large collection of Kwong ligands, CM-Phos (**251**) has seen the most extensive use in palladium-catalyzed cross-coupling reactions, particularly for Suzuki reactions involving aryl tosylates and mesylates [229, 231, 232], but also for Hiyama couplings

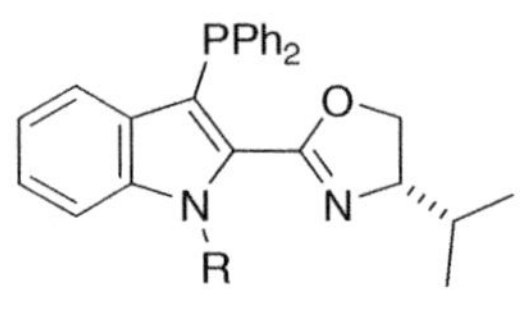

231 (R = Me)
232 (R = CH_2OMe)

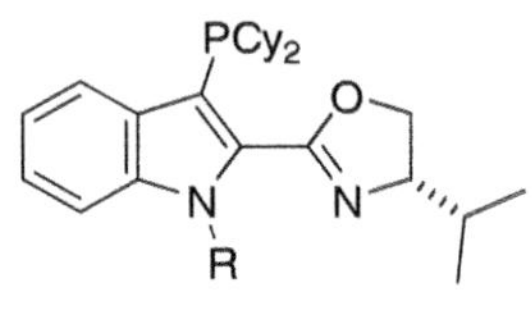

233 (R = Me)
234 (R = CH_2OMe)

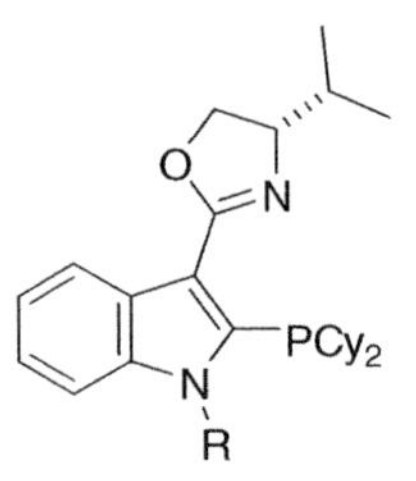

235 (R = Me)
236 (R = CH_2OMe)

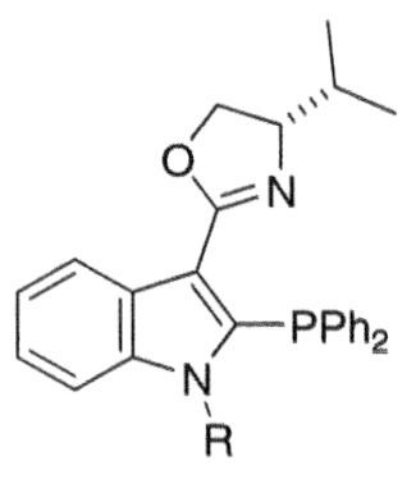

237 (R = Me)
238 (R = CH_2OMe)
239 (R = Bn)
240 (R = $CON(iPr)_2$)
241 (R = Ph)
242 (R = Pr)

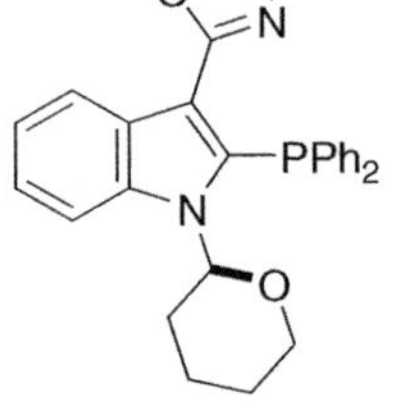

243

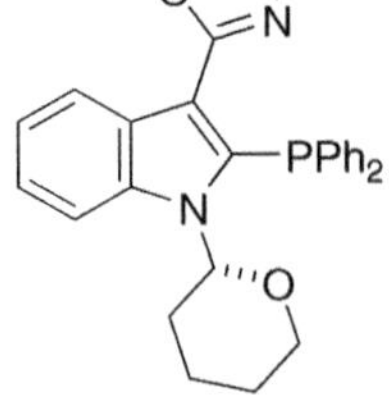

244

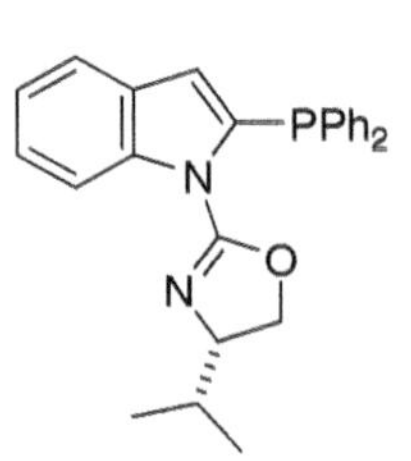

245

Scheme 35 *Franzén Indole-Phosphine Ligands [223–225]*

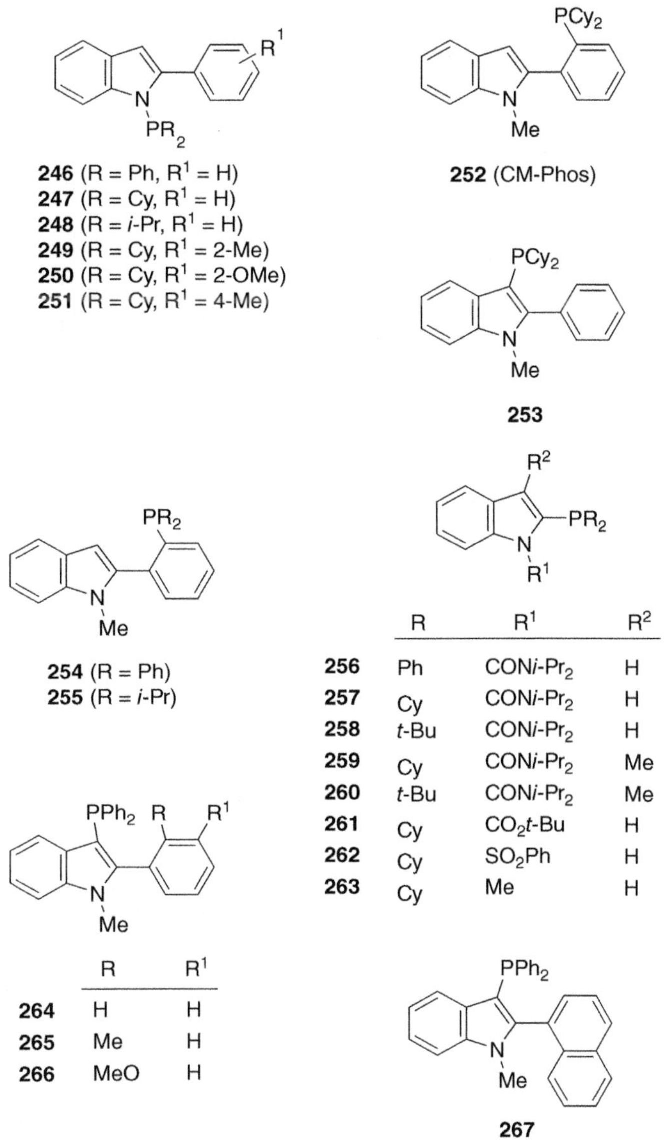

Scheme 36 *Kwong Indole-Phosphine Ligands [226–235]*

[233], amination [227], and cyanation of aryl chlorides in yields up to 96% [234]. Ligand **264** is a member of the new group of indole–phosphine ligands, **264–267**, developed by Kwong [235, 236]. For example, **266** is effective for the palladium-catalyzed borylation of aryl chlorides [235].

The group of Koskinen and Franzén prepared and used a series of indole–oxazoline ligands (Scheme 37). Koskinen and colleagues synthesized **268–273** [237], and Franzén made and evaluated the indole–olefin–oxazoline ligands **274–279** [238]. These last ligands are very effective in the rhodium-catalyzed asymmetric conjugate addition of enones with boron reagents, affording enantioselectivities of up to 94%.

Given the voluminous effort by several groups to design, synthesize, and implement ligands based on indole, it is clear that these remarkable ligands will continue to find important use in asymmetric metal-catalyzed synthesis.

A different type of indole material is Nindigo (**280**), a set of derivatives formed from anilines reacting with indigo [239]. These new materials represent binucleating ligands with twin β-diketiminate-type metal binding sites, illustrated in **281** (Scheme 38). Another novel indole compound is the 1,2-dimethyl-3-sulfonyl group (MIS) **282**, which finds utility as a protecting group for the side chain of arginine [240]. Thus, the corresponding sulfonyl chloride reacts with arginine to give **283**. The MIS group is compatible with tryptophan-containing peptides and is easily removed with mild acid.

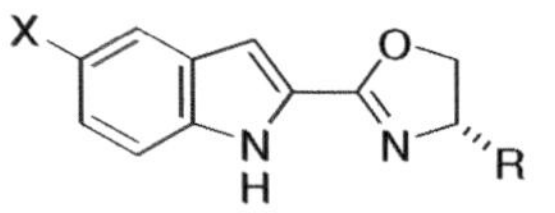

	R	X
268	*t*-Bu	H
269	Bn	H
270	(*R*)-Ph	H
271	*i*-Bu	H
272	*t*-Bu	MeO
273	*t*-Bu	Cl

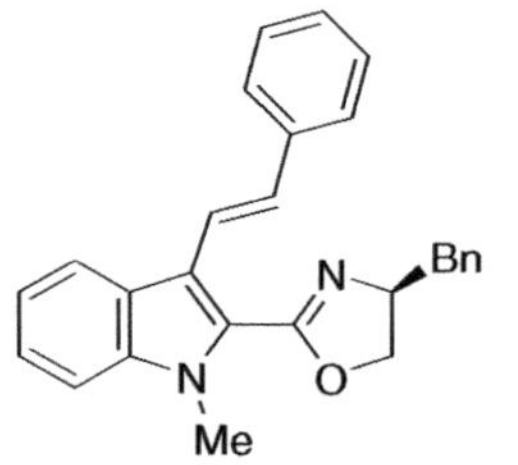

274

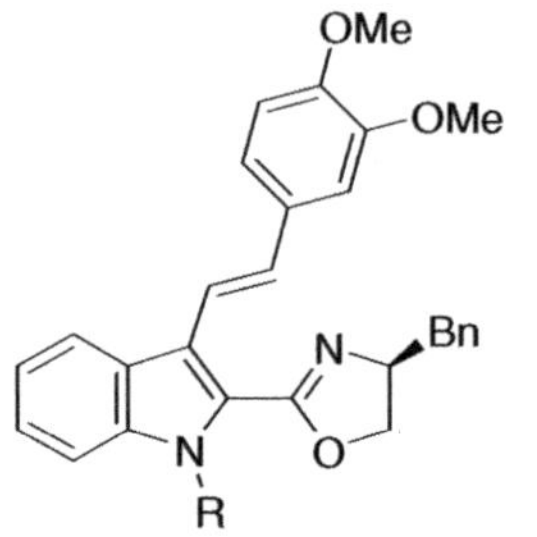

275 (R = Me)
276 (R = OCH_2OMe)

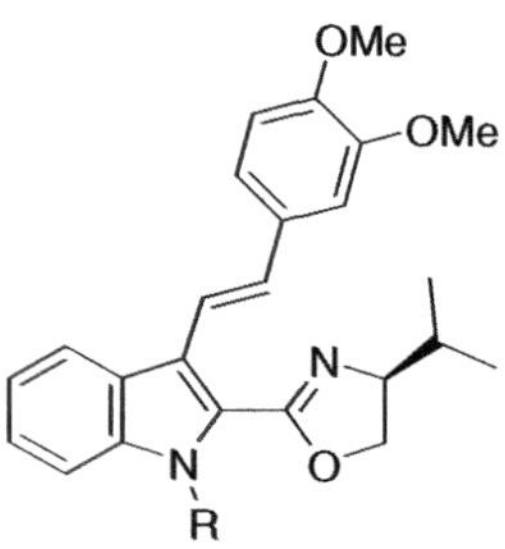

277 (R = Me)
278 (R = OCH_2OMe)

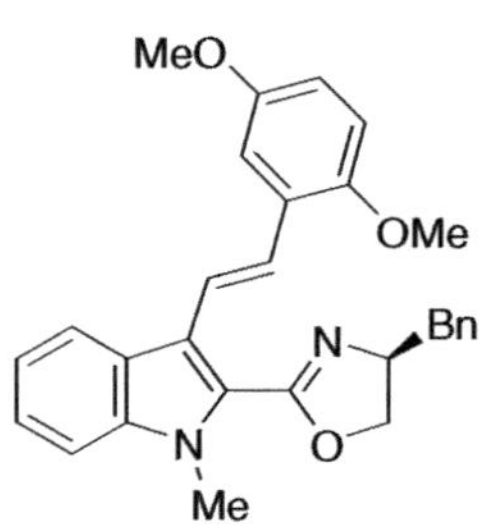

279

Scheme 37 *Indole-Oxazoline Ligands [237, 238]*

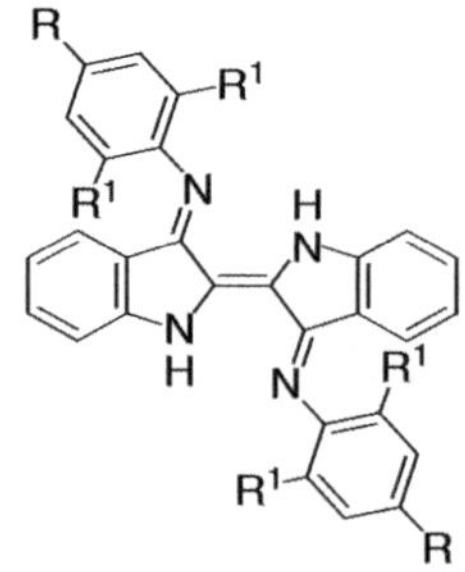

280 (R = H, Me; R^1 = H, Me, *i*-Pr) [239]

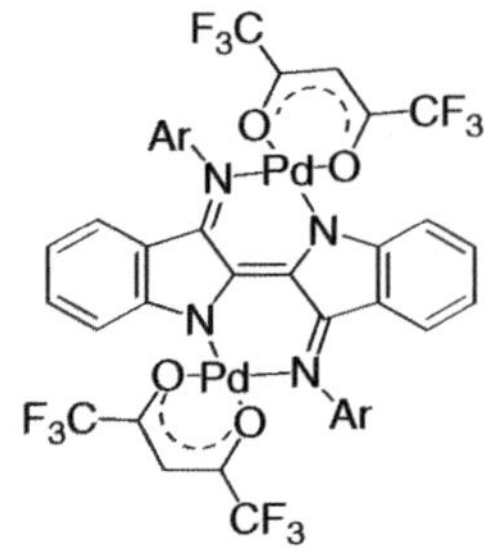

281 [239]

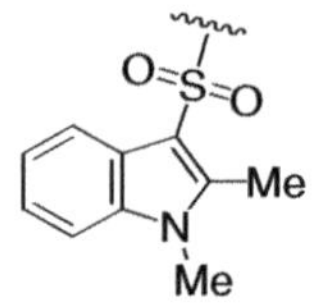

282 [240]

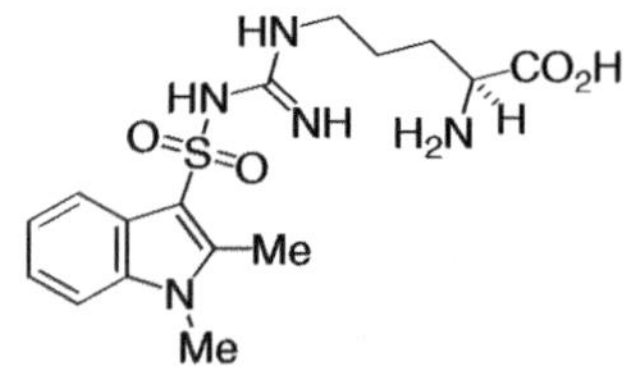

283 [240]

Scheme 38 *Novel Indole Materials*

1.7 Reviews of Indole-Ring Synthesis

Before beginning coverage of the synthesis of the indole ring, it is important to cite the numerous outstanding previous reviews in this area. They are arranged below from general to specialized reviews in descending chronological order. Some reviews of indole-ring analogues follow these. Additional specialized reviews of particular indole ring syntheses are cited within the appropriate chapter.

1.7.1 General Reviews on Indole Ring Synthesis

M. Inman and C.J. Moody, Indole Synthesis – Something Old, Something New, *Chem. Sci.*, 2013, **4**, 29–41. An excellent review that focuses on a monosubstituted arene starting point by a leading practitioner of indole chemistry and synthesis (Moody).

D.F. Taber and P.K. Tirunahari, Indole Synthesis: A Review and Proposed Classification, *Tetrahedron*, 2011, **67**, 7195–7210. A concise, modern review that proposes a useful classification system for constructing the indole ring.

R. Vicente, Recent Advances in Indole Synthesis: New Routes for a Classic Target, *Org. Biomol. Chem.*, 2011, **9**, 6469–6480. A review that covers the most recent developments and innovations.

G.R. Humphrey and J.T. Kuethe, Practical Methodologies for the Synthesis of Indoles, *Chem. Rev.*, 2006, **106**, 2875–2911. A comprehensive review covering both classic and modern indole ring syntheses.

J.A. Joule, Indole and Its Derivatives (2000), in *Science of Synthesis: Houben-Weyl Methods of Molecular Transformations*, Category 2, vol. **10** (ed. E.J. Thomas), George Thieme Verlag, Stuttgart, Germany; Chapter 10.13. An outstanding, comprehensive review written by a leading heterocyclic chemist.

G.W. Gribble, Recent Developments in Indole Ring Synthesis – Methodology and Applications, *J. Chem. Soc., Perkin Trans. 1*, 2000, 1045–1075. A review covering the 1994–1999 literature.

R.J. Sundberg, *Indoles* (1996), Academic Press, London. An excellent review covering the gamut of methods, including selective experimental procedures.

G.W. Gribble, Recent Developments in Indole Ring Synthesis – Methodology and Applications, *Contemp. Org. Synth.*, 1994, 145–172. A review covering the 1990–1993 literature.

1.7.2 Specialized Reviews

1.7.2.1 Solid-Phase Indole Ring Synthesis

S.A Patil, R. Patil, and D.D. Miller, Solid Phase Synthesis of Biologically Important Indoles, *Curr. Med. Chem.*, 2009, **16**, 2531–2565. A recent review with a strong focus on the solid-state synthesis of biologically active indoles.

J. Tois, R. Franzén, and A. Koskinen, Synthetic Approaches Towards Indoles on Solid Phase Recent Advances and Future

Directions, *Tetrahedron*, 2003, **59**, 5395–5405. An excellent complement to the review by Patil, Patil, and Miller.

K. Knepper, R.E. Ziegert, and S. Bräse, Solid Phase Indole Synthesis, *PharmaChem*, 2003, **2**, 4–7. A brief review illustrating the utility of solid-phase synthesis of indoles.

S. Bräse, C. Gil, and K. Knepper, The Recent Impact of Solid-Phase Synthesis on Medicinally Relevant Benzoannelated Nitrogen Heterocycles, *Bioorg. Med. Chem.*, 2002, **10**, 2415–2437. A comprehensive review covering the solid-phase synthesis of not only indoles but also several other nitrogen-containing aromatic heterocycles.

S. Cacchi and G. Fabrizi, Synthesis and Functionalization of Indoles Through Palladium-Catalyzed Reactions – Update 1, *Chem. Rev.*, 2011, **111**, PR215–PR283. An outstanding review and update of their 2005 paper (cited below).

J.J. Song, J.T. Reeves, D.R. Fandrick, *et al.*, Construction of the Indole Nucleus Through C–H Functionalization Reactions, *Arkivoc*, 2010, **1**, 390–449. This very nice review also covers nitrene approaches to indoles.

J. Barluenga, F. Rodríguez, and F.J. Fañanás, Recent Advances in the Synthesis of Indole and Quinoline Derivatives Through Cascade Reactions, *Chem. Asian J.*, 2009, **4**, 1036–1048. This review covers cascade reactions mainly involving the metals palladium, copper, gold, and platinum.

S. Cacchi and G. Fabrizi, Synthesis and Functionalization of Indoles Through Palladium-Catalyzed Reactions, *Chem. Rev.*, 2005, **105**, 2873–2920. The title says it all.

1.7.2.2 Specific Indoles

M. d'Ischia, A. Napolitano, and A. Pezzella, 5,6-Dihydroxyindole Chemistry: Unexplored Opportunities Beyond Eumelanin, *Eur. J. Org. Chem.*, 2011, 5501–5516. A unique microreview on the building blocks of eumelanin polymers.

L.F. Silva, Jr., M.V. Craveiro, and I.R.M. Tébéka, Total Syntheses of Trikentrins and Herbindoles, *Tetrahedron*, 2010, **66**, 3875–3895. Excellent summary of the different syntheses of these novel indole natural products.

O.V. Serdyuk, V.M. Muzalevskiy, and V.G. Nenajdenko, Synthesis and Properties of Fluoropyrroles and Their Analogues, *Synthesis*, 2012, **44**, 2115–2137. An excellent review by a leader of fluorine in heterocycles (Nenajdenko).

V.M. Muzalevskiy, A.V. Shastin, E.S. Balenkova, *et al.*, Synthesis of Trifluoromethyl Pyrroles and Their Benzo Analogues, *Synthesis*, 2009, 3905–3929. A complementary review to Serdyuk, Muzalevskiy, and Nenajdenko (2012) cited above.

D.A. Horton, G.T. Bourne, and M.L. Smythe, The Combinatorial Synthesis of Bicyclic Privileged Structures or Privileged Substructures, *Chem. Rev.*, 2003, **103**, 893–930. A review dealing with indoles and many other aromatic and nonaromatic heterocycles.

A.M. Lobo and S. Prabhakar, Recent Developments in the Synthesis of Biologically Active Indole Alkaloids, *J. Heterocycl. Chem.*, 2002, **39**, 429–436. A short review of indole alkaloid synthesis.

M. Somei, 1-Hydroxyindoles, *Heterocycles*, 1999, **50**, 1157–1211. A nice review by the leading authority on 1-hydroxyindoles.

H.M. Hugel and D.J. Kennaway, Synthesis and Chemistry of Melatonin and of Related Compounds. A Review, *Org. Prep. Proc. Int.*, 1995, **27**, 1–31. An excellent review of this important group of biologically significant compounds.

1.7.2.3 Microwave-Promoted Indole Ring Synthesis

S.A. Patil, R. Patil, and D.D. Miller, Microwave-Assisted Synthesis of Medicinally Relevant Indoles, *Curr. Med. Chem.*, 2011, **18**, 615–637. An excellent review highlighting the utility of microwave heating in indole ring synthesis.

1.7.2.4 Specific Reactions for Indole Ring Synthesis

G. Palmisano, A. Penoni, M. Sisti, *et al.*, Synthesis of Indole Derivatives with Biological Activity by Reactions Between Unsaturated Hydrocarbons and *N*-Aromatic Precursors, *Curr. Org. Chem.*, 2010, **14**, 2409–2441. An excellent recent review focusing primarily on the synthesis of biologically active indoles and starting with a nitrogen-substituted arene.

S.K. Bur and A. Padwa, The Pummerer Reaction: Methodology and Strategy for the Synthesis of Heterocyclic Compounds, *Chem. Rev.*, 2004, **104**, 2401–2432. This review by a leading heterocyclic chemist (Padwa) has a short section on indole alkaloids.

S.P. Gromov, Ring Transformation of Pyridines and Benzo Derivatives Under the Action of C-Nucleophiles, *Heterocycles*, 2000, **53**, 1607–1630. This specialized review covers mostly the excellent indole work of the author, much of which is available only in Russian.

A. Godard, F. Marsais, N. Plé, *et al.*, Connection Between Metalation of Azines and Diazines and Cross-Coupling Strategies for the Synthesis of Natural and Biologically Active Molecules, *Heterocycles*, 1995, **40**, 1055–1091. An excellent account of the work by Quéguiner and his colleagues on the synthesis of azacarbazoles.

1.7.3 Name Reactions

Several outstanding books on name reactions in organic chemistry are available. These typically briefly cover the classic indole name reactions with examples, references, and, in some cases, experimental procedures.

J. Li and J.M. Cook (2005) Indoles, in *Name Reactions in Heterocyclic Chemistry* (ed. J.-J. Li), Wiley-Interscience, Hoboken, New Jersey, Chapter 3, pp. 99–158.

B.P. Mundy, M.G. Ellerd, and F.G. Favaloro, Jr. (2005) *Name Reactions and Reagents in Organic Synthesis*, 2nd Edn., Wiley-Interscience, Hoboken, New Jersey.

E. Kruiswijk (2005) *The Comprehensive e-Book of Name Organic Reactions and Their Mechanisms*, 2nd edn. link.springer.com/content/pdf/10.2478/BF02479284.pdf

J.J. Li (2002) *Name Reactions*, Springer, Berlin.

A. Hassner and C. Stumer (2002) *Organic Syntheses Based on Name Reactions*, 2nd edn, Pergamon, Amsterdam.

A.R. Surrey (1954) *Name Reactions in Organic Chemistry*, Academic Press, New York.

1.7.4 Miscellaneous Reviews

In addition to the references 4–10 in the Preview, other treatments of indole ring synthesis are available.

T.L. Gilchrist, Synthesis of Aromatic Cycles, *J. Chem. Soc., Perkin Trans. 1*, 2001, 2491–2515. This review by a leading heterocyclic chemist has a short section on indoles.

C.J. Moody, "Oxidation by Nitrene Insertion" (1991), in *Comprehensive Organic Synthesis*, vol. **7** (eds. B.M. Trost, I. Fleming, and S.V. Ley), Pergamon Press, Oxford, pp. 21–38.

J.T. Kuethe, A General Approach to Indoles: Practical Applications for the Synthesis of Highly Functionalized Pharmacophores, *Chimia*, 2006, **60**, 543–553.

M. Shiri, M.A. Zolfigol, H.G. Kruger, and Z. Tanbakouchian, Bis- and Trisindolylmethanes (BIMs and TIMs), *Chem. Rev.*, 2010, **110**, 2250–2293.

1.7.5 Synthesis of Carbazoles, Carbolines, and Indolocarbazoles

J. Roy, A.K. Jana, and D. Mal, Recent Trends in the Synthesis of Carbazoles: An Update, *Tetrahedron*, 2012, **68**, 6099–6121. An excellent recent review spanning 2008–2011.

H.-J. Knölker and K.R. Reddy, Isolation and Synthesis of Biologically Active Carbazole Alkaloids, *Chem. Rev.*, 2002, **102**, 4303–4427. An outstanding review by a leading practitioner in the field (Knölker).

J. Bergman, T. Janosik, and N. Wahlström, Indolocarbazoles, *Adv. Heterocycl. Chem.*, 2001, **80**, 1–71. An excellent survey of these (fused) carbazoles, which possess enormous biological activity, by a pioneer in this field (Bergman).

G.W. Gribble and S.J. Berthel A Survey of Indolo[2,3-*a*]carbazoles and Related Natural Products (1993), in *Studies in Natural Products Chemistry,* Volume 12, Structure and Chemistry, Elsevier, Amsterdam, pp. 365–409.

D.P. Chakraborty and S. Roy, Carbazole Alkaloids. III, *Prog. Chem. Org. Nat. Prod.*, 1991, **57**, 71–152. This review and the previous two are excellent accounts of this area of natural products.

U. Pindur, Recent Developments in the Syntheses of Carbazole Alkaloids, *Chimia*, 1990, **44**, 406–412. A short review on the synthesis of biologically active carbazoles.

J.A. Joule, Recent Advances in the Chemistry of 9*H*-Carbazoles, *Adv. Heterocycl. Chem.*, 1984, **35**, 83–198. An excellent review on the chemistry and synthesis of carbazoles.

R.A. Abramovitch and I.D. Spenser, The Carbolines, *Adv. Heterocycl. Chem.*, 1964, **3**, 79–207. A specialized review on these azacarbazoles.

N. Campbell and B.M. Barclay, Recent Advances in the Chemistry of Carbazole, *Chem. Rev.*, 1947, **40**, 359–380; An early review that is a segue to Joule (1984), cited above.

1.7.6 Reviews of Indole Analogues

Although I do not cover the synthesis of indolines, oxindoles, isatins, and azaindoles in this monograph, some excellent reviews on the synthesis of these indole analogues are available [241–243].

All of this rich chemistry on indole ring synthesis puts the question, "Why do we need another review?" My answer is that despite these wonderful voluminous reviews, there is no single monograph covering indole ring synthesis *in toto*. My intent in the present volume is to present all methods for constructing the indole ring that are available to the practicing chemist.

References

[1] A. Baeyer, *Chem. Ber.*, 1880, **13**, 2254–2263.
[2] R.B. Van Order and H.G. Lindwall, *Chem. Rev.*, 1942, **30**, 69–96.
[3] R.J. Sundberg (1970) *The Chemistry of Indoles*, Academic Press, New York.
[4] R.J. Sundberg (1996) *Indoles*, Academic Press, London.
[5] R.J. Sundberg (1984) *Comprehensive Heterocyclic Chemistry*, vol. **4** (eds A.R. Katritzky, C.W. Rees, C.W. Bird, and G.W.H. Cheeseman), Pergamon, Oxford, pp. 313–376.
[6] G.W. Gribble (1996) *Comprehensive Heterocyclic Chemistry II*, vol. **2** (eds A.R. Katritzky, C.W. Rees, E.F.V. Scriven, and C.W. Bird), Pergamon Press, Oxford, pp. 207–257.
[7] J. Bergman and T. Janosik (2008) *Comprehensive Heterocyclic Chemistry III*, vol. **3** (eds A.R. Katritzku, C.A. Ramsden, E.F.V. Scriven, and R.J.K. Taylor), Elsevier, Amsterdam, pp. 269–351.
[8] R.K. Brown (1972) *Indoles*, Part I (ed.W.J. Houlihan), Wiley, New York, pp. 385–396.
[9] G.W. Gribble and J.A. Joule (eds) (2012) *Progress in Heterocyclic Chemistry*, vol. **24**, Chapter 5.2, Pergamon Press, Oxford (and previous volumes in this annual series).
[10] For two excellent introductions to indoles, see (a) J.A. Joule and G.F. Smith (2010) *Heterocyclic Chemistry*, 5th edn, Wiley, New York; (b) T.L. Gilchrist (1997) *Heterocyclic Chemistry*, 3rd edn, Longman, Essex.
[11] M. Ishikura, T. Abe, T. Choshi, and S. Hibino, *Nat. Prod. Rep.*, 2013, **30**, 694–752 and previous reviews in this series.
[12] A.F. Pozharskii, A.T. Soldatenkov, and A.R. Katritzky (2011) *Heterocycles in Life and Society: An Introduction to Heterocyclic Chemistry, Biochemistry, and Applications*, 2nd edn, Wiley, West Sussex.
[13] C.A. Herter, *J. Amer. Med. Assoc.*, 1907, **48**, 985–992.
[14] W. Bergeim, *J. Biol. Chem.*, 1917, **32**, 17–22.
[15] G.S. Clark, *Perf & Flav*, 1995, **20**, 21–31.
[16] J. Meijerink, M.A.H. Braks, A.A. Brack, *et al.*, *J. Chem. Ecol.*, 2000, **26**, 1367–1382.
[17] J.G. Kostelc, G. Preti, P.R. Zelson, *et al.*, *J. Periodontal Res.*, 1980, **15**, 185–192.
[18] A.F. Carey, G. Wang, C.-Y. Su, *et al.*, *Nature*, 2010, **464**, 66–72.
[19] For reviews of halogenated indoles, see (a) G.W. Gribble (2012) Occurrence of halogenated alkaloids, in *The Alkaloids*, vol. **71** (ed. H.-J. Knölker), Elsevier, London, UK, pp. 1–165; (b) G.W. Gribble, *Heterocycles*, 2012, **84**, 157–207.
[20] E.D. Glowacki, G. Voss, L. Leonat, *et al.*, *Israel J. Chem.*, 2012, **52**, 540–551.
[21] U. Pindur and T. Lemster, *Curr. Med. Chem.*, 2001, **8**, 1681–1698.
[22] A. Aygün and U. Pindur, *Curr. Med. Chem.*, 2003, **10**, 1113–1127.
[23] C.-G. Yang, H. Huang, and B. Jiang, *Curr. Org. Chem.*, 2004, **8**, 1691–1720.
[24] W. Gul and M.T. Hamann, *Life Sci.*, 2005, **78**, 442–453.

[25] M. Tadesse, J.N. Tabudravu, M. Jaspars, *et al.*, *J. Nat. Prod.*, 2011, **74**, 837–841.
[26] F. He, Y.-L. Sun, K.-S. Liu, *et al.*, *J. Antibiot.*, 2012, **65**, 109–111.
[27] R. Finlayson, A.N. Pearce, M.J. Page, *et al.*, *J. Nat. Prod.*, 2011, **74**, 888–892.
[28] A.J. Zaharenko, G. Picolo, W.A. Ferreira, Jr., *et al.*, *J. Nat. Prod.*, 2011, **74**, 378–382.
[29] K.Ø. Hanssen, B. Schuler, A.J. Williams, *et al.*, *Angew. Chem. Int. Ed.*, 2012, **51**, 12238–12241.
[30] N-Y. Ji, X-M. Li, L-P. Ding, and B-G. Wang, *Helv. Chim. Acta*, 2007, **90**, 385–391.
[31] I. Zendah, K.A. Shaaban, E. Helmke, *et al.*, *Zeit. Naturforsch.*, 2012, **67b**, 417–420.
[32] H.B. Park, Y.-J. Kim, J.K. Lee, *et al.*, *Org. Lett.*, 2012, **14**, 5002–5005.
[33] L. Zhu, C. Chen, H. Wang, *et al.*, *Chem. Pharm. Bull.*, 2012, **60**, 670–673.
[34] M.S.C. Pedras and E.E. Yaya, *Org. Biomol. Chem.*, 2012, **10**, 3613–3616.
[35] B.L.J. Kindler, H.-J. Krämer, S. Nies, *et al.*, *Eur. J. Org. Chem.*, 2010, 2084–2090.
[36] R.A. Butcher, J.R. Ragains, and J. Clardy, *Org. Lett.*, 2009, **11**, 3100–3103.
[37] A. Bartsch, M. Bross, P. Spiteller, *et al.*, *Angew. Chem. Int. Ed.*, 2005, **44**, 2957–2959.
[38] L.K. Nzowa, R.B. Teponno, L.A. Tapondjou, *et al.*, *Fitoterapia*, 2013, **87**, 37–42.
[39] Q. Zhang, A. Mándi, S. Li, *et al.*, *Eur. J. Org. Chem.*, 2012, 5256–5262.
[40] R.J.R. Jaeger, M. Lamshöft, S. Gottfried, *et al.*, *J. Nat. Prod.*, 2013, **76**, 127–134.
[41] P. Fu, Y. Zhuang, Y. Wang, *et al.*, *Org. Lett.*, 2012, **14**, 6194–6197.
[42] W Maneerat, W. Phakhodee, S. Cheenpracha, *et al.*, *Phytochemistry*, 2013, **88**, 74–78.
[43] V. Sharma, P. Kumar, and D. Pathak, *J. Heterocycl. Chem.*, 2010, **47**, 491–502.
[44] J. Rosén, J. Gottfries, S. Muresan, *et al.*, *J. Med. Chem.*, 2009, **52**, 1953–1962.
[45] M. Segarra-Newnham and T.J. Church, *Annals Pharmacother.*, 2012, **46**, 1678–1687.
[46] I.C. Sutcliffe, *Proc. Natl. Acad. Sci.*, 2012, **109**, 18637–18638.
[47] S. Daly, K. Hayden, I. Malik, *et al.*, *Bioorg. Med. Chem. Lett.*, 2011, **21**, 4720–4723.
[48] S. Samosorn, J.B. Bremner, A. Ball, and K. Lewis, *Bioorg. Med. Chem.*, 2006, **14**, 857–865.
[49] R. Karstad, G. Isaksen, B.-O. Brandsdal, *et al.*, *J. Med. Chem.*, 2010, **53**, 5558–5566.
[50] K. Okamoto, M. Sakagami, F. Feng, *et al.*, *J. Org. Chem.*, 2012, **77**, 1367–1377.
[51] R.J. Worthington, J.J. Richards, and C. Melander, *Org. Biomol. Chem.*, 2012, **10**, 7457–7474.
[52] E. Hedner, M. Sjögren, S. Hodzic, *et al.*, *J. Nat. Prod.*, 2008, **71**, 330–333.
[53] I. Moubax, N. Bontemps-Subielos, B. Banaigs, *et al.*, *Environ. Toxicol. Chem.*, 2001, **20**, 589–596.
[54] S. Tsukamoto, H. Hirota, H. Kato, and N. Fusetani, *Tetrahedron Lett.*, 1993, **34**, 4819–4822.
[55] N. Pénez, G. Culioli, T. Pérez, *et al.*, *J. Nat. Prod.*, 2011, **74**, 2304–2308.
[56] R.J. Deschenes, H. Lin, A.D. Ault, and J.S. Fassler, *Antimicrob. Agents Chemother.*, 1999, **43**, 1700–1703.
[57] S.Y. Ablordeppey, P. Fan, S. Li, *et al.*, *Bioorg. Med. Chem.*, 2002, **10**, 1337–1346.
[58] A.A. Farahat, A. Kumar, M. Say, *et al.*, *Bioorg. Med. Chem.*, 2010, **18**, 557–566.
[59] F. Bouchikhi, F. Anizon, R. Brun, and P. Moreau, *Bioorg. Med. Chem. Lett.*, 2011, **21**, 6319–6321.
[60] www.unaids.org.
[61] T.M. Williams, T.M. Ciccarone, S.C. MacTough, *et al.*, *J. Med. Chem.*, 1993, **36**, 1291–1294.
[62] R. Ragno, A. Coluccia, G. La Regina, *et al.*, *J. Med. Chem.*, 2006, **49**, 3172–3184.
[63] G. La Regina, A. Coluccia, A. Brancale, *et al.*, *J. Med. Chem.*, 2011, **54**, 1587–1598.
[64] Z. Zhao, S.E. Wolkenberg, M. Lu, *et al.*, *Bioorg. Med. Chem. Lett.*, 2008, **18**, 554–559.
[65] A. Gopalsamy, K. Lim, G. Ciszewski, *et al.*, *J. Med. Chem.*, 2004, **47**, 6603–6608.
[66] K. Roy and A.S. Mandal, *J. Enzym. Inhib. Med. Chem.*, 2008, **23**, 980–995.
[67] H. Sano, T. Noguchi, A. Yanatani, *et al.*, *Bioorg. Med. Chem.*, 2005, **13**, 3079–3091.
[68] W. Hu, Z. Guo, X. Yi, *et al.*, *Bioorg. Med. Chem.*, 2003, **11**, 5539–5544.
[69] J. Kaur, A. Bhardwaj, Z. Huang, and E.E. Knaus, *Bioorg. Med. Chem. Lett.*, 2012, **22**, 2154–2159.
[70] T. Kameyama, F. Amanuma, S. Okuyama, *et al.*, *J. Pharmacobio-Dyn.*, 1985, **8**, 477–486.
[71] A.D. Favia, D. Habrant, R. Scarpelli, *et al.*, *J. Med. Chem.*, 2012, **55**, 8807–8826.
[72] M. Arisawa, Y. Kasaya, T. Obata, *et al.*, *J. Med. Chem.*, 2012, **55**, 8152–8163.
[73] S.J. Taylor, A. Abeywardane, S. Liang, *et al.*, *J. Med. Chem.*, 2011, **54**, 8174–8187.
[74] T. Komoda, Y. Shinoda, and S. Nakatsuka, *Biosci. Biotechnol. Biochem.*, 2003, **67**, 659–662.
[75] T. Komoda and S. Nakatsuka, *Heterocycl. Commun.*, 2003, **9**, 119–122.
[76] K.R. Campos, M. Journet, S. Lee, *et al.*, *J. Org. Chem.*, 2005, **70**, 268–274.
[77] K.V. Sashidhara, M. Kumar, R. Sonkar, *et al.*, *J. Med. Chem.*, 2012, **55**, 2769–2779.
[78] J.T. Mihalic, X. Chen, P. Fan, *et al.*, *Bioorg. Med. Chem. Lett.*, 2011, **21**, 7001–7005.
[79] S. Kher, K. Lake, I. Sircar, *et al.*, *Bioorg. Med. Chem. Lett.*, 2007, **17**, 4442–4446.
[80] D. Weber, C. Berger, P. Eickelmann, *et al.*, *J. Med. Chem.*, 2003, **46**, 1918–1930.
[81] N.M. Barnes and T. Sharp, *Neuropharmacology*, 1999, **38**, 1083–1152.
[82] D. Hoyer, J.P. Hannon, and G.R. Martin, *Pharmacol. Biochem. Behav.*, 2002, **71**, 533–554.
[83] M.L. Woolley, C.A. Marsden, and K.C.F. Fone, *Curr. Drug Targets – CNS Neurological Disorders*, 2004, **3**, 59–79.
[84] R.A. Glennon, S.-S. Hong, M. Bondarev, *et al.*, *J. Med. Chem.*, 1996, **39**, 314–322.
[85] D.C. Cole, J.W. Ellingboe, W.J. Lennox, *et al.*, *Bioorg. Med. Chem. Lett.*, 2005, **15**, 379–383.
[86] J. Holenz, R. Mercè, J.L. Díaz, *et al.*, *J. Med. Chem.*, 2005, **48**, 1781–1795.
[87] M.G.N. Russell, R.J. Baker, L. Barden, *et al.*, *J. Med. Chem.*, 2001, **44**, 3881–3895.

[88] Y.-C. Chang, J. Riby, G.H.-F. Chang, *et al.*, *Biochem. Pharmacol.*, 1999, **58**, 825–834.
[89] X. Ge, F.A. Fares, and S. Yannai, *Anticancer Res.*, 1999, **19**, 3199–3204.
[90] J.-R. Weng, C.-H. Tsai, S.K. Kulp, and C.-S. Chen, *Cancer Lett.*, 2008, **262**, 153–163.
[91] H. Mori, K. Niwa, Q. Zheng, *et al.*, *Mutation Res.*, 2001, **480-481**, 201–207.
[92] A.D. Shilling, D.B. Carlson, S. Katchamart, and D.E. Williams, *Toxicol. Appl. Pharmacol.*, 2001, **170**, 191–200.
[93] R. Kronbak, F. Duus, and O. Vang, *J. Agric. Food Chem.*, 2010, **58**, 8453–8459.
[94] W.-S. Li, C.-H. Wang, S. Ko, *et al.*, *J. Med. Chem.*, 2012, **55**, 1583–1592.
[95] V.K. Rao, B.S. Chhikara, A.N. Shirazi, *et al.*, *Bioorg. Med. Chem. Lett.*, 2011, **21**, 3511–3514.
[96] M.-P. Lézé, A. Palusczak, R.W. Hartmann, and M. Le Borgne, *Bioorg. Med. Chem. Lett.*, 2008, **18**, 4713–4715.
[97] D.D. Miller, P. Bamborough, J.A. Christopher, *et al.*, *Bioorg. Med. Chem. Lett.*, 2011, **21**, 2255–2258.
[98] A. Brancale and R. Silvestri, *Med. Res. Rev.*, 2007, **27**, 209–238.
[99] G. De Martino, G. La Regina, A. Coluccia, *et al.*, *J. Med. Chem.*, 2004, **47**, 6120–6123.
[100] G. De Martino, M.C. Edler, G. La Regina, *et al.*, *J. Med. Chem.*, 2006, **49**, 947–954.
[101] G. La Regina, M.C. Edler, A. Brancale, *et al.*, *J. Med. Chem.*, 2007, **50**, 2865–2874.
[102] D. Kaufmann, M. Pojarová, S. Vogel, *et al.*, *Bioorg. Med. Chem.*, 2007, **15**, 5122–5136.
[103] H. Saito, K. Tabata, S. Hanada, *et al.*, *Bioorg. Med. Chem. Lett.*, 2011, **21**, 5370–5373.
[104] J. Qian-Cutrone, S. Huang, Y.-Z. Shu, *et al.*, *J. Am. Chem. Soc.*, 2002, **124**, 14556–14557.
[105] D.W. Nettleton, T.W. Doyle, B. Krishnan, *et al.*, *Tetrahedron Lett.*, 1985, **26**, 4011–4014.
[106] J.A. Bush, B.H. Long, J.J. Catino, *et al.*, *J. Antibiot.*, 1987, **40**, 668–678.
[107] J.S. Sandler, P.L. Colin, J.N.A. Hooper, and D.J. Faulkner, *J. Nat. Prod.*, 2002, **65**, 1258–1261.
[108] D.R. Appleton, M.J. Page, G. Lambert, *et al.*, *J. Org. Chem.*, 2002, **67**, 5702–5704.
[109] D.R. Appleton and B.R. Copp, *Tetrahedron Lett.*, 2003, **44**, 8963–8965.
[110] K. Ohba, H. Watabe, T. Sasaki, *et al.*, *J. Antibiot.*, 1988, **41**, 1515–1519.
[111] S. Ishii, M. Nagasawa, Y. Kariya, *et al.*, *J. Antibiot.*, 1989, **42**, 1713–1717.
[112] N.B. Perry, J.W. Blunt, and M.H.G. Munro, *Tetrahedron*, 1988, **44**, 1727–1734.
[113] R.P. Maskey, I. Grün-Wollny, H.H. Fiebig, and H. Laatsch, *Angew. Chem. Int. Ed.*, 2002, **41**, 597–599.
[114] N. Khorana, K. Changwichit, K. Ingkaninan, and M. Utsintong, *Bioorg. Med. Chem. Lett.*, 2012, **22**, 2885–2888.
[115] P.G. Becher, J. Beuchat, K. Gademann, and F. Jüttner, *J. Nat. Prod.*, 2005, **68**, 1793–1795.
[116] P. Polychronopoulos, P. Magiatis, A.-L. Skaltsounis, *et al.*, *J. Med. Chem.*, 2004, **47**, 935–945.
[117] K.S. MacMillan, J. Naidoo, J. Liang, *et al.*, *J. Am. Chem. Soc.*, 2011, **133**, 1428–1437.
[118] A. Mahapatra, *ACS Chem. Neurosci.*, 2010, **1**, 589.
[119] S. Štolc, V. Šnirc, M. Májeková, *et al.*, *Cell. Mol. Neurobiol.*, 2006, **26**, 1495–1504.
[120] K. Dinnell, G.G. Chicchi, M.J. Dhar, *et al.*, *Bioorg. Med. Chem. Lett.*, 2001, **11**, 1237–1240.
[121] J.L. Stanton and M.H. Ackerman, *J. Med. Chem.*, 1983, **26**, 986–989.
[122] M.-J. Don, D.F.V. Lewis, S.-Y. Wang, *et al.*, *Bioorg. Med. Chem. Lett.*, 2003, **13**, 2535–2538.
[123] M. Jansen and G. Dannhardt, *Eur. J. Med. Chem.*, 2003, **38**, 855–865.
[124] N. Uchiyama, R. Kikura-Hanajiri, and Y. Goda, *Chem. Pharm. Bull.*, 2011, **59**, 1203–1205.
[125] J.L. Wiley, V.J. Smith, J. Chen, *et al.*, *Bioorg. Med. Chem.*, 2012, **20**, 2067–2081.
[126] R.K. Razdan, V.K. Vemuri, A. Makriyannis, and J.W. Huffman (2009) *Cannabinoid Receptor Ligands and Structure-Activity Relationships*, Part 1 (ed. P.H. Reggio), Humana Press, New York, pp. 3–94.
[127] V. Leclerc, S. Yous, P. Delagrange, *et al.*, *J. Med. Chem.*, 2002, **45**, 1853–1859.
[128] Y. Sugiyama, Y. Ito, M. Suzuki, and A. Hirota, *J. Nat. Prod.*, 2009, **72**, 2069–2071.
[129] D.J. Buzard, S. Han, L. Lopez, *et al.*, *Bioorg. Med. Chem. Lett.*, 2012, **22**, 4404–4409.
[130] P. Prasit, M. Belley, M. Blouin, *et al.*, *J. Lipid Mediat.*, 1993, **6**, 239–244.
[131] Z. Diamant, M.C. Timmers, H. van der Veen, *et al.*, *J. Allergy Clin. Immunol.*, 1995, **95**, 42–51.
[132] C.D. Funk, *Nature Rev., Drug Discovery*, 2005, **4**, 664–672.
[133] J.F. Evans, C. Léville, J.A. Manchini, *et al.*, *Mol. Pharmacol.*, 1991, **40**, 22–27.
[134] S. Lucas, M.H. Poulsen, N.G. Nørager, *et al.*, *J. Med. Chem.*, 2012, **55**, 10297-10301.
[135] H. Bräuner-Osborne, J. Egebjerg, E.Ø. Nielsen, *et al.*, *J. Med. Chem.*, 2000, **43**, 2609–2645.
[136] R. Planells-Cases, J. Lerma, and A. Ferrer-Montiel, *Curr. Pharm. Des.*, 2006, **12**, 3583–3596.
[137] S.F. Traynelis, L.P. Wollmuth, C.J. McBain, *et al.*, *Pharmacol. Rev.*, 2010, **62**, 405–496.
[138] M.-J. Don, D.F.V. Lewis, S.-Y. Wang, *et al.*, *Bioorg. Med. Chem. Lett.*, 2003, **13**, 2535–2538.
[139] M. Somei, *Heterocycles*, 2011, **82**, 1007–1027.
[140] M.E. Welsch, S.A. Snyder, and B.R. Stockwell, *Curr. Opin. Chem. Biol.*, 2010, **14**, 347–361.
[141] I.S. Johnson, J.G. Armstrong, M. Gorman, and J.P. Burnett Jr., *Cancer Res.*, 1963, **23**, 1390–1427.
[142] P.G. Goppi, C. Broglia, F. Merli, *et al.*, *Cancer*, 2003, **98**, 2393–2401.
[143] G.M. Cragg, P.G. Grothaus, and D.J. Newman, *Chem. Rev.*, 2009, **109**, 3012–3043.
[144] M. Hussain, U. Vaishampagan, L.K. Heilbrun, *et al.*, *Invest. New Drugs*, 2003, **21**, 465–471.

[145] G.H. Schwartz, A. Patnaik, L.A. Hammond, *et al.*, *Ann. Oncology*, 2003, **14**, 775–782.
[146] T. Hata, Y. Sato, R. Sugawara, *et al.*, *J. Antibiot. Ser. A*, 1956, **9**, 141–146.
[147] M. Tomasz, *Chem. Biol.*, 1995, **2**, 575–579.
[148] C.P. Miller, H.A. Harris, and B.S. Komm, *Drugs Fut.*, 2002, **27**, 117–121.
[149] L.A. Sorbera, J. Castaner, and J.S. Silvestre, *Drugs Fut.*, 2002, **27**, 942–947.
[150] Y.-J. Wu, *Prog. Heterocycl. Chem.*, 2012, **24**, 1–42.
[151] P. Revill, N. Mealy, N Serradell, *et al.*, *Drugs Fut.*, 2007, **32**, 315–322.
[152] H. Huynh and J. Fargnoli, *Drug Fut.*, 2009, **34**, 881–895.
[153] L.A. Sorbera, N. Serradell, E. Rosa, *et al.*, *Drugs Fut.*, 2007, **32**, 577–589.
[154] Y. Wang, *Drugs Fut.*, 2009, **34**, 177–182.
[155] Y. Wang, N. Serradell, J. Bolos, and E. Rosa, *Drugs Fut.*, 2007, **32**, 228–233.
[156] G.W. Gribble and S.J. Berthel, *Stud. Nat. Prod. Chem.*, 1993, **12**, 365–409.
[157] M. Prudhomme, *Curr. Pharm. Des.*, 1997, **3**, 265–290.
[158] P.R. Revill, N. Serradell, J. Bolos, and E. Rosa, *Drugs Fut.*, 2007, **32**, 215–222.
[159] L.A. Sorbera, N. Serradell, J. Bolos, and E. Rosa, *Drugs Fut.*, 2007, **32** 297–309.
[160] C. Dulsat and R. Castaner, *Drugs Fut.*, 2009, **34**, 270–275.
[161] R. Talpir, Y. Benayahu, Y. Kashman, *et al.*, *Tetrahedron Lett.*, 1994, **35**, 4453–4456.
[162] L.-C. Hsu, D.E. Durrant, C.-C. Huang, *et al.*, *Invest. New Drugs*, 2012, **30**, 1379–1388.
[163] M.G. Saulnier, B.H. Long, D.B Frennesson, *et al.*, *J. Med. Chem.*, 2005, **48**, 2258–2261.
[164] J.J. Li, D.S. Johnson, D.D. Sliskovic, and B.D. Roth (2004) Triptans for migraine, in *Contemporary Drug Synthesis*, John Wiley & Sons, Hoboken, NJ, Chapter 12.
[165] J.J. Li (2009) *Triumph of the Heart*, Oxford University Press, New York.
[166] A. Graul, J. Silvestre, and J. Castaner, *Drugs Fut.*, 1999, **24**, 38–44.
[167] M. McCall, S. Merani, C. Toso, and A.M.J. Shapiro, *Drugs Fut.*, 2009, **34**, 618–623.
[168] C.N. Eid, Jr., Y.J. Wu, and G. Kenny (2002) Cognition enhancers, in *Burger's Medicinal Chemistry and Drug Discovery* (ed. D.J. Abraham), John Wiley & Sons, New York, Chapter 12, pp. 779–835.
[169] L.K. Larsen, R.E. Moore, and G.M.L. Pattersen, *J. Nat. Prod.*, 1994, **57**, 419–421.
[170] H.-L. Yin, J.-H. Li, J. Li, *et al.*, *Fitoterapia*, 2013, **84**, 360–365.
[171] R.J. Lake, M.M. Brennan, J.W. Blunt, *et al.*, *Tetrahedron Lett.*, 1988, **29**, 4971–4972.
[172] S. Sagar, M. Kaur, and K.P. Minneman, *Mar. Drugs*, 2010, **8**, 2619–2638.
[173] J.R. Woodworth, E.H. Nyhart, G.L. Brier, *et al.*, *Antimicrob. Agents Chemother.*, 1992, **36**, 318–325.
[174] R.M. Scarborough, *Drugs Fut.*, 1998, **23**, 585–590.
[175] L.A. Sorbera, X. Rabasseda, J. Silvestre, and J. Castaner, *Drugs Fut.*, 2001, **26**, 247–252.
[176] H. Elokdah, M. Abou-Gharbia, J.K. Hennan, *et al.*, *J. Med. Chem.*, 2004, **47**, 3491–3494.
[177] L.A. Sorbera, J. Silvestre, X. Rabasseda, and J. Castaner, *Drugs Fut.*, 2000, **25**, 1017–1026.
[178] R.W. Schevitz, N.J. Bach, D.G. Carlson, *et al.*, *Nature Struct. Biol.*, 1995, **2**, 458–465.
[179] M.C. Oliancis and P. Onali, *Life Sci.*, 1999, **65**, 2233–2240.
[180] S. Heptinstall, D.I. Espinosa, P. Manolopoulos, *et al.*, *Platelets*, 2008, **19**, 605–613.
[181] A.B. Bredt and G.J.D. Girey, *Cancer*, 1982, **50**, 1430–1433.
[182] P.A. Gale, *Chem. Commun.*, 2008, 4525–4540.
[183] L. Panzella, A. Pezzella, M. Arzillo, *et al.*, *Tetrahedron*, 2009, **65**, 2032–2036.
[184] Y. Shiraishi, H. Maehara, and T. Hirai, *Org. Biomol. Chem.*, 2009, **7**, 2072–2076.
[185] G.W. Lee, N.-K. Kim, and K.-S. Jeong, *Org. Lett.*, 2010, **12**, 2634–2637.
[186] L. Wang, X. He, Y. Guo, *et al.*, *Org. Biomol. Chem.*, 2011, **9**, 752–757.
[187] K.N. Skala, K.G. Perkins, A. Ali, *et al.*, *Tetrahedron Lett.*, 2010, **51**, 6516–6520.
[188] A.L. Whiting and F. Hof, *Org. Biomol. Chem.*, 2012, **10**, 6885–6892.
[189] J. Schönhaber, W. Frank, and T.J.J. Müller, *Org. Lett.*, 2010, **12**, 4122–4125.
[190] J.H. Lee, J.-H. So, J.H. Jeon, *et al.*, *Chem. Commun.*, 2011, **47**, 7500–7502.
[191] K.M. Schlitt, A.L. Millen, S.D. Wetmore, and R.A. Manderville, *Org. Biomol. Chem.*, 2011, **9**, 1565–1571.
[192] L. Chen, T.-S. Hu, and Z.-J. Yao, *Eur. J. Org. Chem.*, 2008, 6175–6182.
[193] R.A. Illos, D. Shamir, L.J.W. Shimon, *et al.*, *Tetrahedron Lett.*, 2006, **47**, 5543–5546.
[194] G. Chen, D.J. Yee, N.G. Gubernator, and D. Sames, *J. Am. Chem. Soc.*, 2005, **127**, 4544–4545.
[195] H. Hiyoshi, T. Aoyama, T. Wada, *et al.*, *Heterocycles*, 2009, **77**, 595–602.
[196] H. Tsuji, Y. Yokoi, C. Mitsui, *et al.*, *Chem. Asian J.*, 2009, **4**, 655–657.
[197] T.V. Pho, J.D. Yuen, J.A. Kurzman, *et al.*, *J. Am. Chem. Soc.*, 2012, **134**, 18185–18188.
[198] T. Lei, J.-H. Dou, Z.-J. Ma, *et al.*, *J. Am. Chem. Soc.*, 2012, **134**, 20025–20028.
[199] M. Irimia-Vladu, E.D. Glowacki, P.A. Troshin, *et al.*, *Adv. Mater.*, 2012, **24**, 375–380.
[200] M. Yao, M. Araki, H. Senoh, *et al.*, *Chem. Lett.*, 2010, **39**, 950–952.

[201] E. Wang, Z. Ma, Z. Zhang, *et al.*, *J. Am. Chem. Soc.*, 2011, **133**, 14244–14247.
[202] P. Remón, M Bälter, S. Li, *et al.*, *J. Am. Chem. Soc.*, 2011, **133**, 20742–20745.
[203] J.-F. Morin, S. Beaupré, M. Leclerc, *et al.*, *Appl. Phys. Lett.*, 2002, **80**, 341–343.
[204] J.-F. Morin and M. Leclerc, *Macromolecules*, 2001, **34**, 8413–8417.
[205] B.-Y. Ryu and T. Emrick, *Angew. Chem. Int. Ed.*, 2010, **49**, 9644–9647.
[206] L. Ginnari-Satriani, V. Casagrande, A. Bianco, *et al.*, *Org. Biomol. Chem.*, 2009, **7**, 2513–2516.
[207] J. Barbaric, C. Wanninger-Weiß, and H.-A. Wagenknecht, *Eur. J. Org. Chem.*, 2009, 364–370.
[208] C. Wanninger and H.-A. Wagenknecht, *Synlett*, 2006, 2051–2054.
[209] C.L. Chepanoske, C.R. Langelier, N.H. Chmiel, and S.S. David, *Org. Lett.*, 2000, **2**, 1341–1344.
[210] J.S. Biradar, B.S. Sasidhar, and R. Parveen, *Eur. J. Med. Chem.*, 2010, **45**, 4074–4078.
[211] M. Bandini and A. Eichholzer, *Angew. Chem. Int. Ed.*, 2009, **48**, 9608–9644.
[212] T. Benincori, E. Brenna, F. Sannicolo, *et al.*, *J. Organometal. Chem.*, 1997, **529**, 445–453.
[213] T. Benincori, O. Piccolo, S. Rizzo, and F. Sannicolo, *J. Org. Chem.*, 2000, **65**, 8340–8347.
[214] J. Wassenaar and J.N.H. Reek, *Dalton Trans.*, 2007, **34**, 3750–3753.
[215] J. Wassenaar, M. Kuil, and J.N.H. Reek, *Adv. Synth. Catal.*, 2008, **350**, 1610–1614.
[216] J. Wassenaar, S. van Zutphen, G. Mora, *et al.*, *Organometallics*, 2009, **28**, 2724–2734.
[217] F. Rataboul, A. Zapf, R. Jackstell, *et al.*, *Chem. Eur. J.*, 2004, **10**, 2983–2990.
[218] Y.L. Choi, C.-M. Yu, B.T. Kim, and J.-N. Heo, *J. Org. Chem.*, 2009, **74**, 3948–3951.
[219] T. Mino, S. Komatsu, K. Wakui, *et al.*, *Tetrahedron Asym.*, 2010, **21**, 711–718.
[220] R. Ghosh, N.N. Adarsh, and A. Sarkar, *J. Org. Chem.*, 2010, **75**, 5320–5322.
[221] R. Ghosh and A. Sarkar, A., *J. Org. Chem.*, 2011, **76**, 8508–8512.
[222] R. Ghosh and A. Sarkar, A., *J. Org. Chem.*, 2010, **75**, 8283–8286.
[223] Y. Wang, A. Hämäläinen, J. Tois, and R. Franzén, *Tetrahedron: Asym.*, 2010, **21**, 2376–2384.
[224] Y. Wang, M.J.P. Vaismaa, A.M. Hämäläinen, *et al.*, *Tetrahedron: Asym.*, 2011, **22**, 524–529.
[225] Y. Wang, M.J.P. Vaismaa, K. Rissanen, and R. Franzén, *Eur. J. Org. Chem.*, 2012, 1569–1576.
[226] C.M. So, C.P. Lau, and F.Y. Kwong, *Org. Lett.*, 2007, **9**, 2795–2798.
[227] C.M. So, Z. Zhou, C.P. Lau, and F.Y. Kwong, *Angew. Chem. Int. Ed.*, 2008, **47**, 6402–6406.
[228] H.W. Lee, F.L. Lam, C.M. So., *et al.*, *Angew. Chem. Int. Ed.*, 2009, **48**, 7436–7439.
[229] C.M. So, C.P. Lau, A.S.C. Chan, and F.Y. Kwong, *J. Org. Chem.*, 2008, **73**, 7731–7734.
[230] C.M. So, C.C. Yeung, C.P. Lau, and F.Y. Kwong, *J. Org. Chem.*, 2008, **73**, 7803–7806.
[231] W.K. Chow, C.M. So, C.P. Lau, and F.Y. Kwong, *J. Org. Chem.*, 2010, **75**, 5109–5112.
[232] P.Y. Wong, W.K. Chow, K.H. Chung, *et al.*, *Chem. Commun.*, 2011, **47**, 8328–8330.
[233] C.M. So, H.W. Lee, C.P. Lau, and F.Y. Kwong, *Org. Lett.*, 2009, **11**, 317–320.
[234] P.Y. Yeung, C.M. So, C.P. Lau, and F.Y. Kwong, *Org. Lett.*, 2011, **13**, 648–651.
[235] W.K. Chow, O.Y. Yuen, C.M. So, *et al.*, *J. Org. Chem.*, 2012, **77**, 3543–3548.
[236] C.M. So, W.K. Chow, P.Y. Choi, *et al.*, *Chem. Eur. J.*, 2010, **16**, 7996–8001.
[237] M.J. Oila, J.E. Tois, and M.P. Koskinen, *Synth. Commun.*, 2008, **38**, 361–370.
[238] N. Kuuloja, J. Tois, and R. Franzén, *Tetrahedron: Asym.*, 2011, **22**, 468–475.
[239] S.R. Oakley, G. Nawn, K.M. Woldie, *et al.*, *Chem. Commun.*, 2010, **46**, 6753–6755.
[240] A. Isidro, D. Latassa, M. Giraud, *et al.*, *Org. Biomol. Chem.*, 2009, **7**, 2565–2569.
[241] D. Liu, G. Zhao, and L. Xiang, *Eur. J. Org. Chem.*, 2010, 3975–3984.
[242] F. Zhou, Y.-L. Liu, and J. Zhou, *Adv. Synth. Catal.*, 2010, **352**, 1381–1407.
[243] C. Marti and E.M. Carreira, *Eur. J. Org. Chem.*, 2003, 2209–2219.

PART I

Sigmatropic Rearrangements

Although the term *sigmatropic rearrangement* was unknown to Emil Fischer in 1883, this important rearrangement is the key element in Chapters 2 through 7, and, of course, it is a common mechanism in other aspects of organic chemistry.

Indole Ring Synthesis: From Natural Products to Drug Discovery, First Edition. Gordon W. Gribble.

2

Fischer Indole Synthesis

2.1 Preview

A logical beginning in the presentation of indole ring synthesis is with Emil Fischer and his indole synthesis, which has been in vogue since his discovery in 1883 [1] and subsequent exploration [2–5].

In addition to the general reviews of indole ring synthesis cited in Chapter 1, the Fischer indole synthesis *per se* has been extensively reviewed [6–10]. The excellent documentation (>3000 citations in reference 8) will not be repeated here. Rather in this chapter, I focus on recent applications in drug development, materials discovery, and natural-product synthesis.

The mechanism of the Fischer indole synthesis has been extensively studied [11–38], and the accepted mechanism is shown in Scheme 1. The so-called abnormal Fischer indolization (Scheme 2) has been studied by Ishii and Murakani and their coworkers [39–44], and the interrupted Fischer indolization (Scheme 3) has been evaluated theoretically by Houk, Garg, and colleagues [38].

Several key pieces of the mechanistic puzzle that encompass the Fischer indolization mechanism are depicted in Scheme 4 [45–48]. The reader is advised to consult the relevant references for further examples and details [49].

2.2 Methods

2.2.1 Traditional Methods

Although Fischer and his students explored a few early methods to perform indolization of phenylhydrazones, the methodology has been greatly improved and developed in the intervening 130 years. Table 1 summarizes and compares some of these methods [50–65]. In each case one of the better results, or the best result, is selected. Noteworthy are the uses of ionic liquids (entries 16–18, 28–30), clays (19, 20), and the "green" catalysts bismuth nitrate (entry 23), hydrous zirconia–supported niobium oxide (entry 24), and the novel tartaric acid–dimethylurea melt (entries 25–27), conditions that also allow the synthesis of indolenines in excellent yield.

Several workers have described Fischer indolization without the use of acidic catalysts, the epitome of green chemistry. These thermal syntheses are shown in Table 2 [66–68]. The first practical account of noncatalytic Fischer indole synthesis appears to be that of Fitzpatrick and Hiser (entries 1–3) [66]. Matsumoto and colleagues have effected indolizations with *p*-toluenesulfonic acid in the absence of solvent at 250 °C to afford indoles **1–4** from the corresponding ketones and phenylhydrazine hydrochlorides (Scheme 5) [69]. In some cases, trichloroacetic acid was employed.

As in other areas of organic synthesis, microwave irradiation has been adapted in Fischer indolizations (Table 3) [70–77]. Whereas montmorillonite KSF worked well as reported for entries 1 to 3 [70], it utterly failed as reported for entries 4 to 5 [71]. Entry 9 represents a novel decarboxylation of indole-2-carboxylic acids during microwave heating [73]. For entries 10 and 11, yields are invariably higher under microwave conditions than with conventional heating, and the former represents an excellent synthesis of γ-carbolines [74]. Entries 12 to 15 feature propyl phosphonic acid cyclic anhydride (T3P) as a mild water scavenger [75]. Kremsner and Kappe have studied microwave-assisted organic synthesis in near-critical water at 300 °C,

Indole Ring Synthesis: From Natural Products to Drug Discovery, First Edition. Gordon W. Gribble.

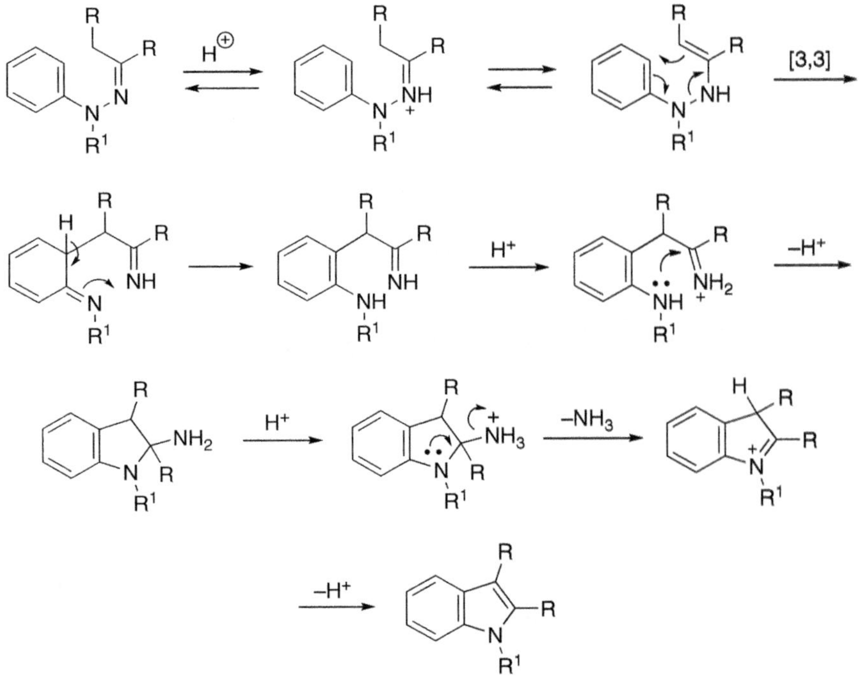

Scheme 1 *Fischer Indole Synthesis Mechanism*

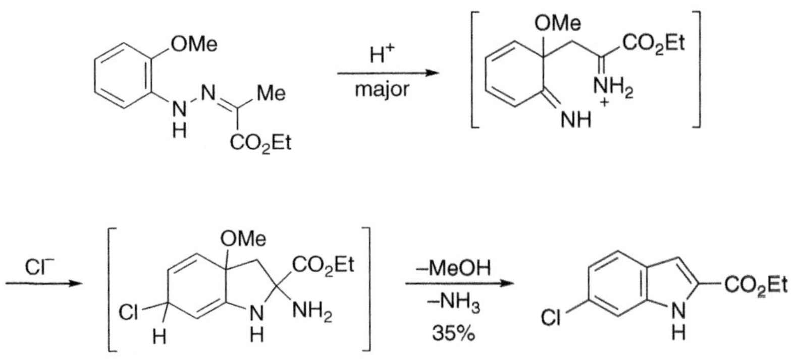

Scheme 2 *Interrupted Fischer Indole Synthesis*

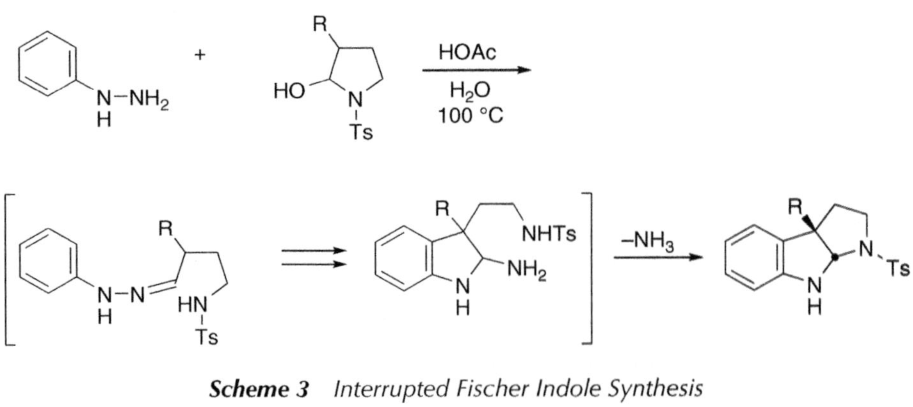

Scheme 3 *Interrupted Fischer Indole Synthesis*

and they demonstrated that Fischer indolization with 2-butanone could be effected to give 2,3-dimethylindole in 64% yield [78].

Several other newer indolization techniques completely unknown to Fischer and his contemporaries are cited here. The synthesis of indoles using continuous flow reactors has been described by the groups of Watts [79] and Cosford [80]. The latter group synthesized a series of biologically active 2-(1*H*-indol-3-yl)thiazoles **5** as shown in Scheme 6. This features a sequential Hantzsch thiazole synthesis, deketalization, and Fischer indolization. A facile preparation of 3*H*-indolium perchlorates **6** is achieved in one pot from suitable phenylhydrazines, α-branched ketones, and perchloric acid in ethanol (Scheme 7). More than two dozen examples are described [81]. The role of 12 Lewis acids and ten solvents was studied with several ketones in an attempt to understand the regiochemistry that is observed with unsymmetrical ketones [82]. Perhaps surprisingly,

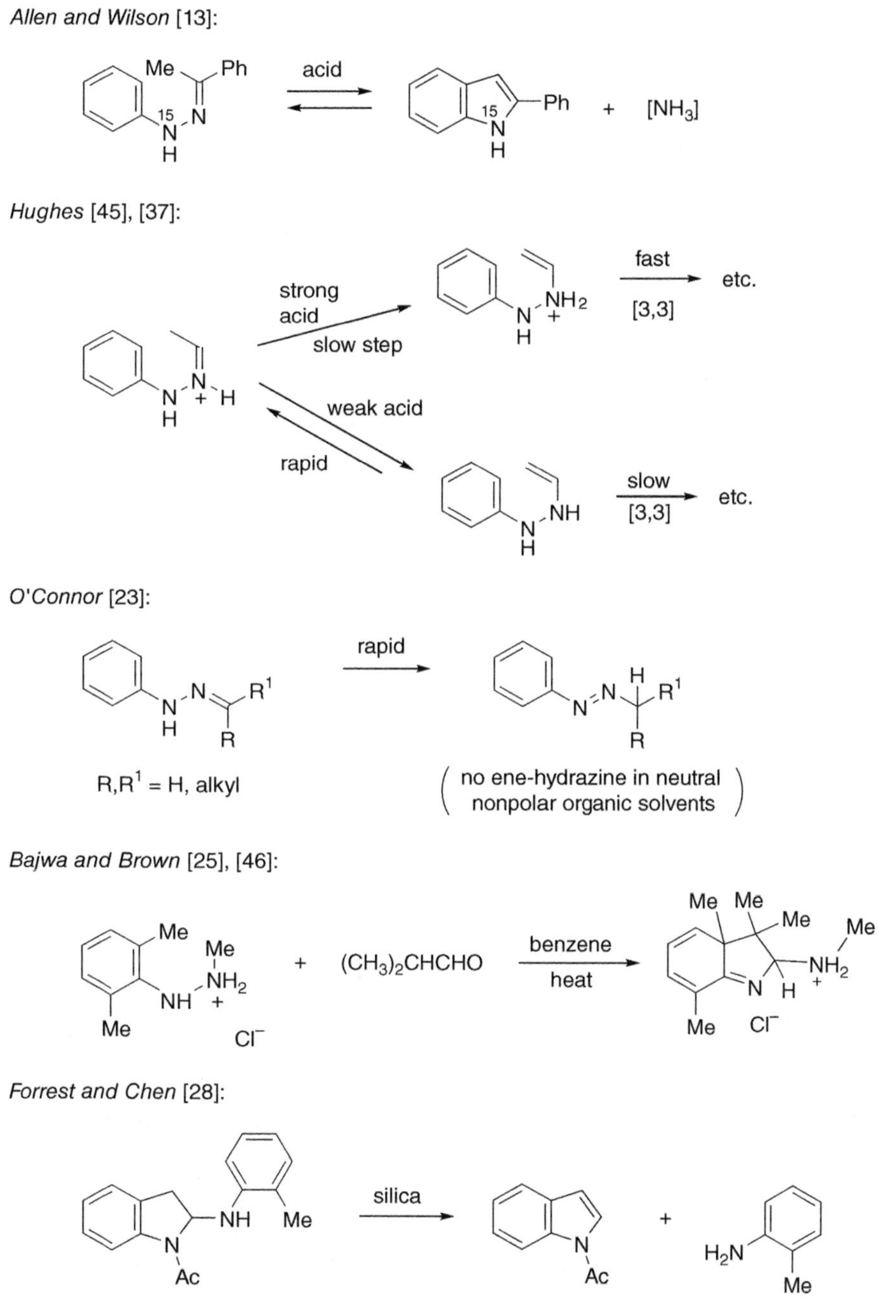

Scheme 4 *Mechanistic Studies of the Fischer Indole Synthesis*

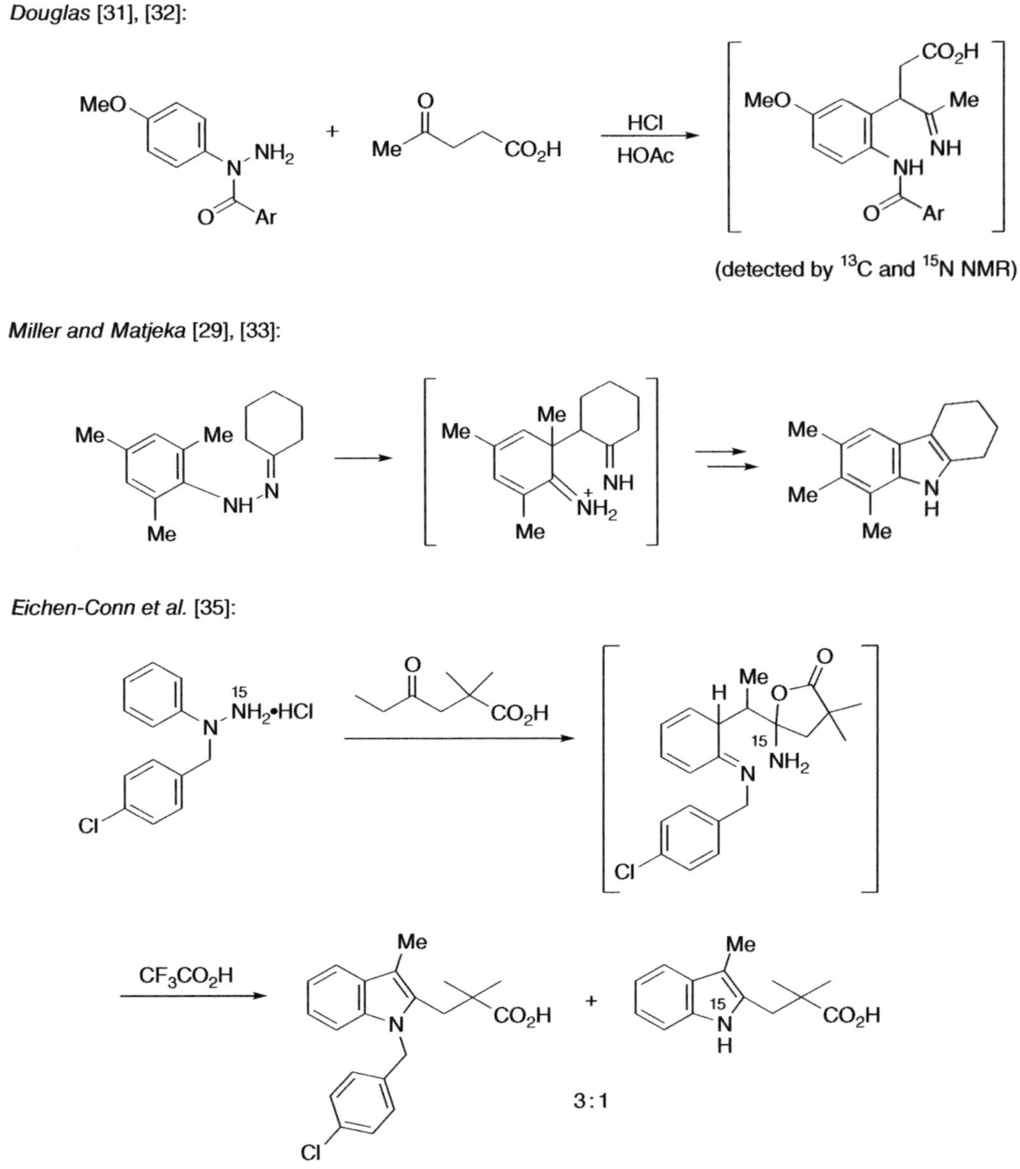

Scheme 4 *(continued)*

whereas the structure of the phenylhydrazone is important, the structure of the Lewis acid has no control over the regiochemistry, and the solvent is only marginally involved. Interestingly, the catalytic system of zeolites in xylene favors the formation of 3-methyl-2-*n*-propylindoles over the isomeric 2,3-diethylindoles (e.g., 9:1) [83, 84]. Similar regiochemistry is observed with phenylhydrazones prepared from unsymmetrical aliphatic ketones. Another relatively new tactic is to employ stabilized arylhydrazines in the Fischer indole synthesis. Thus, Cho and Lim find that stable *N*-Boc arylhydrazines, which are readily available, are smoothly converted to indoles under mild conditions (*p*-TsOH, ethanol, reflux) [85]. The *N*-Boc group is removed during the reaction. Two examples are shown in Scheme 8. These *N*-Boc arylhydrazines can be accessed by palladium-catalyzed cross-coupling; for example, to give enehydrazine **7** and tetrahydrocarbazole **8**.

2.2.2 Metal-Catalyzed Methods

In lieu of the traditional acid catalysts that we saw in the previous section, several groups have explored the use of metals (Rh, Ti, Ru, Au, Zn, Al) to achieve Fischer indolization or to access phenylhydrazines.

Following early work with rhodium by Watanabe [86] to give indoles in low to modest yields (Scheme 9, equations

Garg, Houk [38]:

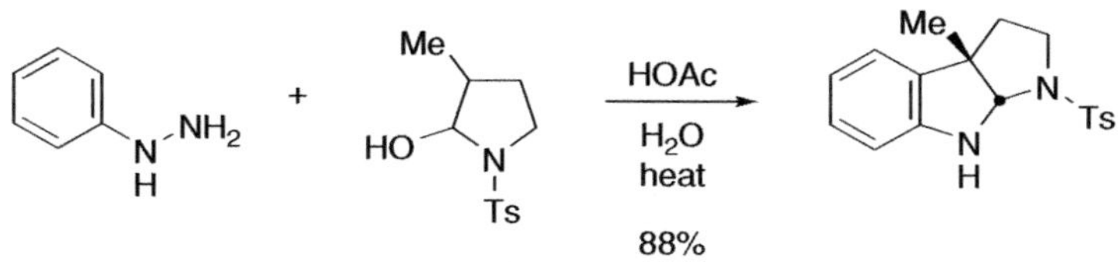

Huisgen [47]:

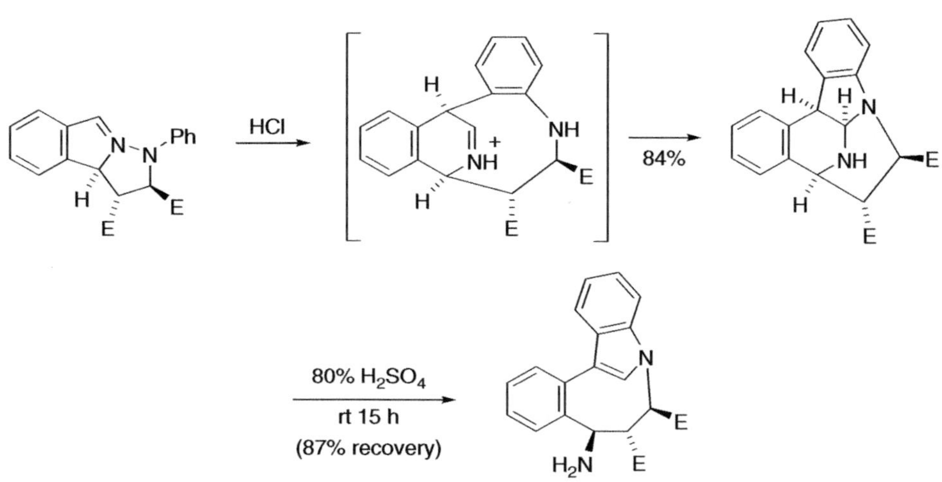

Benke and Brooke [48]:

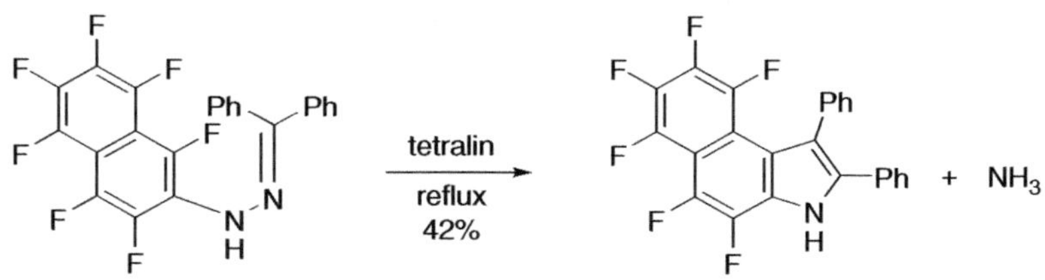

Scheme 4 *(continued)*

1 and 2), a reaction thought to involve *ortho*-metalation of an ene-hydrazine, Sheldon and colleagues [87] described the rhodium-catalyzed hydroformylation of *N*-allylacetamide leading to a synthesis of melatonin (**9**) (Scheme 9, equation 3).

This hydroformylation Fischer indole synthesis was extended and developed by Eilbracht and his research group [88–93]. A small selection of this elegant chemistry is shown in Scheme 10 (equations 1–5). This chemistry has been adapted to the synthesis of tetrahydro-β-carbolines, a process that uses 2,5-dihydropyrroles and arylhydrazines and features spiroindoleninium cation intermediates (equation 5) [92].

Titanium-crafted catalysts have been employed by several groups to effect hydroamination of alkynes with hydrazines to afford hydrazones en route to indoles via Fischer indolization. Odom and coworkers performed this chemistry with 1,1-disubstituted hydrazines; a selection is shown in Table 4 (entries 1–4) [94, 95]. In a similar approach to aryl hydrazones and indoles, Beller and colleagues employed a titanium catalyst to hydroaminate terminal alkynes leading to indoles (Table 4, entries 5–7) [96–98]. Beller's protocol is particularly valuable for the synthesis of tryptamines, tryptophols, and their homologues. Interestingly, Beller found that 1-alkyl-1-phenylhydrazines react with acetylenedicarboxylates in the absence of titanium to give the corresponding indole-2,3-dicarboxylates and 2-arylindole-3-carboxylates after treatment with zinc chloride [99]. This reaction was apparently first discovered by Diels and Reese in 1935 [100] and by others subsequently [101, 102]. Some examples of Beller's work are shown in Scheme 11 (equations 1–3) [99].

Nagase et al. [49]:

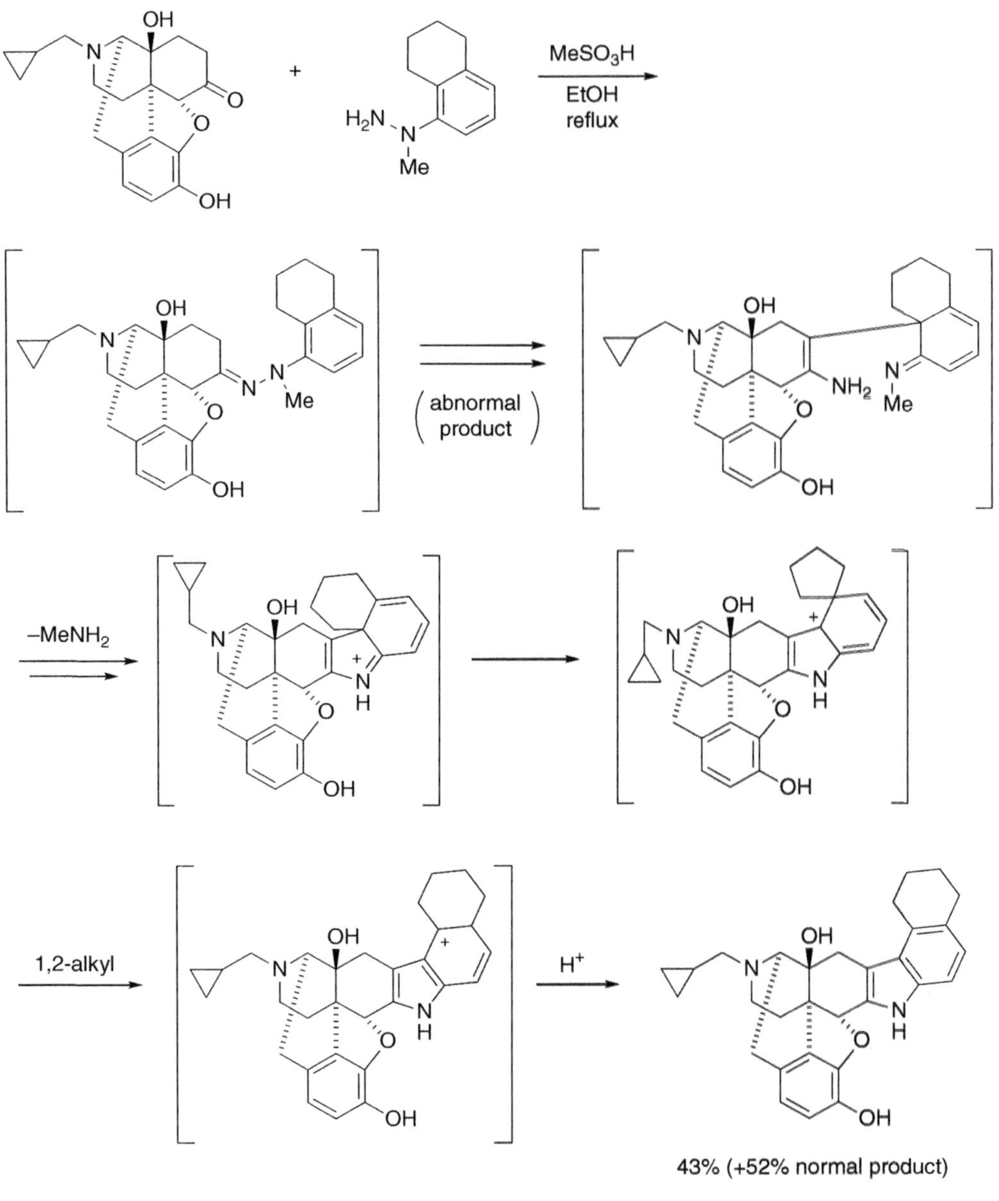

Scheme 4 *(continued)*

Ackermann used $TiCl_4$ with t-$BuNH_2$ as a catalyst for the hydroamination of alkynes with hydrazines, leading to indoles. Three examples of this nice Fischer indolization are shown in equations 4 to 6 [103].

Several other metal-assisted Fischer indole syntheses are summarized in Scheme 12. Yamamoto employed organoaluminum amides to effect regioselective Fischer indolizations [104]. For example, *Z*-hydrazone **10** affords 1,3-dimethyl-2-(2-methylbutyl)indole (equation 1), whereas *E*-hydrazone **11** gives 3-*sec*-butyl-2-ethyl-1-methylindole (equation 2). Both reactions are highly regioselective. The mechanism involves isomerization of the phenylhydrazone to the corresponding enehydrazine followed by the usual [3,3] sigmatropic rearrangement and loss of ammonia. Rasmussen found that the isomerization of allylphenylhydrazines to the corresponding enehydrazines is achieved by a combination of the first-generation Grubbs catalyst and lithium triethylborohydride, which affords indoles under the reaction conditions (Scheme 12, equation 4) [105]. The requisite allylphenylhydrazines are available by allylation of either the appropriate phenylhydrazine or the phenylhydrazide followed by base hydrolysis.

Table 1 *Typical Methods of Fischer Indolization*

Entry	Substrate	Conditions	Product	% Yield	Ref.
1	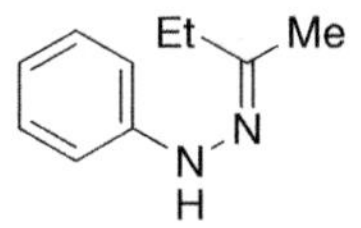	BF_3•Et_2O HOAc reflux	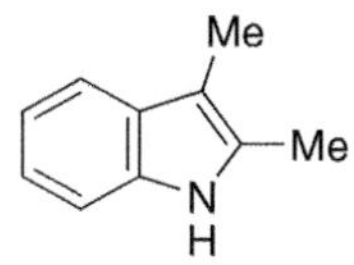	85%	50
2	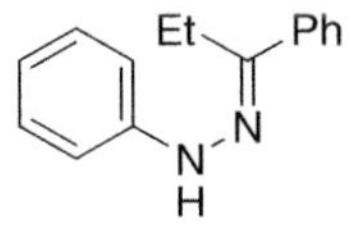	BF_3•Et_2O HOAc reflux	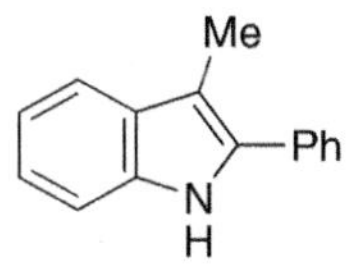	87%	50
3	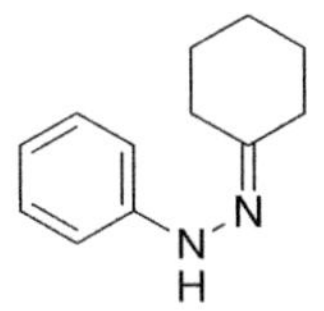	BF_3•Et_2O HOAc reflux	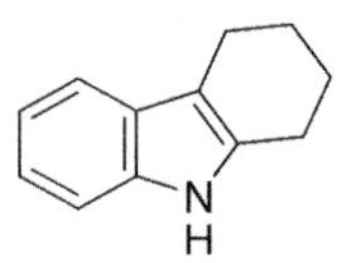	93%	50
4	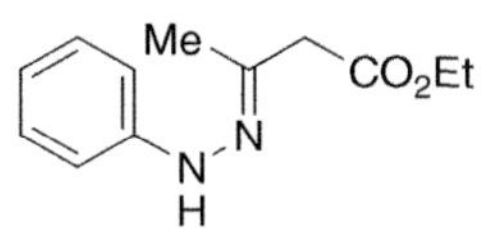	BF_3•Et_2O 127 °C	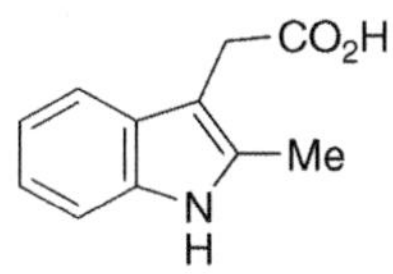	96%	50
5	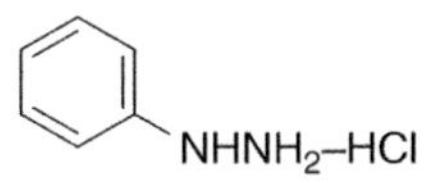	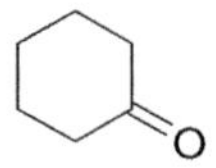 pyridine, reflux	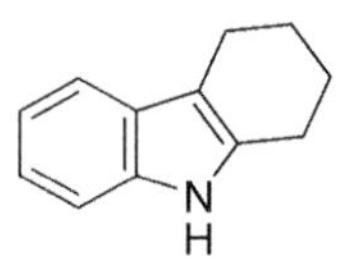	78%	51
6	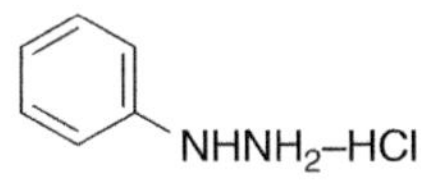	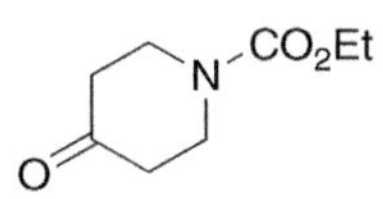 pyridine, reflux	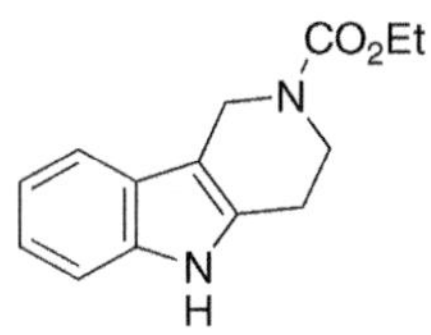	98%	51
7	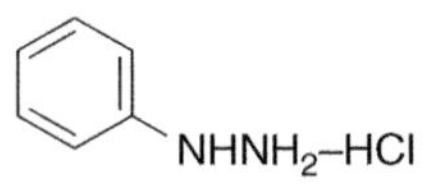	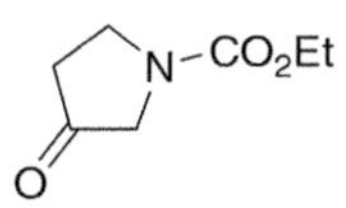 pyridine, reflux	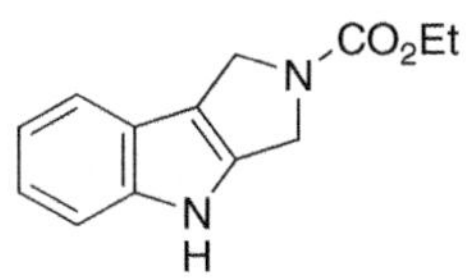	80%	51
8	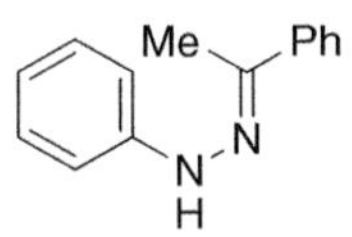	PPA, 130 °C	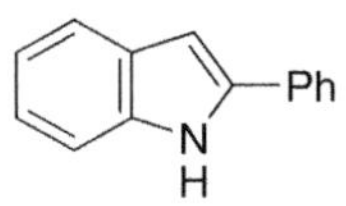	83%	52
9	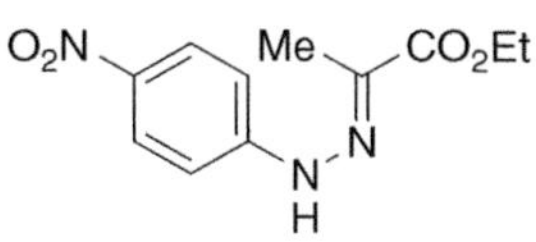	PPA,110 °C	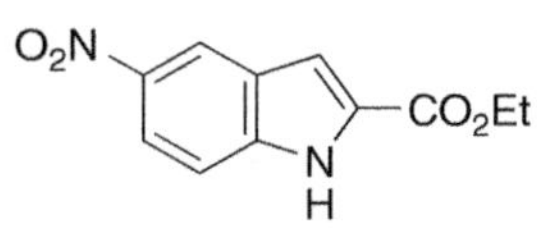	78%	52
10	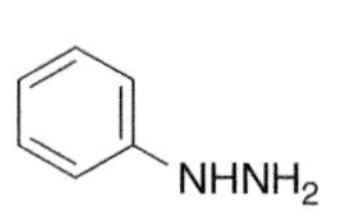	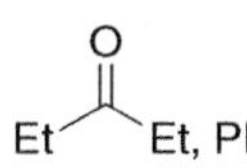 Et, PPSE $(CH_2Cl)_2$, 85 °C 10 min	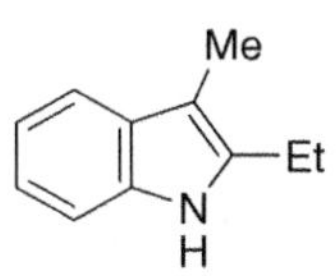	88%	53

(continued overleaf)

Table 1 *(continued)*

Entry	Substrate	Conditions	Product	% Yield	Ref.
11	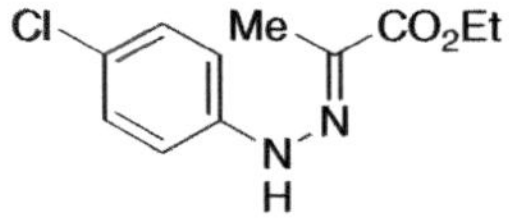	*p*-TsOH, PhH	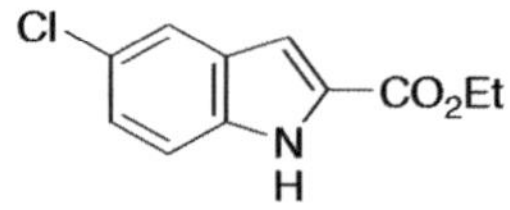	89%	54
12	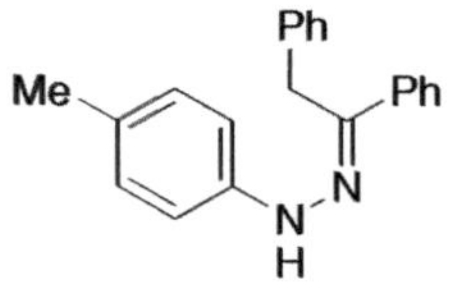	PCl_3, rt CH_2Cl_2	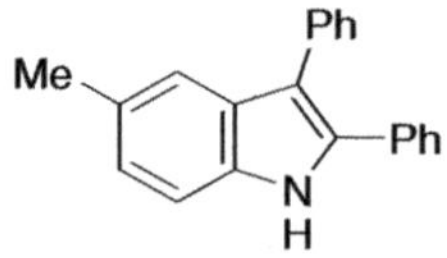	85%	55
13	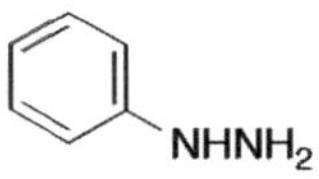	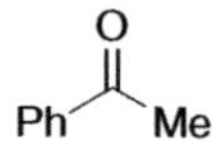 PPA, 180 °C	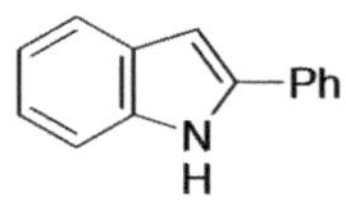	76%	56
14	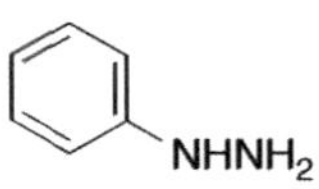	, HOAc reflux	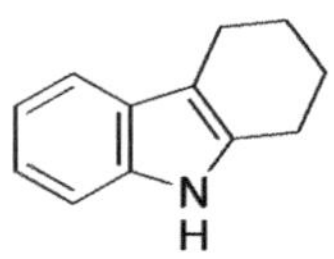	88%	57
15	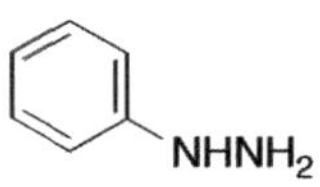	, HCl, EtOH reflux	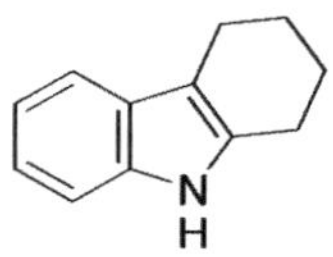	95%	57
16	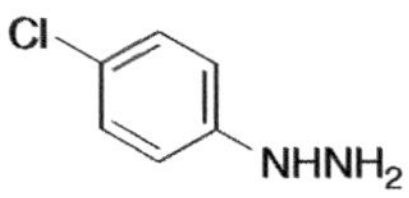	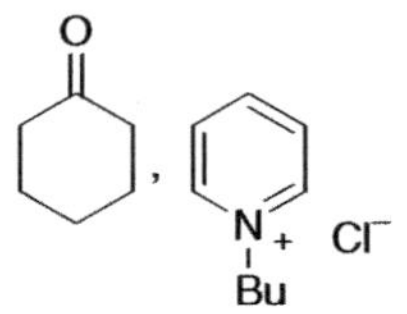 $AlCl_3$ 180–185 °C	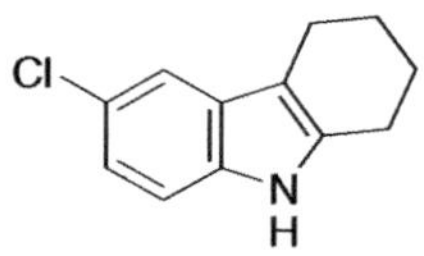	92%	58
17	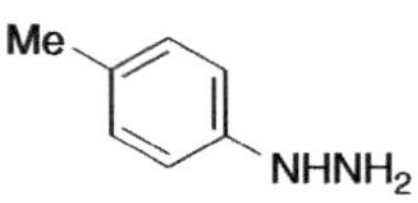	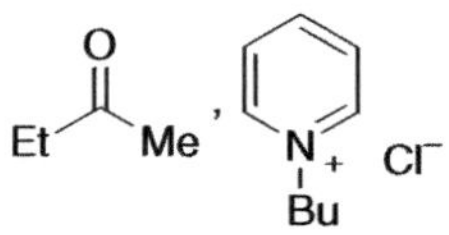 $AlCl_3$ 180–185 °C	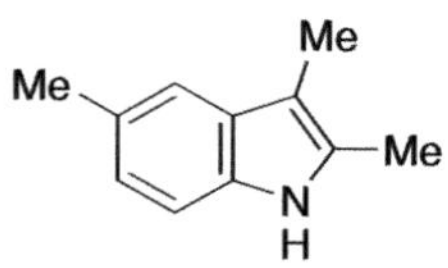	90%	58
18	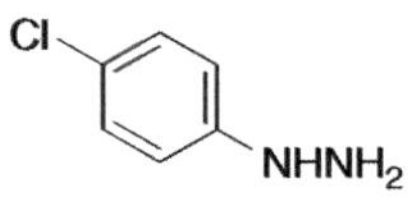	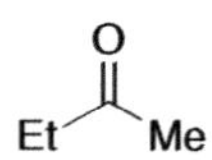, 120 °C choline chloride/$2ZnCl_2$	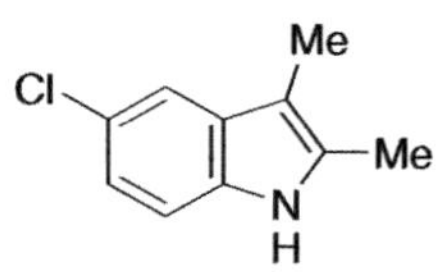	88%	59
19	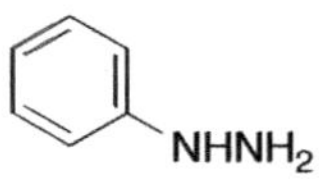	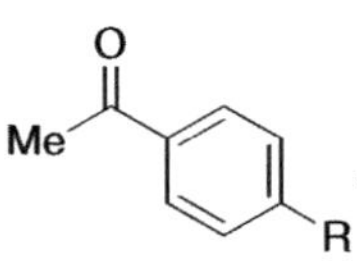, MeOH, 60 °C K-10 montmorillonite	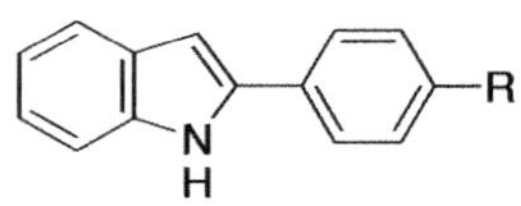 R = H, Cl, Me, NO_2	88–92%	60
20	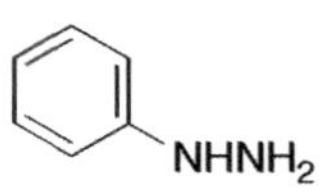	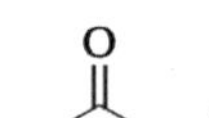, MeOH, 60 °C K-10 montmorillonite	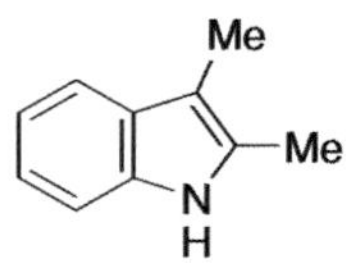	93%	60

Table 1 *(continued)*

Entry	Substrate	Conditions	Product	% Yield	Ref.
21	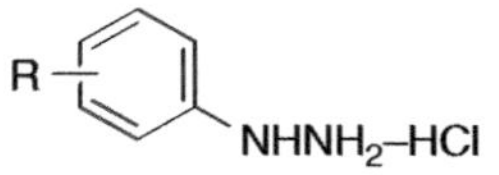 R = H, 4-F, 4-Me, 4-OMe, 2-Me, 2,5-Cl_2	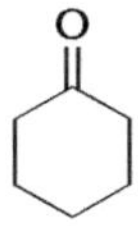 , $Ce(NH_4)_2(NO_2)_6$ MeOH, reflux	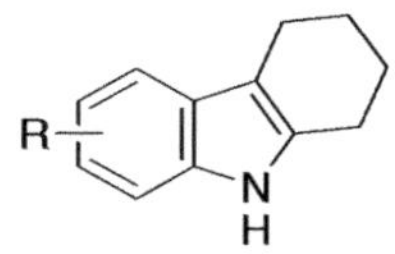	85–95%	61
22	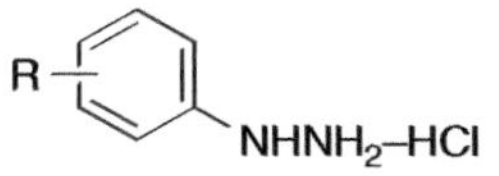 R = H, 4-F, 4-Me, 4-OMe, 2-Me, 2-NO_2, 2,5-Cl_2	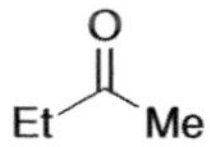 , $Ce(NH_4)_2(NO_2)_6$ MeOH, reflux	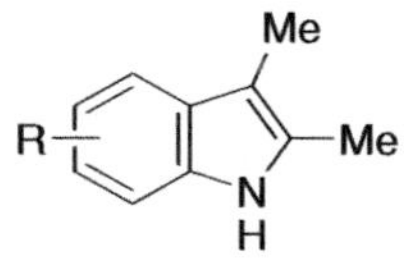	65–95%	61
23	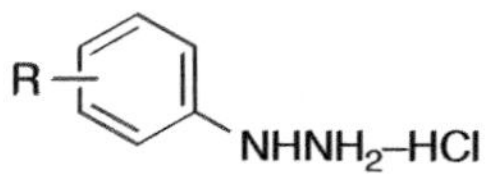 R = H, 4-NO_2, 3-Cl, 3-NO_2, 2-NO_2, 3-Br, 4-Br, 4-F, 3-Cl, 2-F, 4-NO_2	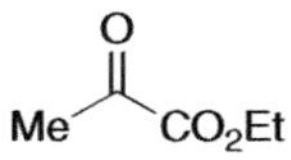 , $BiNO_3$ PPA, EtOH, reflux	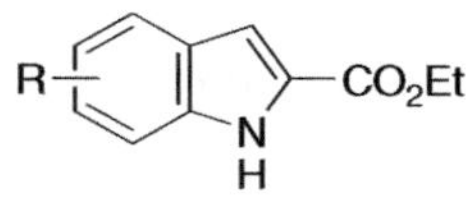	80–90%	62
24	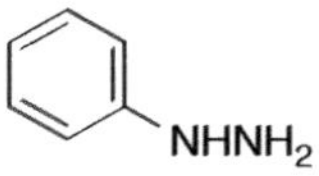	 , $Nb_2O_5/Zr(OH)_2$ *p*-xylene, 140 °C	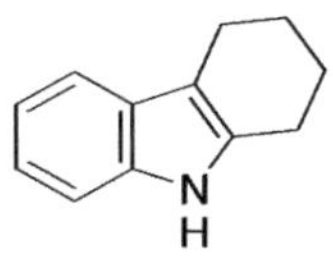	88%	63
25	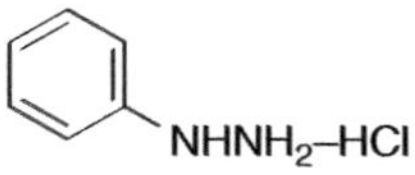	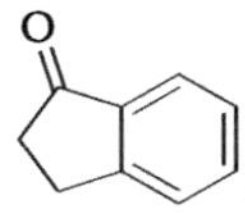 , tartaric acid-dimethylurea melt 90 °C	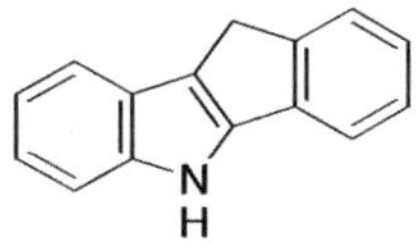	97%	64
26	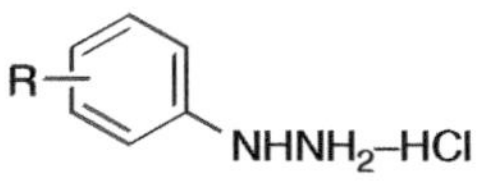 R = 4-OMe, 4-Me, 4-Cl, 2,4-Cl_2	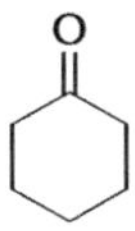 , tartaric acid-dimethylurea, 70–90 °C	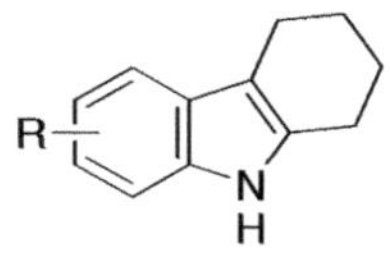	90–97%	64
27	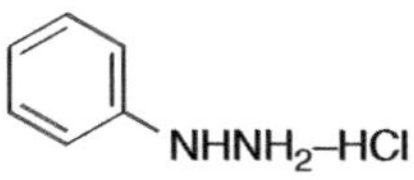	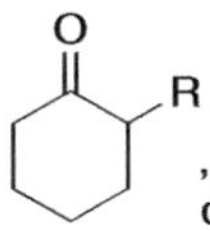 , tartaric acid-dimethylurea, 70 °C	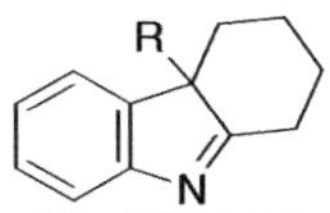 R = Me, $(CH_2)_2CO_2Et$, $(CH_2)_2N_3$, $CH_2CH{=}CH_2$, $(CH_2)_2CN$	90–99%	64
28	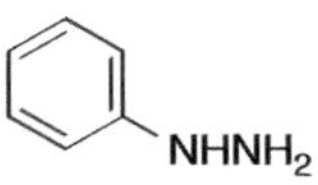	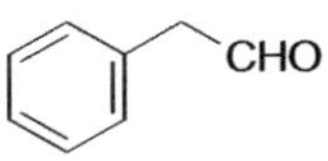 , $BMImHSO_4$ 70 °C	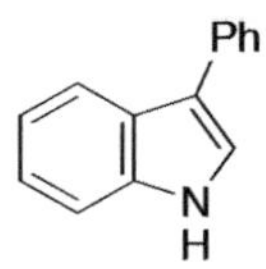	87%	65
29	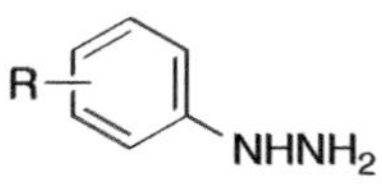 R= H, 2,5-Cl_2, 2-Me	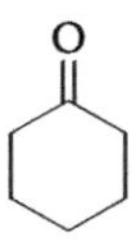 , $BMImHSO_4$ 70 °C	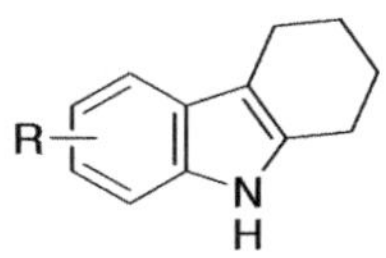	92–96%	65
30	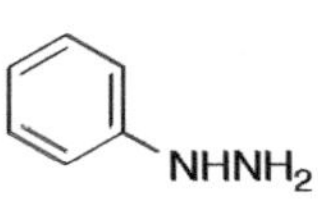	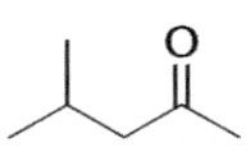 , $BMImHSO_4$ 110 °C	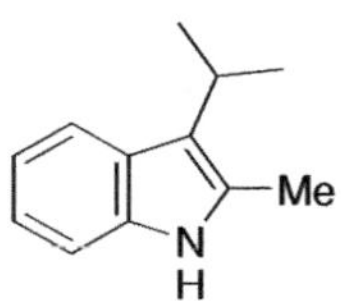	85%	65

Table 2 *Thermal Fischer Indolization*

Entry	Substrate	Conditions	Product	% Yield	Ref.
1	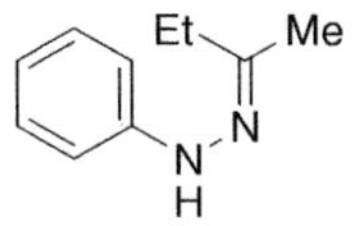	$(HOCH_2)_2$, reflux	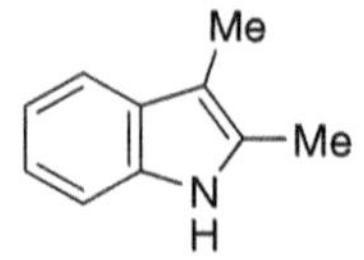	70%	66
2	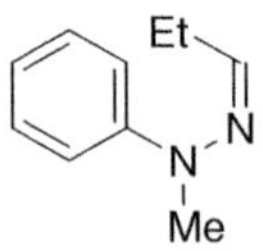	$(HOCH_2)_2$, reflux	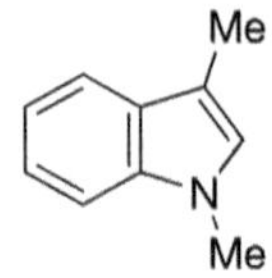	70%	66
3	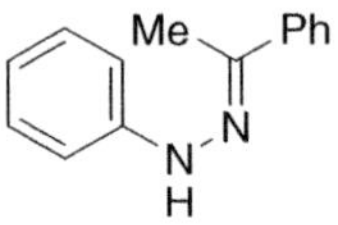	$(HOCH_2)_2$, reflux	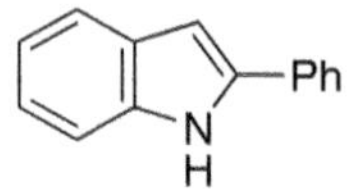	54%	66
4	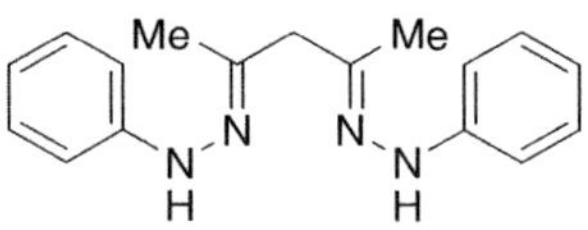	$(HOCH_2)_2$, reflux	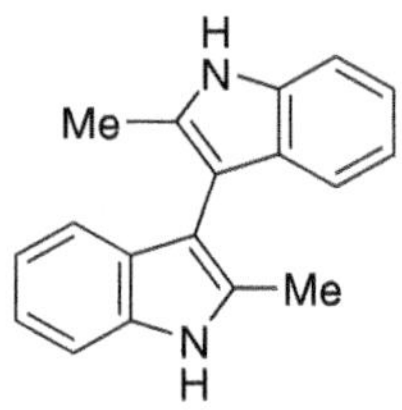	5%	67
5	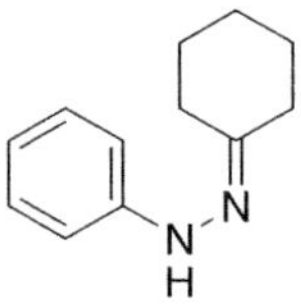	$(HOCH_2)_2O$, reflux	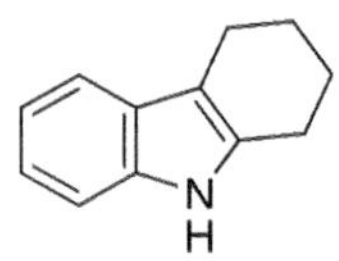	83%	68
6	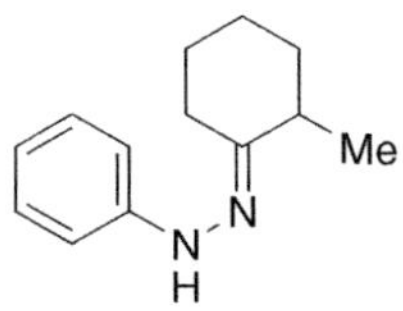	$(HOCH_2)_2O$, reflux	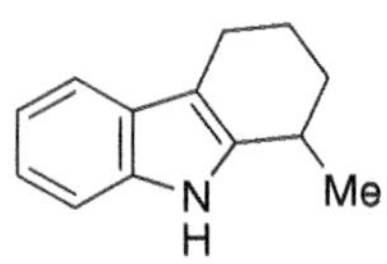	96%	68

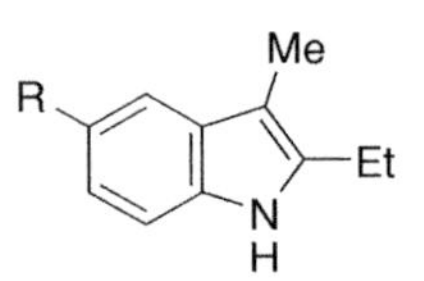

1, R = H, Me, Cl (78–100%)

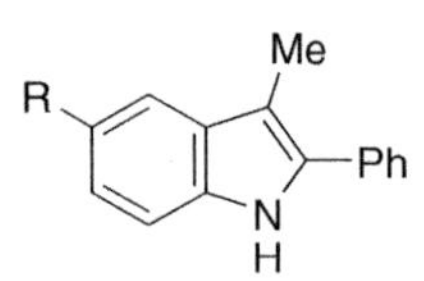

2, R = H, Me, Cl (79–99%)

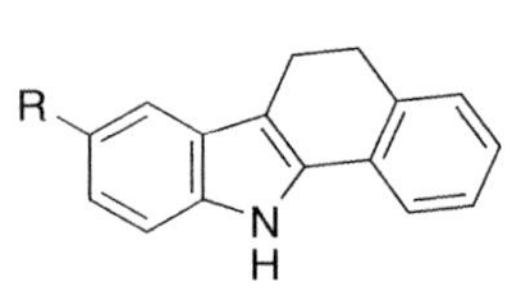

3, R = H, Me, Cl (91–98%)

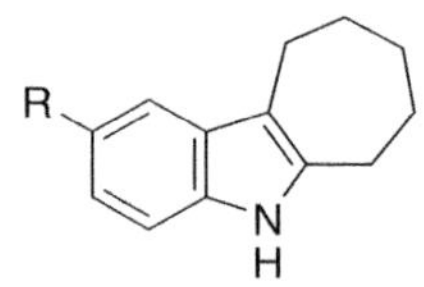

4, R = H, Me, Cl (90–100%)

Scheme 5 *Matsumoto Indole Synthesis [69]*

Knochel and colleagues have developed a zinc-promoted approach to indoles that does not involve arylhydrazones but rather aryldiazonium salts as a route to enehydrazines (Scheme 13) [106, 107]. Aryldiazonium salts react with organozinc reagents to form the aryl azo derivative that isomerizes to the corresponding enehydrazine and thence to indole. Several examples of this unique method of Fischer indolization are shown in Scheme 13. Knochel has synthesized indomethacin and iprindole via this methodology. As shown in equation 2, the crucial enehydrazine intermediate is obtained by Grignard addition to the azobenzene. The trimethylsilyl chloride step effects the final indolization of the enehydrazine. A different use of zinc is that of Beller and colleagues to

Table 3 *Microwave-Facilitated Fischer Indolizations*

Entry	Substrate	Conditions	Product	% Yield	Ref.
1	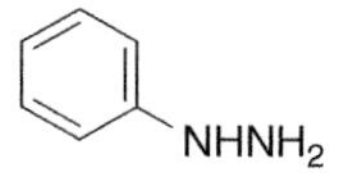	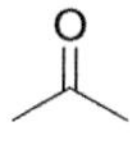 , montmorillonite KSF μν 160W, 5 min	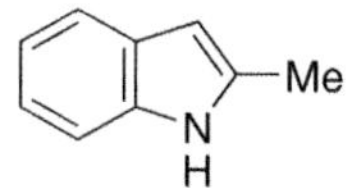	86%	70
2	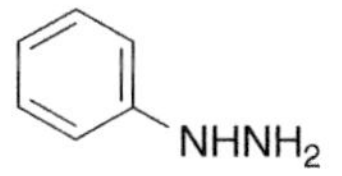	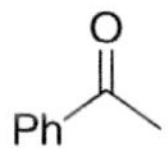 , montmorillonite KSF μν 160W, 5 min	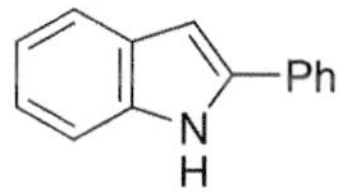	90%	70
3	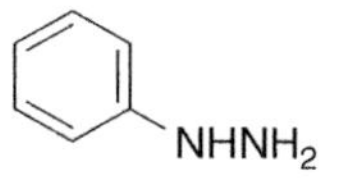	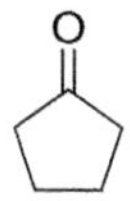 , montmorillonite KSF μν 160W, 5 min	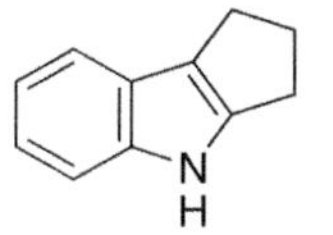	82%	70
4	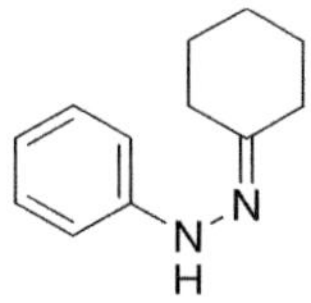	96% HCO_2H, μν 2 min	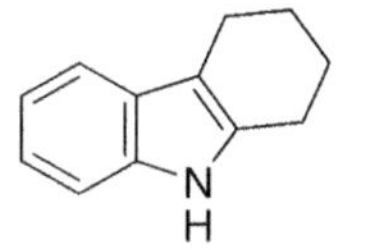	100%	71
5	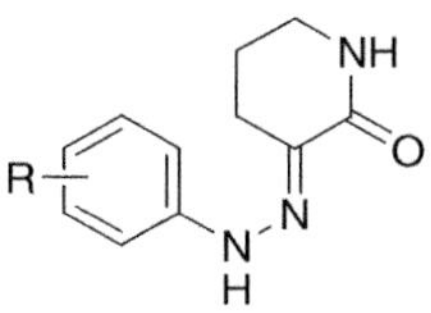 R = H, 2-Me, 4-OMe, 4-Br, 4-NO_2	96% HCO_2H, μν 2 min	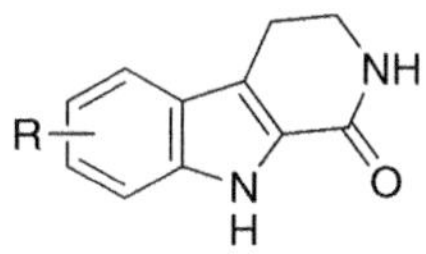	73–88%	71
6	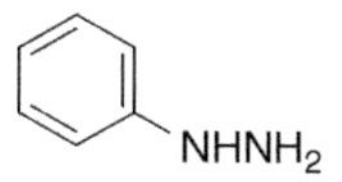	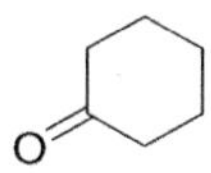 μν, 1–2 min $KHSO_4/H_2O/SiO_2$	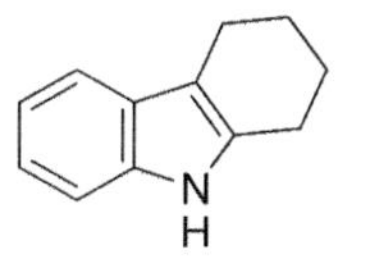	99%	72
7	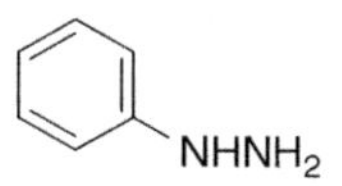	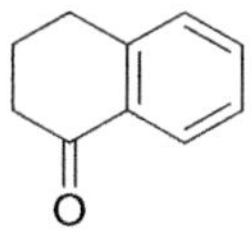 μν, 1–2 min $KHSO_4/H_2O/SiO_2$	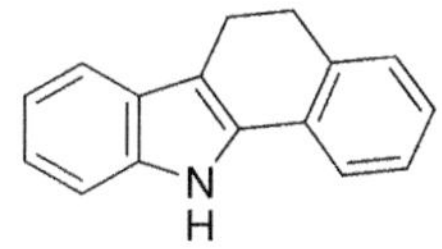	94%	72
8	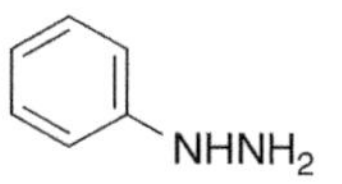	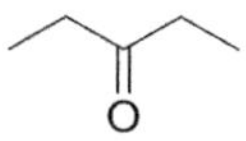 μν, 1–2 min $KHSO_4/H_2O/SiO_2$	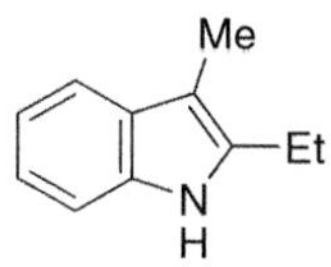	94%	72
9	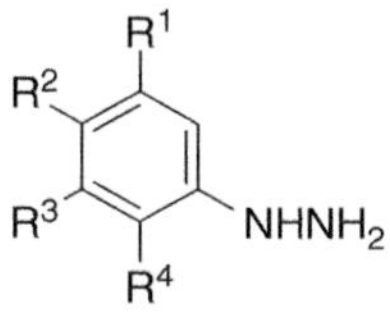 R^1= H, OMe R^2= H, Br, F, NO_2 R^3= H, OMe R^4= H, Me, NO_2, Cl	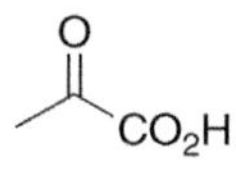 μν, $ZnCl_2$ PCl_5	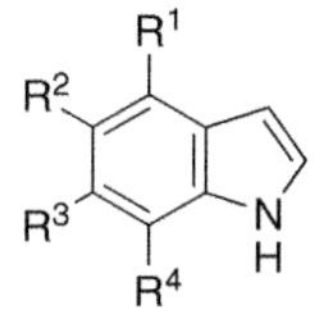	71–88%	73

(continued overleaf)

Table 3 *(continued)*

Entry	Substrate	Conditions	Product	% Yield	Ref.
10	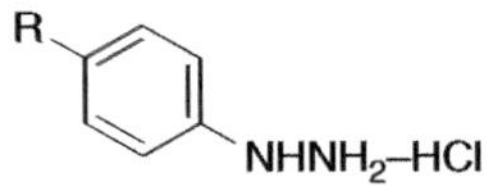 R= H, OMe, Me, F, Cl, Br	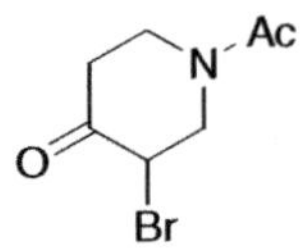 μν, HOAc 230 °C, 5 min	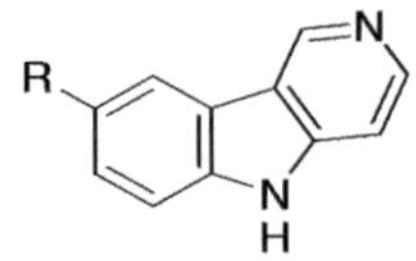	54–83%	74
11	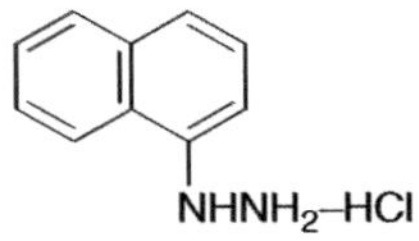	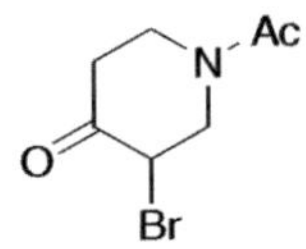 μν, HOAc 230 °C, 5 min	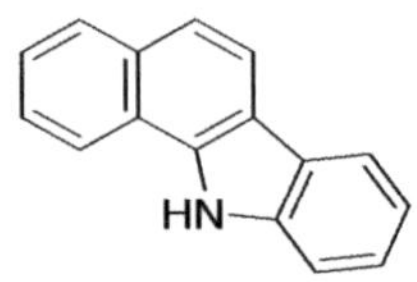	44%	74
12	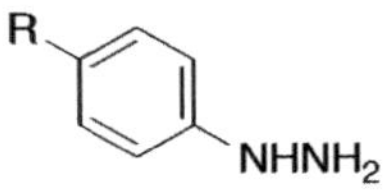 R= H, OMe, Me, Br, F, CF_3	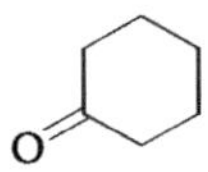 μν, EtOAc T3P	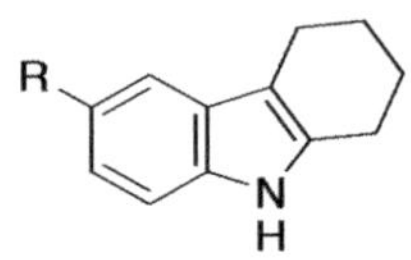	82–93%	75
13	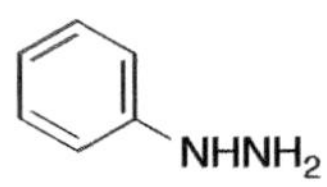	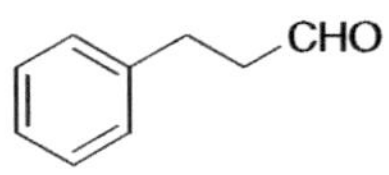 μν, EtOAc T3P	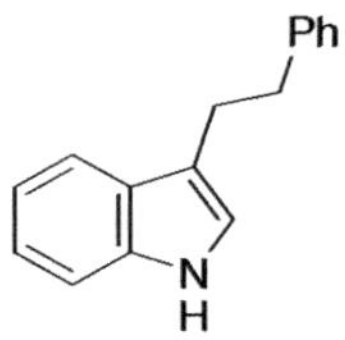	82%	75
14	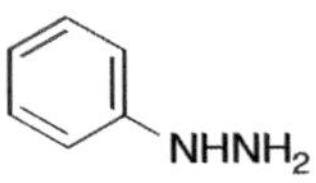	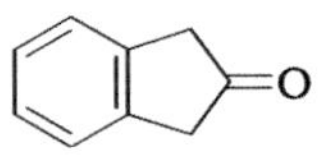 μν, EtOAc T3P	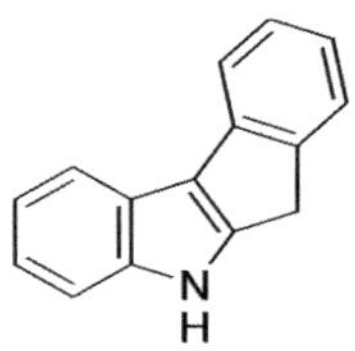	75%	75
15	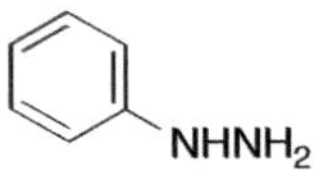	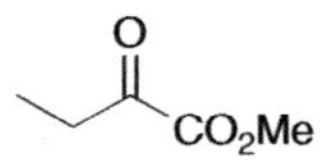 μν, EtOAc T3P	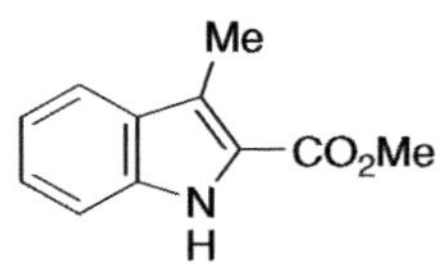	90%	75
16	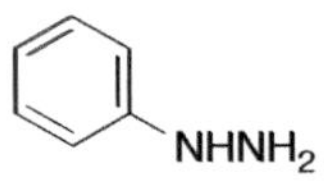	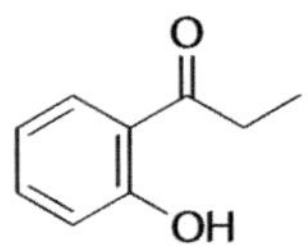 μν, *p*-TSA	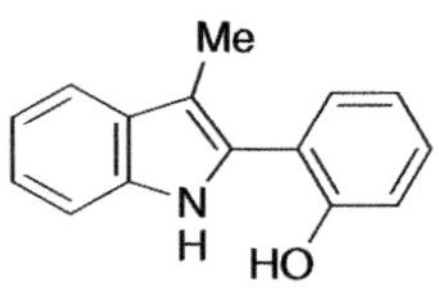	94%	76
17	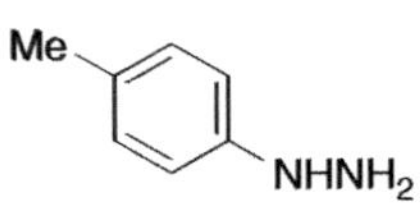 +	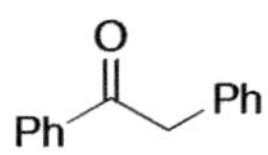 μν, *p*-TSA	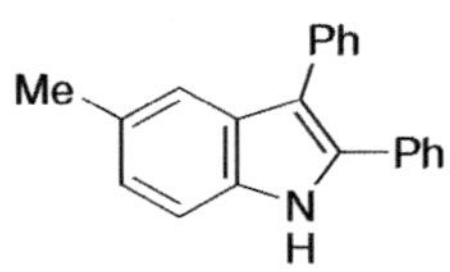	87%	76
18	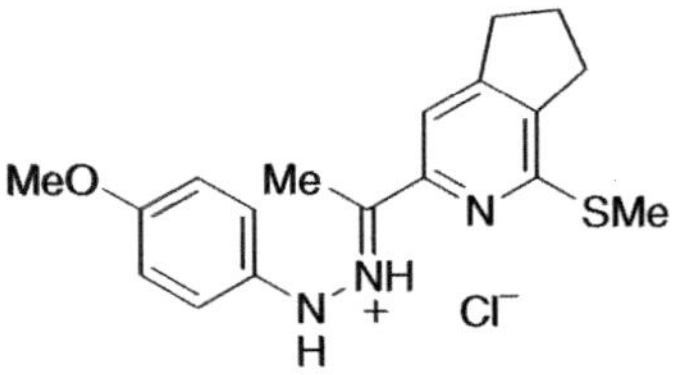	μν, $ZnCl_2$ TGE	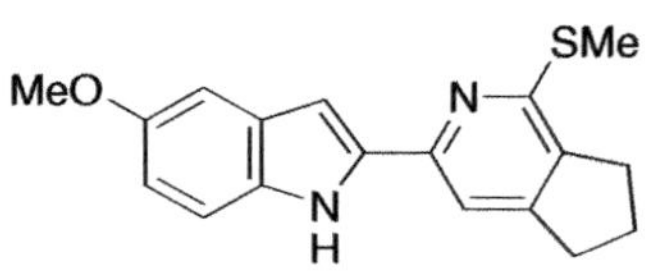	63%	77

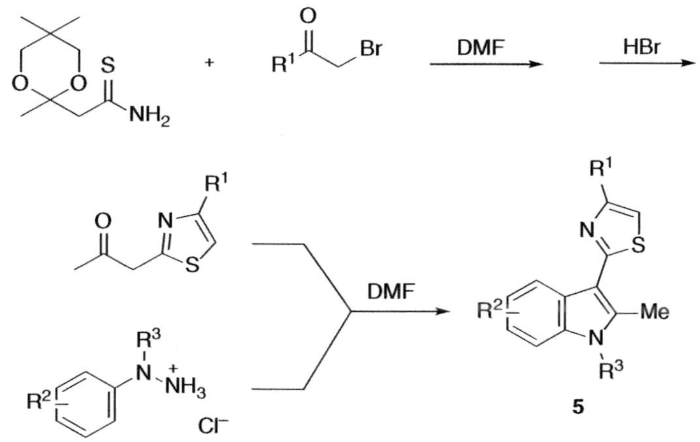

Scheme 6 Cosford Indole Synthesis [80]

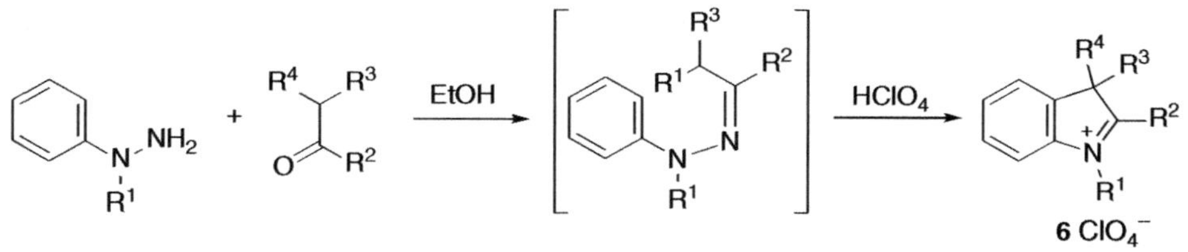

R^1	R^2	R^3	R^4	% yield
Me	Me	Me	Me	83
Mr	Me	Me	Bn	67
i-Pr	Me	Me	Et	49
Me	4-FPh	Me	Me	37
Me	Ph	Me	Me	54

Scheme 7 Zimmermann Indole Synthesis [81]

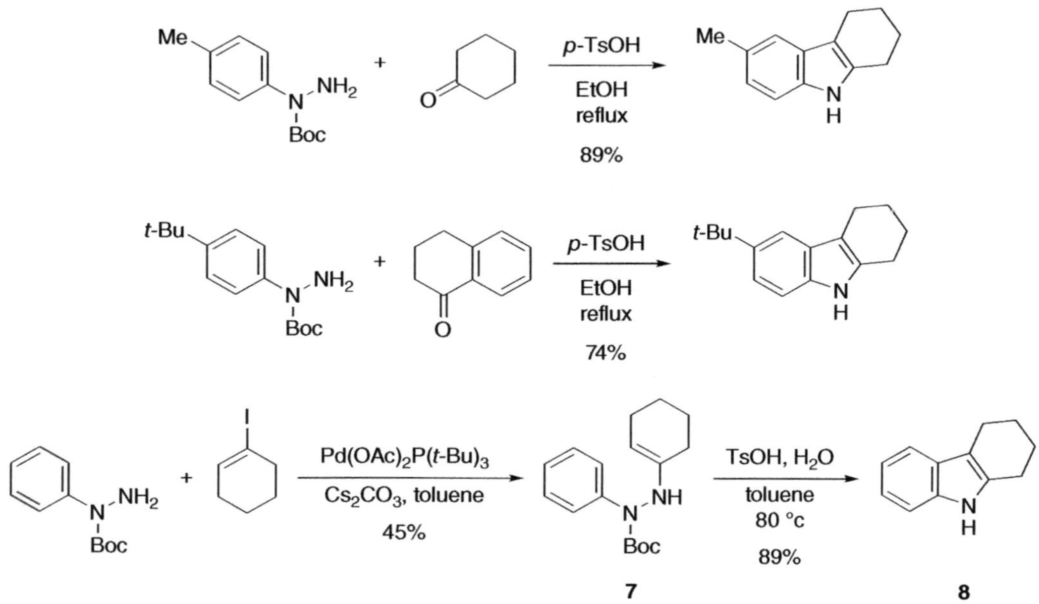

Scheme 8 Cho Indole Synthesis [85]

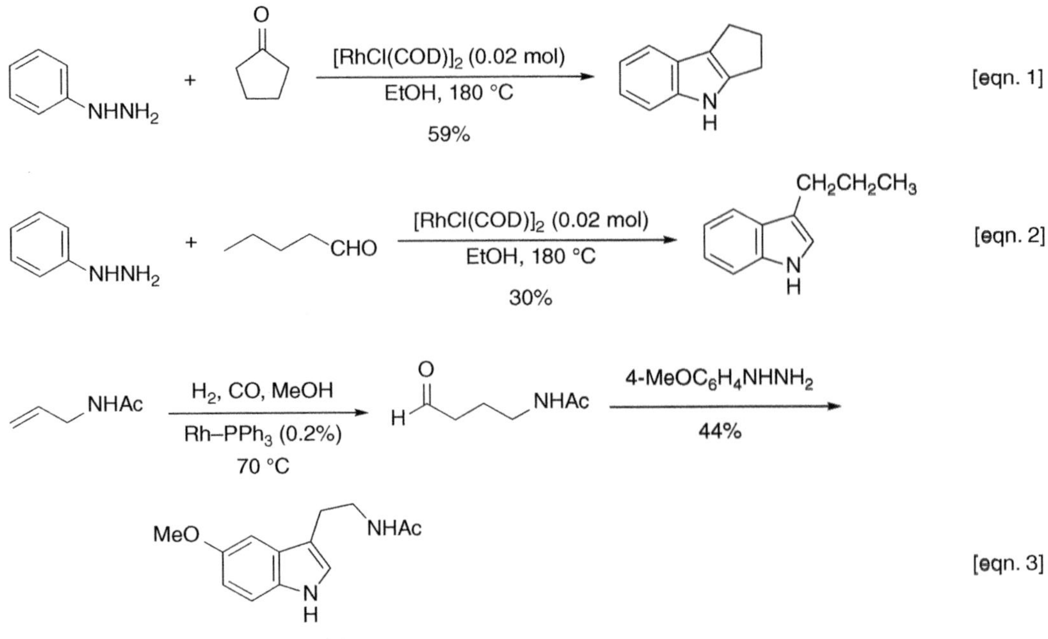

Melatonin (**9**)

Scheme 9 *Watanabe [86] and Sheldon [87] Indole Syntheses*

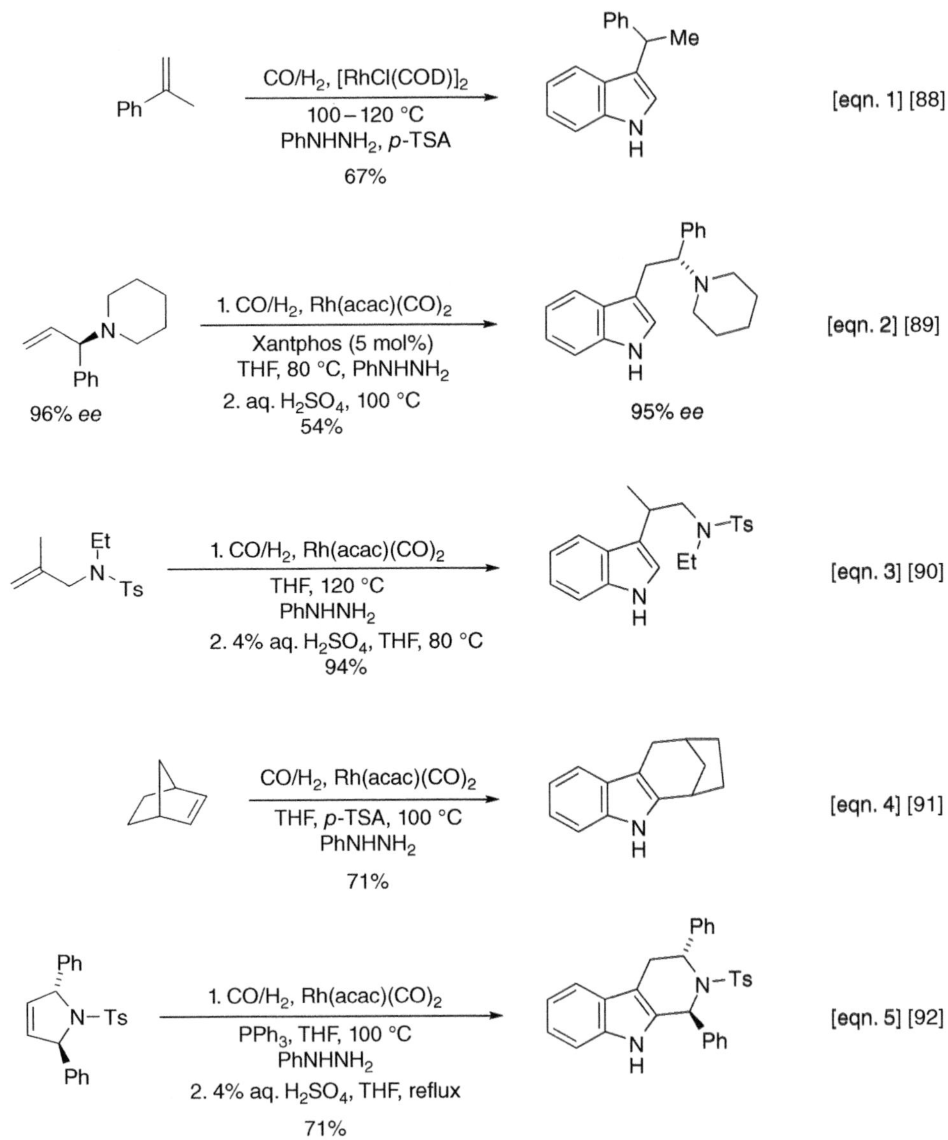

Scheme 10 *Eilbracht Indole Synthesis*

Table 4 *Alkyne Hydroamination and Fischer Indolization*

Entry	Substrate	Conditions	Product	% Yield	Ref.
1	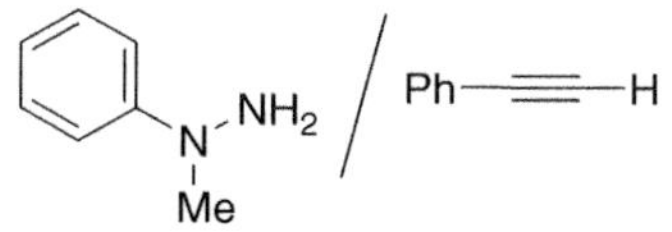	1. $Ti(SC_6F_5)_2(NMe_2)_2(NHMe_2)$ toluene, 100 °C 2. $ZnCl_2$, 100 °C	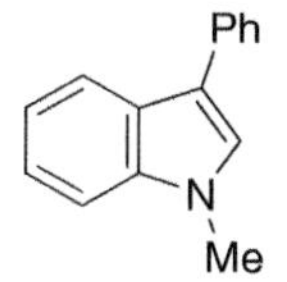	95%	94
2	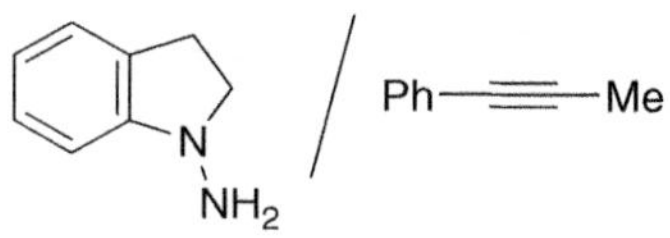	$Ti(SC_6F_5)_2(NMe_2)_2(NHMe_2)$ toluene, 100 °C	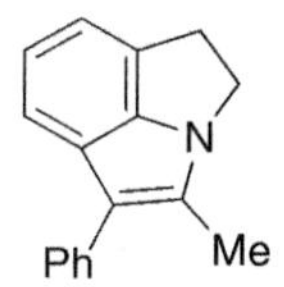	66%	94
3	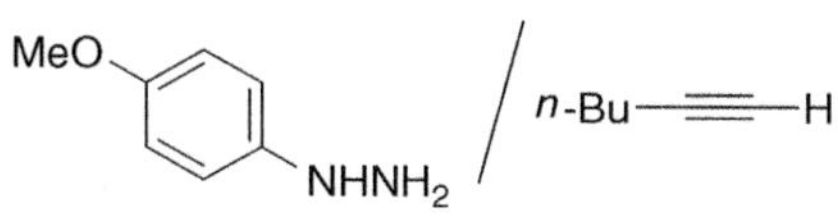	1. $Ti(enp)(NMe_2)_2$ toluene, 80 °C 2. $ZnCl_2$, 100 °C	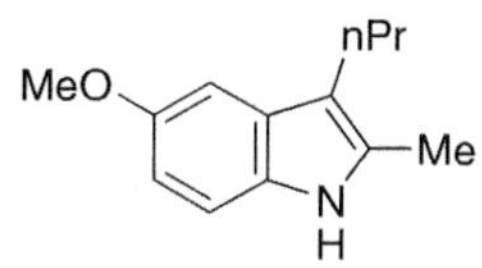	76%	95
4	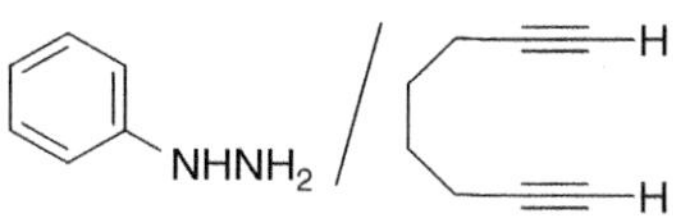	1. $Ti(enp)(NMe_2)_2$ toluene, 80 °C 2. $ZnCl_2$, 100 °C	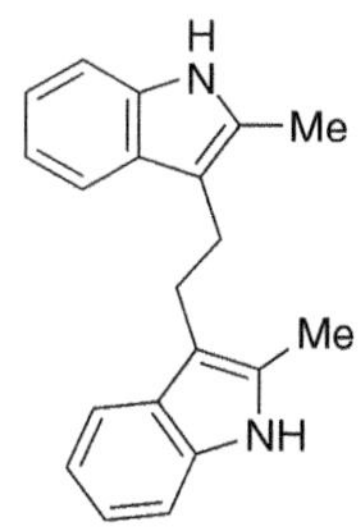	70%	95
5	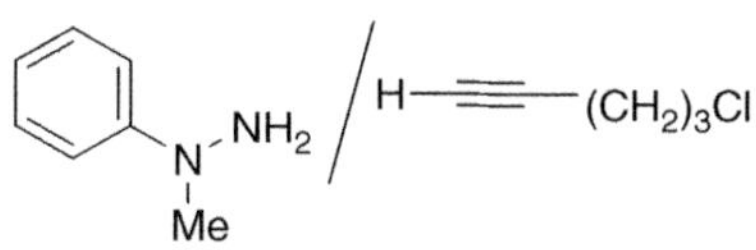	1. 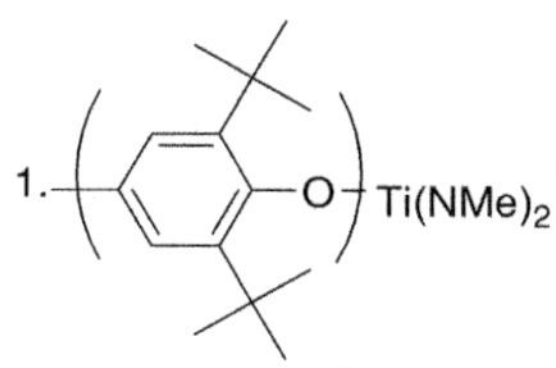 toluene, 100 °C 2. NaOH	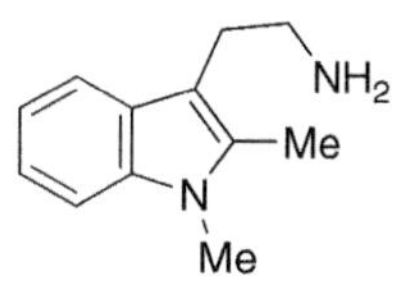	82%	96
6	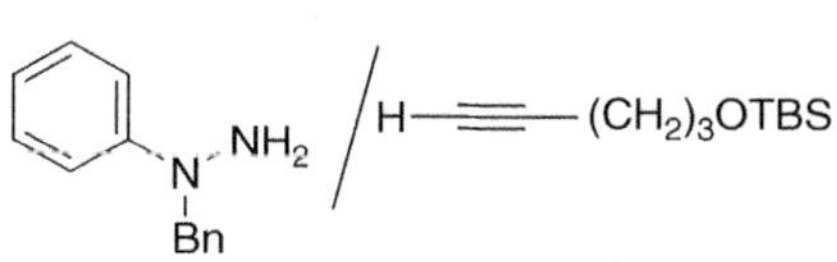	1. $Ti(NEt_2)_4$/L tol, 100 °C 2. $ZnCl_2$ tol, 100 °C 3. TBAF	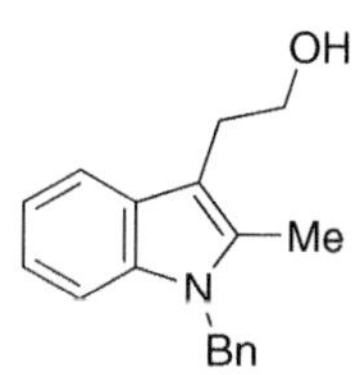	60%	97
7	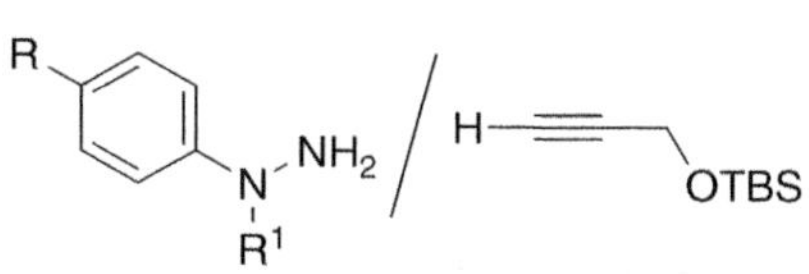 R= H, Br, Cl, F, OMe, Me, SO_2Me R^1= Me, Bn	1. $Ti(NEt_2)_4$/L toluene, 100 °C 2. $ZnCl_2$, 100 °C	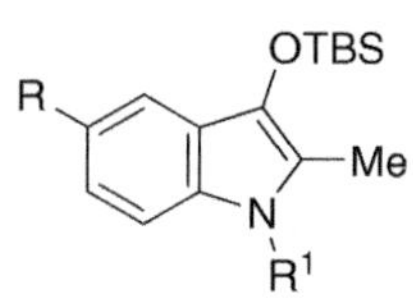	20–65%	98

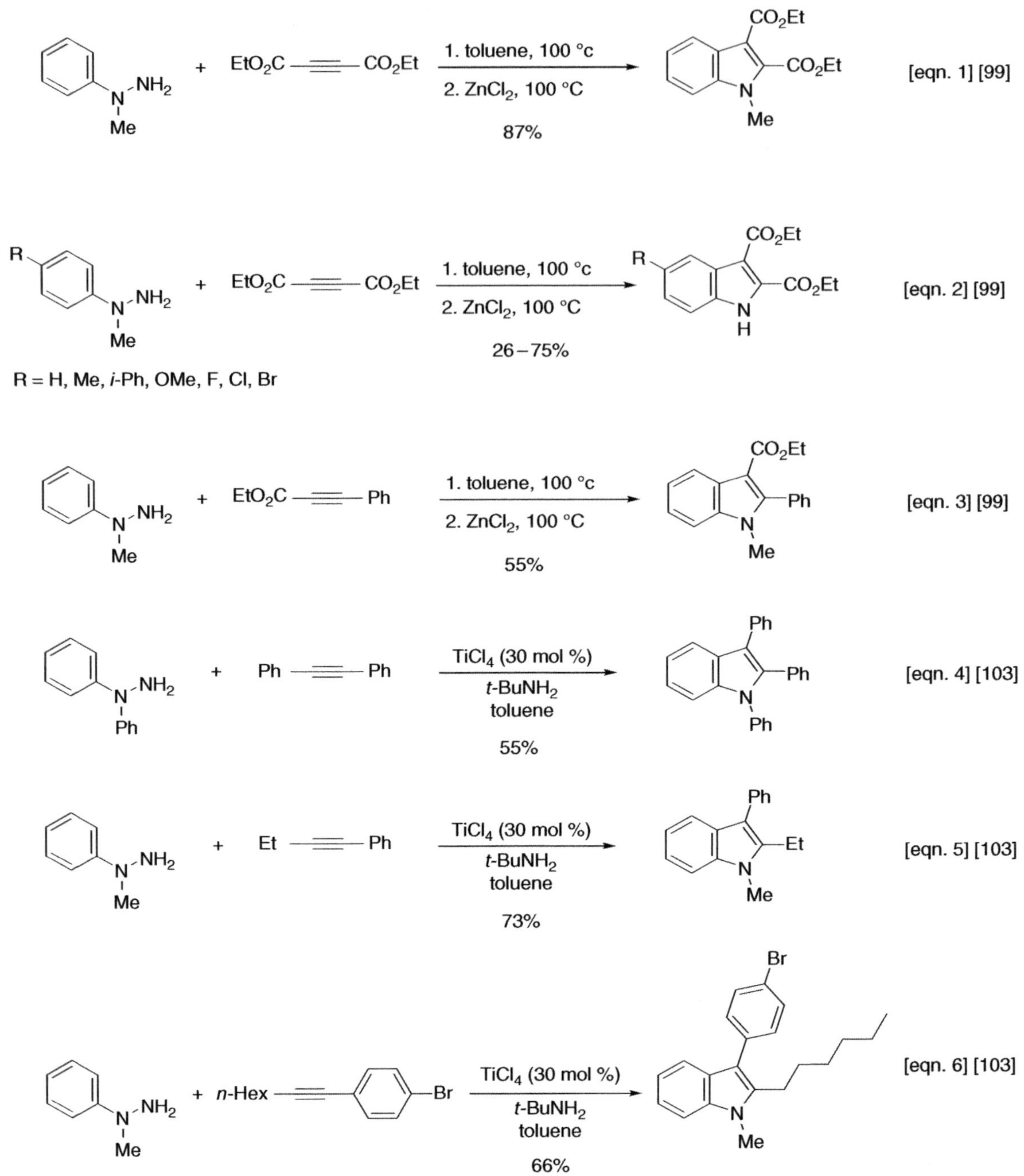

Scheme 11 *Beller and Ackermann Indole Syntheses*

access 3-aminoindoles as shown in Scheme 13 (equation 4) [108]. Compounds **12–14** were synthesized by Beller using this extraordinarily simple process.

Patil and Konala performed hydrohydrazination on alkynes using the gold catalyst $Ph_3PAuNTf_2$/*p*-TSA, eventually affording indoles via the enehydrazine (Scheme 14) [109]. Gade and colleagues found a zirconium-catalyzed route to indoles not involving a Fischer indolization [110]. These are shown in Scheme 14 (equations 3 and 4) along with the zirconium catalyst **15** and the diazazirconacycle **16**.

2.2.3 Solid-Phase Fischer Indolization Method

As we will see in the Applications section to follow, solid-phase synthesis of indoles has assumed an important role in all of synthetic organic chemistry, including Fischer

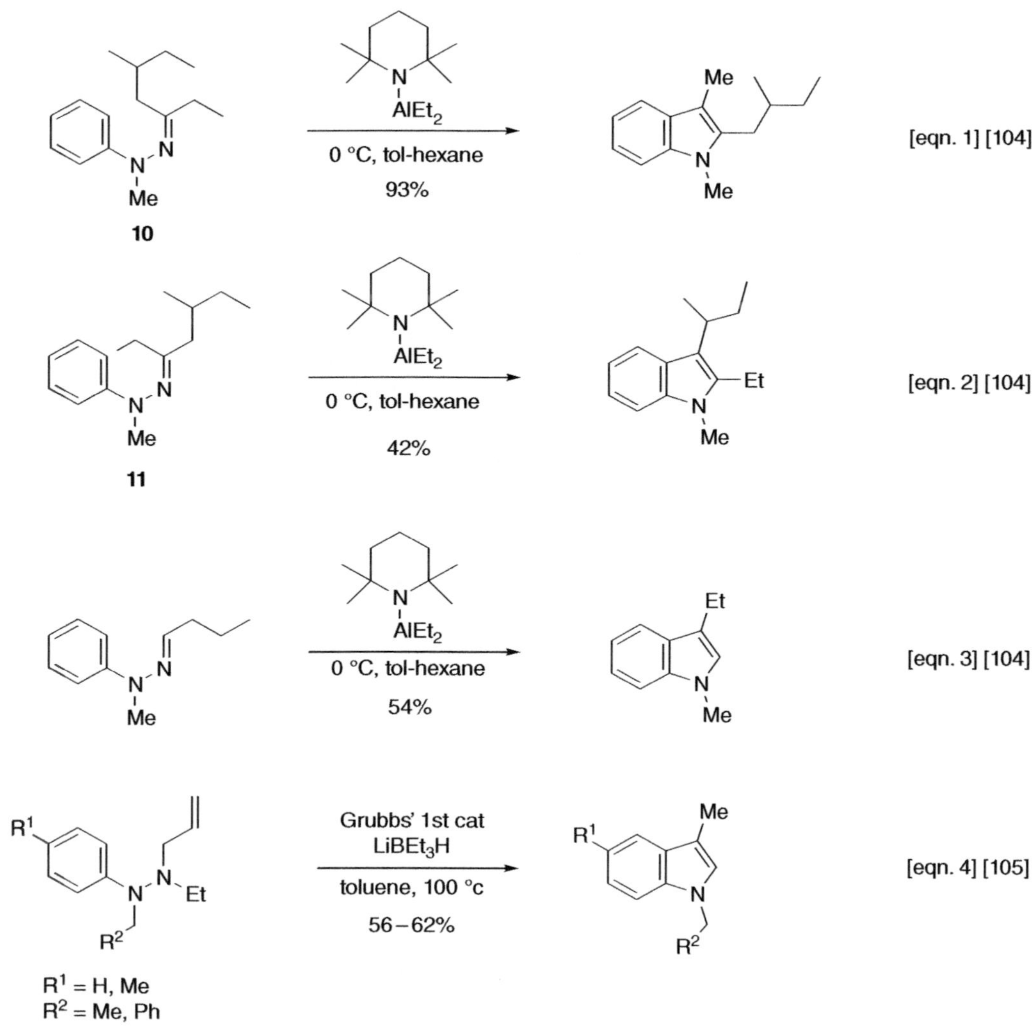

Scheme 12 *Yamamoto and Rasmussen Indole Syntheses*

indolization. A few such examples are mentioned here and depicted in Scheme 15. Yang used a carbamate linker of hydroxymethyl polystyrene **17** to effect Fischer indole synthesis on the solid support and subsequent cleavage (equation 1) [111]. Waldmann's group employed polymer-bound phenylhydrazine on a Merrifield resin (equation 2) [112]. Takahashi and colleagues used a Wang-type linker to prepare a series of naltrindole derivatives that features a one-pot release and cyclization process (equation 3) [113]. Takahashi's group also used solid-supported sulfonylhydroxylamine for the *N*-amination of anilines, giving arylhydrazine for the solid-phase synthesis of naltrindoles [114]. Jeong and coworkers described a traceless silicon linker with which to synthesize 2,3-disubstituted indoles, following cleavage of the carbon–silicon bond with trifluoroacetic acid (equation 4) [115]. Some 16 mostly unsymmetrical 2,3-disubstituted indoles were prepared in this study. Breinbauer and colleagues used a polystyrene sulfonyl chloride linker resin for the synthesis of amino-containing indoles (equation 5) [116]. Several novel tryptamines and homotryptamines were prepared by this group.

2.2.4 Other General Methods

A number of excellent variations on the classic Fischer indolization are known that fall outside of the previous three categories. Like solid-phase synthesis, multicomponent and domino reactions are of major interest to organic chemists. Ganem reported a three-component Fischer indole variation that embraces the union of nitriles or carboxylic acids with organometallics followed by the addition of arylhydrazine. This exquisite method is summarized in Scheme 16 [117]. In both variations, *in situ* formation of the requisite arylhydrazones culminates in indolization. The presence of sufficient arylhydrazine hydrochloride

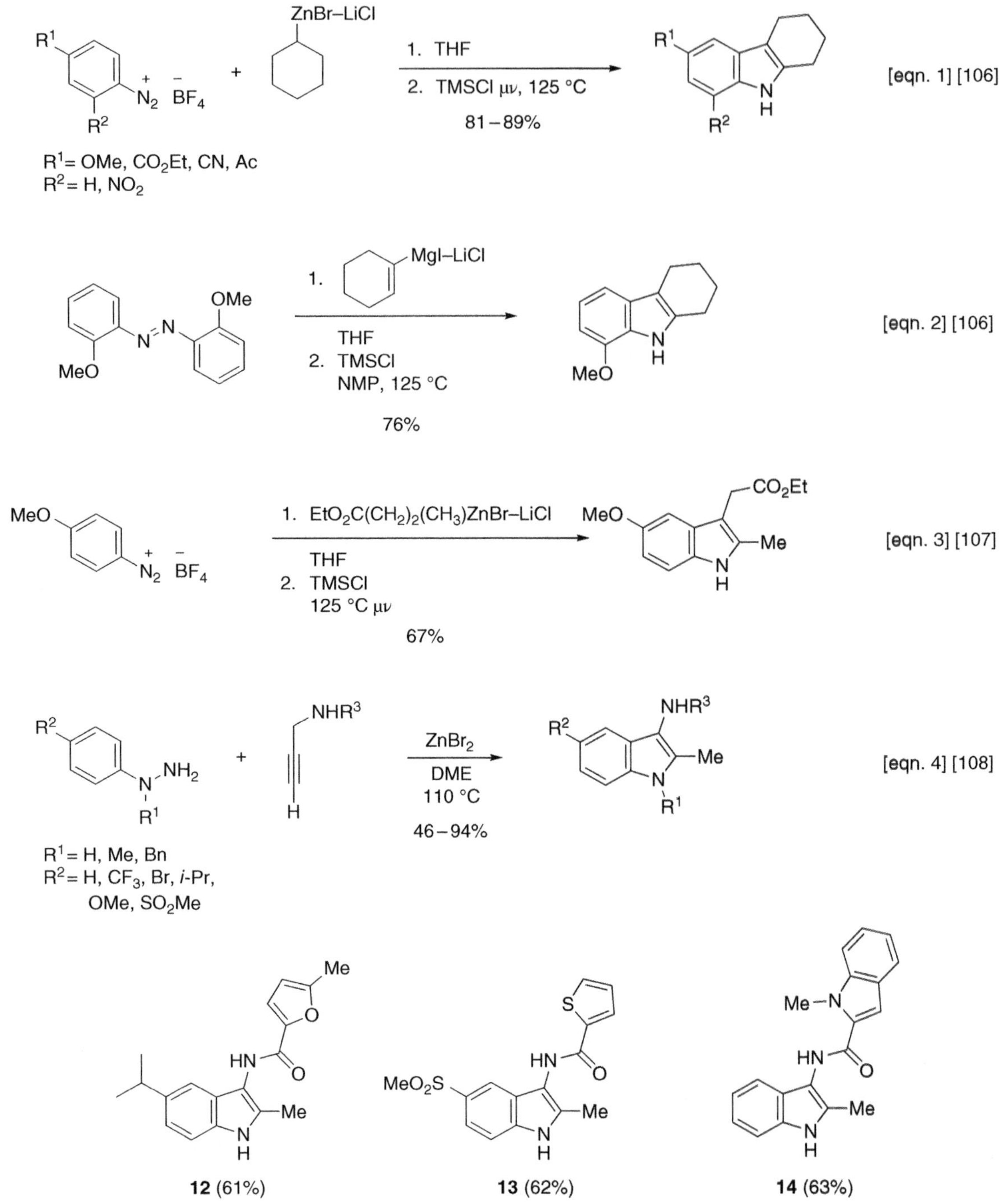

Scheme 13 *Knochel and Beller Indole Syntheses*

ensures generation of the corresponding ketone from the starting carboxylic acid. Ganem and colleagues effected a one-pot indole synthesis from aliphatic nitro compounds and arylhydrazines (Scheme 17) [118]. This Nef-type reaction presumably involves formation of the *aci*-nitro tautomer, which is captured by the arylhydrazine to give the arylhydrazone. Müller's group used a three-component coupling cycloisomerization to synthesize blue-luminescent 5-(3-indolyl)oxazoles via a microwave-assisted Fischer indolization as summarized in Scheme 17, equation 3 [119]. This tandem reaction involves palladium-catalyzed formation of ynone **17**, cycloisomerization to oxazole **18**, hydrazone formation, and final indolization (Scheme 17).

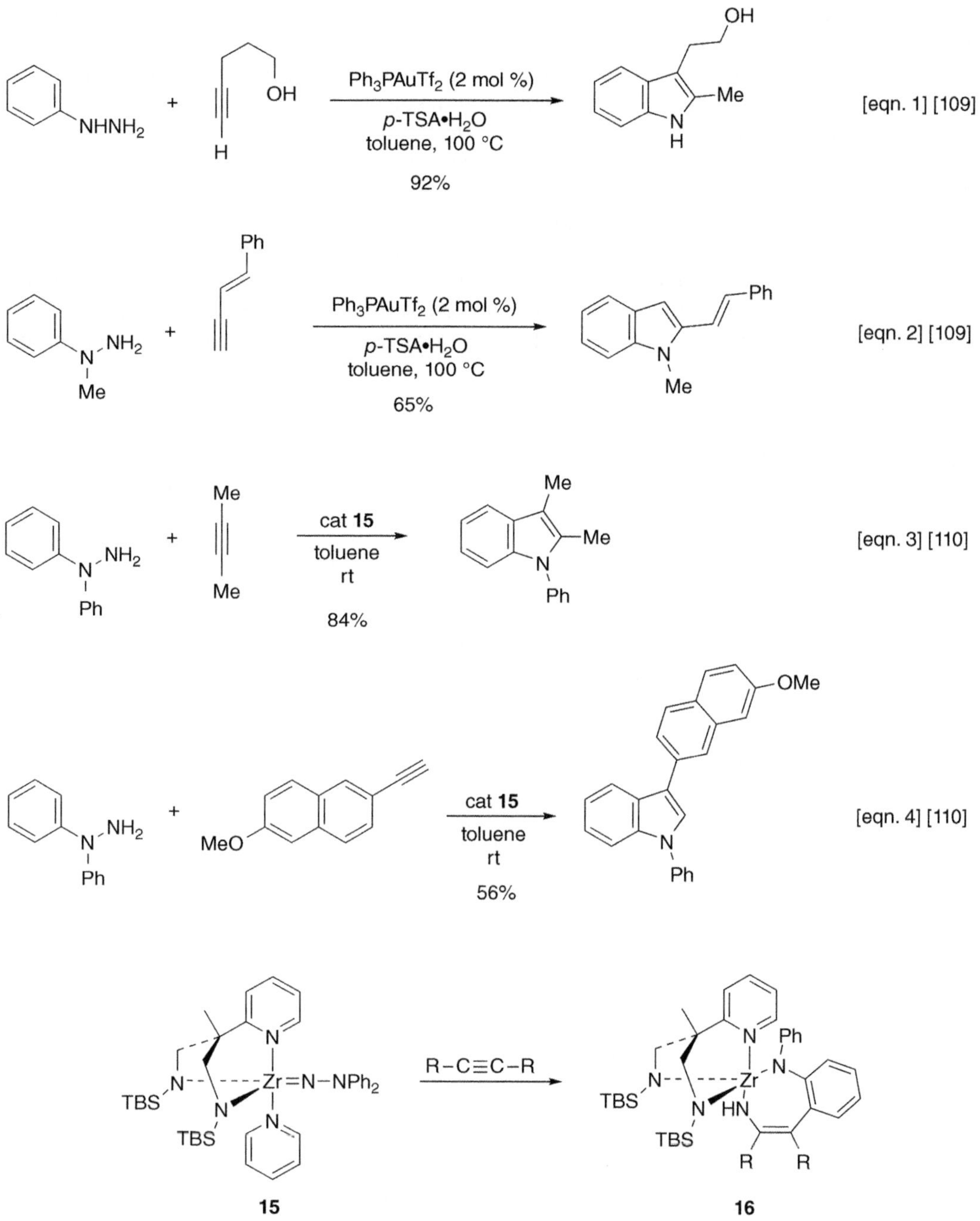

Scheme 14 *Patil and Gade Indole Syntheses*

An extraordinarily simple approach to a Fischer indolization is that of Taylor and Donald [120] involving the transformation of epoxides to indoles via aldehydes or ketones (the Meinwald rearrangement) in the presence of arylhydrazines, and thence to indoles, obviating the need for isolating the carbonyl compound. Selected examples of this simple chemistry are shown in Scheme 18 (equations 1, 2). Of the Lewis acids screened, scandium(III) triflate proved to be superior. Greaney and colleagues employed the highly reactive, unisolable, yet amazingly useful benzyne in a Fischer indole synthesis. Thus, the reaction between benzyne, generated in the typical manner from 2-(trimethylsilyl)phenyl triflate and cesium fluoride, and *N*-tosylhydrazones affords indoles (Scheme 18, entries 3–5) [121]. Presumably, the mechanism involves arylation of the tosylhydrazone, isomerization to the enehydrazine, and acid-catalyzed indole ring formation.

Moody and colleagues described a two-step route to a wide range of indoles that uses haloarenes as the starting point. Halogen–magnesium exchange, reaction with di-*tert*-butyl azodicarboxylate to give the corresponding doubly Boc-protected arylhydrazine, and final reaction with ketones under acidic conditions forms an indole in good to excellent yield [122, 123]. This attractive method is summarized in Scheme 19 (equations 1 and 2), including a selection of indoles prepared, **19–21**.

List and coworkers reported the first asymmetric Fischer indolization as summarized in Scheme 20. This elegant chemistry features a novel spiro chiral phosphoric acid **22** to induce enantioselective indole ring formation, and it culminates in a formal synthesis of *(S)*-ramatroban (**23**), a thromboxane receptor antagonist [124]. Enantiomeric enrichments as high as 98.5:1.5 were achieved. List identified the weakly acidic cation exchange resin Amberlite CG50 for the removal of ammonia, which restores the reactivity of the catalyst. A beautiful piece of chemical engineering!

Another approach to arylhydrazones was devised by Heinrich and Prechter, who found that hydroperoxides (**24**) derived from cyclic ketones react with aryldiazonium salts to give azo carboxylic acids **25**, which are in equilibrium

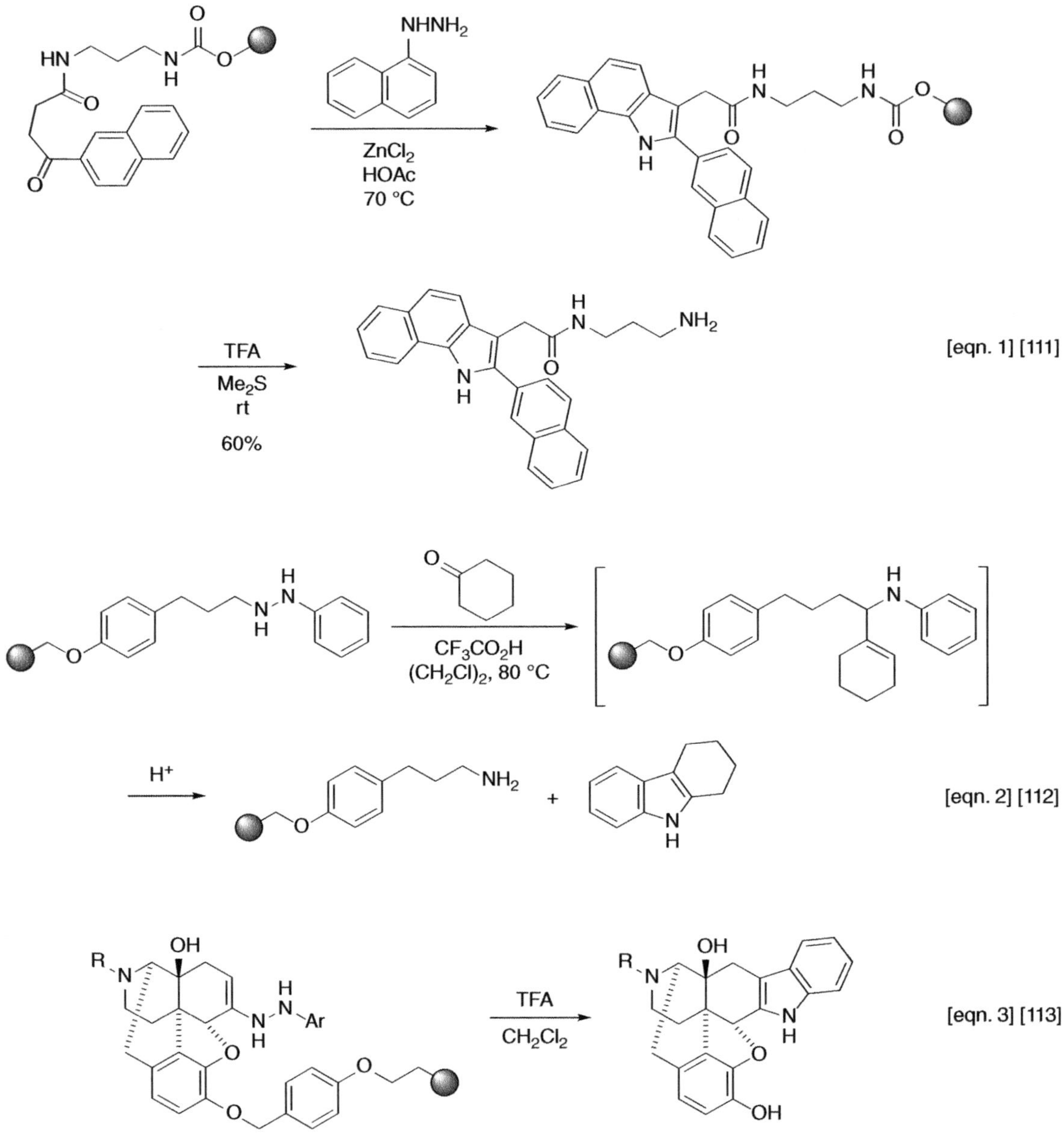

Scheme 15 *Solid Phase Fischer Indole Syntheses*

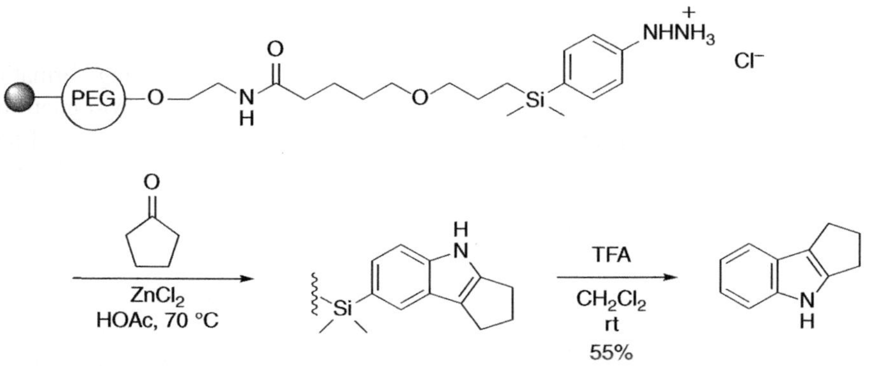

[eqn. 4] [115]

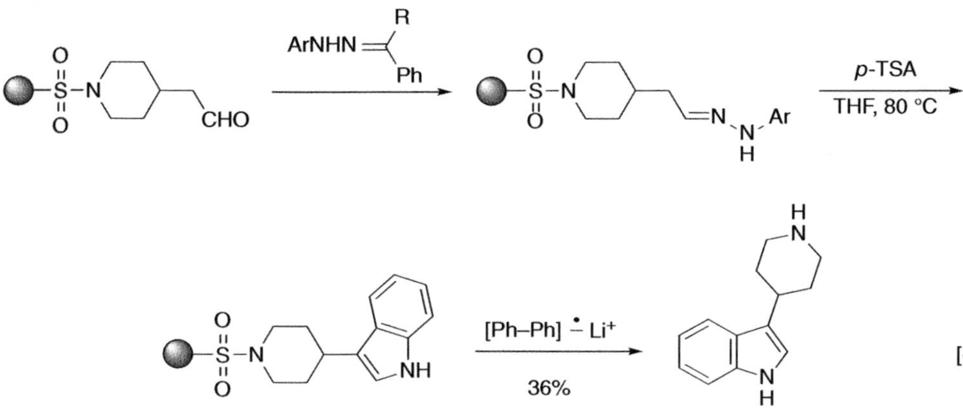

[eqn. 5] [116]

Scheme 15 *(continued)*

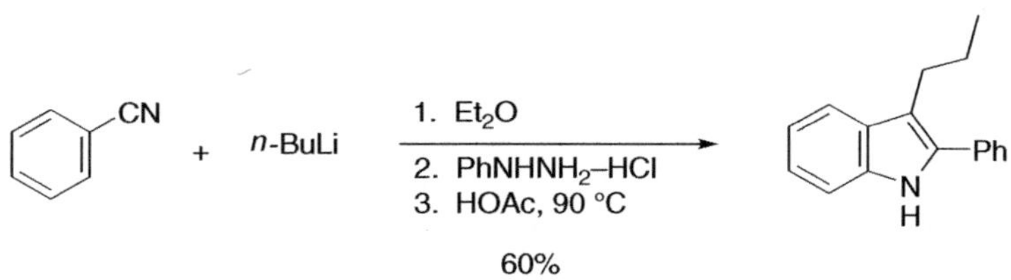

[eqn. 1] [117]

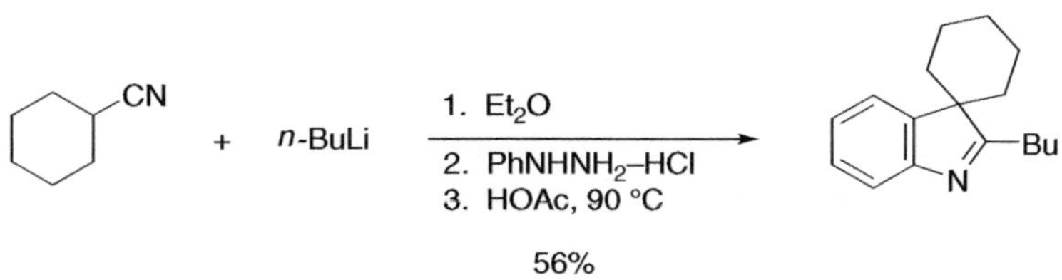

[eqn. 2] [117]

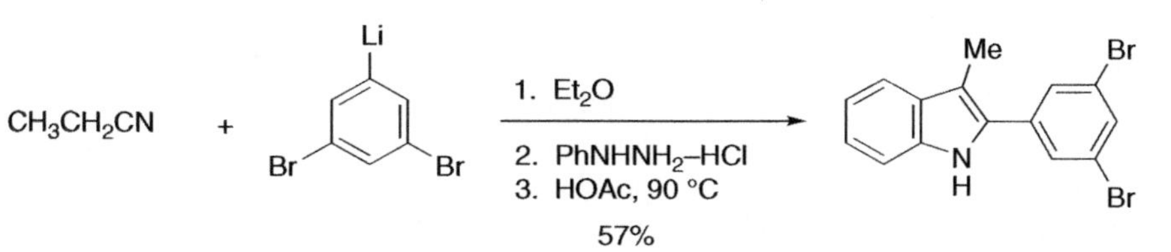

[eqn. 3] [117]

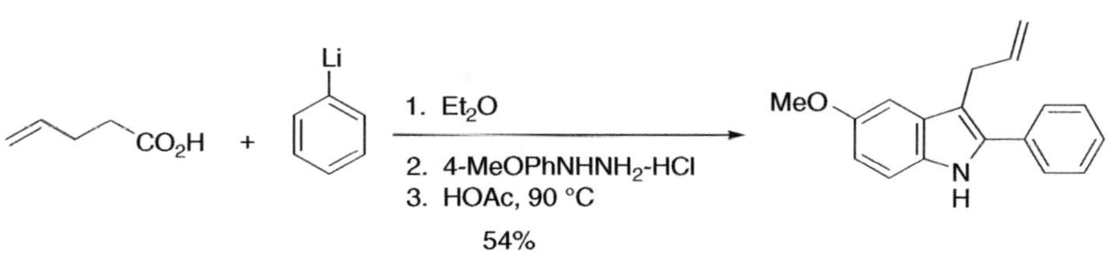

[eqn. 4] [117]

Scheme 16 *Fischer Indole Synthesis Variations*

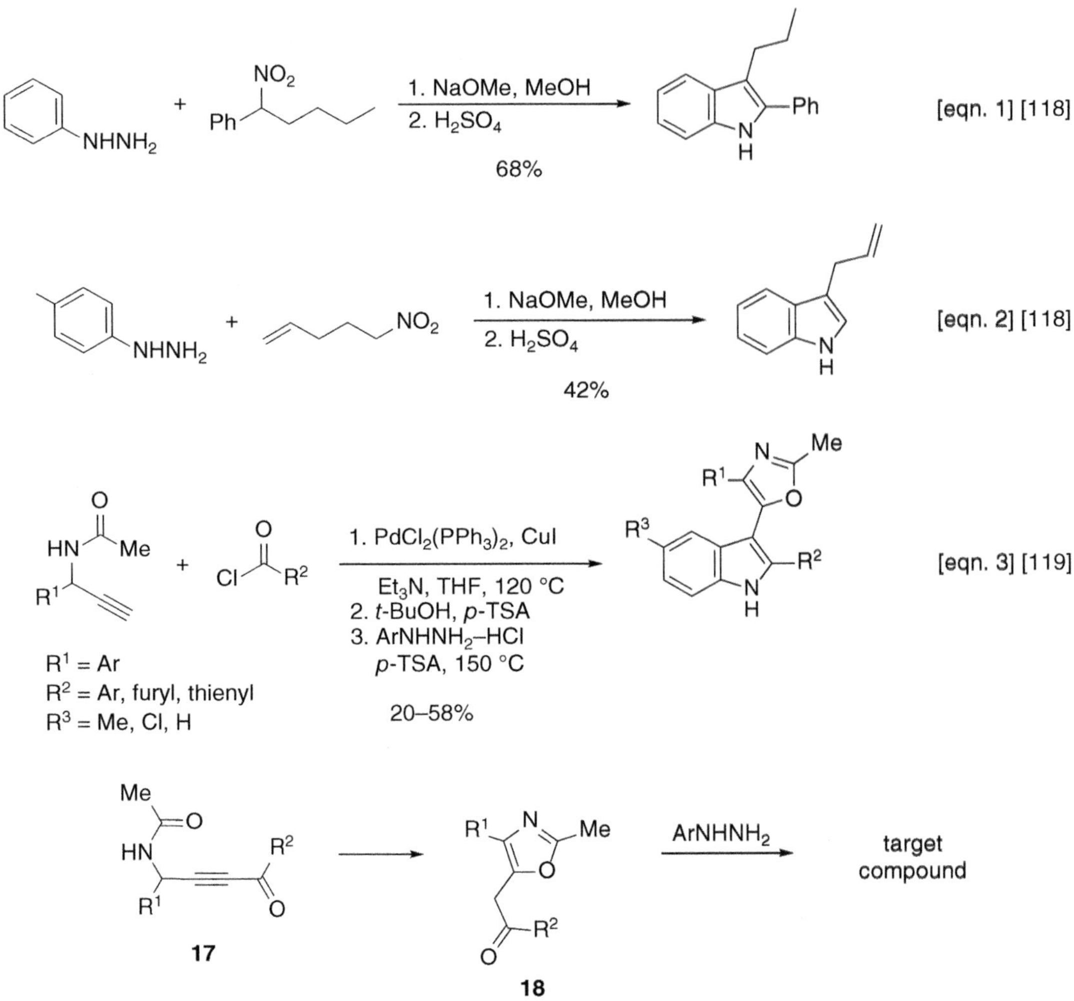

Scheme 17 *Ganem and Müller Indole Syntheses*

with arylhydrazones **26**. In one case, the arylhydrazone (**26**, R=Cl) was converted to the corresponding indole **27** (Scheme 21) [125]. Other azo carboxylic acids prepared in this study are shown (**28–36**). In several isolations of these azo compounds, the corresponding hydrazone was also present. Although this route to indoles has yet to be exploited, it does offer an appealing entry to indole C-3 substituted long-chain carboxylic acids.

We have already alluded to the well-known intervention of 3,3-disubstituted indolenines in the Pictet–Spengler cyclization and other indole reactions, but Liu and colleagues have exploited these ubiquitous, usually invisible, intermediates in a practical synthesis of 2,3-disubstituted indoles (Scheme 22) [126]. Thus, the reaction of phenylhydrazine with α-branched aldehydes affords indolenines that rearrange under acidic conditions to yield the corresponding indoles **37–42**. The novelty of Liu's work is the careful optimization of the reaction, because other conditions ($ZnCl_2$, HOAc; MsOH, toluene; HCl, EtOH; HCl, dioxane; TFA, toluene) afford exclusively or mainly the indolenine. In the case of unsymmetrical α,α′-branched aldehydes, the question of migratory aptitude comes into play (**41**, **42**). Whereas the yields are modest, the process is exceptionally simple.

Perhaps the most serious problem with the Fischer indolization is the abnormal reaction that we discussed in Section 1. Heathcock and Szczepankiewicz addressed this problem as shown in Scheme 23 [127]. Their well-designed conception was to block the abnormal pathway that occurs in **43** (cyclization to the methoxyl carbon), resulting in a poor yield of the desired **44**. Thus, via 1,4-benzoxazine hydrazone **46,** the Fischer indolization to **47** proceeds cleanly. Removal of the benzoxazine gave the desired indole **48** in excellent yield.

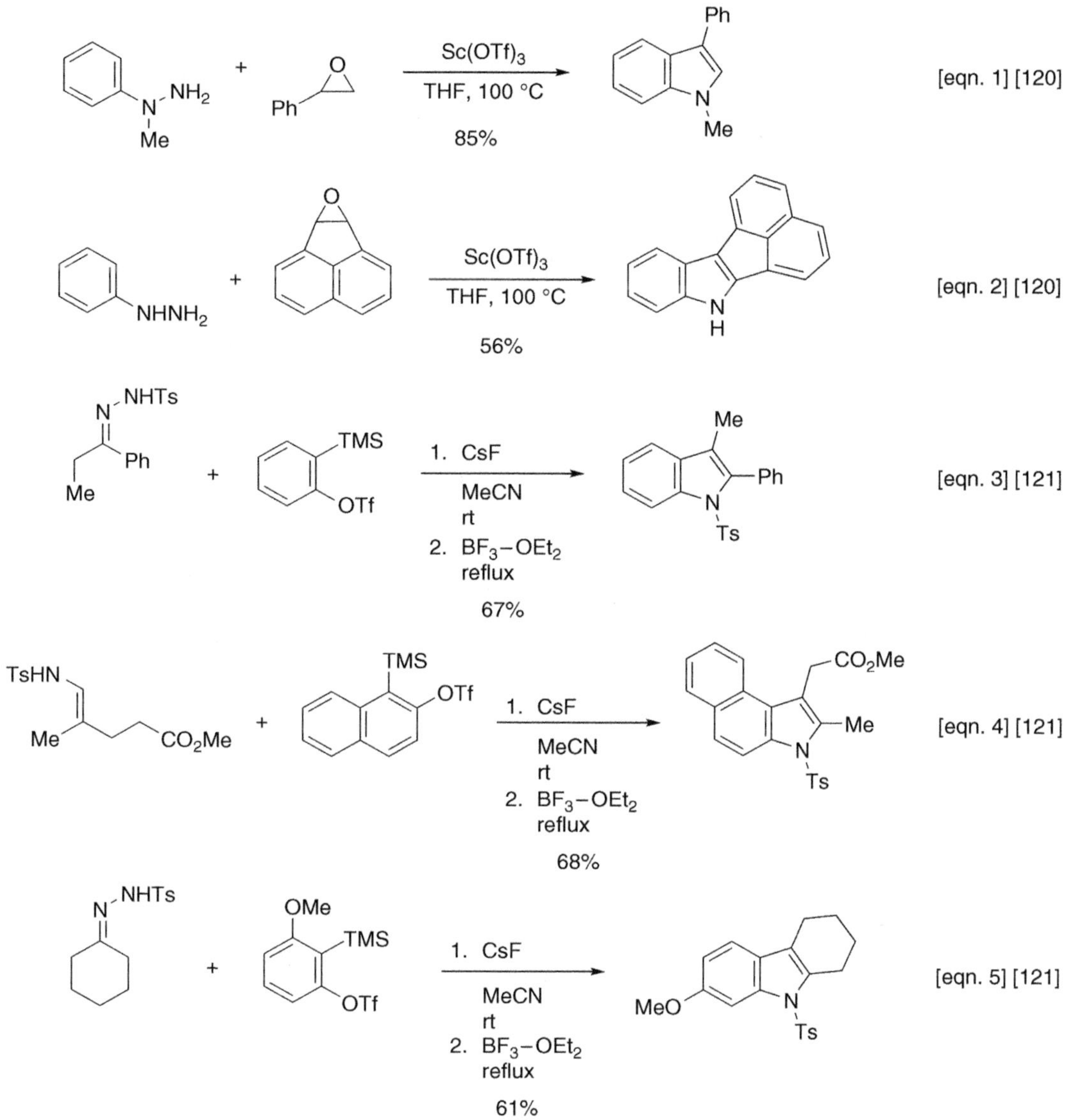

Scheme 18 *Taylor and Greaney Indole Syntheses*

2.2.5 Hydrazones

Obvious to the success of the Fischer indole synthesis is the availability of arylhydrazines and arylhydrazones. Several groups have pursued new methods for these syntheses, which are listed in chronological order in Table 5 [128–134]. Several conditions are typically used depending on the substrate, but only one is shown for each entry. Entry 4 features a reaction scale of 1.6 kg of 4-chlorotoluene, and aryl chlorides also function in this amination protocol, as do heterocyclic substrates (entry 7).

Takamura and colleagues parlayed their hydrazone synthesis (Table 5, entry 5) into an indole synthesis as summarized in Scheme 24 (equation 1) [132, 135]. Beller and colleagues employed their rhodium chemistry (cf. Scheme 10) to a synthesis of arylhydrazones from olefins and hydrazines, which culminates in an *in situ* Fischer indolization (Scheme 24, equations 2 and 3) [136]. Upon screening 30 catalytic systems, Chakraborti and colleagues found that magnesium perchlorate (5 mol %) in dichloromethane at room temperature was the best for preparing arylhydrazones from benzaldehydes, particularly those strongly electron-rich benzaldehydes that are sluggish or nonreactive under noncatalytic conditions (Scheme 24, **49–51**) [137].

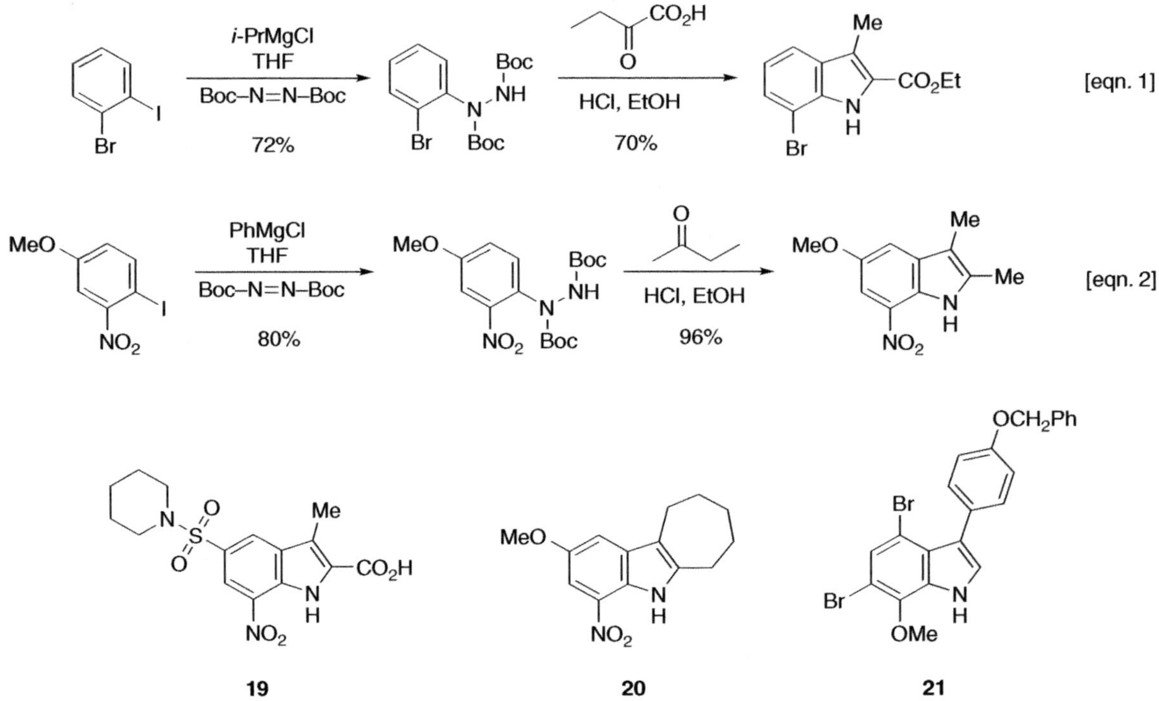

Scheme 19 *Moody Indole Synthesis [122, 123]*

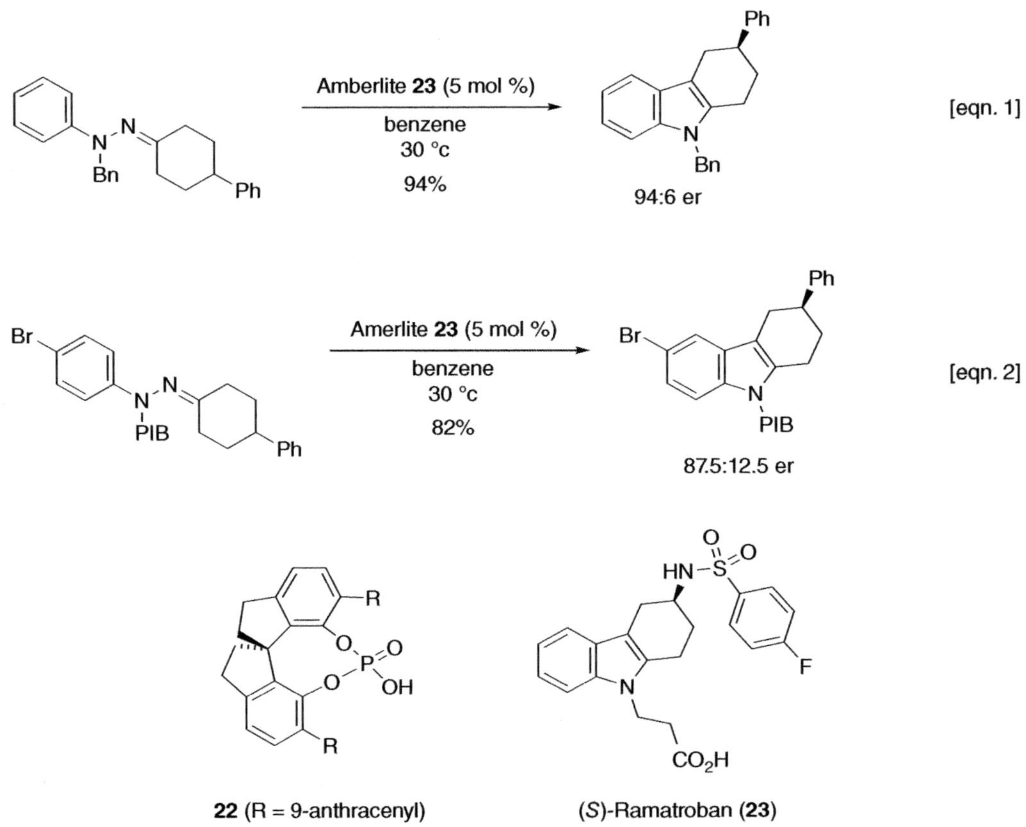

Scheme 20 *List Indole Synthesis [124]*

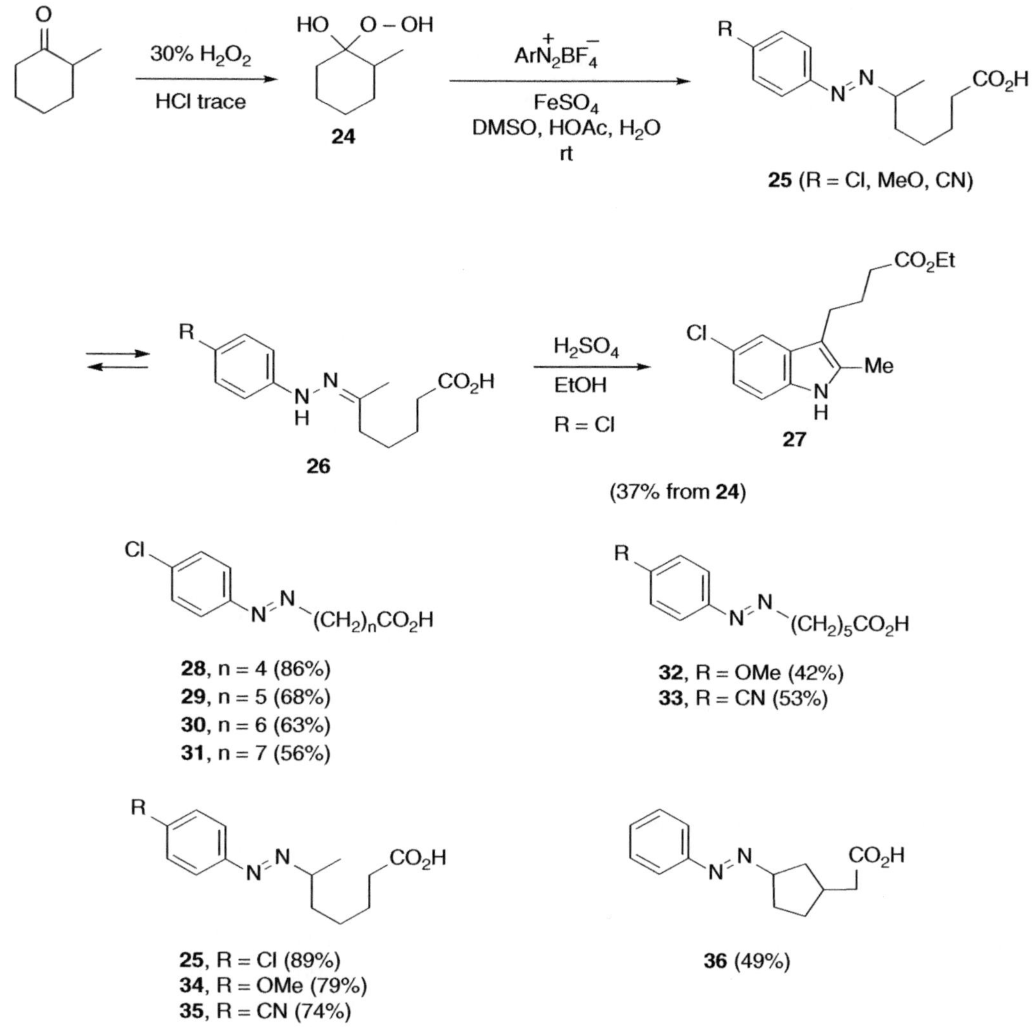

Scheme 21 *Heinrich Indole Synthesis [125]*

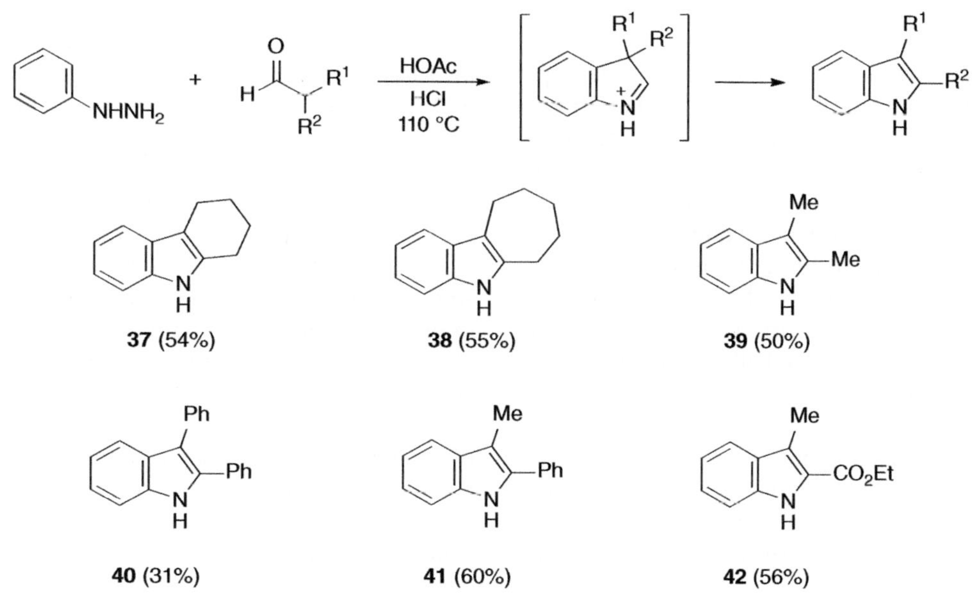

Scheme 22 *Liu Indole Synthesis [126]*

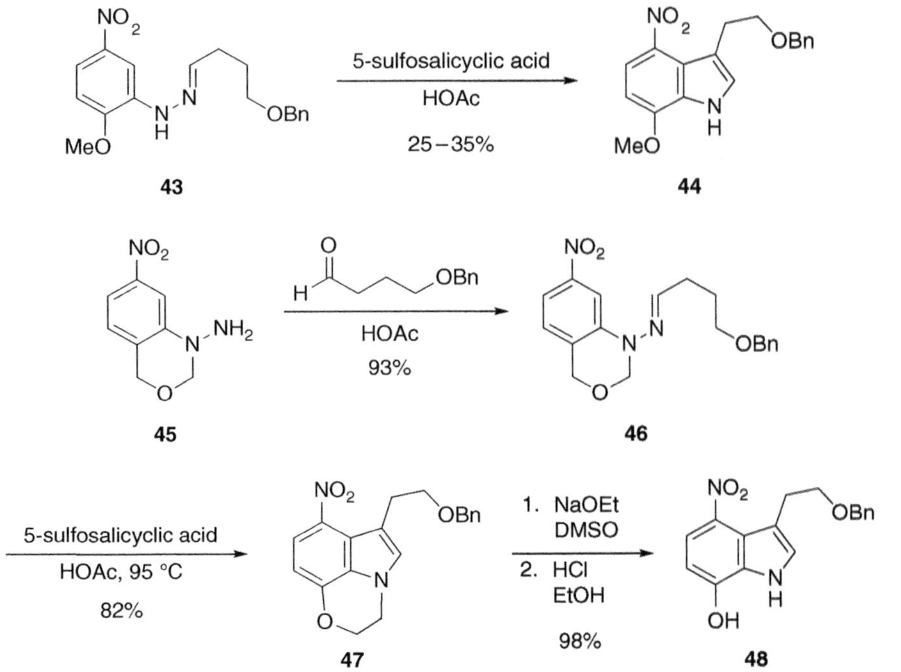

Scheme 23 *Heathcock Indole Synthesis [127]*

2.2.6 Other Variations of Fischer Indole Synthesis

The best known and most significant Fischer indolization variation is that of Japp and Klingemann, described in 1887 [138] and 1888 [139] and reviewed by Phillips in 1959 [140] and Li in 2007 [141]. This reaction involves the union of an aryldiazonium salt with a 1,3-dicarbonyl compound to give the requisite arylhydrazone following loss of one of the carbonyl fragments (Scheme 25). Examples of the Japp–Klingemann reaction are combined with the normal Fischer indolization in the Applications section.

Much less commonly found are a few other Fischer-type variations that will be briefly mentioned. The Piloty–Robinson indole synthesis is a simple variation of the Piloty pyrrole synthesis, where the [3,3] sigmatropic event is triggered by acetic anhydride or methyl iodide (Scheme 26) [142, 143]. An example is shown in equation 2 [143].

The Borsche–Drechsel carbazole synthesis is simply a Fischer indolization involving cyclohexanones, followed by oxidation of the tetrahydrocarbazole to a carbazole [144–147]. For reviews, see Campbell and Barclay [148] and Fultz [149]. Examples appear in the Applications section.

The Bucherer carbazole synthesis is the reaction of aromatic phenols, typically naphthols, with phenylhydrazine to form the corresponding phenylhydrazone via the naphthol tautomer [150, 151]. Bucherer was an early pioneer in studying the enol-keto tautomerism in phenolic rings [151], and the Bucherer reaction is the conversion of naphthols (and certain other phenols prone to tautomerism) to naphthyl amines [152, 153]. Interestingly, in 1901 Japp and Maitland observed the formation of phenyl-α-naphthylcarbazole from α-naphthol and phenylhydrazine hydrochloride [154]. For a short review of the Bucherer carbazole synthesis, see Moore [155]. Some examples are shown in Scheme 27 [155–157].

The most important and widely used variation of the standard Fischer indolization is that of Grandberg and his Russian colleagues. This variation, more of which we will encounter in the Applications section, involves the use of dihydrofurans and dihydropyrans to access tryptophols and homotryptophols [158], respectively, and aminobutanal equivalents to afford tryptamines [159–161]. Campos extended this method to a range of cyclic enol ethers and unsaturated lactones [162], and, somewhat earlier, Soll and colleagues used dihydrofuran in the synthesis of antiinflammatory etodolac derivatives [163, 164]. Selective examples of these methods are shown in Scheme 28. The synthesis of 2-methyltryptamine (equation 5) was carried out on a 19-kg scale of phenylhydrazine for a synthesis of the histone deacetylase inhibitor LBH589 [165]. We will encounter additional examples of these two powerful Grandberg syntheses in the Applications section.

Table 5 *Synthesis of Arylhydrazines/Arylhydrazones/Arylhydrazides*

Entry	Substrate	Conditions	Product	% Yield	Ref.
1	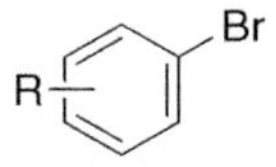 R= 4-MeO, 2-Me, 4-CF_3, 4-Ac, 4-Me	$Ph_2C{=}N{-}NH_2$ $Pd(OAc)_2$, DPPF NaO*t*-Bu or Cs_2CO_3 toluene, 90 °C	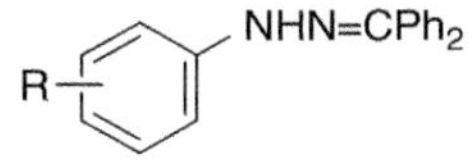	80–99%	128
2	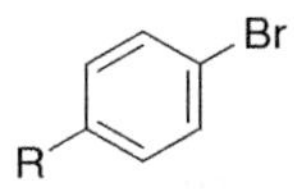 R= CO_2Me, Ac, CN, NO_2, CF_3, COPh, H (others)	$BocNHNH_2$, $Pd_2(dba)_3$ Cs_2CO_3, toluene DPPF, 100 °C	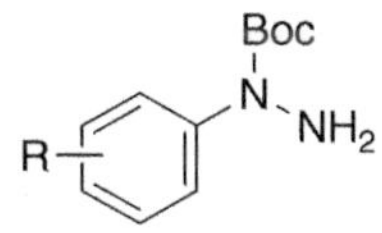	18–84%	129
3	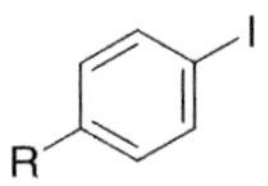 R= H, *i*-Pr, OMe, OH, NH_2, Br, CO_2Et, CN, Ac	$BocNHNH_2$, CuI 1,10-phen, Cs_2CO_3 DMF, 80 °C	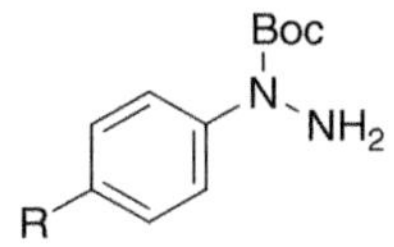	43–97%	130
4	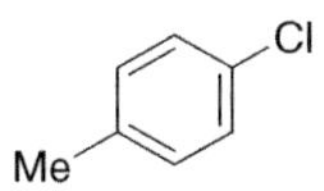	$Ph_2C{=}NNH_2$ $Pd(OAc)_2$, *t*-BuOH MePhos, NaOH	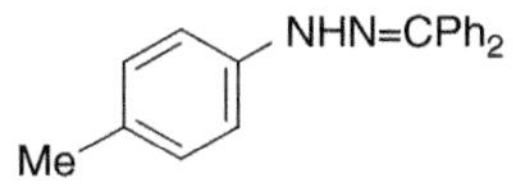	93%	131
5	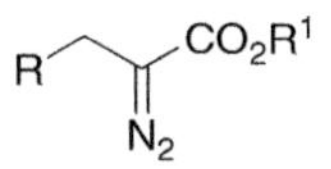 R= Ph, *i*-Pr, CH_2CO_2Et, $MeSCH_2$, BuS, indole R^1= Me, Et	1. PhLi, THF 2. HOAc	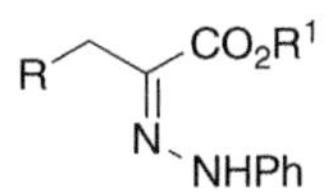	67–97%	132
6	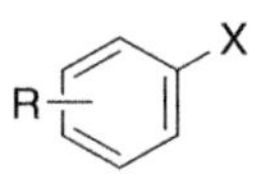 X= Br, I R= 4-Me, 4-OH, 4-NH_2, 4-F, 4-CO_2Me, 4-Br, 4-NO_2, 3-F, 3-NO_2, 3-Me, etc.	$BocNHNH_2$, CuI Cs_2CO_3, DMSO, 50–80 °C	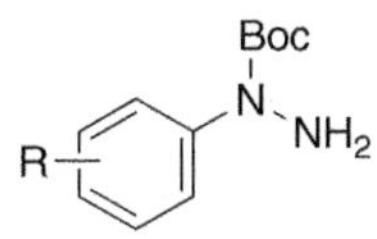	43–92%	133
7	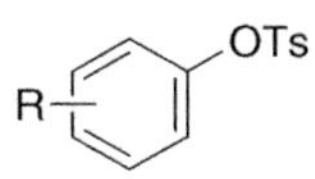	1. $N_2H_4{-}H_2O$, Pd NaO*t*-Bu, 60 °C toluene 2. PhCHO, MeOH	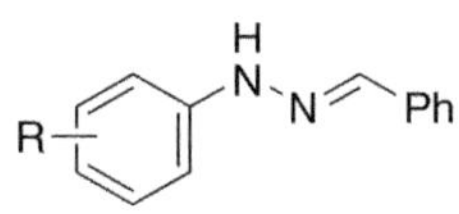	51–97%	134

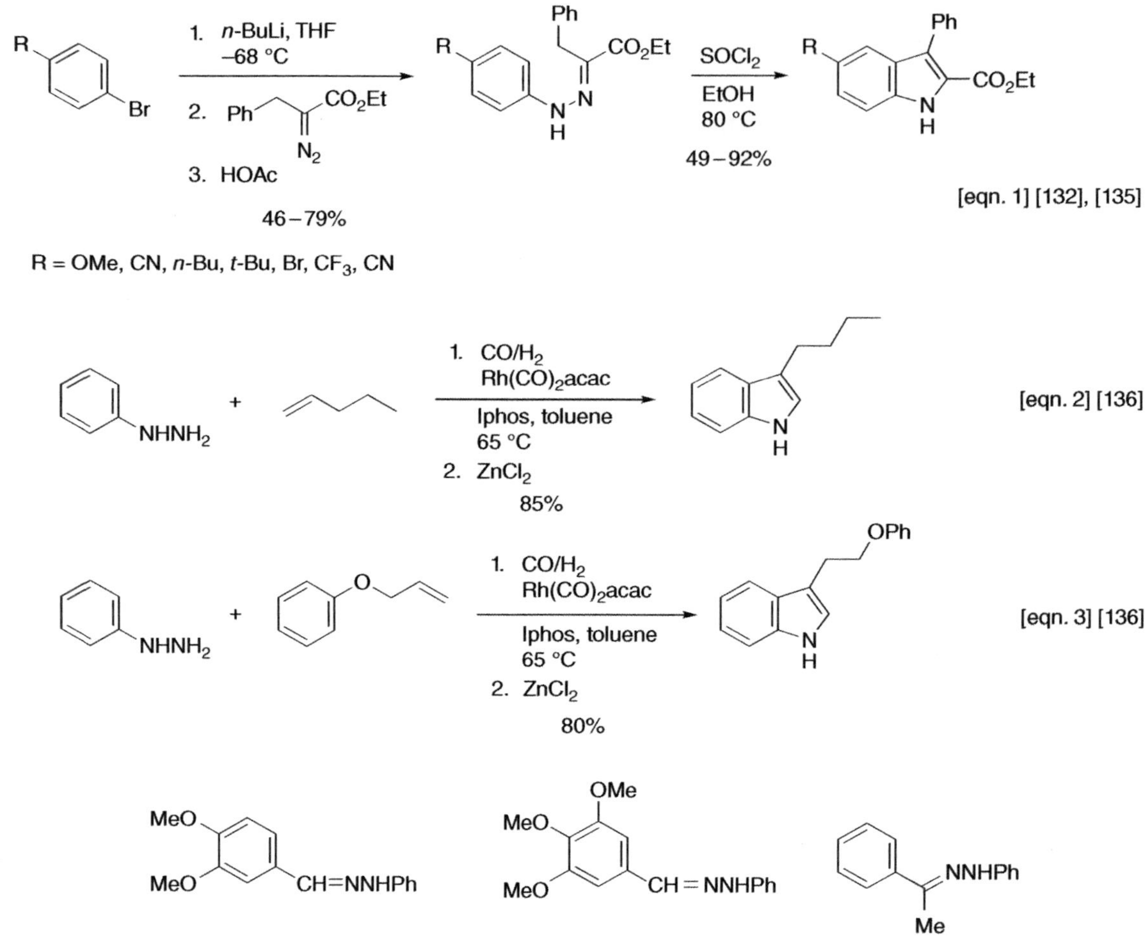

49 (98%) **50** (90%) **51** (95%)

Scheme 24 *Takamura [132, 135] and Beller [136] Indole Syntheses*

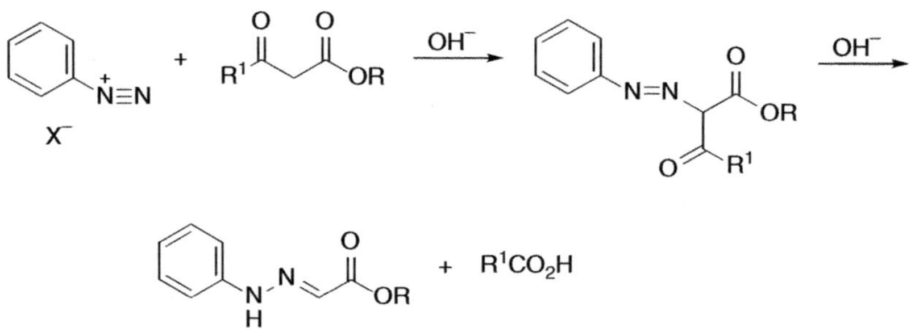

Scheme 25 *Japp–Klingemann Indole Synthesis*

2.3 Applications of Fischer Indolizations

2.3.1 Drug Targets

Perhaps more than any other indole ring synthesis, the Fischer indolization has proved to be of paramount importance in the discovery of new indole-containing drugs. This section illustrates many of these drugs and drug candidates. Recall that we encountered several of these in the Introduction. This section is organized by disease and/or receptor target.

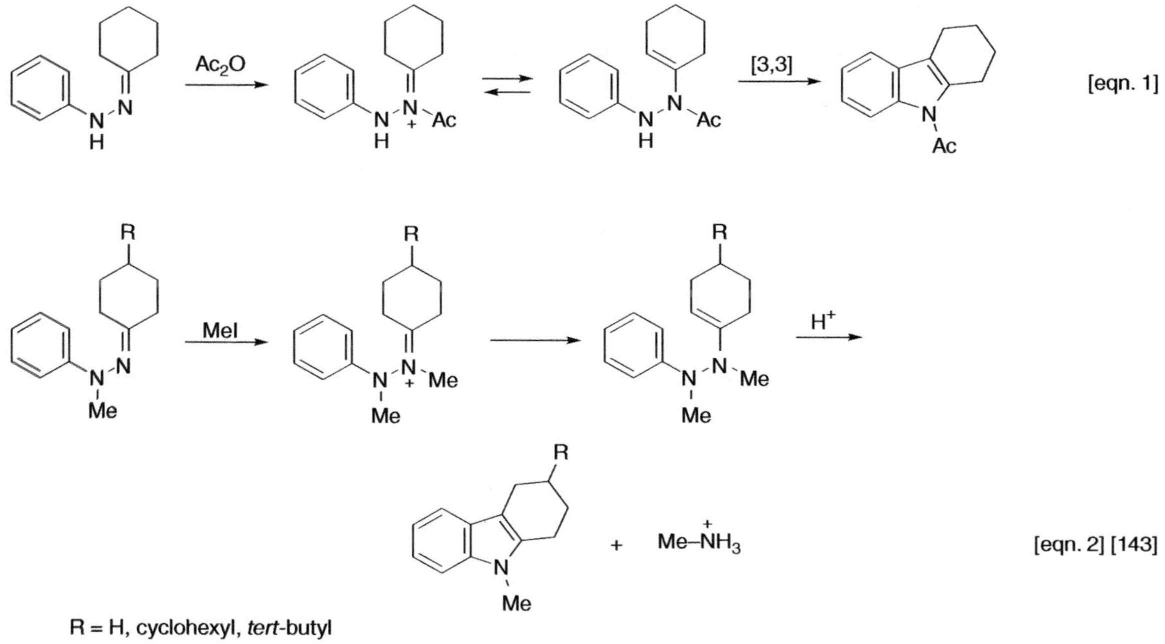

Scheme 26 *Piloty Robinson Indole Synthesis*

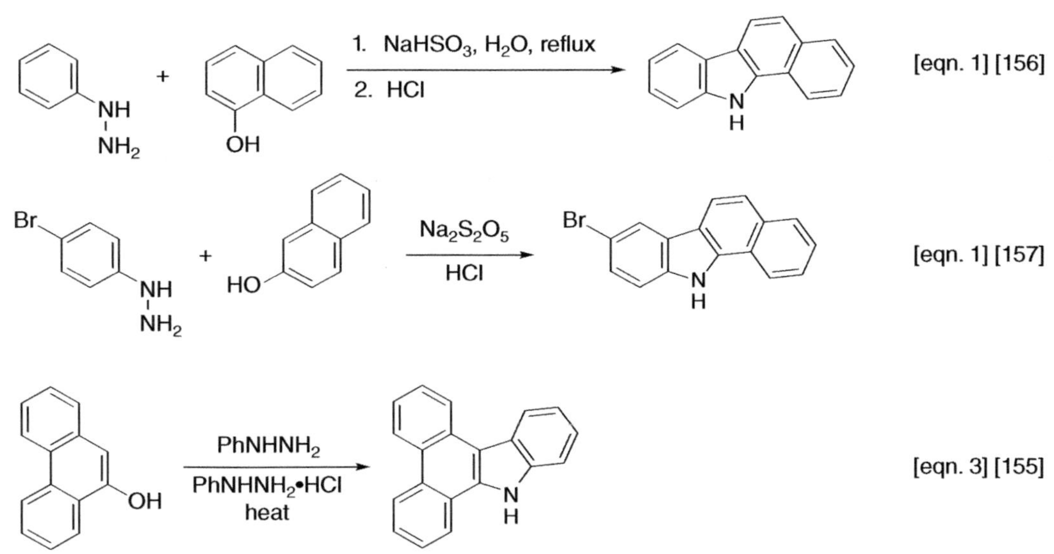

Scheme 27 *Bucherer Carbazole Synthesis*

Serotonin (**53**) (5-hydroxytryptamine [5-HT]) is a neurotransmitter that is involved in controlling several physiological functions. Bufotenine (**54**) (*N*,*N*-dimethyl-5-hydroxytryptamine) is a potent hallucinogen (Scheme 29). As a class, dimethyltryptamines are excellent 5-HT_{1D} agonists that are used to treat migraine headaches. Sumatriptan (**55**) was the first drug discovered to be successful for this malady. As shown in Scheme 29, the Fischer indolization has been employed for the synthesis of other 5-HT agonists. Only the key steps are shown for each

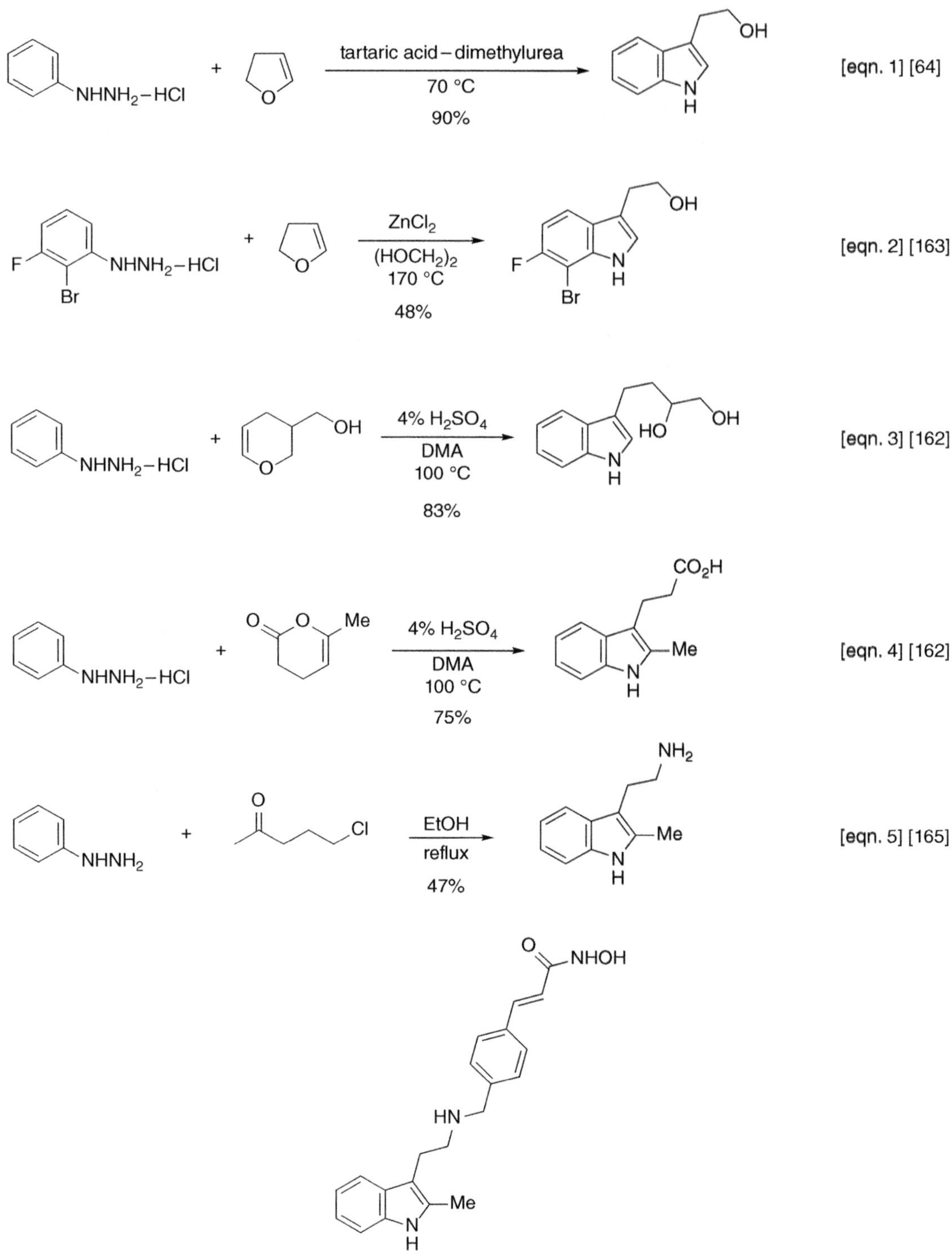

LBH589 (**52**)

Scheme 28 *Grandberg Indole Synthesis*

synthesis. The 1,2,4-oxadiazole L-695,894 (**56**) is active against the 5-HT_{1D} receptor and is a new candidate for migraine treatment (equation 1) [166]. The Grandberg modification was run on a kilogram scale (76% yield). A similar Grandberg variation gave tryptamine amide **57** (equation 2), which is a potent and selective 5-HT_{1F} receptor agonist [167]. The novel cyclobutane indole 4991W93 (**58**) is a partial 5-$HT_{1B/1D}$ receptor agonist and is also highly

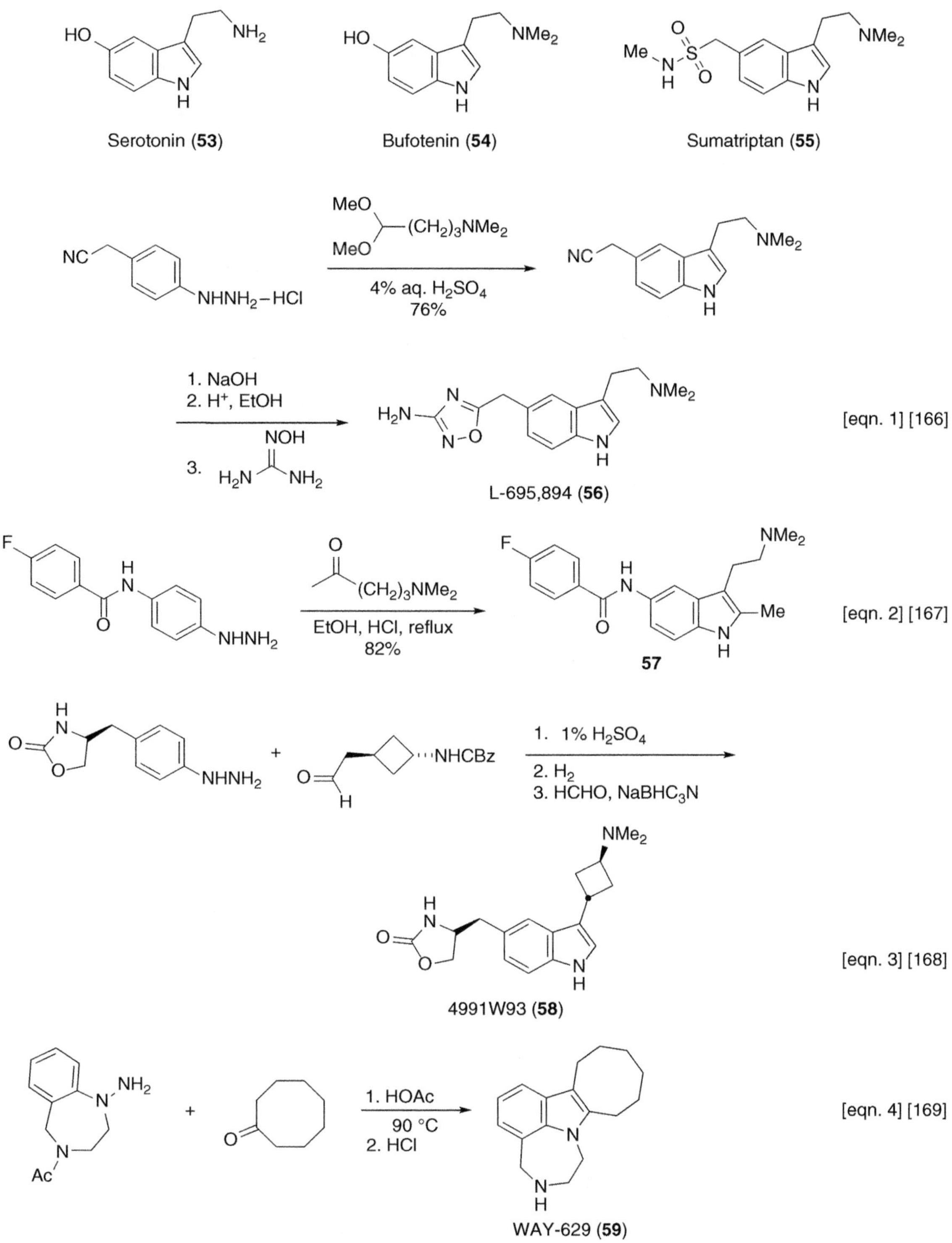

Scheme 29 *Fischer Indole Syntheses of 5-HT Agonists*

active and selective against 5-HT_{1F} receptors (equation 3) [168]. A series of benzodiazepinoindoles are active agonists against the human 5-HT_{2C} receptor; WAY-629 (**59**) shows nanomolar activity (equation 4) [169].

Other 5-HT receptor binders that were synthesized via Fischer indole methodology are shown in Scheme 30 [170–177]. Azepinoindole **62** was formed via Liu's earlier described branched-aldehyde method [172]. Indole **63** was

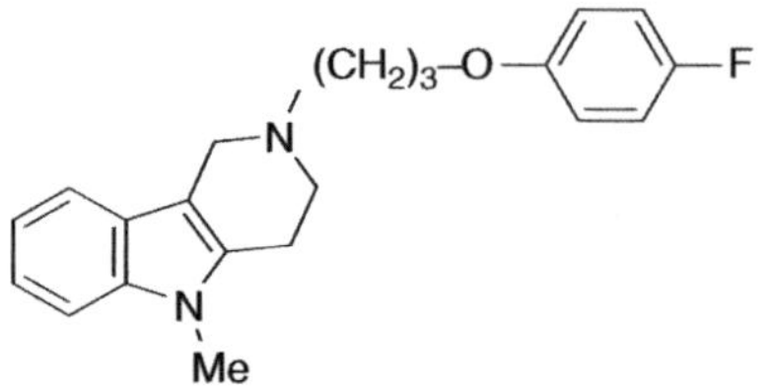

60 (5-HT_{5A}) [170]

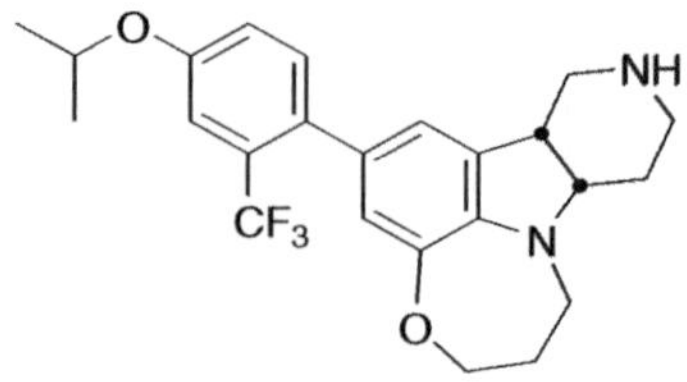

61 (5-HT_{2C}) [171]

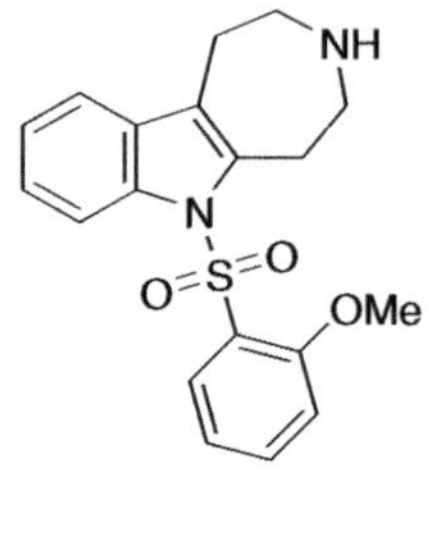

62 (5-HT_6) [172]

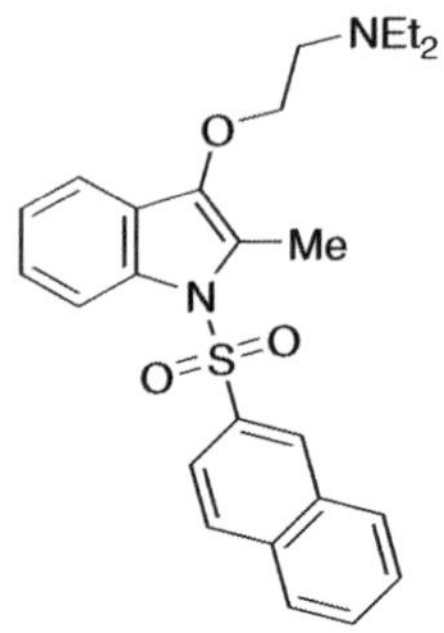

63 (5-HT_6) [173]

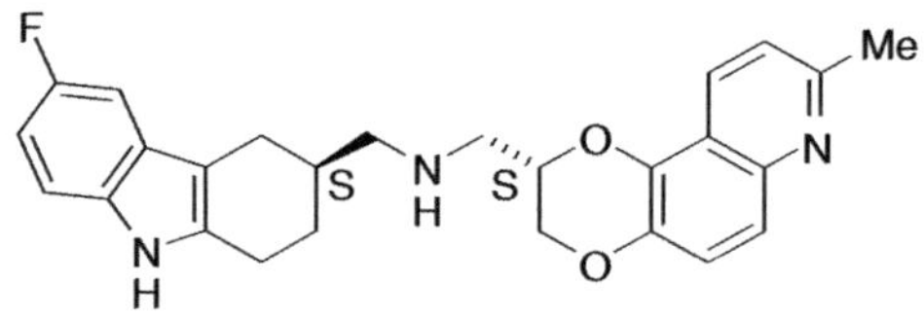

64 (SSRI/5-HT_{1A}) [174]

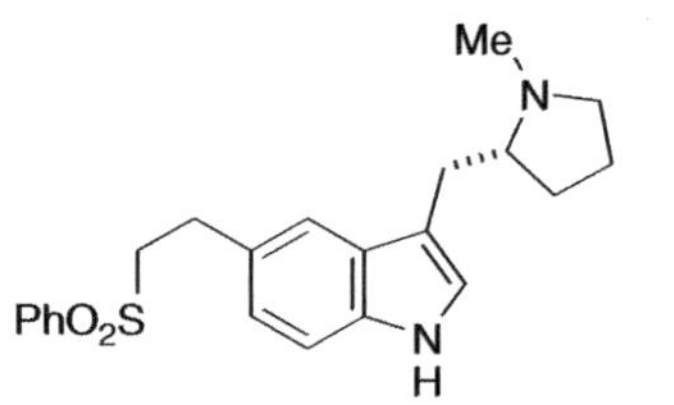

65 (5-HT) [175]

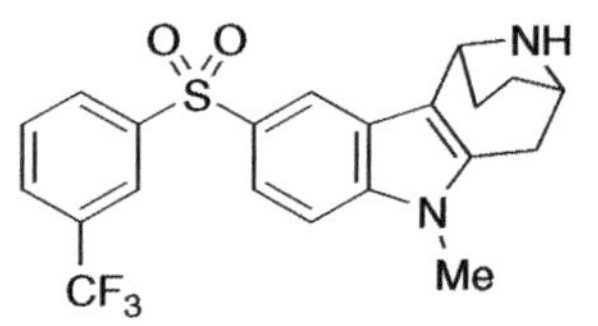

66 (5-HT_6) [176]

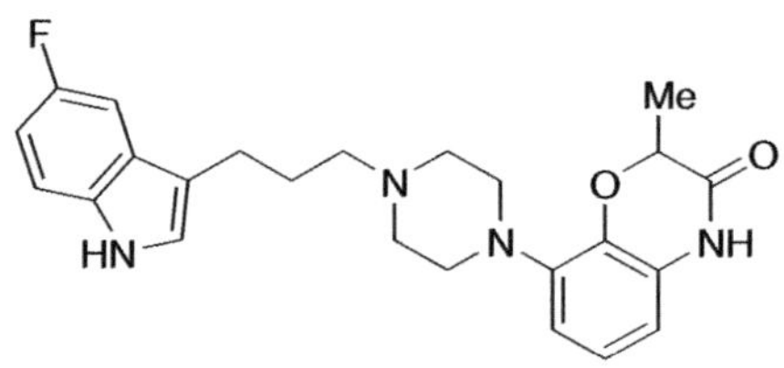

67 (dopamine D_2) [177]

Scheme 30 *Fischer Indole Syntheses of 5-HT Receptor Binders*

prepared using Beller's titanium hydroamination chemistry [172]. A new commercial synthesis of eletriptan (**65**) involves an improved Fischer indolization [175]. A Grandberg synthesis was used to craft the potent dopamine D_2 receptor antagonist **67** (SLV314) [177].

Several *N*-methyl-D-aspartate (NMDA) receptor-active indoles have been synthesized via a Fischer indolization (Scheme 31). For example, a Japp–Klingemann synthesis gave **68**, a potent binder of the glycine site of the NMDA receptor (equation 1) [178, 179]. The related NMDA-type

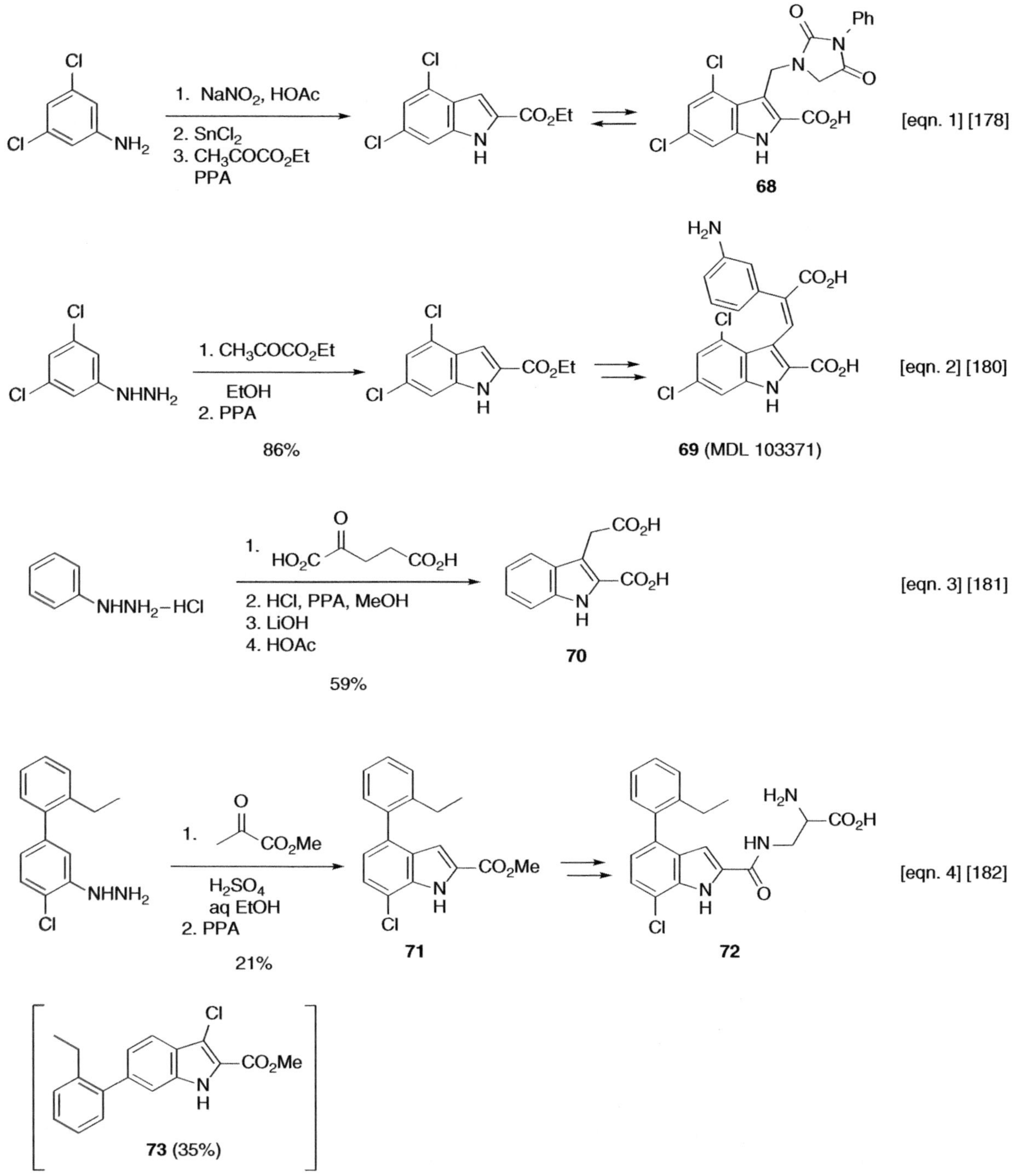

Scheme 31 *Fischer Indole Syntheses of NMDA Receptor Active Indoles*

glycine-receptor antagonist MDL 103371 (**69**) is under development for the treatment of stroke (equation 2) [180]. A standard Fischer indolization led to the kainate analogue **70**, having antagonist activity at ionotropic glutamate receptors (equation 3) [181]. Several partial agonists at the glycine site on the NMDA receptor complex were developed by Koller and colleagues; for example, **72** (equation 4) [182]. In addition to indole **71**, 35% of the abnormal Fischer indole **73** was formed.

A number of antitumor agents have been made available through Fischer indole syntheses (Scheme 32). Cyclopent[*b*]indol-3-ones were used to prepare novel antitumor aziridinyl

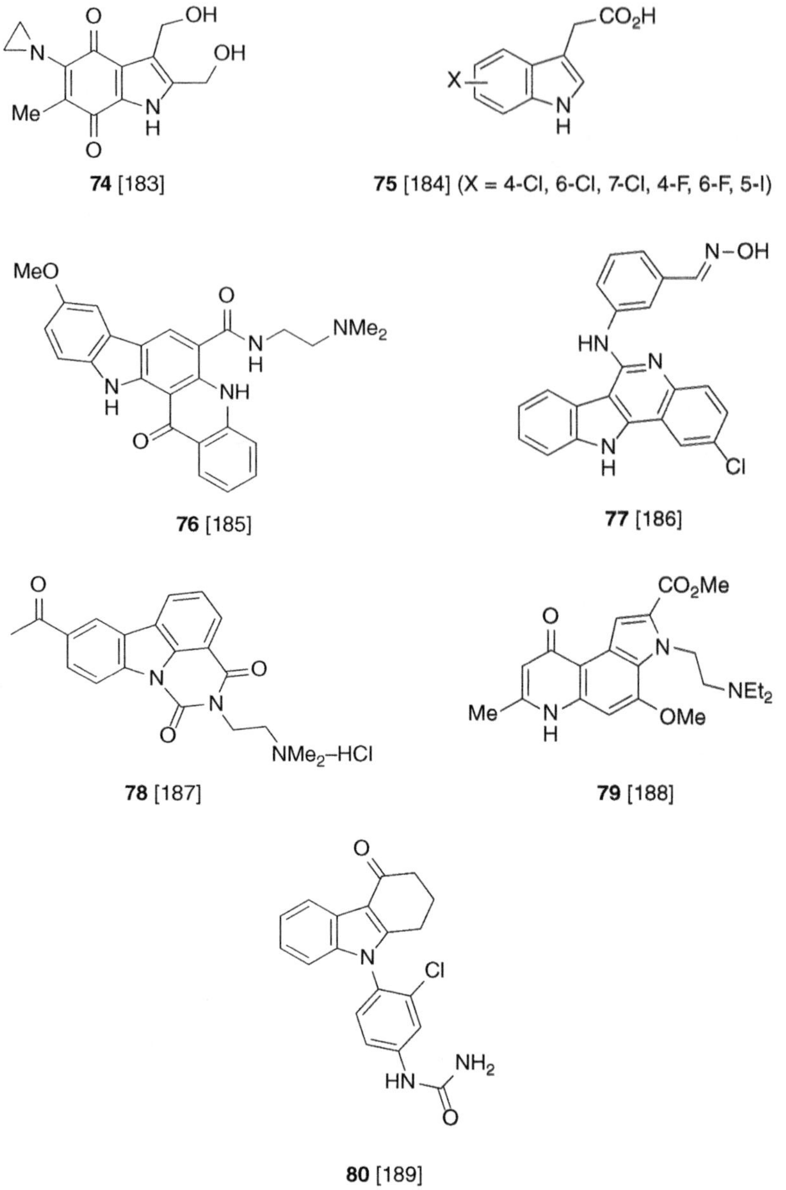

Scheme 32 *Fischer Indole Syntheses of Antitumor Indoles*

quinones (e.g., **74**) [183]. A Fischer indole route to halogenated indole-3-acetic acids produced the cytotoxic **75**, from which 6-chloroindole-3-acetic acid exhibited the highest toxicity upon activation by horseradish peroxidase. This apparently converts the prodrug **75** to a 3-methylene oxindole as the proximate electrophile [184]. Likewise, a classic Fischer indolization affords the weakly antitumor indolo [2,3-*a*]acridine **76**. The somewhat structurally related pyrazolo[3,4,5-*kl*]acridines are significantly more active than **76** [185]. Moreover, indolo[3,2-*c*]quinoline **77** is very active against NCI-H460 lung, MCF7 breast, SF-268 central nervous system, and several other human tumor cell lines at 0.8–2 μM levels. A Borsche–Drechsel carbazole synthesis was employed to obtain the core of **77**, which is targeted for *in vivo* evaluation [186]. Similarly, a Borsche–Drechsel indolization and oxidation starting with cyclohexanone and 2-hydrazinobenzoic acid hydrochloride led to carbazole **78**. This compound, ER-37326 (**78**), was more efficacious than etoposide *in vivo* in the solid tumor M5076 [187]. The water-soluble pyrrolo[3,2-*f*]quinoline **79**, which is very active against melanoma and other cutaneous cancers, was formed via a Fischer indolization between 2-methoxy-4-nitrophenylhydrazine hydrochloride and methyl pyruvate. Subsequent conversion of the resulting methyl

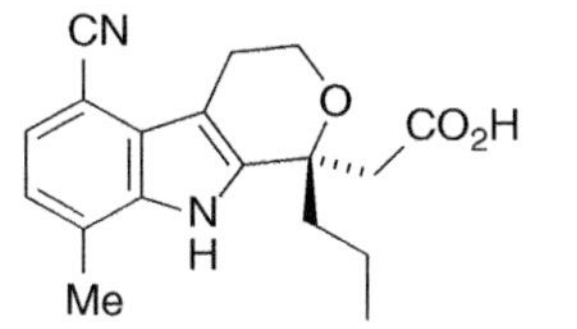

81 [190]

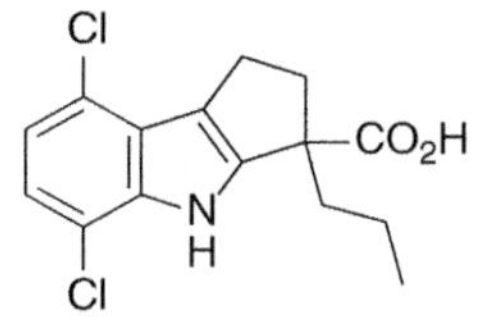

82 [191]

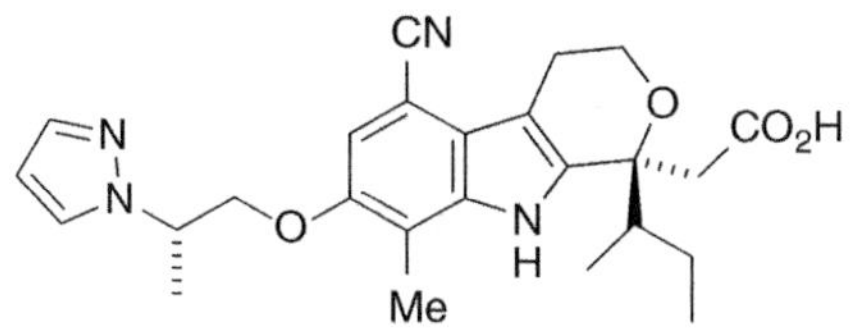

83 [192]

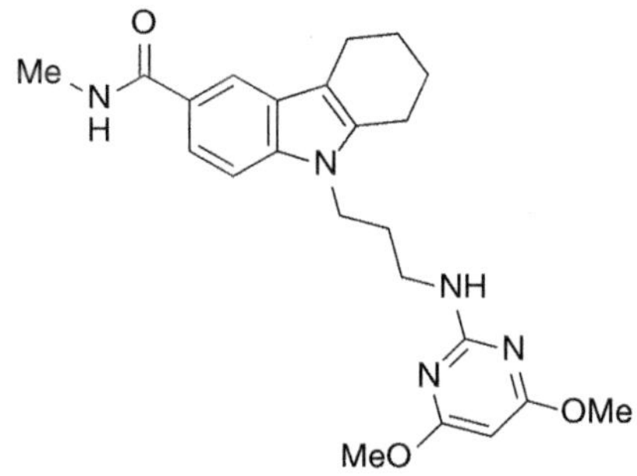

84 [193]

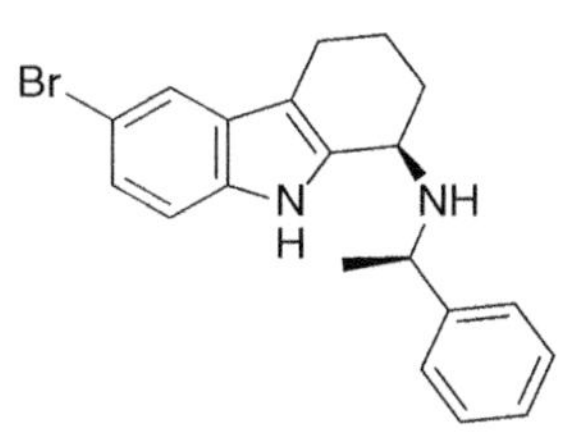

85 [194]

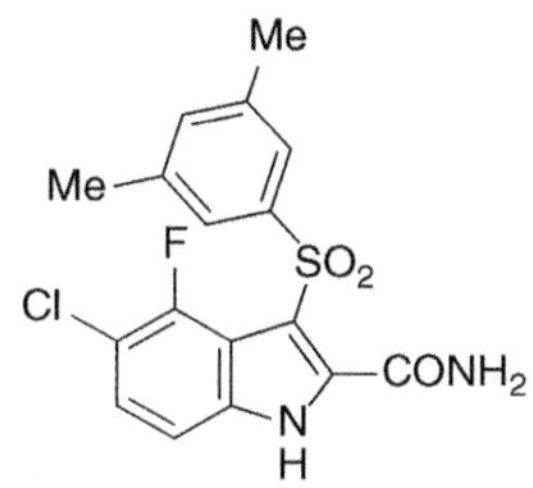

86 [195]

Scheme 33 *Fischer Indole Syntheses of Antiviral Indoles*

7-methoxy-5-nitroindole-2-carboxylate gave **79** [188]. Carbazol-4-one **80** has its origins in the reaction of 1,3-cyclohexanedione and phenylhydrazine in a standard Fischer indolization. This compound is a potent inhibitor of microtubule polymerization, with kinetics similar to those of vinblastine and vincristine and superior to that of the colchicine analogue colcemid. It shows excellent activity against the cell lines K562, MCF-7, SW620, and HT29 [189].

Hepatitis C virus (HCV) is a major global health problem, with an estimated 170 million people worldwide and more than 4 million Americans carrying HCV and up to 12,000 HCV-related deaths annually in the United States. Obviously, new drugs are needed to treat HCV infections. One set of compounds, pyrano[3,4-*b*]indoles, shows excellent and selective activity against HCV NS5B RNA-dependent RNA polymerase, which is essential for replication of this virus (Scheme 33). The most promising indole to emerge from this study is **81**, which was synthesized using the Grandberg tryptophol method from 5-bromo-2-methylphenylhydrazine hydrochloride and 2,3-dihydrofuran [190]. A subsequent study from this research team discovered cyclopent[*b*]indole **82**, which has activity comparable to that of **81**. Indole **82** was synthesized beginning with 5-bromo-2-methylphenylhydrazine and the appropriate cyclopentanone [191]. Another group uncovered a different set of pyrano[3,4-*b*]indoles that are active against the NS5B polymerase, most notably **83** and a related demethyl analogue. A Grandberg Fischer indole variation starting from benzyl 5-bromo-3-hydrazino-2-methylbenzoate and 2,3-dihydrofuran gave the desired tryptophol, which was elaborated to **83**. This pyranoindole is significantly more active (20–600 times) than **81** in three HCV enzyme systems [192]. A class of tetrahydrocarbazoles, such as **84**, is active against HCV entry into normal cells, with activity of $IC_{50} < 1$ nM in a tissue culture HCV infectious assay. The synthesis involved a Borsche–Drechsel cyclohexanone-4-hydrazinobenzoic acid combination followed by methanolic sulfuric acid at reflux [193].

Human papillomaviruses (HPVs) cause a variety of benign and premalignant tumors, and several infect the genitalia. HPV infections are the most widespread sexually transmitted disease worldwide, with nearly 6 million new

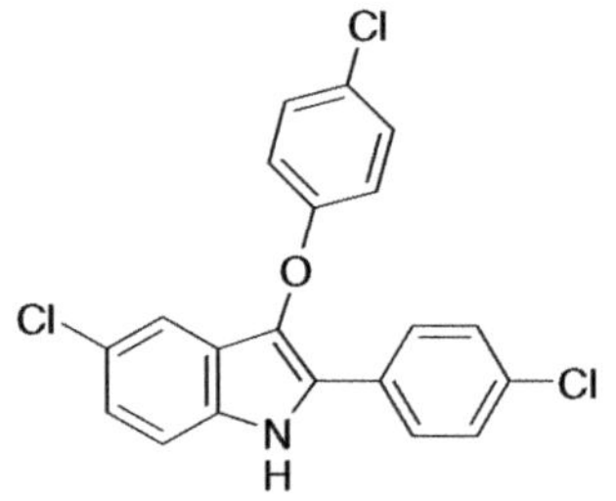

87 [198]

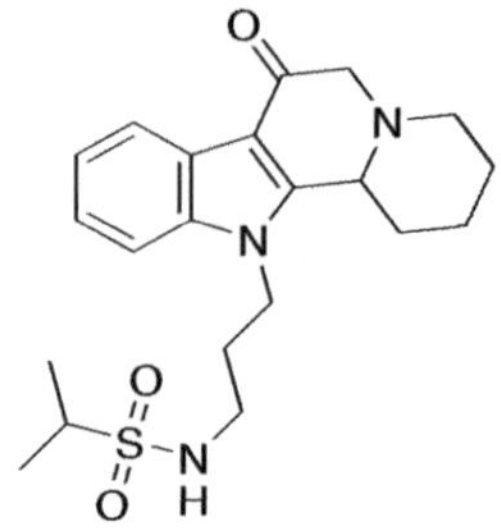

88 [199]

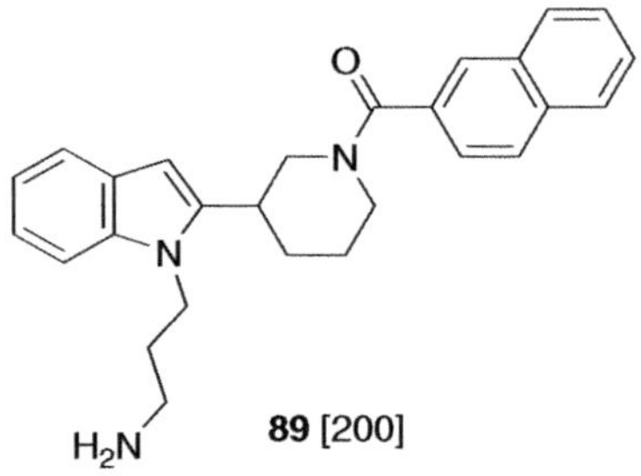

89 [200]

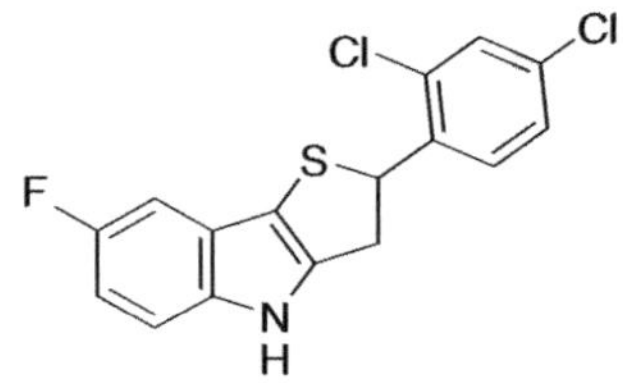

90 [201]

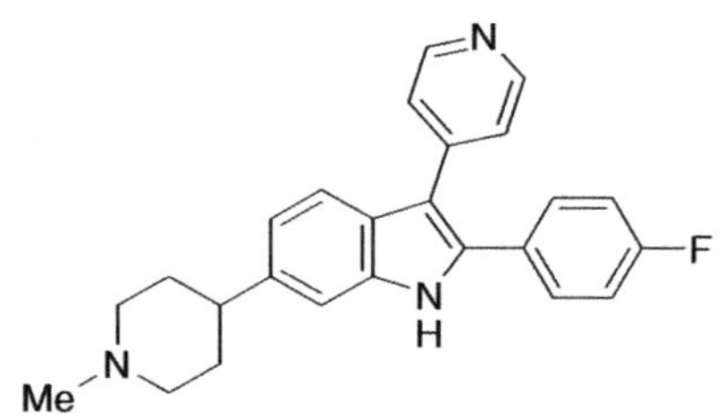

91 [203]

Scheme 34 *Fischer Indole Syntheses of Antibacterial Indoles*

cases annually in the United States and more than 20 million people currently infected in the United States. The incidence of HPV infections is double that caused by herpes simplex. Several tetrahydrocarbazoles are active against HPV, and one compound, **85**, is active against episomal HPV-16 DNA (IC_{50} = 0.03 μM) [194]. The C-1 epimer of **85** is essentially inactive. The synthesis entailed a Japp–Klingemann reaction between 4-bromophenyldiazonium salt and 2-formylcyclohexanone (49%), followed by indolization with HCl/HOAc (88%), to give the expected 1-ketotetrahydrocarbazole [194].

The well-known and ubiquitous human immunodeficiency virus (HIV) has been and continues to be an enormous global menace. Nevertheless, many drugs are available to fight AIDS (acquired immunodeficiency syndrome), the disease caused by HIV. A new collection of potential drugs as HIV-1 non-nucleoside reverse transcriptase inhibitors are indolyl aryl sulfones, such as the highly potent **86**. A Japp–Klingemann protocol fashioned the requisite ethyl 5-chloro-4-fluoroindole-2-carboxylate needed for **86**. Indole **86** is strongly active against HIV-1 wild type (ED_{50} = 0.5 nM) and is a very potent inhibitor of the reverse transcriptases carrying the K103N, Y1811, and L100I mutations [195].

Like the battle against viruses, the war to stop bacterial and parasitic infections has reached the critical stage, with multidrug-resistant bacteria, such as methicillin-resistant *Staphylococcus aureus* (MRSA) and vancomycin-resistant *Enterococcus faecalis* (VRE), killing thousands of people annually in the United States [196]. Likewise, tuberculosis (TB) has killed about 1 billion people worldwide over the past two centuries. Some 3.5 million people in India are currently infected with TB [197].

Magedov and coworkers employed a Fischer indolization between arylhydrazines and acetophenones to prepare a series of 2-aryl-3-aryloxyindoles with potent activity against several gram-positive drug-resistant bacterial strains (Scheme 34). For example, **87** is also comparable in activity to vancomycin and penicillin G. Unexpectedly, **87** is also active against the human cancer cell line HeLa (GI_{50} 8 μM) [198]. A target for antibacterial therapy is the inhibition of the interaction between the prokaryotic tubulin analogue FtsZ and the membrane-anchored protein ZipA, the binding of which is essential for *Escherichia coli* cell division. Overexpression or depletion of ZipA thwarts cell division. Jennings and colleagues discovered two classes of indoles that inhibit the ZipA–FtsZ interaction and display efficacy against both gram-positive and gram-negative bacteria. One compound from each study is shown in **88** [199] and **89** [200]. Both tetracyclic indole **88** and indole **89** originate from phenylhydrazine and 2-acetylpyridine via a polyphosphoric acid (PPA)-catalyzed Fischer indolization, following by reduction to a piperidine ring and further elaboration. Both sets of indoles inhibit the ZipA–FtsZ interaction but at levels deemed unsatisfactory with the current compounds [199, 200]. A novel set of thieno[3,2-*b*] indoles is active against *Mycobacterium tuberculosis* H37Rv (MTB) and multidrug-resistant *M. tuberculosis* (MDR-TB), and **90** shows *in vitro* activity MIC = 0.4 μg/mL against both bacterial strains. A Fischer indolization between 4-fluorophenylhydrazine and 2,4-dichlorophenyl-3(2*H*)-thiophenone cleanly afforded **90** with no trace of the isomeric thieno[3,4-*b*]indole [201]. Importantly, **90** and other thieno[3,2-*b*]indoles in this study were more potent than the currently available anti-TB drugs: isoniazid, rifampicin, ethambutol, and pyrazinamide.

A third major global health problem are parasitic protozoa that cause myriad diseases and afflictions in humans and animals. A serious problem in the poultry industry is coccidiosis, caused by the parasites *Eimeria tenella*, *E. acervulina*, *E. mitis*, and *E. maxima*, which invade the avian intestinal lining [202], leading to illness and death [202]. A new potential set of coccidiostats are 2,3-diarylindoles such as **91**, which has excellent *in vitro* (IC_{50} 0.5 nM) and *in vivo* activity [203]. These indoles were prepared either via a Fischer indolization between arylhydrazines and α-arylacetophenones or via a reductive cyclization of a nitroketone, which is presented in a later chapter.

Several indoles are ligands for estrogen receptors. For example, enzothieno[3,2-*b*]indole **92** is a potent and selective estrogen-receptor ligand for the subtypes ERα and ERβ. The effect of this compound to increase bone density in ovariectomized mice is comparable with that of raloxifene (Scheme 35). A Fischer indolization (HOAc, 80 °C) between 4-methoxyphenylhydrazine and 6-methoxybenzo[*b*]thiophen-3(2*B*)one initiates the synthetic route to **92** [204]. The biindole **93** is a highly potent and selective antagonist of estradiol activity in uterine tissue and MCF-7 cancer cells, with high affinity for the ERα receptor. A standard Fischer indolization between 4-benzyloxyphenylhydrazine hydrochloride and the appropriate ketone leads to other compounds in the series, although **93** itself was prepared from 2-acetyl-5-methoxy-3-methylindole using conventional chemistry [205]. The indole poly(amido)amine conjugate **94** is also a potent ligand for ERα. The synthesis of **94** starts with a Fischer indolization between 4-methoxyphenylhydrazine hydrogen chloride and 4-methoxypropiophenone (HCl, EtOH, 80 °C) [206].

Zhang and colleagues have reported a new indole-based scaffold to mimic the tetracyclic testosterone ring system. These indoles, such as **95**, are strong binders to androgen receptors. Such nonsteroidal androgen-receptor antagonists can be employed in the treatment of prostate and breast cancer, acne, anorexia, alopecia, and hirsutism in women, whereas agonists of the androgen receptor can be engaged in male contraception and sexual performance, cancer treatment, AIDS, and other disorders [207]. The advantage of these nonsteroid indole-androgens is that they may avoid the undesired biological properties of androgens (rapid metabolism, poor oral bioavailability, and activation of nontarget receptors). The synthesis of **95** involves indole ring formation between 4-methoxyphenylhydrazine hydrochloride and 7,8-dihydroquinolin-5(6*H*)-one (HCl, EtOH, reflux; 73%) followed by minor tweaking of the indole product [207].

The excess production of prostaglandins plays a central role in inflammation such as that associated with pain, asthma, fever, and allergic diseases [208]. Several indole-based compounds have an effect on controlling the production of prostaglandin D_2 (PGD_2) by antagonizing the PGD_2 receptor. One such group of compounds is indole-3-acetic acids like the nonsteroidal antiinflammatory drug indomethacin. A newer class is the *N*-sulfonylindole analogues such as **96** (Scheme 36), which is a potent and selective antagonist toward the human CHTH2 receptor on which PGD_2 normally acts. A Fischer indolization between 4-fluorophenylhydrazine hydrochloride and ethyl levulinate led the way to **96** [209]. Similarly, conventional Fischer indole syntheses were used by Merck Frosst chemists to prepare **97** [210] and **98** [211]. Indole **97** is a potent and selective inhibitor of prostaglandin E_2 synthase (mPGES-1), which is the major enzyme involved in cyclooxygenase-2 (COX-2)–mediated prostaglandin E_2 (PGE_2) production, leading to inflammation and pain. Thus, inhibitors of mPGES-1 offer attractive potential antiinflammatory and analgesic drugs. Indole **98** is a potent and highly selective antagonist for the PDG_2 DP-1 receptor and will be developed further. Indole **97** began its journey with the union of *N*-4-chlorobenzyl-*N*-4-methoxyphenylhydrazine hydrochloride with 2,2-diethyl-4-oxohexanoic acid (*t*-BuOH, reflux, 54%) [210], whereas **98** starts with 2-bromo-4-fluorophenylhydrazine hydrochloride and ethyl 2-(2-oxocyclohexyl)acetate (HOAc, reflux, 32%) [211]. Both Fischer indolization yields are only modest but provide richly endowed indoles for the journey to **97** and **98**.

The widely prescribed antiinflammatory drug indomethacin (**99**) [212] inhibits both COX-1 (constitutive) and COX-2 (inducible) enzymes. Overexpression of the latter enzyme results in the production of proinflammatory

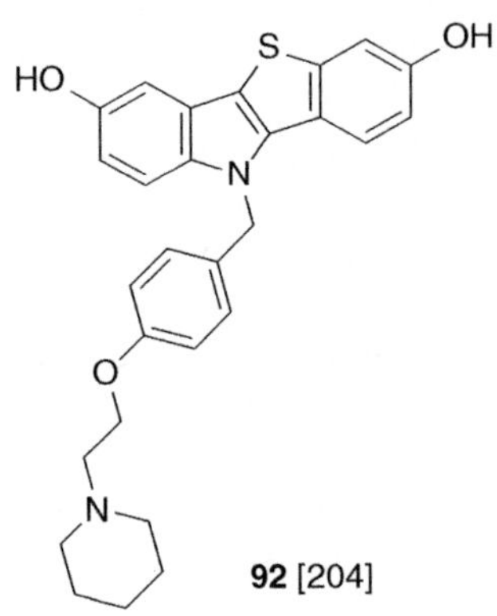

92 [204]

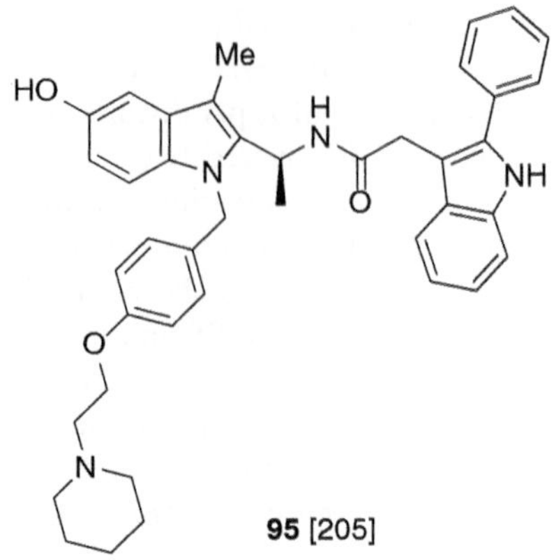

95 [205]

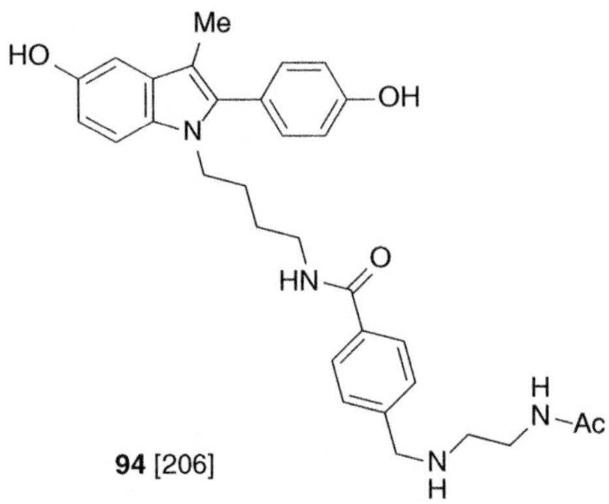

94 [206]

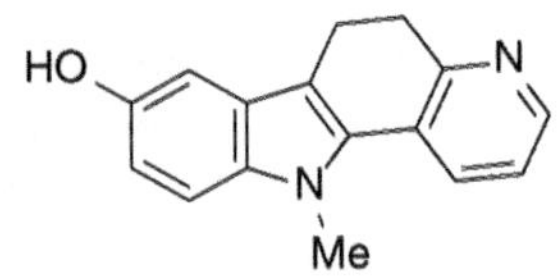

95 [207]

Scheme 35 *Fischer Indole Syntheses of Hormonal Receptors*

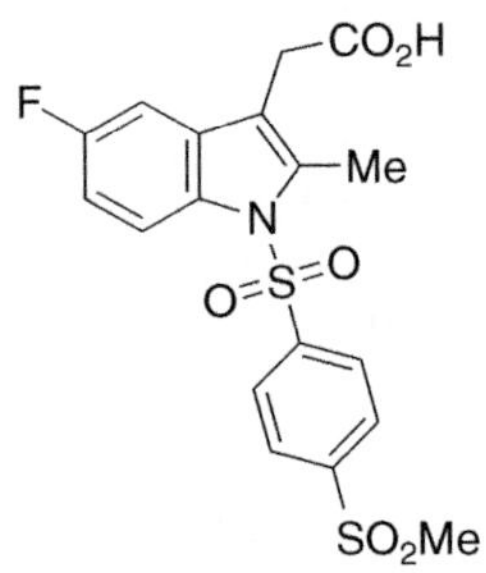

96 [209]

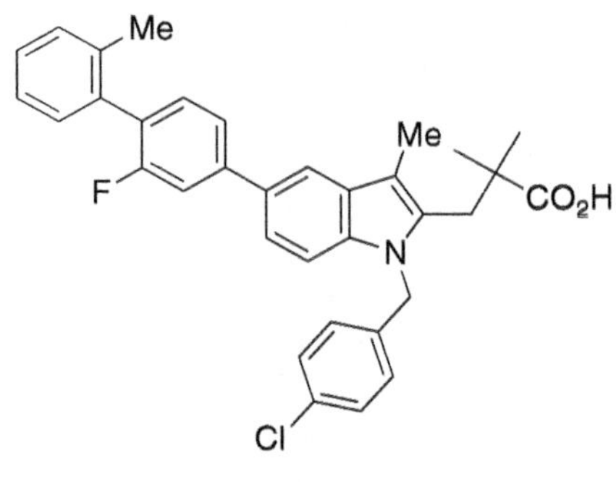

97 [210]

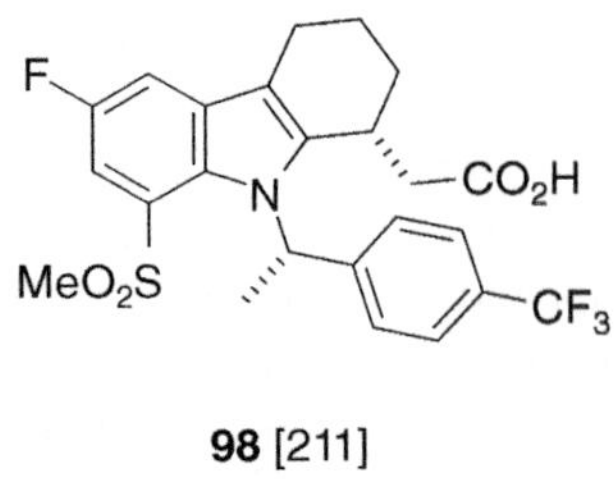

98 [211]

Scheme 36 *Fischer Indole Syntheses of Hormonal Receptors*

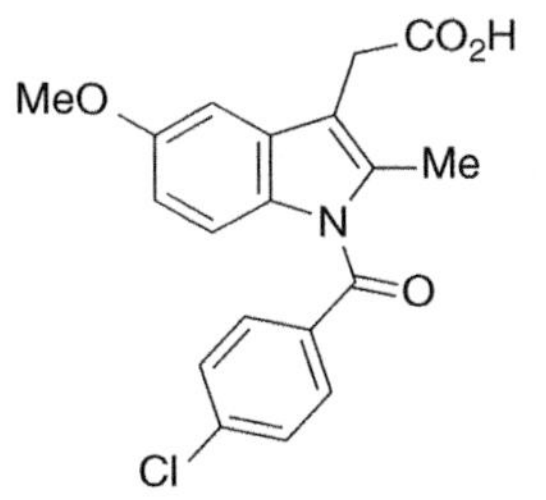
Indomethacin (**99**)

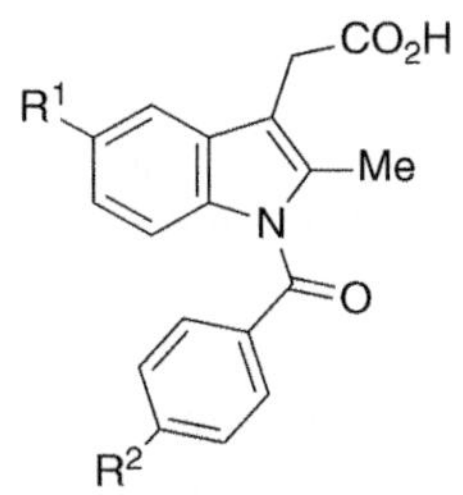
100, R^1 = OMe, R^2 = Br
101, R^1 = H, R^2 = Cl

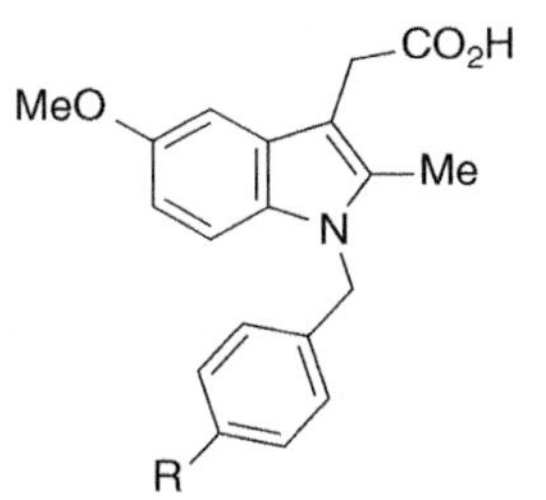
102, R = Cl
103, R = Br
104, R = F

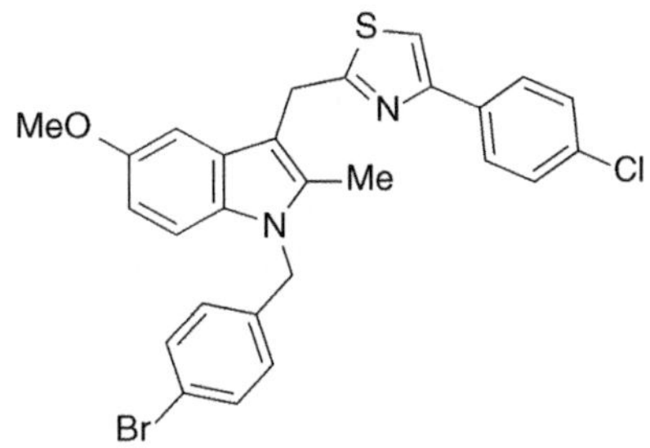
105 [216]

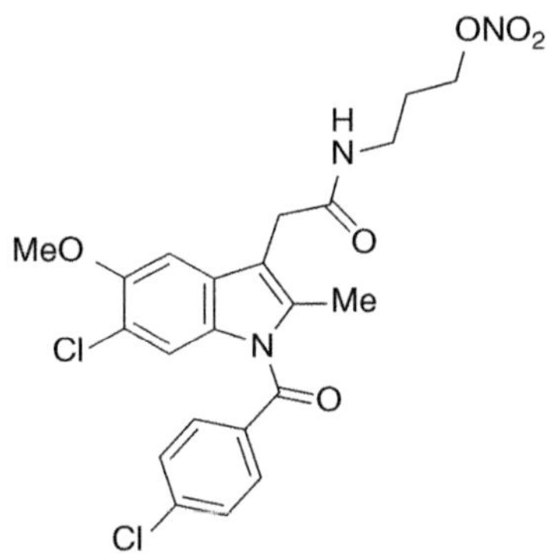
107 [218]

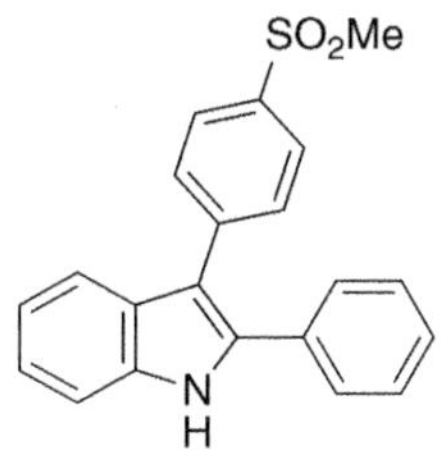
106 [217]

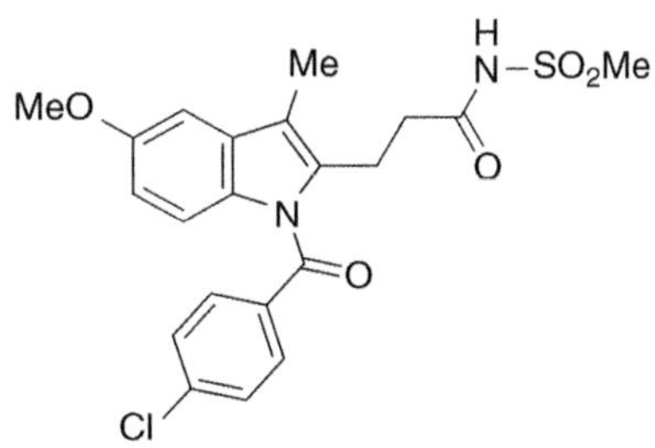
108 [219]

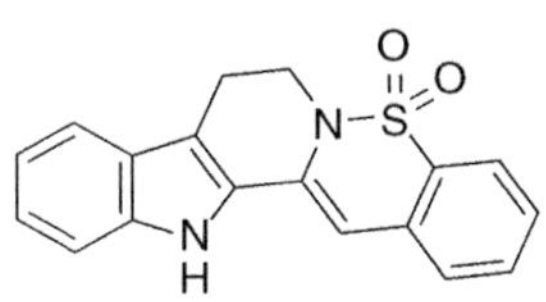
109 [220]

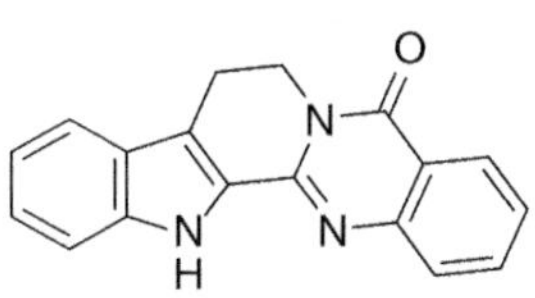
Rutaecarpine (**110**)

Scheme 37 *Fischer Indole Syntheses of COX-2 Inhibitors*

prostaglandins, leading to inflamed tissue, pain, and fever. Therefore, there is a strong need for selective COX-2 inhibitors because COX-1 is essential for the regulation of normal physiological functions. Moreover, the inhibition of COX-1 leads to gastrotoxicity [213, 214]. The Fischer indolization has featured prominently in the synthesis of selective and potent COX-2 inhibitors. Several examples are shown in Scheme 37. Indomethacin analogues **100–104** all show selective COX-2 inhibitory activity [215], as do **105** [216], **106** [217], and **107** [218]. For example, **106** is a potent inhibitor of COX-2 (IC_{50}=0.22 nM) but not of COX-1 (IC_{50} = >10 μM), and it is more potent than celecoxib. Several indomethacin analogues (e.g., **108**) have been constructed via Fischer indolization [219]. Somewhat different is the antiinflammatory rutaecarpine bioisosteric hybrid **109**, which was prepared via a Fischer indole cyclization [220]. Rutaecarpine (**110**) and its homologues are reported to be selective COX-2 inhibitors [221].

Obesity and the often-resulting type 2 diabetes are growing global health problems typically culminating in atherosclerosis, chronic kidney disease, insulin resistance, blindness, and death [222, 223]. The transcription factor PPARγ (peroxisome proliferator-activated receptor gamma) mediates adipocyte differentiation [224], and agonists of PPARγ decrease blood glucose levels in type 2 diabetes patients [225]. One new compound of a large indole set that is a selective and potent PPARγ modulator and that displays antidiabetic activity in mice and rats is indole **111** (Scheme 38) [226]. A retrosynthesis is illustrated for the Fischer indolization. Another enzyme target for diabetes is aldose reductase, the lead protein in the polyol pathway. This cellular pathway is believed to contribute to diabetic complications by depleting NADPH, which in turn reduces glutathione. A reduction in this potent antioxidant leads to oxidative stress and diabetic complications. Indole **112** possesses both aldose reductase inhibitory and antioxidant activity. This simple tetrahydro-β-carboline was fashioned from the 1-(2-phenethyl)-4-piperidone and 4-carboxymethylphenylhydrazine hydrochloride in a Fischer indolization (HCl, MeOH, 45%) [227]. A Grandberg synthesis of various tryptophols was used to prepare 2-(3-(indolyl)ethyloxy)adenosine (**113**) as a novel and potent agonist of the human A_{2B} adenosine receptor [228]. It is known that A_{2B} adenosine receptor antagonists are targeted for asthma and diabetes, and, therefore, agonists like adenosine-indole **113**

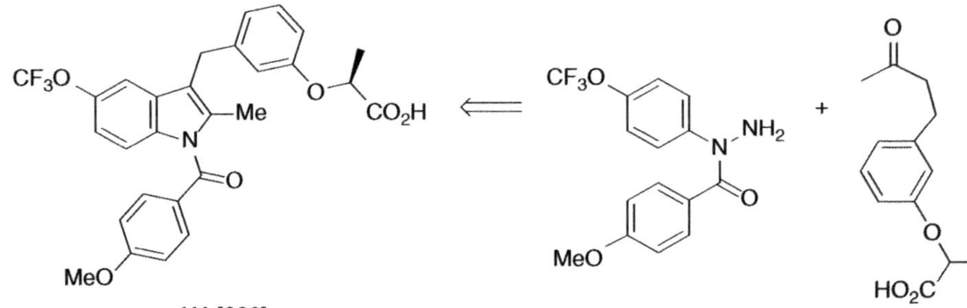

111 [226]

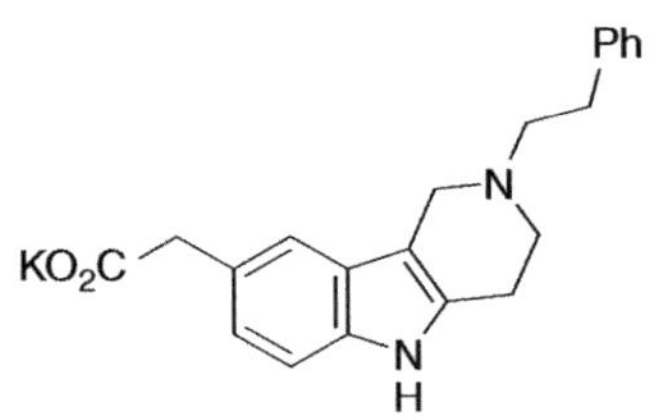

112 [227]

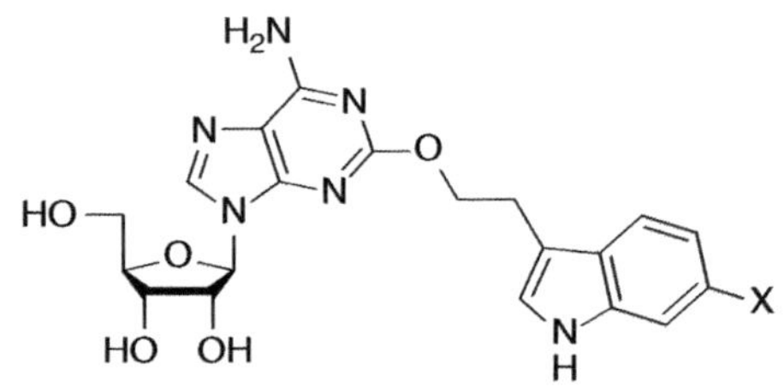

113, X = H [228]
114, X = Br [228]

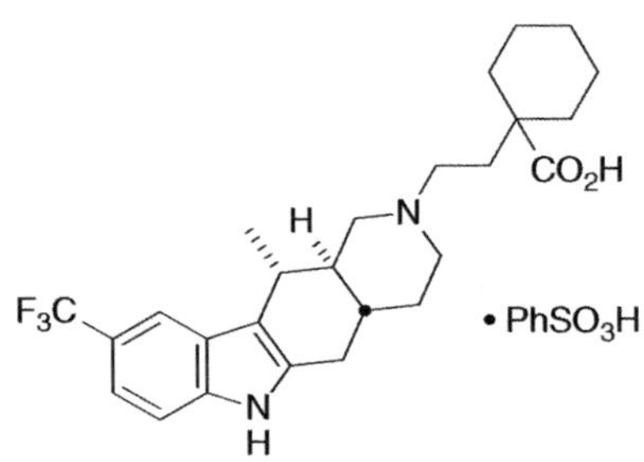

AMG 076 (**115**) [229]

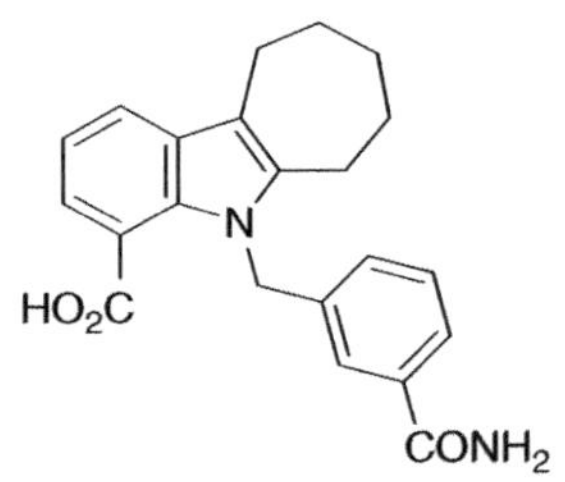

116 [230]

Scheme 38 *Fischer Indole Syntheses of Antiobesity Indoles*

would be a useful pharmacological probe for this receptor. The more active 2-bromo analogue (**114**) was prepared similarly.

The melanin-concentrating hormone (MCH) is involved in mammalian feeding behavior, and the action of this hormone results from its binding to the MCH receptor 1 (MCHr1). Therefore, antagonism of MCHr1 is a reasonable strategy for appetite suppression and increased metabolism. One such potent antagonist is indole AMG 076 (**115**). The preparation of this compound was initiated by a Fischer indolization between 4-trifluoromethylphenylhydrazine and the appropriate octahydroisoquinoline (H_2SO_4, dioxane, 70 °C, 92%) in an 8:1 ratio favoring the linear isomer, isolated as the phosphoric acid salt [229]. Barf and his coworkers found that cyclohept[*b*]indole **116** is a potent and selective inhibitor of A-FABP (fatty acid binding protein) [230]. The implication is that inhibition or reduced expression of A-FABP will lower risk for developing type 2 diabetes and coronary heart disease [231].

Inhibitors of the numerous kinase enzymes are of great importance in the control and treatment of human disease and as physiological probes. Several important kinase inhibitors are indole- or carbazole-based compounds. In addition to its other biological activities, indomethacin **176** (Chapter 1) and analogues are active against various kinases. For example, indole **117** (Scheme 39) represents a new class of angiogenesis-related kinase inhibitors [232]. Aberrant angiogenesis is a crucial step in cancer growth and metastasis, and several specific receptors are involved in this process [233]. Specifically, **117** and related structures inhibit the vascular endothelial growth factors 2 and 3 (VEGFR-2 and -3), the Tie-2 receptor, the fibroblast growth factor receptor 1 (FGFR-1), and the insulin-like growth factor 1 receptor (IGF1R), all of which are implicated in the growth and spread of cancer. Of some 134 indomethacin analogues screened, six were found to inhibit these receptors at low micromolar levels. Indole **117** was synthesized using a solid-phase Fischer indole synthesis [234]. Several indole pyrazoles **118–121** show high *in vivo* potency and oral bioavailability against the aurora kinases A and B, enzymes that have roles in mitosis and are feasible targets for cancer therapy [235]. Indeed, these indole pyrazoles show activity *in vivo* against the HCT116 mouse xenograft model. The synthesis of **117–120** involves a Japp–Klingemann reaction with aniline and the appropriate 2-formylcycloalkanone that gives the expected tricyclic indole ketone that was converted to the pyrazoles.

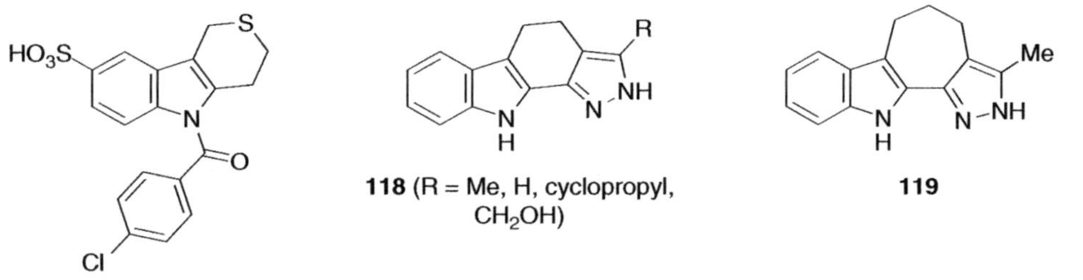

117　**118** (R = Me, H, cyclopropyl, CH_2OH)　**119**

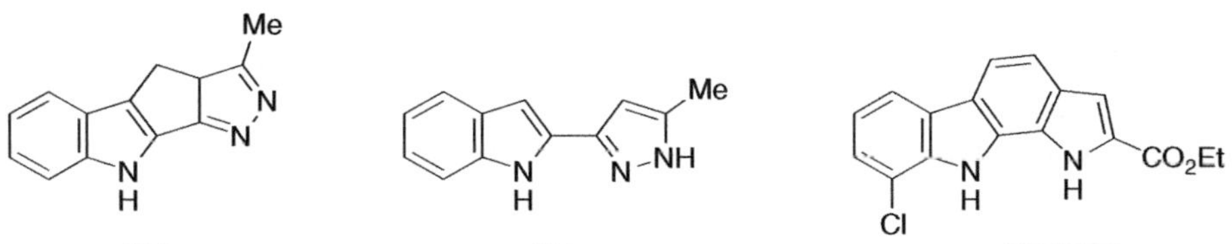

120　**121**　**122** [237]

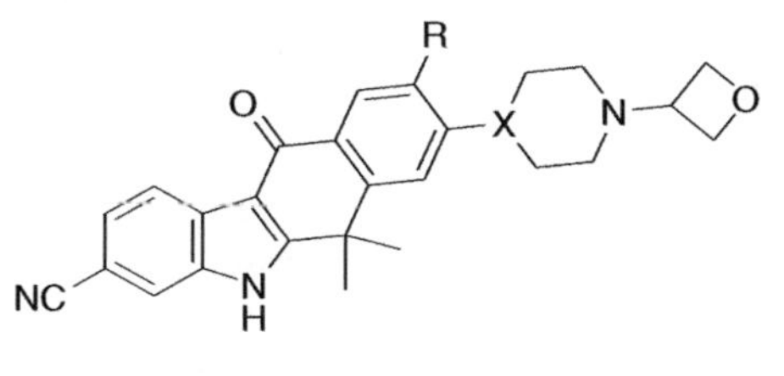

123, R = H, X = CH [238]
124, R = Et, X = N [239]

Scheme 39 *Fischer Indole Synthesis of Kinase Inhibitors*

Another family of serine–threonine kinases are the cyclin-dependent kinases (CDKs), which play a central role in cell cycling, apoptosis, differentiation, and transcription [236]. The normal activity of CDK enzymes is strictly regulated by various processes, but this control can be lost in cancer cells. As a result, CDKs are appealing targets for drug intrusion. In a search for inhibitors of CDK1, Nikolaropoulos and colleagues found that pyrrolo[2,3-*a*] carbazoles are active and compound **122** is especially potent [237]. Using a computational model, these workers found the likely binding of **122** to the ATP-binding groove. A Fischer indolization using PPSE (polyphosphoric acid trimethylsilyl ester) in nitromethane on the hydrazone formed between 2-chlorophenylhydrazine hydrochloride and the appropriate 7-oxotetrahydroindole gave the desired pyrrolocarbazoles. Kinoshita and coworkers discovered several benzocarbazoles with highly potent and selective activity as inhibitors of anaplastic lymphoma kinase (ALK) [238, 239]. Two of these, **123** and **124**, also display antitumor efficacy against lymphoma in mice and strong antiproliferative activity against the KARPAS-299 cell line. The normal function of this receptor tyrosine kinase is currently unknown. A Fischer indolization between 3-cyanophenylhydrazine and the appropriate 2-tetralone began the long journey to **123** and **124**.

A collection of pharmacologically active indoles prepared via Fischer indolization is listed in Table 6 [240–268], along with their putative biological target and the disease(s) that are associated with that target. In several cases the compound shown is selected from a large collection of equally active compounds.

2.3.2 Natural Products

Beyond serotonin and tryptophan, the indole ring is embedded in myriad natural products. This section presents those naturally occurring indoles and related unnatural analogues that have been synthesized via Fischer indolization. In a few cases only the approach to an indole natural product is reported.

Melatonin (**125**) has been synthesized in several ways [269], and a commercial process (Japp–Klingemann) on a 10-kg scale is very efficient (Scheme 40, equation 1) [270]. He and colleagues developed a microwave version of the former method and achieved slightly higher yields of **125** [271]. A thrust in this area has been the synthesis and biological evaluation of melatonin analogues. Thus, a standard Fischer indolization afforded the melatonin-related 2-alkyltryptamines **126** (equation 2) [272]. One isomelatonin **127** was also prepared, in which the reacting enehydrazine favors π-conjugation with the phenyl ring (equation 3). Several fluoro-containing melatonins were accessed via the Grandberg tryptamine method (e.g., 7-fluoromelatonin (**128**), equation 4), as was 7-fluoroserotonin [273]. A one-pot three-component tryptamine (**129**) synthesis was reported by Wang's group (equation 5) [274]. Reaction of 5-methyl-3,4-dihydro-2*H*-pyrrole with an acyl chloride (R^1Cl) and arylhydrazine afforded **129** in good to excellent yields. The method was applied to the synthesis of homotryptamines.

The marine environment is a vast repository of organic compounds, many of which are indoles, and most, if not all, of these possess pronounced biological activity. Fischer indolization features prominently in the synthesis of these indoles.

The indole ring embedded in makaluvamine D (**130**) was fashioned by White and coworkers using a Grandberg tryptophol synthesis (Scheme 41, equation 1) [275]. This pyrroloiminoquinone alkaloid was characterized in the Fijian sponge *Zyzzya* cf. *marsailis* and is a topoisomerase II inhibitor. The indolization reaction afforded a 1.7:1 ratio of the desired to the ("abnormal") undesired tryptophol. In contrast to the aforementioned early-stage Fischer indolization of White's team, Delfourne and colleagues performed a Japp–Klingemann indolization near the final stage of their synthesis of the marine alkaloids arnoamines A and B (**131**) (equation 2) [276]. Arnoamine A is demethyl arnoamine B. A standard Fischer indole cyclization was used by Palermo and Franco to complete their synthesis of iso-meridianin (**132**) (equation 3), which is isomeric to the meridianin family of marine alkaloids from the subantarctic tunicate *Aplidium meridianum* [277]. In contrast to the culminating Fischer indole synthesis of **132**, McWhorter and Liu initiated their elegant synthesis of 8-desbromohinckdentine A (**133**) with the Fischer indolization between phenylhydrazine hydrochloride and 2-bromoacetophenone on a 50-g scale (equation 4) [278]. Hinckdentine A is a marine alkaloid found in the Tasmanian bryozoan (moss animal) *Hincksinoflustra denticulate.*

In an approach to the marine alkaloid dimer dendridine A, which was isolated from the Okinawan sponge *Dictyodendrilla* sp., Sperry and Boyd synthesized 5-bromo-7-methoxytryptamine (**134**) via a Fischer–Grandberg tryptamine synthesis from 4-bromo-2-methoxyphenylhydrazine hydrochloride (Scheme 42, equation 1). Not unexpectedly, the abnormal Fischer product **135** was also obtained, along with a second abnormal product **136** [279]. Cook and co-workers synthesized indole **137** en route to 6-chloro-5-hydroxytryptophan, a key component of the cyclic hexapeptide keramamide A found in the Okinawan sponge *Theonella* sp. The isomeric Japp–Klingemann indole **138** was obtained in lesser yield (Scheme 42, equation 2) [280]. A strain of *Leptosphaeria* sp. that was found on the marine alga *Sargassum tortile* was cultured to yield leptosins, a series of hexahydropyrrolo[2,3-*b*]indoles. Crich and colleagues synthesized the core unit **139** from the appropriate aldehyde via a conventional Fischer indolization (equation 3) [281].

Table 6 *Drug Candidates Synthesized by the Fischer Indolization*

Entry	Candidate	Drug Target	Disease Indication	Ref.
1	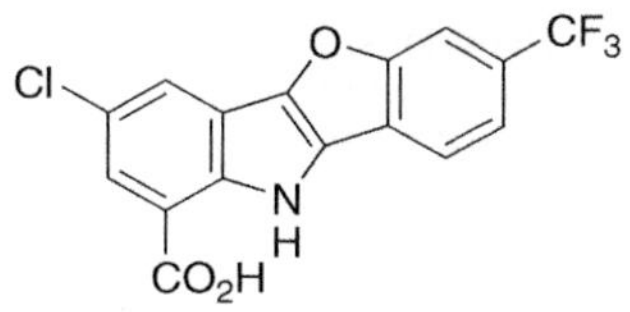	BK_{ca} – Channel Opener	Neuronal damage, Ischemic events Stroke, Convulsions Asthma Hypertension	240
2	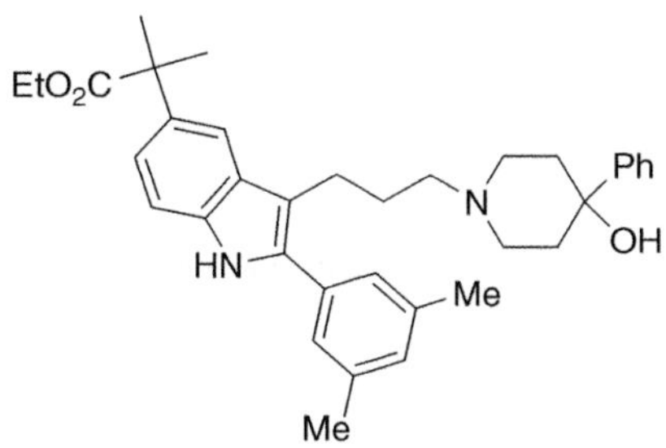	G-Protein coupled receptors		241
3	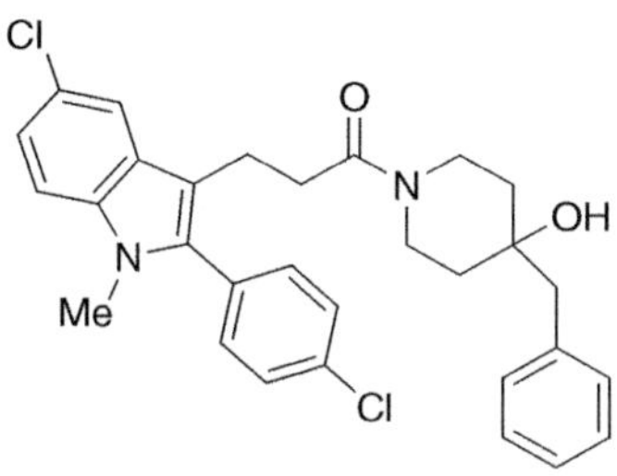	Neurokinin 1		242, 253
4	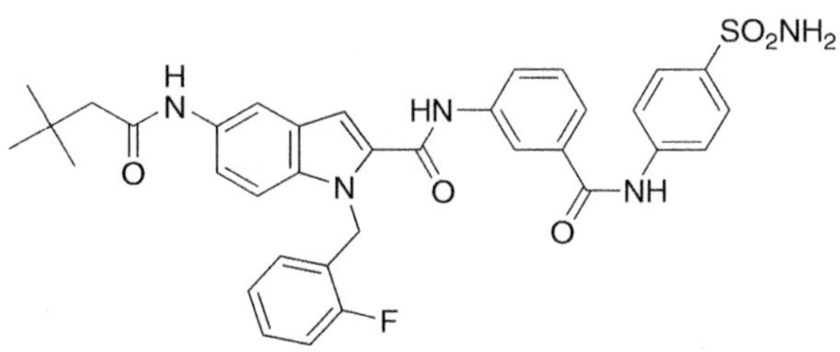	Endothelin-Converting enzyme	Hypertension, Congestive heart failure, Cancer	243
5	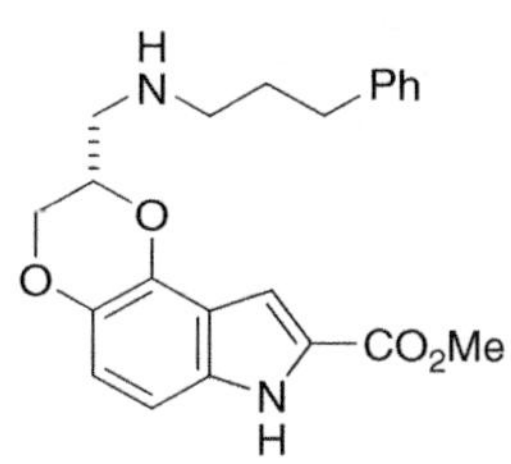	5-HT_{1A} receptor	Hypertension	244
6	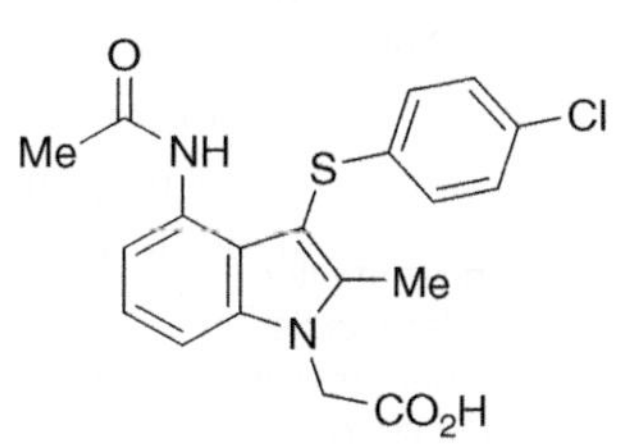	$CRTh_2$ or DP_2	Asthma Allergies	245

(continued overleaf)

Table 6 *(continued)*

Entry	Candidate	Drug Target	Disease Indication	Ref.
7	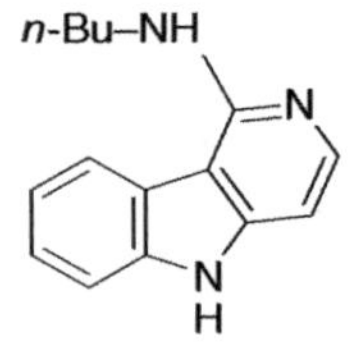	Urotensin-II	Hypertension Heart failure Renal failure Atherosclerosis Diabetes	246
8	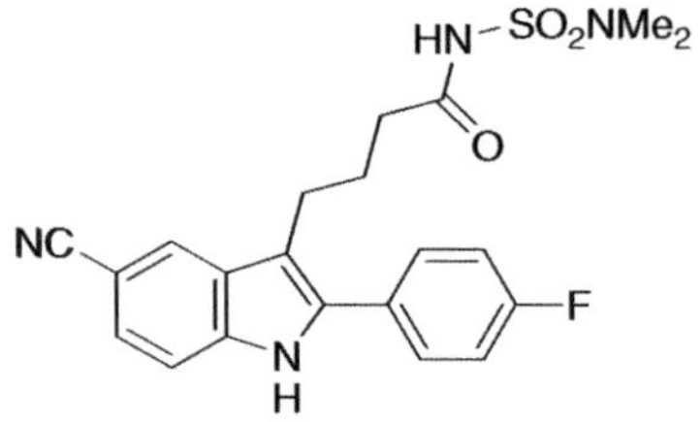	$CXCR_2$	Reperfusion injury	247
9	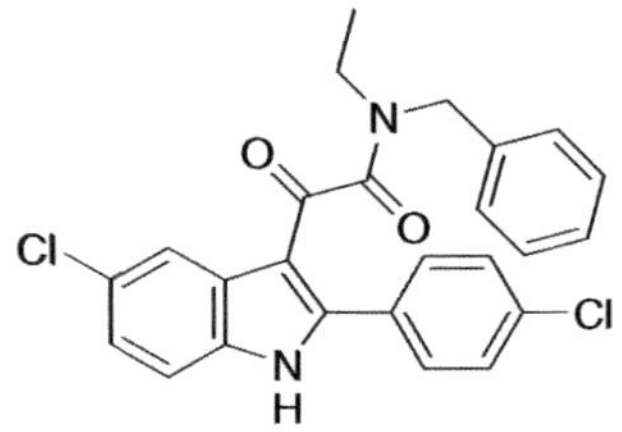	(Peripheral- Benzodiazepine receptor)	Anxiety	248
10	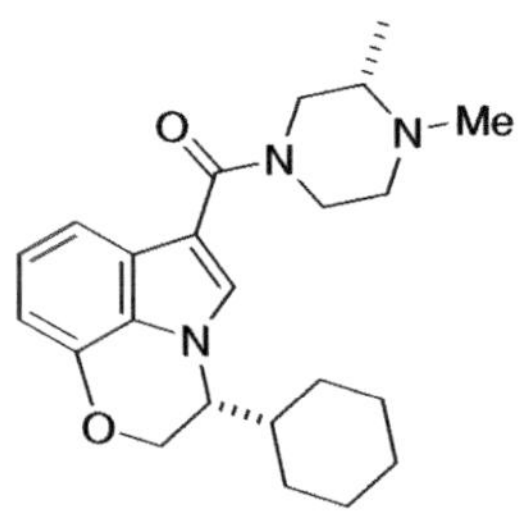	CB1 Cannabinoid Receptor	Pain Glaucoma Multiple sclerosis Antiemetic Appetite stimulants	249
11	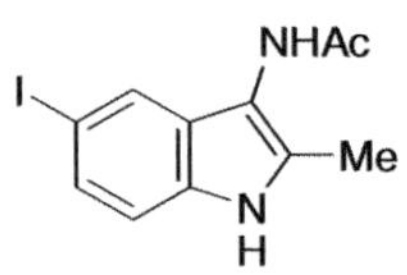	GSK-3β	Bipolar disorders Alzheimer's disease Diabetes	250
12	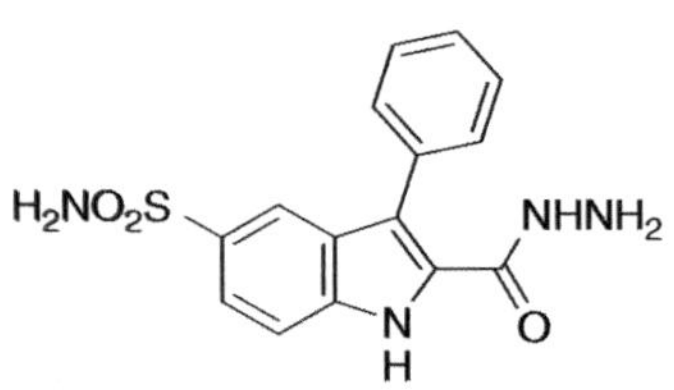	Carbonic anhydrase (several isoforms)	Glaucoma Diuretics	251, 252
13	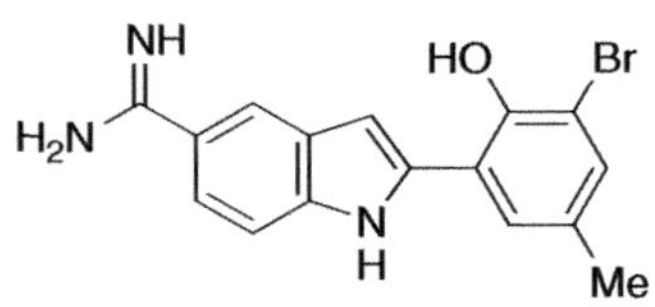	Urokinase-type plasminogen activator Factor X_a	Inflammation Metastasis Cellular invasion Blood coagulation	254

Table 6 *(continued)*

Entry	Candidate	Drug Target	Disease Indication	Ref.
14	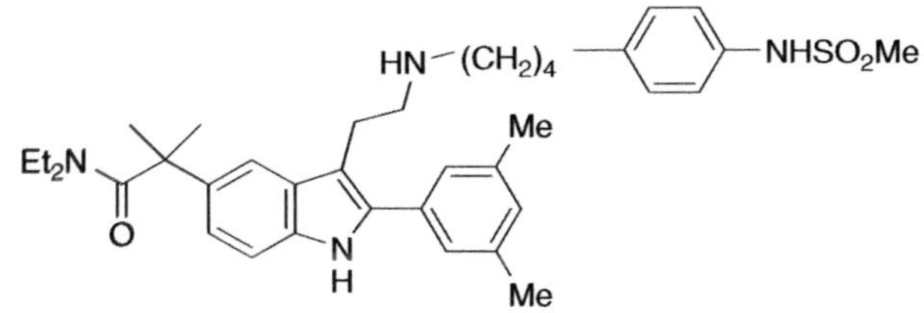	Gonadotropon-releasing hormone	Cancer Endometriosis Uterine fibroids Assisted reproduction	255, 256
15	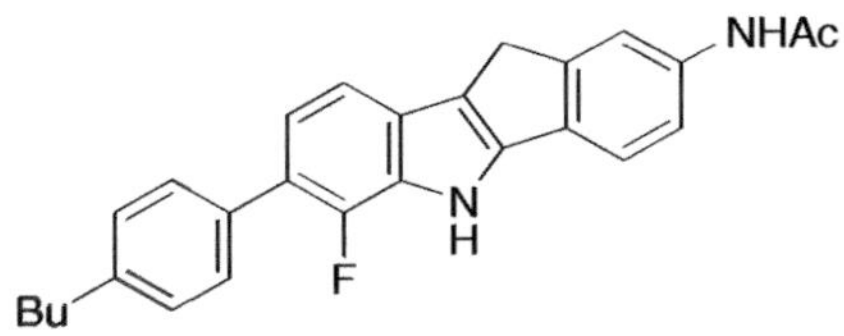	Thrombopoietin receptor	Hepatis C cirrhosis Low platelets	257
16	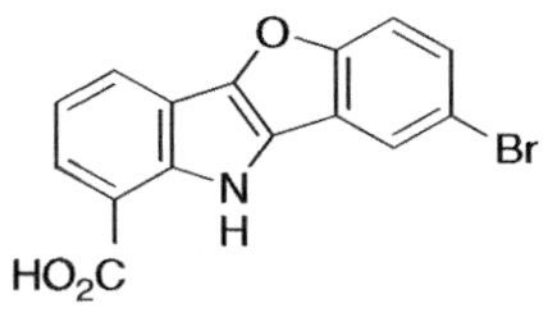	BK_{ea}-Channel Opener	Urge urinary incontinence	258
17	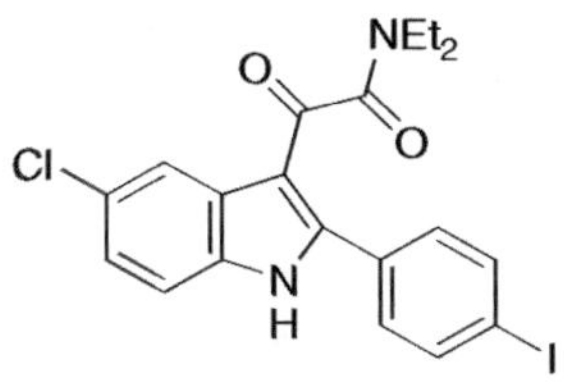	Peripheral benzodiazepine binding sites	Huntington's disease Alzheimers' disease Multiple sclerosis Cancer	259, 260
18	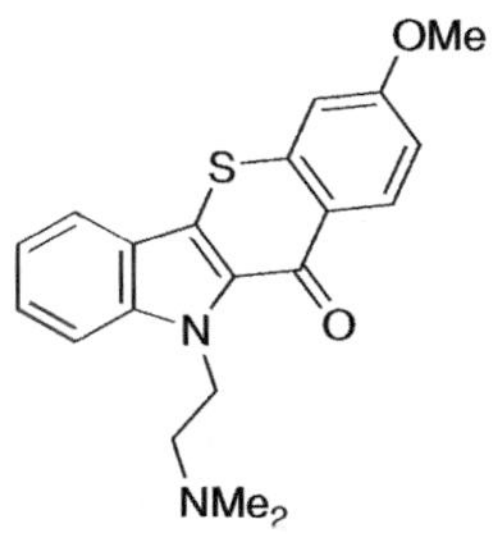	Antiproliferation (HL-60, HeLa)	Cancer	261
19	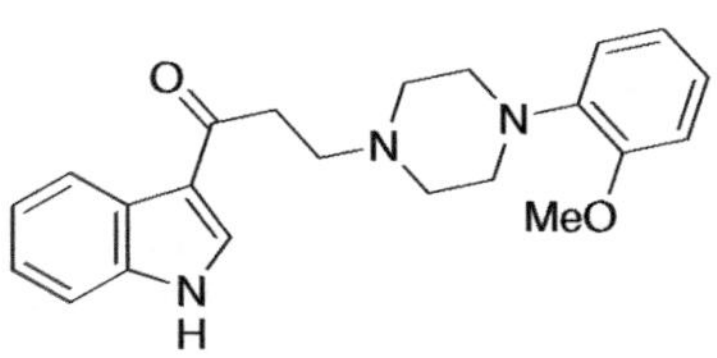	α_1-Adrenergic receptor	Cardiovascular Hypertension Benign prostatic hyperplasia	262
20	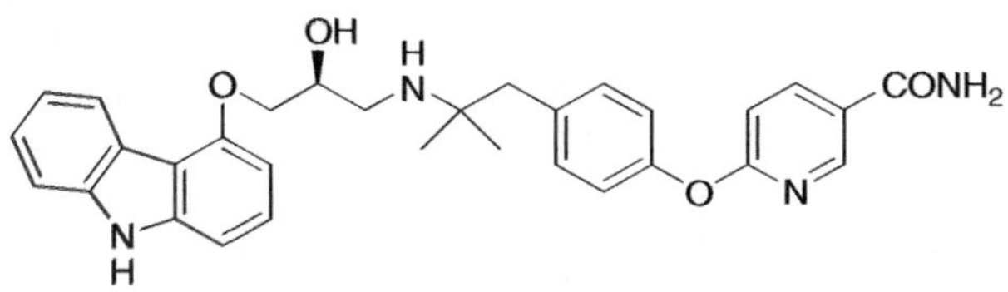	β_3-Adrenergic receptor	Obesity Diabetes	263

(continued overleaf)

Table 6 *(continued)*

Entry	Candidate	Drug Target	Disease Indication	Ref.
21	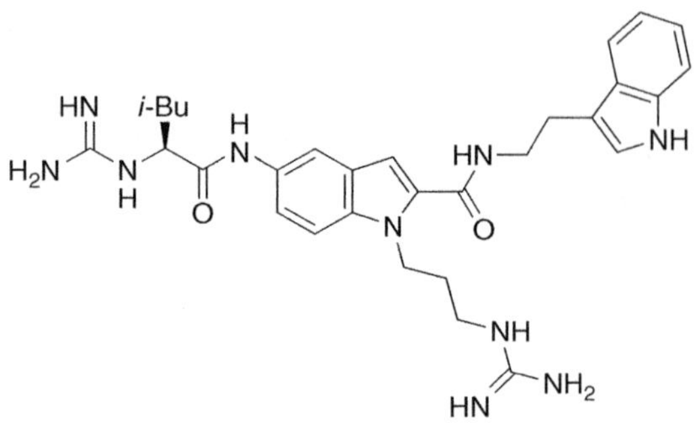	G-protein coupled CXCR 4	Tumor growth Metastasis Rheumatoid arthritis HIV-1	264
22	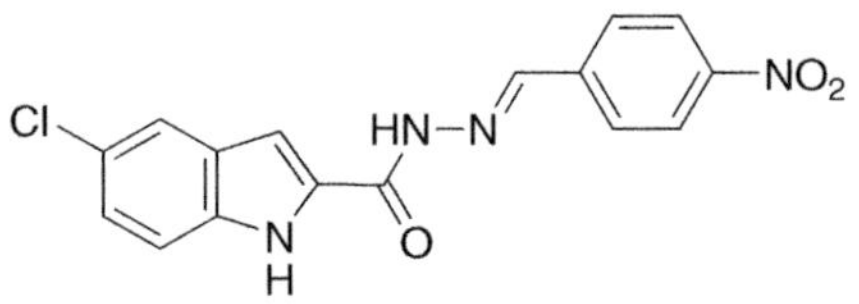	Apoptosis inducer (tubulin polymerization inhibitor)	Cancer	265
23	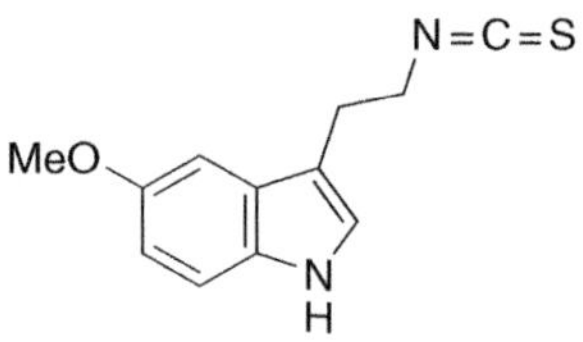	Melatonin receptor	Insomnia Antioxidative Antiproliferative	266
24	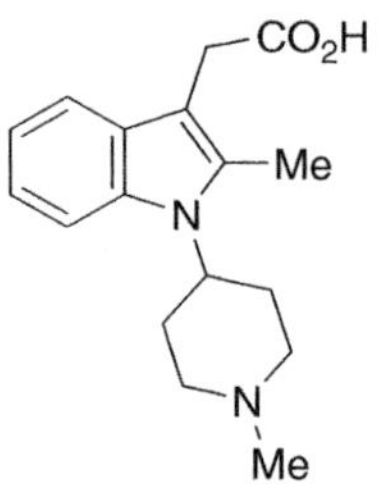	Opiate μ-receptor COX-2	Analgesia Inflammation	267
25	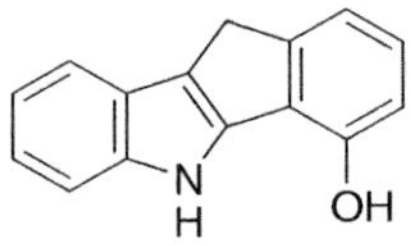	Removal of excess H_2O_2, superoxide anion	Antioxidant Radical scavenger	268

Several β-carboline alkaloids were synthesized via Fischer indole methodology. For example, the blue-green alga *Dichothrix baueriana* alkaloids bauerines A, B, and C were synthesized by two groups via a Japp–Klingemann approach (Scheme 43, equations 1 and 2) [282, 283]. Only the routes to bauerine C (**140**) are shown. Bauerines A and B were prepared in poor yields by reduction of the Fischer indole lactam product and DDQ oxidation of the tetrahydro-β-carbolines. Islam's synthesis [283] of bauerine C is essentially identical to that of Bracher [282], and in roughly comparable yields. A Fischer indolization was employed by Zhidkov and colleagues to synthesize fascaplysin (**141**) (equation 2), a novel biologically active alkaloid found in the marine sponge *Fascaplysinopsis* Bergquist sp [284]. Some β-carbolines **142** were synthesized and evaluated as novel inhibitors of IKK (I_KB kinase complex) (IC_{50} = 1.1 μM) (equation 3) [285]. Other related active β-carbolines (**142**, R = OMe, F, Br, Cl) were prepared other ways. Several paullone derivatives are selective inhibitors of glycogen synthase kinase-3β (GSK-3β) [286] and are new agents against the parasite *Leishmania donovani,* which causes visceral leishmaniasis in tropical and

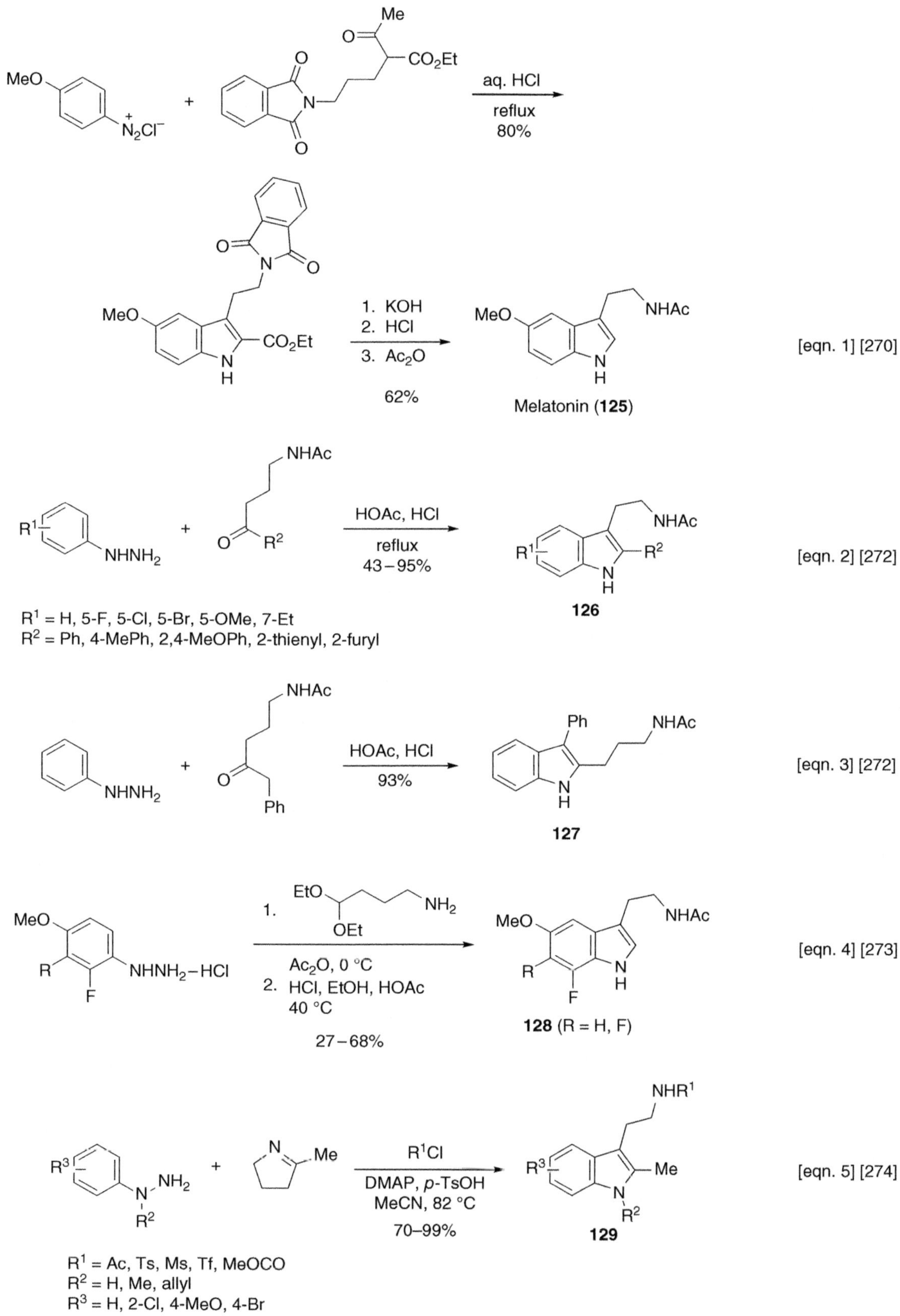

Scheme 40 *Fischer Indole Synthesis of Tryptamines*

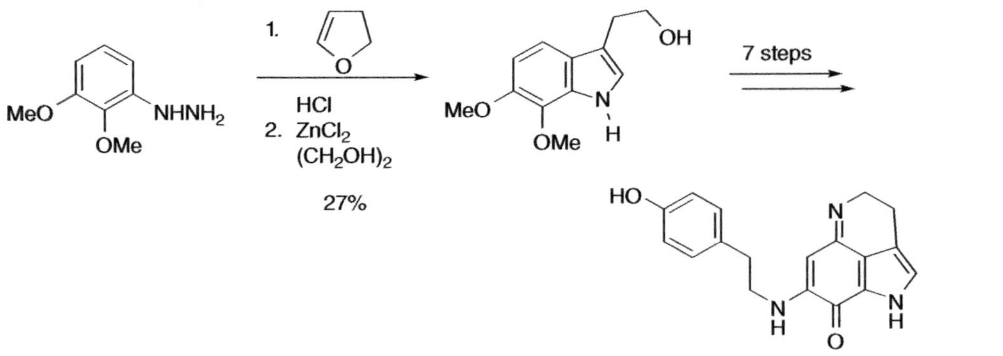

[eqn. 1] [275]

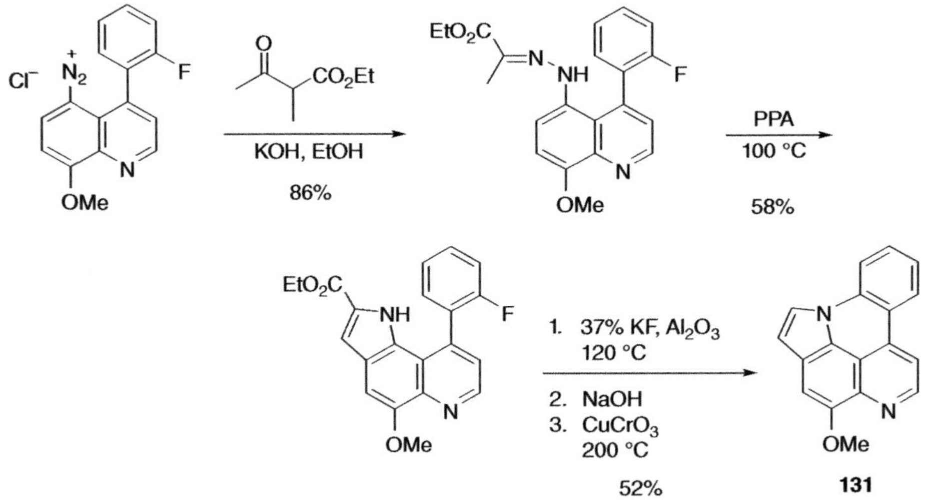

131

[eqn. 2] [276]

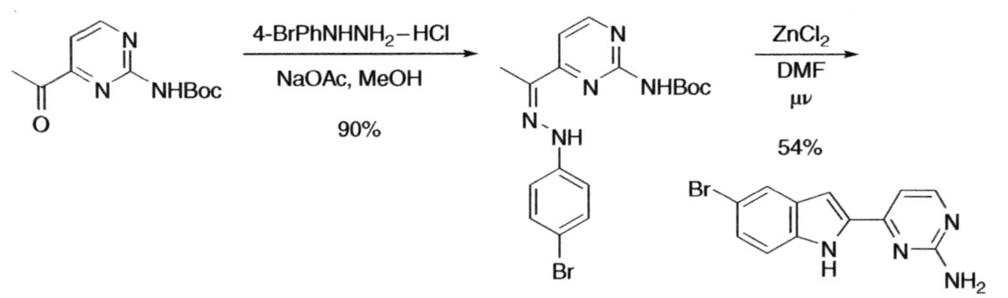

132

[eqn. 3] [277]

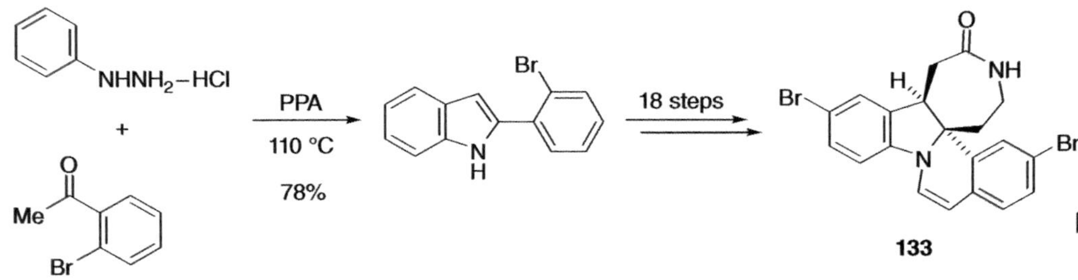

133

[eqn. 4] [278]

Scheme 41 *Fischer Indole Syntheses of Marine Indoles*

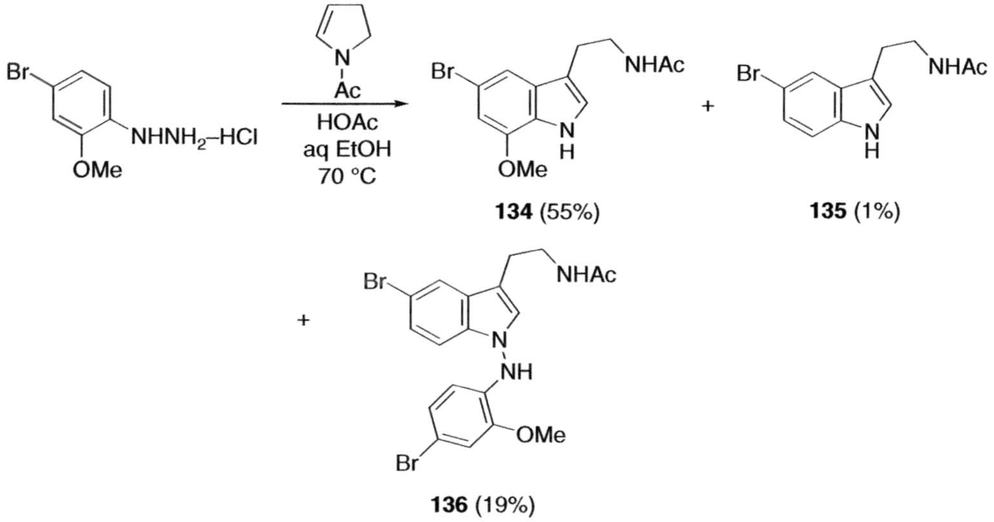

[eqn. 1] [279]

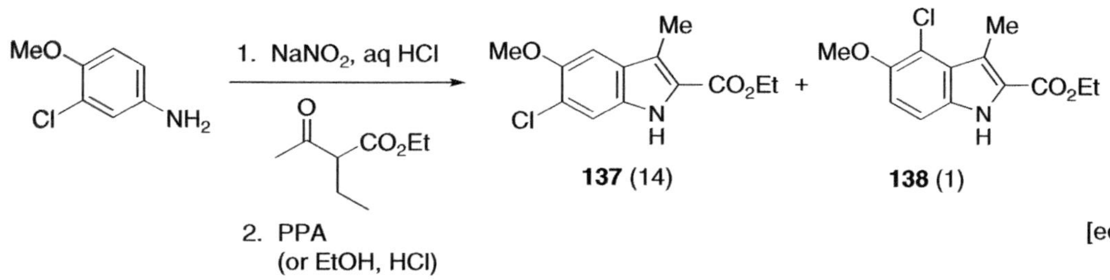

[eqn. 2] [280]

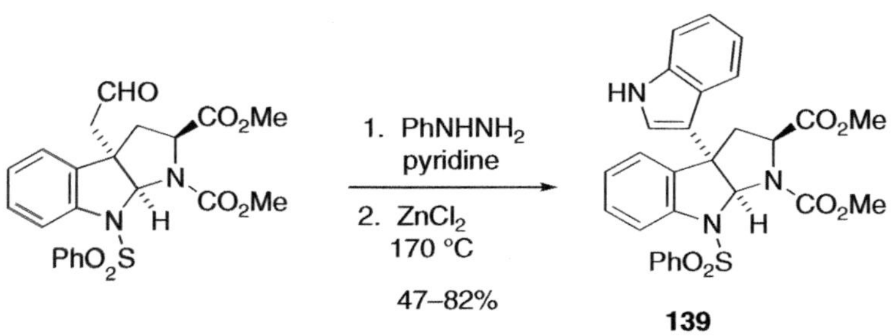

[eqn. 3] [281]

Scheme 42 *Fischer Indole Syntheses of Marine Indoles 2*

subtropical regions. The Fischer indolization is a powerful method in the authors' program for these new compounds. Thus, **143** has excellent activity (IC_{50} = 0.018 μM) against GSK-3β [286], and **144** has excellent antileishmanial activity [287].

The Fischer indole synthesis and its variations are prominently highlighted in the synthesis or partial synthesis of many natural products, many of which are summarized in Table 7 [288–311]. In addition to these examples, Fischer indolization was often applied to the synthesis of pyrrolo[2,3-*b*]indole alkaloids in what Southwick in 1963 coined an arrested Fischer indole synthesis [312], later to be expanded by Garg as an interrupted Fischer indolization [313–317], as featured in his elegant syntheses of physovenine [313, 314], debromoflustramine B [314], (+)-phenserine [316], the core of perophoramidine [317], and (±)-aspidophylline A [315]. This synthesis of the pyrrolo[2,3-*b*]indole alkaloids (±)-physostigmine and (±)-phenserine has also been described by Nishida [318] and Rigby [319], respectively. An example (**145**) from the work of Garg is shown in Scheme 44 (equation 1) [316]. Several indole-fused triterpenoids were prepared via standard Fischer methodology (equation 2 [320], and **147–148**). A betulinic acid analogue shows good activity against the cell lines MIAPaCa, PA-1, and A549 (IC_{50} = 0.67, 3.0, and 3.53 μg/mL, respectively) [321], and the carboxylic acid corresponding to **146** is a good inhibitor of protein tyrosine phosphatase 1B (PTP 1B) (IC_{50} = 0.61 μM) [322]. Inhibition

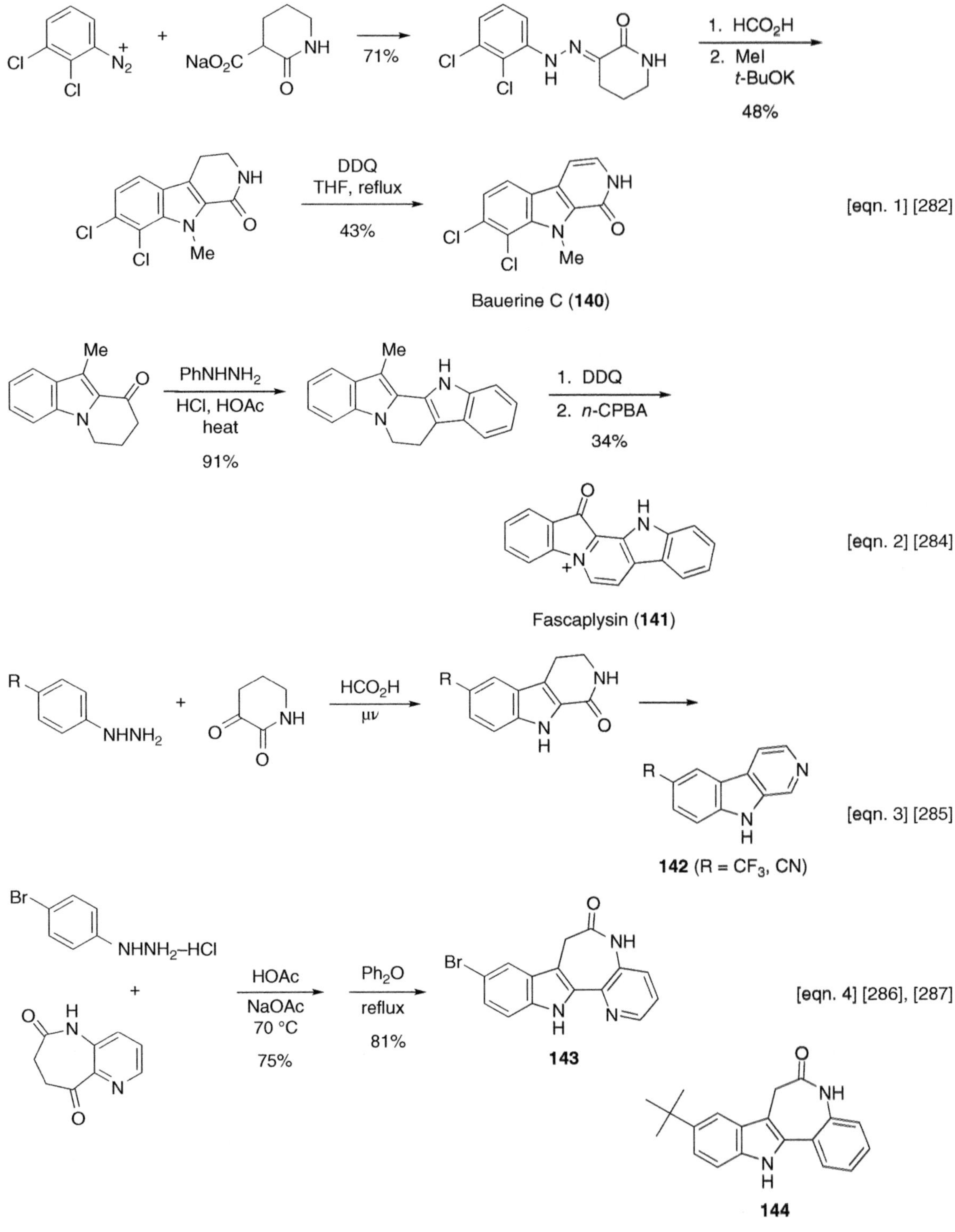

Scheme 43 *Fischer Indole Syntheses of Carbolines*

of this protein may offer a therapy for type 2 diabetes and obesity [323]. The indole derivative **148** of the diterpenoid isosteviol is highly active against α-glucosidase [324]. The Fischer indolization has been recruited to prepare several substituted tryptophans and tryptamines for subsequent use in the synthesis of indole-containing peptides such as chloptosin (6-chlorotryptophan) [325], tryprostatin A (6-methoxytryptophan) [326], and gardnerine and gardnutine

Table 7 *Fischer Indolization of Selected Natural Products*

Entry	Fischer Indolization	%, Yield	Natural product target	Ref.
1	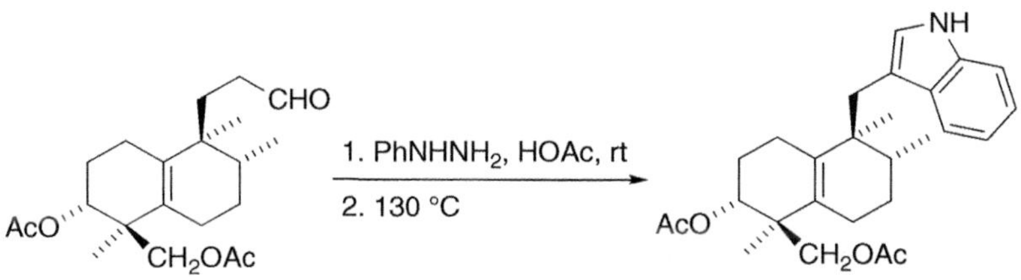	93%	(+)-Thiersindole	288, 298
2	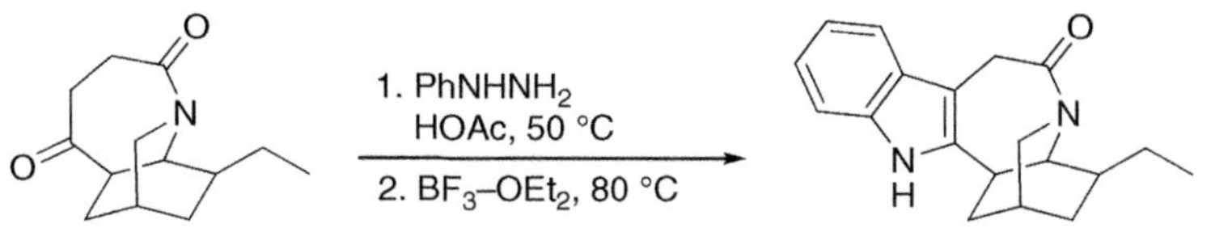	77%	(–)-Ibogamine	289, 290
3	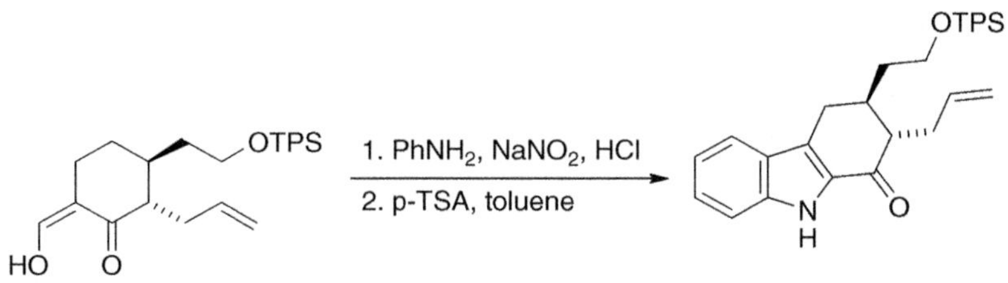	72%	(–)-Gilbertine	291
4	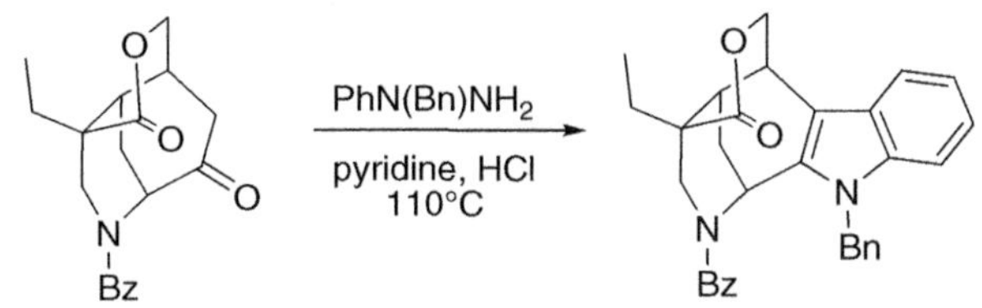		(+)-Scholarisine	292
5	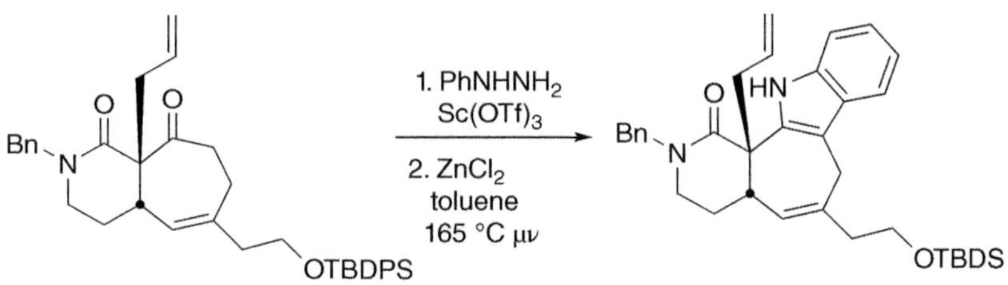	64%	(±)-Actinophyllic acid	293
6	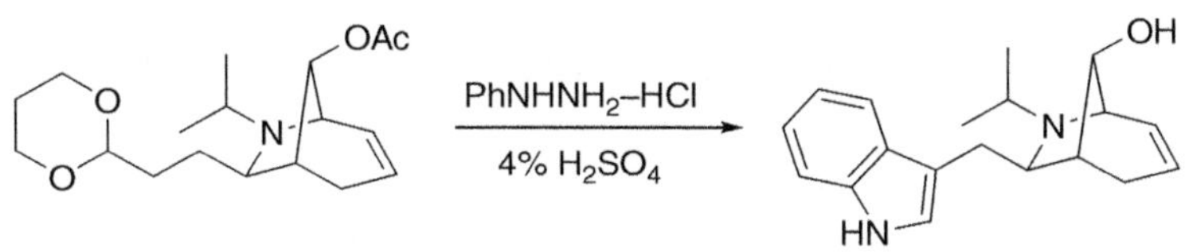	75%	(±)-Peduncularine	294
7	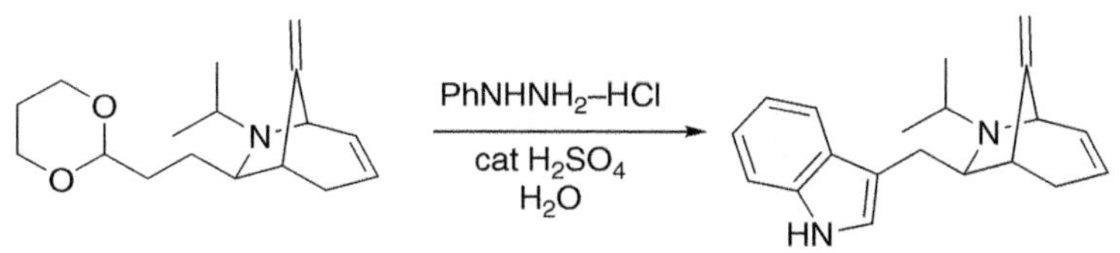	63%	(±)-Peduncularine	295

(continued overleaf)

Table 7 *(continued)*

Entry	Fischer Indolization	%, Yield	Natural product target	Ref.
8	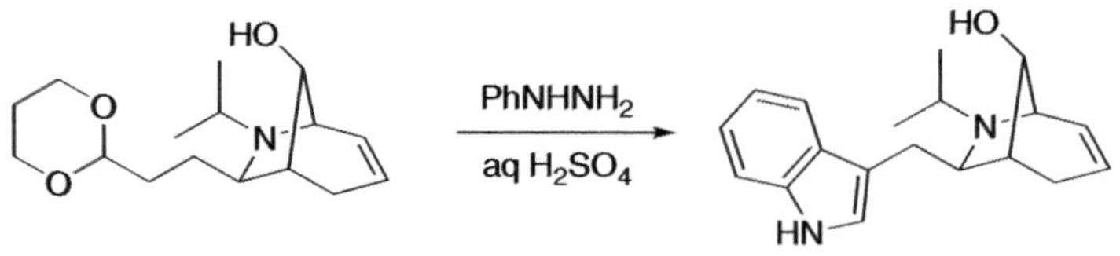	72%	(±)-Peduncularine	296
9	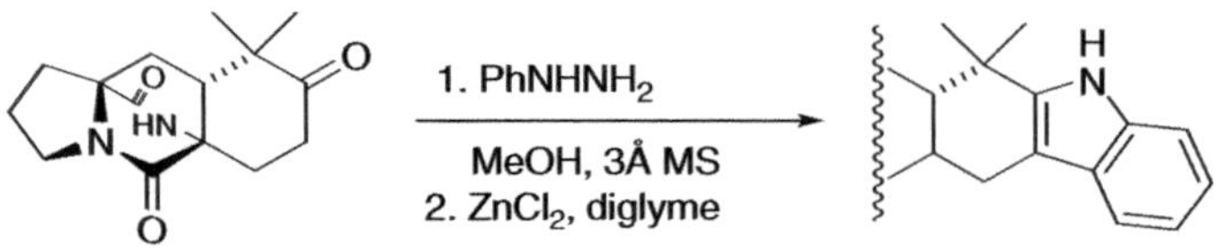	58%	(±)-Brevianamide B	297
10	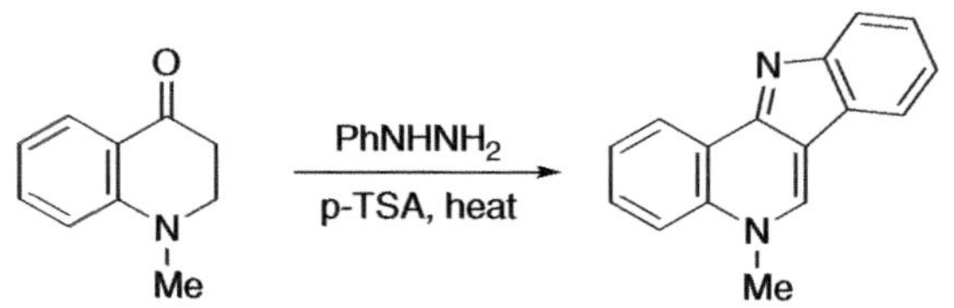	83%	Cryptosanguinolentine	299
11	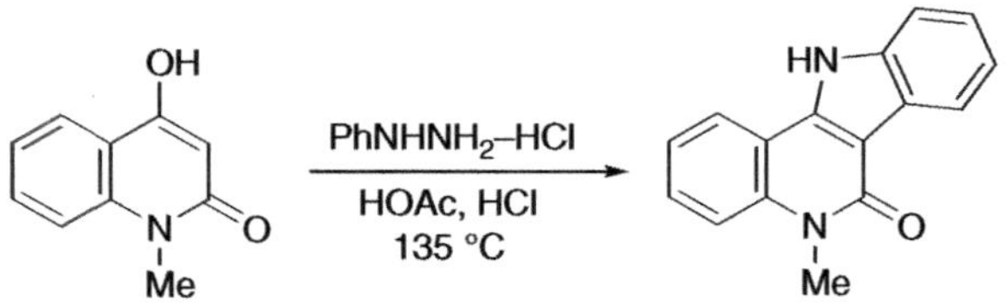	65%	Cryptosanguinolentine	300
12	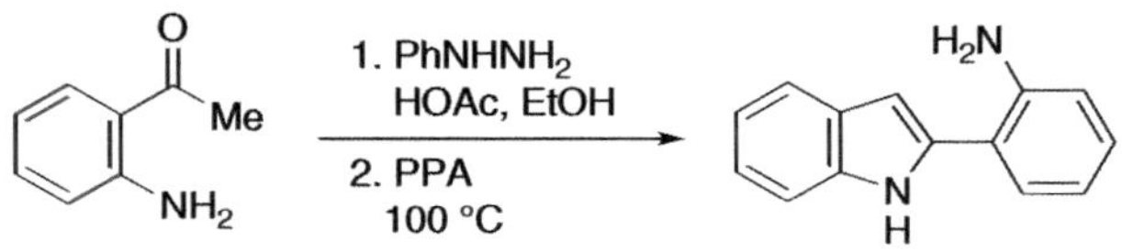	—	Isocryptolepine	301
13	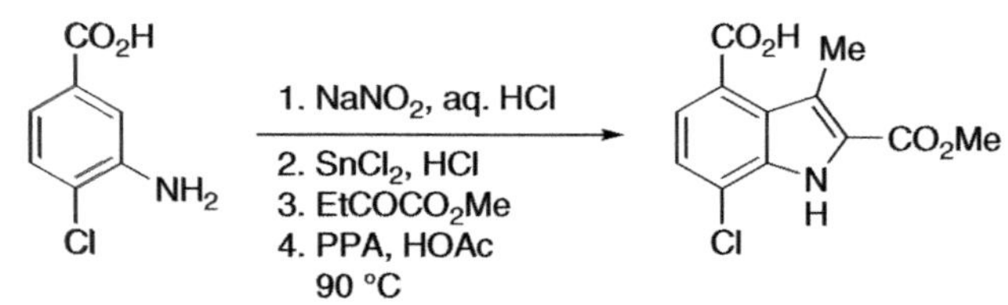	87%	Nosiheptide	302
14	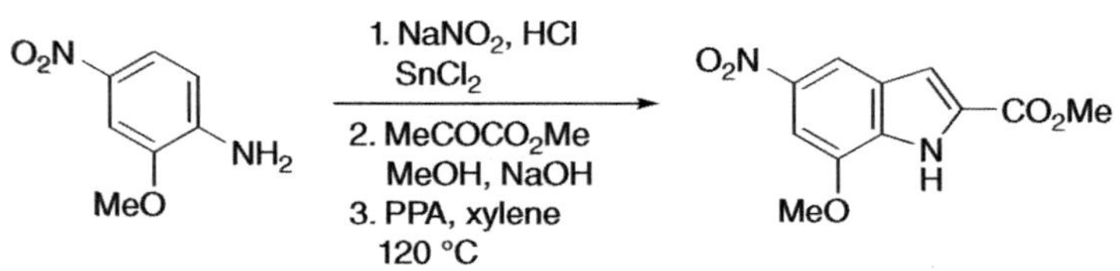	44%	*seco*-Duocarmycin SA	303
15	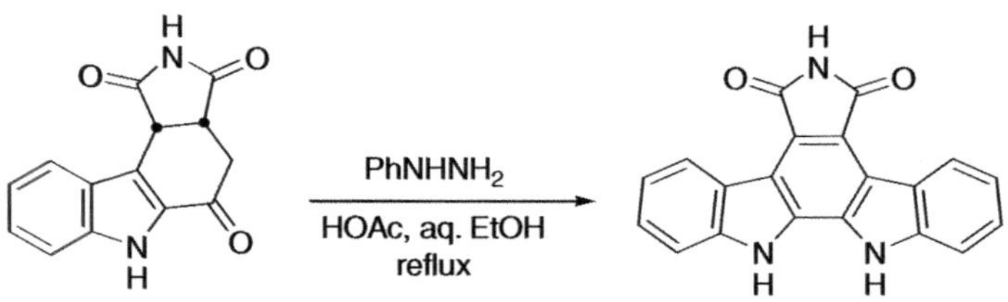	57%	Arcyriaflavin-A	304

Table 7 *(continued)*

Entry	Fischer Indolization	%, Yield	Natural product target	Ref.
16	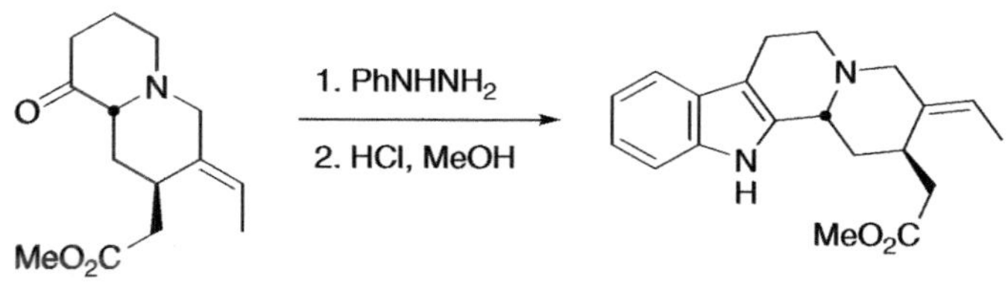	64%	(±)-Deformyl-isogeissoschizine	305
17	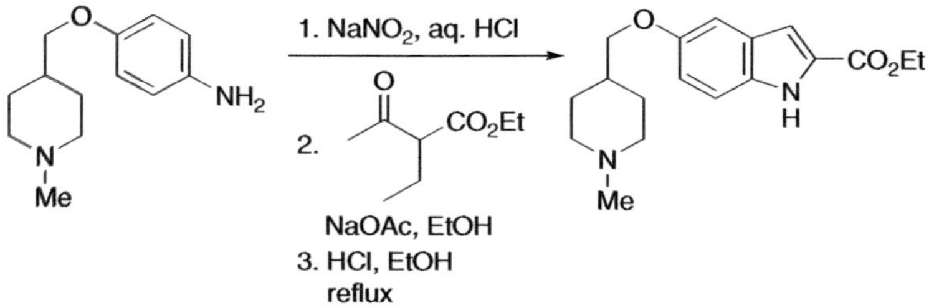	72%	CC-1065 analogues	306
18	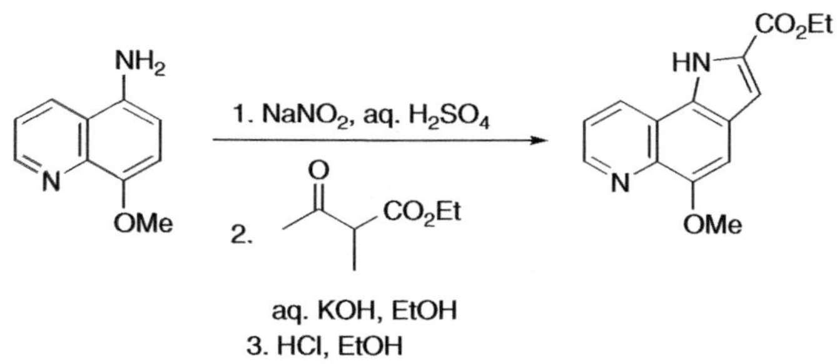	54%	Methoxathine analogue	307
19	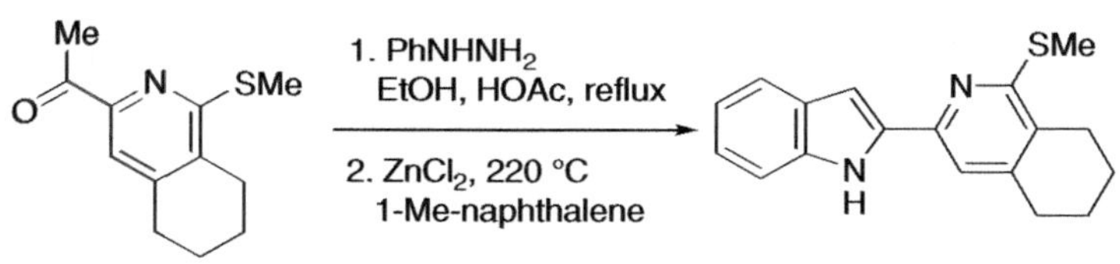	55%	Sempervirine	308, 309, 310, 311

(6-methoxytryptophan) [327]. Moreover, Fischer indole ring syntheses are used to generate intermediates for the synthesis of several macroline/sarpagine alkaloids [328], 19,20-dihydroakuammicine [329], tubifolin and tubifolidine [330], and (–)-majvinine, (–)-10-methoxyaffinisine, (+)-N_a-methylsarpagine, and macralstonidine [331].

As we encountered earlier, Fischer indole synthesis methodology has been used to synthesize the plant alkaloid rutaecarpine (**110**) and many analogues of this COX-2 inhibitor. Several examples are shown in Scheme 45. Equation 1 is a simple synthesis of rutaecarpine itself [332]. This plant alkaloid, found in *Evodia rutaecarpa,* has long been used in Asian folk medicine for treating inflammation. The plant Wu-Chu-Yu has been employed as a remedy for gastrointestinal disorders, headache, dysentery, and postpartum hemorrhage. Several halogenated and other analogues were prepared as shown (equation 2) [333], and other rutaecarpine analogues are depicted (**149–151**) [334, 335].

Although morphine and the related opium alkaloids are not indoles, a great number of analogues are indoles, the prototype being naltrindole (**152**), which is the fused-indole analogue of naltrexone (**153**) (Scheme 46). Naltrindole is a δ-opioid receptor antagonist and κ agonist, and it shows greater selectivity for the δ-opioid receptor than **153** [336–338]. A standard Fischer indolization (phenylhydrazine hydrochloride, HOAc, HCl, MeOH) transformed naltrexone to naltrindole [336]. Since the discovery of naltrindole by Portoghese in 1988, this compound has been of enormous utility as a pharmacological tool. A summary of these newer analogues of naltrindole is shown in Scheme 46.

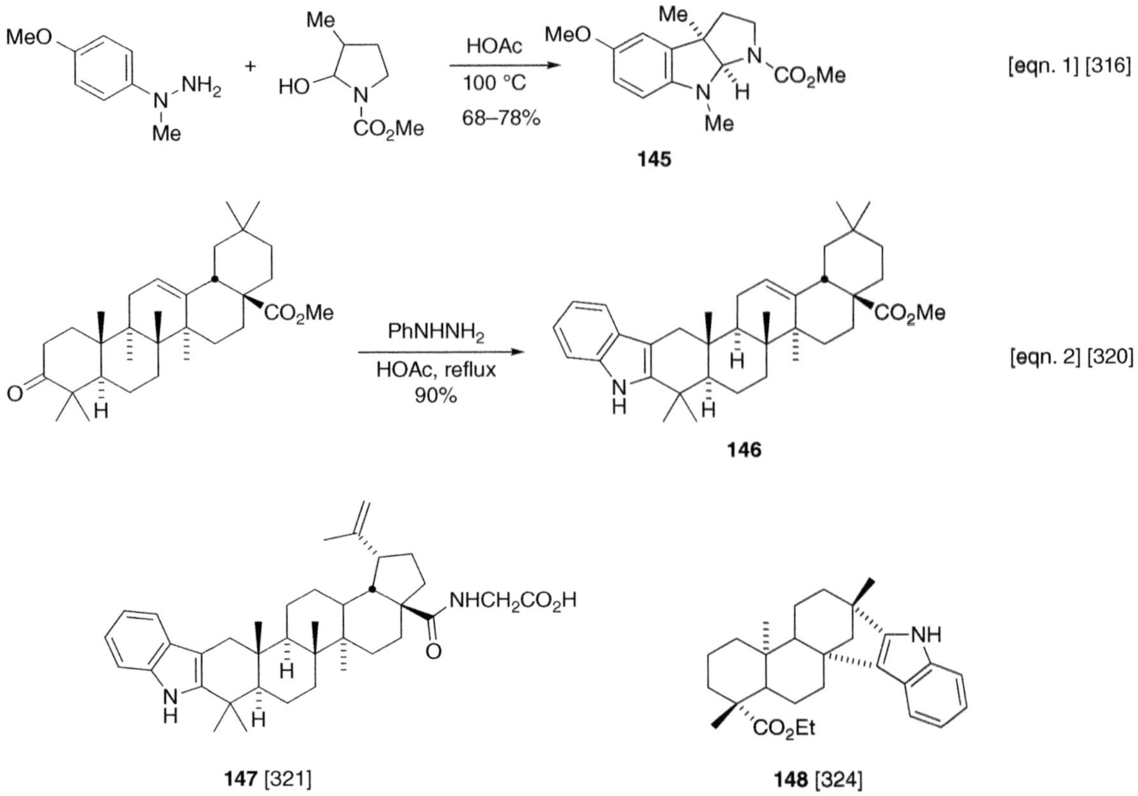

Scheme 44 *Fischer Indole Syntheses of Biologically Active Indoles*

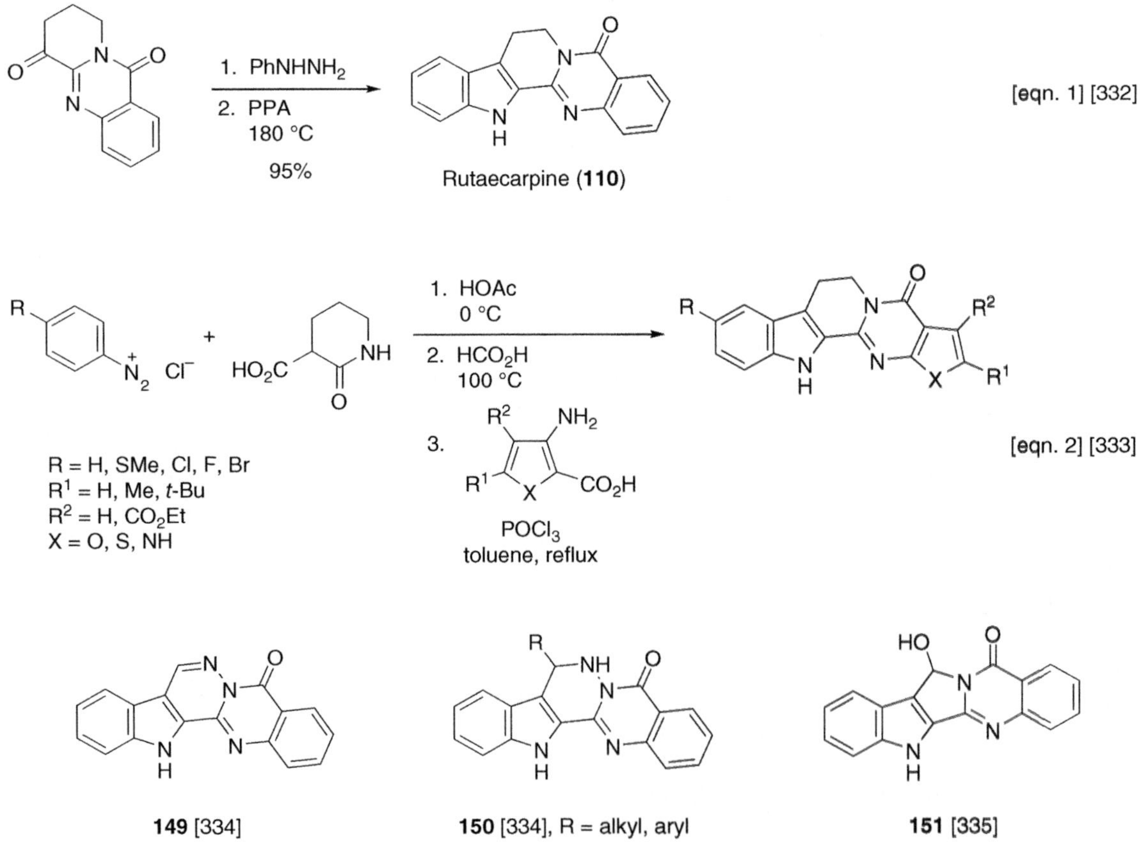

Scheme 45 *Fischer Indole Syntheses of Rutaecarpine and Analogues*

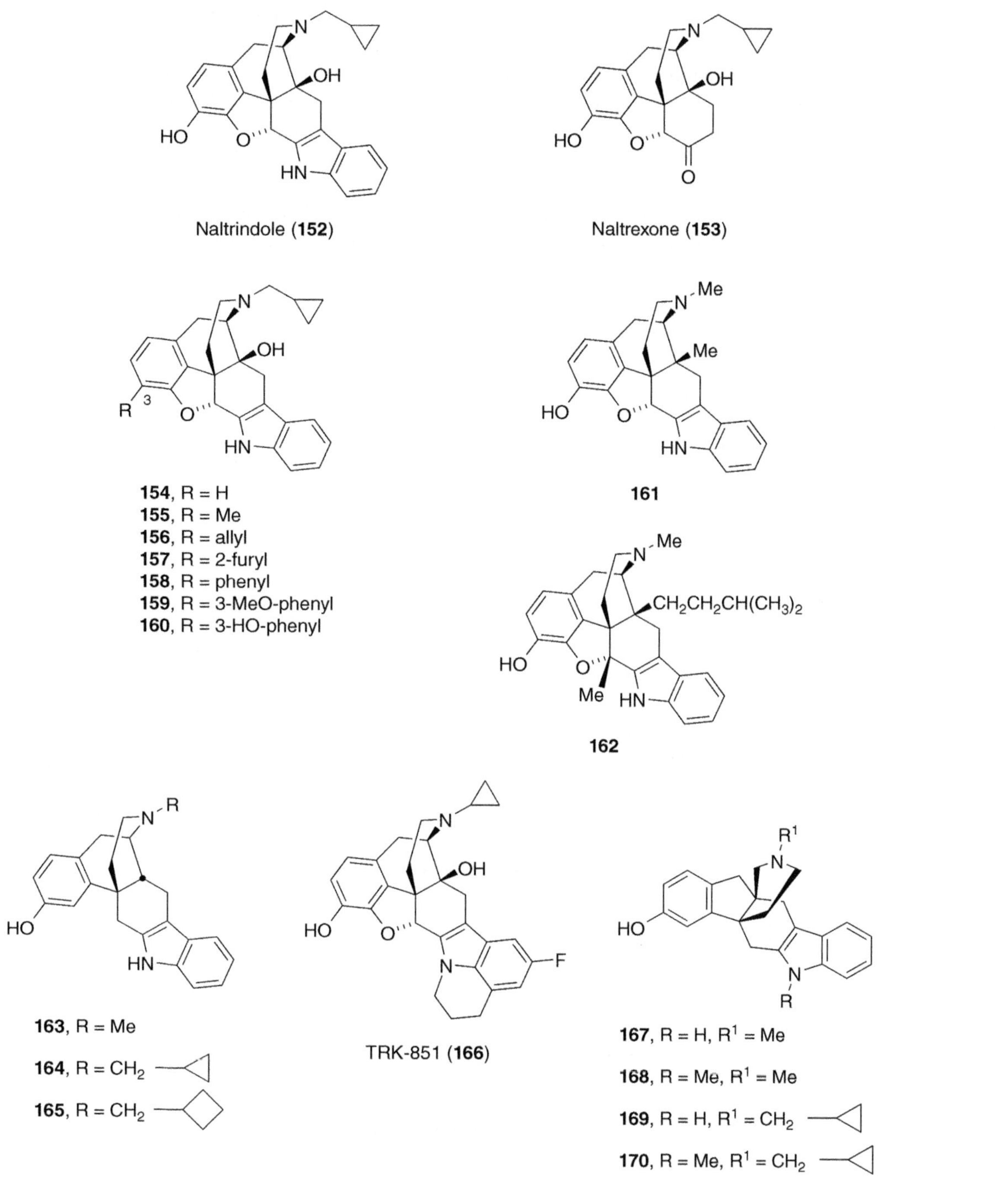

Scheme 46 *Fischer Indole Syntheses of Naltrindole and Analogues*

The C-3 substituted derivatives **154–160** were prepared from the corresponding naltrexone by Fischer indolization ($PhNHNH_2$–HCl, *p*-TSA, EtOH, reflux; 22%–72%). Of these indoles, although only **154** shows comparable selectivity with naltrindole for the δ-opioid receptor, none of these analogues **154–160** possess acceptable affinity for this receptor [339]. Indolomorphinan **161** is a δ-opioid receptor antagonist (K_i=0.7 nM) and has good δ selectivity [340], as does the 5β-methyl-14-isopentyl derivative **162** [341]. Handa and coworkers have synthesized several latex nanoparticles that carry naltrindole derivatives as probes for affinity studies [342]. Neumeyer's group found that indole morphinans **163–165** are potent at the δ receptor, with **163** exhibiting the highest selectivity over κ and μ receptors [343]. Nagase and colleagues synthesized a variety of highly active and novel δ-opioid receptor antagonists [344, 345], one of which, TRK-851 (**166**), is a stable, orally active antitussive agent [345]. A Fischer indolization was employed to synthesize **161–166**. The indolopropellanes **167–170** are a series of rearranged indolomorphinans that show μ-opioid receptor activity and selectivity [346], with **168** exhibiting the highest affinity for the μ receptor ($K_i = 40$ nM).

A relatively new class of indole-containing natural products are the indolocarbazoles and related fused heterocycles. For example, the indolo[2,3-*a*]pyrrolo[3,4-*c*]carbazole alkaloids display enormous biological activity, and several examples are known [347–350]. Bergman and his co-workers have been major players in this area and has developed Fischer indolization methods to synthesize these indolocarbazoles. For example, arcyriaflavin A (**171**) was prepared in good yield by the bis-Fischer indolization shown in Scheme 47 [351]. A Fischer indole ring synthesis has been adopted by another group to prepare a series of open indolocarbazole analogues,

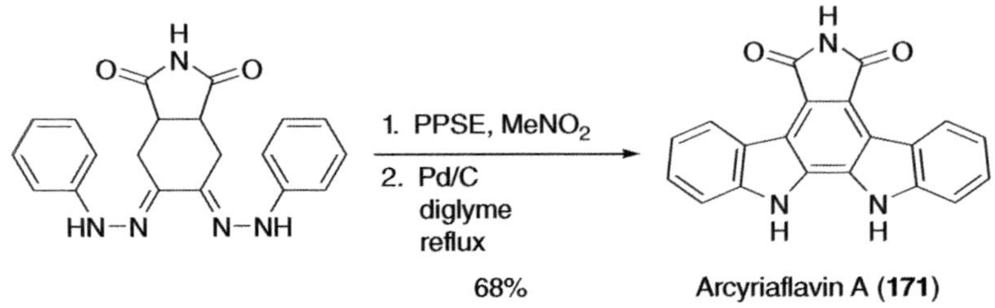

Arcyriaflavin A (**171**)

[eqn. 1] [350]

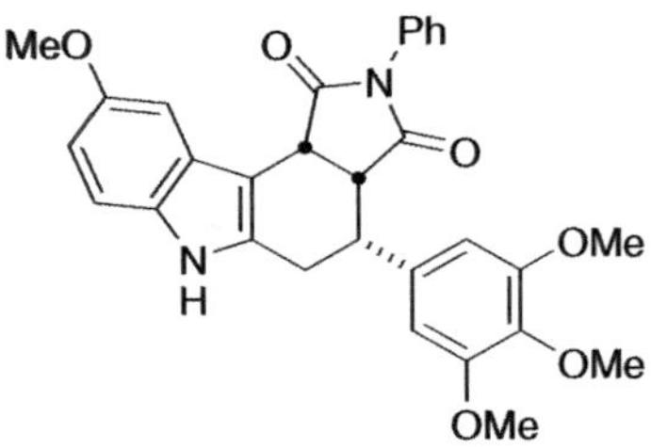

172 [353]

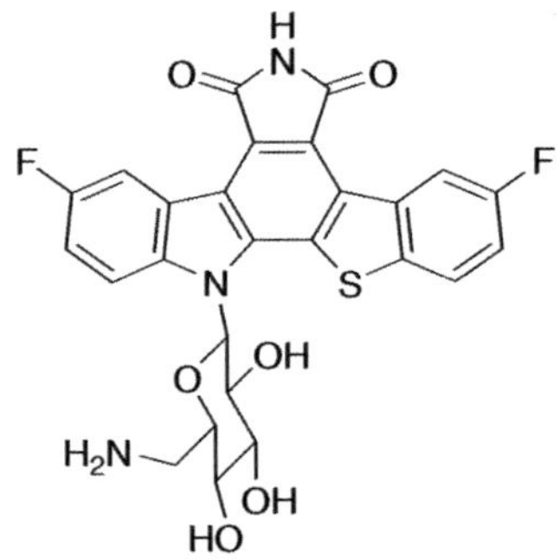

BMS-251873 (**173**) [354]

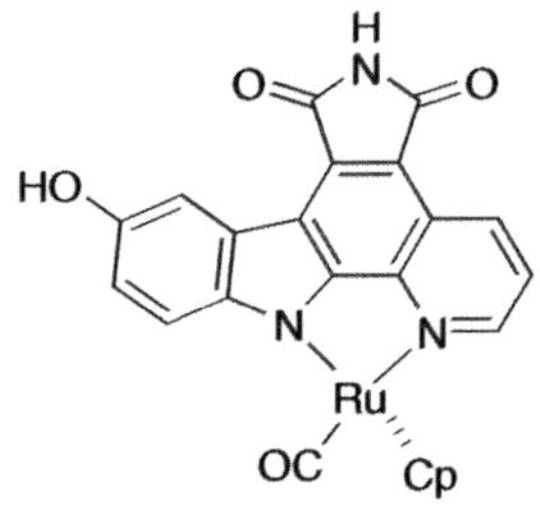

174 [355]

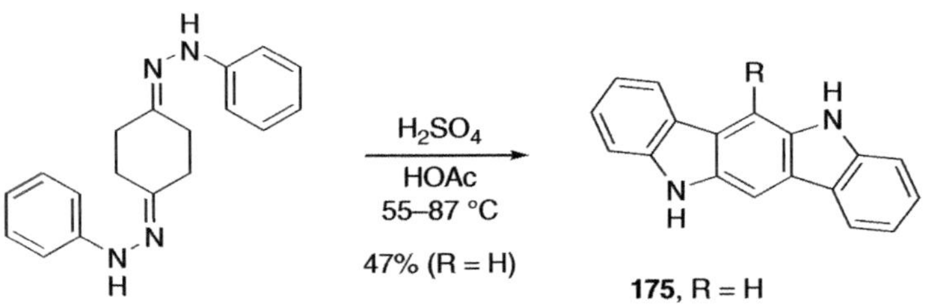

175, R = H
176, R = CHO

[eqn. 2] [356]

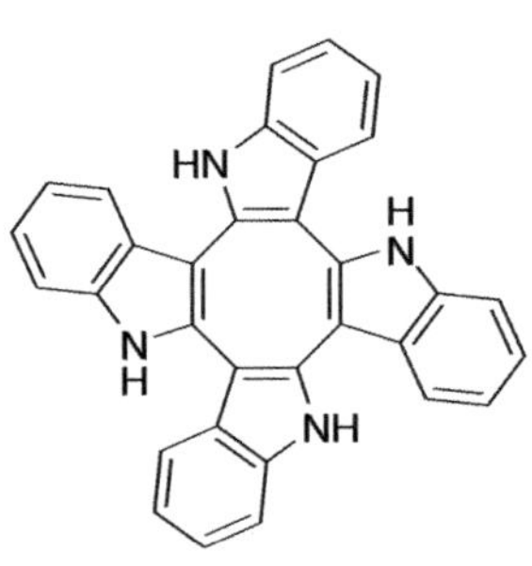

177 [357]

Scheme 47 *Fischer Indole Syntheses of Indolocarbazoles* e.g., **172**, which has good *in vivo* activity against P-388, A-549 (human lung), HT-29 (human colon), and MEL-28 (human melanoma) [352, 353]. A related indole is (slightly less) active against UACC-62 (human melanoma), TK-10 (human kidney), and MCF-7 (human breast). A rebeccamycin analogue BMS-251873 (**173**) is a clinical candidate for cancer [354]. Compounds **172** and **173** were prepared using Fischer indolizations. A series of pyridopyrrolocarbazole–ruthenium complexes are very potent inhibitors for glycogen synthase kinase 3 (GSK-3); for example, **174** shows $IC_{50}=0.3$ nM [355]. Bergman and Yudina have effected a route to several indolo[3,2-*b*]carbazoles, such as **175** (equation 2) via a double Fischer indolization [356]. Such compounds are of biological interest because **175** is formed *in vivo* after the consumption of cruciferous vegetables. It has a strong affinity for the dioxin (2,3,7,8-tetrachlorodibenzo-*p*-dioxin) (TCDD) Ah-receptor. Moreover, the 6-formyl derivative **176** binds more strongly to the Ah-receptor than TCDD itself [357, 358]. Saracoglu and Talaz have transformed cyclooctanone into tetrameric indole **177** by stepwise Fischer indolization steps [359].

Following the work of Stork and Dolfini on the synthesis of (±)-aspidospernine [360] and the earlier (1957) demonstration of spiro-indolenine formation via a Fischer indolization by Georgian (Scheme 48, equations 1–3) [361], many workers have employed Fischer indole methodology in total syntheses of *Aspidosperma* and related indole alkaloids. Indeed, Georgian recognized the potential value of his discovery as providing synthetic entry to both morphine and strychnos alkaloids [361]. Examples of these syntheses are (+)-aspidospermidine [362], (±)-aspidospermidine [363], (–)-aspidospermine [364], and (+)-haplophytine [365], each of which features a Fischer indolization as the final or penultimate step. An example is shown in equation 4 [363].

2.3.3 Materials

An expanding area of opportunity for organic synthesis is materials chemistry. The surge of interest in electronic devices, biomaterials, metal receptors, fullerenes, quantum dots, and more includes applications in indole chemistry.

For a number of years, Thummel fabricated various nitrogen-based heterocyclic receptors using Fischer indolization. For example, urea receptors **178–180** are depicted in Scheme 49 (equations 1 and 2) [365]. Thummel, Wang, and their coworkers synthesized luminescent diphenylboron

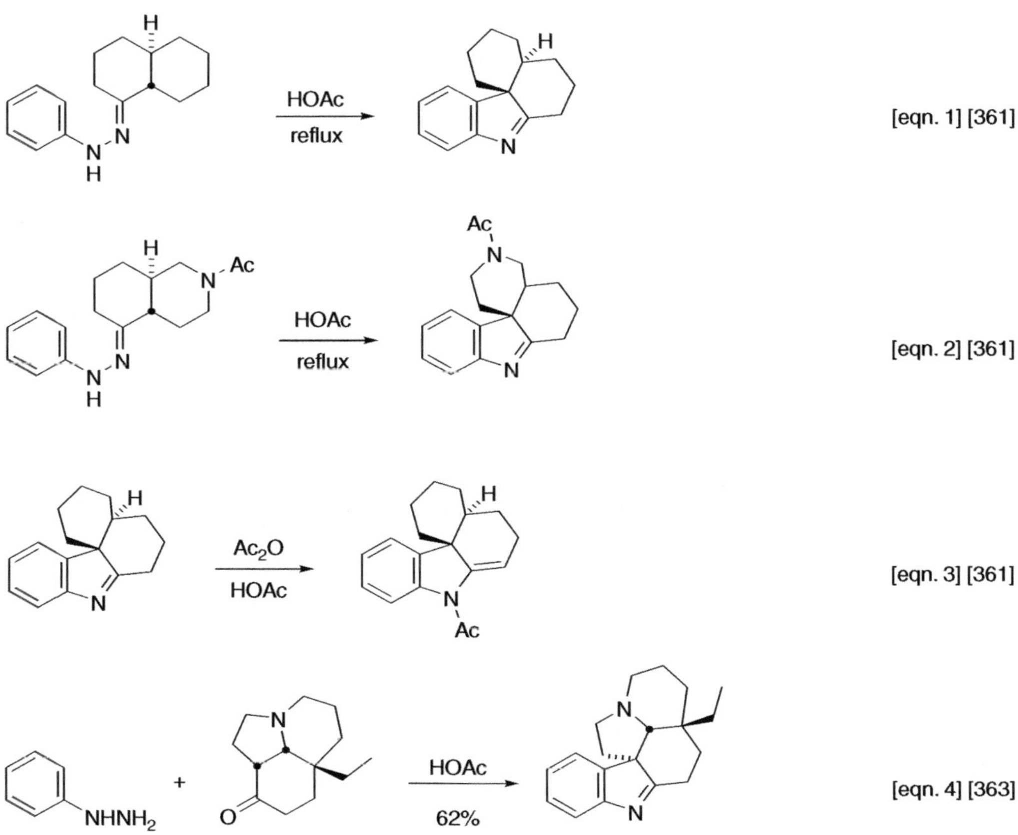

Scheme 48 *Fischer Indole Syntheses of Aspidospermine Analogues*

complexes of 2-(2′pyridyl)indoles (**181–183**) via the Fischer indolization between 2-acetylpyridine and 4-substituted phenylhydrazines followed by treatment with triphenylboron. These three complexes are luminescent, with **181** having the highest emission efficiency and being blue-shifted to 490 nm. An electroluminescent device was built using **181** as the emitting layer, indium–tin oxide as the electron transport layer, and *N,N*′-di-1-naphthyl-*N,N*′-diphenylbenzidine as the hole transport layer in the device [367]. Müller and colleagues synthesized a set of blue-luminescent 5-(3-indolyl) oxazoles **184** by a three-component coupling process that culminated in a Fischer indolization (equation 4). These compounds show large Stokes shifts upon UV irradiation [368].

In their quest to prepare pentaleno[2,1-*b*:5,4-*b*′]-diindoles, Cook and his students effected a bis-Fischer indolization on bicyclo[3.3.0]octan-3,7-dione to yield only diindole **185** (Scheme 50), albeit in low yield. Attempts to oxidize **185** to the target compound were unsuccessful [369]. Beer prepared several new indolo[2,3-*a*]carbazoles **186** and found that they behave as anion sensors (equation 2) [370]. Thus, **186** binds in a 1:1 ratio to these anions in decreasing association: benzoate > hydrogen > phosphate > fluoride > chloride > bisulfate, with 3,8-dibromo **186** showing the strongest binding (log K_a = 5.9) in acetone. Complementary H-bonding may be the explanation for the preferred benzoate complexation (**187**). Functionalized indolo[3,2-*b*]carbazoles **188** act as novel high-performance organic semiconductors suitable for organic thin-film transistor applications. Their synthesis is achieved via a double Fischer indolization of 1,4-cyclohexanedione with phenylhydrazine as shown earlier in Scheme 47, equation 2, followed by the appropriate *N*-alkylation. These indolocarbazoles have excellent environmental stability (oxidative and photochemical) owing to their relatively low-lying HOMOs and large band gaps [371].

Sessler and colleagues performed a double Fischer indolization of bis-hydrazone **189** to build a novel dinaphthoporphycene (not shown) via benzo[*e*]pyrrolo[3,2-*g*]indoles **190** (Scheme 51, equation 1). Twin formylation of naphthobipyrrole **190** followed by McMurry coupling and DDQ oxidation gave the desired porphycenes [372]. Similarly, a double Fischer indolization of bis-hydrazone **191** yielded bis-(2-indolyl)-dibenzofuran and the corresponding carbazole **192** (equation 2). Subsequent twin indole C-3 formulation, and reductive amination with diamines gave the target macrocycles (not shown) [373]. Using the Buchward variation of the Fischer indole ring synthesis, Gmeiner and colleagues prepared indoloparacyclophane **194** from hydrazone **193**, which in turn originated from 2-bromo[2.2]paracyclophane via a palladium-catalyzed coupling with benzophenone hydrazone. Another indoloparacyclophane synthesized by these workers has strong and selective affinity for the dopamine D4 receptor [374].

A series of pyrrolocoumarins was prepared by Yao and colleagues and evaluated as new fluorescent molecules having large Stokes shifts and as potential biological imaging agents. For example, **195** was prepared as shown in Scheme 52, equation 1; it is excited by visible light and emits intensely green fluorescence with a Stokes shift of 113 nm. This small-molecule organic dye may find application in biological FRET (fluorescence resonance energy transfer) devices [375]. The fluorogenic cyclooctyne **197**, which undergoes a 10-fold enhancement in fluorescence quantum yield upon triazole formation using click chemistry, was crafted from Fischer indolization product **196** (equation 2) [376]. The indolyl phosphine ligands **198** that were presented in Chapter 1, part 6, were prepared from the appropriate 2-arylindole available by Fischer indolization (equation 3) [377]. As we saw in Chapter 1, these ligands provide great flexibility and efficiency in achieving transition-metal-catalyzed reactions.

2.3.4 General

This final section on the Fischer indole ring synthesis discusses applications to the synthesis of indoles that were not included in the previous three sections. I have attempted to present mainly current examples of Fischer indolization.

Ironically, indole itself was not obtained via Fischer indolization until the work of Nakazaki and Yamamoto in 1976, who accomplished this preparation from acetaldehyde phenylhydrazone and zinc chloride in a heat combustion tube (Scheme 53, equation 1) [378]. A few other simple indoles were also synthesized. Several authors have observed that *N*-protected arylhydrazines often behave better in Fischer indolizations than do unprotected arylhydrazines. Three examples are shown [379–381]. These protected arylhydrazines are more stable than their unprotected counterparts, and they can be easily synthesized as we saw in the Methods section earlier in this chapter.

In the inaugural generation and Diels–Alder cycloaddition reactions of 4,5-, 5,6-, and 6,7-indole arynes, Buszek and colleagues used Fischer indolization to prepare the prerequisite indoles **199** and **200** (Scheme 54, equations 1 and 2) [382,383]. A series of 2-trifluoromethylindoles **201** was prepared starting with the appropriate trifluoromethyl-ketones (equation 3) [384].

As Table 8 illustrates, the venerable Japp–Klingemann variation of the Fischer indolization continues to attract interest. Entry 1 features a three-component Japp–Klingemann indolization involving diazonium salts, acid chlorides, and amines or alcohols and involving the intermediacy of α-hydrazono carboxylic acid esters and amides prior to indole ring formation [385]. Entry 2 describes a preparation of 5-indolyl-Mannich bases and, hence, an expedient source of 5-(chloromethyl)indoles (acetyl

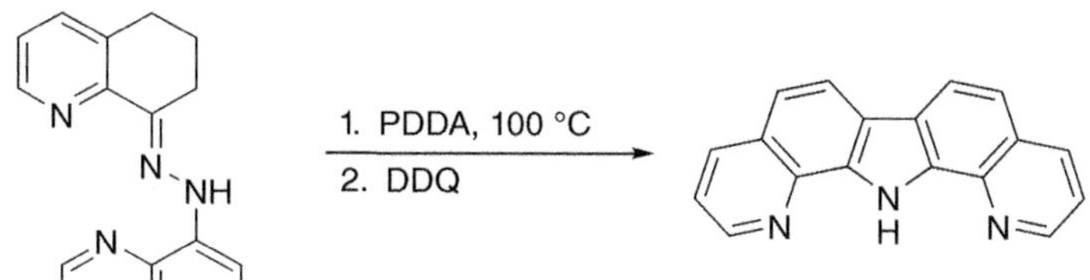

.96 × .99

178

[eqn. 1] [366]

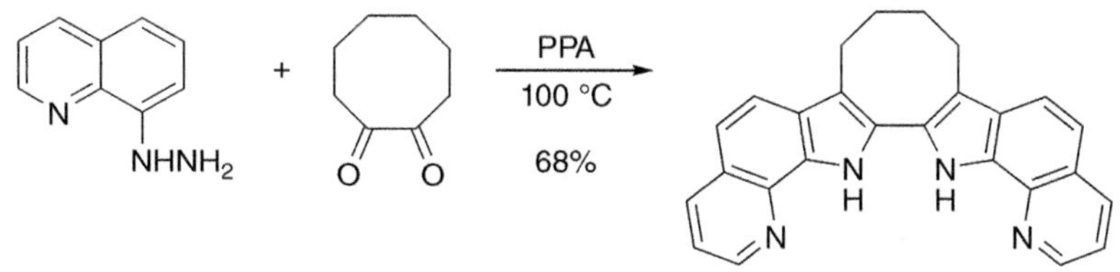

179

[eqn. 2] [366]

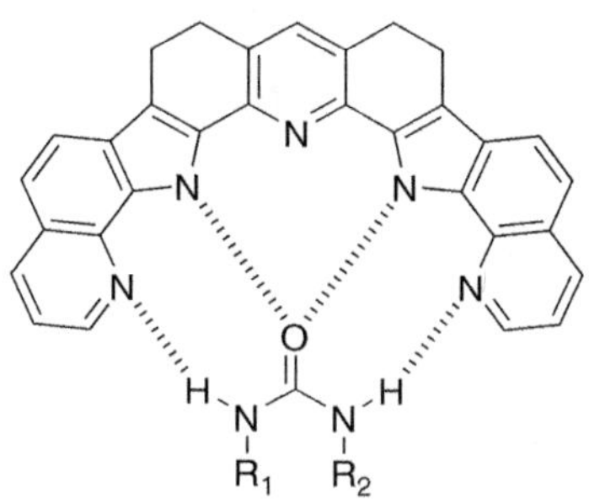

180 [366]

R_1	R_2	K_a	
Me	Me	0.5	(weak binder)
H	*n*-Bu	80	
CH_2–CH_2		3530	(strong binder)

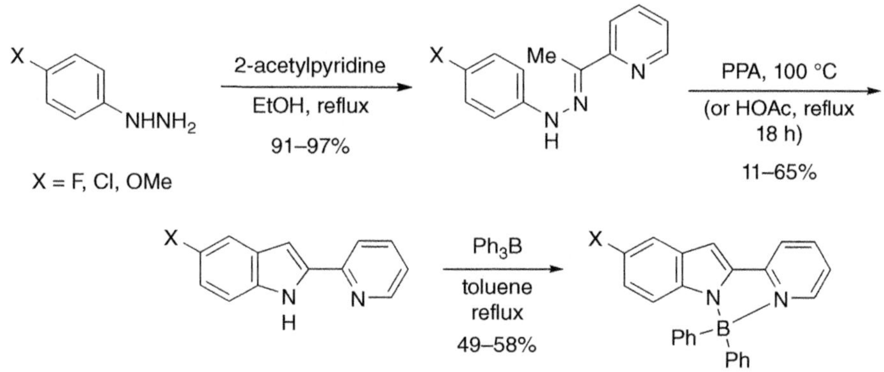

X = F, Cl, OMe

181, X = F
182, X = Cl
183, X = OMe

[eqn. 3] [367]

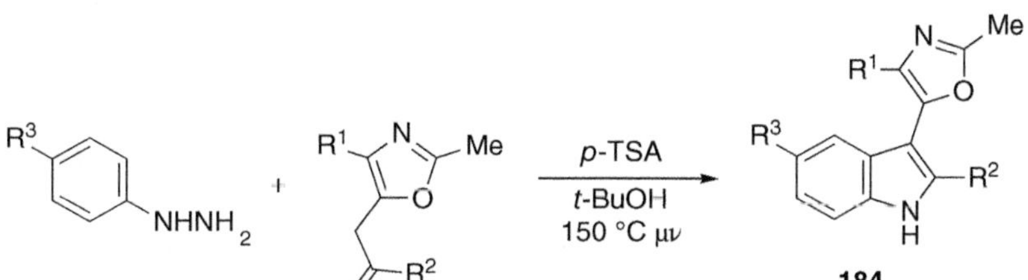

184

[eqn. 4] [368]

R^1 = 4-MePh, 4-ClPh
R^2 = Ar, 2-thienyl, 2-furyl
R^3 = H, Me, Cl

Scheme 49 *Fischer Indole Syntheses of Indole-Based Receptors*

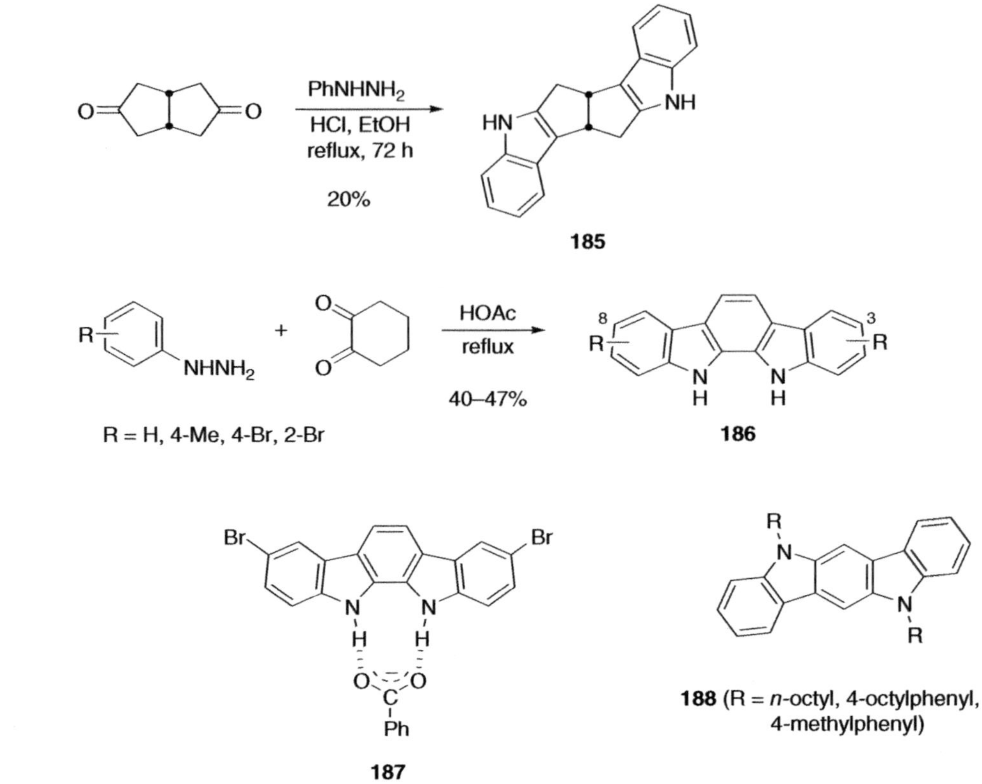

Scheme 50 *Fischer Indole Syntheses of Bisindoles*

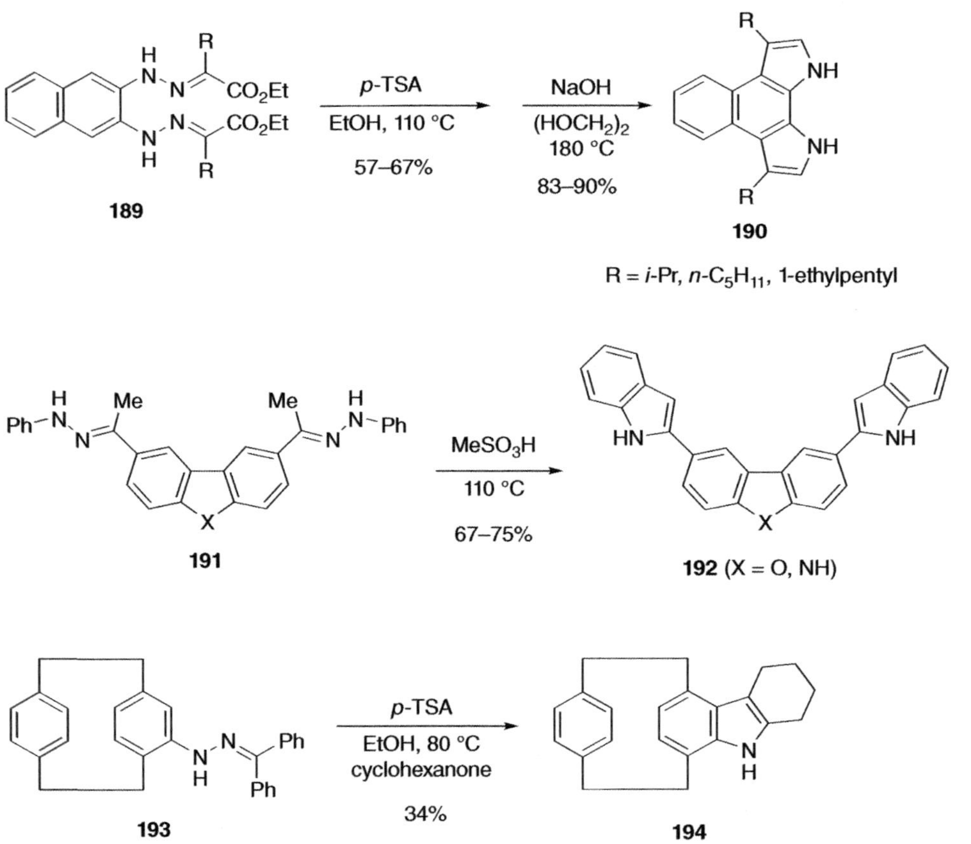

Scheme 51 *Fischer Indole Syntheses of Unusual Indoles*

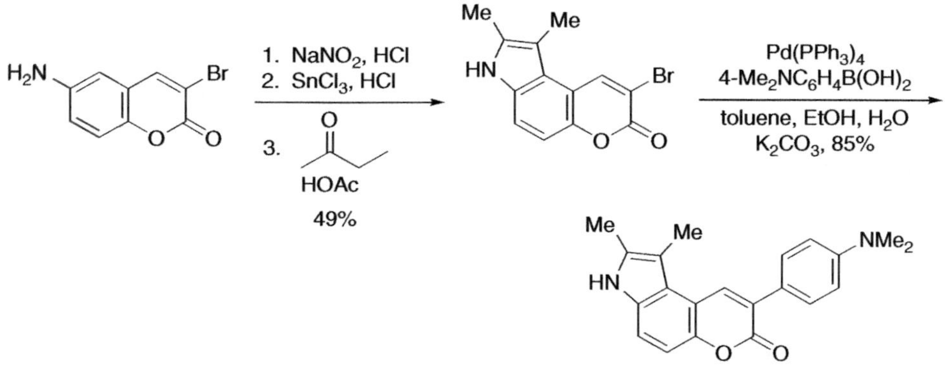

195

[eqn. 1] [375]

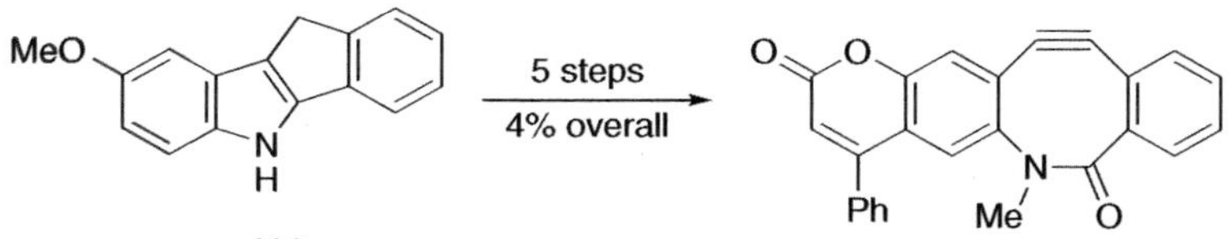

196 **197**

[eqn. 2] [376]

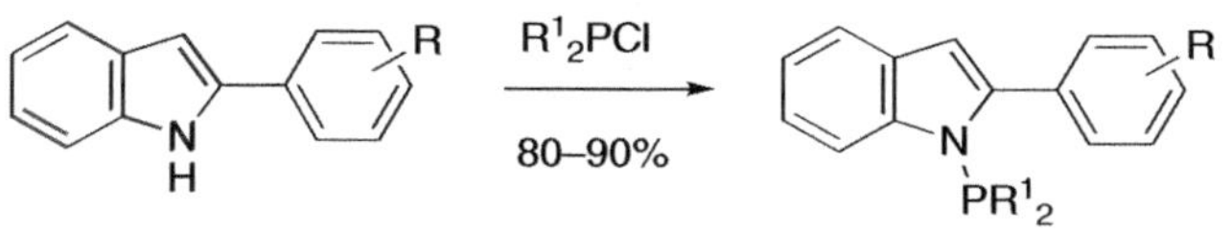

R = H, 2-Me, 4-Me, 2-OMe
R^1 = Ph, cyclohexyl, *i*-Pr

198

[eqn. 3] [377]

Scheme 52 *Fischer Indole Syntheses of Unusual Indoles 2*

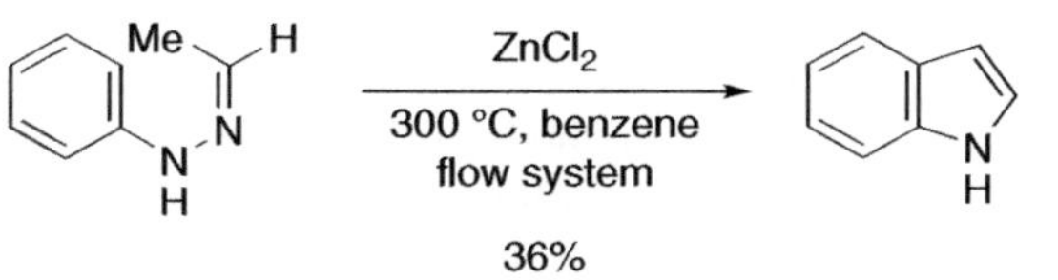

[eqn. 1] [378]

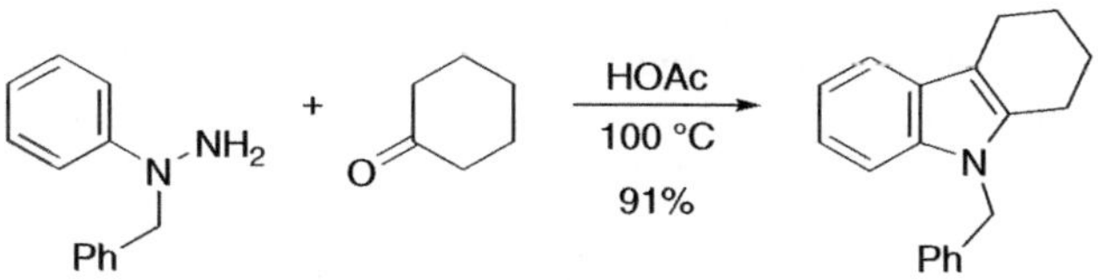

[eqn. 2] [379]

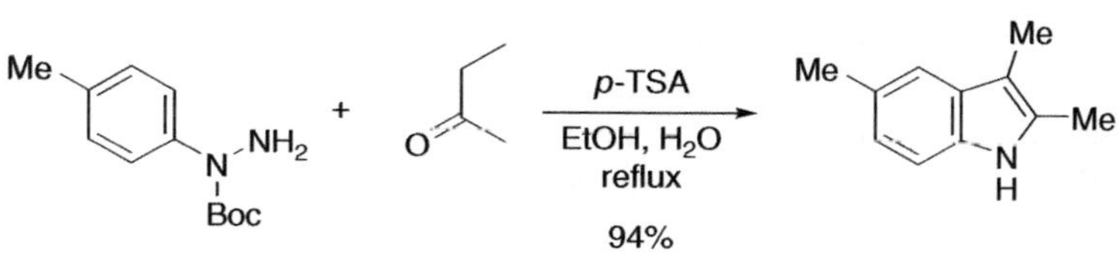

[eqn. 3] [380]

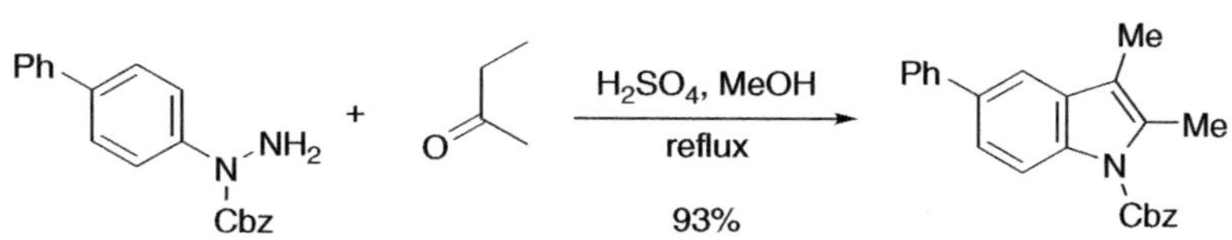

[eqn. 4] [381]

Scheme 53 *Fischer Indole Syntheses of Simple Indoles*

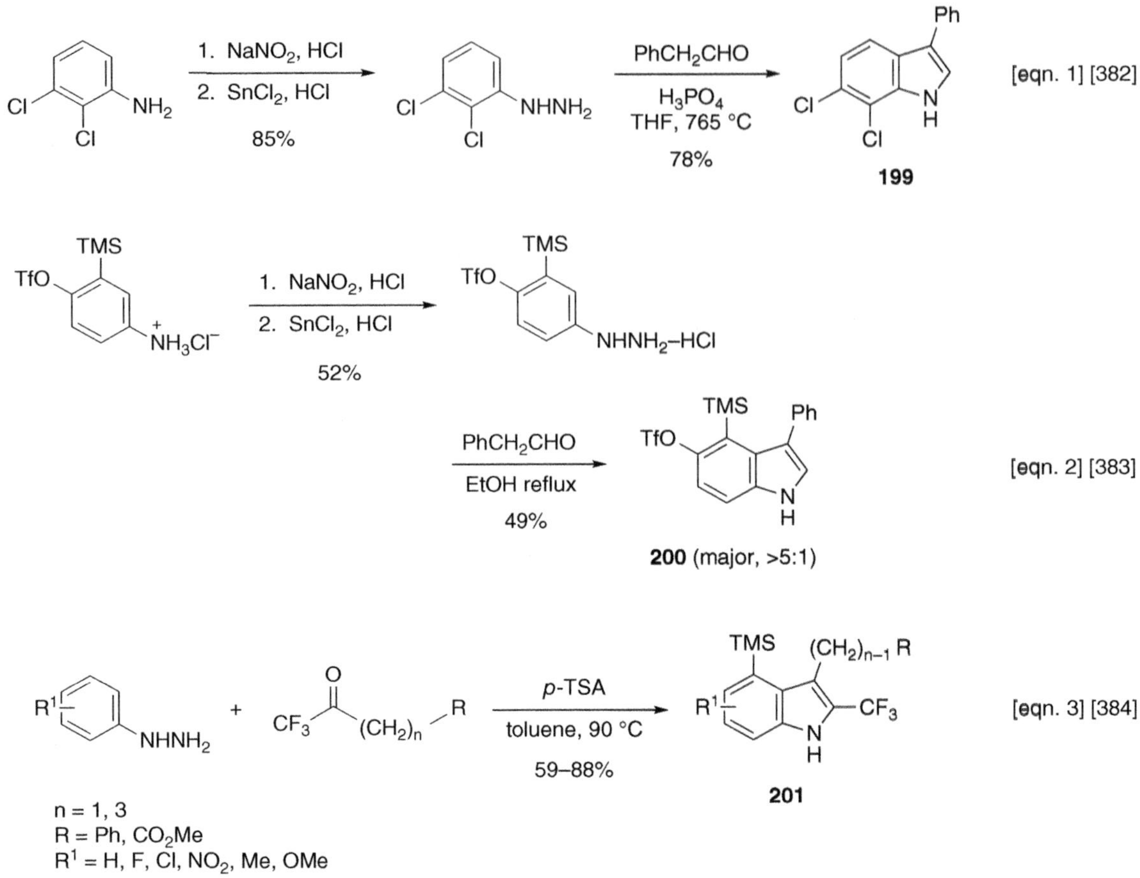

Scheme 54 *Fischer Indole Syntheses of Simple Indoles 2*

chloride, rt, quantitative) for further elaboration [386]. A simple synthesis of 7-oxygenated α-methyltryptamines was discovered by Satyam and colleagues (Entry 3) [387]. Independently, the groups of Zhang (Entry 4) and Chen (Entry 5) employed phase-transfer methods in the course of a Japp–Klingemann indolization [388, 389]. Zhang found that the best phase-transfer agent was dimethyldioctadecyl ammonium chloride (DMDOA), whereas Chen used triethylbenzylammonium chloride to prepare the keto ester phthalimide. Prasad adopted the Japp–Klingemann reaction to construct a suite of indolo[2,3-*b*]carbazol-6-ones as possible Ah receptor ligands (Entry 6) [390]. Basanagoudar and colleagues observed the loss of a methyl group while carrying out the Japp–Klingemann reaction depicted in Entry 7 [391]. Hillier and colleagues synthesized a series of cycloalkyl[*b*]indolones for use in preparing the corresponding cycloalkyl[*b*]indoles, and they used a Japp–Klingemann protocol to accomplish this (Entry 8) [392].

The Grandberg tryptophol-tryptamine Fischer indolization variation has found utility in many areas of indole chemistry. Some examples are illustrated in Scheme 55. Singh and coworkers demonstrated the power of the Grandberg method for the preparation of both tryptophols and homotryptophols (equations 1–3) [393], whereas Bosch applied the Grandberg tryptamine synthesis to the preparation of 5-(sulfamoylmethyl)indoles (equation 4) [394].

Eisenbeis and colleagues crafted a collection of azepino[3,4-*b*]indoles [395] via the so-called Plancher rearrangement, which involves the formation of an indolenium ion and then rearrangement to the final indole product (Scheme 56, equation 1) [396–398]. Cho's group reported a novel intramolecular Fischer indolization leading to tricyclic benzo[*cd*]indoles, and a spectacular example is shown in equation 2 [399]. Alford and colleagues used a Fischer indolization to prepare 2,3′-biindoles (equation 3) [400].

Christoffers and coworkers examined the regioselectivity of the Fischer indolization of bicyclic ketones. Whereas the *trans*-ketone affords the linear isomer, the *cis*-ketone gives the angular isomer (Scheme 57, equations 1 and 2). This same regiochemistry obtains for both the five- and seven-membered ring fused cyclohexanones [401, 402]. Hu and colleagues employed 2-aminocyclohexanones in

Table 8 *Japp-Klingemann Indole Ring Syntheses*

Entry	Substrate	Conditions	Indole	% Yield	Ref.
1	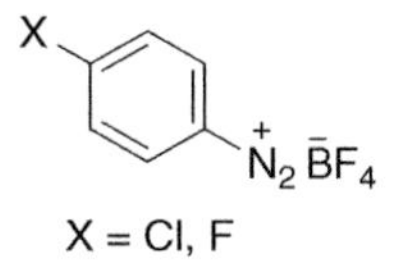 X = Cl, F	1. 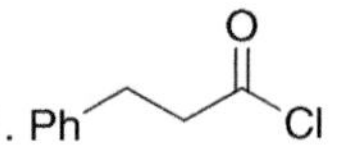 pyridine 2. RXH 3. HOAc, reflux	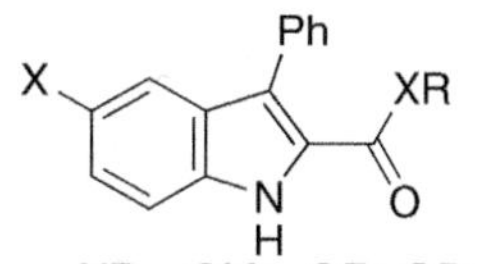 XR = OMe, OEt, OBn, NEt_2, NHCy	34–63%	385
2	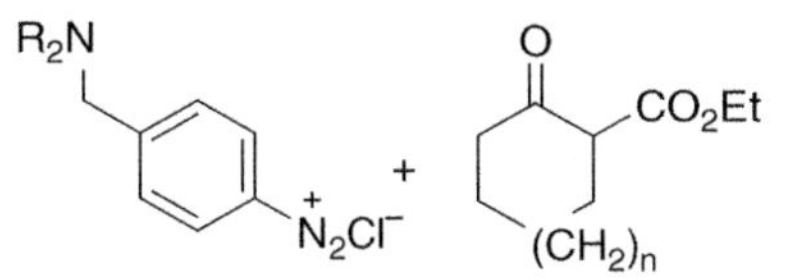 n = 0, 1	1. KOAc 2. HCl, HOAc 100 °C	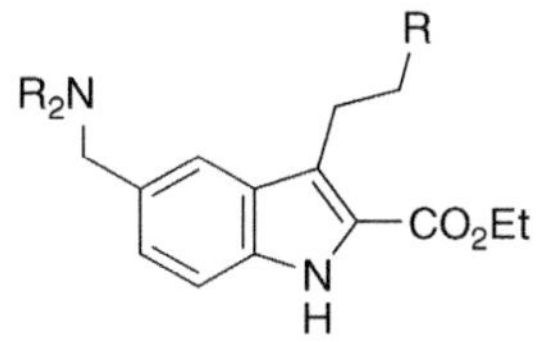 R = CH_2CO_2H, $(CH_2)_2CO_2H$	42–72%	386
3	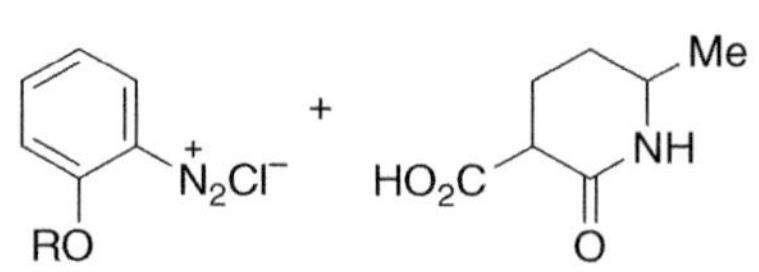 R = Me, Bn	1. NaOAc, pH 4.5 2. HOAc, reflux 3. KOH, EtOH 4. HCl, reflux 5. NaOH	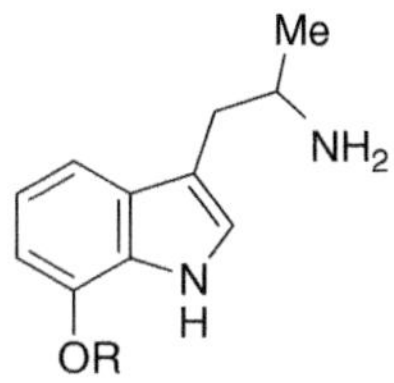	19–37%	387
4	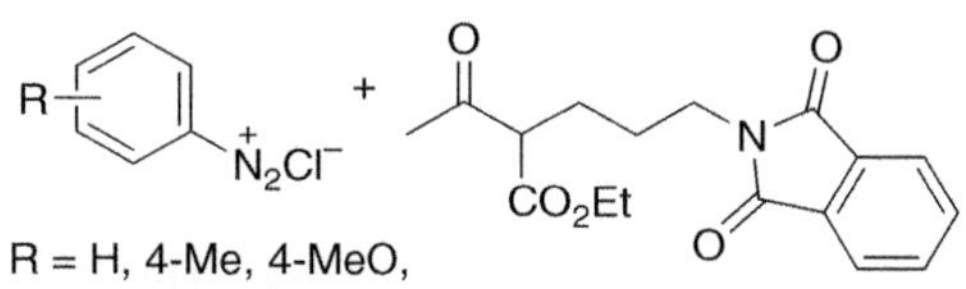 R = H, 4-Me, 4-MeO, 4-OBn, 4-NO_2, 3-OBn, 2-NO_2, 4-Cl, 4-Br, 4-Ac	1. NaOAc, H_2O, PT 2. HCl, EtOH, reflux	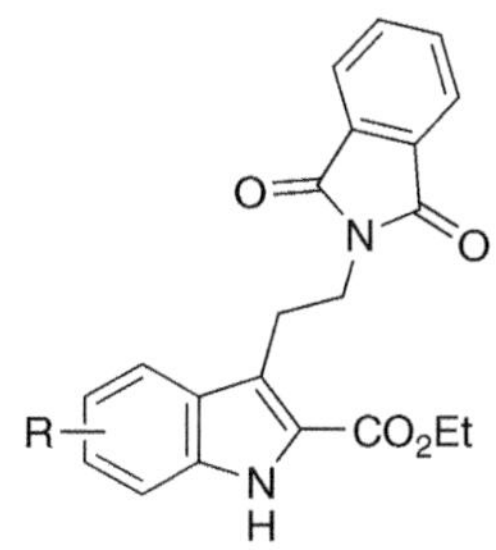	75–87%	388
5	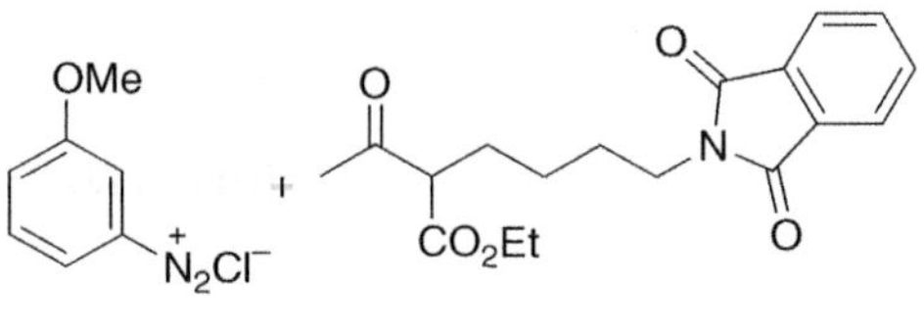	1. NaOAc, MeOH 2. KOH, H_2O 3. HCl, reflux	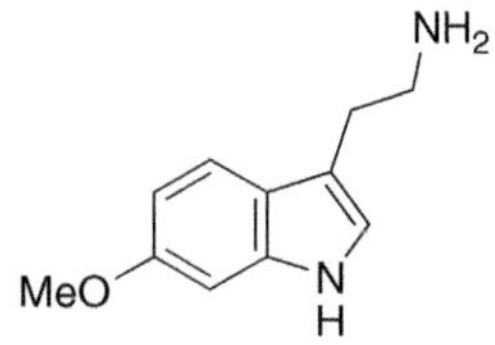	68%	389
6	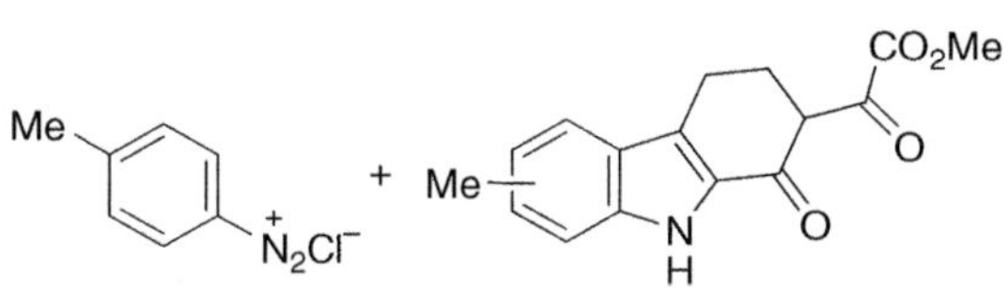	1. NaOAc, MeOH 2. HOAc, HCl, 120 °C	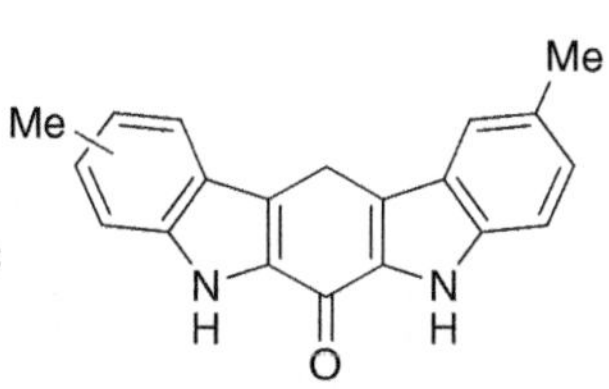	45–49%	390

(continued overleaf)

Table 8 *(continued)*

Entry	Substrate	Conditions	Indole	% Yield	Ref.
7	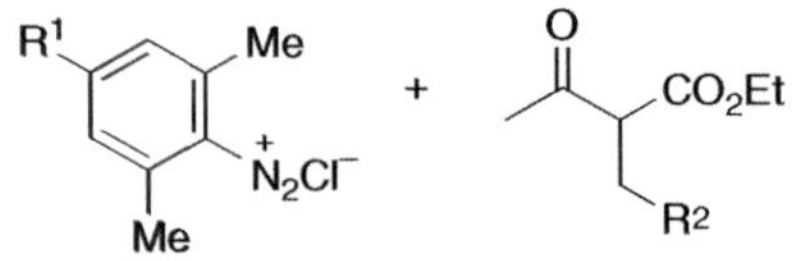 R^1 = H, Me; R^2 = H, Me, Et, Bn	1. NaOAc, NaOH EtOH 2. HCl, EtOH	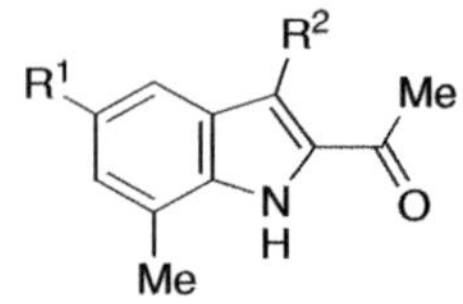	15–35%	391
8	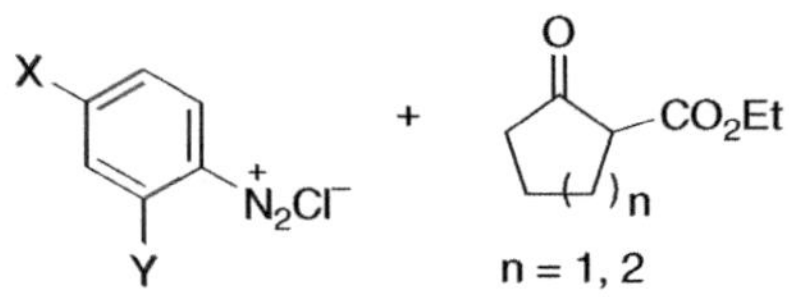 n = 1, 2; X = F, H, OMe; Y = H, H, Br	1. NaOH, H_2O, rt 2. H_2SO_4, MeCN 80 °C	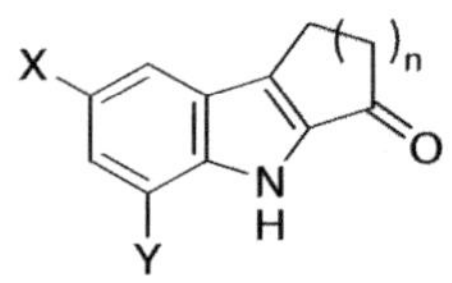	36–90%	392

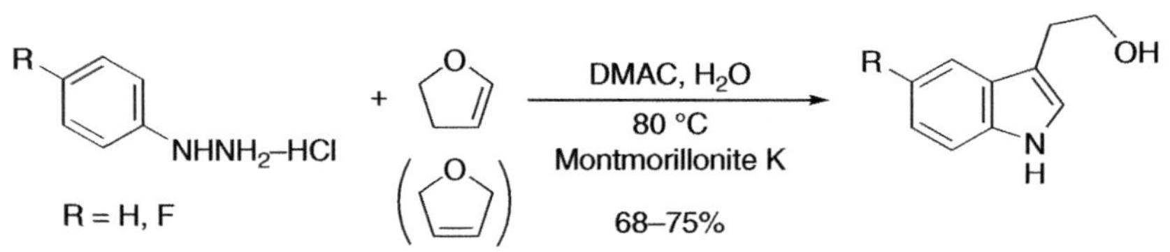

[eqn. 1] [393]

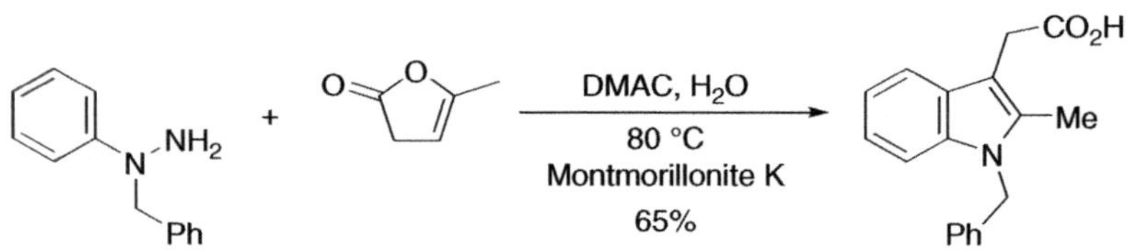

[eqn. 2] [393]

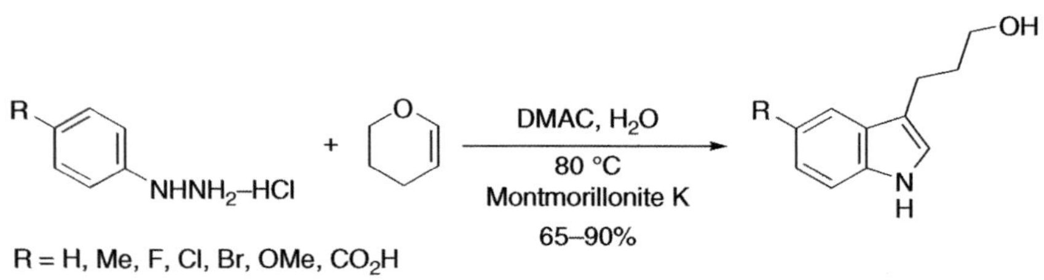

[eqn. 3] [393]

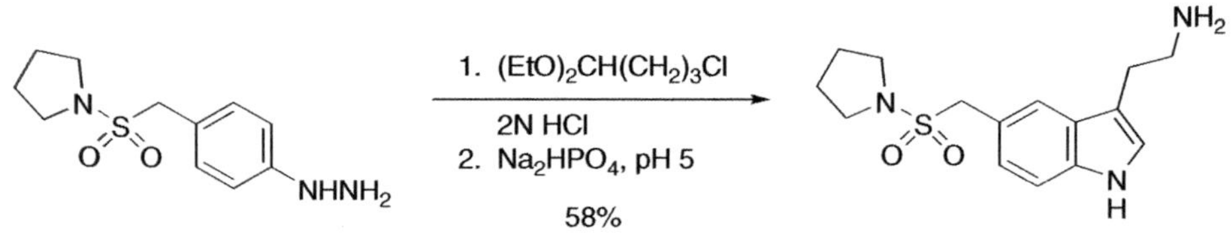

[eqn. 4] [394]

Scheme 55 *Grandberg Tryptophol-Tryptamine Syntheses*

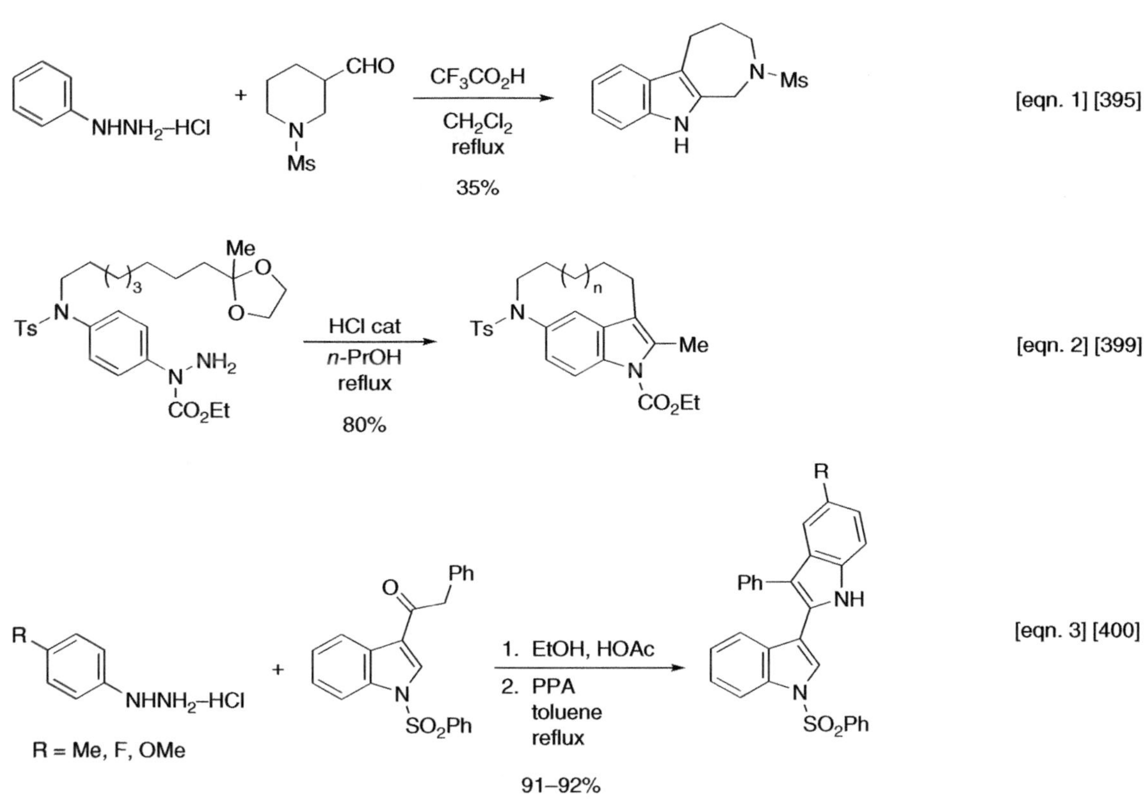

Scheme 56 *Eisenbeis, Cho, and Gribble Indole Syntheses*

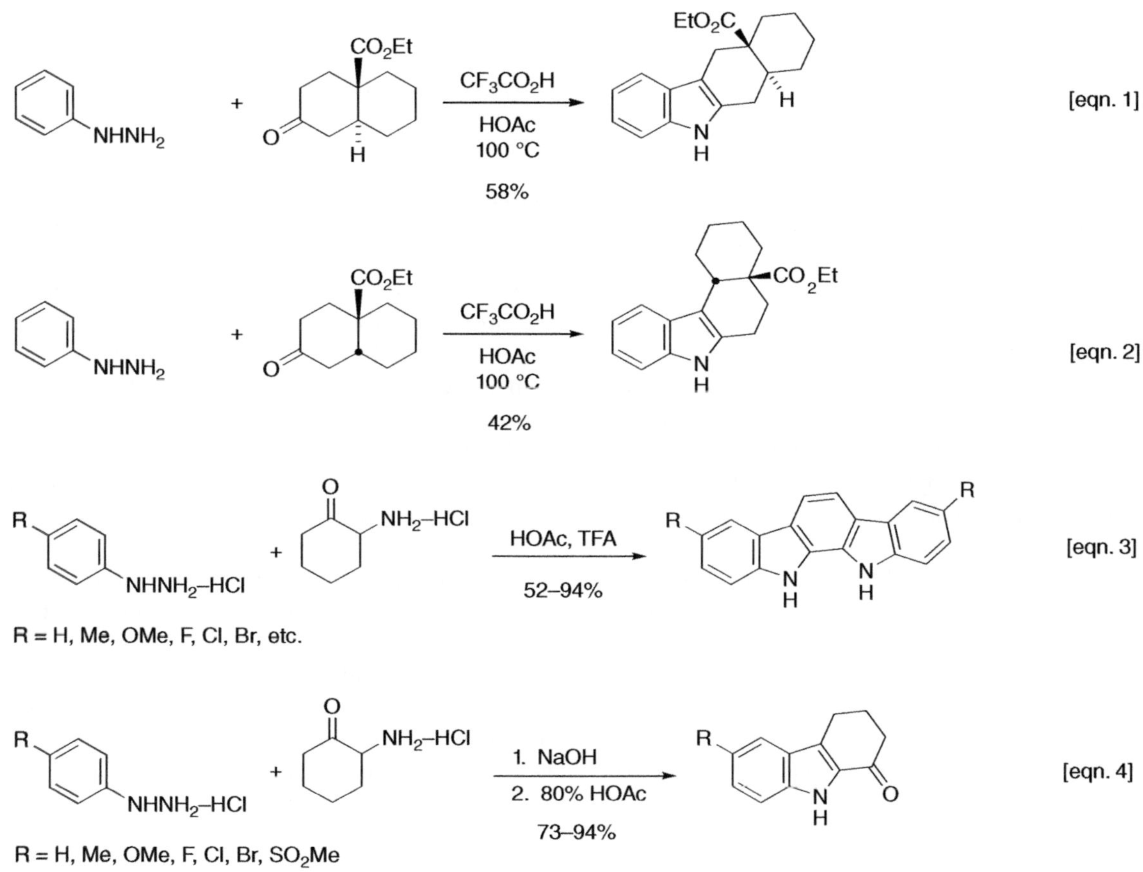

Scheme 57 *Christoffers [401, 401] and Hu [403, 404] Indole Syntheses*

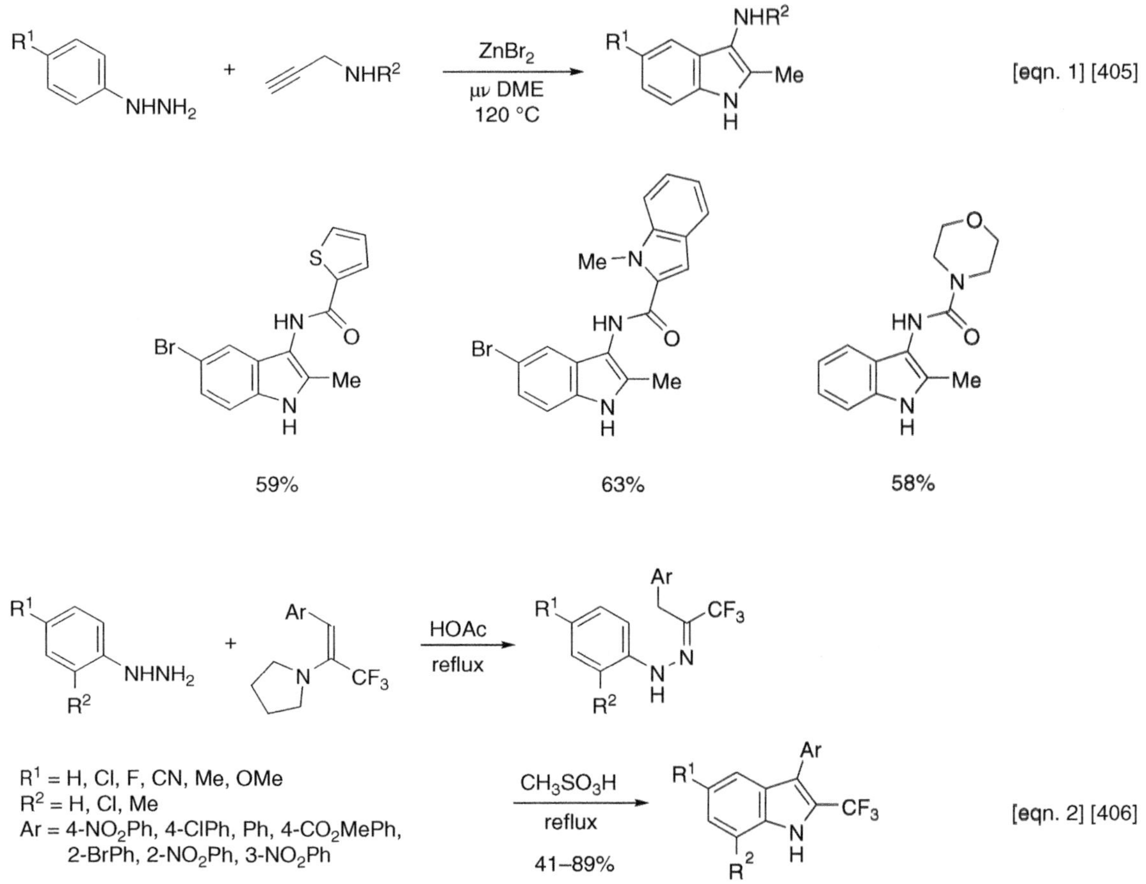

Scheme 58 *Beller [405] and Haufe [406] Indole Syntheses*

Fischer indolizations to give indolo[2,3-*a*]carbazoles (equation 3) [403] and 1-oxo-1,2,3,4-tetrahydrocarbazoles (equation 4) [404].

A fitting conclusion to this chapter is to cite two modern variations of Fischer indolizations, those by Beller [405] and Haufe [406] (Scheme 58, equations 1 and 2). Beller's work provides an excellent route to a range of 3-amidoindoles from arylhydrazines and propargyl amines. Nenajdenko, Haufe, and colleagues use α-trifluoromethyl-β-aryl enamines to access 2-trifluoromethylindoles. The trifluoromethyl group on heterocyclic rings has assumed great importance in medicinal chemistry [407].

Naito and colleagues parlayed the Fischer enehydrazine intermediate into an indole ring synthesis as summarized in Scheme 59 [408–413]. For example, *N′*-*N′*-diphenylcyclopentenyl-*N*-trifluoroacetylenehydrazine **202**, which was prepared by trifluoroacetic anhydride acylation of the corresponding hydrazone (TFAA, 0 °C), is smoothly converted to indoline **203** (equation 1). Interestingly, this [3,3]-sigmatropic rearrangement occurs slowly at room temperature (20 days). Further heating of **203** in xylene gave indole **204** in excellent overall yield [408]. In the case of enehydrazine **205**, no indoline was isolated (equation 2). Acyclic enehydrazines also afford indoles (equation 3). Naito was able to isolate and characterize by x-ray the dienylimine intermediate **208** from the cyclization of enehydrazine **207** (equation 4) [409]. This research group developed a one-pot synthesis of these indoles from arylhydrazines and ketones [410], prepared some naturally occurring indoles [411], published a full account of this methodology [412, 413], and applied it to the synthesis of benzo[*b*]furans[413]. Murphy's team has described the alkylidenation of acylhydrazides to access aryl enehydrazines and to effect an indole ring synthesis (equations 5 and 6) [414].

The Fischer indole ring synthesis, with all of its modifications and variations, may be the single most important method for constructing this heterocycle, but other methods exist, as we shall see.

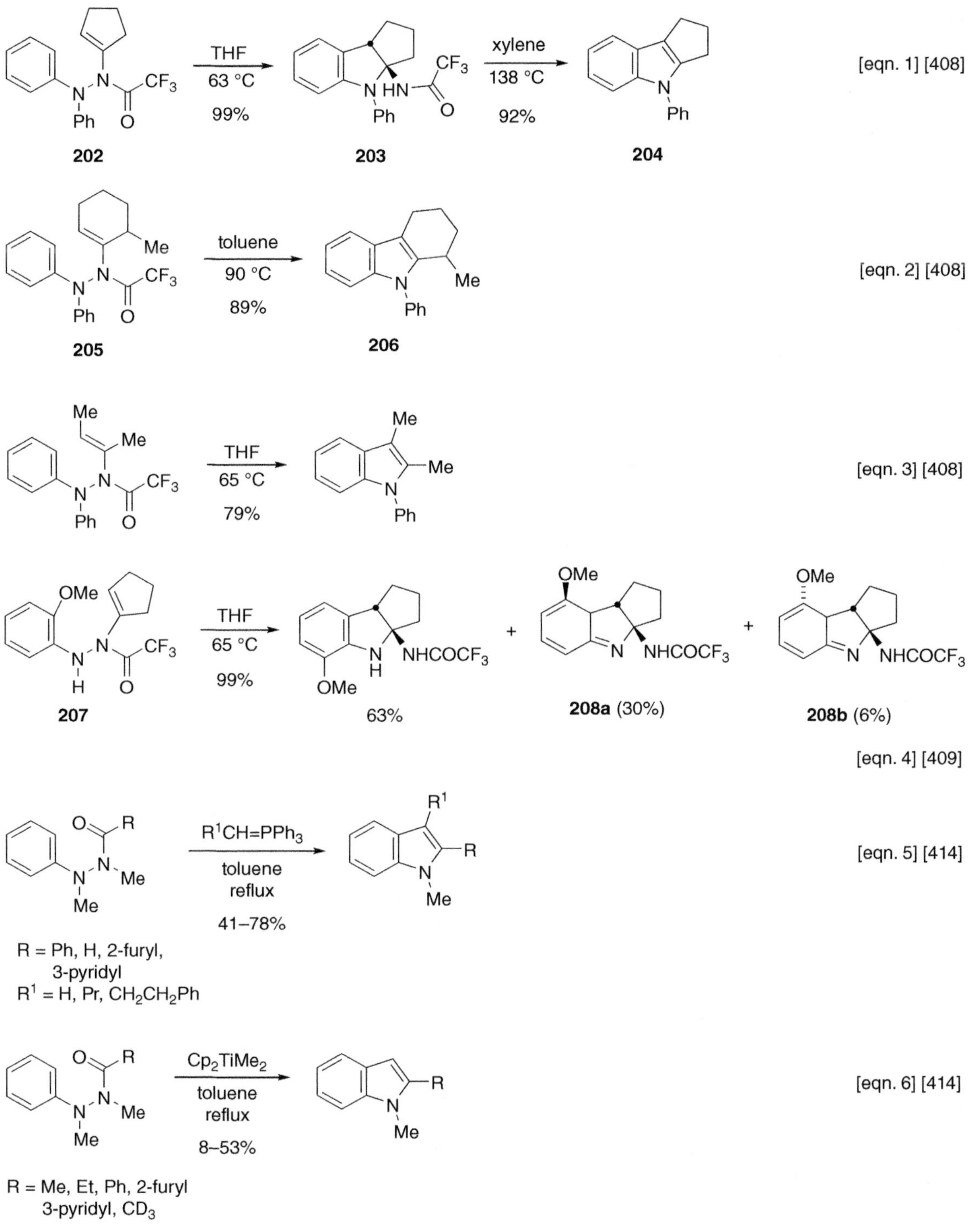

Scheme 59 *Naito [408–413] and Murphy [414] Indole Syntheses*

References

[1] E. Fischer and F. Jourdan, *Ber.*, 1883, **16**, 2241–2245.
[2] E. Fischer and O. Hess, *Ber.*, 1884, **17**, 559–568.
[3] E. Fischer, *Ann.*, 1886, **236**, 116–151.
[4] E. Fischer, *Ber.*, 1886, **19**, 1563–1570.
[5] E. Fischer, *Liebigs Ann. Chem.*, 1886, **236**, 116–126.
[6] B. Robinson, *Chem. Rev.*, 1963, **63**, 373–401.
[7] B. Robinson, *Chem. Rev.*, 1969, **69**, 227–250.
[8] B. Robinson (1982) *The Fischer Indole Synthesis*, John Wiley and Sons, New York.
[9] D.L. Hughes, *Org. Prep. Proc. Int.*, 1993, **25**, 609–632.
[10] I.I. Grandberg and V.I. Sorokin, *Russ. Chem. Rev.*, 1974, **43**, 115–128.
[11] G.M. Robinson and R. Robinson, *J. Chem. Soc.*, 1918, **113**, 639–645.
[12] G.M. Robinson and R. Robinson, *J. Chem. Soc.*, 1924, **125**, 827–840.
[13] C.F.H. Allen and C.V. Wilson, *J. Am. Chem. Soc.*, 1943, **65**, 611–612.
[14] R.B. Carlin and E.E. Fisher, *J. Am. Chem. Soc.*, 1948, **70**, 3421-3424.
[15] K.H. Pausacker and C.I. Schubert, *J. Chem. Soc.*, 1949, 1384–1389.
[16] K.H. Pausacker and C.I. Schubert, *J. Chem. Soc.*, 1950, 1814–1816.
[17] K. Clusius and H.R. Weisser, *Helv. Chim. Acta*, 1952, **35**, 400–406.
[18] R.B. Carlin, J.G. Wallace, and E.E. Fisher, *J. Am. Chem. Soc.*, 1952, **74**, 990–994.
[19] R.B. Carlin, *J. Am. Chem. Soc.*, 1952, **74**, 1077–1078.
[20] R.B. Carlin and G.W. Larson, *J. Am. Chem. Soc.*, 1957, **79**, 934–941.
[21] R.B. Carlin and D.P. Carlson, *J. Am. Chem. Soc.*, 1957, **79**, 3605–3606.
[22] J. McLean, S. McLean, and R.I. Reed, *J. Chem. Soc.*, 1955, 2519–2520.
[23] R. O'Connor, *J. Org. Chem.*, 1961, **26**, 4375–4380.
[24] G.S. Bajwa and R.K. Brown, *Can. J. Chem.*, 1968, **46**, 1927–1938.
[25] G.S. Bajwa and R.K. Brown, *Can. J. Chem.*, 1969, **47**, 785–794.
[26] R.H.C. Elgersma and E. Havinga, *Tetrahedron Lett.*, 1969, 1735–1736.
[27] G.S. Bajwa and R.K. Brown, *Can. J. Chem.*, 1970, **48**, 2293–2299.
[28] T.P. Forrest and F.M.F. Chen, *J. Chem. Soc., Chem. Commun.*, 1972, 1067.
[29] B. Miller and E.R. Matjeka, *Tetrahedron Lett.*, 1977, 131–134.
[30] Yu.P. Kitaev and T.V. Troepol'skaya, *Chem. Heterocycl. Comp.*, 1978, **14**, 807–821.
[31] A.W. Douglas, *J. Am. Chem. Soc.*, 1978, **100**, 6463–6469.
[32] A.W. Douglas, *J. Am. Chem. Soc.*, 1979, **101**, 5676–5678.
[33] B. Miller and E.R. Matjeka, *J. Am. Chem. Soc.*, 1980, **102**, 4772–4780.
[34] N.M. Przheval'skii, L.Yu. Kostromina, and I.I. Grandberg, *Chem. Heterocycl. Compd.*, 1988, **24**, 709–721.
[35] R.S. Eichen Conn, A.W. Douglas, S. Karady, *et al.*, *J. Org. Chem.*, 1990, **55**, 2908–2913.
[36] S.V. Luis and M.I. Burguete, *Tetrahedron*, 1991, **47**, 1737–1744.
[37] D.L. Hughes and D. Zhao, *J. Org. Chem.*, 1993, **58**, 228–233.
[38] N. Çelebi-Ölçüm, B.W. Boal, A.D. Huters, *et al.*, *J. Am. Chem. Soc.*, 2011, **133**, 5752–5755.
[39] H. Ishii, Y. Murakami, K. Hosoya, *et al.*, *Chem. Pharm. Bull*, 1973, **21**, 1481–1494.
[40] H. Ishii, Y. Murakami, T. Furuse, *et al.*, *Chem. Pharm. Bull.*, 1973, **21**, 1495–1505.
[41] H. Ishii, *Acc. Chem. Res.*, 1981, **14**, 275–283.
[42] H. Ishii, T. Sugiura, Y. Akiyama, *et al.*, *Chem. Pharm. Bull.*, 1990, **38**, 2118–2126.
[43] H. Ishii, Y. Murakami, and T. Ishikawa, *Chem. Pharm. Bull.*, 1990, **38**, 597–604.
[44] Y. Murakami, *Proc. Jpn. Acad., Ser. B*, 2012, **88**, 1–17.
[45] D.L. Hughes, *J. Phys. Org. Chem.*, 1994, **7**, 625–628.
[46] G.S. Bajwa and R.K. Brown, *Can. J. Chem.*, 1968, **46**, 3105–3109.
[47] K. Bast, T. Durst, R. Huisgen, *et al.*, *Tetrahedron*, 1998, **54**, 3745–3764.
[48] F.D. Benke and G.M. Brooke, *J. Fluorine Chem.*, 1984, **26**, 77–86.
[49] H. Fujii, A. Mizusuna R. Tanimura, and H. Nagase, *Heterocycles*, 1997, **45**, 2109–2112.
[50] H.R. Snyder and C.W. Smith, *J. Am. Chem. Soc.*, 1943, **65**, 2452–2454.
[51] W.M. Welch, *Synthesis*, 1977, 645–646.
[52] A. Guy and J.-P. Guetté, *Synthesis*, 1980, 222–223.
[53] K. Yamamoto and H. Watanabe, *Chem. Lett.*, 1982, 1225–1228.
[54] Y. Murakami, Y. Yokoyama, T. Miura, *et al.*, *Heterocycles*, 1984, **22**, 1211–1216.
[55] G. Baccolini and E. Marotta, *Tetrahedron*, 1985, **41**, 4615–4620.
[56] H.M. Kissman, D.W. Farnsworth, and B. Witkop, *J. Am. Chem. Soc.*, 1952, **74**, 3948–3949.
[57] C.U. Rogers and B.B. Corson, *J. Am. Chem. Soc.*, 1947, **69**, 2910–2911.
[58] G.L. Rebeiro and B.M. Khadilkar, *Synthesis*, 2001, 370–372.
[59] R.C. Morales, V. Tambyrajah, P.R. Jenkins, *et al.*, *Chem. Commun.*, 2004, 158–159.
[60] A. Dhakshinamoorthy and K. Pitchumani, *Appl. Catal. A*, 2005, **292**, 305–311.
[61] P.P. Varma, B.S. Sherigara, K.M. Mahadevan, and V. Halikal, *Synth. Commun.*, 2009, **39**, 158–165.

[62] A. Sudhakara, H. Jayadevappa, K.M. Mahadevan, and V. Hulikal, *Synth. Commun.*, 2009, **39**, 2506–2515.
[63] L. Zhong and G.-K. Chuah, *Aust. J. Chem.*, 2009, **62**, 1027–1033.
[64] S. Gore, S. Baskaran, and B. König, *Org. Lett.*, 2012, **14**, 4568–4571.
[65] D.-Q. Xu, W.-L. Yang, S.-P. Luo, *et al.*, *Eur. J. Org. Chem.*, 2007, 1007–1012.
[66] J.T. Fitzpatrick and R.D. Hiser, *J. Org. Chem.*, 1957, **22**, 1703–1704.
[67] B. Robinson, *Can. J. Chem.*, 1964, **42**, 2900–2902.
[68] A.H. Kelly, D.H. McLeod, and J. Parrick, *Can. J. Chem.*, 1965, **43**, 296–301.
[69] K. Matsumoto, A. Tanaka, I. Yukio, *et al.*, *Heterocycl. Commun.*, 2003, **9**, 9–12.
[70] D. Villemin, B. Lahiad, and Y. Ouhilal, *Chem. Ind.*, 1989, 607–609.
[71] R.A. Abramovitch and A. Bulman, *Synlett*, 1992, 795–796.
[72] K.K. Kapoor, B.A. Ganai, S. Kumar, and C.S. Andotra, *Synth. Commun.*, 2006, **36**, 2727–2735.
[73] G. Bratulescu, *Tetrahedron Lett.*, 2008, **49**, 984–986.
[74] J. Chen, W. Chen, and Y. Hu, *Synlett*, 2008, 77–82.
[75] M. Desroses, K. Wieckowski, M. Stevens, and L.R. Odell, *Tetrahedron Lett.*, 2011, **52**, 4417–4420.
[76] E.C. Creencia, M. Tsukamoto, and T. Horaguchi, *J. Heterocycl. Chem.*, 2011, **48**, 1095–1102.
[77] T.M. Lipinska and S.J. Czarnocki, *Org. Lett.*, 2006, **8**, 367–370.
[78] J.M. Kremsner and C.O. Kappe, *Eur. J. Org. Chem.*, 2005, 3672–3679.
[79] B. Wahab, G. Ellames, S. Passey, and P. Watts, *Tetrahedron*, 2010, **66**, 3861–3865.
[80] N. Pagano, M.L. Heil, and N.D.P. Cosford, *Synthesis*, 2012, **44**, 2537–2546.
[81] T. Zimmermann, *J. Heterocycl. Chem.*, 2000, **37**, 1571–1574.
[82] M.P. Prochazka and R. Carlson, *Acta Chem. Scand.*, 1989, **43**, 651–659.
[83] M.P. Prochazka, L. Eklund, and R. Carlson, *Acta Chem. Scand.*, 1990, **44**, 610–613.
[84] M.P. Prochazka and R. Carlson, *Acta Chem. Scand.*, 1990, **44**, 614–616.
[85] Y.-K. Lim and C.-G. Cho, *Tetrahedron Lett.*, 2004, **45**, 1857–1859.
[86] T. Watanabe, M. Yamamoto, S.C. Shim, *et al.*, *Chem. Lett.*, 1980, 603–604.
[87] G. Verspui, G. Elbertse, F.A. Sheldon, *et al.*, *Chem. Commun.*, 2000, 1363–1364.
[88] P. Köhling, A.M. Schmidt, and P. Eilbracht, *Org. Lett.*, 2003, **5**, 3213–3216.
[89] A.M. Schmidt and P. Eilbracht, *J. Org. Chem.*, 2005, **70**, 5528–5535.
[90] A.M. Schmidt and P. Eilbracht, *Org. Biomol. Chem.*, 2005, **3**, 2333–2343.
[91] P. Linnepe, A.M. Schmidt, and P. Eilbracht, *Org. Biomol. Chem.*, 2006, **4**, 302–313.
[92] B.P. Bondzić and P. Eilbracht, *Org. Lett.*, 2008, **10**, 3433–3436.
[93] B.P. Bondzic, A. Farwick, J. Liebich, and P. Eilbracht, *Org. Biomol. Chem.*, 2008, **6**, 3723–3731.
[94] C. Cao, Y. Shi, and A.L. Odom, *Org. Lett.*, 2002, **4**, 2853–2856.
[95] S. Banerjee, E. Barnea, and A.L. Odom, *Organometallics*, 2008, **27**, 1005–1014.
[96] V. Khedkar, A. Tillack, M. Michalik, and M. Beller, *Tetrahedron Lett.*, 2004, **45**, 3123–3126.
[97] V. Khedkar, A. Tillack, M. Michalik, and M. Beller, *Tetrahedron*, 2005, **61**, 7622–7631.
[98] N. Schwarz, K. Alex, I.A. Sayyed, *et al.*, *Synlett*, 2007, 1091–1095.
[99] I.A. Sayyed, K. Alex, A. Tillack, *et al.*, *Eur. J. Org. Chem.*, 2007, 4525–4528.
[100] O. Diels and J. Reese, *Liebigs Ann. Chem.*, 1935, **519**, 147–157.
[101] R.M. Acheson and J.M. Vernon, *J. Chem. Soc.*, 1962, 1148–1157.
[102] Y. Miki, K. Matsushita, H. Hibino, and H. Shirokoshi, *Heterocycles*, 1999, **51**, 1585–1592.
[103] L. Ackermann and R. Born, *Tetrahedron Lett.*, 2004, **45**, 9541–9544.
[104] K. Maruoka, M. Oishi, and H. Yamamoto, *J. Org. Chem.*, 1993, **58**, 7638–7639.
[105] S.D. Nielsen, T. Ruhland, and L.K. Rasmussen, *Synlett*, 2007, 443–446.
[106] B.A. Haag, Z.-G. Zhang, J.-S. Li, and P. Knochel, *Angew. Chem. Int. Ed.*, 2010, **49**, 9513–9516.
[107] Z.-G. Zhang, B.A. Haag, J.-S. Li, and P. Knochel, *Synthesis*, 2011, 23–29.
[108] A. Pews-Davtyan and M. Beller, *Org. Biomol. Chem.*, 2011, **9**, 6331–6334.
[109] N.T. Patil and A. Konala, *Eur. J. Org. Chem.*, 2010, 6831–6839.
[110] T. Gehrmann, J.L. Fillol, S.A. Scholl, *et al.*, *Angew. Chem. Int. Ed.*, 2011, **50**, 5757–5761.
[111] L. Yang, *Tetrahedron Lett.*, 2000, **41**, 6981–6984.
[112] C. Rosenbaum, C. Katzka, A. Marzinzik, and H. Waldmann, *Chem. Commun.*, 2003, 1822–1823.
[113] H. Tanaka, H. Ohno, K. Kawamura, *et al.*, *Org. Lett.*, 2003, **5**, 1159–1162.
[114] H. Ohno, H. Tanaka, and T. Takahashi, *Synlett*, 2005, 1191–1994.
[115] H.-S. Mun, W.-H. Ham, and J.-H. Jeong, *J. Comb. Chem.*, 2005, **7**, 130–135.
[116] M. Mentel, A.M. Schmidt, M. Gorray, *et al.*, *Angew. Chem. Int. Ed.*, 2009, **48**, 5841–5844.
[117] C.A. Simoneau and B. Ganem, *Tetrahedron*, 2005, **61**, 11374–11379.
[118] C.A. Simoneau, A.M. Strohl, and B. Ganem, *Tetrahedron Lett.*, 2007, **48**, 1809–1811.

[119] O. Grotkopp, A. Ahmad, W. Frank, and T.J.J. Müller, *Org. Biomol. Chem.*, 2011, **9**, 8130–8140.
[120] J.R. Donald and R.J.K. Taylor, *Synlett*, 2009, 59–62.
[121] D. McAusland, S. Seo, D.G. Pintori, *et al.*, *Org. Lett.*, 2011, **13**, 3667–3669.
[122] M. Inman and C.J. Moody, *Chem. Commun.*, 2011, **47**, 788–790.
[123] M. Inman, A. Carbone, and C.J. Moody, *J. Org. Chem.*, 2012, **77**, 1217–1232.
[124] S. Müller, M.J. Webber, and B. List, *J. Am. Chem. Soc.*, 2011, **133**, 18534–18537.
[125] A. Prechter and M.R. Heinrich, *Synthesis*, 2011, 1515–1525.
[126] K.G. Liu, A.J. Robichaud, J.R. Lo, *et al.*, *Org. Lett.*, 2006, **8**, 5769–5771.
[127] B.G. Szczepankiewicz and C.H. Heathcock, *Tetrahedron*, 1997, **53**, 8853–8870.
[128] J.F. Hartwig, *Angew. Chem. Int. Ed.*, 1998, **37**, 2090–2093.
[129] Z. Wang, R.T. Skerlj, and G.J. Bridger, *Tetrahedron Lett.*, 1999, **40**, 3543–3546.
[130] M. Wolter, A. Klapars, and S.L. Buchwald, *Org. Lett.*, 2001, **3**, 3803–3805.
[131] C.C. Mauger and G.A. Mignani, *Org. Process Res. Dev.*, 2004, **8**, 1065–1071.
[132] E. Yasui, M. Wada, and N. Takamura, *Tetrahedron Lett.*, 2006, **47**, 743–746.
[133] L. Jiang, X. Lu, H. Zhang, *et al.*, *J. Org. Chem.* , 2009, **74**, 4542–4546.
[134] R.J. Lundgren and M. Stradiotto, *Angew. Chem. Int. Ed.*, 2010, **49**, 8686–8690.
[135] E. Yasui, M. Wada, and N. Takamura, *Tetrahedron*, 2009, **65**, 461–468.
[136] M. Ahmed, R. Jackstell, A.M. Seayad, *et al.*, *Tetrahedron Lett.*, 2004, **45**, 869–873.
[137] A.K. Chakraborti, S. Bhagat, and S. Rudrawar, *Tetrahedron Lett.*, 2004, **45**, 7641–7644.
[138] F.R. Japp and F Klingemann, *Chem. Ber.*, 1887, **20**, 2942–2944.
[139] F.R. Japp and F. Klingemann, *Ann.*, 1888, **247**, 190–225.
[140] R.B. Phillips, *Org. React.*, 1959, **10**, 143–178.
[141] J. Li (2007) Japp-Klingemann hydrazone synthesis, in *Name Reactions for Functional Group Transformations* (eds. J.J. Li and E.J. Corey), Wiley & Sons, Hoboken, New Jersey, pp. 630–634.
[142] O. Piloty, *Ber.*, 1910, **43**, 489–498.
[143] H. Posvic, R. Dombro, H. Ito, and T. Telinski, *J. Org. Chem.*, 1974, **39**, 2575–2580.
[144] E. Drechsel, *J. Prakt. Chem.*, 1888, **38**, 65–74.
[145] W. Borsche and M. Feise, *Ber.*, 1907, **40**, 378–386.
[146] W. Borsche, A. Witte, and W. Bothe, *Liebigs Ann. Chem.*, 1908, **359**, 49–80.
[147] W. Borsche and G.A. Kienitz, *Chem. Ber.*, 1910, **43**, 2333–2337.
[148] N. Campell and B.M. Barclay, *Chem. Rev.*, 1947, **40**, 359–380.
[149] M.W. Fultz (2011) Borsche-Drechsel cyclization, in *Name Reactions in Heterocyclic Chemistry – II* (eds. J.J. Li and E.J. Corey), Wiley & Sons, Hoboken, New Jersey, pp. 91–101.
[150] H.T. Bucherer and F. Seyde, *J. Prakt. Chem.*, 1908, **77**, 403–413.
[151] H.T. Bucherer, *J. Prakt. Chem.*, 1904, **69**, 87.
[152] N.L. Drake, *Org. React.*, 1942, **1**, 105–128.
[153] H. Seeboth, *Angew. Chem. Int. Ed.*, 1967, **6**, 307–317.
[154] F.R. Japp and W. Maitland, *Proc. Chem. Soc.*, 1901, **17**, 176–177.
[155] A.J. Moore (2005) Bucherer carbazole synthesis, in *Name Reactions in Heterocyclic Chemistry* (ed. J.J. Li), Wiley & Sons, Hoboken, New Jersey, pp 110–115.
[156] H.T. Bucherer and E.F. Sonnenburg, *J. Prakt. Chem.*, 1910, **81**, 1–48.
[157] J.A. Hill and J.F. Eaddy, *J. Labelled Compd. Radiopharm.*, 1994, **34**, 697–706.
[158] I.I. Grandberg and T.P. Moskvina, *Chem. Heterocycl. Compd.*, 1970, **6**, 875–877.
[159] I.I. Grandberg, *Chem. Heterocycl. Compd.*, 1974, **10**, 501–510.
[160] I.I. Grandberg, *J. Org. Chem. (English Translation)*, 1983, **19**, 2135–2147.
[161] T.I. Bidylo and M.A. Yurovskaya, *Chem. Heterocycl. Compd.*, 2008, **44**, 379–418.
[162] K.R. Campos, J.C.S. Woo, S. Lee, and R.D. Tillyer, *Org. Lett.*, 2004, **6**, 79–82.
[163] B. McKittrick, A. Failli, R.J. Steffan, *et al.*, *J. Heterocycl. Chem.*, 1990, **27**, 2151–2163.
[164] C.A. Demerson, L.G. Humber, A.H. Philipp, and R.R. Martel, *J. Med. Chem.*, 1976, **19**, 391–395.
[165] J. Slade, D. Parker, M. Girgis, *et al.*, *Org. Process Res. Dev.*, 2007, **11**, 721–725.
[166] C. Chen, C.H. Senanoyake, T.J. Bill, *et al.*, *J. Org. Chem.*, 1994, **59**, 3738–3741.
[167] Y.-C. Xu, K.W. Johnson, L.A. Phebus, *et al.*, *J. Med. Chem.*, 2001, **44**, 4031–4034.
[168] K.S. Jandu, V. Barrett, M. Brockwell, *et al.*, *J. Med. Chem.*, 2001, **44**, 681–693.
[169] A.L. Sabb, R.L. Vogel, G.S. Welmaker, *et al.*, *Bioorg. Med. Chem. Lett.*, 2004, **14**, 2603–2607.
[170] N. Khorana, C. Smith, K. Herrick-Davis, *et al.*, *J. Med. Chem.*, 2003, **46**, 3930–3937.
[171] L.A. Hobson, W.A. Nugent, S.R. Anderson, *et al.*, *Org. Process Res. Dev.*, 2007, **11**, 985–995.
[172] K.G. Liu, J.R. Lo, T.A. Comery, *et al.*, *Bioorg. Med. Chem. Lett.*, 2008, **18**, 3929–3931.

[173] K. Alex, N. Schwarz, V. Khedkar, *et al.*, *Org. Biomol. Chem.*, 2008, **6**, 1802–1807.
[174] A. Nikitenko, D. Evrard, A.L. Sabb, *et al.*, *Org. Process Res. Dev.*, 2008, **12**, 76–80.
[175] C.P. Ashcroft, P. Hellier, A. Pettman, and S. Watkinson, *Org. Process Res. Dev.*, 2011, **15**, 98–103.
[176] A.J. Henderson, P.R. Guzzo, A. Ghosh, *et al.*, *Bioorg. Med. Chem. Lett.*, 2012, **22**, 1494–1498.
[177] P. Smid, H.K.A.C. Coolen, H.G. Keizer, *et al.*, *J. Med. Chem.*, 2005, **48**, 6855–6869.
[178] M. Jansen, H. Potschka, C. Brandt, *et al.*, *J. Med. Chem.*, 2003, **46**, 64–73.
[179] M. Jansen and G. Dannhardt, *Eur. J. Med. Chem.*, 2003, **38**, 855–865.
[180] T.J.N. Watson, S.W. Horgan, R.S. Shah, *et al.*, *Org. Process Res. Dev.*, 2000, **4**, 477–487.
[181] X. Shou, R. Miledi, and A.R. Chamberlin, *Bioorg. Med. Chem. Lett.*, 2005, **15**, 3942–3947.
[182] S. Urwyler, P. Floersheim, B.L. Roy, and M. Koller, *J. Med. Chem.*, 2009, **52**, 5093–5107.
[183] E.B. Skibo and C. Xing, *J. Med. Chem.*, 2001, **44**, 3545–3562.
[184] S. Rossiter, L.K. Folkes, and P. Wardman, *Bioorg. Med. Chem. Lett.*, 2002, **12**, 2523–2526.
[185] X. Bu, J. Chen, L.W. Deady, and W.A. Denny, *Tetrahedron*, 2002, **58**, 175–181.
[186] Y.-L. Chen, C.-H. Chung, I.-L. Chen, *et al.*, *Bioorg. Med. Chem.*, 2002, **10**, 2705–2712.
[187] J. Kamata, T. Okada, Y. Kotake, *et al.*, *Chem. Pharm. Bull.*, 2004, **52**, 1071–1081.
[188] M.G. Ferlin, C. Marzano, L.D. Via, *et al.*, *Bioorg. Med. Chem.*, 2005, **13**, 4733–4739.
[189] T.E. Barta, A.F. Barabasz, B.E. Foley, *et al.*, *Bioorg. Med. Chem. Lett.*, 2009, **19**, 3078–3080.
[190] A. Gopalsamy, K. Lim, G. Ciszewski, *et al.*, *J. Med. Chem.*, 2004, **47**, 6603–6608.
[191] A. Gopalsamy, M. Shi, G. Ciszewski, *et al.*, *Bioorg. Med. Chem. Lett.*, 2006, **16**, 2532–2534.
[192] R.W. Jackson, M.G. LaPorte, T. Herbertz, *et al.*, *Bioorg. Med. Chem. Lett.*, 2011, **21**, 3227–3231.
[193] G.K. Mittapalli, A. Jackson, F. Zhao, *et al.*, *Bioorg. Med. Chem. Lett.*, 2011, **21**, 6852–6855.
[194] K.S. Gudmundsson, P.R. Sebahar, L.D. Richardson, *et al.*, *Bioorg. Med. Chem. Lett.*, 2009, **19**, 3489–3492.
[195] G. La Regina, A. Coluccia, F. Piscitelli, *et al.*, *J. Med. Chem.*, 2007, **50**, 5034–5038.
[196] Centers for Disease Control and Prevention, MRSA Statistics. http://www.cdc.gov/mrsa/index.html
[197] World Health Organization, Tuberculosis Fact Sheet 2006. http://www.cdc.gov/tb/default.htm
[198] S. Daly, K. Hayden, I. Malik, *et al.*, *Bioorg. Med. Chem. Lett.*, 2011, **21**, 4720–4723.
[199] L.D. Jennings, K.W. Foreman, T.S. Rush, III, *et al.*, *Bioorg. Med. Chem. Lett.*, 2004, **14**, 1427–1431.
[200] L.D. Jennings, K.W. Foreman, T.S. Rush, III, *et al.*, *Bioorg. Med. Chem.*, 2004, **12**, 5115–5131.
[201] S.V. Karthikeyan, S. Perumal, K.A. Shetty, *et al.*, *Bioorg. Med. Chem. Lett.*, 2009, **19**, 3006–3009.
[202] R.B. Williams, *Int. J. Parasitol.*, 1999, **29**, 1209–1229.
[203] A. Scribner, J.A. Moore III, G. Ouvry, *et al.*, *Bioorg. Med. Chem. Lett.*, 2009, **19**, 1517–1521.
[204] Q. Ji, J. Gao, J. Wang, *et al.*, *Bioorg. Med. Chem. Lett.*, 2005, **15**, 2891–2893.
[205] K.D. Dykstra, L. Guo, E.T. Birzin, *et al.*, *Bioorg. Med. Chem. Lett.*, 2007, **17**, 2322–2328.
[206] B.G. Trogden, S.H. Kim, S. Lee, and J.A. Katzenellenbogen, *Bioorg. Med. Chem. Lett.*, 2009, **19**, 485–488.
[207] X. Zhang, X. Li, G.F. Allan, A. Musto, *et al.*, *Bioorg. Med. Chem. Lett.*, 2006, **16**, 3233–3237.
[208] E. Ricciotti and G.A. FitzGerald, *Arterioscler. Thromb. Vasc. Biol.*, 2011, **31**, 986–1000.
[209] R.E. Armer, M.R. Ashton, E.A. Boyd, *et al.*, *J. Med. Chem.*, 2005, **48**, 6174–6177.
[210] D. Riendeau, R. Aspiotis, D. Ethier, *et al.*, *Bioorg. Med. Chem. Lett.*, 2005, **15**, 3352–3355.
[211] L. Li, C. Beaulieu, M.-C. Carriere, D. Denis, *et al.*, *Bioorg. Med. Chem. Lett.*, 2010, **20**, 7462–7465.
[212] A.L. Blobaum and L.J. Marnett, *J. Med. Chem.*, 2007, **50**, 1425–1441.
[213] J.L. Masferrer, B.S. Zweifel, P.T. Manning, *et al.*, *Proc. Natl. Acad. Sci. U.S.A.*, 1994, **91**, 3228–3232.
[214] K. Seibert, Y. Zhang, K. Leahy, *et al.*, *Proc. Natl. Acad. Sci. U.S.A.*, 1994, **91**, 12013–12017.
[215] A.R. Maguire, S.J. Plunkett, S. Papot, *et al.*, *Bioorg. Med. Chem.*, 2001, **9**, 745–762.
[216] K.W. Woods, R.W. McCroskey, M.R. Michaelides, *et al.*, *Bioorg. Med. Chem. Lett.*, 2001, **11**, 1325–1328.
[217] W. Hu, Z. Guo, X. Yi, *et al.*, *Bioorg. Med. Chem.*, 2003, **11**, 5539–5544.
[218] S.-J. Wey, M.E. Augustyniak, E.D. Cochran, *et al.*, *J. Med. Chem.*, 2007, **50**, 6367–6382.
[219] A.J. Liedtke, K. Kim, D.F. Stec, *et al.*, *Tetrahedron*, 2012 **68**, 10049–10058.
[220] M. Bubenyák, B. Noszál, K. Kóczián, *et al.*, *Tetrahedron Lett.*, 2008, **49**, 5711–5713.
[221] H.W. Chang, S.I. Kim, H. Jung, and Y. Jahng, *Heterocycles*, 2003, **60**, 1359–1366.
[222] J.M. Friedman, *Nature*, 2000, **404**, 632–634.
[223] P.G. Kopelman, *Nature*, 2000, **404**, 635–643.
[224] P. Tontonoz, E. Hu, and B.M. Spiegelman, *Cell*, 1994, **79**, 1147–1156.
[225] B.C.C. Cantello, M.A. Cawthorne, D. Haigh, *et al.*, *J. Med. Chem.*, 1994, **37**, 3977–3985.
[226] J.J. Acton, III, R.M. Black, A.B. Jones, *et al.*, *Bioorg. Med. Chem. Lett.*, 2005, **15**, 357–362.
[227] M. Stefek, V. Snirc, P.-O. Djoubissie, *et al.*, *Bioorg. Med. Chem.*, 2008, **16**, 4908–4920.
[228] H. Adachi, K.K. Palaniappan, A.A. Ivanov, *et al.*, *J. Med. Chem.*, 2007, **50**, 1810–1827.
[229] D. Andersen, T. Storz, P. Liu, *et al.*, *J. Org. Chem.*, 2007, **72**, 9648–9655.

[230] T. Barf, F. Lehmann, K. Hammer, *et al.*, *Bioorg. Med. Chem. Lett.*, 2009, **19**, 1745–1748.
[231] G. Tuncman, E. Erbay, I. De Vivo, *et al.*, *Proc. Natl. Acad. Sci.*, 2006, **103**, 6970–6975.
[232] C. Rosenbaum, P. Baumhof, R. Mazitschek, *et al.*, *Angew. Chem. Int. Ed.*, 2004, **43**, 224–228.
[233] P. Cameliet and R.K. Jain, *Nature*, 2000, **407**, 249–257.
[234] T.E. Rawson, M. Rüth, E. Blackwood, *et al.*, *J. Med. Chem.*, 2008, **51**, 4465–4475.
[235] P. Meraldi, R. Honda, and E.A. Nigg, *Curr. Opin. Genet. Dev.*, 2004, **14**, 29–36.
[236] M. Knockaert, P. Greengard, and L.M. Meijer, *Trends Pharmacol. Sci.*, 2002, **43**, 417–427.
[237] M.A. Fousteris, A. Papakyriakou, A. Koutsourea, *et al.*, *J. Med. Chem.*, 2008, **51**, 1048–1052.
[238] K. Kinoshita, Y. Ono, T. Emura, *et al.*, *Bioorg. Med. Chem. Lett.*, 2001, **21**, 3788–3793.
[239] K. Kinoshita, T. Kobayashi, K. Asoh, *et al.*, *J. Med. Chem.*, 2011, **54**, 6286–6294.
[240] A.E. Gormemis, T.S. Ha, I. Im, *et al.*, *ChemBioChem*, 2005, **6**, 1745–1748.
[241] C.A. Willoughby, S.M. Hutchins, K.G. Rosauer, *et al.*, *Bioorg. Med. Chem. Lett.*, 2002, **12**, 93–96.
[242] D. Shaw, G.G. Chicchi, J.M. Elliott, *et al.*, *Bioorg. Med. Chem. Lett.*, 2001, **11**, 3031–3034.
[243] M. Brands, J.-K. Ergüden, K. Hashimoto, *et al.*, *Bioorg. Med. Chem. Lett.*, 2005, **15**, 4201–4205.
[244] J. Chae and S.L. Buchwald, *J. Org. Chem.*, 2004, **69**, 3336–3339.
[245] T. Luker, R. Bonnert, S. Brough, *et al.*, *Bioorg. Med. Chem. Lett.*, 2011, **21**, 6288–6292.
[246] Y. Wang, Z. Wu, B.F. Guida, *et al.*, *Bioorg. Med. Chem. Lett.*, 2008, **18**, 4936–4939.
[247] M.P. Winters, C. Crysler, N. Subasinghe, *et al.*, *Bioorg. Med. Chem. Lett.*, 2008, **18**, 1926–1930.
[248] F. Da Settimo, F. Simorini, S. Taliani, *et al.*, *J. Med. Chem.*, 2008, **51**, 5798–5806.
[249] T. Kiyoi, M. York, S. Francis, *et al.*, *Bioorg. Med. Chem. Lett.*, 2010, **20**, 4918–4921.
[250] A. Pews-Davtyan, A. Tillack, A.-C. Schmöle, *et al.*, *Org. Biomol. Chem.*, 2010, **8**, 1149–1153.
[251] Ö. Güzel, C. Temperini, A. Innocenti, *et al.*, *Bioorg. Med. Chem. Lett.*, 2008, **18**, 152–158.
[252] Ö. Güzel, A. Innocenti, A. Scozzafava, *et al.*, *Bioorg. Med. Chem.*, 2008, **16**, 9113–9120.
[253] L.C. Cooper, G.G. Chicchi, K. Dinnell, *et al.*, *Bioorg. Med. Chem. Lett.*, 2001, **11**, 1233–1236.
[254] E. Verner, B.A. Katz, J.R. Spencer, *et al.*, *J. Med. Chem.*, 2001, **44**, 2753–2771.
[255] W.T. Ashton, R.M. Sisco, Y.T. Yang, *et al.*, *Bioorg. Med. Chem. Lett.*, 2001, **11**, 1723–1726.
[256] J.P. Simeone, R.L. Bugianesi, M.M. Ponpipom, *et al.*, *Tetrahedron Lett.*, 2001, **42**, 6459–6461.
[257] T.H. Marsilje, P.B. Alper, W. Lu, *et al.*, *Bioorg. Med. Chem. Lett.*, 2008, **18**, 5259–5262.
[258] J.A. Butera, S.A. Antane, B. Hirth, *et al.*, *Bioorg. Med. Chem. Lett.*, 2001, **11**, 2093–2097.
[259] T.P. Homes, F. Mattner, P.A. Keller, and A. Katsifis, *Bioorg. Med. Chem.*, 2006, **14**, 3938–3946.
[260] I. Bennacef, C.N. Haile, A. Schmidt, *et al.*, *Bioorg. Med. Chem.*, 2006, **14**, 7582–7591.
[261] L.D. Via, S.M. Magno, O. Gia, *et al.*, *J. Med. Chem.*, 2009, **52**, 5429–5441.
[262] G. Romeo, L. Materia, V. Pittalà, *et al.*, *Bioorg. Med. Chem.*, 2006, **14**, 5211–5219.
[263] B.A. Czeskis and W.J. Wheeler, *J. Label. Compd. Radiopharm.*, 2005, **48**, 407–419.
[264] S. Ueda, M. Kato, S. Inuki, *et al.*, *Bioorg. Med. Chem. Lett.*, 2008, **18**, 4124–4129.
[265] H.-Z. Zhang, J. Drewe, B. Tseng, *et al.*, *Bioorg. Med. Chem.*, 2004, **12**, 3649–3655.
[266] A. Tsotinis, P.A. Afroudakis, K. Davidson, *et al.*, *J. Med. Chem.*, 2007, **50**, 6436–6440.
[267] R. Vardanyan, G. Vijay, G.S. Nichol, *et al.*, *Bioorg. Med. Chem.*, 2009, **17**, 5044–5053.
[268] O. Talaz, İ. Gülçin, S. Göksu, and N. Saracoglu, *Bioorg. Med. Chem.*, 2009, **17**, 6583–6589.
[269] H.M. Hügel and F. Nurlawis, *Heterocycles*, 2003, **60**, 2349–2354.
[270] C. Prabhakar, N.V. Kumar, M.R. Reddy, *et al.*, *Org. Proc. Res. Dev.*, 1999, **3**, 155–160.
[271] L. He, J.-L. Li, J.-J. Zhang, *et al.*, *Synth. Commun.*, 2003, **33**, 741–747.
[272] V.G. Nenajdenko, E.P. Zakurdaev, E.V. Prusov, and E.S. Balenkova, *Tetrahedron*, 2004, **60**, 11719–11724.
[273] J. Heredia-Moya, Y. Hayakawa, and K.L. Kirk, *J. Fluorine Chem.*, 2006, **127**, 1256–1260.
[274] S.J. Yeo, Y. Liu, and X. Wang, *Tetrahedron*, 2012, **68**, 813–818.
[275] J.D. White, K.M. Yager, and T. Yakura, *J. Am. Chem. Soc.*, 1994, **116**, 1831–1838.
[276] E. Delfourne, C. Roubin, and J. Bastide, *J. Org. Chem.*, 2000, **65**, 5476–5479.
[277] L.H. Franco and J.A. Palermo, *Chem. Pharm. Bull.*, 2003, **51**, 975–977.
[278] Y. Liu and W.W. McWhorter, Jr., *J. Am. Chem. Soc.*, 2003, **125**, 4240–4252.
[279] E.M. Boyd and J. Sperry, *Tetrahedron Lett.*, 2012, **53**, 3623–3626.
[280] P. Zhang, R. Liu, and J.M. Cook, *Tetrahedron Lett.*, 1995, **36**, 7411–7414.
[281] D. Crich, E. Fredette, and W.J. Flosi, *Heterocycles*, 1998, **48**, 545–547.
[282] B. Pohl, T. Lucherhandt, and F. Bracher, *Synth. Commun.*, 2007, **37**, 1273–1280.
[283] Y. Lingam, D.M. Rao, D.R. Bhowmik, and A. Islam, *Synth. Commun.*, 2007, **37**, 4313–4318.
[284] M.E. Zhidkov, O.V. Baranova, N.S. Kravchenko, and S.V. Dubovitskii, *Tetrahedron Lett.*, 2010, **51**, 6498–6499.
[285] A.C. Castro, L.C. Dang, F. Soucy, *et al.*, *Bioorg. Med. Chem. Lett.*, 2003, **13**, 2419–2422.
[286] C. Kunick, K. Lauenroth, M. Leost, *et al.*, *Bioorg. Med. Chem. Lett.*, 2004, **14**, 413–416.

[287] C. Reichwald, O. Shimony, U. Dunkel, *et al.*, *J. Med. Chem.*, 2008, **51**, 659–665.
[288] I.S. Marcos, M.A. Escola, R.F. Moro, *et al.*, *Synlett*, 2007, 2017–2022.
[289] J.D. White and Y. Choi, *Org. Lett.*, 2000, **2**, 2373–2376.
[290] J.D. White and Y. Choi, *Helv. Chim. Acta*, 2002, **85**, 4306–4327.
[291] J. Jiricek and S. Blechert, *J. Am. Chem. Soc.*, 2004, **126**, 3534–3538.
[292] G.L. Adams, P.J. Carroll, and A.B. Smith, III, *J. Am. Chem. Soc.*, 2012, **134**, 4037–4040.
[293] R.G. Vaswani, J.J. Day, and J.L. Wood, *Org. Lett.*, 2009, **11**, 4532–4535.
[294] C.W. Roberson and K.A. Woerpel, *J. Am. Chem. Soc.*, 2002, **124**, 11342–11348.
[295] M. Kitamura, Y. Ihara, K. Uera, and K. Narasaka, *Bull. Chem. Soc. Jpn.*, 2006, **79**, 1552–1560.
[296] D.M. Hodgson, R.E. Shelton, T.A. Moss, and M. Dekhane, *Org. Lett.*, 2010, **12**, 2834–2837.
[297] L.A. Adams, M.W.N. Valente, and R.M. Williams, *Tetrahedron*, 2006, **62**, 5195–5200.
[298] I.S. Marcos, R.F. Moro, I. Costales, *et al.*, *Tetrahedron*, 2009, **65**, 10235–10242.
[299] D. Kumar, M. Kumar, and V.S. Rao, *Chem. Lett.*, 2009, **38**, 156–157.
[300] T. Dhanabal, R. Sangeetha, and P.S. Mohan, *Tetrahedron Lett.*, 2005, **46**, 4509–4510.
[301] P.K. Agarwal, D. Sawant, S. Sharma, and B. Kundu, *Eur. J. Org. Chem.*, 2009, 292–303.
[302] D.J. Bentley, J. Fairhurst, P.T. Gallagher, *et al.*, *Org. Biomol. Chem.*, 2004, **2**, 701–708.
[303] L.F. Tietze, F. Haunert, T. Feuerstein, and T. Herzig, *Eur. J. Org. Chem.*, 2003, 562–566.
[304] D. Alonso, E. Caballero, M. Medarde, and F. Tomé, *Tetrahedron Lett.*, 2005, **46**, 4839–4841.
[305] R.S. Fornicola, K. Subburaj, and J. Montgomery, *Org. Lett.*, 2002, **4**, 615–617.
[306] L.F. Tietze and F. Major, *Eur. J. Org. Chem.*, 2006, 2314–2321.
[307] H. Suzuki, T. Watanabe, Y. Yokoyama, and Y. Murakami, *Heterocycles*, 2002, **56**, 515–518.
[308] A. Rykowski and T. Lipińska, *Synth. Commun.*, 1996, **26**, 4409–4414.
[309] T. Lipinska, *Tetrahedron Lett.*, 2002, **43**, 9565–9567.
[310] T. Lipinska, *Tetrahedron Lett.*, 2004, **45**, 8831–8834.
[311] T.M. Lipinska, *Tetrahedron*, 2006, **62**, 5736–5747.
[312] P.L. Southwick, B. McGrew, R.R. Engel, *et al.*, *J. Org. Chem.*, 1963, **28**, 3058–3065.
[313] B.W. Boal, A.W. Schammel, and N.K. Garg, *Org. Lett.*, 2009, **11**, 3458–3461.
[314] A.W. Schammel, B.W. Boal, L. Zu, *et al.*, *Tetrahedron*, 2010, **66**, 4687–4695.
[315] L. Zu, B.W. Boal, and N.K. Garg, *J. Am. Chem. Soc.*, 2011, **133**, 8877–8879.
[316] A.W. Schammel, G. Chiou, and N.K. Garg, *J. Org. Chem.*, 2012, **77**, 725–728.
[317] A.W. Schammel, G. Chiou, and N.K. Garg, *Org. Lett.*, 2012, **14**, 4556–4559.
[318] R. Tsuji, M. Nakagawa, and A. Nishida, *Heterocycles*, 2002, **58**, 587–593.
[319] J.H. Rigby and S. Sidique, *Org. Lett.*, 2007, **9**, 1219–1221.
[320] H.J. Finlay, T. Honda, and G.W. Gribble, *ARKIVOC*, 2002, **xii**, 38–46.
[321] V. Kumar, N. Rani, P. Aggarwal, *et al.*, *Bioorg. Med. Chem. Lett.*, 2008, **18**, 5058–5062.
[322] W.-W. Qiu, Q. Shen, F. Yang, *et al.*, *Bioorg. Med. Chem. Lett.*, 2009, **19**, 6618–6622.
[323] S. Zhang and Z.Y. Zhang, *Drug Discovery Today*, 2007, **12**, 373–381.
[324] Y. Wu, J.-H. Yang, G.-F. Dai, *et al.*, *Bioorg. Med. Chem.*, 2009, **17**, 1464–1473.
[325] Y.-A. Kim and S.-Y. Han, *Synth. Commun.*, 2004, **34**, 2931–2943.
[326] T. Gan, R. Liu, P. Yu, S. Zhao, and J.M. Cook, *J. Org. Chem.*, 1997, **62**, 9298–9304.
[327] H. Zhou, D. Han, X. Liao, and J.M. Cook, *Tetrahedrorn Lett.*, 2005, **46**, 4219–4224.
[328] D. Gennet, P. Michel, and A. Rassat, *Synthesis*, 2000, 447–451.
[329] K. Ohori, S. Shimizu, T. Ohshima, and M. Shibasaki, *Chirality*, 2000, **12**, 400–403.
[330] Y. Ergün, S. Patir, and G. Okay, *J. Heterocyclic Chem.*, 2002, **39**, 315–317.
[331] S. Zhao, X. Liao, T. Wang, *et al.*, *J. Org. Chem.*, 2003, **68**, 6279–6295.
[332] S.H. Lee, S.I. Kim, J.G. Park, *et al.*, *Heterocycles*, 2001, **55**, 1555–1559.
[333] B. Baruah, K. Dasu, B. Vaitilingam, *et al.*, *Bioorg. Med. Chem.*, 2004, **12**, 1991–1994.
[334] A. Kiss, J. Kökösi, R. Rotter, and I. Hermecz, *Tetrahedron*, 2000, **56**, 7987–7994.
[335] M. Bubenyák, M. Pálfi, M. Takács, *et al.*, *Tetrahedron Lett.*, 2008, **49**, 4937–4940.
[336] P.S. Portoghese, M. Sultana, H. Nagase, and A.E. Takemori, *J. Med. Chem.*, 1988, **31**, 281–282.
[337] P.S. Portoghese, M. Sultana, and A.E. Takemori, *J. Med. Chem.*, 1990, **33**, 1714–1720.
[338] A.E. Takemori, M. Sultana, H. Nagase, and P.S. Portoghese, *Life Sci.*, 1992, **50**, 1491–1495.
[339] H. Kubota, R.B. Rothman, C. Dersch, *et al.*, *Bioorg. Med. Chem. Lett.*, 1998, **8**, 799–804.
[340] H. Yu, T. Prisinzano, C.M. Dersch, *et al.*, *Bioorg. Med. Chem. Lett.*, 2002, **12**, 165–168.
[341] P. Grundt, F. Martinez-Bermejo, J.W. Lewis, and S.M. Husbands, *Helv. Chim. Acta*, 2003, **86**, 793–798.
[342] M. Hasegawa, H. Ohno, H. Tanaka, *et al.*, *Bioorg. Med. Chem. Lett.*, 2006, **16**, 158–161.
[343] A. Zhang, F. Li, C. Ding, *et al.*, *J. Med. Chem.*, 2007, **50**, 2747–2751.

[344] S. Sakami, M. Maeda, K. Kawai, *et al.*, *J. Med. Chem.*, 2008, **51**, 4404–4411.
[345] S. Sakami, K. Kawai, M. Maeda, *et al.*, *Bioorg. Med. Chem.*, 2008, **16**, 7956–7967.
[346] F. Li, L. Gaob, C. Yin, *et al.*, *Bioorg. Med. Chem. Lett.*, 2009, **19**, 4603–4606.
[347] M. Gill and W. Steglich, *Prog. Chem. Org. Nat. Prod.*, 1987, **51**, 216–226.
[348] J. Bergman, *Stud. Nat. Prod. Chem., Part A*, 1988, **1**, 3–30.
[349] G.W. Gribble and S.J. Berthel, *Stud. Nat. Prod. Chem.*, 1993, **12**, 365–409.
[350] U. Pindur, Y.S. Kim, and F. Mehrabani, *Curr. Med. Chem.*, 1999, **6**, 29–69.
[351] J. Bergman and B. Pelcman, *J. Org. Chem.*, 1989, **54**, 824–828.
[352] M. Adeva, H. Sahagún, E. Caballero, *et al.*, *J. Org. Chem.*, 2000, **65**, 3387–3394.
[353] E. Caballero, M. Adeva, S. Calderón, *et al.*, *Bioorg. Med. Chem.*, 2003, **11**, 3413–3421.
[354] B.N. Balasubramanian, D.R. St. Laurent, M.G. Saulnier, *et al.*, *J. Med. Chem.*, 2004, **47**, 1609–1612.
[355] H. Bregman, D.S. Williams, and E. Meggers, *Synthesis*, 2005, 1521–1527.
[356] L.N. Yudina and J. Bergman, *Tetrahedron*, 2003, **59**, 1265–1275.
[357] U. Rannug, A. Rannug, U. Sjöberg, *et al.*, *J. Chem. Biol.*, 1995, **2**, 841–845.
[358] Y. Wei, H. Helleberg, U. Rannug, and A. Rannug, *Chem. Biol. Interact.*, 1998, **110**, 39–55.
[359] O. Talaz and N. Saracoglu, *Tetrahedron*, 2010, **66**, 1902–1910.
[360] G. Stork and J.E. Dolfini, *J. Am. Chem. Soc.*, 1963, **85**, 2872–2873.
[361] V. Georgian, *Chem. Ind.*, 1957, 1124–1125.
[362] R. Iyengar, K. Schildknegt, and J. Aubé, *Org. Lett.*, 2000, **2**, 1625–1627.
[363] L.A. Sharp and S.Z. Zard, *Org. Lett.*, 2006, **8**, 831-834.
[364] Y. Fukuda, M. Shindo, and K. Shishido, *Org. Lett.*, 2003, **5**, 749–751.
[365] H. Ueda, H. Satoh, K. Matsumoto, *et al.*, *Angew. Chem. Int. Ed.*, 2009, **48**, 7600–7603.
[366] V. Hegde, P. Madhukar, J.D. Madura, and R.P. Thummel, *J. Am. Chem. Soc.*, 1990, **112**, 4549–4550.
[367] Q. Liu, M.S. Mudadu, H. Schmider, *et al.*, *Organometallics*, 2002, **21**, 4743–4749.
[368] O. Grotkopp, A. Ahmad, W. Frank, and T.J.J. Müller, *Org. Biomol. Chem.*, 2011, **9**, 8130–8140.
[369] K. Sambasivarao, S. Hollinshead, D. Grubisha, *et al.*, *J. Org. Chem.*, 1990, **55**, 3858–3866.
[370] D. Curiel, A. Cowley, and P.D. Beer, *Chem. Commun.*, 2005, 236–238.
[371] Y. Wu, Y. Li, S. Gardner, and B.S. Ong, *J. Am. Chem. Soc.*, 2005, **127**, 614–618.
[372] V. Roznyatovskiy, V. Lynch, and J.L. Sessler, *Org. Lett.*, 2010, **12**, 4424–4427.
[373] I.F. Sengul, K. Wood, N. Kumar, and D.StC. Black, *Tetrahedron*, 2012,**68**, 9050–9055.
[374] B. Ortner, R. Waibel, and P. Gmeiner, *Angew. Chem. Int. Ed.*, 2001, **40**, 1283–1285.
[375] L. Chen, T.-S. Hu, and Z.-J. Yao, *Eur. J. Org. Chem.*, 2008, 6175–6182.
[376] J.C. Jewett and C.R. Bertozzi, *Org. Lett.*, 2011, **13**, 5937–5939.
[377] C.M. So, C.P. Lau, and F.Y. Kwong, *Org. Lett.*, 2007, **9**, 2795–2798.
[378] M. Nakazaki and K. Yamamoto, *J. Org. Chem.*, 1976, **41**, 1877.
[379] R.B. Perni and G.W. Gribble, *Org. Prep. Proc. Int.*, 1982, **14**, 343–346.
[380] Y.-K. Lim and C.-G. Cho, *Tetrahedron Lett.*, 2004, **45**, 1857–1859.
[381] I.-K. Park, S.-E. Suh, B.-Y. Lim, and C.-G. Cho, *Org. Lett.*, 2009, **11**, 5454–5456.
[382] K.R. Buszek, D. Luo, M. Kondrashov, *et al.*, *Org. Lett.*, 2007, **9**, 4135–4137.
[383] N. Brown, D. Luo, D.V. Vander Velde, *et al.*, *Tetrahedron Lett.*, 2009, **50**, 63–65.
[384] H. Jiang, Y. Wang, W. Wan, and J. Hao, *Tetrahedron*, 2010, **66**, 2746–2751.
[385] L.E. Kaïm, L. Grimaud, and C. Ronsseray, *Synlett*, 2010, 2296–2298.
[386] B. Pete, *Tetrahedron Lett.*, 2008, **49**, 2835–2838.
[387] N.P. Dubash, N.K. Mangu, and A. Satyam, *Synth. Commun.*, 2004, **34**, 1791–1799.
[388] W. He, B.-L. Zhang, Z.-J. Li, and S.-Y. Zhang, *Synth. Commun.*, 2005, **35**, 1359–1368.
[389] Y. Kuang, J. Huang, and F. Chen, *Synth. Commun.*, 2006, **36**, 1515–1519.
[390] A.E. Martin and K.J.R. Prasad, *Synth. Commun.*, 2008, **38**, 1778–1783.
[391] S.J. Maddirala, V.S. Gokak, S.B. Rajur, and L.D. Basanagoudar, *Tetrahedron Lett.*, 2003, **44**, 5665–5668.
[392] M.C. Hillier, J.-F. Marcoux, D. Zhao, *et al.*, *J. Org. Chem.*, 2005, **70**, 8385–8394.
[393] P.R. Singh, M.P. Surpur, and S.B. Patil, *Tetrahedron Lett.*, 2008, **49**, 3335–3340.
[394] J. Bosch, T. Roca, M. Armengol, and D. Fernández-Forner, *Tetrahedron*, 2001, **57**, 1041–1048.
[395] S.A. Eisenbeis, J.R. Phillips, D. Rescek, and Y. Oyola-Cintron, *Tetrahedronn Lett.*, 2010, **51**, 4303–4305.
[396] G. Plancher, *Gazz. Chin. Ital.*, 1898, **28**, II, 374–391.
[397] G. Plancher, *Atti. Accad. Lincei*, 1900, **9**, (I-5), 115–122.
[398] H.S. Boyd-Barrett, *J. Chem Soc.*, 1932, 321–325.
[399] I.-K. Park, J. Park, and C.-G. Cho, *Angew. Chem. Int. Ed.*, 2012, **51**, 2496–2499.
[400] P.E. Alford, L. Fu, J.K. Kubert, *et al.*, *Tetrahedron Lett.*, 2011, **52**, 2642–2644.

[401] J. Christoffers, *Synlett*, 2006, 318–320.
[402] C.L. Diedrich, W. Frey, and J. Christoffers, *Eur. J. Org. Chem.*, 2007, 4731–4737.
[403] Y.-Z. Hu and Y.-Q. Chen, *Synlett*, 2005, 42–48.
[404] R. Sheng, L. Shen, Y.-Q. Chen, and Y.-Z. Hu, *Synth. Commun.*, 2009, **39**, 1120–1127.
[405] A. Pews-Davtyan and M. Beller, *Org. Biomol. Chem.*, 2011, **9**, 6331–6334.
[406] V.M. Muzalevskiy, V.G. Nenajdenko, A.V. Shastin, *et al.*, *Tetrahedron*, 2009, **65**, 7553–7561.
[407] S. Roy, B.T. Gregg, G.W. Gribble, *et al.*, *Tetrahedron*, 2011, **67**, 2161–2195.
[408] O. Miyata, Y. Kimura, K. Muroya, *et al.*, *Tetrahedron Lett.*, 1999, **40**, 3601–3604.
[409] O. Miyata, Y. Kimura, and T. Naito, *Chem. Commun.*, 1999, 2429–2430.
[410] O. Miyata, Y. Kimura, and T. Naito, *Synthesis*, 2001, 1635–1638.
[411] O. Miyata, M. Takeda, and T. Naito, *Heterocycles*, 2002, **57**, 1101–1107.
[412] O. Miyata, N. Takeda, Y. Kimura, *et al.*, *Tetrahedron*, 2006, **62**, 3629–3647.
[413] O. Miyata, N. Takeda, and T. Naito, *Heterocycles*, 2009, **78**, 843–871.
[414] K. Hisler, A.G.J. Commeureuc, S. Zhou, and J.A. Murphy, *Tetrahedron Lett.*, 2009, **50**, 3290–3293.

3

Gassman Indole Synthesis

In 1973, Paul Gassman and his coworkers described a very clever indole ring synthesis that encompasses a [2,3]-sigmatropic rearrangement as the key step. The process is initiated by *N*-chlorination of aniline, nucleophilic displacement of chloride on nitrogen by methylthio-2-propanone to give sulfonium salt **1**. Triethylamine then generates sulfonium ylide **2** that undergoes a [2, 3]-sigmatropic (Sommelet–Hauser) rearrangement to afford amino ketone **3**. Ring closure ensues to form the thiomethyl indole **4**. Raney nickel desulfurization to **5** completes the process (Scheme 1) [1–9]. Gassman's related synthesis of oxindoles will not be covered here [10].

A summary of the indoles prepared by Gassman is depicted in Table 1. A driving force for this reaction is the acidity ($pK_a \sim 15$) of the α-hydrogen between the carbonyl and the sulfonium groups in **1**, leading to ylide **2** [11]. In entry 8, the use of methylthioacetaldehyde allows the isolation of the intermediate acetal from the [2, 3]-sigmatropic rearrangement. Gassman's use of different methylsulfides has greatly expanded the method to the synthesis of 2,3-disubstituted indoles (Table 1, entries 4, 5). The reduction of the 3-thiomethylindolenines in these entries can also be accomplished with sodium borohydride. Another modification is the use of the *S*-chloro sulfonium salt to

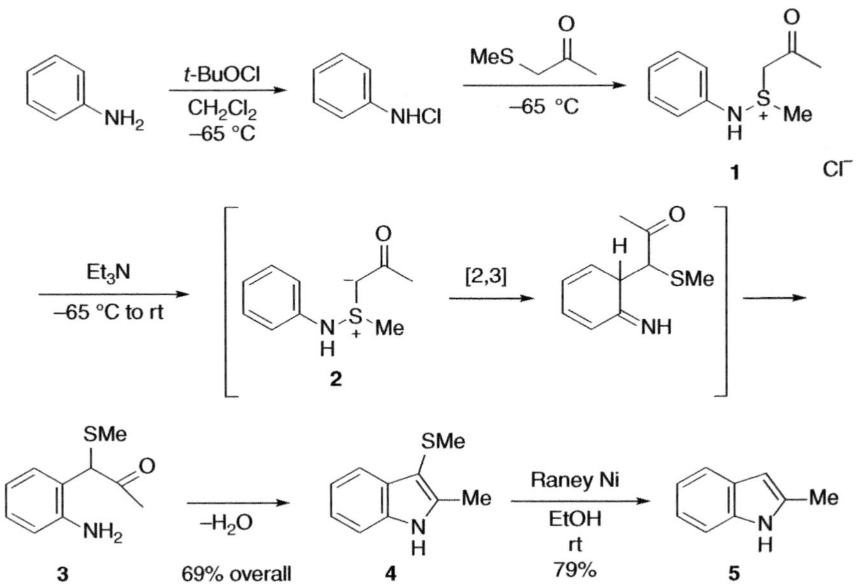

Scheme 1 *Gassman Indole Synthesis*

Indole Ring Synthesis: From Natural Products to Drug Discovery, First Edition. Gordon W. Gribble.

Table 1 *Gassman Synthesis of Indoles*

Entry	Aniline	Conditions	Thioindole	% Yield	Conditions	Indole	% Yield	Ref.
1	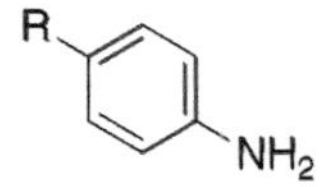 R = OAc, Me, Cl, CO_2Et	1. *t*-BuOCl 2. $MeSCH_2(CO)Me$ 3. Et_3N	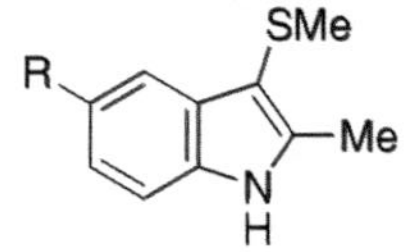	58–72%	Ra-Ni EtOH	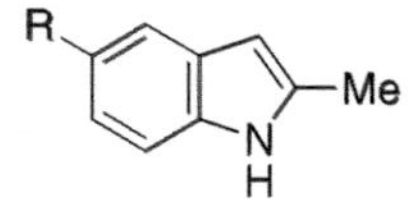	72–83%	1
2	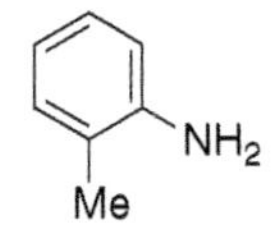	1. *t*-BuOCl 2. $MeSCH_2(CO)Me$ 3. Et_3N	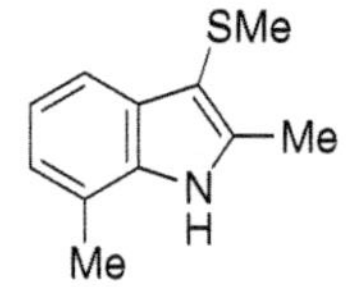	72%	Ra-Ni EtOH	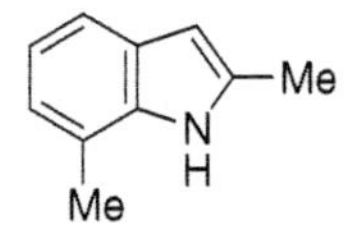	73%	1
3	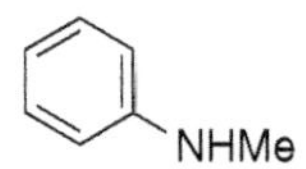	1. *t*-BuOCl 2. $MeSCH_2(CO)Me$ 3. Et_3N	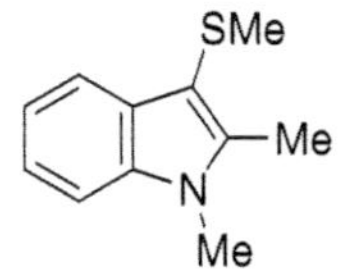	54%	Ra-Ni EtOH	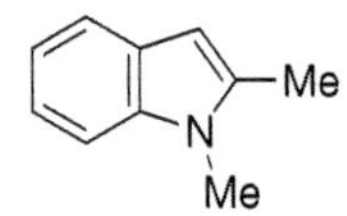	76%	2
4	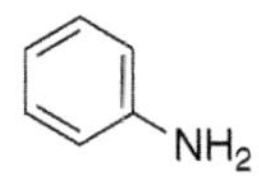	1. *t*-BuOCl 2. $MeSCH_2(CO)Ph$ 3. Et_3N	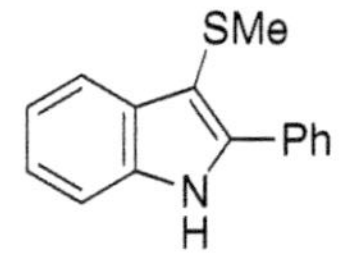	81%	Ra-Ni EtOH	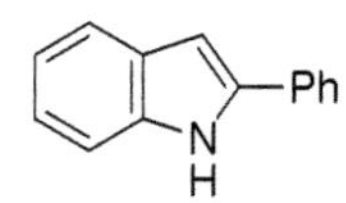	74%	1
5	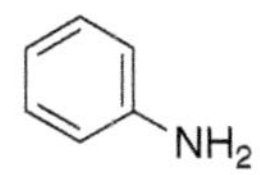	1. *t*-BuOCl 2. $MeSCH_2CHO$ 3. Et_3N	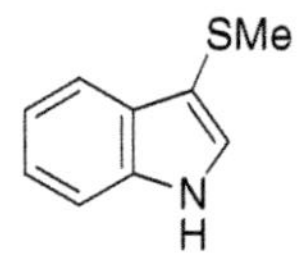	30%	Ra-Ni EtOH	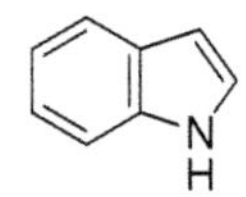	82%	2
6	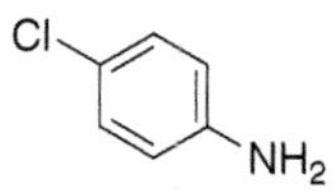	1. *t*-BuOCl 2. $MeSCH_2CHO$ 3. Et_3N	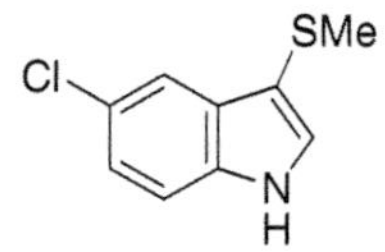	50%	—	—	—	2
7	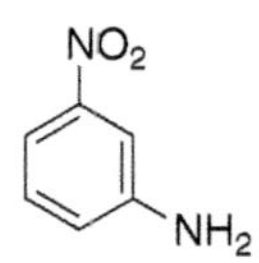	1. *t*-BuOCl 2. $MeSCH_2CHO$ 3. Et_3N	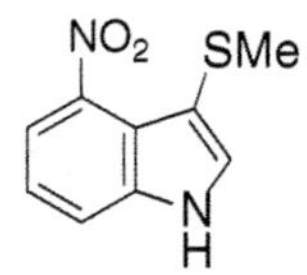	38%	—	—	—	2
8	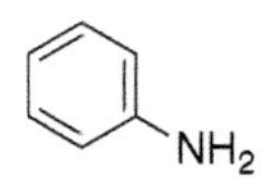	1. *t*-BuOCl 2. $MeSCH_2CH(OMe)_2$ 3. Et_3N	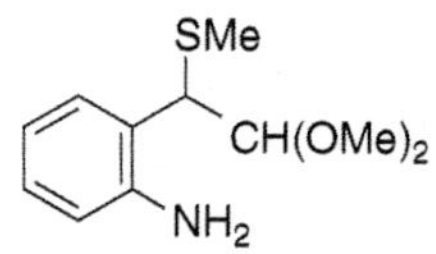	57%	aq HCl	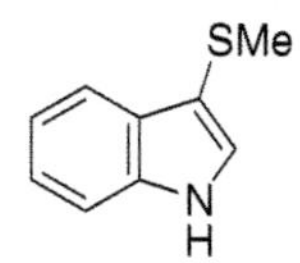	97%	2

(continued overleaf)

Table 1 (continued)

Entry	Aniline	Conditions	Thioindole	% Yield	Conditions	Indole	% Yield	Ref.
9	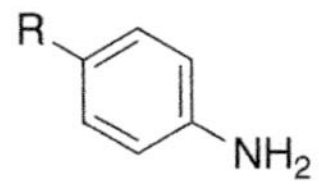 R = Cl, Me	1. *t*-BuOCl 2. $MeSCH_2CH(OMe)_2$ 3. Et_3N	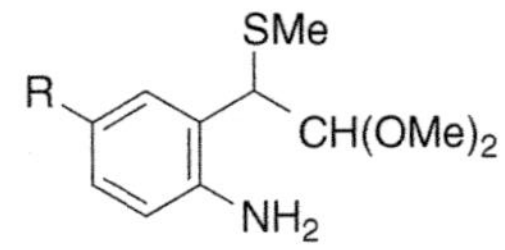 not isolated	—	aq HCl	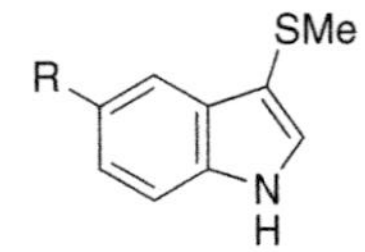	23–39%	2
10	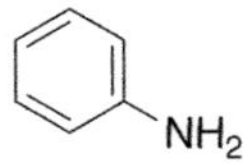	1. *t*-BuOCl 2. 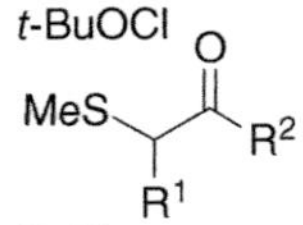 3. Et_3N	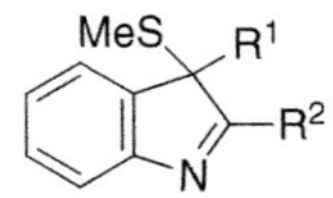 R^1 = Me, Et, Ph R^2 = Me, Et, Ph	—	$LiAlH_4$	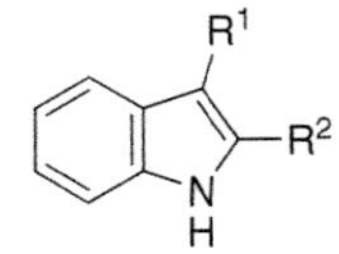	41–85%	4
11	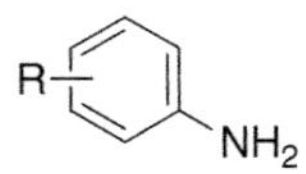 R = H, 4-Me, 4-Cl, 4-CO_2Et, 2-Me, 2-Cl	1. *t*-BuOCl 2. 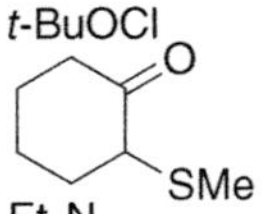 3. Et_3N	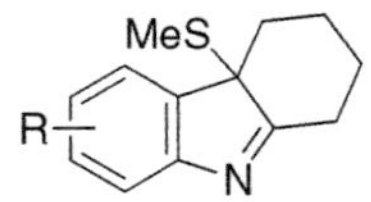	29–70%	Ra-Ni R = H	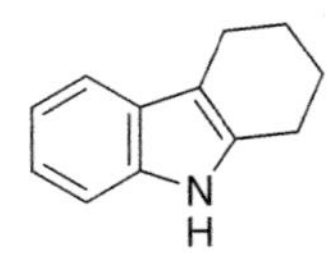	83%	4
12	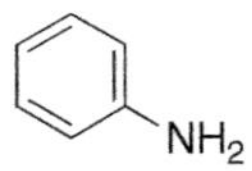	1. *t*-BuOCl 2. 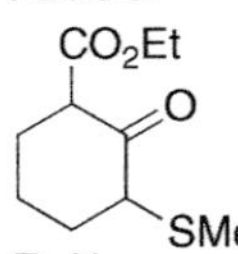 3. Et_3N	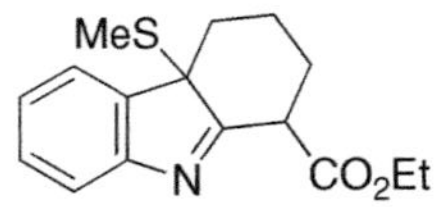	—	Raney Ni	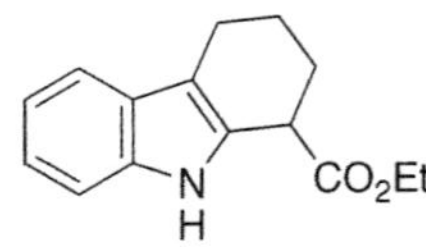	70%	5
13	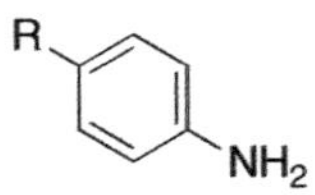 R = H, OMe, Cl, CO_2Et	1. $MeS^{+}(Cl)CH_2COMe$ 2. Et_3N	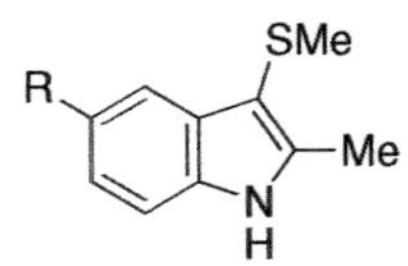	33–68%	Raney Ni	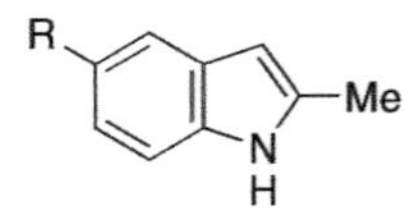	72% R = OMe	6
14	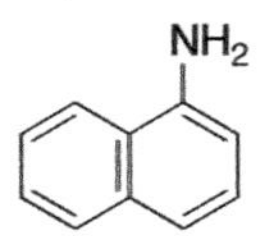	1. *t*-BuOCl 2. $MeSCH_2COMe$ 3. Et_3N	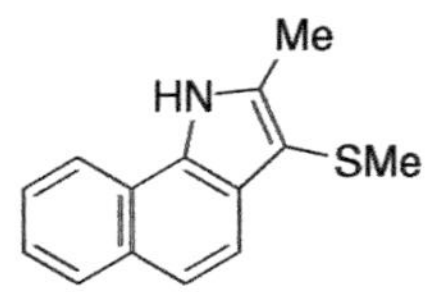	65%	Raney Ni	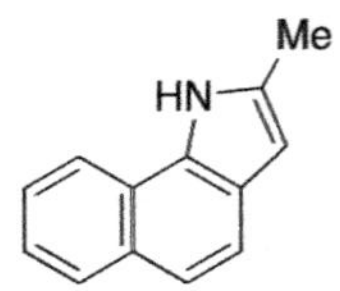	90%	7
15	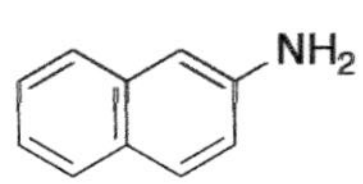	1. *t*-BuOCl 2. $MeSCH_2COMe$ 3. Et_3N	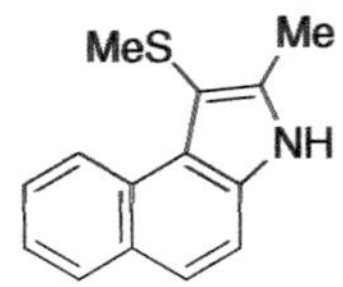	73%	Raney Ni	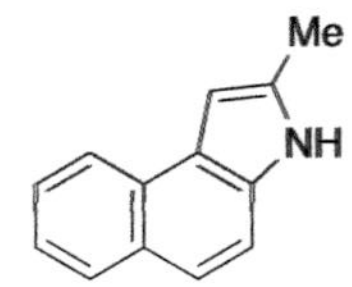	88%	7
16	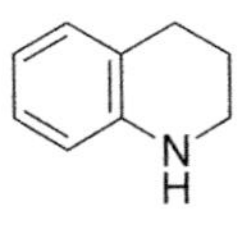	1. 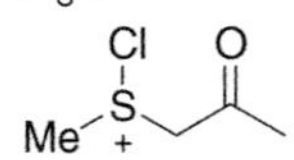 2. Et_3N	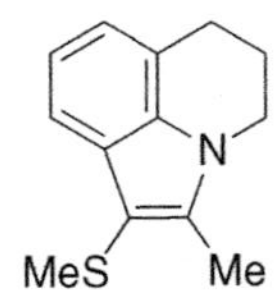	—	—	—	39%	8

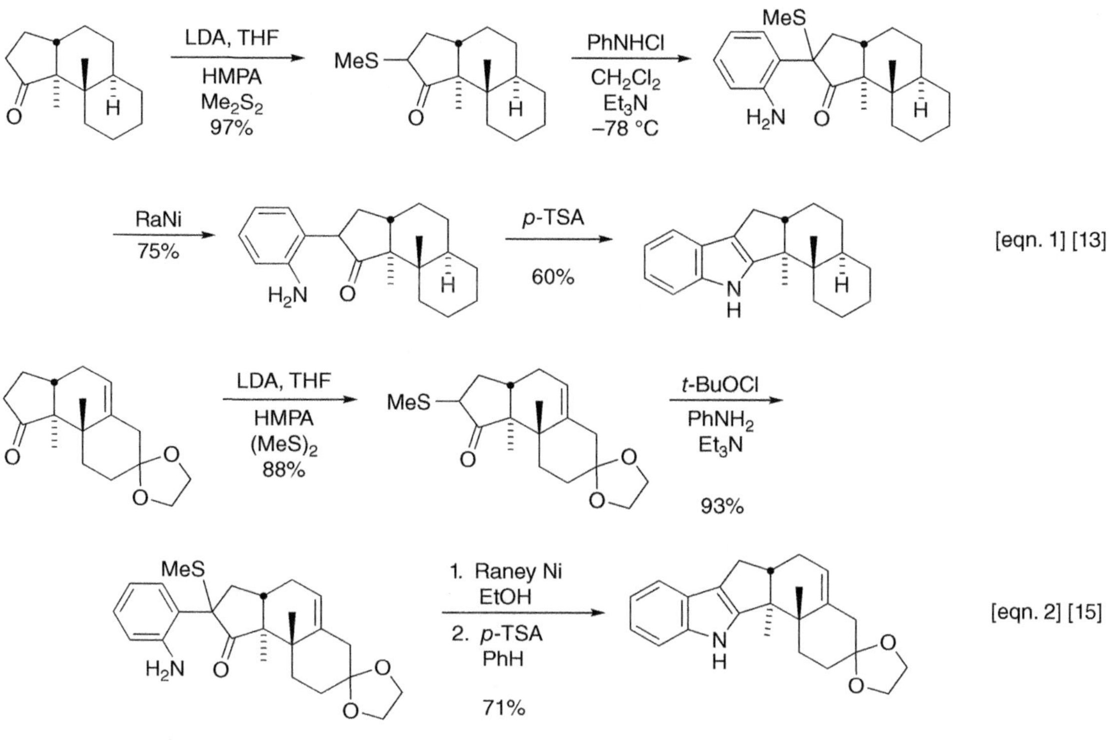

Scheme 2 *Smith's Application of the Gassman Indole Synthesis*

circumvent the use of *t*-butyl hypochloride (Entry 13) [6]. Akin to Entry 15, the parent 3*H*-benz[*e*]indole was obtained in low overall yield from 2-aminonaphthalene using methylthioacetaldehyde [7]. Gassman has published an *Organic Synthesis* procedure on his synthesis of ethyl 2-methylindole-5-carboxylate [9].

Although it has not acquired the popularity of the Fischer indolization, the Gassman indole synthesis has been employed with good success by several researchers. In their syntheses of the indole diterpene alkaloids (−)-paspaline, (+)-paspalicine, and (+)-paspalinine (not shown), Smith and coworkers employed the Gassman indole synthesis to great effect as summarized in Scheme 2 (equations 1 and 2) [12–15]. Inoue and colleagues adapted the Gassman protocol to prepare *N*-tosylindoles (Scheme 3, equation 1) [16, 17]. The conversion of **6** to indole **7** is thought to proceed by way of an episulfonium intermediate.

Van Vrankcn and colleagues synthesized 2,4-dimethyl-7-chloroindole from 2-chloro-5-methylaniline in 36% yield via a standard Gassman sequence [18]. Alper and Nguyen used the Murai variation of the Gassman procedure to prepare methyl 7-chloroindole-4-carboxylate (**11**) (Scheme 3, equation 2) [19]. Sulfuryl chloride was used to chlorinate the sulfide. Methoxide cleavage of pivalate ester **8** afforded the lactone **9** after trifluoromethyl acylation. Oxidation to unsaturated lactone **10** was followed by methanolysis to give indole **11**. A 2:1 mixture of 2-methyl-3-methylthio-6-trifluoromethoxyindole and 2-methyl-3-methylthio-4-trifluoromethoxyindole was obtained in 87% yield using a Gassman indole synthesis. Subsequent Raney nickel reduction and chromatography gave the desired 2-methyl-6-trifluoromethoxyindole in 49% yield [20, 21]. For a synthesis of an antibacterial indoloquinone, Ishikawa's group used the Gassman synthesis to prepare 4,5-difluoro-2-methylindole [22]. This particular indole, like many "Gassman indoles," was obtained in poor yield by Fischer indolization. Dobson and coworkers prepared 7-benzyoxyindole in 23% overall yield from 2-benzyloxyaniline using the standard Gassman conditions [23].

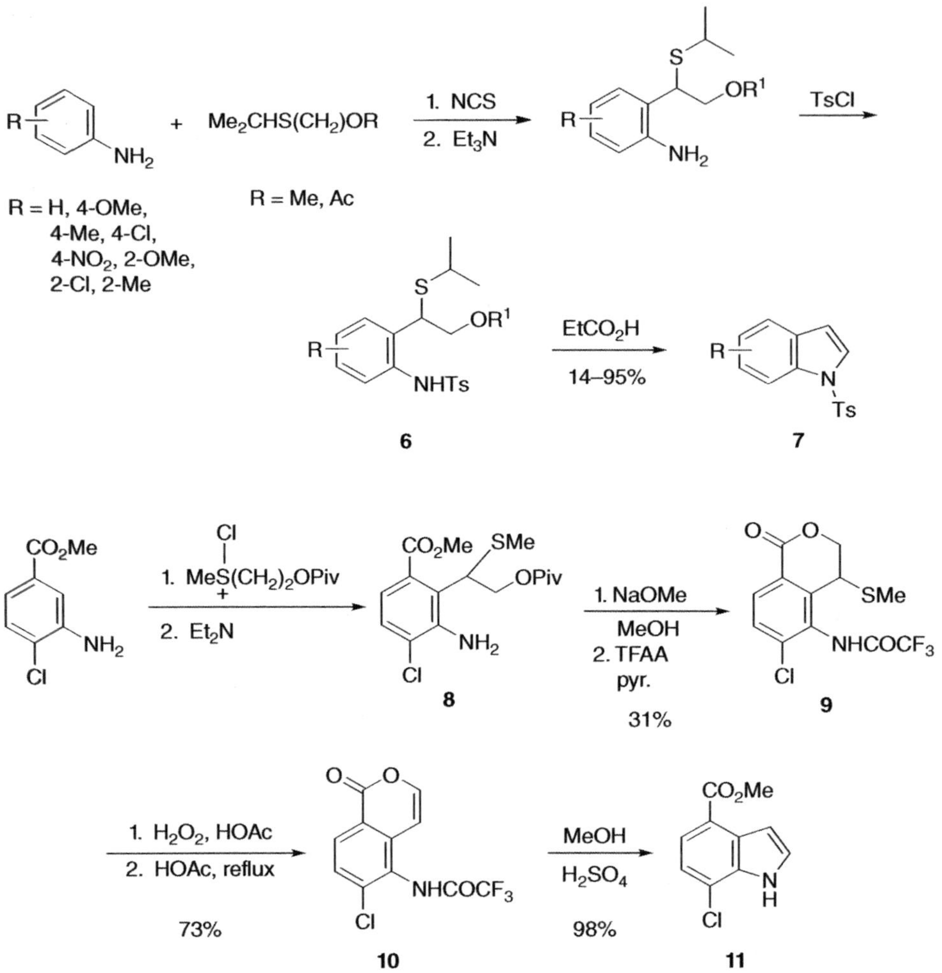

[eqn. 1] [17]

[eqn. 2] [19]

Scheme 3 *Applications of the Gassman Indole Synthesis*

References

[1] P.G. Gassman and T.J. van Bergen, *J. Am. Chem. Soc.*, 1973, **95**, 590–591.
[2] P.G. Gassman and T.J. van Bergen, *J. Am. Chem. Soc.*, 1973, **95**, 591–592.
[3] P.G. Gassman, G. Gruetzmacher, and T.J. van Bergen, *J. Am. Chem. Soc.*, 1973, **95**, 6508–6509.
[4] P.G. Gassman, D.P. Gilbert, and T.J. van Bergen, *J. Chem. Soc., Chem. Comm.*, 1974, 201–202.
[5] P.G. Gassman, T.J. van Bergen, D.P. Gilbert, and B.W. Cue, Jr., *J. Am. Chem. Soc.*, 1974, **96**, 5495–5508.
[6] P.G. Gassman, G. Gruetzmacher, and T.J. van Bergen, *J. Am. Chem. Soc.*, 1974, **96**, 5512–5517.
[7] P.G. Gassman and W.N. Schenk, *J. Org. Chem.*, 1977, **42**, 3240–3243.
[8] P.G. Gassman, J.J. Roos, and S.J. Lee, *J. Org. Chem.*, 1984, **49**, 717–718.
[9] P.G. Gassman and T.J. van Bergen, *Org. Synth. Coll. vol.* **6**, 1988, 601–605.
[10] P.G. Gassman and T.J. van Bergen, *J. Am. Chem. Soc.*, 1974, **96**, 5508–5512.
[11] F.G. Bordwell, private communication; F.G. Bordwell, *Acct. Chem. Res.*, 1988, **21**, 456–463.
[12] A.B. Smith, III and R. Mewshaw, *J. Am. Chem. Soc.*, 1985, **107**, 1769–1771.
[13] R.E. Mewshaw, M.D. Taylor, and A.B. Smith III, *J. Org. Chem.*, 1989, **54**, 3449–3462.
[14] A.B. Smith, T. Sunazuka, T.L. Leenay, and J. Kingery-Wood, *J. Am. Chem. Soc.*, 1990, **112**, 8197–8198.
[15] A.B. Smith, III, J. Kingery-Wood, T.L. Leenay, *et al.*, *J. Am. Chem. Soc.*, 1992, **114**, 1438–1449.
[16] Y. Murai, G. Masuda, S. Inoue, and K. Sato, *Heterocycles*, 1991, **32**, 1377–1386.
[17] Y. Murai, S. Kobayashi, S. Inoue, and K. Sato, *Heterocycles*, 1992, **34**, 1017–1029.
[18] S.J. Stachel, M. Nilges, and D.L. Van Vranken, *J. Org. Chem.*, 1997, **62**, 4756–4762.
[19] P.B. Alper and K.T. Nguyen, *J. Org. Chem.*, 2003, **68**, 2051–2053.
[20] P.E. Maligres, G.R. Humphrey, J.-F. Marcoux, *et al.*, *Org. Process Res. Dev.*, 2009, **13**, 525–534.
[21] J.J. Acton, III, T.E. Akiyama, C.H. Chang, *et al.*, *J. Med. Chem.*, 2009, **52**, 3846–3854.
[22] H. Ishikawa, T. Uno, H. Miyamoto, *et al.*, *Chem. Pharm. Bull.*, 1990, **38**, 2459–2462.
[23] D. Dobson, A. Todd, and J. Gilmore, *Synth. Commun.*, 1991, **21**, 611–617.

4

Bartoli Indole Synthesis

The Bartoli indole synthesis consists of the reaction of 2-substituted nitroarenes with a vinyl Grignard reagent to furnish an indole. This particular indole synthesis is an extension of the reaction of an unsubstituted nitroarene with Grignards to give conjugative ring alkylation, among other products previously discovered by Bartoli and his colleagues [1–3]. In 1989, Bartoli's group found that nitroarenes react with three equivalents of vinylmagnesium bromide to afford indoles after aqueous workup (Scheme 1) [4].

A summary of Bartoli's results is presented in Table 1. Scheme 1 shows a possible mechanism, which has been investigated and put forth by Bartoli [5]. The obvious limitation in this indole synthesis is that only *ortho*-substituted nitroarenes afford useful results. Thus, 3-chloronitrobenzene, 4-bromonitrobenzene, 4-chloronitrobenzene, and 2-nitronaphthalene give the corresponding indoles in only 19%, 12%, 17%, and 17%, respectively. Side products in this reaction are anilines and oxindoles. Bartoli and colleagues extended the chemistry to include the reaction of nitrosoarenes with 1-trimethylsilylmagnesium bromide to give the corresponding 2-trimethylsilylindoles (Table 1, entries 9–10) [6]. The reactions in Entry 10 reveal that these *ortho*-substituted nitrosoarenes are essential for reasonable yields. The use of *n*-dibutyl ether as solvent avoids the formation of azoxy- and azobenzenes, which do form in tetrahydrofuran and diethyl ether. A comprehensive review of this chemistry has been written by Dalpozzo

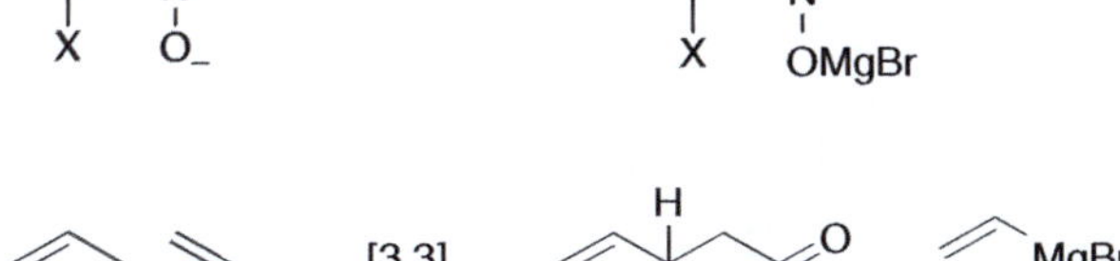

[eqn. 1]

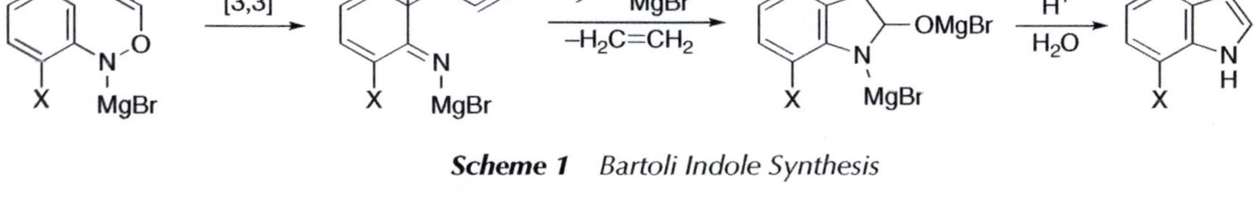

Scheme 1 *Bartoli Indole Synthesis*

Indole Ring Synthesis: From Natural Products to Drug Discovery, First Edition. Gordon W. Gribble.

Table 1 *Bartoli Indole Synthesis*

Entry	Substrate	Conditions	Indole	% Yield	Ref.
1	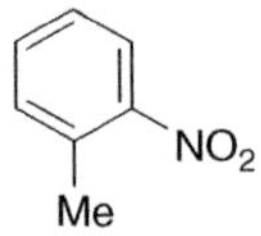	 MgBr THF, –40 °C	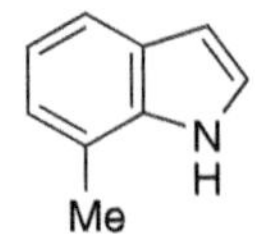	67%	4
2	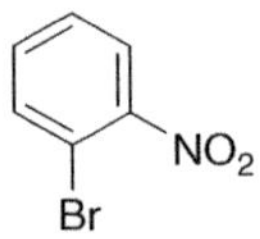	 MgBr THF, –40 °C	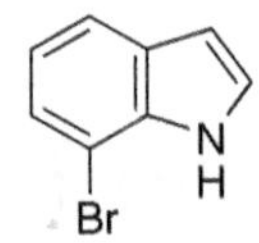	62%	4
3	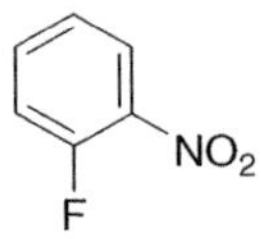	 MgBr THF, –40 °C	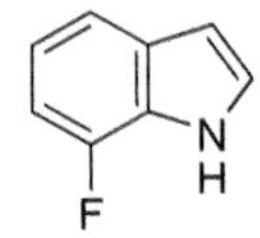	42%	4
4	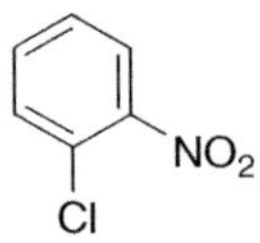	 MgBr THF, –40 °C	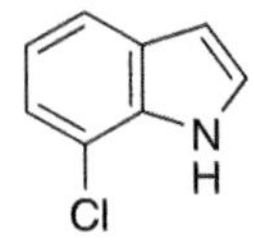	63%	4
5	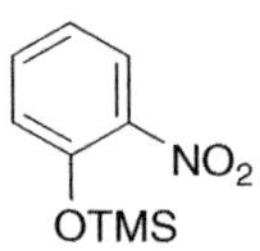	 MgBr THF, –40 °C	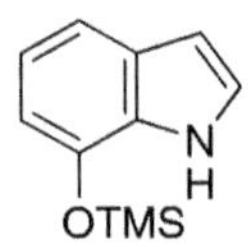	41%	4
6	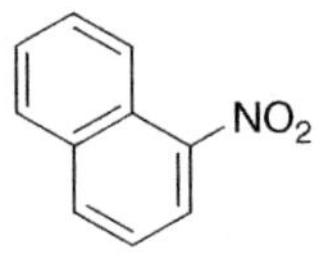	 MgBr THF, –40 °C	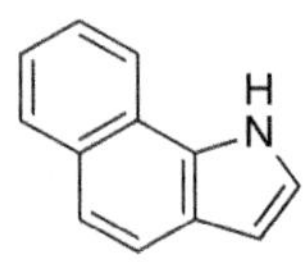	54%	4
7	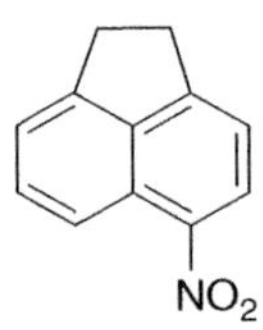	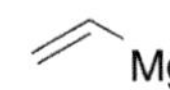 MgBr THF, –40 °C	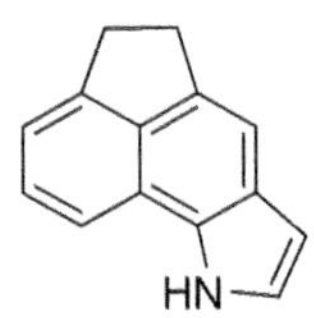	59%	4
8	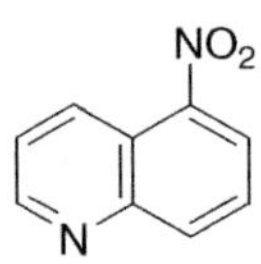	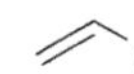 MgBr THF, –40 °C	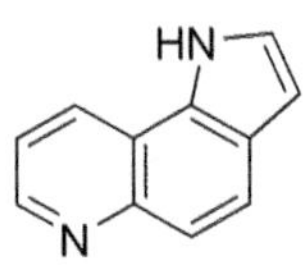	42%	4
9	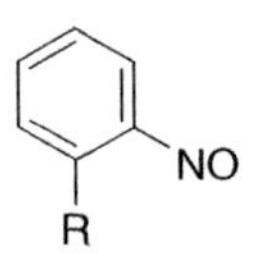 R = Me, Cl, Br, Et	1. 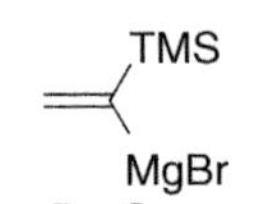 n-Bu_2O –50 °C to 0 °C 2. aq. NH_4Cl	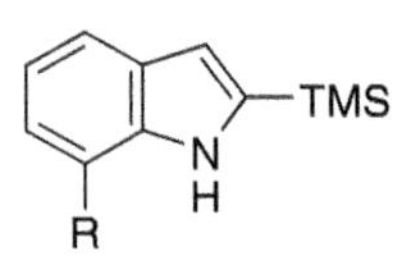	39–50%	6
10	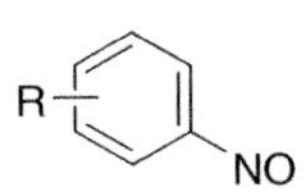 R = 3-Me, 4-Me, 3-Cl, 4-Cl	1. 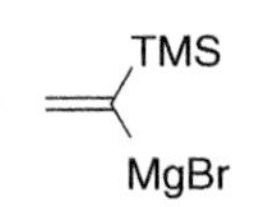 n-Bu_2O –50 °C to 0 °C 2. aq. NH_4Cl	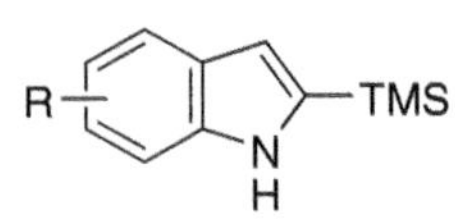	11–35%	6

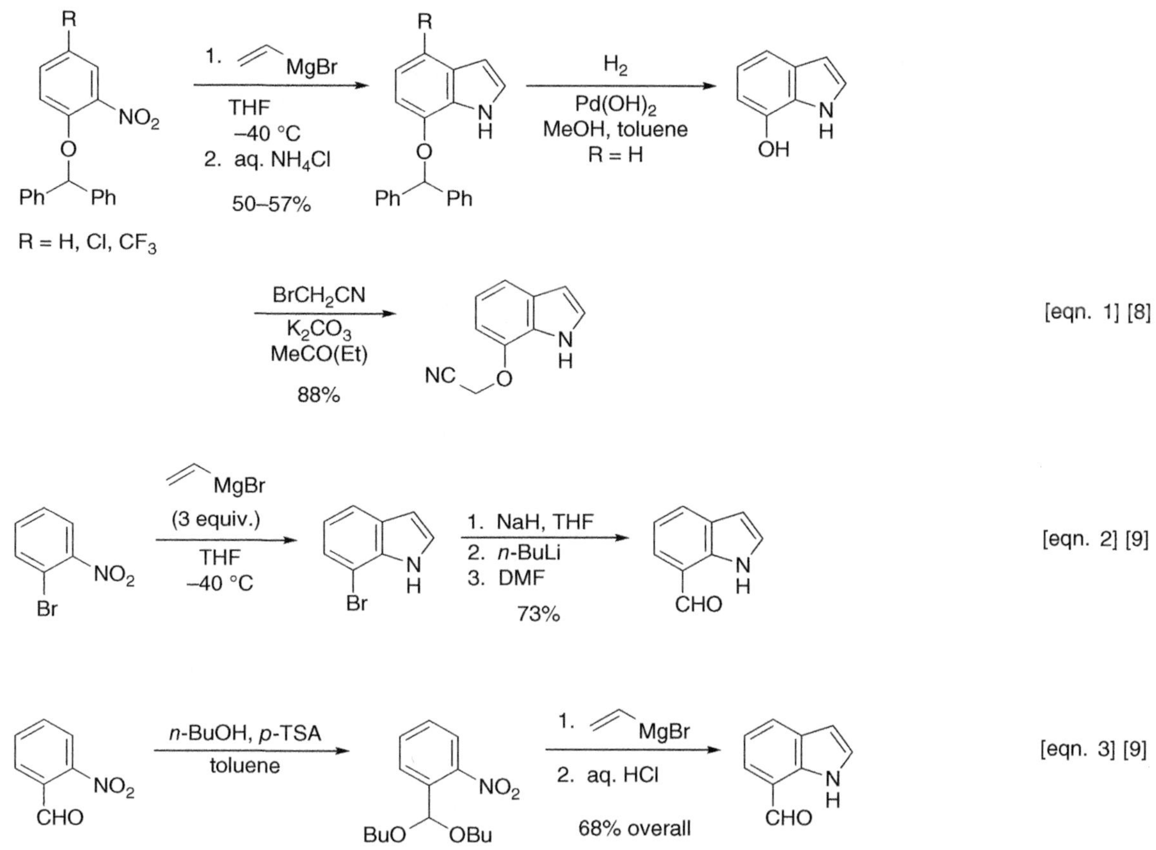

Scheme 2 *Gilmore Application of the Bartoli Indole Synthesis*

and Bartoli [7]. Gilmore and coworkers extended the Bartoli reaction to an efficient synthesis of 7-benzhydryloxyindoles (Scheme 2, equation 1) [8]. The benzhydryl protecting group is superior to benzyl, trityl, and 9-anthracenemethyl. The crude (unstable) 7-hydroxyindole was capped with bromoacetonitrile in good overall yield. This research team also used the Bartoli method to synthesize 7-formylindole (equations 2, 3) [9].

Several groups have used the Bartoli preparation of 7-bromoindole as the starting point in the synthesis of, for example, 7-arylindoles [10], *N*-allyl and *N*-benzyl radical cyclization substrates [11], the pyrrolophenanthridone alkaloid hippadine [12], a series of *N*-(7-indolyl)benzenesulfonamides [13], 7-bromo-D-tryptophan [14], analogues of the proteasome inhibitor TMC-95A [15], and 7-pyridylindoles [16]. Interestingly, the Bartoli indole synthesis is the starting point for the synthesis of the trikentrin indole alkaloids by four separate groups [17–20]. These studies and other applications of the Bartoli methodology are listed in Table 2 [17–23]. Entry 5 reiterates the result found by Bartoli and other workers [7] that *ortho*-alkoxy nitrobenzene behaves badly in this reaction, especially with smaller alkoxy groups. Dobbs reexamined and extended the Bartoli reaction to other nitro aromatics as shown in Entries 6–20 [22, 23]. Under identical conditions, vinylmagnesium chloride gives ~15% lower yield of 7-methylindole (Entries 7 vs. 6). Both 3- and 4-trifluoronitrobenzene give no indole product under the usual conditions, but rather a milieu of other products (azo, azoxy, nitroso, and amine compounds). The yields of indole compounds from 3-bromo- and 3-trimethyl-4-methylnitrobenzene are 34% and 36%, respectively. Dobbs utilized the products in Entries 15 and 17 to synthesize three naturally occurring indoles isolated from the European fungi *Tricholoma virgatum* and *T. sciodes* [23] (2,4-dimethylindole, 4-(hydroxymethyl)-2-methylindole, and 4-(methoxymethyl)-2-methylindole) by solvolysis/methanolysis of the C-4 benzylic bromine and tri-*n*-butyltin debromination of the C-7 bromine.

Following these early investigations on the Bartoli indole synthesis, several research groups made use of this reaction to great effect as a starting point in natural product synthesis despite the inherent low to modest yields. Stolz employed the Bartoli synthesis to prepare 7-benzyloxy-4-bromoindole (33% yield from 2-benzyloxy-5-bromonitrobenzene and vinylmagnesium bromide) in the first total synthesis of the marine bis-indole (±)-dragmacidin D [24].

Table 2 *Application of the Bartoli Indole Synthesis [a]*

Entry	Substrate	Conditions	Indole	% Yield	Ref.
1	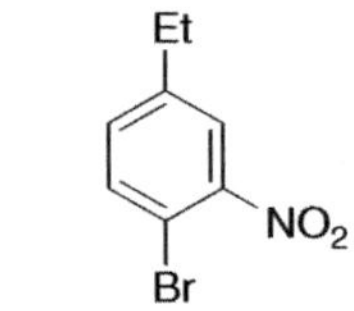		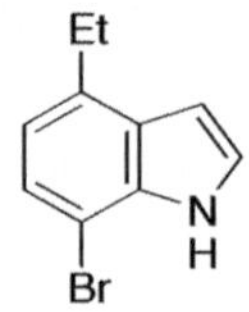	47%, 53%	17, 18
2	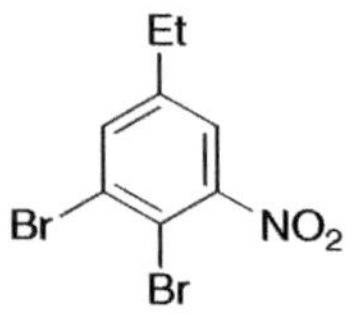		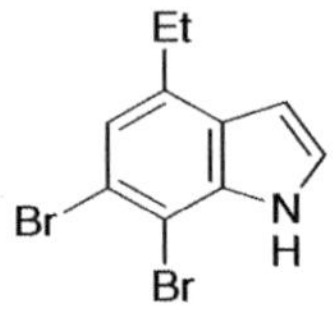	52%	19
3	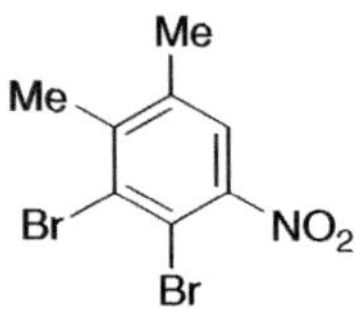	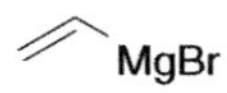	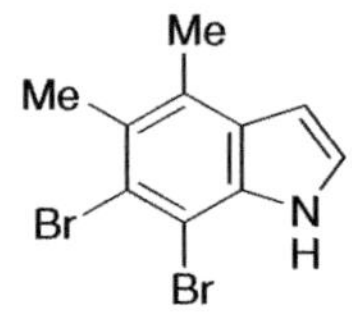	36%	19
4	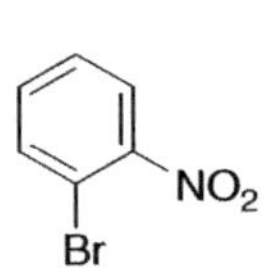	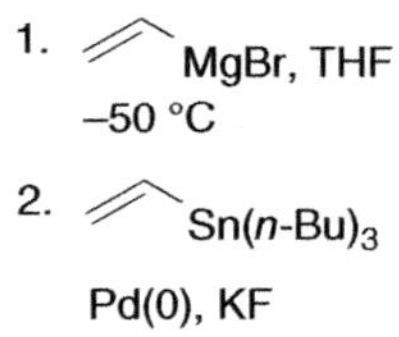	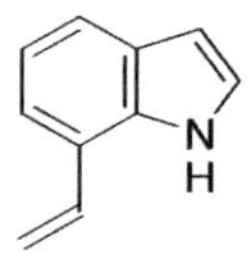	42%	20
5	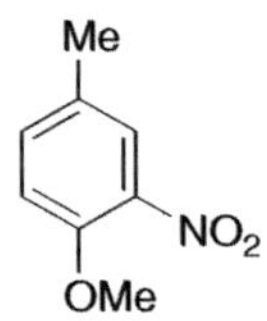	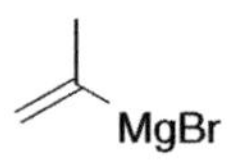	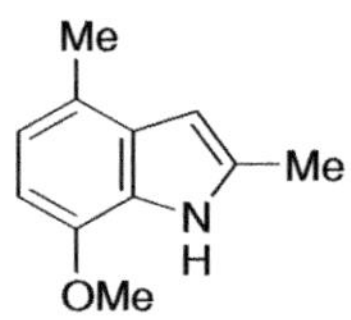	5%	21
6	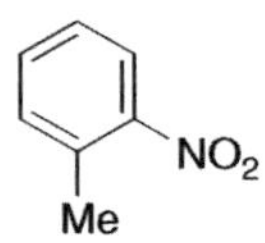		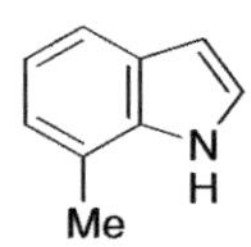	71%	22, 23
7	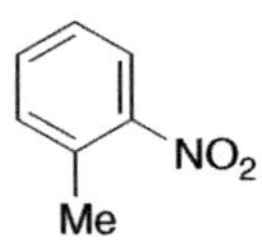	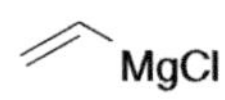	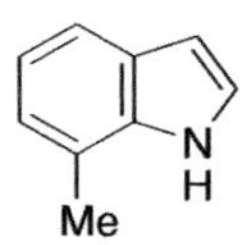	58%	22
8	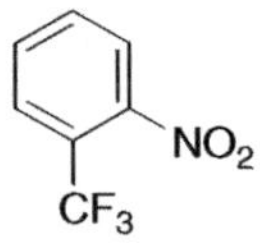	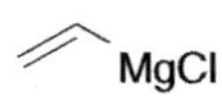	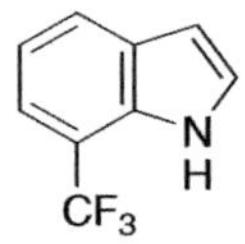	56%	22
9	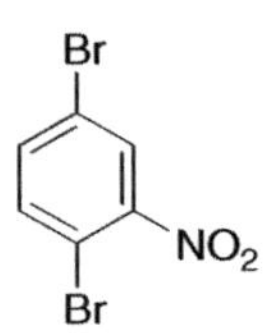	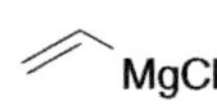	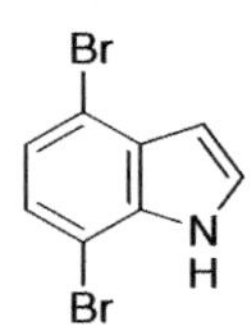	69%	22
10	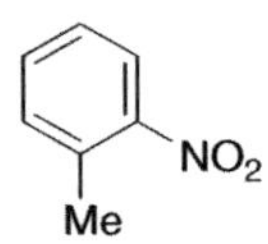	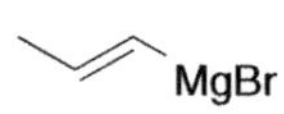	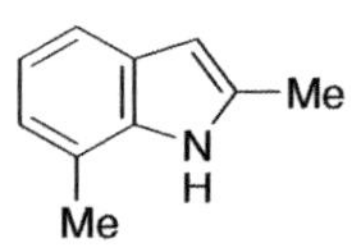	59%	22

Table 2 (continued)

Entry	Substrate	Conditions	Indole	% Yield	Ref.
11	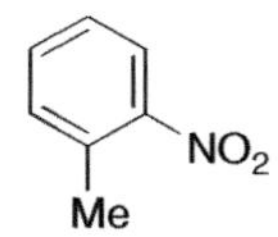	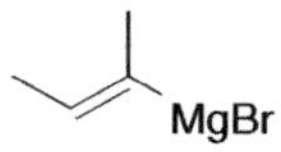	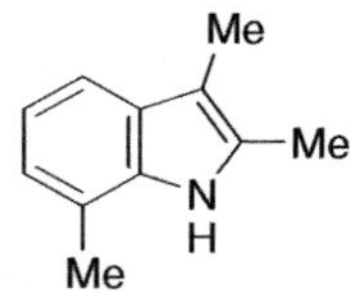	73%	22
12	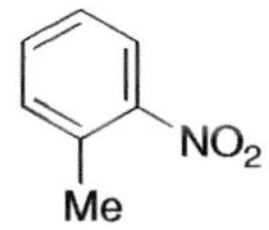	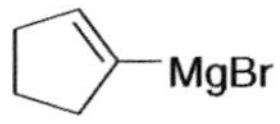	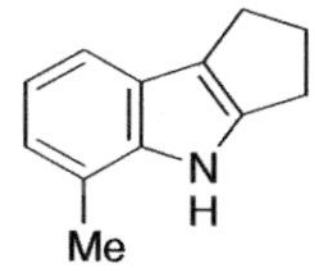	49%	22
13	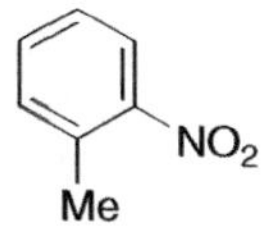	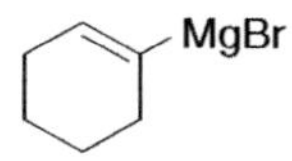	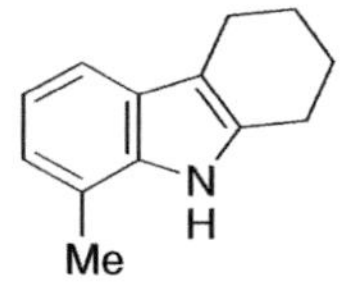	47%	22, 23
14	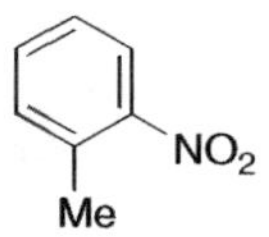	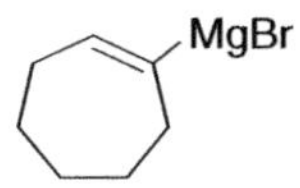	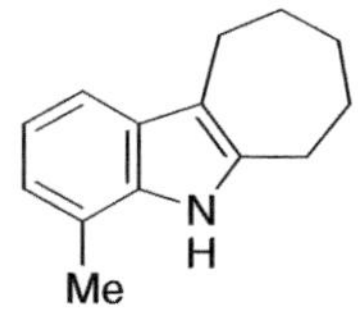	41%	22
15	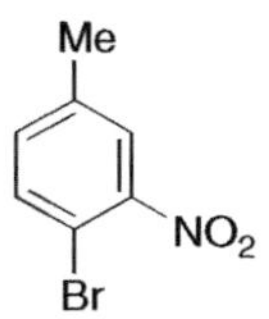	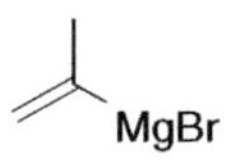	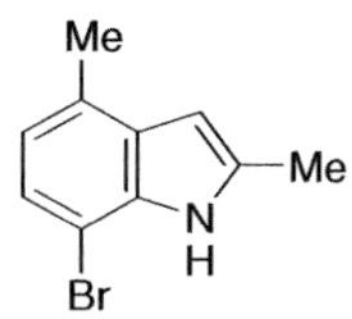	67%	23
16	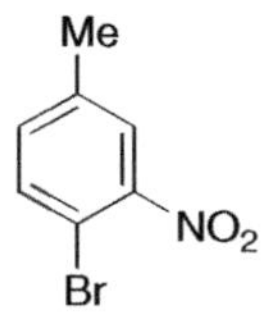		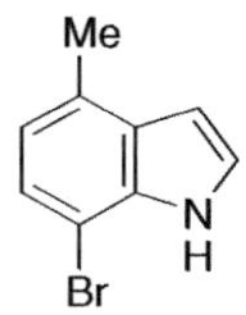	59%	23
17	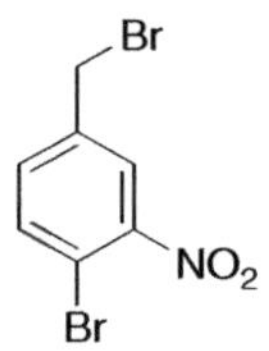	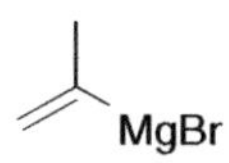	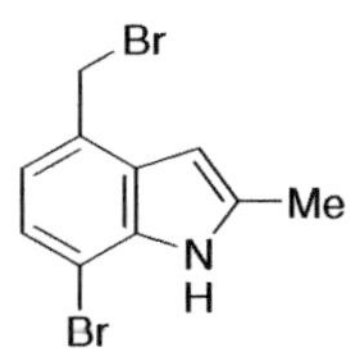	61%	23
18	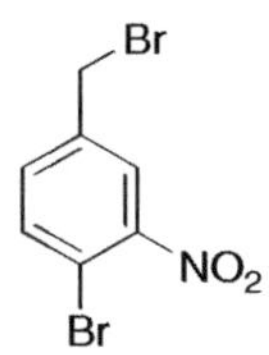		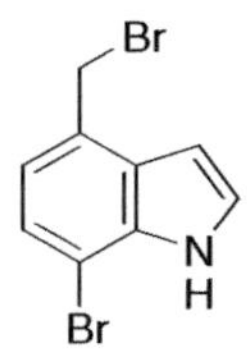	48%	23
19	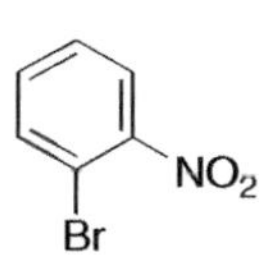	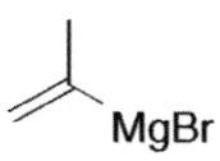	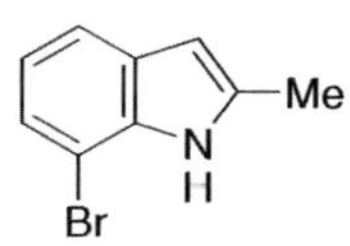	62%	23
20	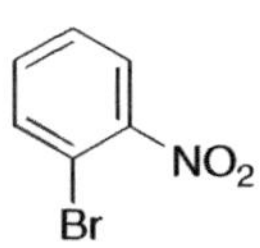	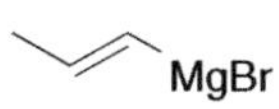	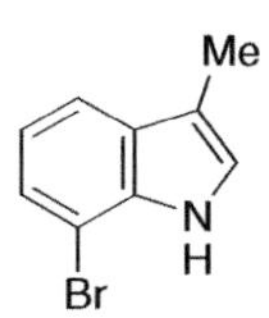	63%	23

[a] Unless otherwise indicated, all reactions were run at –40 °C in THF with three equivalents of Grignard reagent and quenched by the addition of aqueous ammonium chloride.

Table 3 *Applications of the Bartoli Indole Synthesis to Medicinal Agents*

Entry	Substrate	Conditions	Indole	% Yield	Target	Ref.
1	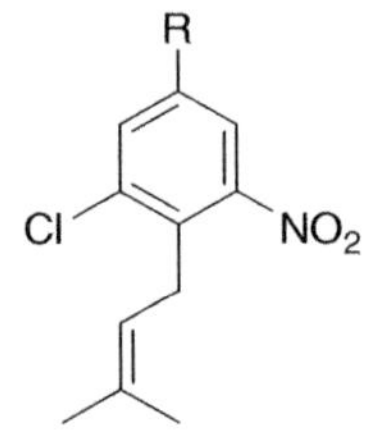	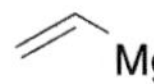 MgBr THF, –48 °C R = H, Cl	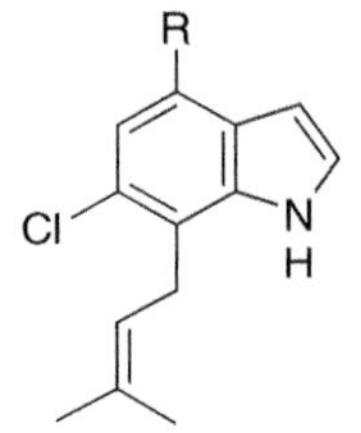	38–39%	Cdc25 phosphatase	27
2	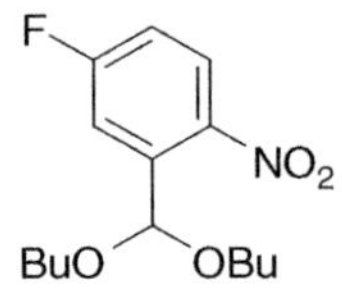	1. 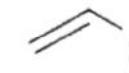 MgBr THF, –40 °C 2. aq. HCl	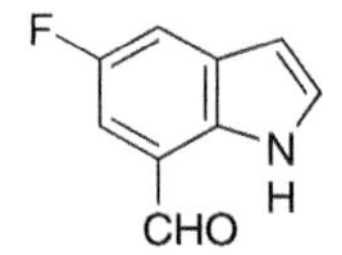	55%	Glycogen Synthase Kinase-3	28, 34
3	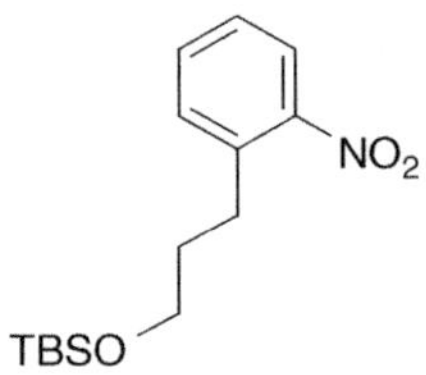	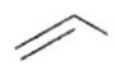 MgBr THF, –45 °C	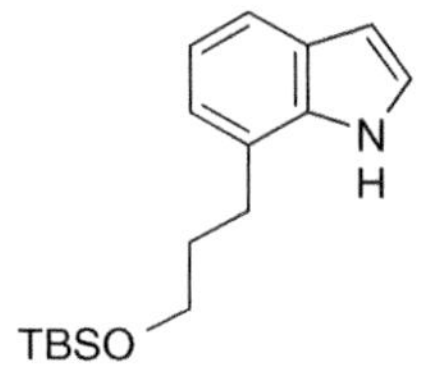	56%	Cyclin D1/CDK4	29
4	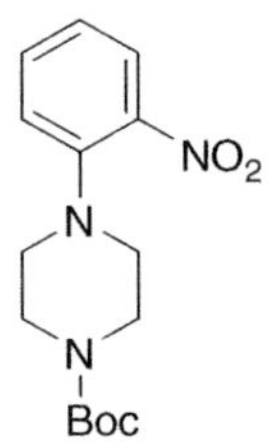	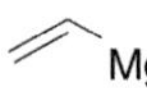 MgBr THF, –40 °C	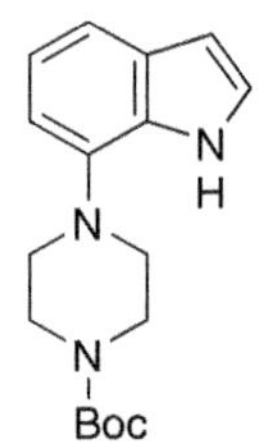	11%	5-HT$_6$ receptor	30
5	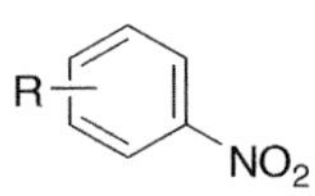 R = 2,4-Cl$_2$, 2,3-Cl$_2$, 2-Br, 2-Cl, 2-Cl-4-OMe	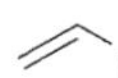 MgBr THF, –40 °C	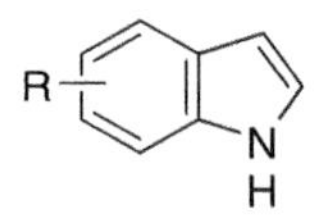	40–65%	Necrosis	31
6	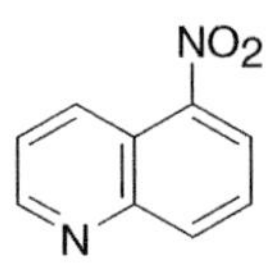	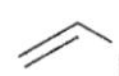 MgBr THF, –40 °C	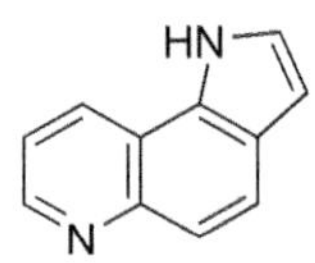	—	5-HT$_{2c}$ receptor	32
7	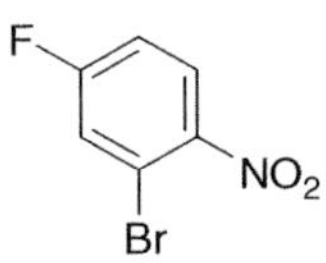	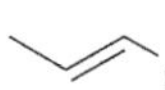 MgBr THF, –50 °C	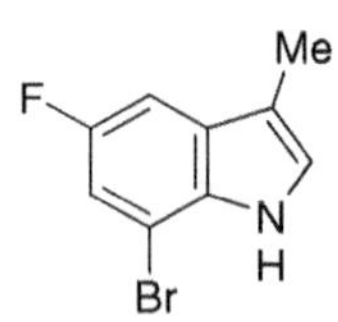	42%	EP$_3$ receptor	33

Table 4 *Applications of the Bartoli Indole Synthesis to Functionalized Indoles*

Entry	Substrate	Conditions	Indole	% Yield	Ref.
1	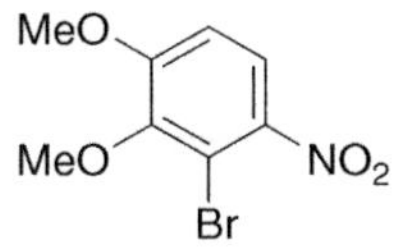	 THF, –40 °C	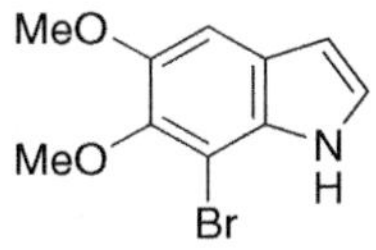	66%	35
2	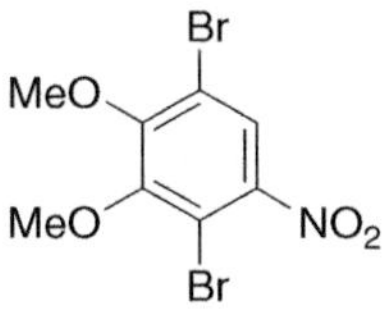	 THF, –40 °C	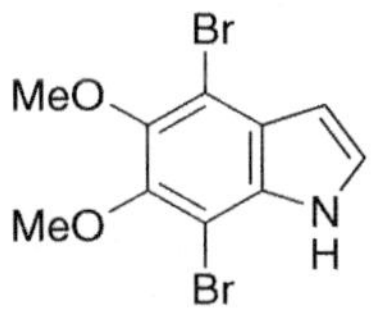	38%	35
3	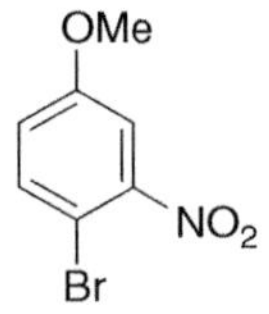	 THF <5 °C	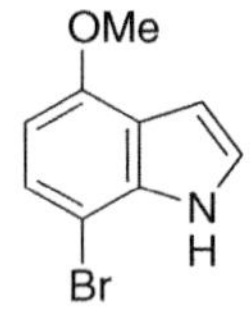	41%	36
4	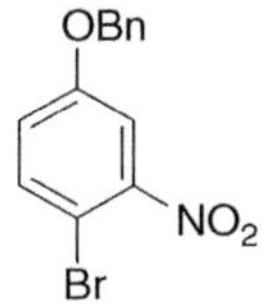	 THF <5 °C	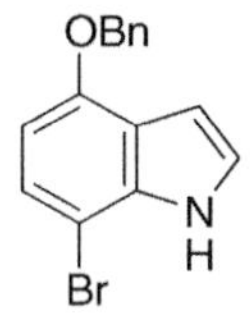	37%	36
5	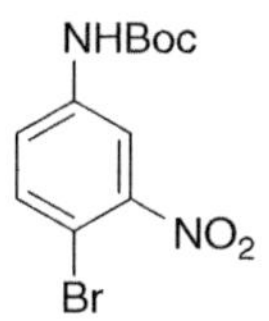	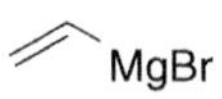 THF/Et_2O –40 °C	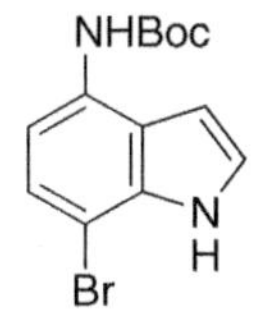	64%	37
6	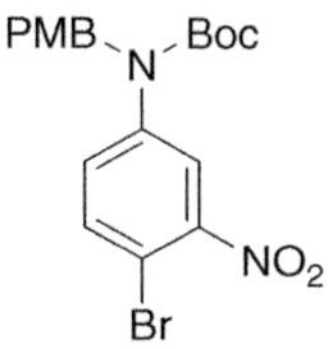	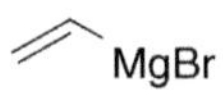 THF/Et_2O –40 °C	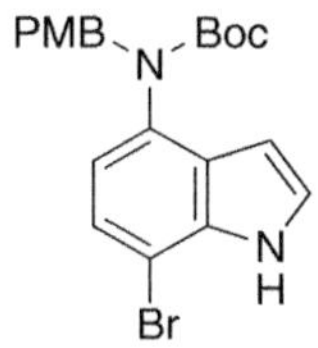	47%	37
7	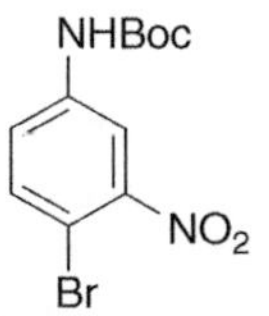	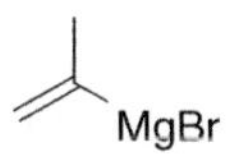 THF–Et_2O –40 °C	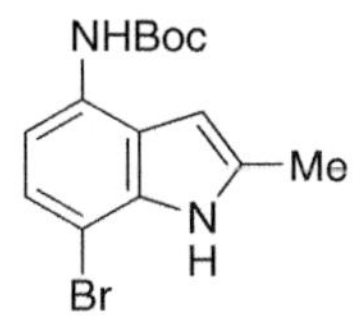	44%	37
8	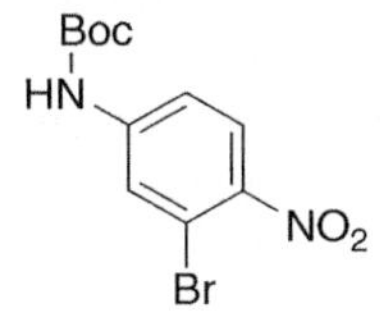	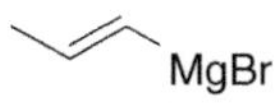 THF–Et_2O –40 °C	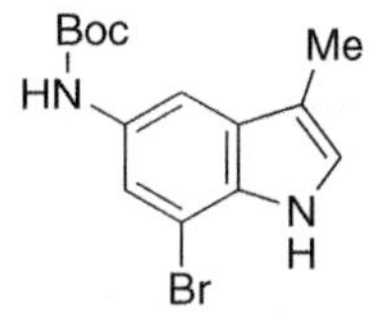	33%	37
9	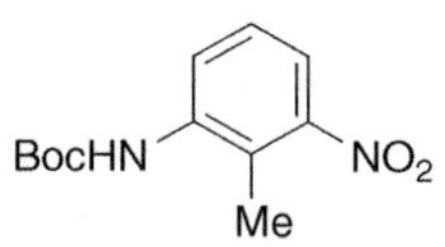	 THF–Et_2O –40 °C	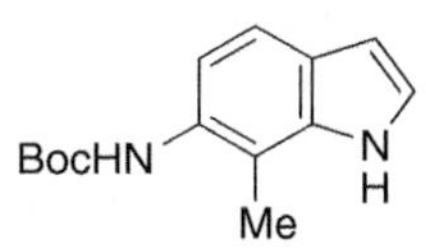	27%	37

(continued overleaf)

Table 4 *(continued)*

Entry	Substrate	Conditions	Indole	% Yield	Ref.
10	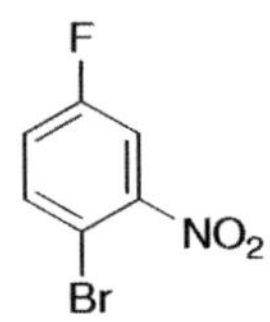	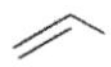 MgBr THF, –40 °C	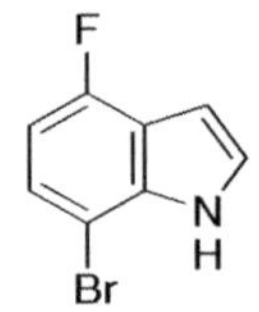	52%	38
11	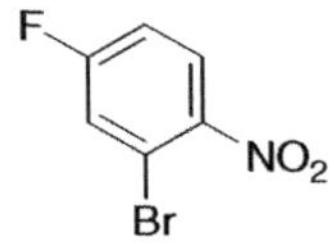	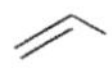 MgBr THF, –40 °C	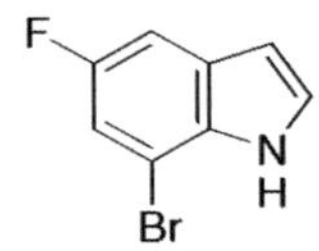	57%	38
12	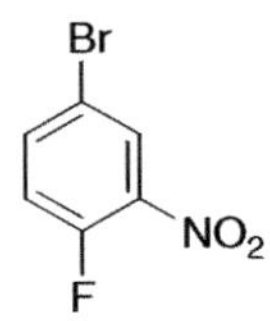	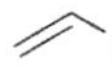 MgBr THF, –40 °C	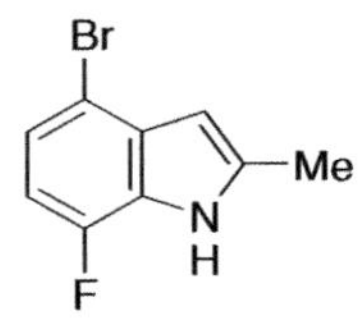	34%	38
13	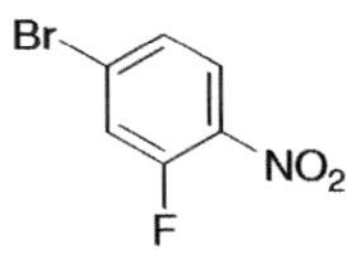	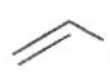 MgBr THF, –40 °C	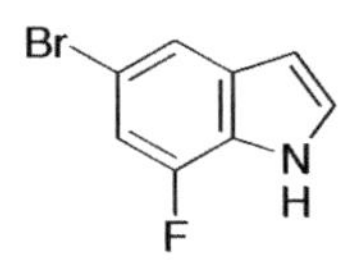	36%	38
14	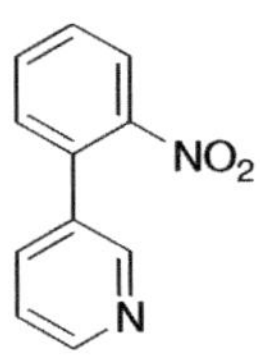	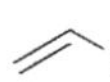 MgBr THF, –40 °C	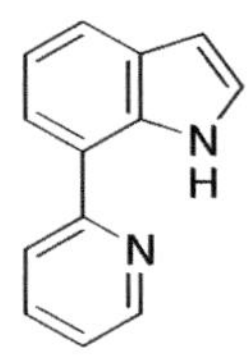	58%	16
15	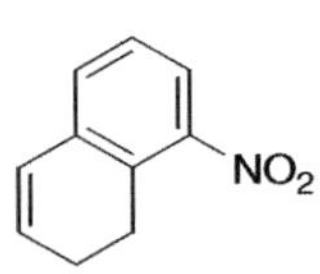	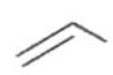 MgBr THF, –45 °C	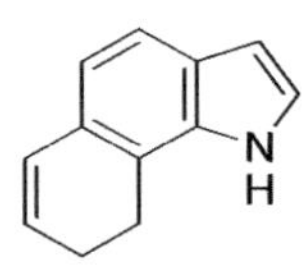	56%	41

Barrett synthesized the anthelmintic pyrrolobenzoxazine terpenoid CJ-12662 that began with a Bartoli synthesis of 7-chloro-3-methylindole (53% yield from 2-chloronitrobenzene and 2-propenylmagnesium bromide) [25]. Sperry prepared 7-benzyloxy-5-bromoindole from 2-benzyloxy-4-bromonitrobenzene and vinylmagnesium bromide (31% yield) in an approach to *hemi*-dendridine A acetate [26]. Sperry and Boyd synthesized this alkaloid using a Fischer indolization (cf. Chapter 2). Table 3 summarizes other uses of the Bartoli indole synthesis, particularly those applied to medicinally active compounds [27–29]. These examples illustrate the great versatility of the Bartoli reaction, even given the modest yields. The example in Entry 4 was also applied to the corresponding nitropyridines to afford azaindoles [30].

The Bartoli indole synthesis has found use in the preparation of highly functionalized indoles, including multi-oxygenated, 3,4,7-trisubstituted, amino-substituted, and fluorinated. A selection of these reactions is shown in Table 4 [31–38]. Entries 5–8 feature the use of diethyl ether as a cosolvent with slight improvement in reaction yield. This cosolvent seems to increase the solubility of the Grignard reagent at low temperature [37]. Nevertheless, several combinations of Grignard reagent and nitrobenzene substrate in some cases failed to afford any indole; for example, the *N*-Boc nitrobenzene in Entry 7 gave no desired product upon reaction with 1-methyl-1-propenylmagnesium bromide and with 2-methyl-2-butenylmagnesium bromide. Likewise, 2-bromo-3-*N*-Boc nitrobenzene failed to afford indole products with any of the four

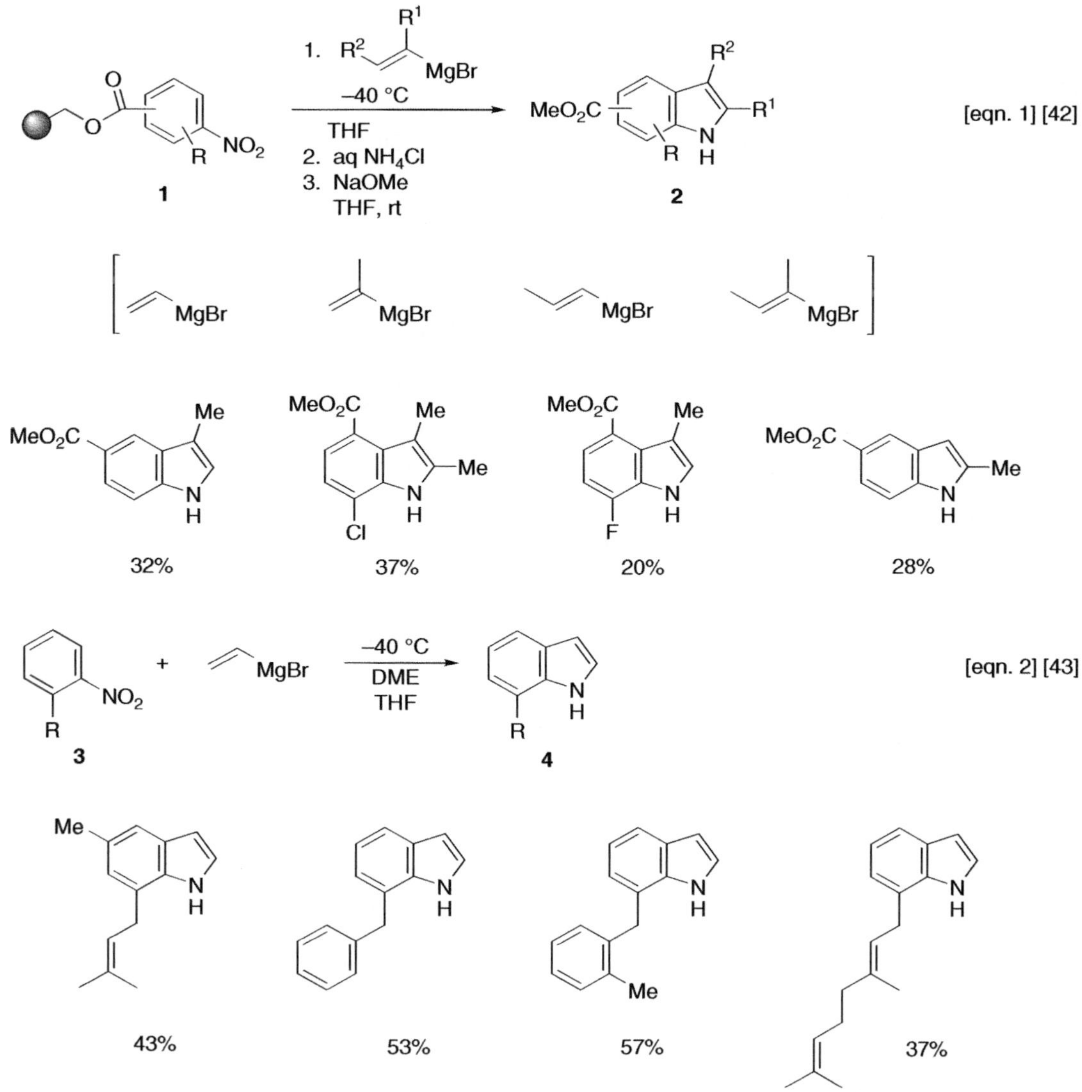

Scheme 3 *Knepper [42] and Pirrung [43] Applications of the Bartoli Indole Synthesis*

vinylmagnesium bromides examined [37]. Murugan and colleagues prepared 4,7-diarylindole iodide-fluorescent sensors using 4,7-dibromoindole prepared in a Bartoli synthesis [39], and this latter indole was used by RajanBabu and coworkers to synthesize enantiopure polycyclic diketopiperazines [40]. As mentioned earlier, 7-(3′-pyridyl)indole was prepared using a Bartoli procedure (Entry 14) [16], and 8,9-dihydro-1*H*-benz[*g*]indole was synthesized from the requisite 1,2-dihydronaphthalene (Entry 15) [41].

As illustrated in Scheme 3 (equation 1), Bräse and Knepper adapted the Bartoli indole synthesis to the solid state [42]. These workers used four vinylmagnesium bromides in the conversion of nitroarene Merrifield resin **1** to more than a dozen carbomethoxyindoles **2**. As we have already seen in the Bartoli reaction, the main products are the corresponding reduced anilines. Pirrung and colleagues described a modification of the Bartoli indolization that entails a six-fold excess of vinylmagnesium bromide in a solvent mixture of THF and dimethoxyethane (DME) and the addition of the nitrobenzene **3** to the Grignard. This method was used to prepare 7-alkylindoles **4** (equation 2) [43].

Even though the yields of indoles from the Bartoli synthesis are usually below 50%–60%, and *ortho*-substitution to the nitro group is mandatory, this methodology is extraordinarily powerful for constructing indoles with substitution in the benzene ring.

References

[1] G. Bartoli and G. Rosini, *Synthesis*, 1976, 270–271.
[2] G. Bartoli, *Acc. Chem. Res.*, 1984, **17**, 109–115.
[3] G. Bartoli, M. Bosco, and R. Dalpozzo, *Tetrahedron Lett.*, 1985, **26**, 115–118.
[4] G. Bartoli, G. Palmieri, M. Bosco, and R. Dalpozzo, *Tetrahedron Lett.*, 1989, **30**, 2129– 2132.
[5] M. Bosco, R. Dalpozzo, G. Bartoli, *et al.*, *J. Chem. Soc., Perkin Trans. 2*, 1991, 657–663.
[6] G. Bartoli, M. Bosco, R. Dalpozzo, *et al.*, *J. Chem. Soc., Perkin Trans. 1*, 1991, 2757–2761.
[7] R. Dalpozzo and G. Bartoli, *Curr. Org. Chem.*, 2005, **9**, 163–178.
[8] D. Dobson, A. Todd, and J. Gilmore, *Synth. Commun.*, 1991, **21**, 611–617.
[9] D.R. Dobson, J. Gilmore, and D.A. Long, *Synlett*, 1992, 79–80.
[10] G.M. Carrera, Jr. and G.S. Sheppard, *Synlett*, 1994, 93–94.
[11] A.P. Dobbs, K. Jones, and K.T. Veal, *Tetrahedron Lett.*, 1997, **38**, 5379–5382.
[12] D.C. Harrowven, D. Lai, and M.C. Lucas, *Synthesis*, 1999, 1300–1302.
[13] T. Owa, T. Okauchi, K. Yoshimatsu, *et al.*, *Bioorg. Med. Chem. Lett.*, 2000, **10**, 1223–1226.
[14] Y. Konda-Yamada, C. Okada, K. Yoshida, *et al.*, *Tetrahedron*, 2002, **58**, 7851–7861.
[15] A. Berthelot, S. Piguel, G. Le Dour, and J. Vidal, *J. Org. Chem.*, 2003, **68**, 9835–9838.
[16] M.S. Mudadu, A. Singh, and R.P. Thummel, *J. Org. Chem.*, 2006, **71**, 7611–7617.
[17] P. Wiedenau, B. Monse, and S. Blechert, *Tetrahedron*, 1995, **51**, 1167–1176.
[18] L.F. Silva, Jr. and M.V. Craveiro, *Org. Lett.*, 2008, **10**, 5417–5420.
[19] K.R. Buszek, N. Brown, and D. Luo, *Org. Lett.*, 2009, **11**, 201–204.
[20] W. Liu, H.J. Lim, and T.V. RajaBabu, *J. Am. Chem. Soc.*, 2012, **134**, 5496–5499.
[21] S.J. Stachel, M. Nilges, and D.L. Van Vranken, *J. Org. Chem.*, 1997, **62**, 4756–4762.
[22] A.P. Dobbs, M. Voyle, and N. Whittall, *Synlett*, 1999, 1594–1596.
[23] A. Dobbs, *J. Org. Chem.*, 2001, **66**, 638–641.
[24] N.K. Garg, R. Sarpong, and B.M. Stoltz, *J. Am. Chem. Soc.*, 2002, **124**, 13179–13184.
[25] C. Didier, D.J. Critcher, N.D. Walshe, *et al.*, *J. Org. Chem.*, 2004, **69**, 7875–7879.
[26] E.M. Boyd and J. Sperry, *Tetrahedron Lett.*, 2012, **53**, 3623–3626.
[27] J. Sohn, B. Kiburz, Z. Li, *et al.*, *J. Med. Chem.*, 2003, **46**, 2580–2588.
[28] T.A. Engler, J.R. Henry, S. Malhotra, *et al.*, *J. Med. Chem.*, 2004, **47**, 3934–3937.
[29] M.M. Faul, T.A. Engler, K.A. Sullivan, *et al.*, *J. Org. Chem.*, 2004, **69**, 2967–2975.
[30] M. Ahmed, M.A. Briggs, S.M. Bromidge, *et al.*, *Bioorg. Med. Chem. Lett.*, 2005, **15**, 4867–4871.
[31] X. Teng, A. Degterev, P. Jagtap, *et al.*, *Bioorg. Med. Chem. Lett.*, 2005, **15**, 5039–5044.
[32] D.R. Adams, J.M. Bentley, K.R. Benwell, *et al.*, *Bioorg. Med. Chem. Lett.*, 2006, **16**, 677–680.
[33] S. Zegar, C. Tokar, L.A. Enache, *et al.*, *Org. Process Res. Dev.*, 2007, 11, 747–753.
[34] N.A. Magnus, C.P. Ley, P.M. Pollock, and J.P. Wepsiec, *Org. Lett.*, 2010, **12**, 3700–3703.
[35] P.B. Huleatt, S.S. Choo, S. Chua, C.L.L. Chai, *Tetrahedron lett.*, 2008, 49, 5309–5311.
[36] S.W. Grant, T.F. Gallagher, M.A. Bobko, *et al.*, *Tetrahedron Lett.*, 2011, **52**, 3376–3378.
[37] L. Wylie, P. Innocenti, D.K. Whelligan, and S. Hoelder, *Org. Biomol. Chem.*, 2012, **10**, 4441–4447.
[38] M. Schlosser, A. Ginanneschi, and F. Leroux, *Eur. J. Org. Chem.*, 2006, 2956–2969.
[39] K.R. Rathikrishnan, V.K. Indirapriyadharshini, S. Ramakrishna, and R. Murugan, *Tetrahedron*, 2011, **67**, 4025–4030.
[40] H.J. Lim, J.C. Gallucci, and T.V. RajanBabu, *Org. Lett.*, 2010, **12**, 2162–2165.
[41] L.F. Silva, Jr., M.V. Craveiro, and M.T.P. Gambardella, *Synthesis*, 2007, 3851–3857.
[42] K. Knepper and S. Bräse, *Org. Lett.*, 2003, **5**, 2829–2832.
[43] M.C. Pirrung, M. Wedel, and Y. Zhao, *Synlett*, 2002, 143–145.

5

Thyagarajan Indole Synthesis

Our fourth example of an indole ring synthesis that features a sigmatropic rearrangement as the key step was discovered in 1974 by Thyagarajan [1–3] and later expanded by Majumdar and extended to other heterocycles. Makisumi and Takada uncovered this same rearrangement independently [4]. The initial discoveries by these two groups are shown in Schemes 1 and 2, respectively. Both results reveal that the protonated amine oxide **3** is stable, but when the neutral amine oxides **4** and **8** are unleashed, they undergo rapid transformation to **5** and **9**, respectively, but sequential 2,3]- and [3,3]-sigmatropic rearrangements (Scheme 3). Although 3-methylene indolines **5** and **9** can be isolated, they are both quite labile and are smoothly converted to indoles **6** and **10**, respectively. A mechanism for this interesting chemistry is shown in Scheme 3. For the formation of indole **2** the nucleophile is *meta*-chlorobenzoate.

A summary of the applications of this rearrangement to indole synthesis by Makisumi and Takada is illustrated in Scheme 4. In the case of reactions involving cyanide (equation 3), when R=H, the 2-cyano-3-methylindoles are also isolated in 7%–9% yield resulting from nucleophilic attack at the C-2 indolenium ion formed by loss of hydroxide from **9**. Likewise, the extensions of Thyagarajan's initial result to the synthesis of other indoles are shown in Scheme 5.

Majumdar and his co-workers have continued the work begun by Thyagarajan and some of these results are shown in Scheme 6. In each case the amine oxide depicted undergoes sequential [2,3]- and [3,3]-sigmatropic rearrangements followed by nucleophilic ambush to give an indole (equation 1). In the absence of a nucleophile such as cyanide, the dimeric product **13** is formed (equation 1) [5, 6]. The amine oxide **14** affords the ten-membered ring cyclic bis-ether **15** under the usual conditions, by the suggested pathway shown. No evidence for the formation of the expected products **16** (5-*endo*-trig) and **17** was found [7, 8]. Several ring substituted amine oxides afford the same bis-ether **15** (4-Me, 4-Br, 4-Cl, 2-Me) [8].

Additional examples of this fascinating chemistry are listed in Table 1 [9, 10]. For Entries 1–3, the initially formed *m*-chlorobenzoate product is methanolized to the methyl ether (not shown). Entry 3 represents an attractive synthesis of pyrrolo[3,2-*d*]pyrimidines [11]. Water is the nucleophile in Entry 4, not *m*-chlorobenzoate, in a sequence that provides pyrrolo[3,2-*c*]coumarins [12]. Entry 5 describes the preparation of pyrrolo[3,2-*c*] [1] benzothiopyran-4-ones, which is a new ring system and where water has captured the intermediate 3-methylene-2-hydroxy indoline [13]. Entry 6 features the synthesis of pyrrolo[3,2-*c*]pyrones [14]. Entry 7 describes a similar rearrangement of *N*-alkyl-*N*-allenylmethylanilines with magnesium monoperoxyphthalate (MMPP) to afford 2-vinylindoles [15]. This reaction presumably follows the one described earlier, although none of the presumed intermediates could be isolated.

Two reviews written by Majumdar are available that cover the aforementioned chemistry and other advances in the aromatic *ortho*-Claisen and aza-Claisen rearrangements [16, 17]. What characterizes the Thyagarajan indole syntheses are the very mild conditions (spontaneous) and the excellent yields.

Indole Ring Synthesis: From Natural Products to Drug Discovery, First Edition. Gordon W. Gribble.

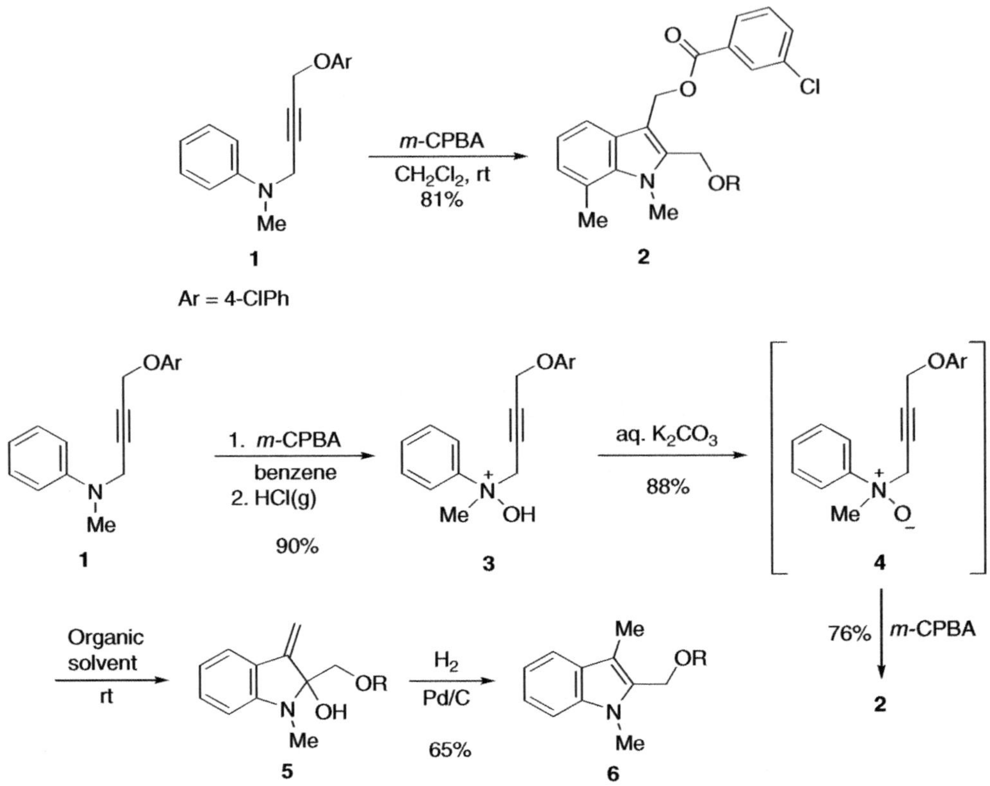

Scheme 1 *Thyagarajan Indole Synthesis*

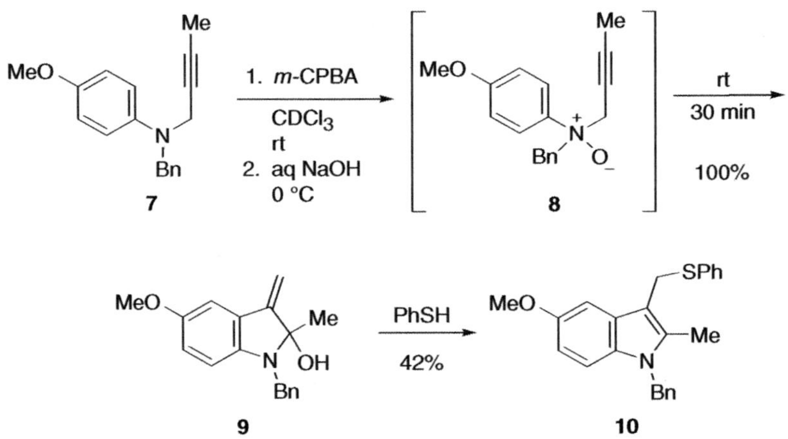

Scheme 2 *Makisumi and Takada Indole Synthesis*

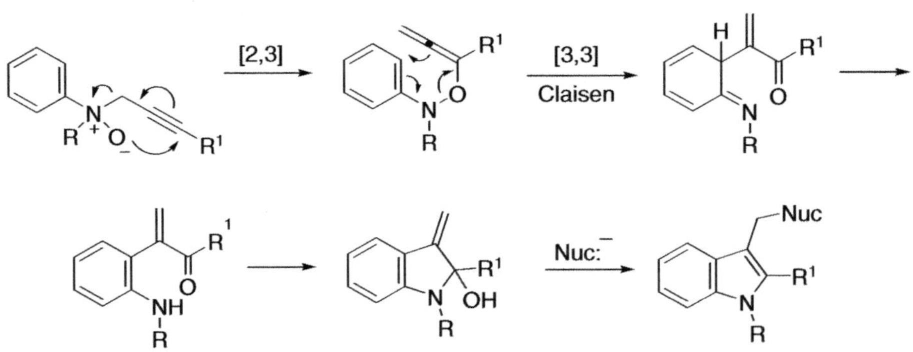

Scheme 3 *Proposed Mechanism of this Indole Ring Synthesis*

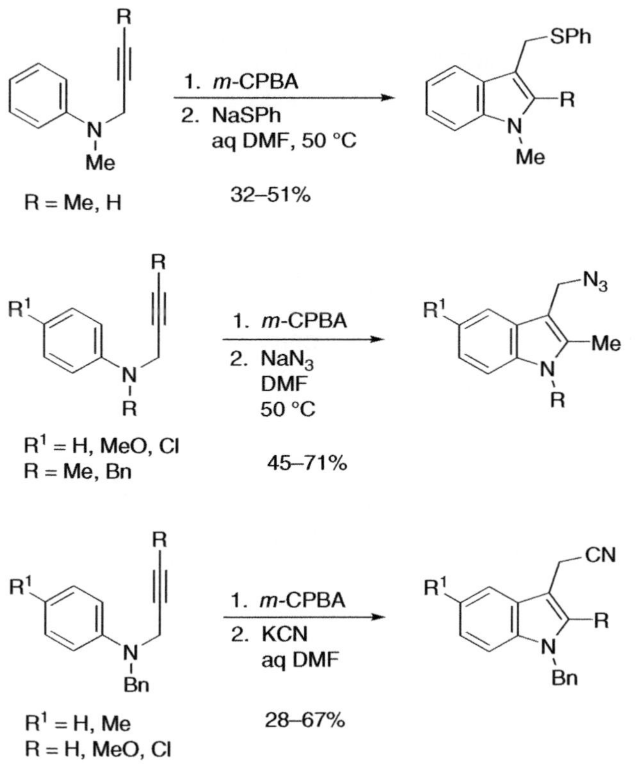

[eqn. 1] [4]

[eqn. 2] [4]

[eqn. 3] [4]

Scheme 4 *Makisumi and Takada Indole Syntheses [4]*

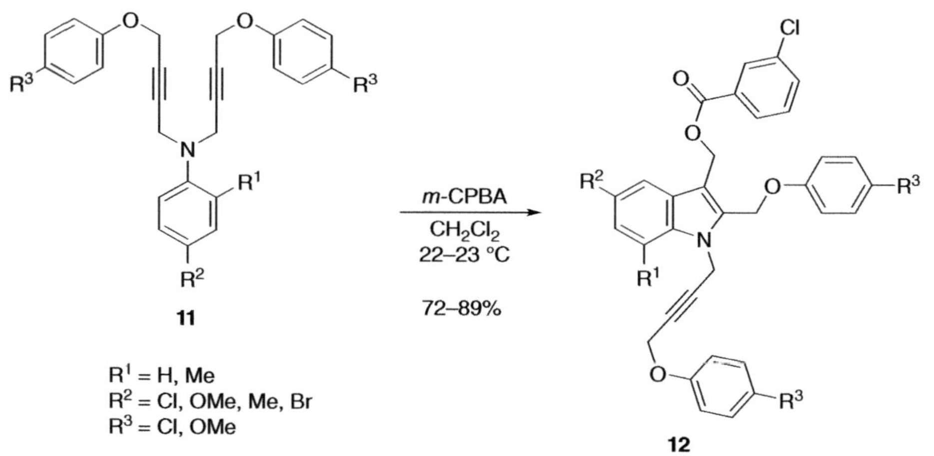

[eqn. 1]

Scheme 5 *Thyagarajan Synthesis of Indole 12 [2, 3]*

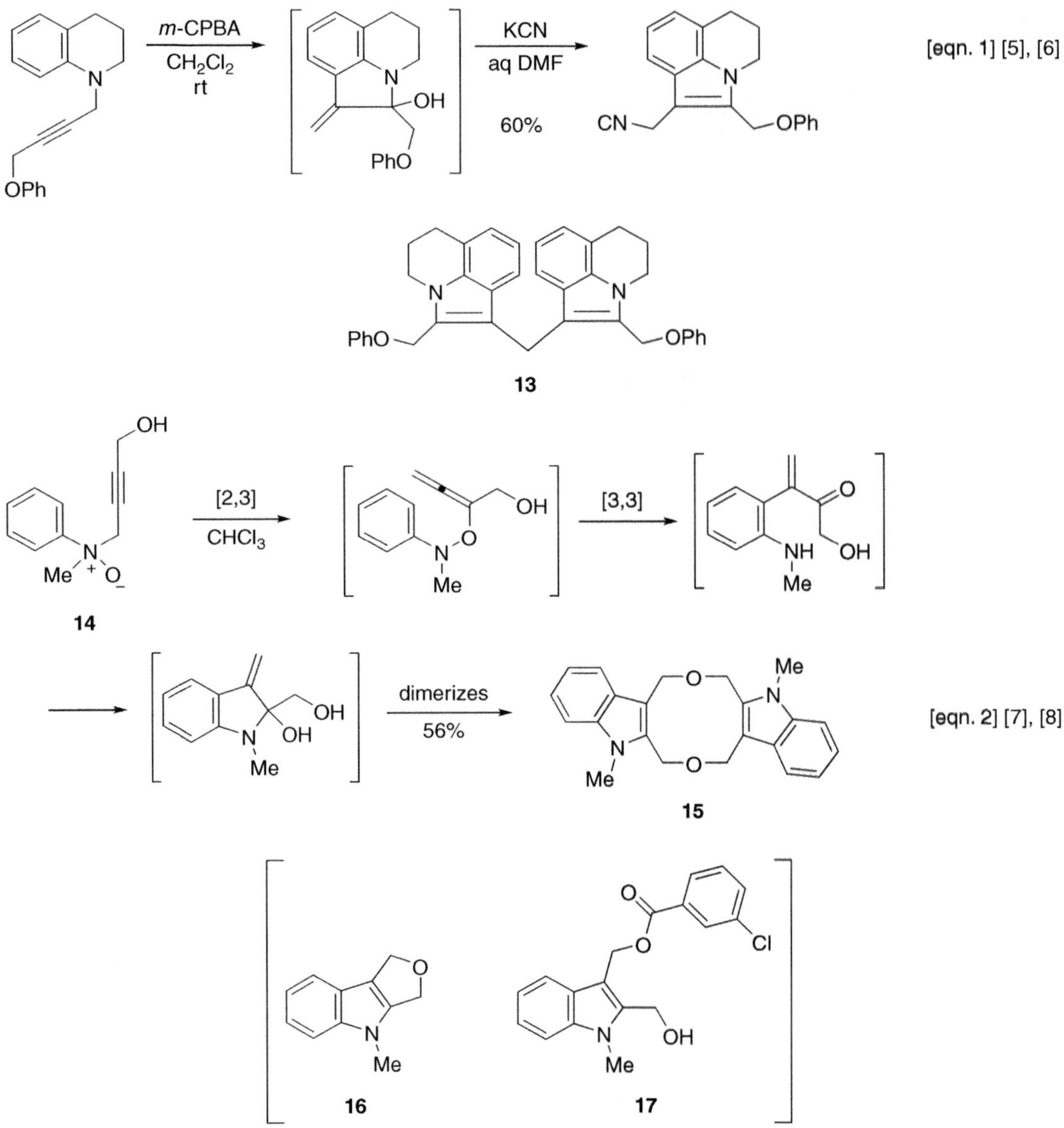

Scheme 6 *Majumdar Indole Syntheses [5–8]*

Table 1 *Examples of the Thyagarajan Indole Synthesis*

Entry	Substrate	Conditions	Product	% Yield	Ref.
1	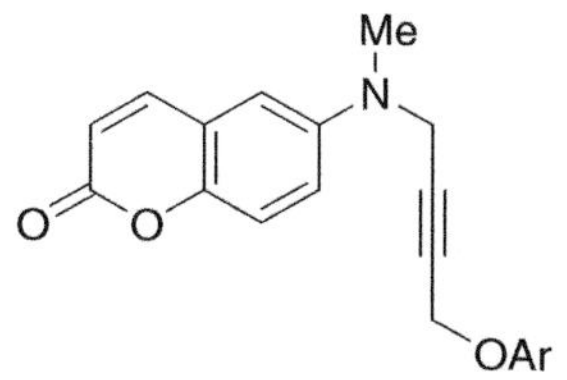 Ar = Ph, 2-ClPh, 4-ClPh, 2-MePh, 4-MePh, 4-BrPh, 4-NO_2Ph, 2,4-Cl_2Ph	*m*-CPBA CH_2Cl_2, 0–5 °C	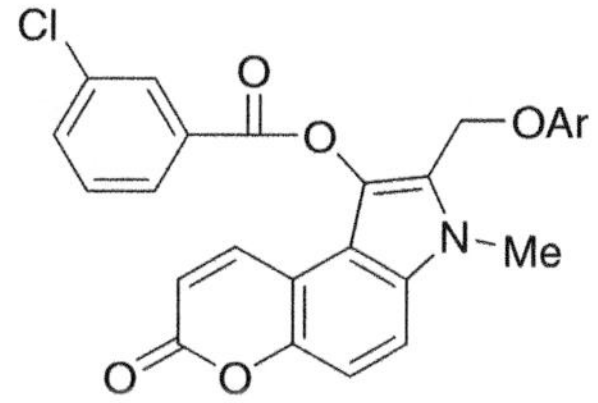	75–97%	9
2	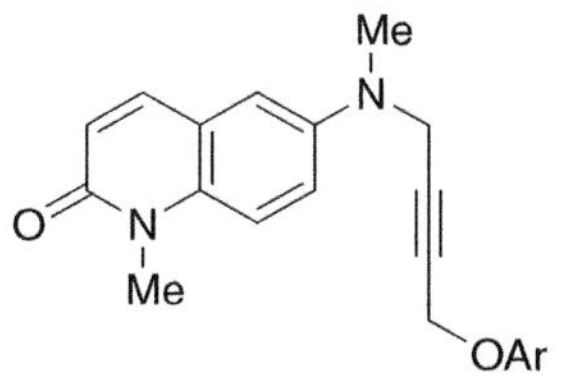 Ar = Ph, 4-ClPh, 4-MePh, 2,4-Cl_2Ph, 2,4-Me_2Ph, 3,5-Me_2Ph, 4-NO_2Ph	*m*-CPBA CH_2Cl_2, rt	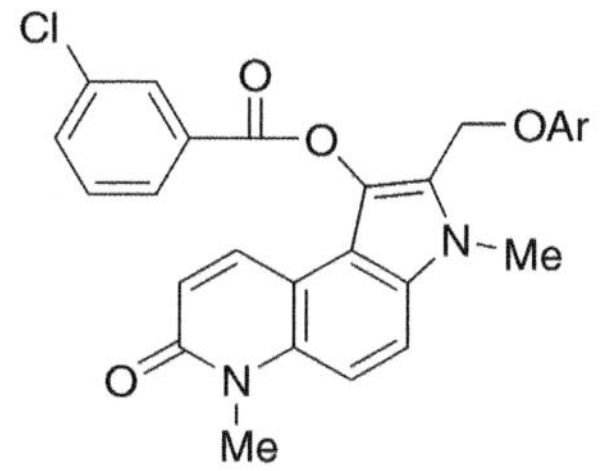	—	10
3	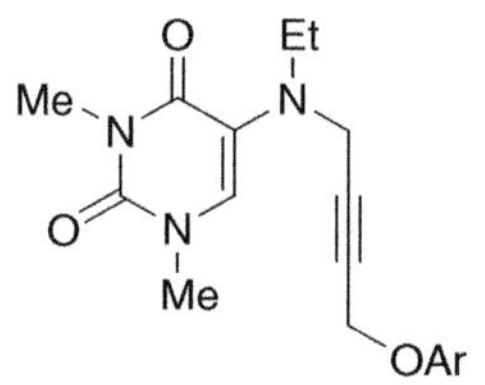 Ar = Ph, 2-ClPh, 4-ClPh, 4-MePh, 4-MeOPh, 2-MePh	*m*-CPBA CH_2Cl_2 0–5 °C	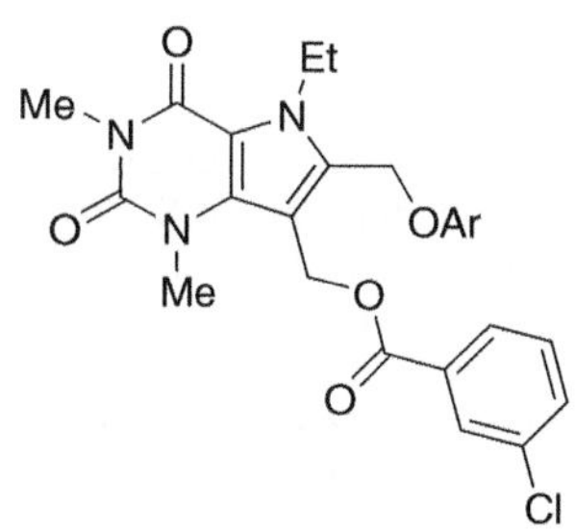	90–95%	11
4	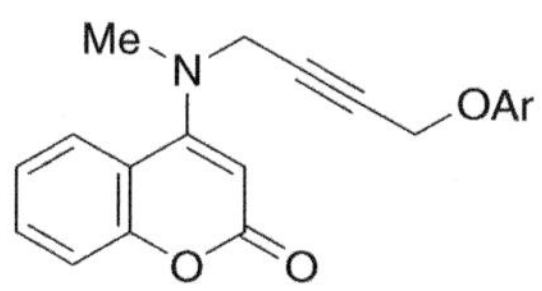 Ar = Ph, 2-ClPh, 4-MePh, 4-MeOPh, 2,4-Cl_2Ph	*m*-CPBA $CHCl_3$ 0–5 °C	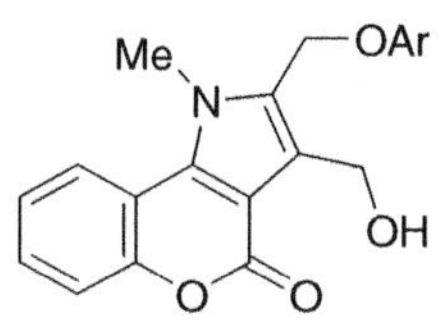	70–75%	12
5	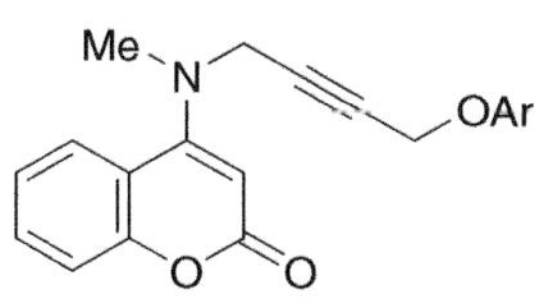 Ar = 4-BrPh, 4-MeOPh, 3-MePh, 4-MePh, 2-BrPh, 3,5-Me_2Ph, 2,4-Me_2Ph	*m*-CPBA CH_2Cl_2 0 °C	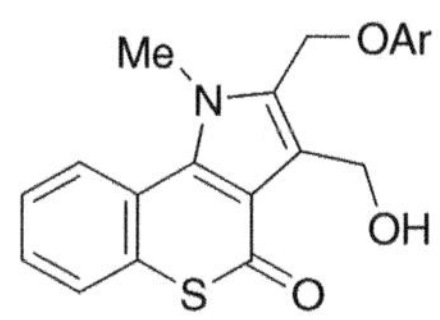	55–72%	13

(continued overleaf)

Table 1 *(continued)*

Entry	Substrate	Conditions	Product	% Yield	Ref.
6	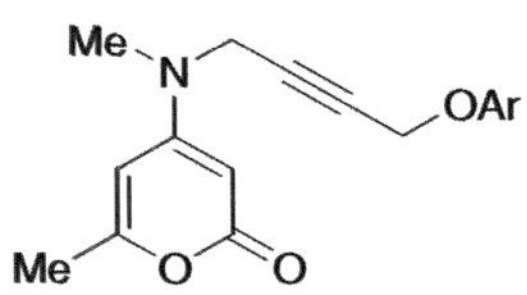 Ar = 4-ClPh, 2-BrPh, 2-ClPh, 2,4-Cl_2Ph, 3,5-Me_2Ph, 3-Me-2ClPh, 4-Cl-2MePh	*m*-CPBA CH_2Cl_2 0 °C	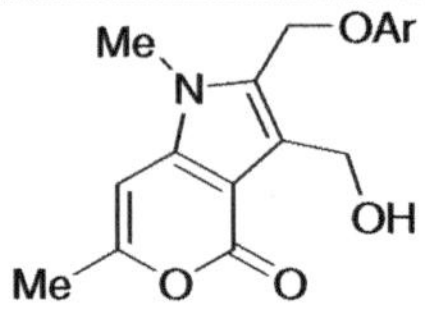	80–85%	14
7	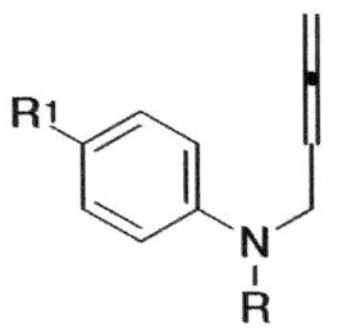 R = Me, Et, Bn, Et R^1 = H, Cl	MMPP aq MeOH rt	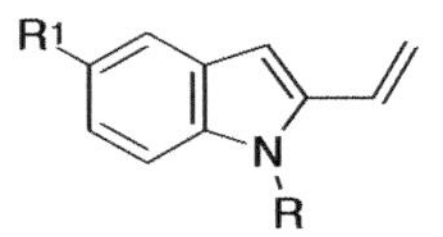	63–82%	15

References

[1] B.S. Thyagarajan, J.B. Hillard, K.V. Reddy, and K.C. Majumdar, *Tetrahedron Lett.*, 1974, 1999–2002.
[2] J. Hillard, K.V. Reddy, K.C. Majumdar, and B.S. Thyagarajan, *J. Heterocycl. Chem.*, 1974, **11**, 369–375.
[3] B.S. Thyagarajan and K.C. Majumdar, *J. Heterocycl. Chem.*, 1975, **12**, 43–47.
[4] Y. Makisumi and S. Takada, *Chem. Pharm. Bull.*, 1976, **24**, 770–777.
[5] K.C. Majumdar and S.K. Chattopadhyay, *J. Chem. Soc., Chem. Commun.*, 1987, 524–525.
[6] K.C. Majumdar, S.K. Chattopadhyay, and A.T. Khan, *J. Chem. Soc., Perkin Trans. 1*, 1989, 1285–1288.
[7] K.C. Majumdar, G.H. Jana, and U. Das, *Chem. Commun.*, 1996, 517–518.
[8] K.C. Majumdar, G.H. Jana, and U. Das, *J. Chem. Soc., Perkin Trans. 1*, 1997, 1229–1231.
[9] K.C. Majumdar and S.K. Ghosh, *J. Chem. Soc., Perkin Trans. 1*, 1994, 2889–2894.
[10] K.C. Majumdar, P. Biwas, and G.H. Jana, *J. Chem. Res. (S)*, 1997, 310–311.
[11] K.C. Majumdar, U. Das, and N.K. Jana, *J. Org. Chem.*, 1998, **63**, 3550–3553.
[12] K.C. Majumdar and S.K. Samant, *Tetrahedron Lett.*, 2002, **43**, 2119–2121.
[13] K.C. Majumdar, S.K. Chattopadhyay, and P.P. Mukhopadhyay, *Synth. Commun.*, 2006, **36**, 1291–1297.
[14] K.C. Majumdar, H. Rahaman, and B. Roy, *Lett. Org. Chem.*, 2012, **2**, 739–741.
[15] T. Balasubramanian and K.K. Balasubramanian, *J. Chem. Soc., Chem. Commun.*, 1994, 1237–1238.
[16] K.C. Majumdar, *Synlett*, 2008, 2400–2411.
[17] K.C. Majumdar, T. Bhattacharyya, B. Chattopadhyay, and B. Sinha, *Synthesis*, 2009, 2117–2142.

6

Julia Indole Synthesis

As with the Thyagarajan indolization, the Julia indole synthesis involves a [3,3]-sigmatropic rearrangement. Moreover, like the former method, the Julia indolization is not widely used despite its simplicity and potential generality [1, 2]. The basic reaction is shown in Scheme 1. The [3,3]-sigmatropic rearrangement of sulfinamide **2** to indole **5** also gives sulfenic acid, which decomposes presumably to sulfur, hydrogen sulfide, sulfur dioxide, and sulfuric acid. Although the yields of sulfinamides **2** from *N*-sulfinyl anilines **1** are excellent (82%–97%), those for the formation of indoles are less so (19%–75%). A few indoles that were prepared by Julia and Baudin are shown in Scheme 1 (combined yields for the two steps). The carbophilic cyclization of the rearrangement product **3** to the unstable indole-2-sulfenic acid **4** could instead involve a thiophilic cyclization to a benzo[c][1,2]thiazine 2-oxide and loss of SO.

The use of triethyloxonium tetrafluoroborate accelerates the [3,3]-sigmatropic rearrangement of sulfinamide **6** to indole **7**, as summarized in Scheme 2.

Other examples of sigmatropic rearrangements leading to indoles are presented in the next chapter, or more appropriately elsewhere in this monograph.

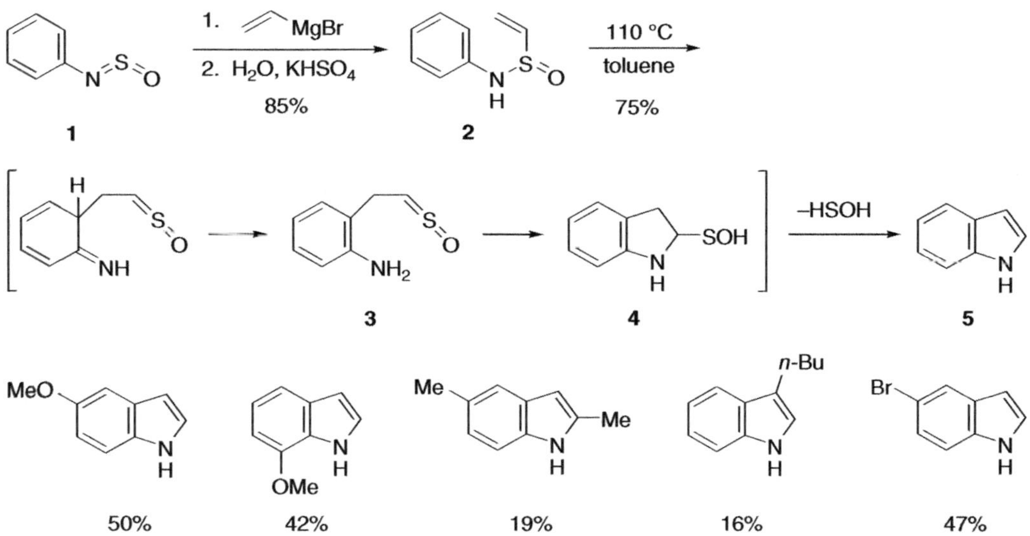

Scheme 1 *Julia Indole Synthesis*

Indole Ring Synthesis: From Natural Products to Drug Discovery, First Edition. Gordon W. Gribble.

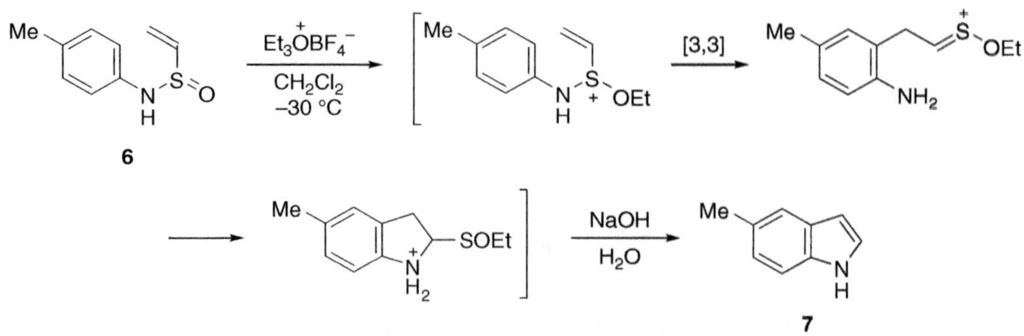

Scheme 2 *Improved Julia Indole Synthesis*

References

[1] J.-B. Baudin and S.A. Julia, *Tetrahedron Lett.*, 1986, **27**, 837–840.
[2] J.-B. Baudin, M.-G. Comménil, S.A. Julia, *et al.*, *Bull. Soc. Chim. Fr.*, 1996, **133**, 329–350.

7

Miscellaneous Sigmatropic Rearrangements

In this chapter I cover other sigmatropic rearrangements that generate the indole ring and were not covered in earlier chapters. While some of these sigmatropic rearrangements could reasonably be assigned as a Name reaction, they are included in this chapter as a group.

Oikawa, Yonemitsu and colleagues have demonstrated that 2-substituted indoles are accessible via the reaction of acyl Meldrum's acids and phenylhydroxylamine. Their overall sequence is shown in Scheme 1 [1, 2]. Reaction of phenylhydroxyamine (**1**) with acylated Meldrum's acids **2** gives *N*-acylacetylphenylhydroxyamines **3** in good to excellent yields. A subsequent reaction of **3** with **2** in refluxing toluene affords indoles **4** and smaller amounts of 5-substituted isoxazolin-3-ones **5**. Likewise, heating *N*-acylphenylhydroxamines **6** with Meldrum's acid **2** (R = Bn) gives indoles **7** along with the intermediate *ortho*-alkylation products **8**. Further heating of **8** (R = Bz) in xylene or with acid affords **7** (R = Bz). Yields of indoles **7** in this reaction were increased to 64%–67% by heating the reaction in the presence of copper powder. The *N*-acylindoles could be quantitatively converted to the corresponding indoles by treatment with sodium carbonate in aqueous methanol at room temperature. The suggested mechanism for this indole synthesis is shown in Scheme 2.

Martin also employed *N*-acylphenylhydroxylamines (hydroxamic acids) to set up a [3,3]-sigmatropic rearrangement leading to indoles, as summarized in Scheme 3 [3]. A palladium-catalyzed vinylation of **9** gives the *o*-vinyl intermediate **10**, which rapidly undergoes the hetero-Cope [3,3]-sigmatropic rearrangement to **11** that cyclizes with loss of water to give *N*-acylindoles **12** [3]. Tricyclic indole **13** was also prepared. In a sequel, Martin applied his method to the synthesis of the marine indoles glossobalol (**14**) and balanoglossol (**15**) (Scheme 3) [4].

A nice synthesis of benz[*g*]indoles comes from the work of Pinna and colleagues. Thus, alpha-tetralone oxime **16** reacts with methyl propiolate to give oxime ether **17**, which upon heating affords pyrrole **18** via a [3,3]-sigmatropic rearrangement. Oxidation with DDQ gives benz[*g*]indole **19** (Scheme 4, equation 1) [5]. Saczewski and colleagues find that the *N*-phenylhydroxylamine dihydroimidazolium **20** reacts with ethyl propiolate to form indole **21** (equation 2) [6].

Parsons and colleagues combined a [2,3]-sigmatropic rearrangement with a Michael addition to convert acetylenic alcohols **22** to indoles **25** via allenes **23** (Scheme 5). The starting alcohols **22** were prepared by ethynylmagnesium chloride addition to the appropriate benzaldehyde or acetophenone in 76%–99% yield. Reaction of **22** with phenylsulfenyl chloride gave **23** and triggered a sulfenate-sulfoxide [2,3]-sigmatropic rearrangement to give **24**, which cyclizes to the indole **25**. The authors note that allene **24** could be detected by TLC analysis but not isolated. Some of the prepared indoles are shown [7].

In work related to that shown in Schemes 3 and 4, Hwu and coworkers synthesized indoles **29** from arylhydroxylamines **26** and activated alkynes (Scheme 6, equation 1). The acetylene adds to both oxygen and nitrogen, giving **27**, which undergoes a [3,3]-sigmatropic rearrangement to **28** and thence to indole **29** [8]. With *N*-benzyl-*N*-arylhydroxlamines, under the same reaction conditions indoles are formed, but by a different pathway that is discussed in a later chapter. Wojciechowski finds that aza-*ortho*-xylylenes **31** are generated by the extrusion of sulfur dioxide from 2,1-benzisothiazoline 2,2-dioxides **30** to form

Indole Ring Synthesis: From Natural Products to Drug Discovery, First Edition. Gordon W. Gribble.

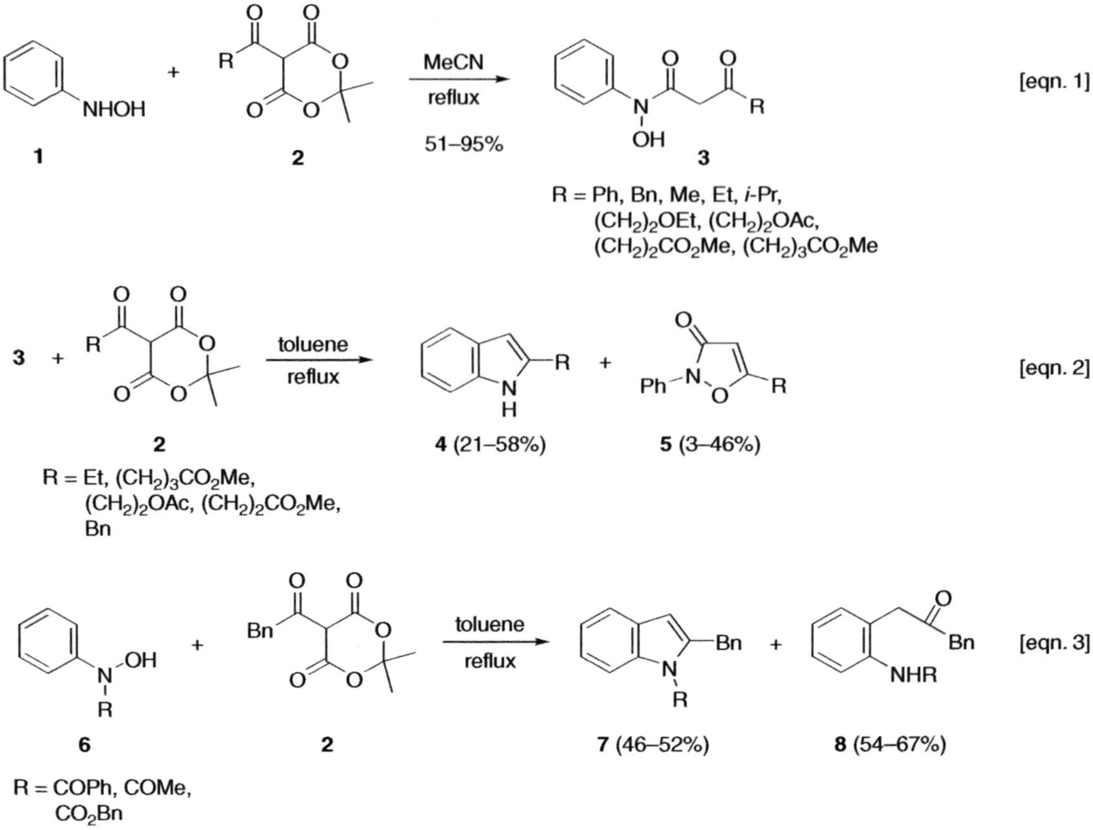

Scheme 1 *Oikawa, Yonemitsu Indole Synthesis*

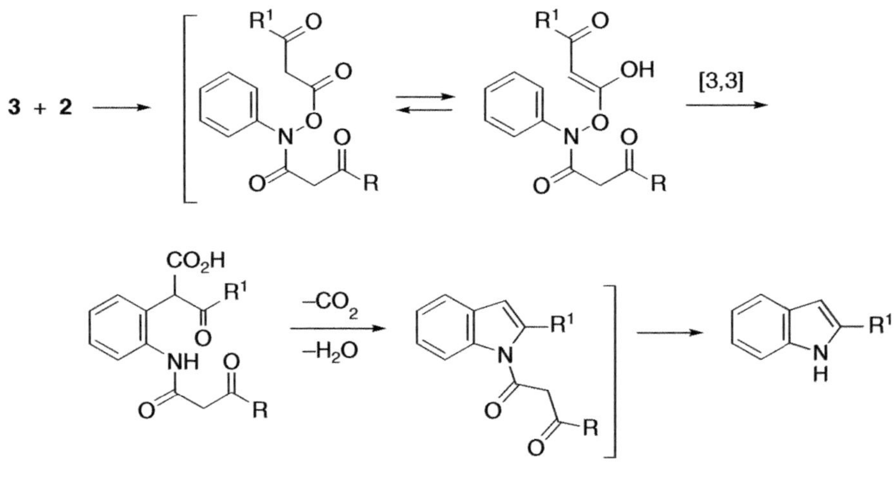

Scheme 2 *Suggested Mechanism for Oikawa, Yonemitsu Indole Synthesis*

indoles **34** (Scheme 6, equation 2) [9]. The key step is a [1,5]-hydrogen shift from **31** to generate **32–33**, which cyclizes to nitroindoles **34**. Interestingly, the major products are the uncyclized vinyl chlorides **32** (40%–53%), whereas the allylic chlorides **33** are the precursors to indoles **34**. This chemist also found that thermolysis of benzosultams **35** gives 2-arylindoles **37**, via the corresponding aza-*ortho*-xylylene **36** [10].

An imino ene reaction was employed by Burger's group to prepare α-fluoromethyl tryptophans **40** from the union of 1-sulfonyl-3-methylene indolines **38** with electrophilic imines **39** (Scheme 7, equation 1) [11].

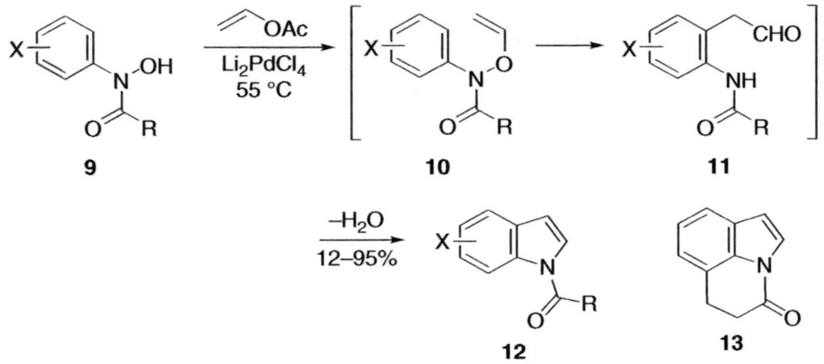

[eqn. 1] [3]

R = Me, OEt, H, Ph, vinyl, CH_2CH_2Cl, CH_2Cl
X = H, 2-Cl, 4-Cl, 2-Me, 3-Me, 4-Me

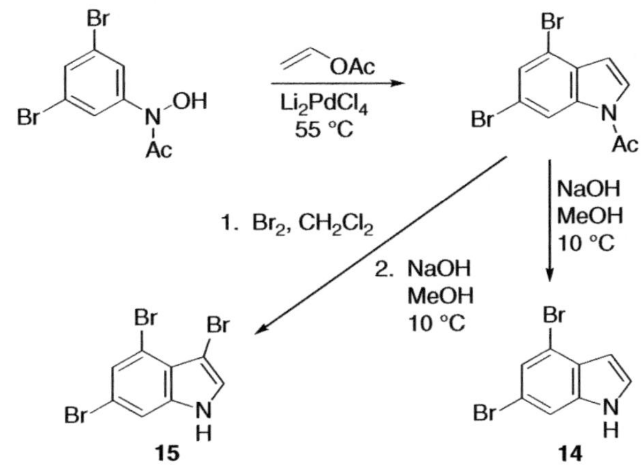

Scheme 3 *Martin Indole Synthesis*

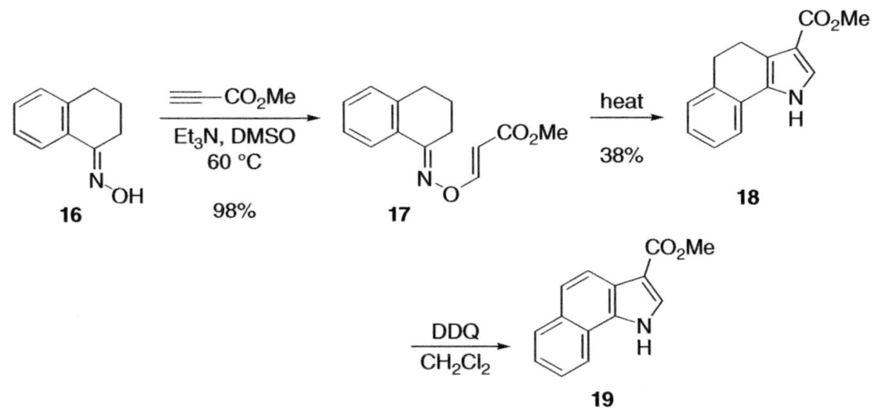

[eqn. 1] [5]

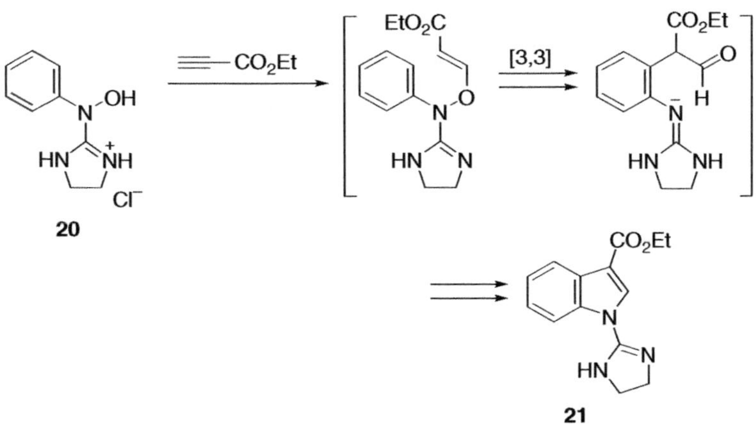

[eqn. 2] [6]

Scheme 4 *Pinna and Saczewski Indole Syntheses*

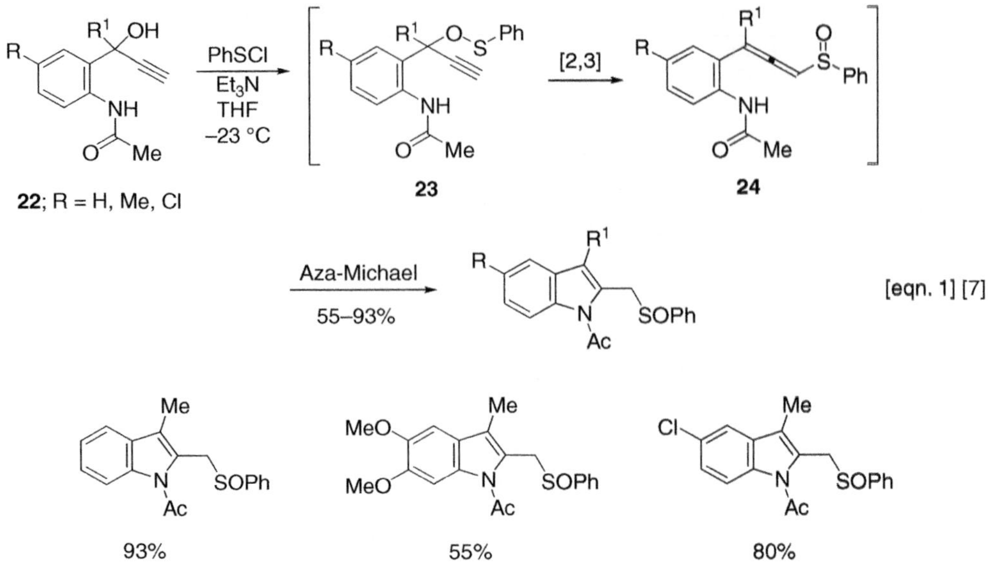

Scheme 5 *Parsons Indole Synthesis*

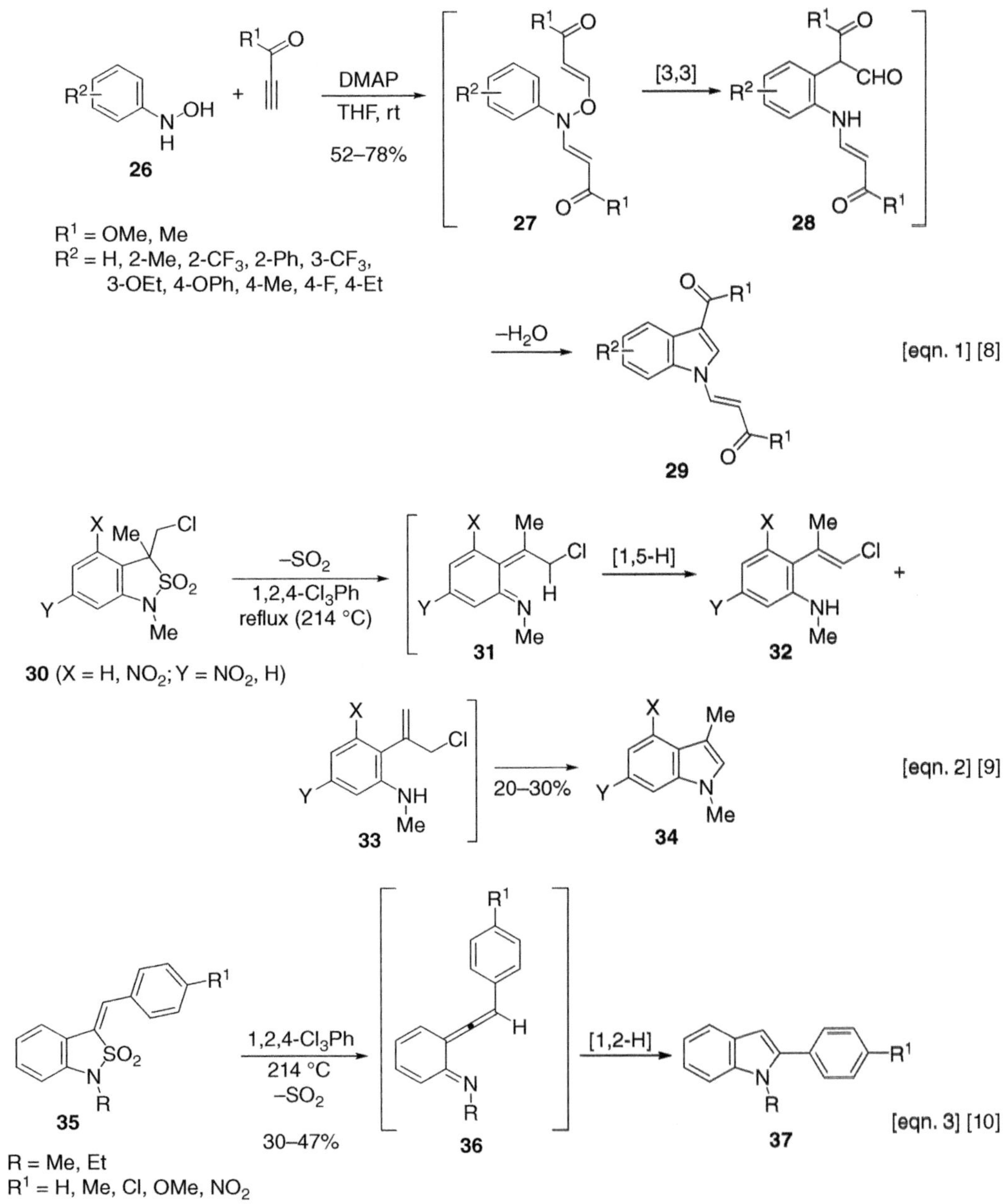

Scheme 6 *Hwu and Wojciechowski Indole Syntheses*

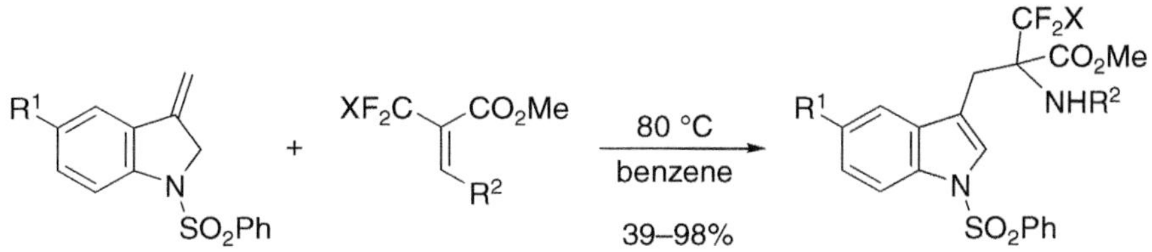

38 **39** **40**

[eqn. 1] [11]

R^1 = H, F
R^2 = SO_2Ph, Ts, SO_2Me, Boc
X = F, Br

Scheme 7 *Burger Indole Synthesis*

References

[1] K. Mohri, Y. Oikawa, K. Hirao, and O. Yonemitsu, *Heterocycles*, 1982, **19**, 515–520.
[2] K. Mohri, Y. Oikawa, K. Hirao, and O. Yonemitsu, *Chem. Pharm. Bull.*, 1982, **30**, 3097–3105.
[3] P. Martin, *Helv. Chim. Acta*, 1984, **67**, 1647–1649.
[4] P. Martin, *Tetrahedron Lett.*, 1987, **28**, 1645–1646.
[5] G.A. Pinna, M.A. Pirisi, and G. Paglietti, *J. Chem. Res. (S)*, 1990, 360–361.
[6] F. Saczewski, T. Debowski, M. Gdaniec, and Z. Gdaniec, *J. Org. Chem.*, 1996, **61**, 5425–5429.
[7] M. Gray, P.J. Parsons, and A.P. Neary, *Synlett*, 1993, 281–282.
[8] J.R. Hwu, H.V. Patel, R.J. Lin, and M.O. Gray, *J. Org. Chem.*, 1994, **59**, 1577–1582.
[9] K. Wojciechowski, *Tetrahedron*, 1993, **49**, 10017–10026.
[10] K. Wojciechowski, *Molecules*, 1996, **1**, 51–56.
[11] S.N. Osipov, N.M. Kobel'kova, A.F. Kolomiets, *et al.*, *Synlett*, 2001, 1287–1289.

PART II
Nucleophilic Cyclization

Nucleophilic cyclization is a common tactic in all branches of organic chemistry, and several indole ring syntheses, which are presented in Chapters 8 through 22, employ this reaction as the central theme in the synthesis. The Sugasawa indole synthesis (Chapter 22) uses both nucleophilic and electrophilic reactions.

Indole Ring Synthesis: From Natural Products to Drug Discovery, First Edition. Gordon W. Gribble.

8

Madelung Indole Synthesis

Twenty-six years after the initial discovery by Mauthner and Suida [1], Madelung explored and developed the cyclization of *ortho*-alkylamides with alkoxy bases at high temperatures to give indoles [2–4]. These initial reactions are shown in Scheme 1, in addition to subsequent applications by other workers. Tyson found that potassium salts such as potassium amide and potassium *t*-butoxide were better bases than the original sodium alkoxides in preparing indole itself (equation 3) [5]. Slightly lower yields were obtained with potassium amide, potassium ethoxide, and potassium methoxide, and virtually no indole was obtained with the corresponding sodium alkoxides. It should be noted that Madelung was unable to produce indole from *ortho*-toluide and sodium alkoxide. Galat and Friedman modified the Tyson discovery by avoiding the hazardous potassium metal by substituting potassium acetate, because the potassium ion is the key factor in these Madelung reactions (equation 4) [6]. In both reactions, the alcohol used to prepare the alkoxides is removed by distillation prior to the high-temperature reactions. A further modification by Tyson involved the generation of sodium or potassium *ortho*-toluide in the presence of carbon monoxide at high temperature and high pressure. Yields of indole under these conditions ranged from 56% to 82% based on uncovered *ortho*-toluidine [7]. An *Organic Synthesis* is available. Pichat and colleagues made use of Tyson's work [5, 7–9] to prepare ^{14}C-2 labeled indole and discussed the mechanism involving potassium formate [10].

Applications of the original Madelung method are tabulated in Table 1 [11–16, 18–25]. As was found earlier, Augustine observed that potassium *tert*-butoxide is superior to sodium methoxide, sodamide, lithium amide, and *n*-butyllithium (Entries 4–6) [13]. The key factor in using potassium *tert*-butoxide is to avoid sublimation of this base before cyclization has occurred. The mild conditions reported by Fuhrer and Gschwend (Entry 8) [15] have also been pursued by Houlihan and colleagues [16] (Entry 9) and earlier by Piozzi and Langella [17], who synthesized a large number of 2-alkylindoles.

The two-step procedure shown in Entry 12 was first described by Clark and colleagues as summarized in Scheme 2 [26, 27]. This procedure is a powerful alternative to the classic Madelung protocol.

A major development in the Madelung indole synthesis is the introduction of electron-withdrawing groups to facilitate formation of a benzylic anion. A Wittig-type solution to this situation is presented in the next section. Bergman and colleagues found that 4-nitroindoles can be prepared via a modified Madelung indole synthesis that employs an oxalate ester functionality to acidify the benzylic hydrogen (Scheme 3) [28–30]. It might be noted that a conventional Madelung reaction on 4-nitro-2-methylacetanilide caused an explosion [31]. The acylation of *ortho*-methylnitrobenzenes (**1** and **3**), base-catalyzed cyclization, and loss of

Indole Ring Synthesis: From Natural Products to Drug Discovery, First Edition. Gordon W. Gribble.

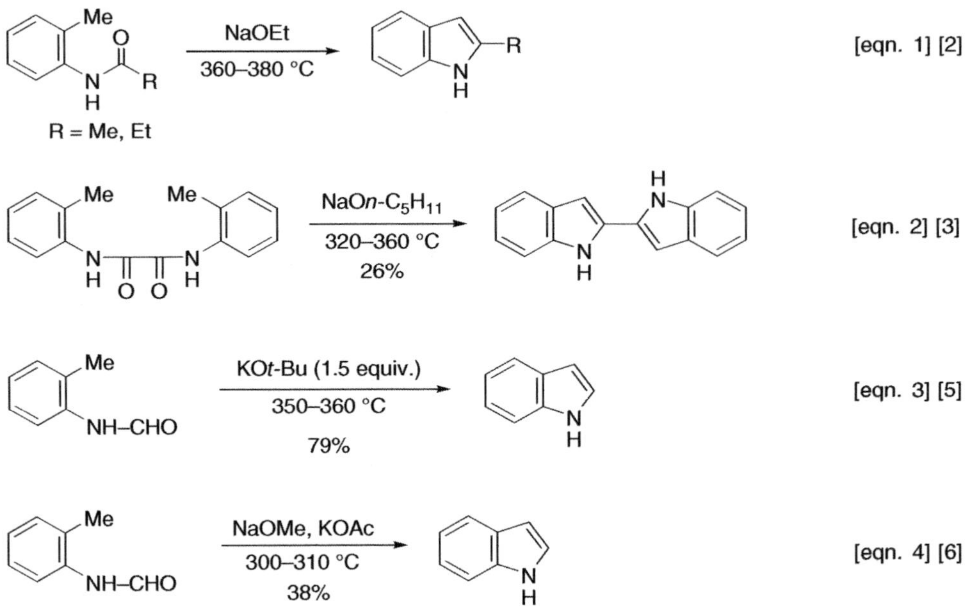

Scheme 1 *Madelung Indole Syntheses*

glyoxylate all occur in the same pot to give the indoles **2** and **4** (equations 1 and 2). Imidates such as **5** also undergo this facile cyclization to afford nitroindoles (equation 3). Melhado and Brodsky employed the Bergman cyclization to prepare 4-nitroindole and derived compounds. Reinhoudt and coworkers used other electron-withdrawing groups to promote a facile Madelung indole ring synthesis [33, 34]. Two examples are shown in Scheme 3 (equations 4 and 5).

Bartoli and colleagues employed *ortho*-trimethylsilylmethyl anilides to effect a Madelung cyclization in an intramolecular Peterson olefination, as shown in Scheme 4 [35, 36]. Whereas a normal Madelung product is obtained in equation 1, a reverse Madelung compound is the result in equation 2, where the 2-R^1 substitution is the result of the ester R^1CO_2Me.

Wacker and Kasireddy described a solid-phase synthesis of 2,3-disubstituted indoles (Scheme 5) [37] via the modified Madelung synthesis reported by Reinhoudt [34, 35]. The basic approach is shown in equation 1, and several synthesized indoles are posted.

A collection of Madelung indole syntheses using electron-withdrawing activation of the benzylic methylene group and other variations is summarized in Table 2 [38–43]. That extreme mildness is possible is shown by the cyclization in Entry 1, and Entry 4 features the synthesis of a new class of D_2 *in vivo* agonists.

Table 1 *Applications of the Classic Madelung Indole Synthesis*

Entry	Substrate	Conditions	Product	% Yield	Ref.
1	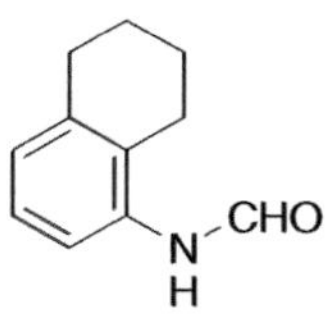	KO*t*-Bu 340–360 °C	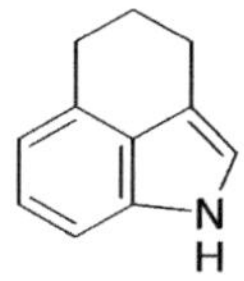	11%	11
2	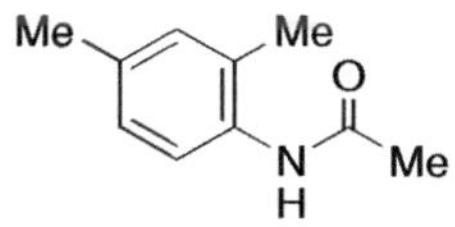	$NaNH_2$ *N,N*-DiEt-aniline 190–200 °C	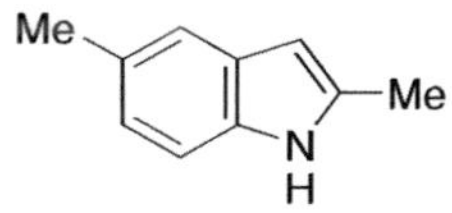	63%	12
3	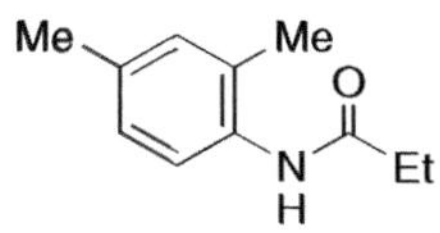	*N,N*-diEt-aniline 200 °C	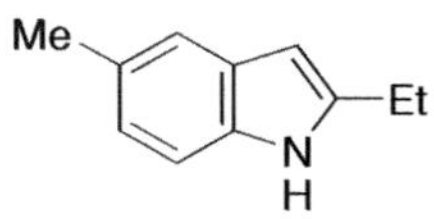	67%	12
4	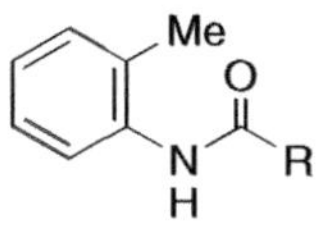 R = Me, Ph, cyclopropyl	KO*t*-Bu 300–325 °C	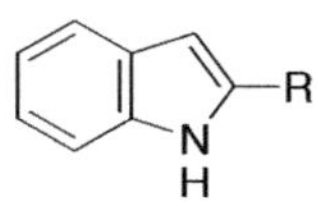	60–86%	13
5	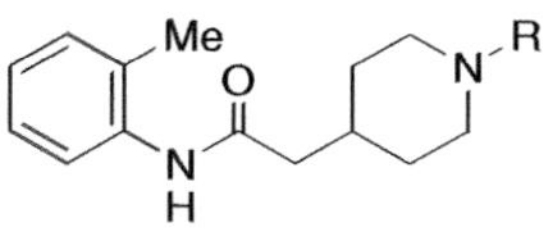 R = Me, Bn	KO*t*-Bu 300–340 °C	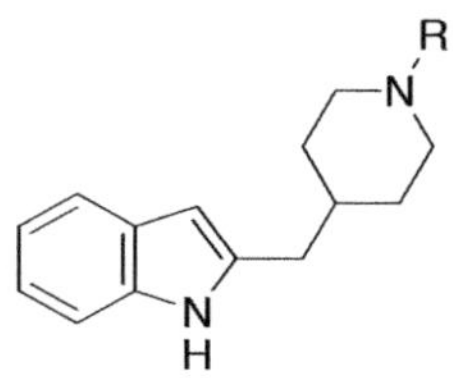	72–92%	13
6	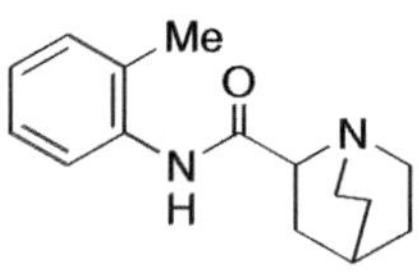	KO*t*-Bu 300 °C	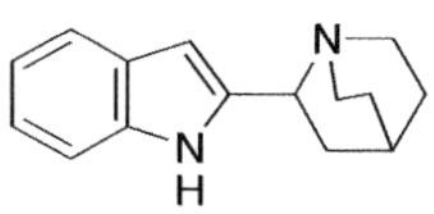	66%	13
7	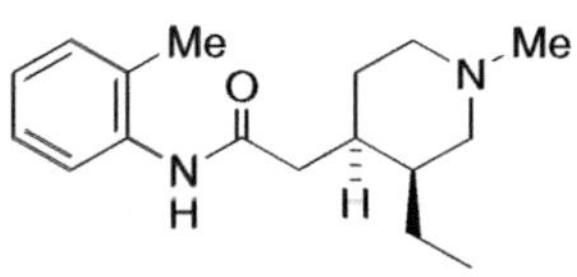	KO*t*-Bu 340 °C	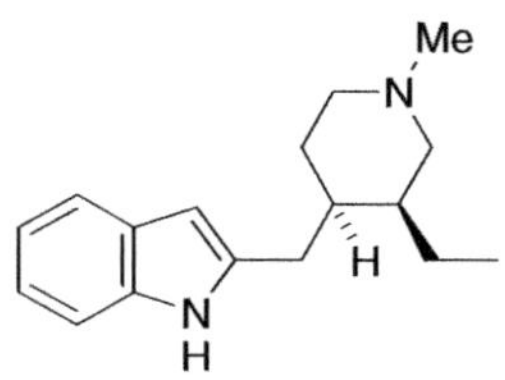	76%	14
8	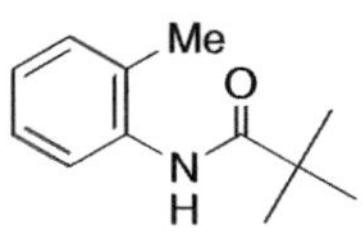	*n*-BuLi (2 equiv.) THF, rt	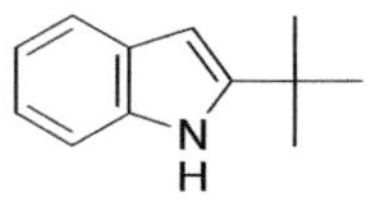	>90%	15
9	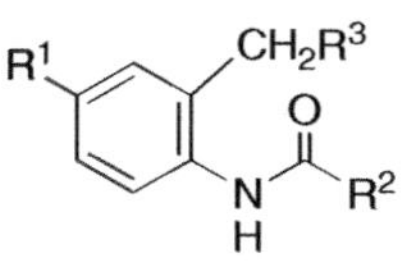 R^1 = H, OMe, Cl R^2 = Ph, *t*-Bu, 1-adamantanyl R^3 = H, Me	*n*-BuLi (2-3 equiv.) THF, rt	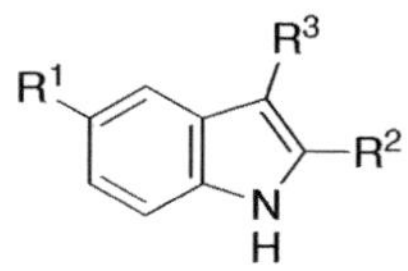	20–90%	16

(continued overleaf)

Table 1 *(continued)*

Entry	Substrate	Conditions	Product	% Yield	Ref.
10	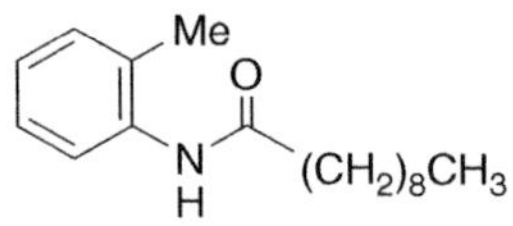	$NaNH_2$ 250 °C	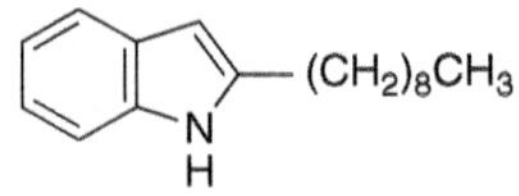	81%	18
11	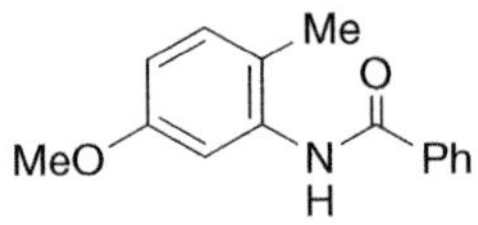	*t*-BuLi THF, rt	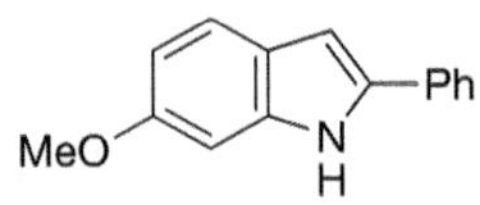	20%	19
12	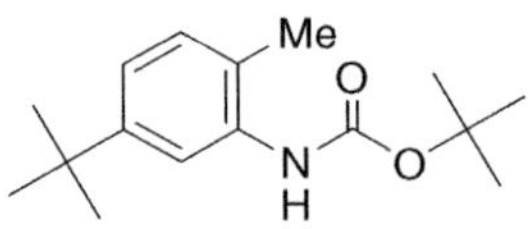	1. *sec*-BuLi, –78 °C to 0 °C 2. DMF	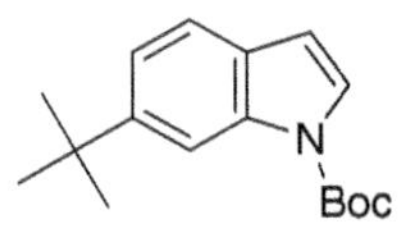	65%	20
13	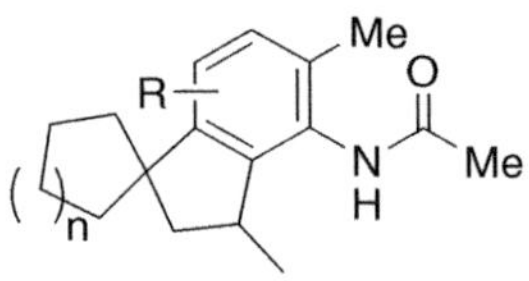 n = 1, 2, 3 R = H, 5-Cl, 4-Me, 5-Me (indole numbering)	$NaNH_2$, DMA reflux	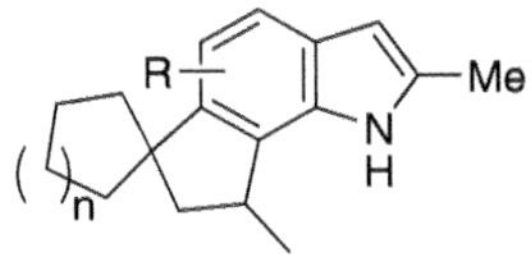	30–94%	21
14	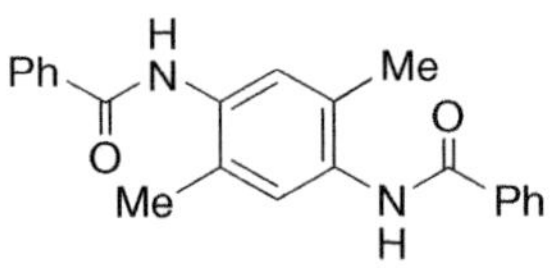	KO*t*-Bu 320–330 °C	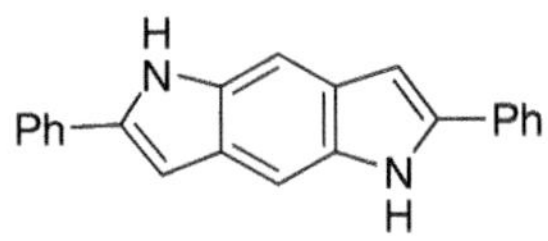	—	22
15	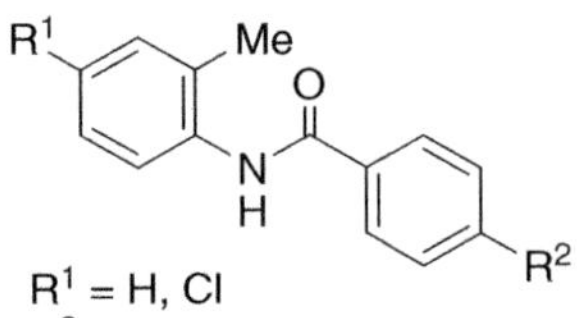 R^1 = H, Cl R^2 = H, Cl, Me	*n*-BuLi THF, 0 °C	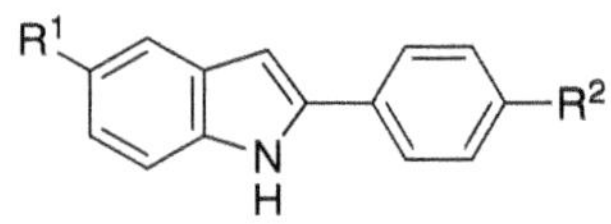	75–88%	23
16	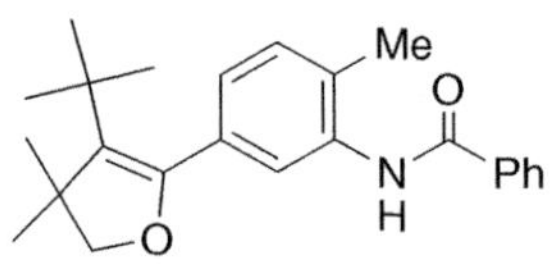	*n*-BuLi THF, 50 °C	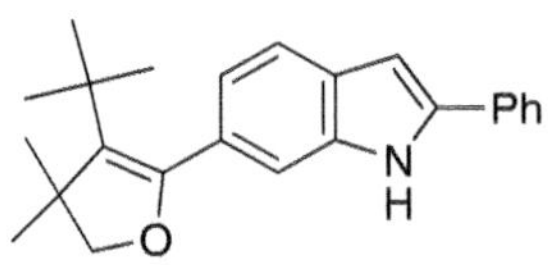	55%	24
17	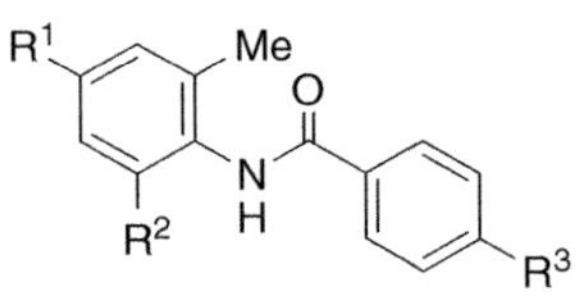 R^1 = H, NO_2, OMe, F, Cl R^2 = H, Cl, Me R^3 = H, F, CF_3, NO_2	*n*-BuLi THF, –20 °C to rt	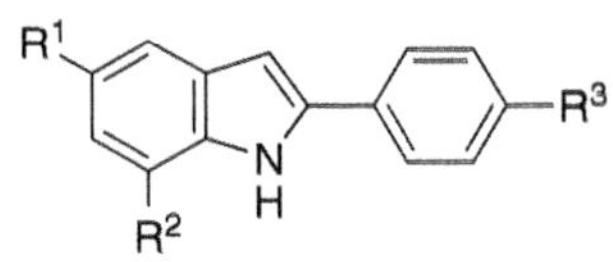	30–85%	25

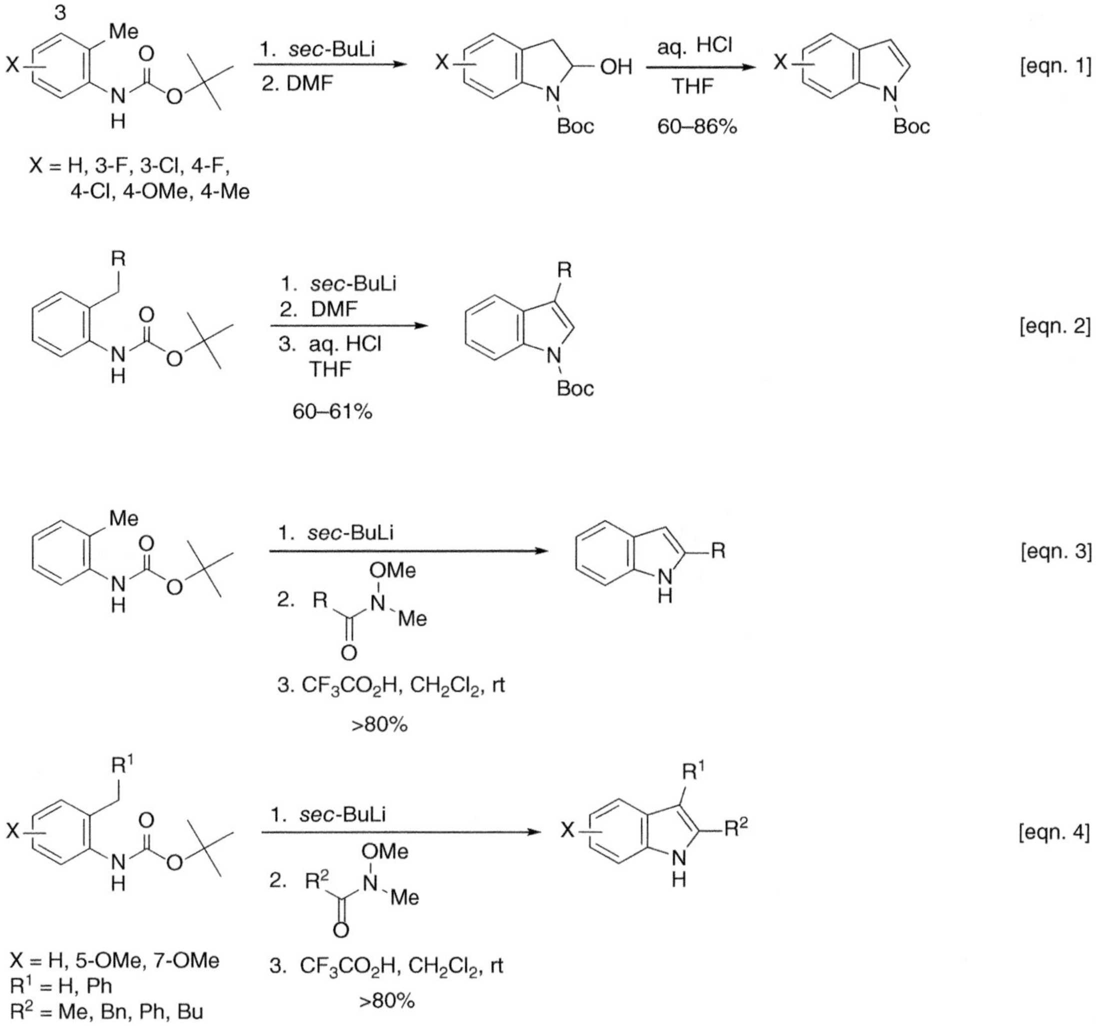

Scheme 2 *Clark Modification of the Madelung Indole Synthesis*

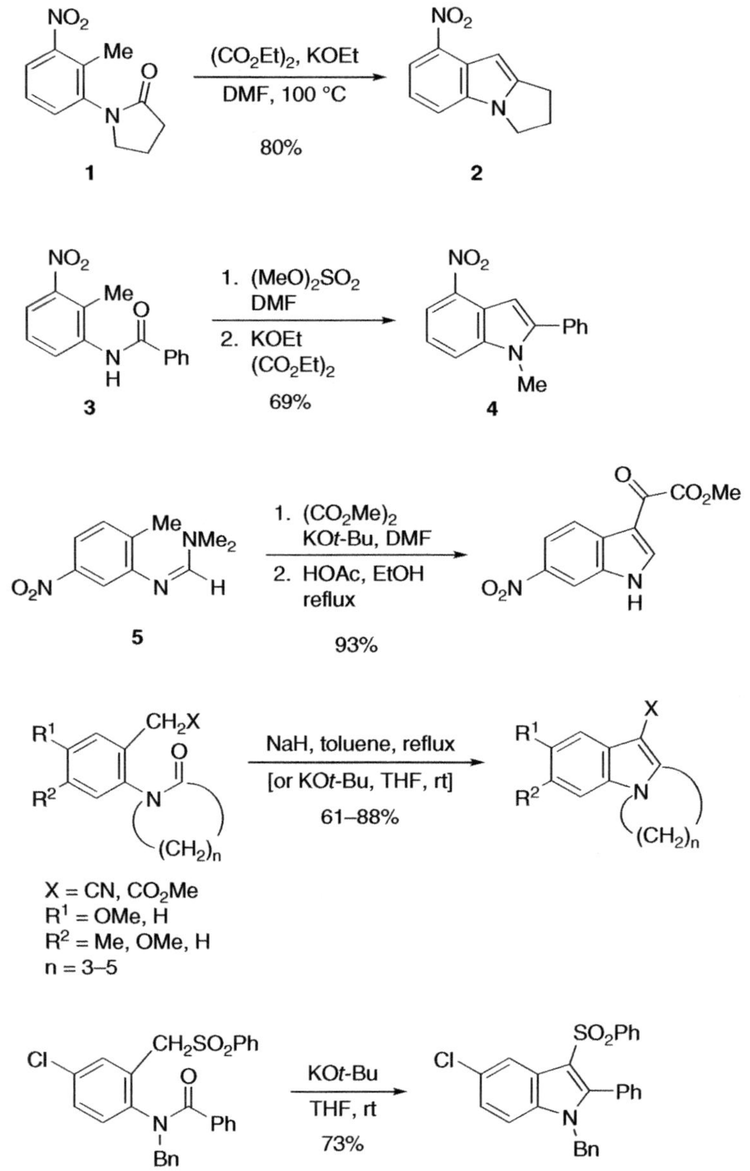

[eqn. 1] [30]

[eqn. 2] [30]

[eqn. 3] [30]

[eqn. 4] [33]

[eqn. 5] [34]

Scheme 3 *Applications of the Madelung Indole Synthesis*

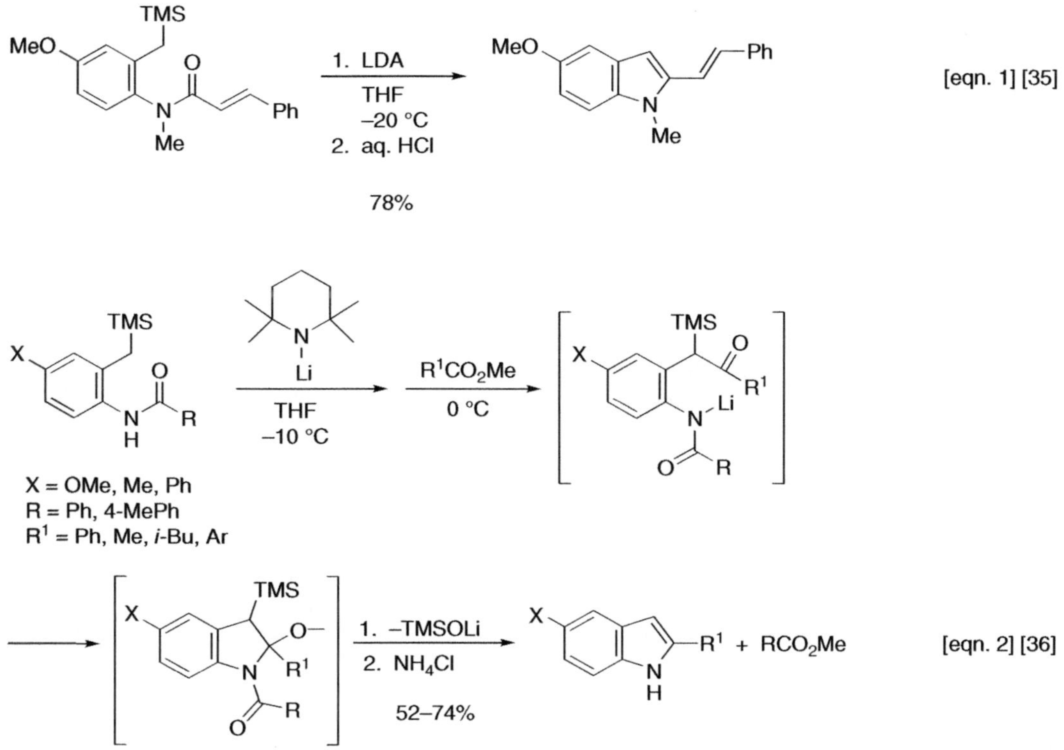

Scheme 4 *Bartoli Modification of the Madelung Indole Synthesis*

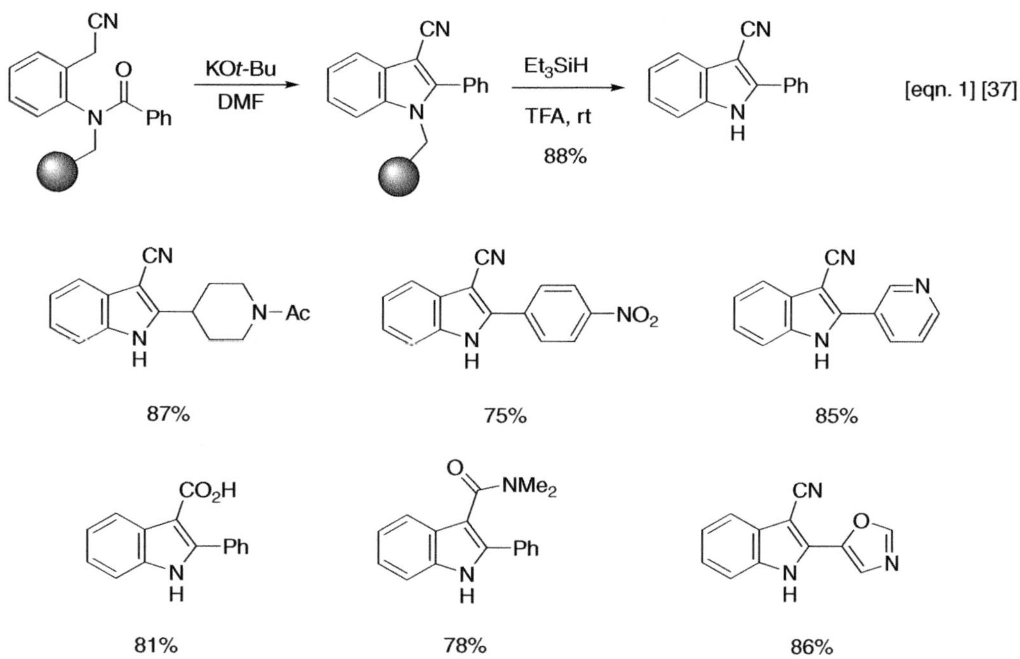

Scheme 5 *Wacker Solid-Phase Madelung Indole Synthesis*

Table 2 *Miscellaneous Applications of Modified Madelung Indole Syntheses*

Entry	Substrate	Conditions	Product	% Yield	Ref.
1	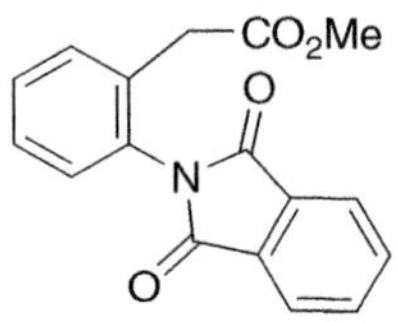	K_2CO_3, DMF 80–90 °C	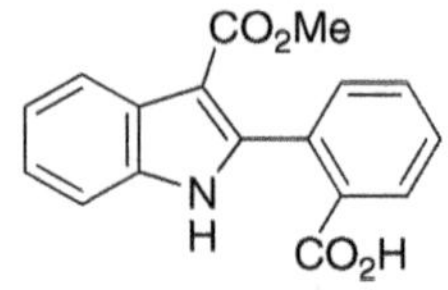	89%	38
2	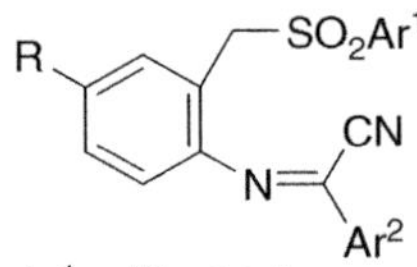 Ar^1 = Ph, 4-tol R = H, Cl Ar^2 = Ph, 4-tol	NaOH, DMSO 80–90 °C	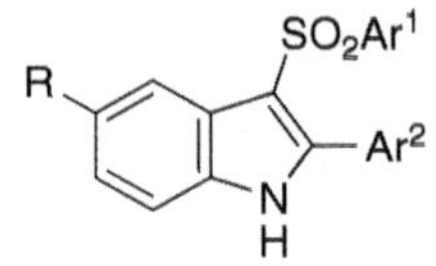	56–83%	39
3	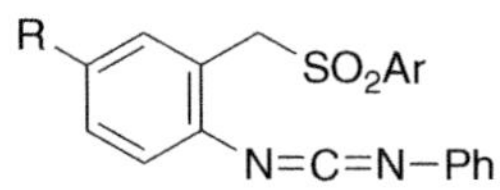 Ar = Ph, 4-tol R = H, Cl	NaOH, DMSO rt	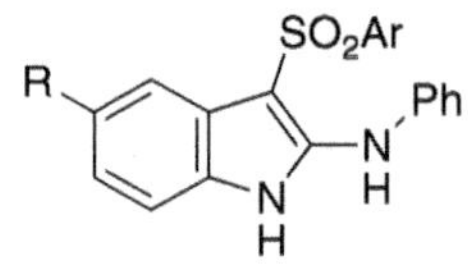	68–83%	39
4	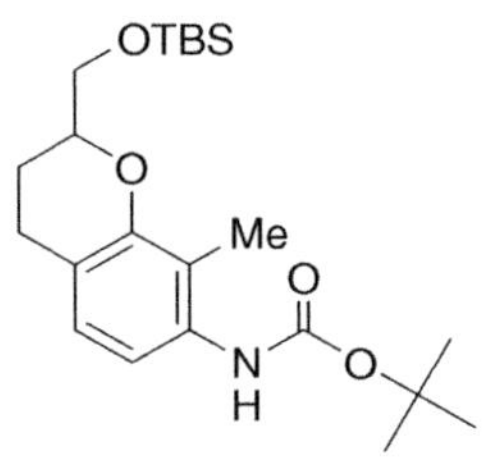	1. *s*-BuLi 2. DMF 3. HCl, THF	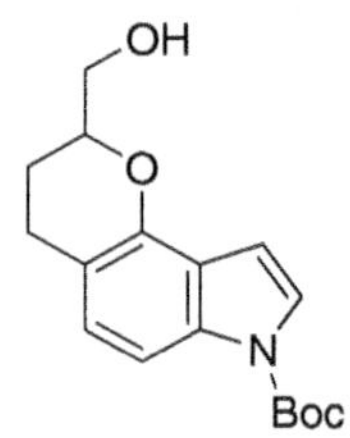	90%	40
5	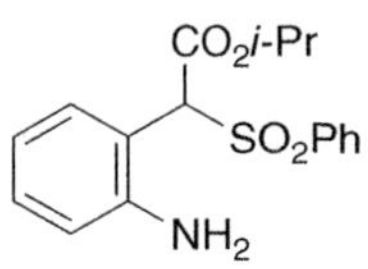	PhCHO MeCN reflux	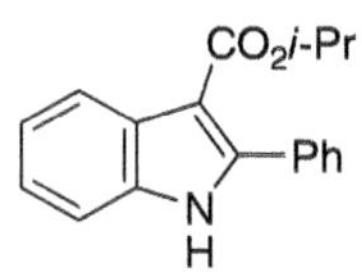	83%	41
6	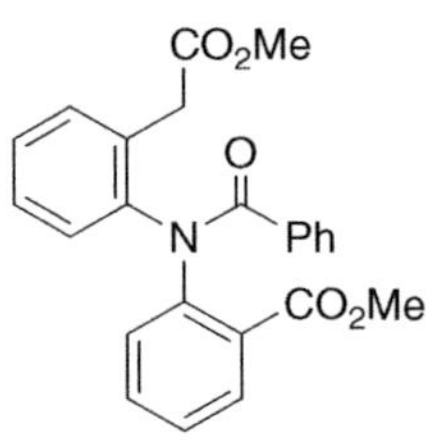	NaOMe benzene reflux	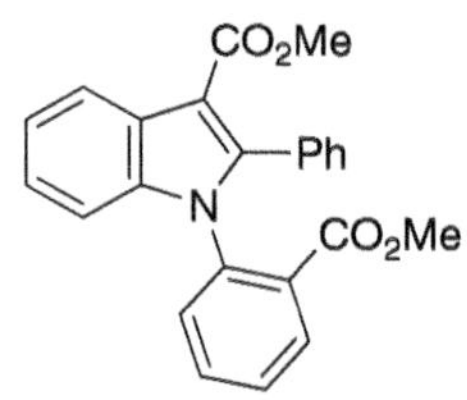	63%	42, 43

References

[1] J. Mauthner and W. Suida, *Monatsh. Chem.*, 1886, **7**, 230–240.
[2] W. Madelung, *Ber.*, 1912, **45**, 1128–1134.
[3] W. Madelung, *Justus Liebigs Ann. Chem.*, 1914, **405**, 58–95.
[4] W. Madelung, *Ber.*, 1913, **45**, 3521–3527.
[5] F.T. Tyson, *J. Am. Chem. Soc.*, 1941, **63**, 2024–2025.
[6] A. Galat and H.L. Friedman, *J. Am. Chem. Soc.*, 1948, **70**, 1280–1281.
[7] F.T. Tyson, *J. Am. Chem. Soc.*, 1950, **72**, 2801–2803.
[8] F.T. Tyson, *Org. Synth.*, 1943, **23**, 42–45.
[9] F.T. Tyson, *Org. Syn. Coll.* Vol. **3**, 1955, 479–482.
[10] L. Pichat, M. Audinot, and J. Monnet, *Bull. Chim. Soc. Fr.*, 1954, **21**, 85–88.
[11] F.C. Uhle, C.G. Vernick, and G.L. Schmir, *J. Am. Chem. Soc.*, 1955, **77**, 3334–3337.
[12] E. Walton, C.H. Stammer, R.F. Nutt, *et al.*, *J. Med. Chem.*, 1965, **8**, 204–208.
[13] R.L. Augustine, A.J. Gustavsen, S.F. Wanat, *et al.*, *J. Org. Chem.*, 1973, **38**, 3004–3011.
[14] A. Wu and V. Snieckus, *Tetrahedron Lett.*, 1975, 2057–2060.
[15] W. Fuhrer and H.W. Gschwend, *J. Org. Chem.*, 1979, **44**, 1133–1136.
[16] W.J. Houlihan, V.A. Parrino, and Y. Uike, *J. Org. Chem.*, 1981, **46**, 4511–4515.
[17] F. Piozzi and M.R. Langella, *Gazz. Chim. Ital.*, 1963, **93**, 1382–1391.
[18] M. Arcari, R. Aveta, A. Brandt, *et al.*, *Gazz. Chim. Ital.*, 1991, **121**, 499–504.
[19] G. Tarzia, G. Diamantini, B. Di Giacomo, and G. Spadoni, *J. Med. Chem.*, 1997, **40**, 2003–2010.
[20] J.V.N.V. Prasad, *Org. Lett.*, 2000, **2**, 1069–1072.
[21] V. Kouznetsov, F. Zubkov, A. Palma, and G. Restrepo, *Tetrahedron Lett.*, 2002, **43**, 4707–4709.
[22] H.Z. Chen, Y.D. Jin, R.S. Xu, *et al.*, *Synth. Metals*, 2003, **139**, 529–534.
[23] G. Primofiore, F. Da Settimo, S. Taliani, *et al.*, *J. Med. Chem.*, 2004, **47**, 1852–1855.
[24] N. Watanabe, M. Ichikawa, A. Ono, *et al.*, *Chem. Lett.*, 2005, **34**, 718–719.
[25] F. Da Settimo, F. Simorini, S. Taliani, *et al.*, *J. Med. Chem.*, 2008, **51**, 5798–5806.
[26] R.D. Clark, J.M. Muchowski, M. Souchet, and D.B. Repke, *Synlett*, 1990, 207–208.
[27] R.D. Clark, J.M. Muchowski, L.E. Fisher, *et al.*, *Synthesis*, 1991, 871–878.
[28] J. Bergman, P. Sand, and U. Tilstam, *Tetrahedron Lett.*, 1983, **24**, 3665–3668.
[29] J. Bergman and P. Sand, *Org. Synth.*, 1987, **56**, 146–149.
[30] J. Bergman and P. Sand, *Tetrahedron*, 1990, **46**, 6085–6112.
[31] W.E. Noland, L.R. Smith, and K.R. Rush, *J. Org. Chem.*, 1965, **30**, 3457–3469.
[32] L.L. Melhado and J.L. Brodsky, *J. Org. Chem.*, 1988, **53**, 3852–3855.
[33] W. Verboom, E.O.M. Orlemens, H.J. Berga, *et al.*, *Tetrahedron*, 1986, **42**, 5053–5064.
[34] E.O.M. Orlemans, A.H. Schreuder, P.G.M. Conti, *et al.*, *Tetrahedron*, 1987, **43**, 3817–3826.
[35] G. Bartoli, M. Bosco, R. Dalpozzo, and P.E. Todesco, *J. Chem. Soc., Chem. Commun.*, 1988, 807–808.
[36] G. Bartoli, G. Palmieri, M. Petrini, *et al.*, *Tetrahedron*, 1990, **46**, 1379–1384.
[37] D.A. Wacker and P. Kasireddy, *Tetrahedron Lett.*, 2002, **43**, 5189–5191.
[38] G. Kim and G. Keum, *Heterocycles*, 1997, **45**, 1979–1988.
[39] M. Takahashi and D. Suga, *Synthesis*, 1998, 986–990.
[40] R.E. Mewshaw, K.L. Marquis, and X. Shi, *Tetrahedron*, 1998, **54**, 7081–7108.
[41] J. Garcia, R. Greenhouse, J.M. Muchowski, and J.A. Ruiz, *Tetrahedron Lett.*, 1985, **26**, 1827–1830.
[42] J.W. Schulenberg, *J. Am. Chem. Soc.*, 1968, **90**, 7008–7014.
[43] J.W. Schulenberg, *J. Am. Chem. Soc.*, 1968, **90**, 1367–1368.

9

Wittig–Madelung Indole Synthesis

Several investigators realized that converting the *ortho*-alkyl group into a phosphonium ylide would greatly facilitate the Madelung indole cyclization, and thus was born the Wittig–Madelung indole synthesis. The inaugural exploration was that of Le Corre and coworkers as shown in Scheme 1 (equation 1) [1, 2]. Several different bases were explored. Zimmer and Crenshaw applied this method to the synthesis of several fused indoles (equations 2, 3) [3], as did Flitsch and colleagues at about the same time (equation 4) [4, 5].

Capuano and coworkers extended the Wittig–Madelung indole synthesis to several 2-acylindoles, 2-indolylcarboxylates, 2,2′-biindolyls, and 1,2-diindolylethylenes as summarized in Scheme 2 [6, 7]. The reaction of 2-(α-ketoacyloxy) amino compounds affords mainly 2-quinolones by cyclization to the more-reactive ketone carbonyl (equation 2). Although the yields are low, the chemistry shown in equations 3 and 4 provides a simple synthesis of 2,2′-biindolyls and 1,2-di(2-indolyl)ethylenes, respectively. Bernauer and Mahboobi applied the Capuano method to the synthesis of methyl indole-2-acetate (**6**) as the starting point in an alkaloid synthesis program. These workers used sodium *tert*-pentylate in refluxing toluene to obtain the indole in 70% yield [8]. Likewise, Danieli's team, for a synthesis of (±)-3-oxovincadifformine ethyl ester, prepared ethyl 2-(indol-2-yl) propanoate (**7**) in 64% yield using potassium *tert*-butoxide in refluxing toluene [9]. Both groups started with the requisite benzylic triphenyl phosphonium bromide *ortho*-anilide.

Pindur and Eitel prepared several 2-vinylindoles via a Wittig–Madelung synthesis (Scheme 3, equation 1), work that included the first synthesis of the parent 2-vinylindole [10]. Imaniski and colleagues prepared a series of 2-trifluoromethylindoles via a Wittig–Madelung synthesis (equations 2, 3) [11, 12]. The sensitivity of the trifluoromethyl group to strong base militates against using a conventional Madelung approach. The yields shown in equation 2 are based on consumed starting material. No reaction occurred with the amide **8** ($R^1 = R^2 = H$; and $R^1 = OMe$, $R^2 = H$). Superior yields were achieved when the phosphonium salts were used directly (equation 3). In neither case was added base employed. As we saw earlier (Scheme 2, equation 2), the reaction of the pyruvinamide derivative afforded mainly the quinolone (62%) rather than the indole (19%). The use of benzyl ether **8** in the thermal Wittig–Madelung process is believed to proceed via the phosphonium salt, in which case methoxide can function as a base, although the authors favor a seven-membered ring intermediate phosphorane that loses triphenylphosphine oxide, giving the indole [12].

Imamoto and coworkers developed a modification of the Wittig–Madelung indole synthesis that involves the reaction of (2-aminobenzyl)triphenylphosphonium salts and acid anhydrides in the presence of triethylamine (Scheme 4, equation 1) [13]. The base 2,6-lutidine allowed the use of acid chlorides (equation 2). Obviously, mild base cleavage of the *N*-acyl group affords the respective indole. The authors present evidence that a *N,N*-diacylated intermediate is the cyclization precursor. Thus, a monoacylated compound does not cyclize to an indole under the amine reaction conditions, whereas an isolable but unstable *N,N*-diacylated derivative does afford the cyclized indole product. Moreover, successful crossover control experiments—the dream of all mechanistic chemists—clearly show the intermediacy of *N,N*-diacylindole **9**, as summarized in equations 3 and 4. Imamoto extended his procedure to a vinylogous Wittig–Madelung reaction to yield

Indole Ring Synthesis: From Natural Products to Drug Discovery, First Edition. Gordon W. Gribble.

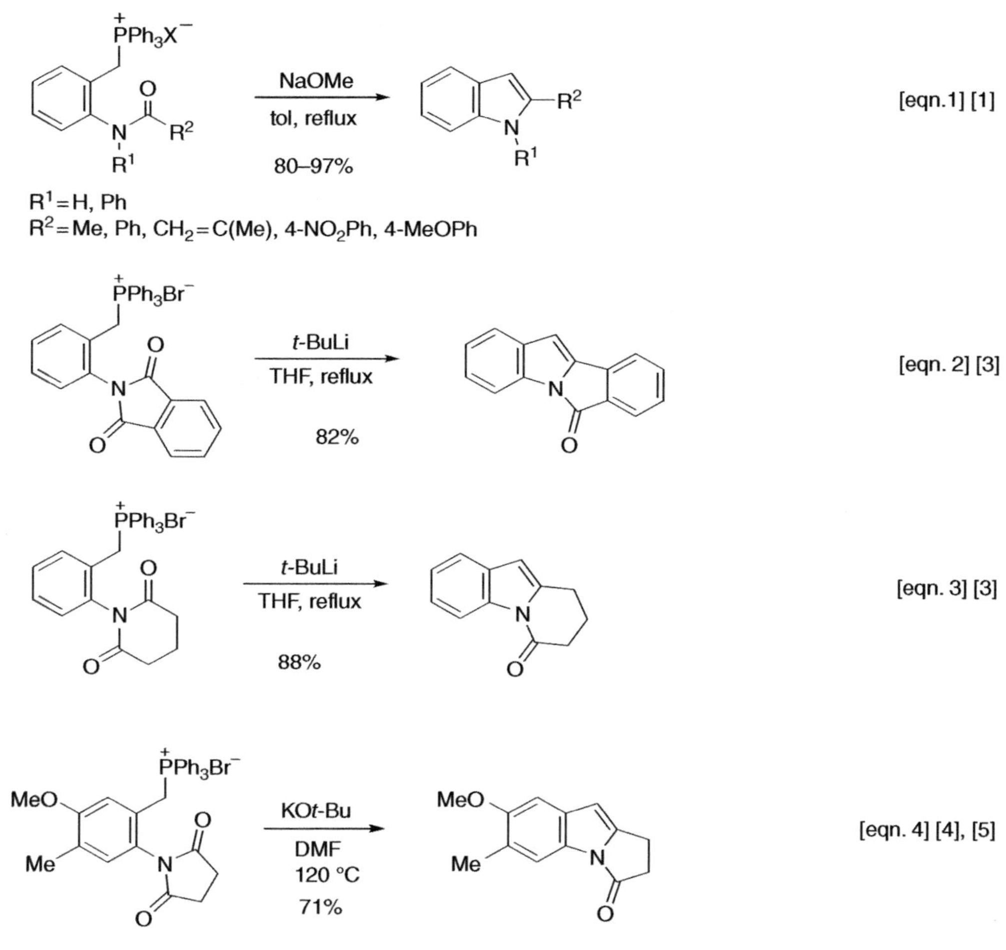

Scheme 1 *Wittig-Madelung Indole Synthesis*

1,3-diacylindoles (equation 5) [14]. Acid chlorides also function well in this process.

Kraus and his colleagues were the first to apply extensively the Wittig–Madelung protocol to the synthesis of indole alkaloids [15–18]. Scheme 5 presents the Kraus synthesis of 2-substituted indoles that culminated in a formal synthesis of arcyriacyanin A. The key feature of this approach is the use of aldehydes and microwave conditions with the commercially available phosphonium salt **11**. The resulting imines undergo base-induced indolization in one pot to give a range of indoles (equation 1) [15]. Indole **12** can be converted to arcyriacyanin A in one step. Likewise, indole **13** is an advanced intermediate in the synthesis of rutaecarpine alkaloid analogues (equation 2) [16]. The simple expediency of employing substituted phosphonium salts **14** provides a direct synthesis of 2,3-disubstituted indoles (equation 3) [16].

Kraus and his group extended their Wittig–Madelung indole synthesis to other activated benzylic systems as shown in Scheme 6 (equations 1–3) [17]. These include phenylsulfonyl, thiomethyl, and cyano, and this team applied this chemistry to a synthesis of the alkaloid isocryptolepine. As we saw twice in the previous chapter, the cyclization of phosphonium salt **15** affords quinolone **16**, which was used to synthesize both neocryptolepine (**17**) and isocryptolepine (**18**).

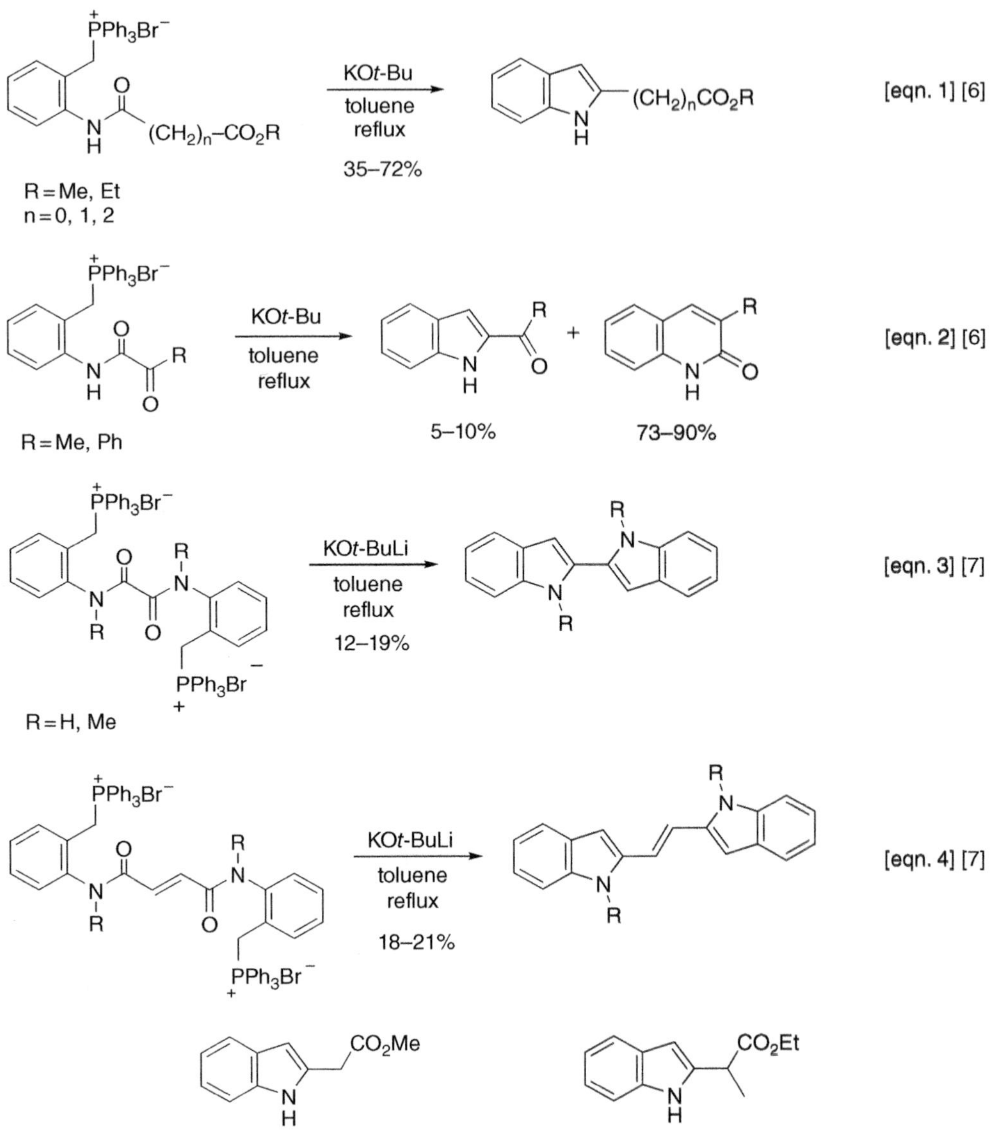

6 (70%) [8]

7 (64%) [9]

Scheme 2 *Applications of the Wittig-Madelung Indole Synthesis*

Lin and coworkers applied the Wittig–Madelung indole synthesis in a manner so as to obtain 3-acyloxyindoles **19** as shown in Scheme 7 (equation 1) [19]. Interesting is the observation that the R^2CO acyl group is not transferred during the reaction but rather undergoes a normal Wittig–Madelung cyclization. Lin's group extended this chemistry to a Michael-addition version involving enone amides **20**, giving rise to vinylogous 3-acyloxyindoles **21** under exceptionally mild conditions (equation 2) [20].

Hughes developed a polymer-bound phosphonium salt support for a solid phase Wittig–Madelung indole synthesis (Scheme 8) [21]. The method uses a commercially available polymer-bound triphenylphosphine **22**.

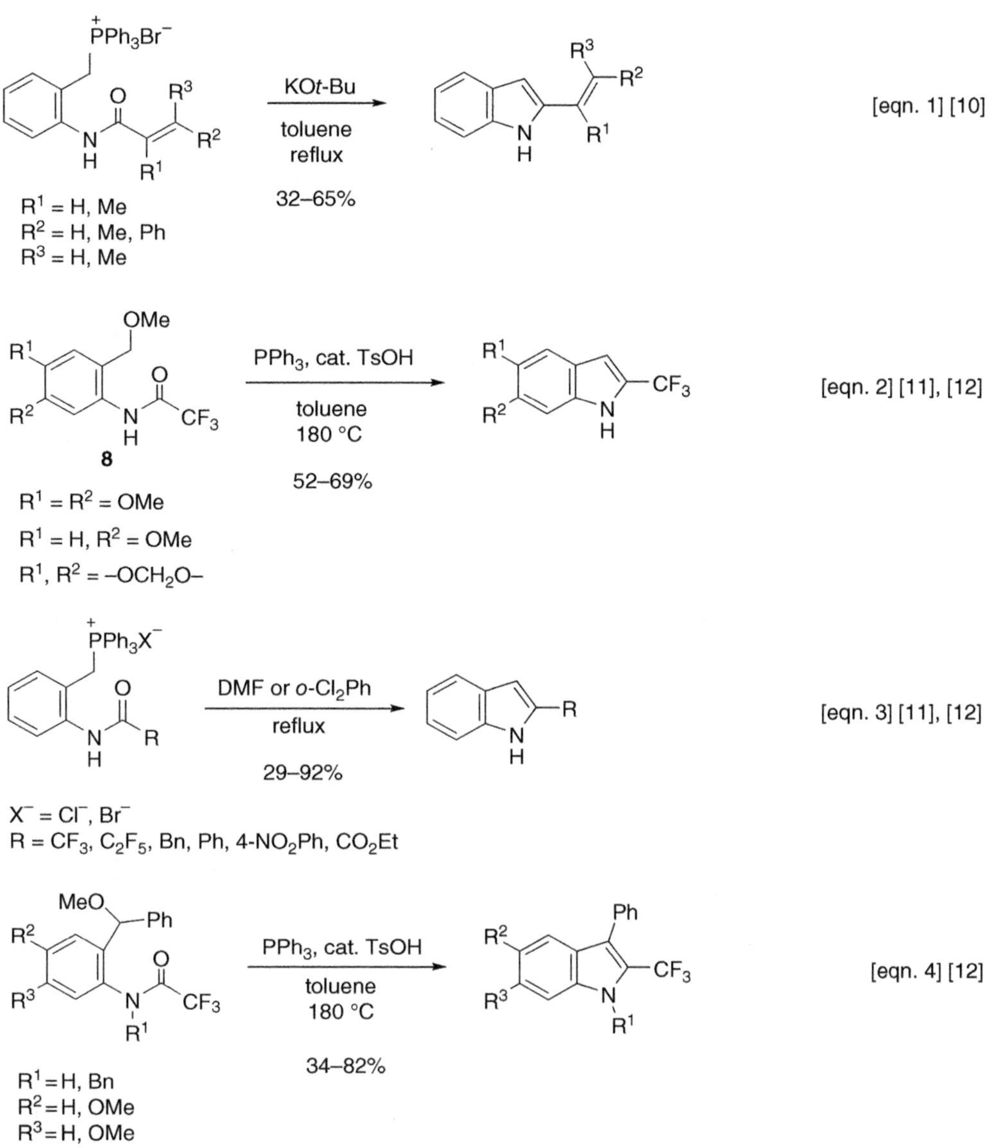

Scheme 3 *Pindur and Imaniski Indole Syntheses*

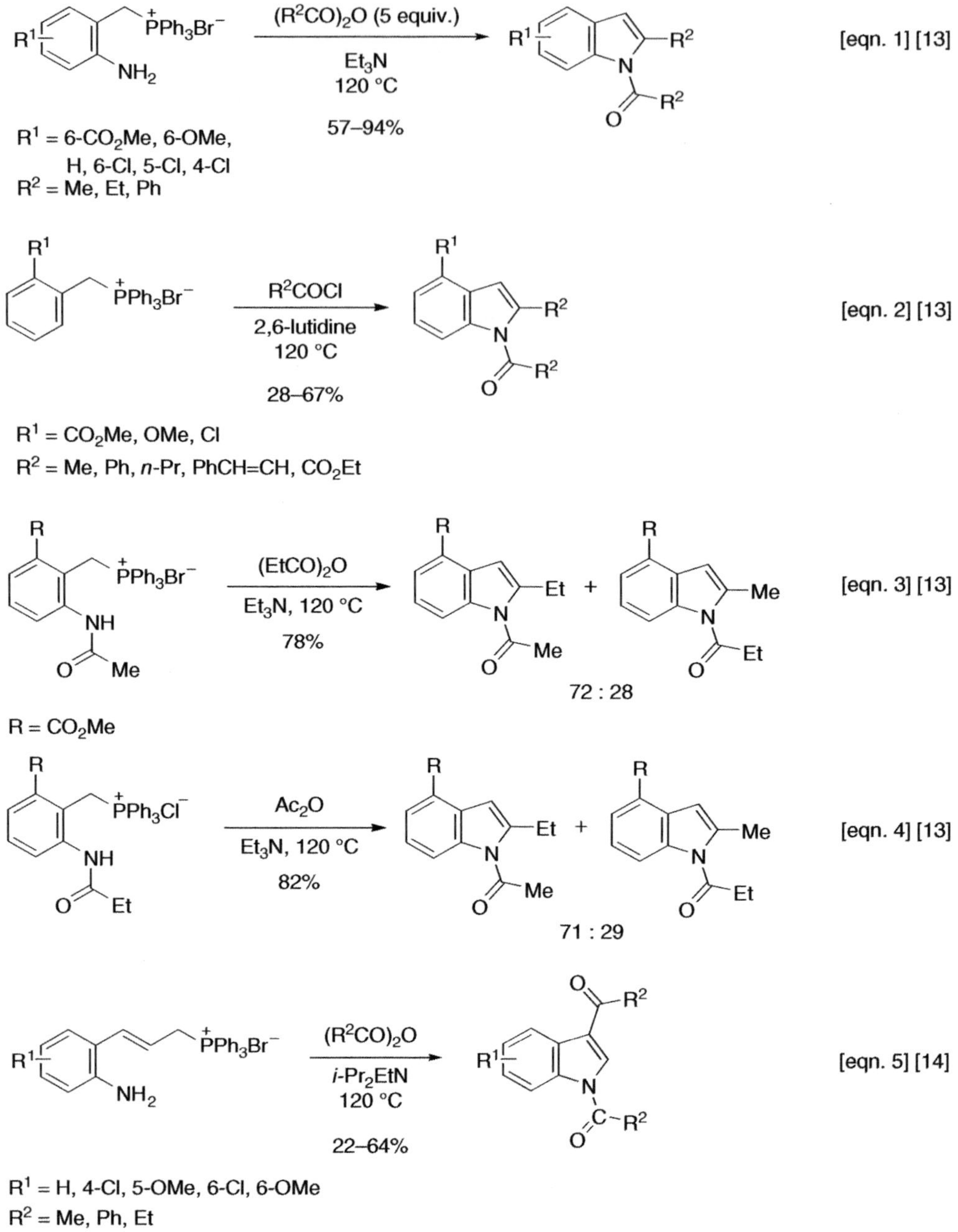

Scheme 4 *Imamoto Indole Synthesis*

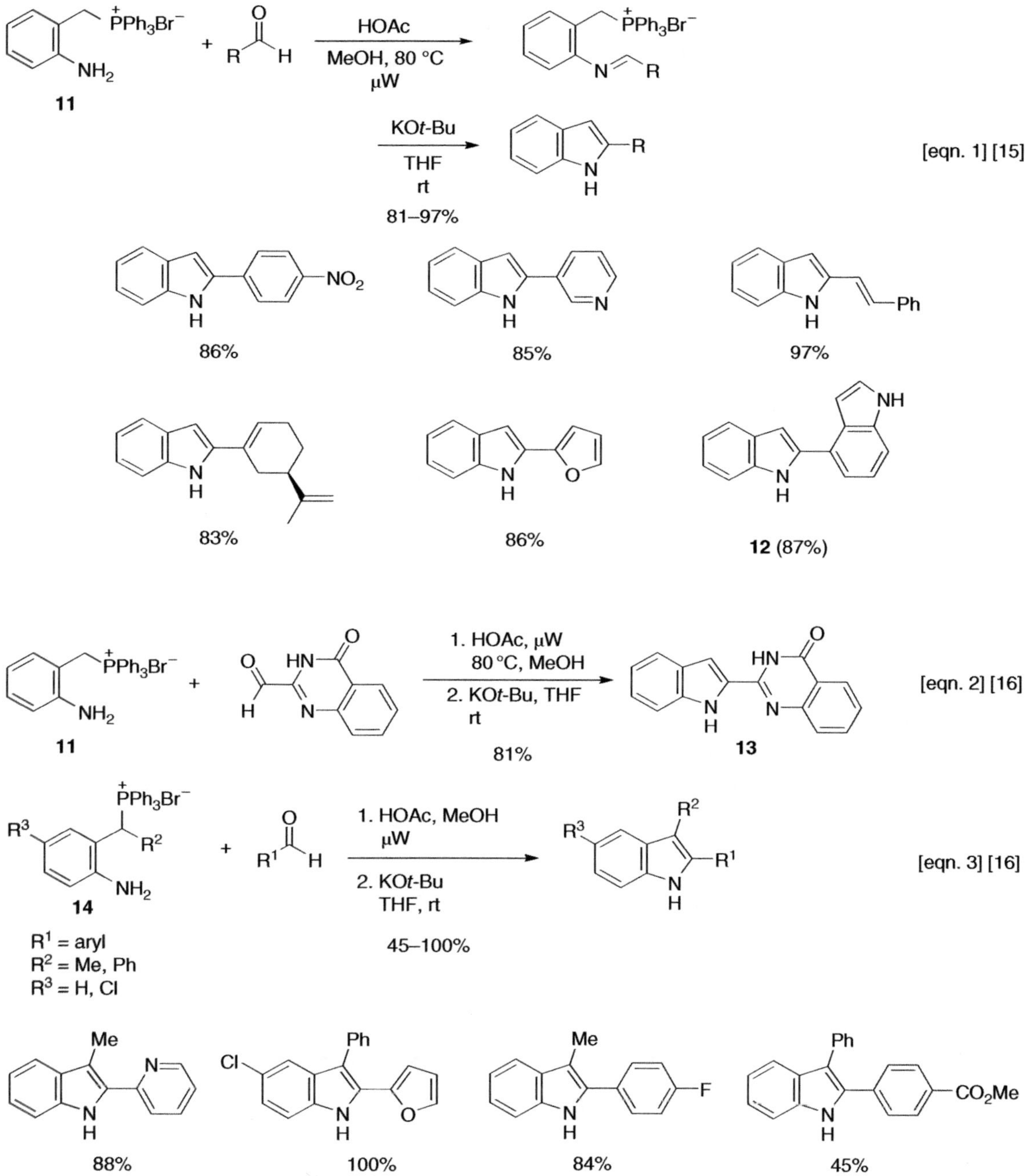

Scheme 5 *Kraus Indole Synthesis*

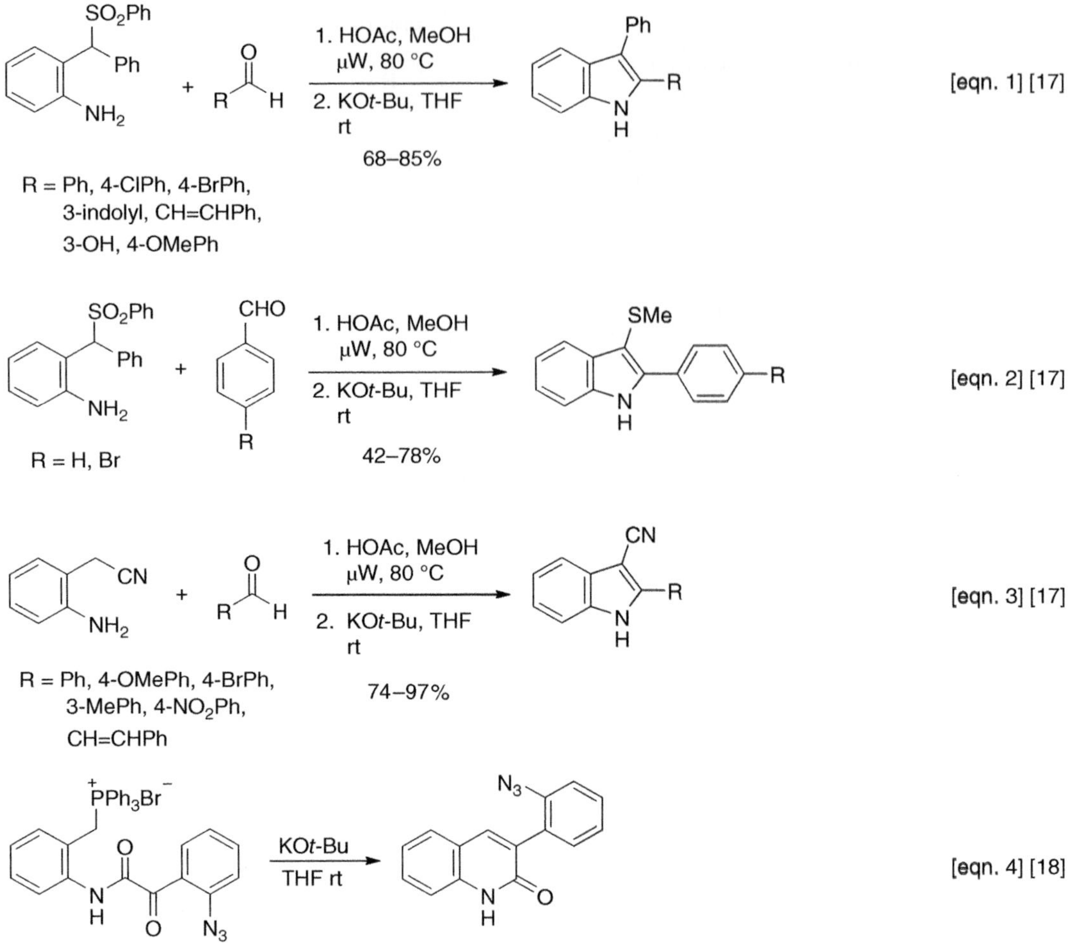

15 **16**

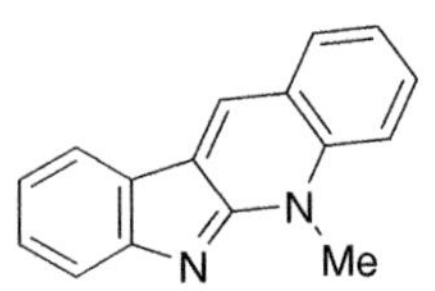

Neocryptolepine (**17**)

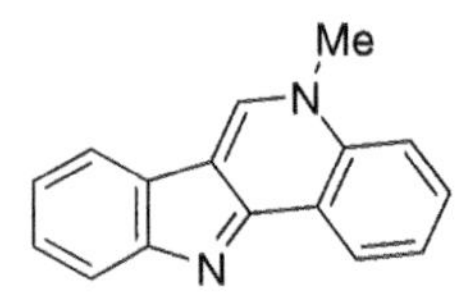

Isocryptolepine (**18**)

Scheme 6 *Kraus Indole Synthesis – 2*

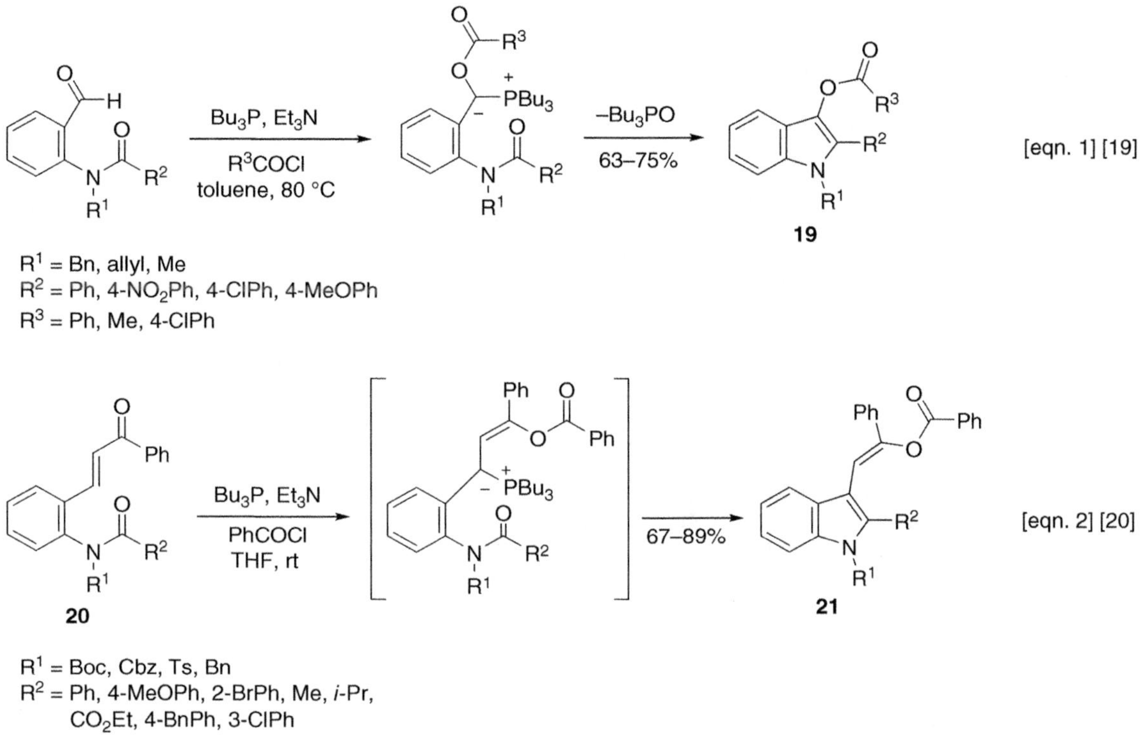

Scheme 7 *Lin Indole Synthesis*

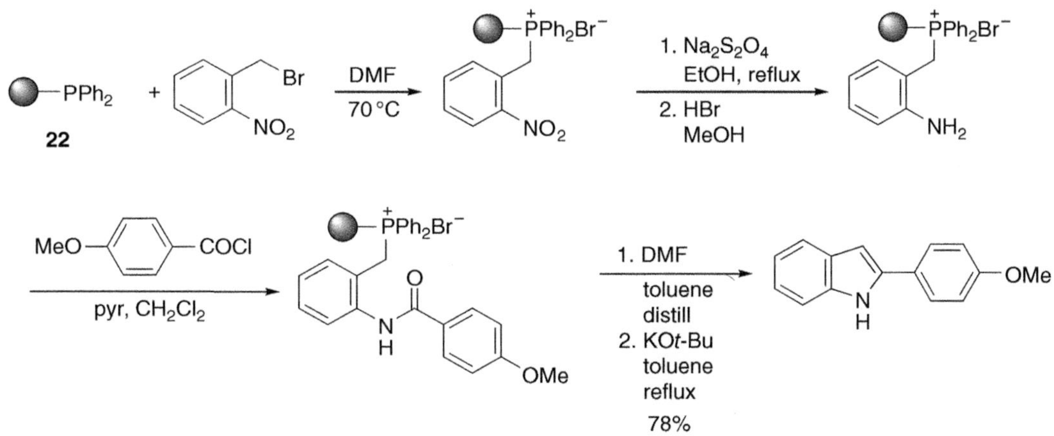

Scheme 8 *Hughes Solid-Phase Madelung Indole Synthesis*

References

[1] M. Le Corre, A. Hercouet, and H. Le Baron, *J. Chem. Soc., Chem. Comm.*, 1981, 14–15.
[2] M. Le Corre, A. Hercouet, Y. Le Stanc, and H. Le Baron, *Tetrahedron*, 1985, **41**, 5313–5320.
[3] M.D. Crenshaw and H. Zimmer, *J. Heterocycl. Chem.*, 1984, **21**, 623–624.
[4] W. Flitsch and P. Rußkamp, *Heterocycles*, 1984, **22**, 541–544.
[5] W. Flkitsch, P. Rußkamp, and W. Langer, *Liebigs Ann. Chem.*, 1985, 1413–1421.
[6] L. Capuano, A. Ahlhelm, and H. Hartmann, *Chem. Ber.*, 1986, **119**, 2069–2074.
[7] L. Capuano, S. Drescher, V. Hammerer, and M. Hanisch, *Chem. Ber.*, 1988, **121**, 2259–2261.
[8] S. Mahboobi and K. Bernauer, *Helv. Chim. Acta*, 1988, **71**, 2034–2041.
[9] B. Danieli, G. Lesma, G. Palmisano, *et al.*, *Tetrahedron*, 1994, **50**, 6941–6954.
[10] M. Eitel and U. Pindur, *Synthesis*, 1989, 364–367.
[11] K. Miyashita, K. Tsuchiya, K. Kondoh, *et al.*, *Heterocycles*, 1996, **42**, 513–516.
[12] K. Miyashita, K. Kondoh, K. Tsuchiya, *et al.*, *J. Chem. Soc., Perkin Trans. 1*, 1996, 1261–1268.
[13] S. Taira, H. Danjo, and T. Imamoto, *Tetrahedron Lett.*, 2002, **43**, 2885–2888.
[14] S. Taira, H. Danjo, and T. Imamoto, *Tetrahedron Lett.*, 2002, **43**, 8893–8896.
[15] G.A. Kraus and H. Guo, *Org. Lett.*, 2008, **10**, 3061–3063.
[16] G.A. Kraus and H. Guo, *J. Org. Chem.*, 2009, **74**, 5337–5341.
[17] G.A. Kraus, H. Guo, G. Kumar, *et al.*, *Synthesis*, 2010, 1386–1393.
[18] G.A. Kraus and H. Guo, *Tetrahedron Lett.*, 2010, **51**, 4137–4139.
[19] S. Syu, Y.-T. Lee, Y.-J. Jang, and W. Lin, *Org. Lett.*, 2011, **13**, 2970–2973.
[20] Y.-T. Lee, Y.-J. Jang, S. Syu, *et al.*, *Chem. Commun.*, 2012, **48**, 8135–8137.
[21] I. Hughes, *Tetrahedron Lett.*, 1996, **37**, 7595–7598.

10

Jones–Schmid Indole Synthesis

In what can be construed as a reverse-Madelung indole synthesis, Jones [1, 2] and Schmid [3] apparently independently discovered the reaction shown in Scheme 1 (equations 1–5), although other workers reported similar reactions, but without subsequent exploitation (equations 6 and 7) [4, 5]. The *N*-alkylated intermediate **1** is normally not isolated, but control experiments show that it is the intermediate (equation 2). Acetal **2** was prepared from the corresponding aniline and converted *in situ* to 2-benzoylindole (equation 3), indicating that acidic conditions can accomplish this indole synthesis. Indeed, the simplicity of this reverse-Madelung cyclization suggests that this indole ring synthesis has been encountered by others well prior to the work of Jones and Schmid. Indeed, as we will see in a later section, Nenitzescu discovered a related acid-induced cyclization leading to *N,O*-diacetylindoxyl. The generality of Jones's second method is seen by the mild base conversion of acyl sulfonamides to indoles (equation 4) [2]. Greuter and Schmid employed stronger basic conditions in their cyclization of amides **3** to indoles **4** (equation 5) [3]. Other examples of this reverse-Madelung indole synthesis are depicted in Table 1 [6–17]. Another early example is shown in Entry 1, where the triketone was prepared *in situ* from the *ortho*-amino ketone and the bromomethyl diketone [6]. Exceptionally mild conditions formed the indole. Molina and coworkers employed the conditions of Schmid to synthesize a range of iminophosphoranes (Entry 2) [7]. This reverse-Madelung indole construction is applicable to the synthesis of 3-aminoindoles, which entails cyclization onto the appropriate nitrile (Entries 4–11) [11–17]. In Entry 6 the labile 3-aminoindoles were capped with acetic anhydride or, in one case, with methyl iodide [13]. Koutentis and Michaelidou (Entry 7) also prepared the 5-nitroindole derivative directly from 5-nitro-2-(tosylamino)benzonitrile and chloroacetonitrile (40%) [14]. Entries 9 and 10 both feature *in situ* alkylation with ethyl bromoacetate. In Entry 10 the minor alternative regioisomer (not shown) is formed in 9% to 23% [16].

An important variation of the Jones–Schmid indole synthesis is the base-mediated cyclization onto esters, which affords 3-hydroxyindoles (indoxyl tautomers) under mild conditions. Three case studies are shown in Scheme 2 (equations 1–3) [18–20]. This chemistry, like that in Table 1 (Entries 4–11), represents a powerful route to C-3 functionalized indoles. Kraus and coworkers use the sterically hindered phosphazine base, P_4-*t*-Bu, to prepare the 5,6-dihydroindolo[2,1-*a*]isoquinoline ring system (equation 4) [21]. Several derivatives display immunosuppressive activity.

Stevens and colleagues extended the base-promoted reverse-Madelung reaction to a vinylogous version (Scheme 3, equation 1) [22]. Among the many 2-acyl-3-indole esters (and carboxylic acids) that were prepared are **5–7**. This clever tactic has been used by Opatz [23, 24] and Webb [25] as summarized in Table 2. Scheme 4 depicts the reaction pathway and includes some of Opatz's examples (**8–10**).

One of the lesser known Nenitzescu name reactions is the cyclization of *N*-acetylphenylglycine-*ortho*-carboxylic acid **11** to *N,O*-diacetylindoxyl **12** (Scheme 5, equation 1) [26, 27]. While this precise reaction has a long history [28], it was Nenitzescu and colleagues who greatly improved the preparation of this important indole. As you can see, the reaction involves an acid-promoted reverse-Madelung cyclization. Some examples of this

Indole Ring Synthesis: From Natural Products to Drug Discovery, First Edition. Gordon W. Gribble.

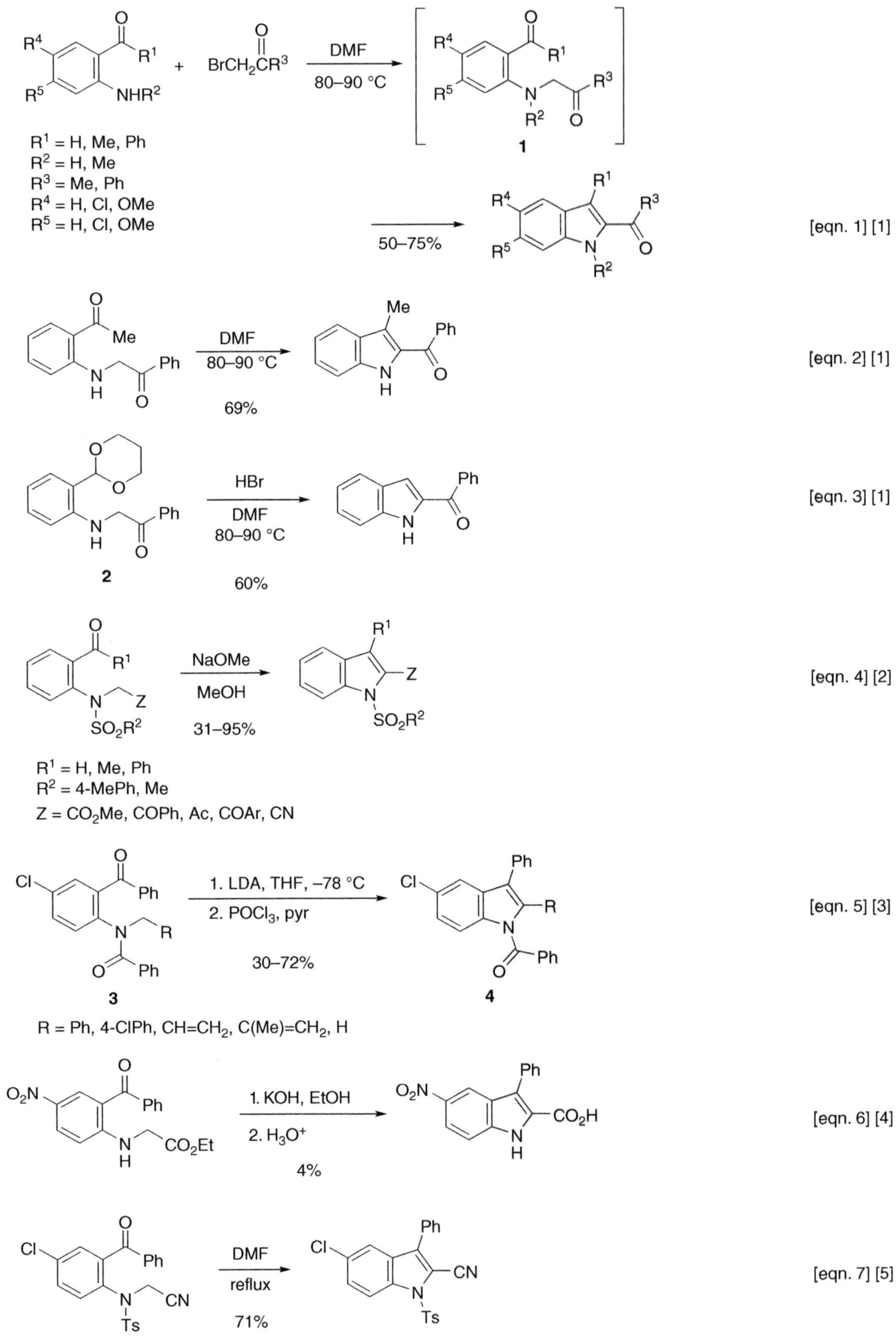

Scheme 1 *Jones-Schmid Indole Synthesis*

Table 1 *Reverse-Madelung Indole Synthesis*

Entry	Substrate	Conditions	Indole	% Yield	Ref.
1	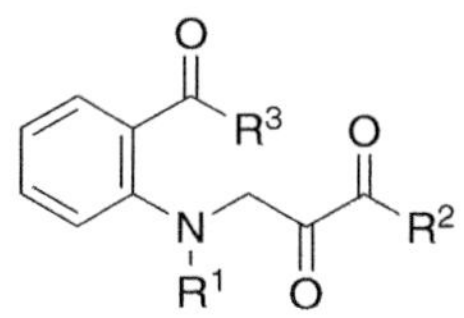 R^1 = Me, Et; R^2 = R^3 = Me, Ph	$NaHCO_3$ 75–80 °C EtOH	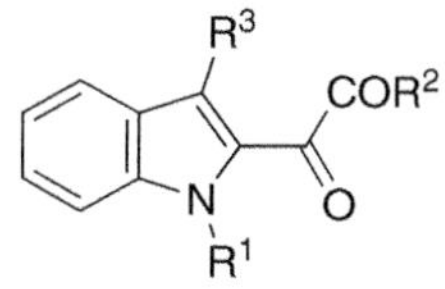	26–39%	6
2	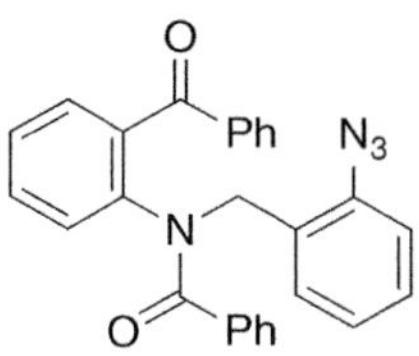	1. LDA, THF −78 °C 2. $SOCl_2$, pyr	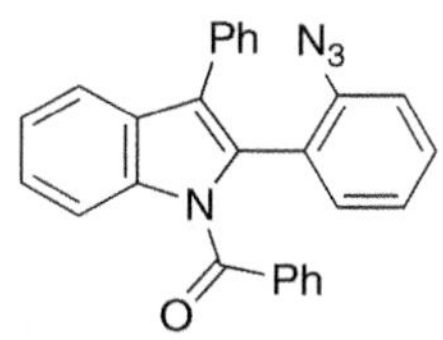	73%	7
3	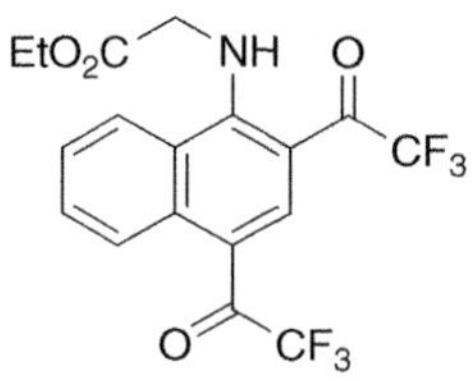	1. Et_3N, MeCN reflux 2. TFA, rt	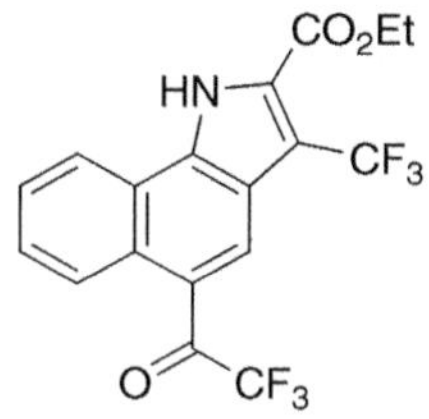	99%	8–10
4	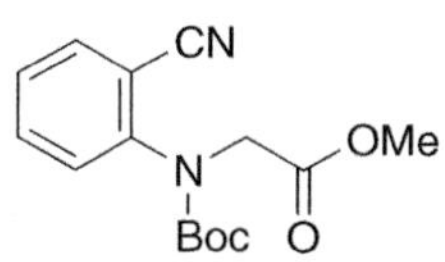	LDA, THF −78 °C	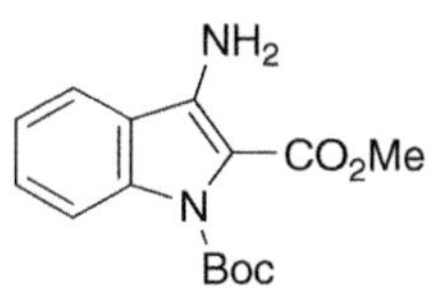	89%	11
5	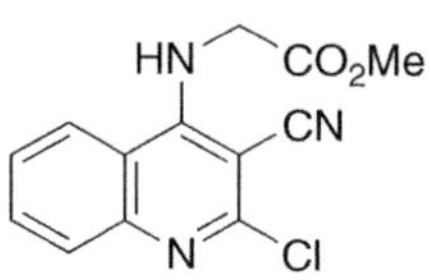	NaOMe MeOH reflux	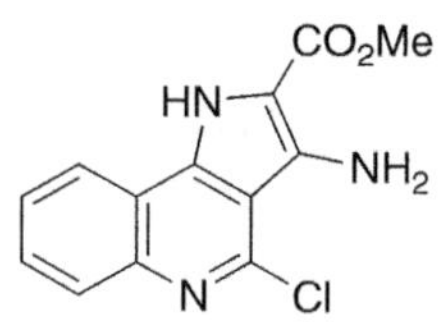	85%	12
6	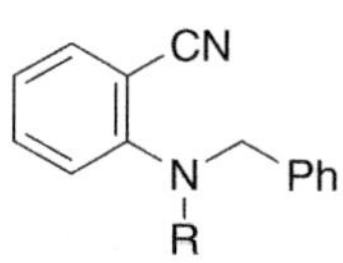 R = Me, Et, Bn, *n*-Pr	1. NaH DMF 80 °C 2. Ac_2O	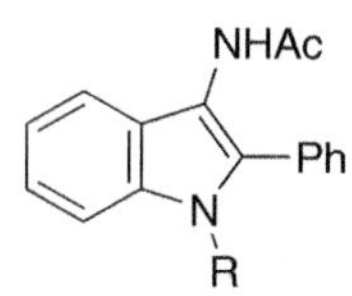	80–90%	13
7	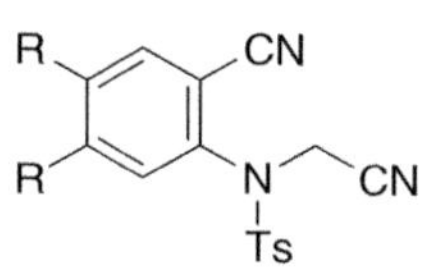 R = H OMe	K_2CO_3, EtOH rt or 78 °C	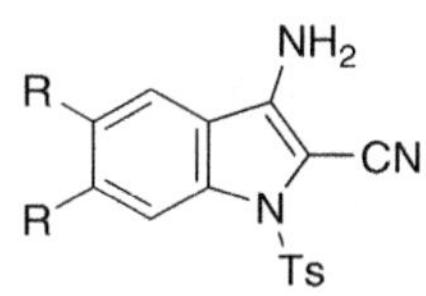	93–96%	14
8	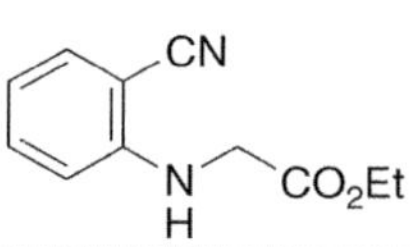	KO*t*-Bu, THF rt	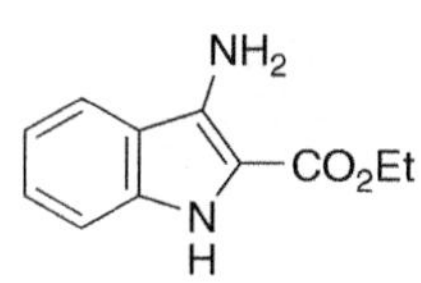	90%	15

(continued overleaf)

Table 1 *(continued)*

Entry	Substrate	Conditions	Indole	% Yield	Ref.
9	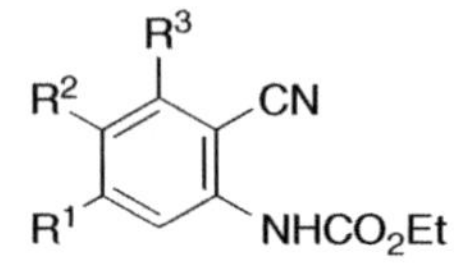 R^1 = H, Me, Cl R^2 = H, Cl, OMe R^3 = H, Cl, Me	NaH, DMF $BrCH_2CO_2Et$ rt	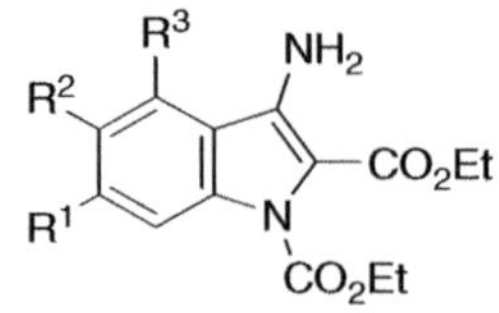	70–96%	15
10	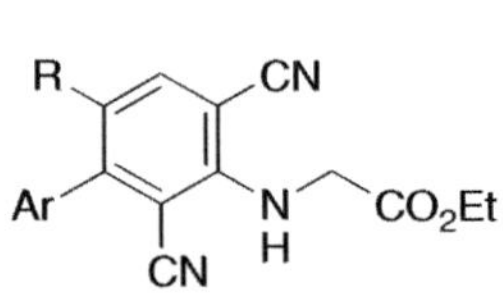 R = Me, *n*-Pr, *n*-Hex Ar = 4-OMePh, 1-naphthyl, 2-thienyl, 2-furanyl, 3,4-(OCH_2O)Ph	KOH MeCN rt	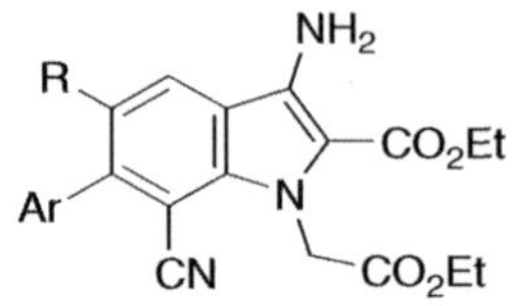	61–85%	16
11	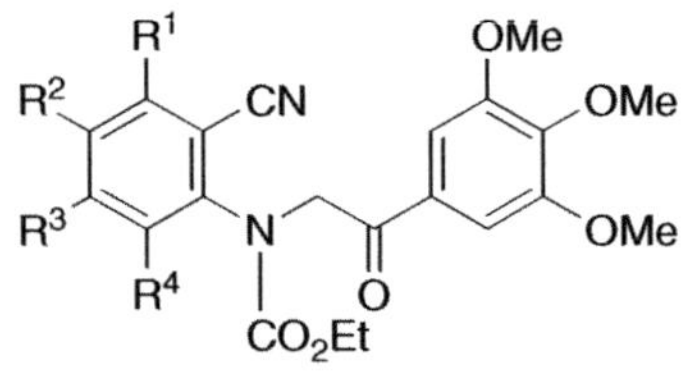 R^1 = H, Me, OMe R^2 = H, OMe, Cl, Me R^3 = H, OMe, Cl, Me R^4 = H, Me	aq NaOH EtOH reflux	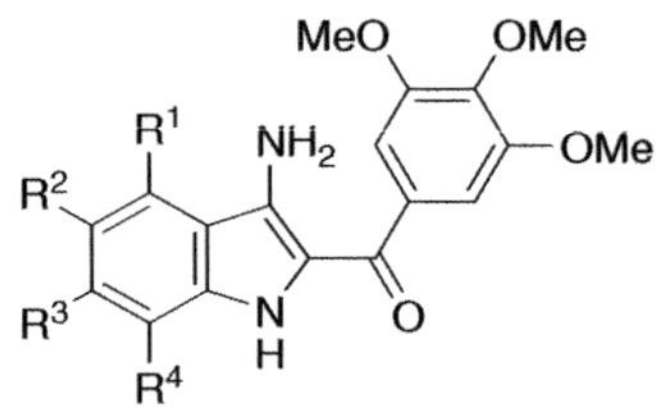		17

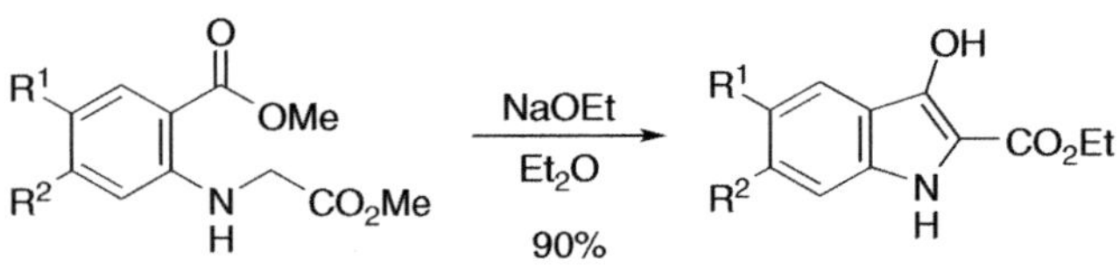

R^1 = H, Cl, Me, Et, *i*-Bu
R^2 = H, Cl, OCF_3

[eqn. 1] [18]

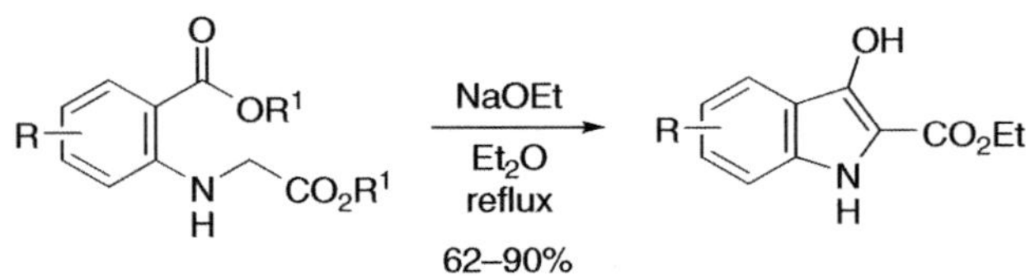

R^1 = Me, Et
R = H, Br, Cl, F, Me, *i*-Pr
(various positions)

[eqn. 2] [19]

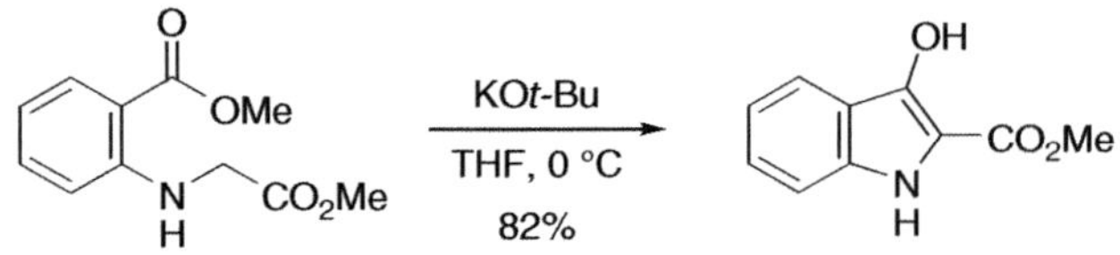

[eqn. 3] [20]

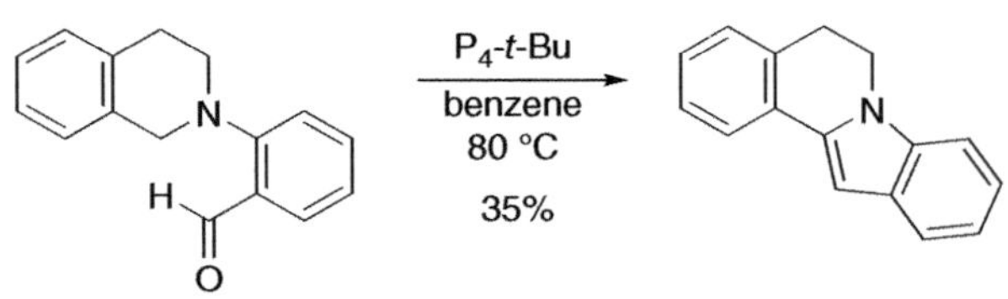

[eqn. 4] [21]

Scheme 2 *Variations of the Jones-Schmid Indole Synthesis*

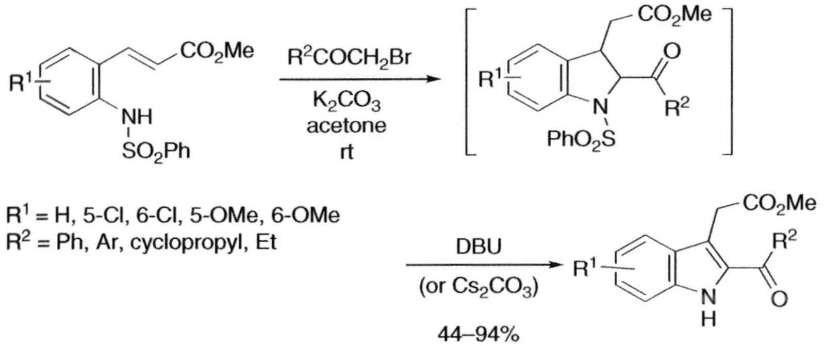

[eqn. 1] [22]

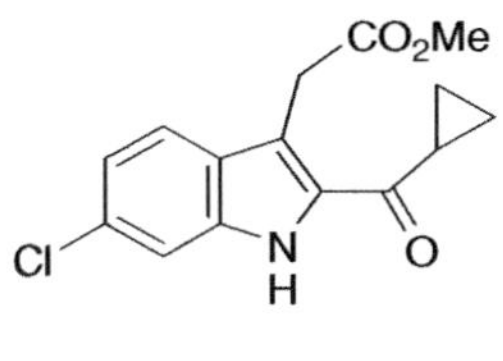 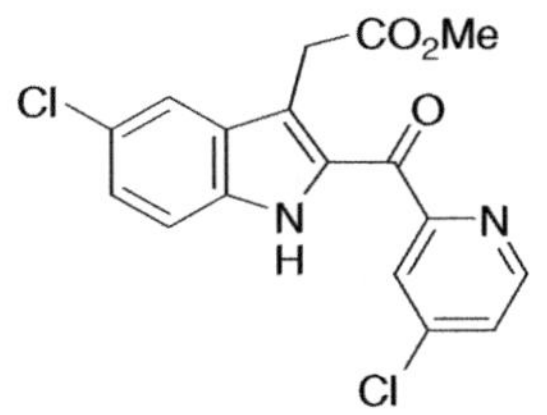 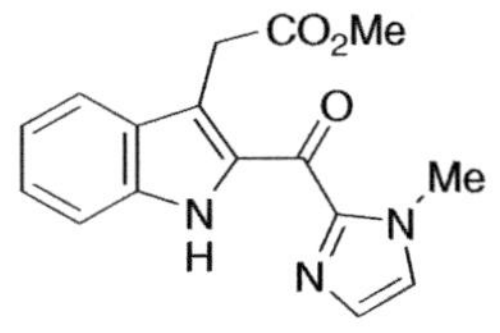

5 (94%) **6** (72%) **7** (57%)

Scheme 3 *Stevens Indole Synthesis*

Table 2 *Stevens-Opatz Modification of the Jones-Schmid Indole Synthesis*

Entry	Substrate	Conditions	Indole	% Yield	Ref.
1	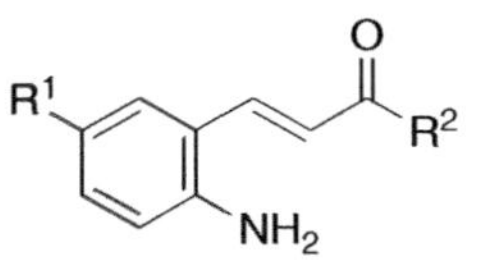 R^1 = H, F; R^2 = OEt, O-*n*-Bu, N(CH$_2$)$_5$; R^3 = 2-naphthyl, 4-Pyr, 3-Pyr, 4-FPh, 4-PhPh, 3-MeOPh, 4-ClPh, 2-thienyl	1. R^{3}CHO, HOAc KCN 2. KO*t*-Bu	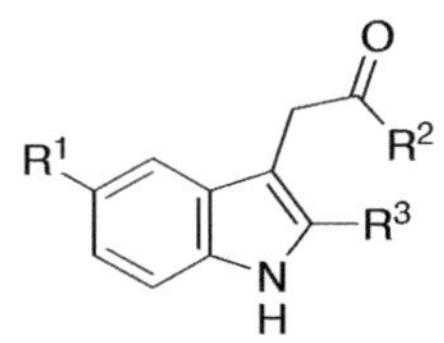	37–95%	23
2	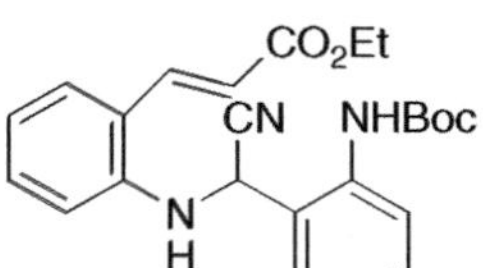	KO*t*-Bu EtOH	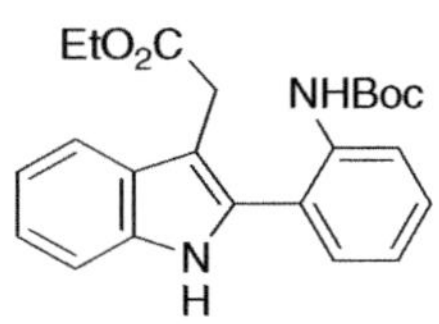	74%	24
3	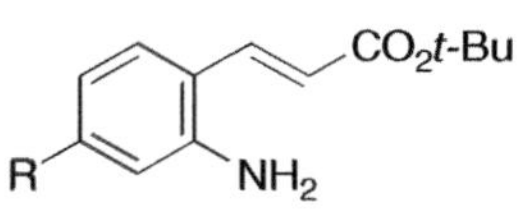 R = H, Cl	1. 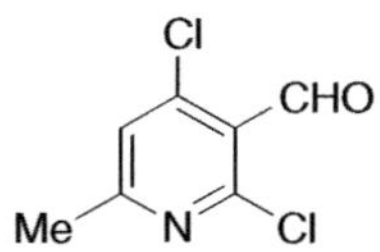 EtOH 2. KCN, HOAc EtOH	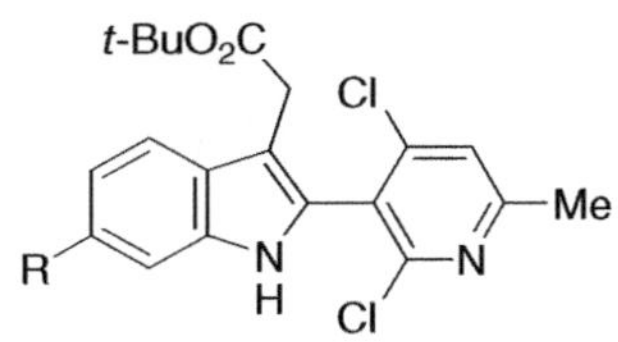	68–73%	25

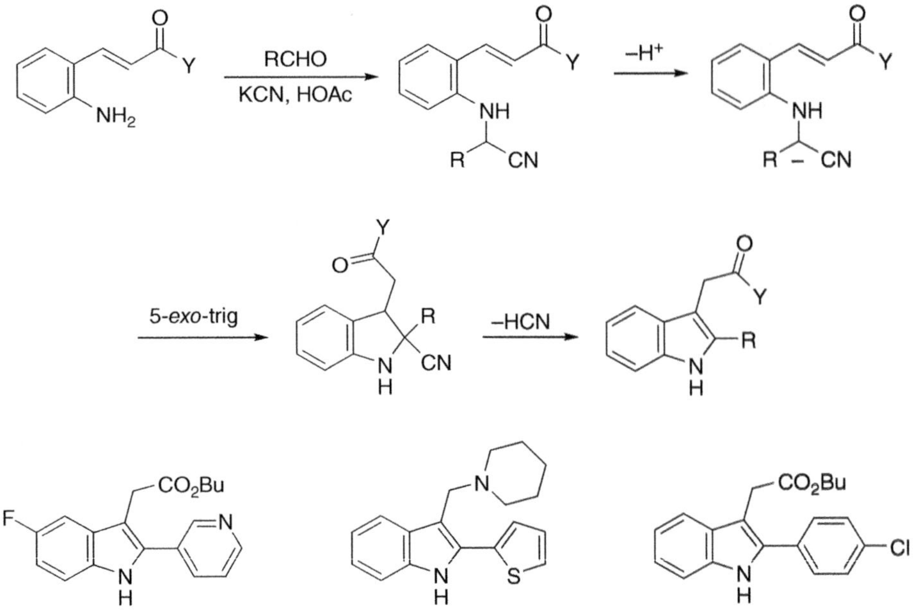

8 (69%) **9** (51%) **10** (81%)

Scheme 4 *Opatz Indole Synthesis*

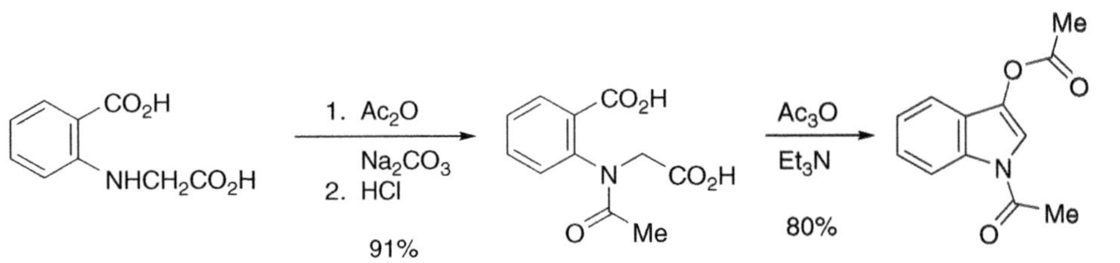

[eqn. 1] [26]

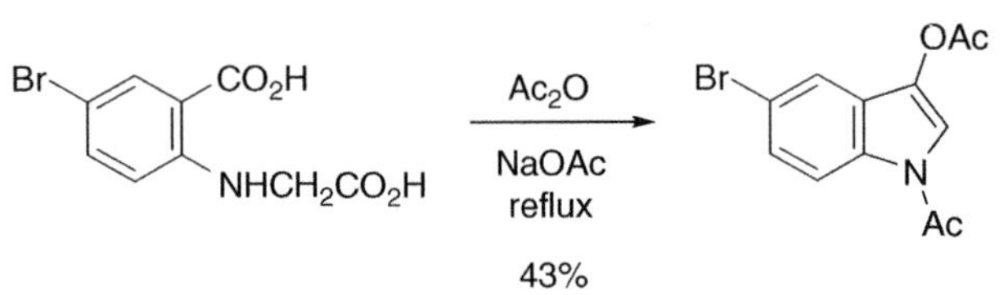

[eqn. 2] [29]

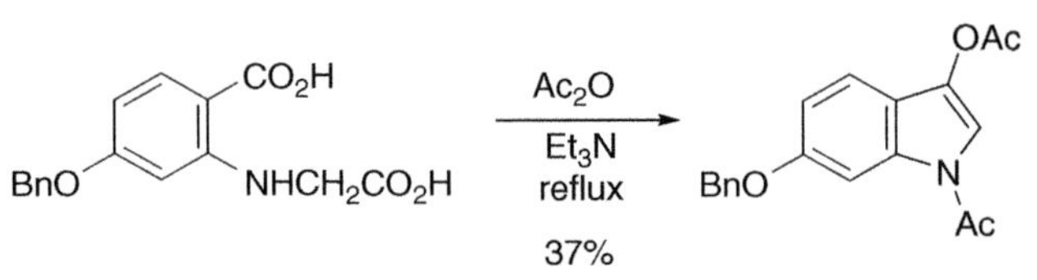

[eqn. 3] [30]

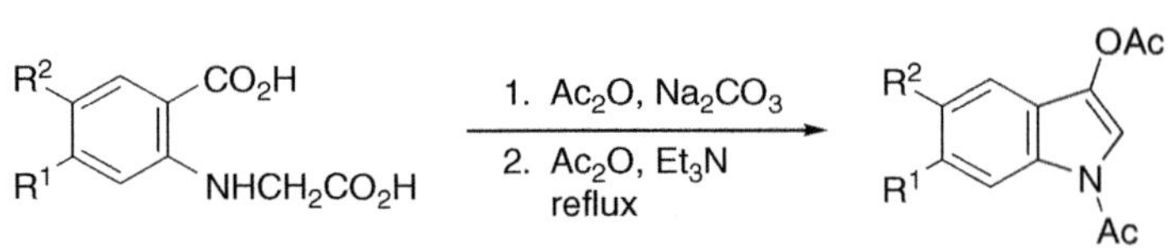

R^1 = H, Cl
R^2 = H, Br

[eqn. 4] [31]

Scheme 5 *Nenitzescu Indole Synthesis*

chemistry are shown in Scheme 5 [29–31]. An important utility of these *N,O*-diacetylindoxyls was demonstrated by Wright's group in their synthesis of cryptolepine antimalarial analogues by condensation with isatins [31, 35]. Some additional examples of this *N,O*-diacetylindoxyl preparation are shown in Table 3 [32–38]. The chemistry by Funk and Huntley in Entry 4 was a key feature in their synthesis of the dragmacidin E core system [36]. Entry 6 reveals that different electrophilic conditions (Vilsmeier), other than acetic anhydride, can accomplish indole ring formation.

Mohanakrishnan and colleagues described a one-pot synthesis of 1-phenylsulfonyl-2-aroylindoles and related indoles that nicely complements the chemistry we have been discussing (Scheme 6, equations 1–3) [39].

A review of the chemistry of anthranilic acid by Wiklund and Bergman includes some of the chemistry in this chapter [40].

Table 3 *Syntheses of* N,O*-Diacetylindoxyls and Related Chemistry*

Entry	Substrate	Conditions	Indole	%Yield	Ref.
1	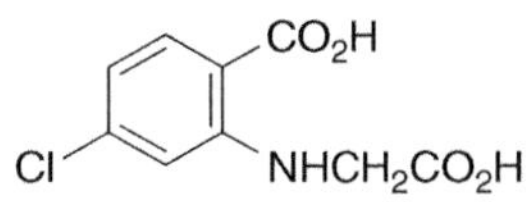	Ac_2O, NaOAc reflux	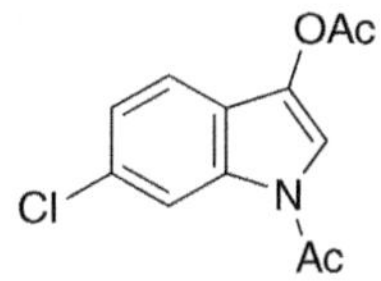	45%	32
2	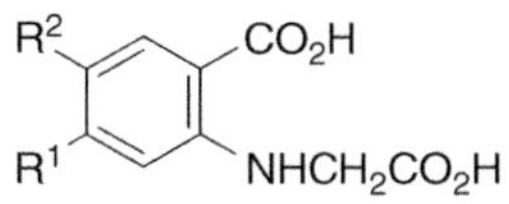 R^1 = H, Cl, NO_2; R^2 = H, Br, Cl, NO_2	Ac_2O, NaOAc reflux	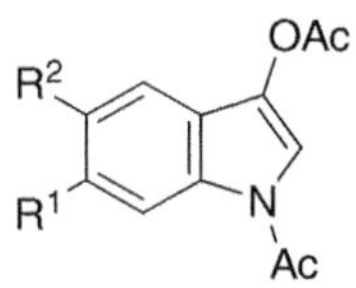	10–70%	33, 34
3	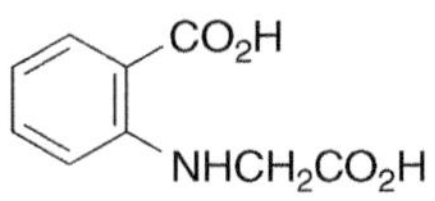	1. aq. Ac_2O, μν 2. Ac_2O, Et_3N	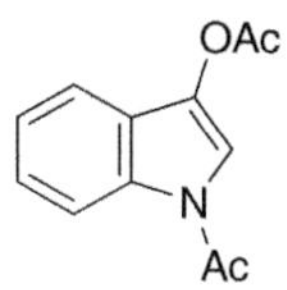	79%	35
4	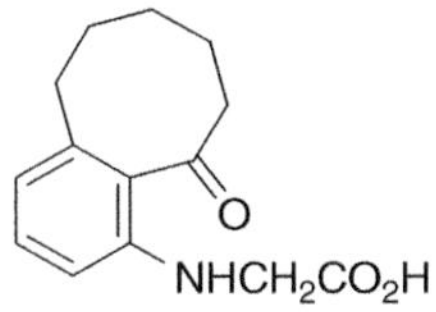	Ac_2O, 130 °C Et_3N	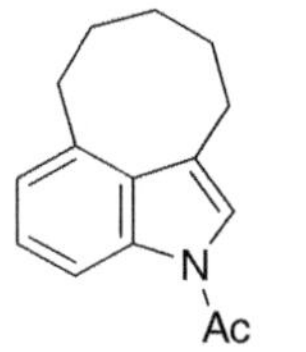	75%	36
5	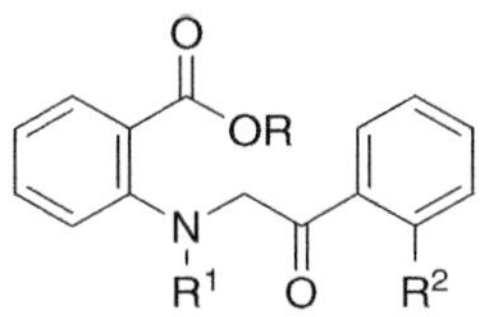 R = Me, H, CH_2Bz; R^1 = H, Me; R^2 = H, OH	NaOMe benzene	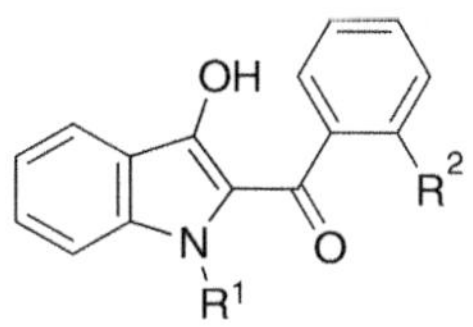	21–59%	37
6	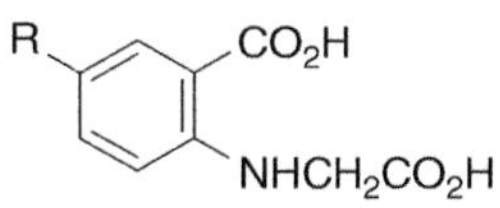 R = H, Me, Cl, Br	DMF, $POCl_3$ 90 °C	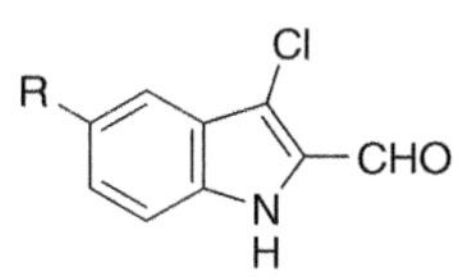	45–75%	38

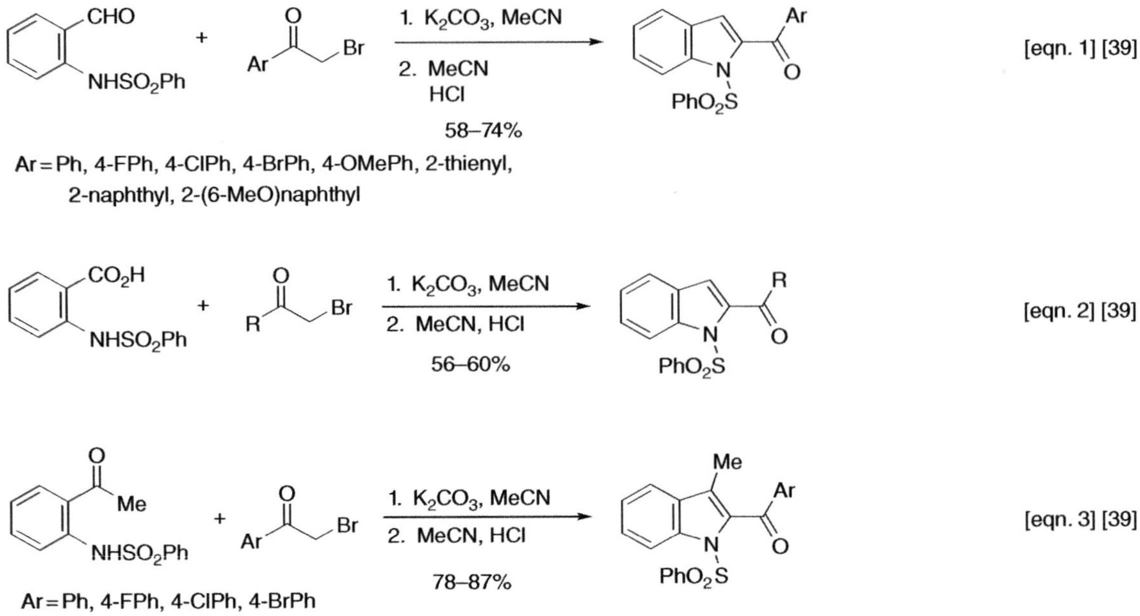

Scheme 6 *Mohanakrishnan Indole Synthesis*

References

[1] C.D. Jones and T. Suárez, *J. Org. Chem.*, 1972, **37**, 3622–3623.
[2] C.D. Jones, *J. Org. Chem.*, 1972, **37**, 3624–3625.
[3] H. Greuter and H. Schmid, *Helv. Chim. Acta*, 1974, **57**, 281–286.
[4] R.I. Fryer, J.V. Earley, and L.H. Sternbach, *J. Org. Chem.*, 1967, **32**, 3798–3803.
[5] M. Oklobdžija, M. Japelj, and T. Fajdiga, *J. Heterocyclic Chem.*, 1972, **9**, 161–163.
[6] G. Kempter and E. Schiewald, *J. Prakt. Chem.*, 1965, **28**, 169–171.
[7] P. Molina, M. Alajarin, and A. Vidal, *Tetrahedron*, 1990, **46**, 1063–1078.
[8] M. Hojo, R. Masuda, E. Okada, and H. Miya, *Synthesis*, 1989, 550–552.
[9] E. Okada and N. Tsukushi, *Synlett*, 1999, 210–212.
[10] E. Okada and N. Tsukushi, *Heterocycles*, 2000, **53**, 127–134.
[11] D.L. Boger and T. Nishi, *Bioorg. Med. Chem.*, 1995, **3**, 67–77.
[12] R.A. Mekheimer, *Synthesis*, 2000, 2078-2084.
[13] C.M. Seong, C.M. Park, J. Choi, and N.S. Park, *Tetrahedron Lett.*, 2009, **50**, 1029–1031.
[14] S.S. Michaelidou and P.A. Koutentis, *Tetrahedron*, 2010, **66**, 3016–3023.
[15] P. Diana, A. Stagno, P. Barraja, *et al.*, *Tetrahedron*, 2011, **67**, 3374–3379.
[16] S.P. Sawargave, A.S. Kudale, J.V. Deore, *et al.*, *Tetrahedron Lett.*, 2011, **52**, 5491–5493.
[17] R. Romagnoli, P.G. Baraldi, T. Sarkar, *et al.*, *J. Med. Chem.*, 2008, **51**, 1464–1468.
[18] J.F. Dropinski, T. Akiyama, M. Einstein, *et al.*, *Bioorg. Med. Chem. Lett.*, 2005, **15**, 5035–5038.
[19] M. Bös, F. Jenck, J.R. Martin, *et al.*, *Eur. J. Med. Chem.*, 1997, **32**, 253–261.
[20] M. Jessing and P.S. Baran, *Heterocycles*, 2011, **82**, 1739–1745.
[21] G.A. Kraus, V. Gupta, M. Kohut, and N. Singh, *Bioorg. Med. Chem. Lett.*, 2009, **19**, 5539–5542.
[22] K. Nakao, Y. Murata, H. Koike, *et al.*, *Tetrahedron Lett.*, 2003, **44**, 7269–7271.
[23] T. Opatz and D. Ferenc, *Org. Lett.*, 2006, **8**, 4473–4475.
[24] T. Opatz and D. Ferenc, *Synthesis*, 2008, 3941–3944.
[25] P.J. Slavish, Q. Jiang, X. Cui, *et al.*, *Bioorg. Med. Chem.*, 2009, **17**, 3308–3316.
[26] D. Răileanu, O. Constantinescu-Simon, E. Moşanu, and C.D. Nenitzescu, *Rev. Roum. Chim.*, 1967, **12**, 105–108.
[27] D. Raileanu, E. Tighineanu, and F. Dumitrascu, *Revista de Chimie*, 1992, **43**, 506–511.
[28] (a)A. Rössing, *Ber. Dtsch. Chem. Ges.*, 1884, **17**, 2988–3010. (b)D. Vorländer, *Ber. Dtsch. Chem. Ges.*, 1902, **35**, 1683–1698.
[29] H.C.F. Su and K.C. Tsou, *J. Am. Chem. Soc.*, 1960, **82**, 1187–1189.
[30] E. Tighineanu, F. Chiraleu, and D. Răileanu, *Tetrahedron*, 1980, **36**, 1385–1397.
[31] O. Onyeibor, S.L. Croft, H.I. Dodson, *et al.*, *J. Med. Chem.*, 2005, **48**, 2701–2709.
[32] A. Balbuzano-Deus, J.C. Rodríguez-Domínguez, A. Fernández-Villalobo, *et al.*, *Org. Prep. Proc. Int.*, 2006, **38**, 87–99.
[33] J.C. Rodríguez-Domínguez, A. Balbuzano-Deus, M.A. López-López, and G. Kirsch, *J. Heterocycl. Chem.*, 2007, **44**, 273–275.

[34] J.C. Rodríguez-Dominguez, X. Gang, and G. Kirsch, *Synthesis*, 2009, 2345–2348.
[35] T.K. Lai, A. Chatterjee, J. Banerji, *et al.*, *Helv. Chim. Acta*, 2008, **91**, 1975–1983.
[36] R.J. Huntley and R.L. Funk, *Org. Lett.*, 2006, **8**, 4775–4778.
[37] K. Görlitzer, *Arch. Pharmaz.*, 1974, **307**, 523–539.
[38] V.J. Majo and P.T. Perumal, *J. Org. Chem.*, 1996, **61**, 6523–6525.
[39] V. Dhayalan, G. Panchapakesan, and A.K. Mohanakrishnan, *Synth. Commun.*, 2012, **42**, 402–411.
[40] P. Wiklund and J. Bergman, *Curr. Org. Synth.*, 2006, **3**, 379–402.

11

Couture Indole Synthesis

The Couture indole synthesis is a modified reverse-Madelung reaction in which the methylene protons adjacent to the aniline nitrogen are acidified by a diphenylphosphinoyl group [1, 2]. Thus, it might be termed a reverse-Horner–Wittig–Madelung indole synthesis. The basic chemistry is illustrated by the examples in Scheme 1. The Horner–Wittig reagents **1** are prepared as shown in two steps from the appropriate anilines (equation 1). Depending on the starting material **1**, one has access to 3-alkyl-, 3-hydroxy, or 3-aminoindoles (equations 2–4). With the base potassium bis(trimethylsilyl)amide followed by acidic workup are provided the simple C-2 unsubstituted indoles (equation 5) [2].

Indole Ring Synthesis: From Natural Products to Drug Discovery, First Edition. Gordon W. Gribble.

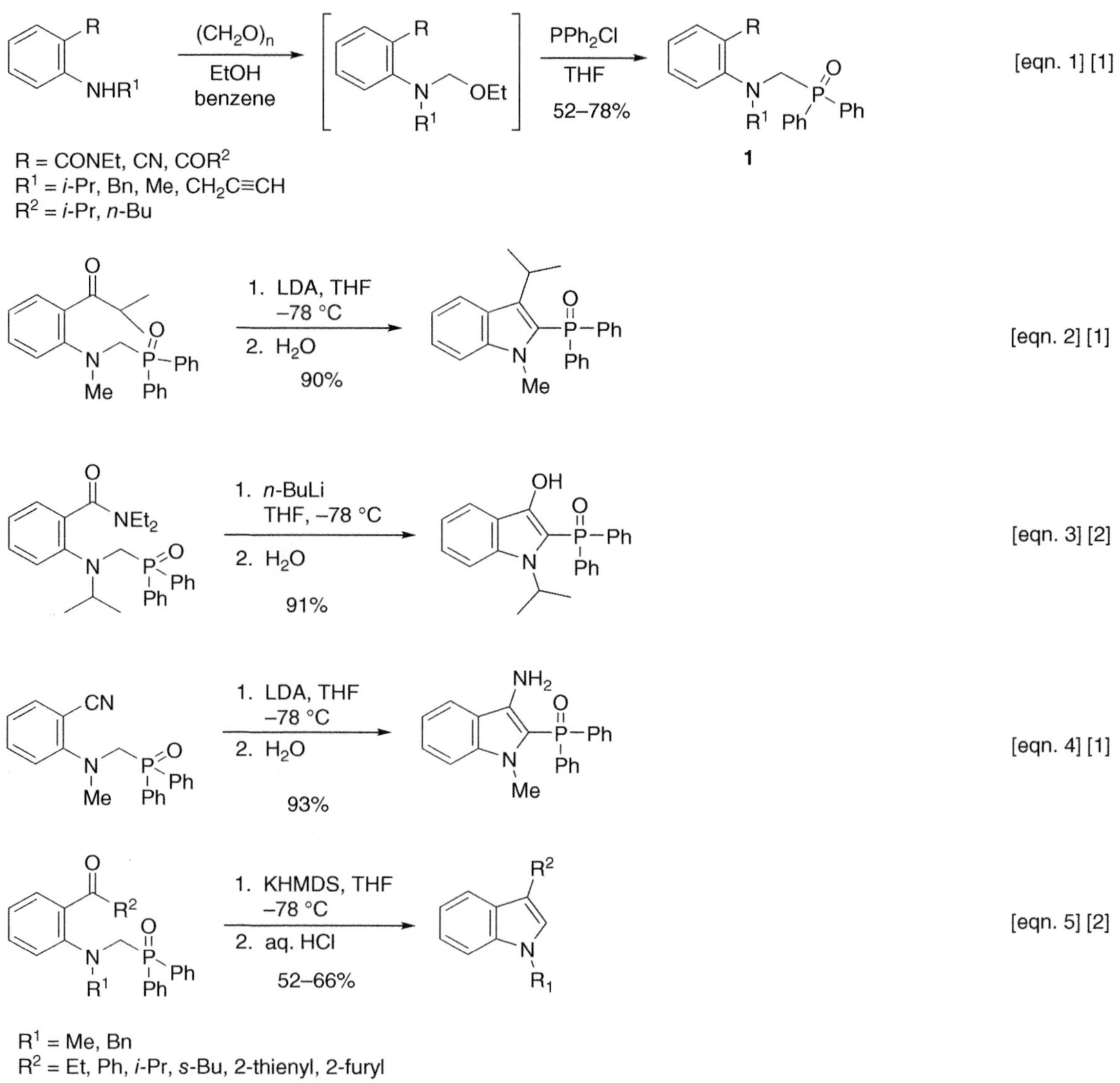

Scheme 1 *Couture Indole Synthesis*

References

[1] A. Couture, E. Deniau, Y. Gimbert, and P. Grandclaudon, *Tetrahedron*, 1993, **49**, 1431–1444.
[2] A. Couture, E. Deniau, Y. Gimbert, and P. Grandclaudon, *J. Chem. Soc., Perkin Trans. 1*, 1993, 2463–2466.

12

Wender Indole Synthesis

Wender and White published a very simple indole ring synthesis that involves the generation of a bis-lithio anion **1** and its reaction with an α-halo carbonyl compound **2**, followed by acid- or base-catalyzed dehydration [1, 2]. The overall transformation is shown in Scheme 1, along with three examples. This chemistry illustrates yet another indole ring synthesis that uses α-halo carbonyl compounds [3]. We will encounter these compounds again with the venerable Bischler indole synthesis in Chapter 23. A summary of several Wender indole ring syntheses is tabulated in Table 1 [1, 2, 4–7]. Entry 5 features a directed lithiation method to the bis-lithio nucleophile [5], a modification also described by Wender and White [2]. Sainsbury and colleagues found that cerium trichloride suppresses enolate formation of the chloroindanone and improves the overall yield of indole product (Entry 6) [6]. In fact, the reaction fails in the absence of cerium(III) chloride. In several of the examples in Table 1, the intermediate chlorohydrin is isolable in what is a formal two-step transformation to the indole product.

Wender and White found that bis-electrophiles other than α-halo carbonyls engage in this process. Two examples are illustrated in Scheme 2. Nicolaou and coworkers have developed a somewhat related tryptamine synthesis involving lithiated Boc-anilines in reactions with *N*-Boc-pyrrolidin-3-one (Scheme 3) [8].

Indole Ring Synthesis: From Natural Products to Drug Discovery, First Edition. Gordon W. Gribble.

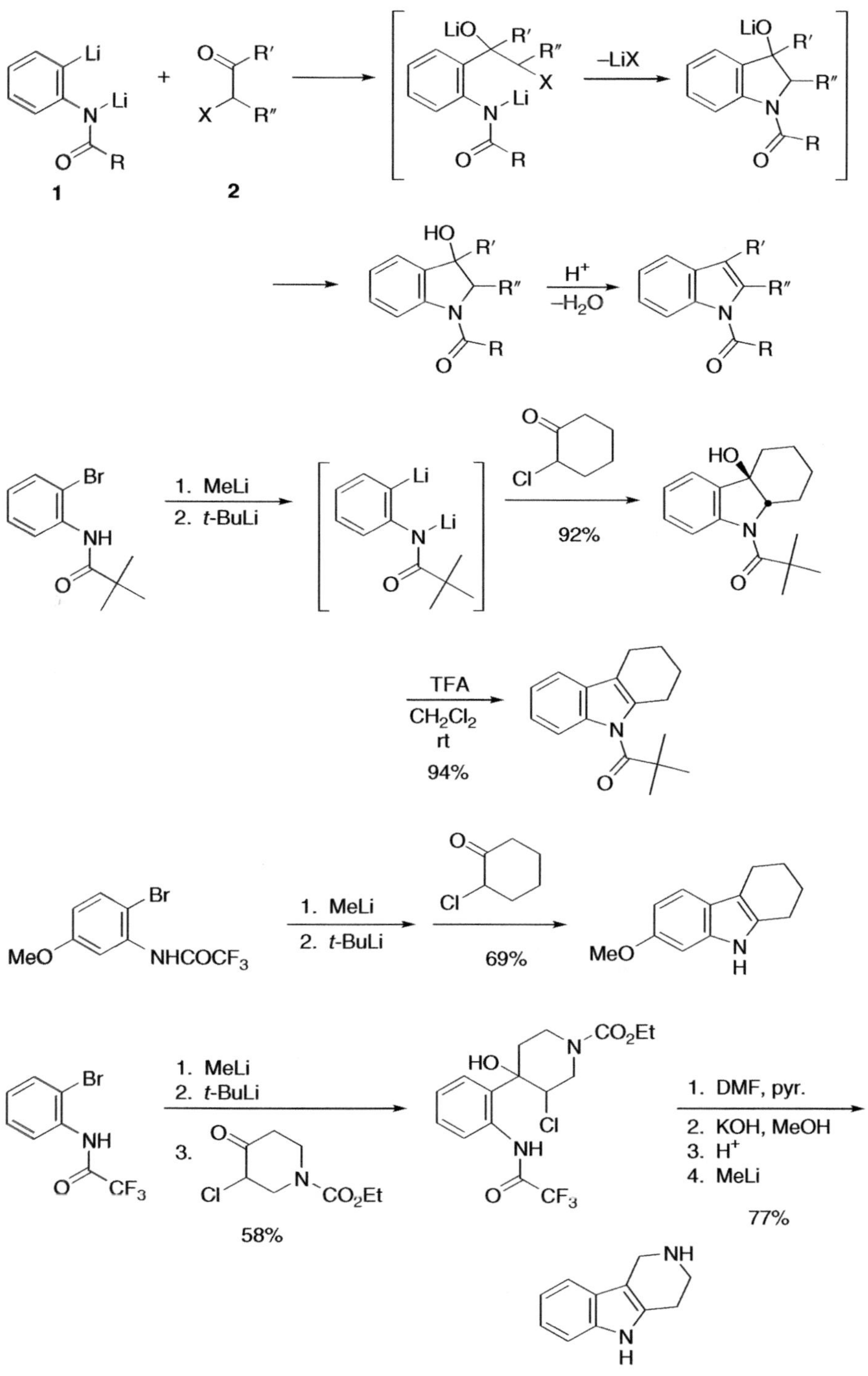

[eqn. 1] [1]

[eqn. 2] [1]

[eqn. 3] [2]

Scheme 1 *Wender Indole Synthesis*

Table 1 *Applications of the Wender Indole Synthesis*

Entry	Substrate	Conditions	Indole	% Yield	Ref.
1	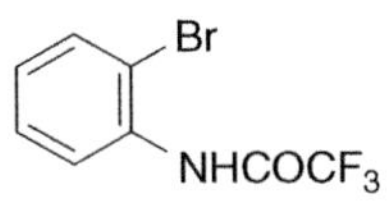	1. MeLi 2. *t*-BuLi 3. 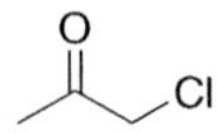 4. HOAc 5. Et_3N 6. KOH, MeOH	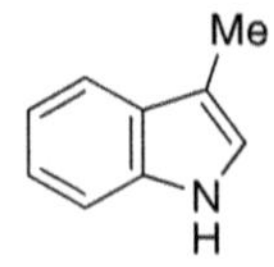	52%	1
2	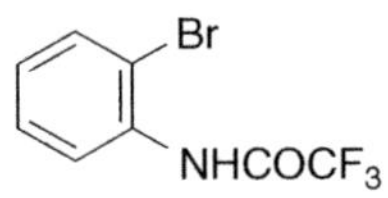	1. MeLi 2. *t*-BuLi 3. 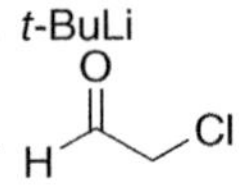 4. HOAc 5. Et_3N 6. TFA	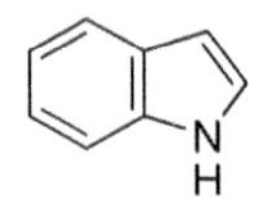	60%	2
3	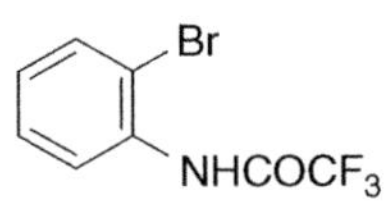	1. MeLi 2. *t*-BuLi 3. 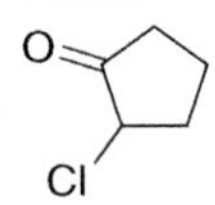 4. KOH, MeOH 5. H^+	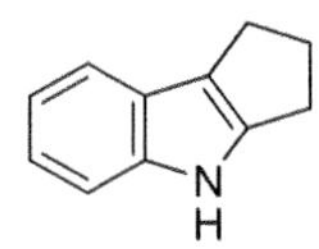	67%	2
4	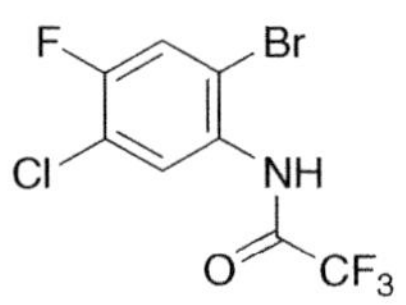	1. MeLi 2. *t*-BuLi −100 °C 3. 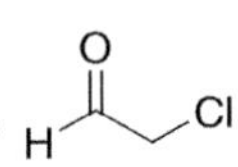 4. *p*-TSA 5. NaOH, MeOH	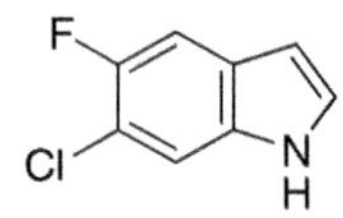	47%	4
5	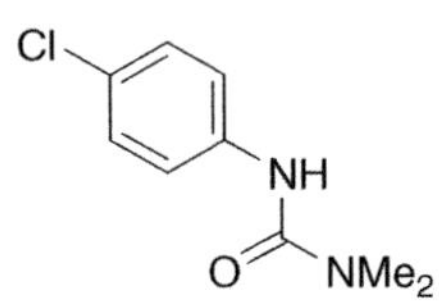	1. *n*-BuLi THF, 0 °C 2. 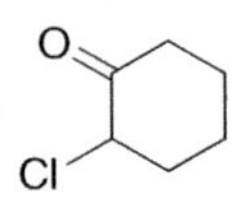 3. H^+ 4. TFA	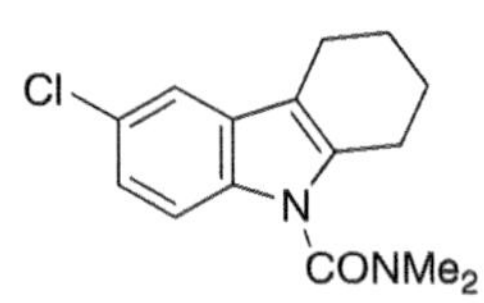	80%	5
6	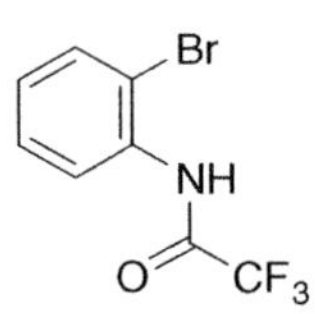	1. MeLi 2. *t*-BuLi 3. 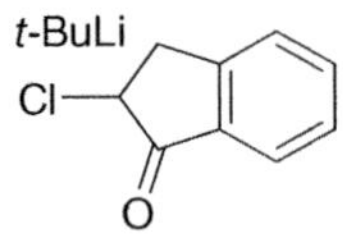 $CeCl_3$ 4. KO*t*-Bu 5. KOH 6. TFA	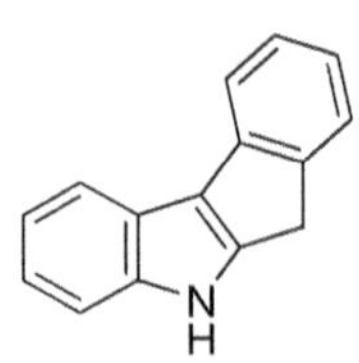	17%	6, 7

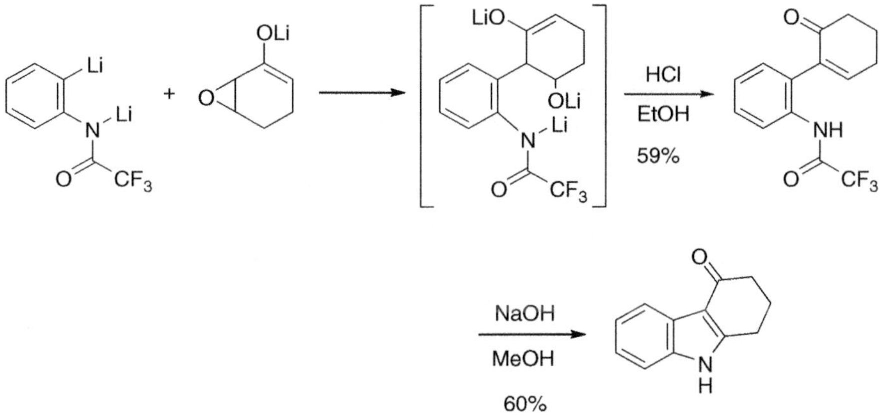

[eqn. 1] [1], [2]

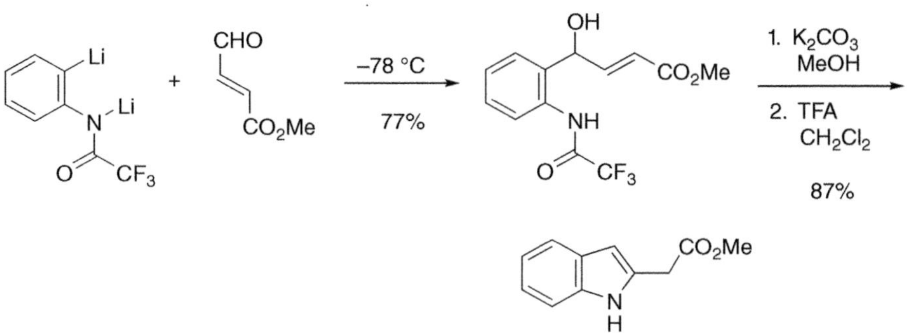

[eqn. 2] [1], [2]

Scheme 2 *Wender Indole Synthesis 2*

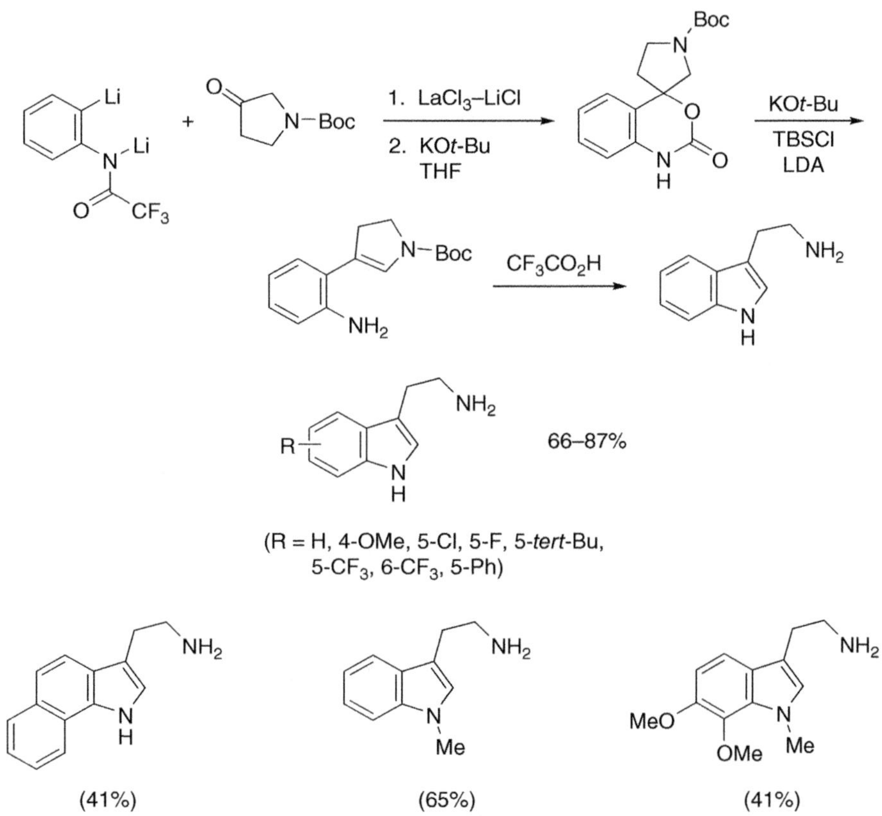

Scheme 3 *Nicolaou Tryptamine Synthesis*

References

[1] P.A. Wender and A.W. White, *Tetrahedron Lett.*, 1981, **22**, 1475–1478.
[2] P.A. Wender and A.W. White, *Tetrahedron*, 1983, **39**, 3767–3776.
[3] For a review, see A.W. Erian, S.M. Sherif, and H.M. Gaber, *Molecules*, 2003, **8**, 793–865.
[4] M. Bos, F. Jenck, J.R. Martin, *et al.*, *J. Med. Chem.*, 1997, **40**, 2762–2769.
[5] K. Smith, G.A. El-Hiti, and A.P. Shukla, *J. Chem. Soc., Perkin Trans.* **11**, 1999, 2305–2313.
[6] J. Graham, A. Ninan, K. Reza, *et al.*, *Tetrahedron*, 1992, **48**, 167–176.
[7] D.W. Brown, P.R. Graupner, M. Sainsbury, and H.G. Shertzer, *Tetrahedron*, 1991, **47**, 4383–4408.
[8] K.C. Nicolaou, A. Krasovskiy, V.É. Trépanier, and D.Y.-K. Chen, *Angew. Chem. Int. Ed.*, 2008, **47**, 4217–4220.

13

Smith Indole Synthesis

Somewhat related to the Madelung and Wender indole syntheses is the method developed by Smith and Visnick, which features the dilithio species from *ortho*-alkyl-*N*-trimethylsilyl anilines reacting with carboxylic acid esters to give 2-substituted and 2,3-disubstituted indoles [1, 2]. The value of this indole synthesis is seen by its numerous applications by Smith and coworkers in the synthesis of indole alkaloids [3–12]. The basic reaction and some examples are shown in Scheme 1. The requisite silylated anilines were prepared by lithiation (*n*-butyllithium, –78 °C) of the aniline followed by quenching with trimethylsilyl chloride. For the synthesis of 2,3-disubstituted indoles an inverse quench is preferred (equation 3). To lithiate *ortho*-ethyl-*N*-methylsilyl aniline, *n*-butyllithium-tetramethylethylenediamine (TMEDA) was required. Indole **1** is an intermediate in the synthesis of (+)-cinchonamine.

The key indole-forming step for each of Smith's alkaloid syntheses is shown in Scheme 2. For the synthesis of a model indole related to penitrem D, indole **2** was an advanced intermediate (equation 1) [3]. Likewise, indole **3** was synthesized as a molecule that will soon embody the A, F, and I rings of penitrem D (equation 2) [4], and indole **4** was employed in the synthesis of (–)-penitrem D (equation 3) [5, 6]. The related indole alkaloid (–)-21-isopentenylpaxilline was approached synthetically via indole **5** (equation 4) [7, 8]. In Smith's syntheses of the nodulisporic acids [9–12], the anticipated amine–ketone indolization did not occur, and it required a separate step. This is illustrated for the synthesis of key indole intermediate **6** (equation 5).

Several other groups have made use of the Smith indole synthesis, and a summary of this work is shown in Table 1. Gribble and coworkers prepared 2-(2-pyridinyl)indoles en route to syntheses of several zwitterionic indolo[2,3-*a*]quinolizine alkaloids (Entry 1) [13], and Husson and Hashimoto synthesized a derivative of the alkaloid goniomitine via a Smith indole synthesis (Entry 2) [14]. Henegar and Hunt reported a very convenient synthesis of 2-trifluoromethylindole (Entry 3) [15], in which TMEDA was found to be necessary. Two bis-indoles were prepared in low yield (Entry 4), but this approach failed for n = 1 [16]. The bridged indole shown in Entry 5 was synthesized following acid-catalyzed cyclization of the initially formed indole [17].

Indole Ring Synthesis: From Natural Products to Drug Discovery, First Edition. Gordon W. Gribble.

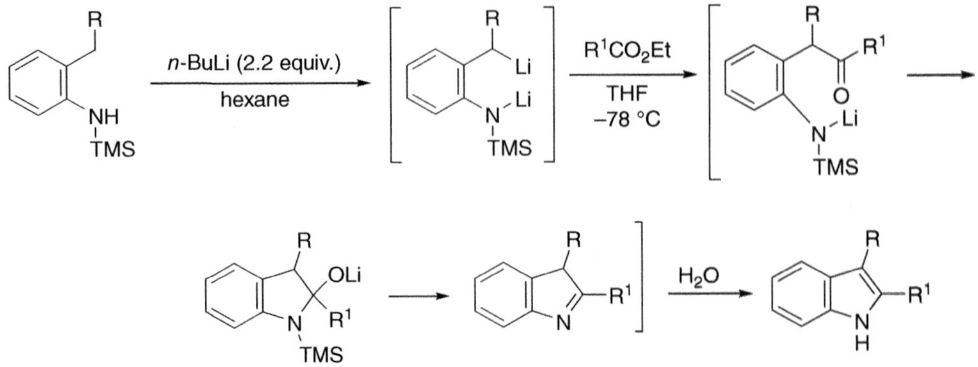

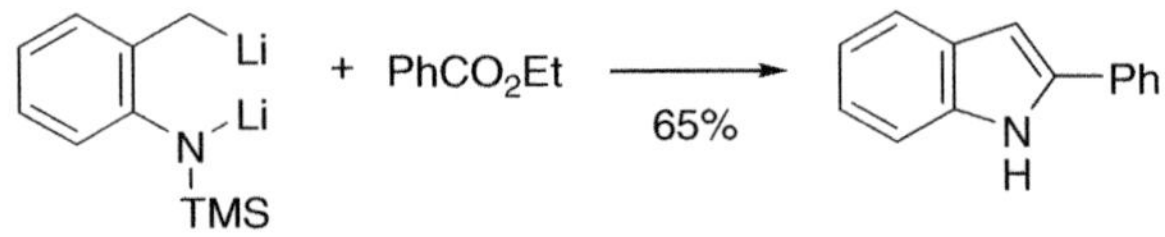

[eqn. 1] [1], [2]

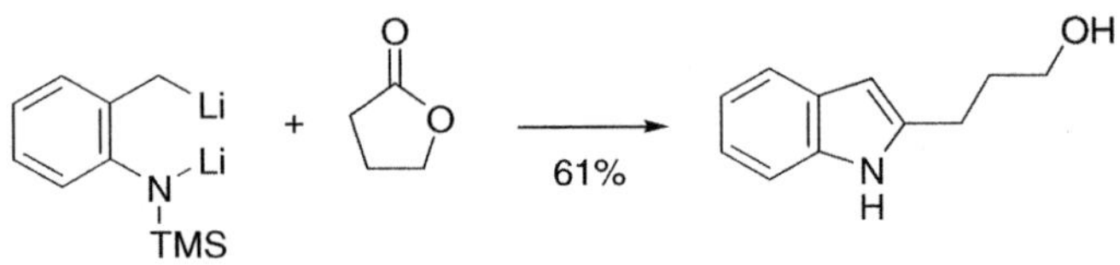

[eqn. 2] [1], [2]

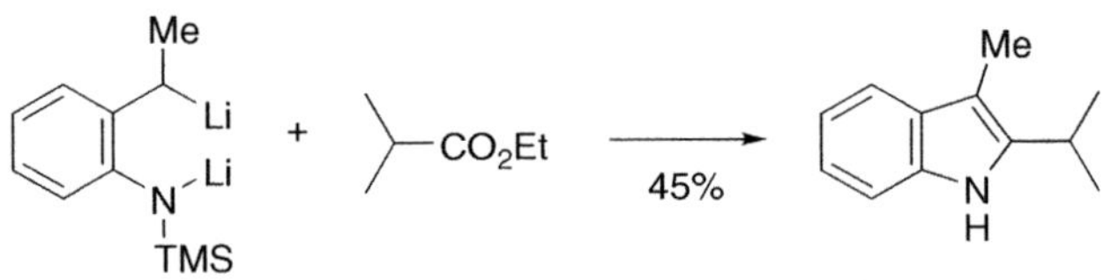

[eqn. 3] [1], [2]

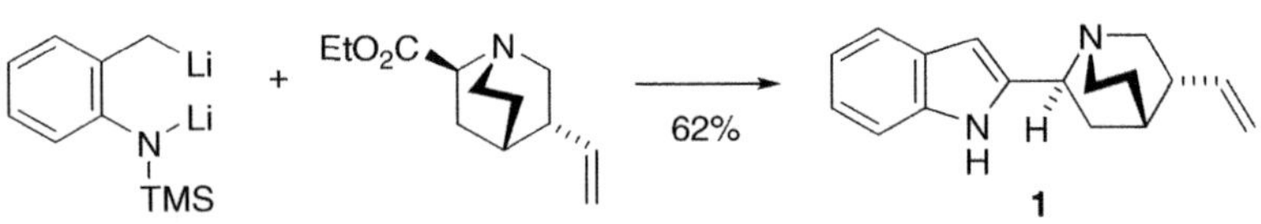

1

[eqn. 4] [2]

Scheme 1 *Smith Indole Synthesis*

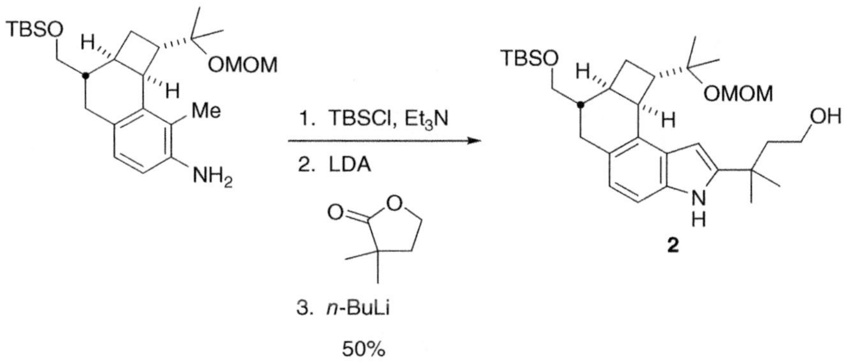

[eqn. 1] [3]

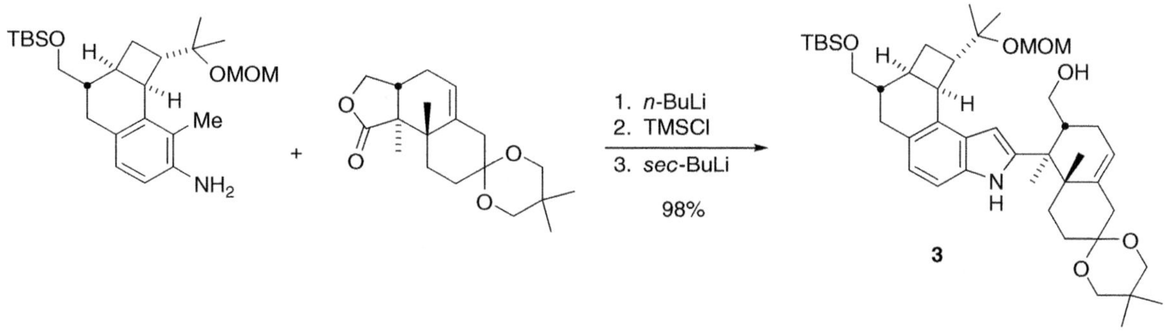

[eqn. 2] [4]

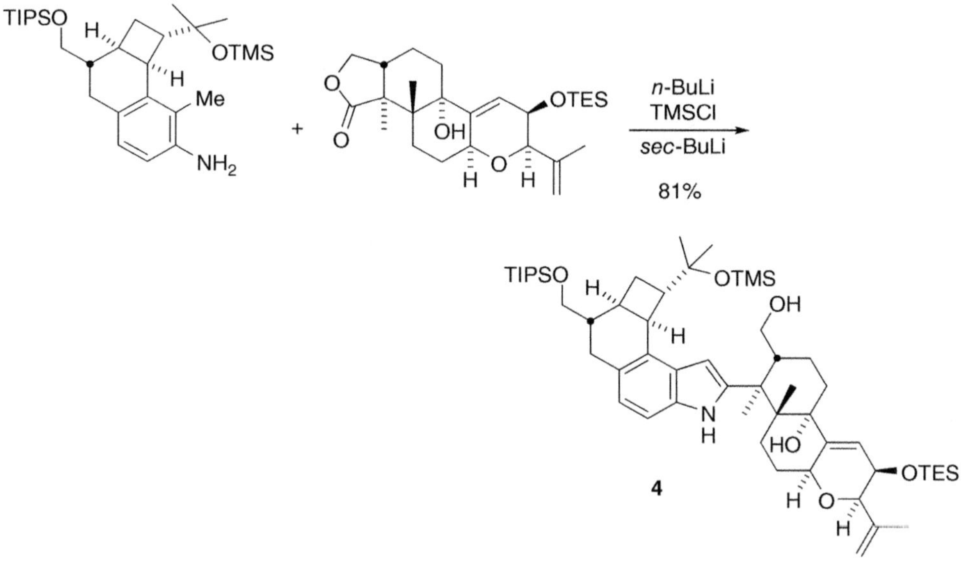

[eqn. 3] [5], [6]

Scheme 2 *Applications of the Smith Indole Synthesis*

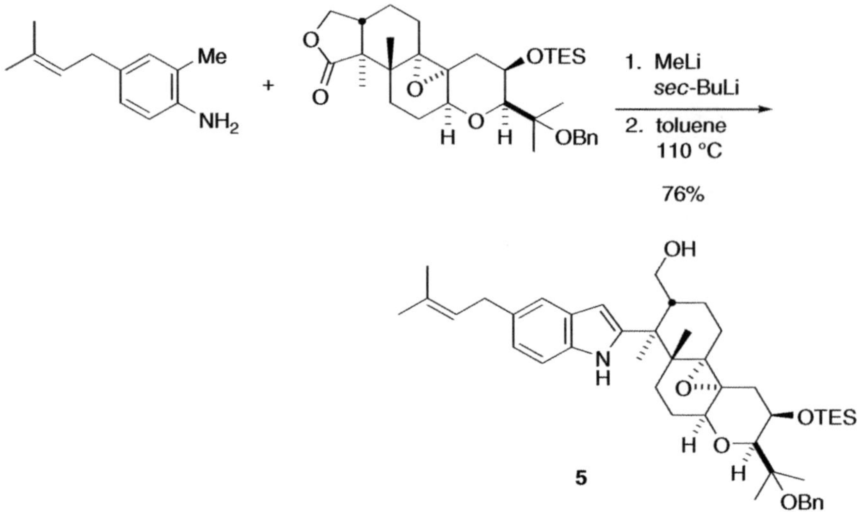

[eqn. 4] [7], [8]

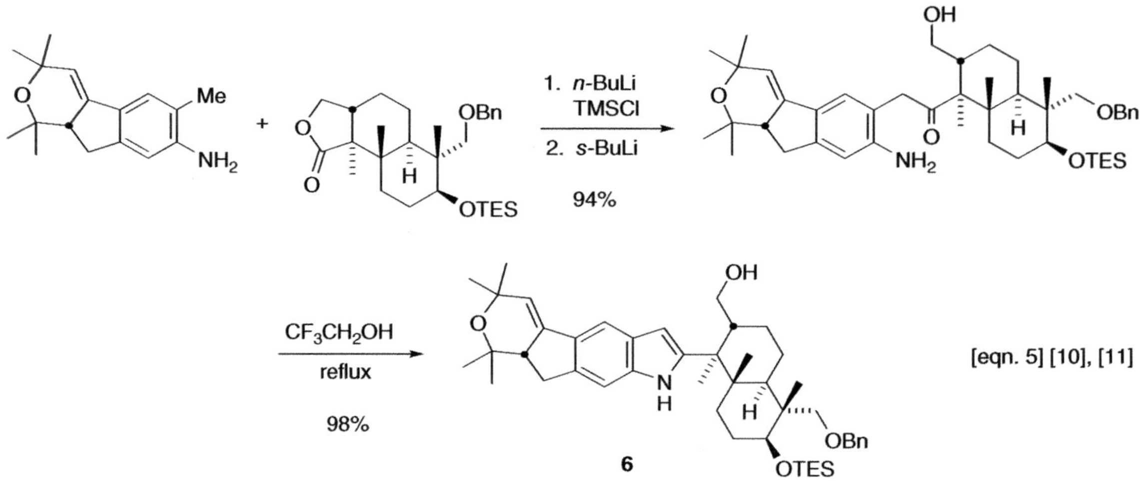

[eqn. 5] [10], [11]

Scheme 2 *(continued)*

Table 1 *Applications of the Smith Indole Synthesis*

Entry	Substrate	Conditions	Indole	% Yield	Ref.
1	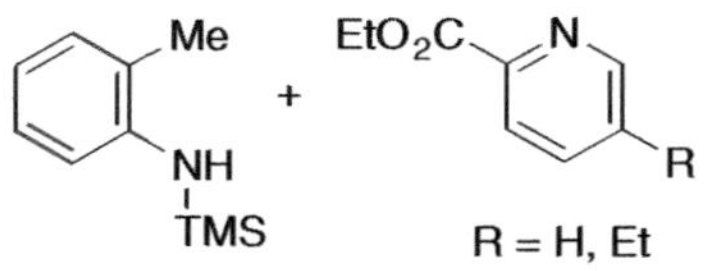	*n*-BuLi THF –78 °C to rt	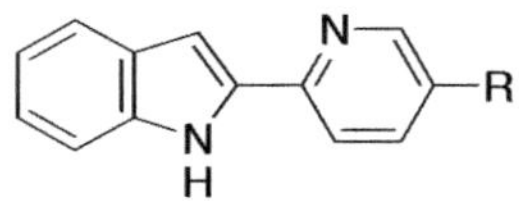	63–67%	13
2	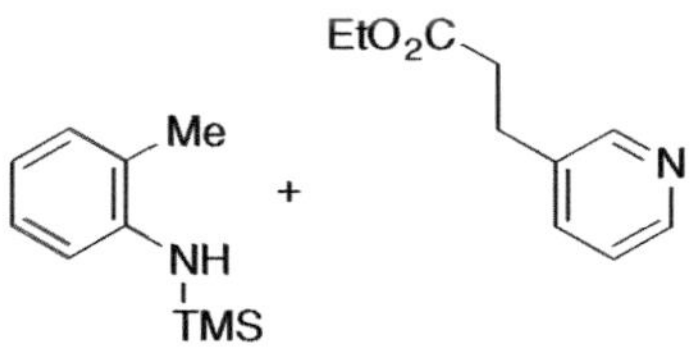	*n*-BuLi hexane reflux	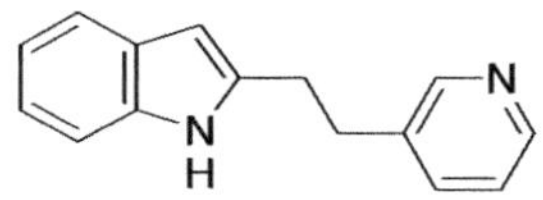	55%	14
3	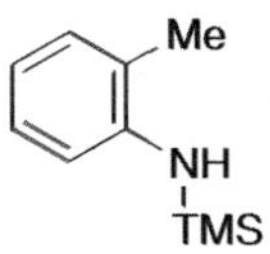 + CF_3CO_2Et	*n*-BuLi TMEDA hexane –78 °C to rt	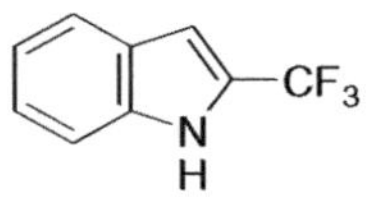	47%	15
4	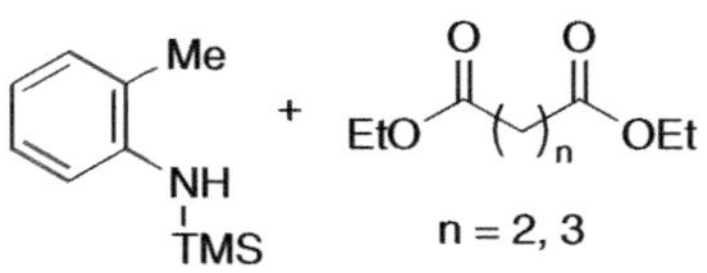	*n*-BuLi hexane –78 °C to rt	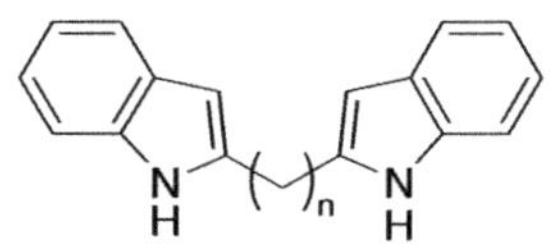	18–22%	16
5	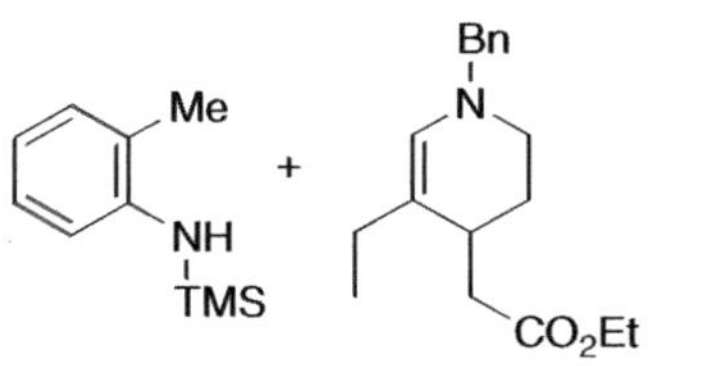	1. *n*-BuLi hexane, THF –78 °C to 25 °C 2. HCl, THF	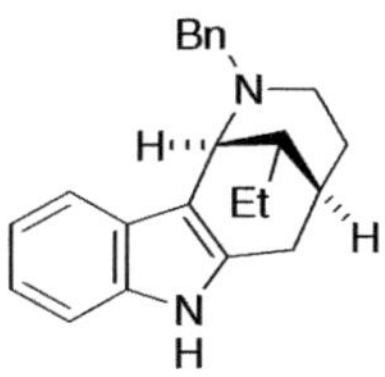	44%	17

References

[1] A.B. Smith, III and M. Visnick, *Tetrahedron Lett.*, 1985, **26**, 3757–3760.
[2] A.B. Smith, III, M. Visnick, J.N. Haseltine, and P.A. Sprengeler, *Tetrahedron*, 1986, **42**, 2957–2969.
[3] A.B. Smith, III, J.N. Haseltine, and M. Visnick, *Tetrahedron*, 1989, **45**, 2431–2449.
[4] A.B. Smith, III, N. Kanoh, N. Minakawa, *et al.*, *Org. Lett.*, 1999, **1**, 1263–1266.
[5] A.B. Smith, III, N. Kanoh, H. Ishiyama, and R.A. Hartz, *J. Am. Chem. Soc.*, 2000, **122**, 11254–11255.
[6] A.B. Smith, III, N. Kanoh, H. Ishiyama, *et al.*, *J. Am. Chem. Soc.*, 2003, **125**, 8228–8237.
[7] A.B. Smith, III and H. Cui, *Org. Lett.*, 2003, **5**, 587–590.
[8] A.B. Smith III and H. Cui, *Helv. Chim. Acta*, 2003, **86**, 3908–3938.
[9] A.B. Smith, III, A.H. Davulcu, and L. Kürti, *Org. Lett.*, 2006, **8**, 1665–1668.
[10] A.B. Smith, III, A.H. Davulcu, and L. Kürti, *Org. Lett.*, 2006, **8**, 1669–1672.
[11] A.B. Smith, III, A.H. Davulcu, Y.S. Cho, *et al.*, *J. Org. Chem.*, 2007, **72**, 4596–4610.
[12] A.B. Smith, III, L. Kürti, A.H. Davulcu, *et al.*, *J. Org. Chem.*, 2007, **72**, 4611–4620.
[13] G.W. Gribble and D.A. Johnson, *Tetrahedron Lett.*, 1987, **28**, 5259–5262.
[14] C. Hashimoto and H.-P. Husson, *Tetrahedron Lett.*, 1988, **29**, 4563–4566.
[15] K.E. Henegar and D.A. Hunt, *Heterocycles*, 1996, **43**, 1471–1475.
[16] S. Mahboobi, T. Burgemeister, S. Dove, *et al.*, *J. Org. Chem.*, 1999, **65**, 8130–8137.
[17] J.M. Aurrecoechea, J.M. Gorgoja, and C. Saornil, *J. Org. Chem.*, 2005, **70**, 9640–9643.

14
Kihara Indole Synthesis

The little-used Kihara indole synthesis is an example of an intramolecular Barbier reaction applied to the synthesis of 3-phenyl- and 3-alkyl-3-hydroxyindolines, which can be dehydrated to the corresponding indoles (Scheme 1) [1]. In some cases, the dehydration of indoline to indole was spontaneous. The major byproduct in the Kihara indole synthesis is deiodination of the starting *ortho*-iodo aniline. Some examples are shown in Scheme 1 (equations 2–5). The ease of acid-catalyzed indoline dehydration presumably reflects the relative stability of the corresponding benzylic carbocation under the silica gel workup conditions. All of the indolines could be dehydrated to indoles with 5% hydrochloric acid.

Although limited in scope at the present time, the Kihara indole synthesis seems particularly well suited for the synthesis of 3-arylindoles.

Indole Ring Synthesis: From Natural Products to Drug Discovery, First Edition. Gordon W. Gribble.

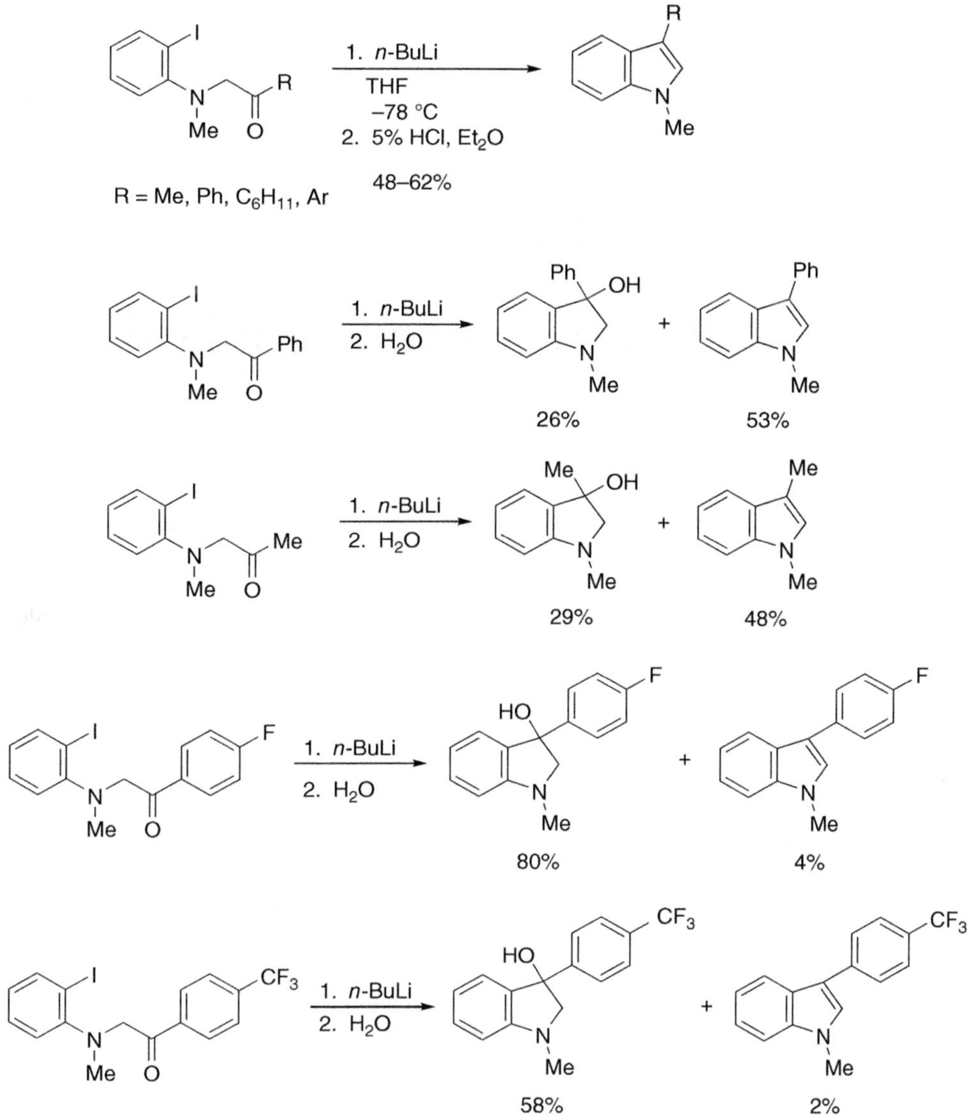

[eqn. 1] [1]

[eqn. 2] [1]

[eqn. 3] [1]

[eqn. 4] [1]

[eqn. 5] [1]

Scheme 1 *Kihara Indole Synthesis*

Reference

[1] M. Kihara, Y. Iwai, and Y. Nagao, *Heterocycles*, 1995, **41**, 2279–2287.

15

Nenitzescu 5-Hydroxyindole Synthesis

Although Nenitzescu is known for several indole construction methods, he is most recognized for the reaction between benzoquinones and ethyl β-aminocrotonate leading to 5-hydroxyindole esters. The prototypical reaction is shown in Scheme 1 [1, 2]. Although this indole synthesis remained relatively obscure for two decades, the recognized biological importance of 5-oxygenated indoles in the 1950's revitalized the Nenitzescu 5-hydroxyindole synthesis. An excellent review is available [3]. Some of the early examples, including two by Nenitzescu, are shown in Scheme 1a. Base hydrolysis and heating of these indole esters effected decarboxylation [4], except for the 2-ethoxyindoles (equation 6). These undergo acid hydrolysis and decarboxylation to the respective oxindoles [5]. Nenitzescu showed that nitriles (β-aminocinnamic nitrile) can be employed (equation 8) [6]. The reaction shown in equation 1 has been performed by Steck's group on a 1.1-kg scale to give ethyl 5-hydroxy-2-methylindole-3-carboxylate in 26% crystallized yield [7]. These workers also prepared a series of 1-substituted indoles **1–4** by using *N*-alkyl β-aminocrotonates Scheme 1b.

Although the combination of reactants (benzoquinone + β-aminocrotonate) gives the 5-hydroxyindole ester + water, the mechanism is far from simple, as early workers realized (Beer [5], Nenitzescu [6, 8], Allen [9–11], Monti [12]). For example, Beer and colleagues recognized that hydroquinone **5** and quinone **6** were conceivable intermediates leading to 5- (but not 6-) hydroxyindoles but were not isolated (Scheme 2). Moreover, the reaction of benzoquinone with ethyl β-amino-α-methylcrotonate afforded a hydrate of the hydroquinone **7** (R = H), which on treatment with hydroxide gave 5-hydroxy-2,3-dimethylbenzofuran (coumarone) **8** (R = H) (confirmed by independent synthesis) (equation 1) [5]. Likewise, the reaction of 2-methoxybenzoquinone and ethyl β-amino-α-methylcrotonate afforded hydroquinone **7** (R = OMe) and benzofuran **8** (R = OMe) under similar conditions. Thus, Beer's work seemed to demonstrate that the initial bond formation involves a Michael addition, and not carbonyl addition, to the benzoquinone. Subsequent work by Nenitzescu, Allen, and Monti confirmed this. Nenitzescu found that ethyl β-aminocinnamate reacts with benzoquinone in refluxing chloroform or benzene to give hydroquinone cinnamate **9** (equation 2) [6]. Allen and his colleagues were the first to show that the isomeric 7-alkyl-5-hydroxyindoles can form in the Nenitzescu indole synthesis [9, 11]. For example, the condensation of 2-methylbenzoquinone with ethyl 3-aminocrotonate actually affords both **10** and **11** in essentially equal amounts after partition chromatography (equation 3). Similar results obtain with ethyl 3-amino-2-methylcrotonate (equation 4) [9, 11]. Allen and Weiss also isolated in low yields the hydroquinones **14–16**, and quinone **17** [10, 11]. The latter quinone and its conjugate acid gave ethyl 5-hydroxy-6-methoxy-2-methylcarboxylate (22%) but only after the addition of the reducing agent sodium hydrosulfite [10, 11]. All of these data led Allen and Nenitzescu to propose the mechanism shown in abbreviated form in Scheme 3.

Following the initial Michael addition and tautomerism, **A** → **B**, oxidation, perhaps by starting benzoquinone, gives **C**. Cyclization of **C** to **D** is followed by loss of water to afford **E**. A final reduction, perhaps by a hydroquinone, yields **F** and then **G**. Monti also provided support for the proposed mechanism. He hydrolyzed the hydroquinone **B** to keto ester **18**, which was cyclized with zinc chloride to the known benzofuran **19** (equation 1) [12]. Subsequent work by Răileanu and collegues substantiated the mechanism shown in Scheme 3 [8].

Indole Ring Synthesis: From Natural Products to Drug Discovery, First Edition. Gordon W. Gribble.

(a)

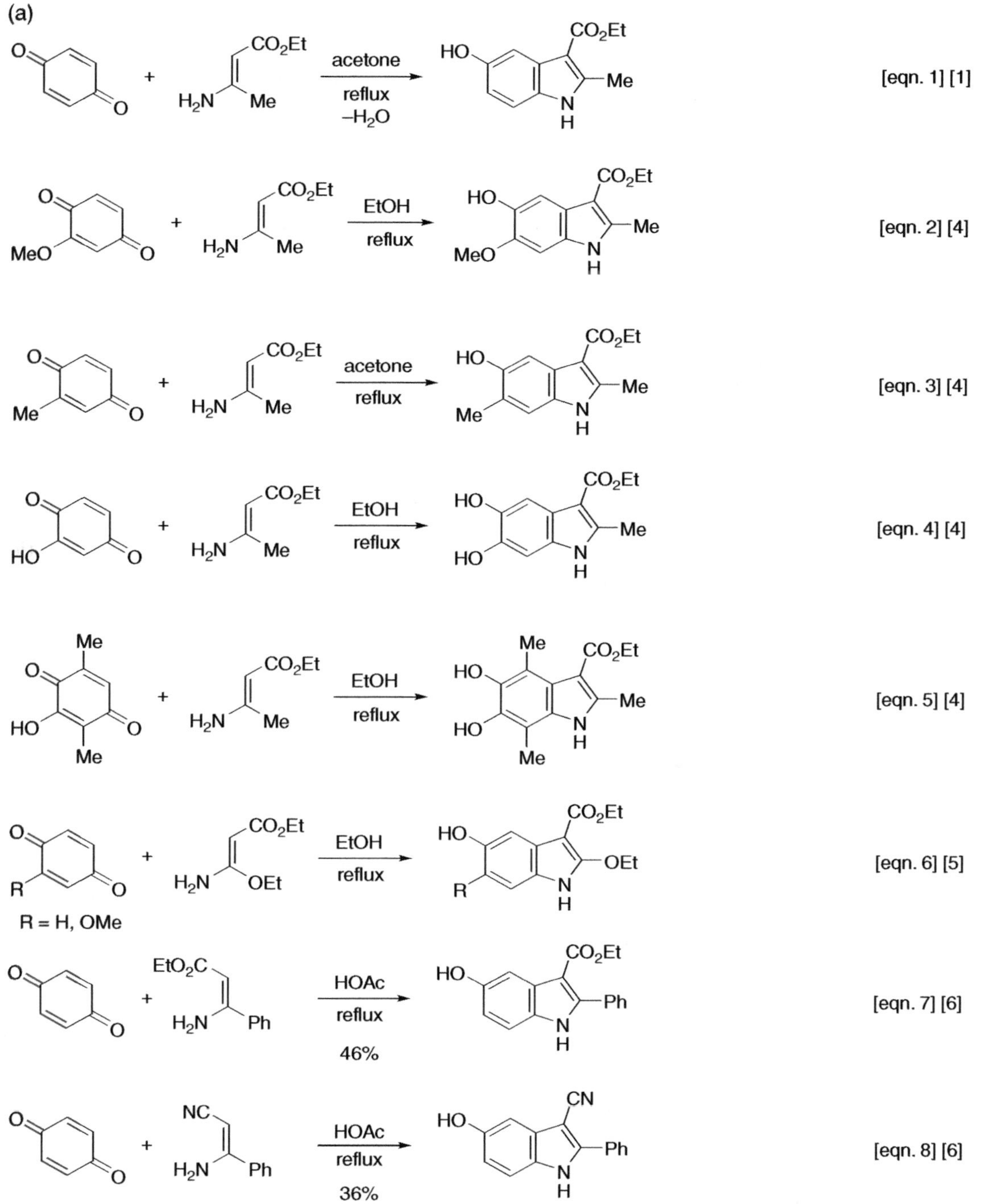

Scheme 1a *The Nenitzescu Indole Synthesis*

(b)

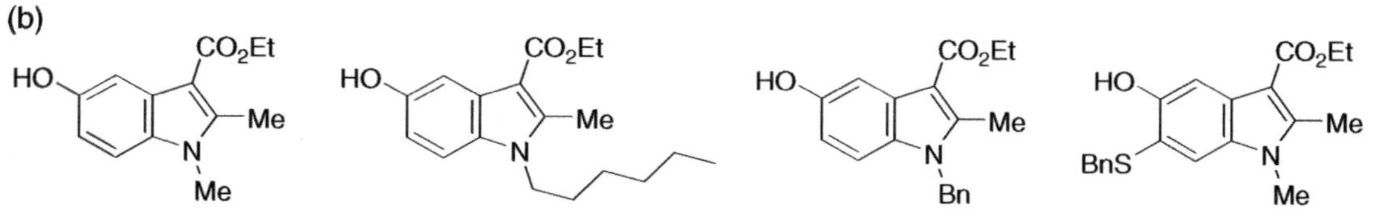

1 (32%) **2** (22%) **3** (22%) **4** (46%)

Scheme 1b *Steck Applications of the Nenitzescu Indole Synthesis*

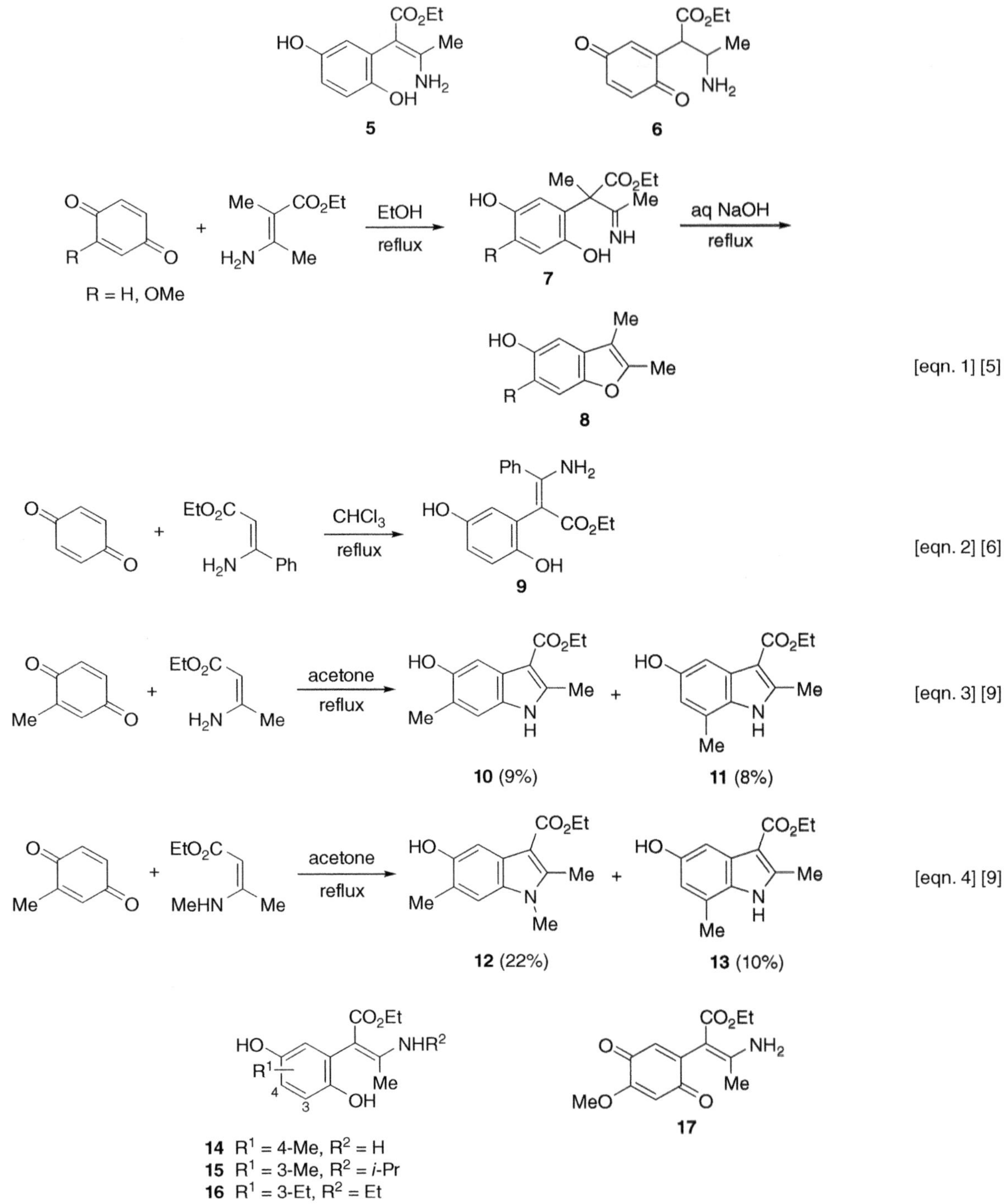

Scheme 2 *Applications of the Nenitzescu Indole Synthesis*

Even though the proposed mechanism does not preclude formation of 4-alkyl-5-hydroxyindoles, no examples were known until Allen and Weiss isolated ethyl 4-carbomethoxy-5-hydroxy-2-methylindole-3-carboxylate (**22**) from the reaction of 2-carbomethoxy-1,4-benzoquinone (**20**) with ethyl 3-aminocrotonate, following treatment of the initially formed, and isolated, Michael adduct **21** with acetic acid and 0.18 equivalents of quinone **20** (Scheme 4, equation 1) [13]. These conditions presumably isomerize **21** to the *E*-isomer that would lead to cyclization and completion of the Nenitzescu indole synthesis. As might be expected, the isocarbostyryl **23**

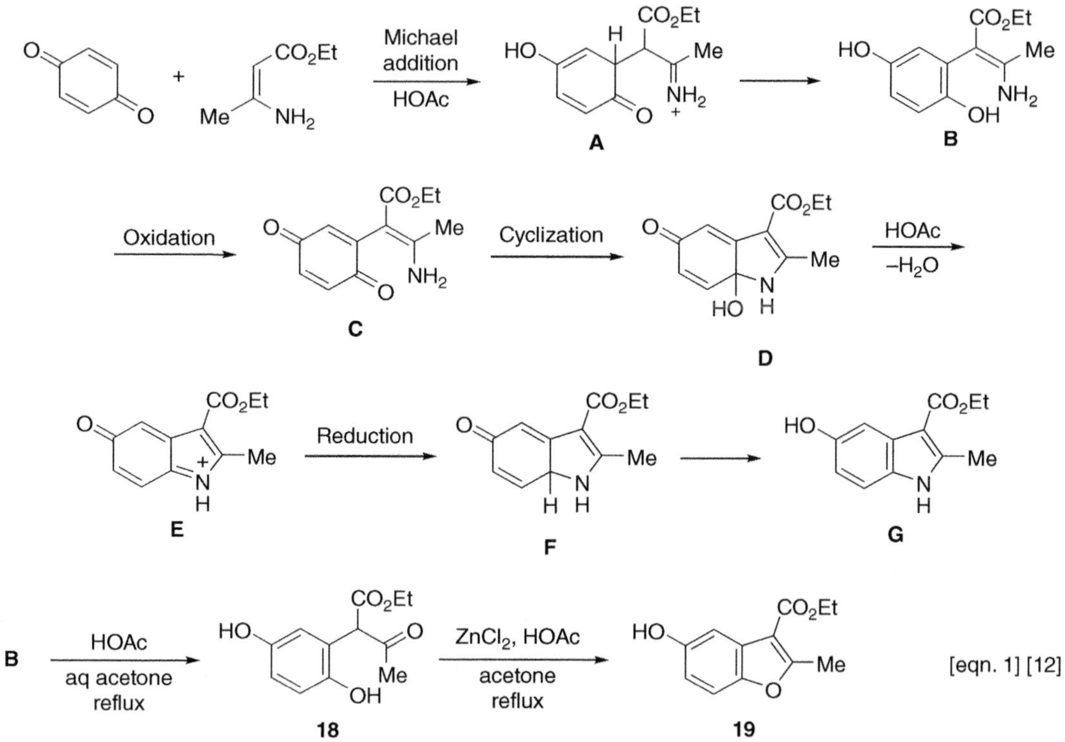

Scheme 3 *Nenitzescu and Allen Mechanism for this Indole Synthesis*

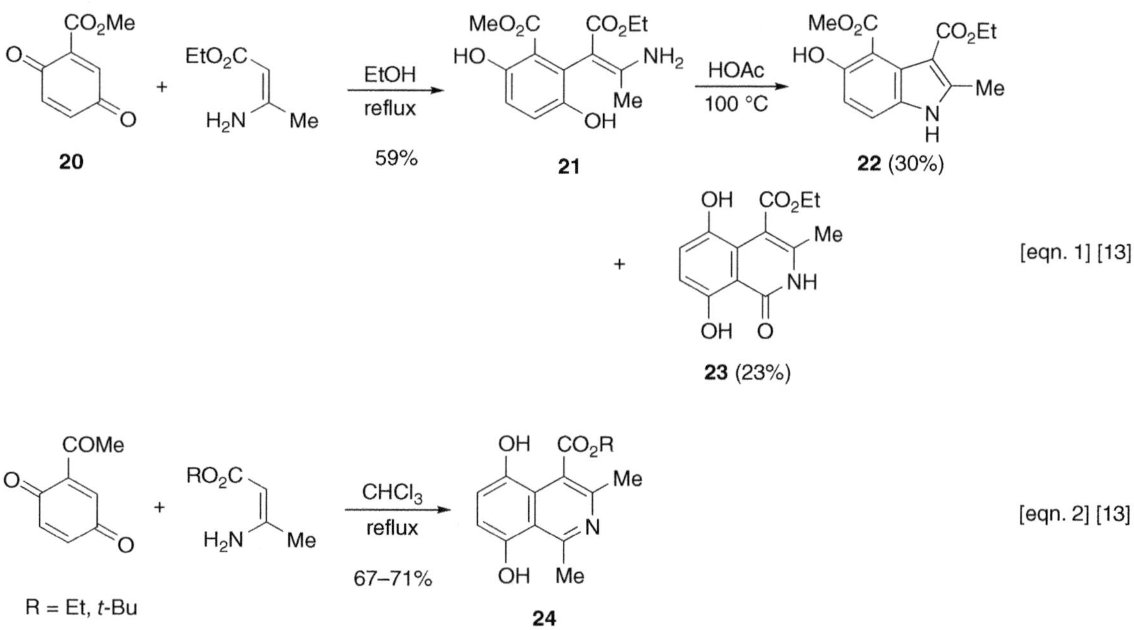

Scheme 4 *Allen and Weiss Applications of the Nenitzescu Indole Synthesis*

also formed in this reaction. Heating **22** with 20% aqueous HCl afforded 5-hydroxy-2-methylindole (60%). Interestingly, 2-acetyl-1,4-benzoquinone gave no indole product with 3-aminocrotonates, but rather only isoquinolines **24** (equation 2).

Allen and colleagues studied the effect of halogen-substitution on the Nenitzescu indolization, and their results are summarized in Table 1 [14, 15]. The reactions of 2-halo-1,4-benzoquinones with ethyl 3-aminocrotonate gave poor to low yields of indoles, with the initial Michael

Table 1 *Effect of Halogenated 1,4-Benzoquinones on the Nenitzescu Indole Synthesis*

Entry	Benzoquinone	Aminocrotonate	Indole Products		% Yield	Ref.
1	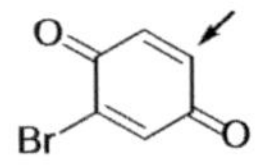	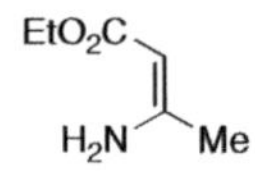	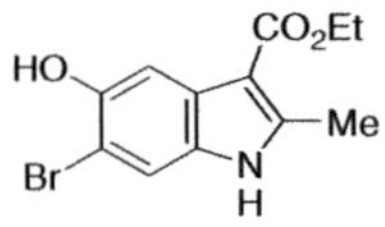 (5%)	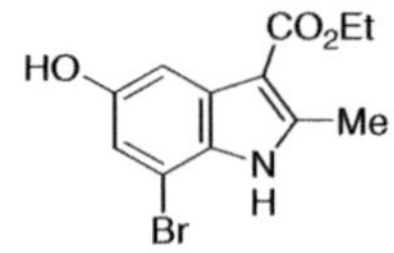 (2%)	7%	14
2	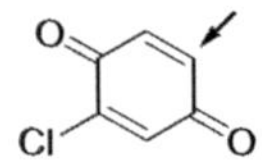	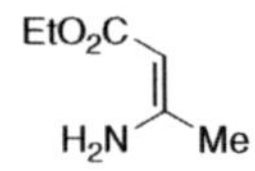	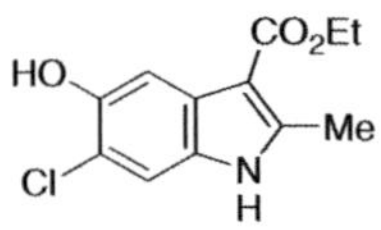 (20%)	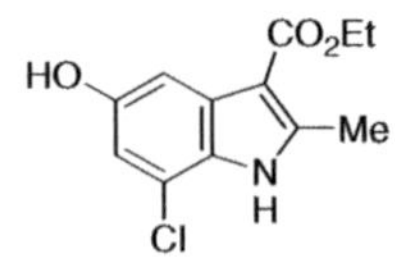 (4%)	24%	14
3	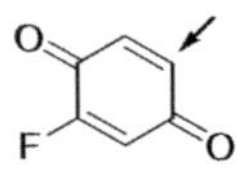	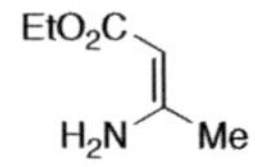	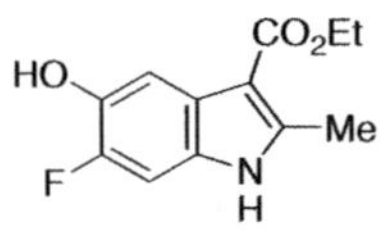 (12%)	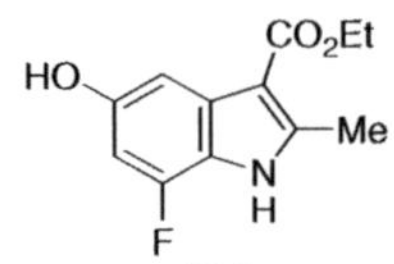 (0%)	12%	14
4	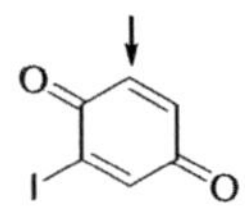	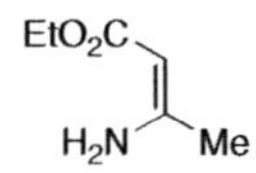	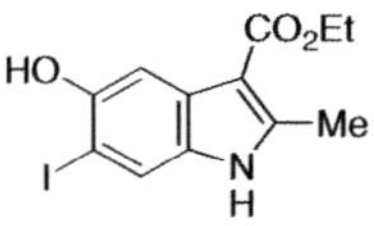 (1%)	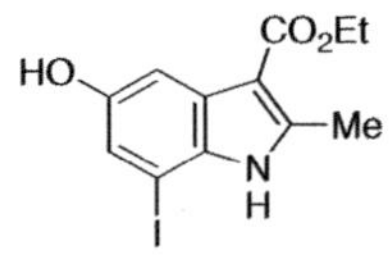 (7%)	8%	14
5	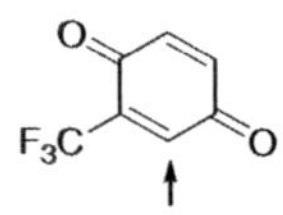	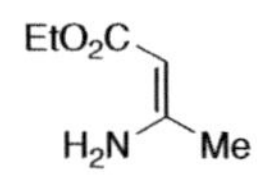	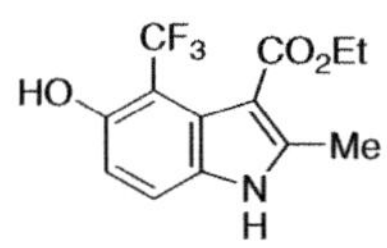 (5%)		54%	14
6	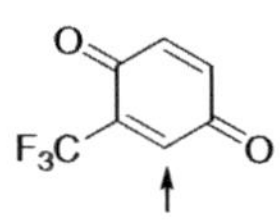	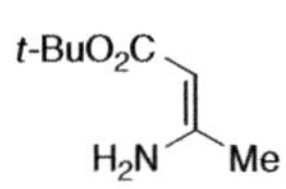	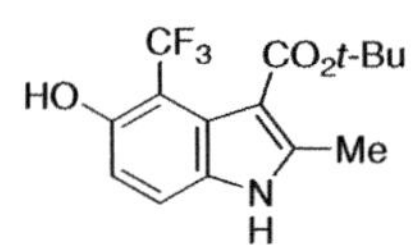		62%	14
7	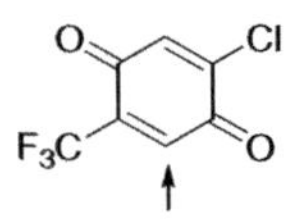	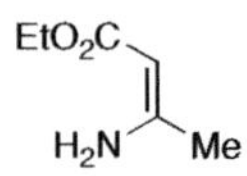	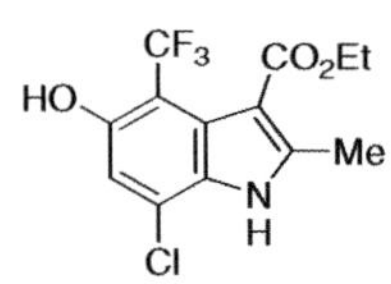		74%	14
8	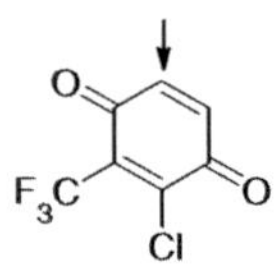	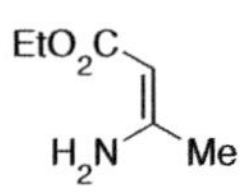	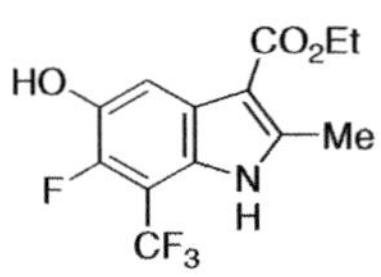		78%	14
9	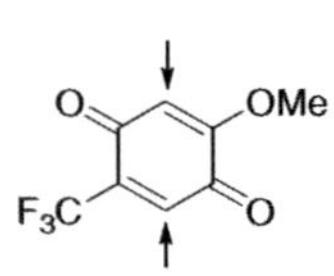	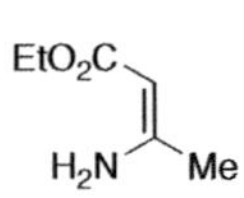	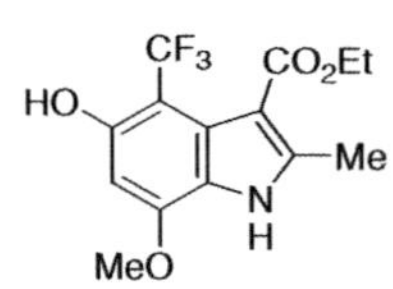 (25%)	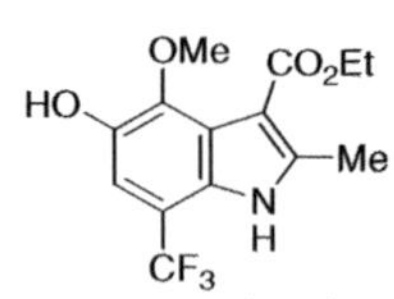 (25%)	50%	14
10	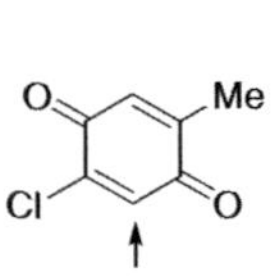	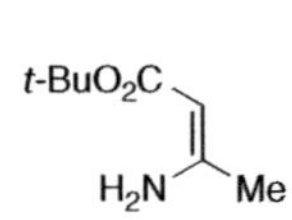	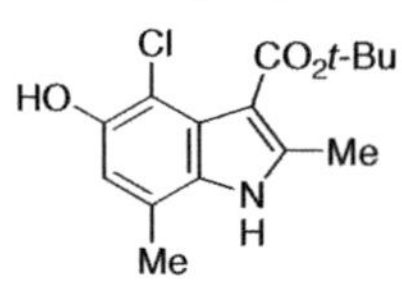		51%	15

addition favoring attack at C-5 of the benzoquinone, giving the 6-haloindole as the major isomer (Entries 1–3); however, the trifluoromethyl group exerts a major inductive-withdrawing influence on the direction and yield of the reaction (Entries 5–7). This powerful trifluoromethyl inductive effect is offset by the equally strong electron-donating effect of the methoxyl group such that an equal mixture of indoles was obtained from the reaction of 2-methoxy-5-trifluoromethyl-1,4-benzoquinone with ethyl 3-aminocrotonate (Entry 9). Further manipulation of these halogenated indoles afforded 5-hydroxyindoles that are difficult to access other ways (Scheme 5) [14, 15].

The major alternative mechanism of the Nenitzescu indole synthesis is that proposed by Steck and colleagues [7], which involves the initial obvious condensation of the aminocrotonate nitrogen with the benzoquinone carbonyl group to give **26**. This mechanism is shown in abbreviated form in Scheme 6. The proposed steps from **26** to **27** to **28** (not shown) make no chemical sense. In any event, Allen and colleagues have shown that compounds of type **26** are not intermediates in the formation of indoles [16, 17]. Thus, whereas the reaction between quinone **29** and enaminone **30** gave amine-carbonyl condensate **31** and cyclized carbinol amine **32**, this latter compound did not furnish indole **33** under conditions of acid-catalysis to **33**, quinone oxidation to **34**, or fragmentation to **29**+**30**. Furthermore, Allen's evidence shows that such intermediates as **26** and **31** are only converted into indoles after reversion to compounds resulting from initial carbon–carbon bond formation as in Scheme 3.

In a series of papers, Kuckländer has provided additional evidence for the Nenitzescu–Allen mechanism, in particular the **D**→**E**→**F**→**G** transformation [18–20]. For example, isolated carbinolamine **35** gives byproducts **36–38** in addition to indole **39** (Scheme 7, equation 1) [18]. The formation of **36** presumably involves reaction of iminium ion **40** with acetate and rearrangement of **41** to **36** as shown (Scheme 7, equation 2) [20]. Kuckländer and Hühnermann found that benzoquinone reacts with β-arylaminocrotonates in propionic acid to afford 6-hydroxyindoles **42** (10%–30%) (equation 3), along with the byproducts 5-acyloxy-4-hydroxyindoles (20%–35%) and arylaminobenzofurans (10%–15%) [21].

Given the typical low to modest yields of the aforementioned Nenitzescu indole syntheses, several improvements have been reported. Patrick and Saunders found that nitromethane solvent greatly improved the efficiency of the condensation between β-aminocrotonates and 1,4-benzoquinones (Scheme 8) [22]. More remarkable is the

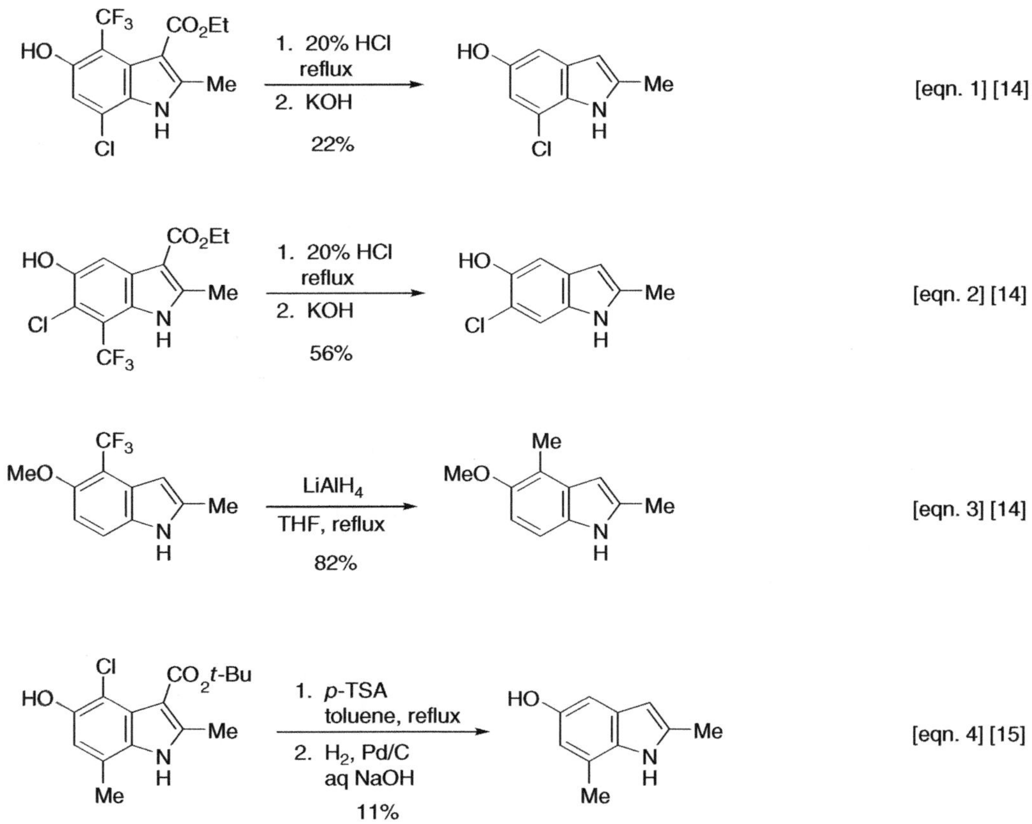

Scheme 5 *Allen and Weiss Applications of the Nenitzescu Indole Synthesis*

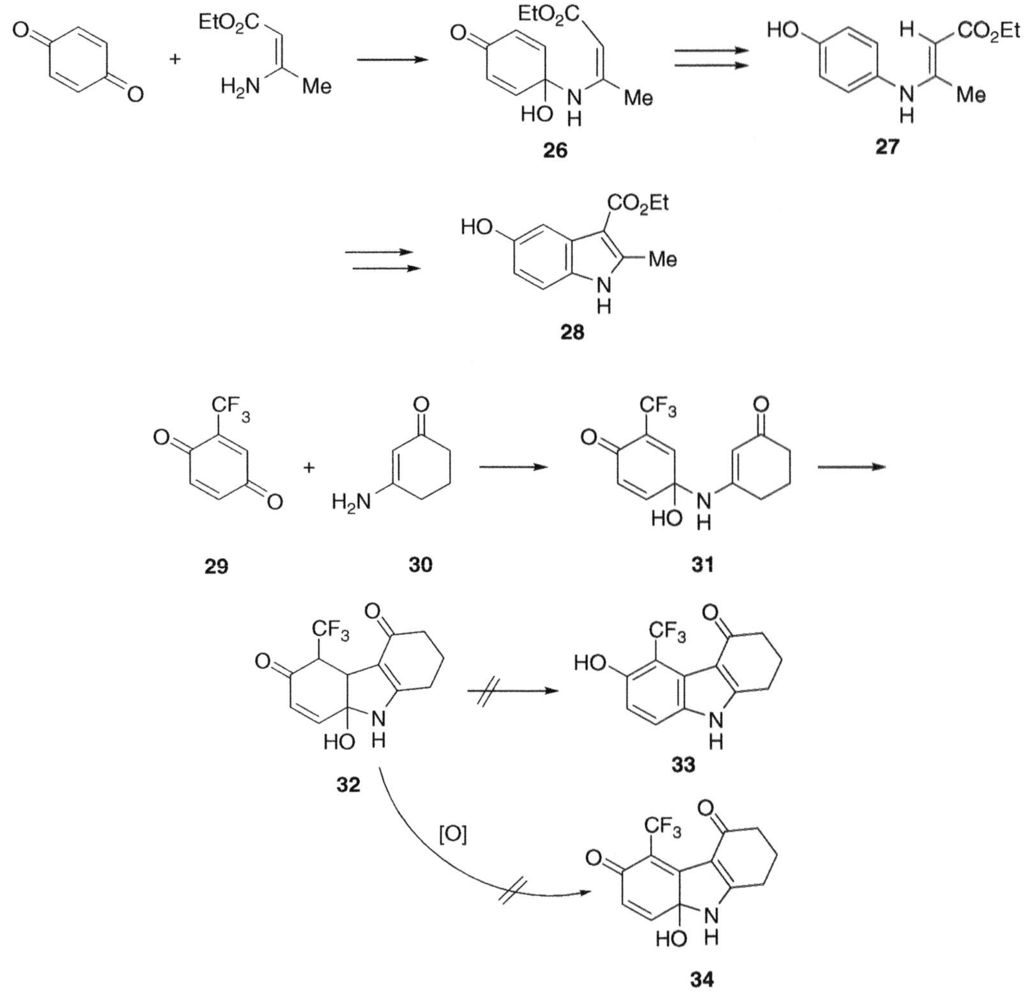

Scheme 6 *Steck's Proposed Mechanism of the Nenitzescu Indole Synthesis*

observation that *methyl* β-aminocrotonates give much higher yields than either ethyl or *tert*-butyl esters. A few examples are shown in Scheme 8. These workers also propose a face-to-face sandwich electron-transfer complex that mediates the oxidation–reduction step (equation 1). Another advantage of nitromethane seems to be that the indole product crystallizes from the reaction mixture as it forms.

Velezheva and coworkers described the Lewis acid–catalyzed Nenitzescu indolization using zinc iodide in dichloromethane (Scheme 9, equation 1 and **43–45**) [23], later improved to the use of zinc chloride (equations 2, 3) [24]. Boruah and colleagues employed a Lewis acid–catalyzed microwave-modified Nenitzescu indolization to prepare 3-amino-5-hydroxybenzo[*g*]indoles (Scheme 9, equation 4, and **46–47**) [25]. Boron trifluoride etherate was superior to other Lewis acids studied ($TiCl_4$, $AlCl_3$, $ZnCl_2$, $InCl_3$). The reaction presumably proceeds by a Michael addition of the carbonyl component, in the form of a β-hydroxyenamine to the naphthoquinone, and the urea serves as a controlled-release source of ammonia.

As with other indole ring syntheses, the Nenitzescu method has been employed to prepare biologically active indoles and drug candidates. A summary of these reports is listed in Table 2 [26–39]. The examples in Entries 4, 6, and 7 feature the use of nitromethane as solvent (cf. Scheme 8). In Entries 8, 9, 13, and 14 the Nenitzescu product is the biologically active target indole. Although the yields are low in Entries 13 and 14, the Troschütz synthesis of highly complex fused indoles is powerful testimony to the value of Nenitzescu indolization [38–41]. Some additional examples from the research of Troschütz and colleagues are shown in Scheme 10. Somewhat earlier, Bernier and colleagues described related syntheses of 6-hydroxycarbazoles and 6-hydroxypyrimido[4,5-*b*]indoles by the condensation of 1,4-benzoquinone with 4-nitroaniline and 1,3-dimethyl-6-aminouracil, respectively (equations 5, 6) [42, 43].

With the invention of several variations on the basic Nenitzescu indole synthesis, both the scope and efficiency of this method have been greatly expanded and

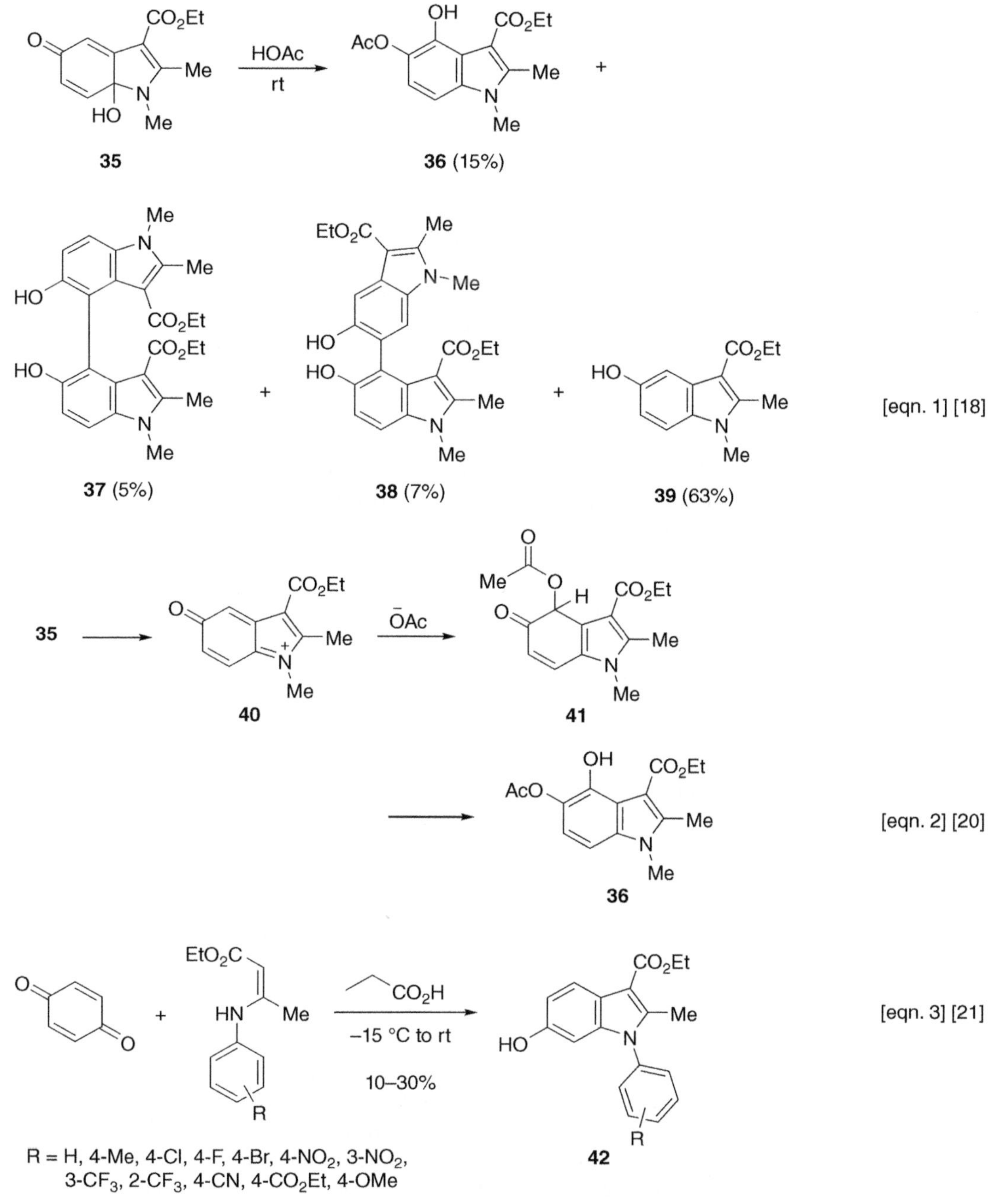

Scheme 7 *Kuckländer Applications of the Nenitzescu Indole Synthesis*

improved. Some examples are presented in the following schemes. Ila, Junjappa, and coworkers used ketene acetals as the enamine components to prepare, for example, 3-nitroindoles (Scheme 11, equations 1, 2) [44]. The reaction is invariably accompanied more or less by the corresponding benzofuran. Mayadeo and Gandhi used substituted enamines to prepare *N*-aryl-3-substituted 5-hydroxyindoles (equation 3) [45]. Granik and coworkers used enehydrazines (e.g., **48**) in the Nenitzescu synthesis to synthesize 1-amino-6-hydroxyindoles (equation 4) [46, 47]. Kheder employed bis-aminocrotonate **49** in a Nenitzescu indolization to give bis-indole **50** (equation 5) [48].

A Nenitzescu route to the medicinally important 2-methyl-4-(trifluoromethyl)indole-5-carbonitrile was adopted by Boros and colleagues to synthesize the intermediate 5-hydroxyindole **53**, which features the *in situ* oxidation of indole **51** to benzoquinone **52** (Scheme 12, equation 1) [49].

Entry	R^1	R^2	R^3	Solvent	% Yield
1a	H	H	Et	$MeNO_2$	23
1b	H	H	Me	$MeNO_2$	82
2a	H	*i*-Pr	Et	$MeNO_2$	5
2b	H	*i*-Pr	Me	$MeNO_2$	95
3a	Me	*i*-Pr	Et	HOAc	18
3b	Me	*i*-Pr	Et	$MeNO_2$	82
4a	Me	*n*-Pr	Et	HOAc	21
4b	Me	*n*-Pr	Et	$MeNO_2$	24
4c	Me	*n*-Pr	Me	$MeNO_2$	75
5a	prenyl	H	Et	$MeNO_2$	0
5b	prenyl	H	Me	$MeNO_2$	39

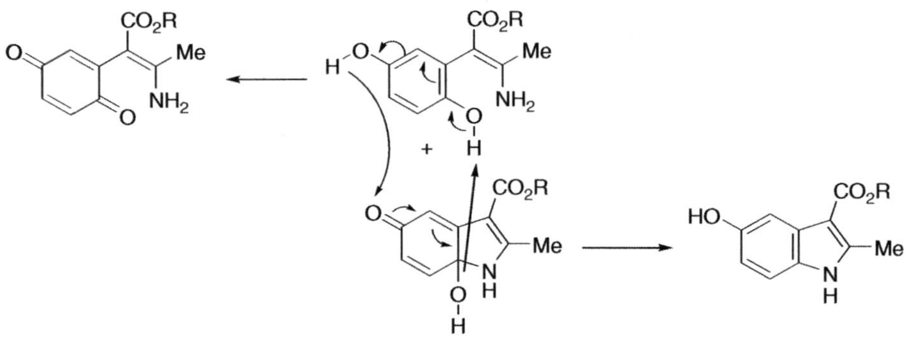

[eqn. 1] [22]

Scheme 8 *Patrick and Saunders Applications of the Nenitzescu Indole Synthesis*

Menéndez and colleagues used a three-component cerium(IV) ammonium nitrate (CAN) domino reaction to effect the synthesis of 5-hydroxybenzo[*g*]indoles and 5-hydroxyindoles in yields up to 96% (equations 2, 3) [50]. The reaction presumably proceeds via enaminone **54** and activation of the quinone by CAN. The diphenylcarbinol unit serves as a source of the C-2 methyl group following loss of benzophenone.

Lin and colleagues expanded the Nenitzescu indole synthesis to include a wide variety of 1,3-diazaheterocycle-fused [1,2-*a*]indoles (Scheme 13, equation 1 and **55–57**) [51]. Parr and Reiss employed the Nenitzescu indole synthesis to prepare several benz[*g*]indoles and 1,2-annulated indoles (equations 2, 3) [52].

Junjappa and Ila effected a Nenitzescu condensation of tetrahydroisoquinoline-derived enaminones with benzoquinones to afford indolo[2,1-*a*]isoquinolines (Scheme 14, equation 1) [53]. Kuckländer and coworkers employed quinone-2,3-dicarboxylates to access benzo[*g*]indoles [54], and they used a similar strategy to synthesize the antitumor 2-hydroxy-5*H*-benzo[*a*]carbazoles (equation 2) [55]. Although the yield is low, a Nenitzescu indolization led to the novel furo[2,3-*g*]indole ring system **58** (equation 3) [56].

Ozaki, Kim, and colleagues fashioned the indole ring from 1,4-cyclohexanedione and *N*-protected α-amino aldehydes in what might be construed as a variation on the Nenitzescu indole synthesis (Scheme 15, equation 1) [57, 58]. In an elegant extension of their original work, Ozaki, Kim, and coworkers carried out the sequence shown in equation 2, wherein the last step is reminiscent of a step in the Nenitzescu mechanism (involving oxidation to a quinone). Yields are good, and this protocol would seem to be a very attractive route to 3-substituted 5-hydroxyindoles. Ketcha and colleagues described a solid-phase Nenitzescu indole synthesis using a Rink-NH_2 support (equation 3) [59]. For a review on the importance of solid-phase synthesis to biologically active benzoannulated nitrogen heterocycles, see Bräse, Gil, and Knepper [60].

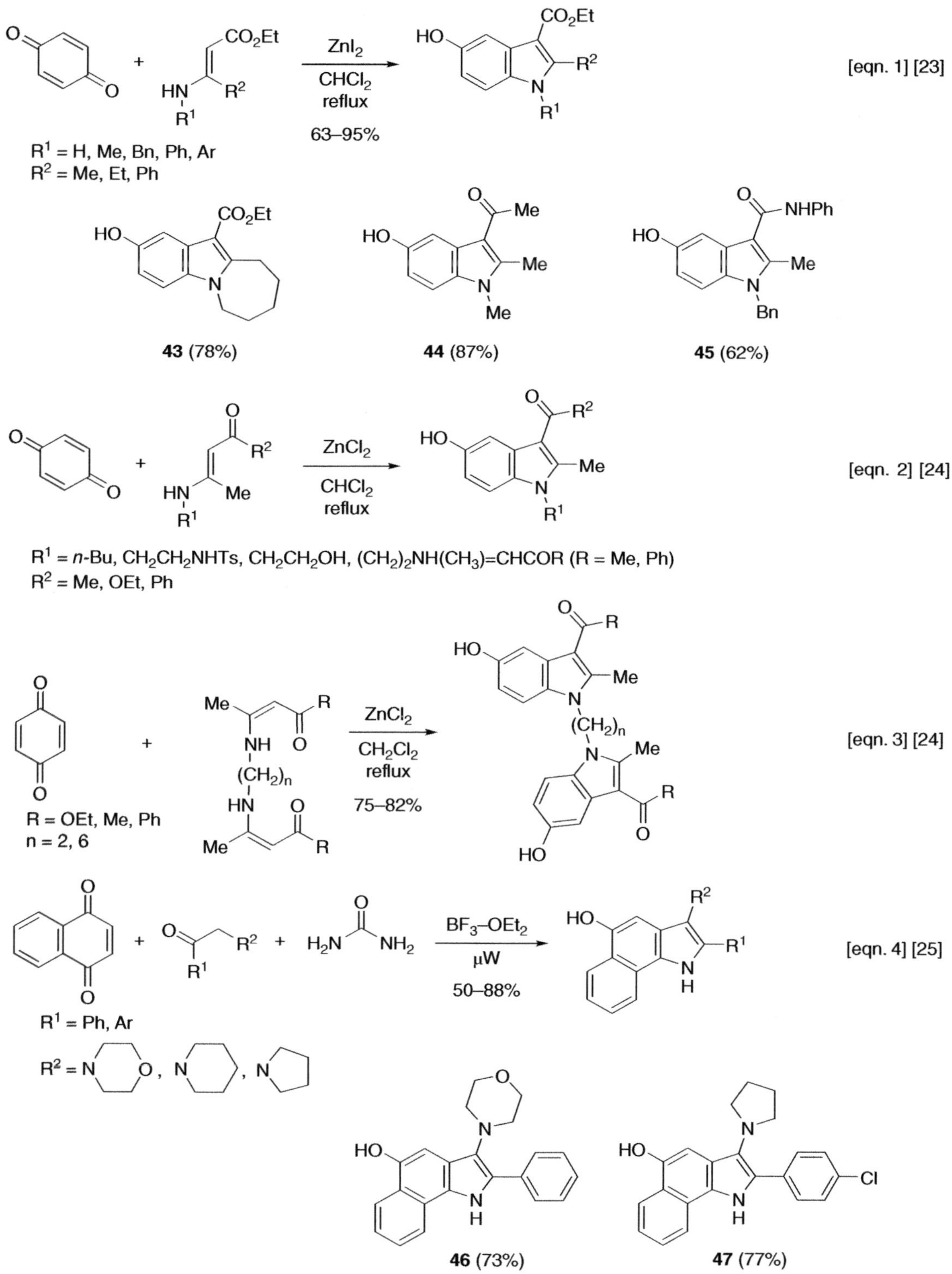

Scheme 9 *Velezheva and Boruah Application of the Nenitzescu Indole Synthesis*

Table 2 *Applications of the Nenitzescu Indole Synthesis to Biological Targets*

Entry	Substrates	Indole	% Yield	Indole Target	Activity	Ref.
1	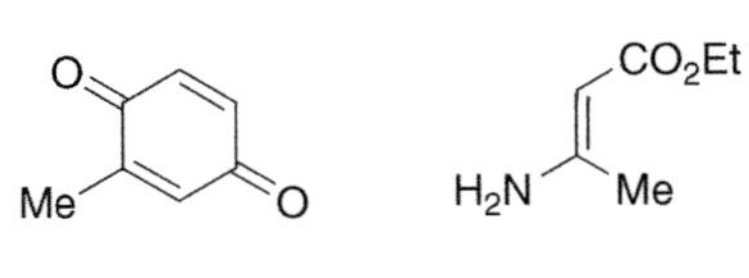	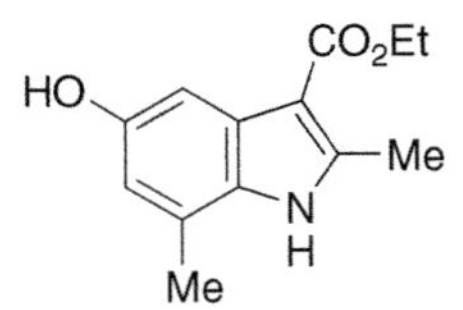	10–15%	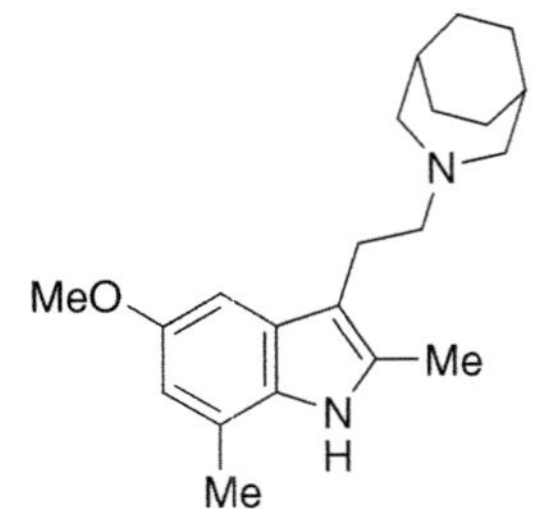	Antiinflammatory	26
2	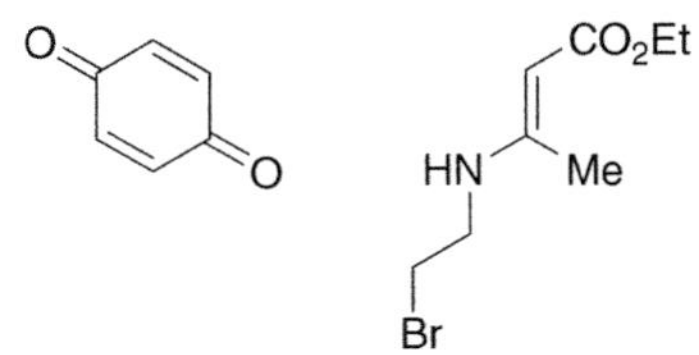	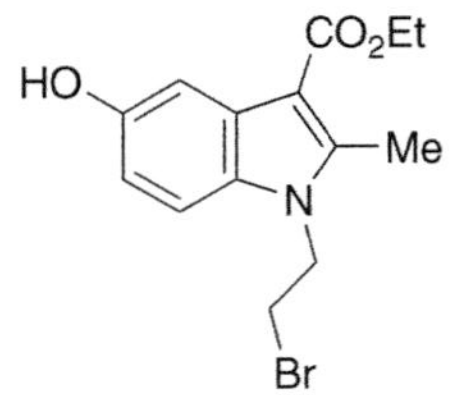	—	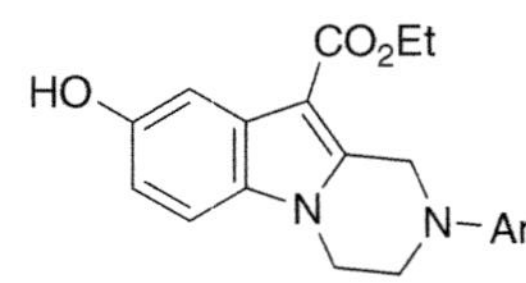	Antidepressant	27
3	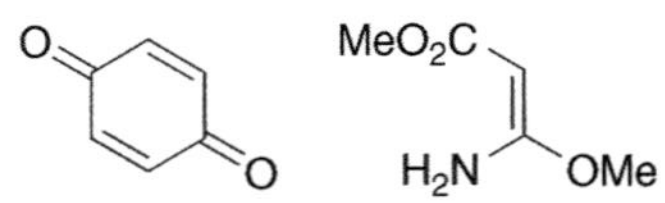	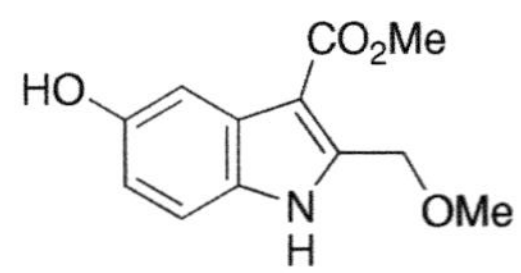	45–54%	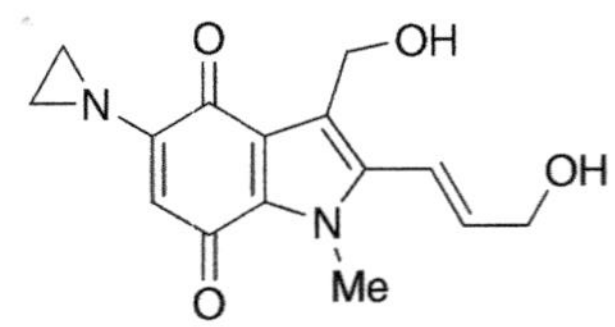 EO9	Anticancer	28
4	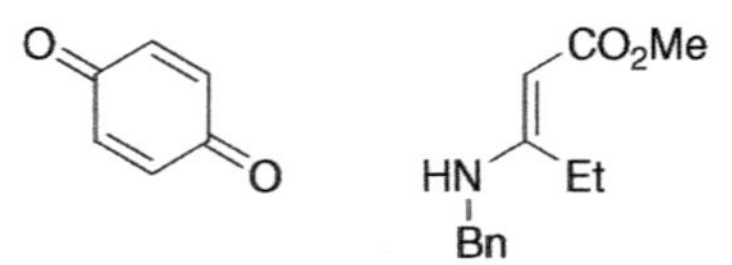	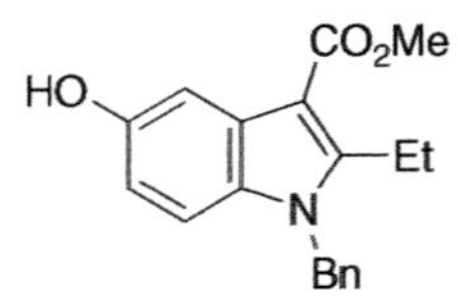	83%	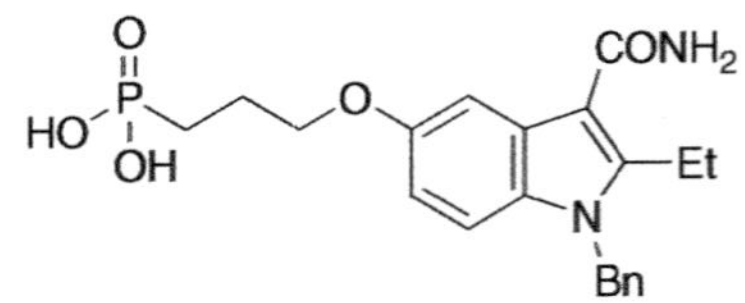 LY311T27	s-PLA_2 Inhibitor	29
5	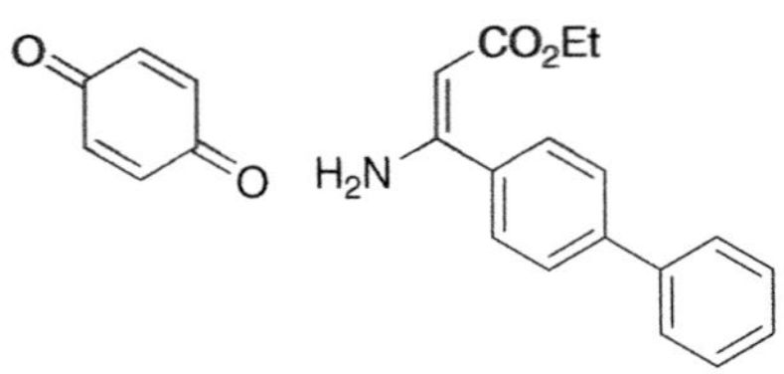	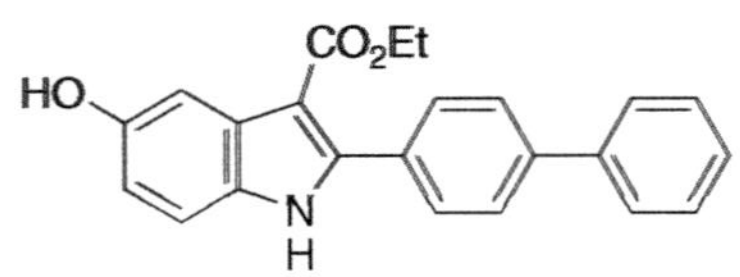	90%	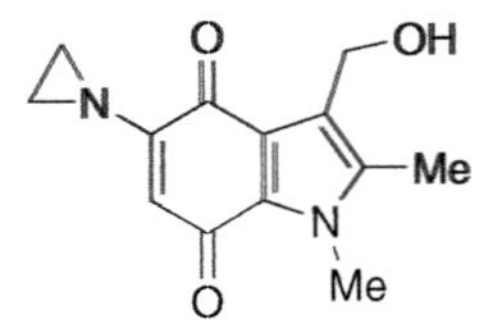	Anticancer	30

6	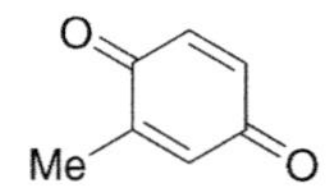	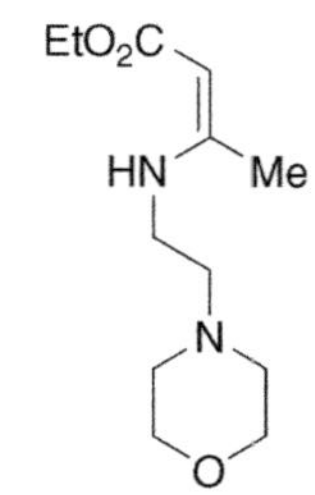	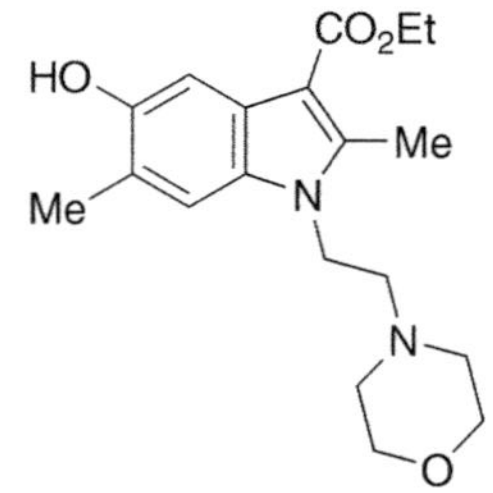	39%	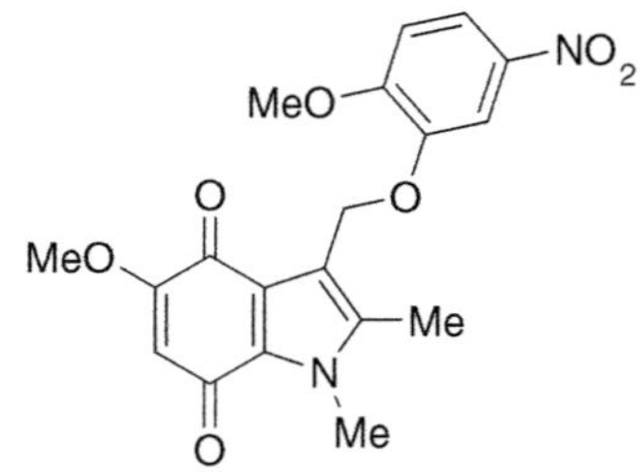	Anticancer	31
7	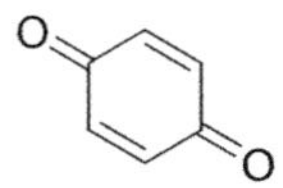	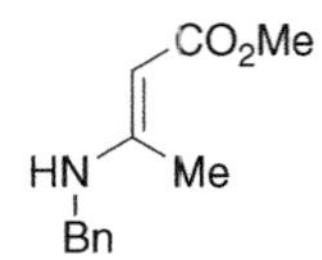	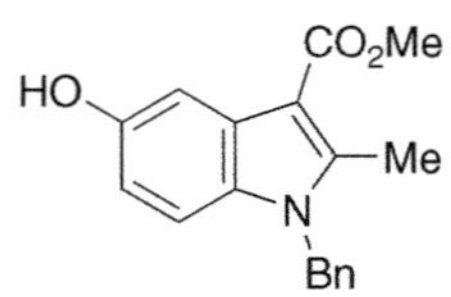	47%	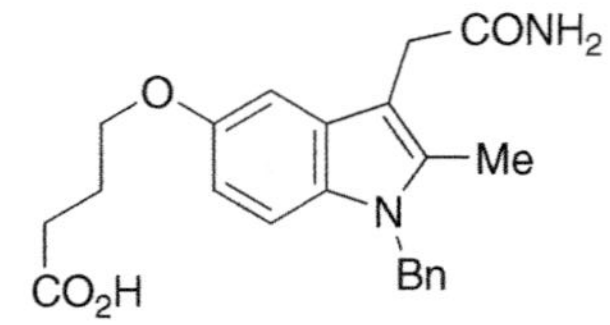	Antiinflammatory	32
8	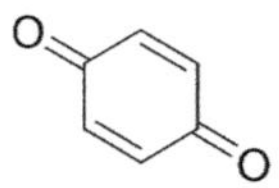	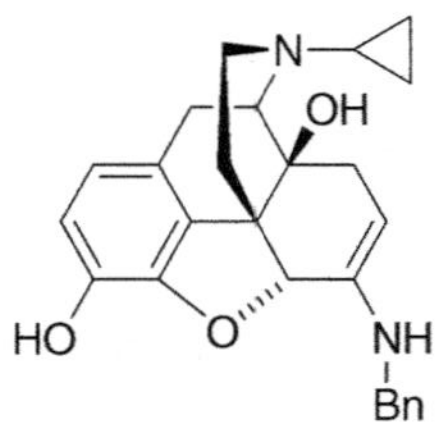	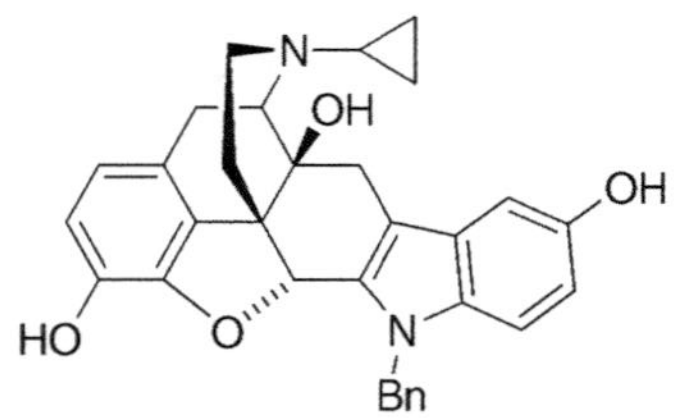	55%	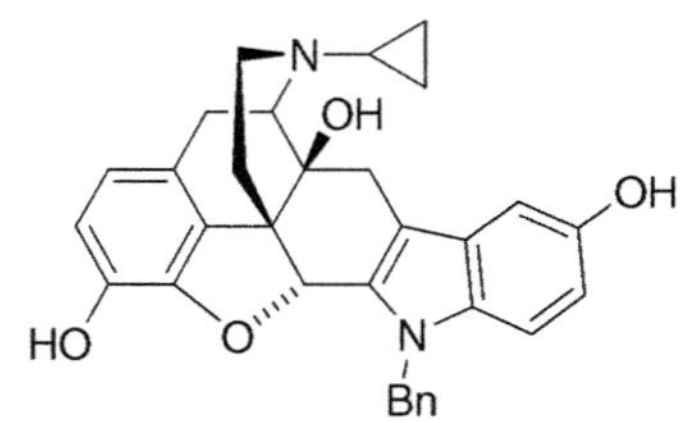	δ-opioid receptor antagonist	33
9	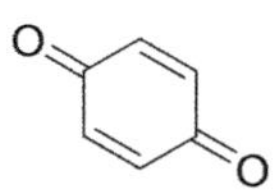	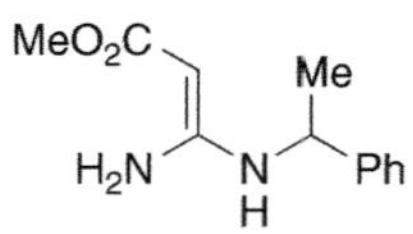	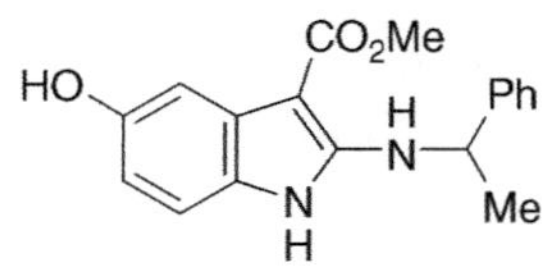	32%	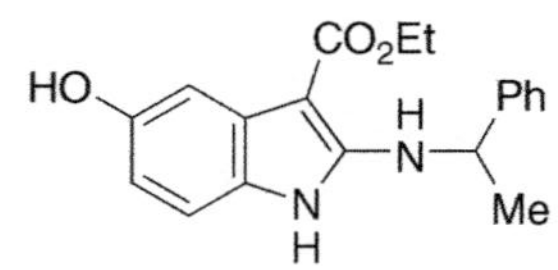	5-lipoxgenase inhibitor	34
10	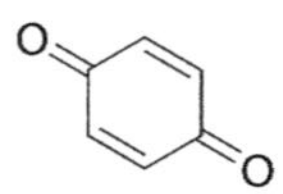	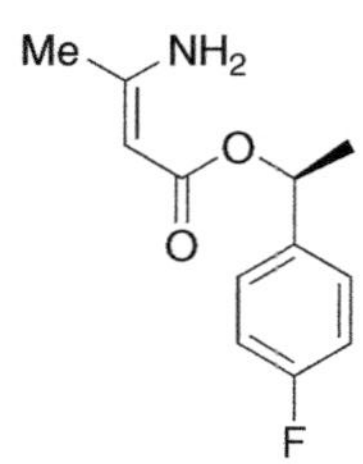	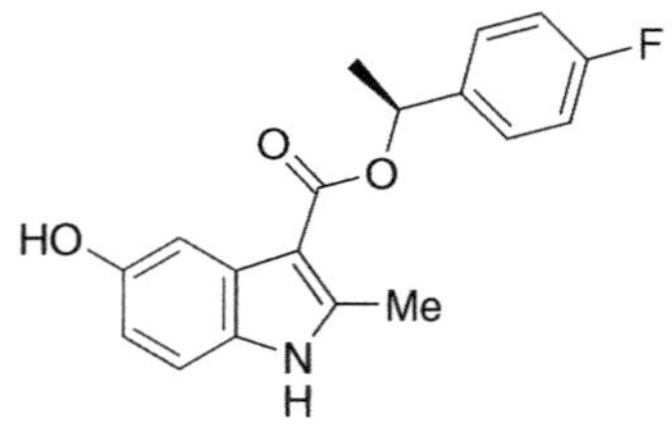	—	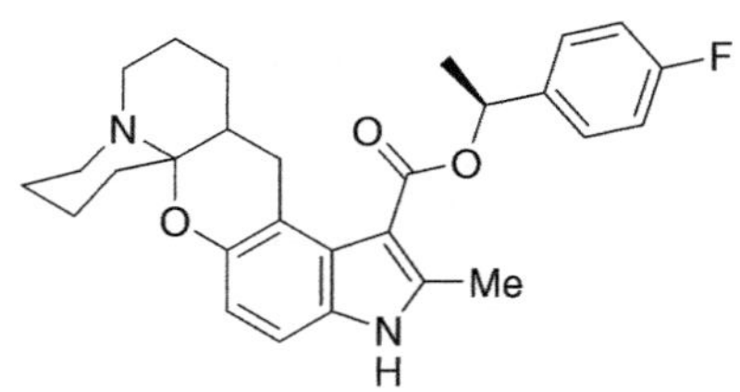	HIV	35

(continued overleaf)

Table 2 *(continued)*

Entry	Substrates	Indole	% Yield	Indole Target	Activity	Ref.
11	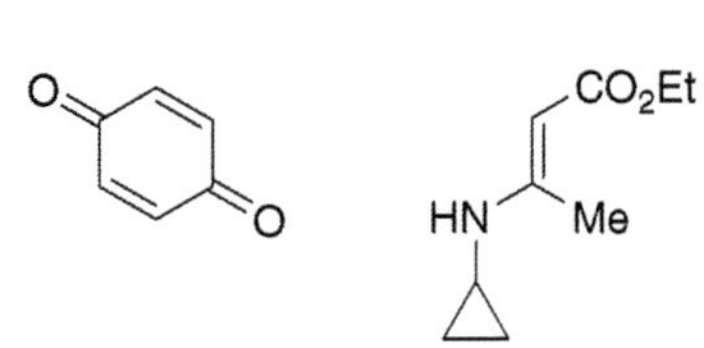	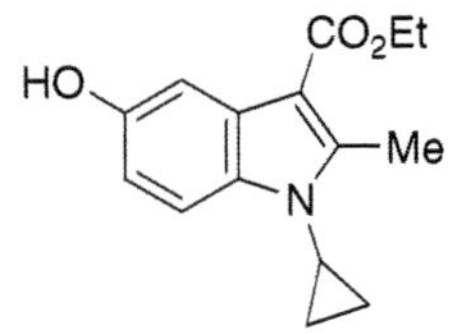	—	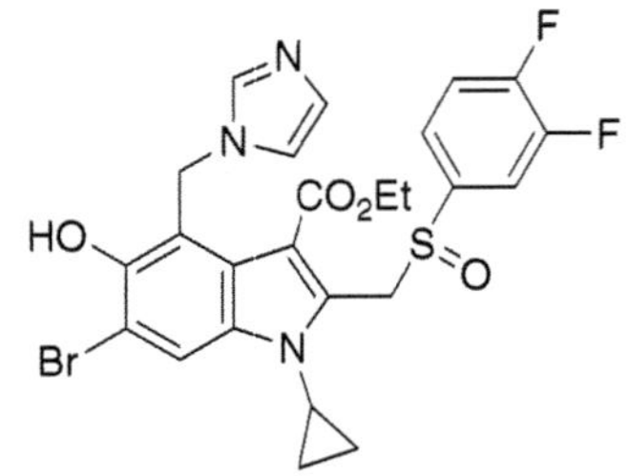	Hepatitis B	36
12	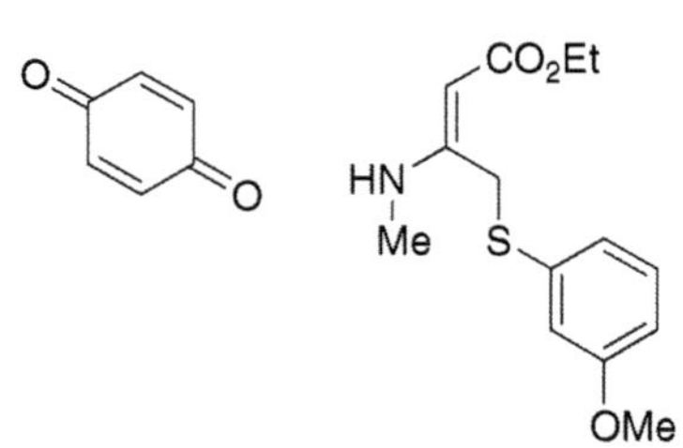	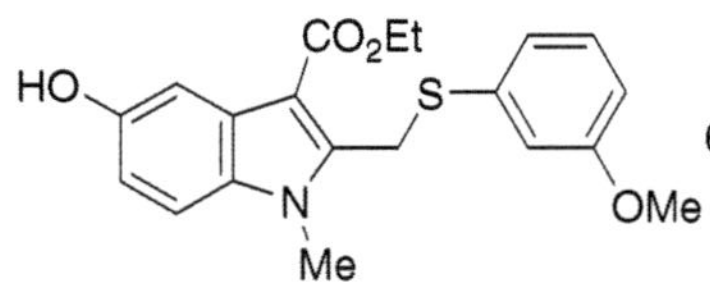	66–68%	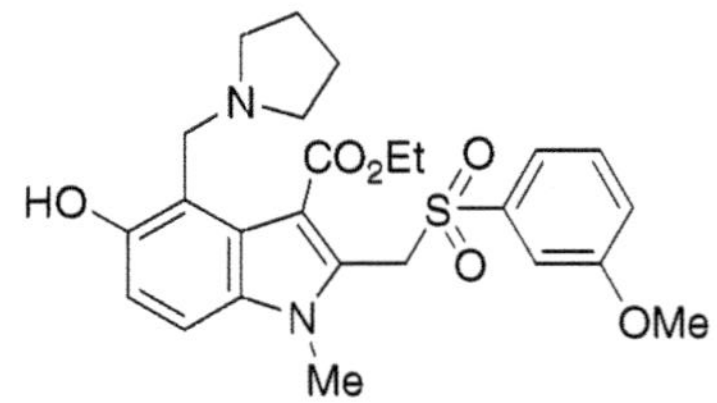	Hepatitis B	37
13	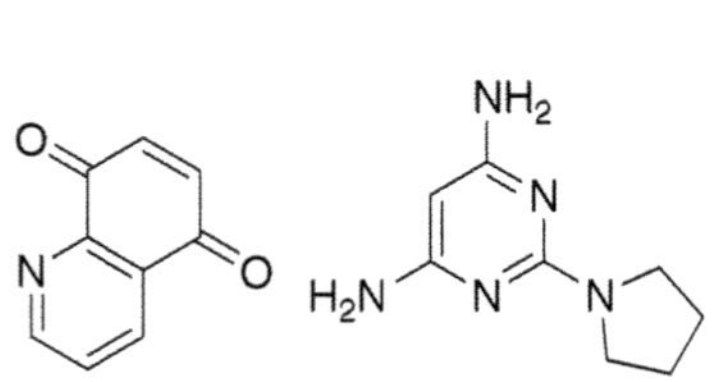	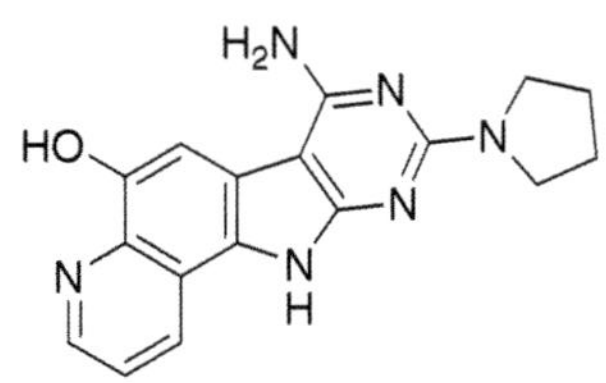	11%	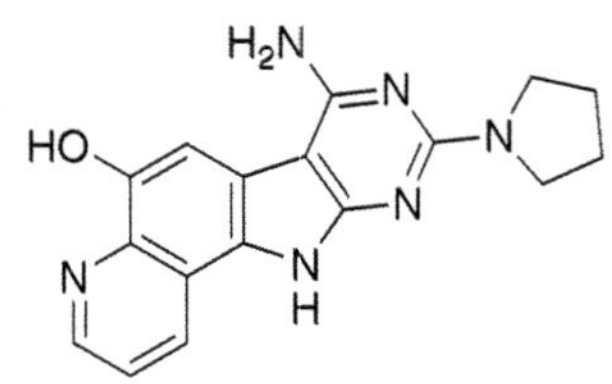	Anticancer	38
14	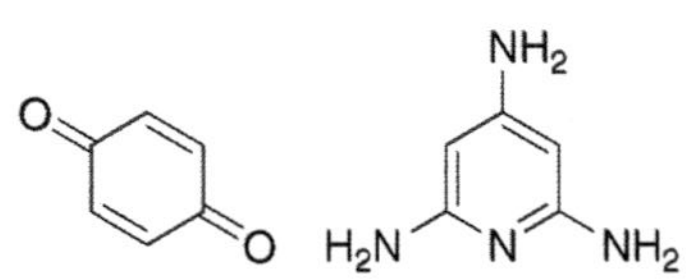	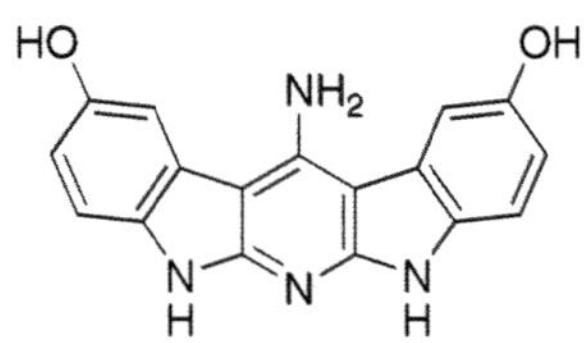	2.2%	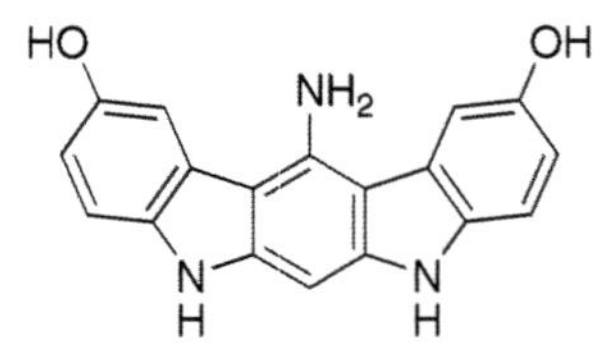	Anticancer	39

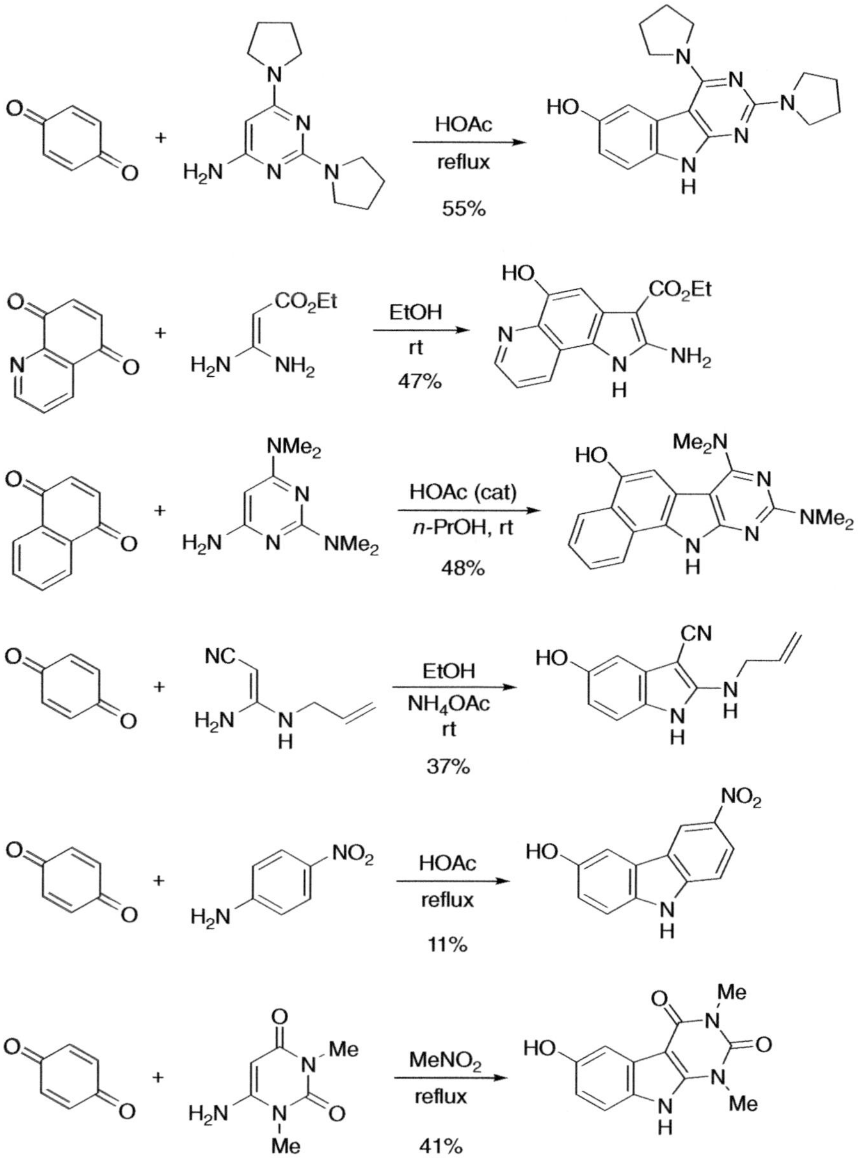

[eqn. 1] [40]

[eqn. 2] [41]

[eqn. 3] [40]

[eqn. 4] [41]

[eqn. 5] [42]

[eqn. 6] [43]

Scheme 10 *Troschütz and Bernier Applications of the Nenitzescu Indole Synthesis*

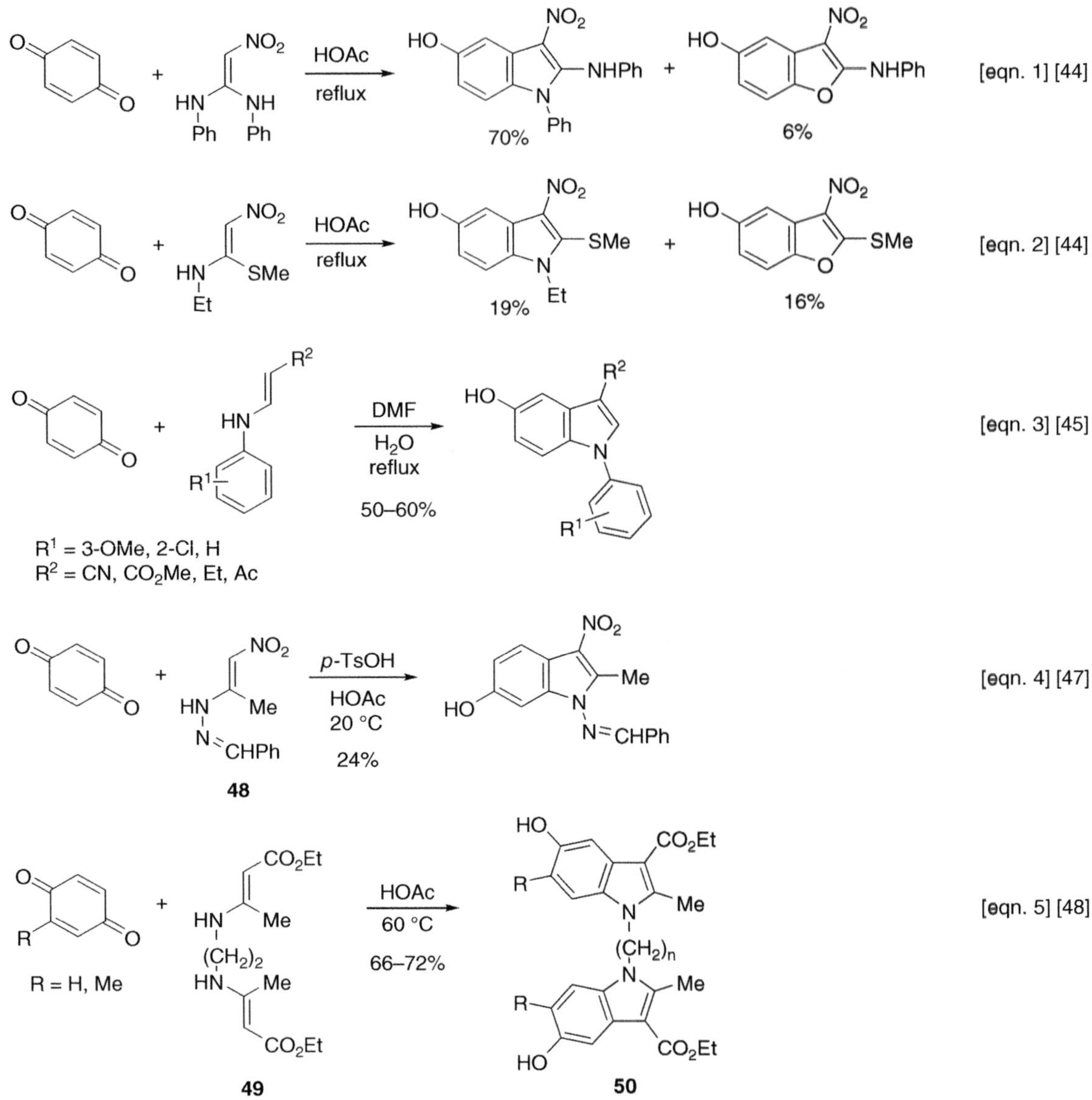

Scheme 11 *Applications of the Nenitzescu Indole Synthesis*

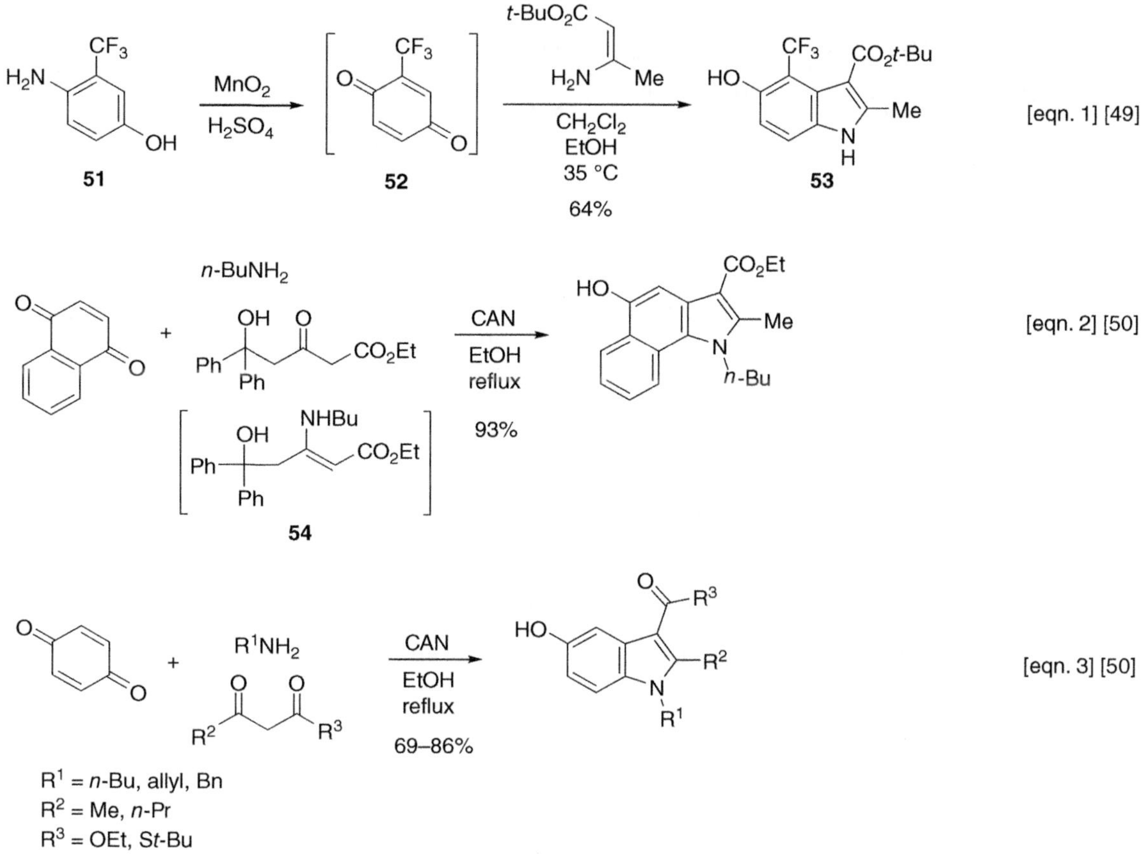

Scheme 12 *Applications of the Nenitzescu Indole Synthesis 2*

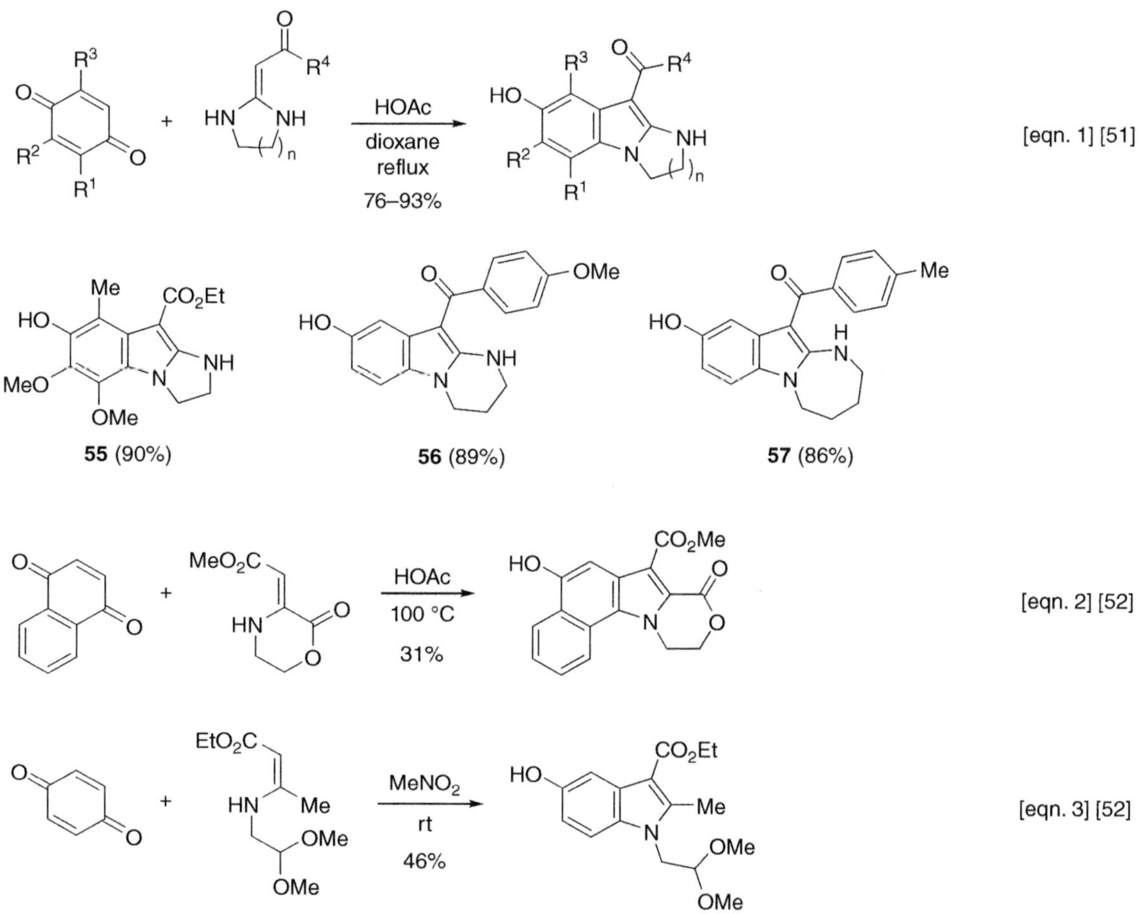

Scheme 13 *Applications of the Nenitzescu Indole Synthesis 3*

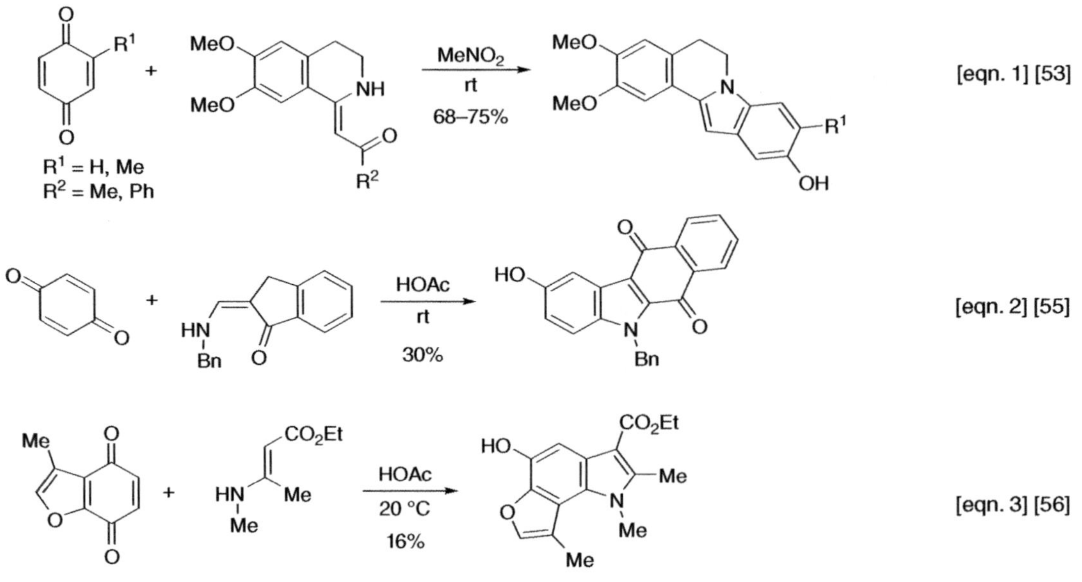

58

Scheme 14 *Applications of the Nenitzescu Indole Synthesis 4*

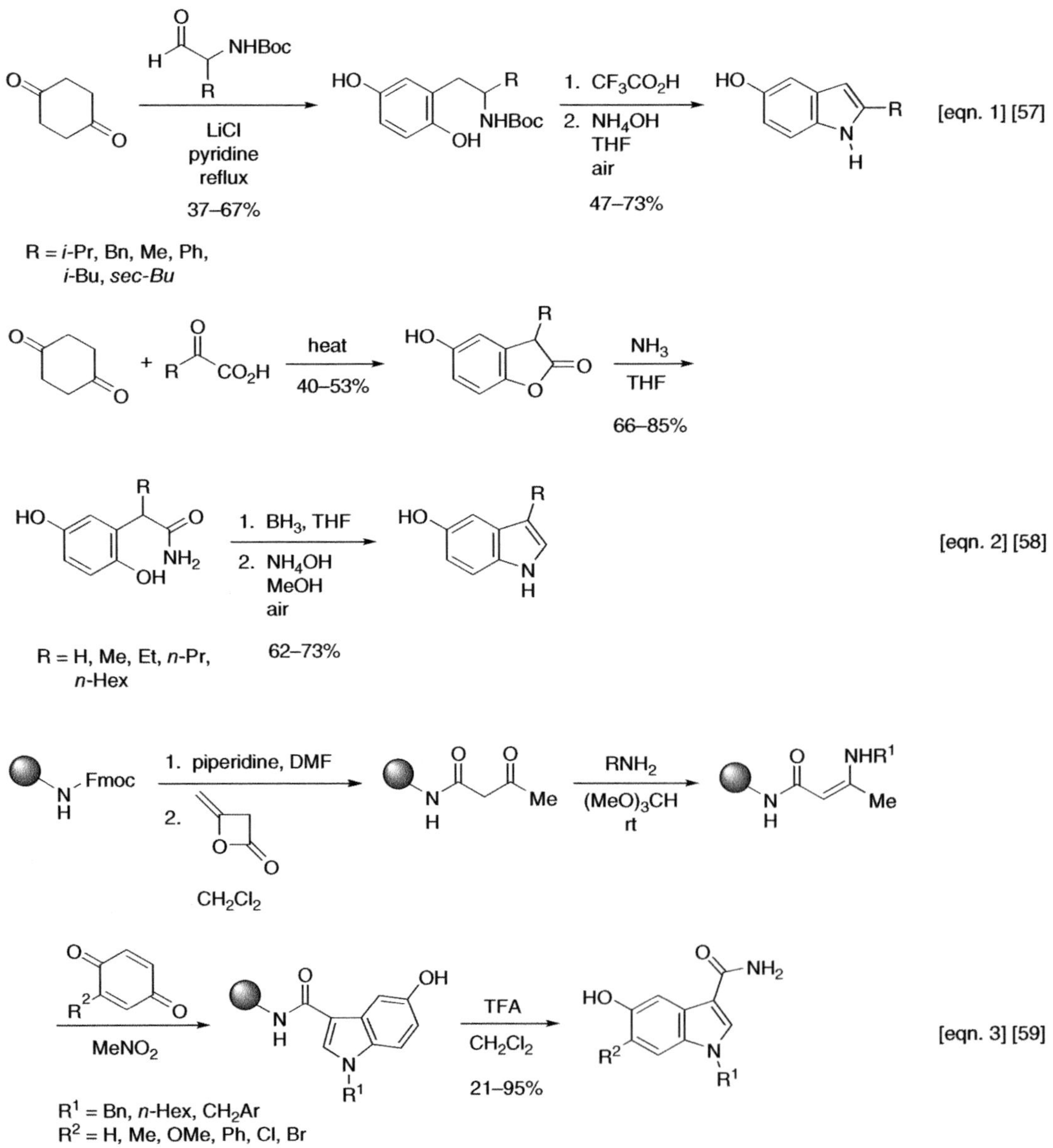

Scheme 15 *Variations of the Nenitzescu Indole Synthesis*

References

[1] C.D. Nenitzescu, *Bull. Soc. Chim. Romania*, 1929, **11**, 37–43.
[2] C.D. Nenitzescu, *Chem. Zentralbl.*, 1929, **11**, 2331–2332.
[3] G.R. Allen, Jr., *Org. React.*, 1973, **20**, 337–454.
[4] R.J.S. Beer, K. Clarke, H.F. Davenport, and A. Robertson, *J. Chem. Soc.*, 1951, 2029–2032.
[5] R.J.S. Beer, H.F. Davenport, and A. Robertson, *J. Chem. Soc., Abst.*, 1953, 1262–1264.
[6] D. Raileanu and C.D. Nenitzescu, *Rev. Roumaine Chim.*, 1965, **10**, 339–353.
[7] E.A. Steck, R.P. Brundage, and L.T. Fletcher, *J. Org. Chem.*, 1959, **24**, 1750–1752.
[8] D. Răileanu, M. Palaghită, and C.D. Nenitzescu, *Tetrahedron*, 1971, **27**, 5031–5047.
[9] G.R. Allen, Jr., C. Pidacks, and M.J. Weiss, *Chem. Ind.*, 1965, 2096–2097.
[10] G.R. Allen, Jr. and M.J. Weiss, *Chem. Ind.*, 1966, 117–118.
[11] G.R. Allen, Jr., C. Pidacks, and M.J. Weiss, *J. Am. Chem. Soc.*, 1966, **88**, 2536–2544.
[12] S.A. Monti, *J. Org. Chem.*, 1966, **31**, 2669–2672.
[13] G.R. Allen, Jr. and M.J. Weiss, *J. Org. Chem.*, 1968, **33**, 198–200.
[14] R. Littell and G.R. Allen, Jr., *J. Org. Chem.*, 1968, **33**, 2064–2069.
[15] J.F. Poletto and M.J. Weiss, *J. Org. Chem.*, 1970, **35**, 1190–1191.
[16] R. Littell, G.O. Morton, and G.R. Allen, Jr., *Chem. Commun.*, 1969, 1144.
[17] R. Littell, G.O. Morton, and G.R. Allen, Jr., *J. Am. Chem. Soc.*, 1970, **92**, 3740–3746.
[18] U. Kuckländer, *Tetrahedron*, 1972, **28**, 5251–5259.
[19] U. Kuckländer, *Tetrahedron*, 1973, **29**, 921–927.
[20] U. Kuckländer, *Tetrahedron*, 1975, **31**, 1631–1639.
[21] U. Kuckländer and W. Hühnermann, *Arch. Pharm.*, 1979, **312**, 515–526.
[22] J.B. Patrick and E.K. Saunders, *Tetrahedron Lett.*, 1979, 4009–4012.
[23] V.S. Velezheva, A.G. Kornienko, S.V. Topilin, *et al.*, *J. Heterocycl. Chem.*, 2006, **43**, 873–879.
[24] V.S. Velezheva, A.I. Sokolov, A.G. Kornienko, *et al.*, *Tetrahedron Lett.*, 2008, **49**, 7106–7109.
[25] M. Borthakur, S. Gogoi, J. Gogoi, and R.C. Boruah, *Tetrahedron Lett.*, 2010, **51**, 5160–5163.
[26] J.F. Poletto, G.R. Allen, Jr., A.E. Sloboda, and M.J. Weiss, *J. Med. Chem.*, 1973, **16**, 757–765.
[27] N.G. Tsyshkova, F.A. Trofimov, V.P. Marinchenko, *et al.*, *Khim.-Farm. Zh.*, 1992, **26**, 70–72.
[28] M. Kinugawa, H. Arai, H. Nishikawa, *et al.*, *J. Chem. Soc., Perkin Trans. 1*, 1995, 2677–2678.
[29] J.M. Pawlak, V.V. Khau, D.R. Hutchison, and M.J. Martinelli, *J. Org. Chem.*, 1996, **61**, 9055–9059.
[30] H.D. Beall, S. Winski, E. Swann, *et al.*, *J. Med. Chem.*, 1998, **41**, 4755–4766.
[31] E. Swann, P. Barraja, A.M. Oberlander, *et al.*, *J. Med. Chem.*, 2001, **44**, 3311–3319.
[32] A.R. Maguire, S.J. Plunkett, S. Papot, *et al.*, *Bioorg. Med. Chem.*, 2001, **9**, 745–762.
[33] Shefali, S.K. Srivastava, S.M. Husbands, and J.W. Lewis, *J. Med. Chem.*, 2005, **48**, 635–638.
[34] J. Landwehr, S. George, E.-M. Karg, *et al.*, *J. Med. Chem.*, 2006, **49**, 4327–4332.
[35] M. Rönn, Q. McCubbin, S. Winter, *et al.*, *Org. Process Res. Dev.*, 2007, **11**, 241–245.
[36] H. Chai, Y. Zhao, C. Zhao, and P. Gong, *Bioorg. Med. Chem.*, 2006, **14**, 911–917.
[37] C. Zhao, Y. Zhao, H. Chai, and P. Gong, *Bioorg. Med. Chem.*, 2006, **14**, 2552–2558.
[38] B. Dotzauer, R. Grünert, P.J. Bednarski, *et al.*, *Bioorg. Med. Chem.*, 2006, **14**, 7282–7292.
[39] C. Willemann, R. Waibel, R. Grünert, *et al.*, *J. Heterocycl. Chem.*, 2008, **45**, 1517–1519.
[40] B. Dotzauer and R. Troschütz, *Synlett*, 2004, 1039–1043.
[41] J. Landwehr and R. Troschütz, *Synthesis*, 2005, 2414–2420.
[42] J.-L. Bernier, J.-P. Hénichart, C. Vaccher, and R. Houssin, *J. Org. Chem.*, 1980, **45**, 1493–1496.
[43] J.-L. Bernier and J.-P. Hénichart, *J. Org. Chem.*, 1981, **46**, 4197–4198.
[44] V. Aggarwal, A. Kumar, H. Ila, and H. Junjappa, *Synthesis*, 1981, 157–158.
[45] M.S. Mayadeo and S.A. Gandhi, *J. Indian Chem. Soc.*, 1994, **71**, 281–282.
[46] V.M. Lyubchanskaya, S.A. Savina, L.M. Alekseeva, *et al.*, *Mendeleev Commun.*, 2004, **14**, 73–75.
[47] V.M. Lyubchanskaya, S.A. Savina, L.M. Alekseeva, *et al.*, *Russ. Chem. Bull.*, 2004, **53**, 2834–2839.
[48] N.A. Kheder, *Heterocycles*, 2009, **78**, 1281–1288.
[49] E.E. Boros, I. Kaldor, and P.S. Turnbull, *J. Heterocycl. Chem.*, 2011, **48**, 733–736.
[50] P.A. Suryavanshi, V. Sridharan, and J.C. Menéndez, *Org. Biomol. Chem.*, 2010, **8**, 3426–3436.
[51] L.-J. Yang, S.-J. Yan, W. Chen, and J. Lin, *Synthesis*, 2010, 3536–3544.
[52] R.W. Parr and J.A. Reiss, *Aust. J. Chem.*, 1984, **37**, 1263–1270.
[53] O. Barun, S. Chakrabarti, H. Ila, and H. Junjappa, *J. Org. Chem.*, 2001, **66**, 4457–4461.
[54] L.W. Schenck, A. Sippel, K. Kuna, *et al.*, *Tetrahedron*, 2005, **61**, 9129–9139.
[55] C. Asche, W. Frank, A. Albert, and U. Kucklaender, *Bioorg. Med. Chem.*, 2005, **13**, 819–837.
[56] V.M. Lyubchanskaya, L.M. Alekseeva, S.A. Savina, *et al.*, *Chem. Heterocycl. Comp.*, 2003, **39**, 872–877.
[57] Y. Ozaki, K. Okamura, A. Hosoya, and S.-W. Kim, *Chem. Lett.*, 1997, 679–680.
[58] Y. Ozaki, Z.S. Quan, K. Watabe, and S.W. Kim, *Heterocycles*, 1999, **51**, 727–731.
[59] D.M. Ketcha, L.J. Wilson, and D.E. Portlock, *Tetrahedron Lett.*, 2000, **41**, 6253–6257.
[60] S. Bräse, C. Gil, and K. Knepper, *Bioorg. Med. Chem.*, 2002, **10**, 2415–2437.

16

Engler-Kita Indole Synthesis

The Engler indole synthesis entails the Lewis acid–promoted reaction of benzoquinone imines with styrenes and enol ethers to give a suite of heterocycles including dihydroindoles and indoles [1–6]. Focusing on indoles, Engler's basic strategy is shown in Scheme 1. Whereas titanium(IV) Lewis acids afford mainly dihydroindoles **1** (equation 1), boron trifluoride yields nearly exclusively dihydrobenzofurans **2** (equation 2) [2]. Other examples of dihydroindoles prepared are shown in equations 3–5. Note that the bisimide **3** yields dihydroindole **4** with boron trifluoride (equation 3). These dihydroindoles are smoothly oxidized to the corresponding indoles with 2,3-dichloro-5,6-dicyanobenzoquinone (DDQ) in excellent yield (equation 5). We will revisit the indoline to indole synthetic route in Chapters 66 and 67. Although direct DDQ oxidation of the 5-hydroxydihydroindoles (e.g., **5**) was not possible, this failure was circumvented as shown in equation 6 [4]. Thus, the Engler indole synthesis is related to the Nenitzescu indolization both in benzoquinone starting materials and 5-hydroxyindole products.

Engler and colleagues extended to the reaction of *N*-phenylsulfonylimino-1,4-benzoquinones with enol ethers derived from 3- and 4-piperidone to give β- and γ-tetrahydrocarbolines (Scheme 2) [5, 6]. Once again the choice of Lewis acid is critical for this chemistry ($TiCl_4$:Ti(O*i*-Pr)$_4$), lest the corresponding benzofurans intervene (BF_3). Carbolines **6** and **7** are formed similarly using quinone bisimide **3** shown in Scheme 1 (equation 3). Conditions for oxidation to the aromatic carbolines are shown in equations 3 and 4. Interestingly, **8** is converted to **9** in one step with excess potassium *tert*-butoxide in THF (71%). Engler parlayed his synthesis of oxygenated carbolines into the preparation of the potential antitumor compounds **10**, which are abbreviated versions of ellipticine alkaloids.

The mechanistic rationale for the dramatic Lewis acid effect on the course of the Engler indole/benzofuran synthesis involving styrenes is illustrated in Scheme 3. Whereas titanium forms a bidentate complex **11** with the C-1 carbonyl and C-alkoxy oxygens, which leads to the dihydroindole (equation 1), boron trifluoride engages in monodentate complexation **12** with the sulfonyl nitrogen, leading to dihydrobenzofuran formation (equation 2) [3, 4]. Likewise, the reaction between **11** and an enol ether derived from 4-piperidone is depicted in equation 3 [5].

While the Engler indole synthesis has not been pursued by others, Kita and coworkers developed a related intramolecular amine cyclization onto benzoquinones leading to 5-oxygenated indoles (Scheme 4) [7]. The yields are excellent, and several examples are shown (yields are overall from the starting quinones or quinone acetal).

Kita's group also described an intermolecular version that resembles the procedure of Engler [8]. This is shown in Scheme 5 and again features phenyliodine(III) bis(trifluoroacetate) (PIFA) reaction with *N*-tosylaniline, followed by nucleophilic attack by an activated olefin, and cyclization to give either *N*-tosylindoline (equation 1) or *N*-tosylindole (equation 2). The latter is the product when phenyl vinyl sulfides are employed. The thiophenol byproduct was oxidized to diphenyl disulfide with excess PIFA. Four indoles that were prepared in this study are also shown.

Indole Ring Synthesis: From Natural Products to Drug Discovery, First Edition. Gordon W. Gribble.

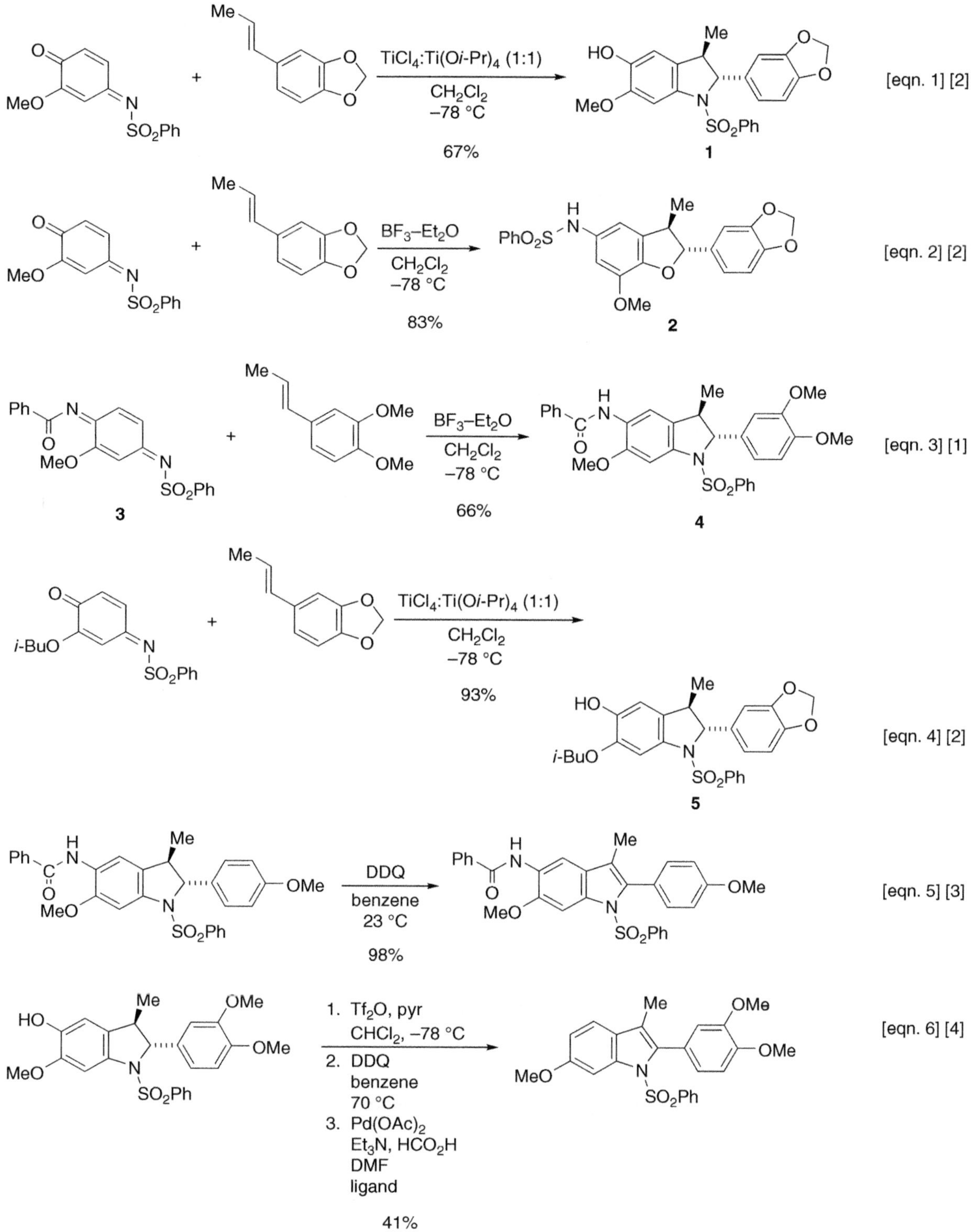

Scheme 1 *Engler Indole Synthesis*

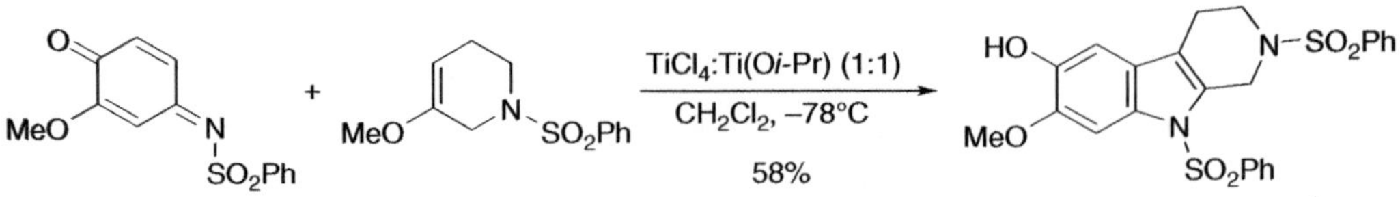

[eqn. 1] [5], [6]

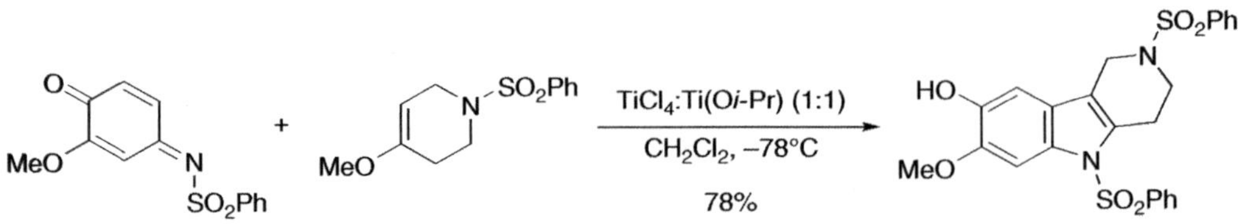

[eqn. 2] [5], [6]

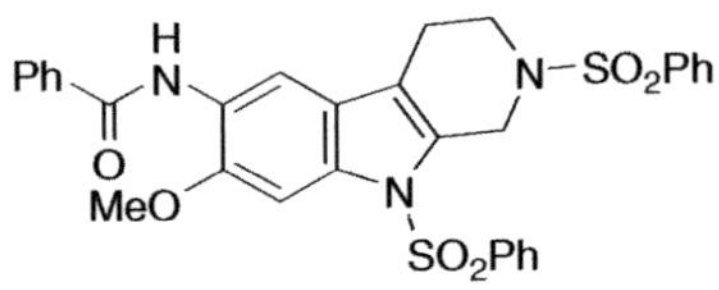

6 (80%)

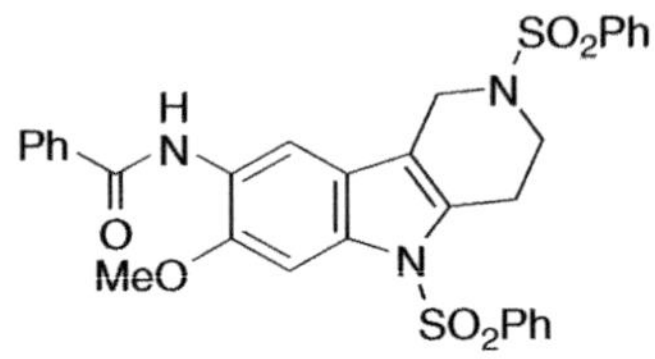

7 (81%)

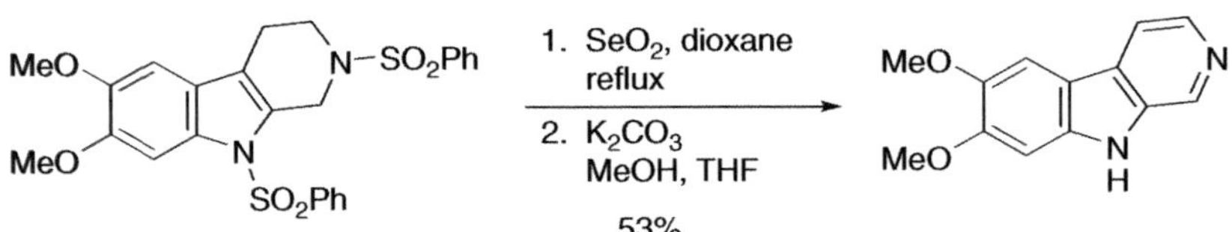

8 **9**

[eqn. 3] [6]

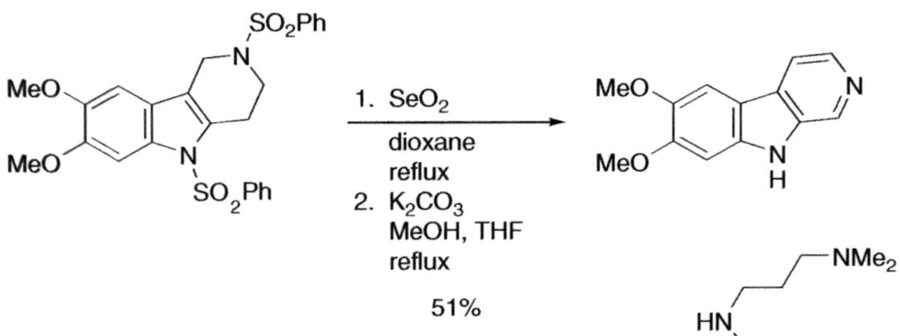

[eqn. 4] [6]

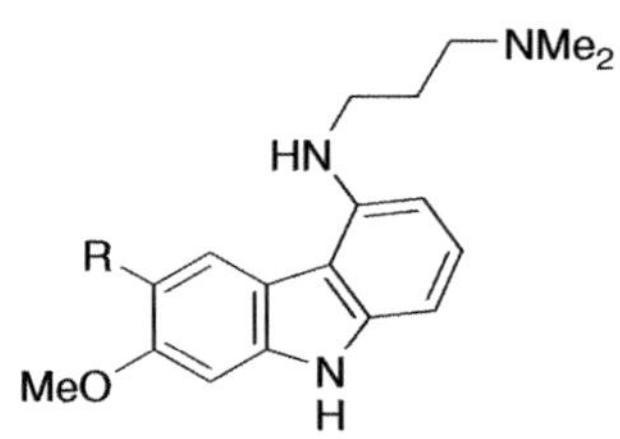

10 (R = H, OMe)

Scheme 2 *Engler Indole Synthesis 2*

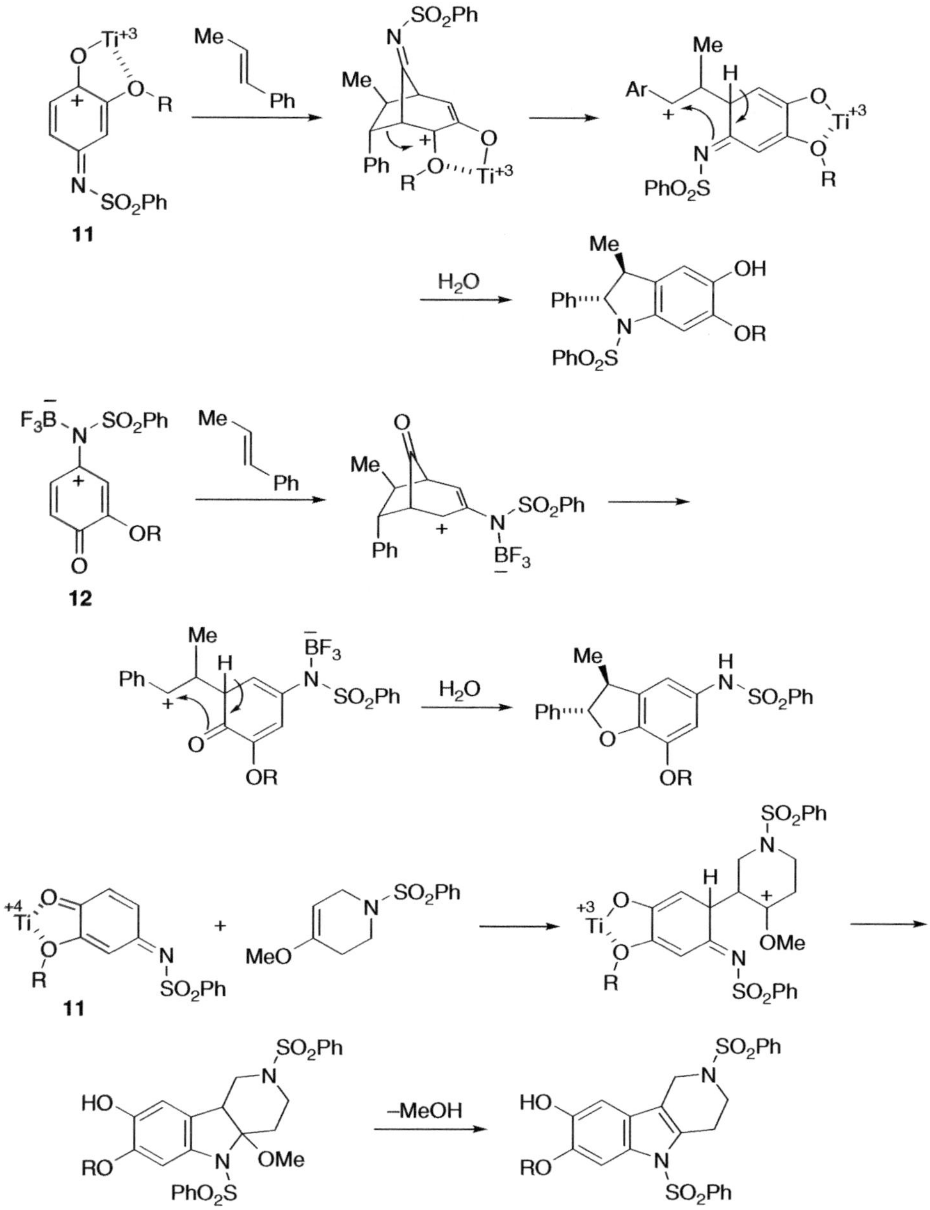

[eqn. 1] [3], [4]

[eqn. 2] [3], [4]

[eqn. 3] [5]

Scheme 3 *Proposed Mechanism of the Engler Indole Synthesis*

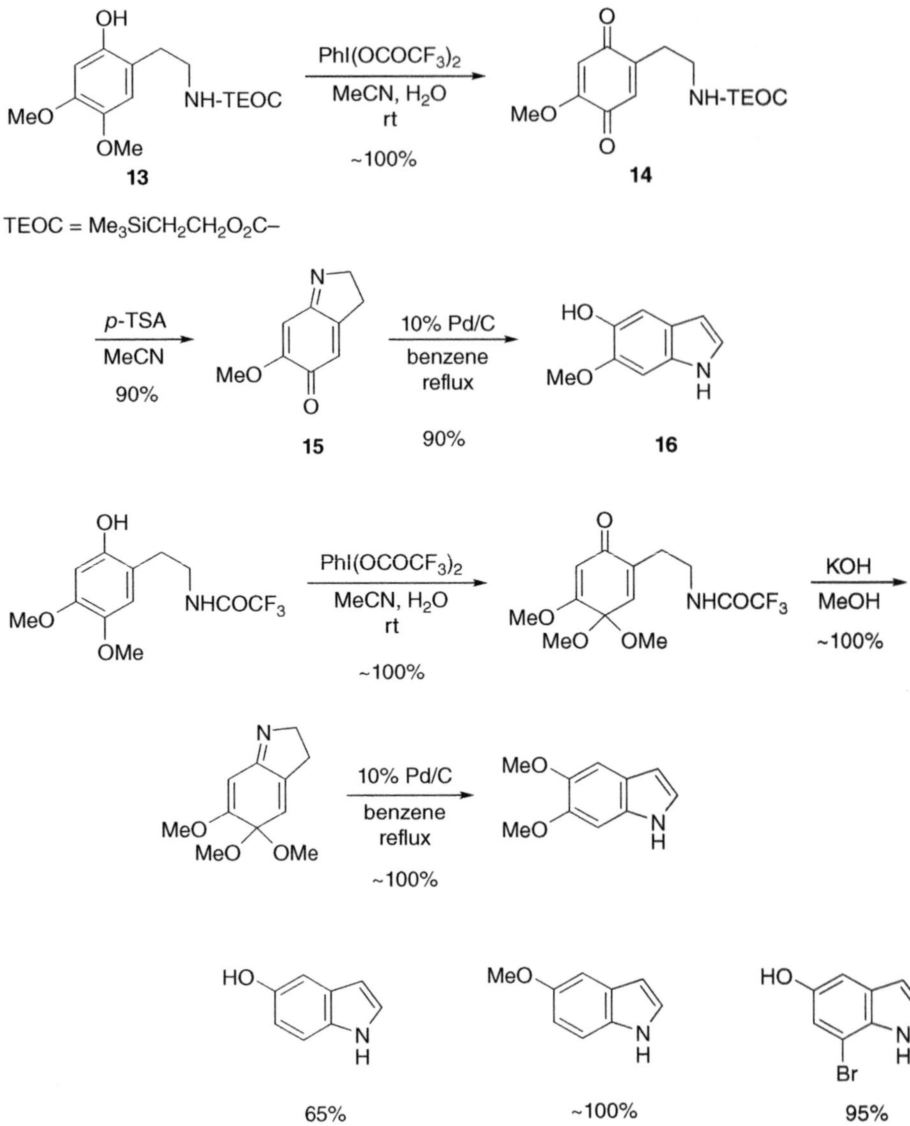

[eqn. 1] [7]

[eqn. 2] [7]

Scheme 4 *Kita Indole Synthesis*

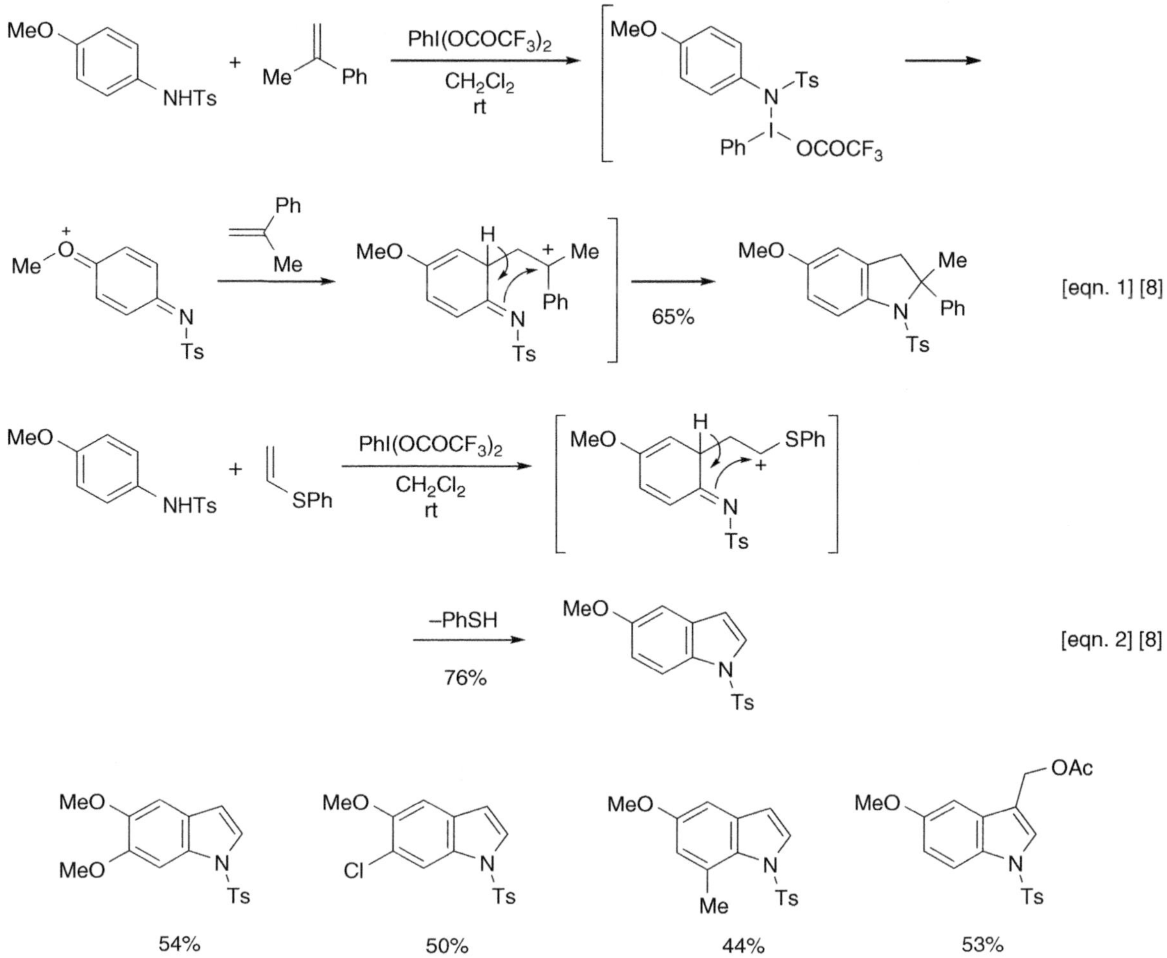

Scheme 5 *Kita Indole Synthesis 2*

Some 40 years earlier, Harley-Mason and Cromartie synthesized 5-hydroxyindole in excellent yield via the oxidation of 2,5-dihydroxyphenylalanine with potassium ferricyanide (Scheme 6, equation 1) [9]. This simple oxidation when applied to 2,3-dihydroxyphenylalanine gave 7-hydroxyindole in 20% yield [9] and 3,4-dihydroxyphenylalanine gave 5,6-dihydroxyindole in 30% yield (but not repeatable) [10]. This simple oxidation procedure with ferricyanide was used to synthesize bufotenine and serotonin [11]. In chemistry related to the Kita indole synthesis, Clive and Stoffman presented a synthesis of 4-halo-5-hydroxyindoles from the corresponding coumarins (Scheme 6, equation 2) [12]. The key step in Clive's sequence is the Hofmann rearrangement of amide **17** to amine **18**, followed by oxidation to quinone **19** with phenyliodine(III) bis(triacetate). In like fashion, 4-bromo-5-hydroxyindole, 5-hydroxy-4-iodoindole, and 5-hydroxy-4-iodo-3-methylindole were prepared in this study. Nishiyama and coworkers adopted an electrochemical oxidation of diaryl amides to carbazoles (equation 3 and **21–22**) [13].

The Engler–Kita indole synthesis, combined with the Nenitzescu indolization, represent powerful and versatile syntheses of 5-hydroxyindoles.

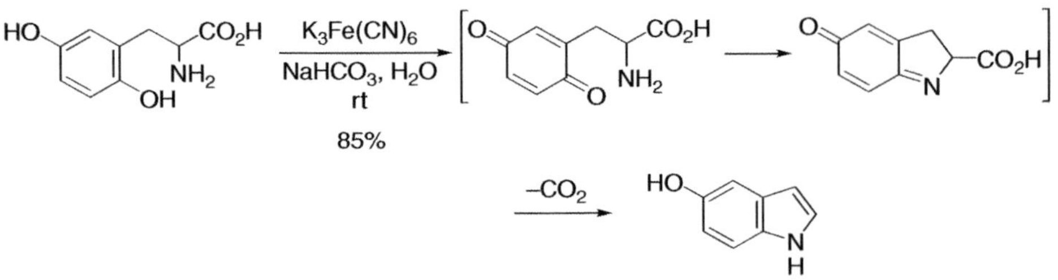

[eqn. 1] [9]

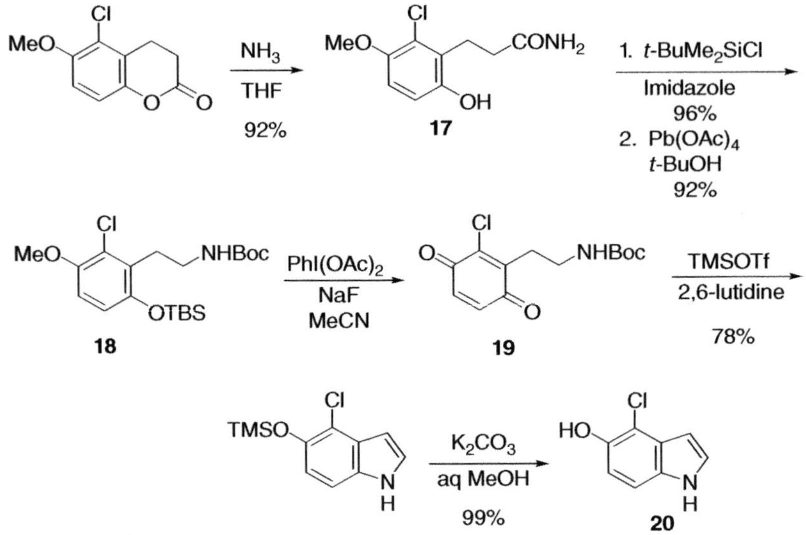

[eqn. 2] [12]

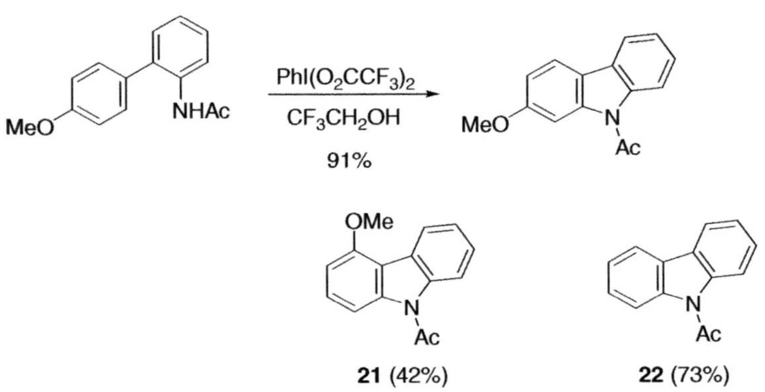

[eqn. 3] [13]

Scheme 6 Applications of the Engler-Kita Indole Synthesis

References

[1] T.A. Engler, K.O. Lynch, Jr., W. Chai, and S.P. Meduna, *Tetrahedron Lett.*, 1995, **36**, 2713–2716.
[2] T.A. Engler, W. Chai, and K.O. Lynch, Jr., *Tetrahedron Lett.*, 1995, **36**, 7003–7006.
[3] T.A. Engler, S.P. Meduna, K.O. LaTessa, and W. Chai, *J. Org. Chem.*, 1996, **61**, 8598–8603.
[4] T.A. Engler, W. Chai, and K.O. LaTessa, *J. Org. Chem.*, 1996, **61**, 9297–9308.
[5] T.A. Engler and J. Wanner, *Tetrahedron Lett.*, 1997, **38**, 6135–6138.
[6] T.A. Engler and J. Wanner, *J. Org. Chem.*, 2000, **65**, 2444–2457.
[7] Y. Kita, H. Tohma, M. Inagaki, and K. Hatanaka, *Heterocycles*, 1992, **33**, 503–506.
[8] H. Tohma, H. Watanabe, S. Takizawa, *et al.*, *Heterocycles*, 1999, **51**, 1785–1788.
[9] R.I.T. Cromartie and J. Harley-Mason, *J. Chem. Soc.*, 1952, 2525–2527.
[10] J.D. Bu'Lock and J. Harley-Mason, *J. Chem. Soc.*, 1951, 2248–2252.
[11] J. Harley-Mason and A.H. Jackson, *J. Chem. Soc.*, 1954, 1165–1171.
[12] E.J.L. Stoffman and D.L.J. Clive, *Org. Biomol. Chem.*, 2009, **7**, 4862–4870.
[13] D. Kajiyama, K. Inoue, Y. Ishikawa, and S. Nishiyama, *Tetrahedron*, 2010, **66**, 9779–9784.

17

Bailey–Liebeskind–O'Shea Indoline–Indole Synthesis

Bailey and Liebeskind independently discovered the anionic cyclization shown in Scheme 1 [1, 2] that leads to indolines and, by subsequent oxidation, to indoles. Both methods are virtually identical, involving bromine–lithium exchange followed by quenching of the aryllithium species with an electrophile (equations 1 and 4). The *N*-allylindole can be deallylated with palladium (equation 2) [3], and the indoline is easily oxidized to the corresponding indole (equation 3). A selection of indoles prepared this way by Liebeskind and Zhang is shown (1–6) [2]. The yields are overall for the two steps. These workers also employed chloranil (*t*-BuOMe, rt) to oxidize indolines to indoles. An excellent review by Bailey and Mealy is available [4]. Bailey's related indoline synthesis via arynic cyclization is presented in the later chapter on Nucleophilic Cyclization of Arynes (Chapter 65).

Several applications of the Bailey–Liebeskind indoline–indole synthesis are tabulated in Table 1 [5–8]. Entry 4 involves the generation of **8**, **9**, and **10** from dibromo compound **7**, and Entry 5 involves the further lithiation, of **10** to **11** (Scheme 2) [9]. Quenching of these species then afforded the products summarized in Table 1 (Entries 4 and 5).

O'Shea and colleagues intensively studied the carbolithiation of *ortho*-aminostyrenes leading to indoles (Scheme 3) [10–14]. A proposed pathway is shown in equation 3 for the reaction with DMF, and a derivative of carbinol **12** was isolated (**12**, R = *t*-Bu) [11].

Furthermore, O'Shea and colleagues reported the enantioselective carbolithiation of *ortho*-aminostyrenes as summarized in Scheme 4 [12, 13]. The chiral intermediate **13** (equation 1) has the *S*-configuration based on the generation of known (*S*)-3-methylheptanoic acid from **14**. Quenching **13** with electrophiles furnished **15** in modest to good yields and in very good enantiorichments (equation 2) [12]. Electrophiles included DMF, PhCN, γ-butyrolactone, *N*-methoxy-*N*-methylacetamide, and 2,2-diethoxypropionitrile. Carbon dioxide afforded the expected oxindole [13]. Debenzylation was effected with Li in NH_3 (equation 3) [13].

These workers expanded their carbolithiation chemistry to *ortho*-aminostibenes (e.g., **16**) in the synthesis of C-3 functionalized indoles (Scheme 5) [14]. Thus, lithiation of **16** in the presence of pentamethyldiethylenetriamine (PMDTA) gave **17** as the major species in equilibrium with the *Z*-isomer. The structure of **17** was established by an x-ray structure determination of the carboxylic acid obtained upon quenching with carbon dioxide. Exposure of **17** to DMF yielded **18** (equation 1), which upon reaction with ethanol/*p*-toluenesulfonic acid, thionyl chloride, and allyltrimethylsilane/BF_3–Et_2O gave **19–21**, respectively [14].

Indole Ring Synthesis: From Natural Products to Drug Discovery, First Edition. Gordon W. Gribble.

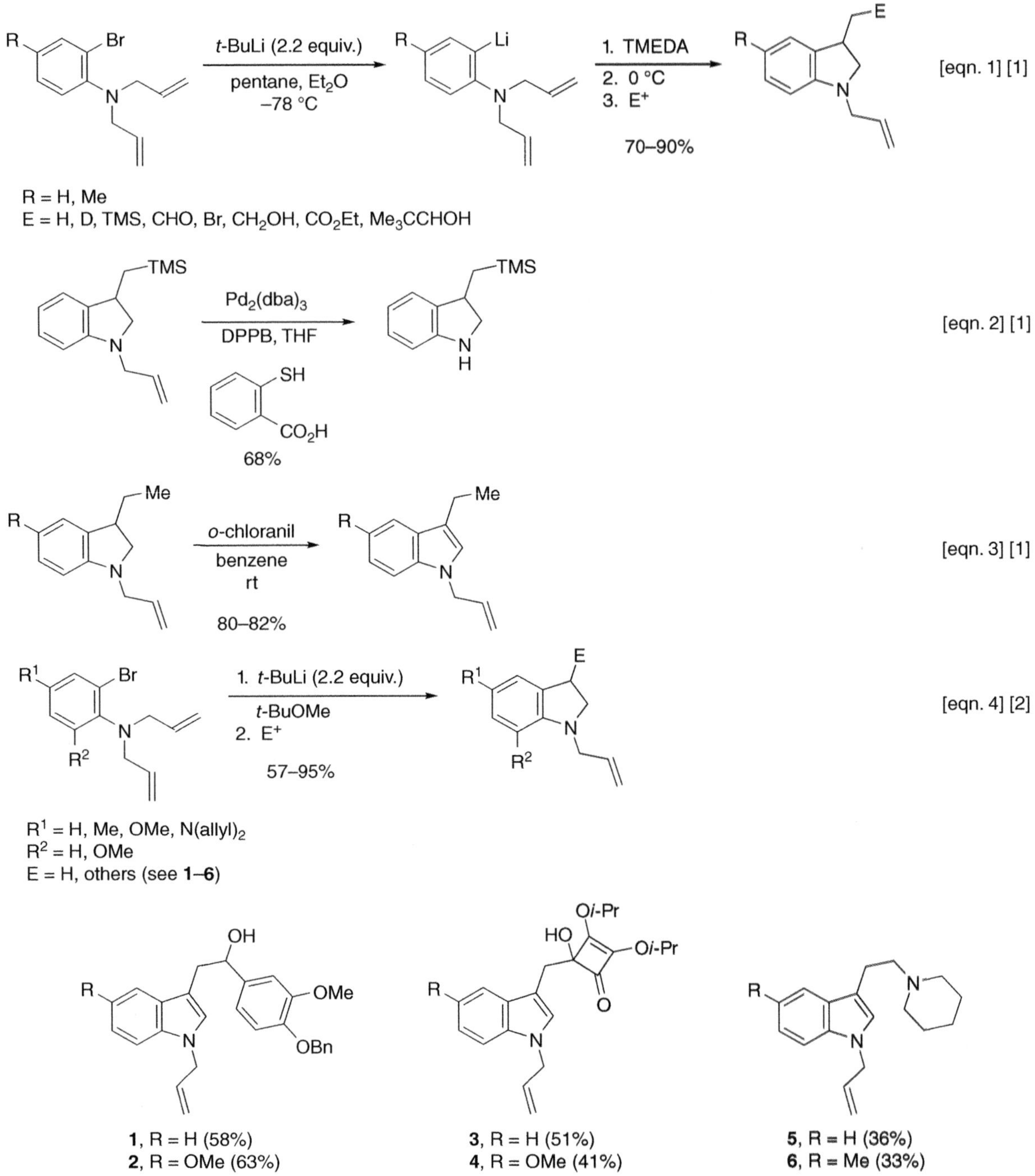

Scheme 1 *Bailey-Liebeskind Indoline-Indole Synthesis*

Table 1 *Applications of the Bailey-Liebeskind Indoline-Indole Synthesis*

Entry	Substrate	Conditions	Indole	% Yield	Ref.
1	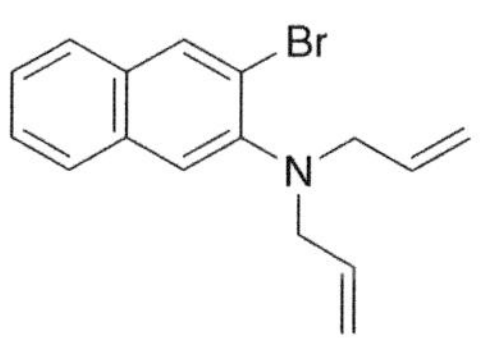	1. *t*-BuLi 2. DMF 3. $(NH_4)_2CO_3$, KCN (49%) 4. Boc_2O, DMAP (89%) 5. Pd(0) (54%) 6. DDQ (51%) 7. LiOH (50%)	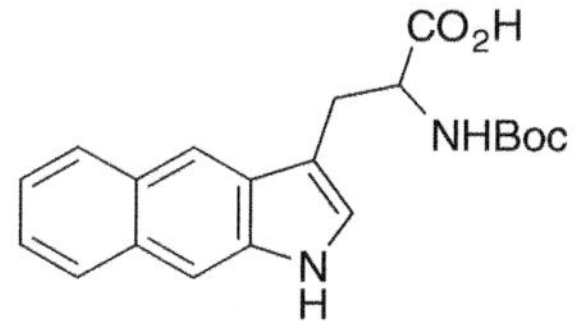	6% (from 2-bromo-3-aminonaphthalene)	5
2	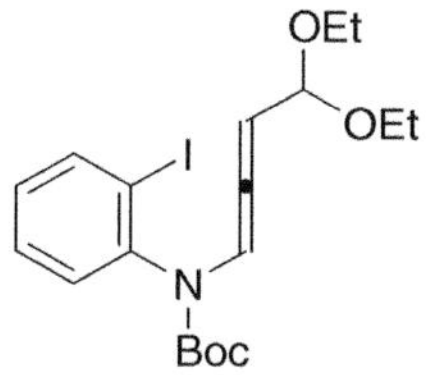	*t*-BuLi, THF –78 °C	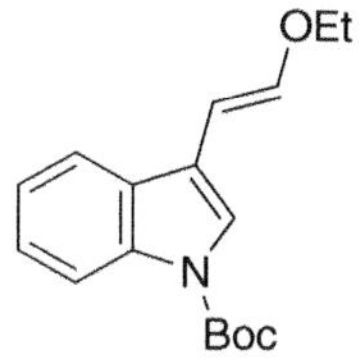 E/Z = 77:23	75%	6
3	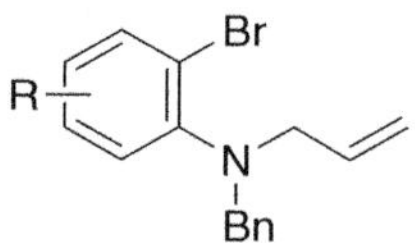 R = H, 4-OBn, 5-OBn, 4-Me, 4-F	1. *t*-BuLi (–)-sparteine toluene –90 °C 2. H_3O^+	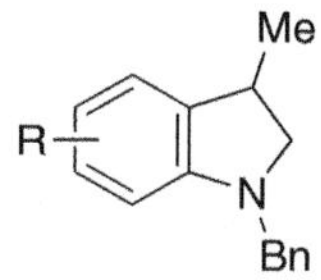	70–90% (85–90% ee)	7, 8
4	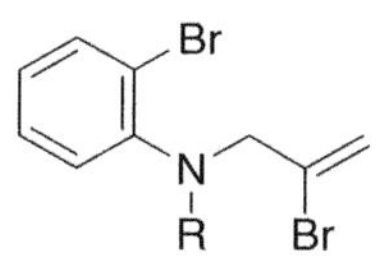 R = Me, Bn, H	1. *t*-BuLi (4 equiv.) –78 °C 2. TMEDA 3. E^+ 4. H_2O	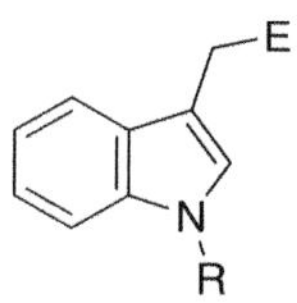 E = D, TMS, SBn, PhNHCO, SPh, Et_2COH, Ph_2COH, 4-MePhCO	51–75%	9
5	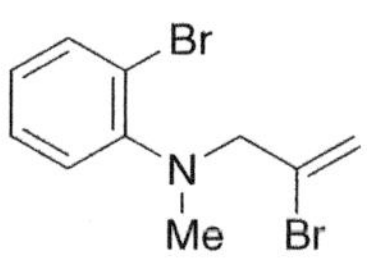	1. *t*-BuLi (5 equiv.) TMEDA, –78 °C 2. 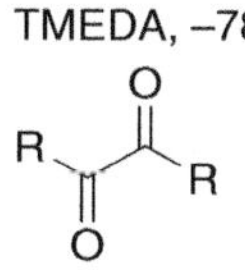	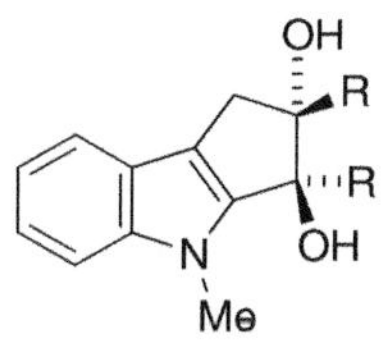 R = Me, Ph	42–47%	9

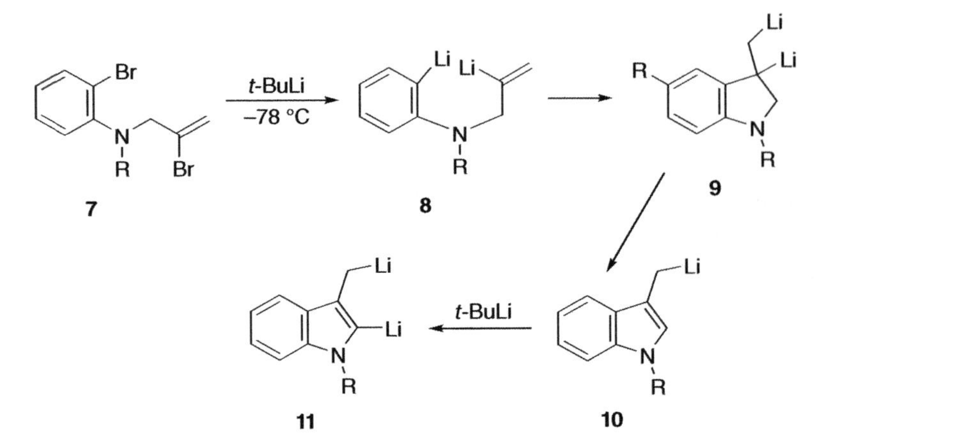

[eqn. 1] [9]

Scheme 2 *Application of the Bailey-Liebeskind Indoline-Indole Synthesis*

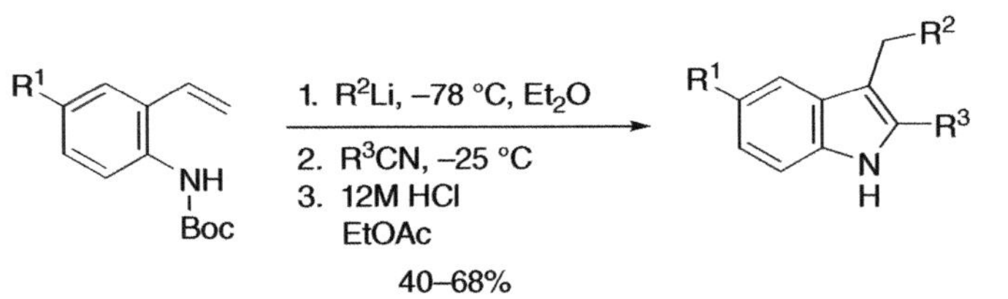

[eqn. 1] [10]

R^1 = H F, OMe
R^2 = *t*-Bu, *n*-Bu
R^3 = Ph, *t*-Bu, Ac, 2-thienyl

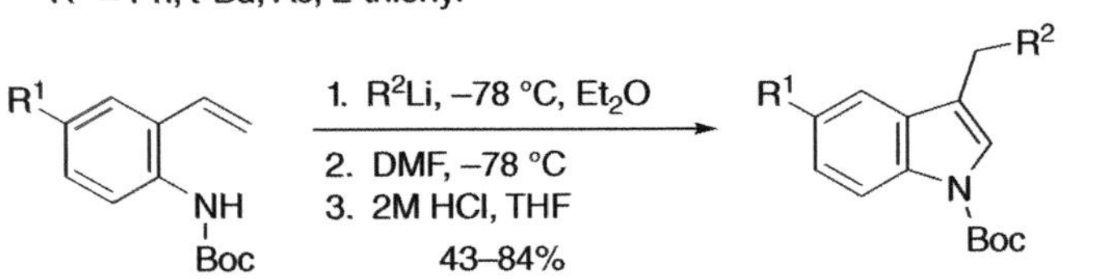

[eqn. 2] [10]

R^1 = H F, OMe
R^2 = *t*-Bu, *n*-Bu, *s*-Bu

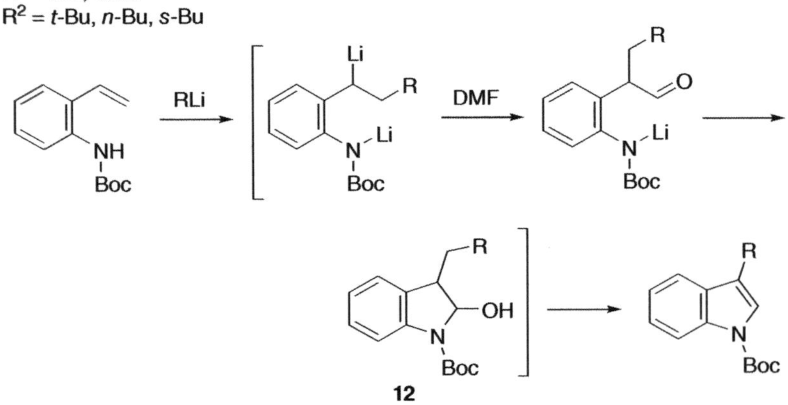

[eqn. 3] [10]

Scheme 3 *O'Shea Indole Synthesis*

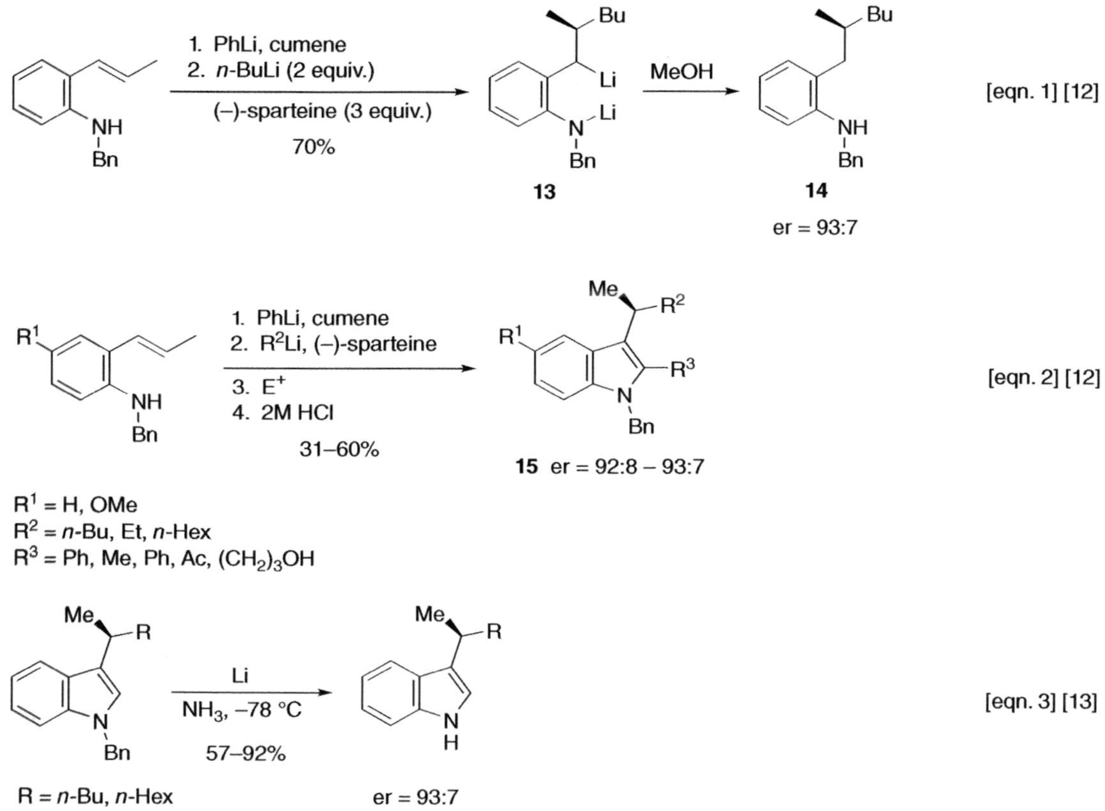

Scheme 4 *O'Shea Indole Synthesis*

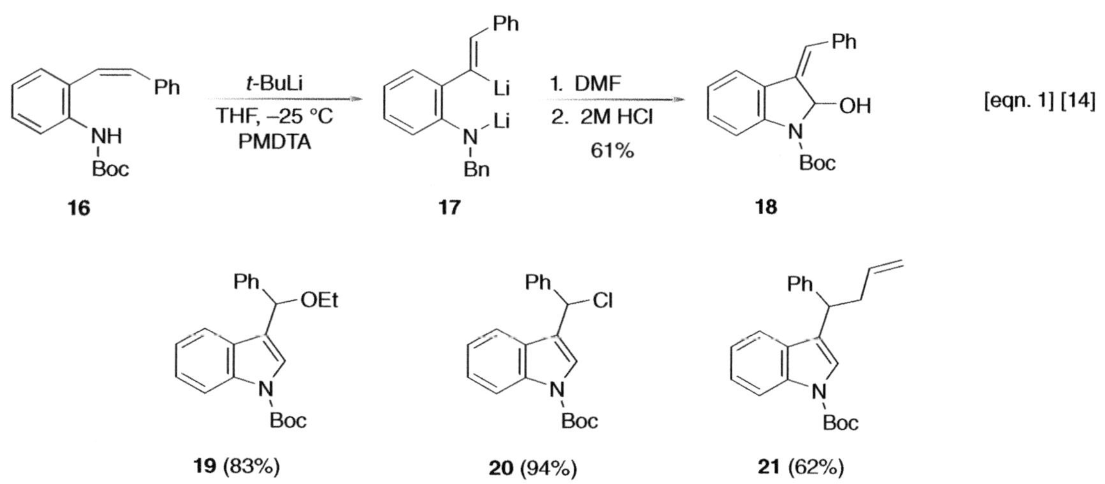

Scheme 5 *O'Shea Indole Synthesis*

References

[1] W.F. Bailey and X.-L. Jiang, *J. Org. Chem.*, 1996, **61**, 2596–2597.
[2] D. Zhang and L.S. Liebeskind, *J. Org. Chem.*, 1996, **61**, 2594–2595.
[3] S. Lemaire-Audoire, M. Savignac, J.P. Genet, and J.-M. Bernard, *Tetrahedron Lett.*, 1995, **36**, 1267–1270.
[4] M.J. Mealy and W.F. Bailey, *J. Organometallic Chem.*, 2002, **646**, 59–67.
[5] T.S. Yokum, P.K. Tungaturthi, and M.L. McLaughlin, *Tetrahedron Lett.*, 1997, **38**, 5111–5114.
[6] F. Le Strat and J. Maddaluno, *Org. Lett.*, 2002, **4**, 2791–2793.

[7] G. Sanz Gil and U.M. Groth, *J. Am. Chem. Soc.*, 2000, **122**, 6789–6790.
[8] M.J. Mealy, M.R. Luderer, W.F. Bailey, and M.B. Sommer, *J. Org. Chem.*, 2004, **69**, 6042–6049.
[9] F.J. Fañanás, A. Granados, R. Sanz, *et al.*, *Chem. Eur. J.*, 2001, **7**, 2896–2907.
[10] C.M. Coleman and D.F. O'Shea, *J. Am. Chem. Soc.*, 2003, **125**, 4054–4055.
[11] A. Kessler, C.M. Coleman, P. Charoenying, and D.F. O'Shea, *J. Org. Chem.*, 2004, **69**, 7836–7846.
[12] A.-M.L. Hogan and D.F. O'Shea, *J. Am. Chem. Soc.*, 2006, **128**, 10360–10361.
[13] A.-M.L. Hogan and D.F. O'Shea, *J. Org. Chem.*, 2008, **73**, 2503–2509.
[14] J. Cotter, A.-M.L. Hogan, and D.F. O'Shea, *Org. Lett.*, 2007, **9**, 1493–1496.

18

Wright Indoline Synthesis

Wright and coworkers reported a novel synthesis of ring-substituted indoline-2,2-dicarboxylates from the tandem bis-alkylation of *ortho*-bromomethyltrifluoroacetanilides (Scheme 1) [1]. Cleavage of the trifluoroacetyl group with Mg/MeOH was accompanied by transesterification (equation 2). Reaction of the anion from **1** with various electrophiles gave **2–6**. Although indoles are not synthesized from these indolines, it seems eminently reasonable to effect a Krapcho decarbomethoxylation and oxidation (DDQ) to afford the corresponding indole-2-carboxylate. This route remains to be seen.

Indole Ring Synthesis: From Natural Products to Drug Discovery, First Edition. Gordon W. Gribble.

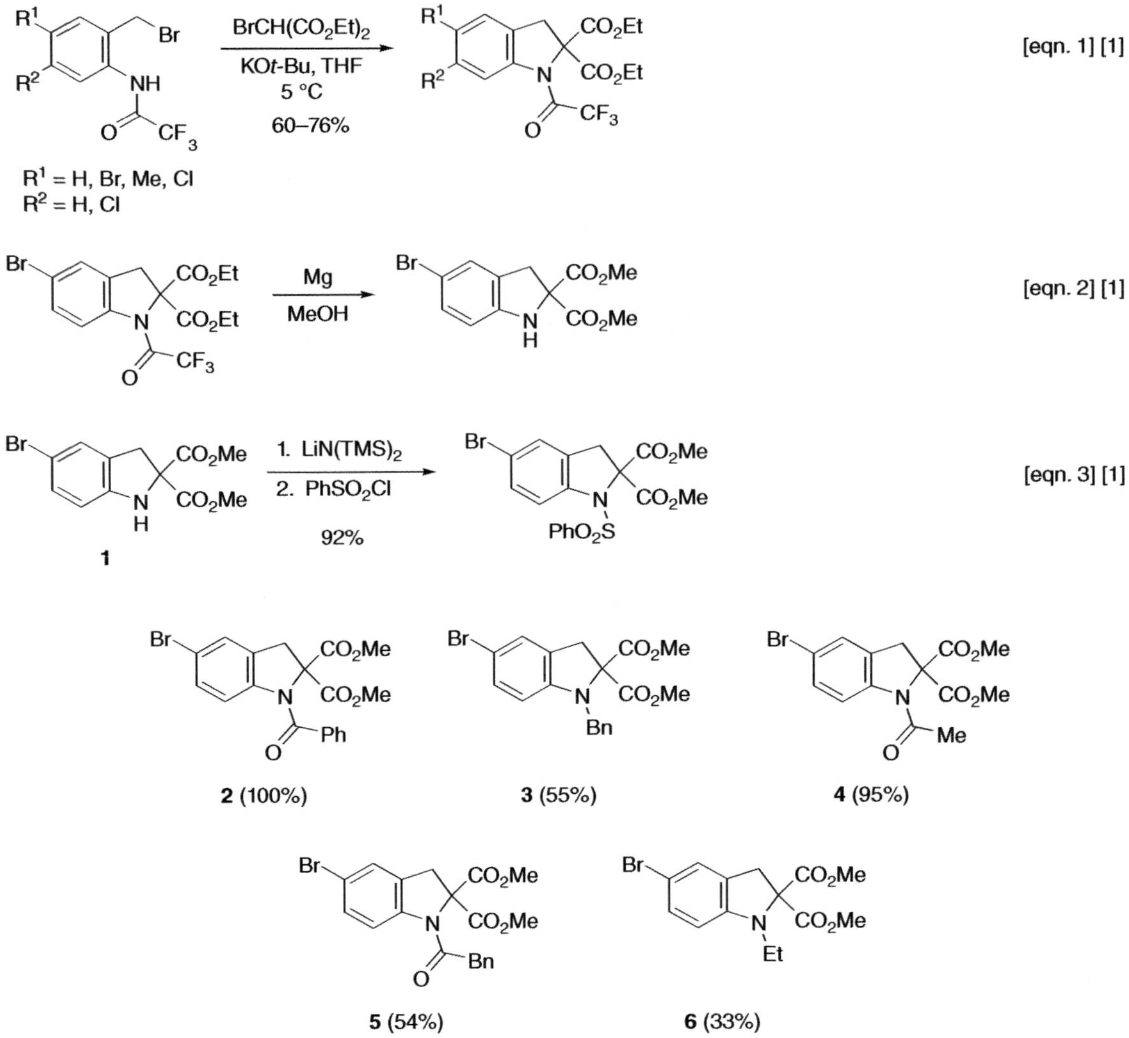

Scheme 1 *Wright Indoline Synthesis*

Reference

[1] S.W. Wright, R.L. Dow, L.D. McClure, and D.L. Hageman, *Tetrahedron Lett.*, 1996, **37**, 6965–6968.

19

Saegusa Indole Synthesis

The Saegusa indole synthesis involves the strong-base promoted cyclization of *ortho*-tolylisocyanides to indoles (Scheme 1) [1–5]. Thus the Saegusa indolization resembles the Madelung indole synthesis. The reaction proceeds by a 5-*endo-dig* cyclization of **2** to form indole (equation 1). The isocyanide anion **2** was alkylated and subsequently cyclized to form 3-substituted indoles (equation 2). The quenching of **2** with epoxides yielded tryptophols (equation 3 and **4** and **5**). Reaction of the ultimately formed *N*-lithioindole with electrophiles gave *N*-substituted indoles (equation 4).

In a variation of their 3-alkylindole synthesis (Scheme 1, equation 2), Saegusa and colleagues found that a one-pot procedure involving the addition of magnesium iodide prior to alkylation gives superior results, presumably via the indolylmagnesium iodide (Scheme 2) [2]. Indoles **6–9** are examples of prepared compounds. Another tweaking of the protocol allowed Saegusa's group to synthesize 3-acylindoles **11** and 2-substituted indoles **12** (equations 2–3) [3]. Although standard acylating agents (acyl halides) failed to give **10**, allyl carboxylates performed well. The cuprous oxide–catalyzed cyclization of **10** to 3-acylindoles **11** is believed to involve organocopper(I) **13** and subsequent insertion to form **14** and then **11** (equation 4).

The reaction of *ortho*-lithiomethylphenyl isocyanide **2** with isocyanates and isothiocyanates produced *ortho*-isocyanophenylacetamides (**15**) and *ortho*-isocyanophenylacetothioamides (**16**), respectively (Scheme 3, equation 1) [4]. Cyclization of **15** and **16** with LDA gave indole-3-carboxamides (**17**) and indole-3-thiocarboxamides (**18**), respectively (equation 1). Whereas *ortho*-isocyanophenylacetamide **19** (**15**, R = Ph) underwent cyclization with cuprous oxide to **20** (equation 2), isocyanide **21** (**15**, R = *n*-Bu) afforded 1,3-benzodiazepin-4-one (**22**). With **15** (R = cyclohexyl), a mixture of indole-3-carboxamide (25%) and 1,3-benzodiazepin-4-one (58%) was obtained. The latter cyclized product probably arises from a nitrogen–copper species. In a full account of their indole syntheses methodology, Saegusa presented more examples of the chemistry already given in this chapter, plus a new application was the synthesis of 2,3-disubstituted indoles **24**. Thus, as shown in equation 4 (Scheme 3) base-induced alkylation of *ortho*-(acylmethyl)phenyl isocyamides **10** prior to cyclization led to 2,3-disubstituted indoles **24** [5].

Saegusa and his colleagues applied their copper-catalyzed cyclization of lithiated phenyl isocyanides to *ortho*-(cyanomethyl)phenyl isocyanide (**25**) and *ortho*-(methoxy carbonylmethyl)phenyl isocyanide (**26**) [6]. Thus, as shown in Scheme 4, lithiation of **25** and **26** with *n*-butyllithium at –78 °C followed by reaction with alkyl halides gave the alkylation products **27** and **28** (equations 1 and 2), respectively. Whereas the copper-catalyzed cyclization of **25** and **26** afforded 3-cyanoindole (**29**) and 3-(methoxycarbonyl) indole (**30**), respectively, in excellent yield (equation 2), cyclization of the alkylated isocyanides **27** and **28** gave indolenines **31** and **32**, respectively, also in good to excellent yield (equation 3). The latter cyclization represents an excellent route to 3-alkyl-3-cyano(methoxycarbonyl)-3*H*-indoles (indolenines). A pathway for the role of copper is suggested to be similar to that shown in Scheme 2 (equation 4).

The Saegusa indole synthesis has been used by several groups as tabulated in Table 1. Indole **34** in Entry 1 arises from a vicarious nucleophilic substitution of hydrogen for chloromethyl phenyl sulfone on *meta*-isocyanonitrobenzene, one of the products of which is isocyanide **33**, which cyclizes to indole **34** [7]. The yield shown is that from the

Indole Ring Synthesis: From Natural Products to Drug Discovery, First Edition. Gordon W. Gribble.

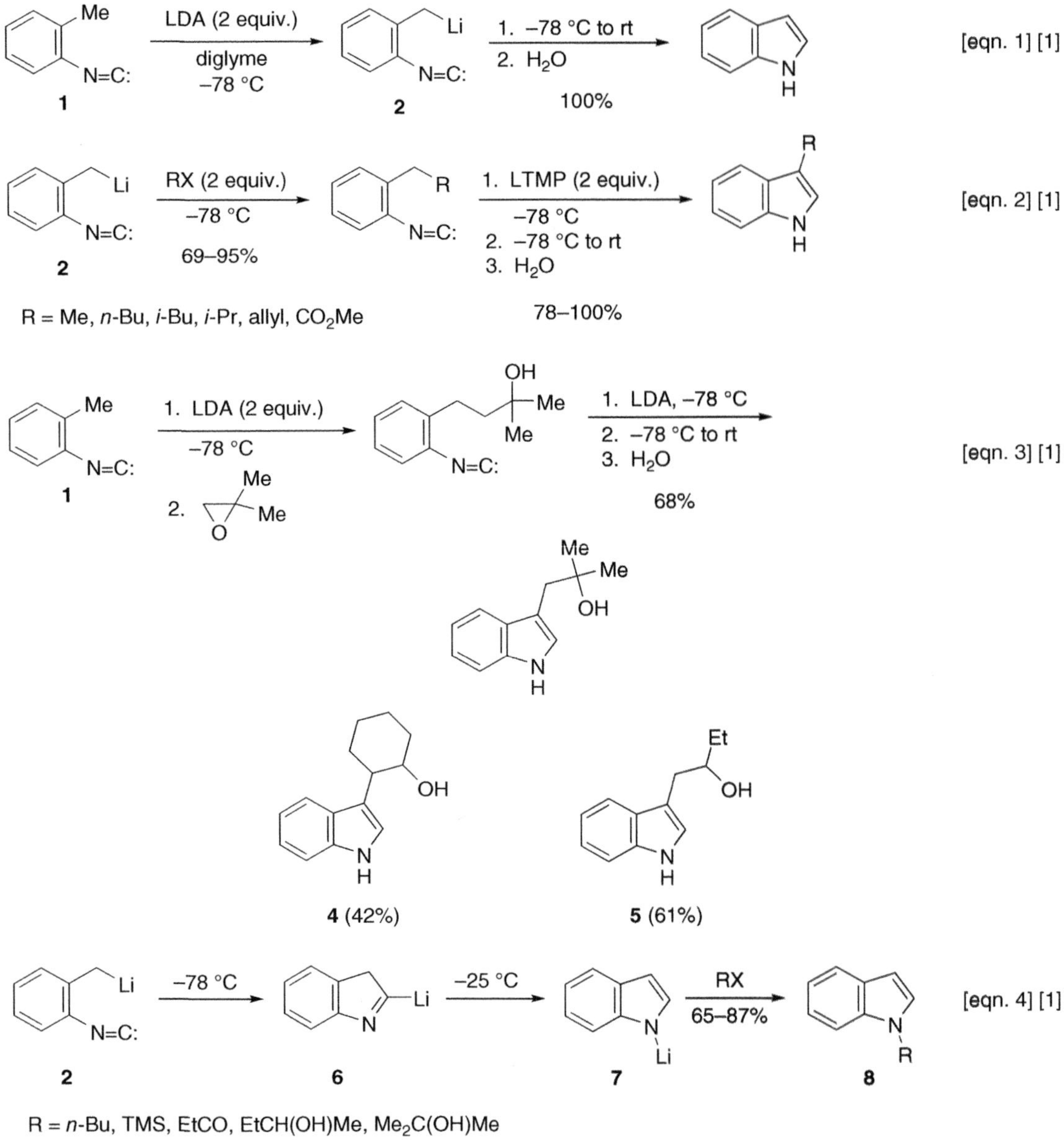

Scheme 1 *Saegusa Indole Synthesis*

vicarious nucleophilic reaction on *meta*-isocyanonitrobenzene. A lesser amount of 4-nitro-3-(phenylsulfonyl)indole (13%) is also obtained. A more accurate yield for the cyclization in Entry 1 is that shown in Entry 2 for a similar base-catalyzed cyclization [8]. For a review of vicarious nucleophilic substitution of hydrogen in the synthesis of heterocycles, see Makosza [9].

The initially formed *N*-sodioindole in Entry 3 is arylated by activated fluorobenzenes to give *N*-aryl-3-(ethoxycarbonyl)indoles in excellent yield [10]. The synthesis of 2-(methylsulfanyl)indoles in Entry 4 involves initial formation of isothiocyanates from the isocyanide, then sodium hydride–induced cyclization to the disodium indol-1-ide-2-thiolates, which are capped by methyl iodide (or other alkyl halide) to give the final product [11]. Alkylation with dibromoalkanes afforded the corresponding tricyclic [1,3]thiazolo-, [1,3]thiazino-, and [1,3]thiazepino[3,2-*a*]indoles. Kobayashi and colleagues in Entry 5 described a reverse Saegusa indolization wherein the organolithium reagent attacks the isocyanide carbon, and the resulting imidoyl anion displaces chloride to form an indolenine. Tautomerization furnishes the indole [12]. Likewise, Entry 6 features the addition of Grignard reagents to the isocyanide carbon, resulting in cyclization onto the ketone carbonyl and formation of 3*H*-indol-3-ols [13]. Moreover, two equivalents of methylmagnesium

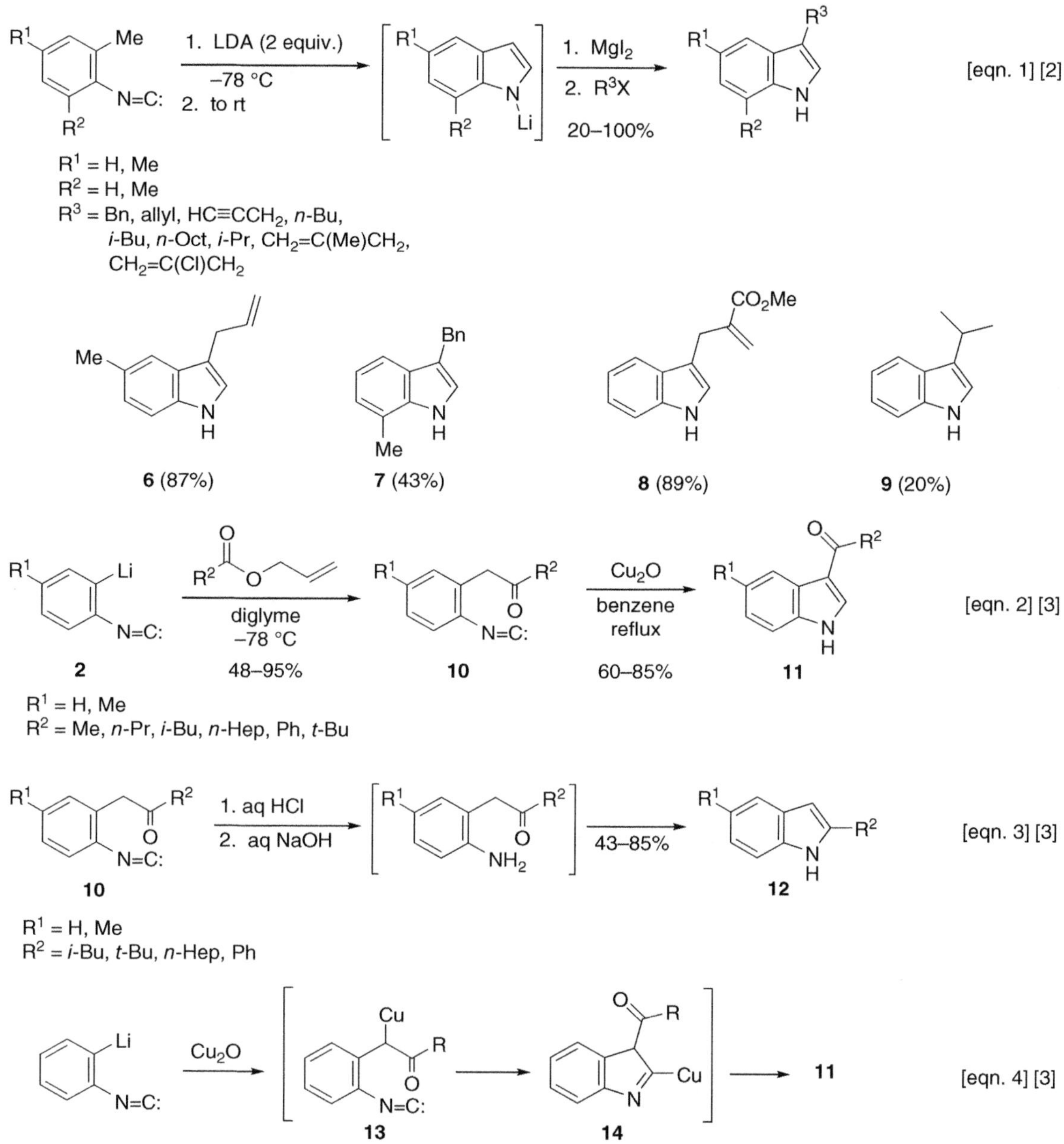

Scheme 2 *Saegusa Indole Synthesis Variations*

bromide led to the 2,2-dimethylindolin-3-ols (41%–48% yield) (not shown).

Mironov and Mokrushin described the extraordinary formation of **35** from the reaction between triethylamine and aryl isocyanides bearing electron-withdrawing groups (Entry 7) [14]. An x-ray crystal determination of **35** ($R^1=R^3=H$, $R^2=R^4=CF_3$) confirmed the structure of this 2-triethylammonium-3-arylaminoindolate. Even more bizarre is the isolation, and x-ray crystal structure, of *N*-lithioindole **36** from the reaction of phenyl isocyanide and $LiCH(TMS)_2$ (Entry 8) [15].

The chemistry described by Saegusa, Kobayashi, and others (Schemes 1 to 4 and Table 1) is reminiscent of the earlier work by Zeeh (Scheme 5) [16–18]. These Lewis acid–catalyzed reactions of isocyanides with ketones are shown in equation 3. The *N-tert*-butyl group in **37** is removable by boron trifluoride etherate. A suggested pathway is illustrated in equation 3 and is initiated by BF_3 activation of the aryl ketone (**38**). Some indoles prepared by Zeeh are **39–41** [17]. The elegant free-radical cyclization of aryl isocyanides developed by Fukuyama is presented in Chapter 49.

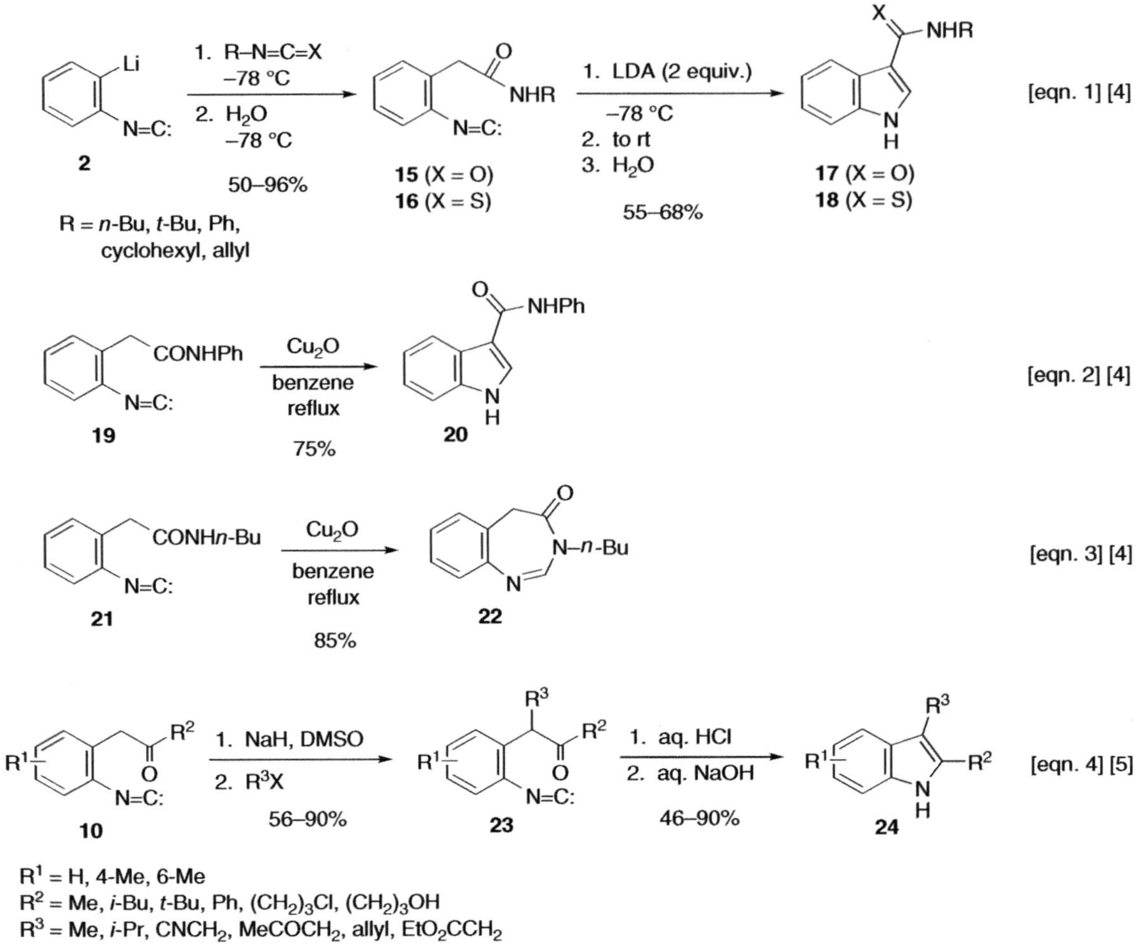

Scheme 3 *Saegusa Indole Synthesis Variations*

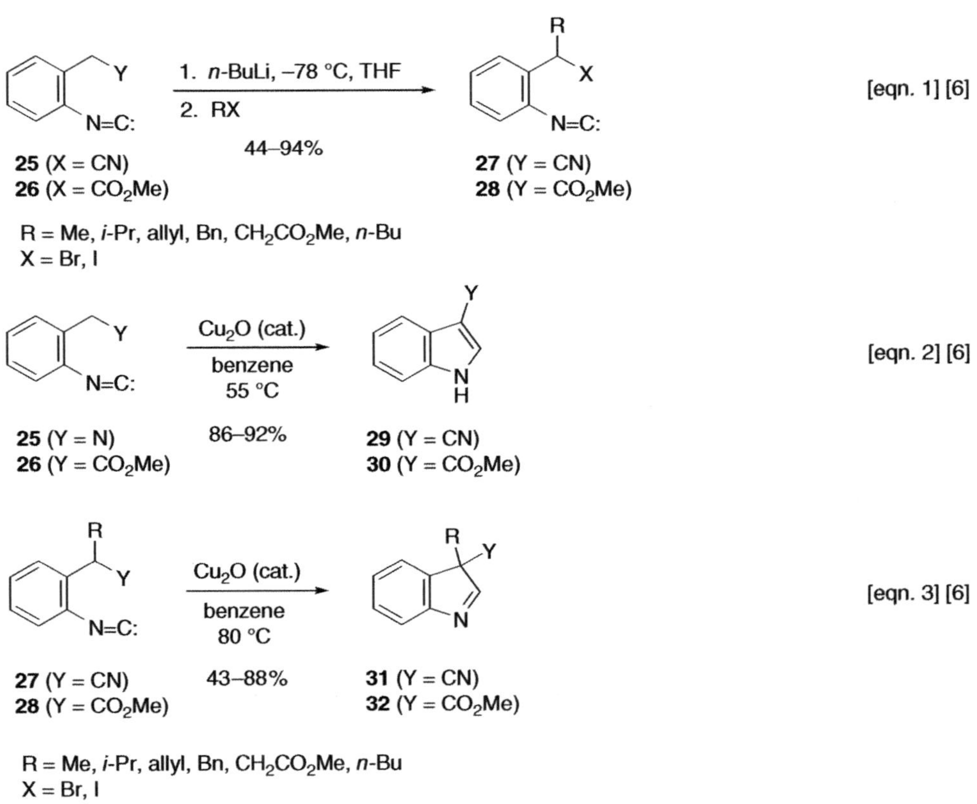

Scheme 4 *Saegusa Indole Synthesis Variations*

Table 1 *Applications of the Saegusa Indole Synthesis*

Entry	Substrate	Conditions	Indole	% Yield	Ref.
1	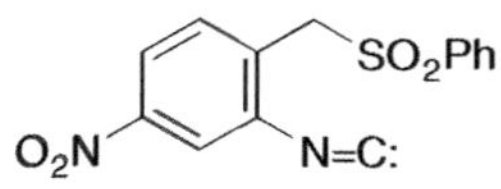 **33**	NaOH, DMSO	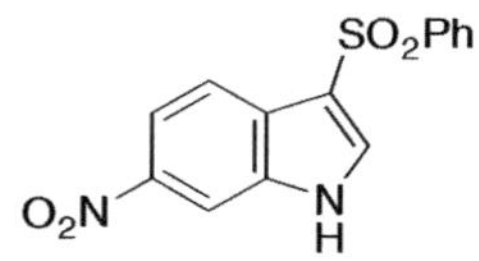 **34**	36%	7
2	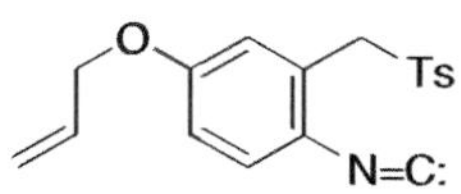	NaOH, DMSO 15–20 °C	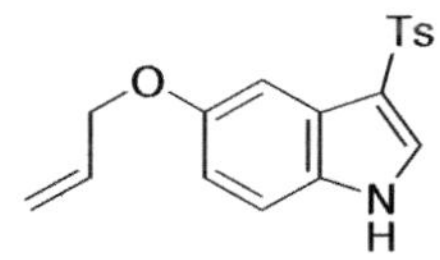	93%	8
3	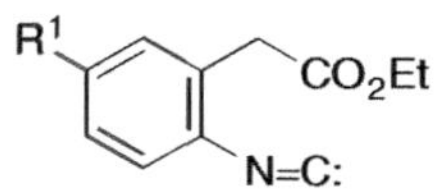 R^1 = H, Me, Cl, OMe R^2 = H, NO_2 R^3 = H, Me, NO_2	1. NaH, DMSO, THF rt 2. 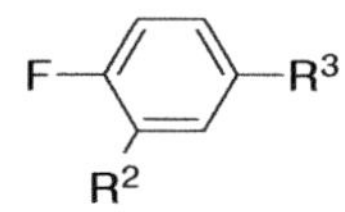	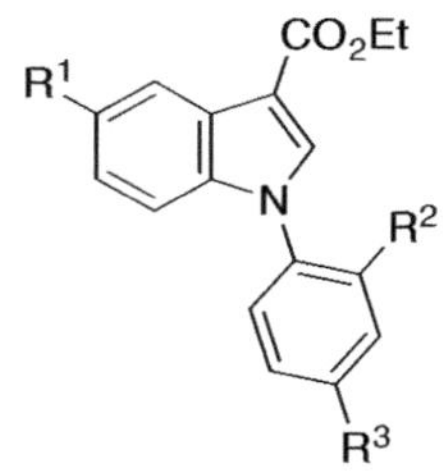	76–91%	10
4	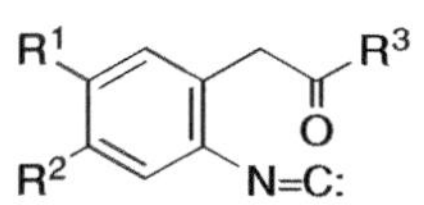 R^1 = H, Cl, Me R^2 = H, Me R^3 = Me, Ph, Et, OEt	1. Et_3N, rt THF, S 2. NaH 3. MeI	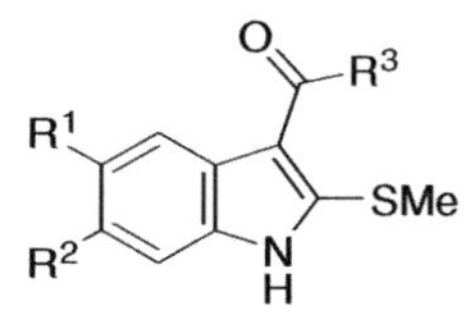	64–91%	11
5	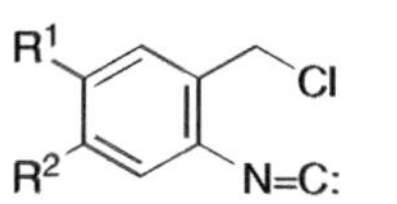 R^1 = H, Me R^2 = H, Cl R^3 = Ph, *p*-Tol, *n*-Bu, *m*-Tol, *o*-Tol, *t*-Bu, 3-ClPh, 4-ClPh, 3-MeOPh, 1-naphthyl	R^3Li THF –78 °C	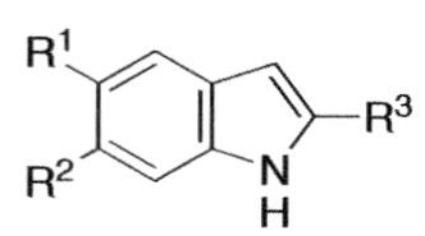	49–74%	12
6	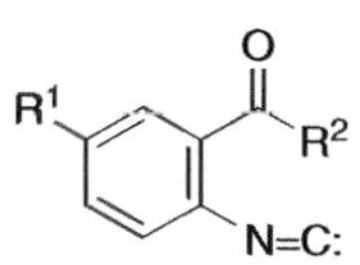 R^1 = H, Cl R^2 = Ph, Et, 4-ClPh, 4-MeOPh R^3 = Me, Ph, Et, 4-Tol, 3-ClPh, 4-ClPh, 3-MeOPh, 3,4(MeO)$_2$Ph	R^3MgBr THF, 0 °C	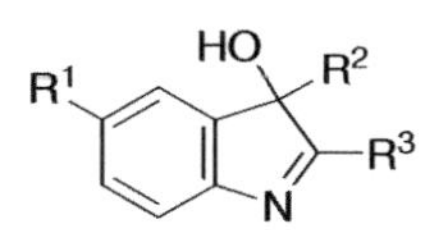	50–98%	13

(continued overleaf)

Table 1 *(continued)*

Entry	Substrate	Conditions	Indole	% Yield	Ref.
7	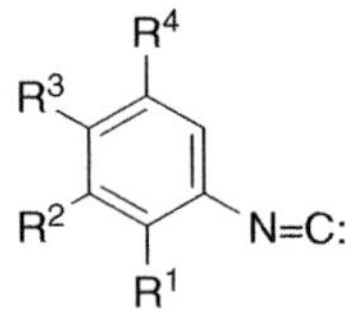 R^1 = H, Br R^2 = H, CF_3, Cl R^3 = H, NO_2, Cl R^4 = H, Cl, CF_3	Et_3N hexane reflux	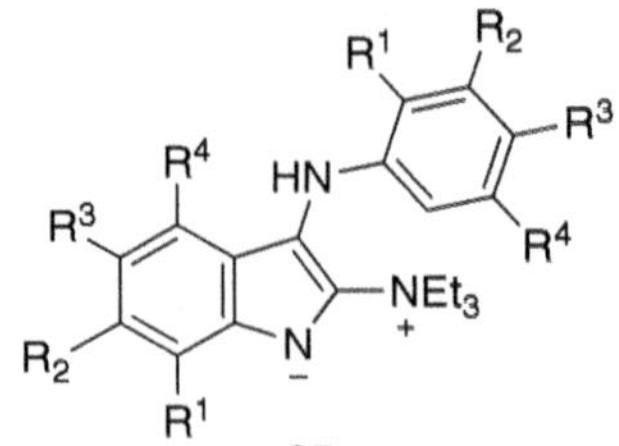 **35**	85% (R^1 = R^3 = H; R^2 = R^4 = CF_3)	14
8	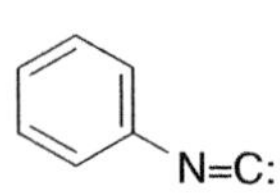	$LiCH(TMS)_2$ 1,2-bis(Me_2N)ethane pentane (tmen) −90 °C to rt	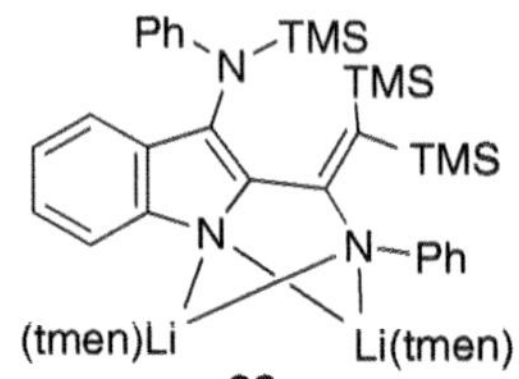 **36**	64%	15

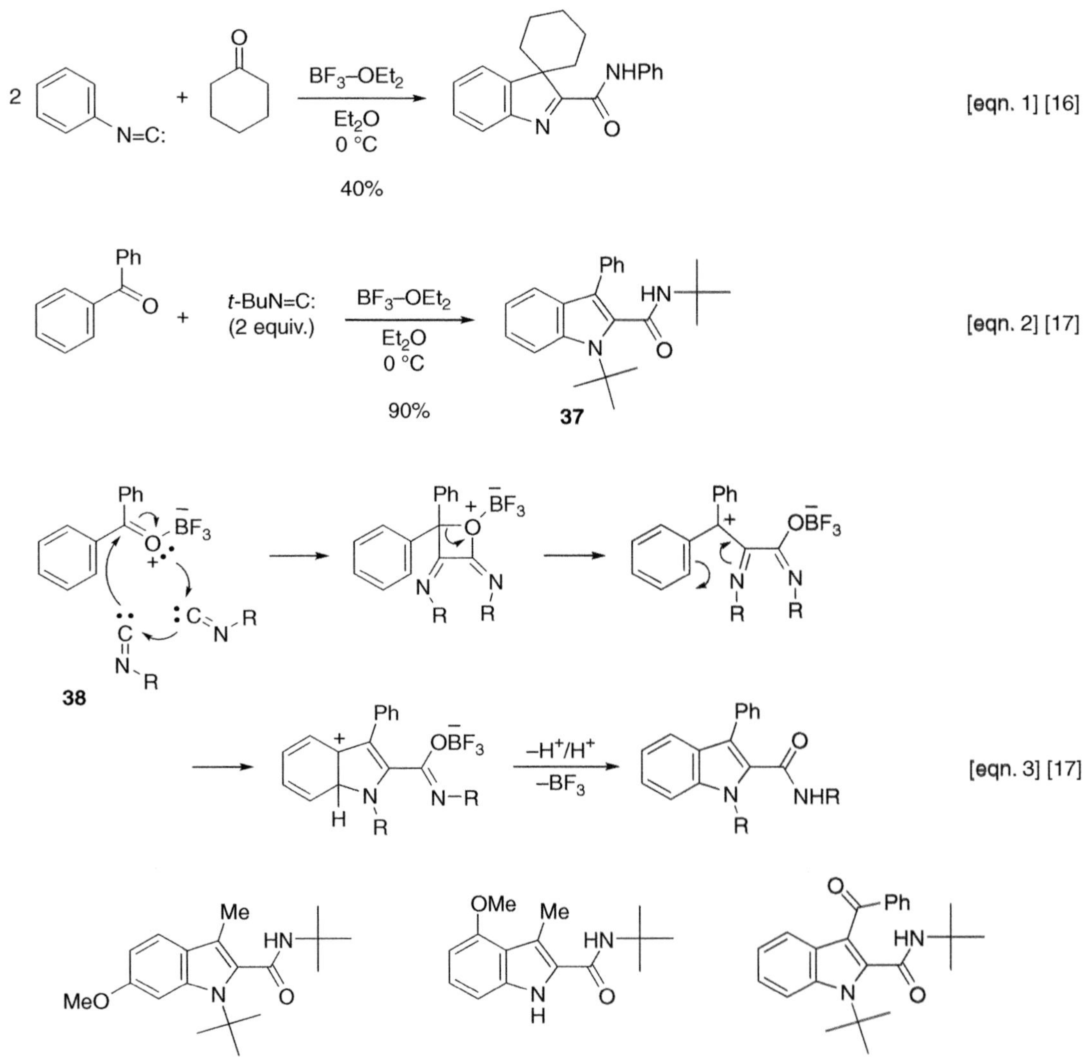

Scheme 5 *Zeeh Indole Synthesis*

References

[1] Y. Ito, K. Kobayashi, and T. Saegusa, *J. Am. Chem. Soc.*, 1977, **99**, 3532–3534.
[2] Y. Ito, K. Kobayashi, N. Seko, and T. Saegusa, *Chem. Lett.*, 1979, 1273–1276.
[3] Y. Ito, K. Kobayashi, and T. Saegusa, *J. Org. Chem.*, 1979, **44**, 2030–2032.
[4] Y. Ito, K. Kobayashi, and T. Saegusa, *Tetrahedron Lett.*, 1979, **20**, 1039–1042.
[5] Y. Ito, K. Kobayashi, N. Seko, and T. Saegusa, *Bull. Chem. Soc. Jpn.*, 1984, **57**, 73–84.
[6] Y. Ito, Y. Inubushi, T. Sugaya, *et al.*, *Bull. Chem. Soc. Jpn.*, 1978, **51**, 1186–1188.
[7] K. Wojciechowski and M. Makosza, *Tetrahedron Lett.*, 1984, **25**, 4793–4794.
[8] M. Makosza, J. Stalewski, K. Wojciechowski, and W. Danikiewicz, *Tetrahedron*, 1997, **53**, 193–214.
[9] M. Makosza, *Synthesis*, 1991, 103–111.
[10] F. Fukamachi, H. Konishi, and K. Kobayashi, *Heterocycles*, 2009, **78**, 161–168.
[11] S. Fukamachi, H. Konishi, and K. Kobayashi, *Synthesis*, 2009, 1786–1790.
[12] K. Kobayashi, D. Iitsuka, S. Fukamachi, and H. Konishi, *Tetrahedron*, 2009, **65**, 7523–7526.
[13] K. Kobayashi, Y. Okamura, S. Fukamachi, and H. Konishi, *Tetrahedron*, 2010, **66**, 7961–7964.
[14] M.A. Mironov and V.S. Mokrushin, *Mendeleev Commun.*, 1998, **8**, 242–243.
[15] M.A. Fernandes, M.F. Lappert, M. Layh, and B. Omondi, *Chem. Commun.*, 2003, 656–657.
[16] B. Zeeh, *Chem. Ber.*, 1968, **101**, 1753–1760.
[17] B. Zeeh, *Chem. Ber.*, 1969, **102**, 678–685.
[18] B. Zeeh, *Chem. Ber.*, 1969, **102**, 1876–1882.

20

Ichikawa Indole Synthesis

The Ichikawa indole synthesis is the base-induced 5-*endo-trig* cyclization of *ortho*-amino-β,β-difluorostyrenes **1** to 2-fluoroindoles **2** (Scheme 1, equation 1) [1–3]. The method is applicable to indolines, which can be oxidized to indoles [4, 5]. This nice chemistry exploits the inherent favorable conversion of a difluorinated sp^2 carbon to an sp^3 carbon, which then expels fluoride to give the aromatic indole. Ichikawa also applied this method to the preparation of 2-fluorobenzo[*b*]furans and 2-fluorobenzo[*b*]thiophenes [1, 2]. The starting difluorostyrenes were synthesized in three steps from trifluoroethyl tosylate, and an *Organic Synthesis* preparation is available [3]. Trifluorostyrene **3** was converted to difluoromethyleneindoline **4** with sodium hydride (equation 2) and to 3-trifluoromethylindoline **5** with DBU [4, 5]. On treatment of **4** ($R^1 = R^2 = H$) with *N*-iodosuccinimide (Et_3N-HF, CH_2Cl_2, −10 °C), bromine (CCl_4, rt), and NaI (TMSCl, H_2O, rt), indoles **6–8** are formed, respectively, each in excellent yield.

Indole Ring Synthesis: From Natural Products to Drug Discovery, First Edition. Gordon W. Gribble.

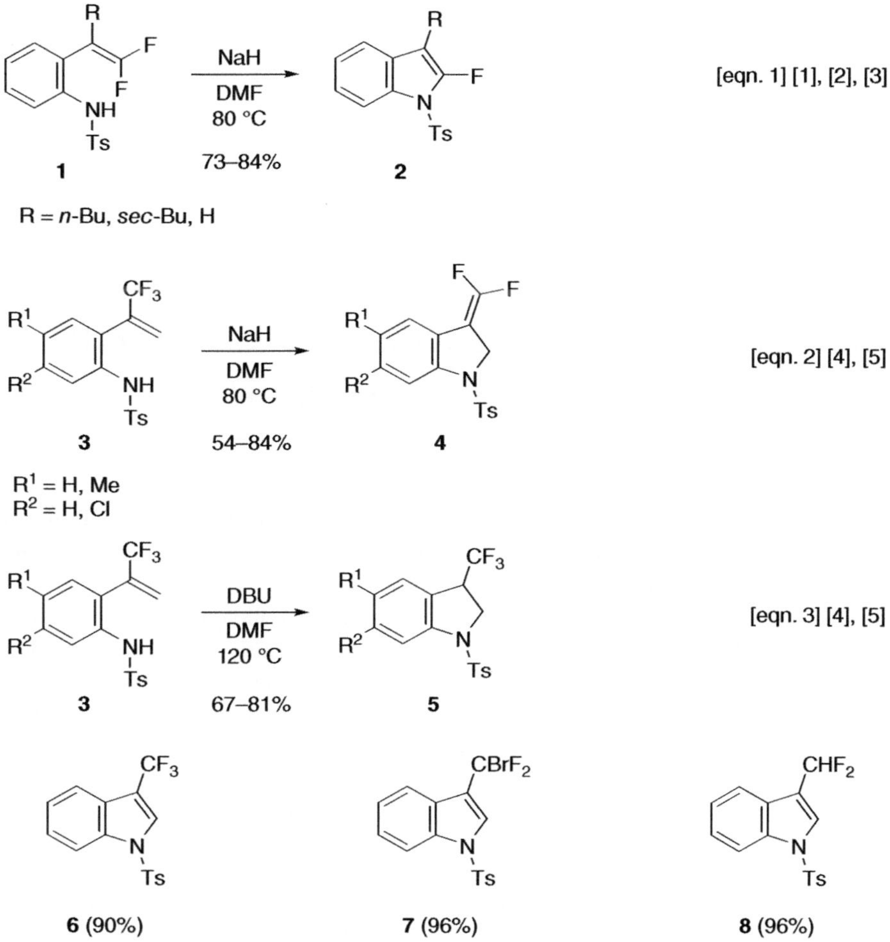

Scheme 1 *Ichikawa Indole Synthesis*

References

[1] J. Ichikawa, Y. Wada, T. Okauchi, and T. Minami, *Chem. Commun.*, 1997, 1537–1538.
[2] J. Ichikawa, Y. Wada, M. Fujiwara, and K. Sakoda, *Synthesis*, 2002, 1917–1936.
[3] J. Ichikawa, R. Nadano, T. Mori, and Y. Wada, *Org. Syn.*, 2006, **83**, 111–120.
[4] J. Ichikawa, T. Mori, and Y. Iwai, *Chem. Lett.*, 2004, **33**, 1354–1355.
[5] J. Ichikawa, Y. Iwai, R. Nadano, *et al.*, *Chem. Asian J.*, 2008, **3**, 393–406.

21

Miscellaneous Nucleophilic Cyclizations that Form the Indole Ring

There exist a plethora of nucleophilic cyclizations leading to the indole ring that are not embraced by the name reactions presented hitherto. Although these indole syntheses are not (yet) graced by the name of their inventor, I will mention the senior author in each case. In many indole ring–forming reactions, the final step is an amino carbonyl cyclization, but the essence of the indolization may place this particular reaction in a separate chapter.

The S_NAr reaction has been employed to prepare several polyfluorinated indoles (Scheme 1). In each case the 5-membered ring cyclization that displaces fluoride from an sp^2 carbon is favorable in this addition–elimination (Meisenheimer) mechanism, and the anion is stabilized by the electron-deficient aromatic ring. Interestingly, Brooke was unable to isolate the presumed Schiff base intermediate leading to indoles **1** and **2** (equations 1, 2) [1, 2], and indole formation failed when 2,3,4,5,6-pentafluorophenylacetaldehyde was employed. Petrov and Barkhash prepared the amino alcohols **3** by the electrochemical reduction of the appropriate 2-nitro-1-pentafluorophenylalkanols in the presence of a carbonyl compound, which introduces R^1 (equation 3) [3]. The reaction can be stopped at the 3-hydroxyindoline stage when sodium bicarbonate is present. For example, 4,5,6,7-tetrafluoro-3-hydroxy-1-methylindoline was isolated in 82% yield. Heating this compound with concentrated hydrochloric acid gave indole **4** (R^1 = Me, R^2 = R^3 = H) in 98% yield.

The enormous reactivity of hexafluorobenzene is displayed in its reaction with enamines as discovered by Wakselman and Blazejewski (equation 4) [4]. Indoles **5** can be oxidized to their respective carbazoles with chloranil (refluxing toluene, 45%–79%). That one of the R groups is lost as RF from intermediate **7** is seen in the isolation of **5** ($R^1 = (CH_2)_4F$) when the pyrrolidine enamine is used. The intermediate C-arylation enamine **6** was observed by ^{19}F-NMR.

Ojima and colleagues found that enamine (imine) **9** from aldehyde **8** readily cyclized to indoles **10** upon treatment with LDA (equation 5) [5, 6]. The *N*-allyl group in **10** was removed by rhodium trichloride isomerization followed by acidic hydrolysis of the *N*-enamine (97% yield). Along similar lines, Filler's team reported an improved synthesis of 4,5,6,7-tetrafluoroindole **13** (equation 6) [7] over the method they had reported earlier. The improvements are in the reduction of nitrile **11** to phenethylamine **12** and the final oxidation to 4,5,6,7-tetrafluoroindole (**13**).

The S_NAr reaction has been featured in other indole syntheses, such as Molina and Fresneda's route to cryptotackieine (Scheme 2, equation 1 [8]) and Orito's approach to indolo[2,1-*a*]isoquinoline alkaloids (equations 2 and 3 [9]), even though neither substrate seems suitable for an S_NAr reaction. The successful indolization in equation 3 precludes a thermal electrocyclization mechanism because the gem-dimethyls block the required imine tautomerism.

Selvakumar and coworkers report a fascinating indole ring synthesis that involves a Krapcho decarbomethoxylation–cyclization sequence (Scheme 3) [10, 11]. Demethoxylation of the *N*-methoxy group is effected by catalytic hydrogenation (90%). As seen by the examples **13–18**, this reaction is an excellent preparation of 3-(methoxycarbonyl)indoles (and 3-cyanoindoles). Dimethyl sulfoxide alone does not effect reaction, and the sodium chloride is catalytic. A possible mechanism for this interesting reaction is shown in equation 3. Decarboxylation affords **19**, which is trapped by the generated methyl chloride to give **20**. Tautomerism of **20** gives **21**, which

Indole Ring Synthesis: From Natural Products to Drug Discovery, First Edition. Gordon W. Gribble.

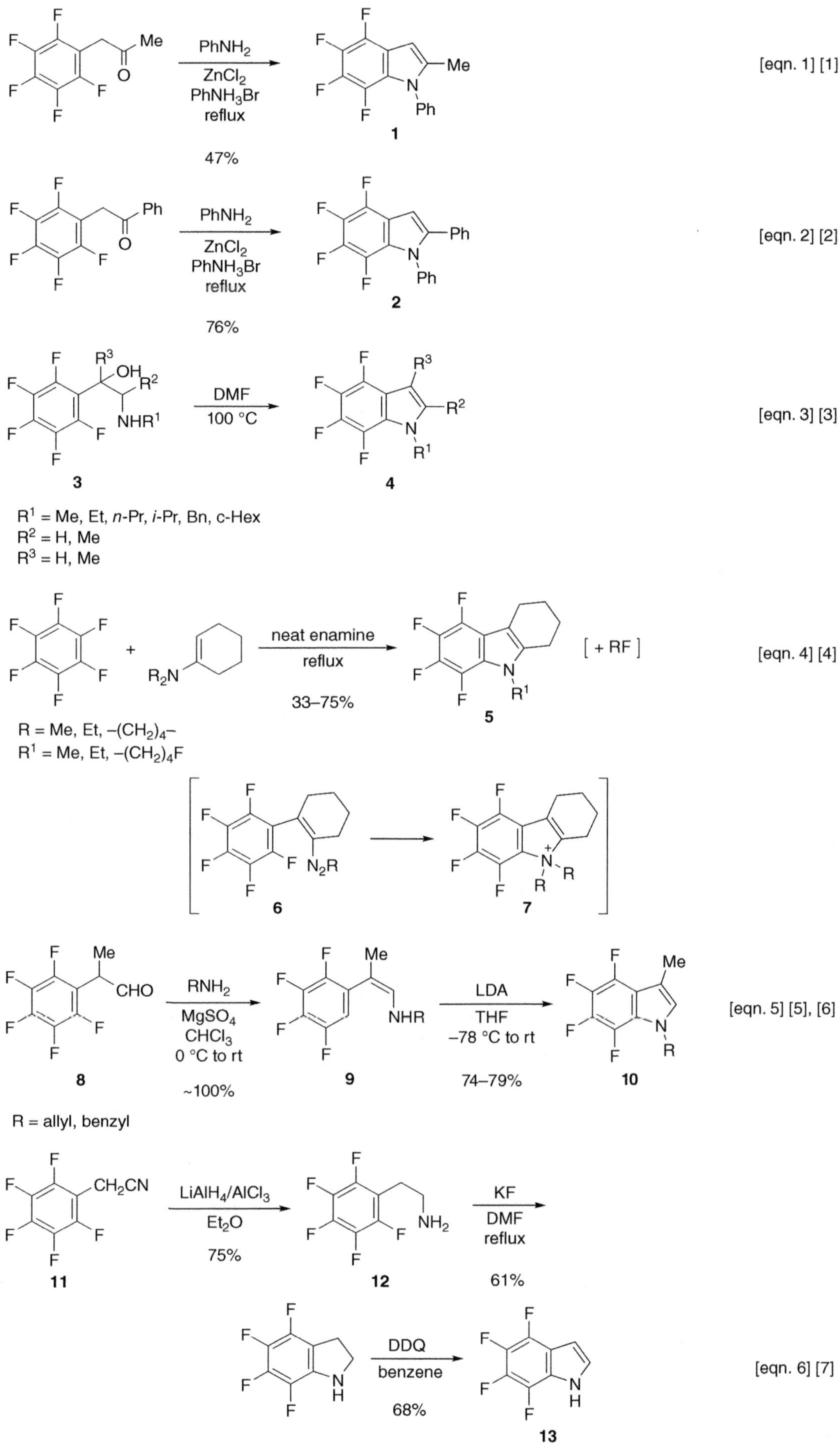

Scheme 1 *Indole Synthesis via Nucleophilic Substitution*

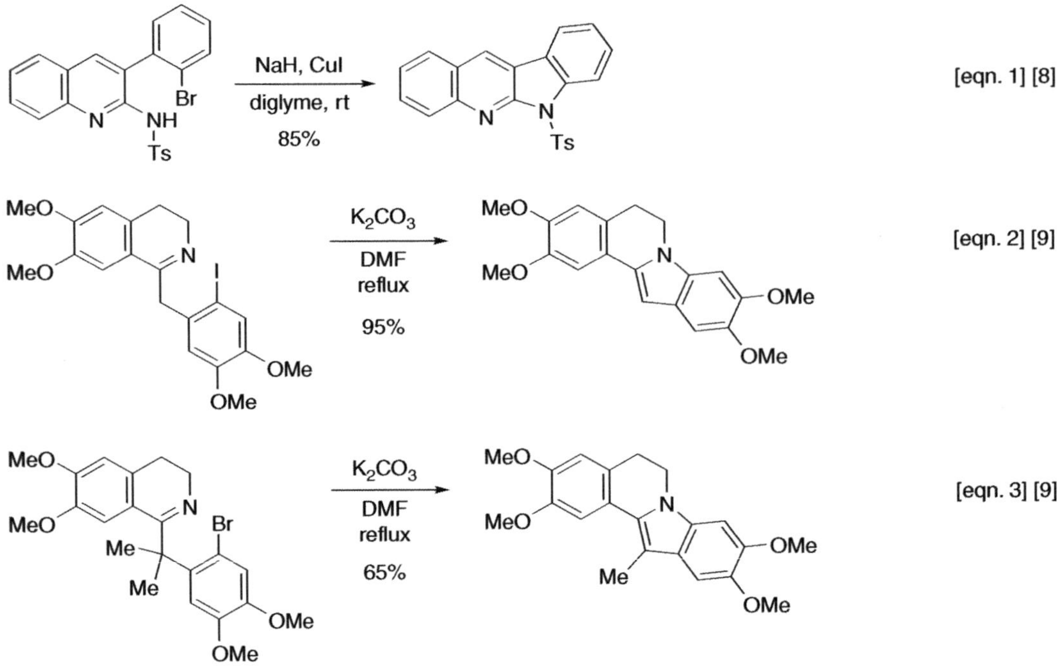

Scheme 2 *Indole Synthesis via Nucleophilic Substitution*

undergoes cyclization to **22**. Formal loss of water yields the indole product.

Ferraccioli, Croce, La Rosa, and coworkers employed sulfonium ylides to access indolines and, subsequently, indoles (Scheme 4 [12–14]). Presumably, in equation 1 the second equivalent of sulfonium ylide functions as a base to eliminate *p*-toluenesulfinic acid from the indoline that initially forms from **25**. The ammonium salts **24** were readily prepared from either *ortho*-aminobenzyl alcohol ($X^-=Cl^-$) or *N,N*-dimethyl-2-aminoaniline ($X^-=I^-$). Sulfonium ylides were generated *in situ* from the sulfonium salts and sodium hydride/dimethyl sulfoxide [12]. The yield of 2-cyanoindole from this method was only 16%. Equations 2 and 3 illustrate the adaptation of the method to the synthesis of indolines **29** and, by subsequent reaction, to indoles **30** and **31** [13]. The benzylic bromides **26** were synthesized from *ortho*-aminobenzophenones, and a second equivalent of ylide **27** served to generate the sulfonamide anion from **28** to effect ring closure to **29**. A related sequence starting with *ortho*-*N*-(phenylsulfonyl)benzaldehyde, and sulfonium ylides also afforded indoles (equation 4) [14]. As before, the second equivalent of ylide presumably effected ring closure of the so-generated sulfonamide anion to the oxirane, **32** to indoline **33**. Dehydration gave the corresponding indole.

Schirok also employed dimethyl sulfonium ylide to synthesize indoles, but where the indole nitrogen is derived from an intermolecular reaction of the styrene epoxide **34** with a primary amine (Scheme 5) [15]. The reaction proceeds by amination of the epoxide **34** to give amino alcohols **35**, which can be isolated at 140 °C (R^1=methyl, R^2=cyclohexyl; 84%). Fluoride is superior to chloride as a leaving group in this S_NAr ring closure, **35** to **36**. Some of the indoles synthesized by Schirok are shown in Scheme 5. The reaction is applicable to the preparation of both *N*-secondary and *N*-tertiary alkylated indoles. Chen and Reamer reported a related indole synthesis (equation 2) [16]. Their procedure uses a sealed tube rather than microwave heating, and the 3-hydroxyindolines can be isolated if desired. Indoles **37–39** were prepared in this study. The starting epoxides were synthesized from the corresponding *ortho*-amino-α-chloroacetophenone by reaction with the appropriate organometallic reagent (ArLi).

The S_NAr reaction is a feature of several other indole ring syntheses, one of which is Sutherland's preparation of indoles **40** involving some unusual chemistry of 1,8-diazabicyclo[5.4.0]undec-8-ene (DBU), which is acting as a carbon-nucleophile (Scheme 6, equation 1) [17]. The initially formed Meisenheimer complex (red color) is oxidized under the reaction conditions to give **40**, which cyclized to indole **41** by displacing nitrite. Another peculiarity of the nitro group is seen in the formation of indoles **43** from the Baylis–Hillman adducts **42** (equation 2) [18]. The optimal conditions for the synthesis of **43** (R^1=Cl, R^2=R^3=H) were three equivalents of KNO_2 in DMF at 0 °C for 1 hour (77% yield).

Michael and other conjugate addition reactions, like the one we just saw in Scheme 6, are key steps in several indole syntheses. For example, Otera and colleagues developed

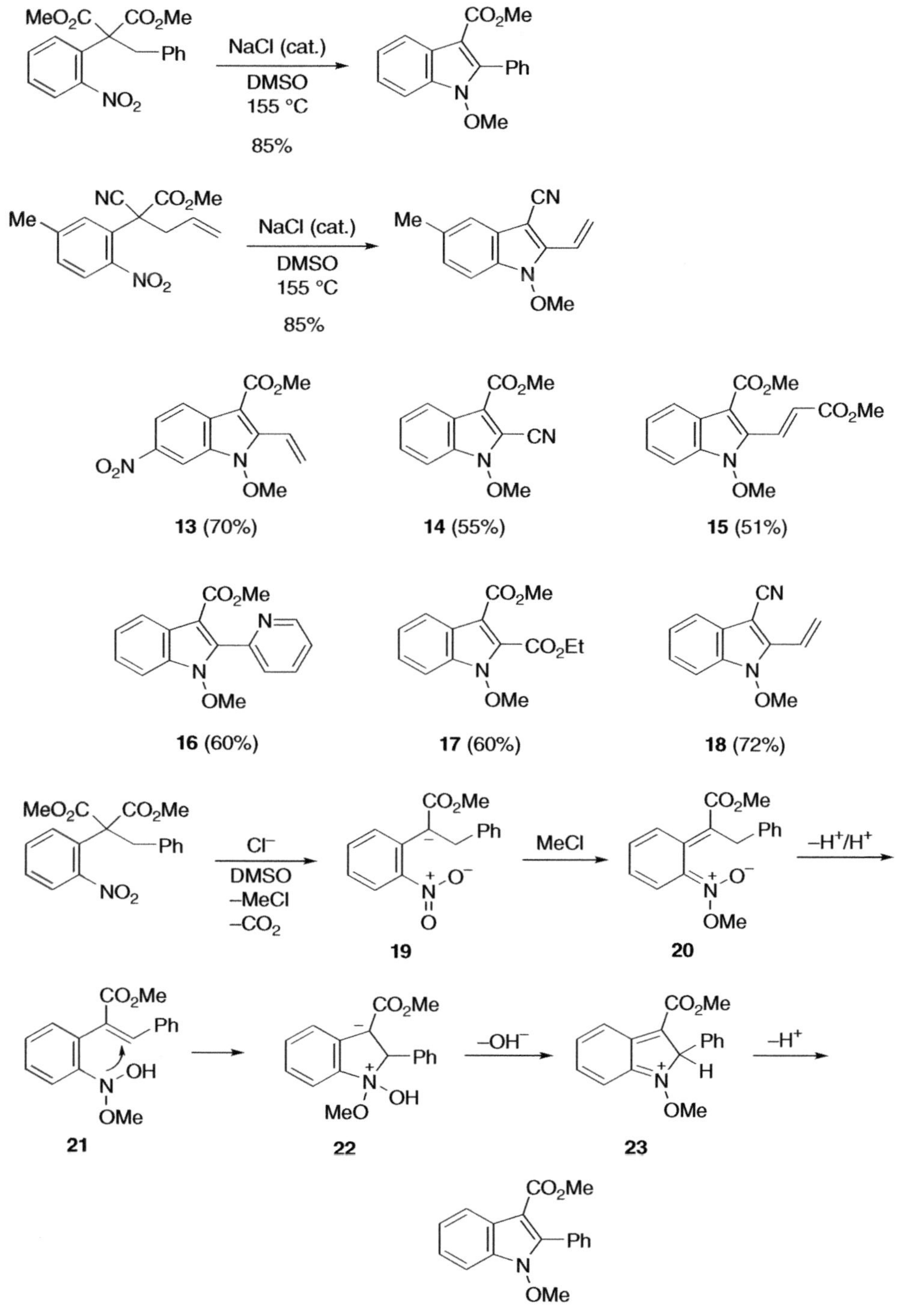

[eqn. 1] [10]

[eqn. 2] [11]

[eqn. 3]

Scheme 3 *Selvakumar Indole Synthesis*

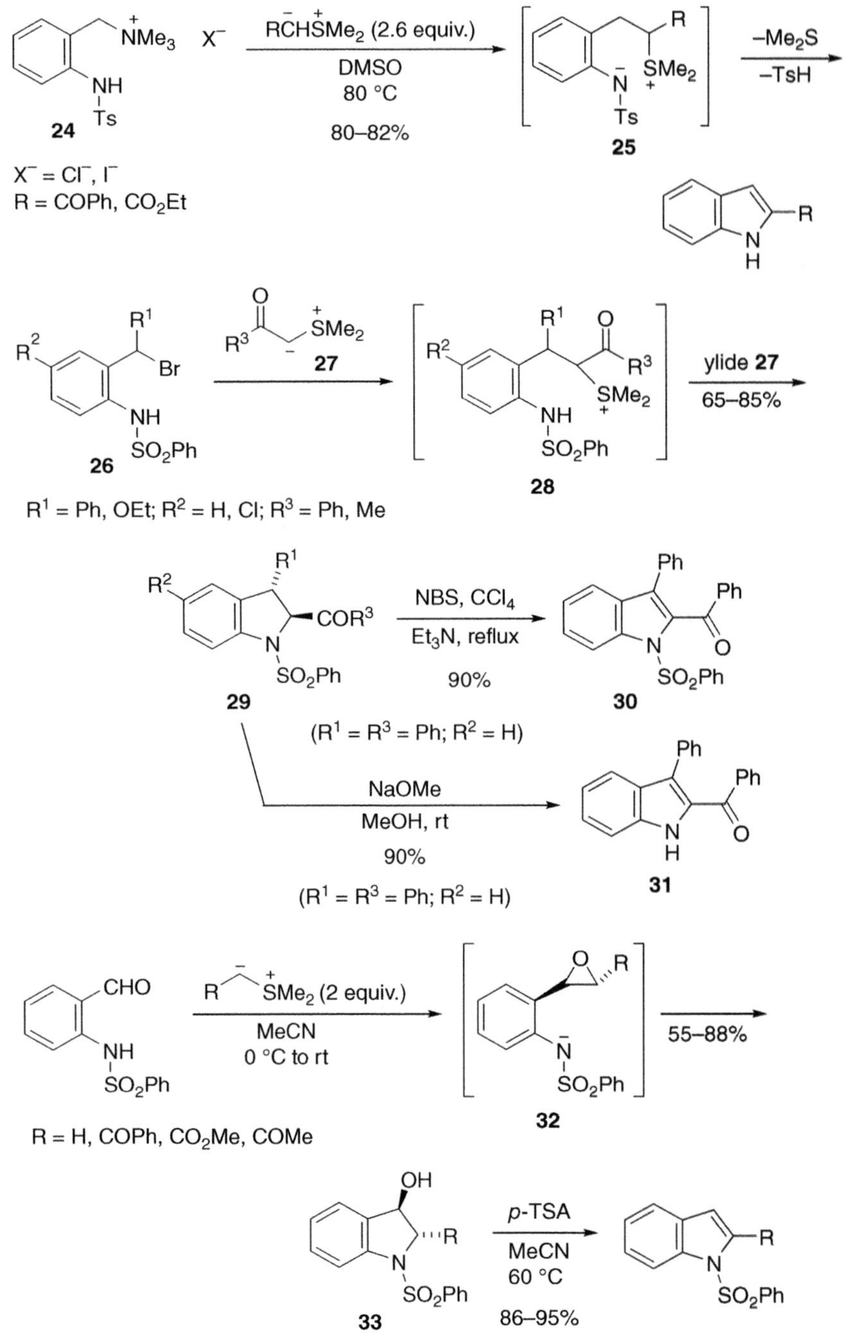

[eqn. 1] [12]

[eqn. 2] [13]

[eqn. 3] [13]

[eqn. 4] [14]

Scheme 4 *Indole Synthesis via Sulfonium Ylides*

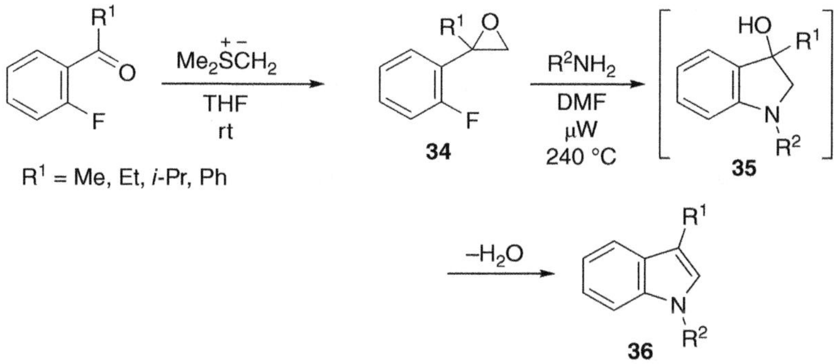

[eqn. 1] [15]

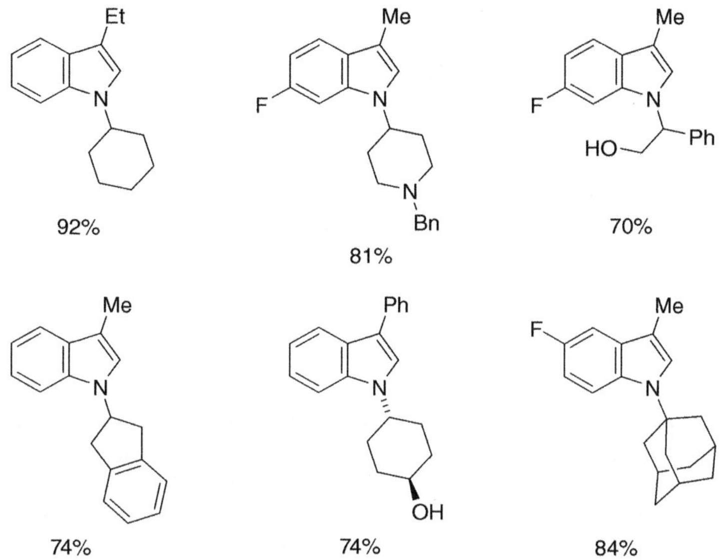

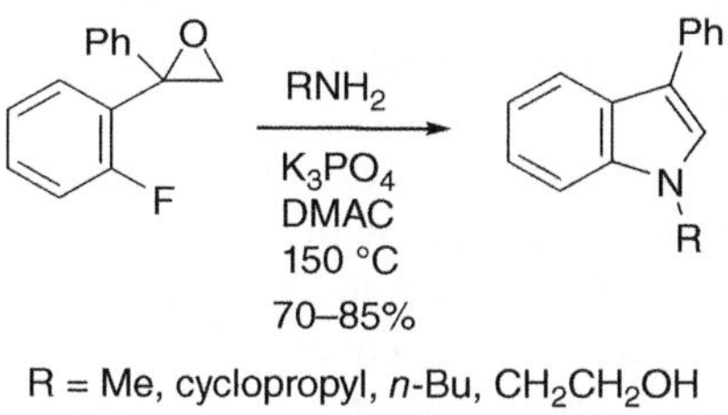

[eqn. 2] [16]

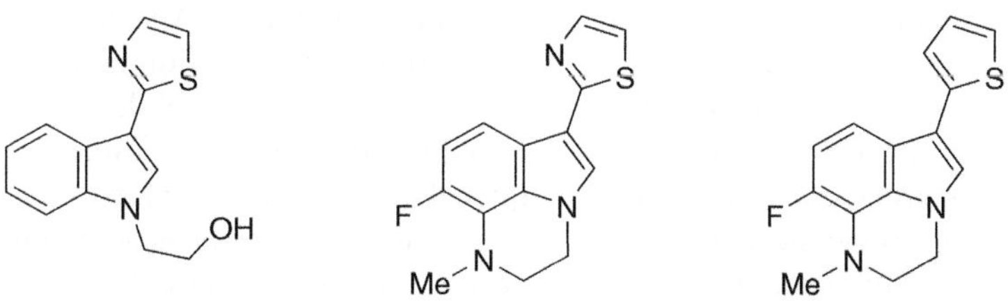

37 (70%) **38** (98%) **39** (84%)

Scheme 5 *Indole Synthesis via Epoxide Ring Opening*

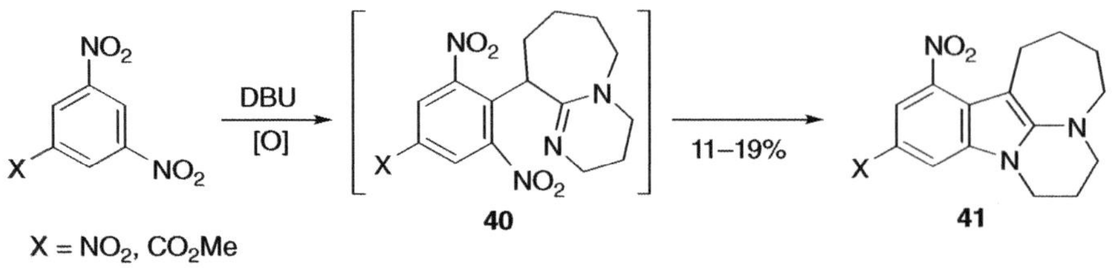

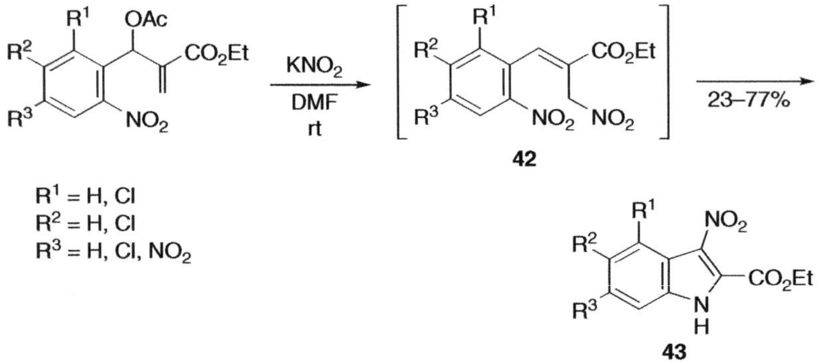

Scheme 6 *Indole Synthesis via S_NAr Chemistry*

the intramolecular cyclization/elimination sequence shown in Scheme 7 (equations 1 and 2) [19]. The starting vinyl sulfones were prepared by condensing the benzyl sulfone anion with the appropriate aldehyde, trapping with diethyl chlorophosphate, and potassium *tert*-butoxide–induced elimination. Equation 2 illustrates an extension to the synthesis of 2,3-disubstituted indoles. Quinones are highly susceptible to conjugate nucleophilic addition, and some examples are shown in equations 3–5 [20–22]. The leuco dyes **46** have potential value as visible or near-infrared color formers [20]. Carbazole quinones **47** show potent antifungal activity [22].

Oxiranes are featured in the indole syntheses shown in Scheme 8. With regard to equation 1, alternatively, the bromohydrin formed in step two was cyclized to 3-hydroxyindoline with sodium hydride (90%) and then dehydrated with *p*-TSA (90%) to the final indole [23]. The reaction of formamides with styrene epoxides (equation 2) was used by Borate's group to synthesize a wide variety of *N*-2-thienylindoles [24]. This sequence again features an intramolecular S_NAr displacement of fluoride. An epoxide intermediate is not involved in equation 3 because that route would lead to 3-substituted indoles. Rather, a [1, 2]-aryl migration was suggested by Pei and colleagues [25]. Likewise, an epoxide may not be involved in equation 4 [26].

The penultimate intermediate in many indole ring formations is an amino carbonyl group (aldehyde or ketone), which then gives the hydroxyindoline and, by dehydration, indole. Scheme 9 illustrates some of these examples that are not covered by one of classic name reactions (e.g., Reissert, Nordlander) [27–30]. Heathcock's elegant cascade (equation 3) was the key step in his total synthesis of (±)-aspidospermidine [29]. The other product (not shown) from the reaction in equation 4 is the cyclized aminal where the hydroxyl group has captured the indolenium ion at C-2 (26%) [30].

Aoyama and colleagues used the reaction of lithium trimethylsilyldiazomethane with *N*-tosyl-*ortho*-acylanilines to give 3-substituted indoles [31] and 2,3-disubstituted indoles [32] (Scheme 10, equations 1 and 2, respectively). Both reactions proceed via the formation of an alkylidene carbene (equation 1). A sequel to this chemistry (equation 2) features a second lithiation at indole C-2 followed by addition of an electrophile [32]. The stronger base *tert*-butyllithium ensures complete C-2 lithiation prior to quenching. Fournier and Levesque have developed a simple synthesis of 3-ethoxycarbonylindoles via the reaction of ethyl diazoacetate (EDA) with 2-aminobenzaldehydes (equation 3) [33]. As was the case in Scheme 8, equation 3, a 1,2-aryl shift is also involved in the present work. Indoles **48–50** were prepared in this study. The appropriate 2-aminoacetophenone gave 2-methylindole **50** in much lower yield.

As we will see in later chapters, anionic and metal-catalyzed nucleophilic additions to alkynes are an important route to indoles. One of the first such examples is that of Johnson and Subramanian, who found that *meta*-anisidine

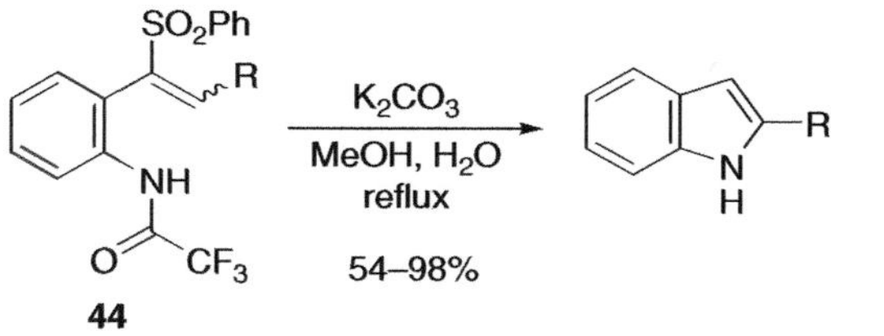

R = Ph, 2-thienyl, PhCH=CH

[eqn. 1] [19]

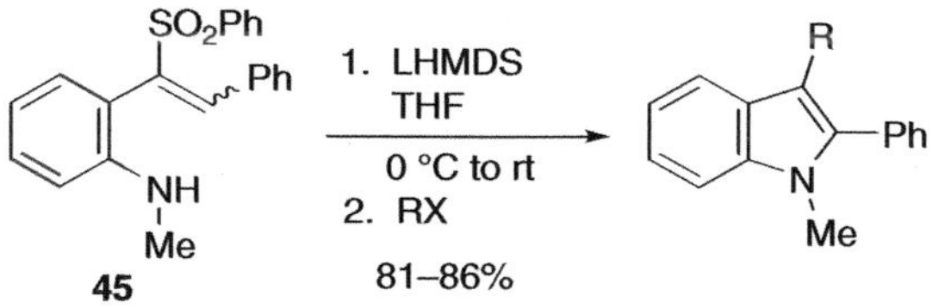

R = Me, Et, n-Pr, allyl, CH_2CO_2Et, COPh, CH_2CONEt_2

[eqn. 2] [19]

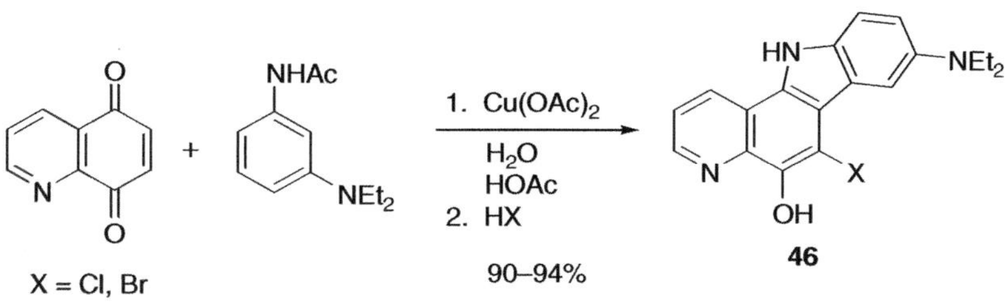

X = Cl, Br

[eqn. 3] [20]

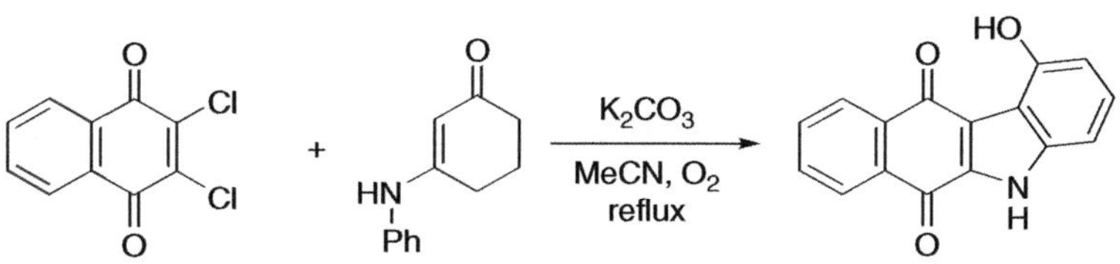

[eqn. 4] [21]

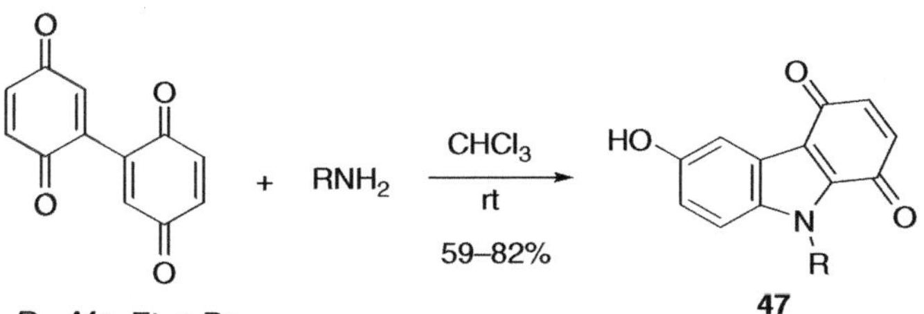

R = Me, Et, n-Pr

[eqn. 5] [22]

Scheme 7 *Indole Synthesis via Conjugate Addition*

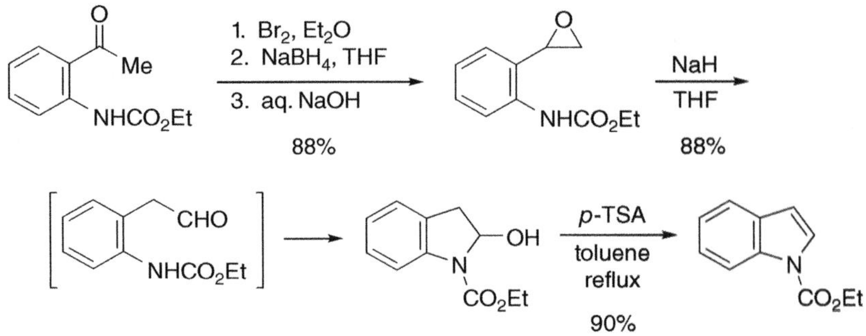

[eqn. 1] [23]

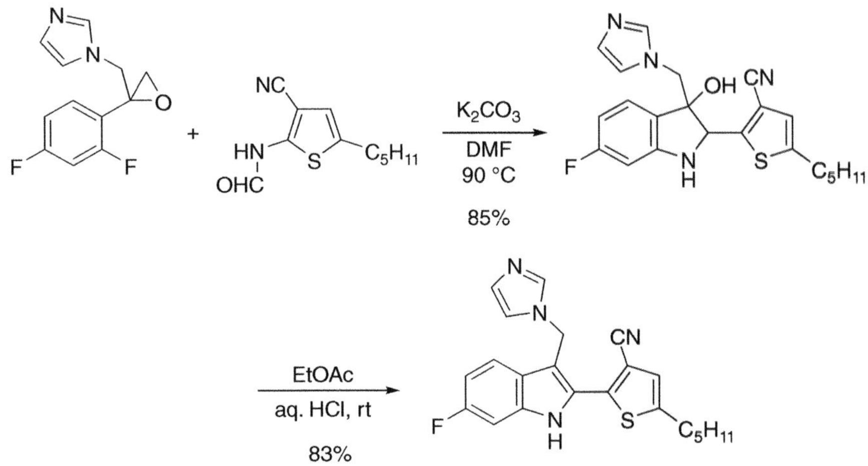

[eqn. 2] [24]

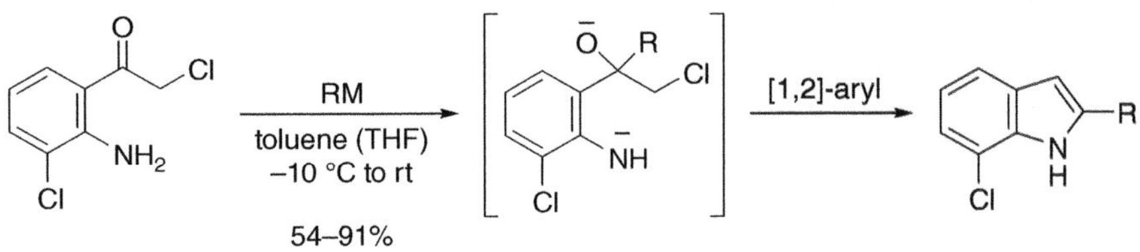

[eqn. 3] [25]

R = allyl, 2-thienyl, cyclopropyl, CH_2TMS, Ph, 2-pyr, 3-pyr, TMSC≡C, *n*-Pen

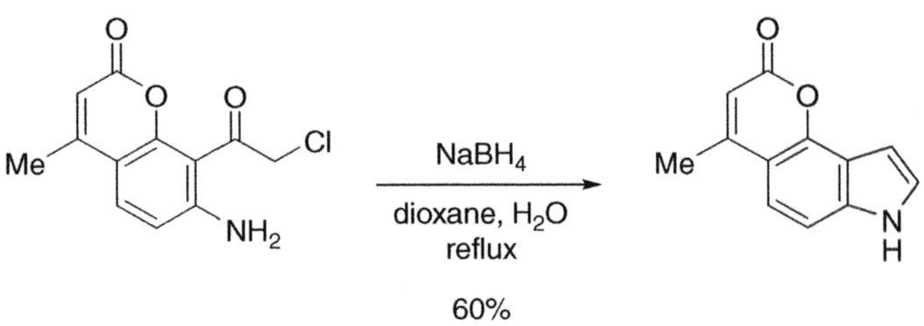

[eqn. 4] [26]

Scheme 8 *Indole Synthesis via Epoxide Ring Opening*

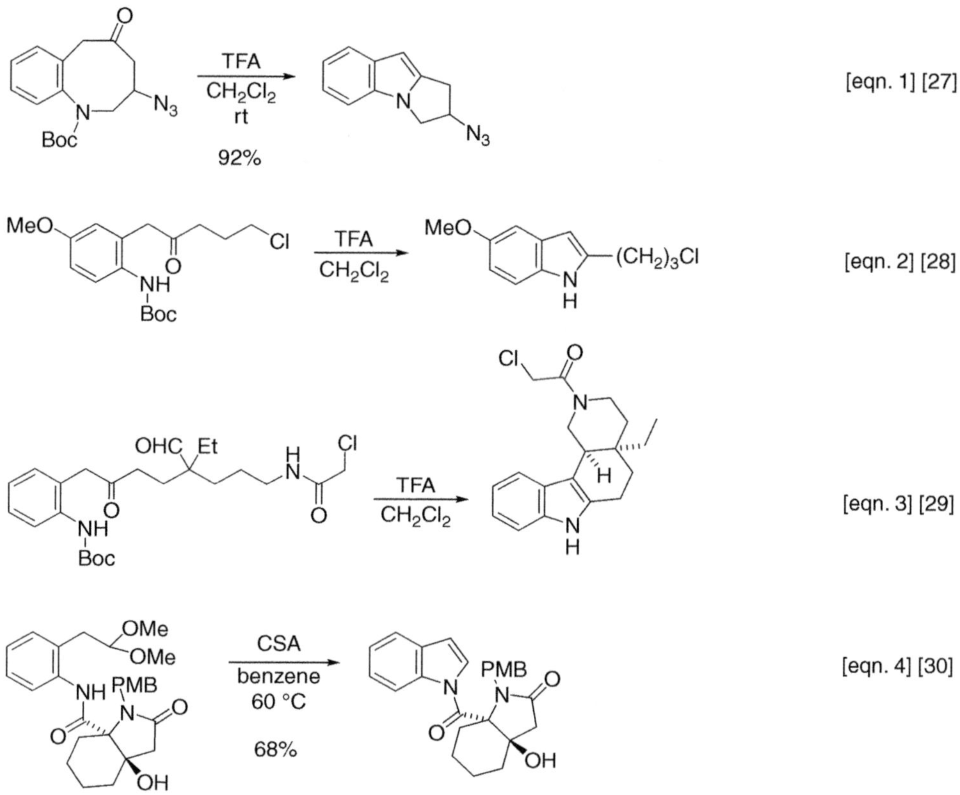

Scheme 9 *Indole Synthesis via Amine Carbonyl Ring Closure*

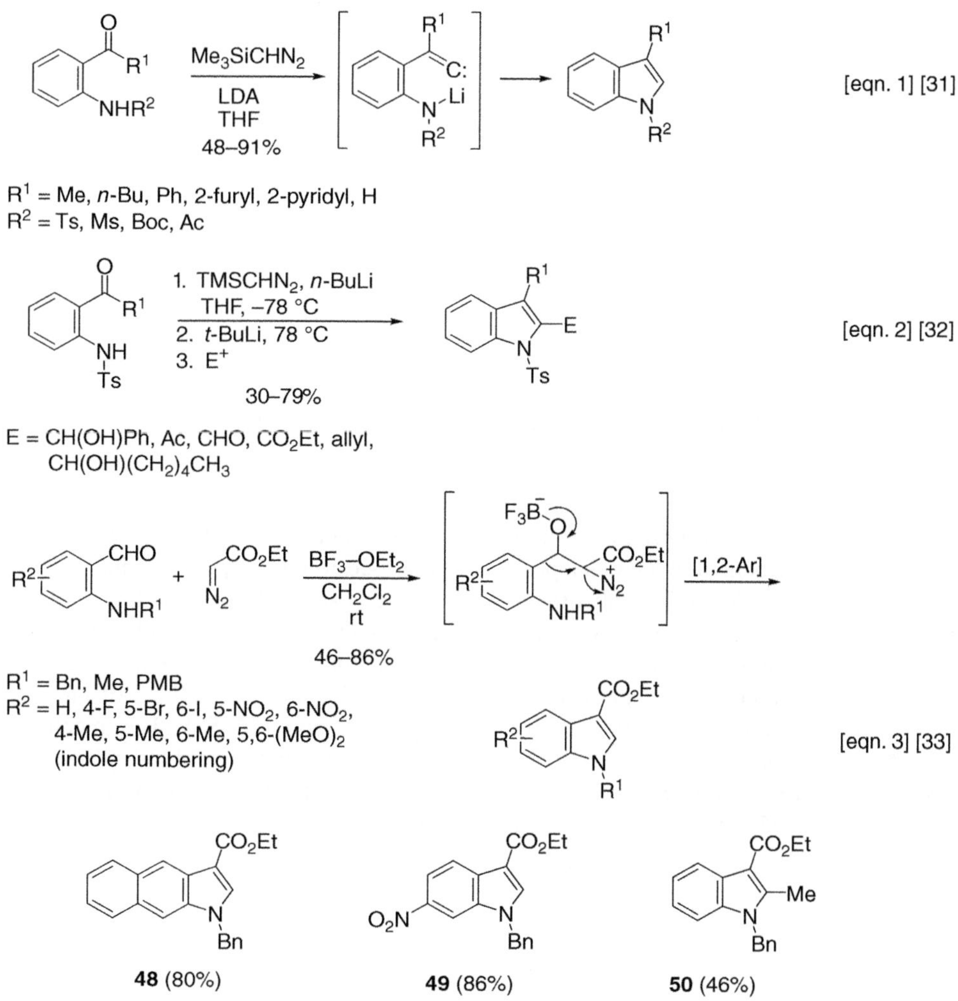

Scheme 10 *Indole Synthesis via Carbenoids*

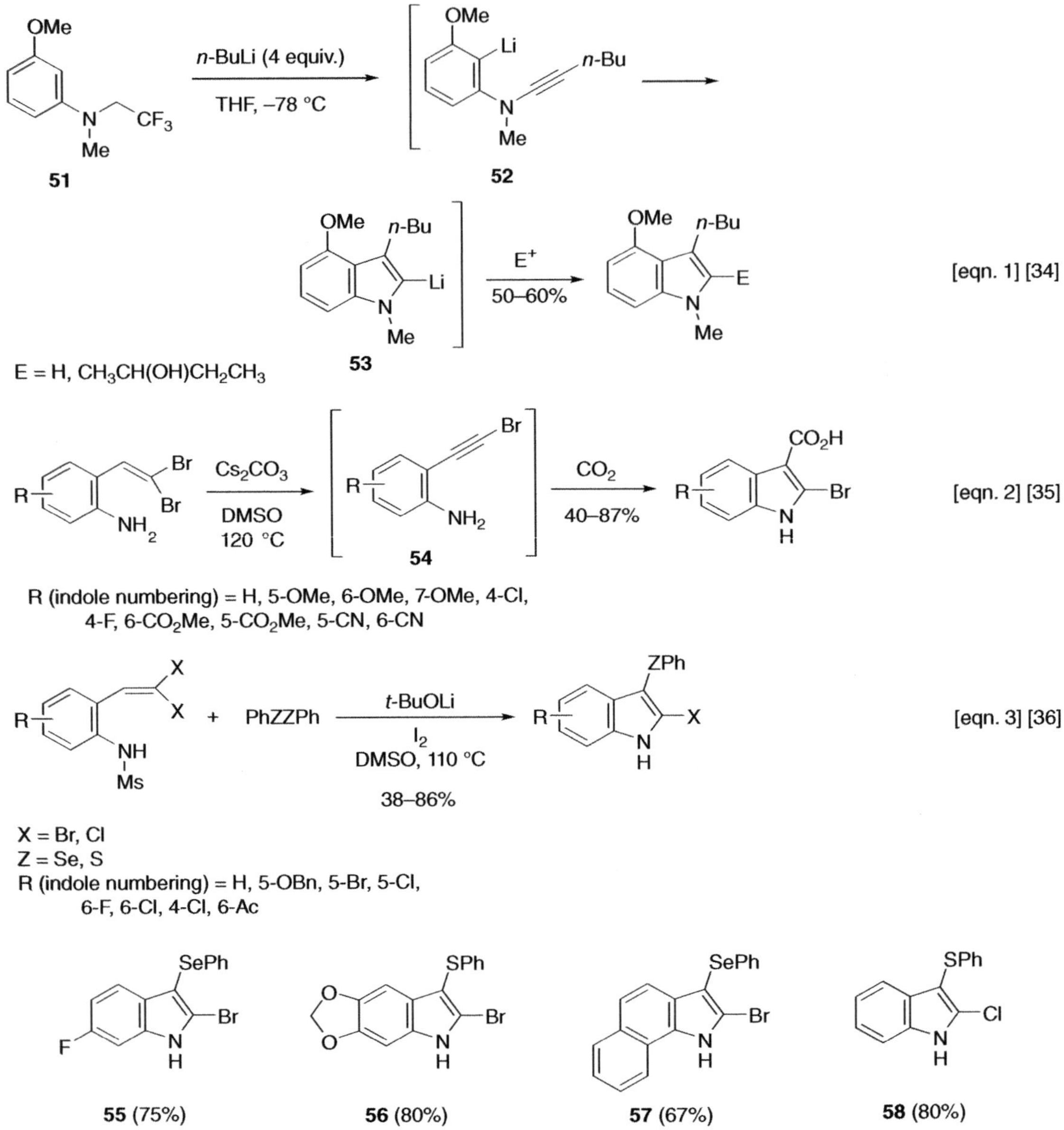

Scheme 11 *Indole Synthesis via Alkyne Cyclization*

51 was converted to **52** and then via 5-*endo-dig* cyclization to **53** with excess *n*-butyllithium. Quenching **53** with a proton source or 1,2-epoxy butane afforded 2-substituted indoles (Scheme 11, equation 1) [34]. Only a few examples were reported in this interesting study. The amine–alkyne nucleophilic addition shown in equation 2 is an efficient synthesis of 2-halo-3-carboxyindoles as reported by Kunzer and Wendt [35]. The reaction also worked with 2,2-dichlorovinyl substrates; for example, 2-chloro-3-carboxyindole was obtained in 95% yield. The reaction generates a bromoacetylene intermediate (**54**) that cyclized to the 3-cesioindole species, which was trapped by *in situ* generated carbon dioxide. At low temperatures **54** (R = H) was isolated. Wang and colleagues described a similar reaction that afforded 2-halo-3-selenyl(sulfenyl)indoles (equation 3) [36]. A small set of indoles prepared in this study is shown (**55–58**). The mechanism involves the formation of 2-haloindole by cyclization of a haloalkyne analogous to **54** (equation 2) and then C-3 electrophilic addition of PhSeI or PhSI.

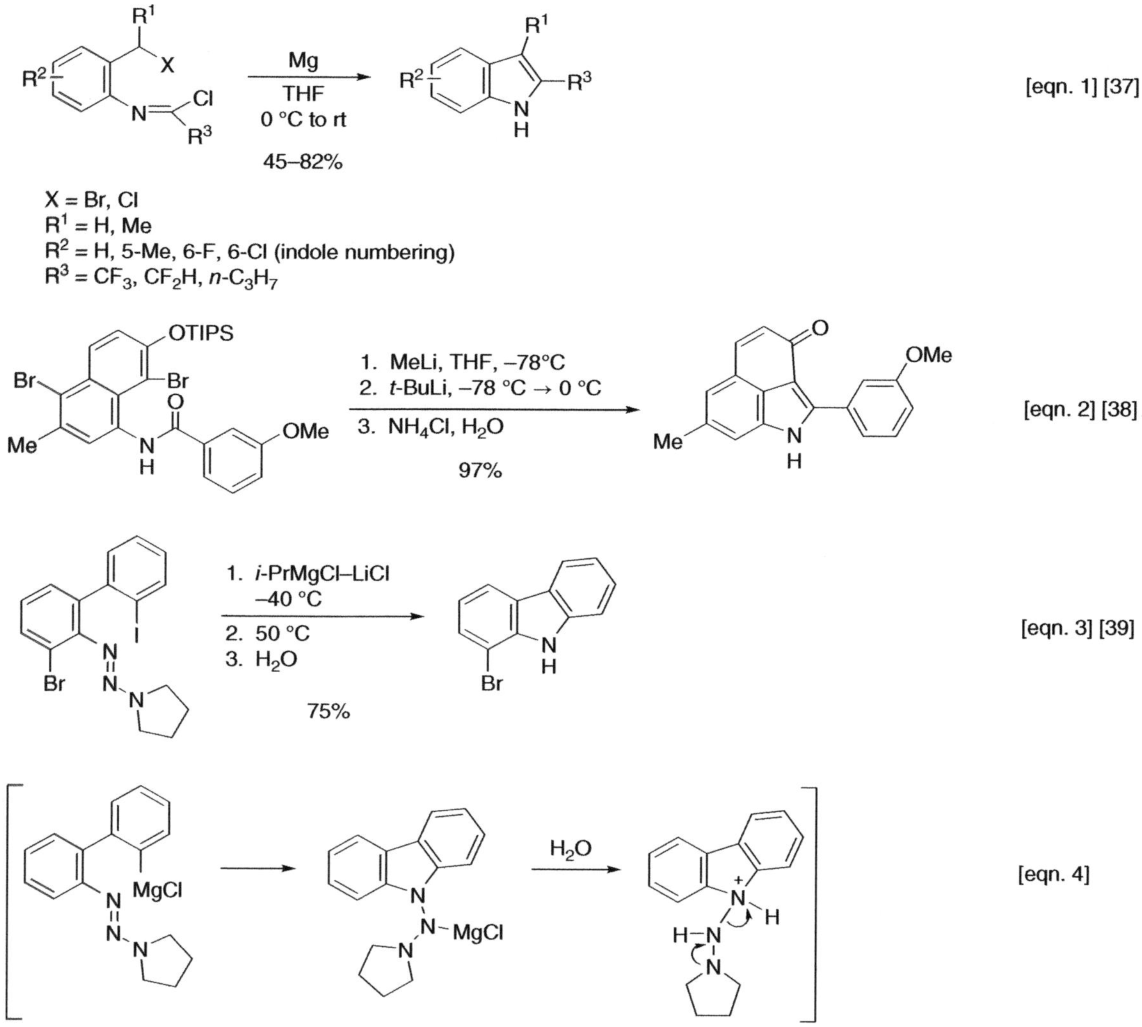

Scheme 12 *Indole Synthesis via Metal-Halogen Exchange*

Metal–halogen exchange is involved in several indole syntheses, and three examples are shown in Scheme 12. Hao and colleagues found that haloalkyl-imidoyl chlorides undergo cyclization upon treatment with Grignard reagents to afford indoles (equation 1) [37]. The starting compounds were prepared from *ortho*-alkylanilines or *ortho*-aminobenzyl alcohol. Kelly and Scopton employed a classic Parham cyclization in their synthesis of HKI 0231B, a *Streptomyces* sp. fermentation product (equation 2) [38]. Knochel and Liu described a new carbazole synthesis that features a novel reaction of triazenes (equation 3) [39]. A pathway is shown in equation 4. These workers also prepared 3-carboethoxycarbazole (70% yield).

This chapter on miscellaneous nucleophilic additions concludes with three interesting, if not yet practical, indole ring formations (Scheme 13). Özcan and Balci reported the reaction of isocyanate **59** with aniline to give the expected urea **60** (equation 1). Treatment with sodium hydride afforded the 2-hydroxyindole **61** [40]. A follow-up paper provides some additional details but precious little on indole ring synthesis because this work is mainly focused on the preparation of indolin-2-ones and isoindolin-1-ones [41]. Knochel and coworkers described a mixed copper–zinc reagent **62** to craft 2-substituted indoles (equation 2) [42]. The reagent was prepared from the corresponding benzyl bromide, which was available in two steps from the appropriate *ortho*-toluidine. Treatment of **62** with an acyl chloride gave the indole by the pathway illustrated. A few of the synthesized indoles are shown, **63–66**.

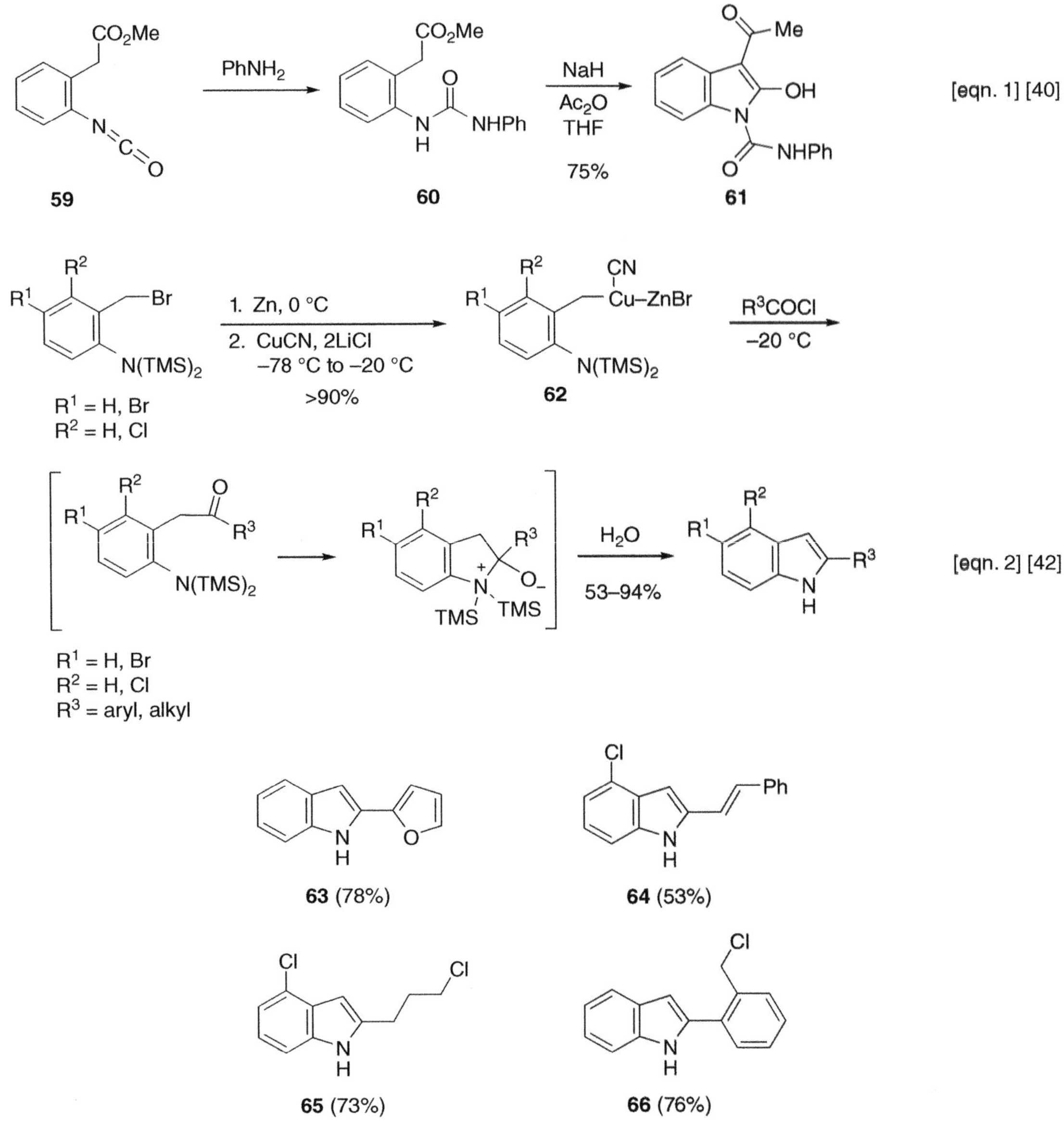

Scheme 13 *Indole Synthesis via Nucleophilic Cyclization*

References

[1] G.M. Brooke, *Tetrahedron Lett.*, 1968, **9**, 4049–4052.
[2] G.M. Brooke, W.K.R. Musgrave, R.J.D. Rutherford, and T.W. Smith, *Tetrahedron*, 1971, **27**, 5653–5658.
[3] V.P. Petrov and V.A. Barkhash, *Chem. Heterocycl. Compd.*, 1970, **6**, 357–360.
[4] C. Wakselman and J.-C. Blazejewski, *J. Chem. Soc. Chem. Commun.*, 1977, 341.
[5] M. Fujita and I. Ojima, *Tetrahedron Lett.*, 1983, **24**, 4573–4576.
[6] I. Ojima, K. Kato, and K. Nakahashi, *J. Org. Chem.*, 1989, **54**, 4511–4522.
[7] R. Filler, W. Chen, and S.M. Woods, *J. Fluorine Chem.*, 1995, **73**, 95–100.
[8] P. Molina, P.M. Fresneda, and S. Delgado, *Synthesis*, 1999, 326–329.
[9] K. Orito, R. Harada, S. Uchiito, and M. Tokuda, *Org. Lett.*, 2000, **2**, 1799–1801.
[10] N. Selvakumar, B.Y. Reddy, A.M. Azhagan, *et al.*, *Tetrahedron Lett.*, 2003, **44**, 7065–7069.
[11] N. Selvakumar, M.K. Khera, B.Y. Reddy, *et al.*, *Tetrahedron Lett.*, 2003, **44**, 7071–7074.
[12] P.D. Croce, R. Ferraccioli, and C. La Rosa, *Heterocycles*, 1996, **43**, 2397–2407.
[13] G. Cremonesi, P.D. Croce, and C. La Rosa, *Heterocycles*, 2005, **66**, 557–562.

[14] G. Cremonesi, P.D. Croce, F. Fontana, and C. La Rosa, *Heterocycles*, 2007, **73**, 873–876.
[15] H. Schirok, *Synthesis*, 2008, 1404–1414.
[16] C. Chen and R.A. Reamer, *Tetrahedron Lett.*, 2009, **50**, 1529–1532.
[17] J.K. Sutherland, *Chem. Commun.*, 1997, 325.
[18] C.R. Horn and M. Perez, *Synlett*, 2005, 1480–1482.
[19] G. Babu, A. Orita, and J. Otera, *Org. Lett.*, 2005, **7**, 4641–4643.
[20] K. Yoshida, N. Oga, M. Kadota, *et al.*, *J. Chem. Soc., Chem. Commun.*, 1992, 1114–1115.
[21] H.-Y. Hu, Y. Liu, M. Ye, and J.-H. Xu, *Synlett*, 2006, 1913–1917.
[22] C.-K. Ryu, S.-Y. Lee, N.Y. Kim, *et al.*, *Bioorg. Med. Chem. Lett.*, 2011, **21**, 427–430.
[23] J.L. Garcia Ruano, C. Pedregal, and J.H. Rodriguez, *Tetrahedron*, 1989, **45**, 203–214.
[24] H.B. Borate, S.P. Sawargave, and S.R. Maujan, *Tetrahedron Lett.*, 2009, **50**, 6562–6566.
[25] (a) T. Pei, D.M. Tellers, E.C. Streckfuss, *et al.*, *Tetrahedron*, 2009, **65**, 3285–3291; (b) T. Pei, C. Chen, P.G. Dormer, and I.W. Davies, *Angew. Chem. Int. Ed.*, 2008, **47**, 4231–4233.
[26] J.C. González, J. Lobo-Antunes, P. Pérez-Lourido, *et al.*, *Synthesis*, 2002, 475–478.
[27] K. Tsuboike, D.J. Guerin, S.M. Mennen, and S.J. Miller, *Tetrahedron*, 2004, **60**, 7367–7374.
[28] R. Peter, P. Waldmeier, and A. Joncour, *Org. Proc. Res. Dev.*, 2005, **9**, 508–512.
[29] M.A. Toczko and C.H. Heathcock, *J. Org. Chem.*, 2000, **65**, 2642–2645.
[30] M. Vamos and Y. Kobayashi, *Tetrahedron*, 2009, **65**, 5899–5903.
[31] T. Miyagi, Y. Hari, and T. Aoyama, *Tetrahedron Lett.*, 2004, **45**, 6303–6305.
[32] Y. Hari, T. Kanie, T. Miyagi, and T. Aoyama, *Synthesis*, 2006, 1249–1252.
[33] P. Levesque and P.-A. Fournier, *J. Org. Chem.*, 2010, **75**, 7033–7036.
[34] F. Johnson and R. Subramanian, *J. Org. Chem.*, 1986, **51**, 5040–5041.
[35] A.R. Kunzer and M.D. Wendt, *Tetrahedron Lett.*, 2011, **52**, 1815–1818.
[36] J. Liu, P. Li, W. Chen, and L. Wang, *Chem. Commun.*, 2012, **48**, 10052–10054.
[37] Z. Wang, F. Ge, W. Wan, *et al.*, *J. Fluorine Chem.*, 2007, **128**, 1143–1152.
[38] A. Scopton and T.R. Kelly, *J. Org. Chem.*, 2005, **70**, 10004–10012.
[39] C.-Y. Liu and P. Knochel, *Org. Lett.*, 2005, **7**, 2543–2546.
[40] S. Özcan and M. Balci, *Tetrahedron*, 2008, **64**, 5531–5540.
[41] A.A. Kilikli, C. Dengiz, S. Özcan, and M. Balci, *Synthesis*, 2011, 3697–3705.
[42] H.G. Chen, C. Hoechstetter, and P. Knochel, *Tetrahedron Lett.*, 1989, **30**, 4795–4798.

22

Sugasawa Indole Synthesis

To bridge the gap between the preceding chapter on nucleophilic cyclization and the forthcoming chapters on electrophilic cyclization, the Sugasawa indole synthesis has features of both pathways (Scheme 1). Thus, Sugasawa and colleagues have effected *ortho* chloroacetylation of anilines to give 2-amino-α-chloroacetophenones **1**, which were reductively cyclized with sodium borohydride in refluxing dioxane (Scheme 1) [1]. At room temperature, reduction of **1** afforded the α-chloromethylbenzyl alcohol, which upon treatment with sodium methoxide or sodium hydride gave indole **2**. The method has been extended to the synthesis of 1-(azacycloalkyl)indoles **3** (equation 2) [2]. Häfelinger and Nimtz used the Sugasawa indole synthesis to prepare a large number of indoles (equation 3) [3], and Wuonola and colleagues employed this indolization in a synthesis of (±)-dragmacidin. Indole **4** had not been previously described, and the Sugasawa method was superior to reduction of the appropriate isatin (33% yield) (equation 4) [4].

Indole Ring Synthesis: From Natural Products to Drug Discovery, First Edition. Gordon W. Gribble.

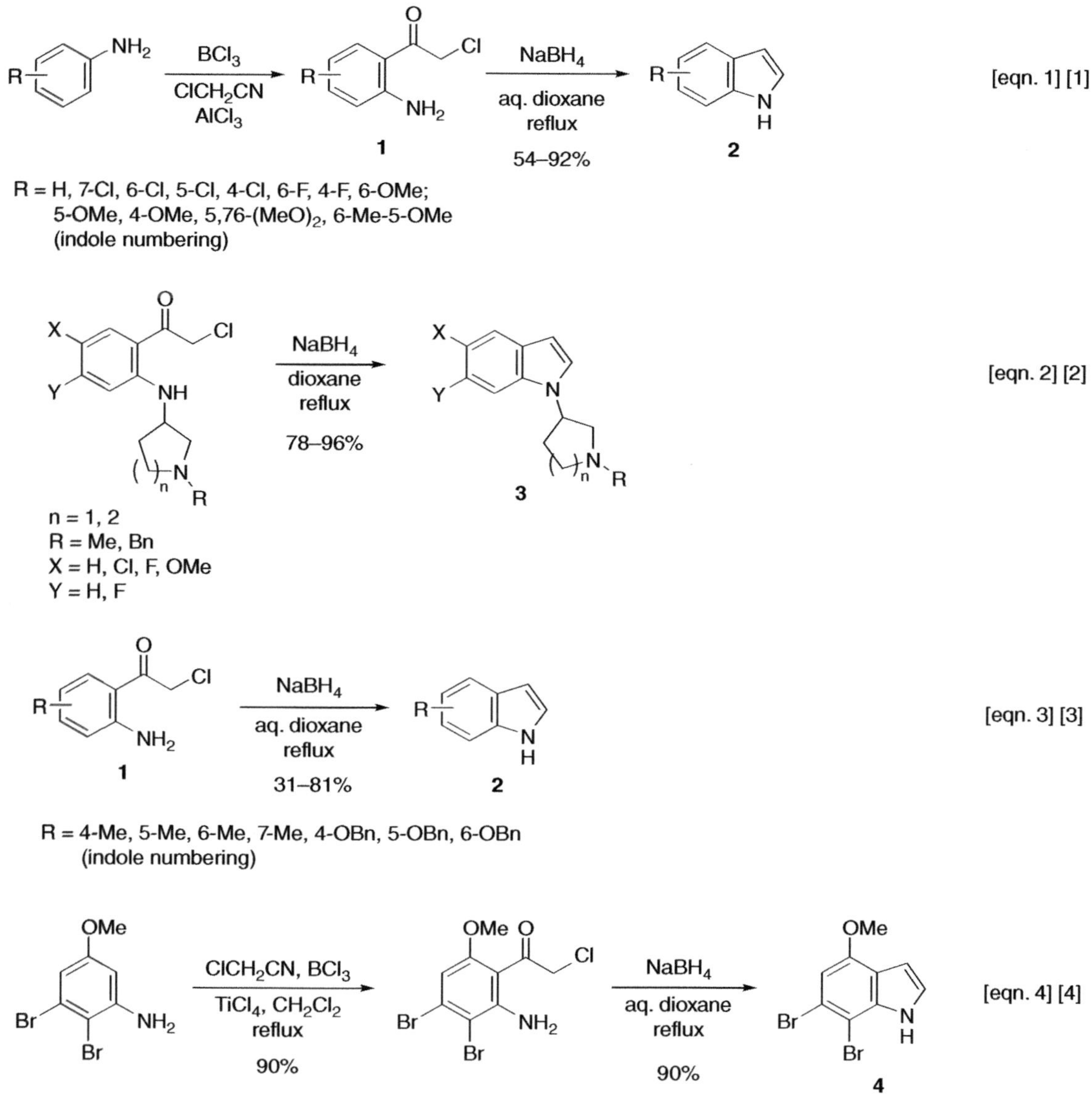

Scheme 1 *Sugasawa Indole Synthesis*

References

[1] T. Sugasawa, M. Adachi, K. Sasakura, and A. Kitagawa, *J. Org. Chem.*, 1979, **44**, 578–586.
[2] K.S. Asakura, M. Adachi, and T. Sugasawa, *Synth. Commun.*, 1988, **18**, 265–273.
[3] M. Nimtz and G. Häfelinger, *Liebigs Ann. Chem.*, 1987, 765–770.
[4] B. Jiang, J.M. Smallheer, C. Amaral-Ly, and M.A. Wuonola, *J. Org. Chem.*, 1994, **59**, 6823–6827.

References

PART III

Electrophilic Cyclization

Electrophilic cyclizations abound in organic synthesis, and several such methods afford the indole nucleus in the key step. In addition to the following chapters, 23 through 38, several other named indole ring syntheses that involve electrophilic cyclization at some stage in their reaction sequence are more appropriately relegated to separate chapters.

Indole Ring Synthesis: From Natural Products to Drug Discovery, First Edition. Gordon W. Gribble.

23

Bischler Indole Synthesis

Like the venerable Fischer, Madelung, and Nenitzescu indole syntheses, the Bischler indolization has stood the test of time and continues to find utility in indole synthesis. This method, which is also known as the Bischler–Möhlau indole synthesis, was discovered by both of these German chemists within the span of a decade [1–4]. Other early pioneers in this area were Nencki and Berlinerblau [5] and Japp and Murray [6], the latter of whom prepared indoles by the reaction of aniline with benzoin in the presence of zinc chloride to give 2,3-diphenylindole [6]. The basic Bischler reaction is shown in Scheme 1 (equation 1). The mechanism is complicated, but one possibility is shown in equation 2 [7–10].

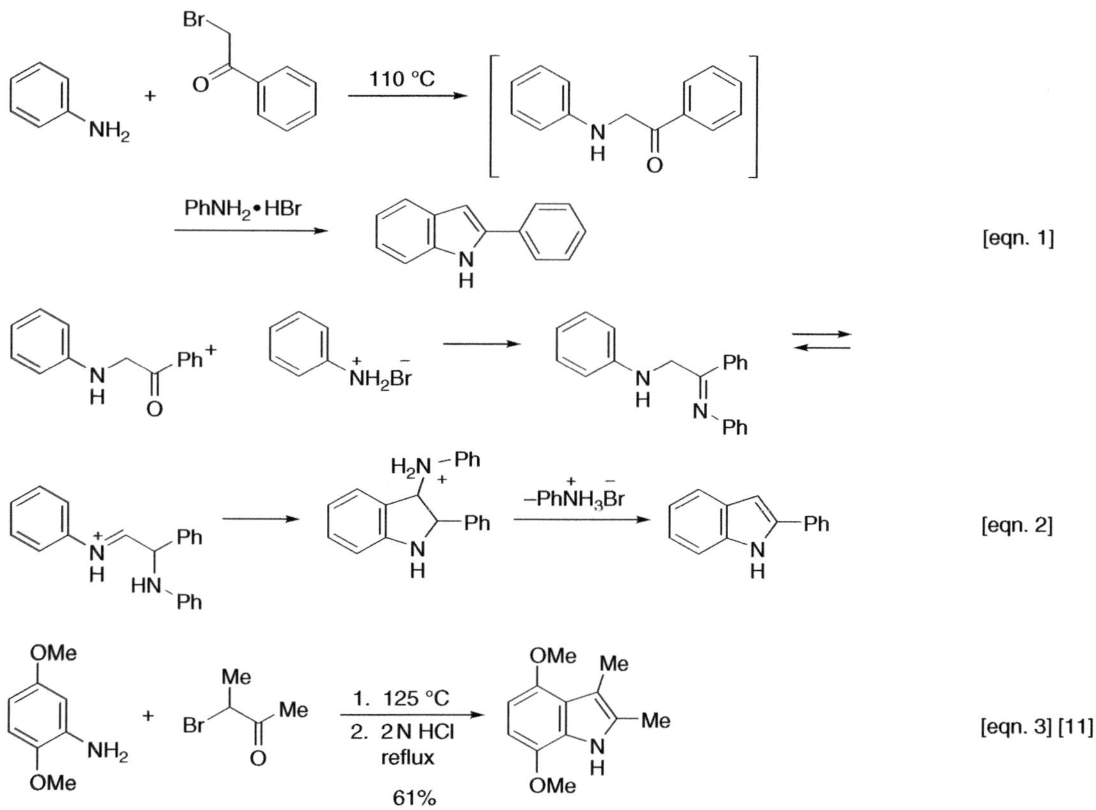

Scheme 1 *Bischler Indole Synthesis*

Indole Ring Synthesis: From Natural Products to Drug Discovery, First Edition. Gordon W. Gribble.

The mechanism is discussed in more detail later in this chapter. An early example of the Bischler indolization is that of Blackhall and Thomson (equation 3) [11].

Despite the vagaries of the mechanism, the Bischler indole synthesis is extremely useful, and several variations are known. Buu-Hoi and colleagues employed a modified version of this indole synthesis to the preparation of 2-arylindoles [12] and polycyclic indoles [13] (Scheme 2, equation 2). Their modification was to heat the ω-arylamino ketone at 230–250 °C in inert silicone oil. Some indoles prepared in these extensive studies are **1–4**. Coïc and Saint-Ruf converted 2-aminophenoxathiin (**5**) to 2-phenylpyrrolo[2,3-*b*]phenoxathiine using the Buu-Hoi method (equation 3) [14]. Thus, heating ketone **6** (0.01 mol) with amine **5** (0.02 mol) and amine hydrobromide (0.005 mol) gave **7** in 70% yield.

For more than 30 years Black and his colleagues made extensive use of the Bischler indole synthesis [15–22]. Some of the highlights of this work are shown in Scheme 3. The conversion of **8** to indoles **9** demonstrates that even an

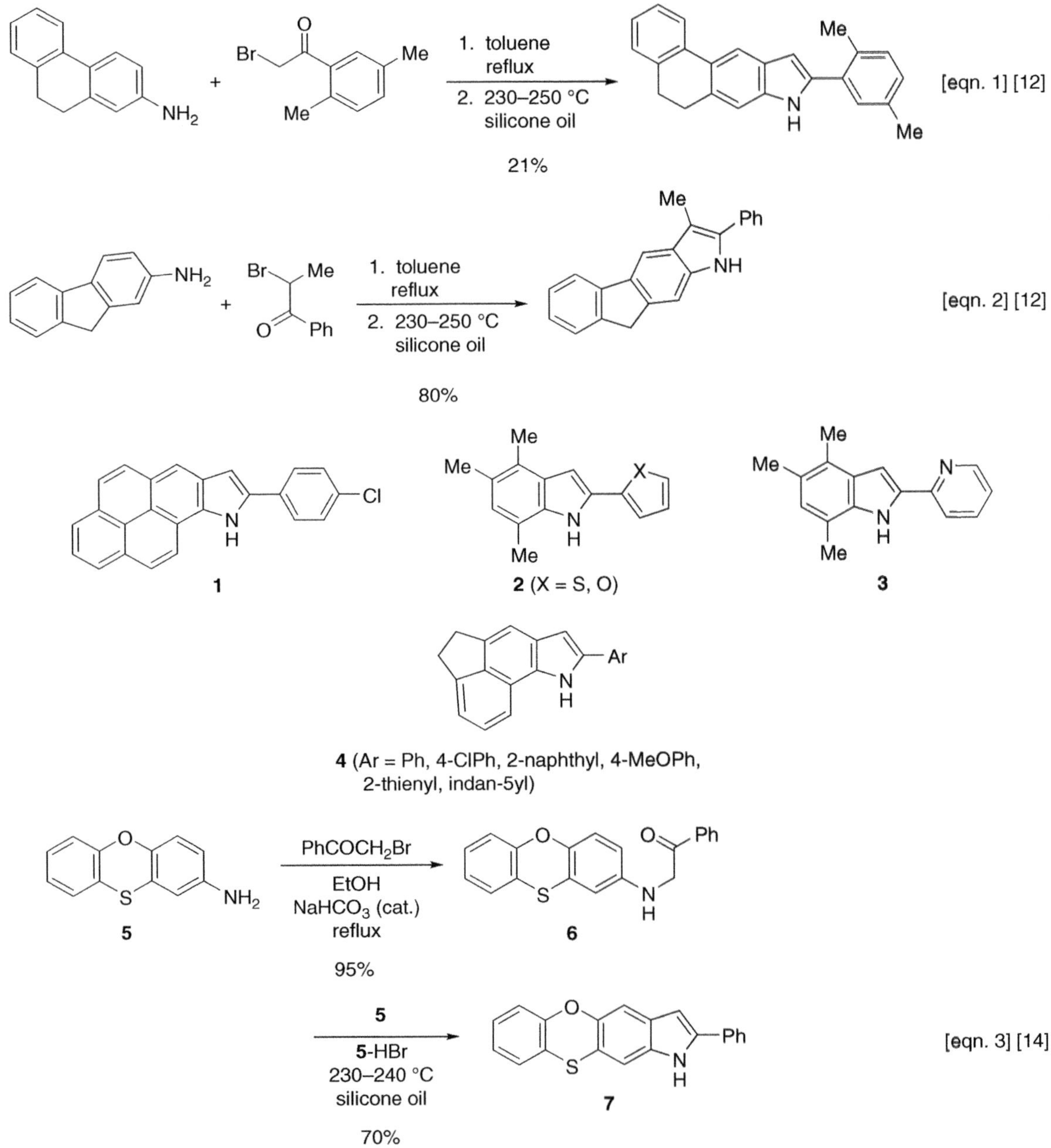

Scheme 2 *Buu-Hoi Variation of the Bischler Indole Synthesis*

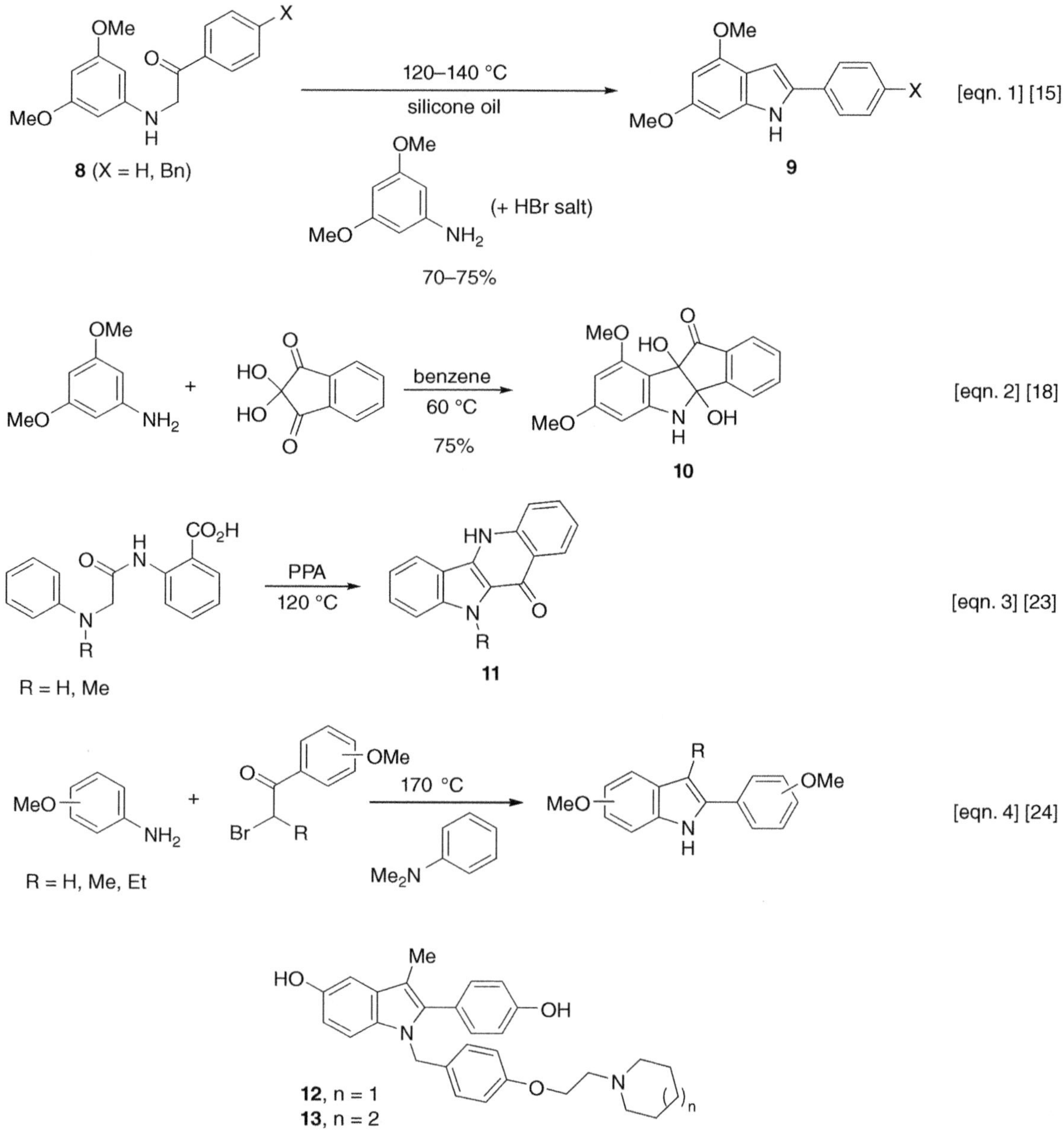

Scheme 3 *Black and other Variations of the Bischler Indole Synthesis*

electron-rich aniline is unable to thwart the normal Bischler equilibrium that favors 2-arylindole formation over 3-arylindole formation [15]. The synthesis of **8** from 3,5-dimethoxyaniline and the appropriate α-bromoacetophenones was achieved in 73% to 75% yield in refluxing ethanol with added sodium bicarbonate. Black and coworkers used 3,5-dimethoxyaniline to prepare a collection of 2-, 3-, and 2,3-substituted 4,6-dimethoxyindoles [16–22]. In some of these studies benzoin was employed rather than a halo ketone. The reaction of 3,5-dimethoxyaniline with ninhydrin afforded compound **10** (equation 2) [18]. The 10*H*-indole[3,2-*b*]quinoline ring system **11** was synthesized by Görlitzer and Weber (equation 3) [23, 24], and von Angerer and colleagues used the Bischler indolization to prepare a large number of hydroxy-substituted 2-phenylindoles via the corresponding methoxylated indoles (equation 4) [25, 26]. Several of these indoles display mammary tumor inhibition. Miller's group discovered similar indole estrogens such as **12** and **13**, which were prepared via Bischler methodology. These compounds have completed phase II clinical trials for hormone-dependent metastatic breast cancer and postmenopausal osteoporosis, respectively [27].

A wide range of applications of the Bischler indole synthesis is tabulated in Table 1 [28–32, 37–43]. In some

Table 1 *Applications of the Bischler Indole Synthesis*

Entry	Substrate	Conditions	Indole	% Yield	Ref.
1	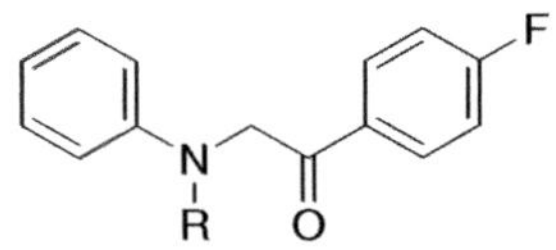 R = H, Me, *i*-Pr	1. $POCl_3$/DMF 80 °C 2. 50% aq. NaOH	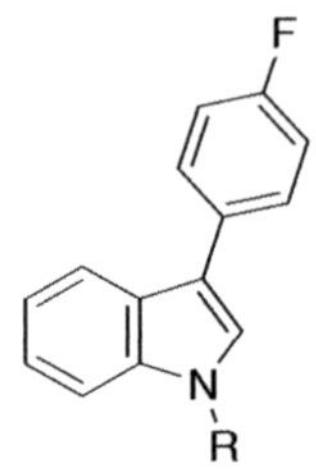	38–86%	28
2	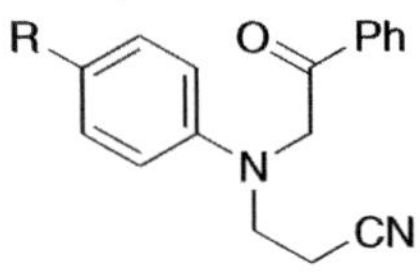 R = H, Me, OMe, OEt	P_2O_5, xylene reflux	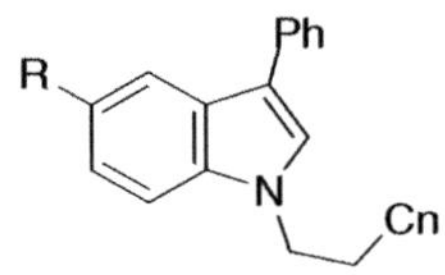	51–57%	29
3	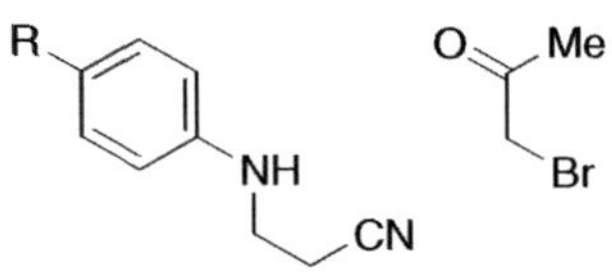 R = H, Me, OMe, OEt	EtOH, reflux	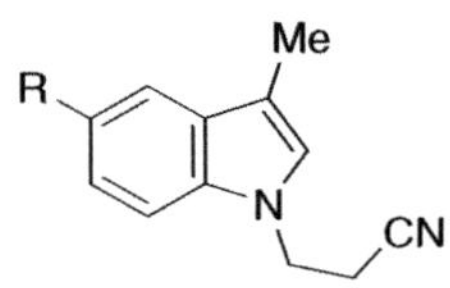	51–55%	29
4	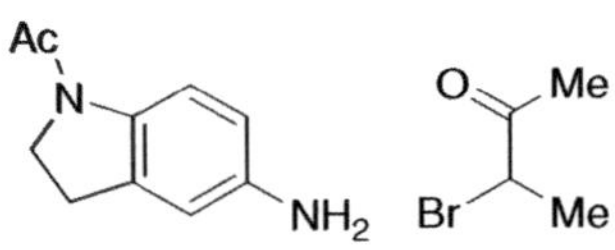	HOAc, *n*-BuOH reflux	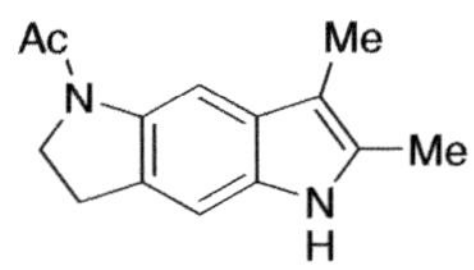	84%	30
5	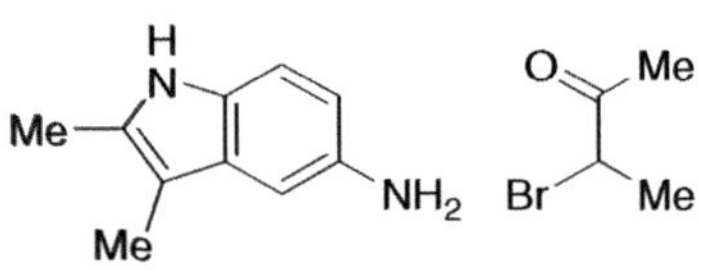	HOAc, *n*-BuOH reflux	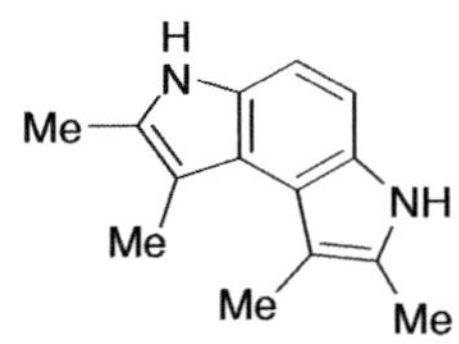	40%	30
6	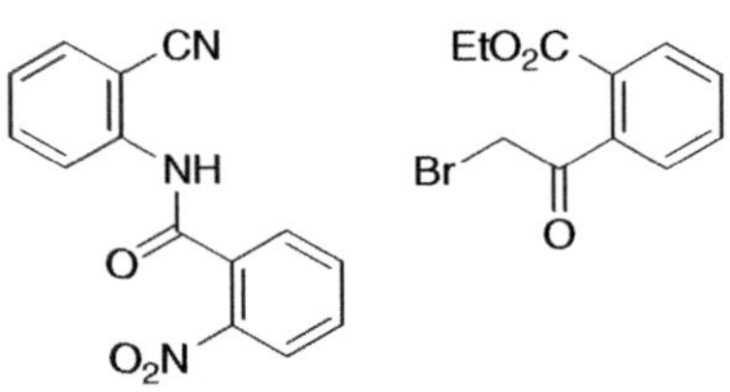	NaH, DMF rt	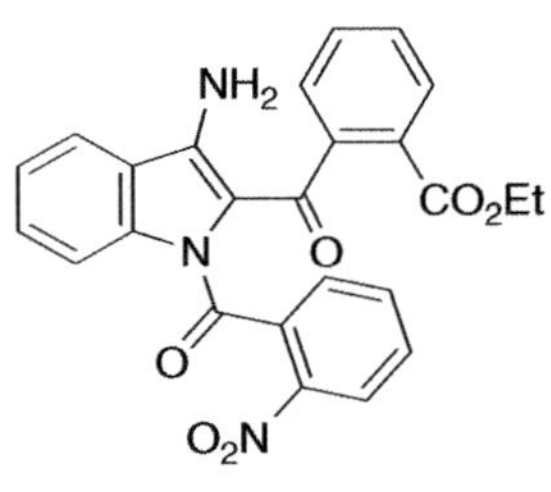	36%	31
7	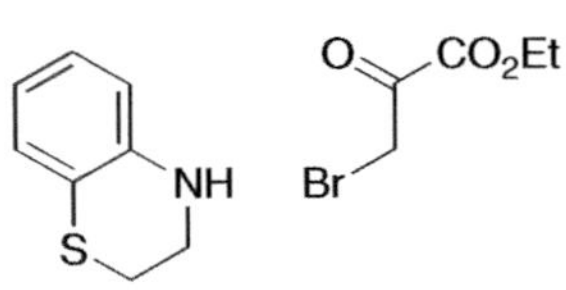	1. THF, rt 2. $MgCl_2$ methylcellosolve 125 °C	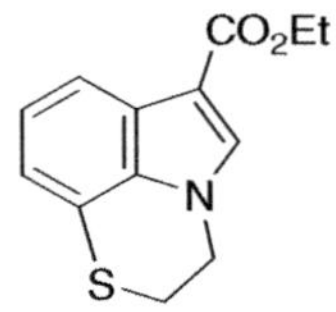	47%	32
8	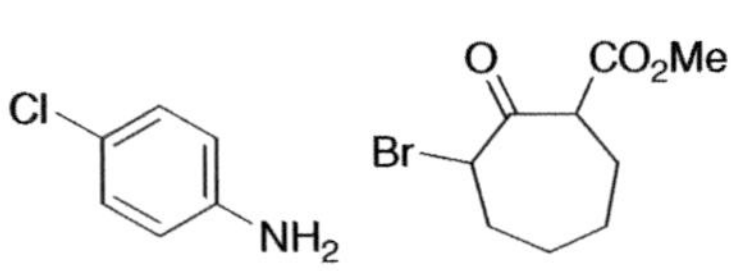	140 °C neat	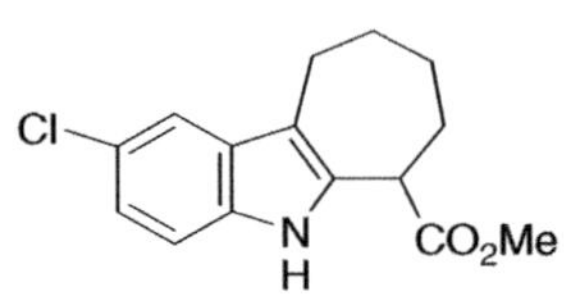	53%	37

Table 1 *(continued)*

Entry	Substrate	Conditions	Indole	% Yield	Ref.
9	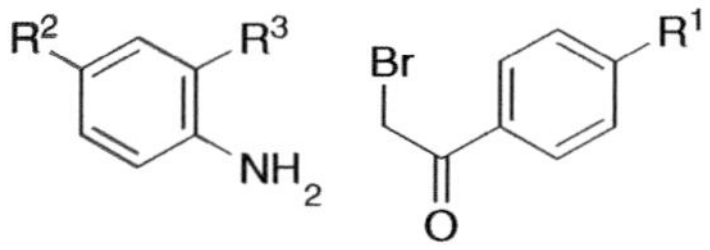 R^1 = H, Me, Cl, OMe, Br R^2 = H, OMe R^3 = H, Me	*N,N*-dimethylaniline 170 °C	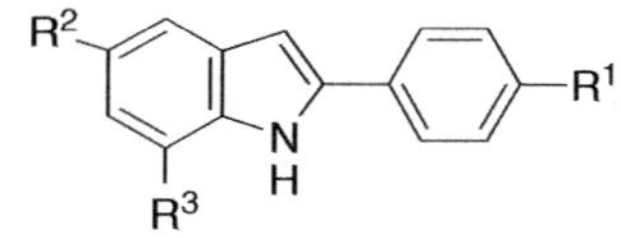	45–62%	38
10	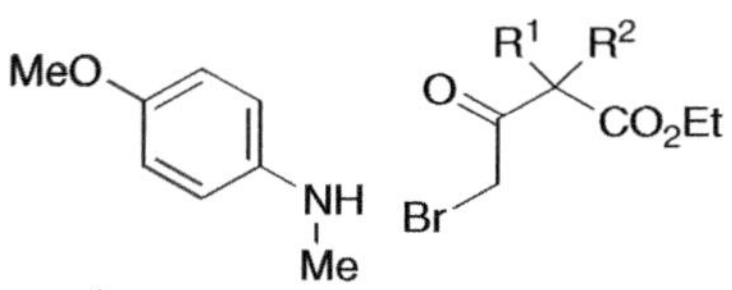 R^1 = Me R^2 = H, Me R^1,R^2 = –$(CH_2)_4$–	1. $ZnCl_2$ 2. NaOH MeOH	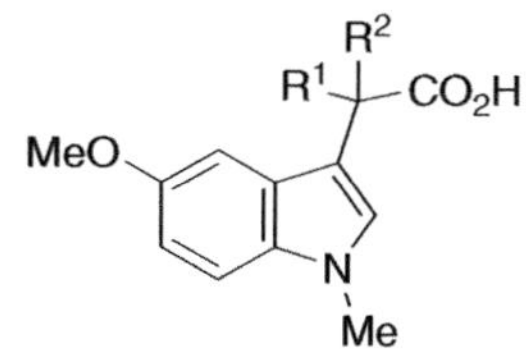	30–54%	39
11	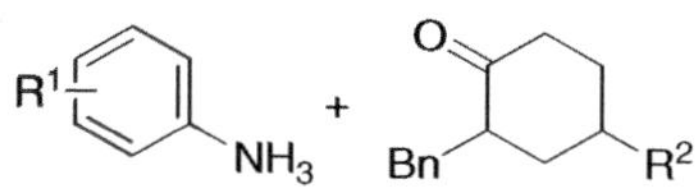 R^1 = H, Me, F, Cl, OMe (different positions) R^2 = H, 4-*t*-Bu, 4-Ph	$EtOCH_2CH_2OH$ μW	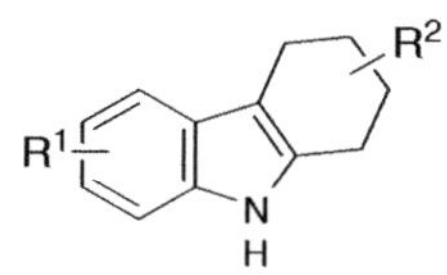	17–84%	40
12	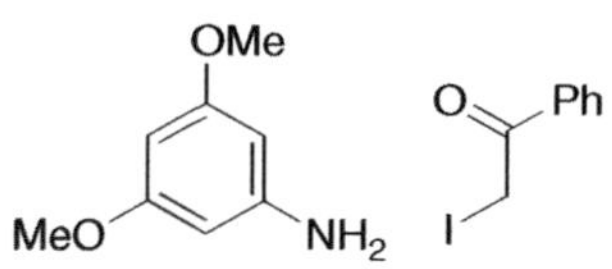	80 °C neat	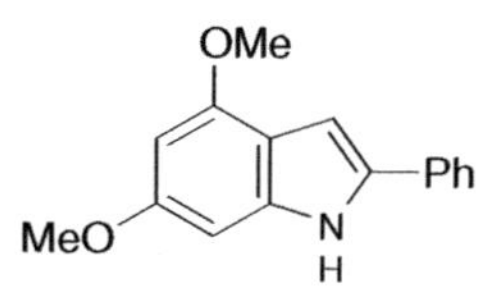	63%	41
13	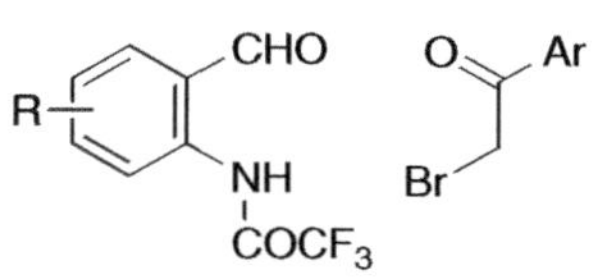 R = H, 5-OBn, 6-Cl, 6-CO_2Me, 5,6-MeO_2 Ar = Ph, 4-MePh, 4-ClPh, 4-FPh, 4-MeOPh, 3-MeOPh, 2-MeOPh, 4-NO_2Ph, 2-BrPh, 2-thienyl, 2-furyl, 2-naphthyl (indole numbering)	K_2CO_3, PEG-400 100 °C	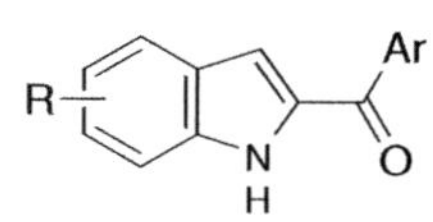	50–85%	42
14	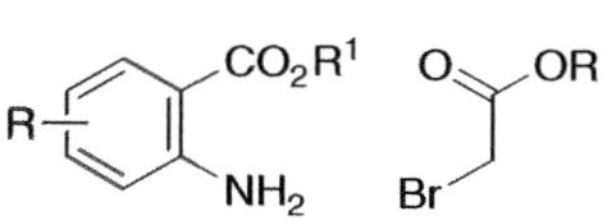 R = H, 7-Br, 6-Cl, 5-F, 5-Cl, 5-Br, 4-F, 4-Br, 4-*i*-Pr, 4-Me (indole numbering)	1. Na_2CO_3 80 °C 2. NaOEt Et_2O, reflux	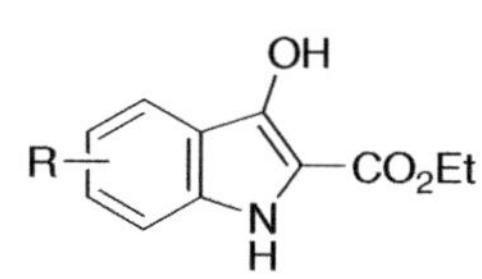	62–90%	43

cases (Entries 1, 2) [28, 29] only the final cyclization is depicted, in a sequence where the two starting materials are the requisite aniline and α-halo ketone. The pair of different cyclized products in Entries 3 and 4 is supported by FMO calculations [30]. The synthesis of 2-acyl-3-aminoindoles by Viti [31] (Entry 6) was exploited by Nettekoven in a combinatorial synthesis [33, 34], by Romagnoli's team in the synthesis of new antitubulin agents [35], and by Narsaiah and coworkers in the synthesis of [1, 3]oxazino[5,4-*b*] indole-6-carbonitriles [36]. The 2-arylindoles in Entry 9 were converted to benz[5,6]azepino[4,3-*b*]indoles [38], and those in Entry 10 were prepared as melatonin analogues to map the melatonin receptor [39]. The work in Entry 11 represents an excellent synthesis of 1,2,3,4-tetrahydrocarbazoles in yields as high as 84% [40]. Zhang and colleagues prepared a series of 2-aroylindoles via the condensation of α-bromoacetophenones with *N*-(2-formylphenyl)trifluoroacetamides (Entry 13) [42]. Bös and coworkers employed anthranilates in a Bischler approach to phenylglycine esters en route to 10-methoxypyrazino[1,2-*a*]indoles (Entry 14) [43]. Dropinski and colleagues used the Bös method to prepare PPARγ partial agonists based on aryl indole-2-carboxylic acids [44].

The novel fluorine-18 labeled tetrahydrocarboline **14**, which was synthesized by Trigg and colleagues as shown in Scheme 4 (equation 1), was further converted to [^{18}F] GE-180 as a new PET tracer for imaging the translocator protein 18 kDa [45]. A vinylogous Bischler indolization was reported by Caron and colleagues that afforded the novel COX-2 inhibitor indole-3-acetic acid **15** (equation 2) [46]. In this method, the sodium hydroxide served to effect the Michael ring closure, the elimination of toluenesulfinate, and the final ester saponification. The two intermediates in this sequence were isolated. Menéndez and colleagues employed a microwave-assisted, solvent-free, one-pot Bischler indole synthesis (equation 3) [47]. As is

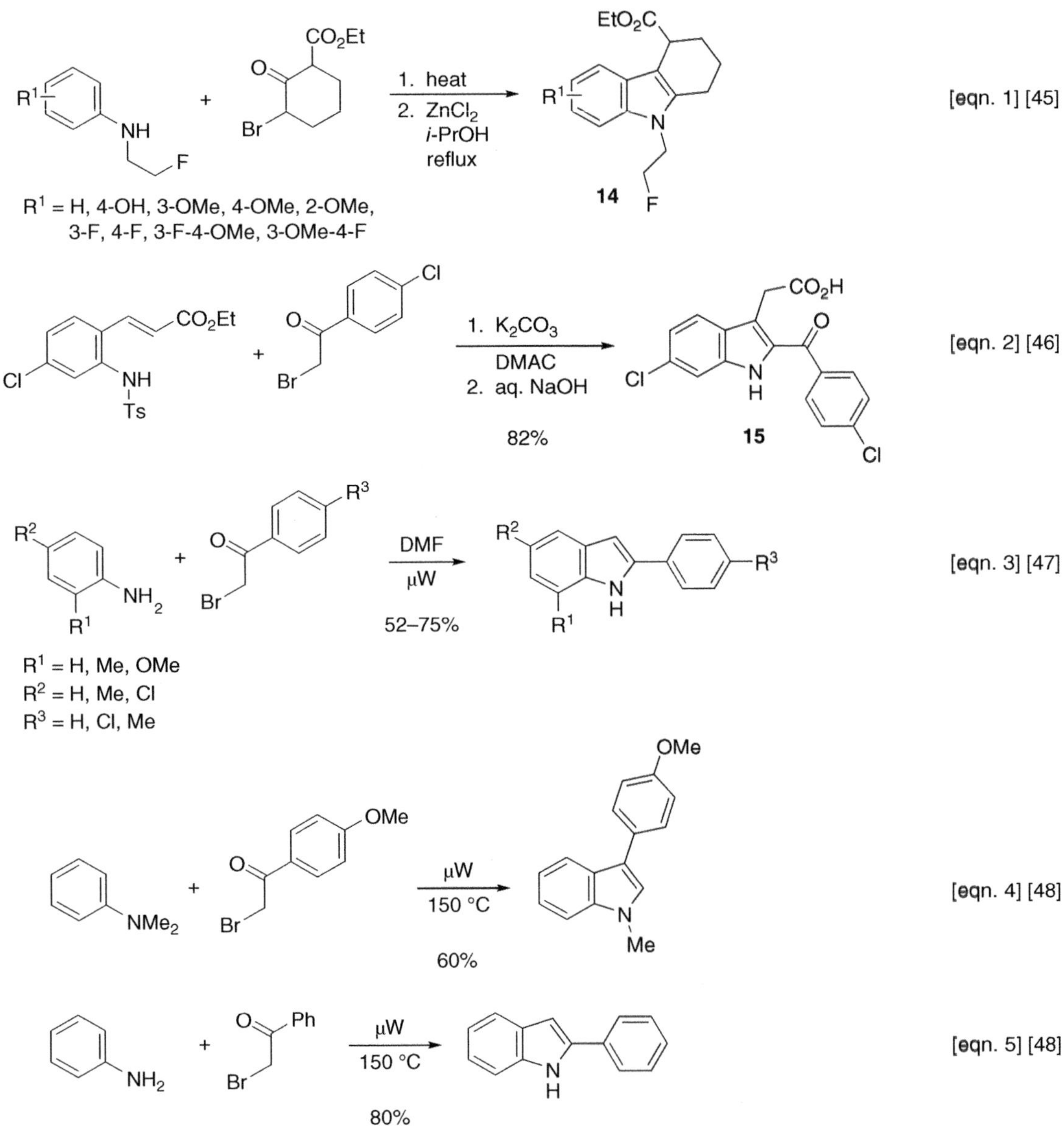

Scheme 4 Applications of the Bischler Indole Synthesis

often customary, the ratio of aniline to phenacyl bromide was 2:1. Cossio and coworkers studied the regiochemistry of the Bischler reaction between aromatic amines and α-bromoketones, both experimentally and computationally [48]. In general, these workers found improved reaction yields under microwave conditions over conventional heating. Only 10 minutes at 150 °C and 20 psi was required to effect the Bischler indolization under microwave irradiation. Moreover, primary anilines (RCH_2NH_2) yielded regiochemistry that was different from that obtained with di- and trisubstituted anilines. Two examples are shown in equations 4 and 5.

Several investigators used the Bischler indole synthesis to craft cryptolepine alkaloids, their analogues, and related fused quinolines. A summary of this chemistry is shown in Table 2. Yamato and colleagues prepared several oxygenated indolo[3,2-*b*]quinolines as potential antitumor compounds (Entry 1) [49–51]. Substituted anilines were also employed in this work [51]. Bierer's group synthesized a wide range of cryptolepine analogues using a Bischler indolization (Entry 2). Some of these compounds derived from the Bischler indoloquinolones possess antihyperglycemic properties [52]. This same synthetic method was used by Wright and colleagues to prepare a series of halogenated cryptolepine analogues [53, 54] and by Moreira and colleagues to synthesize cryptoleptine derivatives having a basic side chain [55]. Ablordeppey and colleagues likewise used this same Bischler indolization to obtain antifungal indoloquinolines [56, 57]. One example is shown in Entry 3.

Some other modifications of the Bischler indole synthesis have been developed. One such modification is that of Moody and coworkers, who employed rhodium to effect N-H insertion of an α-diazo-β-ketoester to an aniline (Scheme 5, equations 1 and 2) [58–60]. The so-formed arylamino β-ketoester **16** was smoothly converted to indole **17** with the ion-exchange resin Amberlyst 15 or in somewhat lower yield with boron trifluoride etherate [59]. These workers extended the method to the synthesis of *N*-unsubstituted indoles using the novel *N*-protecting groups, *N*-(2-ethoxycarbonylethyl)

Table 2 *Applications of the Bischler Indole Synthesis to Cryptolepine and Related Indoles*

Entry	Substrate	Conditions	Indole	% Yield	Ref.
1	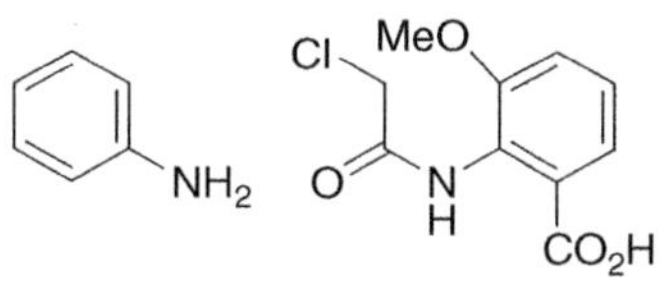	1. DMF 80–90 °C 2. PPA 20 °C	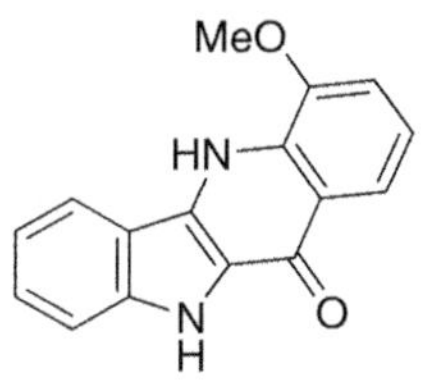	28%	49–51
2	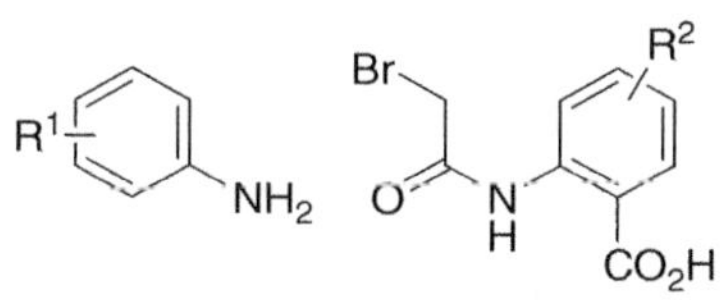 R^1 = H, 2-F, 4-OMe, 2-Cl, 1-Cl; R^2 = H, 7-Br-8-Cl, 6-Cl-7-Br, 6-F, 8-F, 7-Ph, 7-F, 9-F (quindolone numbering)	1. DMF 120 °C 2. PPA 130 °C	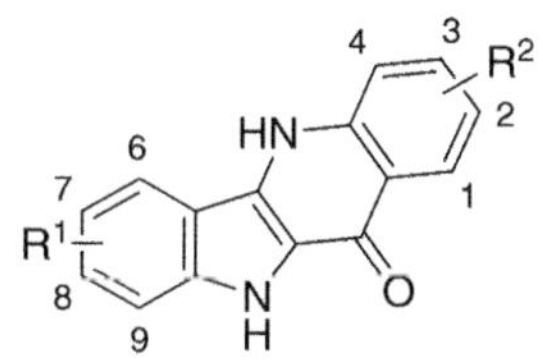	10–50%	52
3	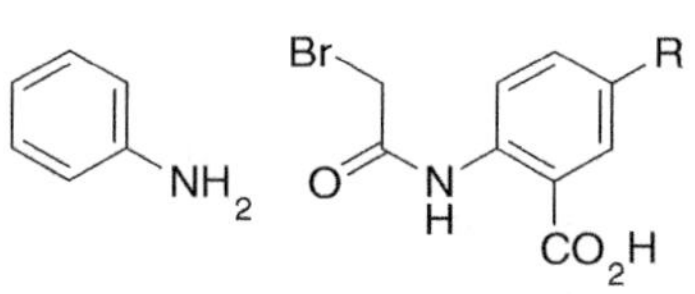 R = H, F, Cl, Br, I, MeSO, $MeSO_2$, PhSO, $PhSO_2$, CN, NO_2	1. DMF 2. PPA 130 °C 3. $POCl_3$ 4. H_2, Pd/C	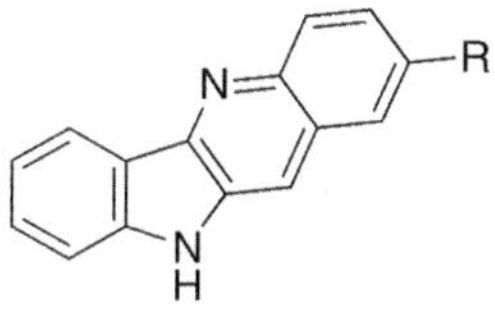	—	56, 57

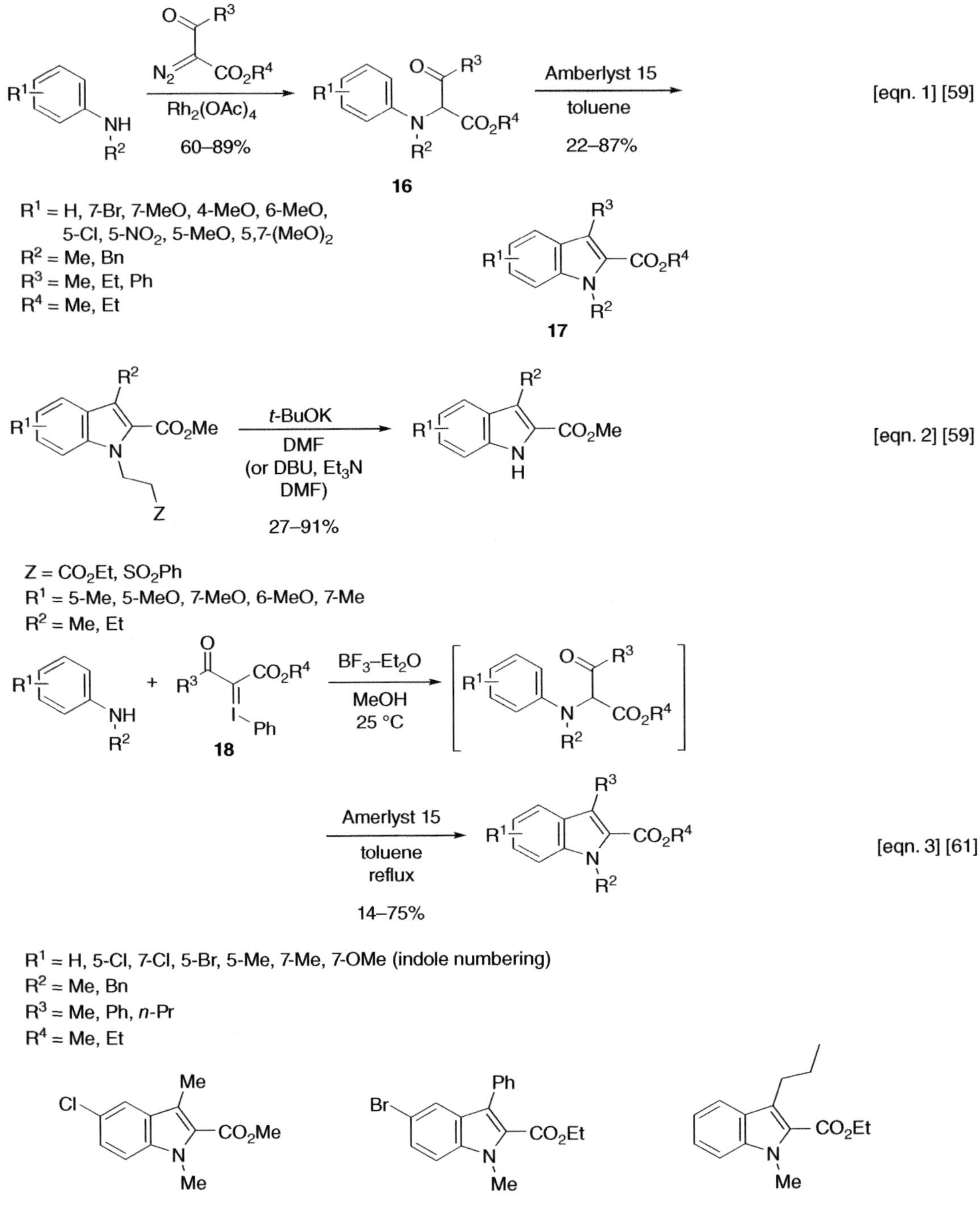

Scheme 5 *Moody and Yu Variations of the Bischler Indole Synthesis*

and *N*-(2-phenylsulfonylethyl), the latter which is cleavable by base (equation 2) [58–60]. Yu's team found that arylamino β-ketoesters **16** can also be prepared using phenyliodonium ylides **18** and BF_3-Et_2O (equation 3) [61]. This synthesis of indoles via **16** can be performed in one pot if desired. Yu's team also found that anilines react with ylides **18** under photochemical conditions to afford indoles in comparable yields to the thermal method. For example, indoles **19–21** were synthesized in this fashion.

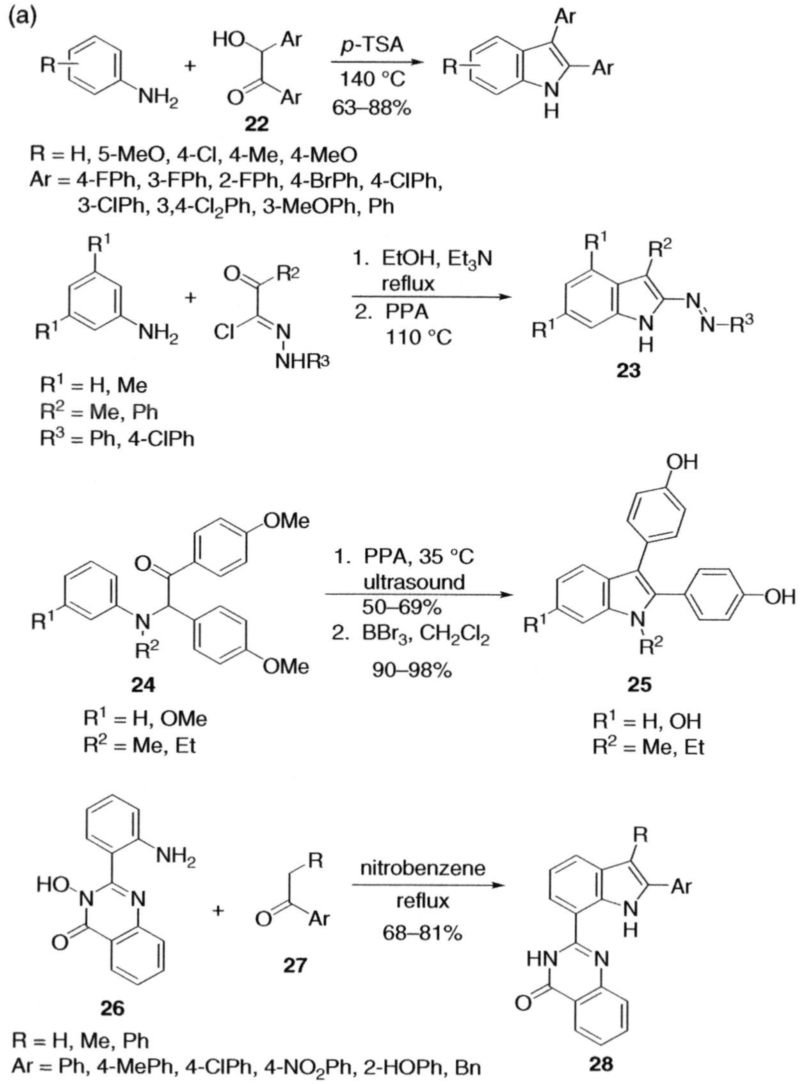

[eqn. 1] [62]

[eqn. 2] [63]

[eqn. 3] [64]

[eqn. 4] [65]

Scheme 6a *Variations of the Bischler Indole Synthesis*

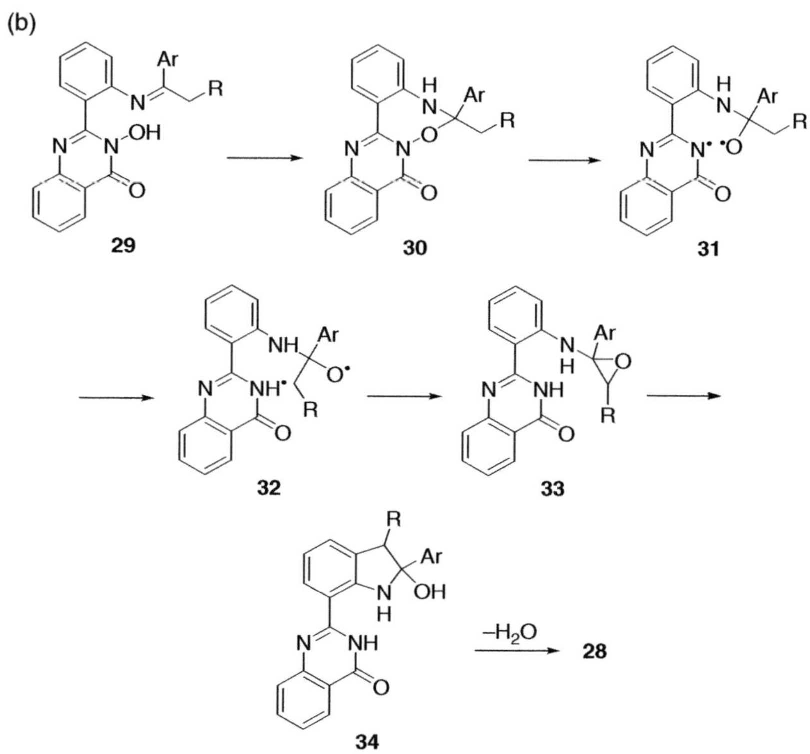

[eqn. 5] [65]

Scheme 6b *Reddy Variation of the Bischler Indole Synthesis*

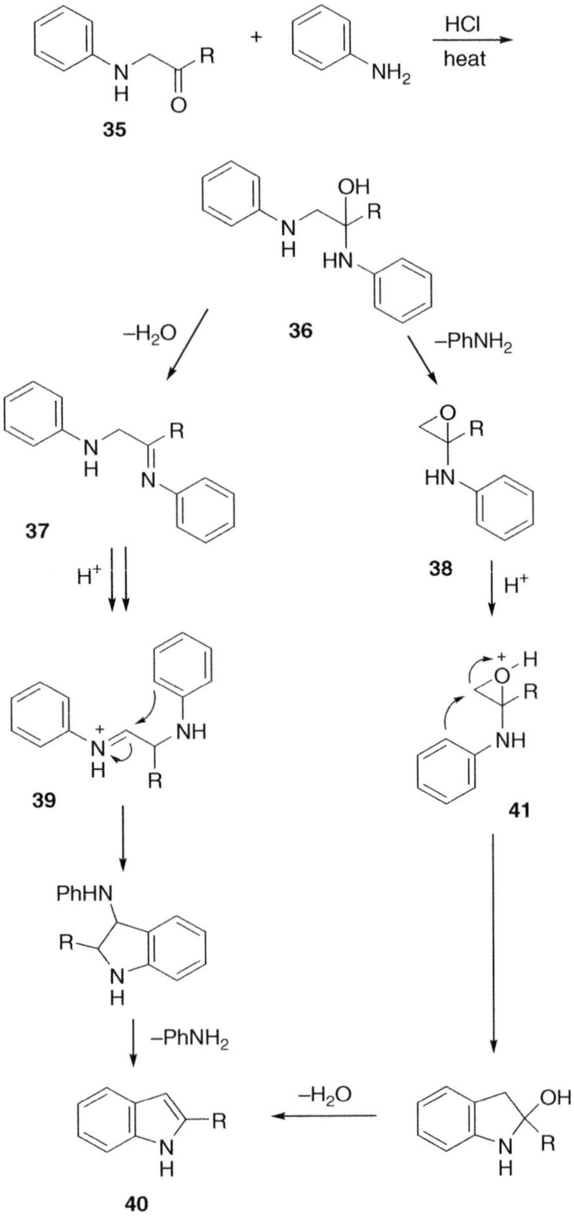

Scheme 7 *Proposed Mechanism for the Bischler Indole Synthesis*

An important variation of the Bischler indolization is to employ benzoins in place of α-halocarbonyls. For example, Yao and colleagues developed a solvent-free Bischler synthesis of 2,3-diarylindoles (Scheme 6, equation 1) [62]. The benzoins **22** were prepared from the appropriate aryl aldehydes by coupling the latter with an *N*-heterocyclic carbene. The conversion to indoles is effected in one pot. Sannicolo and Benincori synthesized several 2-(arylhydrazo)indoles **23** via a Bischler protocol (equation 2) [63], and Haroutounian and Koulocheri employed ultrasound to facilitate the preparation of 2,3-bis(4-hydroxyphenyl)indoles **25** as new fluorescent probes for detecting estrogen receptors in cells (equation 3) [64]. The starting anilino ketones **24** were synthesized from the appropriate anilines and 2-bromo-1,2-bis (4-methoxyphenyl)ethanone. Reddy and colleagues discovered a novel preparation of indolylquinazolinones **28** from the reaction of hydroxamic acid **26** and acetophenones **27** (equation 4) [65]. This novel Bischler-like reaction is proposed to involve cyclization of amino–ketone adduct **29** to quinazolino[3,2-*d*][1, 3, 4]benzoxadiazepin-9-one **30**, which is transformed via **31** and **32** to epoxide **33**. Bischler-type cyclization leads to **34** and then to indole **28** (equation 5).

Based on the available data, the mechanism of the Bischler indole synthesis is generally believed to be that shown in Scheme 7, although, as we have seen, it is strongly governed by substituents on the reacting partners and the reaction conditions. The initially formed amino ketone **35** undergoes addition by a second molecule of aniline to give carbinol amine **36**. This can either lose water to give imine **37** or lose aniline to form epoxide **38**. Acid-catalyzed isomerization of **37** to **39** followed by cyclization leads to the 2-substituted indole **40**. Alternatively, cyclization of protonated epoxide **41** leads to indole **40**. If the less-likely direct cyclization of the iminium ion of **37** transpires, then the 3-substituted indole obtains. For an excellent discussion of the early mechanistic studies and conclusions see Brown [66] and Sumpter and Miller [67].

References

[1] R. Möhlau, *Ber.*, 1881, **14**, 171–175.
[2] R. Möhlau, *Ber.*, 1882, **15**, 2480–2490.
[3] A. Bischler and H. Brion, *Chem. Ber.*, 1892, **25**, 2860–2879.
[4] A. Bischler and P. Fireman, *Chem. Ber.*, 1893, **26**,1336–1349.
[5] M. Nencki and J. Berlinerblau, *Chem. Ber.*, 1887, **20**, 753.
[6] F.R. Japp and T.S. Murray, *J. Chem. Soc.*, 1894, **65**, 889–899.
[7] A.F. Crowther, F.G. Mann, and D. Purdie, *J. Chem. Soc.*, 1943, 58–68.
[8] F. Brown and F.G. Mann, *J. Chem. Soc.*, 1948, 847–858.

[9] R.K. Brown (1972) *Indoles*, Part I (ed. W.J. Houlihan), Wiley, New York, pp. 317–385.
[10] R.M. Cowper and T.S. Stevens, *J. Chem. Soc.*, 1947, 1041–1045.
[11] A. Blackhall and R.H. Thomson, *J. Chem. Soc.*, 1954, 3916–3919.
[12] N.P. Buu-Hoï, G. Saint-Ruf, D. Deschamps, *et al.*, *J. Chem. Soc. (C)*, 1971, 2606–2609.
[13] P. Bigot, G. Saint-Ruf, and N.P. Buu-Hoï, *J. Chem. Soc., Perkin Trans. 1*, 1972, 2573–2576.
[14] J.-P. Coïc and G. Saint-Ruf, *J. Heterocycl. Chem.*, 1978, **15**, 1367–1371.
[15] D.St.C. Black, B.M.K.C. Gatehouse, F. Théobald, and L.C.H. Wong, *Aust. J. Chem.*, 1980, **33**, 343–350.
[16] D.St.C. Black, N. Kumar, and L.C.H. Wong, *Aust. J. Chem.*, 1986, **39**, 15–20.
[17] D.St.C. Black, M.C. Bowyer, P.K. Bowyer, *et al.*, *Aust. J. Chem.*, 1994, **47**, 1741–1750.
[18] D.St.C. Black, M.C. Bowyer, G.C. Condie, *et al.*, *Tetrahedron*, 1994, **50**, 10983–10994.
[19] D.St.C. Black, N. Kumar, and D.B. McConnell, *Tetrahedron*, 2001, **57**, 2203–2211.
[20] K. Pchalek, A.W. Jones, M.M.T. Wekking, and D.St.C. Black, *Tetrahedron*, 2005, **61**, 77–82.
[21] K.A. Clayton, D.St.C. Black, and J.B. Harper, *Tetrahedron*, 2007, **63**, 10615–10621.
[22] S. Rajput, C. Leu, K. Wood, *et al.*, *Tetrahedron Lett.*, 2011, **52**, 7095–7098.
[23] K. Görlitzer and J. Weber, *Arch. Pharm.*, 1981, **314**, 852–861.
[24] K. Görlitzer, R. Stockmann, and R.D. Walter, *Pharmazie*, 1994, **49**, 231–235.
[25] E. von Angerer, J. Prekajac, and J. Strohmeier, *J. Med. Chem.*, 1984, **27**, 1439–1447.
[26] R. Gastpar, M. Goldbrunner, D. Marko, and E. von Angerer, *J. Med. Chem.*, 1998, **41**, 4965–4972.
[27] C.P. Miller, M.D. Collini, B.D. Tran, *et al.*, *J. Med. Chem.*, 2001, **44**, 1654–1657.
[28] R.E. Walkup and J. Linder, *Tetrahedron Lett.*, 1985, **26**, 2155–2158.
[29] L.D. Basanagoudar, C.S. Mahajanshetti, and S.B. Dambal, *Ind. J. Chem.*, 1991, **30B**, 1018–1022.
[30] G.K.B. Prasad, A. Burchat, G. Weeratunga, *et al.*, *Tetrahedron Lett.*, 1991, **32**, 5035–5038.
[31] G. Viti, D. Giannotti, R. Nannicini, *et al.*, *J. Heterocycl. Chem.*, 1991, **28**, 379–384.
[32] I. van Wijingaarden, D. Hamminga, R. van Hes, *et al.*, *J. Med. Chem.*, 1993, **36**, 3693–3699.
[33] M. Nettekoven, *Tetrahedron Lett.*, 2000, **41**, 8251–8254.
[34] M. Nettekoven, *Bioorg. Med. Chem. Lett.*, 2001, **11**, 2169–2171.
[35] R. Romagnoli, P.G. Baraldi, M.K. Jung, *et al.*, *Bioorg. Med. Chem. Lett.*, 2005, **15**, 4048–4052.
[36] D. Maitraie, G.V. Reddy, V.V.V.N.S.R. Rao, *et al.*, *Tetrahedron*, 2005, **61**, 3999–4008.
[37] A.D. Napper, J. Hixon, T. McDonagh, *et al.*, *J. Med. Chem.*, 2005, **48**, 8045–8054.
[38] M. Nyerges, Á. Pintér, A. Virányi, *et al.*, *Tetrahedron Lett.*, 2005, **46**, 377–380.
[39] A. Tsontinis, M. Vlachou, D.P. Papahajis, *et al.*, *J. Med. Chem.*, 2006, **49**, 3509–3519.
[40] J. Chen and Y. Hu, *Synth. Commun.*, 2006, **36**, 1485–1494.
[41] T.J. Donohoe, M.A. Kabeshov, A.H. Rathi, and I.E.D. Smith, *Synlett*, 2010, 2956–2958.
[42] Y. Zhao, D. Li, L. Zhao, and J. Zhang, *Synthesis*, 2011, 873–880.
[43] M. Bös, F. Jenck, J.R. Martin, *et al.*, *Eur. J. Med. Chem.*, 1997, **32**, 253–261.
[44] J.F. Dropinski, T. Akiyama, M. Einstein, *et al.*, *Bioorg. Med. Chem. Lett.*, 2005, **15**, 5035–5038.
[45] H. Wadsworth, P.A. Jones, W.-F. Chau, *et al.*, *Bioorg. Med. Chem. Lett.*, 2012, **22**, 1308–1313.
[46] S. Caron, E. Vazquez, R.W. Stevens, *et al.*, *J. Org. Chem.*, 2003, **68**, 4104–4107.
[47] V. Sridharan, S. Perumal, C. Avendaño, and J.C. Menéndez, *Synlett*, 2006, 91–95.
[48] Y. Vara, E. Aldaba, A. Arrieta, *et al.*, *Org. Biomol. Chem.*, 2008, **6**, 1763–1772.
[49] M. Yamato, Y. Takeuchi, M. Chang, and K. Hashigaki, *Chem. Pharm. Bull.*, 1992, **40**, 528–530.
[50] M. Yamato, Y. Takeuchi, M. Chang, *et al.*, *Chem. Pharm. Bull.*, 1990, **38**, 2048–3052.
[51] M. Chang, Y. Takeuchi, K. Hashigaki, and M. Yamato, *Heterocycles*, 1992, **33**, 147–152.
[52] D.E. Bierer, L.G. Dubenko, P. Zhang, *et al.*, *J. Med. Chem.*, 1998, **41**, 2754–2764.
[53] C.W. Wright, J. Addae-Kyereme, A.G. Breen, *et al.*, *J. Med. Chem.*, 2001, **44**, 3187–3194.
[54] S.R. Gouni, S. Carrington, and C.W. Wright, *J. Heterocycl. Chem.*, 2006, **43**, 171–175.
[55] J. Lavrado, G.G. Cabal, M. Prudencio, *et al.*, *J. Med. Chem.*, 2011, **54**, 734–750.
[56] S.Y. Ablordeppey, P. Fan, S. Li, *et al.*, *Bioorg. Med. Chem.*, 2002, **10**, 1337–1346.
[57] L.G. Mardenborough, X.Y. Zhu, P. Fan, *et al.*, *Bioorg. Med. Chem.*, 2005, **13**, 3955–3963.
[58] C.J. Moody and E. Swann, *Synlett*, 1998, 135–136.
[59] K.E. Bashford, A.L. Cooper, P.D. Kane, *et al.*, *J. Chem. Soc., Perkin Trans. 1*, 2002, 1672–1687.
[60] K.E. Bashford, A.L. Cooper, P.D. Kane, and C.J. Moody, *Tetrahedron Lett.*, 2002, **43**, 135–137.
[61] X. Wang, B. Han, Y. Wang, and W. Yu, *Org. Biomol. Chem.*, 2010, **8**, 3865–3867.
[62] C. Yao, D. Wang, J. Lu, *et al.*, *Tetrahedron Lett.*, 2011, **52**, 6162–6165.
[63] T. Benicori and F. Sannicolò, *J. Org. Chem.*, 1988, **53**, 1309–1312.
[64] S.D. Koulocheri and S.A. Haroutounian, *Eur. J. Org. Chem.*, **2001**, 1723–1729.
[65] D.S. Reddy, P.P. Reddy, and P.S.N. Reddy, *Synthesis*, 2000, 1217–1218.
[66] R.K. Brown (1972) *Indoles, The Chemistry of Heterocyclic Compounds*, vol. **25**, Part 1, Chapter II (ed. W.J. Houlihan), Wiley-Interscience, New York, pp. 317–385.
[67] W.C. Sumpter and F.M. Miller (1954) Heterocyclic compounds with indole and carbazole, in *The Chemistry of Heterocyclic Compounds* (ed. A. Weissberger), Interscience, New York, pp. 12–15.

24

The Nordlander Indole Synthesis

A closely related synthesis to the Bischler indolization is that developed by Nordlander [1]. Quite simply, it involves the cyclization of *N*-(trifluoroacetyl)-2-anilino acetals **1** to the corresponding *N*-(trifluoroacetyl)indoles **2** under the action of trifluoroacetic acid anhydride (TFAA) and trifluoroacetic acid (TFA) (Scheme 1). Base cleavage (KOH, MeOH, rt) of the trifluoroacetyl group furnishes the indole in essentially quantitative yield. The starting acetals are easily prepared in two steps from anilines, and the method is particularly useful for the synthesis of 2,3-unsubstituted indoles, unlike the Bischler indole synthesis *per se*. The importance of this method is that previous similar cyclizations were only applicable to *N*-substituted indoles. For example, Uff and colleagues found that *N*-methyl(ethyl)indole can be prepared from the anilinodiethylacetals and a BF_3-TFAA reagent (equation 2) [2]. This particular indole synthesis can be traced back to the studies of Räth [3], Keimatsu and Izumi [4], and König and Buchheim [5]. Nevertheless, the Nordlander procedure is clearly superior. To illustrate this, the cyclization of *N*-methyl-3-methoxyanilinoacetaldehyde diethyl acetal using dilute HCl in refluxing dioxane afforded a mixture of 4- and 6-methoxyindoles in a combined 43% yield [6]. Similar work by this group to cyclize the *N*-tosyl analogues shows little or no improvement; thus, 4,5,6-trimethoxyindole was obtained from 3,4,5-trimethoxyaniline in very low overall yield [7]. The Sugasawa indole synthesis (Chapter 22) proved to be the method of choice for accessing 4,5,6-trimethoxyindole [7]. Sundberg and Laurino found that titanium tetrachloride effected the cyclization of *N*-(methylsulfonyl)-2-anilino acetals (equation 3) [8]. Dmitrienko and coworkers crafted pyrrolo[2,3-*f*]indole (**1**) using the Nordlander method (equation 4) [9].

Shizuri and his team synthesized the topoisomerase II inhibitor BE 10988 (**2**) in a sequence that featured a zinc chloride cyclization of an *N*-methylanilinoacetal (Scheme 2, equation 1) [10]. As might be expected, this cyclization was only modest. Zhang and Sui synthesized estrieno[2,3-*b*]-(**3**) and estrieno[3,4-*c*]pyrroles using typical Bischler cyclization conditions (equation 2) [11]. The two regioisomers are obtained in a 2:1 ratio (70%) with the major isomer shown. A direct comparison of the original Nordlander procedure to the Sundberg modification found the latter to be superior (51% vs. 25%) (equation 3) [12].

Several additional examples of the Nordlander indole synthesis are tabulated in Table 1 [13–19]. It is obvious that some of these entries could have easily been included in the Bischler chapter because both reactions involve similar if not identical precursor synthesis and acid-catalyzed cyclizations. For example, the substrate in Entry 5 was prepared from the corresponding ethyl bromopyruvate. The example in Entry 3 is one of several similar Nordlander indolizations that yield heavily substituted nitroindoles. Likewise, the chemistry depicted in Entry 6 was applied to the preparation of several 3-trifluoromethylindoles.

Black and colleagues employed Nordlander–Bischler cyclizations to prepare several 3-substituted-4,6-dimethoxyindoles (Scheme 3, equation 1) [20, 21]. In similar fashion 1,3,5-tris(indolyl)benzene **4** was synthesized [21].

Indole Ring Synthesis: From Natural Products to Drug Discovery, First Edition. Gordon W. Gribble.

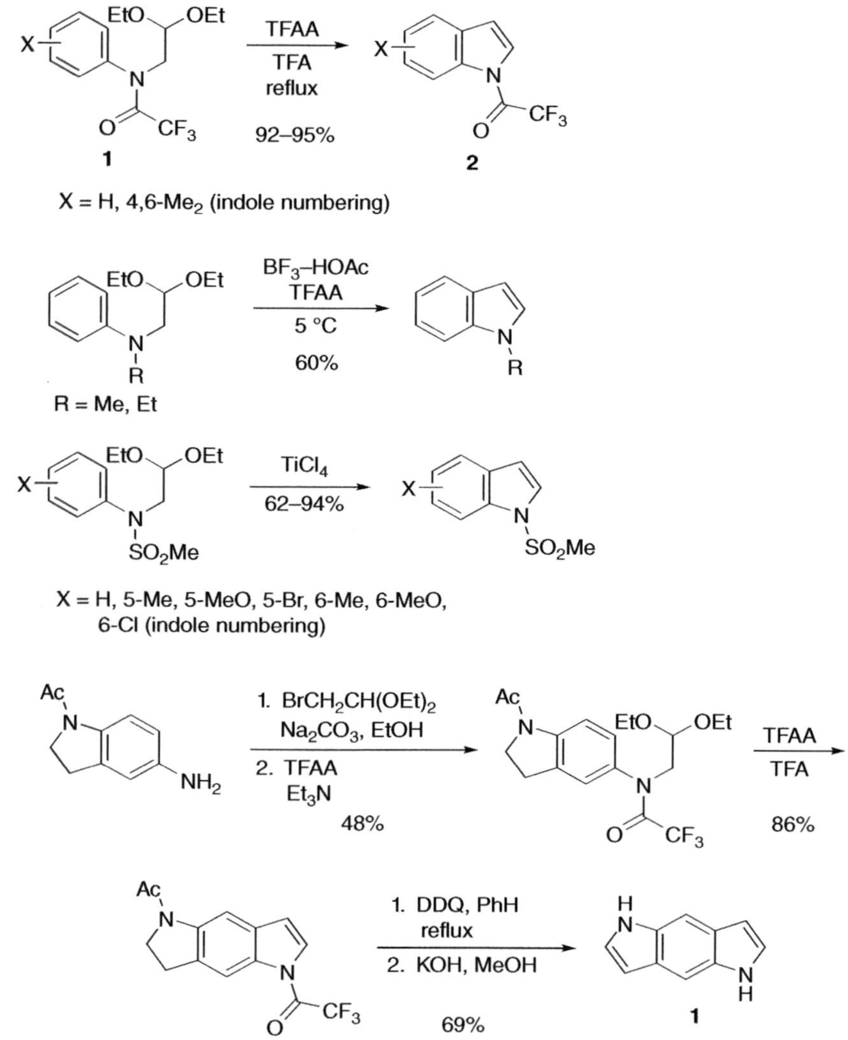

[eqn. 1] [1]

[eqn. 2] [2]

[eqn. 3] [8]

[eqn. 4] [9]

Scheme 1 *Nordlander Indole Synthesis*

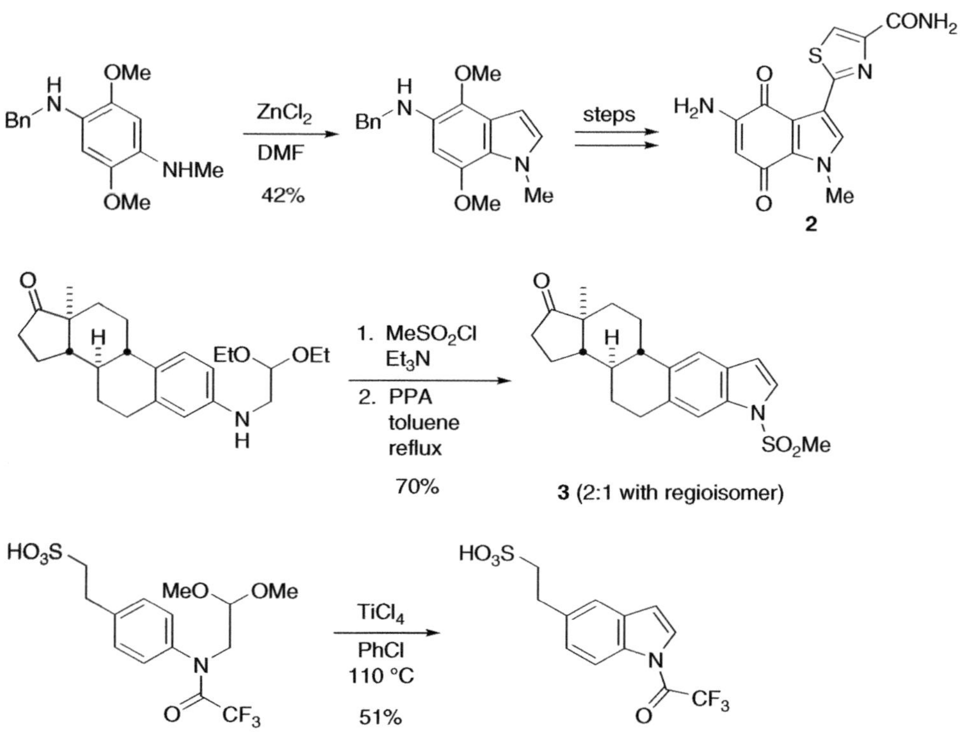

[eqn. 1] [10]

[eqn. 2] [11]

[eqn. 3] [12]

Scheme 2 *Applications of the Nordlander Indole Synthesis*

Table 1 *Applications of the Nordlander Indole Synthesis*

Entry	Substrate	Conditions	Indole	% Yield	Ref.
1	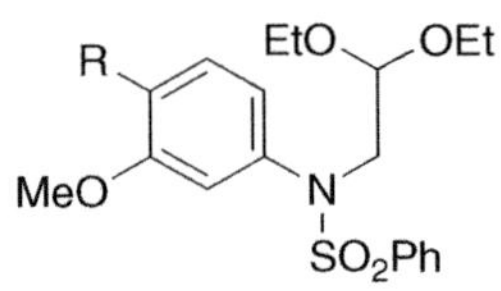 R = H, OMe	$BF_3•Et_2O$ CH_2Cl_2 rt	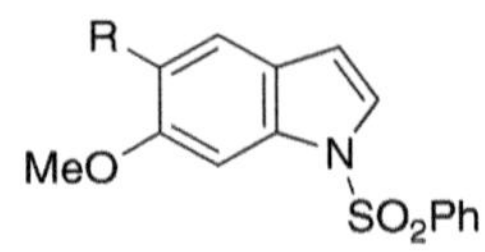	80–97%	13, 14
2	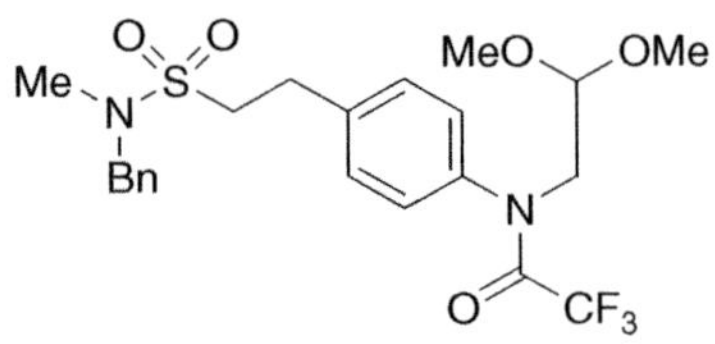	1. $TiCl_4$ PhCl 2. KOH MeOH	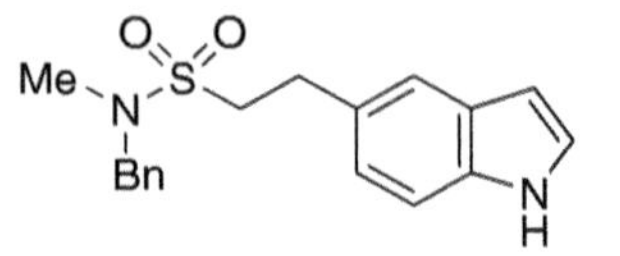	—	15
3	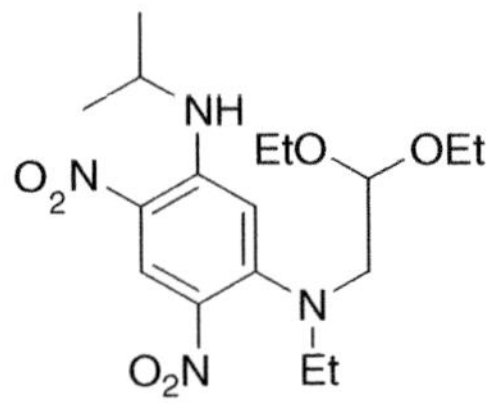	1 M HCl THF rt	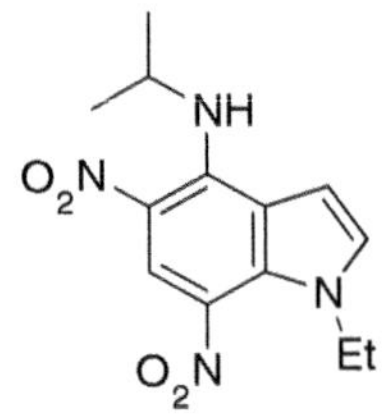	92%	16
4	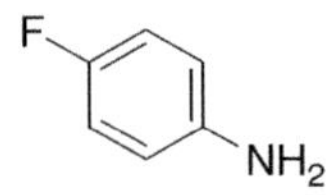	1. NaOEt, EtOH $ClCH_2CH(OEt)_2$ 2. PPA, 100 °C PhCl	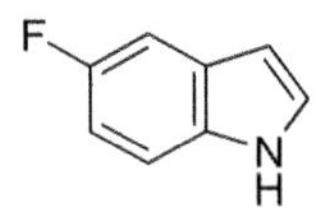	—	17
5	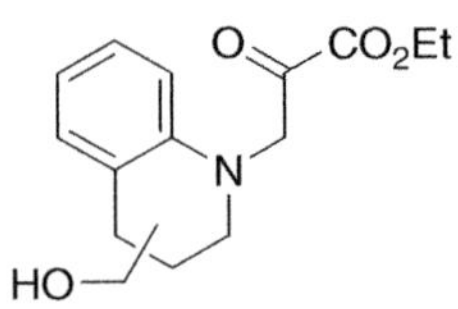	$MgCl_2$ $MeOCH_2CH_2OH$ 125 °C	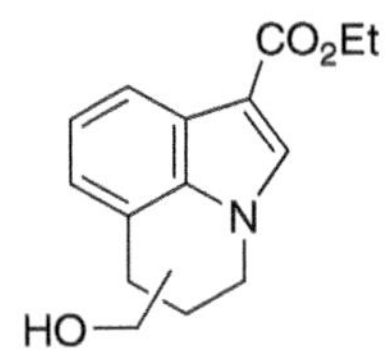	—	18
6	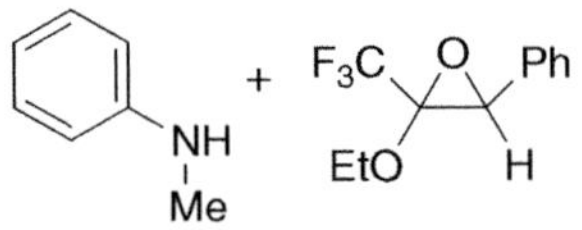	1. $(CF_3)_2CHOH$ rt 2. $SOCl_2$	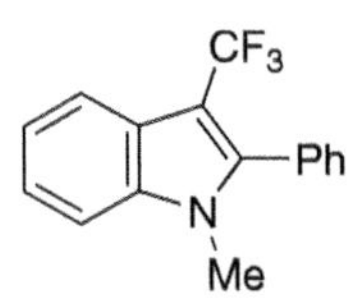	73%	19

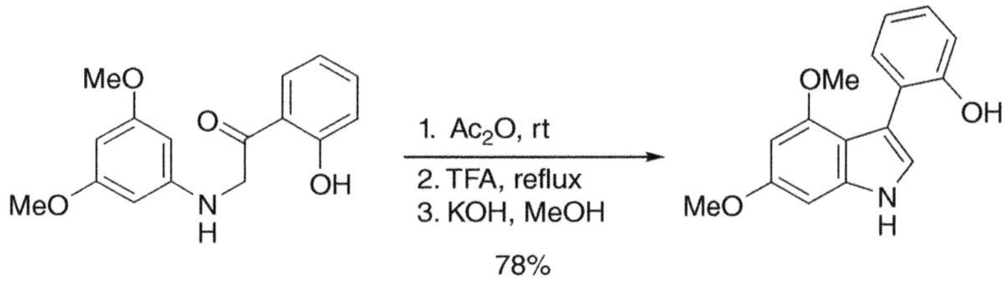

[eqn. 1] [21]

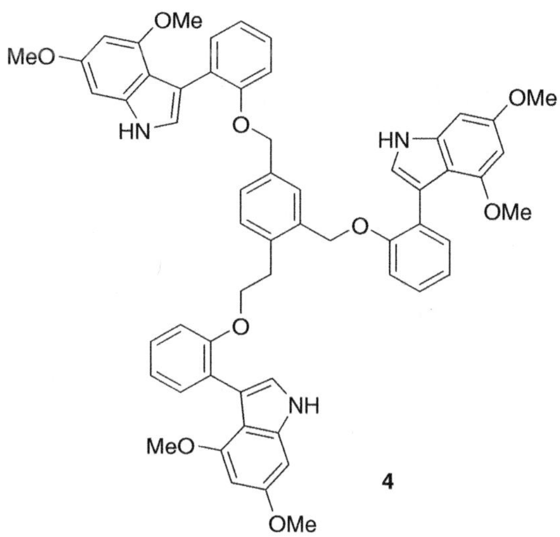

4

Scheme 3 *Black Application of the Nordlander Indole Synthesis*

References

[1] J.E. Nordlander, D.B. Catalane, K.D. Kotian, *et al.*, *J. Org. Chem.*, 1981, **46**, 778–782.
[2] M.J. Bevis, E.J. Forbes, N.N. Naik, and B.C. Uff, *Tetrahedron*, 1971, **27**, 1253–1259.
[3] C. Räth, *Chem. Ber.*, 1924, **57B**, 715–718.
[4] K. Keimatsu and M. Inoue, *J. Pharm. Soc. Japan*, 1925, **518**, 351–354.
[5] W. König and R. Buchheim, *Chem. Ber.*, 1925, **58B**, 2868–2870.
[6] A.H. Jackson, P.R. Jenkins, and P.V.R. Shannon, *J. Chem. Soc., Perkin Trans. 1*, 1977, 1698–1704.
[7] M.J.E. Hewlins, A.H. Jackson, A.-M. Oliveira-Campos, and P.V.R. Shannon, *J. Chem. Soc., Perkin Trans. 1*, 1981, 2906–2911.
[8] R.J. Sundberg and J.P. Laurino, *J. Org. Chem.*, 1984, **49**, 249–254.
[9] G.K.B. Prasad, A. Burchat, G. Weeratunga, *et al.*, *Tetrahedron Lett.*, 1991, **32**, 5035–5038.
[10] H. Suda, M. Ohkubo, K. Matsunaga, *et al.*, *Tetrahedron Lett.*, 1993, **34**, 3797–3798.
[11] X. Zhang and Z. Sui, *Tetrahedron Lett.*, 2003, **44**, 3071–3073.
[12] B. Pete, G. Simig, L. Poszávácz, and L. Töke, *Heterocycles*, 2003, **60**, 2441–2455.
[13] G. Dupeyre, G.G. Chabot, S. Thoret, *et al.*, *Bioorg. Med. Chem.*, 2006, **14**, 4410–4426.
[14] N. Ty, G. Dupeyre, G.G. Chabot, *et al.*, *Bioorg. Med. Chem.*, 2008, **16**, 7494–7503.
[15] L. Poszávácz, G. Simig, J. Fetter, and F. Bertha, *Heterocycles*, 2006, **68**, 713–719.
[16] K. Liu and D. Yin, *Org. Lett.*, 2009, **11**, 637–639.
[17] B.N. Balasubramanian, D.R. St. Laurent, M.G. Saulnier, *et al.*, *J. Med. Chem.*, 2004, **47**, 1609–1612.
[18] G. Zhu, S.E. Conner, X. Zhou, *et al.*, *Bioorg. Med. Chem. Lett.*, 2004, **14**, 3057–3061.
[19] I. Rodrigues, D. Bonnet-Delpon, and J.-P. Bégué, *J. Org. Chem.*, 2003, **66**, 2098–2103.
[20] D.St.C. Black, M.A. Bowyer, P.K. Bowyer, *et al.*, *Aust. J. Chem.*, 1994, **47**, 1741–1750.
[21] D.St.C. Black, N. Kumar, and D.B. McConnell, *Tetrahedron*, 2001, **57**, 2203–2211.

25

Nitrene Cyclization

This chapter on nitrene cyclization is a segue to the following several chapters that employ this tactic in powerful and widely used indole ring syntheses. The use of metals, such as palladium, rhodium, and ruthenium, to generate nitrenes or their equivalent and effect indole ring construction is discussed in later chapters. Söderberg has reviewed the synthesis of heterocycles via the generation and cyclization of nitrene intermediates [1]. Included in this excellent review is coverage of the relevant chapters to follow.

Although their work is of marginal preparative value, Klemm, Horaguchi, and colleagues effected the synthesis of carbazole and related fused indoles by subjecting primary arylamines to calcium oxide at high temperatures

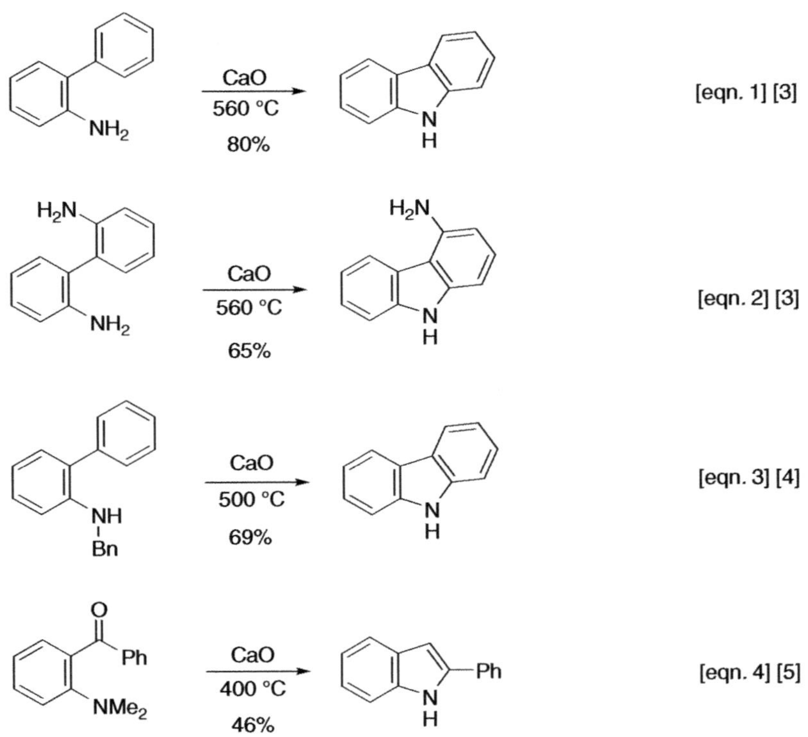

Scheme 1 *Klemm-Horaguchi Indole Synthesis via Nitrene Cyclization*

Indole Ring Synthesis: From Natural Products to Drug Discovery, First Edition. Gordon W. Gribble.

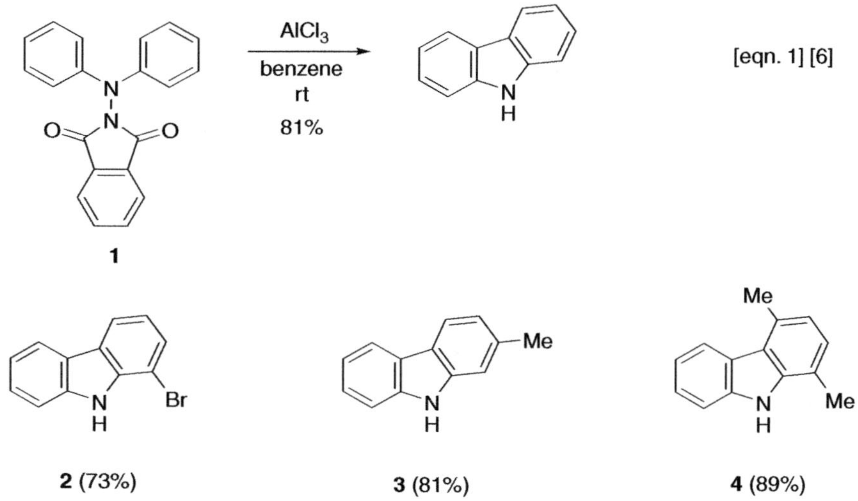

Scheme 2 *Kikugawa Indole Synthesis*

(400–700 °C) [2–5]. A selection of this vapor phase dehydrocyclization is shown in Scheme 1. The mechanism of these reactions is unknown, although it may involve nitrogen radicals.

In contrast to the previous study, Kikugawa and colleagues described an excellent synthesis of carbazoles via the generation of diarylnitrenium ions and subsequent cyclization (Scheme 2) [6]. Thus, reaction of *N*-(*N*, *N*-diphenylanino)phthalimide (1) with aluminum chloride generates diphenylnitrenium ion that cyclizes to carbazole in high yield (equation 1). Other carbazoles prepared in this fashion are **2–4**.

References

[1] B.C.G. Söderberg, *Curr. Org. Chem.*, 2000, **4**, 727–764.
[2] T. Horaguchi, L.H. Klemm, and E.S. Norris, *J. Heterocycl. Chem.*, 1995, **32**, 797–802.
[3] T. Horaguchi and T. Oyanagi, *J. Heterocycl. Chem.*, 2004, **41**, 1–6.
[4] E.C. Creencia and T. Horaguchi, *J. Heterocycl. Chem.*, 2006, **43**, 1441–1446.
[5] E.C. Creencia, A. Takahashi, and T. Horaguchi, *Heterocycles*, 2009, **78**, 1549–1556.
[6] Y. Kikugawa, Y. Aoki, and T. Sakamoto, *J. Org. Chem.*, 2001, **66**, 8612–8615.

26

Cadogan–Sundberg Indole Synthesis

The first widely used indole synthesis involving at least the formal manipulation of aryl nitro compounds with trialkyl phosphites was discovered by Cadogan [1–7] and extended shortly thereafter by Sundberg [8, 9]. Sundberg also explored the reaction of *ortho*-alkylnitrobenzenes with trialkyl phosphites [10]. His indole synthesis involving the thermolysis of *ortho*-azidostyrenes is the subject of Chapter 27. A selection of the Cadogan–Sundberg reactions is presented in Scheme 1. As seen in equation 3, 2-nitrosobiphenyl was converted to carbazole under very mild conditions with triethyl phosphite [3, 5], and *trans*-2-nitrostilbene afforded 2-phenylindole in 58% yield; the *cis* isomer gave this product in 85% yield (equation 4) [4]. Sundberg studied the conversion of β-alkyl-*o*-nitrostyrenes and β-acyl-*o*-nitrostyrenes with triethyl phosphite to form 2-alkylindoles and 2-acylindoles, respectively, the latter in much lower yield [8]. Whereas other trivalent phosphorus reagents also deoxygenate nitrostyrenes to indoles, triethyl phosphite is superior to triphenylphosphine and tri-*n*-butylphosphine. Sundberg also found that these *o*-nitrostyrene reactions could lead to rearrangements (equation 5), ring expansions (equation 6), and *N*-alkylation by the trialkyl phosphite [9]. Nevertheless, this route to indoles has found great utility in synthesis, as we will see in this chapter.

It is generally accepted that the mechanism of this trialkyl phosphite deoxygenation of nitroarenes involves initial transformation to the nitrosoarene and trialkyl phosphate followed by a second deoxygenation to the corresponding nitrene. Support for this pathway is seen by the fact that matching aryl azides afforded similar product ratios under thermolysis conditions [10–12]. Indeed, prior to the work of Cadogan and Sundberg, several workers described indole syntheses from aryl azide pyrolysis, as described in Chapter 27. An early example of the deoxygenation of 2-nitrobiphenyl to carbazole involved ferrous oxalate dihydrate (63%) [13]. Puskas and Fields synthesized several polymethylcarbazoles using triethyl phosphite and nitropolymethylbiphenyls (27%–71%) [14]. Kametani and colleagues employed the Cadogan conditions to prepare a series of β-carbolines [15, 16]. Other applications of the trialkyl phosphite deoxygenation of nitroarenes to indoles are tabulated in Table 1 [17–23]. In Entries 1 and 2, Suschitzky and coworkers describe a new synthesis of the benzo[*b*]thieno[3,2-*b*]indole ring system [17]. Kametani discovered a novel rearrangement as shown in Entry 3, which afforded the [1,3]oxazino[3,4-*a*]indol-1-one ring system [18]. The corresponding ketone in Entry 4 failed to indolize under these conditions [19], and Entries 6–8 illustrate functional group compatibility under these conditions [21–23].

Access to carbazoles has benefitted greatly from the Cadogan–Sundberg indole synthesis. Making use of *t*-butyl blocking groups, Tashiro and Yamato reported an efficient synthesis of 4-methylcarbazole from 2-methyl-4,4′-di-*t*-butylbiphenyl that employed a subsequent Cadogan–Sundberg protocol [24]. The ethylation of carbazoles under typical triethyl phosphite conditions was studied by Kuroki and Tsunashima [25]. An important application of this indole ring synthesis is the construction of functionalized carbazoles as monomers for the synthesis of conjugated polymers. Leclerc and colleagues have been very active in this regard, particularly in synthesizing 2,7-disubstituted carbazoles [26–30]. Thus, these 2,7-disubstituted carbazoles were prepared from the requisite *ortho*-nitrobiphenyl in refluxing triethyl phosphite: 2,7-dichloro- (60%) [26],

Indole Ring Synthesis: From Natural Products to Drug Discovery, First Edition. Gordon W. Gribble.

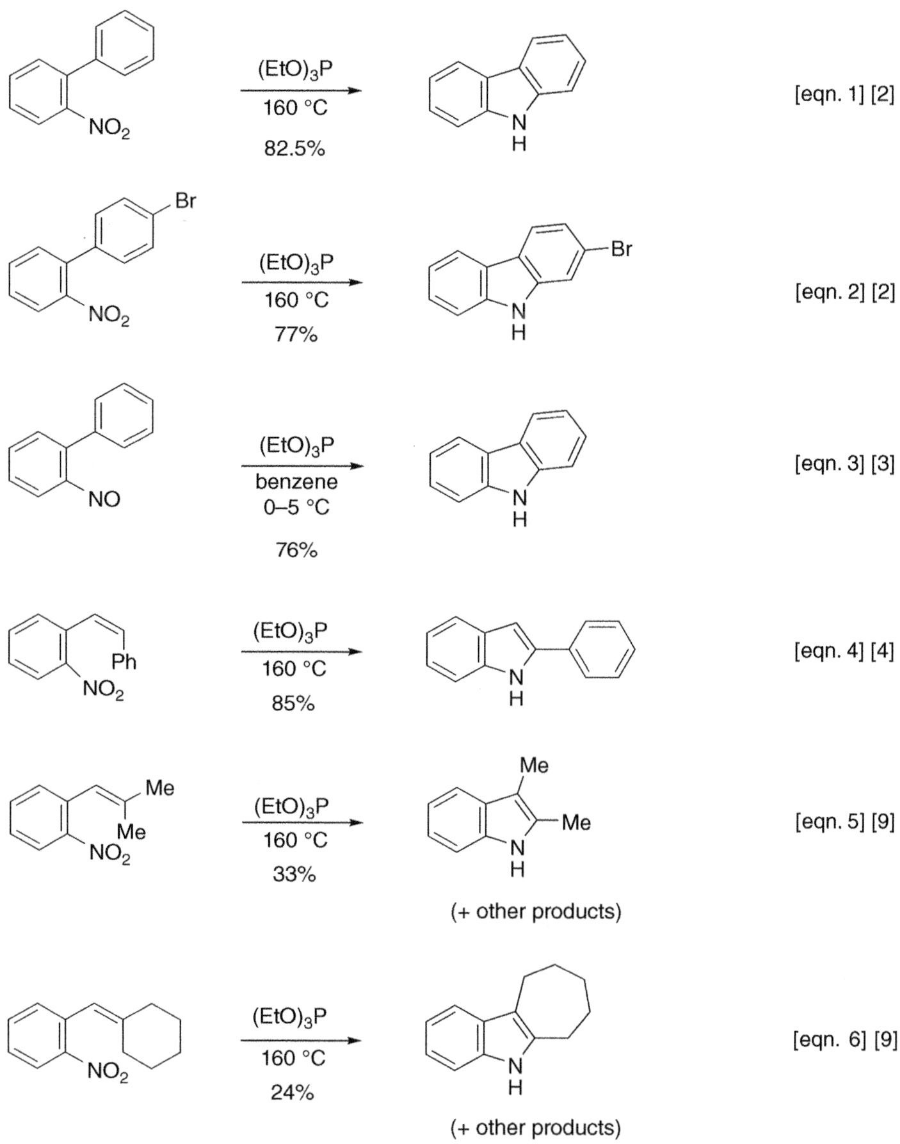

Scheme 1 *Cadogan-Sundberg Indole Synthesis*

2,7-dimethoxy- (60%) [27], and 2,7-bis(methyltrityloxy) carbazole (60%) [28]. In addition, Leclerc and coworkers prepared the carbazoles **1–4** shown in Scheme 2 [29–32]. In each case an arrow indicates the site of ring formation in the triethyl phosphite deoxygenation. Leclerc and Wakim described their research in an account, "Ladder Oligo(*p*-aniline)s" [30]. The research groups of Mullen [33], Bryce [34], and Aaron [35] also synthesized monomeric carbazoles [33, 34] and thieno[3,2-*b*]indoles [35] en route to novel conducting polymers. Scheme 2 includes examples of these Cadogan–Sundberg products from each group (**5–7**).

The range of indoles synthesized via the Cadogan–Sundberg method is enormous, as summarized in Table 2 [36–44]. Several groups employed *ortho*-nitrostyrenes to access 2-substituted indoles. Entry 5 features the use of triphenylphosphine, the so-called Freeman variation, which we discuss later in this chapter.

The Cadogan–Sundberg indole synthesis has been particularly valuable in preparing biindoles, as shown in Scheme 3 [45–47]. In addition to the 2,3′-biindole **8** shown in equation 3, Kuethe and colleagues synthesized a number of 2,2′-biindoles [47, 48], including the marine indolo[2,3-*a*] carbazoles tjipanazoles B, D, E, and I [47]. Furthermore, as shown in Scheme 4 [49–59], this method of synthesis was employed to prepare several indole-fused heterocycles such as **9–19** from the corresponding nitro compounds in refluxing triethyl phosphite. In each case the site of C–N

Table 1 *Early Applications of the Cadogan-Sundberg Indole Synthesis*

Entry	Substrate	Conditions	Indole	% Yield	Ref.
1	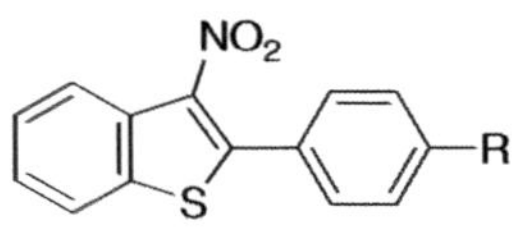 R = H, Me, Cl, Br	$(EtO)_3P$ 160 °C	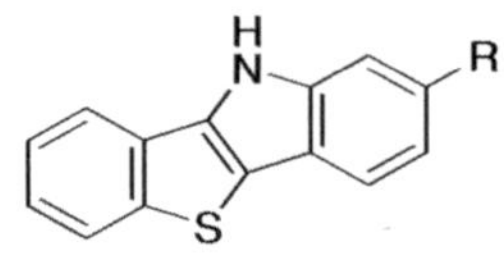	27–57%	17
2	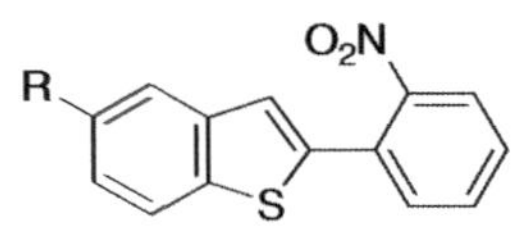 R = H, Me	$(EtO)_3P$ 160 °C	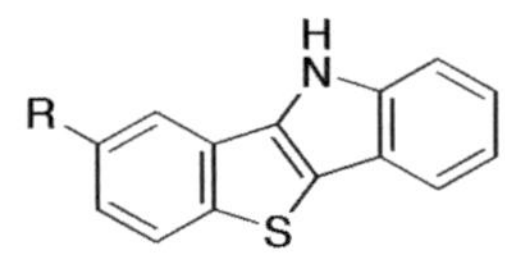	53–60%	17
3	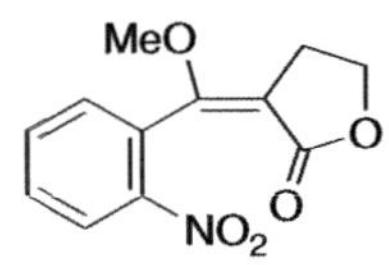	$(EtO)_3P$ 160–170 °C 17 h	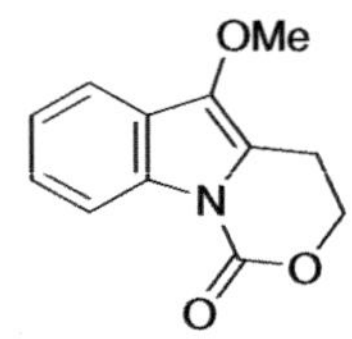	45%	18
4	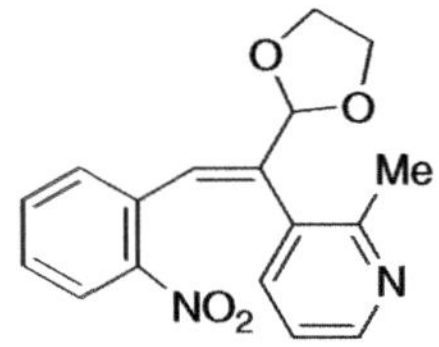	$(EtO)_3P$ 160 °C 5 h	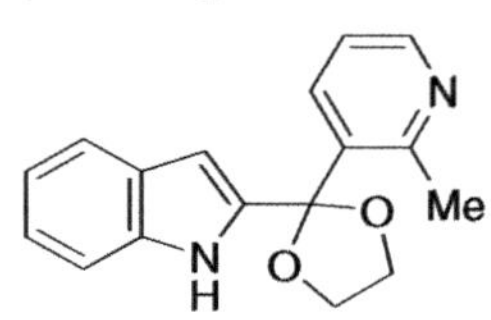	52%	19
5	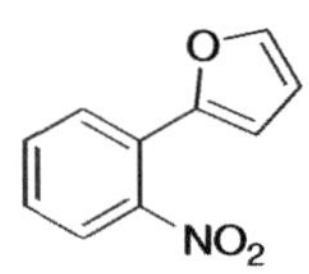	$(EtO)_3P$ 160 °C 5 h	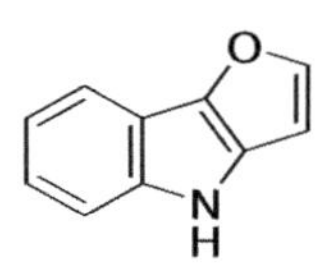	34%	20
6	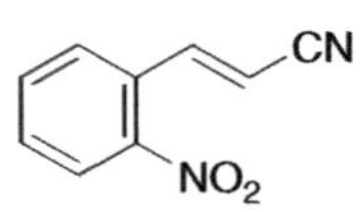	$(EtO)_3P$ mesitylene 160 °C 4 h	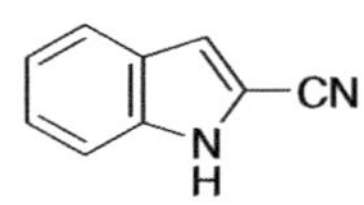	69%	21
7	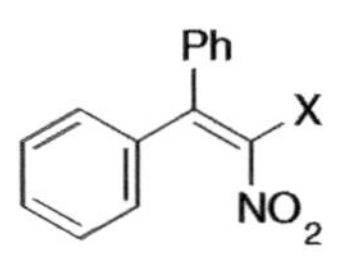 X = H, PhS, *t*-BuS, NO_2	$(EtO)_3P$ 150 °C	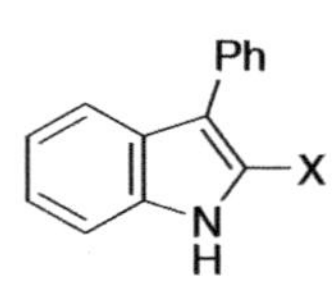	33–99%	22
8	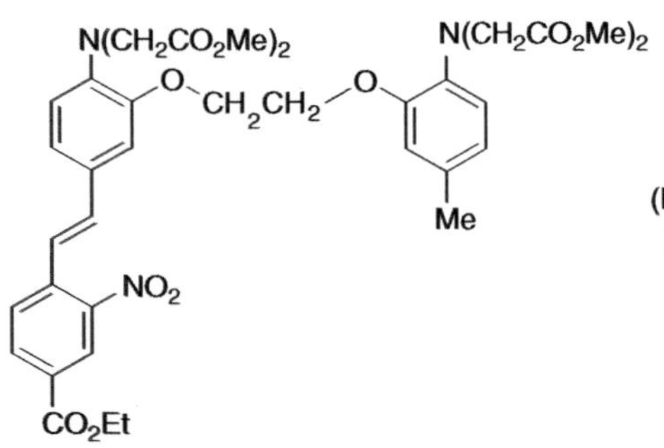	$(EtO)_3P$ reflux	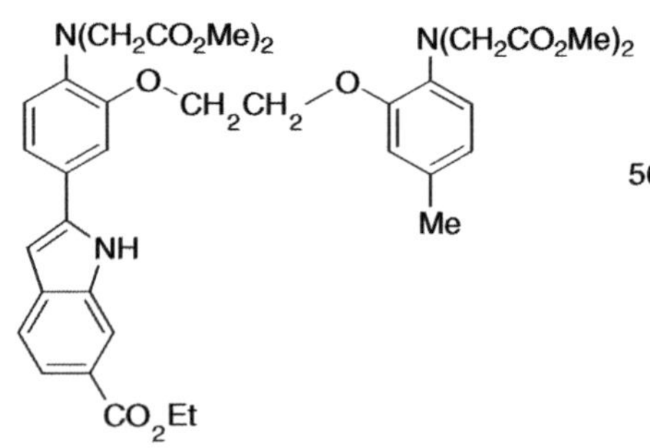	56%	23

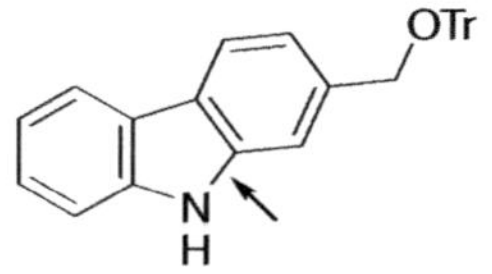

1 (65%) [28]

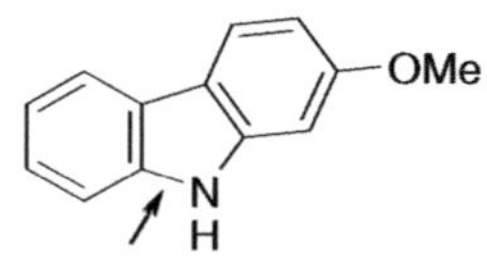

2 (51%) [31]

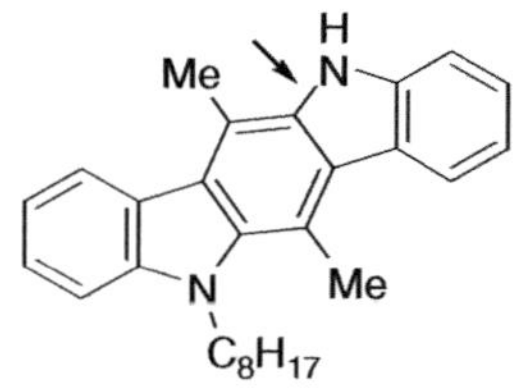

3 (40%) [32]

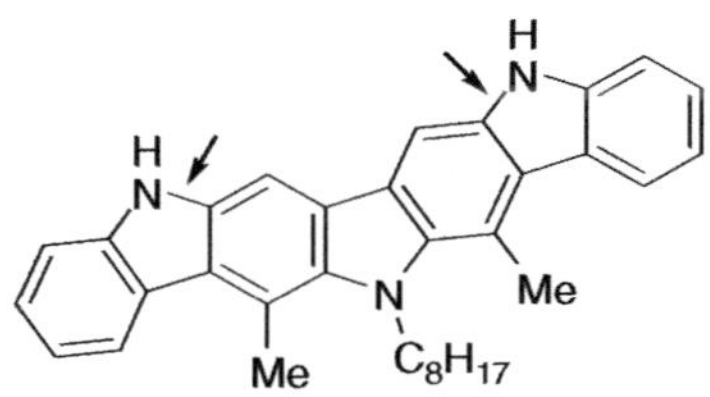

4 (30%) [29]

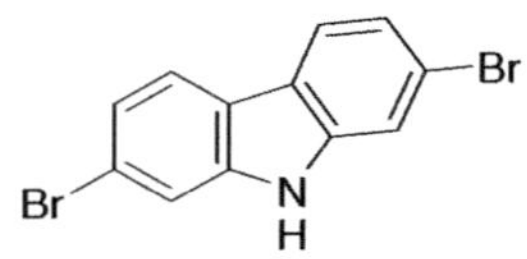

5 (56%) [33]

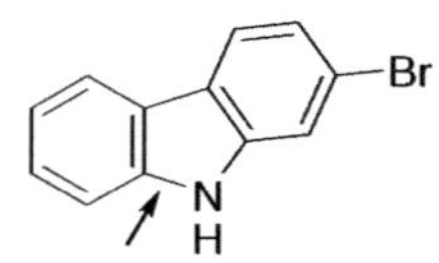

6 (57%) [34]

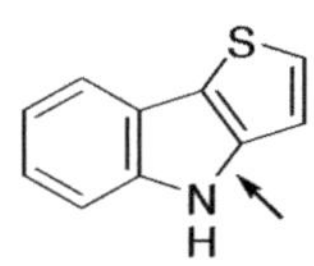

7 (47%) [35]

Scheme 2 *Applications of the Cadogan-Sundberg Indole Synthesis*

bond formation is indicated. Carbazole **12** was subsequently converted to the alkaloid calothrixin B [52], and carbazole **13** was transformed into ellipticine [53]. Indoloquinolines **14** are immediate precursors to the corresponding cryptolepines [54]. The thiazolo[5,4-*b*]indole **15** was prepared using xylene as a cosolvent [55], unlike the majority of Cadogan–Sundberg cyclizations. Furo[3,2-*a*] carbazole **17** is the alkaloid furostifoline *(Murraya euchrestifolia)* [57]. The yield of **17** using the ferrous oxalate method was only 26%. The corresponding sulfur analogue to **18** was obtained in 50% yield [58]. Indole **19** is the immediate precursor to arcyriaflavin-A via a Fischer indole synthesis [59].

An interesting variation of the standard Cadogan–Sundberg procedure involves the use of dialkyl phosphites and alkyl phenylphosphonites to give 1-hydroxyindole-3-phosphonic esters and 1-hydroxyindole-3-phosphinic esters, respectively (Scheme 5, equations 1 and 2) [60]. An earlier report on the synthesis of indole-2-carboxylates using the Cadogan–Sundberg procedure [61] was corrected to show that the products in this reaction are actually indole-3-carboxylates (equation 3) [62]. Thus, the rearranged products result from the more favorable migration of ethoxycarbonyl rather than the methyl group.

Several modifications of the original Cadogan–Sundberg triethyl phosphite reaction conditions have been reported. The aforementioned Freeman variation involves using triphenylphosphine (refluxing *ortho*-dichlorobenzene) in place of triethyl phosphite [63]. Examples are shown in Scheme 6 with a comparison to the classic Cadogan–Sundberg conditions. Advantages of this variation are the avoidance of the *N*-ethylation byproduct, the relative ease of product purification, and generally higher yields than with triethyl phosphite at reflux (156 °C). Výprachtický and coworkers used the Freeman variation to prepare 2,7-diiodocarbazole in 65% yield versus 45% using triethyl phosphite [64]. Moreover, the latter conditions also led to some *N*-ethylation (4%). The Freeman method was also used by Carter and colleagues to synthesize a series of chlorinated carbazoles (65%–87%) [65], and by Liu and colleagues in their preparation and study of indolo[3,2-*b*] benzo[*b*]thiophenes having novel photophysical properties [66]. In combination with microwave heating, Creencia and colleagues used triphenylphosphine to construct some indoles and carbazoles in yields higher than using triethyl phosphite and microwave heating [67].

May and colleagues used triphenyl phosphite, which was found to be superior to both trimethyl and triethyl phosphite, to prepare carbazole precursors to the secretory phospholipase A_2 inhibitors LSN433771 and LSN426891 (Scheme 7, equation 1) [68]. Huleatt and Chai and their coworkers employed the Freeman variation and microwave

Table 2 *Selective Indole Ring Syntheses via the Cadogan-Sundberg Method*

Entry	Substrate	Conditions	Indole product	% Yield	Ref.
1	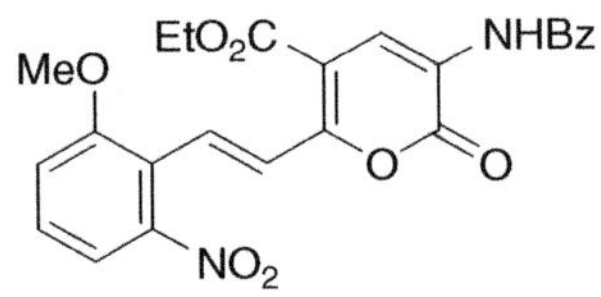	$(EtO)_3P$ 160 °C 4 h	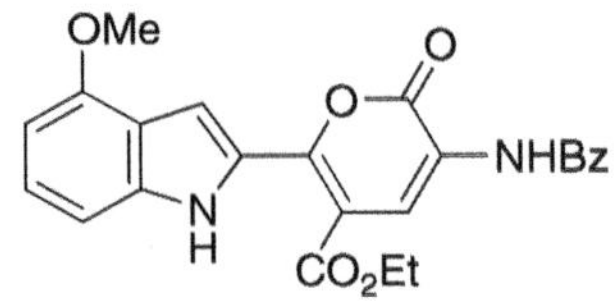	84%	36
2	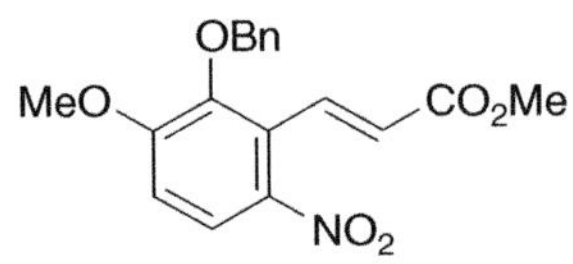	$(EtO)_3P$ 156 °C 3 h	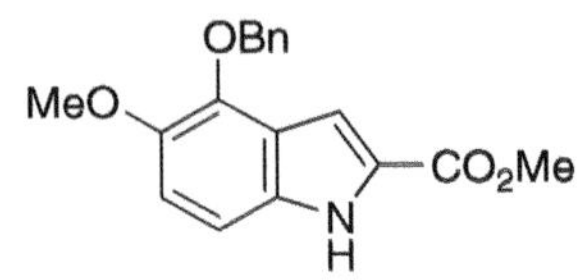	54%	37
3	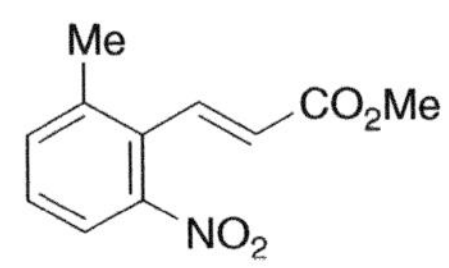	$(EtO)_3P$ 156 °C 20 h	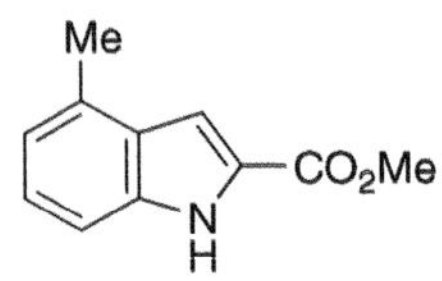	89%	38
4	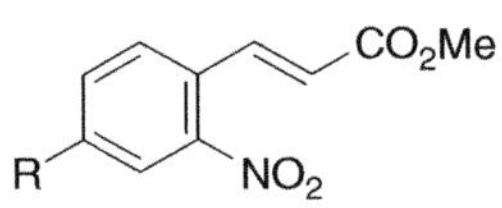 R = H, OMe, *i*-Pr	$(EtO)_3P$ 160 °C	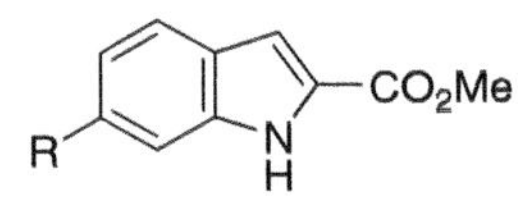	67–81%	39
5	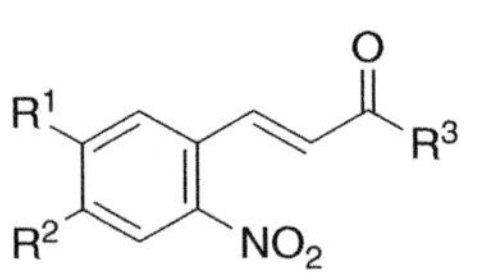 R^1 = H, OMe; R^2 = H, OMe; R^1,R^2 = $-OCH_2O-$; R^3 = Me, Ph	Ph_3P, Ph_2O reflux	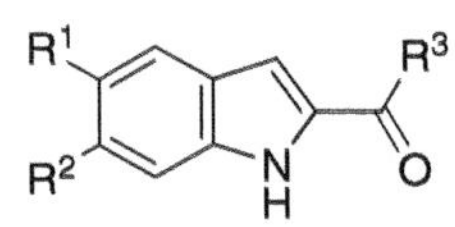	42–57%	40
6	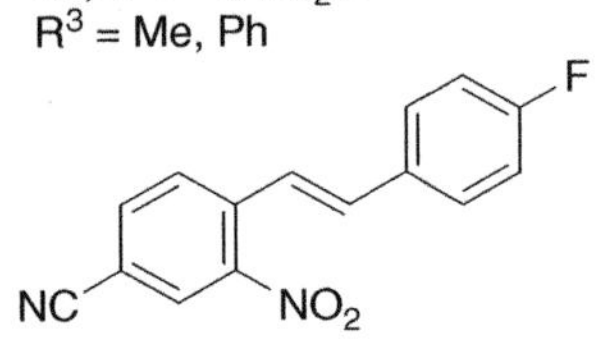	$(EtO)_3P$ 150–160 °C 1.5 h	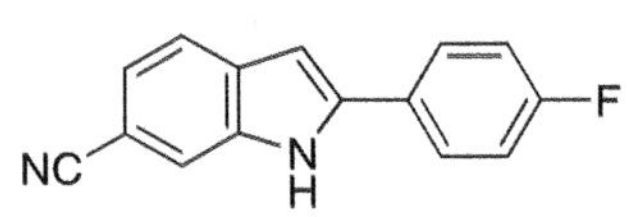	45%	41
7	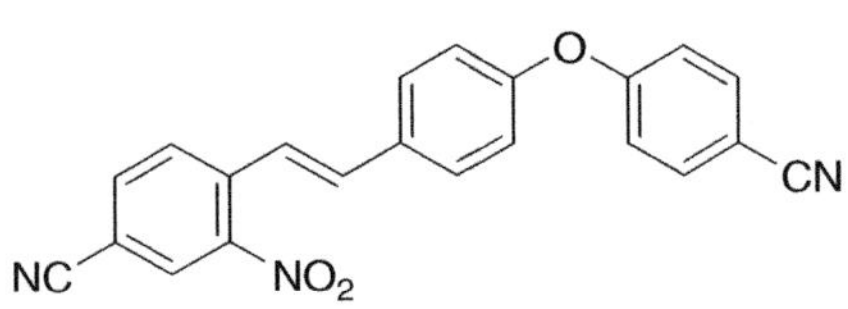	$(EtO)_3P$ 160 °C 13 h	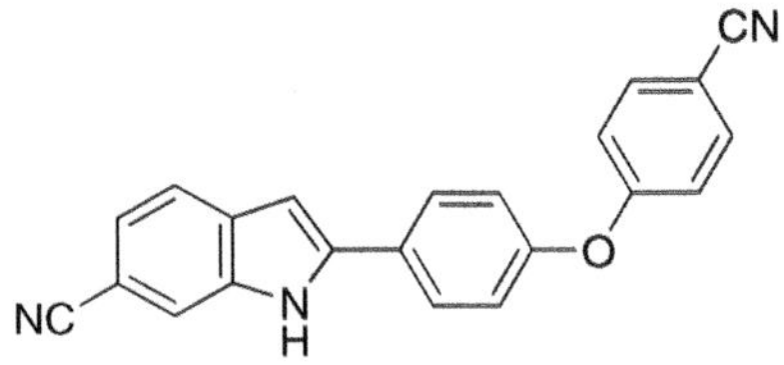	46%	42
8	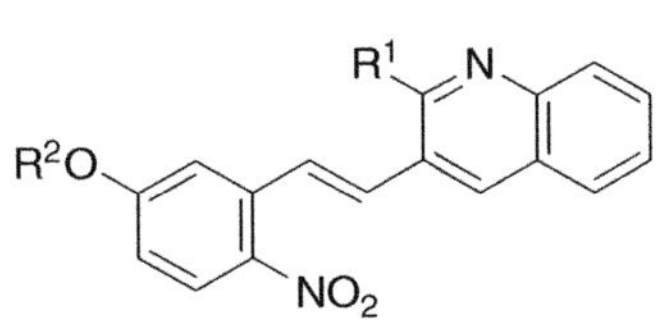 R^1 = OMe, Cl; R^2 = $(CH_2)_2OMe$, $(CH_2)_2NC_5H_{10}$, $(CH_2)_2NMe(CH_2)_2OMe$, others	$(EtO)_3P$ reflux	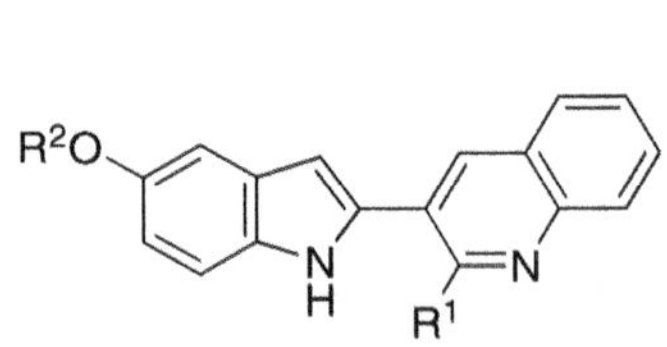	63–76%	43, 44

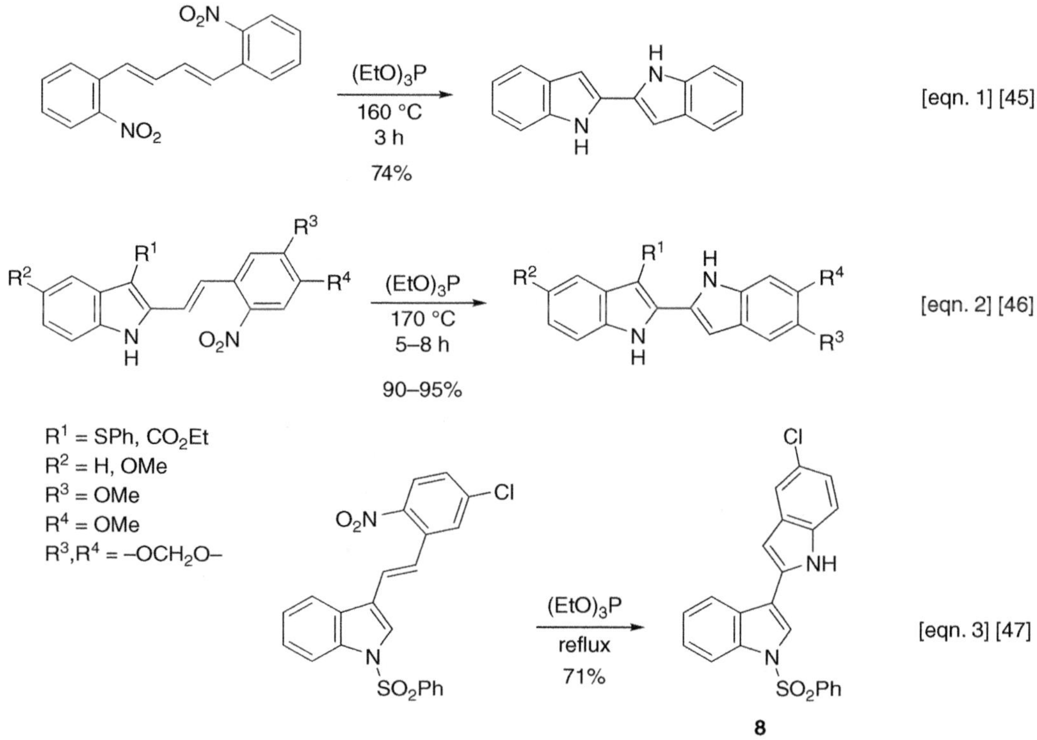

Scheme 3 *Applications of the Cadogan-Sundberg Indole Synthesis to Biindoles*

heating in the presence of $MoO_2Cl_2(dmf)_2$ to access a series of brominated 5,6-dimethoxyindoles from the corresponding *o*-nitrostyrenes in 63% to 92% yields (equation 2) [69]. A series of carbazoles and δ-carbolines was synthesized by Yang and colleagues using 1,2-bis(diphenylphosphino) ethane (DPPE) in place of the more typical triethyl phosphite or triphenylphosphine (equations 3, 4) [70]. In at least one case the yield of a δ-carboline using DPPE was about twice that using triphenylphosphine. A few other examples are shown (**20–22**) with the site of ring closure indicated [70]. Peters and colleagues described an electrochemical reduction of *ortho*-nitrostibenes to give 2-phenylindoles [71]. Conditions are a carbon cathode, dimethylformamide, room temperature, tetramethylammonium tetrafluoroborate, and a proton donor (phenol or 3-oxobutanoate); yields of 2-phenylindoles were 38% to 82%.

Application of the Cadogan–Sundberg indolization continues to be applied to carbazole synthesis, particularly to carbazoles having photophysical properties (e.g., diindenocarbazoles [72], 4-(4-pyridyl)phenylcarbazoles [73], pyrrolo[3,2,1-*jk*]carbazole [74], and indolo[3,2,1-*jk*]carbazole) [74]. Dehaen and coworkers found that microwave heating greatly enhanced both a Suzuki palladium-catalyzed cross-coupling synthesis of *ortho*-nitrobiphenyls and a subsequent Cadogan–Sundberg cyclization (Scheme 8, equations 1, 2) [75]. Some additional carbazoles that were prepared in this fashion are **23–25**. Kuethe and Childers also effected this tandem Suzuki–Cadogan/Sundberg route to afford a wide collection of diverse carbazoles (**26–31**), including the alkaloid glycoborine (**31**) isolated from *Glycosmis arborea* [76]. Cimrová and colleagues reported the synthesis of *N*-alkyl-2,7-dihalocarbazoles by the simultaneous Cadogan–Sundberg ring formation and *N*-alkylation (equation 2) [77]. The *N*-alkylation was facilitated by the additive 4-nitrotoluene. Although the nature of the alkylating species is currently unknown, it may simply be trialkyl phosphate.

Although the Cadogan–Sundberg reaction of nitroarenes to form indoles with triethyl phosphite is generally believed to involve sequential deoxygenation to nitrosoarenes and then to nitrenes followed by insertion into the proximate double bond, other products are often formed. For example, β-nitrostyrene reacted with triethyl phosphite at room temperature to produce **32–34** (Scheme 9, equation 1) [78]. Of greater interest was the formation of *N*-ethoxyindoles, which Peet and coworkers investigated. Using ^{18}O labeled *ortho*-nitrostilbene, Peet and colleagues found that the oxygen atom in 2-aryl-*N*-ethoxyindoles originated from the nitro group and not the triethyl phosphite (equation 2) [79]. Thus, heating nitrostilbene **35** with limited amounts

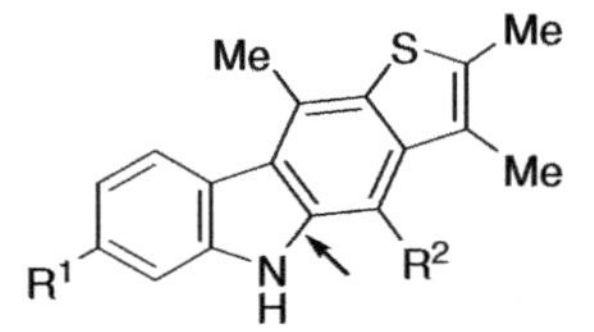

9 (40–70%) [49]
R^1 = OMe, CF_3
R^2 = H, Me

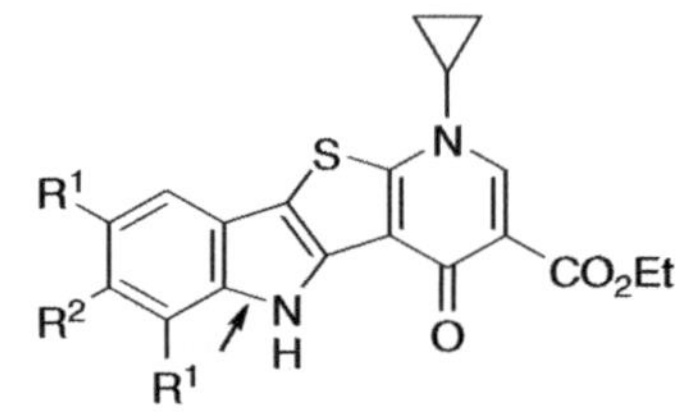

10 (40–63%) [50]
R^1 = H, Me
R^2 = H, Me, OMe, F

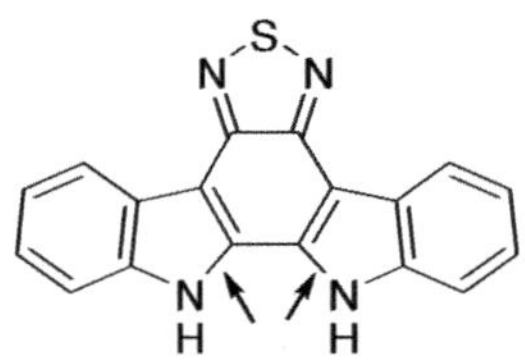

11 (51%) [51]

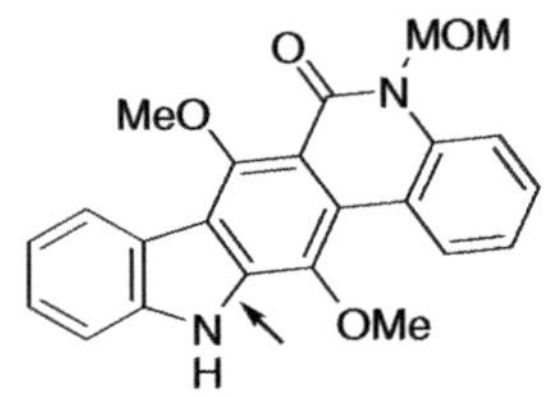

12 (89%) [52]

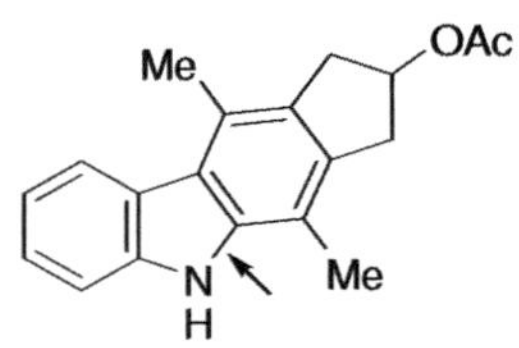

13 (73%) [53]

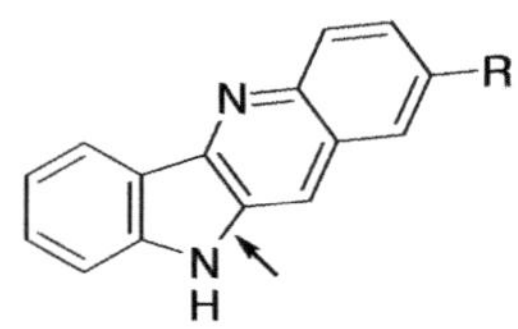

14 (68–75%) [54]
R = H, Me, Br, I

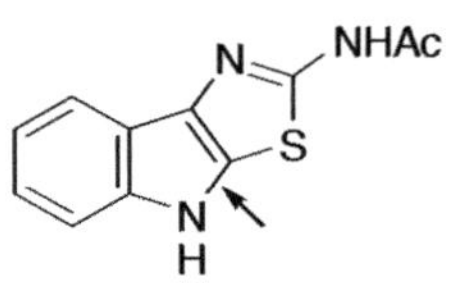

15 (61%) [55]

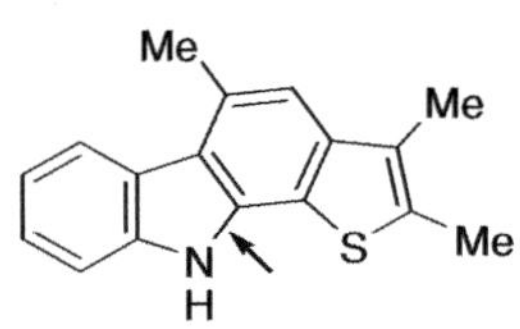

16 (53%) [56]

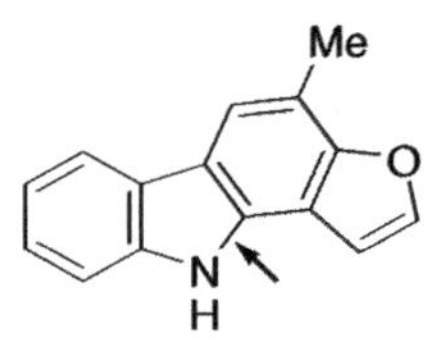

17 (42%) [57]

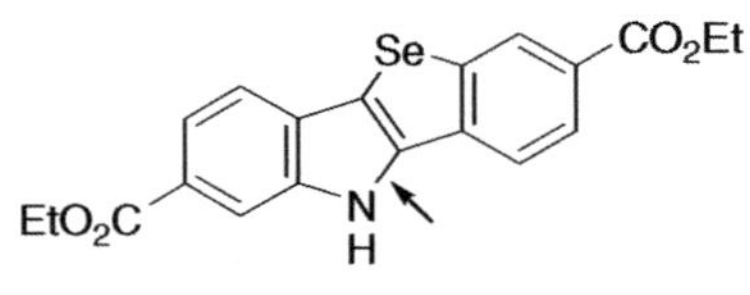

18 (18%) [58]

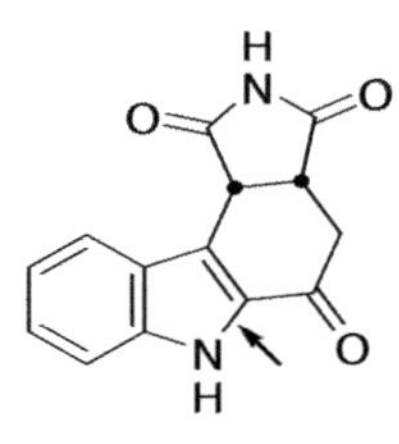

19 (75%) [59]

Scheme 4 *Applications of the Cadogan-Sundberg Indole Synthesis to Fused Indoles*

of triethyl phosphite (150 °C, 2 h) afforded both *N*-ethoxyindoles **36** and indoles **37**, with increased amounts of **36** with only two equivalents of triethyl phosphite. Davies, Houk, and Guner presented theoretical evidence to suggest that the cyclization of *ortho*-nitrosostilbenes, -styrenes, and -biphenyls may involve a 1,5-electrocyclization followed by a 1,5-H shift to afford *N*-hydroxyindoles (equation 3) [80].

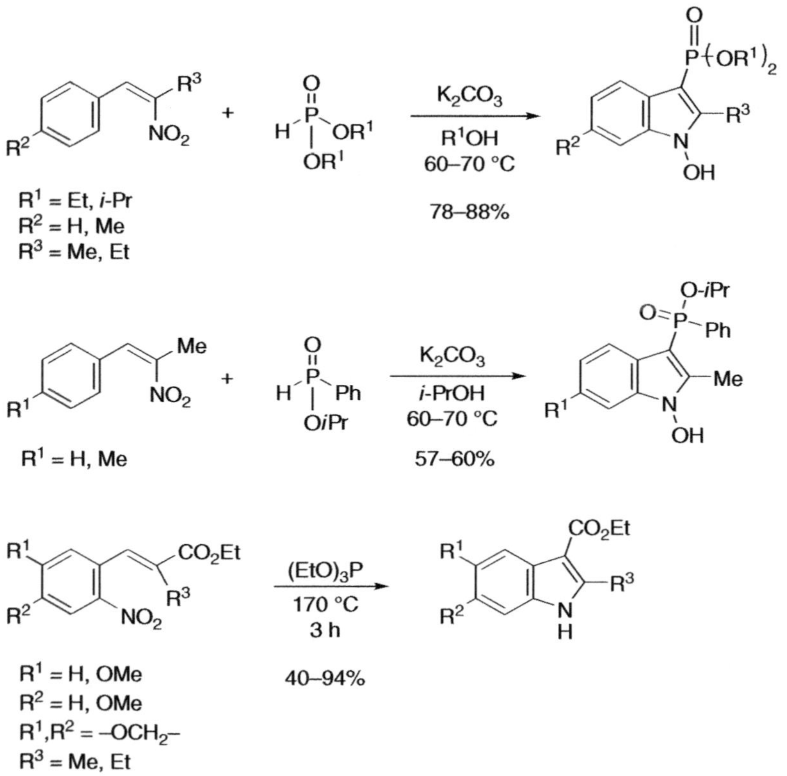

[eqn. 1] [60]

[eqn. 2] [60]

[eqn. 3] [62]

Scheme 5 *Applications of the Cadogan-Sundberg Indole Synthesis Using Dialkyl Phosphites*

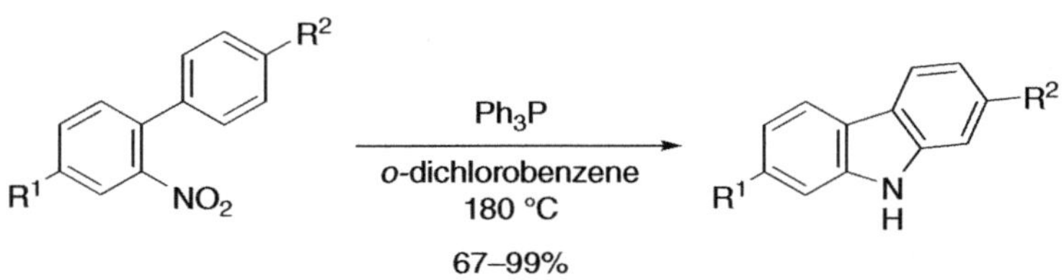

[eqn. 1] [63]

R^1	R^2	%, Yield [63]	% Yield^a
H	*i*-Bu	67	65
MeO	H	91	22
Br	Br	75	56
F	H	91	54
CF_3	H	85	45
CN	H	75	49
CHO	H	78	34
PhCO	H	99	85
CO_2Me	H	90	53
COMe	H	81	82

^{a}Triethyl phosphite, reflux (from the literature) [63]

Scheme 6 *Freeman Variation of the Cadogan-Sundberg Indole Synthesis*

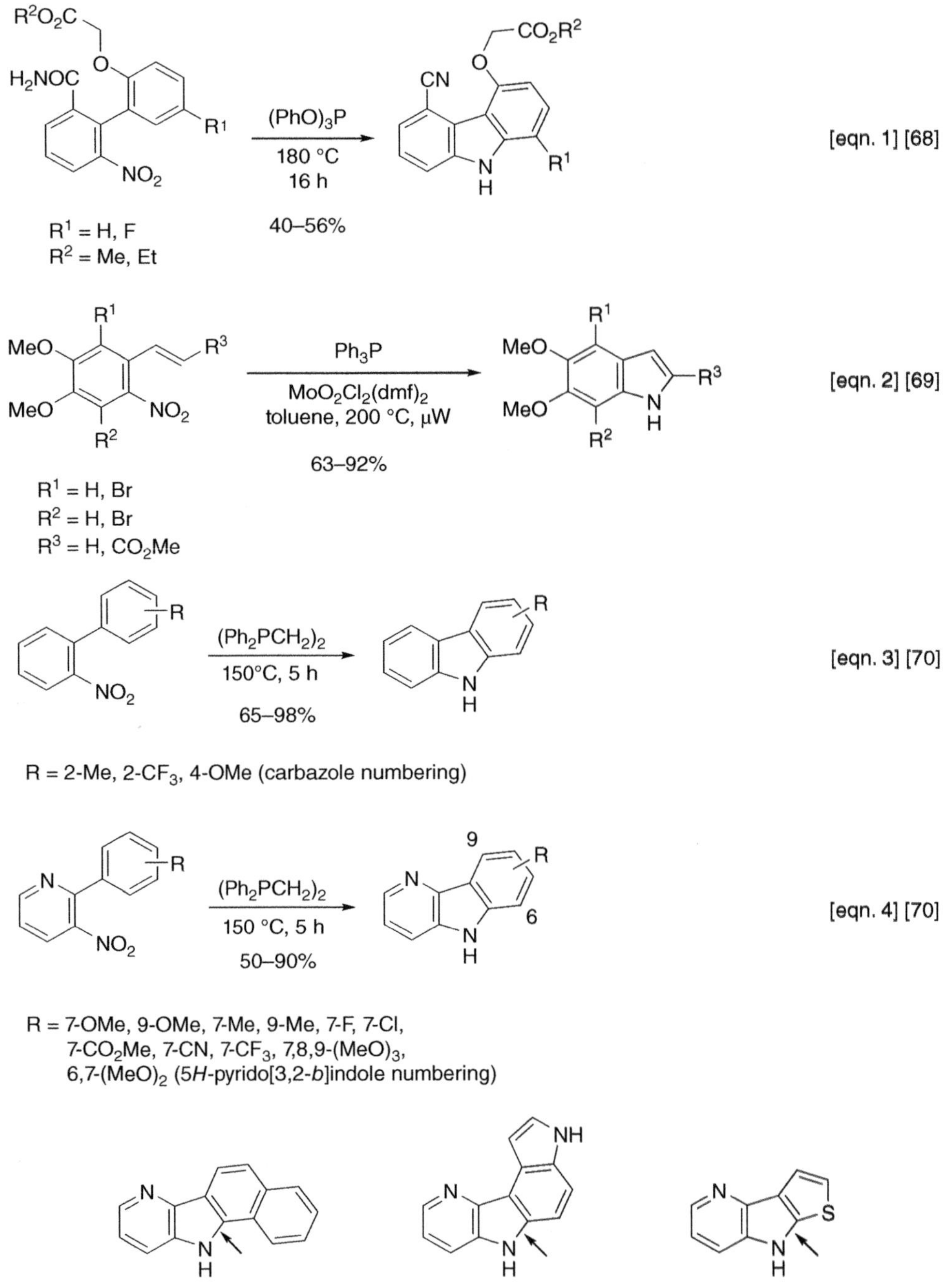

20 (90%) **21** (50%) **22** (60%)

Scheme 7 *Variations of the Cadogan-Sundberg Indole Synthesis*

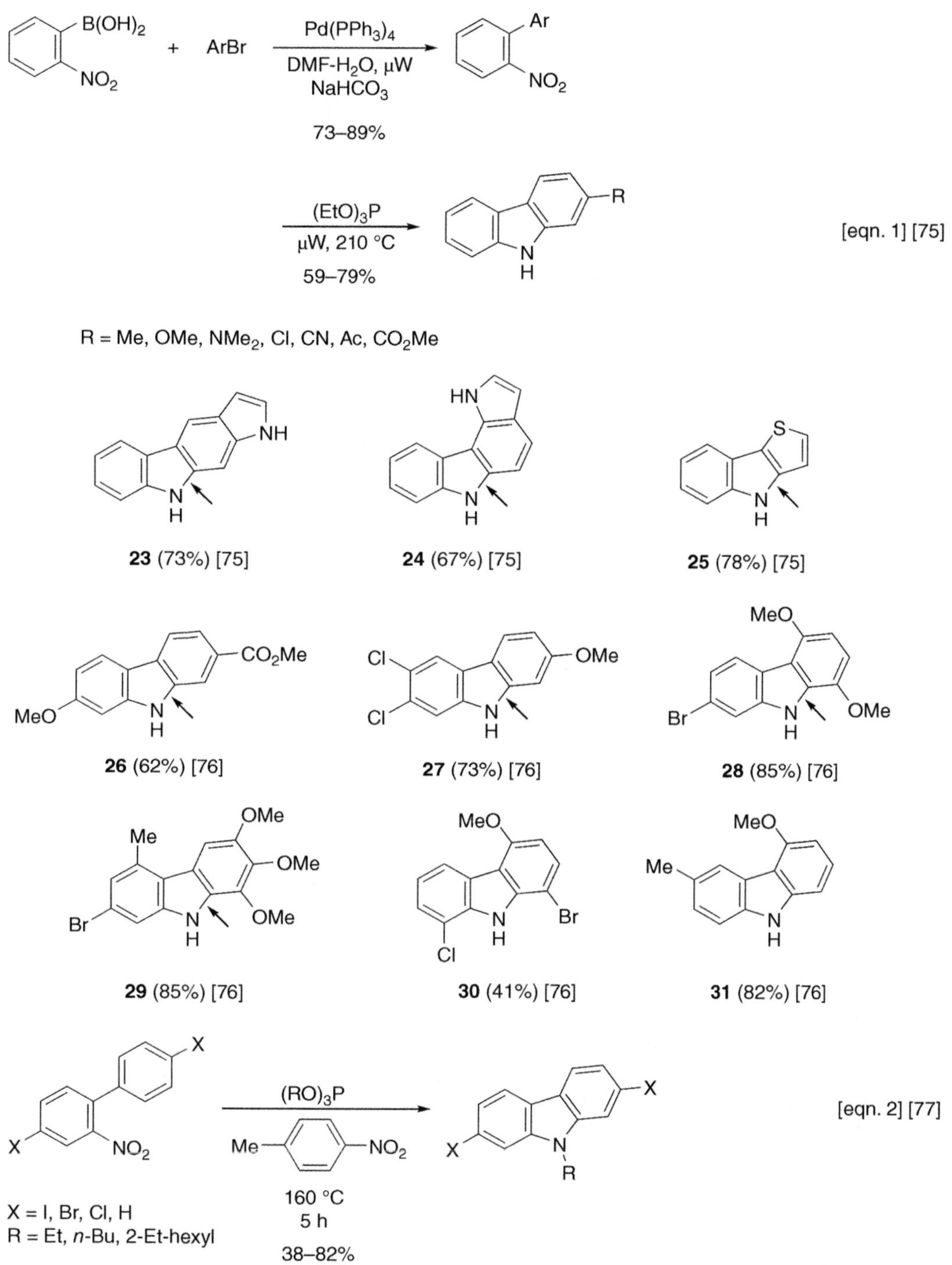

Scheme 8 *The Cadogan-Sundberg Indole Synthesis Applied to Carbazoles*

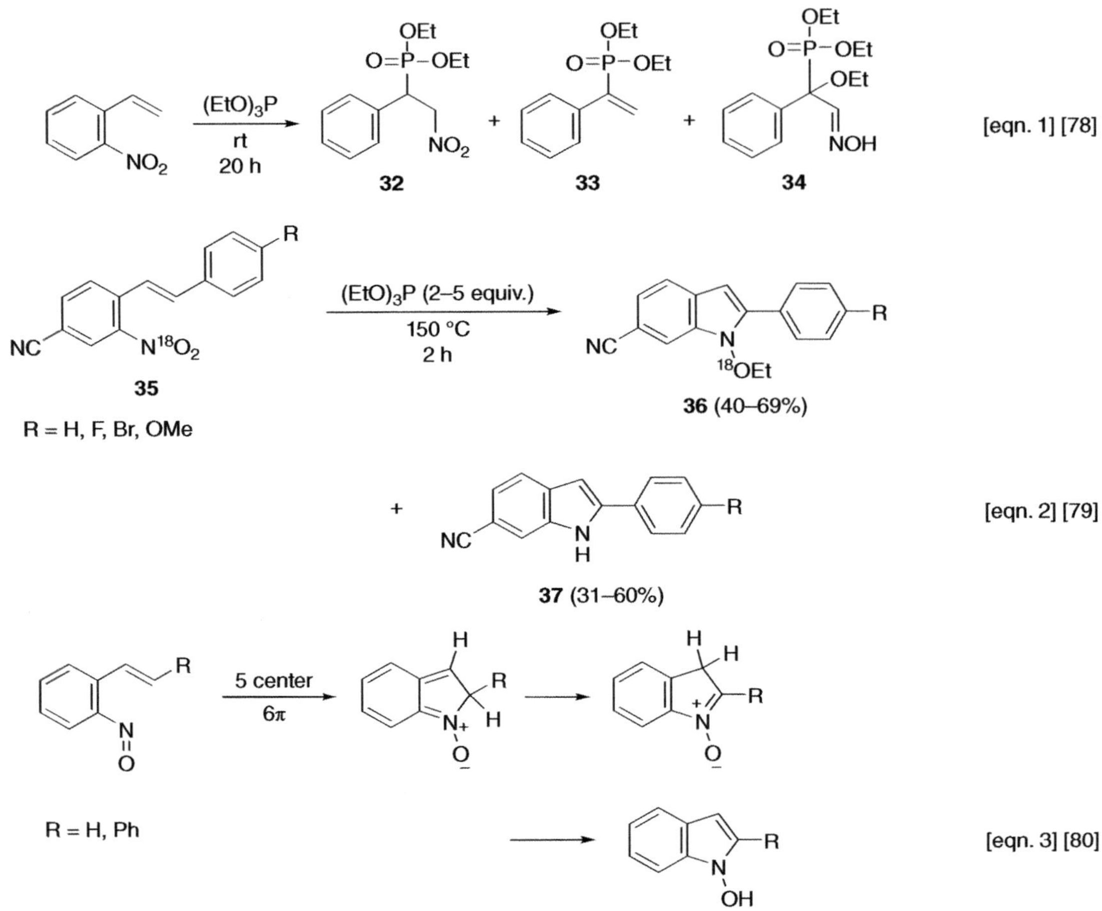

Scheme 9 *Side Reactions in the Cadogan-Sundberg Indole Synthesis*

References

[1] P.J. Bunyan and J.I.G. Cadogan, *Proc. Chem. Soc.*, 1962, 78.
[2] J.I.G. Cadogan and M. Cameron-Wood, *Proc. Chem. Soc.*, 1962, 361.
[3] P.J. Bunyan and J.I.G. Cadogan, *J. Chem. Soc.*, 1963, 42–49.
[4] J.I.G. Cadogan, M. Cameron-Wood, R.K. Mackie, and R.J.G. Searle, *J. Chem. Soc.*, 1965, 4831–4837.
[5] J.I.G. Cadogan and A. Cooper, *J. Chem. Soc. (B)*, 1969, 883–885.
[6] J.I.G. Cadogan, *Synthesis*, 1969, 11–17.
[7] J.I.G. Cadogan and M.J. Todd, *J. Chem. Soc. (C)*, 1969, 2808–2813.
[8] R.J. Sundberg, *J. Org. Chem.*, 1965, **30**, 3604–3610.
[9] R.J. Sundberg and T. Yamazaki, *J. Org. Chem.*, 1967, **32**, 290–294.
[10] R.J. Sundberg, *J. Am. Chem. Soc.*, 1966, **88**, 3781-3789.
[11] R.A. Odum and M. Brenner, *J. Am. Chem. Soc.*, 1966, **88**, 2074–2075.
[12] G. Smolinsky and B.I. Feuer, *J. Org. Chem.*, 1966, **31**, 3882–3884.
[13] H.C. Waterman and D.L. Vivian, *J. Org. Chem.*, 1949, **14**, 289–297.
[14] I. Puskas and E.K. Fields, *J. Org. Chem.*, 1968, **33**, 4237–4242.
[15] T. Kametani, K. Ogasawara, and T. Yamanaka, *J. Chem. Soc. (C)*, 1968, 1006–1007.
[16] T. Kametani, K. Ogasawara, and T. Yamanaka, *J. Chem. Soc. (C)*, 1969, 138–140.
[17] K.E. Chippendale, B. Iddon, and H. Suschitzky, *J. Chem. Soc. (D), Chem. Commun.*, 1971, 203–204.
[18] T. Kametani, F.F. Ebetino, and K. Fikumoto, *J. Chem. Soc., Perkin Trans. 1*, 1974, 861–863.
[19] G.W. Gribble, *J. Org. Chem.*, 1973, **39**, 4074–4075.
[20] A. Tanaka, K. Yakushijin, and S. Yoshina, *J. Heterocycl. Chem.*, 1977, **14**, 975–979.
[21] R.A. Abramovitch and B.W. Cue, Jr., *J. Org. Chem.*, 1980, **45**, 5316–5319.
[22] G.A. Russell, C.-F. Yao, H.I. Tashtoush, *et al.*, *J. Org. Chem.*, 1991, **56**, 663–669.
[23] G. Grynkiewicz, M. Poenie, and R.Y. Tsien, *J. Biol. Chem.*, 1985, **260**, 3440–3450.

[24] M. Tashiro and T. Yamato, *Synthesis*, 1979, 48–50.
[25] Y. Tsunashima and M. Kuroki, *J. Heterocycl. Chem.*, 1981, **18**, 315–318.
[26] J.F. Morin and M. Leclerc, *Macromolecules*, 2001, **34**, 4680–4682.
[27] G. Zotti, G. Schiavon, S. Zecchin, *et al.*, *Macromolecules*, 2002, **35**, 2122–2128.
[28] J. Morin, N. Drolet, Y. Tao, and M. Leclerc, *Chem. Mater.*, 2004, **16**, 4619–4626.
[29] J. Bouchard, S. Wakim, and M. Leclerc, *J. Org. Chem.*, 2004, **69**, 5705–5711.
[30] S. Wakim and M. Leclerc, *Synlett*, 2005, 1223–1234.
[31] M. Belletête, M. Bédard, M. Leclerc, and G. Durocher, *Synth. Met.*, 2004, **146**, 99–108.
[32] S. Wakim, J. Bouchard, M. Simard, *et al.*, *Chem. Mater.*, 2004, **16**, 4386–4388.
[33] F. Dierschke, A.C. Grimsdale, and K. Müllen, *Synthesis*, 2003, 2470–2472.
[34] M. Tavasli, S. Bettington, M.R. Bryce, *et al.*, *Synthesis*, 2005, 1619–1624.
[35] M. Mézlová, J.J. Aaron, J. Svcoboda, *et al.*, *J. Electroanalyt. Chem.*, 2005, **581**, 93–103.
[36] M.L. Gelmi, D. Pocar, and F. Vago, *J. Chem. Soc., Perkin Trans. 1*, 1993, 969–973.
[37] A.S. Cotterill, C.J. Moody, and J.R.A. Roffey, *Tetrahedron*, 1995, **51**, 7223–7230.
[38] D.J. Bentley, J. Fairhurst, P.T. Gallagher, *et al.*, *Org. Biomol. Chem.*, 2004, **2**, 701–708.
[39] C.W. Holzapfel and C. Dwyer, *Heterocycles*, 1998, **48**, 1513–1518.
[40] R.S. Mali, S.G. Tilve, and V.G. Desai, *J. Chem. Res., Synopses*, 2000, 8-9, (*M*), 150–158.
[41] B. Li, R. Pai, S.C. Cardinale, *et al.*, *J. Med. Chem.*, 2010, **53**, 2264–2276.
[42] J.S. Lazo, R. Nunes, J.J. Skoko, *et al.*, *Bioorg. Med. Chem.*, 2006, **14**, 5643–5650.
[43] J.T. Kuethe, A. Wong, and I.W. Davies, *Org. Lett.*, 2003, **5**, 3975–3978.
[44] J.T. Kuethe, A. Wong, C. Qu, *et al.*, *J. Org. Chem.*, 2005, **70**, 2555–2567.
[45] D.J. Koza and W.B. Euler, *Heterocycl. Commun.*, 1999, **5**, 399–402.
[46] K. Jesudoss and P.C. Srinivasan, *Synth. Commun.*, 1994, **24**, 1701–1708.
[47] J.T. Kuethe, A. Wong, and I.W. Davies, *Org. Lett.*, 2003, **5**, 3721–3723.
[48] J.T. Kuethe and I.W. Davies, *Tetrahedron Lett.*, 2004, **45**, 4009–4012.
[49] I.C.F.R. Ferreira, M.-J.R.P. Queiroz, and G. Kirsch, *Tetrahedron Lett.*, 2003, **44**, 4327–4329.
[50] S.A. Al-Trawneh, M.M. El-Abadelah, J.A. Zahra, *et al.*, *Bioorg. Med. Chem.*, 2011, **19**, 2541–2548.
[51] G. Balaji, W.L. Shim, M. Parameswaran, and S. Valiyaveettil, *Org. Lett.*, 2009, **11**, 4450–4453.
[52] P.H. Bernardo, W. Fitriyanto, and C.L.L. Chai, *Synlett*, 2007, 1935–1939.
[53] T.-L. Ho and S.-Y. Hsieh, *Helv. Chim. Acta*, 2006, **89**, 111–116.
[54] B. Dutta, S. Some, and J.K. Ray, *Tetrahedron Lett.*, 2006, **47**, 377–379.
[55] J. Breinholt, C.B. Jeppesen, S. Branner, *et al.*, *J. Heterocycl. Chem.*, 2001, **38**, 569–577.
[56] I.C.F.R. Ferreira, M.-J.R.P. Queiroz, and G. Kirsch, *J. Heterocycl. Chem.*, 2001, **38**, 749–754.
[57] T. Soós, G. Timári, and G. Hajós, *Tetrahedron Lett.*, 1999, **40**, 8607–8609.
[58] P. Kaszynski and D.A. Dougherty, *J. Org. Chem.*, 1993, **58**, 5209–5220.
[59] D. Alonso, E. Caballero, M. Medarde, and F. Tomé, *Tetrahedron Lett.*, 2005, **46**, 4839–4841.
[60] R. Zhang, X. Liao, and Z. Gao, *Synthesis*, 1990, 801–802.
[61] R.S. Mali and V.J. Yadav, *Synthesis*, 1984, 862–865.
[62] M.S. Wadia, R.S. Mali, S.G. Tilve, and V.J. Yadav, *Synthesis*, 1987, 401–404.
[63] A.W. Freeman, M. Urvoy, and M.E. Criswell, *J. Org. Chem.*, 2005, **70**, 5014–5019.
[64] D. Výprachtický, I. Kmínek, P. Pavlačková, and V. Cimrová, *Synthesis*, 2011, 1472–1476.
[65] M.R. Naffziger, B.O. Ashburn, J.R. Perkins, and R.G. Carter, *J. Org. Chem.*, 2007, **72**, 9857–9865.
[66] C. Du, J. Chen, Y. Guo, *et al.*, *J. Org. Chem.*, 2009, **74**, 7322–7327.
[67] E.C. Creencia, M. Kosaka, T. Muramatsu, *et al.*, *J. Heterocycl. Chem.*, 2009, **46**, 1309–1317.
[68] S.A. May, T.M. Wilson, and A.L. Fields, *Tetrahedron Lett.*, 2006, **47**, 1351–1353.
[69] P.B. Huleatt, J. Lau, S. Chua, *et al.*, *Tetrahedron Lett.*, 2011, **52**, 1339–1342.
[70] H. Peng, X. Chen, Y. Chen, *et al.*, *Tetrahedron*, 2011, **67**, 5725–5731.
[71] P. Du, J.L. Brosmer, and D.G. Peters, *Org. Lett.*, 2011, **13**, 4072–4075.
[72] L. Rong, Q. Liu, J. Tang, and Z. Chi, *Heterocycles*, 2010, **81**, 977–984.
[73] C.J. Kelley, K. Ansu, W. Budisusetyo, *et al.*, *J. Heterocycl. Chem.*, 2001, **38**, 11–23.
[74] S.I. Wharton, J.B. Henry, H. McNab, and A.R. Mount, *Chem. Eur. J.*, 2009, **15**, 5482–5490.
[75] P. Appukkuttan, E. Van der Eycken, and W. Dehaen, *Synlett*, 2005, 127–133.
[76] J.T. Kuethe and K.G. Childers, *Adv. Synth. Catal.*, 2008, **350**, 1577–1586.
[77] D. Výprachtický, I. Kmínek, V. Pokorná, and V. Cimrová, *Tetrahedron*, 2012, **68**, 5075–5080.
[78] G.A. Russell and C.-F. Yao, *J. Org. Chem.*, 1992, **57**, 6508–6513.
[79] H. Majgier-Baranowska, J.D. Williams, B. Li, and N.P. Peet, *Tetrahedron Lett.*, 2012, **53**, 4785–4788.
[80] I.W. Davies, V.A. Guner, and K.N. Houk, *Org. Lett.*, 2004, **6**, 743–746.

27

Sundberg Indole Synthesis

In this chapter we will see that aryl azides, like aryl nitro compounds, can form aryl nitrenes, or nitrene equivalents, under suitable conditions, leading to indoles.

Although the cyclization of nitrenes to indoles via *ortho*-azidostyrenes justifiably is associated with Sundberg, several earlier workers described and studied this reaction. Notably, the researches of Smith [1–5], Smolinsky [6–8], Abramovitch [9], Coffin and Robbins [10], and Isomura colleagues [11] describe the pyrolysis of *ortho*-azidobiphenyls to carbazoles and, in a few cases, *ortho*-azidostyrenes to indoles. A sampling of these pre-Sundberg reactions from the work of Smith and coworkers is shown in Scheme 1. The typical solvent used was kerosene, and the conversion of azides to carbazoles was also accomplished photochemically, albeit in lower yield. Other carbazoles prepared thermally were **1–4** in excellent yields. These workers also synthesized 4*H*-thieno[3,2-*b*]indole (**5**) from 2-(*o*-azidophenyl)thiophene [3].

Based on kinetic studies, a lack of a significant solvent effect, and a poor correlation with the Hammett equation, Smith and Hall conclude that an aryl nitrene is formed and undergoes cyclization [4]. This mechanism is supported by the work of Smolinsky [6–8], who also found that 2-azido-1,1-diphenylethylene gave 3-phenylindole on pyrolysis (equation 2) [8]. A similar reaction of 1-azido-2-phenylpropene gave 3-methylindole in 80% yield. Abramovitch and colleagues converted 3-azido-2-phenylpyridine to δ-carboline, which constituted the first unambiguous synthesis of this carboline [9]. Coffin and Robbins employed the solvent diphenyl ether to pyrolyze 2-azido-2′-methylbiphenyl to 4-methylcarbazole (91% yield) and not to the anticipated dihydrophenanthridine that would have arisen by hydrogen abstraction [10]. Independently from Sundberg, Isomura and colleagues found that β-azidostyrenes form indoles upon pyrolysis in refluxing *n*-hexadecane (287 °C) (equation 3) [11]. Likewise, both *cis*- and *trans*-β-azidostyrene gave indole and phenylacetonitrile in equal amounts (85% combined yield) under these conditions. At lower temperatures (100 °C) or with photolysis, the corresponding 2-phenyl-2*H*-azirines form (e.g., equation 4). Pyrolysis of these azirines gave the corresponding indoles, although 2-phenyl-2*H*-azirine itself is very labile in air at room temperature.

Notwithstanding these impressive early studies, Sundberg is most appropriately associated with the pyrolysis of aryl azides to form indoles [12, 13]. A summary of this chemistry is shown in Scheme 2 (equation 1 and **6–7**). Also shown are some early applications of the Sundberg indole synthesis, by Petersen and Lakowitz (equation 2) [14] and Suschitzky and colleagues (equation 3) [15]. Ning and colleagues examined the pyrolysis of 3-azido-4-(2-pyridyl)carbostyrils [16], and both Swenton [17] and Meth-Cohn [18, 19] explored the photolysis of *ortho*-azidobiphenyls as a route to carbazoles. Interestingly, Meth-Cohn's study showed that the carbazole yields were increased when the photochemical reaction temperature is raised. For example, the yield of 4-methylcarbazole from the photolysis of 2-azido-2′-methylbiphenyl at 107 °C (in chlorobenzene) was 90%, whereas at 15 °C (in methylene chloride) the yield was 62%. Thermal decomposition of this azide at 107 °C was negligible [18, 19]. A brief summary of Swenton's work is shown in equation 4 [17]. These two studies demonstrate that the singlet nitrene led to carbazoles, whereas the triplet nitrene afforded phenanthridines (where a methyl group is available) and azo-2-biphenyls.

A tabulation of selected applications of the Sundberg indole synthesis appears in Table 1 [20–26, 28, 29]. In contrast to Entry 4, the corresponding *(Z)-o*-azidocinnamonitriles in refluxing toluene gave the respective tetrazolo[1,5-*a*]

Indole Ring Synthesis: From Natural Products to Drug Discovery, First Edition. Gordon W. Gribble.

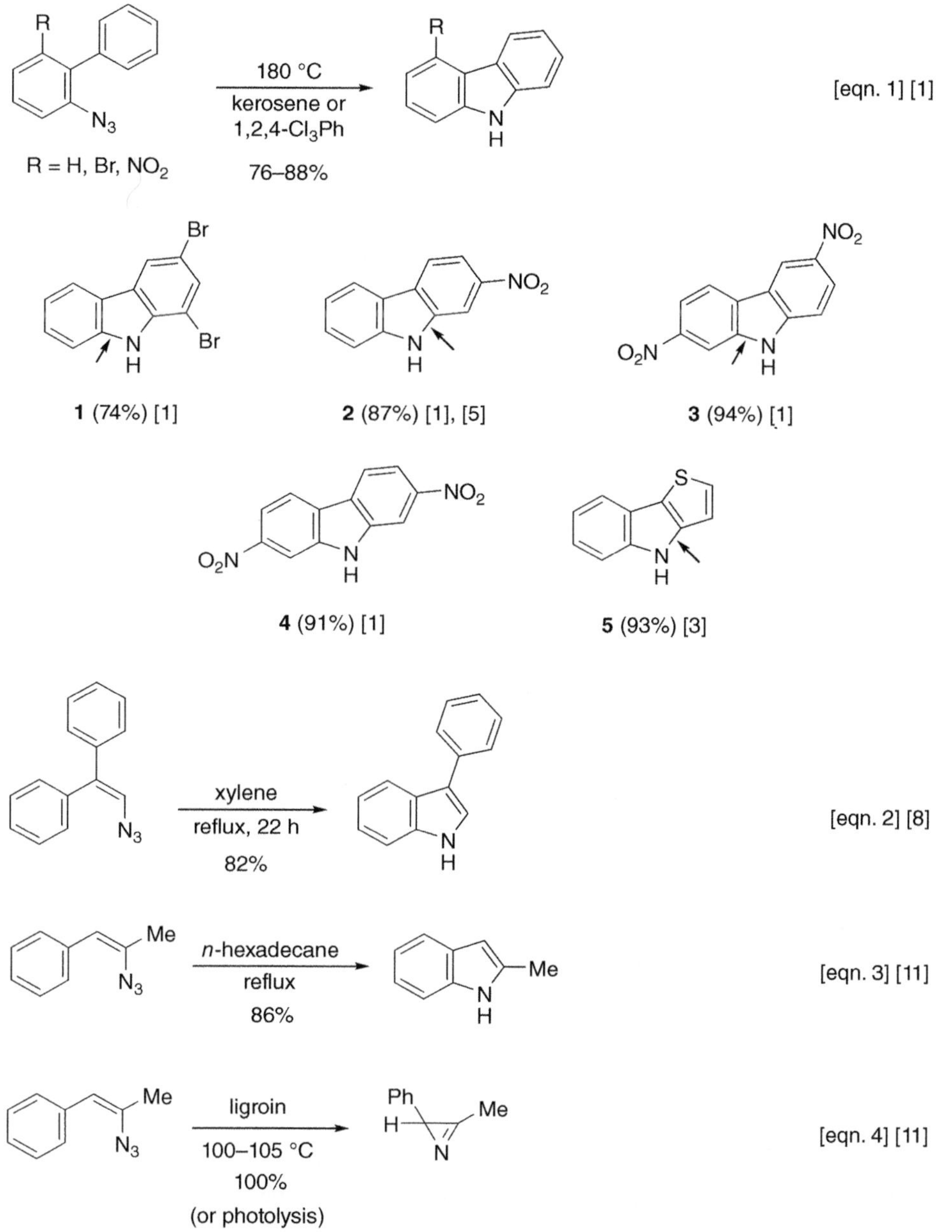

Scheme 1 *Synthesis of Indoles via Azide Pyrolysis*

quinolines in 75% to 83% yield (intramolecular 1,3-dipolar cycloaddition) [23]. A small amount of the mono-cyclization indole was obtained from the reaction shown in Entry 5 [24]. The first synthesis of 2-nitroindole was reported by Pelkey and Gribble via a Sundberg reaction (Entry 6). Similarly, 5,6-dimethoxy-2-nitroindole was prepared in 45% yield [25]. The synthesis of 4,6-dinitroindole (Entry 7) began with 2,4,6-trinitrostyrene and sodium azide followed by *in situ* thermolysis of the 2-azido-4,6-dinitrostyrene [26]. Rozhkov and colleagues prepared several novel 2-aryl-4-alkylsulfonyl-6-nitroindoles from the corresponding aryl azides in 89% to 93% yields, which were prepared *in situ* from the precursor 1-(2-alkylsulfonyl-4,6-dinitrophenyl)-2-arylethenes [27]. A double Sundberg synthesis was employed to prepare the indole in Entry 8 [28]. The reaction in Entry 9 also afforded an isonitrile (4-isocyano-1-(2-TIPS-1*H*-pyrrol-3-yl)but-2-enenitrile, 20% yield), and a similar reaction with 3-azido-4-(thien-3-yl)pyridine gave pyrido[3,4-*b*] thienopyrrole in 29% yield [29].

As we saw with the early work of Smith and others, the pyrolysis of *ortho*-azidobiphenyls is a powerful route to carbazoles. A recent synthesis of 2,7-dinitrocarbazole

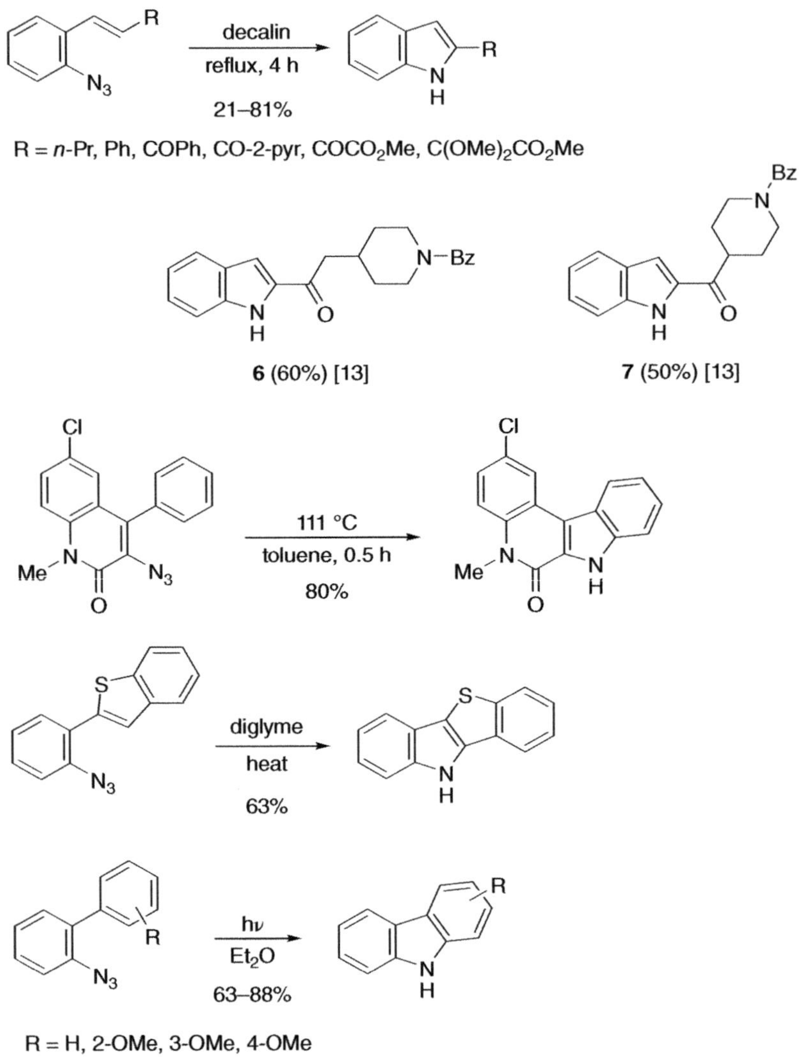

[eqn. 1] [12], [13]

[eqn. 2] [14]

[eqn. 3] [15]

[eqn. 4] [17]

Scheme 2 *Sundberg Indole Synthesis*

made use of the Smith protocol and led to novel conjugated polymers derived from 2,7-dihalocarbazoles [30]. Other examples of this strategy are shown in Scheme 3. Tour and Jian prepared several carbazoles (equation 1) en route to "surface-bound electric field–driven molecular motors" [31]. The requisite azides used by McNulty and coworkers in equation 2 were synthesized via Hartwig palladium-catalyzed amination ($Zn[N(TMS)_2]_2$) followed by diazotization ($NaNO_2$, HOAc, NaN_3) [32]. In complementary fashion, Sapi and colleagues used a Suzuki coupling of arylboronic acids with azido-2-bromobenzene to give the substrates shown in equation 3. Standard pyrolysis gave the desired carbazoles [33]. A small amount (0%–5%) of the isomeric pyrrolo[2,3-*b*]carbazole seemed to be present from the reaction shown in equation 4. A key step in the syntheses of dictyodendrins A–E by Tokuyama and colleagues is a Sundberg reaction (equation 5) [34, 35].

The Sundberg azide pyrolysis methodology was featured in a large number of fused indoles containing additional heterocyclic rings, such as carbolines, indoloquinolines, pyrazinoindoles, and similar ring systems. A sampling of these syntheses is shown in Table 2. Methylation of the indolo[3,2-*c*]quinoline in Entry 1 afforded the alkaloid cryptosanguinolentine [36], and methylation of the indolo[2,3-*c*] quinoline in Entry 2 yielded isoneocryptolepine [37]. The yield shown for the latter reaction is overall from the amine used to prepare the starting azide. The isomeric ring system, indolo[3,2-*c*]isoquinoline, was prepared from the *N*-oxide shown in Entry 3 [38]. The presumed intermediate *N*-oxide was not isolated. This deviation was necessary because the parent azide gave the pyrazole shown in Entry 4 [39]. Similarly, Entry 5 features the synthesis of the novel indazolo[3,2-*a*]-β-carboline ring system [40]. The new indolo[2,3-*c*]isoquinoline was prepared in good yield from

Table 1 *Applications of the Sundberg Indole Synthesis*

Entry	Azide	Conditions	Indole	% Yield	Ref.
1	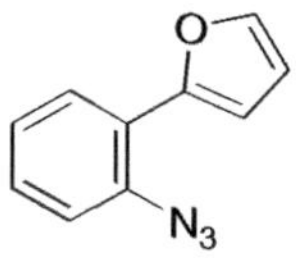	*o*-dichlorobenzene reflux, 1 h	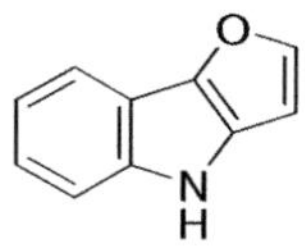	60%	20
2	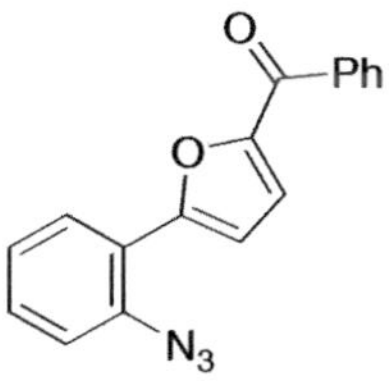	*o*-dichlorobenzene reflux, 1 h	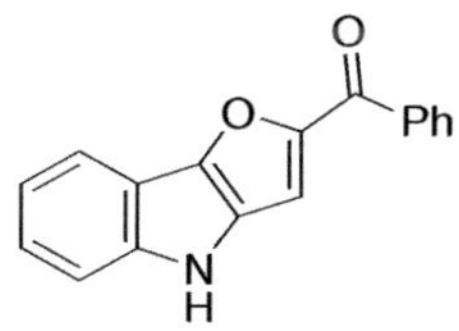	70%	21
3	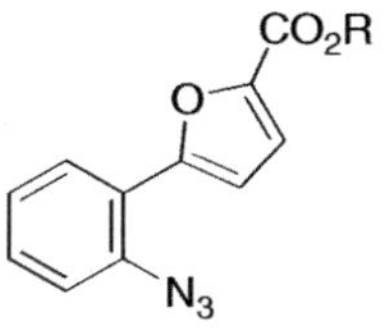 R = Me, Et	*o*-dichlorobenzene reflux, 1 h	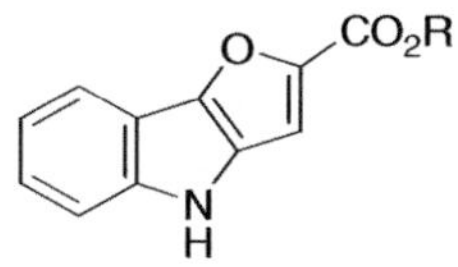	76–77%	22
4	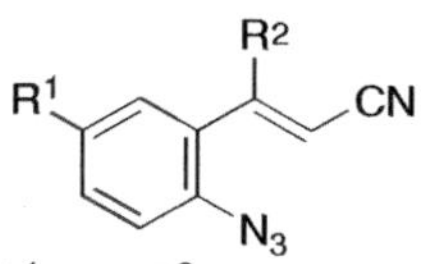 R^1 = H, R^2 = H; R^1 = Cl, R^2 = Ph; R^1 = H, R^2 = Me	DMSO 140 °C 1–2 h	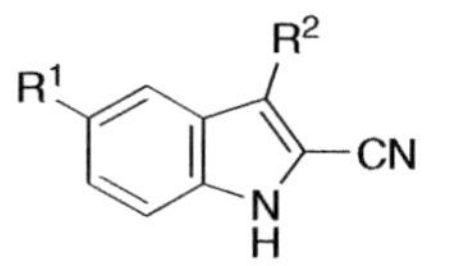	31–60%	23
5	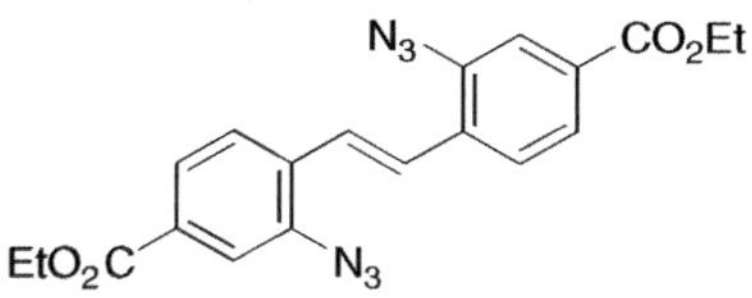	*o*-Cl_2Ph 155 °C	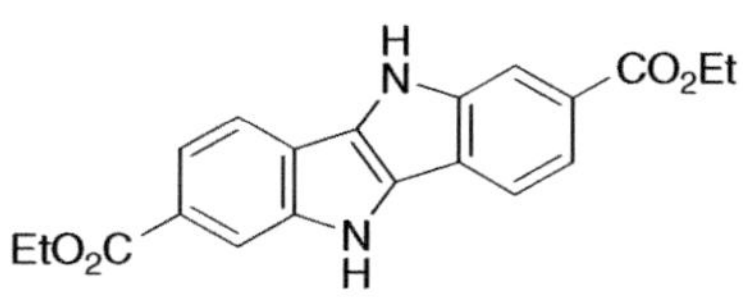	13%	24
6	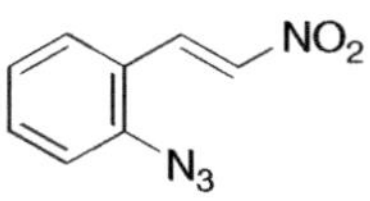	xylene 140 °C 12 h	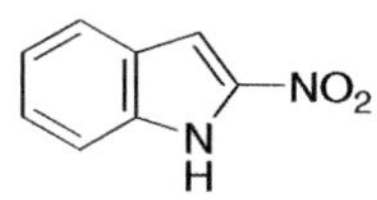	54%	25
7	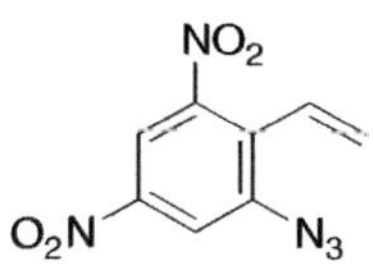	*i*-PrOH reflux	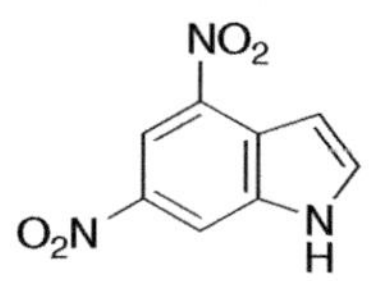	54%	26
8	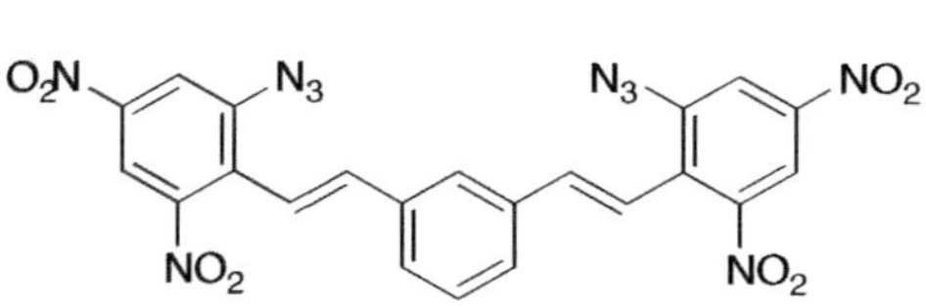	$PhNO_2$ 165 °C 1 h	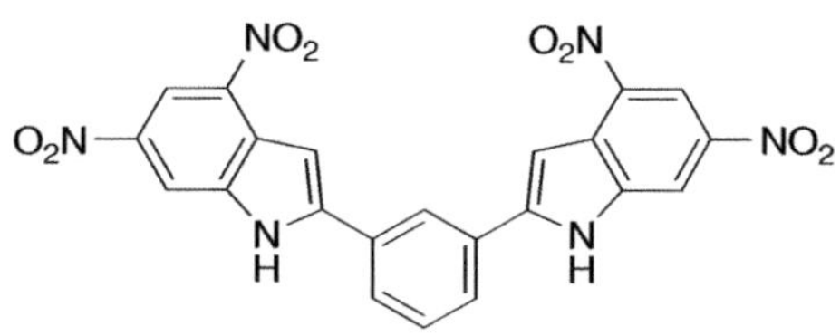	71%	28
9	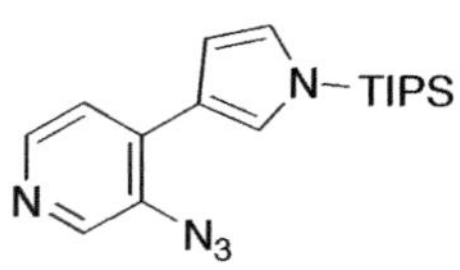	*n*-decane reflux 30 min	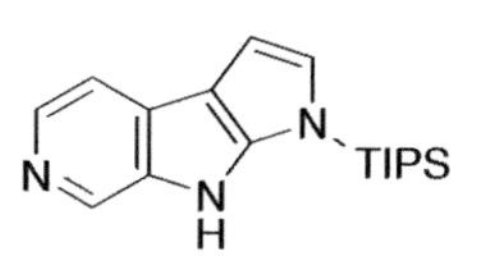	71%	29

(a)

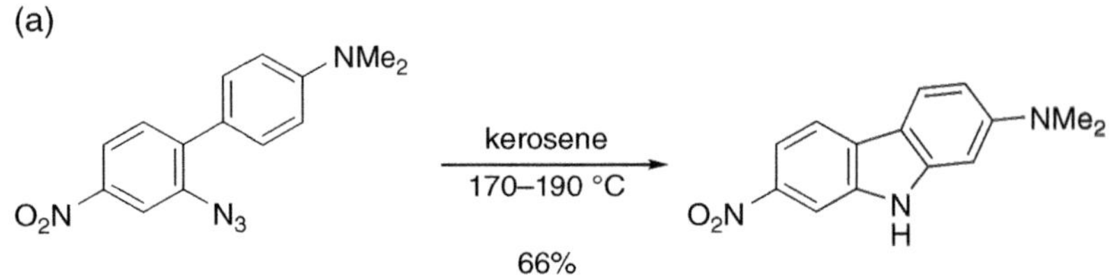

[eqn. 1] [31]

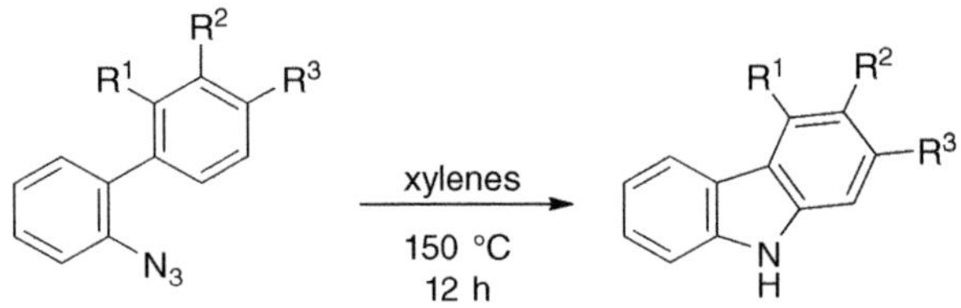

[eqn. 2] [32]

R^1	R^2	R^3	%
H	H	H	92
Me	H	H	88
H	H	OMe	90
OMe	OMe	OMe	82
H	H	Ac	96

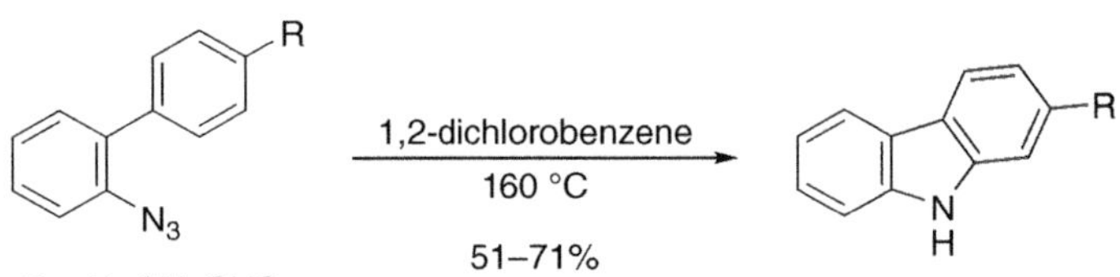

R = H, CN, CHO

[eqn. 3] [33]

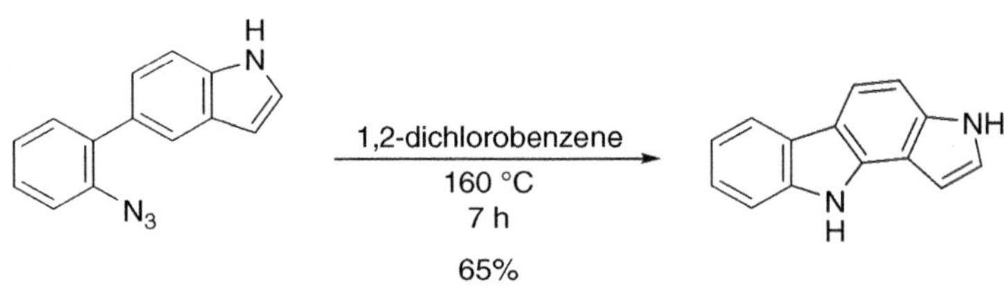

[eqn. 4] [33]

Scheme 3a *Sundberg Indole Synthesis of Carbazoles*

(b)

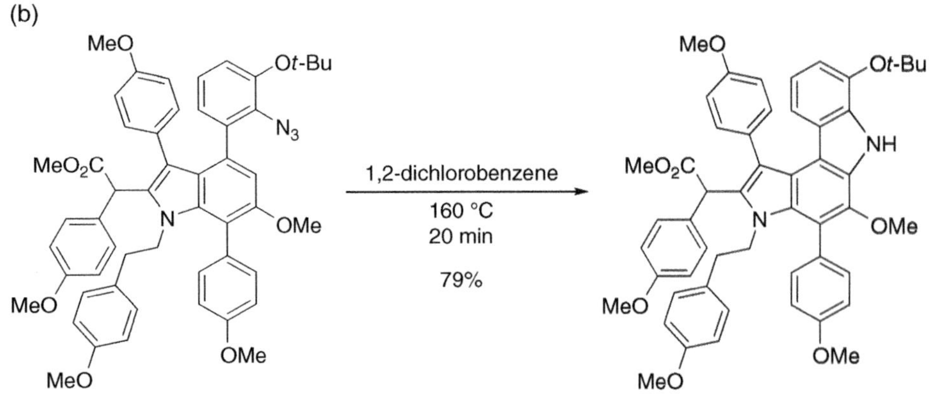

[eqn. 5] [34]

Scheme 3b *Tokuyama Synthesis of Dictyodendrins*

Table 2 *Applications of the Sundberg Indole Synthesis to Fused Indole Heterocycles*

Entry	Azide	Conditions	Indole	% Yield	Ref.
1	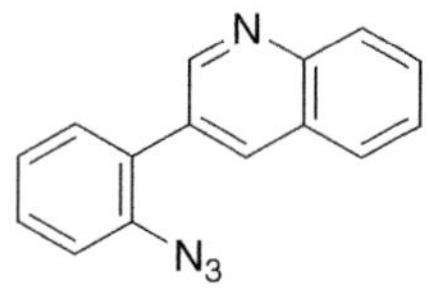	*o*-dichlorobenzene 180 °C, 5 h	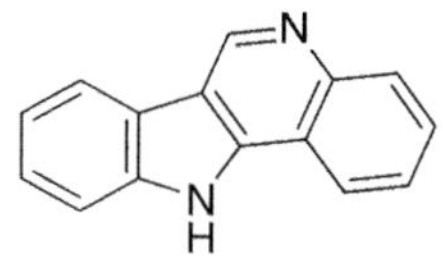	75%	36
2	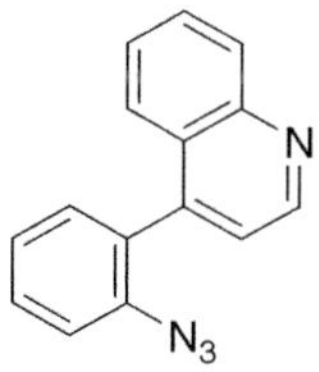	*o*-dichlorobenzene 180 °C, 3 h	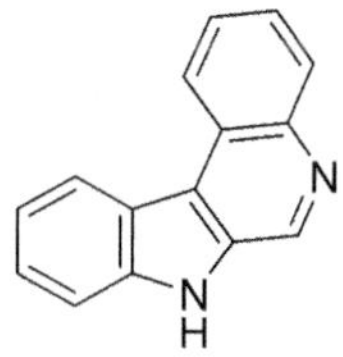	88%	37
3	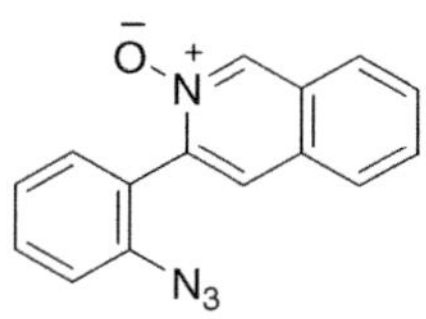	*o*-dichlorobenzene 150 °C, 6 h	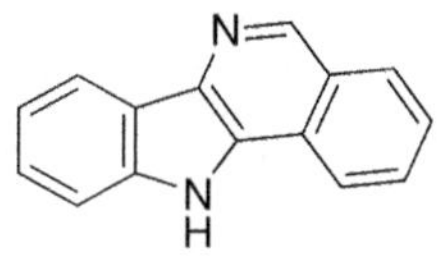	43%	38
4	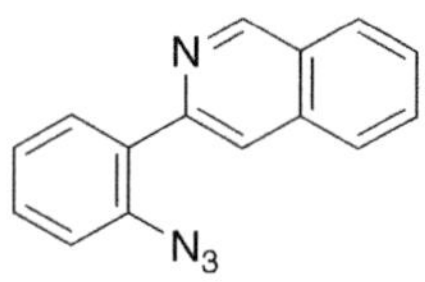	*o*-dichlorobenzene 180 °C	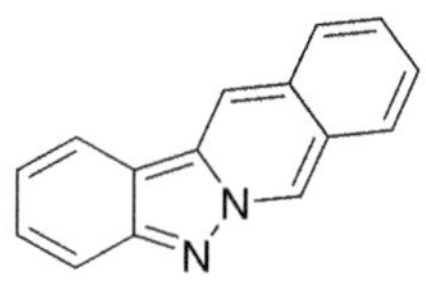	68%	39
5	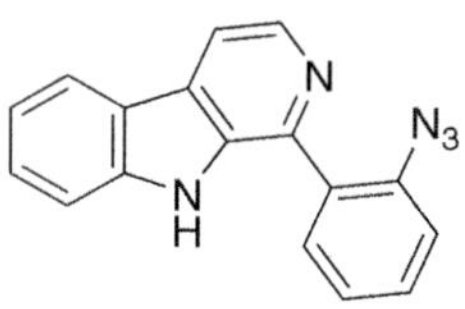	*o*-dichlorobenzene 180 °C	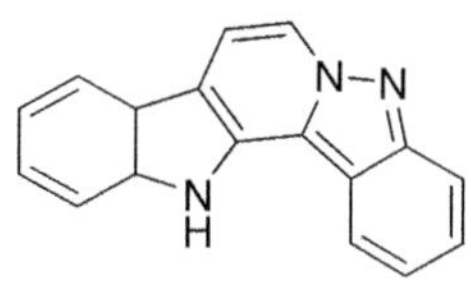	58%	40
6	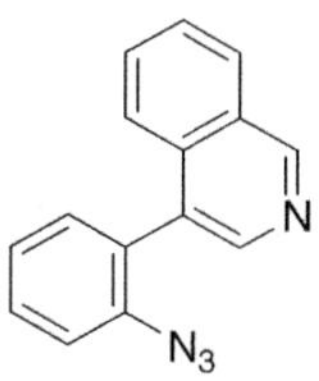	*o*-dichlorobenzene 180 °C, 3 h	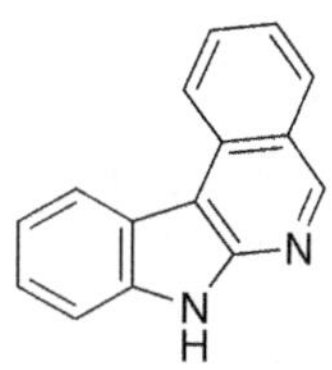	77%	41
7	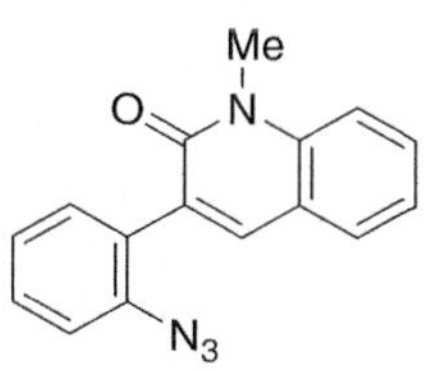	*o*-xylene 150 °C 20 h	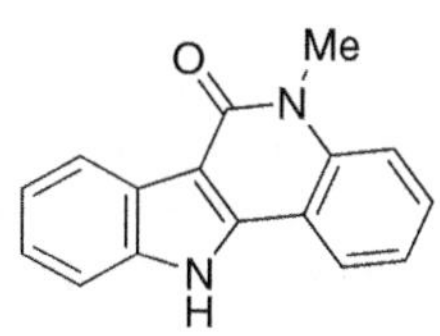	82%	42, 43
8	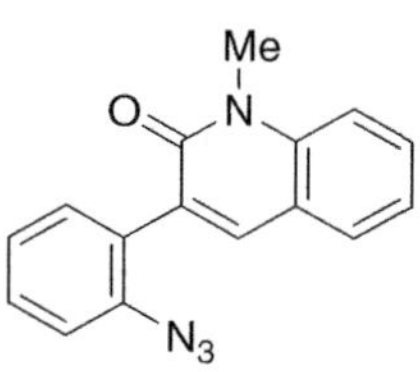	Me_3P, µW nitrobenzene 180 °C	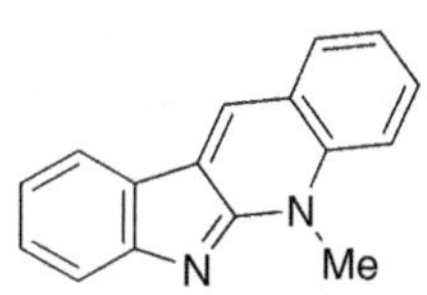	40%	42, 43

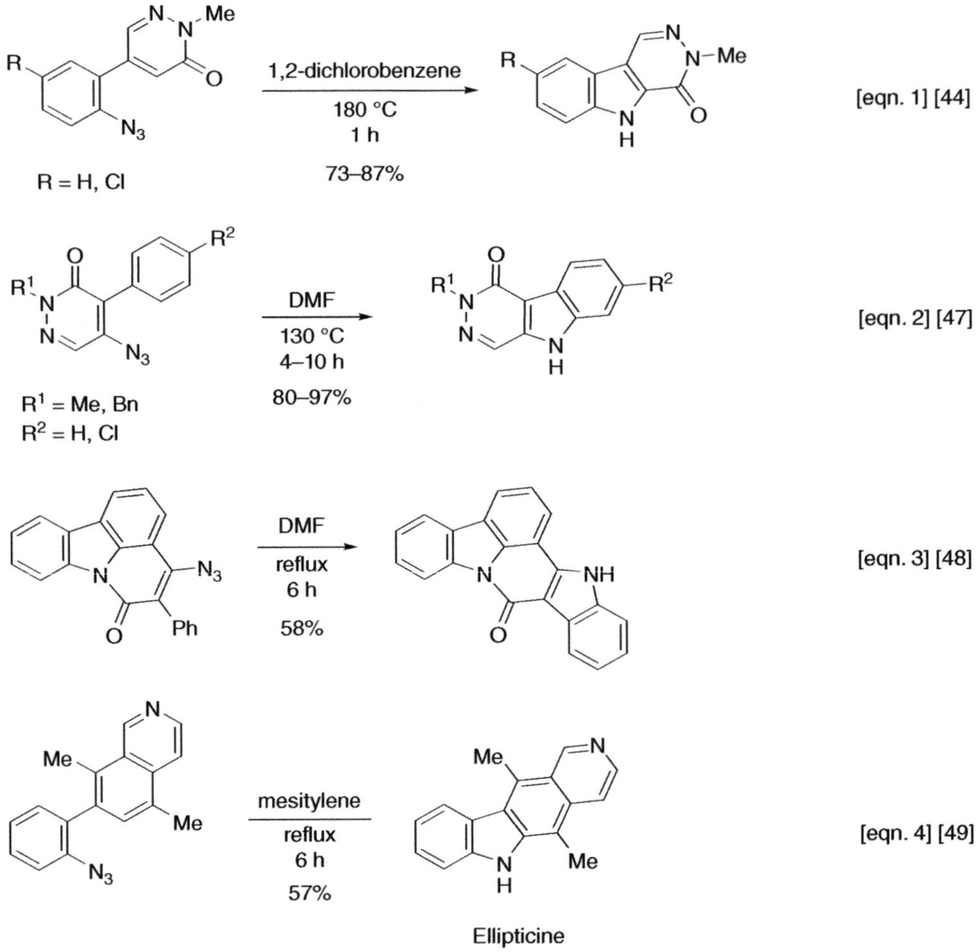

Scheme 4 *Sundberg Indole Synthesis of Fused Heterocyclic Indoles*

the appropriate azide (Entry 6) [41]. The isomeric ring system pyrido[3,4,5-*kl*]acridine, where nitrene addition occurred at C-5 of the isoquinoline ring, was not observed. Another approach to the indolo[3,2-*c*]quinoline ring system and cryptosanguinolentine is shown in Entry 7 as the indoloquinolone was reduced with Red-Al to give the alkaloid [42, 43]. Interestingly, heating this azide with trimethylphosphine gives cryptotackieine (Entry 8) [42, 43].

Several groups have prepared pyridazinoindoles using the Sundberg method, as summarized in Scheme 4 (equations 1–2). In this fashion Hajós and colleagues prepared pyridazino[4,5-*b*]indol-4-ones in very good yield (equation 1) [44]. None of the other possible cyclization product was found. Maes, Mátyus, and coworkers described a similar synthesis of this ring system [45, 46]. Haider and Wobus synthesized the isomeric pyridazino[4,5-*b*]indol-1-ones by switching the site of the azide (equation 2) [47]. The indolo[2,3:4,5]pyrido[3,2,1-*jk*]carbazolone ring system was prepared in good yield from the appropriate azide by Stadlbauer and colleagues (equation 3) [48]. Knochel and Liu extended the Sundberg azide pyrolysis to the synthesis of several pyridocarbazoles, one example of which is depicted in equation 4 [49]. In addition to ellipticine, this group fashioned 9-methoxyellipticine, isoellipticine, and 7-carbethoxyisoellipticine.

As a complement to the basic Sundberg indole synthesis, several important variations should be discussed (Scheme 5). Spagnolo and Zanirato reported that the Sundberg indolization can be effected by boron trichloride or trifluoride [50]. For example, treatment of 2-azido-3-nitrobiphenyl with BF_3 (benzene, 50 °C) gave 1-nitrocarbazole in 75% yield (equation 1). The authors proposed an intermediate nitrenium–trihalogenborane complex that undergoes cyclization and proton and BX_3 loss. Borden and colleagues investigated the photochemistry of *ortho*-azidobiphenyl by laser flash photolysis and employed time-resolved infrared experiments and calculations to detect the so-formed transient intermediates [51]. The initially formed singlet *ortho*-biphenylnitrene relaxes to the lower-energy triplet nitrene, which is persistent at 77 K in a 3-methylpentane glass. At higher temperatures the photolysis led almost immediately to benzazirine **9** and isocarbazole **11**, which isomerized to

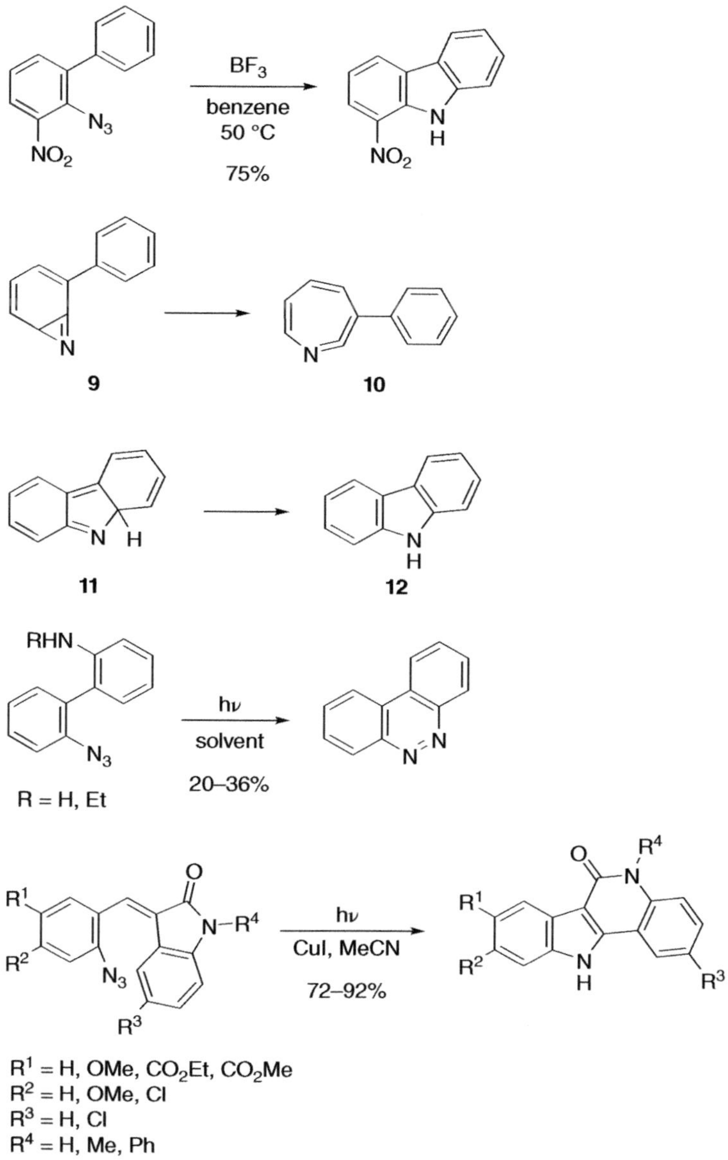

[eqn. 1] [50]

[eqn. 2] [51]

[eqn. 3] [51]

[eqn. 4] [52]

[eqn. 5] [53]

Scheme 5 *Variations of the Sundberg Indole Synthesis*

azacycloheptatraene **10** and carbazole (**12**), respectively (equations 2, 3). Isocarbazole has a lifetime of 70 nanoseconds in pentane at ambient temperature. Tomioka and colleagues studied the photolysis of 2-amino-2′-azidobiphenyls and found that the major product (20%–36%) was the azo compound benzo[*c*]cinnoline (equation 4) [52]. 4-Aminocarbazoles were formed in much lower yields (0%–12%) as were 4,10-dihydroazepino[2,3-*b*]indoles. A more productive photochemical study, in terms of carbazole synthesis, is that of Zhang and colleagues, who synthesized a series of indolo[3,2-*c*]quinolin-6-ones and related heterocycles by the photolysis of the appropriate azides (equation 5) [53]. The addition of copper(I) iodide increased the yields.

Molina and coworkers synthesized 2-(2-azidoethyl) indole by thermolysis of both (*E*)- [54] and (*Z*)-4-(2-azidophenyl)-3-butenylazide [55] (61% yield; toluene, 160 °C). Feldman, López, and colleagues discovered that the thermolysis or photolysis of 2-(allenyl)phenyl azides afforded cyclopentannelated indoles [56–58]. As seen in the work of Zhang [53], the addition of copper(I) iodide enhanced the photochemical process [58].

As with several indole-ring syntheses to be discussed, transition metals have been adapted to the Sundberg azide indole–carbazole synthesis. These include rhodium, ruthenium, palladium, and iron. Rather than discuss these elegant methods in the present chapter, I have relegated them to the respective chapters on metal-promoted indole synthesis. Two excellent reviews discuss the synthesis of nitrogen heterocycles via azides [59] and nitrenes [60].

References

[1] P.A.S. Smith and B.B. Brown, *J. Am. Chem. Soc.*, 1951, **73**, 2435–2437.
[2] P.A.S. Smith and B.B. Brown, *J. Am. Chem. Soc.*, 1951, **73**, 2438–2441.
[3] P.A.S. Smith and J.H. Boyer, *J. Am. Chem. Soc.*, 1951, **73**, 2626–2629.
[4] P.A.S. Smith and J.H. Hall, *J. Am. Chem. Soc.*, 1962, **84**, 480–485.
[5] G.D. Mendenhall and P.A.S. Smith, *Org. Syn. Coll. Vol. V*, 1973, 829–833.
[6] G. Smolinsky, *J. Am. Chem. Soc.*, 1960, **82**, 4717–4719.
[7] G. Smolinsky, *J. Am. Chem. Soc.*, 1961, **83**, 2489–2493.
[8] G. Smolinsky and C.A. Pryde, *J. Org. Chem.*, 1968, **33**, 2411–2416.
[9] R.A. Abramovitch, K.A.H. Adams, and A.D. Notation, *Can. J. Chem.*, 1960, **38**, 2152–2160.
[10] B. Coffin and R.F. Robbins, *J. Chem. Soc.*, 1965, 1252–1257.
[11] K. Isomura, S. Kobayashi, and H. Taniguchi, *Tetrahedron Lett.*, 1968, 3499–3502.
[12] R.J. Sundberg, L.-S. Lin, and D.E. Blackburn, *J. Heterocycl. Chem.*, 1969, **6**, 441.
[13] R.J. Sundberg, H.F. Russell, W.V. Ligon, Jr., and L.-S. Lin, *J. Org. Chem.*, 1972, **37**, 719–724.
[14] J.B. Petersen and K.H. Lakowitz, *Acta Chem. Scand.*, 1969, **23**, 971–974.
[15] K.E. Chippendale, B. Iddon, and H. Suschitzky, *J. Chem. Soc. (D), Chem. Commun.*, 1971, 203–204.
[16] R.Y. Ning, P.B. Madan, and L.H. Sternbach, *J. Org. Chem.*, 1973, **38**, 3995–3998.
[17] J.S. Swenton, T.J. Ikeler, and B.H. Williams, *J. Am. Chem. Soc.*, 1970, **92**, 3103–3109.
[18] J.M. Lindley, I.M. McRobbie, O. Meth-Cohn, and H. Suschitzky, *Tetrahedron Lett.*, 1976, 4513–4516.
[19] J.M. Lindley, I.M. McRobbie, O. Meth-Cohn, and H. Suschitzky, *J. Chem. Soc., Perkin Trans. 1*, 1977, 2194–2204.
[20] A. Tanaka, K. Yakushijin, and S. Yoshina, *J. Heterocycl. Chem.*, 1977, **14**, 975–979.
[21] A. Tanaka, K. Yakushijin, and S. Yoshina, *J. Heterocycl. Chem.*, 1978, **15**, 123–125.
[22] A. Tanaka, K. Yakushijin, and S. Yoshina, *J. Heterocycl. Chem.*, 1979, **16**, 785–788.
[23] L. Garanti and G. Zecchi, *J. Org. Chem.*, 1980, **45**, 4767–4769.
[24] P. Kaszynski and D.A. Dougherty, *J. Org. Chem.*, 1993, **58**, 5209–5220.
[25] E.T. Pelkey and G.W. Gribble, *Tetrahedron Lett.*, 1997, **38**, 5603–5606.
[26] V.V. Rozhkov, A.M. Kuvshinov, and S.A. Shevelev, *Org. Prep. Proc. Int.*, 2000, **32**, 94–96.
[27] V.V. Rozhkov, A.M. Kuvshinov, and S.A. Shevelev, *Synth. Commun.*, 2002, **32**, 1465–1474.
[28] V.V. Rozhkov, A.M. Kuvshinov, V.I. Gulevskaya, *et al.*, *Synthesis*, 1999, 2065–2070.
[29] V. Stockmann, J.M. Bakke, P. Bruheim, and A. Fiksdahl, *Tetrahedron*, 2009, **65**, 3668–3672.
[30] J.F. Morin and M. Leclerc, *Macromolecules*, 2001, **34**, 4680–4682.
[31] J. Jian and J.M. Tour, *J. Org. Chem.*, 2003, **68**, 5091–5103.
[32] E. Ullah, J. McNulty, and A. Robertson, *Eur. J. Org. Chem.*, 2012, 2127–2131.
[33] M. Pudlo, D. Csanyi, F. Moreau, *et al.*, *Tetrahedron*, 2007, **63**, 10320–10329.
[34] K. Okano, H. Fujiwara, T. Noji, *et al.*, *Angew. Chem. Int. Ed.*, 2010, **49**, 5925–5929.
[35] H. Tokuyama, K. Okano, H. Fujiwara, *et al.*, *Chem. Asian J.*, 2011, **6**, 560–572.
[36] G. Timári, T. Soós, and G. Hajós, *Synlett*, 1997, 1067–1068.
[37] S. Hostyn, B.U.W. Maes, L. Pieters, *et al.*, *Tetrahedron*, 2005, **61**, 1571–1577.
[38] M. Béres, G. Timári, and G. Hajós, *Tetrahedon Lett.*, 2002, **43**, 6035–6038.
[39] G. Timári, T. Soós, G. Hajós, *et al.*, *Bioorg. Med. Chem. Lett.*, 1996, **6**, 2831–2835.
[40] G. Hajós, Z. Riedl, G. Timári, *et al.*, *Molecules*, 2003, **8**, 480–487.
[41] G. Van Baelen, C. Meyers, G.L.F. Lemière, *et al.*, *Tetrahedron*, 2008, **64**, 11802–11809.
[42] M. Fresneda, P. Molina, and S. Delgado, *Tetrahedron*, 2001, **57**, 6197–6202.
[43] P.M. Fresneda, P. Molina, and S. Delgado, *Tetrahedron Lett.*, 1999, **40**, 7275–7278.
[44] G. Krajsovszky, P. Mátyus, Z. Riedl, *et al.*, *Heterocycles*, 2001, **55**, 1105–1111.
[45] P. Tapolcsányi, B.U.W. Maes, K. Monsieurs, *et al.*, *Tetrahedron*, 2003, **59**, 5919–5926.
[46] P. Tapolcsányi, G. Krajsovszky, R. Andó, *et al.*, *Tetrahedron*, 2002, **58**, 10137–10143.
[47] N. Haider and A. Wobus, *Heterocycles*, 2006, **68**, 2549–2561.
[48] H.V. Dang, B. Knobloch, N.S. Habib, *et al.*, *J. Heterocycl. Chem.*, 2005, **42**, 85–91.
[49] C.-Y. Liu and P. Knochel, *J. Org. Chem.*, 2007, **72**, 7106–7115.
[50] P. Spagnolo and P. Zanirato, *J. Chem. Soc., Perkin Trans. 1*, 1988, 2615–2620.
[51] M.-L. Tsao, N. Gritsan, T.R. James, *et al.*, *J. Am. Chem. Soc.*, 2003, **125**, 9343–9358.
[52] S. Murata, H. Tsuji, and H. Tomioka, *Bull. Chem. Soc. Jpn.*, 1994, **67**, 895–897.
[53] Z. Shi, Y. Ren, B. Li, *et al.*, *Chem. Commun.* 2010, **46**, 3973–3975.
[54] P. Molina, J. Alcantara, and C. Lopez-Leonardo, *Tetrahedron Lett.*, 1995, **36**, 953–956.
[55] P. Molina, J. Alcántara, and C. López-Leonardo, *Tetrahedron*, 1996, **52**, 5833–5844.
[56] K.S. Feldman, M.R. Iyer, and D.K. Hester, II, *Org. Lett.*, 2006, **8**, 3113–3116.
[57] C.S. López, O.N. Faza, K.S. Feldman, *et al.*, *J. Am. Chem. Soc.*, **129**, 7638–7646.
[58] K.S. Feldman, D.K. Hester II, C.S. López, and O.N. Faza, *Org. Lett.*, 2008, **10**, 1665–1668.
[59] E.F.V. Scriven and K. Turnbull, *Chem. Rev.*, 1988, **88**, 297–368.
[60] B.C.G. Söderberg, *Curr. Org. Chem.*, 2000, **4**, 727–764.

28

Hemetsberger Indole Synthesis

The Hemetsberger indole synthesis involves the thermolysis of α-azidocinnamate esters in xylene to give the corresponding indole-2-carboxylates via nitrene intermediates (Scheme 1) [1, 2]. The starting azides are readily prepared by an aldol condensation between aryl aldehydes and α-azidoacetates [3]. A suggested pathway is shown along with some Hemetsberger results (equations 2, 3). Reversible formation of 2*H*-azirine may also be involved. Indeed, at lower temperature (*n*-heptane, 98 °C) the 2*H*-azirine was formed in 80% yield. Unlike the Cadogan–Sundberg and Sundberg syntheses, the Hemetsberger synthesis leads only to indoles.

A summary of the wide variety of indoles that are accessible via the Hemetsberger synthesis is shown in Table 1 [4, 5, 7–16]. Provided the necessary aryl aldehyde is available, the indolization of the precursor α-azidocinnamate is practically failsafe. In addition to the indoles in Entries 1 and 2 [4, 5], Kim and colleagues synthesized pyrano[3,2-*f*]- and pyrano[2,3-*g*]indoles [6]. The chemistry reported by Rees and coworkers in Entry 5 was employed in a synthesis of the bacterial coenzyme methoxatin [9,10]. Moody uncovered the novel Claisen-type rearrangement following the Hemetsberger indole synthesis in Entry 6 [11]. The indole amide shown in Entry 8 was used to prepare 2-cyano-4-hydroxyindole, a useful intermediate for the synthesis of cyanopindolol [14]. Knittel isolated the azirines shown in Entry 10 (cyclohexane, 80 °C) from their respective azidocinnamates and subjected the azirines to thermolysis in xylene (reflux) to afford indoles in excellent yield [16]. In addition to the work cited above, Moody and colleagues have been major players in the application of the Hemetsberger reaction to the synthesis of indoles and other heterocycles [17, 18].

If the nitrene intermediate has a choice of two cyclization sites, one often gets mixtures. For example, an inseparable 1:1 mixture of two indoles was obtained when indole **13** was subjected to the Hemetsberger synthesis (equation 1, Scheme 2) [17]. Also shown in Scheme 2 [19–23]are some of the natural products (**14–18**) that feature as a key step the Hemetsberger indole synthesis. Besides Boger and his contributions to the synthesis of (+)-duocarmycin A (**18**) and related antitumor antibiotics [24, 25], the groups of Tietze [26] and Terashima [27] employed Hemetsberger tactics to prepare indoles in early stages of their syntheses. Syntheses of the highly complex and potent antitumor antibiotic CC-1065 and related model compounds have relied on Hemetsberger chemistry [28–30].

The Hemetsberger indole synthesis has been extensively used by medicinal chemists to prepare indole-based drugs and drug candidates. A selection of these compounds is shown in Scheme 3 [31–42]. The preparation of each indole involved a Hemetsberger ring formation, and the drug target is indicated. In most cases the lead or most biologically active compound is shown. Obviously, in several cases a decarboxylation step at the indole C-2 position was involved. Indolequinone **27** demonstrates antiproliferative activity against human pancreatic cancer cells (MIA PaCa-2) [39]. The Hemetsberger route was superior to a Reissert indole synthesis, which is presented in Chapter 40.

In addition to the natural products **14–18** shown in Scheme 2, the Hemetsberger indole ring construction played a pivotal role in the syntheses of analogues of the plant growth hormone 3-indoleacetic acid [43], the indole-containing ABCD ring of the insecticidal alkaloid nodulisporic acid [44], the marine alkaloids makaluvamine D and discorhabin C [45], the marine indole

Indole Ring Synthesis: From Natural Products to Drug Discovery, First Edition. Gordon W. Gribble.

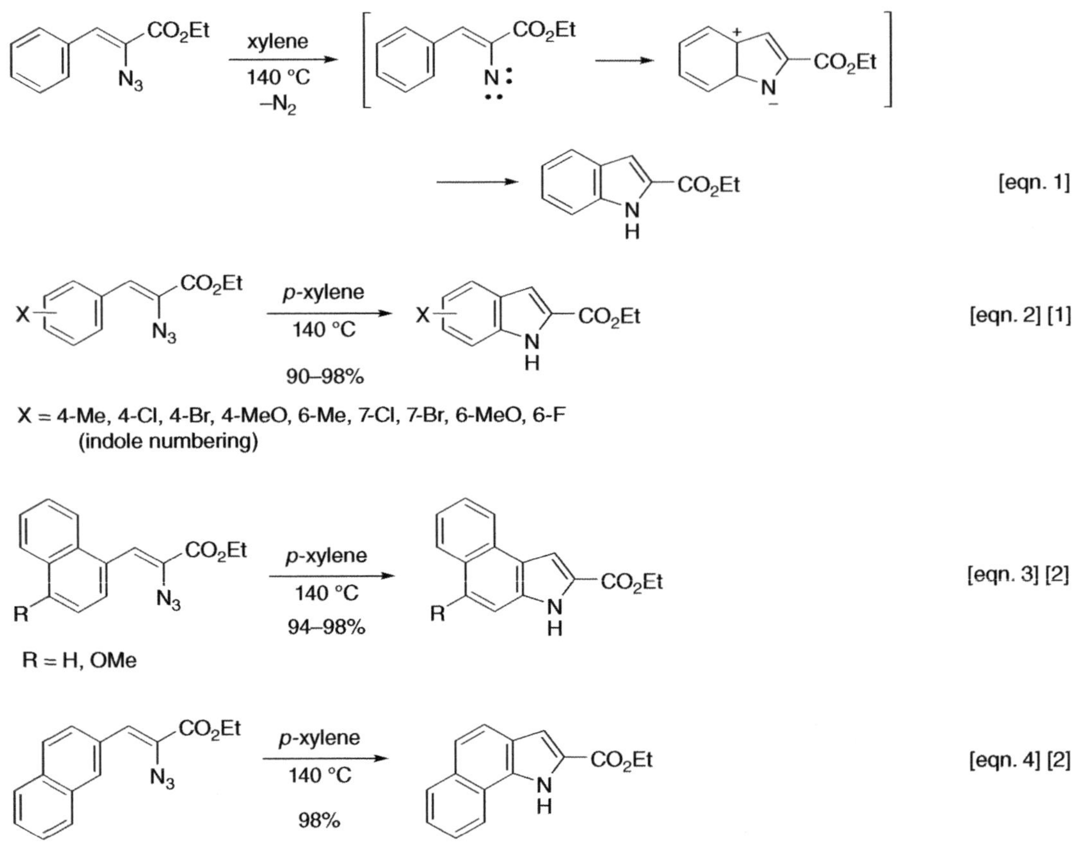

Scheme 1 *Hemetsberger Indole Synthesis*

alkaloids meridianins A, C–E [46], one of the two core indole structures of the bis-indole alkaloid conophylline [47], indole neurotensin mimetics [48, 49], the β-carboline alkaloids roeharmine and (–)-1,2,3,4-tetrahydroroeharmine [50], and a formal synthesis of the antitumor antibiotic CC-1065 [51]. In nearly all of these studies, the Hemetsberger indole synthesis served as the crucial starting point in the synthesis. However, in Bentley's synthesis of the indole fragments of the thiopeptide antibiotics nosiheptide and glycothiohexide, a Hemetsberger approach was less successful than either a Fischer or a Cadogan–Sundberg indole synthesis [52], as we saw for this study in Chapter 26 and earlier in this chapter, respectively. Cook and coworkers synthesized several indole alkaloids via 2-carbomethoxy-6-methoxyindole, which they prepared via the Hemetsberger route on a 300-gram scale in 89% yield [53]. Further applications of the Hemetsberger synthesis to an array of indole targets are shown in Table 2. The reaction in Entry 1 needed to be performed in dilute solution for maximum results. A small amount of methyl 2-(2-benzyloxy-3-methoxyphenyl)-2-cyanoacetate (10%–15%) was also obtained, which arose via the known process from the corresponding 2*H*-azirine [54]. This indole was used to construct novel mitomycin antitumor agents. The isomeric methyl 8*H*-thieno[3,2-*g*]indole-7-carboxylate was also prepared (50% yield) by De and colleagues (Entry 2), but neither indole could be hydrolyzed and decarboxylated [55]. The fluoroindoles in Entry 3 were employed to prepare fluorinated [56] analogues of the hallucinogenic tryptamines. Engels and colleagues prepared several fluorinated indole nucleosides in a Hemetsberger sequence that yielded 4-fluoro-, 6-fluoro-, and 4,6-difluoroindoles [57]. The presence of bromine in Entry 4 prevented the alternative nitrene cyclization that would result from the azide without the blocking atom; that is, a tetrahydro[1,4]oxazino[2,3-*f*]indole would be obtained instead of a tetrahydro[1,4]oxazino[3,2-*g*]indole that is obtained (Entry 4) [58]. In addition to the desired thieno[3,2-*f*]indole, De and Ghosh obtained a thieno[2,3-*h*]benz-1,2-oxazine (Entry 5) [59]. This novel nitrene insertion onto a methoxyl group also resulted from vinyl azides derived from various methoxynaphthaldehydes and 2,6-dimethoxybenzaldehyde. In concert with the Bartoli indole synthesis (Chapter 4), Chai and coworkers employed a Hemetsberger strategy to craft a series of mono- and dibrominated 5,6-dimethoxyindoles and the corresponding 5,6-dihydroxyindole building blocks that are found in

Table 1 *Applications of the Hemetsberger Indole Synthesis*

Entry	Azide	Conditions	Indole	% Yield	Ref.
1	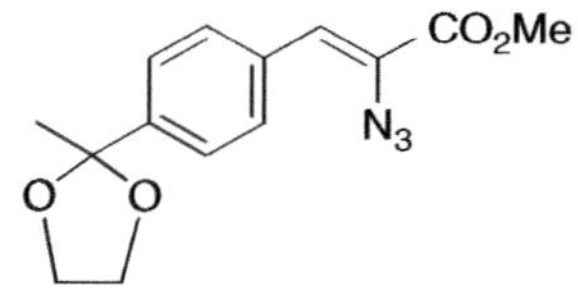	xylene 140 °C 2 h	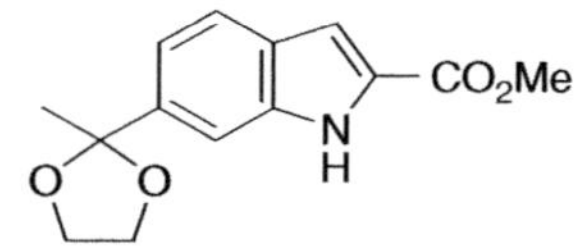	78%	4
2	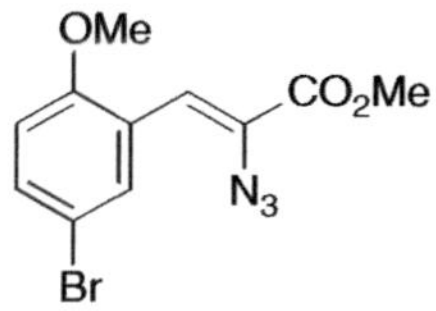	xylene 140 °C 3 h	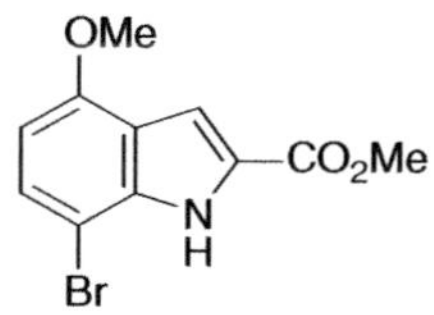	51%	5
3	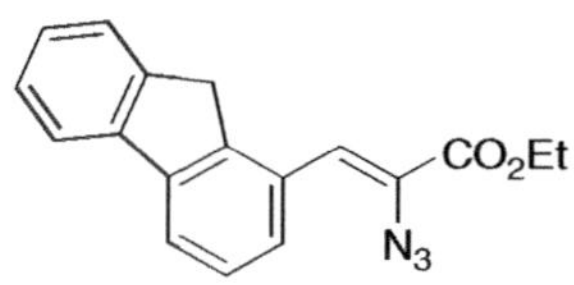	xylene 140 °C 1 h	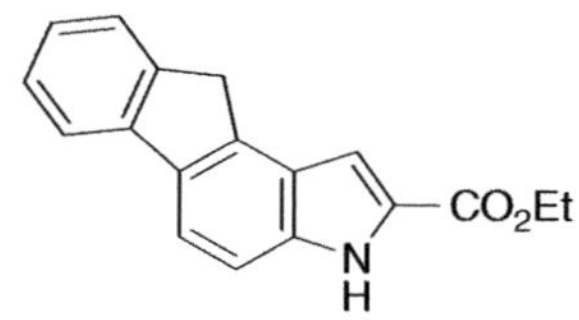	90%	7
4	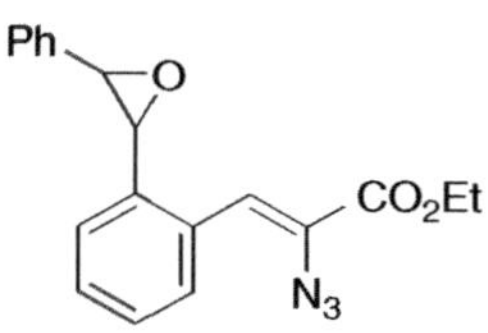	toluene 110 °C	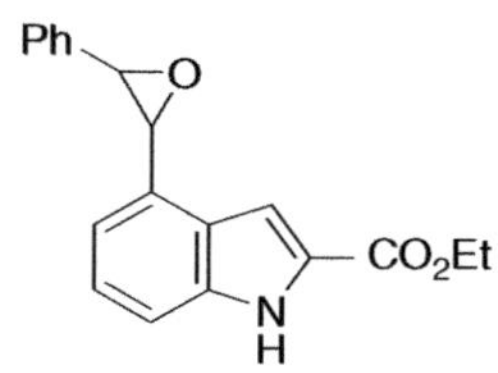	80%	8
5	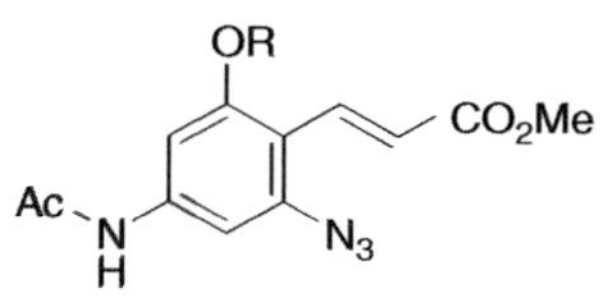 R = Me, Bn`	xylene 140 °C 4 h	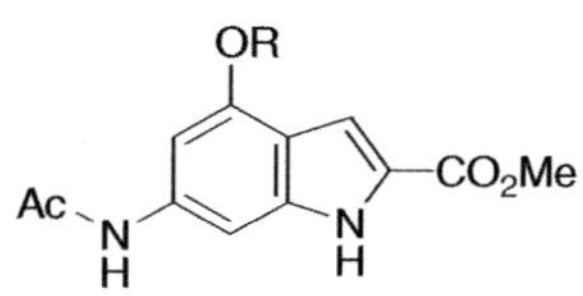	82–87%	9, 10
6	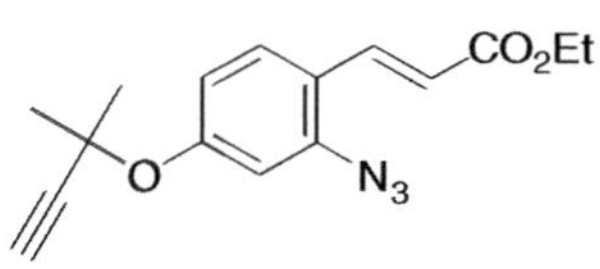	toluene 110 °C 3 h	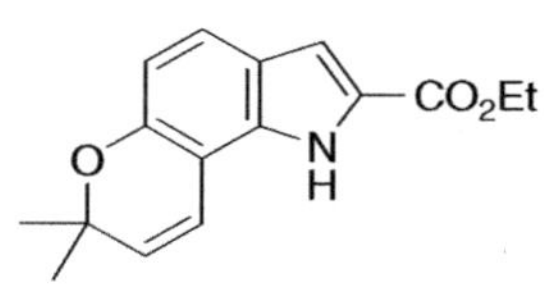	94%	11
7	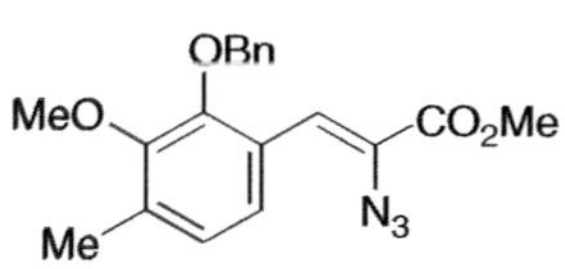	xylene 140 °C 1 h	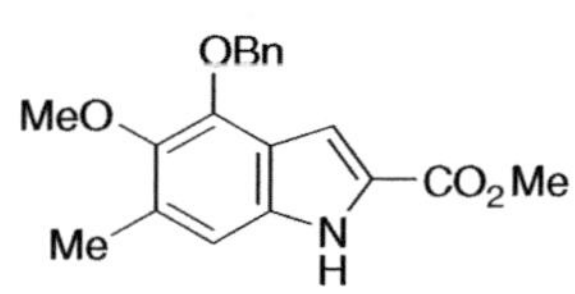	96%	12, 13
8	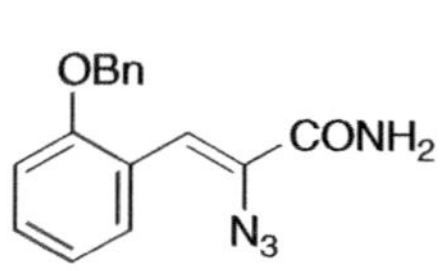	toluene 110 °C	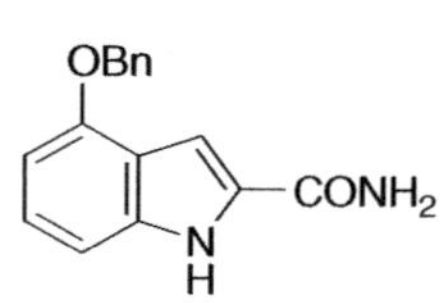	71%	14

(Continued overleaf)

Table 1 *(continued)*

Entry	Azide	Conditions	Indole	% Yield	Ref.
9	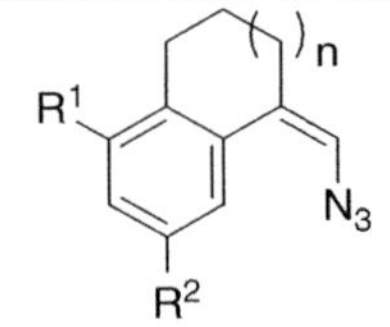 n = 1, 2; R^1 = H, OMe; R^2 = H, OMe	mesitylene reflux	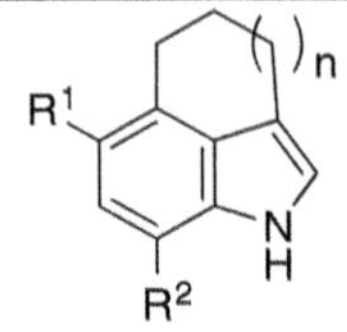	67–72%	15
10	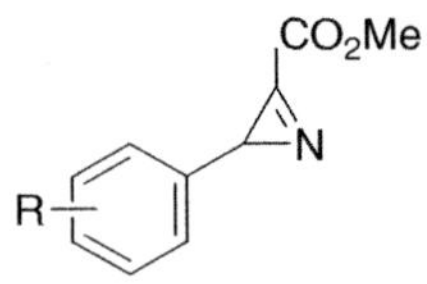 R = H, 6-Me, 6-OMe, 6-Cl (indole numbering)	xylene 140 °C	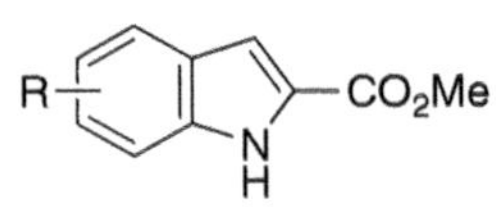	90–97%	16

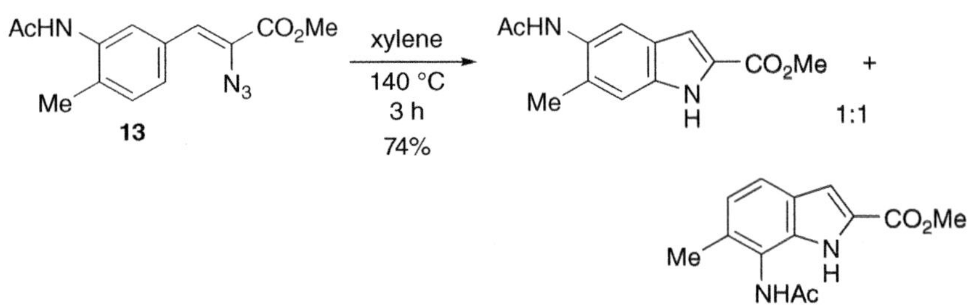

[eqn. 1] [13]

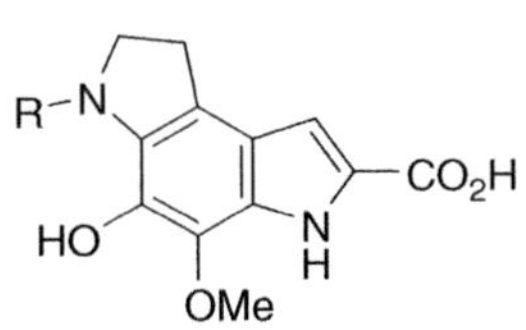

14, R = $CONH_2$ (PDE-1) [19], [20]
15, R = COMe (PDE-2) [19], [20]

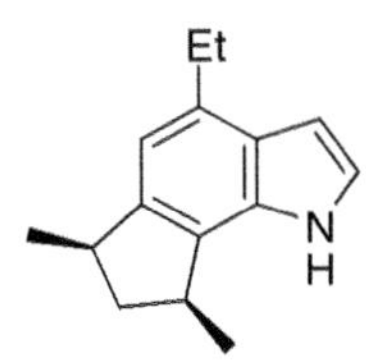

17 (Trikentrin A) [23]

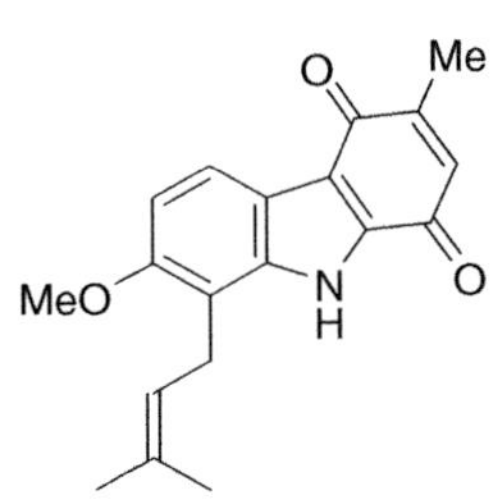

16 (Murrayaquinone-B) [21], [22]

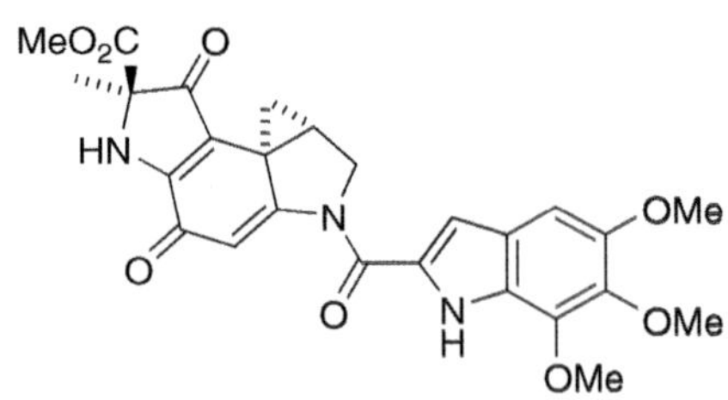

18 (Duocarmycin A) [24]

Scheme 2 *Applications of the Hemetsberger Indole Synthesis*

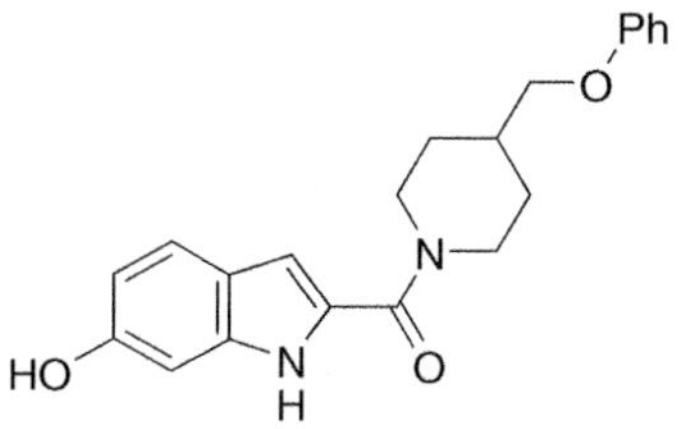

19 (NR2B selective NMDA receptor antagonist) [31]

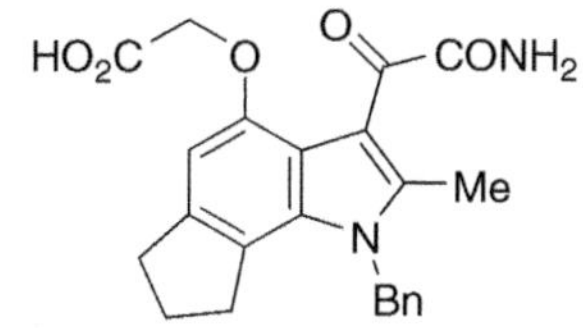

20 (Inhibitor of human nonpancreatic secretory phospholipase A_2) [32]

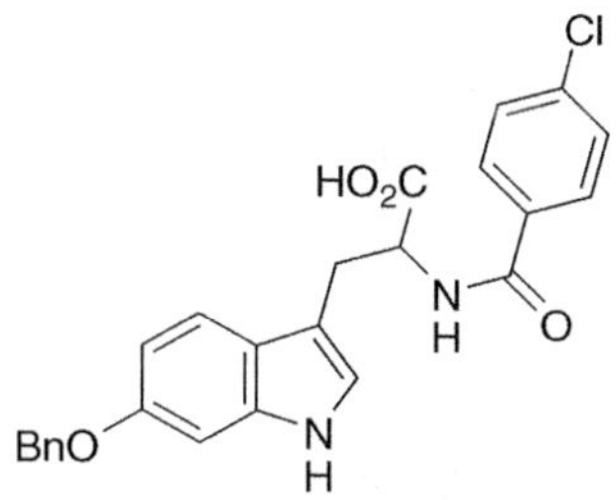

21 (Peroxisome proliferator-activated receptor ligand) [33]

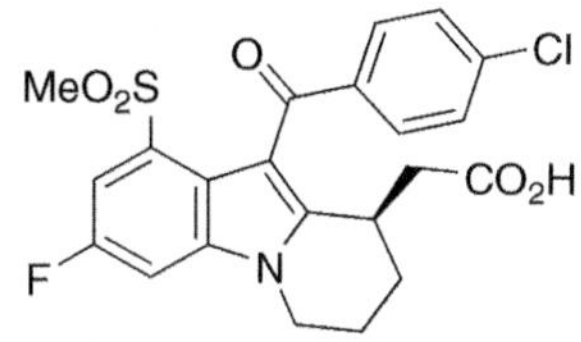

22 (Prostaglandin D_2 receptor subtype 1 antagonist) [34]

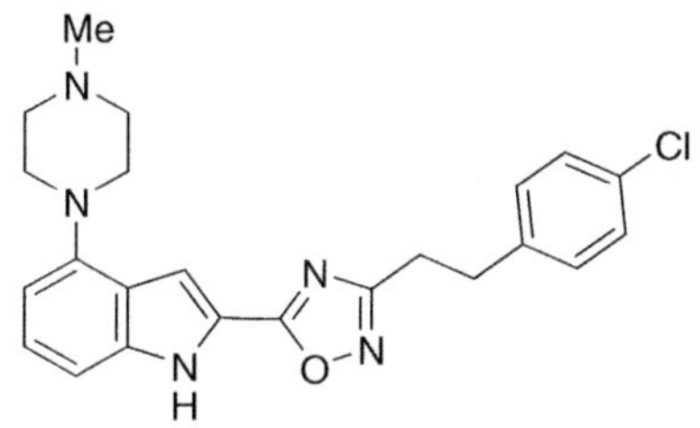

23 (Nociceptin/orphanin FQ (N/OFQ) receptor antagonist) [35]

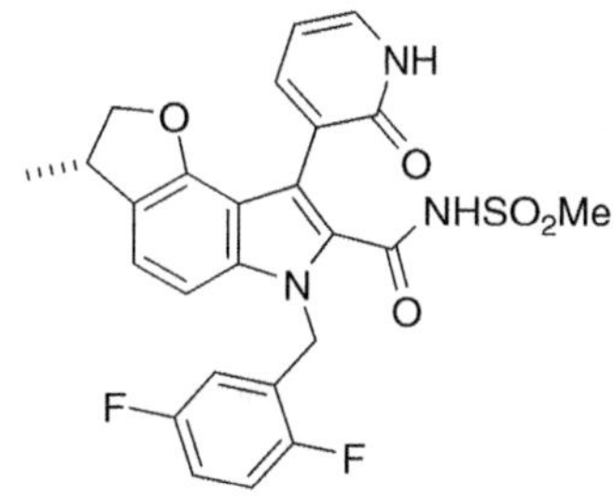

24 (HCV NS5B polymerase inhibitor) [36]

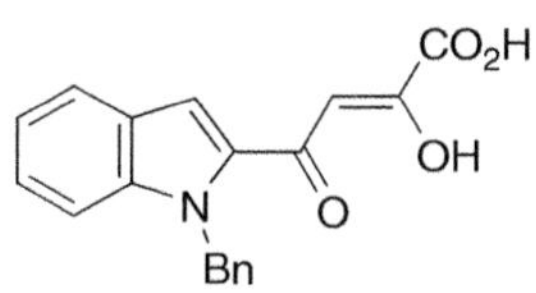

25 (HIV-1 integrase inhibitor) [37]

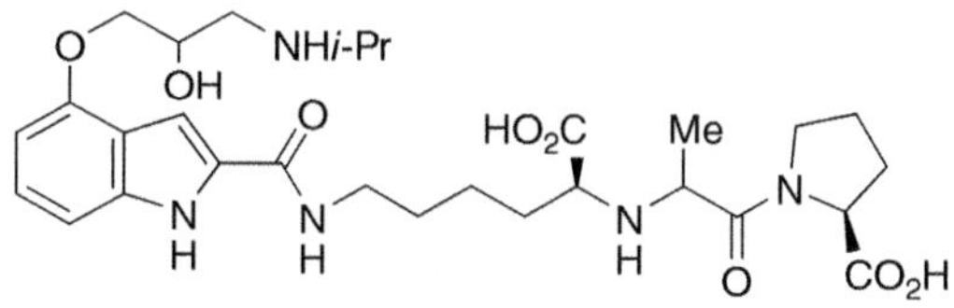

26 (A575C) (Angiotensin converting enyme inhibitor and beta-adrenoceptor blocker) [38]

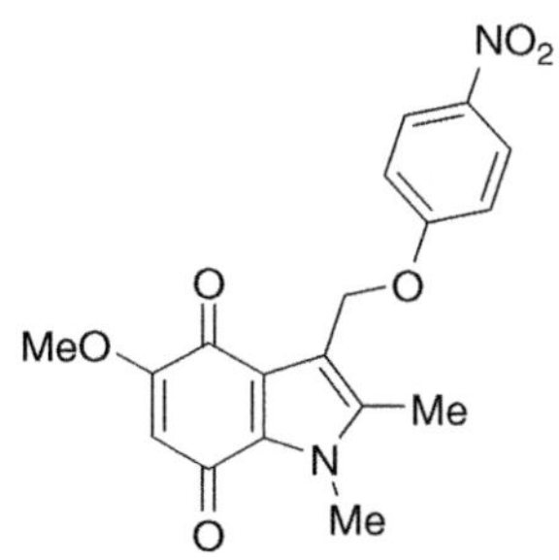

27 (NaD(P)H: Quinone oxidoreductase 1 inhibitor) [39]

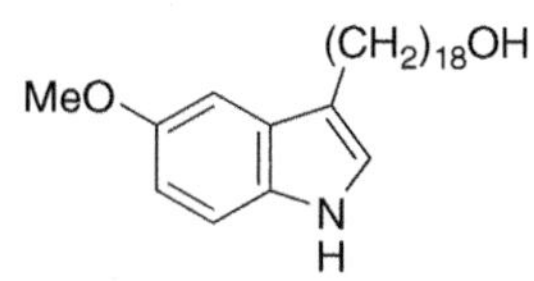

28 (Neural stem cell derived neurosphere differentiation inducer) [40]

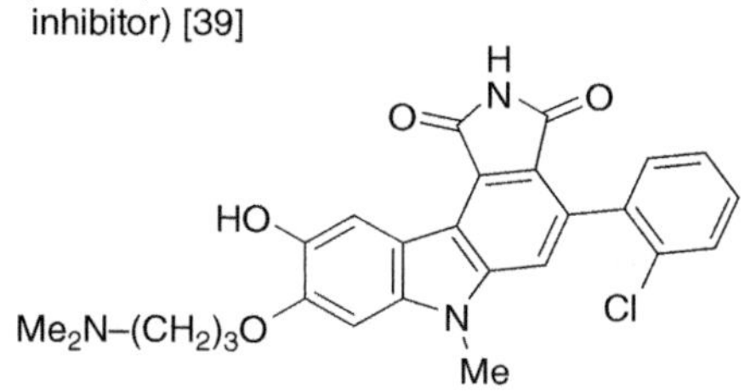

29 (Chk1 checkpoint kinase inhibitor) [41]

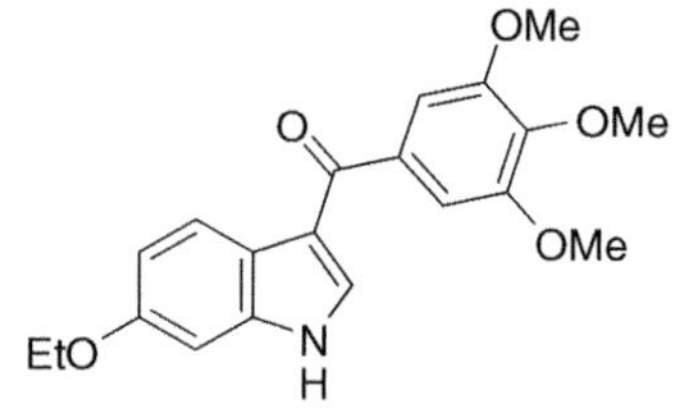

30 (KB antitumor) [42]

Scheme 3 *Applications of the Hemetsberger Indole Synthesis to Drug Synthesis*

Table 2 *Further Applications of the Hemetsberger Indole Synthesis*

Entry	Azide	Conditions	Indole	% Yield	Ref.
1	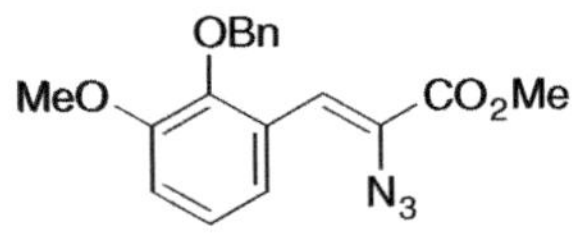	xylene 140 °C 1.75 h	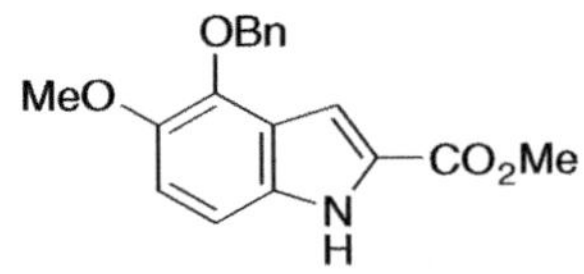	84%	54
2	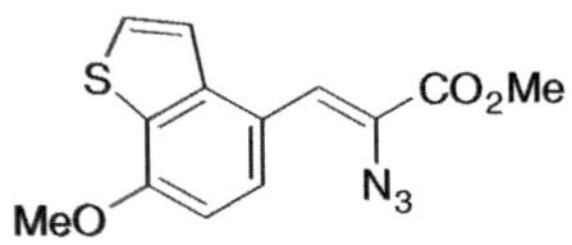	xylene 140 °C 3 h	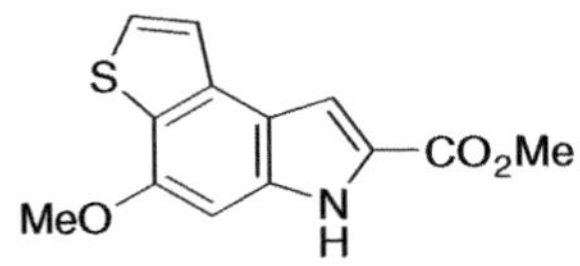	81%	55
3	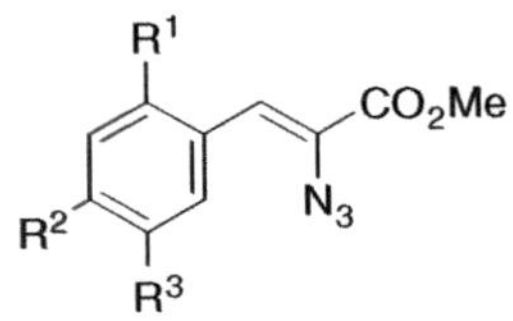 R^1 = OBn, R^2 = F, R^3 = H R^1 = OBn, R^2 = H, R^3 = F R^1 = H, R^2 = F, R^3 = H	xylene 140 °C 3–6 h	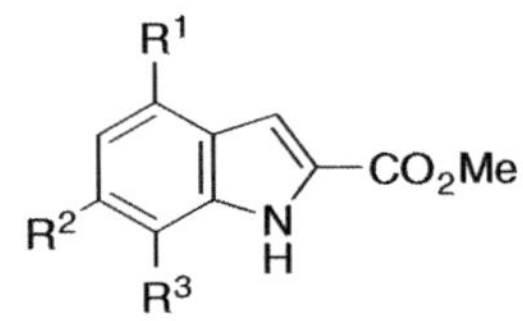	43–54%	56
4	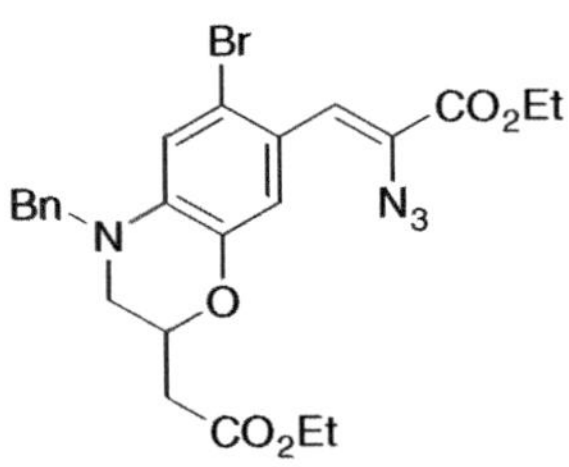	xylene 140 °C	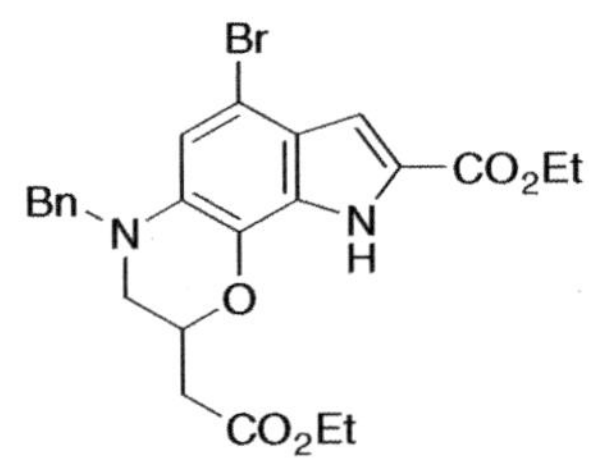	58%	58
5	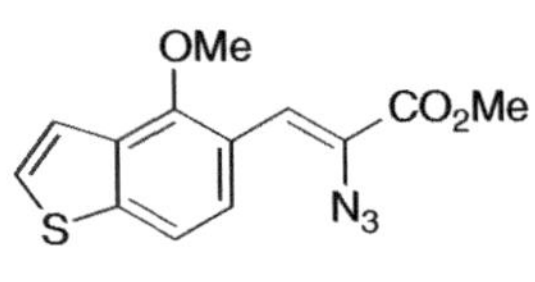	xylene 140 °C 2.5 h	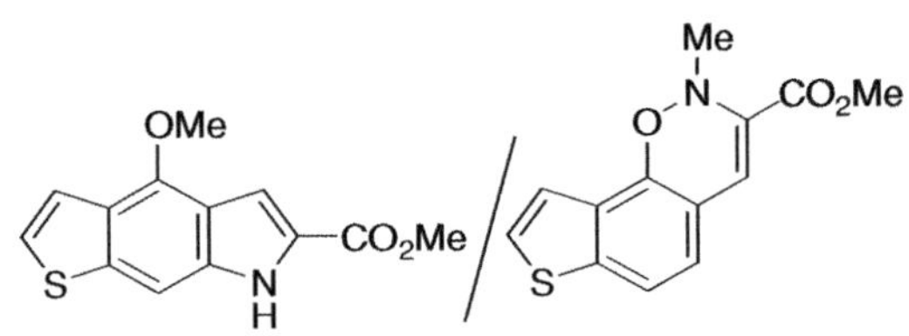	55%/45%	59
6	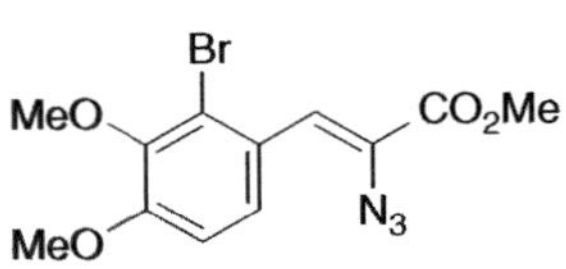	xylene 140 °C 2 h	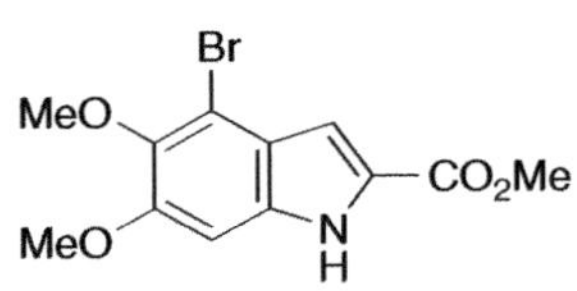	62%	60
7	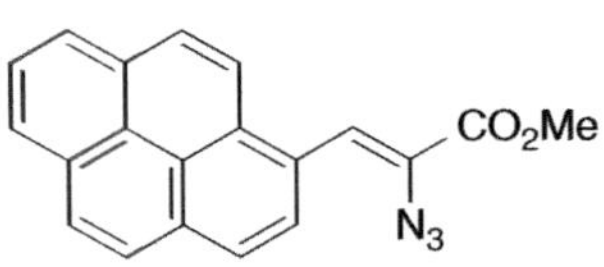	toluene 110 °C 4 h	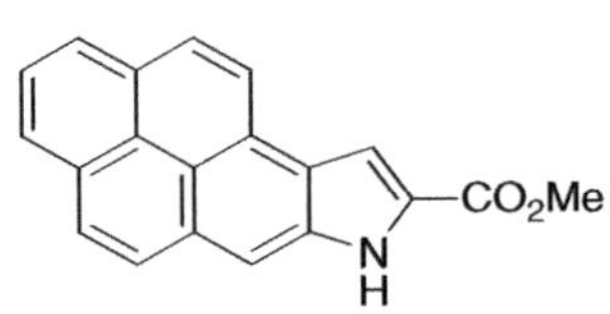	92%	61
8	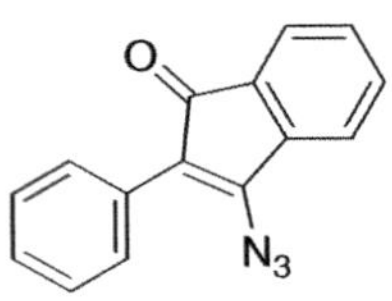	EtOH 70 °C 1 h	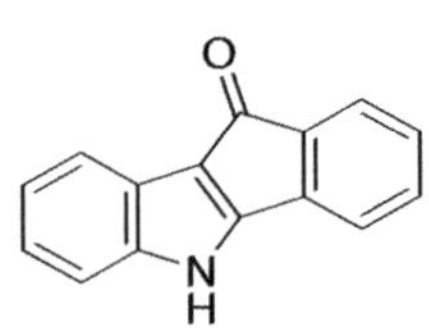	51%	62

Table 2 *(continued)*

Entry	Azide	Conditions	Indole	% Yield	Ref.
9	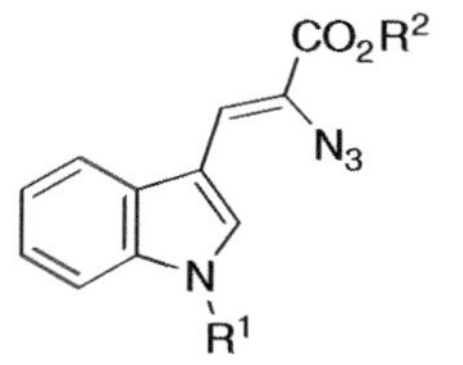 R^1 = Me, Bn, CH_2OMe R^2 = Et, Me	toluene 110 °C 1–2 h	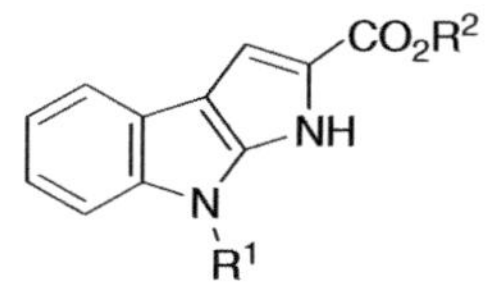	70–94%	64
10	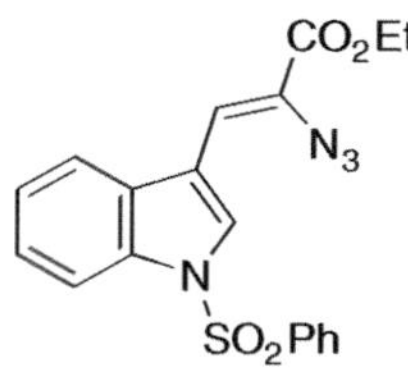	xylene 140 °C 2 h	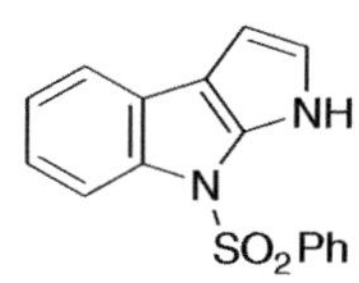	43%	65
11	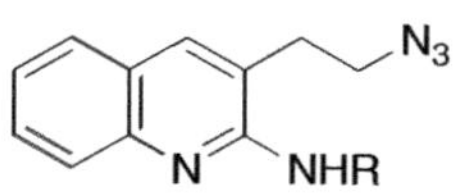 R = Ph, 4-Tol, 4-BrPh	230–240 °C 4 min	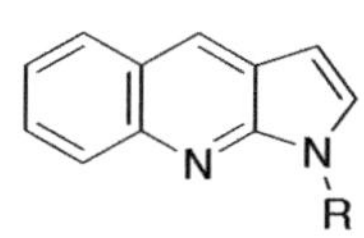	39–70%	66

natural products and biopolymers (Entry 6) [60]. Entry 7 features the synthesis of a novel pyreno[2,1-*b*]pyrrole, which was converted to the novel indigoid analogue, among other derivatives. These pyreno[2,1-*b*]pyrroles exhibit intense fluorescence in both solution and in crystalline form [61]. The 5*H*-indeno[1,2-*b*]indol-10-one prepared in Entry 8 [62] has a role in the enzyme activity of antioxidants [63]. Entries 9 and 10 illustrate the new ring system, pyrrolo[2,3-*b*]indole [64, 65]. The interesting synthesis of pyrrolo[2,3-*b*]quinolines (Entry 11) may involve formation of an aziridine from a nitrene, followed by nucleophilic cyclization and loss of ammonia [66].

Although the Hemetsberger reaction is clearly a powerful synthesis of indole-2-carboxylates, some problems are encountered in addition to the occasional mixture of nitrene cyclization products. Indeed, often the aldolization route to the 2-azidocinnamates proceeds poorly. This can be obviated at lower temperatures (–30 °C) to stop the reaction at the azido alcohol stage. Subsequent dehydration with thionyl chloride in pyridine or triethylamine afforded the 2-azidocinnamates. This two-step process was superior to the conventional one-step aldolization/dehydration reaction, as found by Murakami and coworkers [67]. Tercel's group also observed that a low temperature (–78 °C) aldol condensation, isolation of the azido alcohol, and subsequent dehydration (mesyl chloride/triethylamine) gave the 2-azidocinnamate in much higher yield than the classic methoxide aldolization [68]. As shown in Scheme 4, equation 1, Nair and George reported a new synthesis of 2-azidocinnamates [69]. A 1:1 mixture of *E*- and *Z*-isomers was obtained. Equation 2 depicts a synthesis of β-azidostyrenes when decarboxylation occurs. The intermediate 2-azido-3-nitrato ester was isolated after the first step with cerium(IV) ammonium nitrate, but pyrolysis studies were not reported. Murakami's team used *tert*-butyl azidoacetate to form the cinnamates [70]. The advantages were that the aldol reaction was greatly improved for the less-reactive aryl aldehydes, and the method provided a simple route to the important *tert*-butyl indole-2-carboxylates. As before [67], a two-step synthesis of 2-azidocinnamates was preferred. As shown in equation 3, a microwave-assisted Hemetsberger synthesis afforded very high yields of indoles [71]. Seeberger and colleagues described a continuous flow thermolysis of 2-azidocinnamates to give indoles, including a synthesis of the D-amino acid oxidase (DAAO) inhibitor 4*H*-furo[3,2-*b*]pyrrole-5-carboxylic acid [72].

An important review of the synthesis and utility of azides is that of Scriven and Turnbull [73].

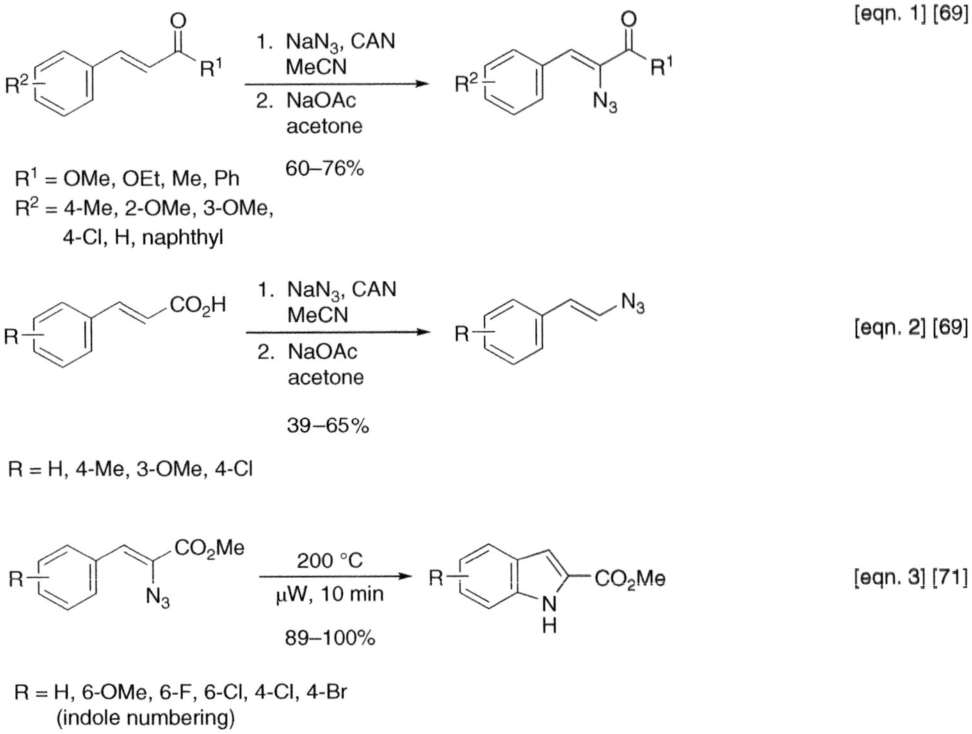

Scheme 4 *Modifications of the Hemetsberger Indole Synthesis*

References

[1] H. Hemetsberger, D. Knittel, and H. Weidmann, *Monatsh.*, 1970, **101**, 161–165.
[2] H. Hemetsberger and D. Knittel, *Monatsh. Chem.*, 1972, **103**, 194–204.
[3] H. Hemetsberger, D. Knittel, and H. Weidmann, *Monatsh. Chem.*, 1969, **100**, 1599–1603.
[4] P. T. Kim, R. Guilard, P. Dodey, and R. Sornay, *J. Heterocycl. Chem.*, 1981, **18**, 1365–1371.
[5] P.T. Kim, R. Guilard, S. Samreth, and R. Sornay, *J. Heterocycl. Chem.*, 1981, **18**, 1373–1377.
[6] P.T. Kim, R. Guilard, and P. Renaut, *Can. J. Chem.*, 1982, **60**, 2093–2098.
[7] D.M.B. Hickey, C.J. Moody, and C.W. Rees, *Chem. Commun.*, 1982, 3–4.
[8] D.M.B. Hickey, C.J. Moody, and C.W. Rees, *J. Chem. Soc., Chem. Commun.*, 1982, 1419–1421.
[9] A.R. MacKenzie, C.J. Moody, and C.W. Rees, *J. Chem. Soc., Chem. Commun.*, 1983, 1372–1373.
[10] A.R. MacKenzie, C.J. Moody, and C.W. Rees, *Tetrahedron*, 1986, **42**, 3259–3268.
[11] C.J. Moody, *J. Chem. Soc., Perkin Trans. 1*, 1984, 1333–1337.
[12] G.B. Jones and C.J. Moody, *J. Chem. Soc., Chem. Commun.*, 1989, 186–187.
[13] G.B. Jones and C.J. Moody, *J. Chem. Soc., Perkin Trans. 1*, 1989, 2455–2462.
[14] R.E. Adams, J.B. Press, and E.G. Deegan, *Synth. Commun.*, 1991, **21**, 675–681.
[15] C.J. Moody, A.L. Beck, and W.J. Coates, *Tetrahedron Lett.*, 1989, **30**, 4017–4018.
[16] D. Knittel, *Synthesis*, 1985, 186–188.
[17] L. Henn, D.M.B. Hickey, C.J. Moody, and C.W. Rees, *J. Chem. Soc., Perkin Trans. 1*, 1984, 2189–2196.
[18] D.M.B. Hickey, A.R. MacKenzie, C.J. Moody, and C.W. Rees, *J. Chem. Soc., Perkin Trans. 1*, 1987, 921–926.
[19] R.E. Bolton, C.J. Moody, C.W. Rees, and G. Tojo, *J. Chem. Soc., Chem. Commun.*, 1985, 1775–1776.
[20] R.E. Bolton, C.J. Moody, C.W. Rees, and G. Tojo, *J. Chem. Soc., Perkin Trans. 1*, 1987, 931–935.
[21] T. Martin and C.J. Moody, *J. Chem. Soc., Chem. Commun.*, 1985, 1391–1392.
[22] T. Martin and C.J. Moody, *J. Chem. Soc., Perkin Trans. 1*, 1988, 241–246.
[23] J.K. MacLeod and L.C. Monahan, *Aust. J. Chem.*, 1990, **43**, 329–337.
[24] M.S. Tichenor, J.D. Trzupek, D.B. Kastrinsky, *et al.*, *J. Am. Chem. Soc.*, 2006, **128**, 15683–15696.
[25] K.S. MacMillan, T. Nguyen, I. Hwang, and D.L. Boger, *J. Am. Chem. Soc.*, 2009, **131**, 1187–1194.
[26] L.F. Tietze and F. Major, *Eur. J. Org. Chem.*, 2006, 2314–2321.
[27] Y. Fukuda, Y. Itoh, K. Nakatani, and S. Terashima, *Tetrahedron*, 1994, **50**, 2793–2808.
[28] D.L. Boger and R.S. Coleman, *J. Org. Chem.*, 1984, **49**, 2240–2245.
[29] D.L. Boger, T. Ishizaki, H. Zarrinmayeh, *et al.*, *J. Am. Chem. Soc.*, 1990, **112**, 8961–8971.
[30] D.L. Boger, L.R. Cerbone, and D. Yohannes, *J. Org. Chem.*, 1988, **53**, 5163–5166.

[31] I. Borza, S. Kolok, A. Gere, *et al.*, *Bioorg. Med. Chem. Lett.*, 2003, **13**, 3859–3861.
[32] J.S. Sawyer, D.W. Geight, E.C.R. Smith, *et al.*, *J. Med. Chem.*, 2005, **48**, 893–896.
[33] X. Dong, Z. Zhang, R. Wen, *et al.*, *Bioorg. Med. Chem. Lett.*, 2006, **16**, 5913–5916.
[34] C. Beaulieu, D. Guay, Z. Wang, *et al.*, *Bioorg. Med. Chem. Lett.*, 2008, **18**, 2696–2700.
[35] Y. Sugimoto, A. Shimizu, T. Kato, *et al.*, *Bioorg. Med. Chem. Lett.*, 2006, **16**, 3569–3573.
[36] F. Velázquez, S. Venkatraman, C.A. Lesburg, *et al.*, *Org. Lett.*, 2012, **14**, 556–559.
[37] M. Sechi, M. Derudas, R. Dallocchio, *et al.*, *J. Med. Chem.*, 2004, **47**, 5298–5310.
[38] G.W. Hardy, D. Bull, H.T. Jones, *et al.*, *Tetrahedron Lett.*, 1988, **29**, 799–802.
[39] M.A. Colucci, P. Reigan, D. Siegel, *et al.*, *J. Med. Chem.*, 2007, **50**, 5780–5789.
[40] D. Coowar, J. Bouissac, M. Hanbali, *et al.*, *J. Med. Chem.*, 2004, **47**, 6270–6282.
[41] J.B. Smaill, H.H. Lee, B.D. Palmer, *et al.*, *Bioorg. Med. Chem. Lett.*, 2008, **18**, 929–933.
[42] T.-S. Wu, M.S. Coumar, J.-Y. Chang, *et al.*, *J. Med. Chem.*, 2009, **52**, 4941–4945.
[43] F.A.F. da Rosa, R.A. Rebelo, and M.G. Nascimento, *J. Braz. Chem. Soc.*, 2003, **14**, 11–15.
[44] P. Magnus and T.E. Mansley, *Tetrahedron Lett.*, 1999, **40**, 6909–6912.
[45] E.V. Sadanandan, S.K. Pillai, M.V. Lakshmikantham, *et al.*, *J. Org. Chem.*, 1995, **60**, 1800–1805.
[46] P.M. Fresneda, P. Molina, and J.A. Bleda, *Tetrahedron*, 2001, **57**, 2355–2363.
[47] S. Ando, Y. Okamoto, K. Umezawa, and M. Otsuka, *J. Heterocycl. Chem.*, 2008, **45**, 1803–1808.
[48] A.P. Kozikowski, D.S. Dodd, J. Zaidi, *et al.*, *J. Chem. Soc., Perkin Trans. 1*, 1995, 1615–1621.
[49] F. Hong, J. Zaidi, B. Cusack, and E. Richelson, *Bioorg. Med. Chem.*, 2002, **10**, 3849–3858.
[50] M.S. Reddy and J.M. Cook, *Tetrahedron Lett.*, 1994, **35**, 5413–5416.
[51] R.E. Bolton, C.J. Moody, M. Pass, *et al.*, *J. Chem. Soc., Perkin Trans 1*, 1988, 2491–2499.
[52] D.J. Bentley, J. Fairhurst, P.T. Gallagher, *et al.*, *Org. Biomol. Chem.*, 2004, **2**, 701–708.
[53] M.S. Allen, L.K. Hamaker, A.J. La Loggia, and J.M. Cook, *Synth. Commun.*, 1992, **22**, 2077–2102.
[54] A.S. Cotterill, P. Hartopp, G.B. Jones, *et al.*, *Tetrahedron*, 1994, **50**, 7657–7674.
[55] S.S. Samanta, S.C. Ghosh, and A. De, *J. Chem. Soc., Perkin Trans. 1*, 1997, 3673–3677.
[56] J.B. Blair, D. Kurrasch-Orbaugh, D. Marona-Lewicka, *et al.*, *J. Med. Chem.*, 2000, **43**, 4701–4710.
[57] J. Božilović, J.W. Bats, and J.W. Engels, *Can. J. Chem.*, 2007, **85**, 283–292.
[58] S. Mayer and J.-Y. Mérour, *Eur. J. Org. Chem.*, 2002, 1646–1653.
[59] S.C. Ghosh and A. De, *Chem. Commun.*, 2000, 979–980.
[60] P.B. Huleatt, S.S. Choo, S. Chua, and C.L.L. Chai, *Tetrahedron Lett.*, 2008, **49**, 5309–5311.
[61] S. Selvi, S.-C. Pu, Y.-M. Cheng, *et al.*, *J. Org. Chem.*, 2004, **69**, 6674–6678.
[62] W. Stadlbauer and M. Fischer, *J. Heterocycl. Chem.*, 2002, **39**, 131–135.
[63] P.R. Graupner, M.F. Mahon, A. Ninan, *et al.*, *Tetrahedron Lett.*, 1995, **36**, 5827–5730.
[64] C.J. Moody and J.G. Ward, *J. Chem. Soc., Perkin Trans. 1*, 1984, 2903–2909.
[65] E.T. Pelkey, L. Chang, and G.W. Gribble, *Chem. Commun.*, 1996, 1909–1910.
[66] P. Molina, J. Alcántara, and C. López-Leonardo, *Tetrahedron*, 1997, **53**, 3281–3286.
[67] Y. Murakami, T. Watanabe, H. Suzuki, *et al.*, *Chem. Pharm. Bull.*, 1997, **45**, 1739–1744.
[68] M. Tercel, M.A. Gieseg, W.A. Denny, and W.R. Wilson, *J. Org. Chem.*, 1999, **64**, 5946–5953.
[69] V. Nair and T.G. George, *Tetrahedron Lett.*, 2000, **41**, 3199–3201.
[70] K. Kondo, S. Morohoshi, M. Mitsuhashi, and Y. Murakami, *Chem. Pharm. Bull.*, 1999, **47**, 1227–1231.
[71] F. Lehmann, M. Holm, and S. Laufer, *Tetrahedron Lett.*, 2009, **50**, 1708–1709.
[72] A.G. O'Brien, F. Lévesque, and P.H. Seeberger, *Chem. Commun.*, 2011, **47**, 2688–2690.
[73] E.F.V. Scriven and K. Turnbull, *Chem. Rev.*, 1988, **88**, 297–368.

29

Taber Indole Synthesis

Related to both the Sundberg and Hemetsberger syntheses is the variation presented by Taber and Tian [1]. The Taber indole synthesis made use of the same azirines that are intermediates in the aforementioned syntheses. Thus, as summarized in Scheme 1, equation 1, the azirines were available in two steps from a ketone via the corresponding oxime, followed by a Neber reaction [2] to give the azirine. Thermolysis afforded the corresponding indole. Interestingly, the azirine could not be isolated before it rearranged to the cyclopent[*b*]indole in equation 3. In the case of the unsymmetrical diaryl ketone **1**, the more electron-rich aromatic ring ambushed the nitrene to afford the major indole shown in equation 4. This was consistent with preferred electrophilic attack on the electron-rich benzene ring.

Prior to Taber and Tian, other groups discovered this azirine-to-indole synthesis, but Taber was the first to recognize this reaction as a potentially significant new indole ring construction. For example, Isomura and coworkers studied the kinetics of the thermal rearrangement of optically active 3-methyl-2-phenyl-2*H*-azirine to 2-methylindole [3]. Their important result was that racemization was much faster than indole formation. Whereas racemization occurred below 120 °C in decalin, indolization of the vinyl nitrene required heating above 185 °C in decalin. Du, Zhao, and colleagues used phenyliodine(III) diacetate (PIDA) to prepare 2*H*-azirines (equation 1), which upon heating were smoothly converted to indoles (equation 2) (Scheme 2) [4]. A suggested pathway is shown in equation 3.

Several metal-catalyzed transformations of 2*H*-azirines to indoles (e.g., with palladium, rhodium, and iron) are discussed later in their respective chapters. Some excellent reviews of 2*H*-azirine chemistry are available [5–7].

Indole Ring Synthesis: From Natural Products to Drug Discovery, First Edition. Gordon W. Gribble.

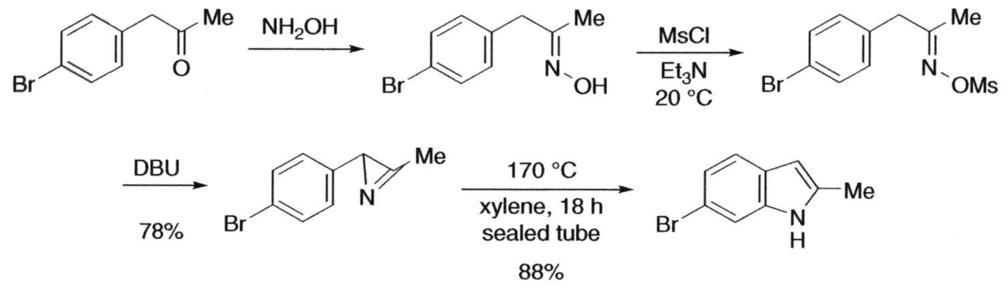

[eqn. 1] [1]

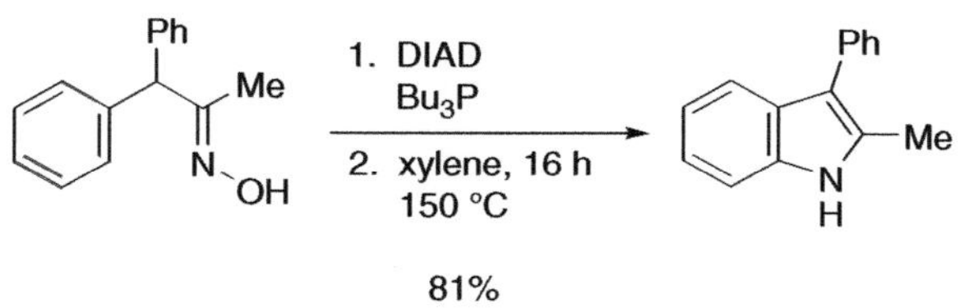

[eqn. 2] [1]

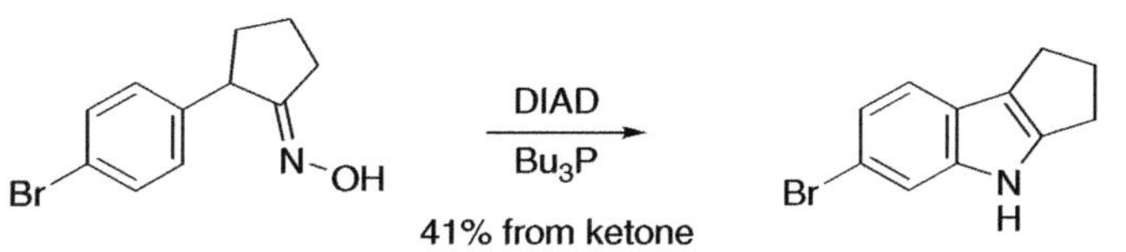

[eqn. 3] [1]

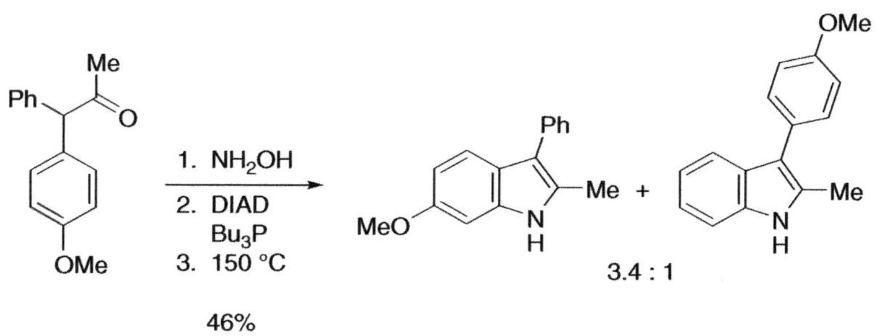

[eqn. 4] [1]

Scheme 1 *Taber Indole Synthesis*

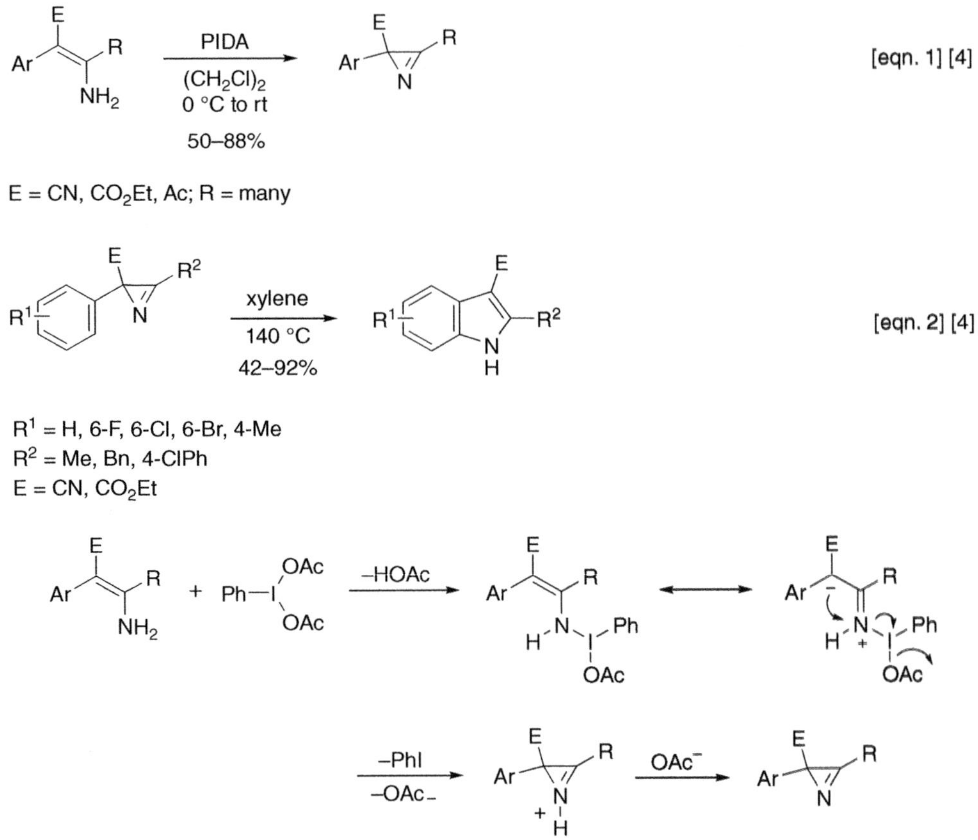

Scheme 2 *Related Azirine to Indole Synthesis*

References

[1] D.F. Taber and W. Tian, *J. Am. Chem. Soc.*, 2006, **128**, 1058–1059.
[2] W.P. Berkowitz, *Org. React.*, 2012, **78**, 321–410.
[3] K. Isomura, G.-I. Ayabe, S. Hatano, and H. Taniguchi, *J. Chem. Soc., Chem. Comm.*, 1980, 1252–1253.
[4] X. Li, Y. Du, Z. Liang, *et al.*, *Org. Lett.*, 2009, **11**, 2643–2646.
[5] F. Palacío, A.M. Ochoa de Retana, E. Martínez de Marigorta, and J. Manuel de los Santos, *Org. Prep. Proc. Int.*, 2002, **34**, 219–269.
[6] F. Palacío, A.M. Ochoa de Retana, E. Martínez de Marigorta, and J. Manuel de los Santos, *Targets in Heterocyclic Systems*, 2003, **7**, 207.
[7] A.F. Khlebnikov and M.S. Novikov, *Tetrahedron*, 2013, **69**, 3363–3401.

30

Knochel Indole Synthesis

The relatively newly described Knochel indole synthesis involves using phenylmagnesium chloride as a reagent to convert nitroarenes into the corresponding aryl nitrenes, which then can undergo cyclization to indoles [1]. The proposed reaction sequence is shown in Scheme 1. The usual addition of the Grignard reagent to the nitrosoarene **1** (cf. Bartoli indole synthesis in Chapter 4) does not occur in the present case owing to steric hindrance imparted by the *ortho*-vinyl group. This abnormal site of attack is amplified by the presence of electron-withdrawing groups in the benzene ring. The resulting aryl nitrene **2** cyclized to an indole. As might be expected from the related Cadogan–Sundberg and Sundberg indolizations, carbazole was formed in low yield (24%) when 2-nitrobiphenyl was treated with phenylmagnesium chloride.

Some examples of this chemistry are shown in Scheme 2. It might be noted that *o*-nitro-substituted amidines and imines were converted to benzimidazoles with phenylmagnesium chloride.

This new indole ring synthesis should find utility for the preparation of 2-arylindoles with electron-withdrawing substituents in the indole ring. Furthermore, as we see in equations 1–3, the bromine atom is impervious to attack by the Grignard reagent. The starting *ortho*-nitrostilbenes were prepared in the standard fashion from the appropriate *ortho*-nitrotoluene and aryl aldehyde with piperidine as base.

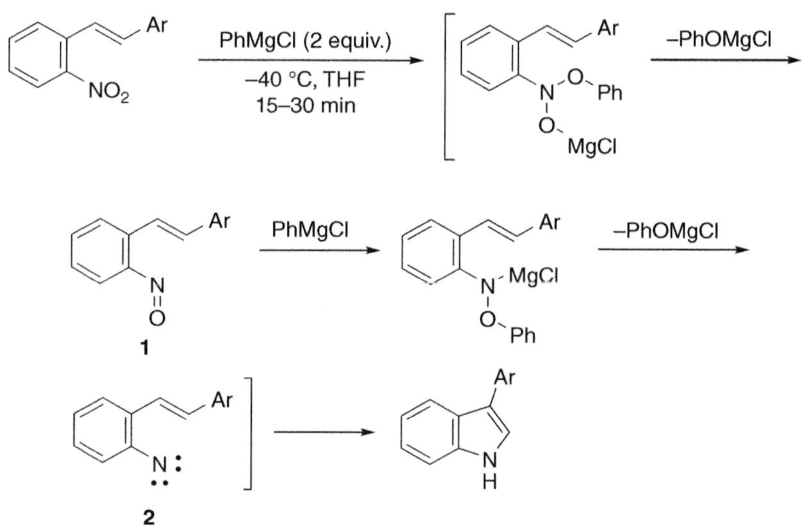

Scheme 1 *Knochel Indole Synthesis*

Indole Ring Synthesis: From Natural Products to Drug Discovery, First Edition. Gordon W. Gribble.

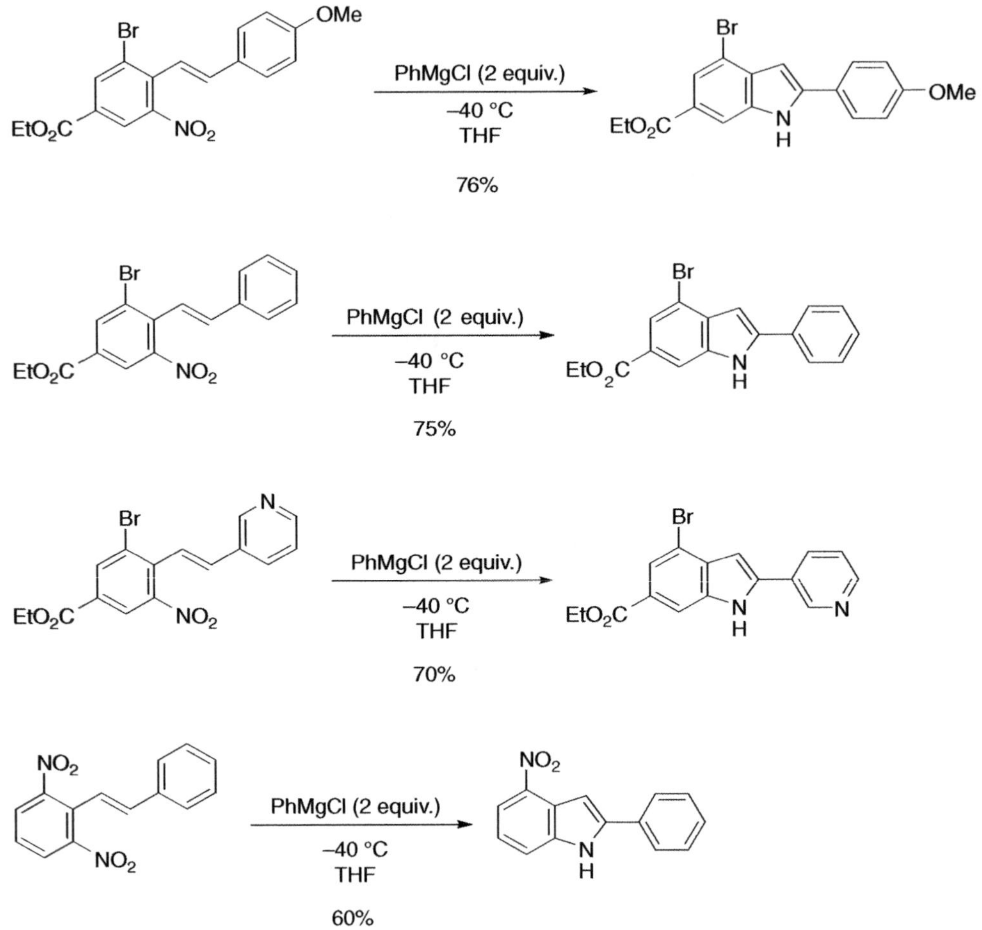

[eqn. 1]

[eqn. 2]

[eqn. 3]

[eqn. 4]

Scheme 2 *Knochel Indole Synthesis*

Reference

[1] W. Dohle, A. Staubitz, and P. Knochel, *Chem. Eur. J.*, 2003, **9**, 5323–5331.

31

Täuber Carbazole Synthesis

The infrequently used carbazole synthesis discovered by Ernst Täuber in 1890 involves the acid-promoted, high-temperature conversion of 2,2′-diaminobiphenyls to carbazoles. Täuber's original observation was the reaction of *meta*-diaminobenzidine with 18% hydrochloric acid at 180–190 °C to afford 2,7-diaminocarbazole (Scheme 1, equation 1) [1–3]. In a second paper, in which carbazole itself was prepared (equation 2), Täuber suggested the intermediacy of the aminophenol **1** followed by loss of water, perhaps via keto tautomer **2**, to give carbazole (equation 3) [2].

The first person to explore the utility of the Täuber synthesis in any depth was Leditschke, who prepared several carbazoles using concentrated phosphoric acid (Scheme 2,

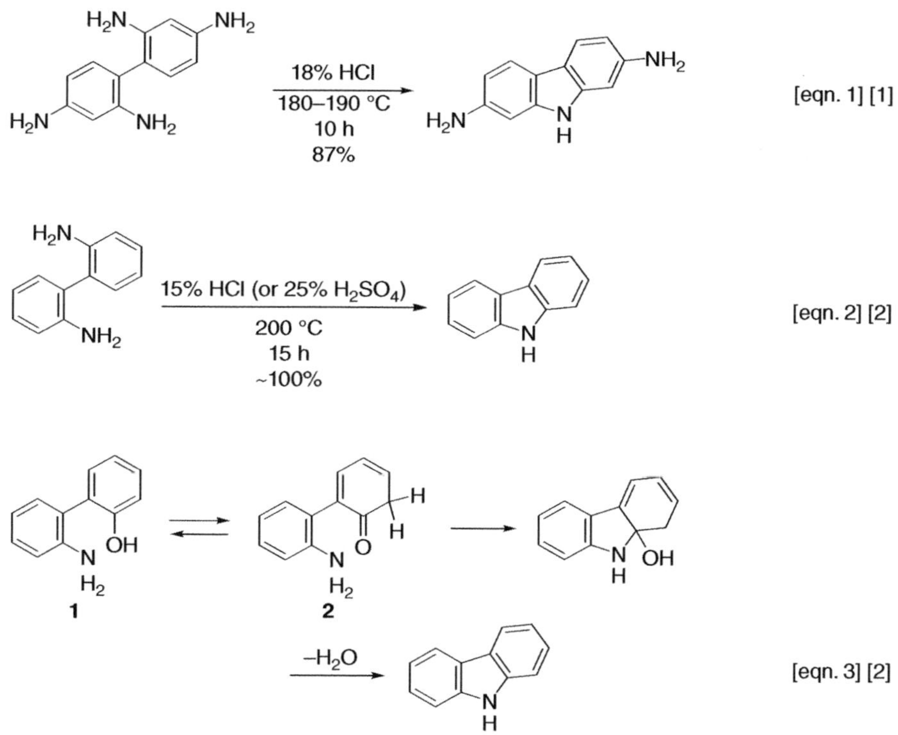

Scheme 1 *Täuber Carbazole Synthesis*

Indole Ring Synthesis: From Natural Products to Drug Discovery, First Edition. Gordon W. Gribble.

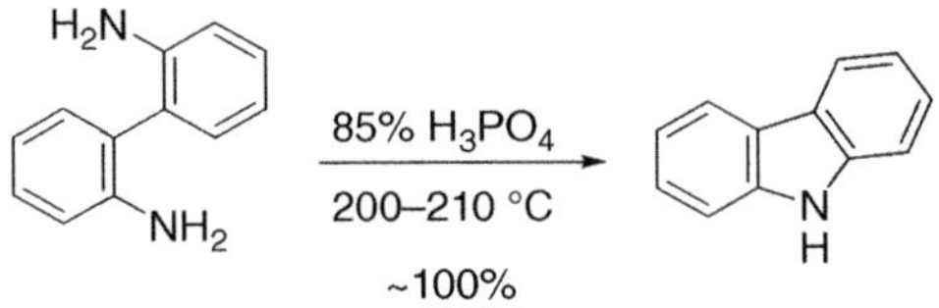

[eqn. 1] [4]

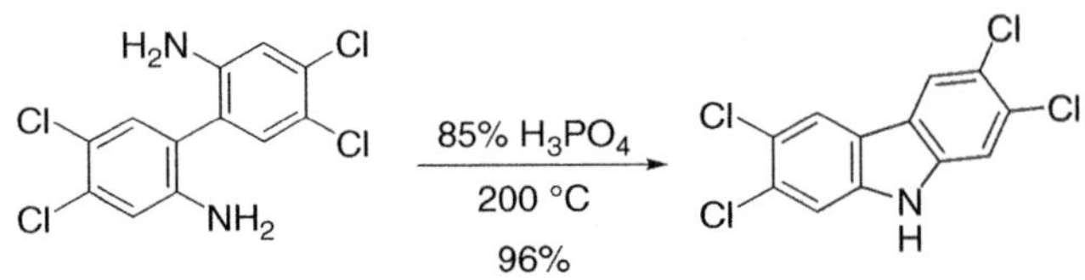

[eqn. 2] [4]

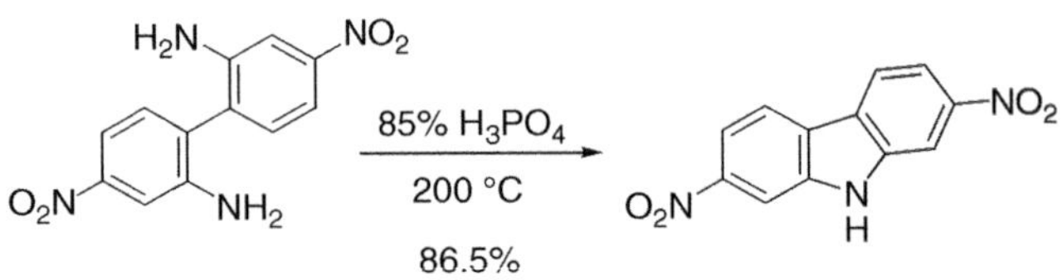

[eqn. 3] [4]

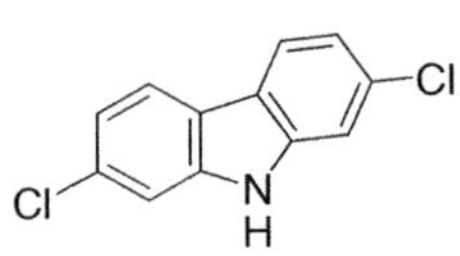

3

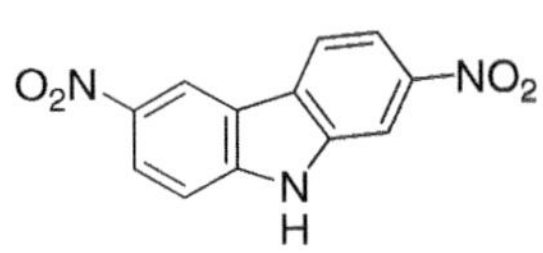

4

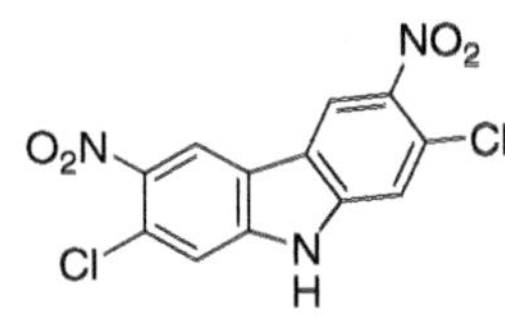

5

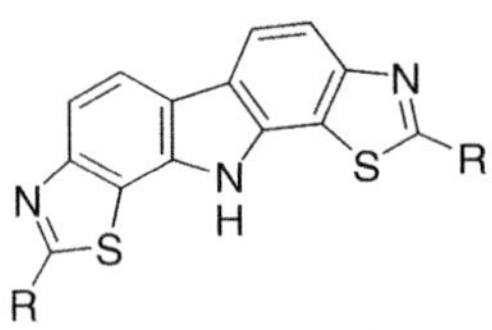

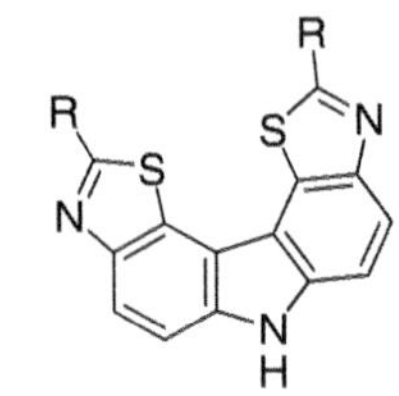

6, R = H, Me (59, 72%) **7**, R = H, Me (65, 76%)

Scheme 2 *Leditschke Variation of the Täuber Carbazole Synthesis*

equations 1–3, and **3–5**) [4]. King and King converted 2,2′-diamino-5-acetamidobiphenyl to 3-acetamidocarbazole with concentrated hydrochloric acid in a sealed tube (200 °C, 15 h), albeit impure and in low yield (35%) [5]. Tidwell and colleagues synthesized 2,7-dibromocarbazole in 85% yield (85% phosphoric acid, 190–200 °C, 26 h) from 2,2′-diamino-4,4′-dibromobiphenyl [6]. Similar conditions, but only heating for 1.5 h at 200 °C, allowed Racané and coworkers to prepare the novel dithiazolocarbazoles **6** and **7** from the corresponding diaminodibenzothiazoles [7]. Tashiro and colleagues studied the cyclization (and de-*tert*-butylation) of 2,2′-diamino-4,4′-di-*tert*-butylbiphenyl to carbazole with 85% phosphoric acid (220 °C, 64% yield) [8].

The first major modification of the original Täuber-Leditschke reaction conditions was that of Olah and coworkers (Scheme 3) [9]. This group found that the perfluorinated sulfonic acid resin Nafion-H, in a suitable solvent, sufficed to convert 2,2′-diaminobiphenyls to carbazoles. The reaction was optimized for solvent using 4,4′-di-*tert*-butyl-2,2′-diaminobiphenyl (equation 1). The carbazoles prepared in this study are **8–12**, the yields shown are gas chromatographic yields, and the solvent was 4-*t*-butyl-*o*-xylene at reflux. The only other product in these reactions was the starting 2,2′-diaminobiphenyl. Cho and coworkers found that a Täuber carbazole synthesis effected the cyclization of 2,2′-diamino-1,1′-biaryls to benzo[*c*]carbazoles under mild conditions and in excellent yields (equation 2) [10]. This latter result may announce a resurrection of the Täuber carbazole synthesis.

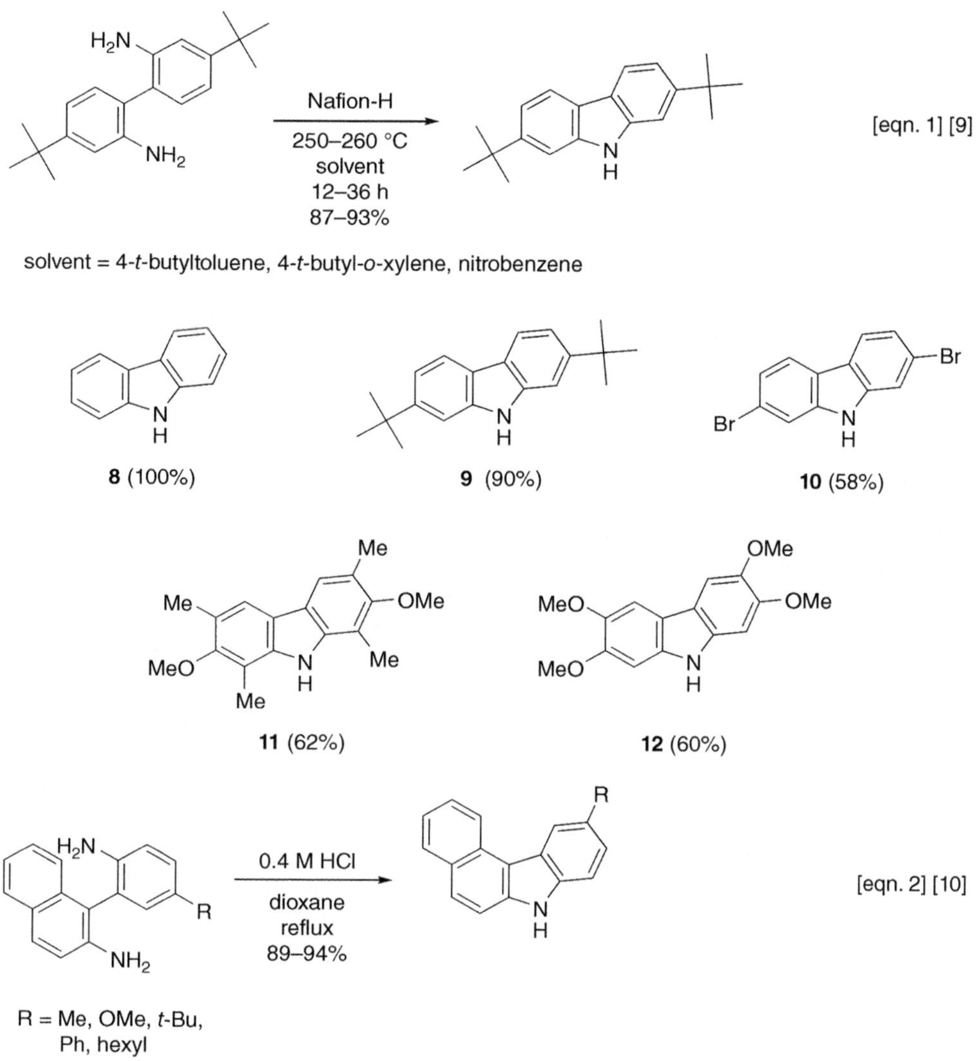

Scheme 3 *Olah and Cho Modifications of the Täuber Carbazole Synthesis*

References

[1] E. Täuber, *Ber.*, 1890, **23**, 3266–3269.
[2] E. Täuber, *Ber.*, 1891, **24**, 197–201.
[3] E. Täuber and R. Loewenherz, *Ber.*, 1891, **24**, 1033–1036.
[4] H. Leditschke, *Chem. Ber.*, 1953, **86**, 522–524.
[5] F.E. King and T.J. King, *J. Chem. Soc.*, 1945, 824–826.
[6] D.A. Patrick, D.W. Boykin, W.D. Wilson, *et al.*, *Eur. J. Med. Chem.*, 1997, **32**, 781–793.
[7] L. Racané, H. Čičak, Z. Mihalić, *et al.*, *Tetrahedron*, 2011, **67**, 2760–2767.
[8] M. Tashiro, Y. Fukuda, and T. Yamato, *Heterocycles*, 1981, **16**, 771–774.
[9] T. Yamato, C. Hideshima, K. Suehiro, *et al.*, *J. Org. Chem.*, 1991, **56**, 6248–6250.
[10] B.-Y. Lim, M.-K. Choi, and C.-G. Cho, *Tetrahedron Lett.*, 2011, **52**, 6015–6017.

32
Quéguiner Azacarbazole Synthesis

For several decades, Guy Quéguiner and his colleagues pursued the lithiation of π-deficient heterocycles. This elegant chemistry has led to an efficient synthesis of all four carbolines and related azacarbazoles (Scheme 1) [1, 2]. The starting azabiphenyls **1** were prepared by palladium-catalyzed cross-coupling between *ortho*-haloiodopyridines and (2-pivaloylaminobenzene)boronic acid. Heating **1** in pyridinium chloride at the boiling point, followed by workup with ammonium hydroxide afforded the carbolines **2–5** [1]. This cyclization was extended to the synthesis of α-substituted β-carbolines **6** [2] and α-substituted δ-carbolines **7** [3].

Concurrent with the chemistry of Quéguiner was the synthesis of α-carbolines by Achab and colleagues [4]. This work is summarized in Scheme 2 and culminated in the synthesis of the tetracyclic ring system **8** embodied in the marine anticancer alkaloids grossularines (equation 1). Several other α-carbolines (**9**) were prepared by Achab using a Stille coupling between an *ortho*-haloacetanilide and a 2,6-dichloro-3-stannylpyridine, followed by base-induced cyclization. Using a similar strategy, Quéguiner and coworkers synthesized 6-hydroxyharman (**10**) and 6-hydroxynorharman (**11**) [5], fascaplysin (**12**) [6], and the lavendamycin skeleton **13** [7, 8]. The points of cyclization are indicated by an arrow. Quéguiner has written a summary of his excellent carboline syntheses [9]. In continuation of Quéguiner's seminal work, Arzel's team synthesized a series of benzo-δ-carbolines **14** (equation 2), including extension to the cryptolepines [10]. Some of these new benzo-δ-carbolines have antimalarial activity.

Indole Ring Synthesis: From Natural Products to Drug Discovery, First Edition. Gordon W. Gribble.

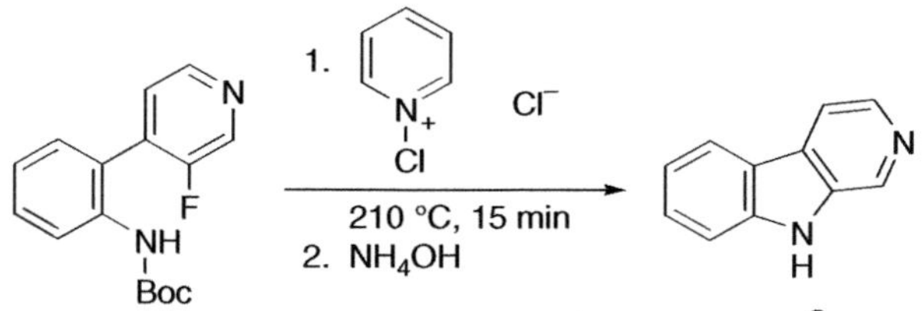

1 86% **2**

[eqn. 1] [1]

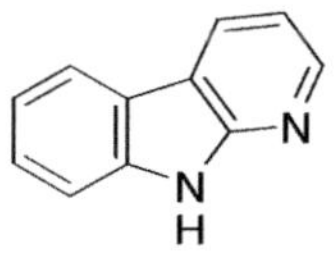

3 (82%) [1]

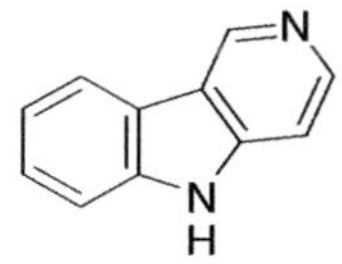

4 (88%) [1]

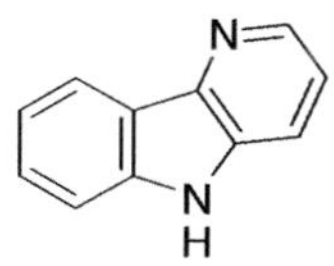

5 (78%) [1]

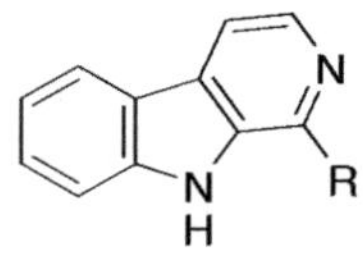

R = Ph, 2-pyridyl, Me, Et, CN, 2-thienyl, 2-quinolyl, vinyl

6 [2] (68–84%)

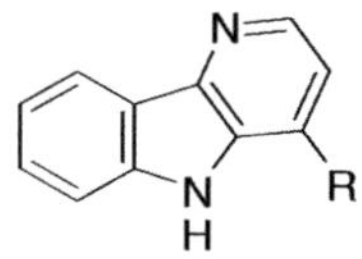

R = Me, Et, Ph, 2-NH_2Ph, 2-pyridyl, 2-thienyl, 2-quinolyl

7 [3] (72–98%)

Scheme 1 *Quéguiner Azacarbazole Synthesis*

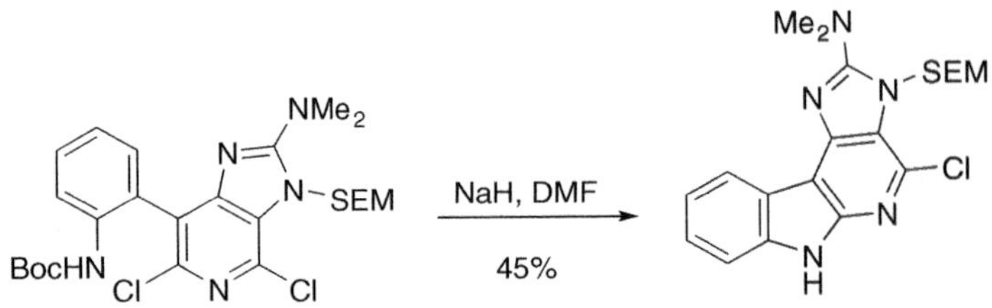

8

[eqn. 1] [4]

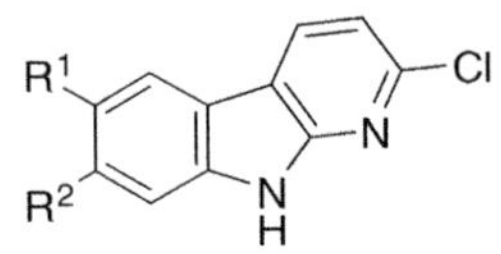

9

R^1	R^2	% Yield
H	H	82%
H	NO_2	75%
F	H	63%
H	Me	73%
H	CF_3	70%
H	OMe	78%

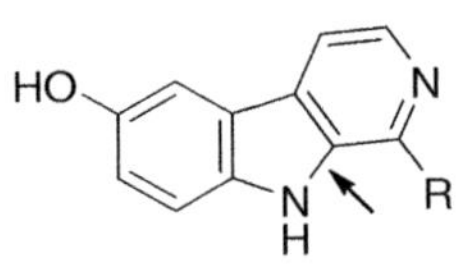

10, R = Me (81%) [5]
11, R = H (89%) [5]

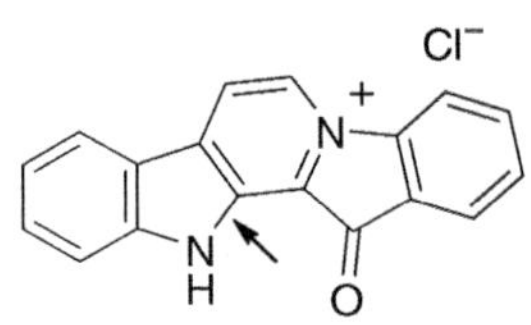

12 (82%) [6]

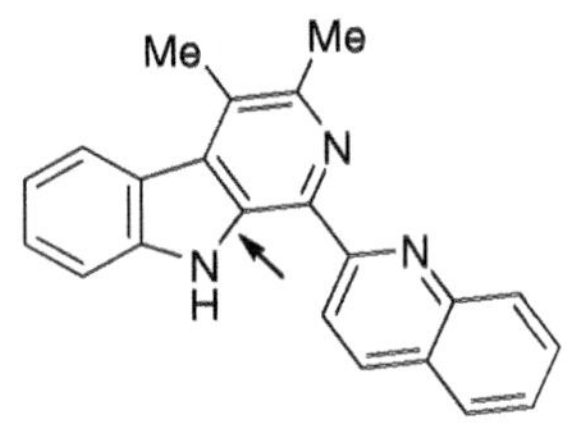

13 (91%) [7]

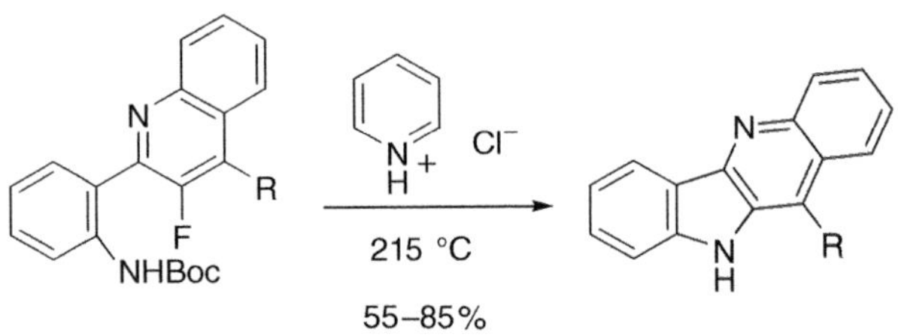

[eqn. 2] [10]

R = H, Me, Et, *i*-Pr, Ph, Cl

Scheme 2 *Applications of the Quéguiner Azacarbazole Synthesis*

References

[1] P. Rocca, F. Marsais, A. Godard, and G. Queguiner, *Tetrahedron*, 1993, **49**, 49–64.
[2] P. Rocca, F. Marsais, A. Godard, and G. Queguiner, *Tetrahedron*, 1993, **49**, 3325–3342.
[3] E. Arzel, P. Rocca, F. Marsais, *et al.*, *J. Heterocycl. Chem.*, 1997, **34**, 1205–1210.
[4] S. Achab, M. Guyot, and P. Potier, *Tetrahedron Lett.*, 1993, **34**, 2127–2130.
[5] P. Rocca, F. Marsais, A. Godard, and G. Quéguiner, *Tetrahedron Lett.*, 1994, **35**, 2003–2004.
[6] P. Rocca, F. Marsais, A. Godard, and G. Quéguiner, *Tetrahedron Lett.*, 1993, **34**, 7917–7918.
[7] P. Rocca, F. Marsais, A. Godard, and G. Quéguiner, *Tetrahedron Lett.*, 1993, **34**, 2937–2940.
[8] A. Godard, P. Rocca, J.-M. Fourquez, *et al.*, *Tetrahedron Lett.*, 1993, **34**, 7919–7922.
[9] A. Godard, F. Marsais, N. Plé, *et al.*, *Heterocycles*, 1995, **40**, 1055–1091.
[10] E. Arzel, P. Rocca, P. Grellier, *et al.*, *J. Med. Chem.*, 2001, **44**, 949–960.

33

Iwao Indole Synthesis

The Iwao indole synthesis involves *ortho*-directed lithiation of *N-tert*-butoxycarbonylanilines, quenching of this lithiated species with 1-trialkylsilyl-1-phenylsulfinylethene, and a thermal sila-Pummerer rearrangement to give 2-thioindolines **1**. Oxidation of **1** with *meta*-chloroperbenzoic acid furnished indoles **2** Scheme 1) [1]. Deprotection of the *tert*-butoxycarbonyl group was facile [2–5]. The sequence can also be carried out on the corresponding *N*-pivaloylanilines, particularly when the lithiation of *N-tert*-butoxycarbonyl-3-anilines is poor.

While the Iwao indole synthesis has yet to grab the attention of the synthetic community, Fukuyama and Chen used a related Pummerer reaction, **3**–**4**, in their synthesis of the marine alkaloid (–)-hapalindole G, as shown in Scheme 2 [6].

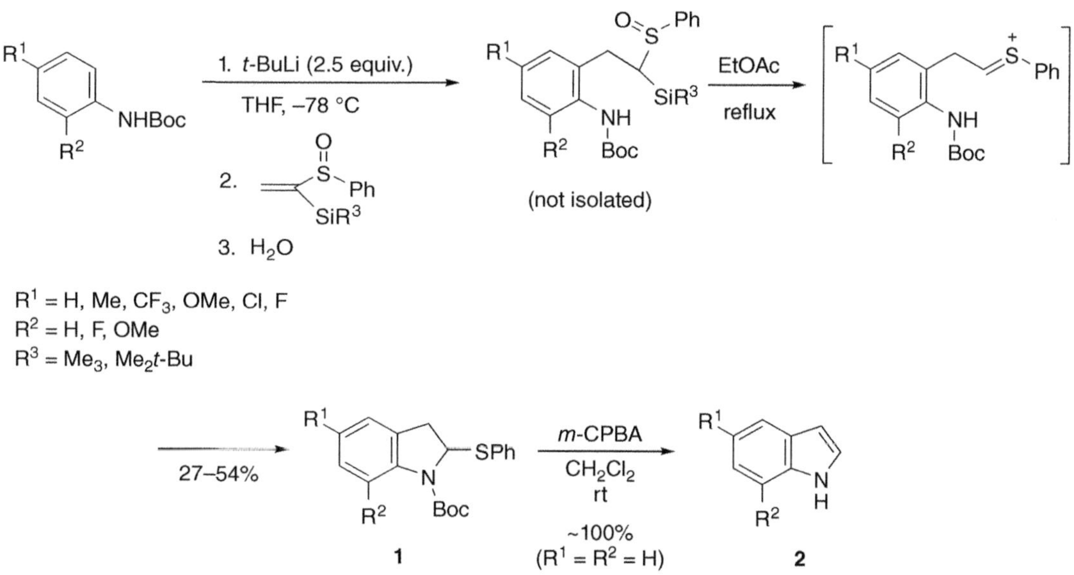

Scheme 1 *Iwao Indole Synthesis*

Indole Ring Synthesis: From Natural Products to Drug Discovery, First Edition. Gordon W. Gribble.

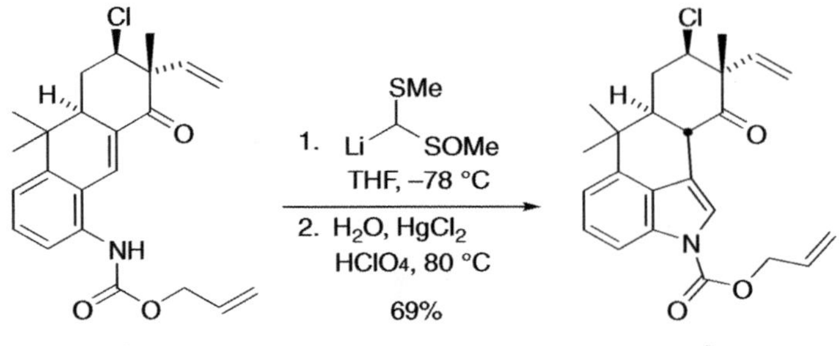

Scheme 2 *Fukuyama Synthesis of Hapalindole G*

References

[1] M. Iwao, *Heterocycles*, 1994, **38**, 45–50.
[2] R.D. Clark, J.M. Muchowski, L.E. Fisher, *et al.*, *Synthesis*, 1991, 871–878.
[3] I. Hasan, E.R. Marinelli, L.-C.C. Lin, *et al.*, *J. Org. Chem.*, 1976, **41**, 163–165.
[4] V.H. Raval and M.C. Cava, *Tetrahedron Lett.*, 1985, **26**, 6141–6142.
[5] D.L. Boger and S.M. Sakya, *J. Org. Chem.*, 1992, **57**, 1277–1284.
[6] T. Fukuyama and X. Chen, *J. Am. Chem. Soc.*, 1994, **116**, 3125–3126.

34

Hewson Indole Synthesis

Closely related to the Iwao indole synthesis (Chapter 33) is the discovery by Hewson and colleagues, who found that *ortho*-sulfonamide ketene dithioacetal *S*-oxides **1** readily cyclize to the corresponding 2-thiomethylindoles **2** upon treatment with hydrochloric acid in the presence of hydrogen sulfide (Scheme 1) [1]. The hydrogen sulfide prevents the formation of 3-chloroindoles, and the starting benzaldehydes were prepared from the appropriate methyl anthranilates by a sequence of reduction ($LiAlH_4$) and oxidation (pyridinium chlorochromate). The presumed pathway is protonation of **1** followed by cyclization and loss of methanesulfinic acid to give **2**.

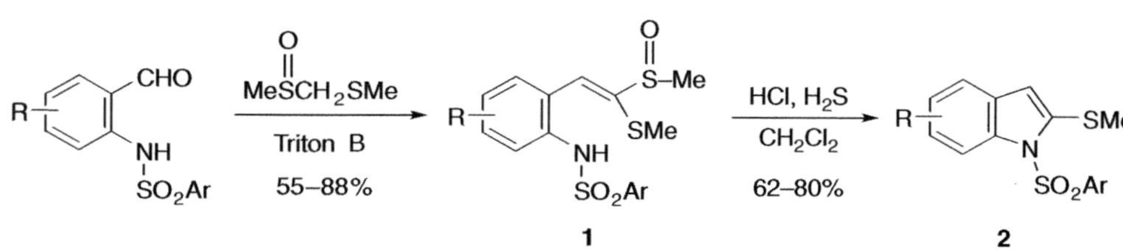

R = H, 5-Me, 6-Cl, 4-Br, 5-Br (indole numbering)
Ar = Ph, 4-MePh

Scheme 1 *Hewson Indole Synthesis*

Reference

[1] A.T. Hewson, K. Hughes, S.K. Richardson, *et al.*, *J. Chem. Soc., Perkin Trans. 1*, 1991, 1565–1569.

Indole Ring Synthesis: From Natural Products to Drug Discovery, First Edition. Gordon W. Gribble.

35

Magnus Indole Synthesis

Magnus and Mitchell discovered a simple indole ring synthesis involving the methanesulfonic acid treatment of terminal triisopropylsilylprop-2-ynylanilines (e.g., **1**) to give 3-methylindoles (e.g., **2**) (Scheme 1) [1]. Several examples are shown of the indoles that were obtained in this fashion, **3–4**. A control experiment revealed that the corresponding ketone was not an intermediate formed before cyclization. However, substrates **5–6** failed to give indoles, but instead afforded ketone products.

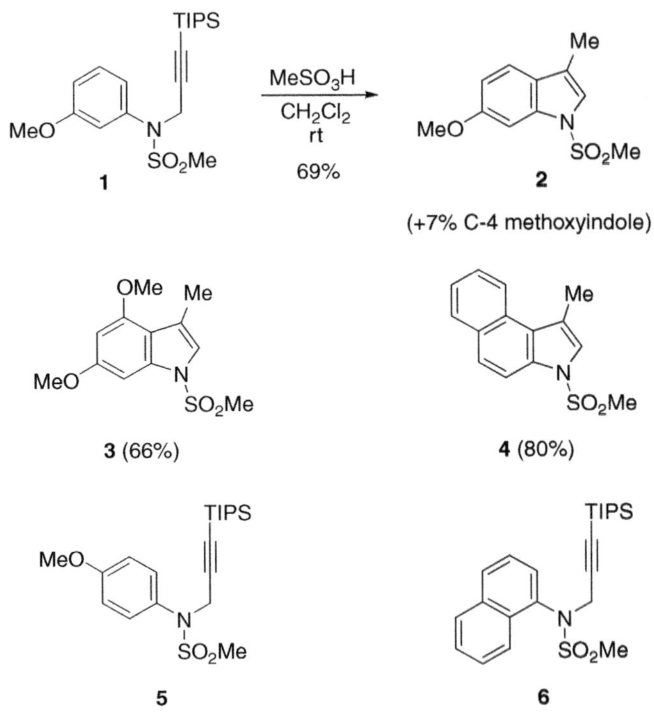

Scheme 1 *Magnus Indole Synthesis*

Reference

[1] P. Magnus and I.S. Mitchell, *Tetrahedron Lett.*, 1998, **39**, 4595–4598.

Indole Ring Synthesis: From Natural Products to Drug Discovery, First Edition. Gordon W. Gribble.

36

Feldman Indole Synthesis

The Feldman indole synthesis describes the reaction between lithiated *N*-phenyl-*p*-toluenesulfonamide (**1**) and phenyl(propynyl)iodonium triflate (**2**) to form indole **3** (Scheme 1, equation 1) [1, 2]. The lithiated species **1** was generated from the corresponding tosylanilide with *n*-butyllithium (THF, –78 °C). Applications are limited thus far to the synthesis of indoles shown in equation 2.

The mechanism of this interesting intermolecular indole synthesis as suggested by Feldman is depicted in Scheme 2 and presumably involves vinyl carbene **6** that cyclized to *N*-tosylindole [2]. The method is applicable to the synthesis of pyrroles and dihydropyrroles.

There is a structural limitation to the Feldman indole ring synthesis. Thus, *o*-methyltosylanilide (**7**) failed to afford indole **9**, apparently due to the unfavorable conformation **8** (Scheme 3, equation 1). Interestingly, the reactions shown in Scheme 1, equations 1 and 2, and Scheme 3, equation 1, gave rise to the azaazulene by product **10** (Scheme 3, equation 2), arising from indole **7**. Thus, vinylcarbene **8** cyclized into the phenylsulfonyl ring as shown. Attempts to thwart this side reaction by changing the *N*-tosyl group to mesityl (**11**) or trisyl (**12**) were unsuccessful, and trifyl (**13**) afforded small amounts of indole (~30%) along with the solvent (THF) insertion product **14**.

Despite the aforementioned obstacles that would need to be overcome or suppressed, the Feldman indole synthesis is a very promising novel route to 2-substituted indoles.

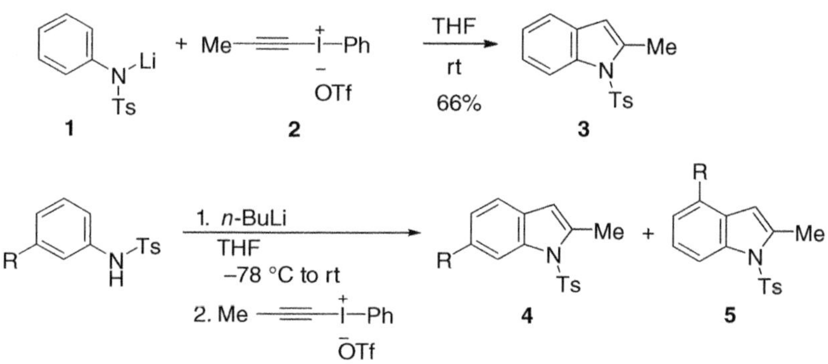

[eqn. 1] [1], [2]

[eqn. 2] [1], [2]

R	% yield	4 : 5
Me	59	1 : 1
OMe	61	1.4 : 1
CO_2Me	51	1.2 : 1
CO_2t-Bu	46	1.3 : 1

Scheme 1 *The Feldman Indole Synthesis*

Indole Ring Synthesis: From Natural Products to Drug Discovery, First Edition. Gordon W. Gribble.

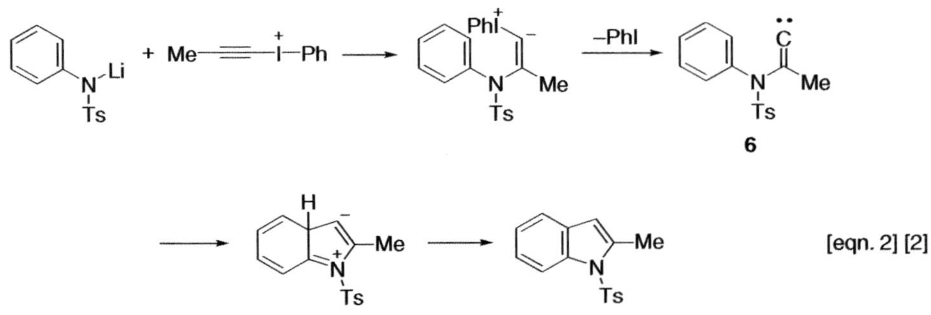

Scheme 2 *Proposed Mechanism for the Feldman Indole Synthesis*

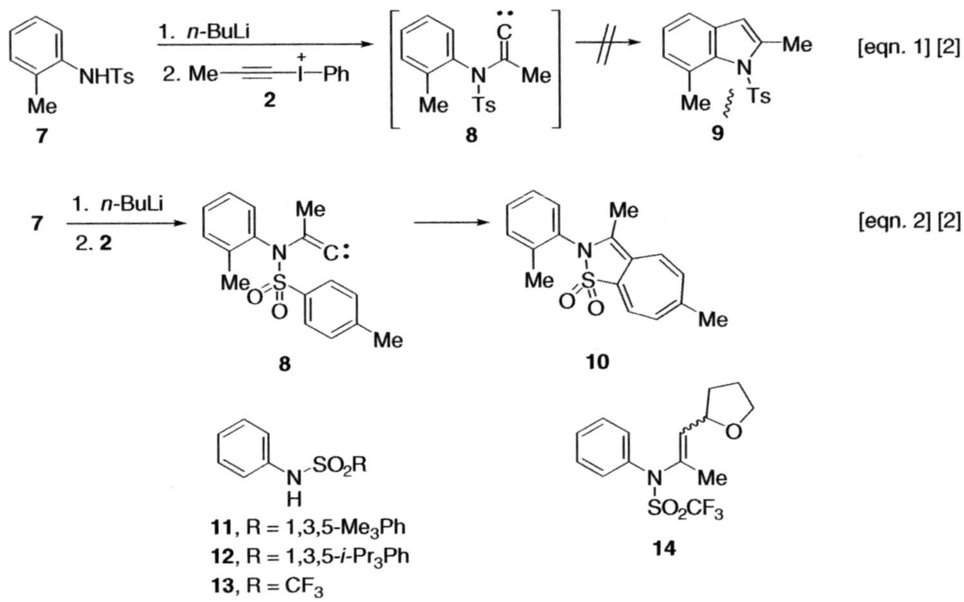

Scheme 3 *Aberrant Reactions in the Feldman Indole Synthesis*

References

[1] K.S. Feldman, M.M. Bruendl, and K. Schildknegt, *J. Org. Chem.*, 1995, **60**, 7722–7723.

[2] K.S. Feldman, M.M. Bruendl, K. Schildknegt, and A.C. Bohnstedt, *J. Org. Chem.*, 1996, **61**, 5440–5452.

37

Butin Indole Synthesis

The Butin indole ring synthesis is a new variation of several amino-carbonyl indole-forming reactions related most closely to the Reissert reaction (Chapter 40). The fundamental reaction is illustrated in Scheme 1 [1, 2]. Alkylation of suitable furans **2** with the benzylic alcohols **1** afforded the *o*-tosylaminobenzylfurans **3** in good yield. A conventional acid hydrolysis of the furan ring, which serves as a 1,4-dicarbonyl unit, led to indoles **5** via enol tautomer **4**.

Displaying remarkable molecular inventiveness, Butin and colleagues parlayed their basic reaction into a number of novel indole ring constructions (Scheme 2). The novel azulenio[7,8-*b*]indole salts **7** can also be obtained from the NH anilines corresponding to acetanilides **6** [3, 4]. Bis-furans **6** were synthesized from *o*-nitrobenzaldehydes and two equivalents of furans [4]. In a variation of the prototypical synthesis, Butin found that the installation of

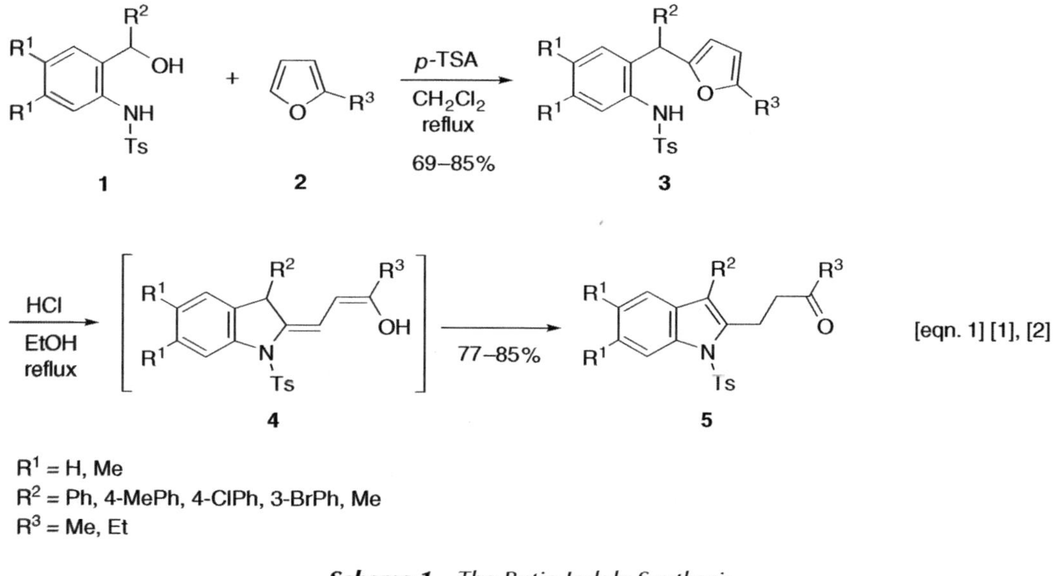

Scheme 1 *The Butin Indole Synthesis*

Indole Ring Synthesis: From Natural Products to Drug Discovery, First Edition. Gordon W. Gribble.

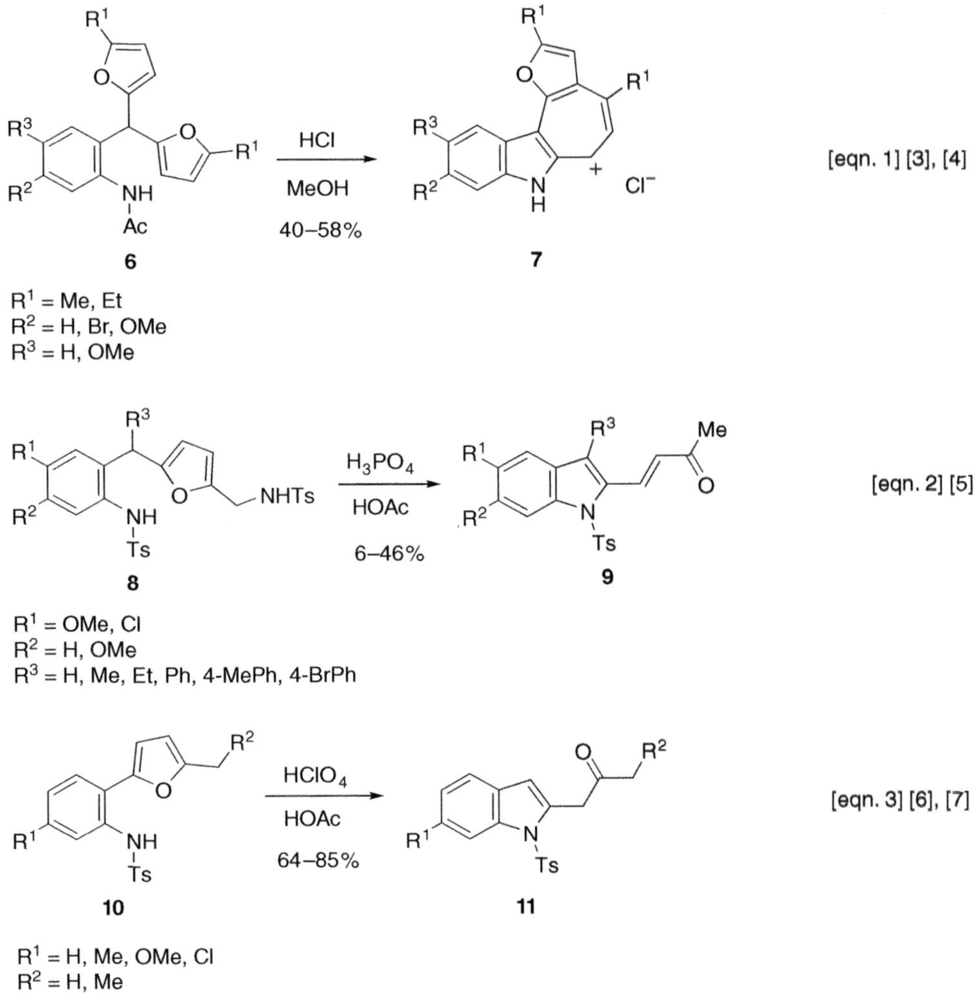

Scheme 2 *Applications of the Butin Indole Synthesis*

the tosylamide leaving group in **8** gave the 4-(2-indolyl)-3-buten-2-ones **9** in modest yields (equation 2) [5]. By attaching the furan ring directly to the tosylanilide (**10**), Butin and colleagues developed a route to indole ketones **11** (equation 3) [6, 7].

Other applications of the Butin indolization, all of which involve acid-catalyzed furan ring opening, are tabulated in Table 1. Note that the only difference between Entries 1 and 2 is the *N*-protecting group and the reaction temperature. The corresponding free anilines also result in the hydrolytic ring opening of both furan rings [8, 9]. The thieno[2,3-*b*] indole ring construction is particularly elegant (Entry 3) [10]. This mechanism presumably involves Lewis acid complexation of the sulfur atom of the isothiocyano group, followed by furan attack on the isothiocyano carbon atom and subsequent unraveling of the furan ring. The reaction of 2-(2-furyl)anilines with aryl aldehydes (Entry 4) and heterocyclic aldehydes (Entry 5) is an excellent route to the respective 3-(2-acylvinyl)-2-(hetero)arylindoles [11]. Indeed, Butin and his group applied this latter chemistry to a synthesis of indolo[3,2-*c*]quinolines **12** and isocryptolepine (**13**) (Scheme 3) [12].

Table 1 *Applications of the Butin Indole Ring Synthesis*

Entry	Substrate	Conditions	Indole	% Yield	Ref.
1	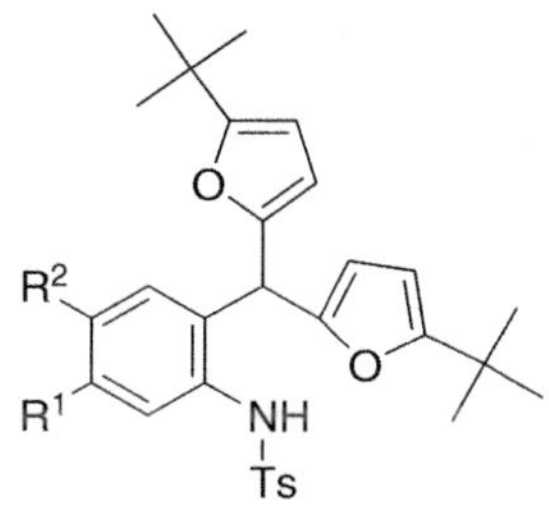 R^1 = H, OMe, –OCH_2O–, –$O(CH_2)_2O$– R^2 = H, OMe	HCl EtOH reflux	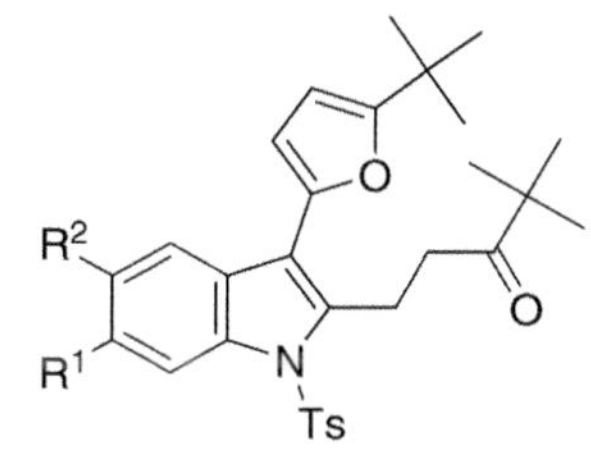	59–65%	8, 9
2	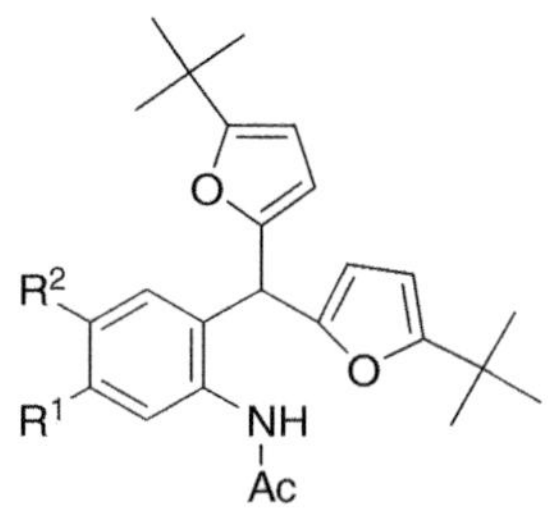 R^1= H, OMe, –OCH_2O–, –$O(CH_2)_2O$– R^2 = H, OMe	HCl EtOH rt	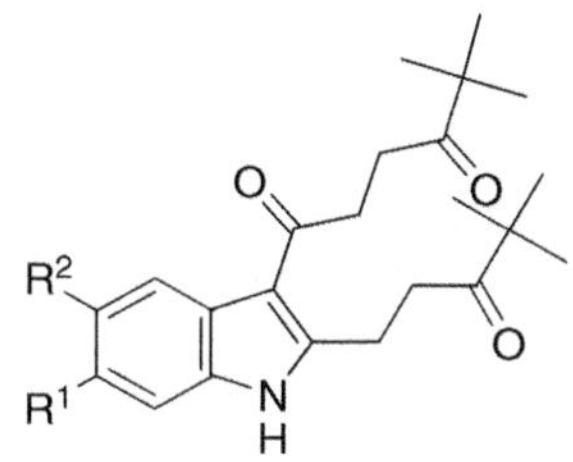	42–56%	8, 9
3	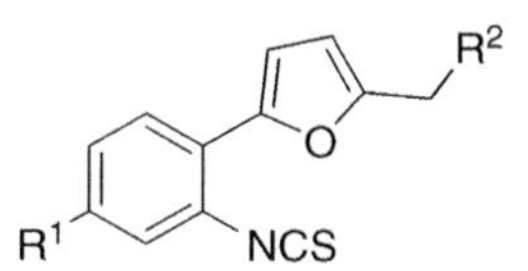	$AlCl_3$ $(ClCH_2)_2$ 50 °C	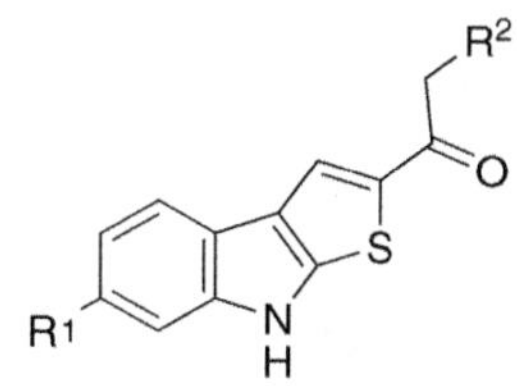	60–83%	10
4	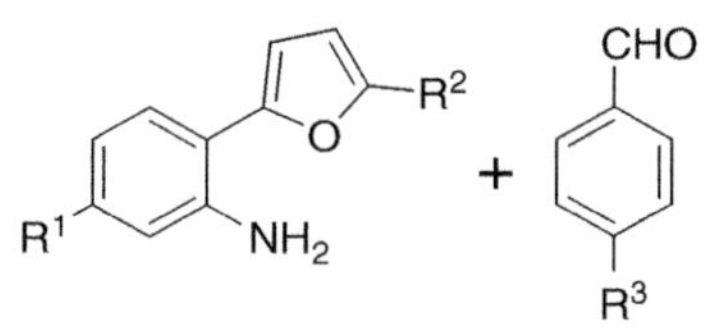 R^1 = H, Cl, Me R^2 = Me, Et R^3 = H, 4-F, 3-NO_2, 4-OEt, 2-OH-5-Cl	aq HCl HOAc 30–35 °C	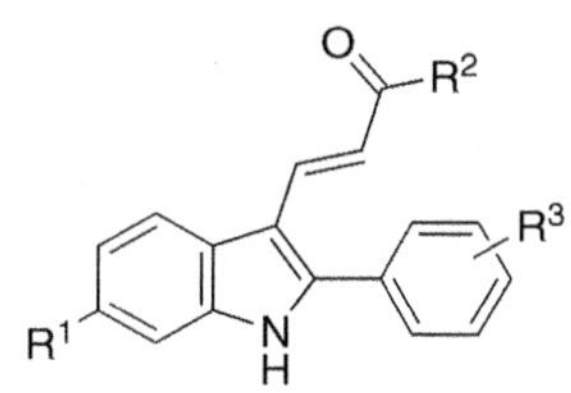	53–79%	11
5	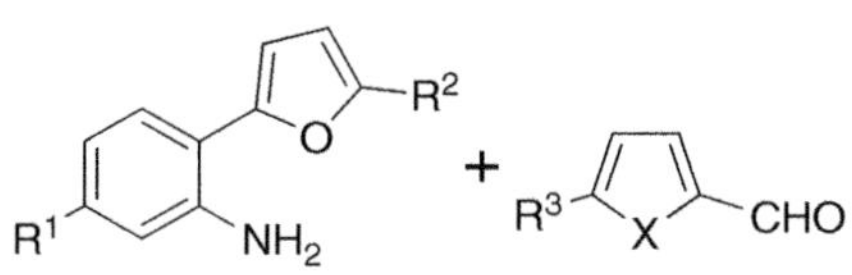 R^1 = H, Cl R^2 = Me, Et X = O, S	aq HCl HOAc 30–35 °C	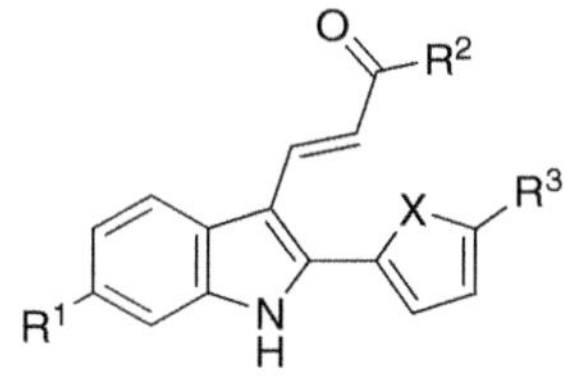	43–80%	11

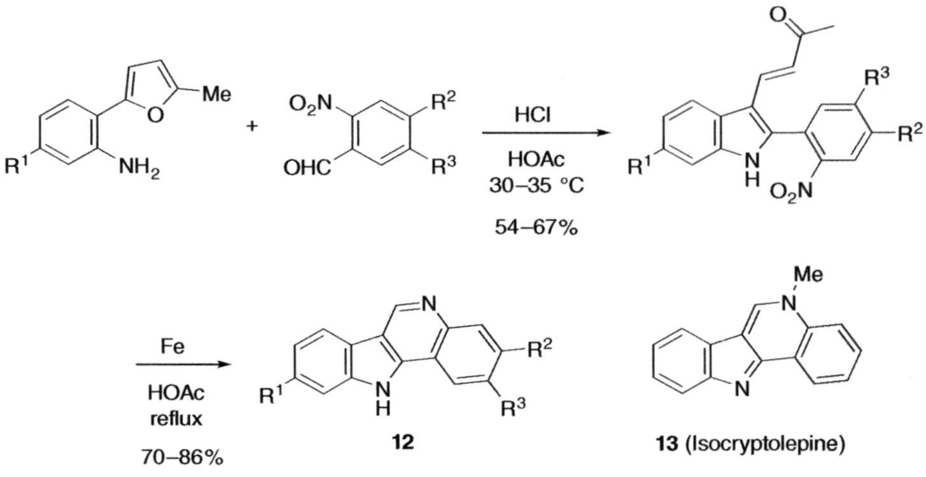

R^1 = H, Cl, Me
R^2 = H, OMe
R^3 = H, OMe, Cl, Br
R^2,R^3 = $-OCH_2-$, $-O(CH_2)_2O-$

Scheme 3 *Butin Synthesis of Indolo[3,2-c]quinolines*

References

[1] A.V. Butin, T.A. Stroganova, I.V. Lodina, and G.D. Krapivin, *Tetrahedron Lett.*, 2001, **42**, 2031–2033.
[2] A.V. Butin, S.K. Smirnov, T.A. Stroganova, *et al.*, *Tetrahedron*, 2007, **63**, 474–491.
[3] S.K. Smirnov, A.V. Butin, T.A. Stroganova, and A.V. Didenko, *Chem. Heterocycl. Cmpd.*, 2005, **41**, 929–931.
[4] A.V. Butin, S.K. Smirnov, and T.A. Stroganova, *J. Heterocycl. Chem.*, 2006, **43**, 623–628.
[5] A.V. Butin and S.K. Smirnov, *Tetrahedron Lett.*, 2005, **46**, 8443–8445.
[6] A.V. Butin, *Tetrahedron Lett.*, 2006, **47**, 4113–4116.
[7] A.S. Pilipenko, V.V. Mel'chin, I.V. Thrushkov, *et al.*, *Tetrahedron*, 2012, **68**, 619–627.
[8] A.V. Butin, S.K. Smirnov, and I.V. Trushkov, *Tetrahedron Lett.*, 2008, **49**, 20–24.
[9] A.V. Butin, S.K. Smirnov, F.A. Tsiunchik, *et al.*, *Synthesis*, 2008, 2943–2952.
[10] A.V. Butin, F.A. Tsiunchik, V.T. Abaev, and V.E. Zavodnik, *Synlett*, 2008, 1145–1148.
[11] A.V. Butin, M.G. Uchuskin, A.S. Pilipenko, *et al.*, *Eur. J. Org. Chem.*, 2010, 920–926.
[12] M.G. Uchuskin, A.S. Pilipenko, O.V. Serdyuk, *et al.*, *Org. Biomol. Chem.*, 2012, **10**, 7262–7265.

38

Miscellaneous Electrophilic Cyclizations

As is the case with miscellaneous nucleophilic cyclizations (Chapter 21), several electrophilic cyclizations leading to the indole ring do not appropriately fit into any of the preceding chapters in this category. As with the aforementioned chapter, I will mention the author of each new electrophilic indolization because future work might elevate that reaction to Name status. As with all ionic reactions, there is an electrophilic and nucleophilic component. Therefore, the inclusion of electrophilic indole ring syntheses in this chapter can be viewed as somewhat arbitrary, as was the case in Chapter 21 (Miscellaneous Nucleophilic Cyclizations).

Suzuki and colleagues found that *o*-aminostyryl ethers, which were prepared from the corresponding *o*-aminoaryl halides via palladium-catalyzed cross coupling with suitable boranes, gave indoles upon acidic hydrolysis (Scheme 1, equation 1) [1]. A similar acid-catalyzed cyclization was used by Ladlow and colleagues to prepare several phase-switching indoles **1** in a parallel synthesis program (equation 2) [2]. These *N*-acylindoles **1** were converted to their respective benzoic acid derivatives on treatment with nucleophiles. Kobayashi and coworkers synthesized a series of 3-aryl-1-thioacylindoles **2** from the corresponding α-aryl-2-isothiocyanato-β-methoxystyrenes (equation 3) [3]. A similar protocol led to 1-thiocarbamoylindoles **3** in a one-pot reaction (equation 4).

Okuma and colleagues converted *o*-alkenylanilides to indoles with dimethyl(methylthio)sulfonium trifluoromethanesulfonate (DMTST) (Scheme 2, equation 1) [4, 5]. This reaction presumably involves initial formation of the 3-methylthioindoline **4,** because this was isolated when $R_1 = R_2 = H$, $R_3 = 4$-Tol (77%). Moreover, the DMTST could be generated *in situ* from a solution of dimethyl disulfide and methyl triflate. A different electrophilic cyclization was described by Tamariz and colleagues that involves an intramolecular Friedel–Crafts reaction of enaminones **6** to indoles **7** (equation 2) [6]. This indolization was accomplished either from **6** or in one pot from the 2-anilinocarbonyl compounds **5** on treatment with *N,N*-dimethylformamide dimethyl acetal (DMF-DMA). Both thermal and microwave heating were explored for both options. Although aluminum chloride was the best catalyst, zinc chloride was preferred for methoxyl-substituted substrates. The substrates **5** were easily prepared by alkylation of the corresponding anilines. Qi, Zhao, and colleagues effected a synthesis of *N*-aryl-3-cyanoindoles using a zinc acetate–catalyzed cyclization of *in situ* halogenated 2-aryl-3-arylamino-2-alkenenitriles (equation 3) [7]. Indoles **8–10** were prepared in this study.

As part of a larger investigation on the synthesis of pyrroles via Lewis acid–catalyzed Mukaiyama–Michael reactions between enol silyl ethers and 1,2-diaza-1,3-butadienes, Filippone's group prepared 1-aminoindole **12** from Michael adduct **11** (Scheme 3, equation 1) [8]. A spontaneous dehydrogenation is involved to afford indole **12**. Serra and Fuganti employed a benzannulation strategy to craft substituted carbazoles **14** (equation 2) [9]. The precursor enynoic acids **13** were prepared via a Wittig coupling. During an attempt to prepare the oxime of 2-phenylcyclohept-2-enone, Ginsburg and Pappo reported the unusual formation of cyclohept[*b*]indole (**15**) (equation 3) [10]. They speculate that an isomer of the oxime led to **15** upon thermal dehydration.

The classic Vilsmeier reagent is featured in two indole syntheses. Meth-Cohn and colleagues converted

Indole Ring Synthesis: From Natural Products to Drug Discovery, First Edition. Gordon W. Gribble.

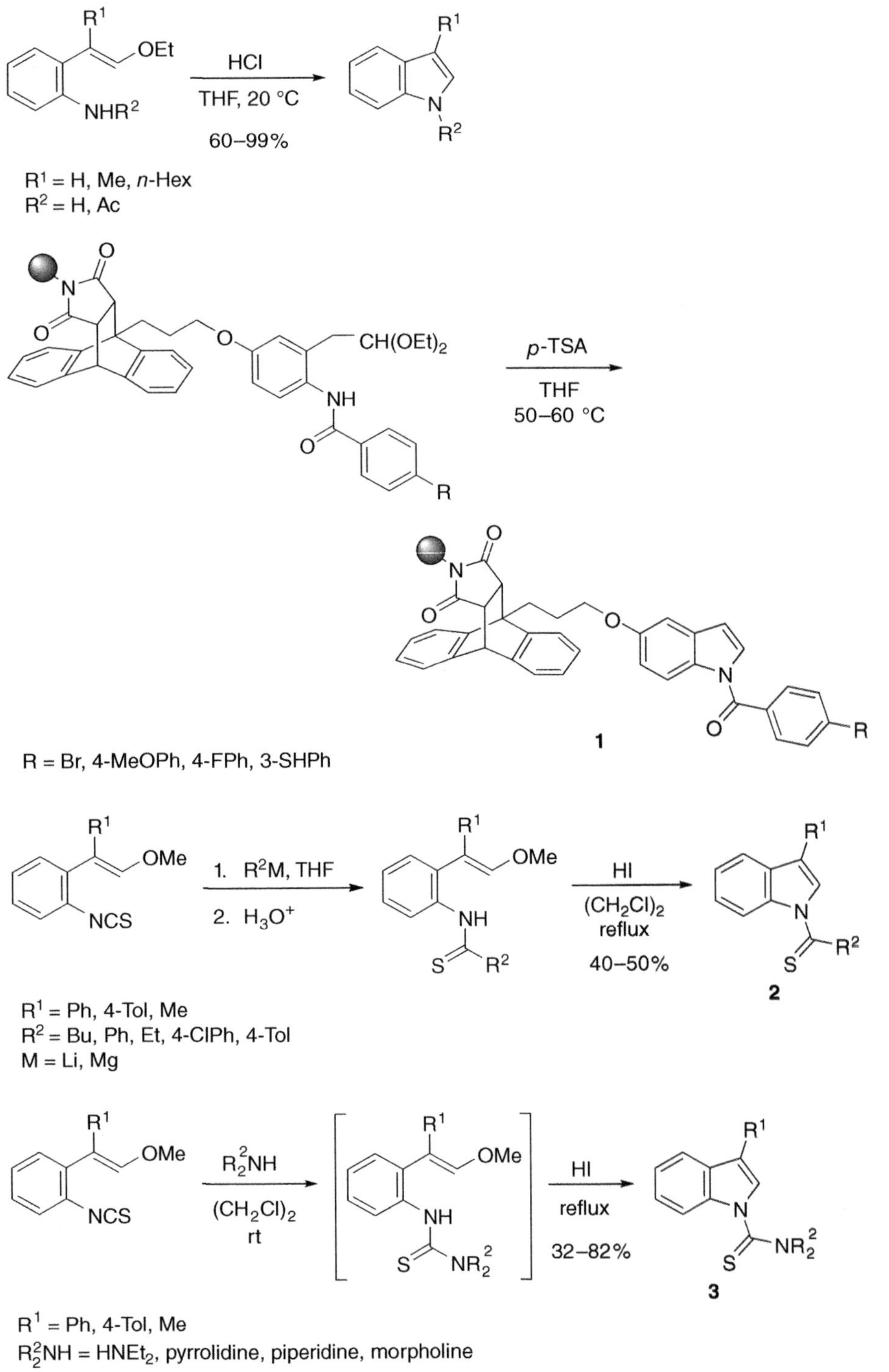

[eqn. 1] [1]

[eqn. 2] [2]

[eqn. 3] [3]

[eqn. 4] [3]

Scheme 1 *Miscellaneous Electrophilic Cyclizations*

N-methylformanilides **16** to tetramers **17**, in addition to other products such as isatins (Scheme 4, equation 1) [11]. Ishikawa's team applied an abnormal Bischler–Napieralski cyclization of Vilsmeier reagents to give unusual azazulenes and related ring systems (equation 2) [12]. Proctor and McDonald cyclized a series of *N*-(2-chloroallyl)anilines with polyphosphoric acid (100 °C) to give the expected 2-methylindoles (equation 3) [13].

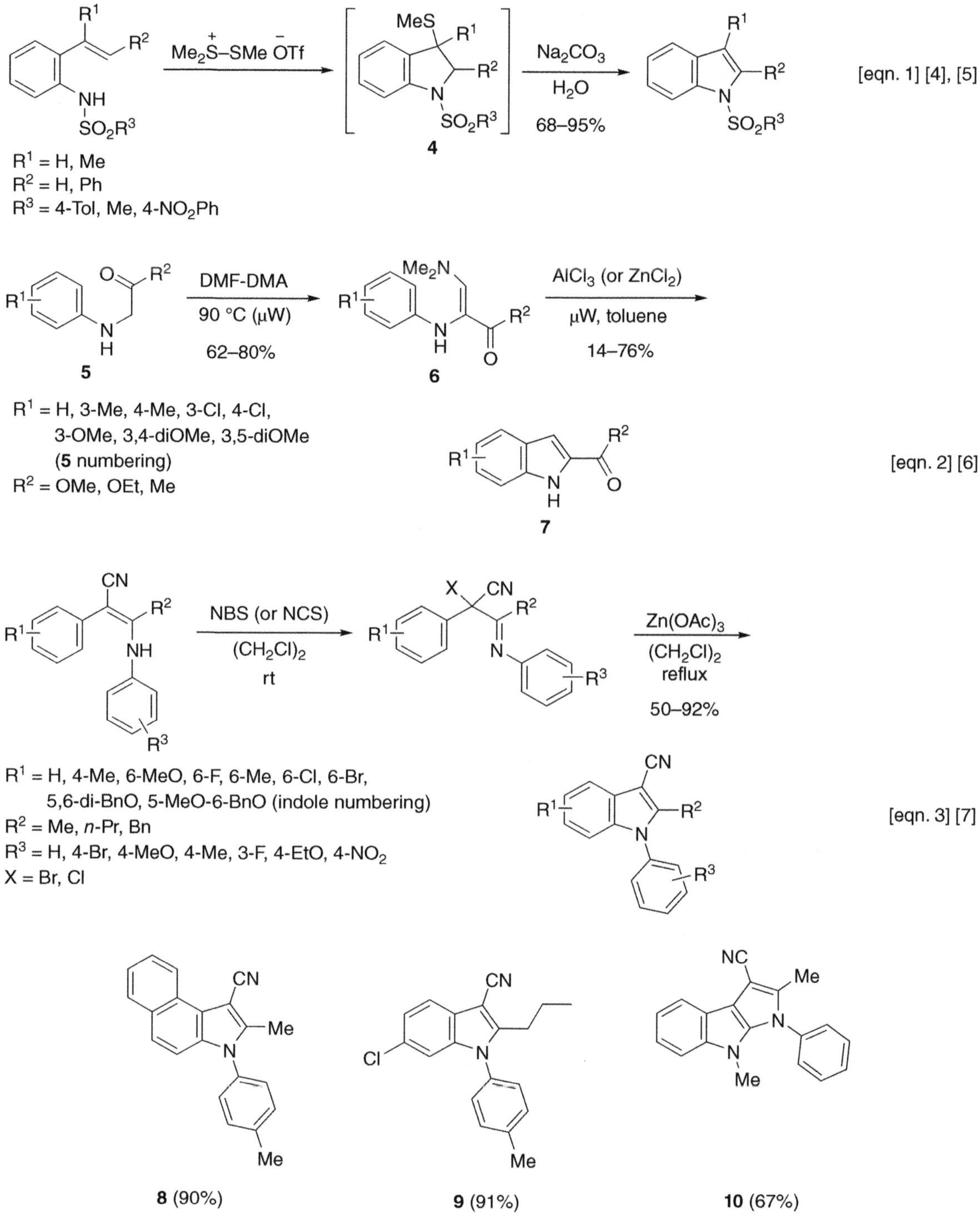

Scheme 2 *Miscellaneous Electrophilic Cyclizations*

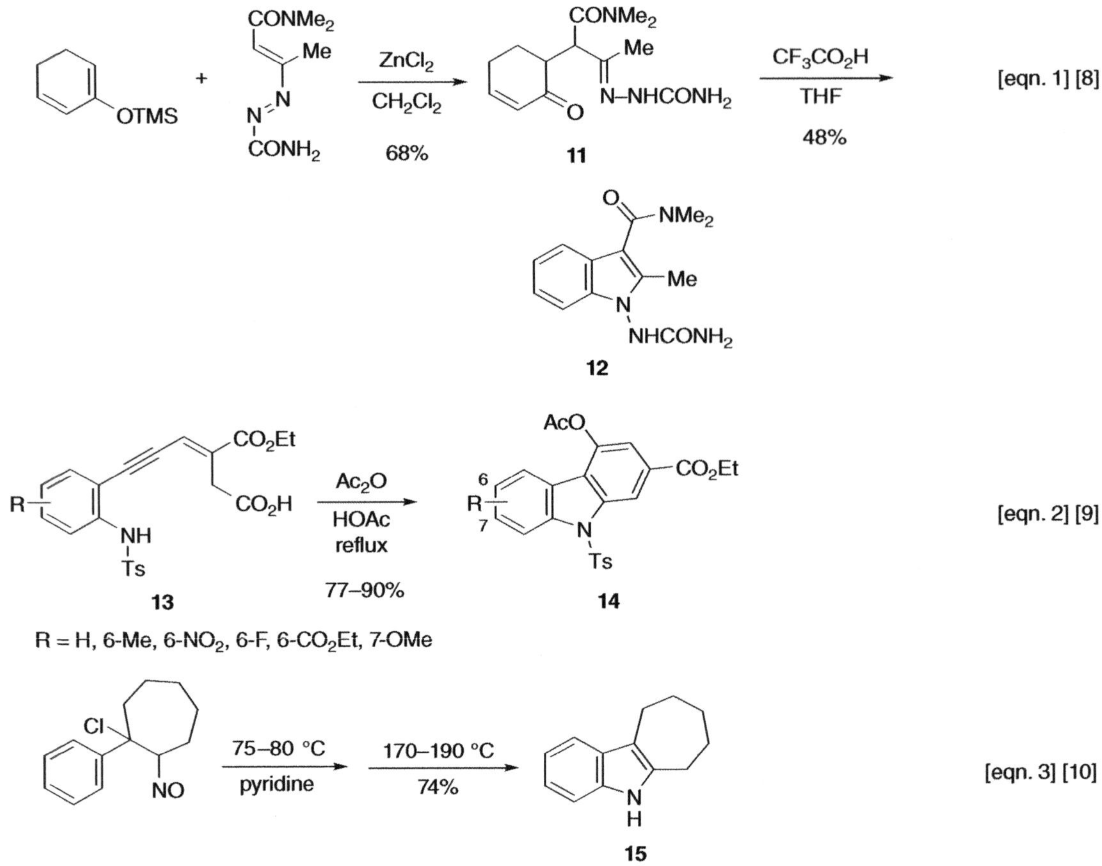

Scheme 3 *Miscellaneous Electrophilic Cyclizations*

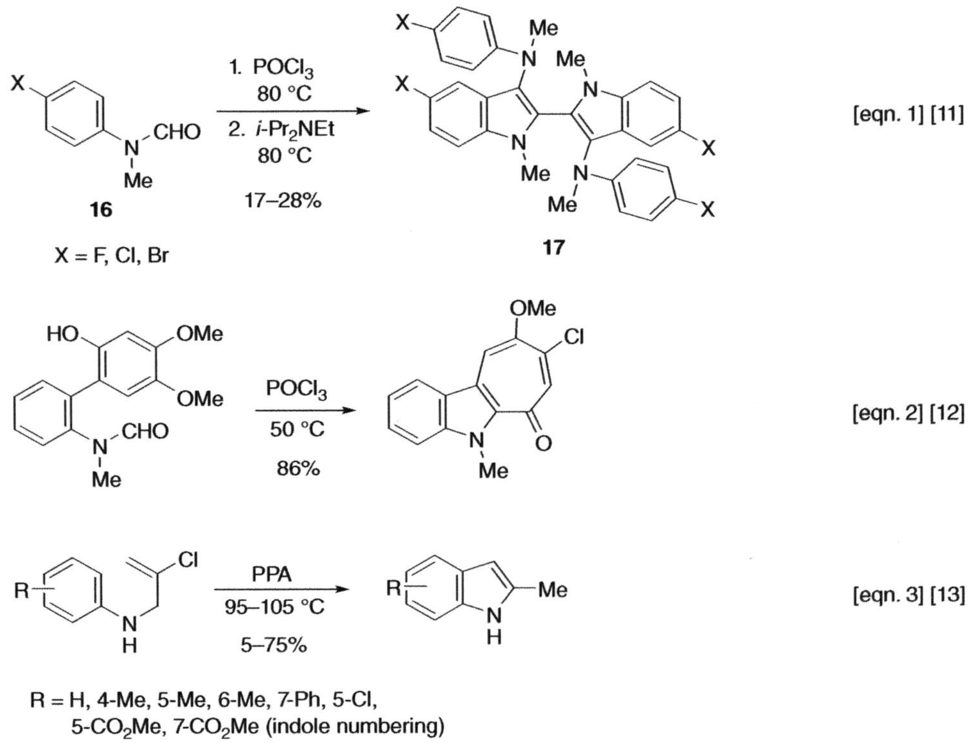

Scheme 4 *Miscellaneous Electrophilic Cyclizations*

References

[1] M. Satoh, N. Miyaura, and A. Suzuki, *Synthesis*, 1987, 373–377.
[2] X. Li, C. Abell, and M. Ladlow, *J. Org. Chem.*, 2003, **68**, 4189–4194.
[3] S. Fukamachi, S. Fujita, K. Murahashi, *et al.*, *Synthesis*, 2010, 2985–2989.
[4] K. Okuma, I. Takeshita, T. Yasuda, and K. Shioja, *Chem. Lett.*, 2006, **35**, 1122–1123.
[5] K. Okuma, T. Yasuda, I. Takeshita, *et al.*, *Tetrahedron*, 2007, **63**, 8250–8254.
[6] M.d.C. Cruz, F. Jiménez, F. Delgado, and J. Tamariz, *Synlett*, 2006, 749–755.
[7] Q. Yan, J. Luo, D. Zhang-Negrerie, *et al.*, *J. Org. Chem.*, 2011, **76**, 8690–8697.
[8] O.A. Attanasi, G. Favi, P. Filippone, *et al.*, *Adv. Synth. Catal.*, 2007, **349**, 907–915.
[9] S. Serra and C. Fuganti, *Synlett*, 2005, 809–812.
[10] D. Ginsburg and R. Pappo, *J. Am. Chem. Soc.*, 1953, **75**, 1094–1097.
[11] Y. Cheng, S. Goon, and O. Meth-Cohn, *J. Chem. Soc., Perkin Trans. 1*, 1998, 1619–1625.
[12] T. Ishikawa, K. Shimooka, T. Narioka, *et al.*, *J. Org. Chem.*, 2000, **65**, 9143–9151.
[13] B.G. McDonald and G.R. Proctor, *J. Chem. Soc., Perkin Trans. 1*, 1975, 1446–1450.

PART IV
Reductive Cyclization

We have already encountered some reduction methods en route to indole ring construction. Several classical indole name reactions feature reduction as the crucial step in indolization, and these are covered now.

Indole Ring Synthesis: From Natural Products to Drug Discovery, First Edition. Gordon W. Gribble.

39

Nenitzescu *o*,β-Dinitrostyrene Reductive Cyclization

Although Nenitzescu is best known for his 5-hydroxyindole synthesis (Chapter 15), he and his colleagues pioneered the reduction and cyclization of *ortho*-β-dinitrostyrenes to indoles. This classic reaction paved the way for other indole ring syntheses involving the reduction of nitroarenes. Nenitzescu described the inaugural reaction in 1925 as shown in Scheme 1, equation 1 [1]. The *o*,β-dinitrostyrene was prepared by the base-catalyzed condensation of *o*-nitrobenzaldehyde with nitromethane (now known as the Henry reaction). Independently and simultaneously with Nenitzescu's discovery, van der Lee described the same reduction of *o*,β-dinitrostyrene to indole [2]. van der Lee used iron/acetic acid, zinc/acetic acid, and aluminum-mercury amalgam. The yields were only 9% to 11%, and no further work was done, so this reaction has normally not included the name of van der Lee, but his independent discovery must be recognized. One of the early practitioners of this Nenitzescu indole synthesis was Beer, who employed it to prepare hydroxyindoles (equation 2 and **1–8**) [3, 4]. The 2-methylindoles (**2–4**, **6**, **8**) were prepared using nitroethane in the benzaldehyde condensation step. Huebner and coworkers synthesized 5,6-dimethoxyindole on a large scale en route to harman analogues (equation 3) [5]. Another early use of the Nenitzescu indolization was that of Witkop and Ek, who prepared 5- and 7-hydroxyindoles via the corresponding benzyloxyindoles [6].

The widespread utility of the Nenitzescu indole synthesis from *o*,β-dinitrostyrenes is illustrated by the examples in Table 1. Entry 1 features the synthesis of 5,6,7-trimethoxyindole, a possible intermediary metabolite of the alkaloid mescaline [7]. Of the seven routes pursued by Benington and colleagues, only the Nenitzescu method proved practical. The conventional iron/acetic acid method was poor for the substrate in Entry 2, and catalytic hydrogenation in the presence of acetic acid was incomplete [8]. Entries 3–6 afforded indoles that were subsequently used to synthesize the corresponding tryptamines and related compounds of biological interest [9–12]. The preparation of 5,6-diacetoxyindole (Entry 7) required milder hydrogenation conditions than normally employed [13], and catalytic hydrogenation in methanol was necessary to give pure 5,6-dihydroxyindole in reasonable yield (Entry 8) [14].

Augustine and coworkers were first to extend the Nenitzescu *o*,β-dinitrostyrene reaction to the preparation of 2-substituted indoles other than 2-methylindoles [15]. Thus, as shown in Scheme 2 (equation 1), 2-ethylindole and 2-benzylindole, along with 2-methylindole, were synthesized in good overall yield from *o*-nitrobenzaldehyde. Interestingly, hydrogenation of α,2-dinitrochalcone gave 3-amino-2-phenylquinoline and no indole product (equation 2) [15]. The first fundamental change in the reduction method was described by Cava and colleagues, who employed ammonium formate as the hydrogen source (equation 3) [16]. The yields of indoles were excellent and were superior to either the Fe/HOAc or the catalytic hydrogenation methods as shown. Other new reduction methods are titanium trichloride (equation 4) [17] and electrolysis (equation 5) [18].

Despite these other reduction methods—notably the excellent Cava variation [16]—the traditional iron/acetic acid and catalytic hydrogenation conditions continue to be preferred. For example, the indoles **9–22** were prepared using the iron/acetic acid conditions (Scheme 3 [19–32]). Indole **10** was converted to 5-hydroxytryptamine-4,7-dione, the neurocytotoxic product of 5,7-dihydroxytryptamine [20], and indoles **11** and **12** were recruited for

Indole Ring Synthesis: From Natural Products to Drug Discovery, First Edition. Gordon W. Gribble.

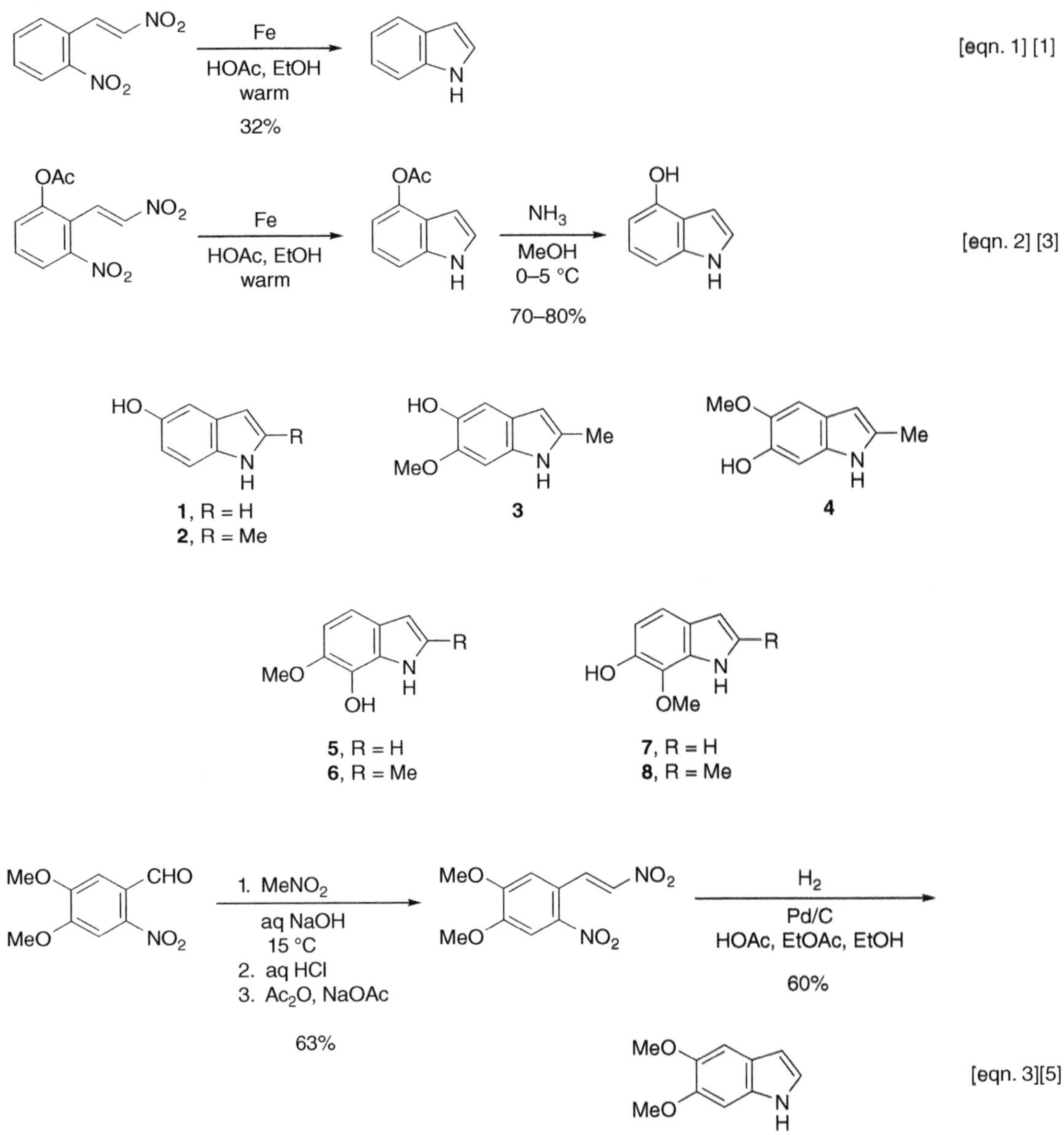

Scheme 1 *Nenitzescu Indole Synthesis*

the synthesis of two natural indole-4,7-quinones found in the muricid gastropod *Drupella fragum* [21]. Indole **13** was a starting point for syntheses of hydroxyellipticines [22], and Delgado and Clardy prepared **14** in a total synthesis of (–)-ovatolide [23]. Magnus and Westlund manufactured the indole alkaloids (±)-lahadinine B and (±)-11-methoxykopsilongine from 6,7-dimethoxyindole (**16**) [25]. A second investigation included syntheses of other indole alkaloids from **16** [26]. Indole **17** was fashioned into 5,5′,6,6′-tetrahydroxy-3,3′-biindolyl, the suggested antioxidant present in beetroot (*Beta vulgaris*) [27]. Indole **18** is one step removed from a potent anticancer 3-aroylindole [28], and 5,6-methylenedioxyindole (**19**) was featured in the synthesis of novel plant hormone indoleacetic acids [29]. Liou and colleagues prepared 5,6,7-trimethoxyindole for a molecular mimic of combretastatin [30], and indole **21** became the indole unit in rapalexin B in a synthesis of this cruciferous phytoalexin by Pedras and colleagues [31]. Indole **22** was needed for a biosynthetic study of the paraherquamide natural products [32], and Corey's team employed 6,7-dimethoxyindole (**16**) (71%) in their synthesis of aspidophytine [33].

Table 1 *Applications of the Nenitzescu Indole Synthesis from o,β-Dinitrostyrenes*

Entry	Substrate	Conditions	Indole	% Yield	Ref.
1	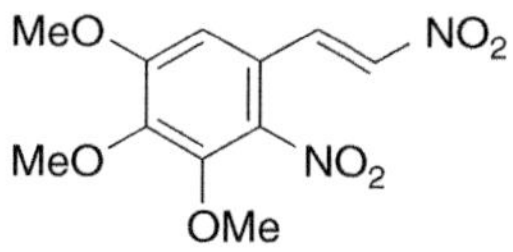	Fe powder HOAc, EtOH	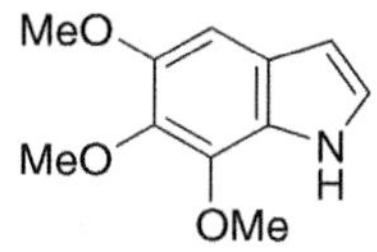	49%	7
2	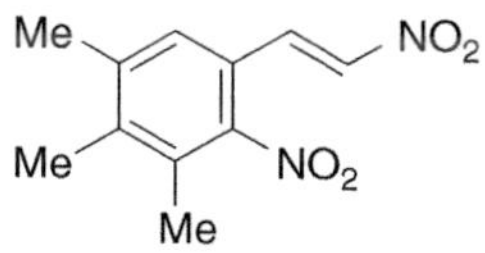	H_2, 10% Pd/C EtOAc EtOH 50 psi	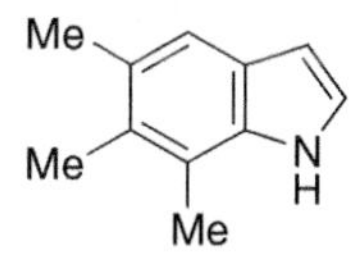	43%	8
3	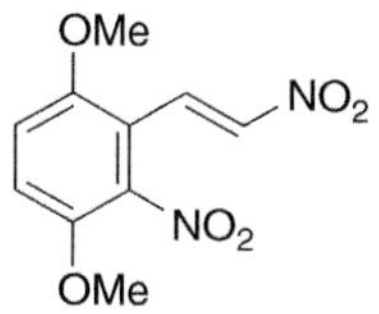	Fe HOAc EtOH	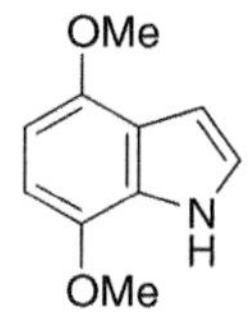	58%	9
4	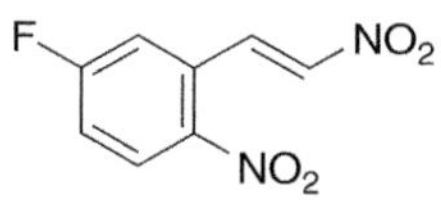	H_2, 10% Pd/C EtOAc, HOAc 40–50 °C 60 psi	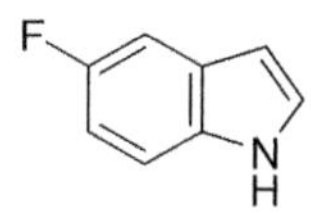	83%	10
5	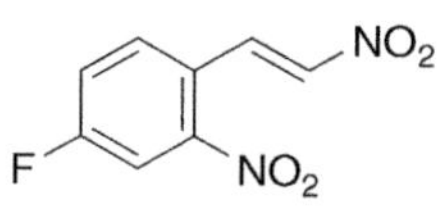	H_2, 10% Pd/C EtOAc, HOAc 60 psi	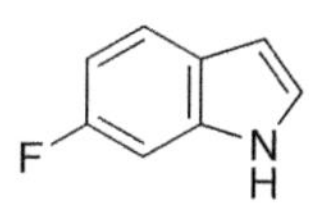	62%	11
6	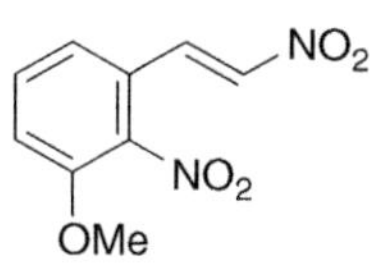	H_2, 10% Pd/C EtOAc, HOAc 4 atm	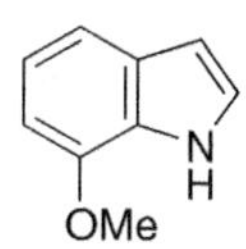	68%	12
7	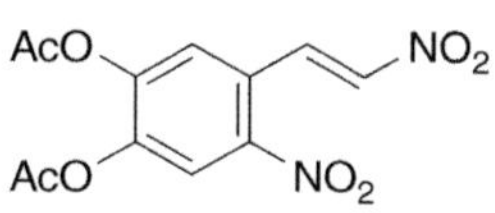	H_2, 5% Pt/C HOAc 15 psi	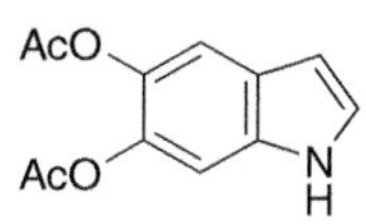	70%	13
8	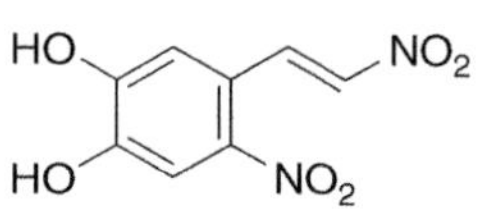	H_2, 10% Pd/C MeOH 50 psi	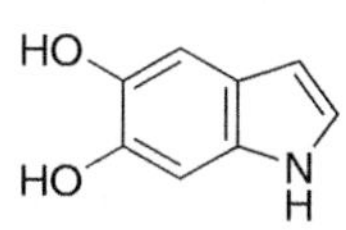	53–76%	14

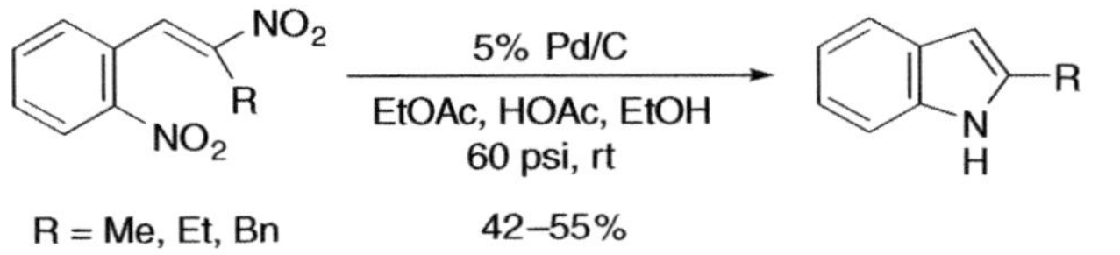

[eqn. 1] [15]

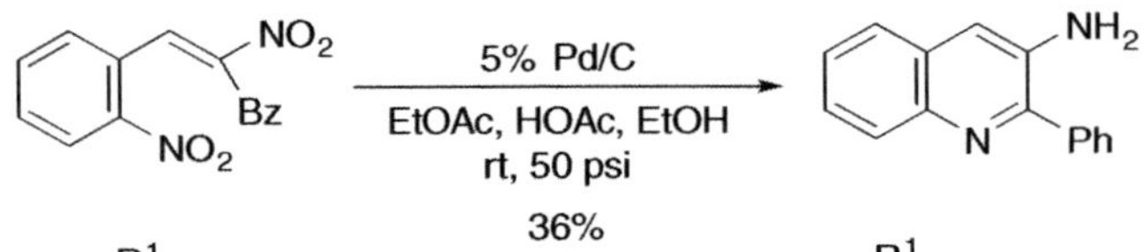

[eqn. 2] [15]

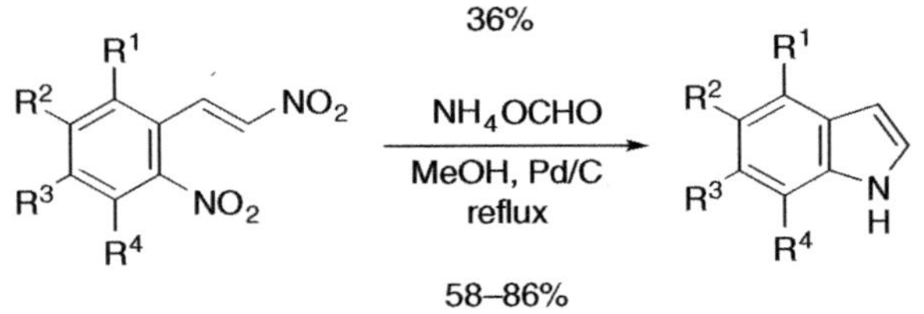

[eqn. 3] [16]

R^1	R^2	R^3	R^4	% Yield [16]	% Yield
H	H	H	H	60%	–
OMe	OMe	H	H	58%	50%[1]
OMe	H	H	OMe	86%	26%[1]
H	OMe	OMe	H	60%	50%[2]
H	H	OMe	OMe	73%	52%[1], 23%[2]
H	H	OH	OMe	73%	44%[1]
H	H	OPh	OMe	74%	15%[1]
H	–OCH_2O–		H	82%	33%[1]

[1]Fe/HOAc procedure (cited in [16])
[2]Catalytic hydrogenation procedure (cited in [16])

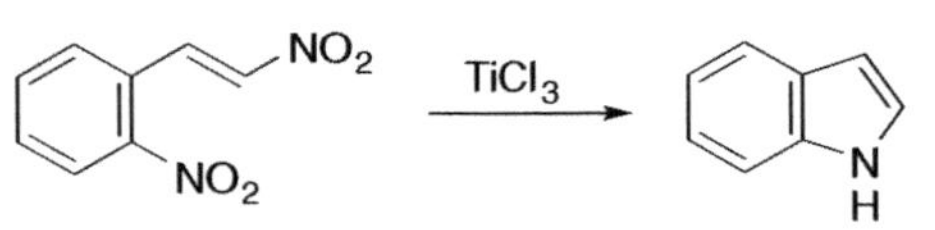

[eqn. 4] [17]

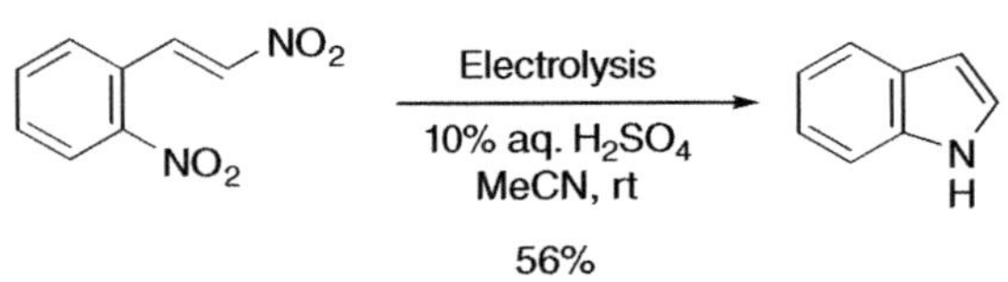

[eqn. 5] [18]

Scheme 2 *Applications of the Nenitzescu Indole Synthesis*

Indoles that were prepared using the catalytic hydrogenation of the *o*,β-dinitrostyrene strategy, in addition to those already discussed, are tabulated in Scheme 4 [31, 34–41]. 4,7-Dimethoxyindole **23** was prepared on a commercial scale [34], and 7-methoxyindole was the starting indole in a synthesis of the potent insect antifeedant dithyreanitrile [35]. The core structure of conophylline made use of indole **25** [36]. Note that the yield of 4,7-dimethoxyindole was significantly higher using catalytic hydrogenation [39] (84% vs. 64% in Schemes 4 (**30**) and 3 (**9**), respectively). Indole **31** was used in a synthesis of an antivascular 3-aroylindole [40], and 5,6-methylenedioxyindole (**32**) was featured in a synthesis of rutaecarpine analogues [41].

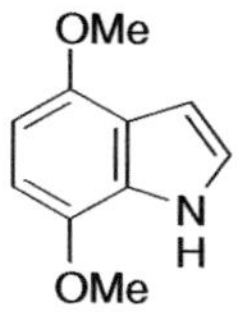

9 (71%) [19]

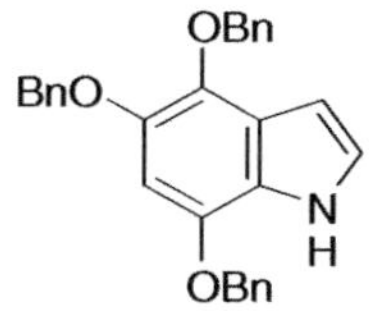

10 (92%) [20]

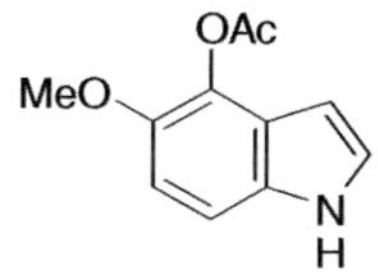

11 (96%) [21]

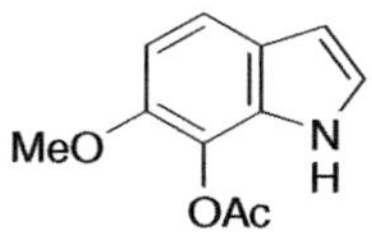

12 (29% from ArCHO) [21]

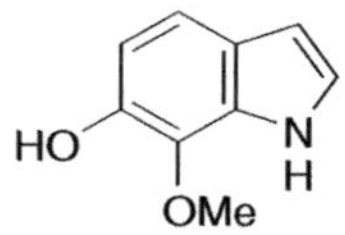

13 [22]

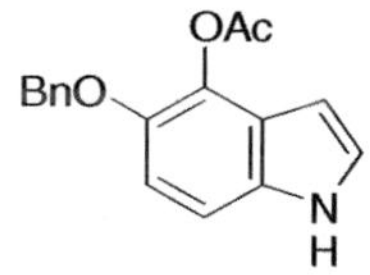

14 (64%) [23]

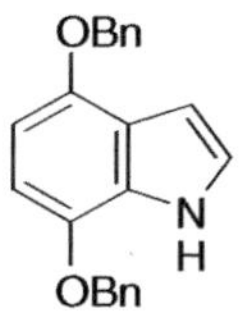

15 (90%) [24]

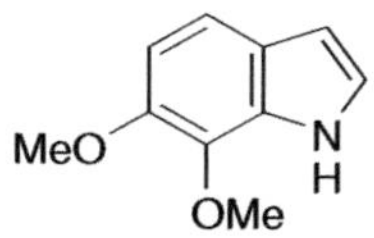

16 (85%) [25], [26]

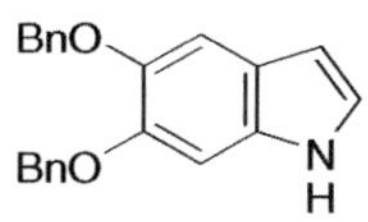

17 (61%) [27]

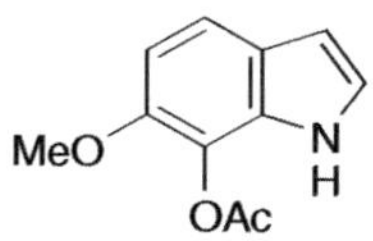

18 (31%) [28]

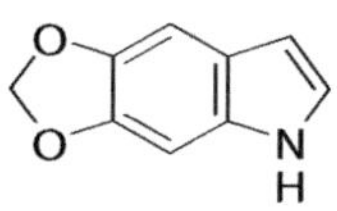

19 (72%) [29]

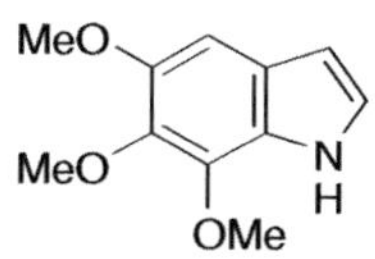

20 (30%) [30]

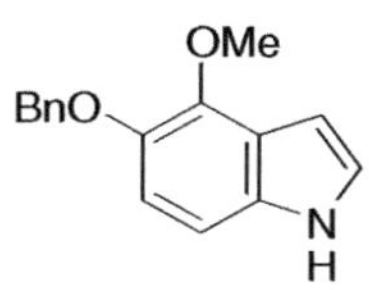

21 (75%) [31]

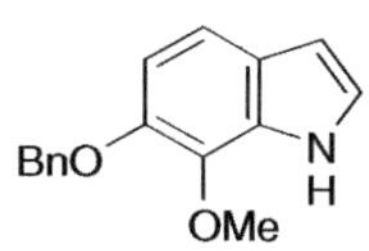

22 (92%) [32]

Scheme 3 *Indoles Prepared by Iron/Acetic Acid in the Nenitzescu Indole Synthesis*

As illustrated by the numerous indoles in Schemes 3 and 4, these two main reduction methods are firmly entrenched in the weaponry of the synthetic chemist. However, a new reduction method was reported by Prota and colleagues [42] and used by Meredith and Atkinson [43], as shown in Scheme 5, equations 1–2. This use of dithionite, zinc sulfate, and 0.1 M phosphate buffer was superior to the two conventional reduction methods for these particular indoles. Yao and coworkers employed the Nenitzescu Fe/HOAc indolization in an interesting fashion to give a series of 3,3′-biindoles (equation 3) and **33** [44], in yields of 79% to 84% [44]. The starting dinitroindole was prepared via a Michael addition of indole to *o*,β-dinitrostyrene.

For an overview of the three distinct Nenitzescu indole ring syntheses, see Ráileanu [45].

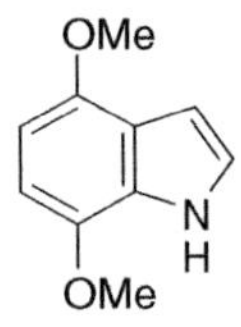

23 (75%) [34]

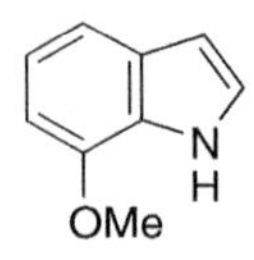

24 (61%) [35]

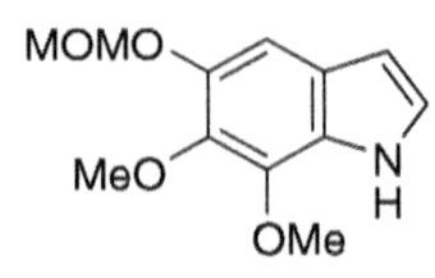

25 (72%) [36]

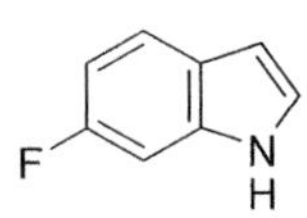

26 (63%) [37]

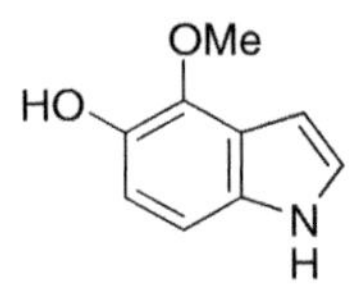

27 (29%) [31]

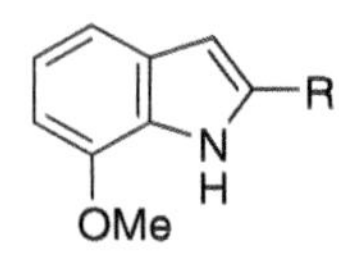

28 (32–62%) [38]
R = H, Me, Et, *n*-Pr

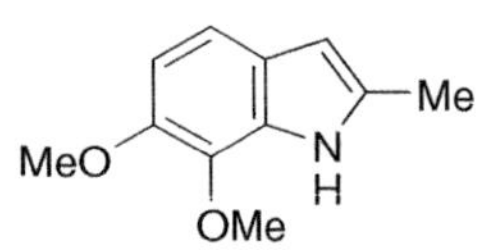

29 (81%) [38]

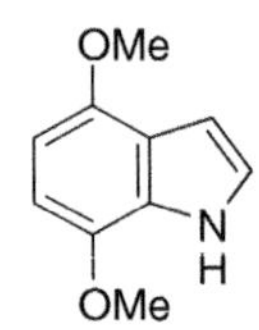

30 (84%) [39]

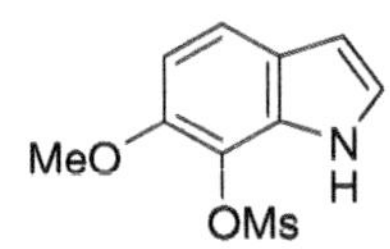

31 (54%) [40]

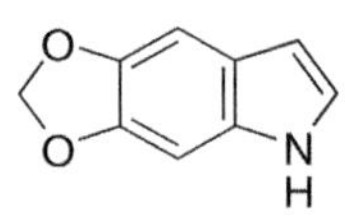

32 (94%) [41]

Scheme 4 *Indoles Prepared by Catalytic Hydrogenation in the Nenitzescu Indole Synthesis*

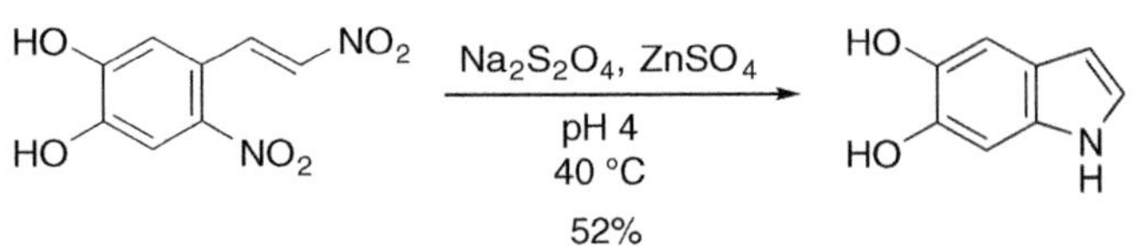

[eqn. 1] [42]

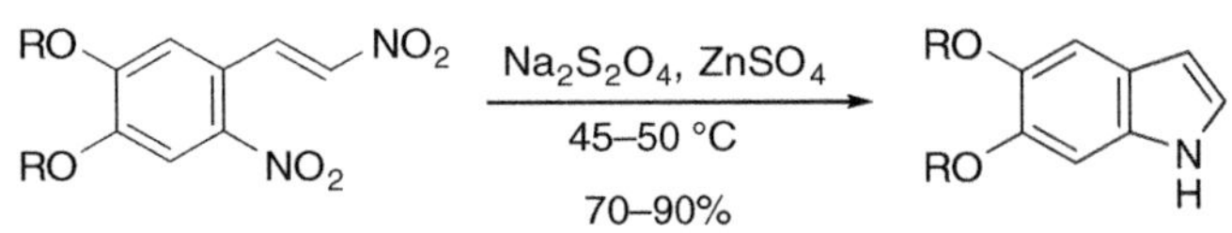

[eqn. 2] [42], [43]

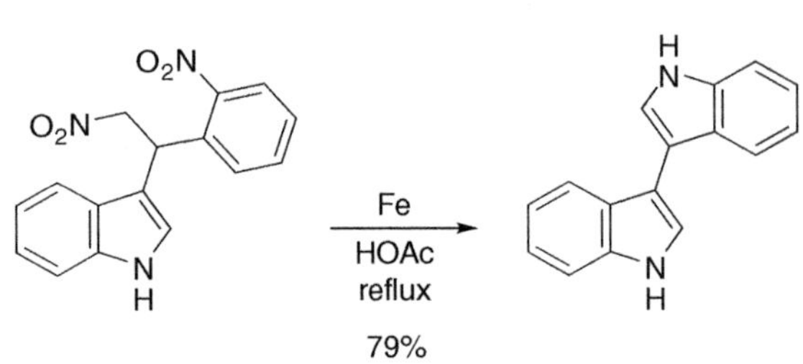

[eqn. 3] [44]

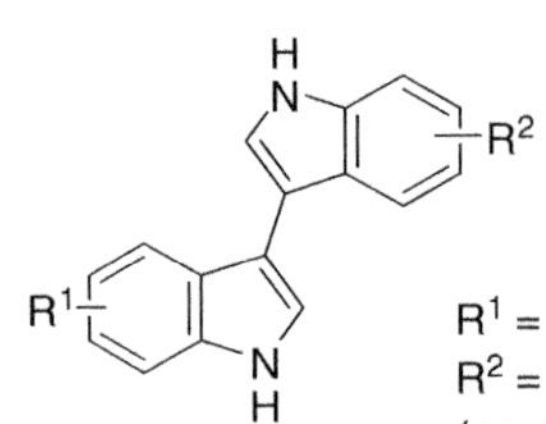

33

R^1 = OMe, Me, Et, F, Cl, Br
R^2 = F, Cl, Br
(several positions on each indole ring)

Scheme 5 *Applications of the Nenitzescu Indole Synthesis*

References

[1] C. Nenitzescu, *Chem. Ber.*, 1925, **58B**, 1063–1064.
[2] J. van der Lee, *Rec. Trav. Chim.*, 1925, **44**, 1089–1092.
[3] R.J.S. Beer, K. Clarke, H.G. Khorana, and A. Robertson, *J. Chem. Soc.*, 1948, 1605–1609.
[4] R.J.S. Beer, K. Clarke, H.F. Davenport, and A. Robertson, *J. Chem. Soc.*, 1951, 2029–2032.
[5] C.F. Huebner, H.A. Toxell, and D.C. Schroeder, *J. Am. Chem. Soc.*, 1953, **75**, 5887–5890.
[6] A. Ek and B. Witkop, *J. Am. Chem. Soc.*, 1954, **76**, 5579–5588.
[7] R.D. Morin, F. Benington, and L.C. Clark, Jr., *J. Org. Chem.*, 1957, **22**, 331–332.
[8] F. Benington, R.D. Morin, and L.C. Clark, Jr. *J. Org. Chem.*, 1960, **25**, 1542–1547.
[9] G. Rodighiero, G. Malesani, and U. Fornasiero, *Gazz. Chim. Ital.*, 1961, **91**, 742–749.
[10] Z. Pelchowicz, A. Kaluszyner, and M. Bentov, *J. Chem. Soc.*, 1961, 5418–5421.
[11] M. Bentov, A. Kaluszyner, and Z. Pelchowicz, *J. Chem. Soc.*, 1962, 2825–2827.
[12] A. Kalir, D. Balderman, H. Edery, and G. Porath, *Isr. J. Chem.*, 1967, **5**, 129–136.
[13] B.P. Murphy, *J. Org. Chem.*, 1985, **50**, 5873–5875.
[14] B.P. Murphy and T.M. Schultz, *J. Org. Chem.*, 1985, **50**, 2790–2791.
[15] R.L. Augustine, A.J. Gustavsen, S.F. Wanat, *et al.*, *J. Org. Chem.*, 1973, **38**, 3004–3011.
[16] S. Rajeswari, K.J. Drost, and M.P. Cava, *Heterocycles*, 1989, **29**, 415–418.
[17] A.S. Ijaz and J. Parrick, *Sci. Int. (Lahore)*, 1989, **1**, 364–365.
[18] S.C. Mishra and R.A. Mishra, *J. Electrochem. Soc. India*, 1990, **39**, 51–53.
[19] A.S. Ijaz, J. Parrick, and A. Yahya, *J. Chem. Res. (S)*, 1990, 116.
[20] A.K. Sinhababu and R.T. Borchardt, *J. Heterocycl. Chem.*, 1988, **25**, 1155–1159.
[21] Y. Fukuyama, C. Iwatsuki, M. Kodama, *et al.*, *Tetrahedron*, 1998, **54**, 10007–10016.
[22] P.M. Dharmasena and P.V.R. Shannon, *Tetrahedron Lett.*, 1994, **35**, 7119–7122.
[23] A. Delgado and J. Clardy, *J. Org. Chem.*, 1993, **58**, 2862–2866.
[24] H.-J. Knölker and K. Hartmann, *Synlett*, 1993, 755–757.
[25] P. Magnus and N. Westlund, *Tetrahedron Lett.*, 2000, **41**, 9369–9372.
[26] P. Magnus, L. Gazzard, L. Hobson, *et al.*, *Tetrahedron*, 2002, **58**, 3423–3443.
[27] S.P.H. Mee, V. Lee, J.E. Baldwin, and A. Cowley, *Tetrahedron*, 2004, **60**, 3695–3712.
[28] T.-S. Wu, M.S. Coumar, J.-Y. Chang, *et al.*, *J. Med. Chem.*, 2009, **52**, 4941–4945.
[29] F.A.F. da Rosa, R.A. Rebelo, and M.G. Nascimento, *J. Braz. Chem. Soc.*, 2003, **14**, 11–15.
[30] H.-Y. Lee, J.-Y. Chang, L.-Y. Chang, *et al.*, *Org. Biomol. Chem.*, 2011, **9**, 3154–3157.
[31] M.S.C. Pedras, Q.-A. Zheng, and R.S. Gadagi, *Chem. Commun.*, 2007, 368–370.
[32] K. Sommer and R.M. Williams, *Tetrahedron*, 2009, **65**, 3246–3260.
[33] F. He, Y. Bo, J.D. Alton, and E.J. Corey, *J. Am. Chem. Soc.*, 1999, **121**, 6771–6772.
[34] H.D.H. Showalter and G. Pohlmann, *Org. Prep. Proc. Int.*, 1992, **34**, 484–488.
[35] E.K. Mantus and J. Clardy, *Tetrahedron Lett.*, 1993, **34**, 1085–1086.
[36] S. Ando, Y. Okamoto, K. Umezawa, and M. Otsuka, *J. Heterocycl. Chem.*, 2008, **45**, 1803–1808.
[37] A. Kalir and S. Szara, *J. Med. Chem.*, 1963, **6**, 716–719.
[38] B.-C. Chen, J. Hynes, Jr., C.R. Pandit, *et al.*, *Heterocycles*, 2001, **55**, 951–960.
[39] V. Bénéteau and T. Besson, *Tetrahedron Lett.*, 2001, **42**, 2673–2676.
[40] N. Ty, G. Dupeyre, G.G. Chabot, *et al.*, *Bioorg. Med. Chem.*, 2008, **16**, 7494–7503.
[41] L-M. Yang, C-F. Chen, and K-H. Lee, *Bioorg. Med. Chem. Lett.*, 1995, **5**, 465–468.
[42] L. Novellino, M. d'Ischia, and G. Prota, *Synthesis*, 1999, 793–796.
[43] S. Atkinson and P. Meredith, *Synlett*, 2003, 1853–1855.
[44] C. Ramesh, V. Kavala, C.-W. Kuo, *et al.*, *Eur. J. Org. Chem.*, 2010, 3796–3801.
[45] D. Raileanu, E. Tighineanu, and F. Dumitrascu, *Revista de Chimie*, 1992, **43**, 506–511.

40

Reissert Indole Synthesis

Given the ease with which the nitro group undergoes reduction to the amino group, and the alacrity of amine–carbonyl condensation, it is not surprising that a very early indole ring synthesis makes use of this transformation. Thus, more than one hundred years ago, Arnold Reissert found that *o*-nitrophenylpyruvic acid was reduced with ferrous sulfate and ammonia to give the corresponding amine, which cyclized without isolation to indole-2-carboxylic acid. Heating the latter above its melting point resulted in decarboxylation to indole (Scheme 1, equation 1) [1–3]. The starting pyruvic acid was readily prepared by acid hydrolysis of ethyl *o*-nitrophenylpyruvate, which was obtained via base-catalyzed condensation of *o*-nitrotoluene with diethyl oxalate. For an excellent discussion of the early work on this reaction, see Brown [4]. Thus, several other base condensation methods and reduction conditions were employed by the early followers of Reissert. For example, potassium ethoxide in dry ether is a stronger base than sodium ethoxide in ethanol, and iron powder in acetic acid/ethanol, iron filings in hydrochloric acid, zinc dust in acetic acid, and sodium dithionite have all been employed for the reduction step [4].

In addition to the several early examples cited by Brown [4], Scheme 2 illustrates some applications of the Reissert indole synthesis. Attempts by Beer and colleagues to improve the decarboxylation step in their synthesis of 5-hydroxyindole (equation 2) were unsuccessful, but these workers were able to synthesize 6- and 7-hydroxyindole via a Reissert method [5]. Uhle described an excellent preparation of 4-chloroindole (equation 2) [6], and Singer and Shive synthesized 7-indolecarboxylic acid (equation 3) [7] and 5-indolecarboxylic acid (equation 4) [8] via a Reissert strategy. As seen, both cyanation steps resulted in decarboxylation and cyanide replacement prior to nitrile

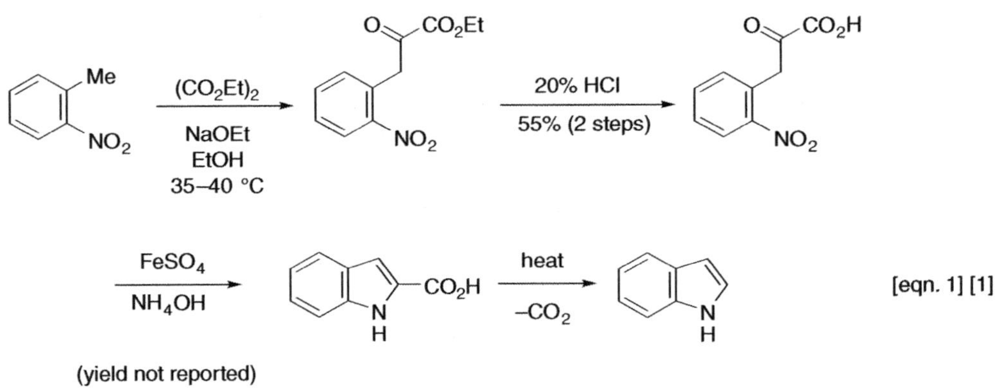

Scheme 1 *Reissert Indole Synthesis*

Indole Ring Synthesis: From Natural Products to Drug Discovery, First Edition. Gordon W. Gribble.

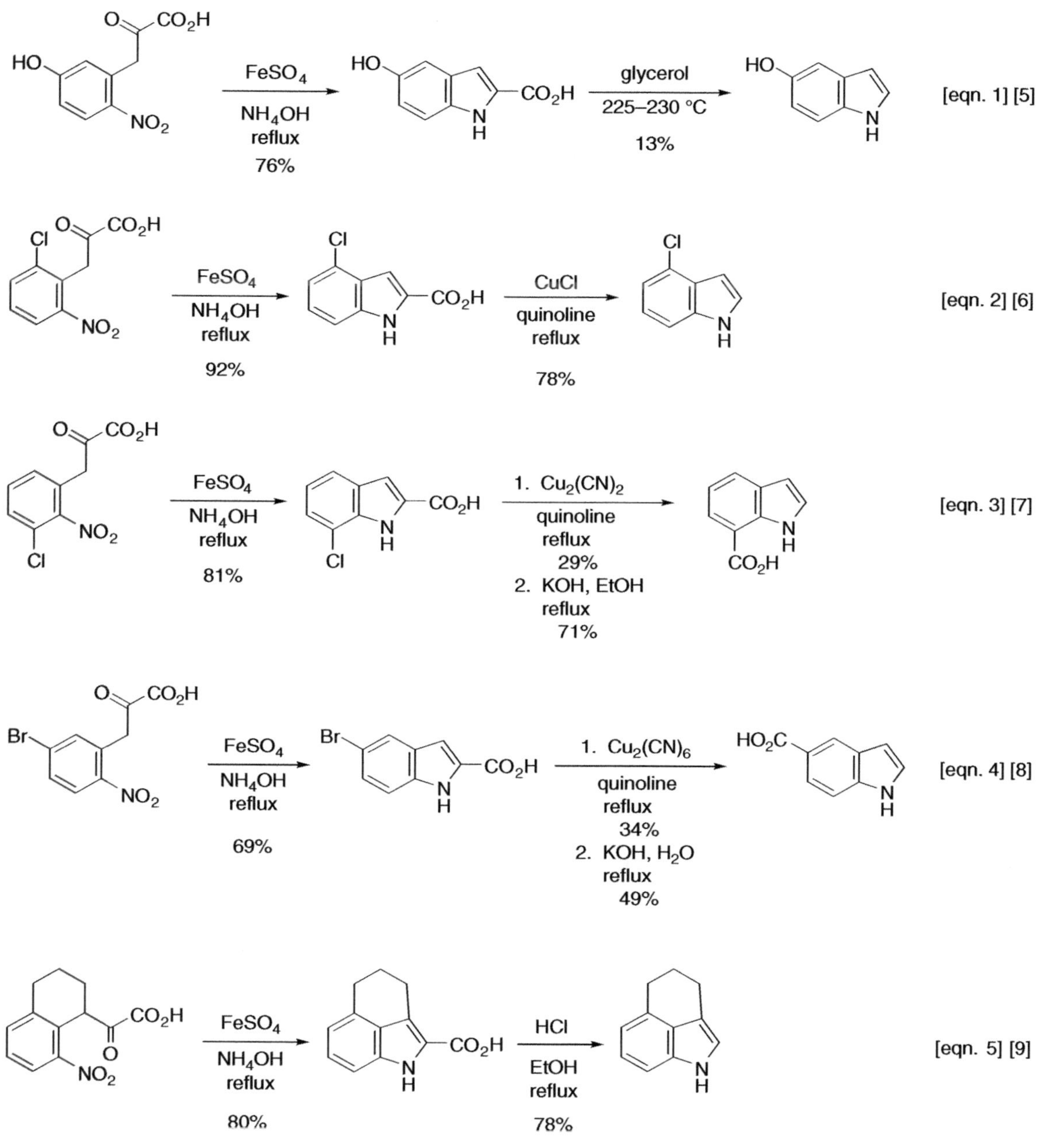

Scheme 2 *Applications of the Reissert Indole Synthesis*

hydrolysis. Uhle and coworkers synthesized 1,3,4,5-tetrahydrobenz[*cd*]indole (equation 5), an important starting point in the synthesis of ergot alkaloids [9].

Stoll and colleagues prepared 4-, 5-, and 6-hydroxytryptamines from the corresponding hydroxyindoles, which were synthesized via a Reissert protocol [10], Plieninger's group used this method to access 4- and 6-bromoindoles [11], and Bergmann and Pelchowicz realized both 5- and 6-fluoroindoles from the respective *o*-nitrophenylpyruvic acids [12]. In addition to these examples, several other Reissert indole syntheses are tabulated in Table 1 [13, 15, 17–22]. Piers and Brown investigated the decarboxylation of indole-2(and 3)-carboxylic acids and found that catalytic amounts of the copper-carboxylate were superior to the use of copper chromite [14]. Noland and Baude illustrate the synthesis of indole-2-carboxylates (entry 2) [15]. Hirata and colleagues employed this latter method to prepare several 5-oxyindole-2-carboxylates [16]. Entry 3 features an indole used in synthetic studies on mitomycins [17], and Entry 6 describes an indole used in the

Table 1 *Applications of the Reissert Indole Synthesis*

Entry	Substrate	Conditions	Indole	% Yield	Ref.
1	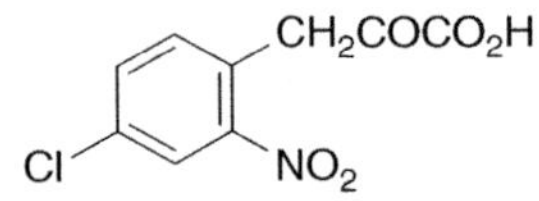	1. $Fe(OH)_2$ (65%) 2. Cu chromite quinoline (66%)	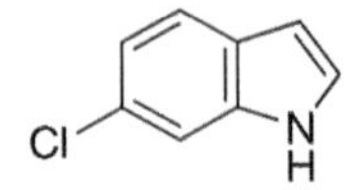	43%	13
2	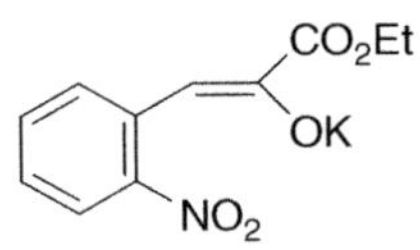	H_2, Pt HOAc	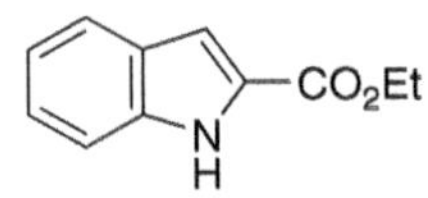	66%	15
3	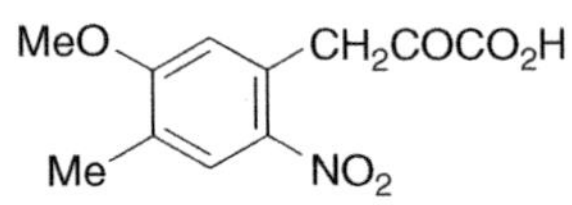	$Fe(NH_4)_3SO_4$	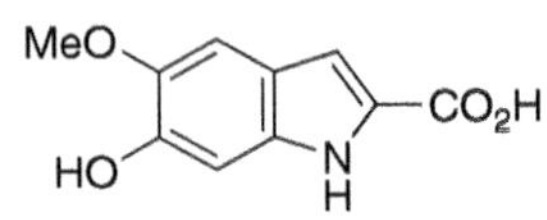	42% (from 2,5-dimethyl-4-nitroanisole)	17
4	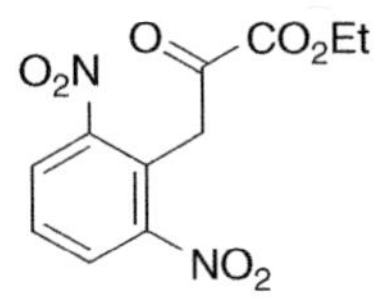	H_2/Pd/C	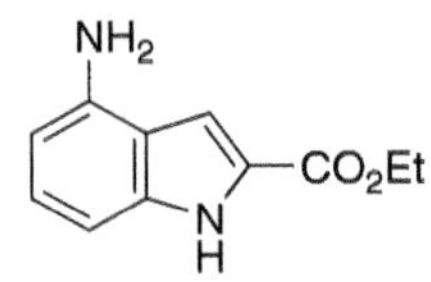	—	18
5	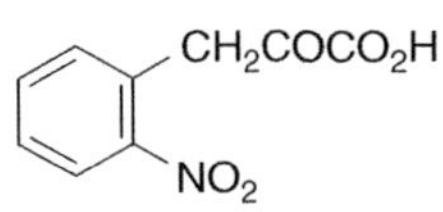	electrolysis aq HOAc, H_2SO_4	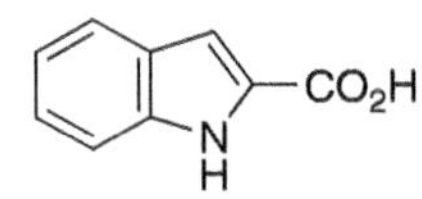	45%	19
6	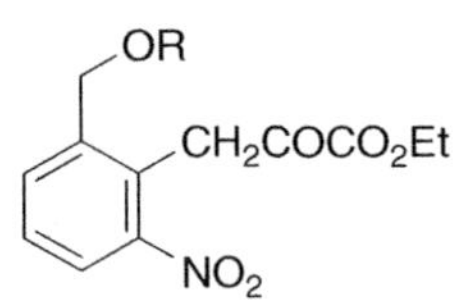 R = CH_2OMe, $CH_2O(CH_2)_2OMe$, 2-THP	H_2 Pd/C EtOH	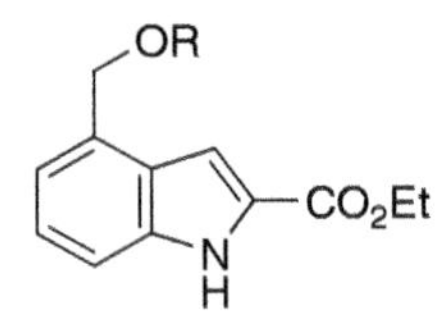	65–70%	20
7	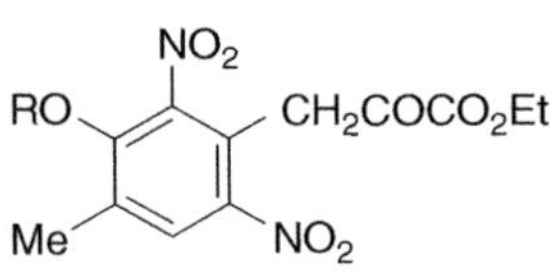 R = Me, Et	Zn, HCl EtOH, Et_2O reflux	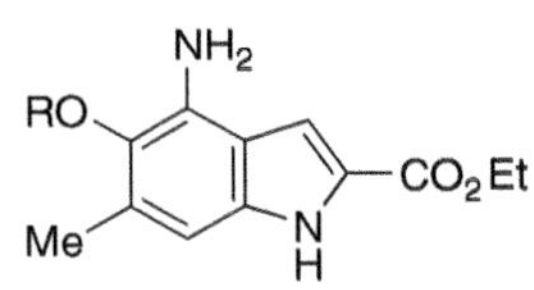	93% (75–82% large scale)	21
8	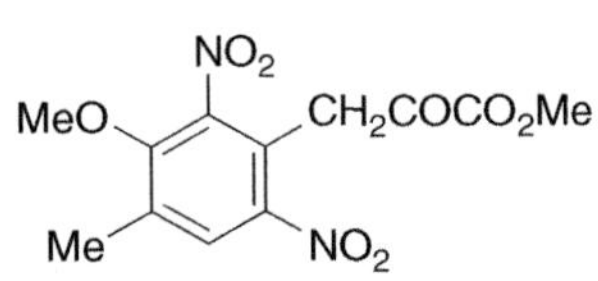	$SnCl_2$, MeOH	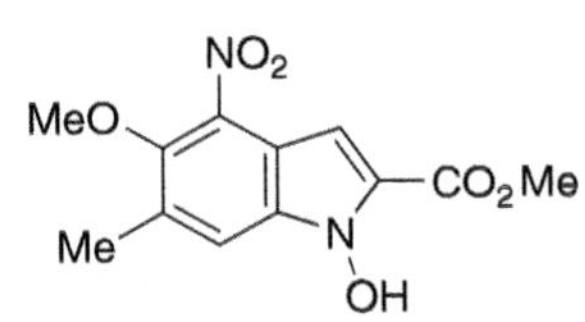	63%	22

synthesis of fragment E of nosiheptide, a *Streptomyces actuosus* antibiotic [20]. Entries 7 and 8 also feature the preparation of mitosene analogues by Jimenez and coworkers [21, 22].

The Reissert indolization continues to be used in modern organic synthesis, and some examples are depicted in Scheme 3. Prasad prepared 2-carbethoxy-6-*tert*-butylindole (equation 1) for use in the synthesis of novel HIV protease

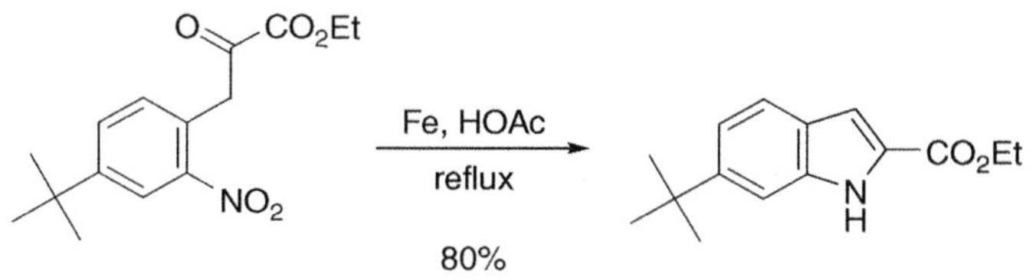

[eqn. 1] [23]

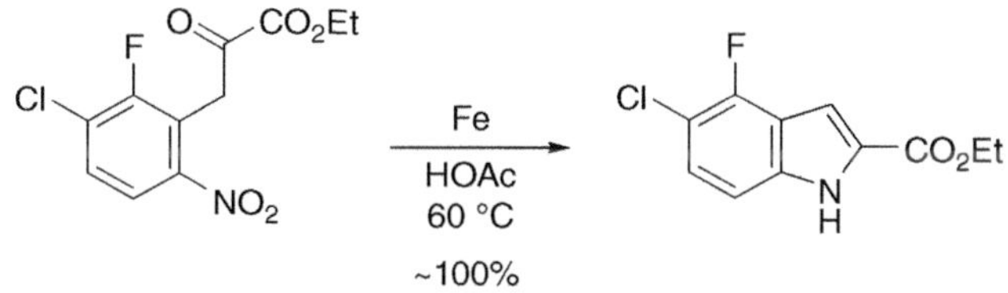

[eqn. 2] [24]

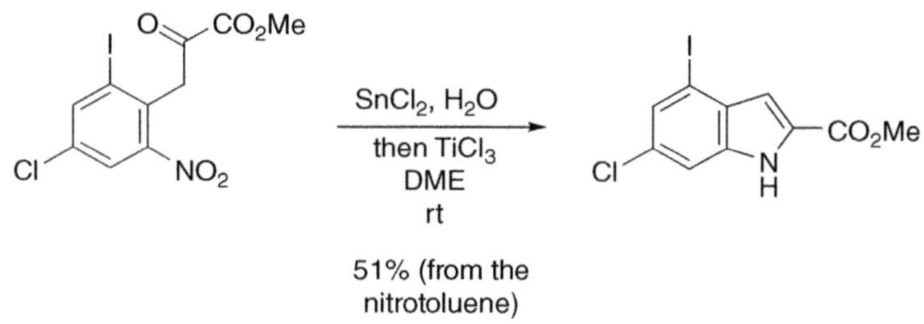

[eqn. 3] [25]

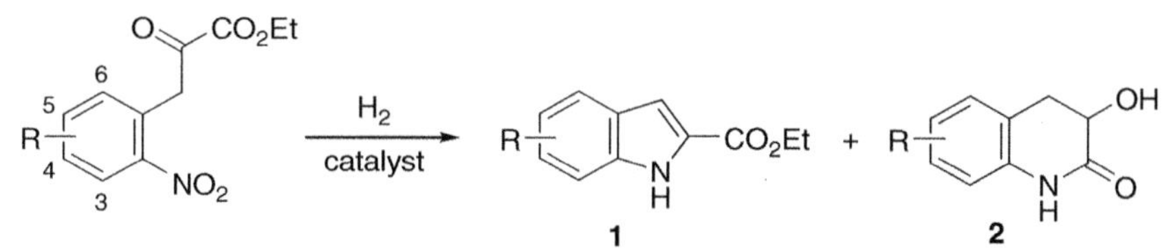

[eqn. 4] [26]

R	Conditions	Total % Yield	Ratio **1**:**2**
H	5% Pd/C, EtOH	83%	100:0
H	PtO_2, EtOH	72%	93:7
6-MeO	5% Pd/C, EtOH	80%	92:8
6-MeO	PtO_2, EtOH	62%	98:2
3-MeO	5% Pd/C, EtOH	85%	94:6
3-MeO	PtO_2, EtOH	86%	16:84
3-*i*-PrO	5% Pd/C, EtOH	81%	90:10
3-*i*-PrO	PtO_2, EtOH	88%	7:93

Scheme 3 *Modern Reissert Indole Syntheses*

inhibitors [23], and Silvestri and colleagues employed a Reissert reaction to make ethyl 5-chloro-4-fluoroindole-2-carboxylate (equation 2), which is an indole embodied in a potent HIV-1 reverse transcriptase inhibitor [24]. This method is superior to a previous Fischer indole approach used by this group. Nagata and colleagues synthesized methyl 6-chloro-4-iodoindole-2-carboxylate (equation 3), en route to some potent NMDA-glycine antagonists [25]. The intermediate *N*-hydroxyindole was reduced by titanium trichloride to the indole product. Murakami's team found that certain reduction conditions can divert the Reissert indolization to give quinolones instead (equation 4) [26]. This is particularly true for 7-substituted indoles and PtO_2 in ethanol.

Nicolaou and coworkers modified the Reissert indole synthesis to incorporate a nucleophilic conjugate step leading to C-3 *N*-hydroxyindoles (Scheme 4) [27–29]. A selection of the *N*-hydroxyindoles that were obtained is

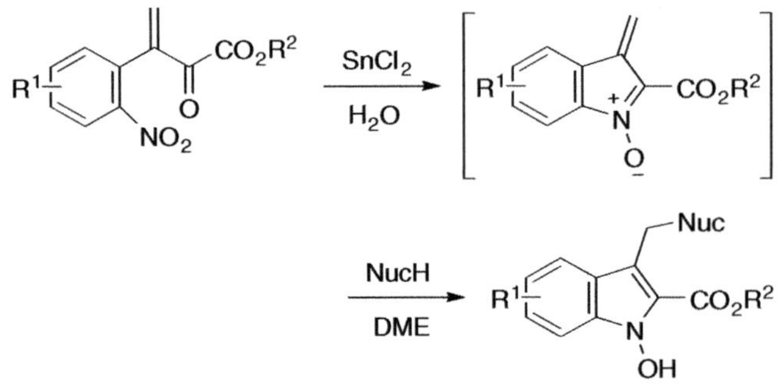

[eqn. 1] [27], [28] [29]

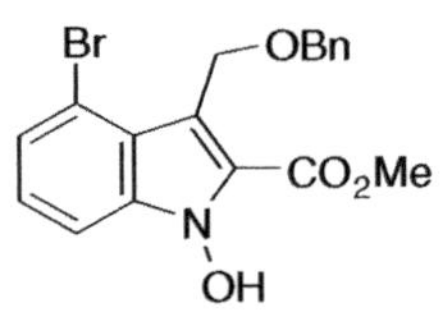

3 (60%) [27]

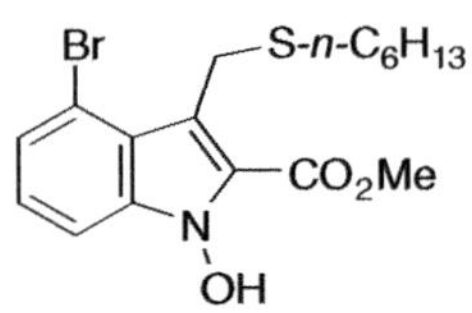

4 (75%) [27]

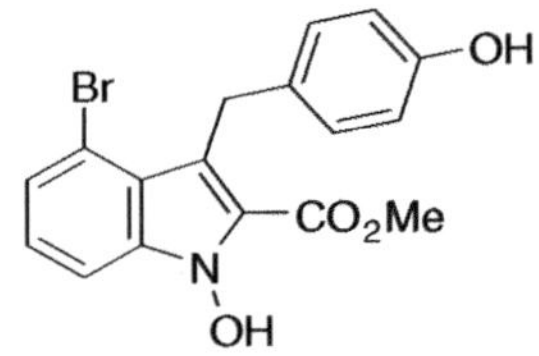

5 (40%) [27]

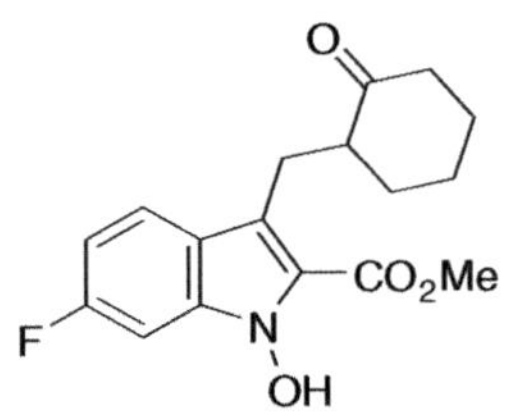

6 (45%) [28]

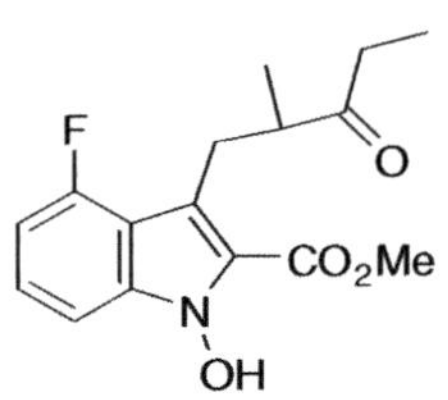

7 (61%) [28]

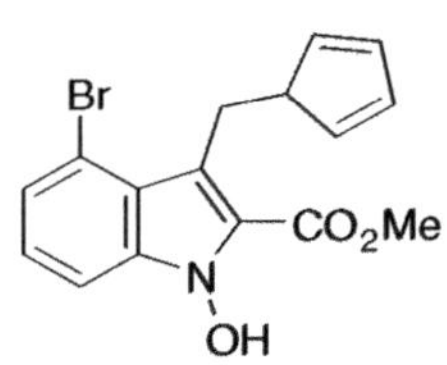

8 (57%) [28]

Scheme 4 *Nicolaou Synthesis of* N*-Hydroxyindoles*

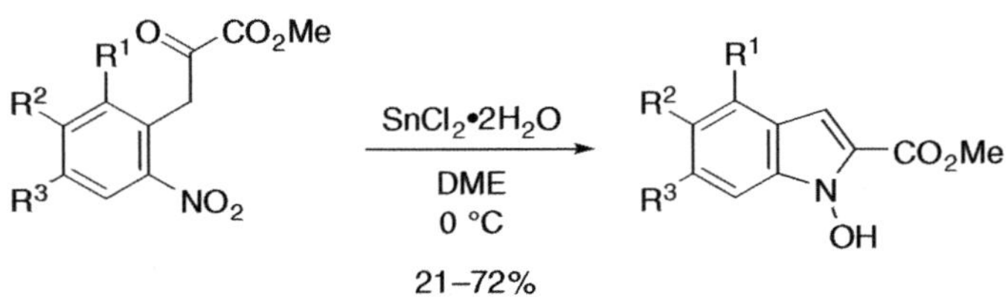

[eqn. 1] [30]

R^1 = H, Me, CF3, Cl, Br, Ph
R^2 = H, Ph
R^3 = H, Br, Ph, tetrazolo

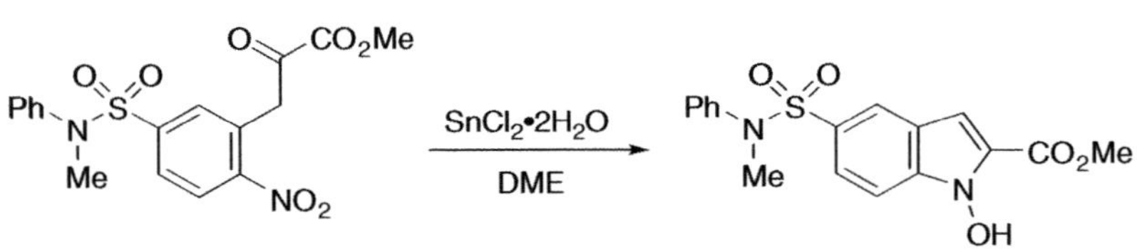

[eqn. 2] [31]

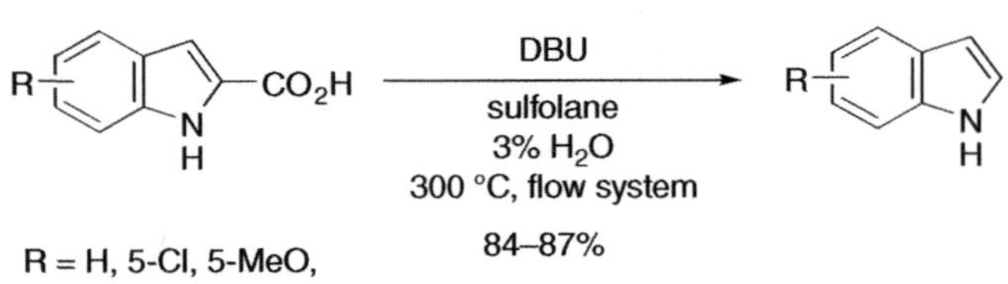

[eqn. 3] [32]

R = H, 5-Cl, 5-MeO, and *N*-Me

Scheme 5 *Minutolo Synthesis of* N*-Hydroxyindoles*

shown (**3–8**), and this work is geared towards model systems of the complex nocathiacin I antibiotic, which contains an *N*-hydroxyindole.

The work of Minutolo and colleagues also featured the Reissert indolization, leading to *N*-hydroxyindoles, which, in the present case, afforded inhibitors of human lactate dehydrogenase isoform A (Scheme 5) [30, 31]. In the case of the R^3=tetrazolo reaction, sodium hypophosphite over Pd/C was the reducing agent. Because decarboxylation of the Reissert indole-2-carboxylic acids was often the ultimate goal in the aforementioned investigations, it should be noted that Tilstam reported a novel method for this conversion (equation 3) [32].

This chapter has focused solely on the classic Reissert reductive cyclization of *o*-nitrophenylpyruvic acids (and pyruvates) to indoles. The myriad other reductive cyclizations of nitro carbonyl compounds to give indoles are relegated to subsequent chapters.

References

[1] A. Reissert, *Ber.*, 1897, **30**, 1030–1053.
[2] A. Reissert, *Ber.*, 1896, **29**, 639–665.
[3] A. Reissert and J. Scherk, *Ber.*, 1898, **31**, 387–397.
[4] R.K. Brown (1972) in *Indoles, Part 1* (ed. W.J. Houlihan), Wiley, New York, pp. 397–413.
[5] R.J.S. Beer, K. Clarke, H.G. Khorana, and A. Robertson, *J. Chem. Soc.*, 1948, 1605–1609.
[6] F.C. Uhle, *J. Am. Chem. Soc.*, 1949, **71**, 761–766.
[7] H. Singer and W. Shive, *J. Am. Chem. Soc.*, 1955, **77**, 5700–5702.
[8] H. Singer and W. Shive, *J. Org. Chem.*, 1955, **20**, 1458–1460.
[9] F.C. Uhle, C.G. Vernick, and G.L. Schmir, *J. Am. Chem. Soc.*, 1955, **77**, 3334–3337.
[10] A. Stoll, F. Troxler, J. Peyer, and A. Hofmann, *Helv. Chim. Acta*, 1955, **38**, 1452–1472.
[11] H. Plieninger, *Chem. Ber.*, 1955, **88**, 370–376.
[12] E.D. Bergmann and Z. Pelchowicz, *J. Chem. Soc.*, 1959, 1913–1914.
[13] F. Benington, R.D. Morin, and L.C. Clark, Jr. *J. Org. Chem.*, 1960, **25**, 1542–1547.
[14] E. Piers and R.K. Brown, *Can. J. Chem.*, 1962, **40**, 559–561.
[15] W.E. Noland and F.J. Baude, *Org. Syn. Coll. Vol. V*, 1973, 567–571.
[16] T. Hirata, Y. Yamada, and M. Matsui, *Tetrahedron Lett.*, 1969, **10**, 19–22.
[17] G. Leadbetter, D.L. Fost, N.N. Ekwuribe, and W.A. Remers, *J. Org. Chem.*, 1974, **39**, 3580–3583.
[18] J. Bergman, P. Sand, and U. Tilstam, *Tetrahedron Lett.*, 1983, **24**, 3665–3668.
[19] S.C. Mishra and R.A. Mishra, *J. Electrochem. Soc. India*, 1990, **39**, 51–53.
[20] C. Shin, Y. Yamada, K. Hayashi, *et al.*, *Heterocycles*, 1996, **43**, 891–898.
[21] Z. Wang and L.S. Jimenez, *J. Org. Chem.*, 1996, **61**, 816–818.
[22] W. Dong and L.S. Jimenez, *J. Org. Chem.*, 1999, **64**, 2520–2523.
[23] J.V.N.V. Prasad, *Org. Lett.*, 2000, **2**, 1069–1072.
[24] F. Piscitelli, G. La Regina, and R. Silvestri, *Org. Prep. Proc. Int.*, 2008, **40**, 204–208.
[25] S. Katayama, N. Ae, and R. Nagata, *J. Org. Chem.*, 2001, **66**, 3474–3483.
[26] H. Suzuki, H. Gyoutoku, H. Yokoo, *et al.*, *Synlett*, 2000, 1196–1198.
[27] K.C. Nicolaou, S.H. Lee, A.A. Estrada, and M. Zak, *Angew. Chem. Int. Ed.*, 2005, **44**, 3736–3740.
[28] K.C. Nicolaou, A.A. Estrada, S.H. Lee, and G.C. Freestone, *Angew. Chem. Int. Ed.*, 2006, **45**, 5364–5368.
[29] K. C. Nicolaou, A. A. Estrada, G. C. Freestone, *et al., Tetrahedron,* 2007, **63**, 6088-6114.
[30] C. Granchi, S. Roy, C. Giacomelli, *et al.*, *J. Med. Chem.*, 2011, **54**, 1599–1612.
[31] C. Granchi, S. Roy, M. Mottinelli, *et al.*, *Bioorg. Med. Chem. Lett.*, 2011, **21**, 7331–7336.
[32] U. Tilstam, *Org. Process Res. Dev.*, 2012, **16**, 1449–1454.

41

Leimgruber–Batcho Indole Synthesis

The relatively new Leimgruber–Batcho indole synthesis has rapidly emerged as a key weapon in the arsenal of the synthetic organic chemist. In 1971, Willy Leimgruber and Andrew Batcho reported at an international meeting the indole synthesis that now bears their name (Scheme 1) [1], followed soon thereafter in patents [2], and then in *Organic Syntheses* [3]. An excellent review is available [4]. Subsequent to the two main methods for reducing the β-aminostyrene **2** to indole **3** (catalytic hydrogenation with palladium and Raney nickel–hydrazine), several other conditions were explored. The nitroenamine **2** can often be isolated and purified prior to reduction, but it need not be. The addition of pyrrolidine (or piperidine) is typically beneficial as it generates a more reactive formylpyrrolidine acetal or a mixed pyrrolidine dimethylamine aminal. Leimgruber and Batcho synthesized a large number of indoles and employed several reduction conditions with and without the isolation of the nitroenamine **2** (Scheme 1).

A general pathway is depicted in Scheme 2. The initial step takes advantage of the enhanced acidity of the benzylic methyl hydrogens, similar to what was observed in the Reissert reaction. Heating *N,N*-dimethylformamide dimethyl acetal (**4**) generates iminium ion **5** and methoxide, which deprotonates *o*-nitrotoluene to give anion **6**. Subsequent condensation of **6** with **5** and elimination of methoxide (methanol) affords nitroenamine **7**. The addition of pyrrolidine can intervene in this pathway at this point or earlier. Reduction of **7** leads to aminoenamine **8**, which cyclizes to 2-aminoindoline **10**, presumably via iminium ion **9**. Final elimination of dimethylamine gives indole.

The Leimgruber–Batcho indole synthesis was quickly adopted by several groups for the synthesis of 2,3-unsubstituted indoles. A tabulation of indoles prepared in this fashion is shown in Scheme 3 [5–8, 10–24]. The yields are for two steps from the respective *o*-nitrotoluene. Indole **11** was obtained in 45% yield in one step without the isolation of the nitroenamine using Raney nickel as the hydrogenation catalyst [6]. The synthesis of indoles **12** and **13** employed Fe/HOAc as the reducing agent, whereas **14** was prepared with $SnCl_2$/HCl in the reduction step [8]. 4-Benzyloxyindole (**15**) featured a Raney nickel–hydrazine reduction [10]. The amine group in **16** and **17** arose from the corresponding dinitrotoluene starting material [12]. The synthesis of 6-aminoindole (**18**) entailed the reduction–cyclization of β-(*N,N*-dimethylamino)-2,4-dinitrostyrene with aluminum amalgam [13]. Both Fe/HOAc and Raney nickel–hydrazine were used to prepare 6-chloro-5-fluoroindole (**19**) and 5-fluoro-6-iodoindole [15, 16]. An excellent synthesis of 4-aminoindole (**20**) began with 2,6-dinitrotoluene using either H_2/Pd or $TiCl_3$ for the twin reduction [19,20]. Indole **21** was used to [21] prepare a novel 5-HT_7 receptor antagonist. Indole **22** was the immediate precursor to a synthesis of the ascidian metabolite arnoamine B, which employed zinc and acetic acid in the second step [23]. As seen in Scheme 3, a very large range of indoles is available via the Leimgruber–Batcho indolization.

As mentioned earlier, the incorporation of a higher boiling secondary amine, such as pyrrolidine or piperidine, can result in significant improvement in both shorter reaction time and increased indole yield. The examples shown in Scheme 4 [25–27, 29–35, 40–48] involve the use of pyrrolidine as an amine additive, and, unless otherwise noted, the yields are overall from the appropriate *o*-nitrotoluene, usually in one pot. Rapoport's syntheses of four bromoindoles **25** used zinc in acetic acid in the second step [26], whereas Ayer used titanium trichloride to obtain 6-bromoindole

Indole Ring Synthesis: From Natural Products to Drug Discovery, First Edition. Gordon W. Gribble.

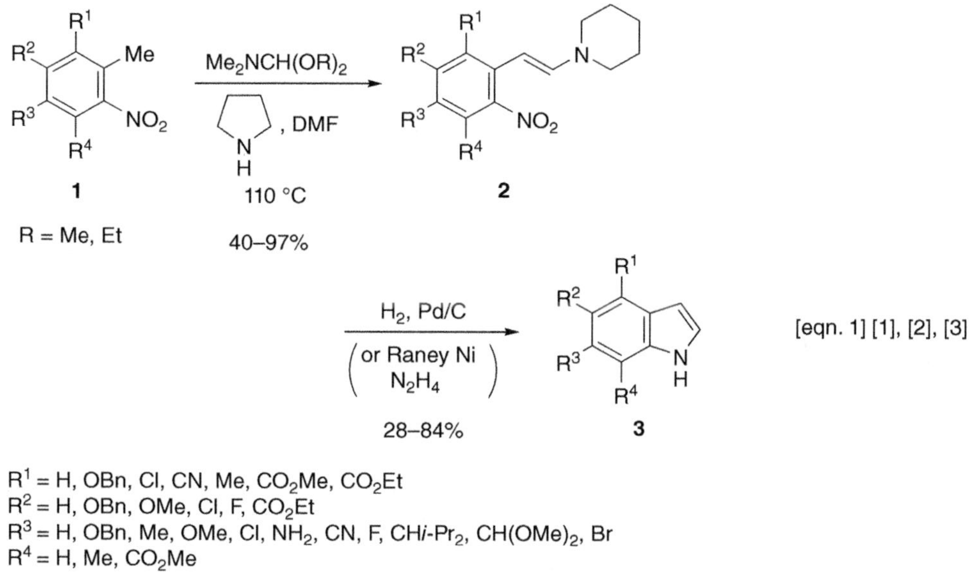

Scheme 1 *Leimgruber-Batcho Indole Synthesis*

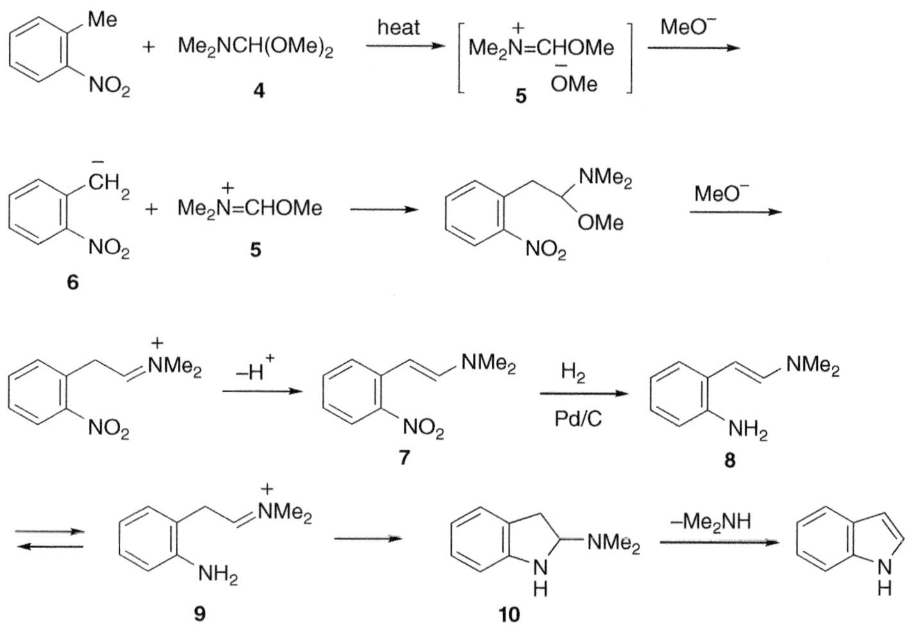

Scheme 2 *Proposed Mechanism for the Leimgruber-Batcho Indole Synthesis*

(**27**) [28]. Carrera and Sheppard employed H_2/Raney nickel to give **27** in 74% yield [29]. To prepare 5- and 6-benzyloxyindoles **29** Ohkubo and colleagues used nickel boride–hydrazine to reduce the nitroenamines [31]. Ziegler and Belema synthesized indole **31** for use in crafting the antitumor agent FR-900482 core structure [33], and Batt and colleagues parlayed 4-arylindoles **32** in a study toward novel analogues of Brequinar and new immunosuppressive agents [34]. Indole **33** was prepared to confirm the structure of 6-bromo-5-hydroxyindole, a marine alkaloid from the gastropod *Drupella fragum* [35]. The β-carboline marine ascidian alkaloids from *Didemnum* sp., didemnolines A and C, were synthesized from 6-bromoindole, prepared in 77% yield in a pyrrolidine–titanium trichloride

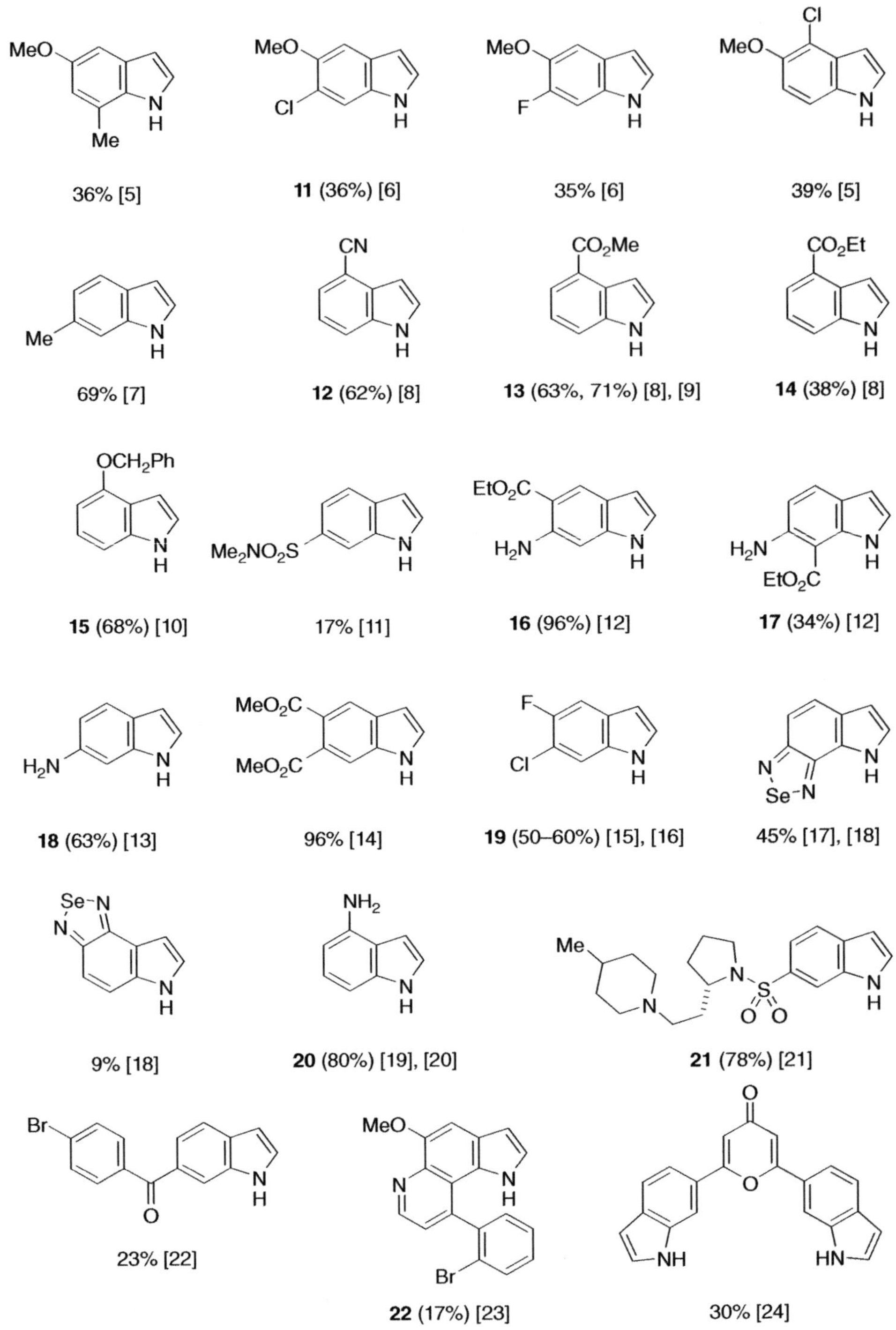

Scheme 3 *Examples of the Leimgruber-Batcho Indole Synthesis*

version of the Leimgruber–Batcho indolization [36]. A pyrrolidine–zinc acetic acid variation led to 6-bromoindole (57%) for use in the synthesis of naturally occurring 6-bromo-D-tryptophan-based peptides [37]. The pyrrolidine–titanium trichloride protocol was adopted by Pedras and colleagues in a synthesis of 6-methoxyindole (73%–78%) en route to 6-methoxycamalexin [38]. Magnus and coworkers used pyrrolidine–zinc/acetic acid for their Leimgruber–Batcho synthesis of 4-bromoindole (70%) in studies toward diazonamide A and B [39]. Indoles **34** and

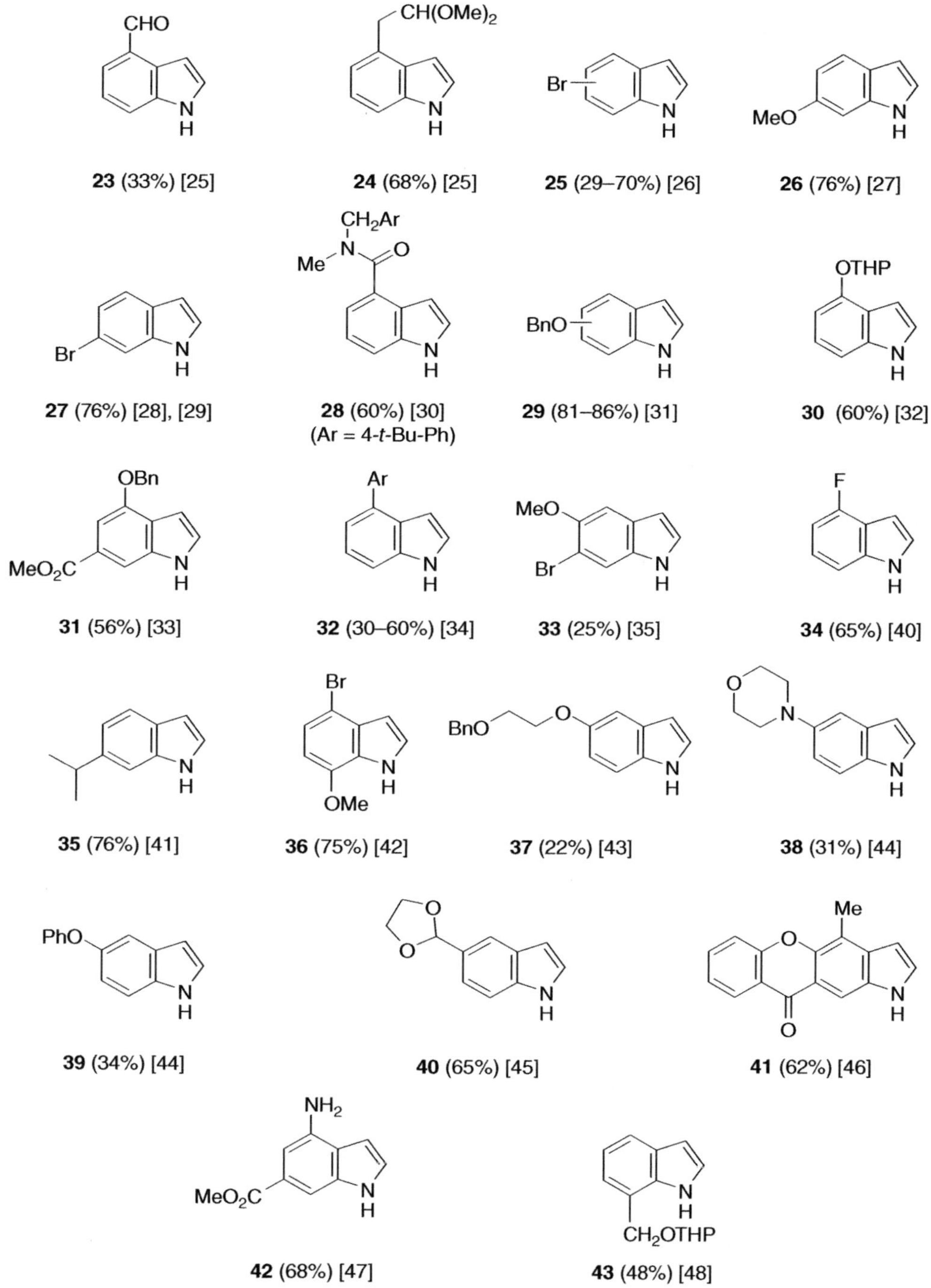

Scheme 4 *Leimgruber-Batcho Indole Syntheses Using Pyrrolidine*

35 were crafted by employing catalytic hydrogenation in the reduction of the intermediate *o*-nitro-β-pyrrolinylstyrenes [40, 41], and indole **36** was prepared with the original Raney nickel–hydrazine reduction method and used in a synthesis of a segment of dragmacidin D [42]. The preparation of indoles **37–42** entailed catalytic hydrogenation for the reduction step, and the corresponding 2,6-dinitrotoluene was the starting point for the preparation of **42** [47]. Indole **43** was synthesized as an alternative to 7-hydroxymethylindole, which could only be realized in 22% yield [48]. Indole **43** was quantitatively converted to 7-hydroxymethylindole with acidic methanol.

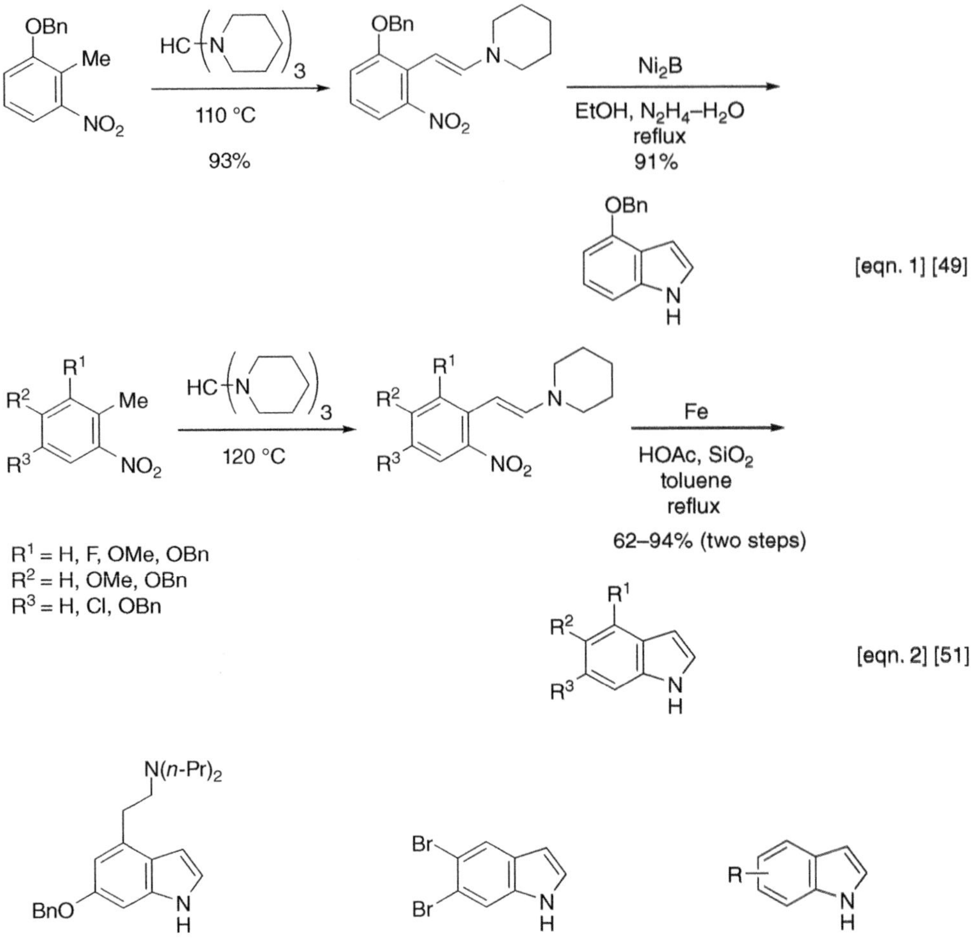

44 (75%) [52] **45** (61%) [53] **46** (40–64%) [54]

R = CO_2Me, CO_2Et, OEt

Scheme 5 *Leimgruber-Batcho Indole Syntheses Using Piperidine*

To a lesser extent than pyrrolidine, piperidine has been used to amplify the Leimgruber–Batcho indole synthesis, and some examples are shown in Scheme 5. Equation 1 reveals that nickel boride can be a good alternative to Raney nickel in the reduction of the piperidinostyrene, which is easily prepared from the appropriate toluene and tripiperidinomethane [49]. Nichols and Lloyd also found that titanium trichloride is an excellent medium for reducing these *o*-nitro-β-piperidinostyrenes, particularly those with oxygen substitution in the benzene ring [50]. Borchardt and colleagues also discovered the power of this simple variation (equation 2) [51]. Indoles **44–46** were prepared using this general methodology with tripiperidinomethane-substituted *o*-nitrostyrenes [52–54]. Similarly, Kruse discovered that *tris*(*N*,*N*-dimethylamino)methane, which dissociates into a stronger base (dimethylamide) than does *N*,*N*-dimethylformamide dimethyl acetal (methoxide), functions in the Leimgruber–Batcho reaction to convert *o*-nitrotoluenes to indoles following catalytic hydrogenation [55]. Nagata's team also employed *tris*(dimethylamino) methane for their preparation of 6-iodo-4-trifluoromethylindole by reduction ($TiCl_3$, aq. NH_4OAc) of the *o*-nitro-β-*N*,*N*-dimethylaminostyrene (83%) [56].

Kruse also reported that the intervention of a semicarbazone derivative prepared from the β-dimethylaminostyrene often led to superior yields of the resulting indole. This modification is illustrated in Scheme 6 (equation 1). Kruse's rationale is that the low solubility of the semicarbazone minimizes polymerization reactions and effectively increases the catalyst loading [55]. This modification was employed by the groups of Leonard [57], Corrie [58], and Harada [59] to synthesize indoles **47**, **48**, and **49**, respectively. 4-Aminoindole (**47**) was prepared from 2,6-dinitrotoluene, 4-*N*,*N*-dimethylaminoindole (**48**) started from

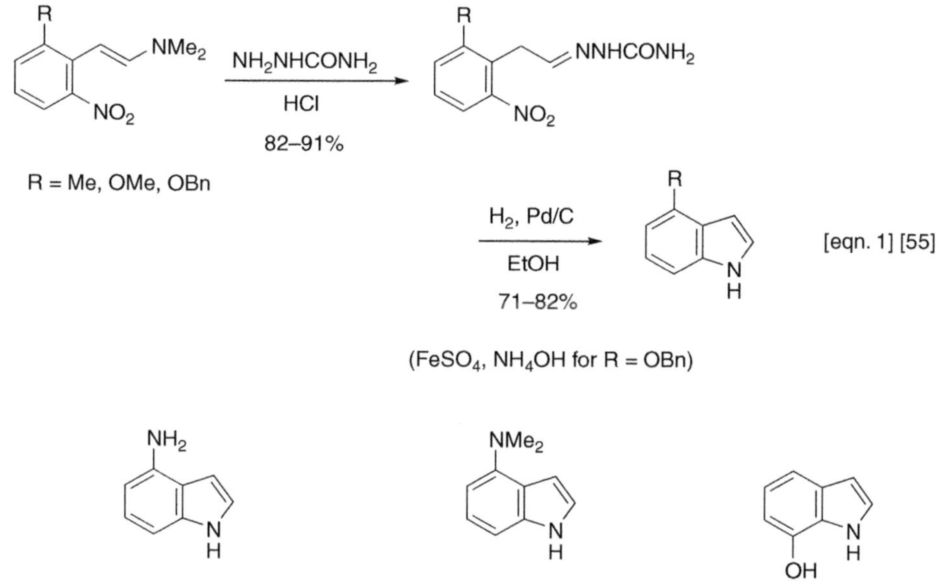

47 (81%) [57] **48** (86%) [58] **49** (80%) [59]

Scheme 6 *Kruse Modification of the Leimgruber-Batcho Indole Synthesis*

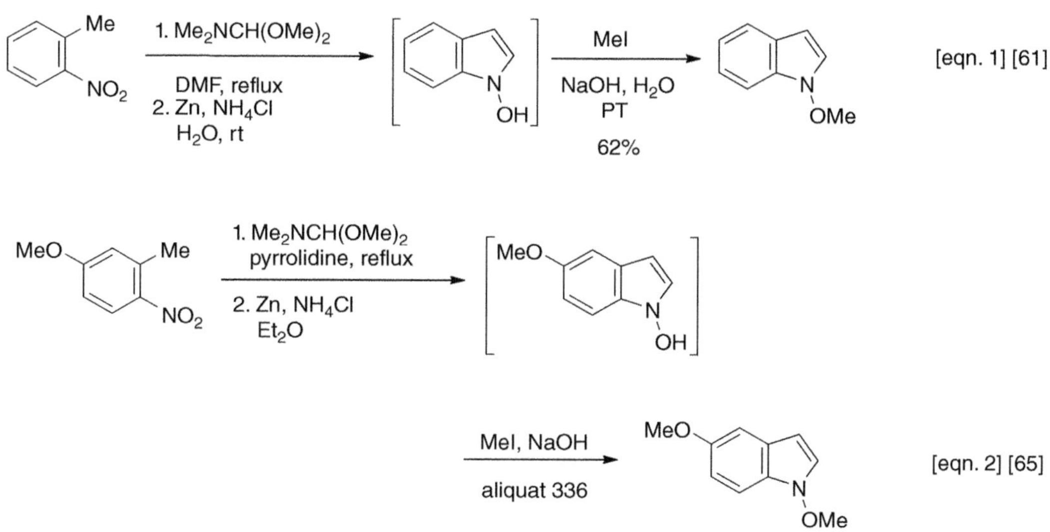

Scheme 7 *Somei 1-Hydroxyindole Synthesis*

N,N-dimethyl-2-methyl-3-nitroaniline, and the synthesis of 7-hydroxyindole **49** began with 3-benzyloxy-2-nitrotoluene. Reduction of the semicarbazone of the latter compound with Fe, NH_4Cl, EtOH gave 7-benzyloxyindole in 76% yield [59].

As for many reductions of the nitro group, the reaction often stops at an intermediate stage (NO, NHOH). Somei was the first to observe this in the Leimgruber–Batcho indole synthesis [60–62]. Although 1-hydroxyindole itself is very unstable, Somei and his coworkers were able to isolate 1-hydroxyindoles bearing an electron-withdrawing group at C-4 (nitro, carbomethoxy), and could otherwise trap unstable 1-hydroxylindoles by acetylation or methylation. Somei's method involved titanium or zinc reduction of the *o*-nitro-β-dimethylaminostyrene, as illustrated in Scheme 7 (equation 1) [61]. Quenching the reaction with acetic anhydride afforded 1-acetoxyindole in 61% yield. Clark and Repke investigated the Somei indole synthesis in detail with regard to catalyst conditions and ring substitution. The yield of 1-hydroxyindoles is only significant when electron-withdrawing groups are in the benzene ring [63]. For example, using catalytic

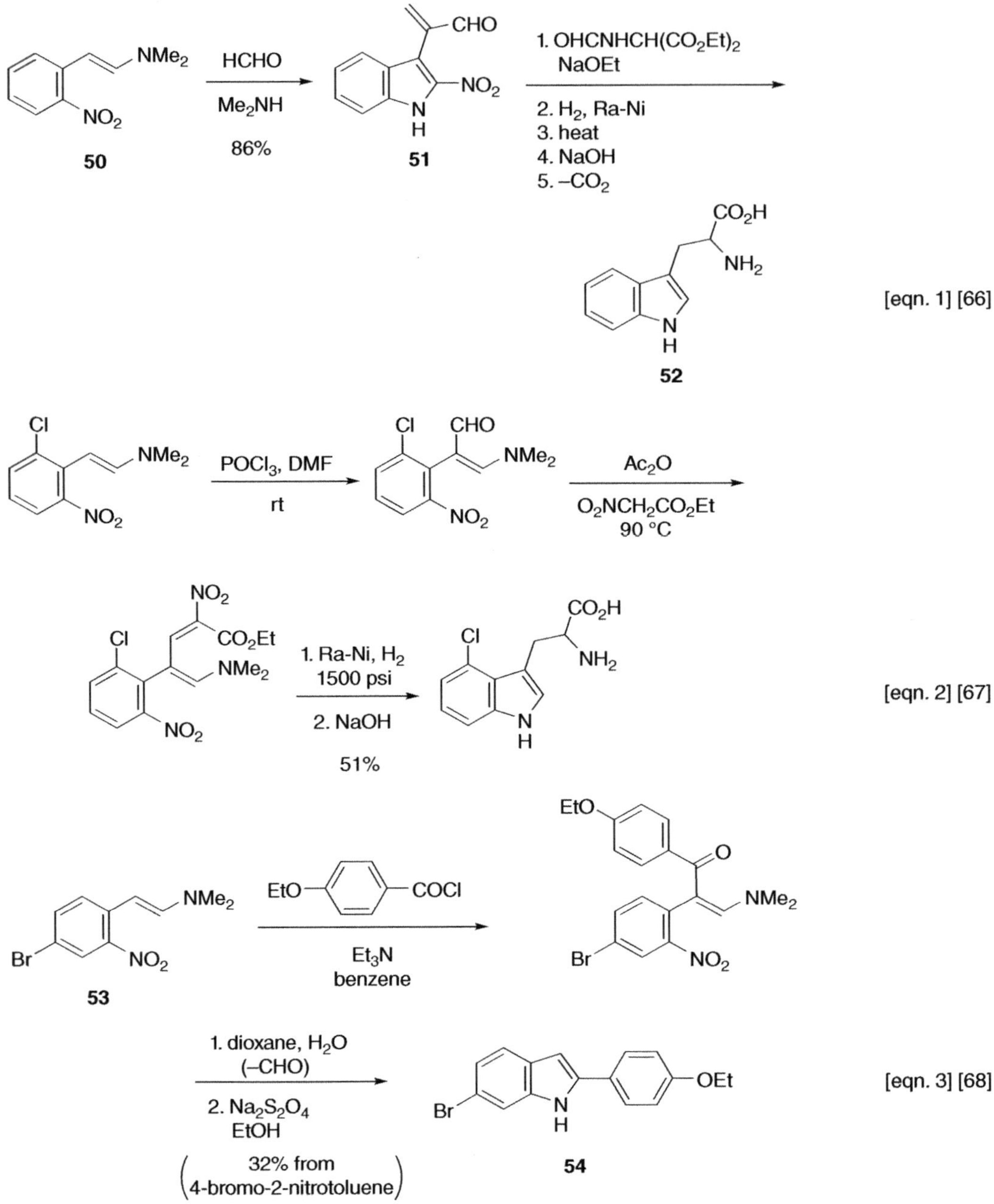

Scheme 8 *Extensions of the Leimgruber-Batcho Indole Synthesis*

hydrogenation under optimized conditions, they prepared 4-cyano-1-hydroxyindole in 75% yield. Preobrazhenskaya and colleagues used the Somei 1-hydroxyindole synthesis to prepare 1-alkoxyneoascorbigen analogues [64], as did Tsotinis and colleagues for their synthesis of 1,5-dimethoxyindole, a precursor to the natural product 1,5-dimethoxygramine found in *Gymnocrantheria paniculata* (equation 2) [65].

An obvious extension of the Leimgruber–Batcho indole synthesis is to modify the intermediate (nucleophilic) aminoenamine prior to reduction. Surprisingly, this tactic has only been minimally explored. For example, Hengartner and colleagues formylated aminoenamines **50** into indoles **51** for a new synthesis of tryptophans (**52**) [66] (Scheme 8, equation 1). Related chemistry gave 6-chloro- and 6-methyltryptophan in 40% to 43% yield from the appropriate *o*-nitrotoluenes. The formulation reaction of **50** could also be effected with the Vilsmeier–Haack reagent (DMF/$POCl_3$). Indeed, these conditions were used by Thiruvikraman's team to synthesize 4-chlorotryptophan via the Hengartner

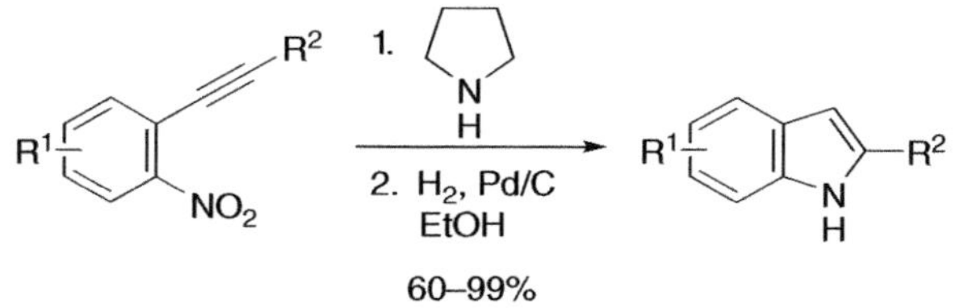

[eqn. 1] [69]

R^1 = H, 6-Me, 6-OMe, 4-MeO-6-CO_2Me
R^2 = H, *n*-Bu, *t*-Bu, Ph, 4-NH_2-6-Me, 6-NH_2, 6-Cl (indole numbering)

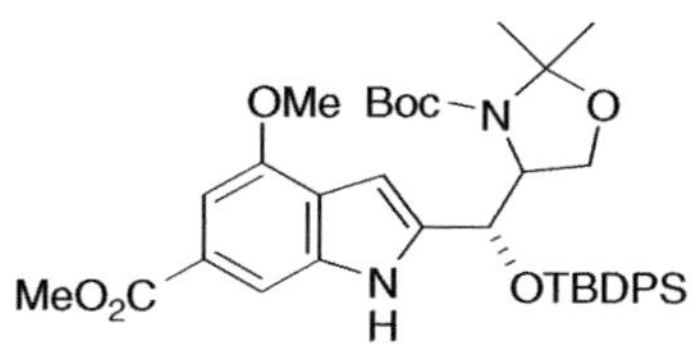

55

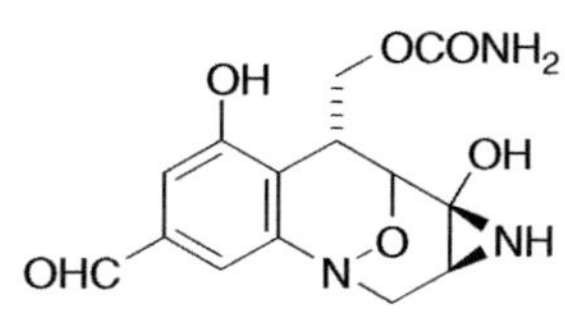

FR-900482 (**56**)

Scheme 9 *Fukuyama Modification of the Leimgruber-Batcho Indole Synthesis*

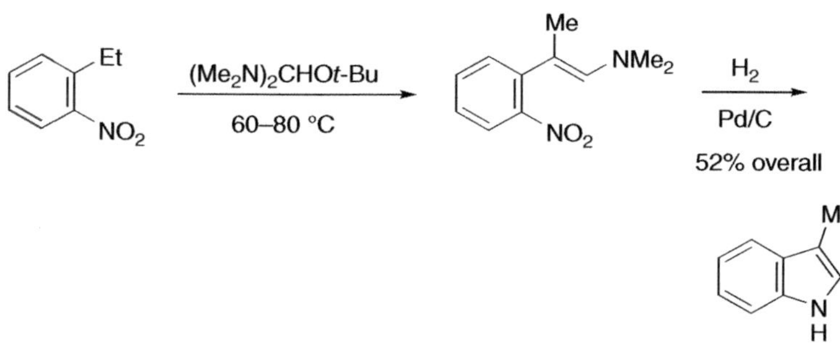

[eqn. 1] [70]

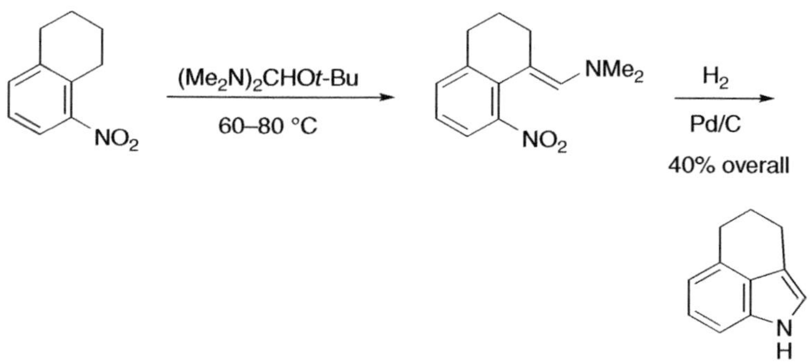

57

[eqn. 2] [70]

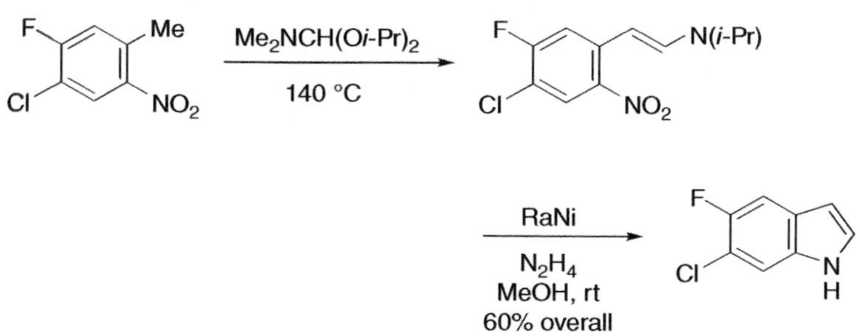

[eqn. 3] [74]

Scheme 10 *Modifications of the Leimgruber-Batcho Indole Synthesis*

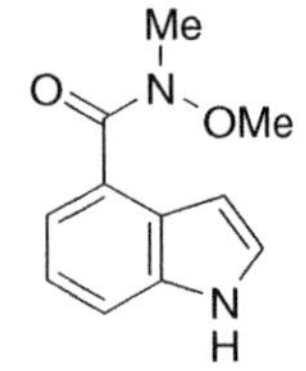

58 (79%) [75]

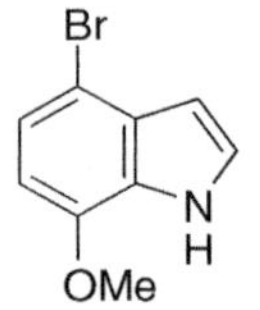

59 (73%) [76]

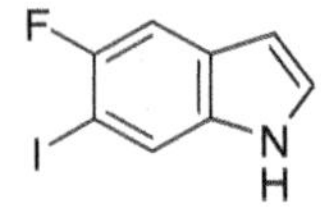

60 [77]

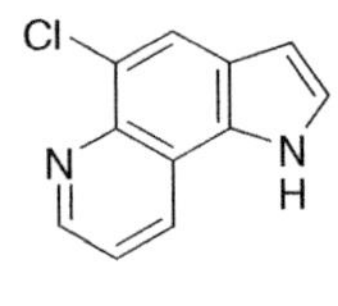

61 (55%) [78]

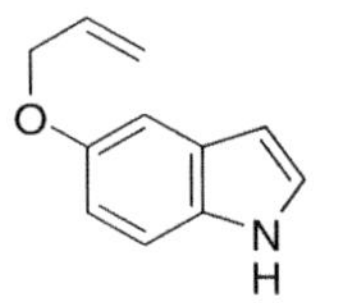

62 (75%) [79]

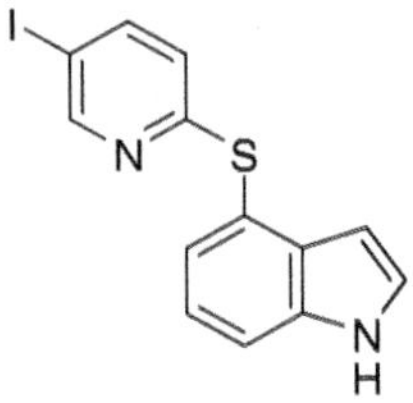

63 (36%) [80]

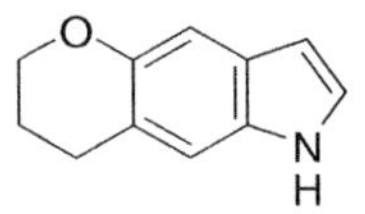

64 [81]

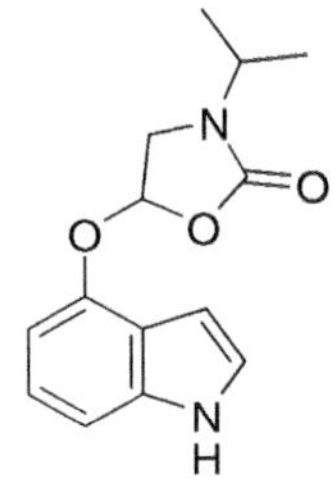

65 (70–75%) [82]

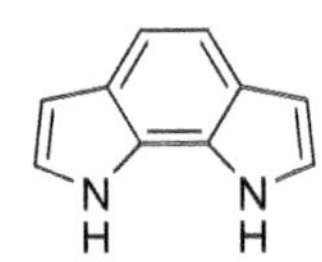

66 (53%) [83]

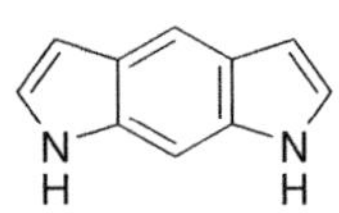

67 (56%) [83]

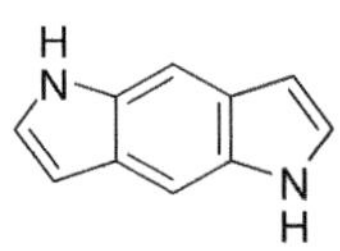

68 (39%) [83]

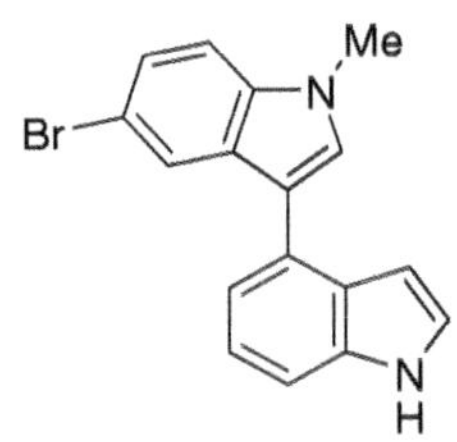

69 (56%) [84]

Scheme 11 *Indoles Synthesized by the Leimgruber-Batcho Indole Synthesis*

protocol (equation 2) [67]. White and colleagues synthesized 2-arylindole **54**, as a precursor to a new DNA minor groove binder, using the acylation of aminoenamine **53** (equation 3) [68], **54** being one of the rare examples of a 2-substituted indole synthesized by means of Leimgruber–Batcho methodology.

Fukuyama and colleagues modified the Leimgruber–Batcho indolization by synthesizing the *o*-nitro-β-aminostyrene in a novel fashion. Thus, the addition of secondary amines to *o*-nitrophenylacetylenes afforded aminoenamines that upon reduction gave indoles (Scheme 9) [69]. Pyrrolidine was superior to other amines (piperidine, morpholine, diethylamine, benzylamine), and several reduction methods were explored. For example, dechlorination was completely prevented using Fe/$FeCl_3$ in the synthesis of 2-*n*-butyl-6-chloroindole. Indole **55** was synthesized as part of a program targeted toward the antitumor antibiotic FR-900482 (**56**). If desired, the intermediate aminoenamine can be hydrolyzed to the corresponding nitro ketone, which is readily reductively cyclized to indole, a reaction we will encounter in later chapters.

Another modification of the Leimgruber–Batcho indole synthesis is the use of *tert*-butoxy-*bis*(dimethylamino) methane (Bredereck's reagent) to craft 3-substituted indoles (Scheme 10, equations 1 and 2) [70]. In these cases, the reaction with DMF dimethyl acetal was sluggish even at 140–170 °C. The well-known Uhle ketone, related to indole **57**, was also synthesized in this study. Akao and colleagues found that catalytic hydrogenation using rhodium on carbon with nickel, iron, or cobalt additives is safe and effective on a large scale for the reduction of *o*-nitro-β-pyrrolindinostyrenes to give indoles [71, 72]. The conditions are suitable for aromatic halides, aldehydes, and benzyl ethers, which are unaffected. Ley's group reported an extensive study of microwave heating for the condensation of *o*-nitrotoluenes with DMF dimethyl acetal and introduced an encapsulated nanoparticulate palladium catalysis (HCO_2H, Et_3N) for the reduction step [73]. Duncton and coworkers were able to circumvent the unwanted displacement of fluoride by methoxide from 3-chloro-4-fluoro-6-methylnitrobenzene by employing the hindered DMF-acetal *N,N*-dimethylformamide di-*iso*-propyl acetal in the preparation of 6-chloro-5-fluoroindole (equation 3) [74]. Gupton and coworkers found that Gold's reagent ($Me_2NCH{=}NCH{=}NMe_2Cl$) with *o*-nitrotoluene (sodium hydride, *N*-methylmorpholine) afforded nitroenamine **7** in 50% yield [85].

To close this presentation of a very important indole ring synthesis, I have depicted in Scheme 11 some of the other indoles that were prepared using the Leimgruber–Batcho indole synthesis. Weinreb amide indole **58** was made for a study of indolmycin analogues [75], and 4-bromo-7-methoxyindole (**59**) was used in a synthesis of the left-hand piece of the marine alkaloid dragmacidin D [76]. 5-Fluoro-6-iodoindole (**60**) was used for the construction of a new glycogen synthase kinase 3β inhibitor [77], and pyrrolo[2,3-*f*]quinoline **61** is a precursor for new 5-HT_{2C} receptor agonists [78]. A total synthesis of (–)-serotobenine featured 5-allyloxyindole (**62**), and a subsequent regioselective Claisen rearrangement gave 4-allyl-5-hydroxyindole for the synthesis of this safflower alkaloid [79]. Indole **63** was synthesized as a serotonin transporter–imaging agent [80]. A Claisen rearrangement was also involved in the preparation of the serotonin analogues pyrano[2,3-*f*]indole (**64**) and pyrano[3,2-*e*]indole [81]. Indole **65** served as the precursor to the three isomeric pindolols [82]. The three benzodipyrroles **66–68** were forged from the appropriate dinitroxylenes [83]. Biindole **69** is a synthetic analogue of the naturally occurring rivularins from the blue-green alga *Rivularia firma* [84].

The myriad chemistry shown in this chapter clearly demonstrates the tremendous importance and enormous utility of the Leimgruber–Batcho indole synthesis.

References

[1] W. Leimgruber and A.D. Batcho (1971) *Third International Congress of Heterocyclic Chemistry*, Tohuku University, Sendai, Japan, August 23–27, Abstracts, p. 462.
[2] A.D. Batcho and W. Leimgruber, U.S. Patent 3732245, 1973; German Offen., 2057840, 1971; U.S. Patent 3976639, 1976; *Chem. Abstr.*, 1971, **75**, P63605v; 1977, **86**, 29624.
[3] A.D. Batcho and W. Leimgruber, *Org. Synth.*, 1985, **63**, 214–225; *Org. Synth. Coll. Vol.*, 1990, **VII**, 34–41.
[4] R.D. Clark and D.B. Repke, *Heterocycles*, 1984, **22**, 195–221.
[5] F. Benington, R.D. Morin, and R.J. Bradley, *J. Heterocycl. Chem.*, 1976, **13**, 749–751.
[6] M.E. Flaugh, T.A. Crowell, J.A. Clemens, and B.D. Sawyer, *J. Med. Chem.*, 1979, **22**, 63–69.
[7] U. Hengartner, D. Valentine, Jr., K.K. Johnson, *et al.*, *J. Org. Chem.*, 1979, **44**, 3741–3747.
[8] G.S. Ponticello and J.J. Baldwin, *J. Org. Chem.*, 1979, **44**, 4003–4005.
[9] A.P. Kozikowski, H. Ishida, and Y.-Y. Chen, *J. Org. Chem.*, 1980, **45**, 3350–3352.
[10] D.B. Repke and W.J. Ferguson, *J. Heterocycl. Chem.*, 1982, **19**, 845–848.
[11] R. Cournoyer, D.H. Evans, S. Stroud, and R. Boggs, *J. Org. Chem.* 1991, **56**, 4576–4579.
[12] H.D.H. Showalter, L. Sun, A.D. Sercel, *et al.*, *J. Org. Chem.*, 1996, **61**, 1155–1158.
[13] F.D. Toste and I.W.J. Still, *Org. Prep. Proc. Int.*, 1995, **27**, 576–579.
[14] S.B. Kalindjian, I.M. Buck, J.M.R. Davies, *et al.*, *J. Med. Chem.*, 1996, **39**, 1806–1815.
[15] J.M. Bentley, D.R. Adams, D. Bebbington, *et al.*, *Bioorg. Med. Chem. Lett.*, 2004, **14**, 2367–2370.
[16] J.M. Bentley, J.E. Davidson, M.A.J. Duncton, *et al.*, *Synth. Commun.*, 2004, **34**, 2295–2300.
[17] M. Edin and S. Grivas, *Arkivoc*, 2000, 1–5.
[18] M. Edin and S. Grivas, *Arkivoc*, 2001, 144–153.
[19] A.I. Suárez, M.C. García, and R.S. Compagnone, *Synth. Commun.*, 2004, **34**, 523–531.
[20] M.G. Ferlin, G. Chiarelotto, V. Gasparotto, *et al.*, *J. Med. Chem.*, 2005, **48**, 3417–3427.
[21] I.T. Forbes, S. Douglas, A.D. Gribble, *et al.*, *Bioorg. Med. Chem. Lett.*, 2002, **12**, 3341–3344.
[22] M.-P. Lézé, A. Palusczak, R.W. Hartmann, and M. Le Borgne, *Bioorg. Med. Chem. Lett.*, 2008, **18**, 4713–4715.
[23] S. Nakahara, A. Kubo, Y. Mikami, and H. Mitani, *Heterocycles*, 2007, **71**, 1801–1806.
[24] A. Shahrisa, Z. Ghasemi, and M. Saraei, *J. Heterocycl. Chem.*, 2009, **46**, 273–277.
[25] H. Maehr and J.M. Smallheer, *J. Org. Chem.*, 1981, **46**, 1752–1755.
[26] M.P. Moyer, J.F. Shiurba, and H. Rapoport, *J. Org. Chem.*, 1986, **51**, 5106–5110.

[27] P.L. Feldman and H. Rapoport, *Synthesis*, 1986, 735–737.
[28] W.A. Ayer, P.A. Craw, Y. Ma, and S. Miao, *Tetrahedron*, 1992, **48**, 2919–2924.
[29] G.M. Carrera, Jr. and G.S. Sheppard, *Synlett*, 1994, 93–94.
[30] P. Stanetty and H. Koller, *Arch. Pharm. (Weinheim)*, 1992, **325**, 433–437.
[31] M. Ohkubo, T. Nishimura, H. Jona, *et al.*, *Tetrahedron*, 1996, **52**, 8099–8112.
[32] M. Brenner, G. Mayer, A. Terpin, and W. Steglich, *Chem. Eur. J.*, 1997, **3**, 70–74.
[33] F.E. Ziegler and M. Belema, *J. Org. Chem.*, 1997, **62**, 1083–1094.
[34] D.G. Batt, J.J. Petraitis, S.R. Sherk, *et al.*, *Bioorg. Med. Chem. Lett.*, 1998, **8**, 1745–1750.
[35] M. Ochi, K. Kataoka, S. Ariki, *et al.*, *J. Nat. Prod.*, 1998, **61**, 1043–1045.
[36] R.W. Schumacher and B.S. Davidson, *Tetrahedron*, 1999, **55**, 935–942.
[37] Y. Konda-Yamada, C. Okada, K. Yoshida, *et al.*, *Tetrahedron*, 2002, **58**, 7851–7861.
[38] M.S.C. Pedras, F.L. Okanga, I.L. Zaharia, and A.Q. Khan, *Phytochemistry*, 2000, **53**, 161–176.
[39] F. Chan, P. Magnus, and E.G. McIver, *Tetrahedron Lett.*, 2000, **41**, 835–838.
[40] U. Laban, D. Kurrasch-Orbaugh, D. Marona-Lewicka, and D.E. Nichols, *Bioorg. Med. Chem. Lett.*, 2001, **11**, 793–795.
[41] T. Sasaki, Y. Igarashi, M. Ogawa, and T. Furumai, *J. Antibiot.*, 2002, **55**, 1009–1012.
[42] C.-G. Yang, J. Wang, and B. Jiang, *Tetrahedron Lett.*, 2002, **43**, 1063–1066.
[43] K. Leonard, W. Pan, B. Anaclerio, *et al.*, *Bioorg. Med. Chem. Lett.*, 2005, **15**, 2679–2684.
[44] M.C. Van Zandt, M.L. Jones, D.E. Gunn, *et al.*, *J. Med. Chem.*, 2005, **48**, 3141–3152.
[45] J. Fetter, F. Bertha, L. Poszávácz, and G. Simig, *J. Heterocycl. Chem.*, 2005, **42**, 137–139.
[46] S.E. Watson, *Synth. Commun.*, 2005, **35**, 2695–2701.
[47] H.A. Braun, A. Zall, M. Brockhaus, *et al.*, *Tetrahedron Lett.*, 2007, **48**, 7990–7993.
[48] V.R. Uchil, M. Gund, and A. Satyam, *Synth. Commun.*, 2006, **36**, 1051–1056.
[49] D.H. Lloyd and D.E. Nichols, *J. Org. Chem.*, 1986, **51**, 4294–4295.
[50] D.H. Lloyd and D.E. Nichols, *Tetrahedron Lett.*, 1983, **24**, 4561–4562.
[51] M. Kawase, A.K. Sinhababu, and R.T. Borchardt, *J. Heterocycl. Chem.*, 1987, **24**, 1499–1501.
[52] J.P. Mayer, J.M. Cassady, and D.E. Nichols, *Heterocycles*, 1990, **31**, 1035–1039.
[53] N. Chandrasoma, N. Brown, A. Brassfield, *et al.*, *Tetrahedron Lett.*, 2013, **54**, 913–917.
[54] Y.-H. Ge, Y.-M. Wu, and Z.-J. Xue, *Chin. J. Org. Chem.*, 2006, **26**, 563–567.
[55] L.I. Kruse, *Heterocycles*, 1981, **16**, 1119–1124.
[56] W.E. Hume, T. Tokunaga, and R. Nagata, *Tetrahedron*, 2002, **58**, 3605–3611.
[57] L.L. Melhado and N.J. Leonard, *J. Org. Chem.*, 1983, **48**, 5130–5133.
[58] G. Papageorgiou and J.E.T. Corrie, *Tetrahedron*, 2000, **56**, 8197–8205.
[59] H. Harada, A. Fujii, and S. Kato, *Synth. Commun.*, 2003, **33**, 507–514.
[60] M. Somei, S. Inoue, S. Tokutake, *et al.*, *Chem. Pharm. Bull.*, 1981, **29**, 726–738.
[61] M. Somei and T. Shoda, *Heterocycles*, 1981, **16**, 1523–1525.
[62] M. Somei, *Chem. Pharm. Bull.*, 1986, **34**, 4109–4115.
[63] R.D. Clark and D.B. Repke, *J. Heterocycl. Chem.*, 1985, **22**, 121–125.
[64] M.N. Preobrazhenskaya, A.M. Korolev, I.I. Rozhkov, *et al.*, *Il Farmaco*, 1999, **54**, 265–274.
[65] A. Tsotinis, A. Eleutheriades, K. Hough, and D. Sugden, *Chem. Commun.*, 2003, 382–383.
[66] U. Hengartner, A.D. Batcho, J.F. Blount, *et al.*, *J. Org. Chem.*, 1979, **44**, 3748–3752.
[67] S.V. Thiruvikraman, Y. Sakagami, M. Katayama, and S. Marumo, *Tetrahedron Lett.*, 1988, **29**, 2339–2342.
[68] C.I. Clark, J.M. White, D.P. Kelly, *et al.*, *Aust. J. Chem.*, 1998, **51**, 243–247.
[69] H. Yokuyama, T. Makido, Y. Han-ya, and T. Fukuyama, *Heterocycles*, 2007, **72**, 191–197.
[70] W. Haefliger and H. Knecht, *Tetrahedron Lett.*, 1984, **25**, 285–288.
[71] A. Akao, N. Nonoyama, T. Mase, and N. Yasuda, *Org. Proc. Res. Dev.*, 2006, **10**, 1178–1183.
[72] A. Akao, K. Sato, N. Nonoyama, *et al.*, *Tetrahedron Lett.*, 2006, **47**, 969–972.
[73] J. Siu, I.R. Baxendale, and S.V. Ley, *Org. Biomol. Chem.*, 2004, **2**, 160–167.
[74] J.M. Bentley, J.E. Davidson, M.A.J. Duncton, *et al.*, *Synth. Commun.*, 2004, **34**, 2295–2300.
[75] D.R. Witty, G. Walker, J.H. Bateson, *et al.*, *Tetrahedron Lett.*, 1996, **37**, 3067–3070.
[76] M. Ikoma, M. Oikawa, and M. Sasaki, *Tetrahedron Lett.*, 2008, **49**, 7197–7199.
[77] I.N. Gaisina, F. Gallier, A.V. Ougolkov, *et al.*, *J. Med. Chem.*, 2009, **52**, 1853–1863.
[78] D.R. Adams, J.M. Bentley, K.R. Benwell, *et al.*, *Bioorg. Med. Chem. Lett.*, 2006, **16**, 677–680.
[79] Y. Koizumi, H. Kobayashi, T. Wakimoto, *et al.*, *J. Am. Chem. Soc.*, 2008, **130**, 16854–16855.
[80] E. Srisook and D.Y. Chi, *Bull. Korean Chem. Soc.*, 2004, **25**, 895–899.
[81] J.E. Macor, K. Ryan, and M.E. Newman, *Tetrahedron*, 1992, **48**, 1039–1052.
[82] Y. Tsuda, K. Yoshimoto, and T. Nishikawa, *Chem. Pharm, Bull.*, 1981, **29**, 3593–3600.
[83] A. Berlin, S. Bradamante, R. Ferraccioli, *et al.*, *J. Chem. Soc., Chem. Commun.*, 1987, 1176–1177.
[84] H. Maehr and J.M. Smallheer, *J. Org. Chem.*, 1984, **49**, 1549–1553.
[85] J. T. Gupton, M. J. Lizzi, and D. Polk, *Synth. Commun.*, 1982, **12**, 939–946.

42

Pschorr–Hoppe Indole Synthesis

In 1910 Robert Pschorr and Gerhard Hoppe reported the synthesis of indole via the reaction of *o*-nitrophenylacetonitrile with tin foil and concentrated hydrochloric acid to give initially *o*-aminophenylacetonitrile. Subsequent treatment with sodium in boiling ethanol afforded indole (Scheme 1, equation 1) [1]. The starting *o*-nitrophenylacetonitrile was synthesized from *o*-nitrophenylacetic acid in a few steps. Although this simple synthesis of indole is first attributed to Pschorr and Hoppe, it was Makosza who greatly extended it by means of his vicarious nucleophilic substitution (VNS) of hydrogen as an efficient route to the prerequisite *o*-nitrophenylacetonitriles, as presented in Chapter 43.

Without reference to Pschorr and Hoppe, Steven in 1925, in the context of a new aldehyde synthesis, prepared indole in 55% yield from *o*-nitrophenylacetonitrile using anhydrous stannous chloride and gaseous hydrogen chloride (Scheme 2, equation 1) [2]. The reduction of the nitrile proceeds to form initially the aldimine stannic chloride complex ($[RCH{=}NH \cdot HCl]_2SnCl_4$), which can be isolated. Aqueous hydrolysis furnishes the aldehyde and then indole by cyclodehydration. Walker first used catalytic hydrogenation in the Pschorr–Hoppe indolization to prepare 5,6-dimethoxyindoles (Scheme 2, equations 2–4) [3]. With four moles of H_2, ammonia is liberated (equations 2, 3), but with three moles of H_2 the product retains the 2-amino group (equations 4, 5) [3]. Independently from Walker, Plieninger and Nógrádi synthesized indole and 5,6-dimethoxyindole from the respective *o*-nitrophenylacetonitriles using Raney nickel hydrogenation and/or sodium in amyl alcohol [4].

Like Walker, Snyder and colleagues found that catalytic hydrogen conditions with palladium-on-carbon improves the yield of indoles and are superior to the tin reduction of Stephen (Scheme 3) [5]. Snyder suggests that the mechanism of this indole synthesis involves initial reduction of the *o*-nitrobenzylcyanide to the imine **1** followed by cyclization and loss of ammonia and tautomerization of indolenine **2** to indole (equation 3), rather than Walker's mechanism via 2-aminoindole.

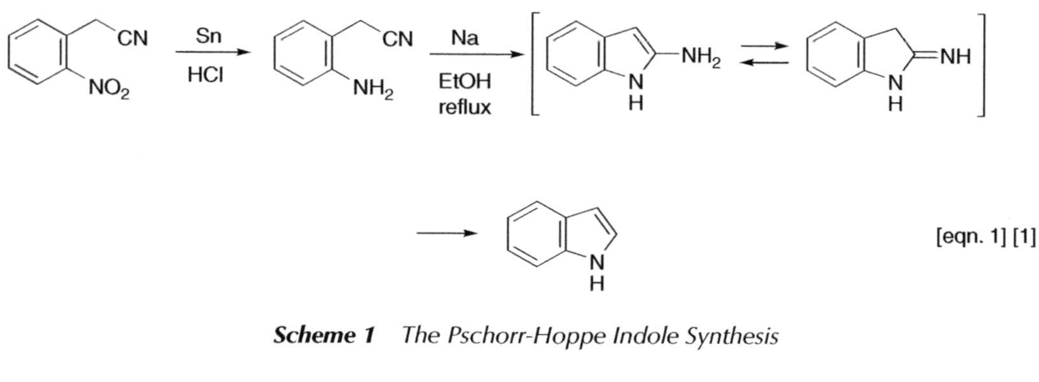

Scheme 1 *The Pschorr-Hoppe Indole Synthesis*

Indole Ring Synthesis: From Natural Products to Drug Discovery, First Edition. Gordon W. Gribble.

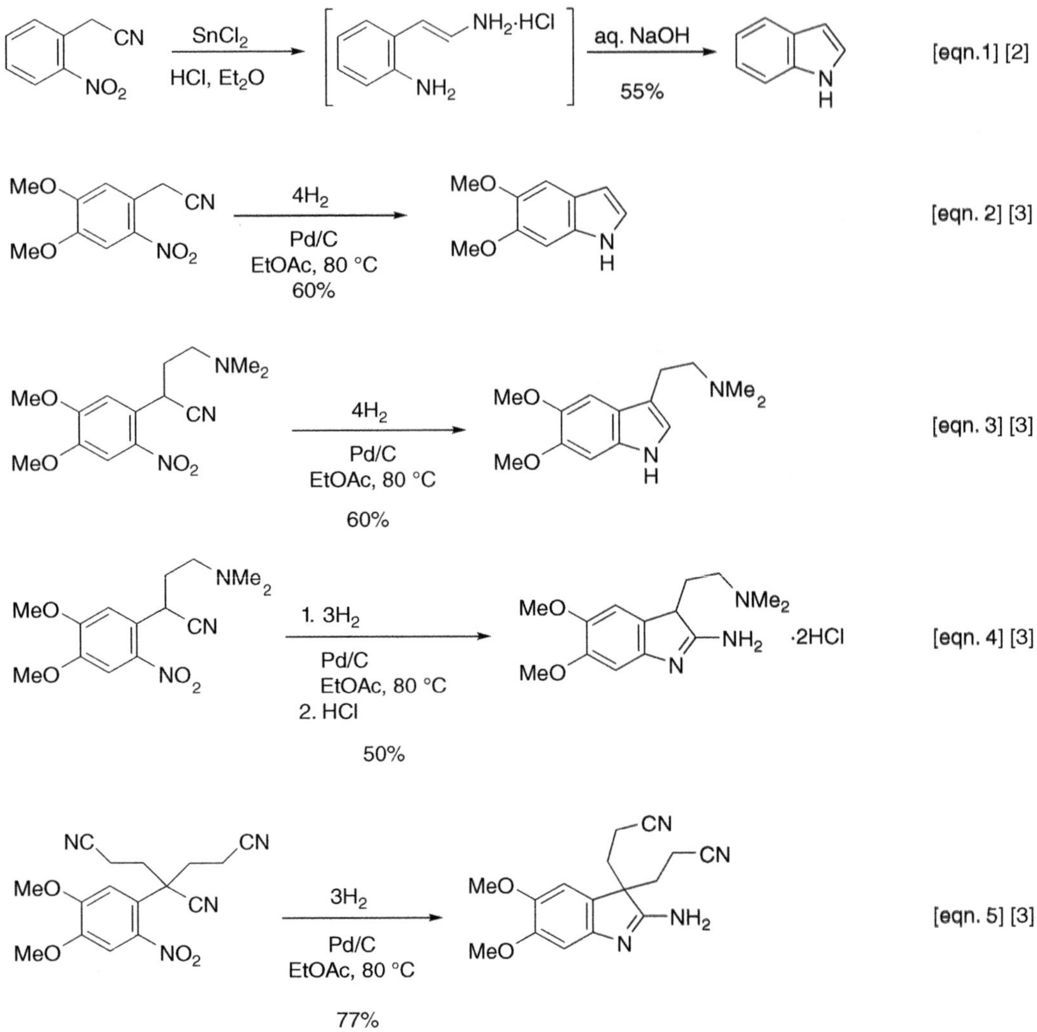

Scheme 2 *Stephen and Walker Modification of the Pschorr-Hoppe Indole Synthesis*

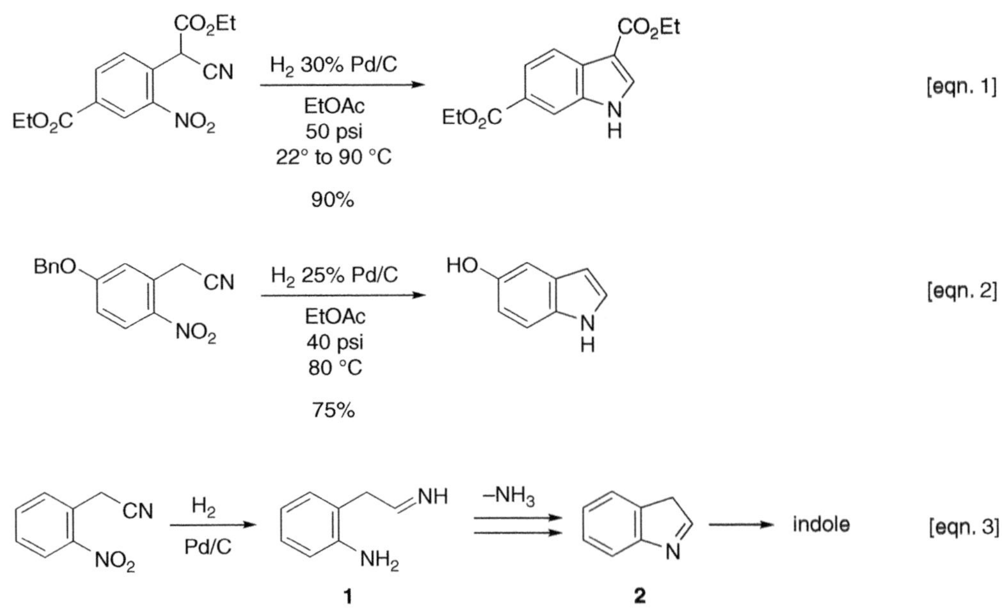

Scheme 3 *Snyder Modification of the Pschorr-Hoppe Indole Synthesis*

Table 1 *Applications of the Pschorr-Hoppe Indole Synthesis*

Entry	Nitrile	Conditions	Indole	% Yield	Ref.
1	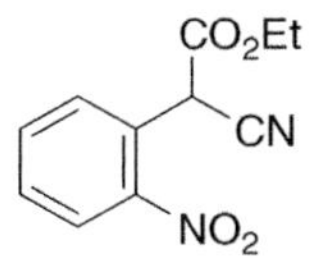	Zn, HOAc 80–100 °C	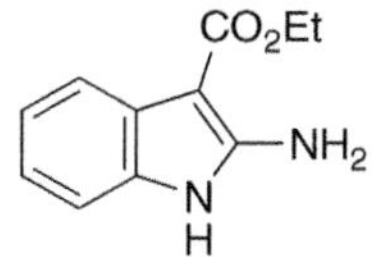	70%	6
2	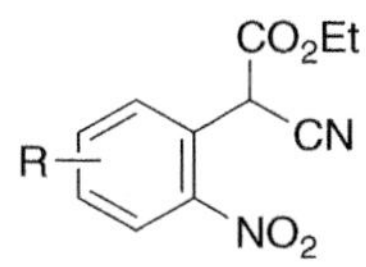 R = H, 5-Me, 6-Me, 6-OMe 6-Cl, 6-NHAc (indole numbering)	Zn, HOAc 15–20 °C	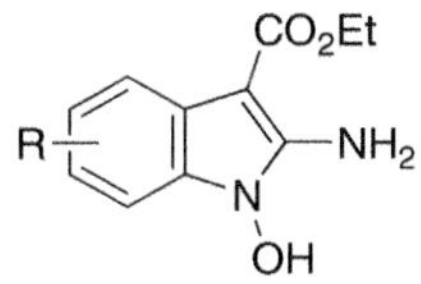	32–65%	7
3	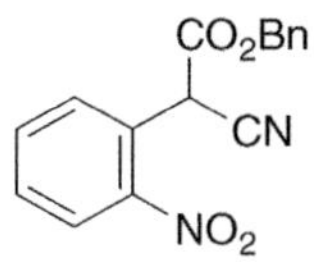	Zn, HOAc 80 °C	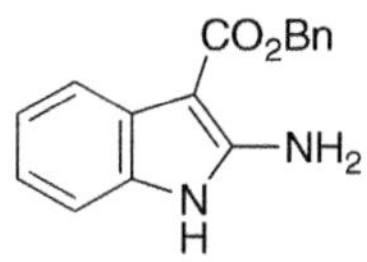	78%	8
4	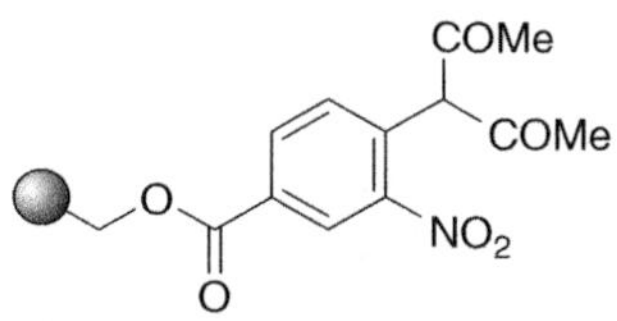	1. $SnCl_2$, NMP 20 °C 2. TFA, CH_2Cl_2	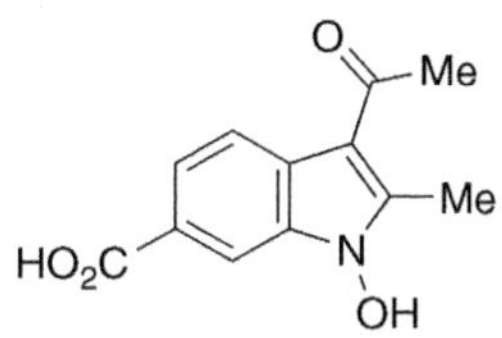	74%	9
5	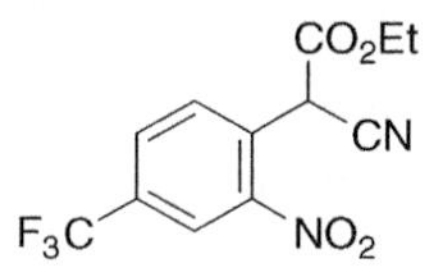	H_2, Pd/C 1 atm EtOH, HOAc rt	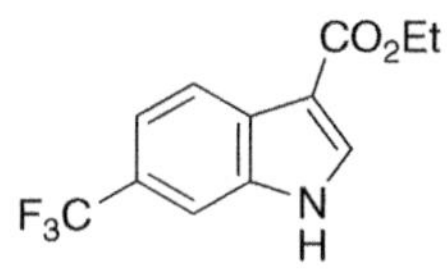	81%	11
6	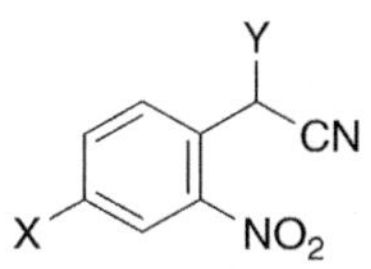 X = H, CF_3, OMe, F, NHAc, SO_2Me Y = CO_2Et, SO_2Me, SO_2Ph, CN	H_2, 10% Pd/C $Pd(Ph_3P)_4$ EtOAc, HOAc rt	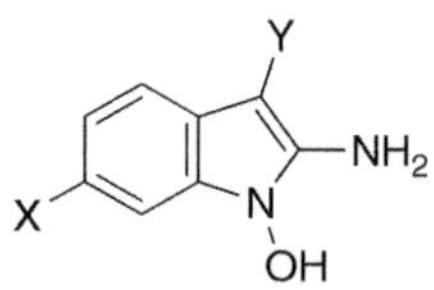	54–94%	12
7	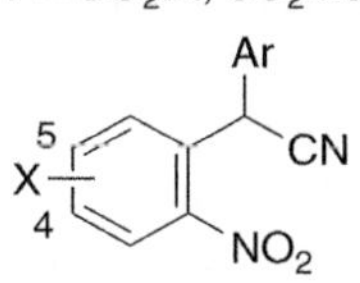 X = 4-CF_3, 4-Cl, 4-F, 5-OBn Ar = Ph, 4-ClPh, 4-MeOPh, 4-CF_3Ph, 4-CNPh, 4-EtPh,	H_2, 10% Pd/C $Pd(Ph_3P)_4$ EtOAc, HOAc	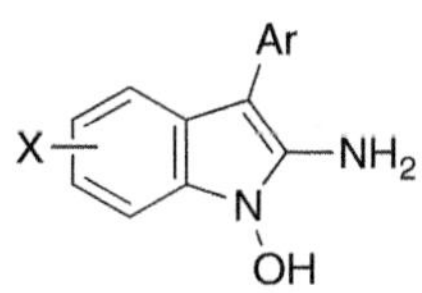	70–95%	13
8	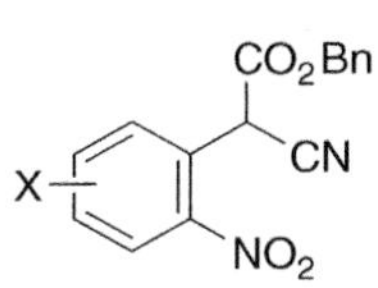 X = 6-CF_3, 5-CF_3, H, 6-Me, 6-OMe, 5-Cl (indole numbering)	H_2, Pd/C 50 psi EtOH cat HOAc	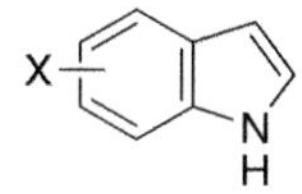	55–88%	14

(continued overleaf)

Table 1 *(continued)*

Entry	Nitrile	Conditions	Indole	% Yield	Ref.
9	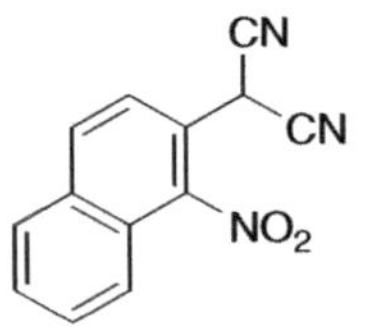	$Na_2S_2O_4$, H_2O DMF, $NaHCO_3$ rt	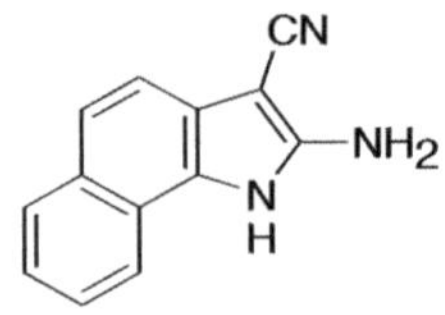	12%	16
10	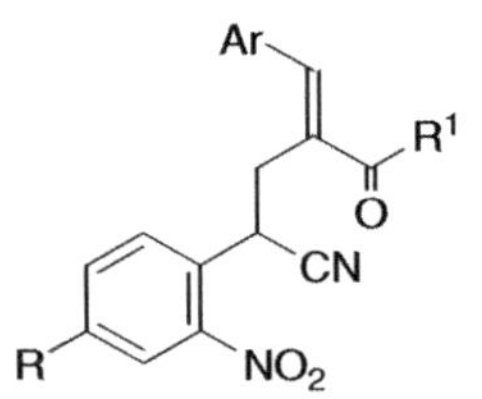 R = H, Me, OMe R^1 = Me, Et Ar = Ph, 4-Tol, 4-ClPh, 4-BrPh, 3-BrPh, 3-OMePh, 2-ClPh, 2-BrPh, 2-Tol	Fe, HOAc reflux	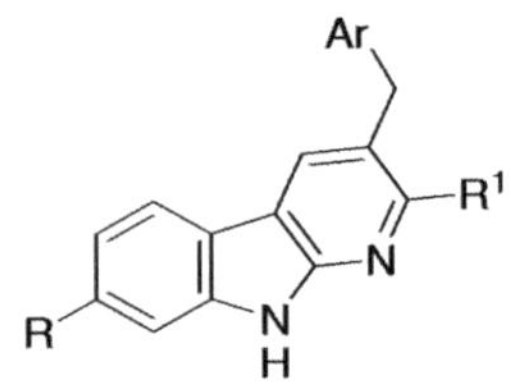	48–67%	17

Several applications of the aforementioned indole syntheses involving reductive cyclization of *o*-nitrobenzylcyanides are tabulated in Table 1. At lower temperatures, the mild zinc–acetic acid reduction method of Grob and Weissbach [6] gives the corresponding intermediate *N*-hydroxyindole as found by de Souza's group (entry 2) [7]. The *o*-nitrophenylcyanoacetates were easily prepared by an S_NAr displacement of chloride from *o*-chloronitrobenzene with cyanoacetate anion. Thompson and coworkers used *o*-fluoronitrobenzene in this protocol to synthesize (entry 3) and then convert benzyl 2-aminoindole-3-carboxylate to the rare imidazo[1,2-*a*]indole ring system [8]. Zaragoza has employed the stannous chloride method in the solid-phase preparation of *N*-hydroxyindoles (entry 4) [9]. Showalter and colleagues found that at 55 °C the zinc–acetic acid reduction of ethyl 2-cyano-2(5-methoxy-2-nitrophenyl)acetate gives a 1:1 mixture of ethyl 2-amino-5-methoxyindole-3-carboxylate and the 1-hydroxy analogue [10].

Belley and colleagues reported the application of the catalytic hydrogenation methodology to access a wide variety of indoles and *N*-hydroxyindoles (entries 5-7) [11–13]. The use of $Pd(Ph_3P)_4$ in entry 6 leads to *N*-hydroxy-2-aminoindoles by decreasing the reduction rate of the intermediate *N*-hydroxylamine prior to nitrile cyclization [12]. Additional examples of reductive cyclization leading to *N*-hydroxyindoles are presented in Chapter 45. Similarly, the corresponding 3-arylindoles are synthesized (entry 7) [13]. Walkington's method (entry 8) was employed to prepare 6-trifluoromethylindole on a 100-kg scale [14]. This reaction was based on the synthesis of 6-trifluoromethylindole by Kalir and Pelah [15]. Although the yield of 2-aminobenzo[*g*]indole-3-carbonitrile is low (entry 9), the use of sodium thiosulfite as the reductant in this procedure gives 2-aminoindole-3-carbonitrile in 57% yield [16]. Basavaiah and Reddy parlayed the Pschorr–Hoppe indolization using iron–acetic acid into a one-pot synthesis of α-carbolines (entry 10) [17].

References

[1] R. Pschorr and G. Hoppe, *Chem. Ber.*, 1910, **43**, 2543–2552.
[2] H. Stephen, *J. Chem. Soc.*, 1925, **127**, 1874–1877.
[3] G.N. Walker, *J. Am. Chem. Soc.*, 1955, **77**, 3844–3850.
[4] H. Plieninger and I. Nógrádi, *Chem. Ber.*, 1955, **88**, 1961–1963.
[5] H.R. Snyder, E.P. Merica, C.G. Force, and E.G. White, *J. Am. Chem. Soc.*, 1958, **80**, 4622–4625.
[6] C.A. Grob and O. Weissbach, *Helv. Chim. Acta*, 1961, **44**, 1748–1753.
[7] K.L. Munshi, H. Kohl, and N.J. de Souza, *J. Heterocycl. Chem.*, 1977, **14**, 1145–1146.
[8] I.T. Forbes, H.K.A. Morgan, and M. Thompson, *Synth. Commun.*, 1996, **26**, 745–754.
[9] H. Stephensen and F. Zaragoza, *Tetrahedron Lett.*, 1999, **40**, 5799–5802.

[10] H.D.H. Showalter, A.J. Bridges, H. Zhou, *et al.*, *J. Med. Chem.*, 1999, **42**, 5464–5474.
[11] M. Belley, J. Scheigetz, P. Dubé, and S. Dolman, *Synlett*, 2001, 222–225.
[12] M. Belley, E. Sauer, D. Beaudoin, *et al.*, *Tetrahedron Lett.*, 2006, **47**, 159–162.
[13] M. Belley, D. Beaudoin, P. Duspara, *et al.*, *Synlett*, 2007, 2991–2994.
[14] A. Walkington, M. Gray, F. Hossner, *et al.*, *Synth. Commun.*, 2003, **33**, 2229–2233.
[15] A. Kalir and Z. Pelsh, *Israel J. Chem.*, 1966, **4**, 155–159.
[16] C. Willermann, R. Grünert, P.J. Bednarski, and R. Troschütz, *Bioorg. Med. Chem.*, 2009, **17**, 4406–4419.
[17] D. Basavaiah and D.M. Reddy, *Org. Biomol. Chem.*, 2012, **10**, 8774–8777.

43

Mąkosza Indole Synthesis

In 1978 Mąkosza and Goliński uncovered a new aromatic nucleophilic substitution reaction: vicarious nucleophilic substitution of hydrogen (VNS) [1]. In the intervening more than three decades, this powerful new reaction has seen many synthetic applications, as reviewed several times by Mąkosza [2–7]. Notably, the synthesis of indoles via VNS has received the most attention by Mąkosza and others, and the first such report in 1984 by Wojciechowski and Mąkosza is shown in Scheme 1, equation 1 [8]. Although the yields are only modest in equation 1, Mąkosza later greatly improved and expanded this indolization as summarized in equation 2 for a synthesis of 3-sulfonylindoles [9]. The starting nitrosulfones **1** are prepared by the VNS substitution of the hydrogen *ortho* to the nitro group in the corresponding arene. A minor modification of the sequence that entails the condensation of aminosulfone **2** with aryl aldehydes leads to 2-arylindoles (equation 3) [10].

A selection of other Mąkosza indolizations is presented in Scheme 2. In each case a VNS reaction was employed to prepare the immediate indole precursor. The cyanoalkylation–hydrogenation sequence in equation 1 was also applied to the preparation of 4-, 5-, 6-, and 7-hydroxy and -methoxyindoles [11]. The mechanism for the interesting cyclization in equation 2 is unclear [12]. Indeed, strongly basic conditions (5 M NaOH, MeOH) afford quinoline *N*-oxides, and intermediate basic conditions (0.1 M NaOH, MeOH) lead to *N*-hydroxyindoles devoid of the C-2 hydroxymethyl group! The synthesis of *N*-hydroxy-2-vinylindoles is achieved under the action of chlorotrimethylsilane and triethylamine (equation 3), via a pathway suggested by Mąkosza to involve an anionic quinoid intermediate that cyclizes to the vinylindole [13]. It should be mentioned that *N*-hydroxyindoles are easily reduced to indoles; for example, 3-cyano-4-vinylindole is obtained in 69% yield with zinc/acetic acid/reflux from the *N*-hydroxyindole [13].

A collection of Mąkosza's and other authors' indolizations that involve a VNS synthesis of the indole precursor are tabulated in Table 1 [14–20]. Entry 3 features a novel route to the indole used by previous workers to synthesize makaluvamine D and discorhabdin C [15], and 6-methoxy-1,3,4,5-tetrahydrobenz[*cd*]indole-4-amine (entry 4), an important biologically active ergot analogue possessing serotonin- and dopamine-receptor activity [16]. The *N*-hydroxyindoles in entry 6 were efficiently reduced to the respective indoles with zinc–acetic acid [17]. Mąkosza has extended his chemistry to the preparation of 4- and 6-nitroindoles (entries 7–9) [18–20]. The regiochemistry in entries 7 and 8 probably results from a steric effect of the nitrile anions [18]. The condensation between 3-nitroaniline and ketones in entry 9 can also form the corresponding 6-nitroindoles in lesser amounts (13%–30%) [19, 20]. The mechanism for this novel 4-nitroindole synthesis involves VNS of the ketone enolate between the amino and nitro groups and not initial ketone carbonyl condensation with the amino group, as shown by control experiments. Thus, the imine synthesized independently from 3-nitroaniline and acetophenone does not yield the expected 4-nitro-2-phenylindole upon treatment with *t*-BuOK/DMSO. Mąkosza believes that oxygen is the oxidant in this oxidative nucleophilic substitution of hydrogen reaction [19, 20].

The first outside group to use the Mąkosza indole synthesis was Macor and colleagues [21–23]. A key feature of this chemistry is the use of a Mitsunobu reaction to functionalize the benzylnitrile as illustrated in Scheme 3 (equation 1) [21, 22]. Equation 2 presents Macor's synthesis of the

Indole Ring Synthesis: From Natural Products to Drug Discovery, First Edition. Gordon W. Gribble.

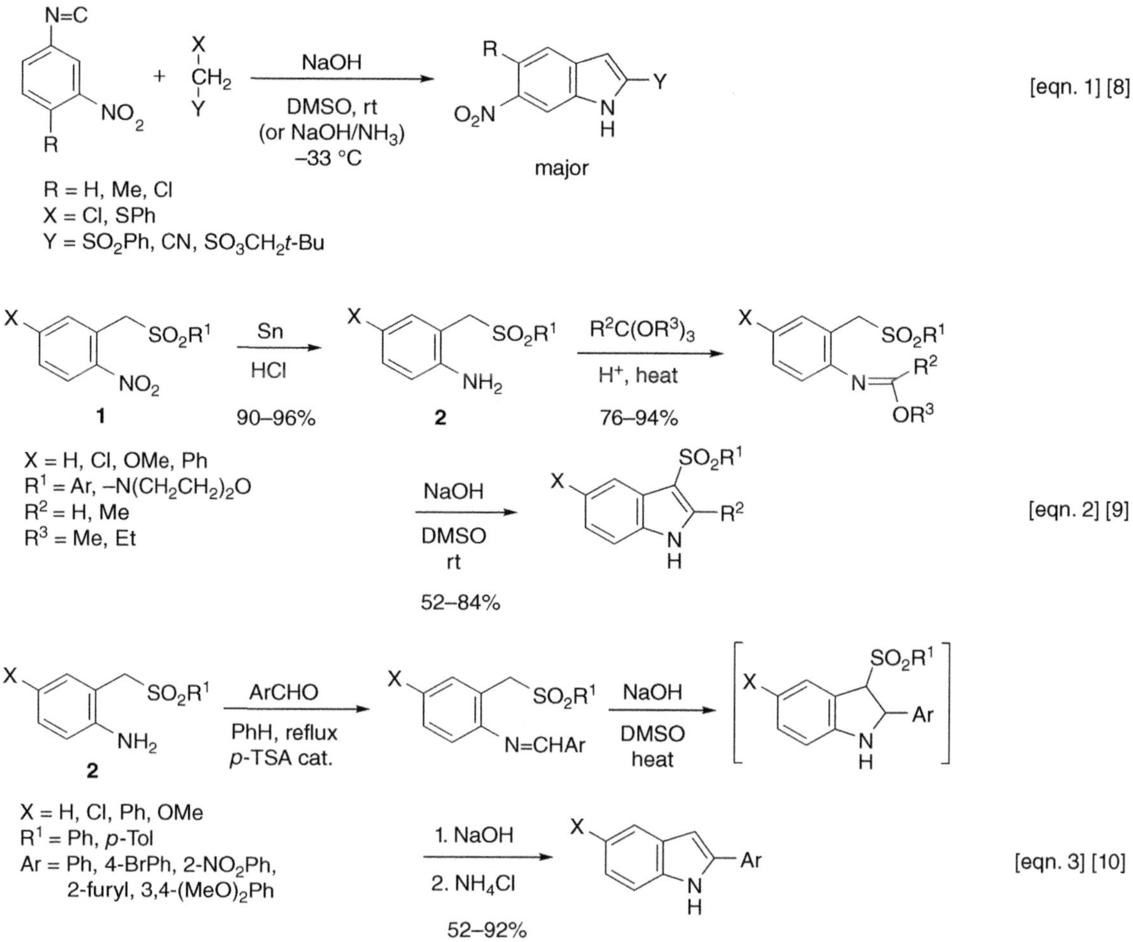

Scheme 1 *The Makosza Indole Synthesis*

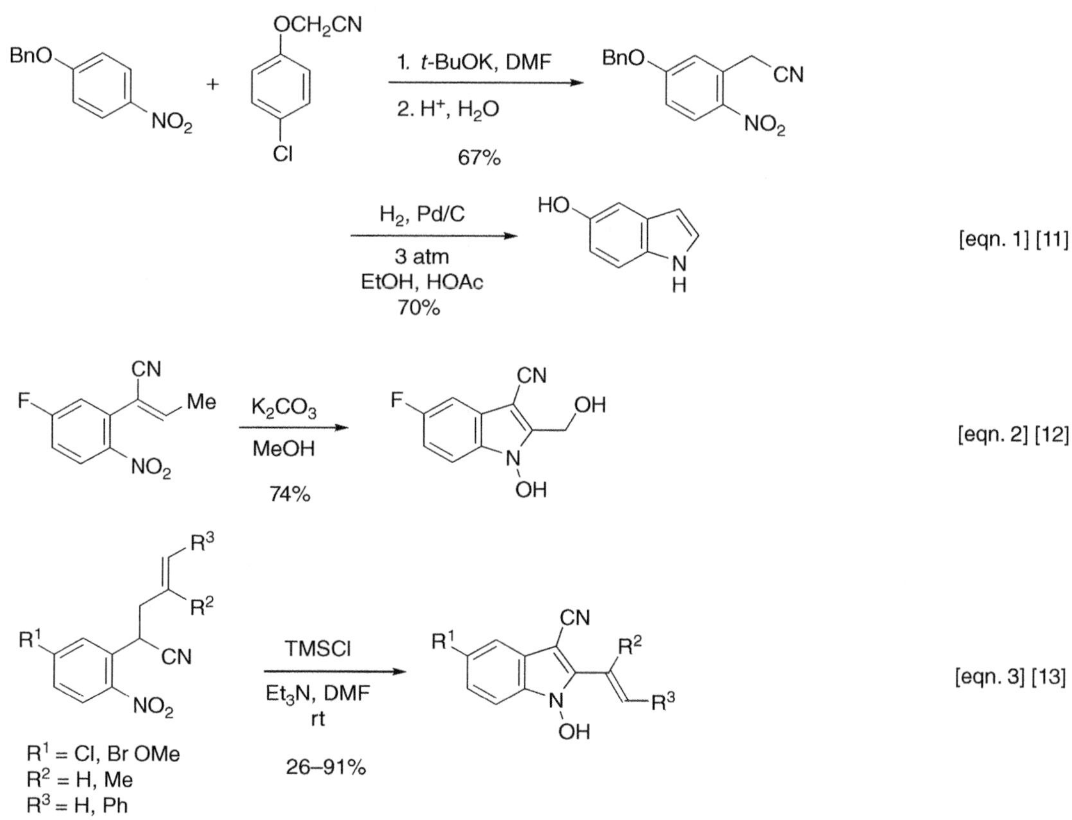

Scheme 2 *The Makosza Indole Syntheses*

Table 1 *Makosza Indole Syntheses*

Entry	Substrate	Conditions	Indole	% Yield	Ref.
1	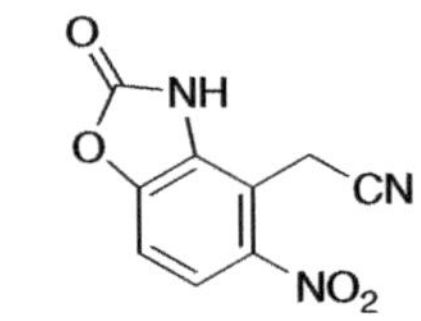	H_2, Pd/C EtOH, HOAc	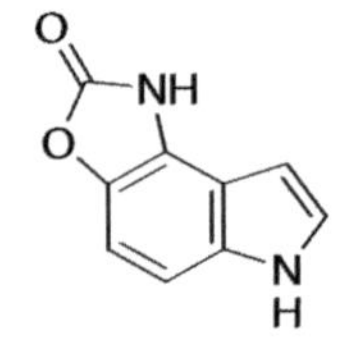	80%	14
2	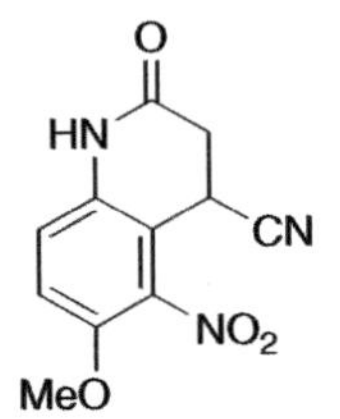	H_2, Pd/C EtOH, HOAc	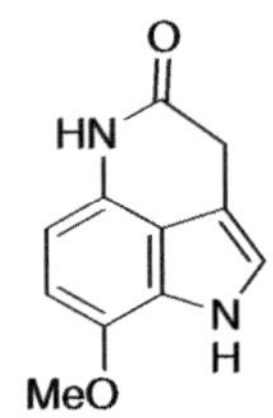	58%	14
3	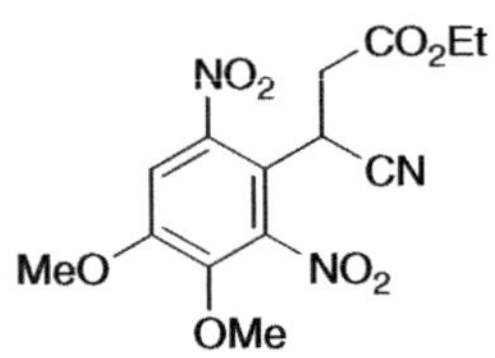	H_2, $PdCl_2$, Fe EtOH, HOAc	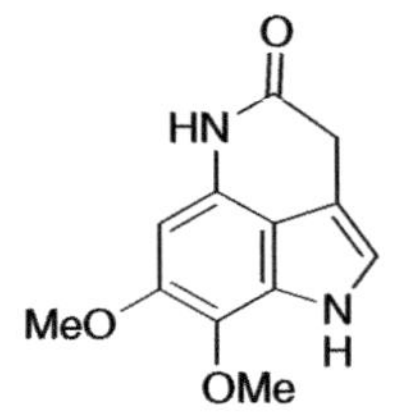	48%	15
4	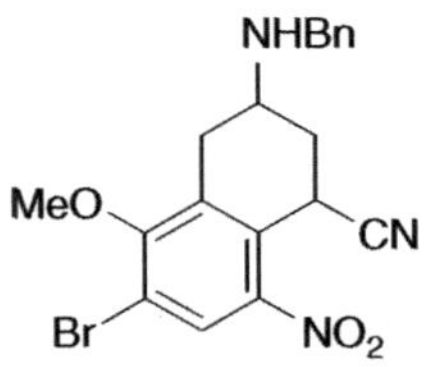	H_2, Pd/C EtOH, Me_2NH	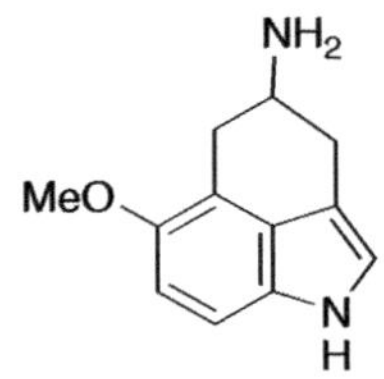	57%	16
5	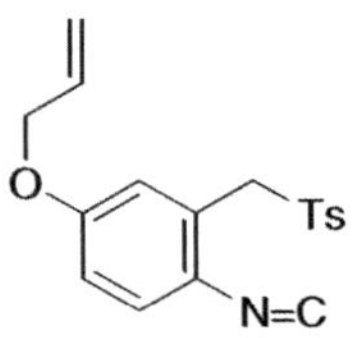	NaOH, DMSO rt	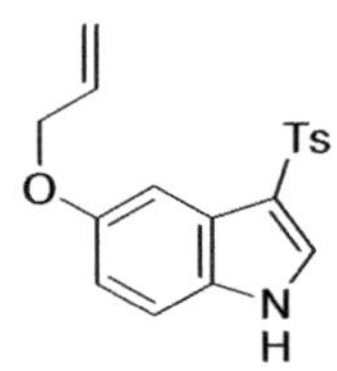	93%	16
6	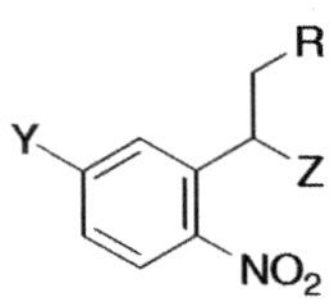 Y = Cl, Br, OMe Z = CN, CO_2*t*-Bu, Ts, SO_2Ph R = CH=CH_2, C(Me)=CH_2, CH=CHPh, CO_2Et, 3,4-Cl_2Ph, CO_2*t*-Bu, Ph, Et	TMSCl, Et_3N DMF, rt (other conditions)	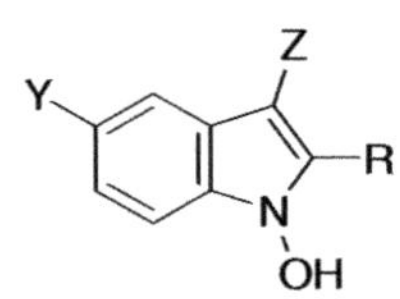	25–91%	17
7	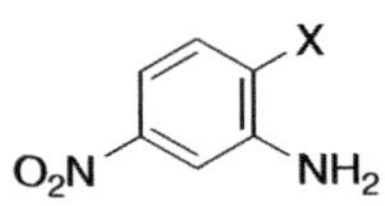 X = H, F, Cl R = Me, Et, Ph, 1-naphthyl	RCH_2CN *t*-BuOK, DMSO	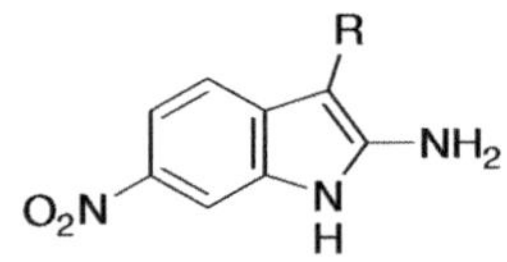	32–97%	18

Table 1 *(continued)*

Entry	Substrate	Conditions	Indole	% Yield	Ref.
8	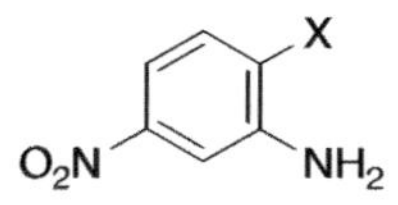 X = H, F, Cl	CH_3CN KOH, DMSO	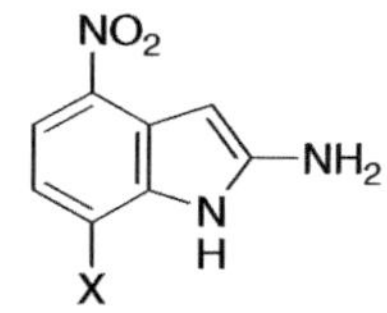	28–65%	18
9	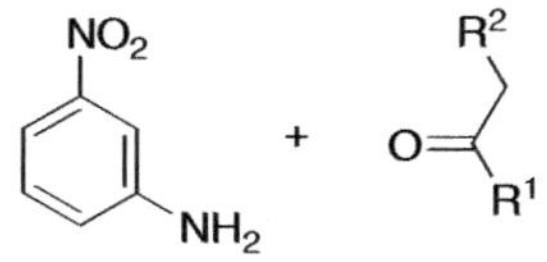 R^1 = Ph, Me, Et, *t*-Bu, 2-pyr, 2-thienyl, 2-furyl R^2 = H, Me R^1,R^2 = $-(CH_2)_4-$	*t*-BuOK DMSO	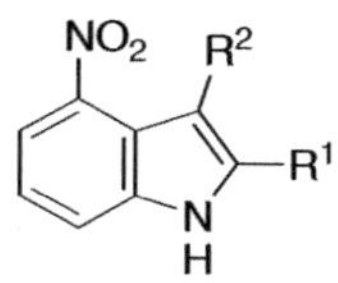	32–67%	19,20

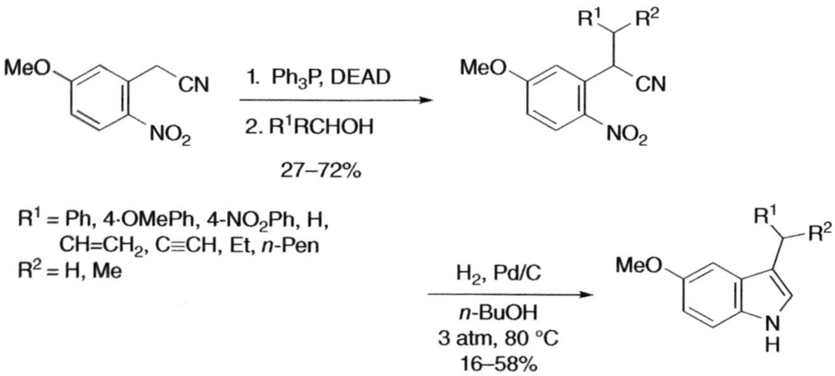

[eqn. 1] [21],[22]

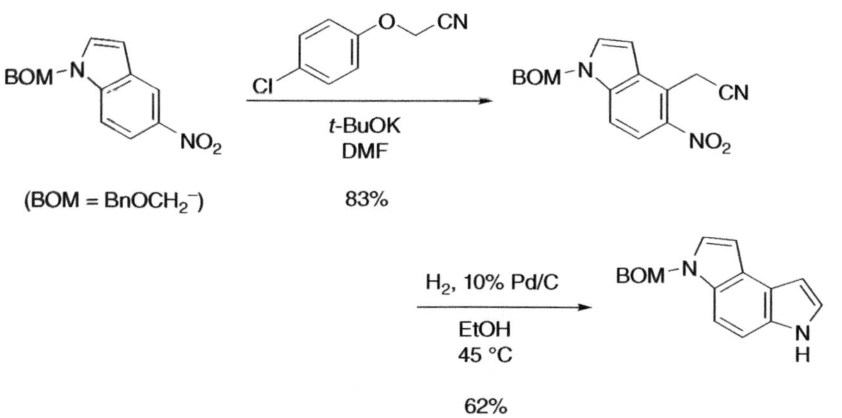

[eqn. 2] [23]

Scheme 3 *Macor Applications of the Makosza Indole Synthesis*

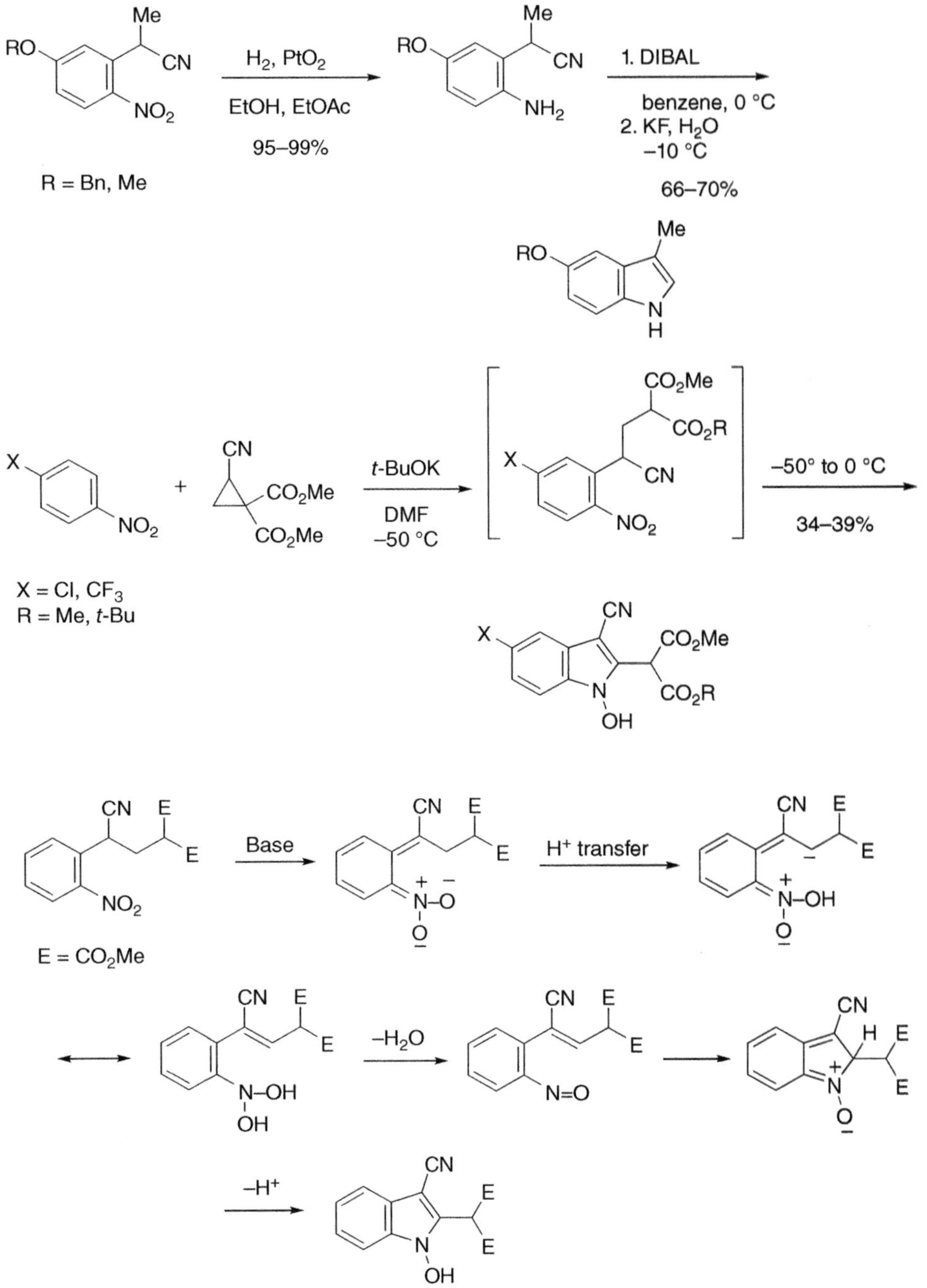

[eqn. 1] [24]

[eqn. 2] [25]

[eqn. 3] [17],[25]

Scheme 4 *Modifications of the Makosza Indole Synthesis*

pyrrolo[3,2-*e*]indole ring system, a fundamental scaffold in several anticancer natural products and drugs such as CC-1065 [23].

In a route to the biologically important 6-alkoxyindoles, Marino and Hurt discovered a noteworthy improvement on the conventional Mąkosza hydrogenation step (Scheme 4) [24]. Partial reduction of the *o*-nitrobenzylnitriles with PtO_2 gave in quantitative yield the corresponding *o*-aminobenzylnitriles (equation 1). Subsequent reductive cyclization with DIBAL (diisobutylaluminum hydride) affords the desired indoles in excellent overall yields. Stalewski presents an interesting twist on the VNS reaction leading to *N*-hydroxyindoles (equation 2) [24]. The final ring closure to *N*-hydroxyindoles proceeds by a fascinating mechanism proposed earlier for entry 6 in Table 1 [17] and is depicted in equation 3 [17, 25].

Table 2 *Applications of the Makosza Indole Synthesis*

Entry	Substrate	Conditions	Indole	% Yield	Ref.
1	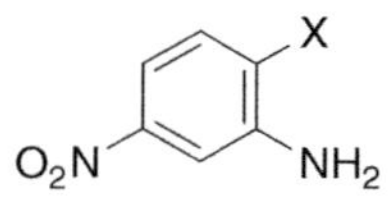 X = F, Cl; R = CN, 2-pyr, 2-NMe-benzimidazole, 2-benzothiazole	RCH_2CN, K_2CO_3, DMF, reflux	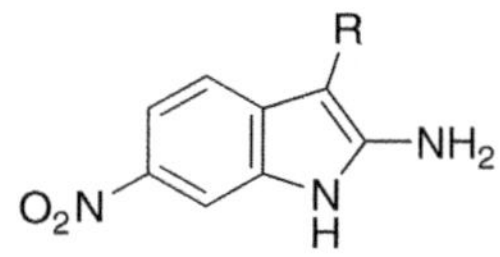	50–65%	26
2	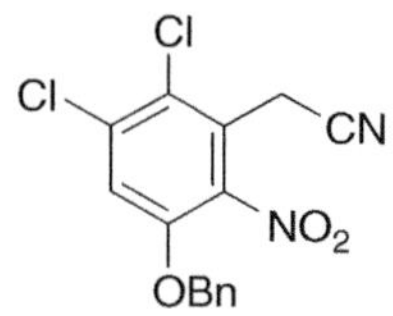	H_2, PtO_2, EtOH	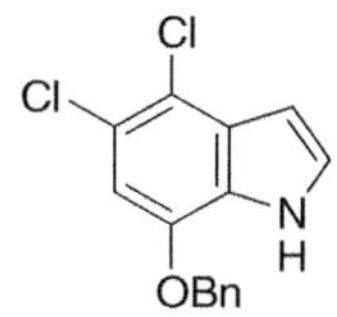	35%	27
3	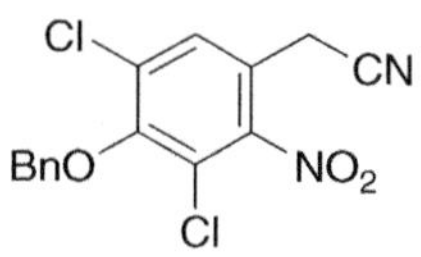	H_2, PtO_2, EtOH	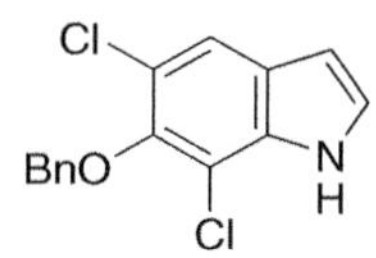	33%	27
4	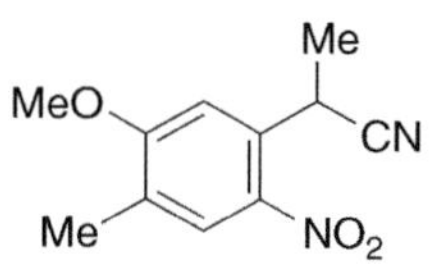	Pd/C, NH_4CHO, HOAc, 80 °C	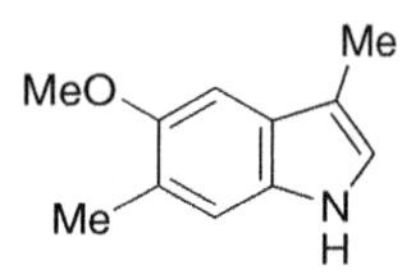	78%	28
5	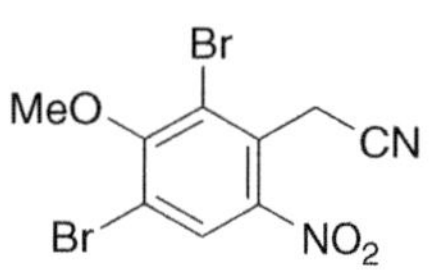	1. DIBAL, CH_2Cl_2, –78 °C 2. Fe, $FeCl_2$, HOAc, EtOH, reflux	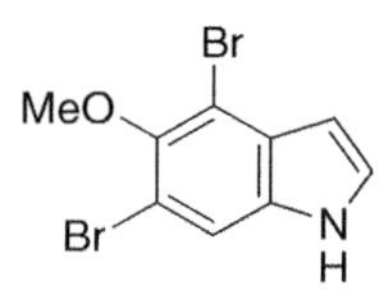	76%	29
6	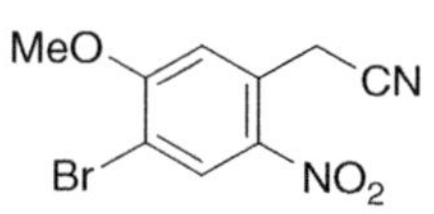	H_2, Rh/C, EtOH	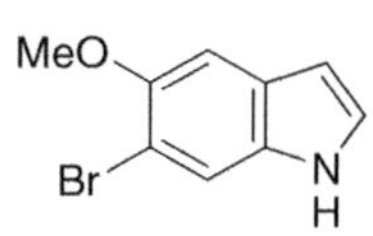	52%	29

Some final examples of the Mąkosza indole synthesis are tabulated in Table 2. Volovenko and colleague report a simple synthesis of 3-substituted-2-amino-5-nitroindoles (entry 1), which were transformed into pyrimido[1,2-*a*] indoles upon reaction with β-dicarbonyl compounds [26]. A conventional VNS method was used by Lerman and colleagues to craft 6- and 7-hydroxyindoles via sacrificial chlorine atoms that serve to increase the electrophilicity of the benzene ring toward cyanomethylation. Subsequent transfer hydrogenation (Pd/C/HCO_2NH_2) gives the respective hydroxyindoles (entries 2, 3) [27]. The preparation of 3,6-dimethyl-5-methoxyindole by Skibo and coworkers (entry 4) was the starting point in a synthesis and study of a cyclopropyl indolequinone methide as a novel reductive alkylating agent [28]. Fukuyama's team employed a mild two-step reduction sequence to avoid the partial debromination of 4,6-dibromo-5-methoxyindole that occurred under catalytic hydrogenation conditions (entry 5) [29]. In the synthesis of 5-bromo-6-methoxyindole (entry 6) these workers found that the bulky 2,6-dichlorophenoxyacetonitrile was superior to the conventional Mąkosza reagent, 4-chlorophenoxyacetonitrile, for the desired regioselective cyanomethylation of 2-bromo-4-nitroanisole [29]. Both of these bromoindoles were employed in the synthesis of the marine alkaloids (–)-eudistomin E and C.

References

[1] J. Goliński and M. Mąkosza, *Tetrahedron Lett.*, 1978, **19**, 3495–3498.
[2] M. Mąkosza and J. Winiarski, *Acc. Chem. Res.*, 1987, **20**, 282–294.
[3] M. Mąkosza, *Synthesis*, 1991, 103–111.
[4] M. Mąkosza and K. Wojciechowski, *Liebigs Ann. Recueil*, 1997, 1805–1816.
[5] M. Mąkosza and K. Wojciechowski, *Heterocycles*, 2001, **54**, 445–474.
[6] M. Mąkosza and K. Wojciechowski, *Chem. Rev.*, 2004, **104**, 2631–2666.
[7] M. Mąkosza, *Chem. Soc. Rev.*, 2010, **39**, 2855–2868.
[8] K. Wojciechowski and M. Mąkosza, *Tetrahedron Lett.*, 1984, **25**, 4793–4794.
[9] K. Wojciechowski and M. Mąkosza, *Synthesis*, 1986, 651–653.
[10] K. Wojciechowski and M. Mąkosza, *Bull. Soc. Chim. Belg.*, 1986, **95**, 671–673.
[11] M. Mąkosza, W. Danikiewicz, and K. Wojciechowski, *Liebigs Ann. Chem.*, 1988, 203–208.
[12] Z. Wróbel and M. Mąkosza, *Tetrahedron*, 1993, **49**, 5315–5326.
[13] Z. Wróbel and M. Mąkosza, *Synlett*, 1993, 597–598.
[14] M. Mąkosza and J. Stalewski, *Tetrahedron*, 1995, **51**, 7263–7276.
[15] M. Mąkosza, J. Stalewski, and O.S. Maslennikova, *Synthesis*, 1997, 1131–1133.
[16] M. Mąkosza, J. Stalewski, K. Wojciechowski, and W. Danikiewicz, *Tetrahedron*, 1997, **53**, 193–214.
[17] Z. Wróbel and M. Mąkosza, *Tetrahedron*, 1997, **53**, 5501–5514.
[18] N. Moskalev and M. Mąkosza, *Heterocycles*, 2000, **53**, 533–536.
[19] N. Moskalev and M. Mąkosza, *Tetrahedron Lett.*, 1999, **40**, 5395–5398.
[20] N. Moskalev, M. Barbasiewicz, and M. Mąkosza, *Tetrahedron*, 2004, **60**, 347–358.
[21] J.E. Macor and J.M. Wehner, *Tetrahedron Lett.*, 1991, **32**, 7195–7198.
[22] J.E. Macor and J.M. Wehner, *Heterocycles*, 1993, **35**, 349–365.
[23] J.E. Macor, J.T. Forman, R.J. Post, and K. Ryan, *Tetrahedron Lett.*, 1997, **38**, 1673–1676.
[24] J.P. Marino and C.R. Hurt, *Synth. Commun.*, 1994, **24**, 839–848.
[25] J. Stalewski, *Tetrahedron Lett.*, 1998, **39**, 9523–9526.
[26] Yu.M. Volovenko and T.A. Volovnenko, *Chem. Heterocycl. Cpd.*, 2001, **37**, 1092–1095.
[27] L. Lerman, M. Weinstock-Rosin, and A. Nudelman, *Synthesis*, 2004, 3043–3046.
[28] O. Khdour, A. Ouyang, and E.B. Skibo, *J. Org. Chem.*, 2006, **71**, 5855–5863.
[29] H. Yamagishi, K. Matsumoto, K. Iwasaki, *et al.*, *Org. Lett.*, 2008, **10**, 2369–2372.

44

Rawal Indole Synthesis

A relatively new example of a reductive cyclization indole synthesis is that of Rawal, who developed it for the pursuit and successful syntheses of *Aspidosperma* alkaloids [1–3]. Taking advantage of previous chemistry described by Kuehne [4], Koser [5], and RajanBabu [6, 7], Rawal and coworkers effected the *o*-nitrophenylation of enol ether **1** with (*o*-nitrophenyl)phenyliodonium fluoride (**2**) to give nitro ketone **3** in excellent yield. Reduction of **3** with $TiCl_3$/NH_4OAc affords tetracyclic indole **4** in high yield (Scheme 1, equation 1) [1, 3]. This nitro group reduction was first described by Sachs and Sichel in 1904 [8], and it was optimized by Ho and Wong 70 years later [9]. As we saw in the Leimgruber–Batcho indole synthesis (Chapter 41) and, as we shall see in later chapters, the Ti(III) reductive cyclization of *o*-nitro carbonyls is a powerful route to the indole ring. Iodonium salt **2** is prepared in good yield from *o*-iodonitrobenzene in two steps: (1) oxidation with potassium persulfate in the presence of benzene to give (*o*-nitrophenyl)phenyliodonium iodide after treatment with potassium iodide (87%) and (2) iodide ion exchange with silver fluoride (78%) [2]. The iodonium iodide can also be obtained via oxidation with chromium trioxide (67%) [3].

Subsequent to their synthesis of tabersonine from intermediate **4** (Scheme 1), Rawal and colleagues demonstrated the utility of their new indolization method in the regiocontrolled preparation of indoles from cyclic ketones via

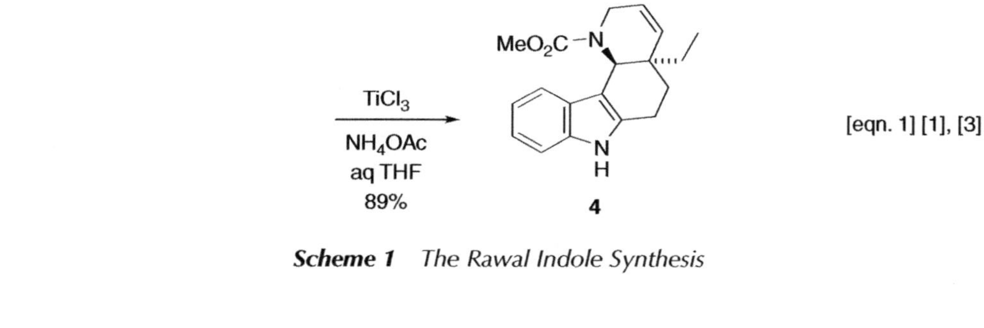

Scheme 1 *The Rawal Indole Synthesis*

Indole Ring Synthesis: From Natural Products to Drug Discovery, First Edition. Gordon W. Gribble.

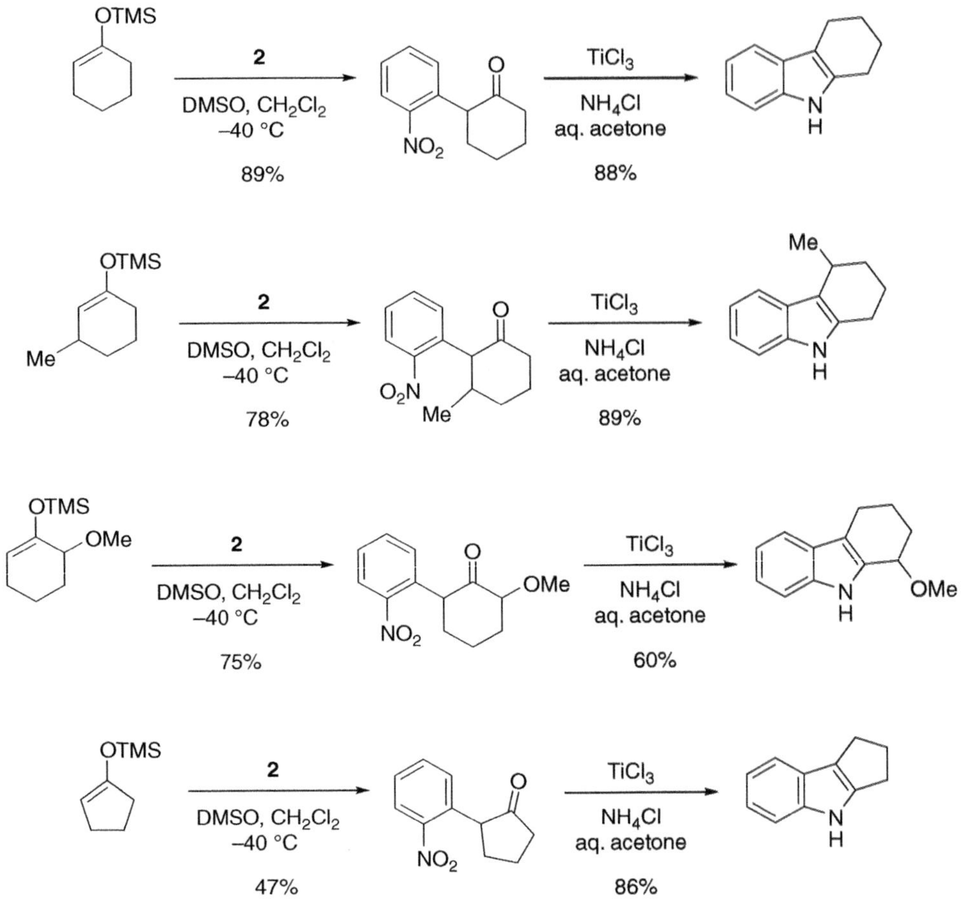

Scheme 2 *Rawal Indole Syntheses*

silyl enol ethers as summarized in Scheme 2 [2]. This method circumvents the lack of regiocontrol often encountered in the Fischer indole synthesis from cyclic ketones. For example, the Fischer indolization of 3-methylcyclohexanone affords 2-methyl-1,2,3,4-tetrahydrocarbazole in preference to 4-methyl-1,2,3,4-tetrahydrocarbazole (10:1) [10].

Two related approaches to indoles, also involving enol ethers, are those of Kuehne [4] and Banwell, which are presented in later chapters.

References

[1] S.A. Kozmin and V.H. Rawal, *J. Am. Chem. Soc.*, 1998, **120**, 13523–13524.
[2] T. Iwama, V.B. Birman, S.A. Kozmin, and V.H. Rawal, *Org. Lett.*, 1999, **1**, 673–676.
[3] S.A. Kozmin, T. Iwama, Y. Huang, and V.H. Rawal, *J. Am. Chem. Soc.*, 2002, **124**, 4628–4641.
[4] M.E. Kuehne, *J. Am. Chem. Soc.*, 1962, **84**, 837–847.
[5] K. Chen and G.F. Koser, *J. Org. Chem.*, 1991, **56**, 5764–5767.
[6] T.V. RajanBabu, G.S. Reddy, and T. Fukunaga, *J. Am. Chem. Soc.*, 1985, **107**, 5473–5483.
[7] T.V. RajanBabu, B.L. Chenard, and M.A. Petti, *J. Org. Chem.*, 1986, **51**, 1704–1712.
[8] F. Sachs and E. Sichel, *Ber.*, 1904, **37**, 1861–1874.
[9] T.L. Ho and C.M. Wong, *Synthesis*, 1974, 45.
[10] D. Stoemer and C.H. Heathcock, *J. Org. Chem.*, 1993, **58**, 564–568.

45

The Baeyer–Jackson Indole Synthesis and Miscellaneous Reductive Cyclization Indole Syntheses

The leading figure in the discovery and early chemistry of indole is Adolf Baeyer, who first synthesized indole by heating oxindole with zinc powder [1] and by reducing 2,3-dichloroindole with zinc/KOH [2]. For an excellent review of the early indole chemistry, see Sumpter and Miller [3]. Given Baeyer's seminal history with indole, it was fitting that he and Jackson discovered in 1880 the first practical synthesis of indoles: the Baeyer–Jackson indole synthesis (Scheme 1, equation 1) [4, 5], which entails the reductive cyclization of *o*-nitrobenzyl carbonyl compounds. Since 1880, this simple indolization has been used extensively, as seen both in this chapter and in previous chapters. Scheme 1 includes some additional early indole syntheses of this type (equations 2-4) [6–8]. For a discussion of the early syntheses of indoloindoles (cf. equation 4), see Samsoniya and Trapaidze [9].

Three other early indole syntheses involving reductive cyclization are depicted in Scheme 2. Weerman effected a Hofmann rearrangement of *o*-nitrocinnamide that was followed by reduction of the nitro group with Fe/HOAc and cyclization with KOH to afford indole (Scheme 2, equation 1) [10]. Baeyer and Emmerling [11] and Beilstein and Kuhlberg [12] separately described the reductive cyclization of *o*-nitrocinnamic acid with Fe/KOH (equation 2). Neither indolization method has received much attention. Ruggli discovered an interesting synthesis of indolo[3,2-*b*]indole via the reduction of *o*,*o*′-dinitrophenylacetylene and eventual cyclization to the diindole (equation 3) [13, 14].

With the knowledge promulgated by Baeyer, Jackson, Weerman, Reissert, Heller, Beilstein, Ruggli, and others, reductive cyclization routes to indole have been explored, utilized, and modified countless times in the intervening decades. In view of the immense literature covering this indolization, the balance of this chapter presents only selected examples that represent unique reaction conditions or afford indoles in acceptable yields. For a review on the reductive cyclization in the synthesis of 5-membered heterocycles, see Tsoungas and Diplas [15].

The reductive cyclization of *o*-nitrophenylacetaldehydes presents the obvious problem in that the formyl group is susceptible to reduction. Nevertheless, several such successful indole ring formations are known. Moreover, the intermediate hydroxylamine can cyclize onto the formyl group to give an *N*-hydroxyindole.

Nevertheless, Ucciani and Bonfand find that a rhodium catalyst effects the desired indole synthesis in high yield from *o*-nitrostyrene via hydroformylation and subsequent reduction and cyclodehydration to give 3-methylindole (Scheme 3, equation 1) [16]. In a project involving the large group of pyrroloiminoquinone natural products, Kraus and Selvakumar hydrogenated nitro aldehyde **1** to tricyclic indole **2** (equation 2) [17]. Aldehyde **1** was prepared from the corresponding ester using DIBAL (92%). A similar reductive cyclization using Zn/HOAc gives indole **3** in 51% yield (equation 3) [18].

A Meerwein arylation protocol was employed by Raucher and Koolpe to construct the requisite *o*-nitro carbonyls for indolization. Thus, arylation of *o*-nitrophenyldiazonium chlorides with vinyl acetate or vinyl bromide followed by reductive cyclization of the adducts affords indoles (Scheme 4, equation 1) [19]. The Raucher indole synthesis has been adopted by Magnus on a multigram scale to

Indole Ring Synthesis: From Natural Products to Drug Discovery, First Edition. Gordon W. Gribble.

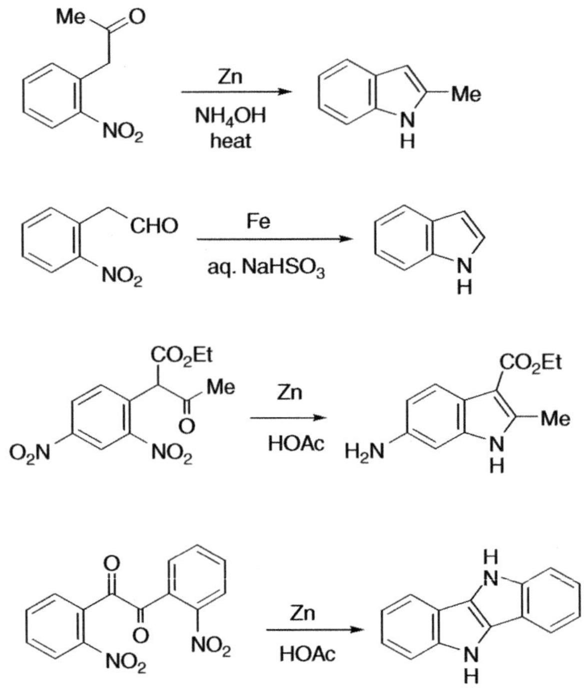

[eqn. 1] [4]

[eqn. 2] [6]

[eqn. 3] [7]

[eqn. 4] [8]

Scheme 1 *The Baeyer-Jackson Indole Synthesis*

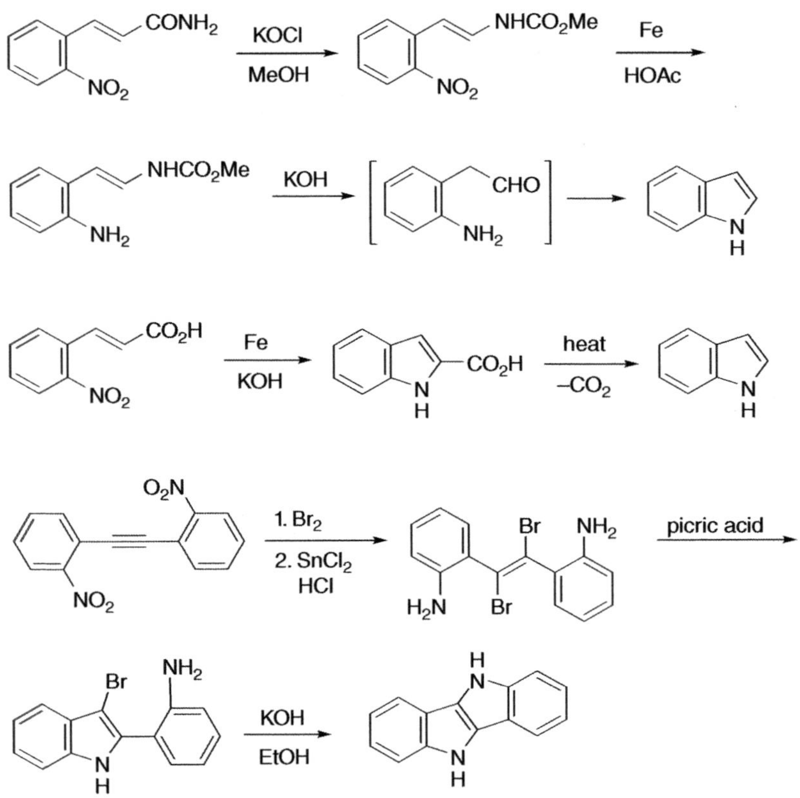

[eqn. 1] [10]

[eqn. 2] [11], [12]

[eqn. 3] [13], [14]

Scheme 2 *Early Reductive Cyclization Indole Syntheses*

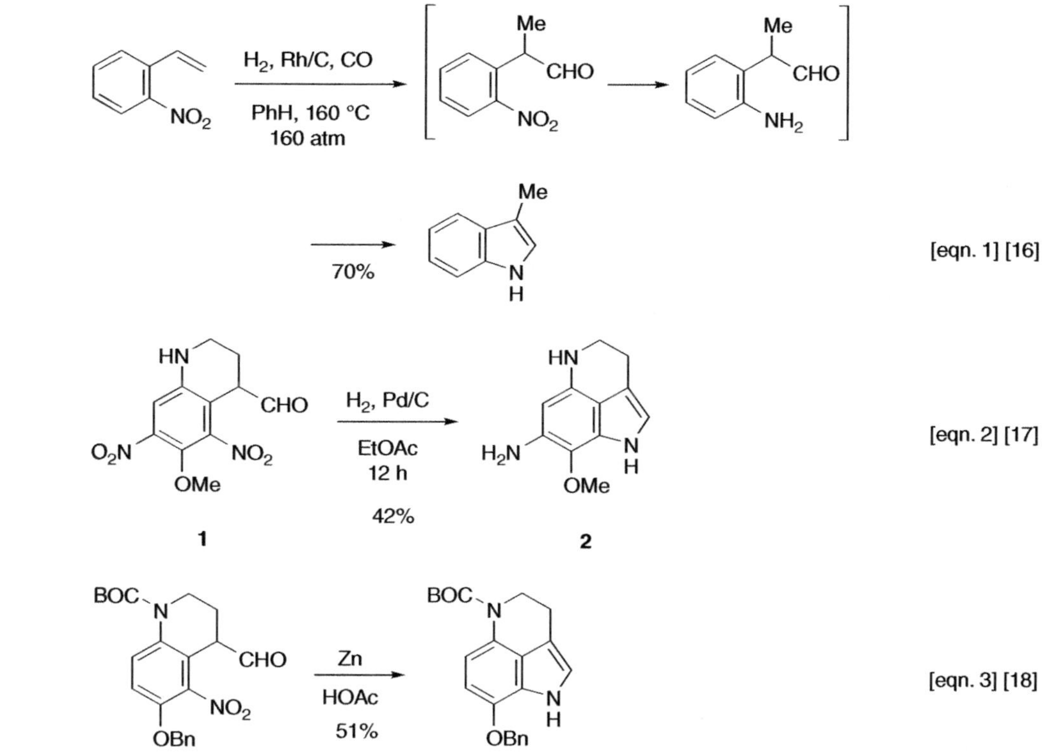

Scheme 3 *Reductive Cyclization of* o-*Nitrophenylacetaldehydes*

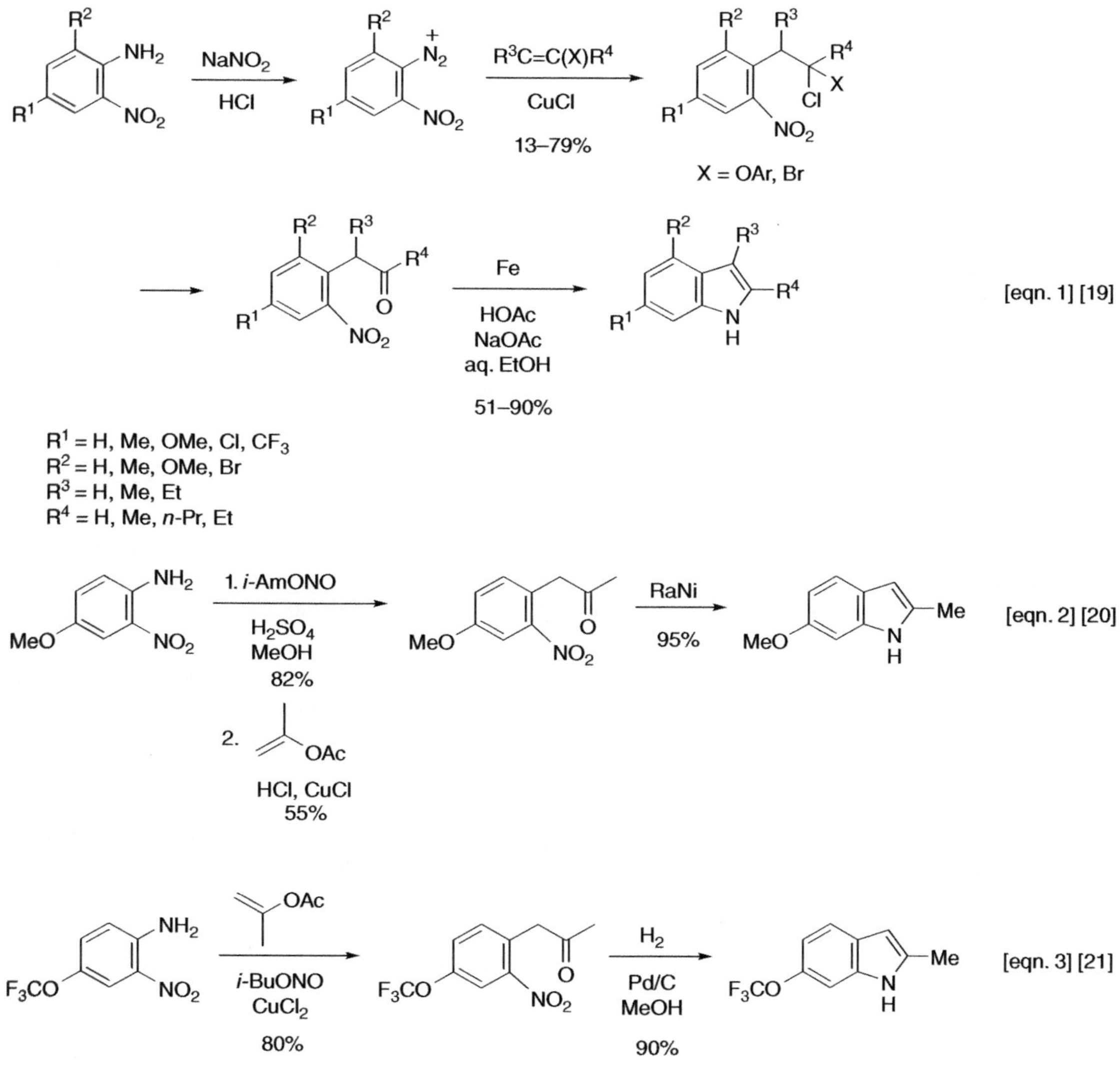

Scheme 4 *The Raucher Indole Synthesis*

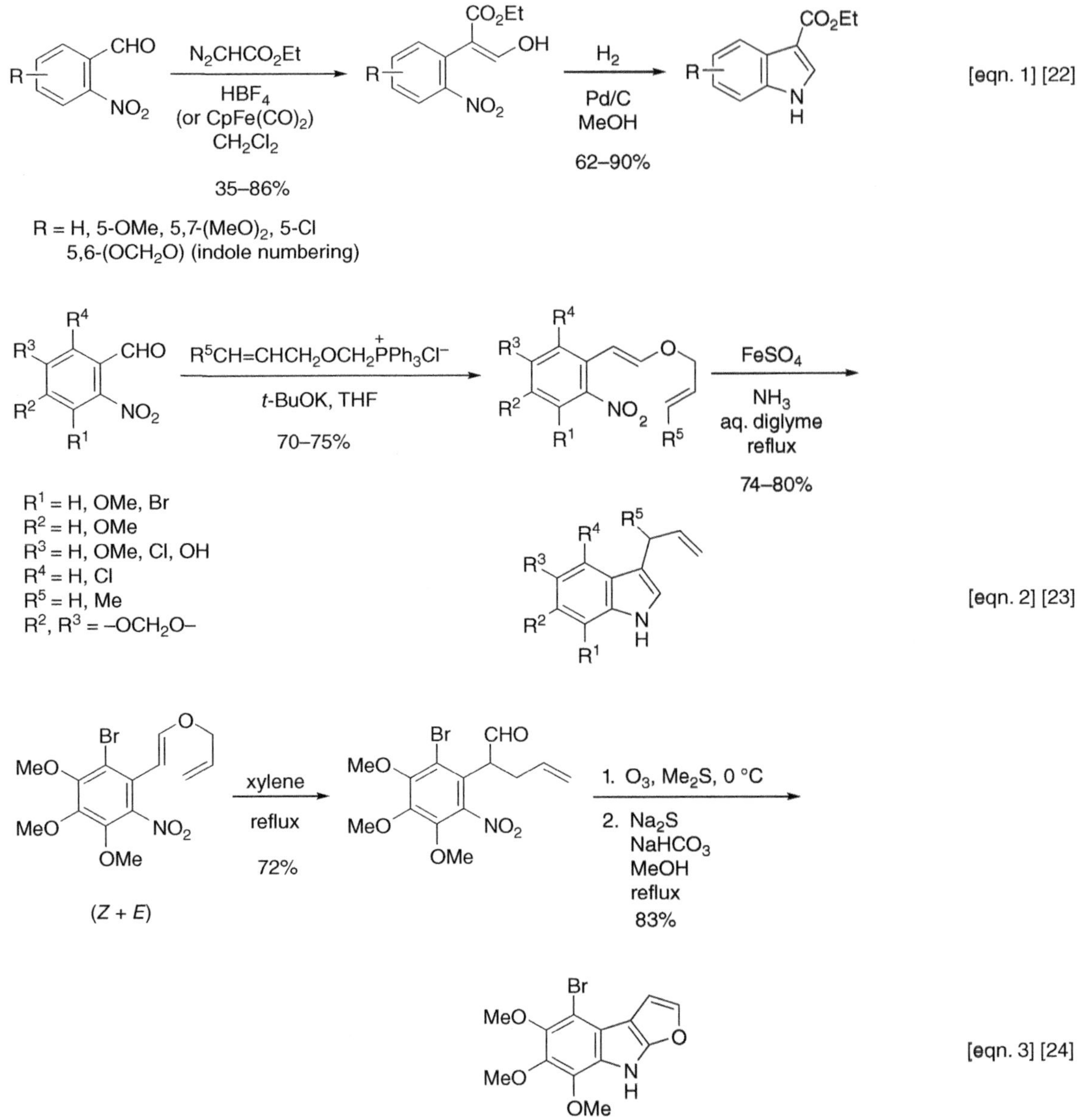

Scheme 5 *The Hossain and Kulkarni Indole Syntheses*

prepare 6-methoxy-2-methylindole for use in a synthesis of 16-methoxytabersonine (equation 2) [20], and it has been adopted by Maligres, Humphrey, and colleagues on a kilogram scale to access 2-methyl-6-(trifluoromethoxy)indole en route to a selective PPARγ modulator (equation 3) [21].

Hossain and colleagues parlayed the reductive cyclization of *o*-nitrophenylacetaldehydes, in the form of 3-hydroxypropenoic acid esters, into a slick synthesis of 3-ethoxycarbonyl indoles (Scheme 5, equation 1) [22] in what might be construed as an isomeric Reissert indole synthesis. Kulkarni, Davawala, and colleagues have developed a simple synthesis of 3-allylindoles that features a Wittig olefination–Claisen rearrangement followed by reductive cyclization (equation 2) [23]. A modification of this methodology affords a concise route to furo[2,3-*b*] indoles (equation 3) [24]. Several examples are described in this study in yields comparable to that shown. A novel feature is the use of NaSH *(in situ)* reduction of the nitro group.

The inherent lability of aldehydes toward reducing agents has led many investigators to employ acetals as an aldehyde equivalent in reductive cyclization to indoles. In a series of papers, Joule and Alvarez employed reductive cyclization of both nitro-aldehydes and nitro-acetals as an entry to the pyrrolo[4,3,2-*de*]quinoline alkaloids (e.g., damirones, batzellines, etc.) [25–30], as shown in Scheme 6, equations 1-3. In all cases, the two-step reductive cyclization of the nitro-acetal (equations 1 and 3) was higher yielding than the nitro-aldehyde (equation 2). Obviously, ring strain in the indole

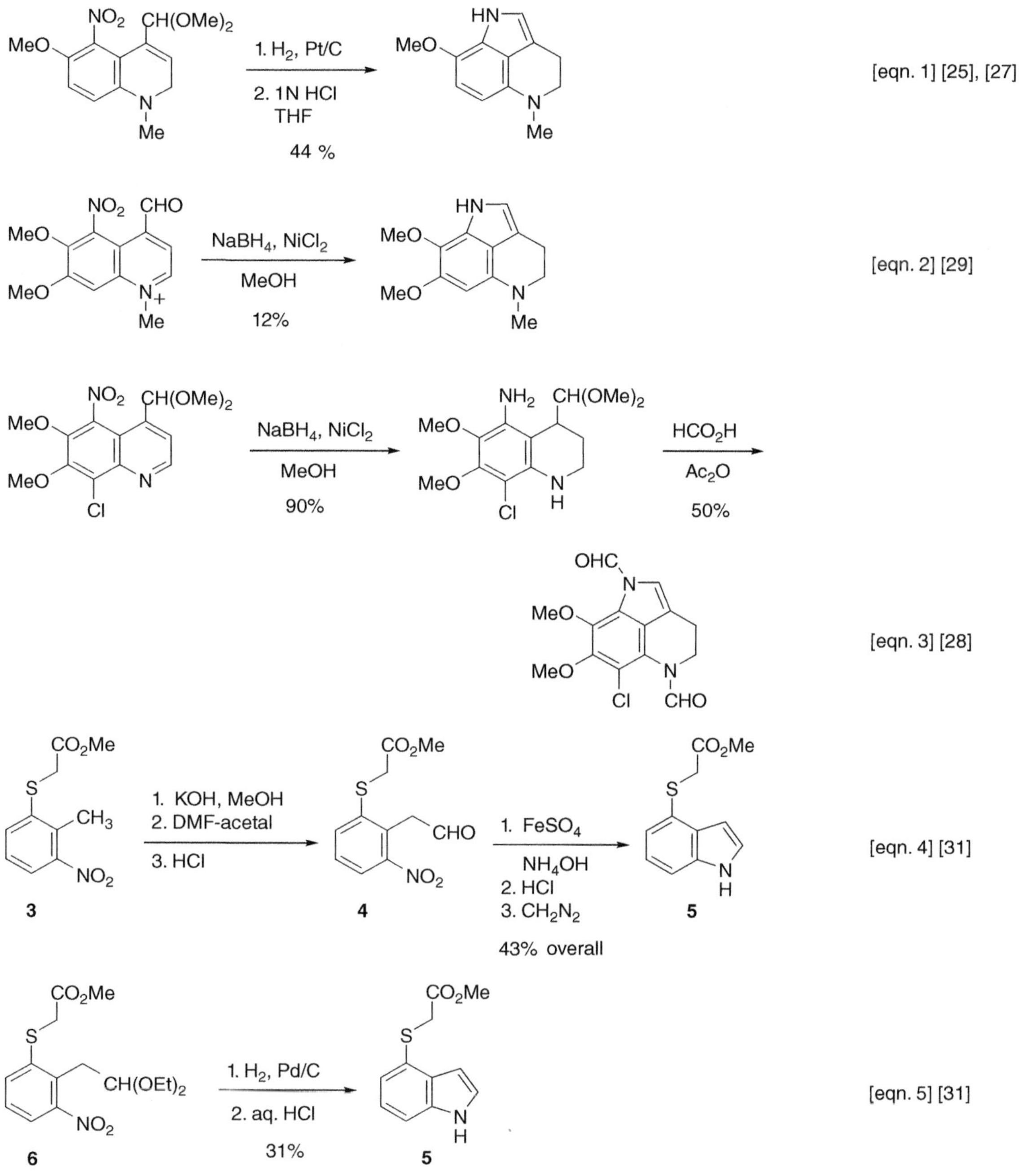

Scheme 6 *Reductive Cyclization of o-Nitrophenylacetaldehyde Acetals*

product is pronounced. An interesting reversal of the preceding situation was observed by Kozikowski and coworkers (equations 4 and 5) in their synthesis of the antibiotic alkaloid chuangxinmycin [31]. Thus, the Leimgruber–Batcho route from **3** to nitro aldehyde **4** followed by reductive cyclization to indole **5** (equation 4) is superior to reductive cyclization of nitro-acetal **6** to **5** (equation 5).

The reductive cyclization of *o*-nitrophenylacetaldehyde dialkyl acetals is a powerful indole synthesis as shown by many investigators (Table 1). The reaction presented in entry 1 begins with *o*-nitrobromobenzene followed by a Sonogashira coupling (TMSC≡CH, $Pd(PPh_3)_2Cl_2$), acetal formation (NaOEt, EtOH), and a two-step reductive cyclization [32, 33]. The lactol ether in entry 2 was prepared in similar fashion by Yamanaka and colleagues from ethyl 2-bromo-3-nitrobenzoate (Sonogashira coupling, then NaOEt/EtOH). Subsequent reductive cyclization affords ethyl 4-indolecarboxylate in 64% yield [32, 33]. Yamanaka also applied this simple indolization to azaindoles (pyrrolopyridines). Related chemistry described by RajanBabu

Table 1 *Reductive Cyclization of o-Nitrophenylacetaldehyde Dialkyl Acetals*

Entry	Substrate	Conditions	Indole	% Yield	Ref.
1	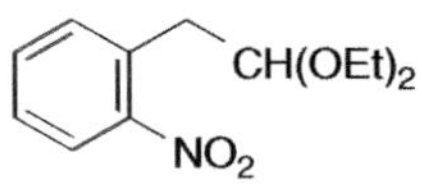	1. H_2, Pd/C EtOH 2. HCl, EtOH	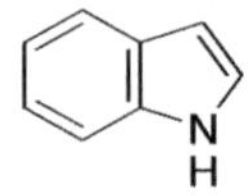	34% (from *o*-nitro-bromobenzene)	32
2	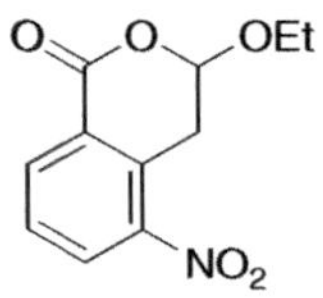	1. H_2, Pd/C EtOH 2. NaOEt, EtOH rt	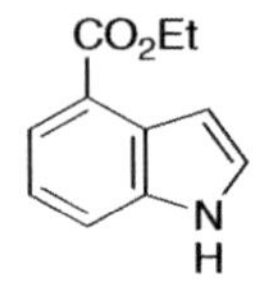	64%	32, 33
3	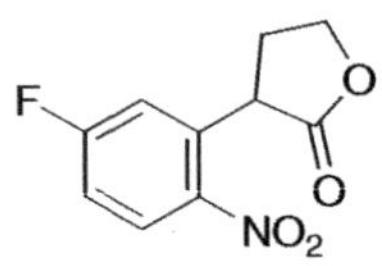	1. DIBAL 2. H_2, Pd/C	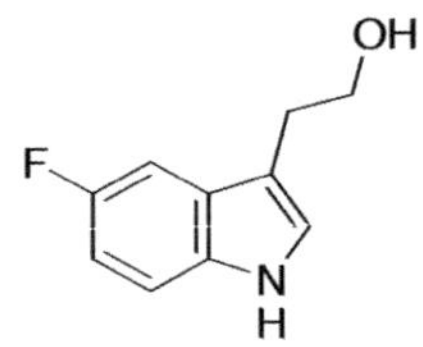	77%	34
4	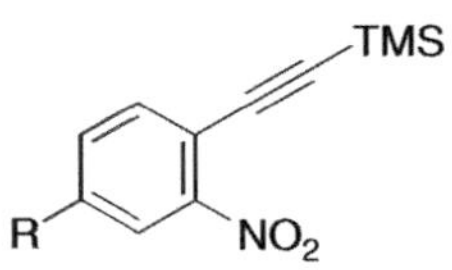 R = CO_2Et, COPh, Ph, CF_3, Me	1. KOH, MeOH 2. HOAc 3. H_2, Pd/C 4. HCl, H_2O, 60 °C	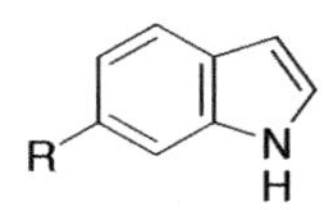	38–86% (overall from *o*-halonitrobenzene)	35
5	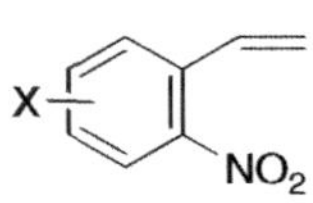 X = H, 4-OMe, 5-OMe, 4-Me, 4-CO_2Me, 4-Cl, 6-Me R = Me, Et, Bu, *t*-Bu	1. RONO, Na_2PdCl_4 O_2, MeOH 2. Fe, HOAc, HCl EtOH, 70–80 °C	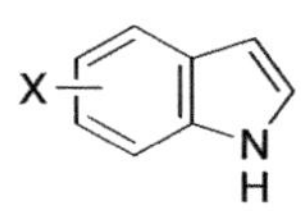	8–70%	36
6	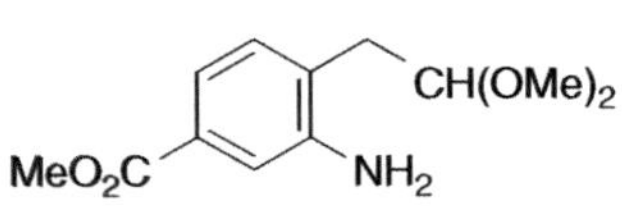	1. *N*-Me-4-piperidone Na_2SO_4 2. $NaBH(OAc)_3$, HOAc 3. HCl, MeOH, reflux	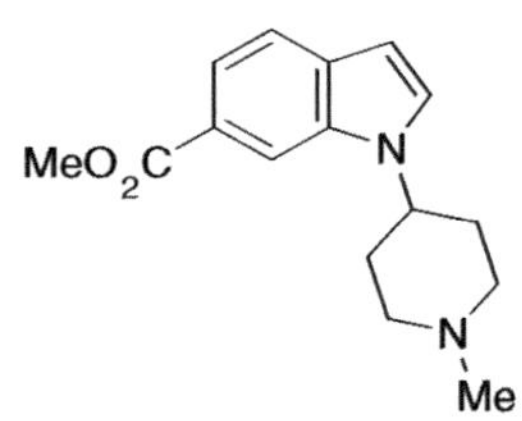	96%	37
7	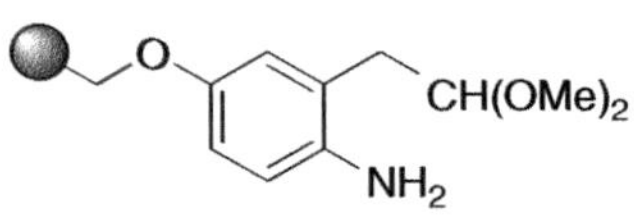 R = 4-I-phenyl, $(CH_2)_4Ph$	1. RCOCl (RCO_2H) 2. PPTS, tol 50 °C	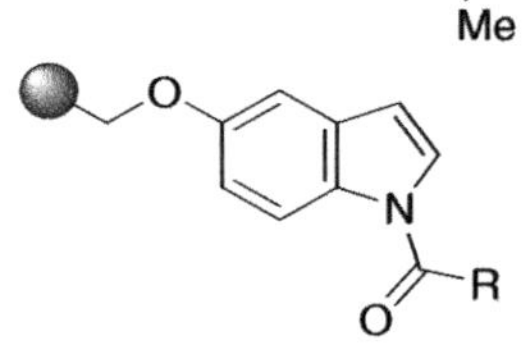	—	39
8	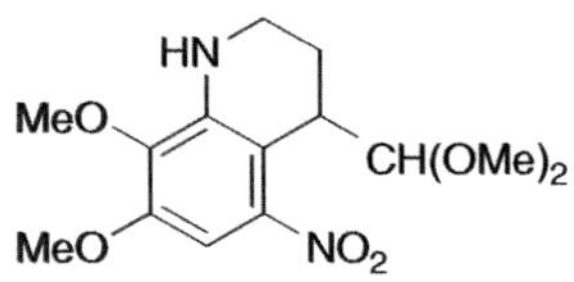	1. $NaBH_4$, $NiCl_2$ MeOH, rt 2. 1N HCl, THF 40 °C	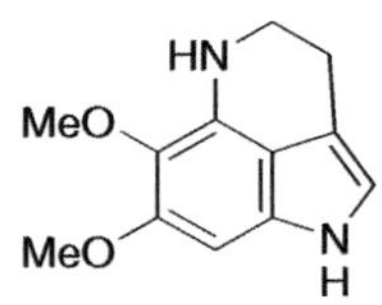	43%	40

Table 1 *(continued)*

Entry	Substrate	Conditions	Indole	% Yield	Ref.
9	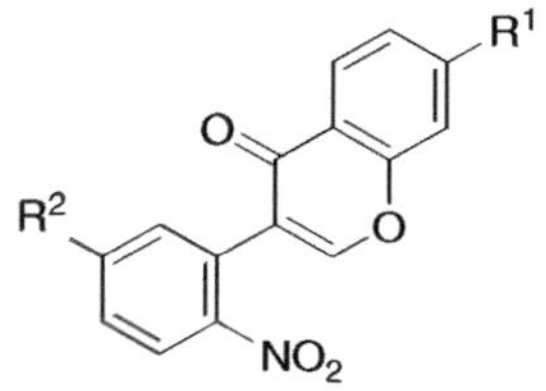 R^1 = Me, Cl, F, H; R^2 = H, OH	H_2, Pd/C EtOH reflux (Zn, HOAc reflux)	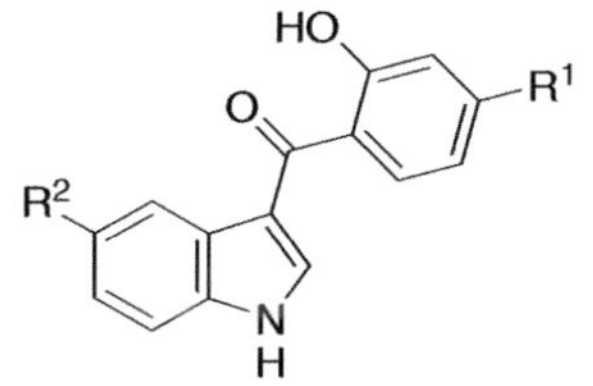	23–54%	41
10	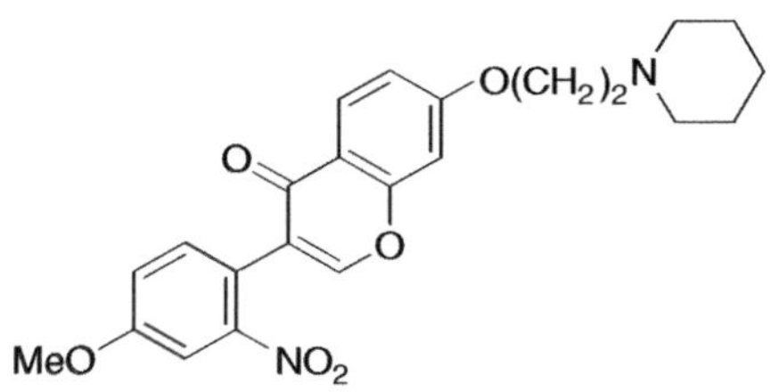	Zn, HOAc reflux	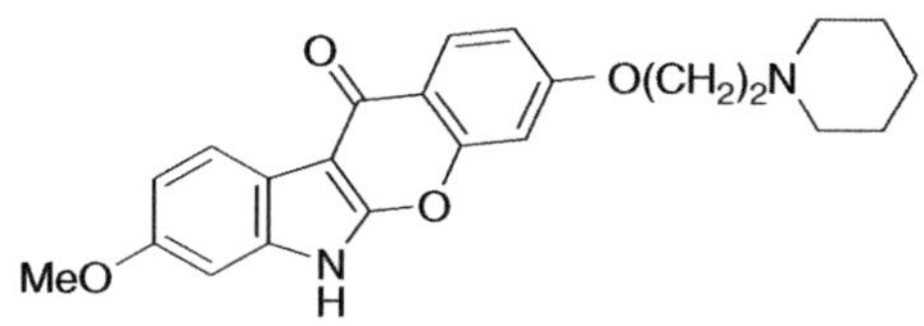	3%	41
11	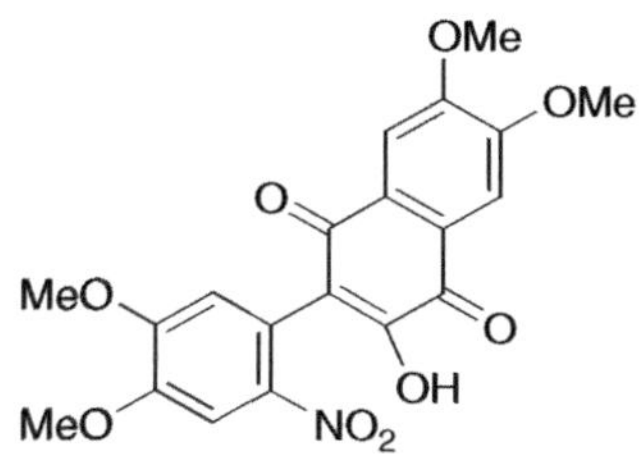	KBH_4, *i*-PrOH rt	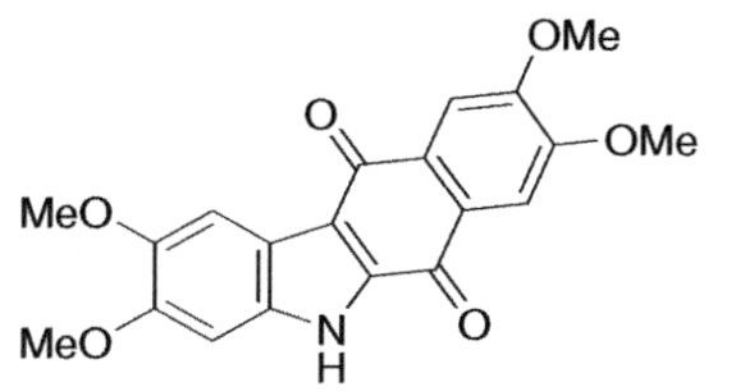	92%	42, 44, 45, 46
12	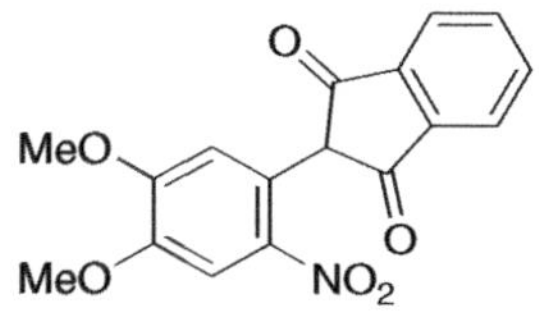	H_2, Pd/C	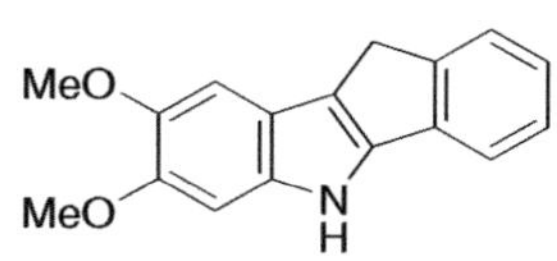	96%	42, 43

gives 5-fluorotryptophol (entry 3) [34]. The same sequential reduction of the corresponding chlorolactone (DIBAL) followed by the nitro group (H_2, Pd/C) provides 1-hydroxy-5-chlorotryptophol (29% yield) [34]. We present additional examples of 1-hydroxyindoles later in this chapter. The *o*-nitro lactones were prepared from the appropriate nitrobenzene and 4,5-dihydro-2-(trimethylsiloxy)furan. RajanBabu also synthesized several 1,2,3,4-tetrahydrocarbazoles in similar fashion. Thus, 6-chloro-1,2,3,4-tetrahydrocarbazole is obtained from α-(2-nitro-5-chlorophenyl)cyclohexanone by reduction with Fe/HOAc (20%), and 6,8-dichloro-1,2,3,4-tetrahydrocarbazole is realized from 2-(3,5-dichloro-2-nitrophenyl)cyclohexanone by hydrogenation with sulfided Pt/C (54%). In both cases, 1-(trimethylsiloxy)cyclohexene was coupled with the appropriate chlorinated nitrobenzene. Tischler and Lanza independently described the same synthesis of 6-substituted indoles from *o*-halonitrobenzenes (entry 4) [35]. The *o*-trimethylsilylethynylnitrobenzene starting material is prepared from an *o*-halonitrobenzene and ethynyltrimethylsilane and $Pd(Ph_3P)_2Cl_2/Et_3N$ (75%–96% yield). Treatment with KOH/MeOH transforms these *o*-nitroethynes to the dimethyl acetals suitable for reductive cyclization. Izumi's group achieved a similar result from *o*-nitrostyrenes and their oxidative conversion to *o*-nitrophenylacetaldehyde dialkyl acetals with nitrous acid alkyl esters and Na_2PdCl_4 (entry 5) [36]. Minor amounts of the corresponding aldehyde are formed in the palladium-catalyzed acetalization step. Coe and colleagues have modified the Leimgruber–Batcho indole synthesis (Chapter 41) to prepare *N*-substituted indoles via reductive amination of ketones and then acid-catalyzed cyclization to form the *N*-substituted indole (entry 6) [37]. Several cyclic ketones, 3-pentanone, and one aldehyde were employed with success in this indolization. The use of sodium sulfate greatly facilitates the alkylation of the aniline prior to cyclization. Moder and coworkers described a similar route to the protein kinase C inhibitor LY317615 using 1-(2-pyridinylmethyl)-4-piperidone in the reductive amination step, similar to entry 6, and trifluoroacetic acid to cyclize the resulting *o*-aminophenylacetaldehyde dimethyl acetal [38]. Abell and colleagues developed a solid-phase synthesis of *N*-acetylindoles using a Leimgruber–Batcho approach to the precursor *o*-aminophenylacetaldehyde acetals (entry 7) [39]. The *N*-acylindoles were subsequently cleaved to form esters and amides. Following Joule's work [29], Hénichart and colleagues synthesized 6,7-dimethoxy-1,3,4,5-tetrahydropyrrolo[4,3,2-*de*] quinoline by reductive cyclization of the appropriate acetal (64% yield) (entry 8) [40]. An interesting variation on the enol ether reductive cyclization to indoles was presented by Löwe's team [41]. Thus, a series of 2′-nitroisoflavones are reductively cyclized to their respective 3-salicyloylindoles (entry 9), although in one case a benzopyrano[2,3-*b*] indol-11-one was isolated as the only product in low yield (entry 10). In a series of investigations Castedo, Estévez, and their coworkers reported a synthesis of benzo[*b*]carbazoles (entry 11) and indeno[1,2-*b*]indoles (entry 12) that features reductive cyclization [42–46]. Fernández also used the chemistry depicted in entry 11 to prepare benzo[*b*]carbazole-6,11-diones and the isomeric indolo[1,2-*b*]isoquinoline-6,11-diones [47].

The sheer number of *o*-nitrophenyl ketone reductive cyclizations to form indoles necessitates an abbreviated section of this material. It is obvious that the challenge in this methodology is the preparation of the precursor nitro ketone because its cyclization is *fait accompli*. Scheme 7 [48–75] illustrates a range of indoles synthesized by

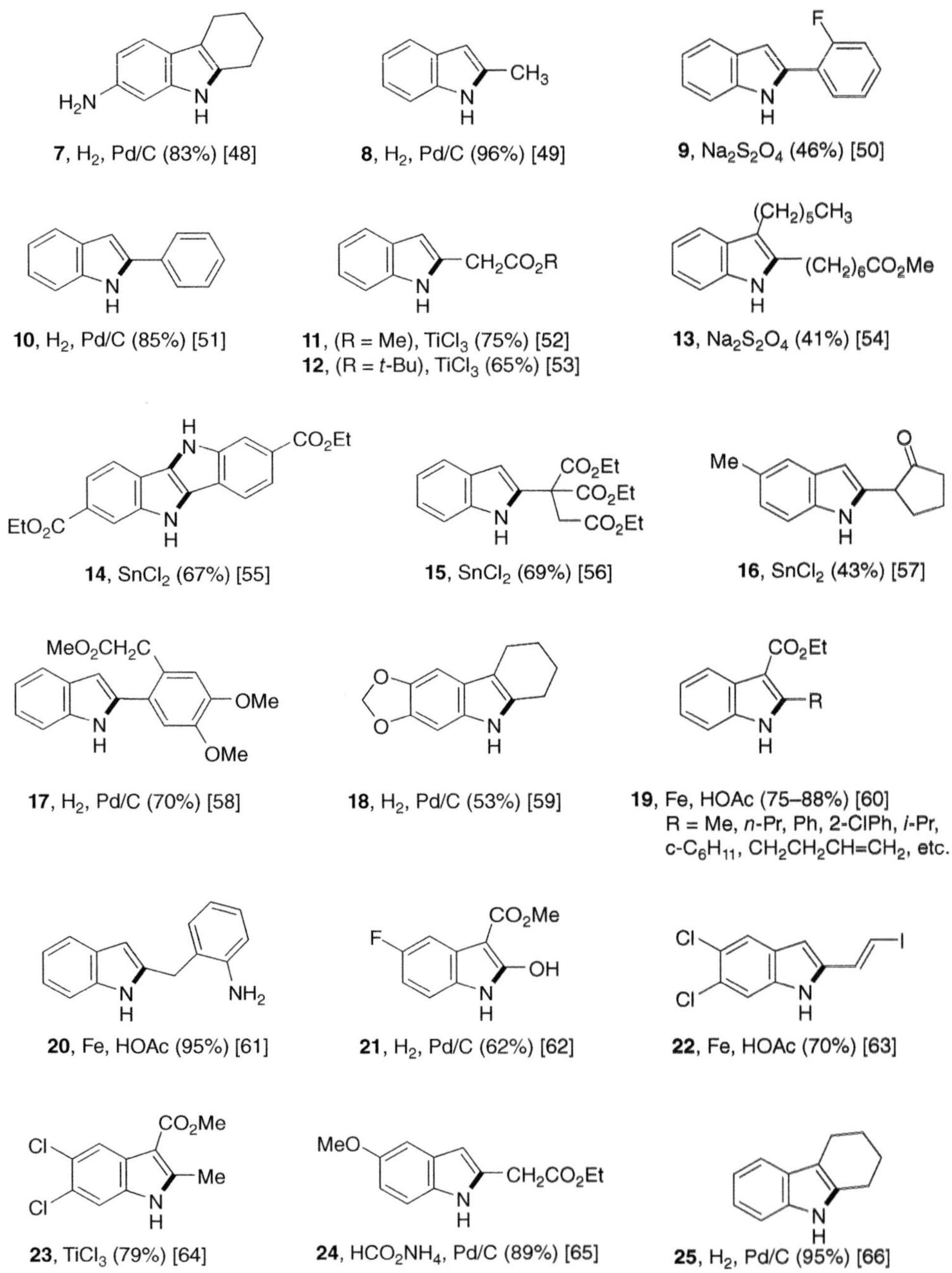

Scheme 7 *Indoles Formed by Reductive Cyclization of Nitro Ketones*

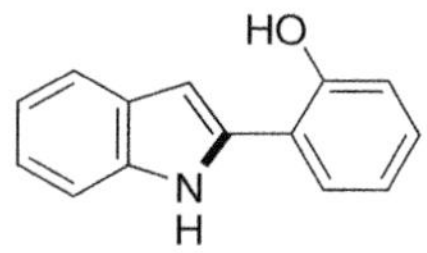

26, H_2, Pd/C (72%) [67]

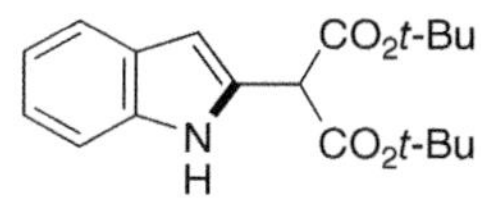

27, H_2, Pd(OH) (69%) [68]

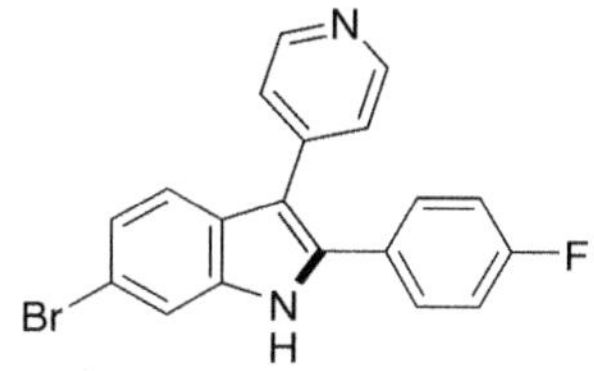

28, $TiCl_3$ [69]

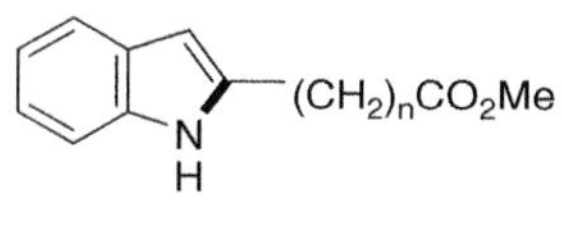

29, (n = 2-4), Fe, HOAc (90–95%)[70]

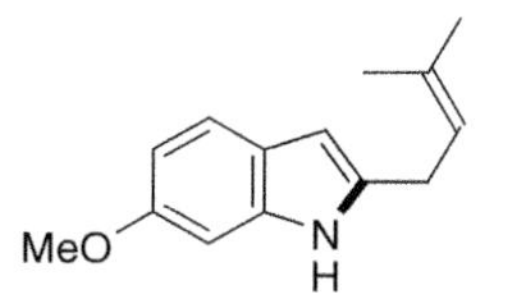

30, Zn, TFA (59%) [71]

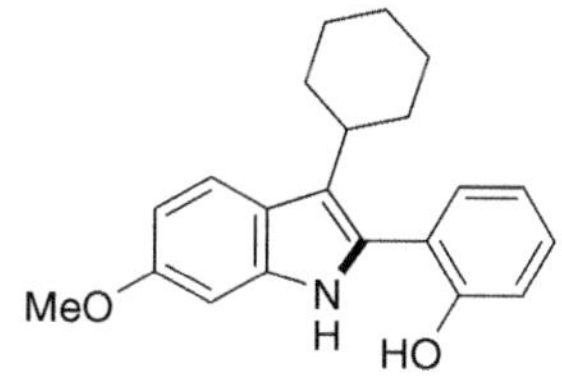

31, H_2, Pd/C (74%) [72]

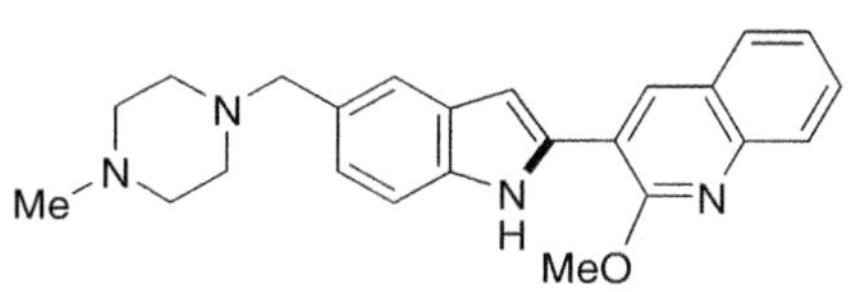

32, H_2, RaNi (90%) [73], [74]

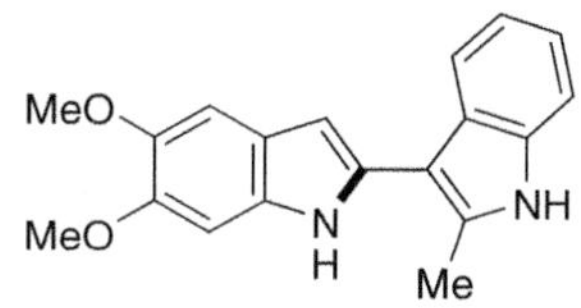

33, $Na_2S_2O_4$ (54%) [75]

Scheme 7 *(continued)*

reductive cyclization of nitro ketones (in chronological sequence). The carbon–nitrogen bond so formed between the nitro group and the ketone carbonyl is shown in bold. The precursor to 7-amino-1,2,3,4-tetrahydrocarbazole (**7**) is 2-(2,4-dinitrophenyl)cyclohexanoney [48]. The route to 2-(2-fluorophenyl)indole (**9**) was also used to synthesize *N*-(2-indolylmethyl)phthalimide in 50% yield with sodium hydrosulfite or in 65% yield with Raney nickel [50]. The preparation of 2-(6-methoxycarbonylhexyl)-3-hexylindole (**13**) was performed on a multigram scale [54]. The precursor to indolo[3,2-*b*]indole diester **14** is the corresponding 2,2′-dinitrobenzil [55], as we saw earlier with the work of Heller (Scheme 1, equation 4). Ho and Jou started their preparation of 2-(2-aminobenzyl)indole (**20**) with 1,3-bis(2-nitrophenyl)propan-2-one [61]. Presumably, 2-hydroxyindole **21** exists as the enol tautomer rather than the corresponding keto tautomer (oxindole). Conde and colleagues, in addition to preparing **22**, synthesized 5,6-dichloro-2-ethynylindole from the respective *o*-nitrophenyl ketone. This indole was used to prepare SB-242784, an osteoclast inhibitor [63]. Likewise Hamprecht's group employed dichloroindole **23** in a synthesis of a selective 5-HT_{2C} receptor antagonist [64]. A transfer hydrogenation method was used by Zlotos and colleagues to craft indole **24**, which eventually led to the development of melatonin receptor ligands [65]. The precursor nitro ketone to indole **26** was prepared by Snape via a Truce–Smiles rearrangement of the appropriate *o*-nitro diaryl ether [67]. Overman and coworkers enlisted indole keto diester **27** in the first total synthesis of the indole alkaloid (±)-actinophyllic acid [68]. Of several reductive cyclization conditions examined (Fe/HOAc, $SnCl_2$, $Na_2S_2O_4$, Lindlar, Zn/HOAc), Fukuyama found that only zinc–trifluoroacetic acid gave an acceptable yield of 2-prenylindole **30**. Most of the aforementioned conditions decomposed the starting nitro ketone. Indole **30** was used to synthesize tryprostatin A and its epimer [71]. The synthesis of indole **31** was performed on a 50-kilogram scale and involved a Truce–Smiles rearrangement to access the requisite *o*-nitrophenyl ketone [72] as developed by Snape [67]. A similar reductive cyclization was used by Kuethe and coworkers to prepare other analogues of **32** as novel KDR kinase inhibitors [73, 74].

Buchwald and his colleagues have made extensive use of the titanium reductive ring closure in an indolization as summarized in Scheme 8, equations 1 and 2 [76]. A small selection of indoles prepared using these methodologies is shown.

Gallou and coworkers described a new general synthesis of 2-arylindole-6-carboxylic acids that features a S_NAr preparation of the precursor nitro ketones (Scheme 9) [77].

A convenient synthesis of carbazol-4-ones and 2,3-disubstituted indoles in general has been developed by Yao and colleagues (Scheme 10) [78]. This protocol duplicates several of the previously presented reductive cyclizations (e.g., Table 1, entries 9–12).

Several unusual reductive cyclizations of nitro ketones are tabulated in Scheme 11. Tilve's group employed a double reductive cyclization to synthesize 6*H*-indolo[2,3-*b*]

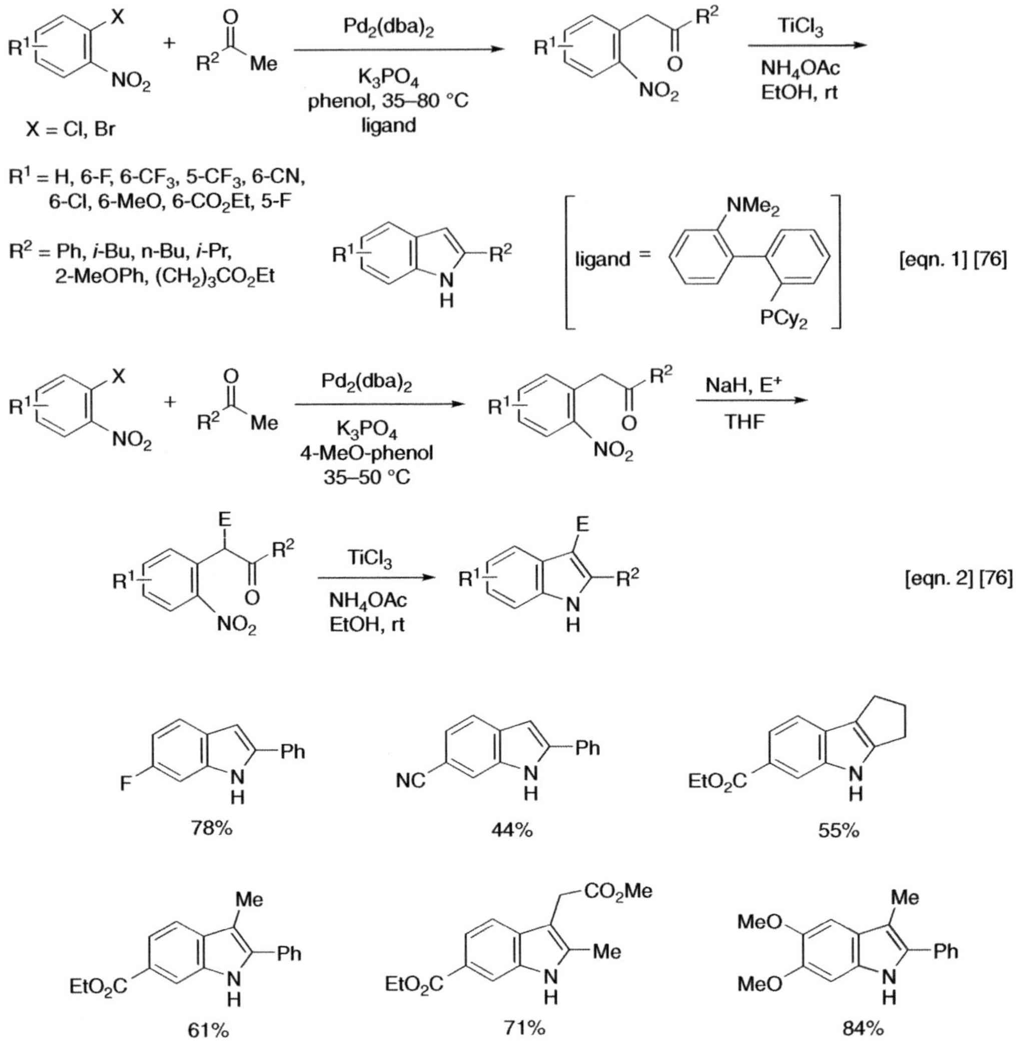

Scheme 8 *Buchwald Indole Ring Synthesis*

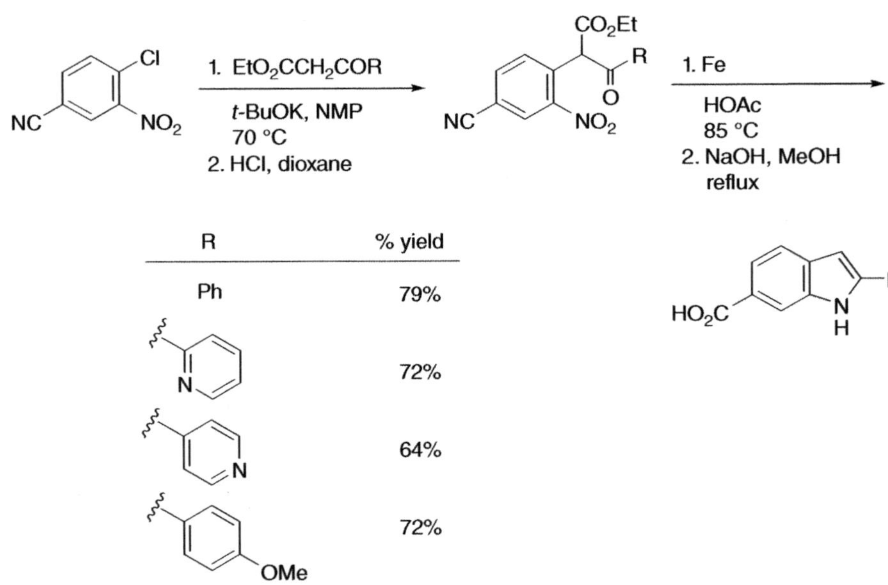

Scheme 9 *Gallou Indole Ring Synthesis*

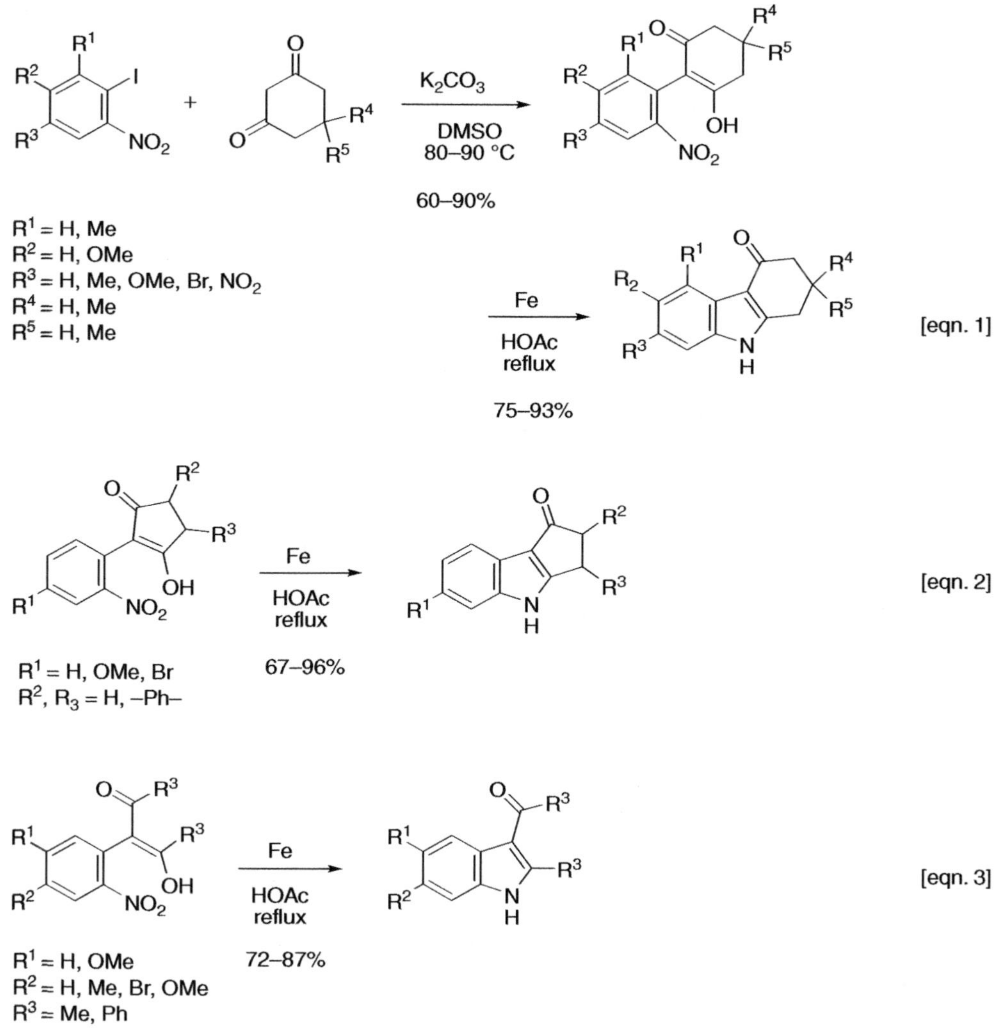

Scheme 10 *Yao Carbazolone Synthesis*

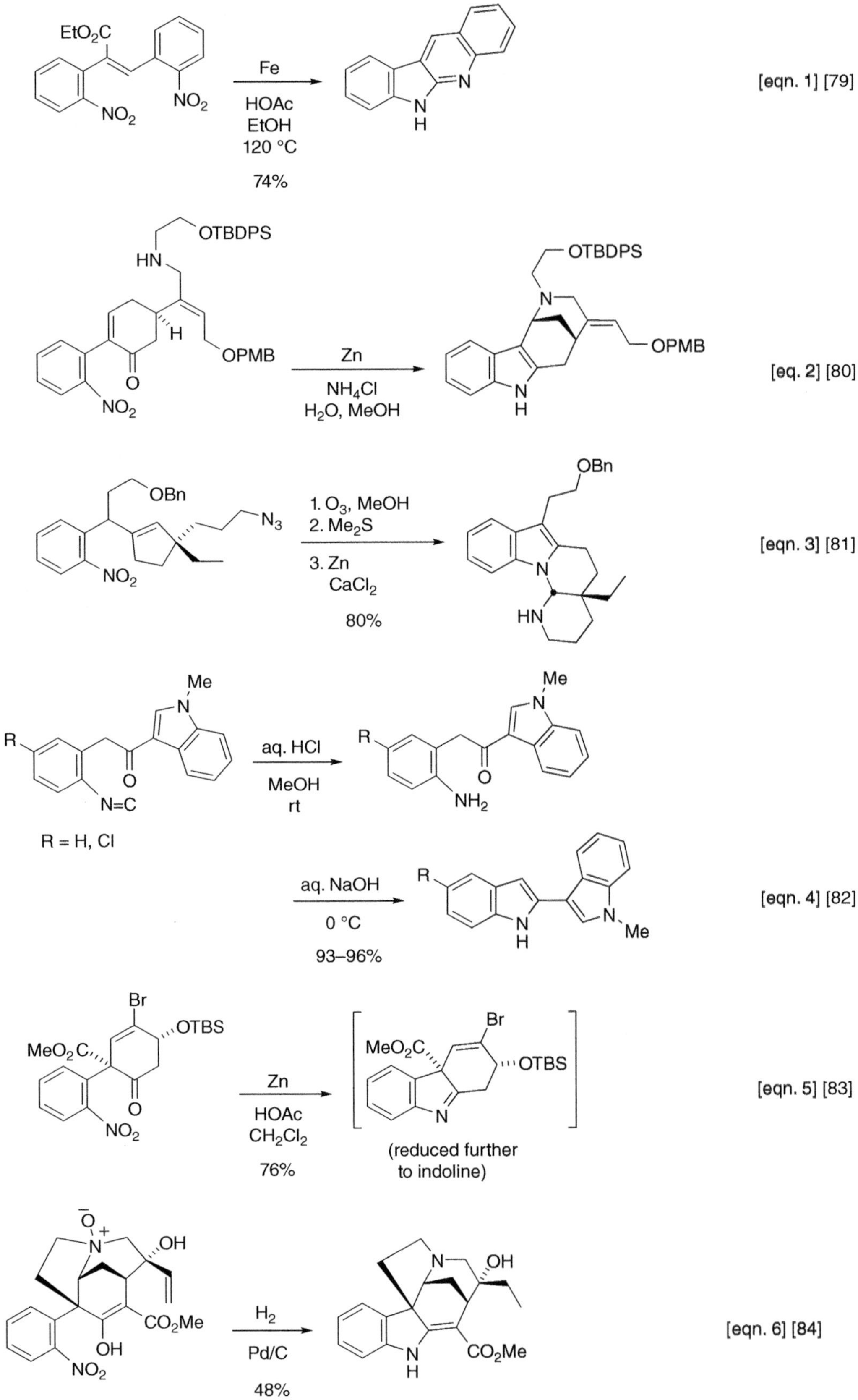

Scheme 11 *Miscellaneous Reductive Cyclization Indole Syntheses*

quinoline (equation 1) in a synthesis of cryptotackieine by subsequent methylation [79]. During the course of Shibasaki's synthesis of (–)-strychnine, the reductive cyclization shown in equation 2 was effected [80]. An equally complex indole alkaloid, (±)-goniomitine, was synthesized by Zhu and colleagues in a sequence involving the reductive cyclization shown in equation 3 [81]. Kobayashi's synthesis of 2,3′-biindolyls features the ring closure of amino ketones derived from isocyano ketones (equation 4) [82]. Several reductive cyclizations of nitro ketones that are interrupted at the imine stage are known (equations 5 and 6) [83, 84]. Bonjoch, Bosch, and colleagues achieved a total synthesis of 20-epilochneridine as shown in equation 6 via a reductive cyclization [84].

As alluded to earlier, the reductive cyclization of nitro ketones often leads to *N*-hydroxyindoles when the intermediate reduction product, a hydroxylamine, undergoes cyclization. This path can be made dominant if desired. Thus, as shown in Scheme 12, several researchers have developed such a methodology. Acheson may have been the first chemist to achieve this reaction. A two-phase zinc reduction of *o*-nitrophenylacetaldehyde yields the unstable *N*-hydroxyindole, which was trapped as the more stable (distillable) *N*-acetoxyindole (equation 1) [85]. Wong and colleagues use a lead-promoted reductive cyclization approach to *N*-hydroxyindoles in excellent yields (equation 2 and **34-37**) [86]. Wojciechowski's team employed a VNS synthesis of the *N*-hydroxyindole precursors (equation 3) [87].

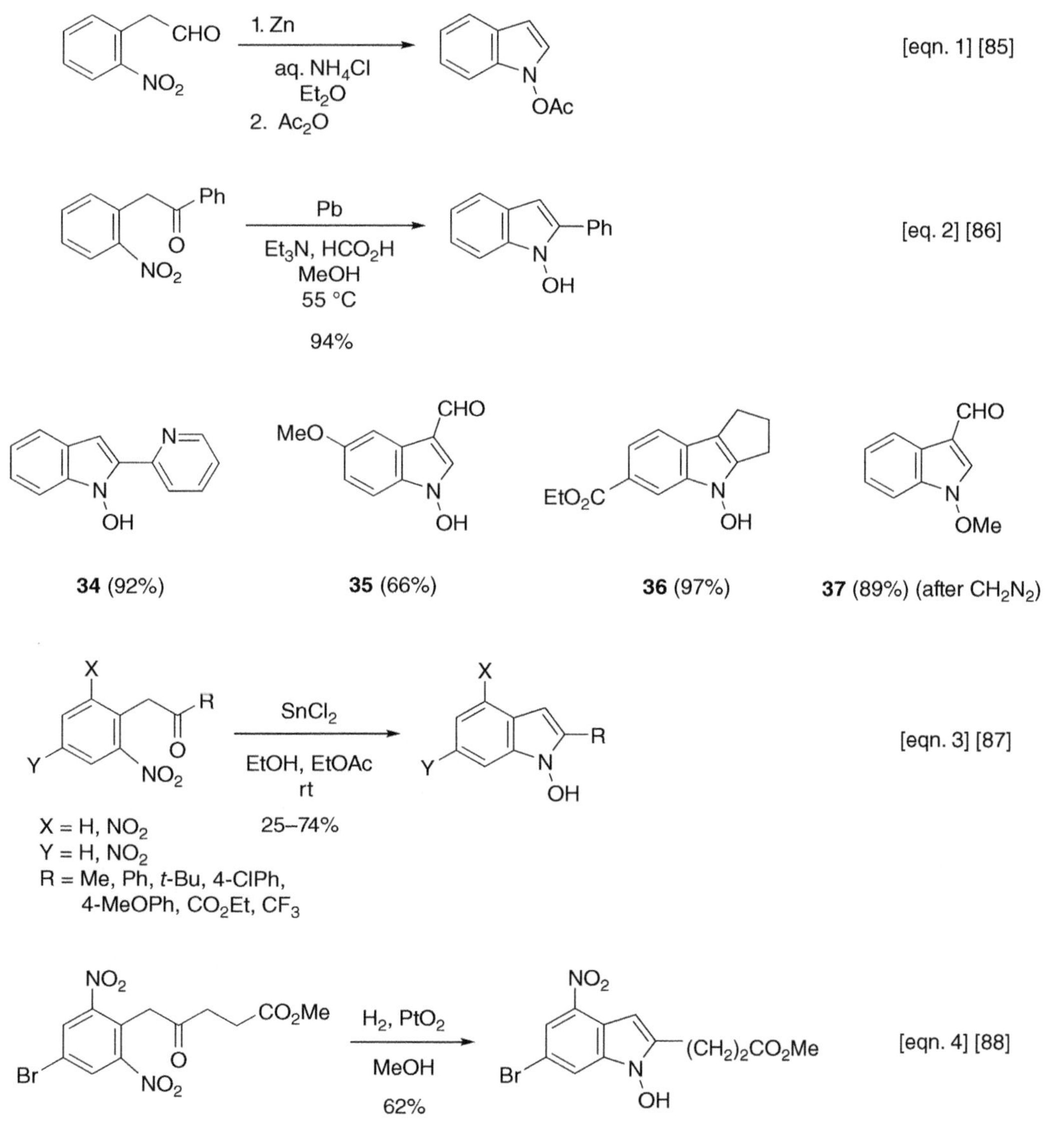

Scheme 12 *Synthesis of* N*-Hydroxyindoles by Reductive Cyclization*

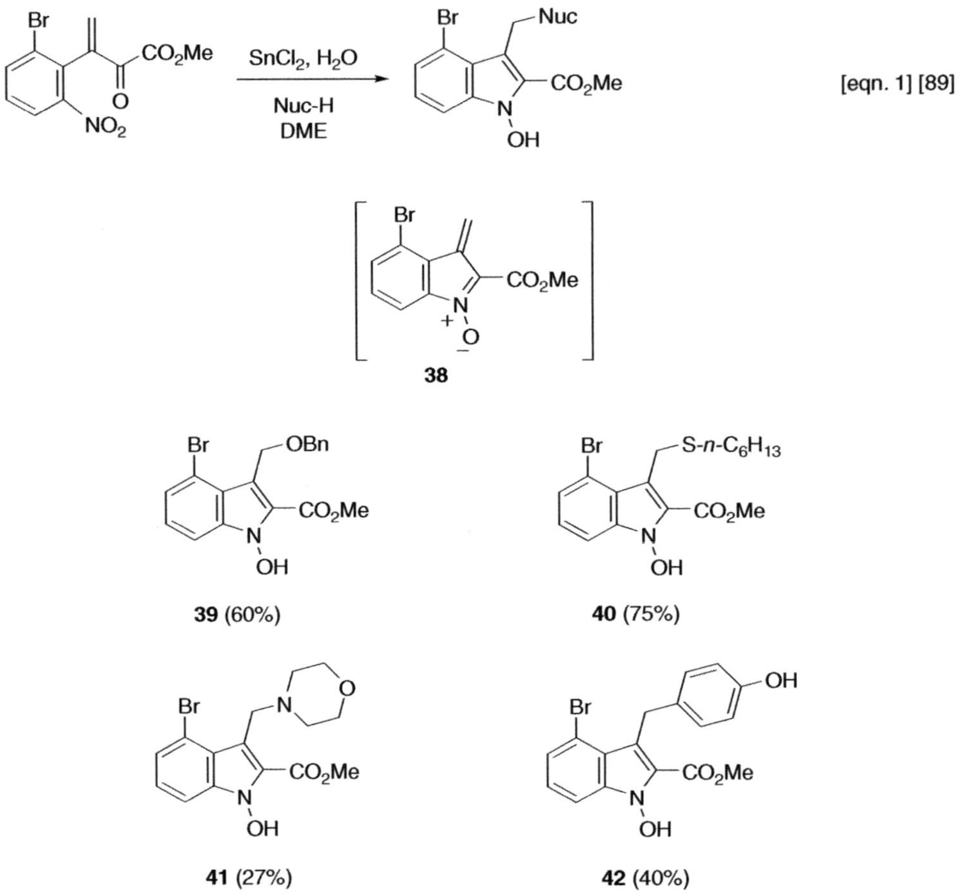

Scheme 13 *Nicolaou* N-*Hydroxyindole Synthesis*

Abbott and colleagues found that catalytic hydrogenation achieves their desired goal (equation 4) in a project targeted at new PDK1 inhibitors [88].

In a program aimed at the synthesis of the *N*-hydroxyindole-containing potent antibiotic nocathiacin, Nicolaou and coworkers developed a clever synthesis of *N*-hydroxyindoles (Scheme 13) [89]. The key intermediate is the α,β-unsaturated nitrone **38** that undergoes nucleophilic addition to *N*-hydroxyindole (equation 1). Several indoles thus prepared are shown **39-42**.

Belley and colleagues demonstrated a general catalytic hydrogenation method for the preparation of *N*-hydroxyindoles (Scheme 14, equations 1-6) [90], which can be converted to *N*-methoxyindoles upon treatment with diazomethane. The reactions in equation 6 afford unstable 1,2-dihydroxyindoles, which are capped with diazomethane. The function of the palladium [$(Ph_3P)_4Pd$] co-catalyst may be to retard reduction of the hydroxylamine intermediate, because without this co-catalyst 1*H*-indoles are obtained.

Two solid-phase syntheses of *N*-hydroxy(alkoxy)indoles have been reported by Roy (equation 1) [91] and Wu (equation 2) [92] respectively (Scheme 15). The *N*-hydroxyindoles in equation 2 were alkylated regioselectively to give a library of 1,2-dialkoxyindoles after cleavage from the SynPhase lanterns [92]. In a program toward the synthesis of the alkaloids avrainvillamide and stephacidin B, Myers and Herzon synthesized 3-alkylidene-3*H*-indole 1-oxide (equation 3), which is the keystone unit of these novel alkaloids [93].

Given the stark simplicity of the final act—cyclodehydration of an amino ketone or aldehyde to form an indole—it is remarkable how diverse and useful this indolization method has become since the early days of Baeyer and his contemporaries.

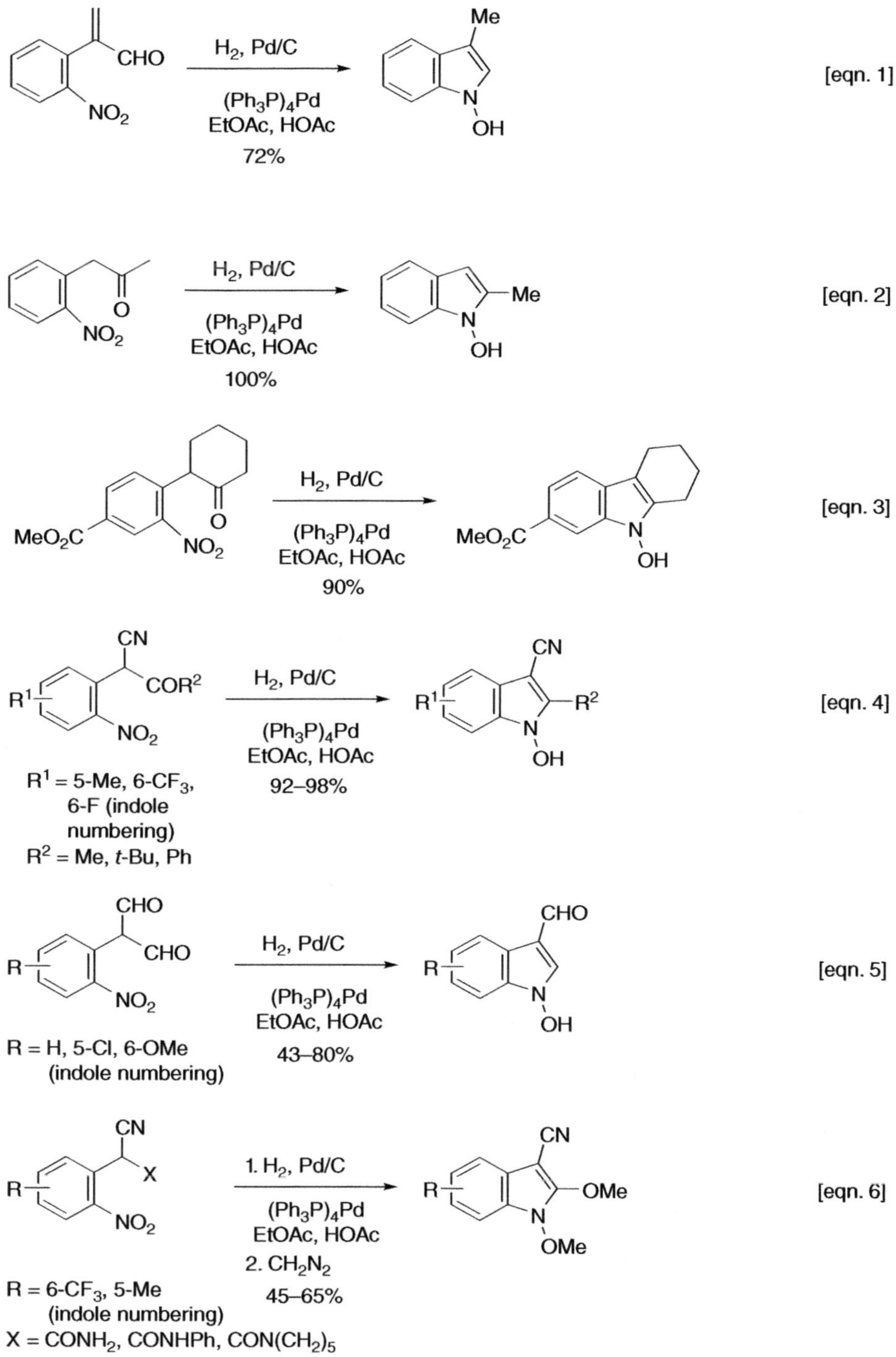

Scheme 14 *Belley N-Hydroxyindole Synthesis*

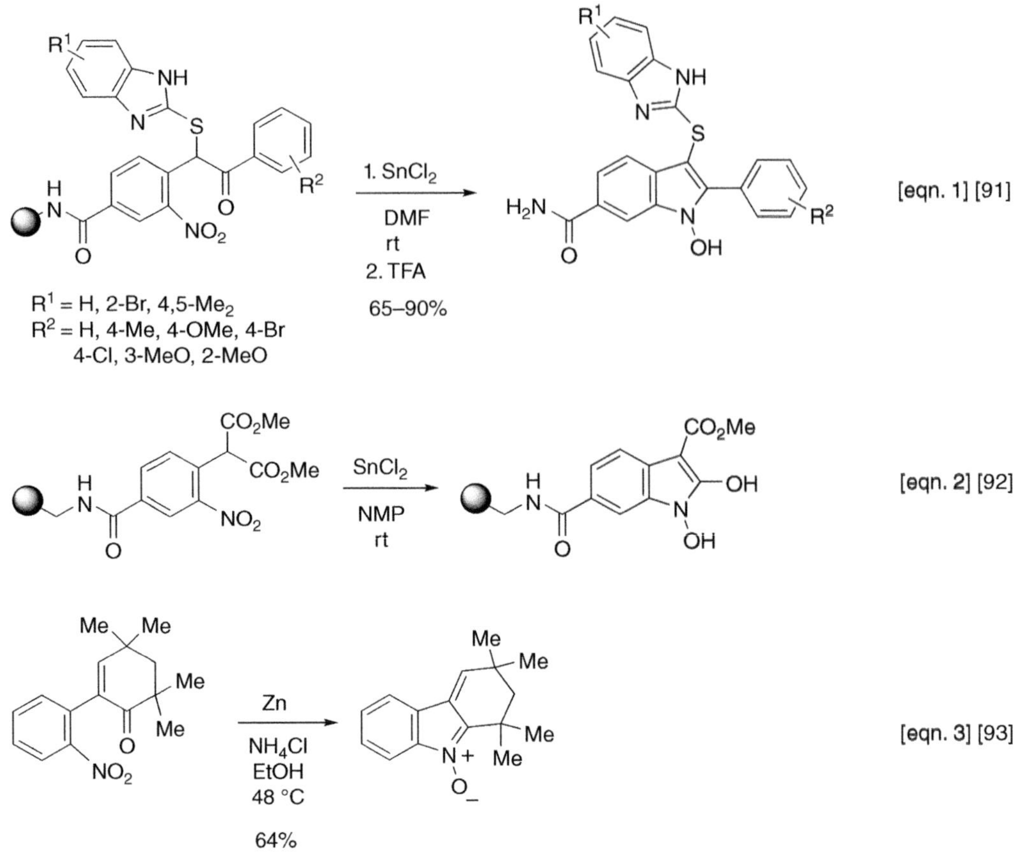

Scheme 15 *Solid Phase Reductive Cyclization Indole Synthesis and Myers 3-Alkylidene-3H-indole 1-Oxide Synthesis*

References

[1] A. Baeyer, *Liebigs Ann. Chem.*, 1866, **140**, 295–296.
[2] A. Baeyer, *Chem. Ber.*, 1879, **12**, 456–461.
[3] W.C. Sumpter and F.M. Miller (1954) Heterocyclic compounds with indole and carbazole systems, in *The Chemistry of Heterocyclic Compounds*, Chapter I (ed. A. Weissberger), Interscience Publishers, Inc., New York.
[4] A. von Baeyer and O.R. Jackson, *Chem. Ber.*, 1880, **13**, 187–189.
[5] O.R. Jackson, *Chem. Ber.*, 1881, **14**, 879–888.
[6] R.A. Weerman, *Recl. Trav. Chim Pay-Bas*, 1910, **29**, 18–21.
[7] A. Reissert and H. Heller, *Chem. Ber.*, 1904, **37**, 4364–4379.
[8] G. Heller, *Chem. Ber.*, 1817, **50**, 1202–1203.
[9] Sh.A. Samsoniya and M.V. Trapaidze, *Russ. Chem. Rev.*, 2007, **76**, 313–326.
[10] R.A. Weerman, *Liebigs Ann. Chem.*, 1914, **401**, 1–20.
[11] A. von Baeyer and A. Emmerling, *Chem. Ber.*, 1869, **2**, 679–682.
[12] F. Beilstein and A. Kuhlberg, *Liebigs Ann. Chem.*, 1872, **163**, 121–143.
[13] P. Ruggli, *Chem. Ber.*, 1917, **50**, 883–893.
[14] P. Ruggli and O. Schmid, *Helv. Chim. Acta*, 1935, **18**, 1215–1228.
[15] P.G. Tsoungas and A.I. Diplas, *Curr. Org. Chem.*, 2004, **8**, 1579–1606.
[16] E. Ucciani and A. Bonfand, *J.C.S. Chem. Comm.*, 1981, 82–83.
[17] G.A. Kraus and N. Selvakumar, *Synlett*, 1998, 845–846.
[18] K. Inoue, Y. Ishikawa, and S. Nishiyama, *Org. Lett.*, 2010, **12**, 436–439.
[19] S. Raucher and G.A. Koople, *J. Org. Chem.*, 1983, **48**, 2066–2069.
[20] K. Cardwell, B. Hewitt, M. Ladlow, and P. Magnus, *J. Am. Chem. Soc.*, 1988, **110**, 2242–2248.
[21] P.E. Maligres, G.R. Humphrey, J.-F. Marcoux, *et al.*, *Org. Proc. Res. Dev.*, 2009, **13**, 525–534.
[22] M.S. Islam, C. Brennan, Q. Wang, and M.M. Hossain, *J. Org. Chem.*, 2006, **71**, 4675–4677.
[23] M.G. Kulkarni, S.I. Davawala, A.P. Dhondge, *et al.*, *Tetrahedron Lett.*, 2006, **47**, 1003–1005.

[24] M.G. Kulkarni, S.W. Chavhan, M.P. Desai, *et al.*, *Tetrahedron Lett.*, 2010, **51**, 4494–4496.
[25] L. Venemalm, C. Estéves, M. Alvarez, and J.A. Joule, *Tetrahedron Lett.*, 1993, **34**, 5495–5496.
[26] D. Roberts, L. Venemalm, M. Alvarez, and J.A. Joule, *Tetrahedron Lett.*, 1994, **35**, 7857–7860.
[27] C. Estévez, L. Venemalm, M. Alvarez, and J.A. Joule, *Tetrahedron*, 1994, **50**, 7879–7888.
[28] D. Roberts, M. Alvarez, and J.A. Joule, *Tetrahedron Lett.*, 1996, **37**, 1509–1512.
[29] D. Roberts and J.A. Joule, *J. Org. Chem.*, 1997, **62**, 568–577.
[30] M. Alvarez, M.A. Bros, and J.A. Joule, *Tetrahedron Lett.*, 1998, **39**, 679–680.
[31] A.P. Kozikowski, M.N. Greco, and J.P. Springer, *J. Am. Chem. Soc.*, 1982, **104**, 7622–7626.
[32] T. Sakamoto, Y. Kondo, and H. Yamanaka, *Heterocycles*, 1984, **22**, 1347–1350.
[33] T. Sakamoto, Y. Kondo, and H. Yamanaka, *Chem. Pharm. Bull.*, 1986, **34**, 2362–2368.
[34] T.V. RajanBabu, B.L. Chenard, and M.A. Petti, *J. Org. Chem.*, 1986, **51**, 1704–1712.
[35] A.N. Tischler and T.J. Lanza, *Tetrahedron Lett.*, 1986, **27**, 1653–1656.
[36] T. Izumi, M. Soutome, and T. Miura, *J. Heterocycl. Chem.*, 1992, **29**, 1625–1629.
[37] J.W. Coe, M.G. Vetelino, and M.J. Bradlee, *Tetrahedron Lett.*, 1996, **37**, 6045–6048.
[38] S. Boini, K.P. Moder, R.K. Vaid, *et al.*, *Org. Proc. Res. Dev.*, 2006, **10**, 1205–1211.
[39] M.H. Todd, S.F. Oliver, and C. Abell, *Org. Lett.*, 1999, **1**, 1149–1151.
[40] S. Chackal, F. Dudouit, R. Houssin, and J.-P. Hénichart, *Heterocycles*, 2003, **60**, 615–622.
[41] W. Löwe, S. Witzel, S. Tappmeyer, and R. Albuschat, *J. Heterocycl. Chem.*, 2004, **41**, 317–326.
[42] J.C. Estévez, R.J. Estévez, and L. Castedo, *Tetrahedron Lett.*, 1993, **34**, 6479–6480.
[43] F.J. Reboredo, M. Treus, J.C. Estévez, *et al.*, *Synlett*, 2002, 999–1001.
[44] J. Cruces, E. Martínez, M. Treus, *et al.*, *Tetrahedron*, 2002, **58**, 3015–3019.
[45] J.C. Barcia, J. Cruces, J.C. Estévez, *et al.*, *Tetrahedron Lett.*, 2002, **43**, 5141–5144.
[46] C.O. Salas, F.J. Reboredo, J.C. Estévez, *et al.*, *Synlett*, 2009, 3107–3110.
[47] M. Fernández, *Synthesis*, 2009, 3051–3060.
[48] M.E. Kuehne, *J. Am. Chem. Soc.*, 1962, **84**, 837–847.
[49] R.L. Augustine, A.J. Gustavsen, S.F. Wanat, *et al.*, *J. Org. Chem.*, 1973, **38**, 3004–3011.
[50] E.E. Garcia and R.I. Fryer, *J. Heterocycl. Chem.*, 1974, **11**, 219–221.
[51] J. Bakke, *Acta Chem. Scand. B*, 1974, **28**, 134–135.
[52] C.J. Moody and K.F. Rahimtoola, *J. Chem. Soc., Perkin. Trans.* **1**, 1990, 673–679.
[53] C.J. Moody and A.I. Morrell, *J. Indian Chem. Soc.*, 1994, **71**, 309–314.
[54] M. Arcari, R. Aveta, A. Brandt, *et al.*, *Gazz. Chim. Ital.*, 1991, **121**, 499–504.
[55] P. Kaszynski and D.A. Dougherty, *J. Org. Chem.*, 1993, **58**, 5209–5220.
[56) M. Taga, H. Ohtsuka, I. Inoue, *et al.*, *Heterocycles*, 1996, **42**, 251–263.
[57] T. Grandi, F. Sparatore, and A. Sparatore, *Il Farmaco*, 1999, **54**, 479–485.
[58] J. Cruces, J.C. Estévez, R.J. Estévez, and L. Castedo, *Heterocycles*, 2000, **53**, 1041–1050.
[59] R.P. Santos, R.S.C. Lopes, and C.C. Lopes, *Synthesis*, 2001, 845–848.
[60] R.A. Bunce, M.H. Randall, and K.G. Applegate, *Org. Prep. Proc. Int.*, 2002, **34**, 483–547.
[61] T.L. Ho and D.G. Jou, *Helv. Chim. Acta*, 2002, **85**, 3823–3827.
[62] N. Selvakumar, A. Malar Azhagan, D. Srinivas, and G. Gopi Krishna, *Tetrahedron Lett.*, 2002, **43**, 9175–9178.
[63] J.J. Conde, M. McGuire, and M. Wallace, *Tetrahedron Lett.*, 2003, **44**, 3081–3084.
[64] D. Hamprecht, F. Micheli, G. Tedesco, *et al.*, *Bioorg. Med. Chem. Lett.*, 2007, **17**, 424–427.
[65) M.I. Attia, D. Güclü, B. Hertlein, *et al.*, *Org. Biomol. Chem.*, 2007, **5**, 2129–2137.
[66] T.L. Scott, N. Burke, G. Carrero-Martínez, and B.C.G. Söderberg, *Tetrahedron*, 2007, **63**, 1183–1190.
[67) T.J. Snape, *Synlett*, 2008, 2689–2691.
[68] C.L. Martin, L.E. Overman, and J.M. Rohde, *J. Am. Chem. Soc.*, 2008, **130**, 7568–7569.
[69] A. Scribner, J.A. Moore III, G. Ouvry, *et al.*, *Bioorg. Med. Chem. Lett.*, 2009, **19**, 1517–1521.
[70] R.A. Bunce and B. Nammalwar, *J. Heterocycl. Chem.*, 2009, **46**, 172–177.
[71) T. Yamakawa, E. Ideue, Y. Iwaki, *et al.*, *Tetrahedron*, 2011, **67**, 6547–6560.
[72) A.D. Alorati, A.D. Gibb, P.R. Mullens, and G.W. Stewart, *Org. Proc. Res. Dev.*, 2012, **16**, 1947–1952.
[73] J.T. Kuethe, A. Wong, C. Qu, *et al.*, *J. Org. Chem.*, 2005, **70**, 2555–2567.
[74) A. Wong, J.T. Kuethe, I.W. Davies, and D.L. Hughes, *J. Org. Chem.*, 2004, **69**, 7761–7764.
[75] T.E. Young, *J. Org. Chem.*, 1962, **27**, 507–510.
[76] J.L. Rutherford, M.P. Rainka, and S.L. Buchwald, *J. Am. Chem. Soc.*, 2002, **124**, 15168–15169.
[77] F. Gallou, N. Yee, F. Qiu, *et al.*, *Synlett*, 2004, 883–885.
[78] D. Janreddy, V. Kavala, J.W.J. Bosco, *et al.*, *Eur. J. Org. Chem.*, 2011, 2360–2365.
[79] P.T. Parvatkar, P.S. Parameswaran, and S.G. Tilve, *Tetrahedron Lett.*, 2007, **48**, 7870–7872.
[80] T. Ohshima, Y. Xu, R. Takita, and M. Shibasaki, *Tetrahedron*, 2004, **60**, 9569–9588.
[81] Z. Xu, Q. Wang, and J. Zhu, *Angew. Chem. Int. Ed.*, 2013, **52**, 3272–3276.
[82] K. Kobayashi, D. Nakai, S. Fukamachi, and H. Konishi, *Heterocycles*, 2009, **78**, 2769–2776.
[83] N.T. Tam, E.-J. Jung, and C.-G. Cho, *Org. Lett.*, 2010, **12**, 2012–2014.
[84] J. Bonjoch, D. Solé, S. García-Rubio, and J. Bosch, *J. Am. Chem. Soc.*, 1997, **119**, 7230–7240.
[85] R.M. Acheson, P.G. Hunt, D.M. Littlewood, *et al.*, *J. Chem. Soc., Perkin Trans. 1*, 1978, 1117–1125.

[86] A. Wong, J.T. Kuethe, and I.W. Davies, *J. Org. Chem.*, 2003, **68**, 9865–9866.
[87] R. Bujok, Z. Wróbel, and K. Wojciechowski, *Synlett*, 2012, **23**, 1315–1320.
[88] M. Brzozowski, N.J. O'Brien, D.J. Wilson, and B.M. Abbott, *Tetrahedron*, 2014, **70**, 318–326.
[89] K.C. Nicolaou, S.H. Lee, A.A. Estrada, and M. Zak, *Angew. Chem. Int. Ed.*, 2005, **44**, 3736–3740.
[90] M. Belley, D. Beaudoin, and G. St-Pierre, *Synlett*, 2007, 2999–3002.
[91] A.D. Roy, S. Sharma, R.K. Grover, *et al.*, *Org. Lett.*, 2004, **6**, 4763–4766.
[92] Z. Wu and N.J. Ede, *Org. Lett.*, 2003, **5**, 2935–2938.
[93] A.G. Myers and S.B. Herzon, *J. Am. Chem. Soc.*, 2003, **125**, 12080–12081.

PART V

Oxidative Cyclization

In contrast to the voluminous reductive cyclization routes to indoles that we have seen in the previous chapters, the number of oxidative cyclization methods pales by comparison. Furthermore, several metal-catalyzed oxidation indole syntheses, which are discussed in later chapters, could well have been included here.

Indole Ring Synthesis: From Natural Products to Drug Discovery, First Edition. Gordon W. Gribble.

46

Watanabe Indole Synthesis

Following their discovery of the ruthenium-catalyzed *N*-alkylation of anilines with alcohols to give secondary and tertiary amines (Scheme 1, equation 1) [1], Watanabe and his colleagues adapted their chemistry to a synthesis of indoles as shown in equations 2 and 3 [2–5]. The reaction of *N*-methylaniline with propylene glycol under typical conditions affords 1,2-dimethylindole and 1,3-dimethylindole in a 1:1 ratio (50% yield), whereas aniline plus styrene glycol gives only 2-phenylindole (43% yield) [4]. The best yield was 89% for the preparation of 5-chloro-2,3-dimethylindole (equation 2).

Although Zamyshlyaeva and her colleagues in 1970 described a low yield of indole (6%–7%) by the catalytic cyclization of *N*-(β-hydroxyethyl)aniline with ThO_2 and Al_2O_3 at 300–380 °C [6, 7], it was Watanabe who elevated this reaction to a bona fide indole synthesis (Scheme 2, equations 1-4) [3, 5]. Indoles are also obtained from 2-nitrophenethyl alcohols using this ruthenium catalyst (equation 4). A mechanism proposed by Watanabe involves hydroxyl coordinating to the Ru center, oxidation to the corresponding aldehyde (loss of H_2), and cyclodehydration to indole [5]. For a detailed mechanistic discussion see Watanabe *et al* [1].

Although applications of the Watanabe indole synthesis are sparse, many investigators have sought to improve upon the original Watanabe catalyst. A review discusses nitrogen heterocycle syntheses involving catalytic hydrogen transfer reactions [8]. A listing of these catalytic systems is shown in Table 1 [9–12, 16–21] in approximately chronological order. The major product in Entry 3 is aniline [11]. In addition to the ruthenium–tin catalytic system used by Shim and colleagues with triethanolamine (Entry 4) [12], this group finds that ethylene glycol with this catalyst is equally suitable for preparing indoles, albeit in lower yields [13], and that dioxane–water (9:1) is a suitable solvent for indole ring formation from anilines reacting with triethanolamine (to give indoles) and triisopropanol (to give 2-methylindoles) [14, 15]. Guo's group found that Raney nickel is an excellent catalyst for obtaining indole on a multigram scale (Entry 7) [18]. The overall yield of indole from *o*-nitrotoluene is 78%. Raney nickel was also used to reduce 2-(2-nitrophenyl)ethanol on a large scale (97%). Guo's synthesis would seem to provide an excellent industrial route to indole. A copper-catalyst system has been developed by Muldoon that affords indole in high yield (90%) along with substituted indoles (Entry 10) [21]. In addition to the examples in Table 1, the research teams of Vaccari (ZrO_2/SiO_2) [22], Xing (CdS) [23], and Shi (Ag/SiO_2-ZnO and Cu/SiO_2-Al_2O_3) [24, 25] describe several catalytic vapor-phase syntheses of indoles from aniline and glycols. Based on a careful study of the reaction products formed between aniline and ethylene glycol at 340–350 °C in a continuous flow reactor over CdS, Xing and Liu propose the mechanism depicted in Scheme 3 [23]. The gas chromatographic analysis of the products is shown. Ethylene glycol is rapidly converted to glycolaldehyde at these temperatures.

In one of the first significant modifications of the Watanabe ruthenium catalyst for indolization, Zhang, Ding, and coworkers used $[Ru(CO)_2Xantphos]_2$ (**1**) to synthesize indoles from anilines and vicinal symmetrical and unsymmetrical diols both (Scheme 4) [26]. Madsen and his colleagues employed a $RuCl_3$/phosphine catalyst to prepare 2,3-disubstituted indoles from anilines and 1,2-diols, and they also found that a $[Cp^*IrCl_2]_2$/MsOH catalyst is equally effective in this reaction [27]. Iridium-based

Indole Ring Synthesis: From Natural Products to Drug Discovery, First Edition. Gordon W. Gribble.

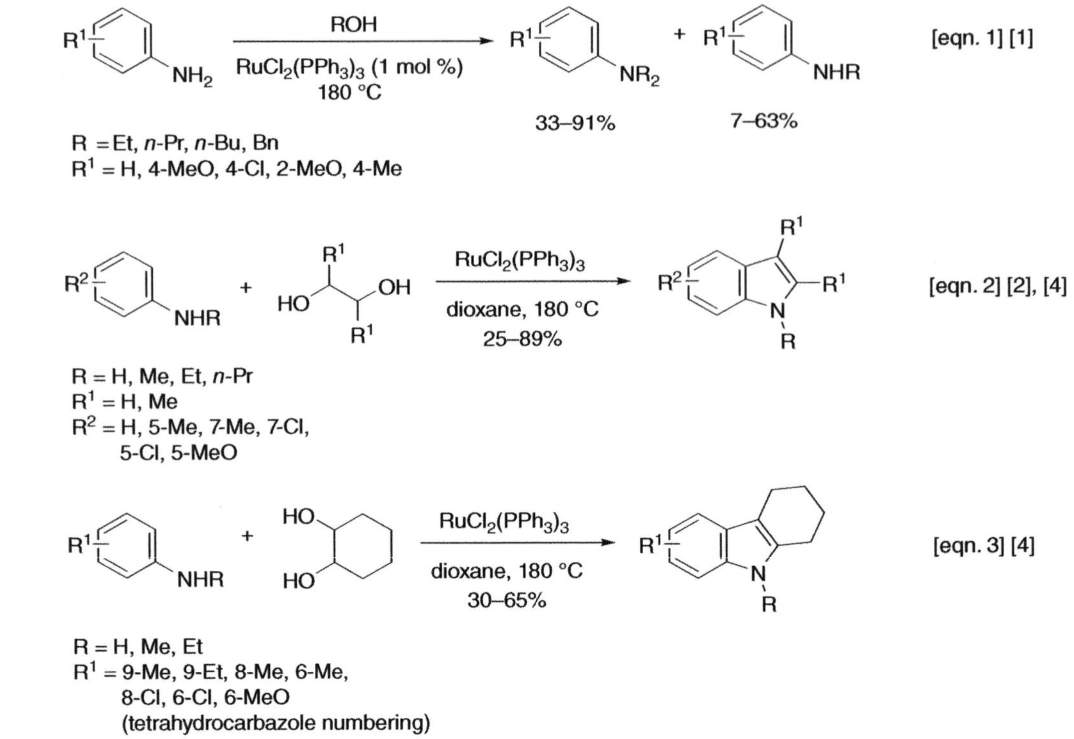

Scheme 1 *The Watanabe Indole Synthesis*

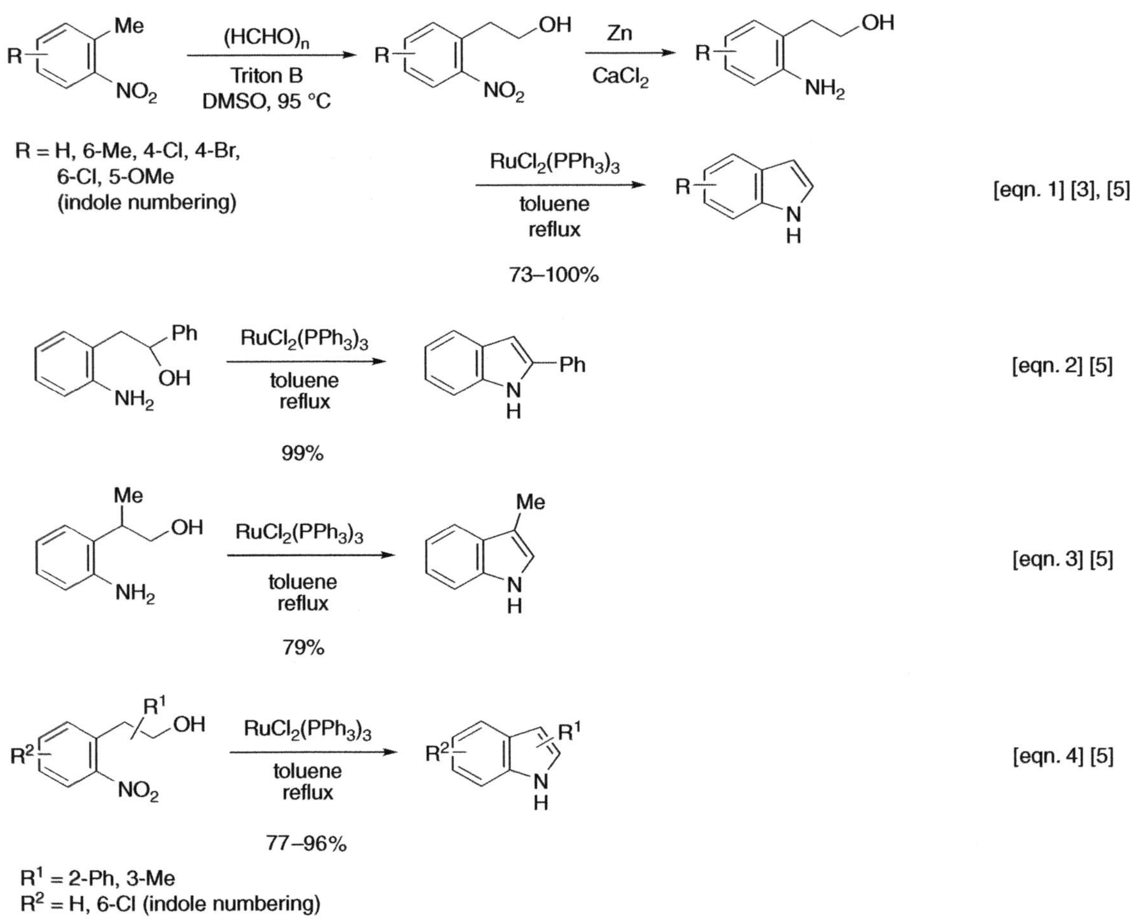

Scheme 2 *Watanabe N-Heterocyclization Indole Synthesis*

Table 1 *New Catalytic Systems Based on the Watanabe Indole Synthesis*

Entry	Substrate		Conditions	Indole	% Yield	Ref.
1	$PhNH_2$	$HOCH_2CH_2OH$	PbI_2–BaI_2 290 °C	indole	76%	9
2	$PhNH_2$	$HOCH_2CH_2OH$	$CaBr_2$–$_3KBr$ 330 °C	indole	76%	10
3	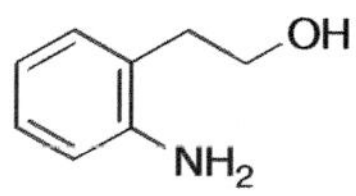	—	$AlPO_4$, 750 °C	indole	2%	11
4	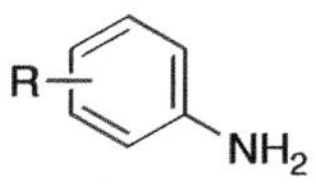 R = H, 7-Me, 5-Me, 4/6-Me 5-MeO, 7-Cl, 5-Cl, 5-*i*-Pr, 6,7-Me_2, 4,6-Me_2, 4,7-Me_2 (indole numbering)	$N(CH_2CH_2OH)_3$	$RuCl_3$, Ph_3P $SnCl_2$•$2H_2O$ dioxane, 180 °C	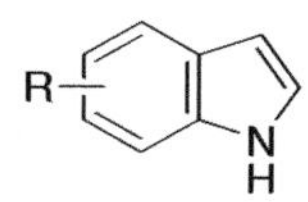	9–99%	12
5	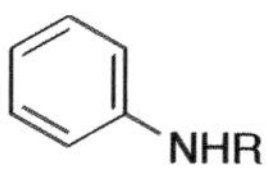 R = Me, Et, *n*-Pr, *n*-Bu, Ph, Bn	$N(CH_2CH_2OH)_3$	$RuCl_2(Ph_3P)_3$ dioxane, 180 °C	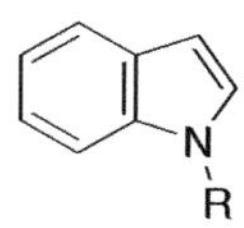	63–78%	16
6	$PhNH_2$	$HOCH_2CH_2OH$	Cu/SiO_2•MgO 325 °C	indole	94%	17
7	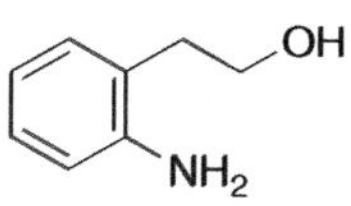	—	Raney Ni 220 °C paraffin	indole	88%	18
8	$PhNH_2$	$HOCH_2CH(OH)CH_2OH$	Cu/SiO_2–Al_2O_3 240 °C	3-methylindole	40%	19
9	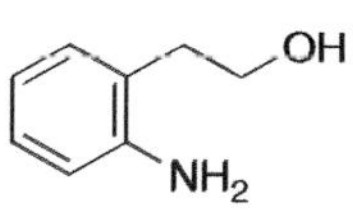	—	Ru/CeO_2 mesitylene 140 °C	indole	99%	20
10	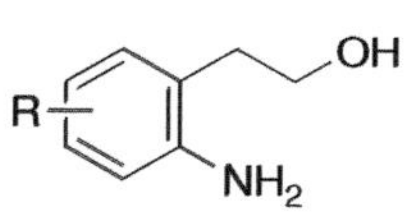 R = H, 4-Cl, 5-OMe, 6-F (indole numbering)	—	$Cu(OTf)_2$ 2,2'-Bispyr TEMPO, DBU MeCN, 60 °C *N*-methylimidazole	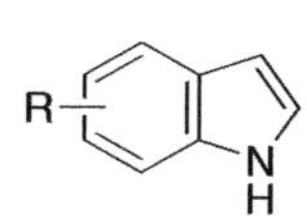	49–90%	21

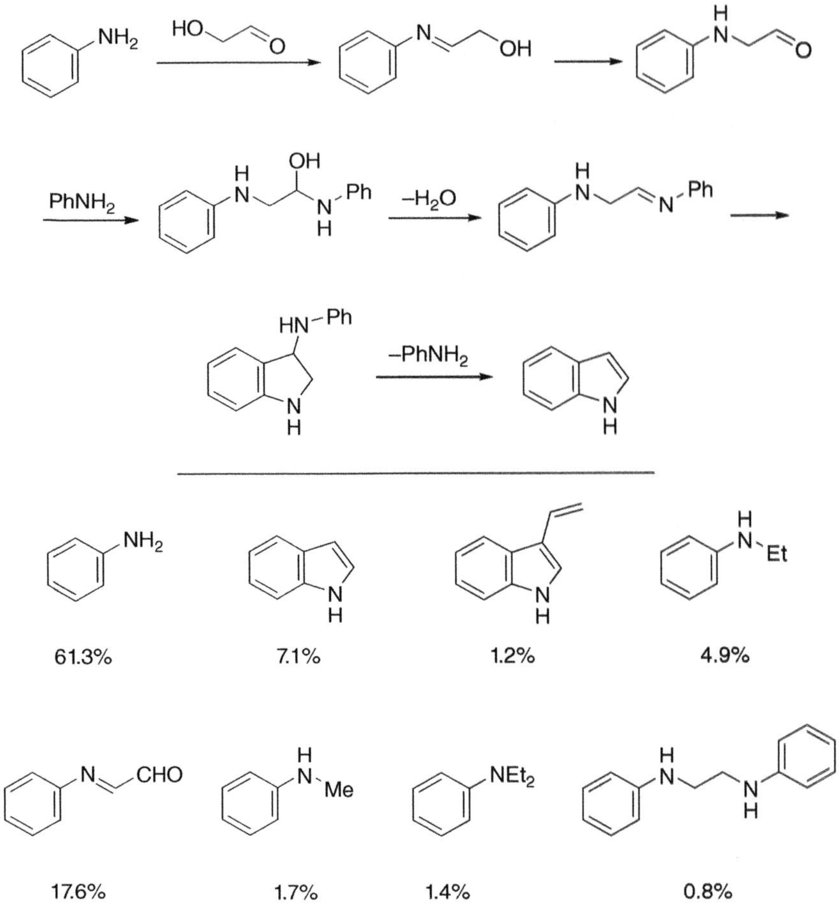

Scheme 3 *Xing and Liu's Reaction between Aniline and Ethylene Glycol*

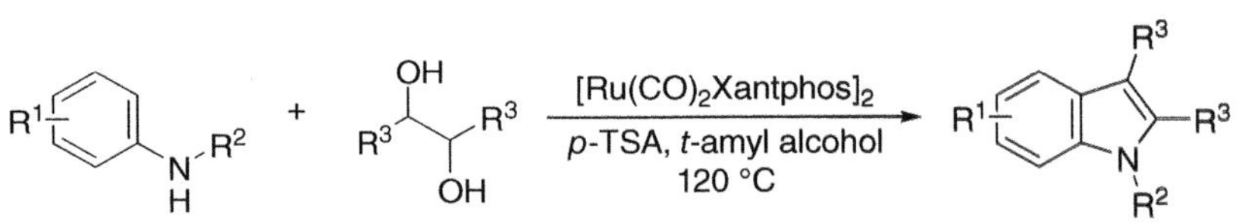

[eqn. 1] [26]

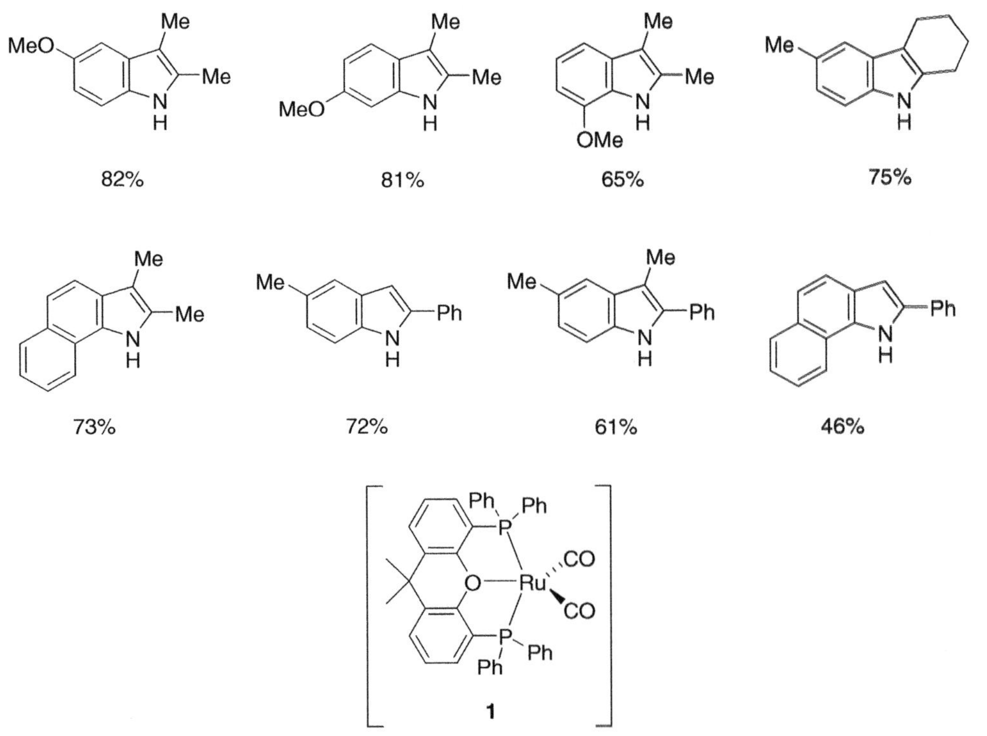

Scheme 4 *Zhang and Ding Indole Synthesis*

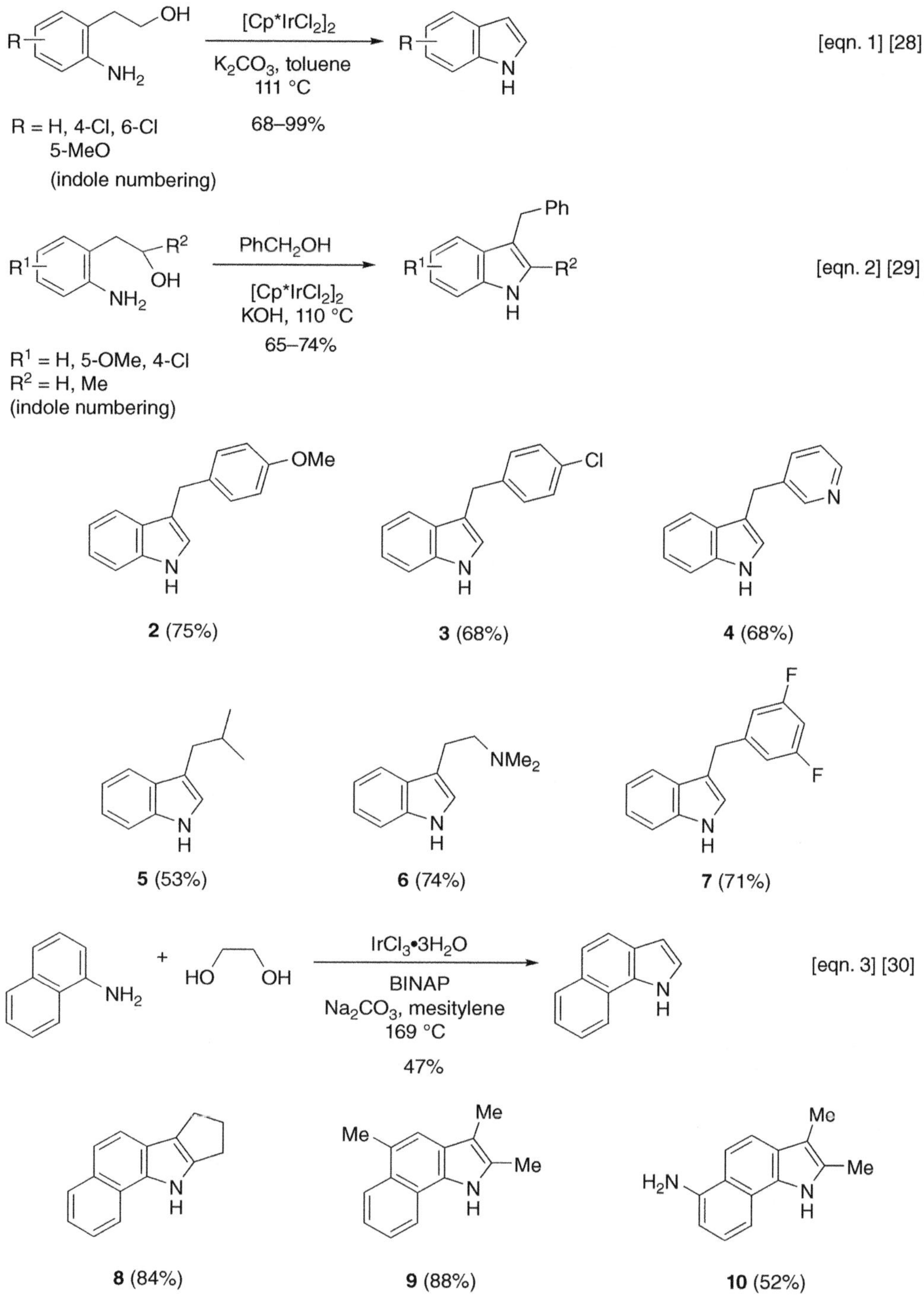

Scheme 5 *Iridium-Based Watanabe Indole Synthesis*

catalysts were used by Fujita and Yamaguchi [28], Grigg [29], and Ishii [30] to synthesize indoles by a Watanabe reaction, of which a selection is shown in Scheme 5. Both 2- and 3-methylindole were also synthesized in the reaction shown in equation 1. Indole is formed from 2-nitrophenethyl alcohol (69%) with this catalyst [28]. The Grigg protocol can be extended to substituted alcohols so as to give 3-substituted indoles **2–6**. Likewise, Ishii extended his methodology to the synthesis of a range of indoles derived from 1-naphthylamines (e.g., **8–10**) [30].

In addition to the above diverse and imaginative variations on the original Watanabe reaction conditions, Table 2

Table 2 *Extensions of the Watanabe Indole Synthesis*

Entry	Substrate	Conditions	Indole	% Yield	Ref.
1	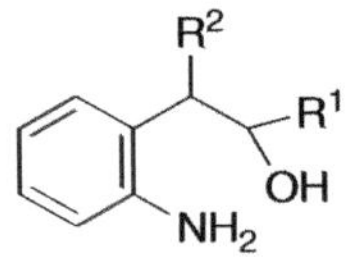 R^1 = H, Me R^2 = Et, Ph, *i*-Pr, *n*-Pr, cyclohexyl, phenethyl	$Pd(PPh_3)_4$, K_2CO_3 mesityl bromide DMF, 150 °C	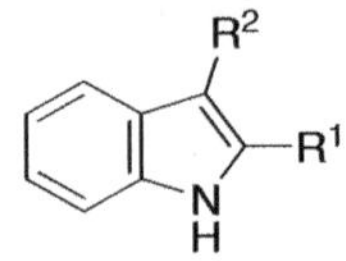	62–98%	31, 32
2	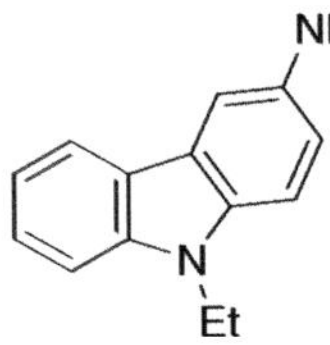	$HOCH_2CH_2OH$, $RuCl_3$ $SnCl_2{\bullet}2H_2O$, toluene 120 °C dppe	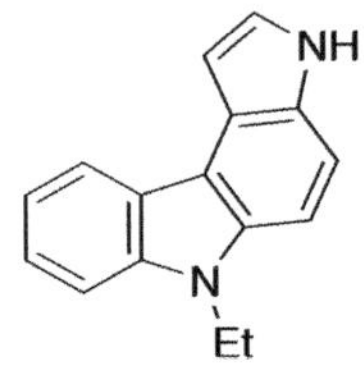	73%	33
3	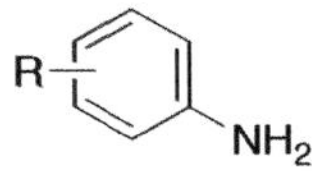 R = H, 4-Me, 3-Me, 4-OMe, 4-Cl, 4-*s*-Bu, 3,5-Me_2, 2,5-Me_2, 2,3-Me_2, 2,5-$(MeO)_2$ (aniline numbering)	$BrCH_2CH_2Br$ $RuCl_3$, Ph_3P dioxane, 180 °C	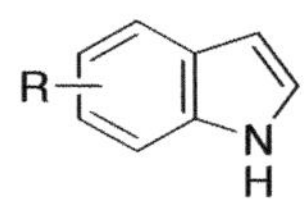	20–96%	34
4	 R^1 = H, Me, Et, Bn, Ph R^2 = H, Me	$Al_2O_3/SiO_2/Na_2O$ 250–400 °C	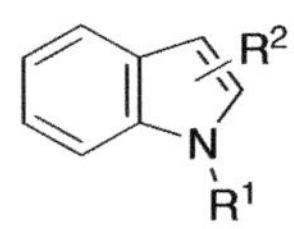	21–43%	35
5	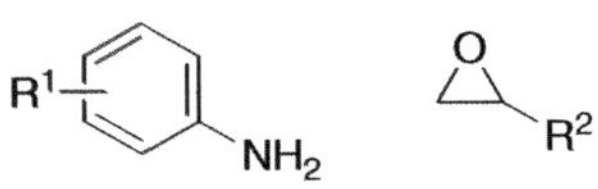 R^1 = H, 4-Me, 3-Me, 2-Me, 4-OMe, 2-OMe, 4-Cl, 4-Bu 4-*s*-Bu, 2,5-Me_2, 3,5-Me_2 (aniline numbering) R^2 = Me, *n*-Bu, Bn, Ph	$RuCl_3/SnCl_2$ PPh_3, dioxane 180 °C	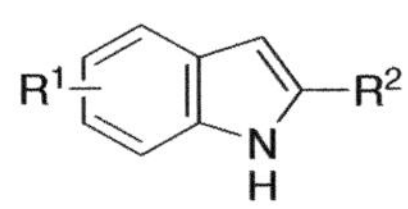	22–98%	36
6	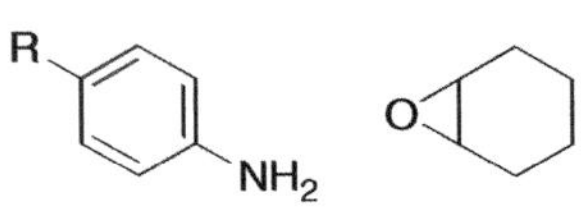	$[Ru_3(CO)_{12}]$ dppf *p*-TSA, dioxane 150 °C	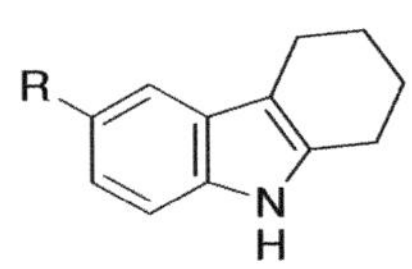	43–85%	37
7	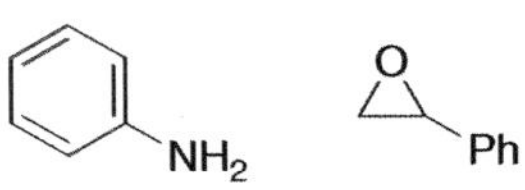	1. $Zn(OTf)_2$, rt 2. $[Ru_3(CO)_{12}]$ dppf *p*-TSA, dioxane, 150 °C	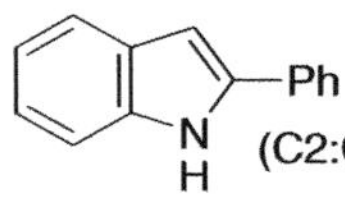 (C2:C3) 99:1	77%	37
8	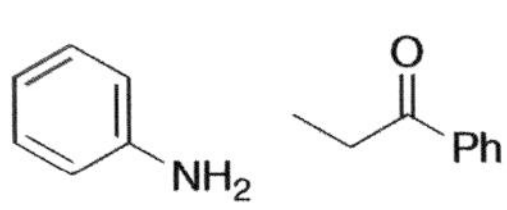	360–380 °C K-16, benzene	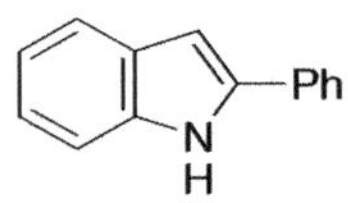	83%	38

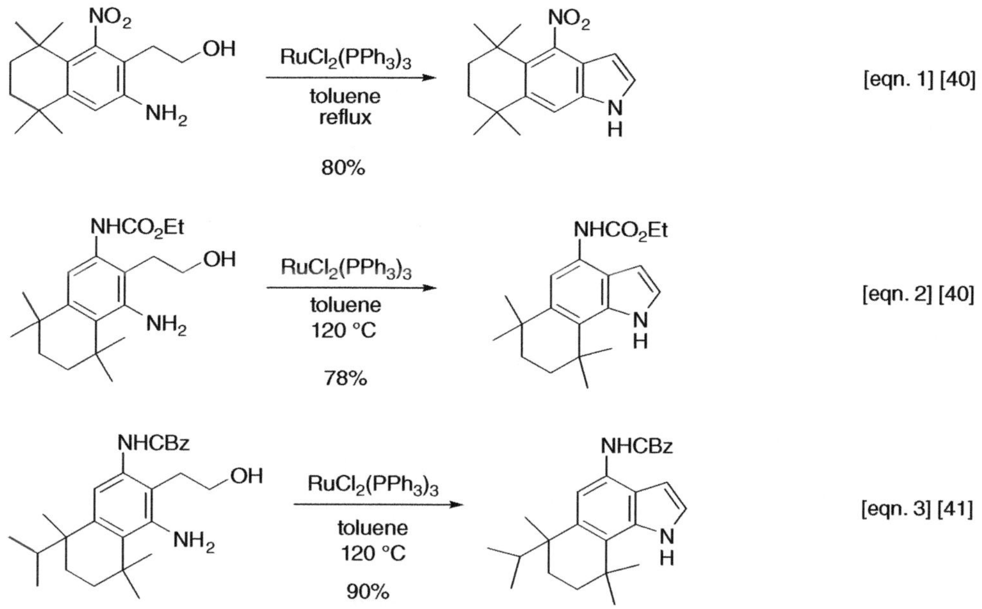

Scheme 6 *Synthesis of Teleocidin Analogues using the Watanabe Indole Synthesis*

tabulates further extensions of what has become a powerful indole synthesis. Ohta and Takeya and their colleagues use a palladium catalyst to effect indolization of a series of amino alcohols (Entry 1) [31, 32]. For example, 1,2,3,4-tetrahydrocarbazole is formed in 92% yield [32]. Nagarajan and Viji adapted the $RuCl_3/SnCl_2$ conditions of Shim [12, 13] to prepare a series of pyrrolo[2,3-*c*]carbazoles (Entry 2) and, subsequently, indolo[2,3-*c*]carbazoles from the former [33]. Cho and Shim's team find that ethylene dibromide can replace ethylene glycol in the Watanabe indolization (Entry 3) [34]. Interestingly, both 1,2-dibromopropane and 1,2-dibromohexane under the same conditions give only 4,7-dimethylindole with 2,5-dimethylaniline (98% and 96%, respectively), reflecting C–C bond cleavage during the reaction. Several groups report that epoxides undergo the Watanabe indolization (entries 4-6) [35–37]. Nishida's vapor phase conditions (Entry 4) were also applied to several 1,2-diols [35]. The typical conditions of Cho and Shim's group with epoxides display excellent regioselectivity in giving 2-substituted indoles (Entry 5) [36]. Beller and coworkers use $[Ru_3(CO)_{12}]$ to prepare indoles from anilines and epoxides (Entry 6) [37]. Moreover, this group employs a two-step sequence to access 2-substituted indoles (Entry 7). This latter protocol provides a series of 2-arylindoles with high regioselectivity (C2>C3) in good yields. 1-Phenylpropylene oxide affords 3-methyl-2-phenylindoles in 46% to 76% yield and 91:9-99:1 C2 regioselectivity. In a not-unrelated reaction to the epoxide substrates, propiophenone was used by Prostakov and colleagues in a 2-phenylindole synthesis (Entry 8) [38]. This interesting product results from methyl cleavage of the initially formed 3-methyl-2-phenylindole.

The Watanabe indole synthesis has been applied to the synthesis of indole natural products. A preparation of L-4-chlorotryptophan, which is found in seeds of the plant *Pisium sativum*, entailed the preparation of 4-chloroindole from 2-amino-5-chlorophenethyl alcohol and $RuCl_2(PPh_3)_3$ in refluxing toluene (80%) [39]. Two groups employed the Watanabe methodology to the syntheses of teleocidin analogues Scheme 6, equations 1 and 2) [40, 41]. Although without experimental details, Hendry's team reported the synthesis of 5- and 6-substituted [2-^{14}C] indoles using [^{14}C] paraformaldehyde to introduce the radioactive label into the substituted *o*-nitrotoluene, followed by a standard catalytic hydrogenation and ruthenium-catalyzed cyclization [42].

What began as almost a curiosity, the Watanabe indolization has developed into a remarkably useful and general indole synthesis from *o*-nitrotoluene substrates.

References

[1] Y. Watanabe, Y. Tsuji, H. Ige, *et al.*, *J. Org. Chem.*, 1984, **49**, 3359–3363.
[2] Y. Tsuji, K. Huh, and Y. Watanabe, *Tetrahedron Lett.*, 1986, **27**, 377–380.
[3] Y. Tsuji, K.-T. Huh, Y. Yokoyama, and Y. Watanabe, *J. Chem. Soc., Chem. Commun.*, 1986, 1575–1576.
[4] Y. Tsuji, K.-T. Huh, and Y. Watanabe, *J. Org. Chem.*, 1987, **52**, 1673–1680.

[5] Y. Tsuji, S. Kotachi, K. Huh, and Y. Watanabe, *J. Org. Chem.*, 1990, **55**, 580–584.
[6] L.I. Zamyshlyaeva, N.N. Mitina, and N.N. Suvorov, *Chem. Heterocycl. Compd.*, 1970, **6**, 707–709.
[7] N.N. Suvorov, N.N. Mitina, and L.I. Zamyshlyaeva, *Chem. Heterocycl. Compd.*, 1970, **6**, 703–706.
[8] R. Yamaguchi, K. Fujita, and M. Zhu, *Heterocycles*, 2010, **81**, 1093–1140.
[9] T. Seto and M. Imanari, *Bull. Chem. Soc. Jpn.*, 1994, **67**, 3139–3141.
[10] T. Seto, K. Kujira, H. Iwane, and M. Imanari, *Bull. Chem. Soc. Jpn.*, 1995, **68**, 3665–3670.
[11] J. Afxantidis, N. Bouchry, and J.-P. Aune, *J. Mol. Catal. A. Chem.*, 1995, **102**, 49–58.
[12] C.S. Cho, H.K. Lim, S.C. Shim, *et al.*, *Chem. Commun.*, 1998, 995–996.
[13] C.S. Cho, M.J. Lee, S.C. Shim, and M.C. Kim, *Bull. Korean Chem. Soc.*, 1999, **20**, 119–121.
[14] C.S. Cho, J.H. Kim, and S.C. Shim, *Tetrahedron Lett.*, 2000, **41**, 1811–1814.
[15] C.S. Cho, J.H. Kim, T.-J. Kim, and S.C. Shim, *Tetrahedron*, 2001, **57**, 3321–3329.
[16] S.C. Shim, Y.Z. Youn, D.Y. Lee, *et al.*, *Synth. Commun.*, 1996, **26**, 1349–1353.
[17] L. Shi, X. Huo, S. Ren, and J. Xu, *Chin. J. Catal.*, 2005, **26**, 449–450.
[18] X. Guo, Z. Peng, S. Jiang, and J. Shen, *Synth. Commun.*, 2011, **41**, 2044–2052.
[19] W. Sun, D.-Y. Liu, H.-Y. Zhu, *et al.*, *Catal. Commun.*, 2010, **12**, 147–150.
[20] S. Shimura, H. Miura, K. Wada, *et al.*, *Catal. Sci. Technol.*, 2011, **1**, 1340–1346.
[21] J.C.A. Flanagan, L.M. Dornan, M.G. McLaughlin, *et al.*, *Green Chem.*, 2012, **14**, 1281–1283.
[22] M. Campanati, S. Franceschini, O. Piccolo, and A. Vaccari, *J. Catal.*, 2005, **232**, 1–9.
[23] J. Xing and Z. Liu, *Adv. Mater. Res.*, 2011, **287–290**, 120–123.
[24] D. Liu, X. Huo, J. Liu, *et al.*, *Reac. Kinet. Mech. Cat.*, 2010, **99**, 335–343.
[25] Z. Wang, X. Li, Y. Zhang, *et al.*, *Chin. J. Catal.*, 2012, **33**: 1139–1145.
[26] M. Zhang, F. Xie, X. Wang, *et al.*, *RSC Adv.*, 2013, **3**, 6022–6029.
[27] M. Tursky, L.L.R. Lorentz-Petersen, L.B. Olsen, and R. Madsen, *Org. Biomol. Chem.*, 2010, **8**, 5576–5582.
[28] K.-I. Fujita, K. Yamamoto, and R. Yamaguchi, *Org. Lett.*, 2002, **4**, 2691–2694.
[29] S. Whitney, R. Grigg, A. Derrick, and A. Keep, *Org. Lett.*, 2007, **9**, 3299–3302.
[30] H. Aramoto, Y. Obora, and Y. Ishii, *J. Org. Chem.*, 2009, **74**, 628–633.
[31] Y. Aoyagi, T. Mizusaki, and A. Ohta, *Tetrahedron Lett.*, 1996, **37**, 9203–9206.
[32] Y. Aoyagi, M. Shishikura, T. Mizusaki, *et al.*, *Heterocycles*, 2008, **75**, 1055–1059.
[33] M. Viji and R. Nagarajan, *Tetrahedron*, 2012, **68**, 2453–2458.
[34] C.S. Cho, D.C. Park, and S.C. Shim, *Bull, Korean Chem. Soc.*, 2007, **28**, 832–834.
[35] T. Nishida, Y. Tokuda, and M. Tsuchiya, *J. Chem. Soc., Perkin Trans. 2*, 1995, 823–830.
[36] C.S. Cho, J.H. Kim, H.-J. Choi, *et al.*, *Tetrahedron Lett.*, 2003, **44**, 2975–2977.
[37] M. Peña-López, H. Neumann, and M. Beller, *Chem. Eur. J.*, 2014, **20**, 1818–1824.
[38] N.S. Prostakov, B.N. Anisimov, A.A. Obynochnyi, and A.V. Varlamov, *J. Org. Chem., USSR*, 1991, **27**, 541–544.
[39] Y. Sakagami, K. Manabe, T. Aitani, *et al.*, *Tetrahedron Lett.*, 1993, **34**, 1057–1060.
[40] R.B. Webb, II, M.C. Venuti, and C. Eigenbrot, *J. Org. Chem.*, 1991, **56**, 4706–4713.
[41] J. Pu, K. Deng, J. Butera, *et al.*, *Tetrahedron*, 2010, **66**, 1963–1972.
[42] D. Hendry, N.S. Nixon, B.S. Roughley, *et al.* (2004) The synthesis of carbon-14 labelled indoles, in *Synthesis and Applications of Isotopically Labelled Compounds*, vol. **8** (eds D.C. Dean, C.N. Filer, and KE. McCarthy), John Wiley & Sons, Ltd, New York, pp. 25–28.

47

Knölker Carbazole Synthesis

An exceedingly versatile oxidative cyclization route to carbazoles was described by Knölker beginning in 1989 with a synthesis of carbazomycin A [1]. This excellent chemistry and myriad applications to naturally occurring carbazoles is discussed in several reviews [2–11]. The overall route is illustrated in Scheme 1 for a synthesis of carbazomycin A (**1**) [12]. The requisite tricarbonyl(η^5-cyclohexadienylium)iron cation **2** is prepared from 1,3-cyclohexadiene and $Fe(CO)_5$ followed by treatment with triphenylmethyl carbocation tetrafluoroborate as shown. Subsequent oxidative cyclization of iron complex **3** with commercial manganese dioxide affords the iron-complexed dihydrocarbazol-3-one **4**.

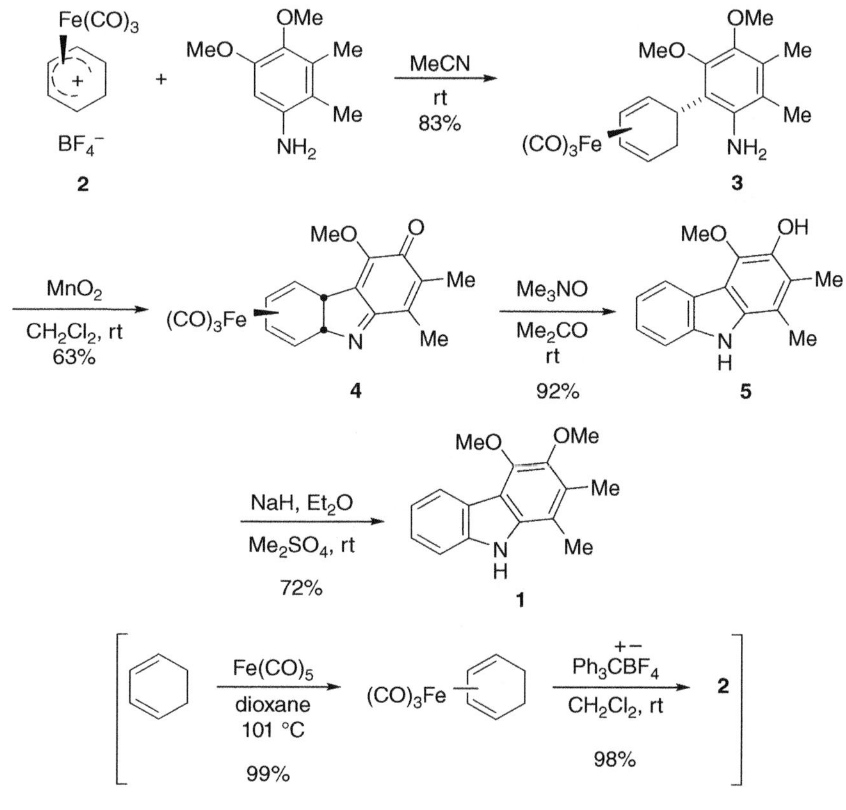

Scheme 1 *Knölker Synthesis of Carbazomycin A (1)*

Indole Ring Synthesis: From Natural Products to Drug Discovery, First Edition. Gordon W. Gribble.

Demetalation of **4** with trimethylamine *N*-oxide yields carbazole **5**. Methylation gives carbazomycin A (**1**) [12]. In addition to carbazoles, Knölker and his students applied this iron-mediated chemistry to the preparation of dihydroindoles [13, 14], (anhydrolycorine), tetrahydroindoles [15], dihydrocarbazoles [16], perhydroacenaphthenes [17], and azaspiroannelated ring systems [18–21]. Several shorter accounts of Knölker's carbazole syntheses [22–25] and the overall utility of tricarbonyl(η^4-diene)iron complexes in organic synthesis are available [26–32].

A reasonably complete list of naturally occurring and related carbazoles synthesized by Knölker and colleagues is as follows: carbazomycins A(**1**) [1, 12, 33, 34], B [12, 33, 34], C [35], D [35], G [36, 37], H [36, 37], mukonine [38–41], murrayanine [38, 39], koenoline [38, 39], murrayafoline A [39], murrayaquinone A [39], mukoeic acid [39], mukonidine [40, 41], pyrido[3,2,1-*jk*]carbazoles [40], 4-deoxycarbazomucin B [42], 3-methoxycarbazole [43], 2-methoxy-3-methylcarbazole [44], 3-hydroxycarbazoles [45], hyellazole [46, 47], isohyellazole [47], 6-chlorohyellazole [48], carazostatin [49, 50], *O*-methylcarazostatin [50], neocarazostatin B [51, 52], mukonal [53], 1,1′-bis(2-hydroxy-3-methylcarbazole) [53], girinimbine [54], murrayacine [54], dihydroxygirinimbine [54], carquinostatin A [51, 55–57, 61], furostifoline [23, 58, 59], carbazoquinocin C [60], indolo[2,3-*b*]carbazole [62, 63], furoclausine-A [23, 64, 65], lavanduquinocin [66–68], carbazomycinal (= carbazomycin E) [69], 7-methoxy-*O*-methylmukonal [70, 71], clausine H (= clauszoline-C) [70, 71], clausine K (= clauszoline-J) [70], clausine O [70, 71], clausine K (= clauszoline-J) [71], antiostatins A [72], B [72], *O*-methylmurrayamine A [73], 7-methoxymurrayacine [73], *trans*-dihydroxygirinimbine [73], streptoverticillin [74], oxydimurrayafoline [75], *O*-methylsiamenol [61], and *O*-methylmicromeline [61]. A selection of these natural carbazoles is shown in Scheme 2.

Several representative syntheses are depicted in Scheme 3, equations 1–3, which illustrate the various conditions employed by Knölker, particularly in the oxidative cyclization step. Molybdenum complexes may also be employed (equation 3).

Interestingly, the initially formed product between cation **2** and *p*-anisidine (**6**) is the isolable *N*-alkylated **7**, which rearranges easily to **8** (Scheme 4, equation 1) [2].

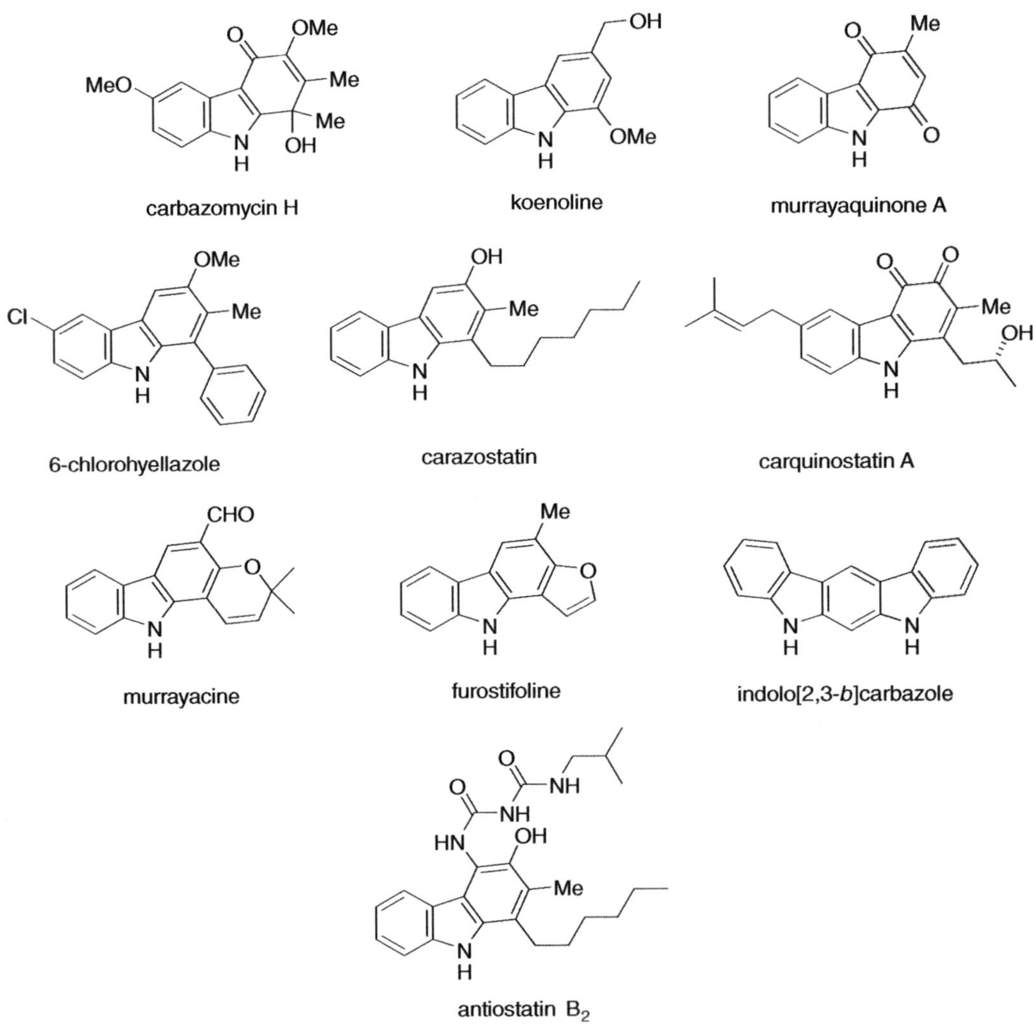

Scheme 2 *Carbazoles Synthesized by Knölker*

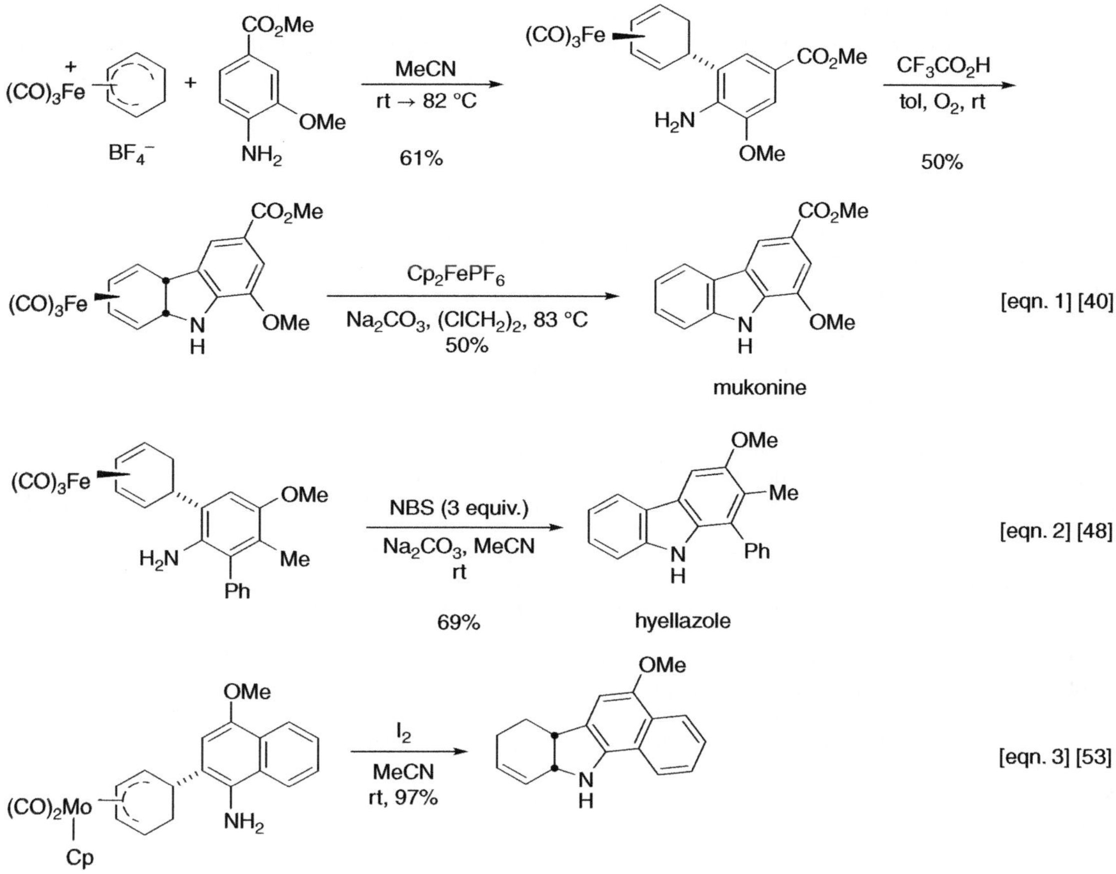

Scheme 3 Knölker Carbazole Synthesis

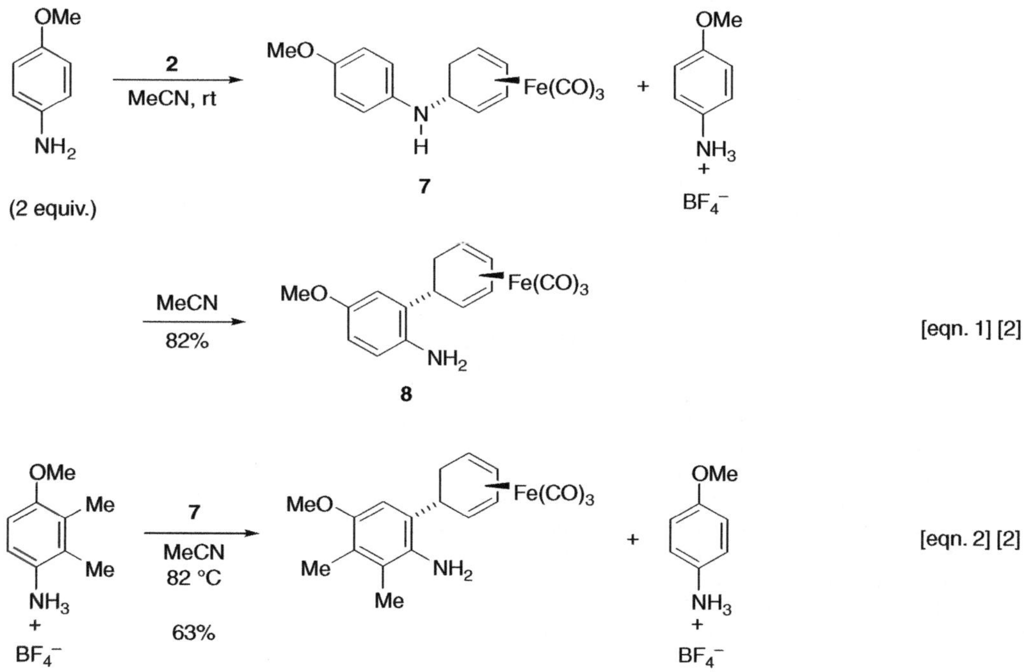

Scheme 4 Pathway of Knölker Carbazole Synthesis and Crossover Experiment

The intermolecular nature of this rearrangement was confirmed by a crossover experiment (Scheme 4, equations 1, 2) [2, 43].

Utilizing previous chemistry of iron-complexed η^5-cyclohexadienylium cations, Knölker and colleagues have developed an extraordinarily versatile carbazole synthesis.

References

[1] H.-J. Knölker, M. Bauermeister, D. Bläser, *et al.*, *Angew. Chem. Int. Ed. Engl.*, 1989, **28**, 223–225.
[2] H.-J. Knölker, *Synlett*, 1992, 371–387.
[3] H.-J. Knölker, *Chem. Rev.*, 2000, **100**, 2941–2961.
[4] H.-J. Knölker, A. Braier, D.J. Bröcher, *et al.*, *Pure Appl. Chem.*, 2001, **73**, 1075–1086.
[5] H.-J. Knölker, *Curr. Org. Synth.*, 2004, **1**, 309–331.
[6] S. Agarwal, S. Cämmerer, S. Filali, *et al.*, *Curr. Org. Chem.*, 2005, **9**, 1601–1614.
[7] H.-J. Knölker, *Top. Curr. Chem.*, 2005, **244**, 115–148.
[8] R. Forke, K.K. Gruner, K.E. Knott, *et al.*, *Pure Appl. Chem.*, 2010, **82**, 1975–1991.
[9] K.K. Gruner and H.-J. Knölker (2011) Carbazoles and acridines, in *Heterocycles in Natural Product Synthesis*, first ed. (eds K.C. Majumdar and S.K. Chattopadhyay), Wiley-VCH Verlag GmbH & Co. KGaA, pp. 341–376.
[10] H.-J. Knölker (2013) Organoiron chemistry, in *Organometallics in Synthesis*, (ed. M. Schlosser), John Wiley & Sons, Hoboken, NJ, pp. 545–776.
[11] A.W. Schmidt, K.R. Reddy, and H.-J. Knölker, *Chem. Rev.*, 2012, **112**, 3193–3328.
[12] H.-J. Knölker and M. Bauermeister, *J. Chem. Soc., Chem. Commun.*, 1989, 1468–1470.
[13] H.-J. Knölker, R. Boese, and K. Hartmann, *Angew. Chem. Int. Ed. Engl.*, 1989, **28**, 1678–1679.
[14] H.-J.; Knölker and S. Filali, *Synlett*, 2003, 1752–1754.
[15] H.-J. Knölker, A.-A. El-Ahl, and G. Weingärtner, *Synlett*, 1994, 194–196.
[16] H.-J. Knölker, P.G. Jones, J.-B. Pannek, and A. Weinkauf, *Synlett*, 1991, 147–150.
[17] H.-J. Knölker, E. Baum, and M. Heininger, *Tetrahedron Lett.*, 1997, **38**, 8021–8024.
[18] H.-J. Knölker, R. Boese, and K. Hartmann, *Tetrahedron Lett.*, 1991, **32**, 1953–1956.
[19] H.-J. Knölker and K. Hartmann, *Synlett*, 1991, 428–430.
[20] H.-J. Knölker, E. Baum, H. Goesmann, *et al.*, *Angew. Chem. Int. Ed.*, 2000, **39**, 781–784.
[21] H.-J. Knölker, E. Baum, and M. Kosub, *Synlett*, 2004, 1769–1771.
[22] H.-J. Knölker, G. Baum, and J.-B. Pannek, *Tetrahedron*, 1996, **52**, 7345–7362.
[23] W. Fröhner, M.P. Krahl, K.R. Reddy, and H.-J. Knölker, *Heterocycles*, 2004, **63**, 2393–2407.
[24] T.A. Choi, R. Czerwonka, R. Forke, *et al.*, *Med. Chem. Res.*, 2008, **17**, 374–385.
[25] H.-J. Knölker, *Chem. Lett.*, 2009, **38**, 8–13.
[26] H.-J. Knölker, G. Baum, N. Foitzik, *et al.*, *Eur. J. Inorg. Chem.*, 1998, 993–1007.
[27] H.-J. Knölker, M. Graf, and U. Mangei, *J. prakt. Chem.*, 1998, **340**, 530–535.
[28] H.-J. Knölker, E. Baum, P. Gonser, *et al.*, *Organometallics*, 1998, **17**, 3916–3925.
[29] H.-J. Knölker, *Chem. Soc. Rev.*, 1999, **28**, 151–157.
[30] H.-J. Knölker, H. Goesmann, and R. Klauss, *Angew. Chem. Int. Ed.*, 1999, **38**, 702–705.
[31] H.-J. Knölker, H. Hermann, and D. Herzberg, *Chem. Commun.*, 1999, 831–832.
[32] H.-J. Knölker, B. Ahrens, P. Gonser, *et al.*, *Tetrahedron*, 2000, **56**, 2259–2271.
[33] H.-J. Knölker and M. Bauermeister, *Helv. Chim. Acta*, 1993, **76**, 2500–2514.
[34] H.-J. Knölker and W. Fröhner, *Tetrahedron Lett.*, 1999, **40**, 6915–6918.
[35] H.-J. Knölker and G. Schlechtingen, *J. Chem. Soc., Perkin Trans. 1*, 1997, 349–350.
[36] H.-J. Knölker and W. Fröhner, *Tetrahedron Lett.*, 1997, **38**, 4051–4054.
[37] H.-J. Knölker, W. Fröhner, and K.R. Reddy, *Eur. J. Org. Chem.* 2003, 740–746.
[38] H.-J. Knölker and M. Bauermeister, *J. Chem. Soc., Chem. Commun.*, 1990, 664–665.
[39] H.-J. Knölker and M. Bauermeister, *Tetrahedron*, 1993, **49**, 11221–11236.
[40] H.-J. Knölker and M. Wolpert, *Tetrahedron Lett.*, 1997, **38**, 533–536.
[41] H.-J. Knölker and M. Wolpert, *Tetrahedron*, 2003, **59**, 5317–5322.
[42] H.-J. Knölker, M. Bauermeister, and J.-B. Pannek, *Tetrahedron*, 1993, **49**, 841–862.
[43] H.-J. Knölker, M. Bauermeister, and J.-B. Pannek, *Chem. Ber.*, 1992, **125**, 2783–2793.
[44] H.-J. Knölker and M. Bauermeister, *J. Indian Chem. Soc.*, 1994, **71**, 345–353.
[45] H.-J. Knölker, M. Bauermeister, J.-B. Pannek, and M. Wolpert, *Synthesis*, 1995, 397–408.
[46] H.-J. Knölker, E. Baum, and T. Hopfmann, *Tetrahedron Lett.*, 1995, **36**, 5339–5342.
[47] H.-J. Knölker, E. Baum, and T. Hopfmann, *Tetrahedron*, 1999, **55**, 10391–10412.
[48] H.-J. Knölker, W. Fröhner, and R. Heinrich, *Synlett*, 2004, 2705–2708.
[49] H.-J. Knölker and T. Hopfmann, *Synlett*, 1995, 981–983.
[50] H.-J. Knölker and T. Hopfmann, *Tetrahedron*, 2002, **58**, 8937–8945.
[51] R. Czerwonka, K.R. Reddy, E. Baum, and H.-J. Knölker, *Chem. Commun.*, 2006, 711–713.
[52] H.-J. Knölker, W. Fröhner, and A. Wagner, *Tetrahedron Lett.*, 1998, **39**, 2947–2950.

[53] H.-J. Knölker, H. Goesmann, and C. Hofmann, *Synlett*, 1996, 737–740.
[54] H.-J. Knölker and C. Hofmann, *Tetrahedron Lett.*, 1996, **37**, 7947–7950.
[55] H.-J. Knölker and W. Fröhner, *Synlett*, 1997, 1108–1110.
[56] H.-J. Knölker, E. Baum, and K.R. Reddy, *Tetrahedron Lett.*, 2000, **41**, 1171–1174.
[57] W. Fröhner, K.R. Reddy, and H.-J. Knölker, *Heterocycles*, 2007, **74**, 895–912.
[58] H.-J. Knölker and W. Fröhner, *Tetrahedron Lett.*, 1996, **37**, 9183–9186.
[59] H.-J. Knölker and W. Fröhner, *Synthesis*, 2000, 2131–2136.
[60] H.-J. Knölker and W. Fröhner, *Tetrahedron Lett.*, 1997, **38**, 1535–1538.
[61] C. Thomas, O. Kataeva, A.W. Schmidt, and H.-J. Knölker, *Org. Biomol. Chem.*, 2014, **12**, 872–875.
[62] H.-J. Knölker and K.R. Reddy, *Tetrahedron Lett.*, 1998, **39**, 4007–4008.
[63] H.-J. Knölker and K.R. Reddy, *Tetrahedron*, 2000, **56**, 4733–4737.
[64] H.-J. Knölker and M.P. Krahl, *Synlett*, 2004, 528–530.
[65] M.P. Krahl, A.W. Schmidt, and H.-J. Knölker, *Heterocycles*, 2012, **86**, 357–370.
[66] H.-J. Knölker and W. Fröhner, *Tetrahedron Lett.*, 1998, **39**, 2537–2540.
[67] H.-J. Knölker, E. Baum, and K.R. Reddy, *Chirality*, 2000, **12**, 526–528.
[68] W. Fröhner, K.R. Reddy, and H.-J. Knölker, *ARKIVOC*, 2012, 330–342.
[69] H.-J. Knölker and M. Bauermeister, *Heterocycles*, 1991, **32**, 2443–2450.
[70] O. Kataeva, M.P. Krahl, and H.-J. Knölker, *Org. Biomol. Chem.*, 2005, **3**, 3099–3101.
[71] M.P. Krahl, O. Kataeva, A.W. Schmidt, and H.-J. Knölker, *Eur. J. Org. Chem.*, 2013, 59–64.
[72] K.E. Knott, S. Auschill, A. Jäger, and H.-J. Knölker, *Chem. Commun.*, 2009, 1467–1469.
[73] K.K. Gruner, T. Hopfmann, K. Matsumoto, *et al.*, *Org. Biomol. Chem.*, 2011, **9**, 2057–2061.
[74] C. Thomas, O. Kataeva, and H.-J. Knölker, *Synlett*, 2011, 2663–2666.
[75] C. Börger, M.P. Krahl, M. Gruner, *et al.*, *Org. Biomol. Chem.*, 2012, **10**, 5189–5193.

48

Miscellaneous Oxidative Cyclizations

Several oxidative cyclizations to form the indole ring not involving a familiar name reaction are covered in this chapter.

Indeed, the biosynthesis of the biopolymer melanin involves the oxidative cyclization of dihydroxyphenylalanine (DOPA) to phenylalanine-3,4-quinone (dopaquinone), which eventually forms 5,6-dihydroxyindole (DHI). Polymerization of DHI affords melanin [1]. Lim and Patil have exploited this biochemical transformation using commercial mushroom tyrosinase in a synthesis of 5,6-dihydroxyindoles protected as the diacetates (**2**) (Scheme 1) [2]. The parent indole ($R^1 = R^2 = H$) is obtained in less than 10% yield. Carpender has reported a similar oxidative cyclization using manganese dioxide to give **2** ($R^1 = Me$) in 80% overall yield from epinine (**1**, $R^1 = Me$, $R^2 = H$) [3]. Other oxidants (H_2O_2, $H_2O_2/FeSO_4$, O_2, NaOCl, $NaClO_3/V_2O_5$) gave little or no product. Choi, Nam, and colleagues have effected an electrochemical oxidation of dopamine (**1**, $R^1 = R^2 = H$) to 5,6-dihydroxyindole that polymerizes to form films of polydopamine suitable for neural attachment and function [4].

Kita has described an excellent synthesis of 5-oxygenated indoles via the phenyliodine (III) bis(trifluoroacetate) (PIFA) oxidation of 2-aminoethyl oxygenated phenols (Scheme 2, equations 1 and 2) [5]. This method was extended to the preparation of 5-hydroxyindole, 5-methoxyindole, and 7-bromo-5-hydroxyindole in 65% to 100% yield from the requisite quinone or quinone acetal.

The versatile reagents phenyliodine (III) bis(trifluoroacetate) (PIFA) and phenyliodine (III) diacetate (PIDA) have found additional utility in the oxidative cyclization of suitable substrates to indoles. Zhao and colleagues used both reagents to effect the oxidative cyclization of β-aminostyrenes to indoles and *N*-aryl enamines to indoles (Scheme 3, equations 1–5) [6–9]. To rationalize the lack of a pronounced electronic effect from the benzene ring in equation 1, the authors proposed a radical mechanism and a single-electron transfer (SET) pathway rather than the more obvious nitrenium ion generation and cyclization [6]. A large number of substituted 3-cyanoindoles were prepared in this study. A conventional electrophilic cyclization path is suggested for the indole formation in equations 3 to 5, possibly via a nitrene [7–9]. The reaction shown in equation 5 has been extended to a synthesis of tetrahydrocycloheptaindolones [9].

The Zhao group found that $FeCl_3$-mediated cyclization of 3-alkoxyimino-2-arylalkylnitriles to 3-cyanoindoles is superior to the use of PIFA (Scheme 4, equation 1) [10]. Some specific examples are shown. The authors propose the formation of an *N*-radical and oxidation to a nitrenium ion that cyclizes eventually to yield the indole product.

Li and colleagues effected the cyclization of *N*-aryl enamines to indoles with iodine itself (Scheme 5, equation 1) [11]. A suggested mechanism is shown, which involves oxidation of substrate by hyperiodide (generated *in situ* from I_2 and NBS) to give *N*-iodo **3** and thence radical **4**. Cyclization and oxidation of **5** affords the indole.

Two groups have employed PIDA to oxidatively cyclize 2-alkynylaniline derivatives to indoles (Scheme 6) [12–16]. Yanada and coworkers employed 2-alkynylbenzamides in this chemistry (equation 1) [12]. The interesting cyclic bis(indoles) **6** were constructed using this methodology. Fan and coworkers also used PIDA to prepare indoles (equations 2–5) [13–16]. This elegant chemistry features

Indole Ring Synthesis: From Natural Products to Drug Discovery, First Edition. Gordon W. Gribble.

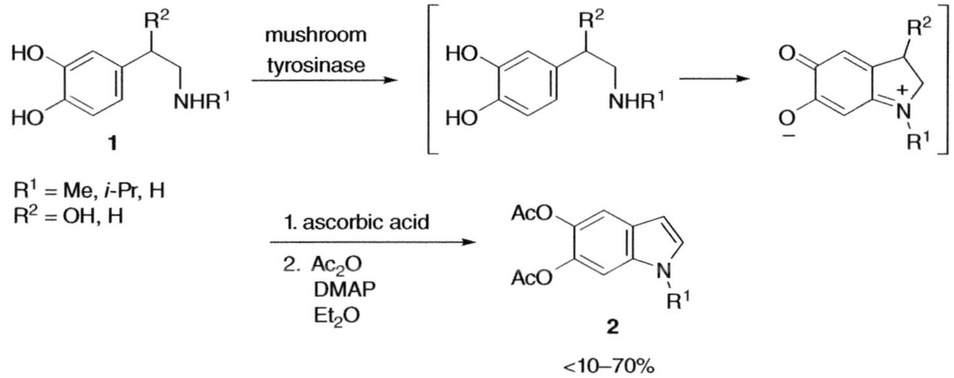

Scheme 1 *Lim and Patil Synthesis of 5,6-Dihydroxyindole Derivatives*

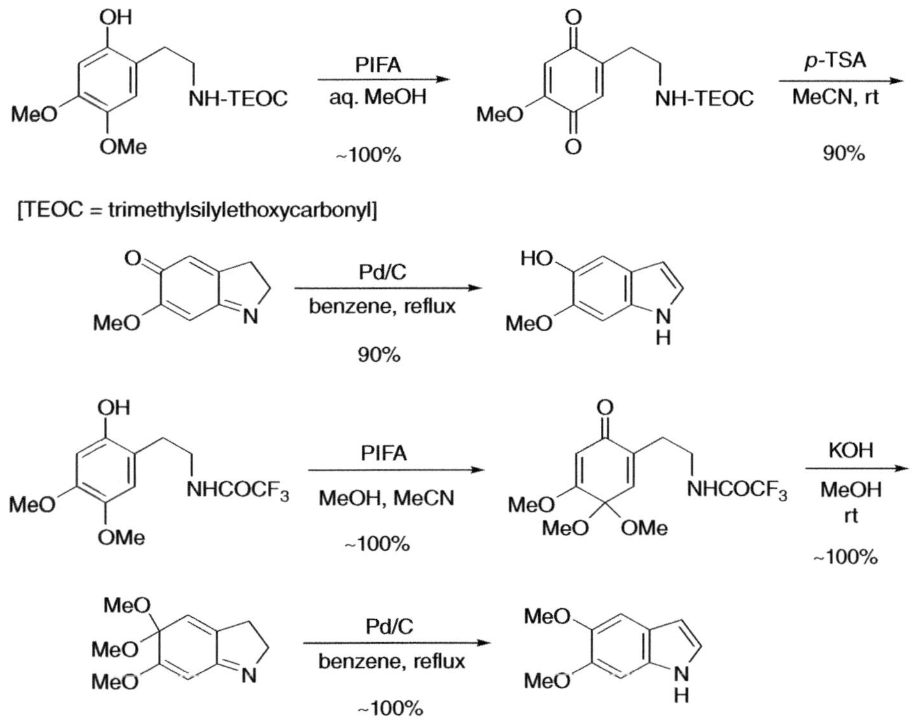

[eqn. 1] [5]

[eqn. 2] [5]

Scheme 2 *Kita Synthesis of 5-Oxygenated Indoles*

nucleophilic trapping of an intermediate cation to give highly substituted indoles. Many examples are described in this work.

Antonchick and colleagues made use of PIDA to craft a series of 1-arylcarbazoles (Scheme 7, equation 1) [17], and Muñiz's team employed iodosobenzene and 2,4,5-tris-isopropybenzene sulfonic acid (TIPBSA) as a modified Koser reagent to synthesize indoles (equation 2) [18].

Jenekhe and coworkers found that DDQ converts anthrazoline **7** to heptacyclic bis(indolo[1,2-*a*]quinolone) **8** (equation 3) [19]. Youn and Jang used DDQ to convert *N*-Ts-2-alkenylanilines into indoles via, it is proposed, a radical cation generated by SET (equation 4) [20]. In the case of β,β-disubstituted 2-alkenylanilines, the 2,3-disubstituted indole product is derived from a C-2 to C-3 aryl migration via a phenonium ion (equation 5).

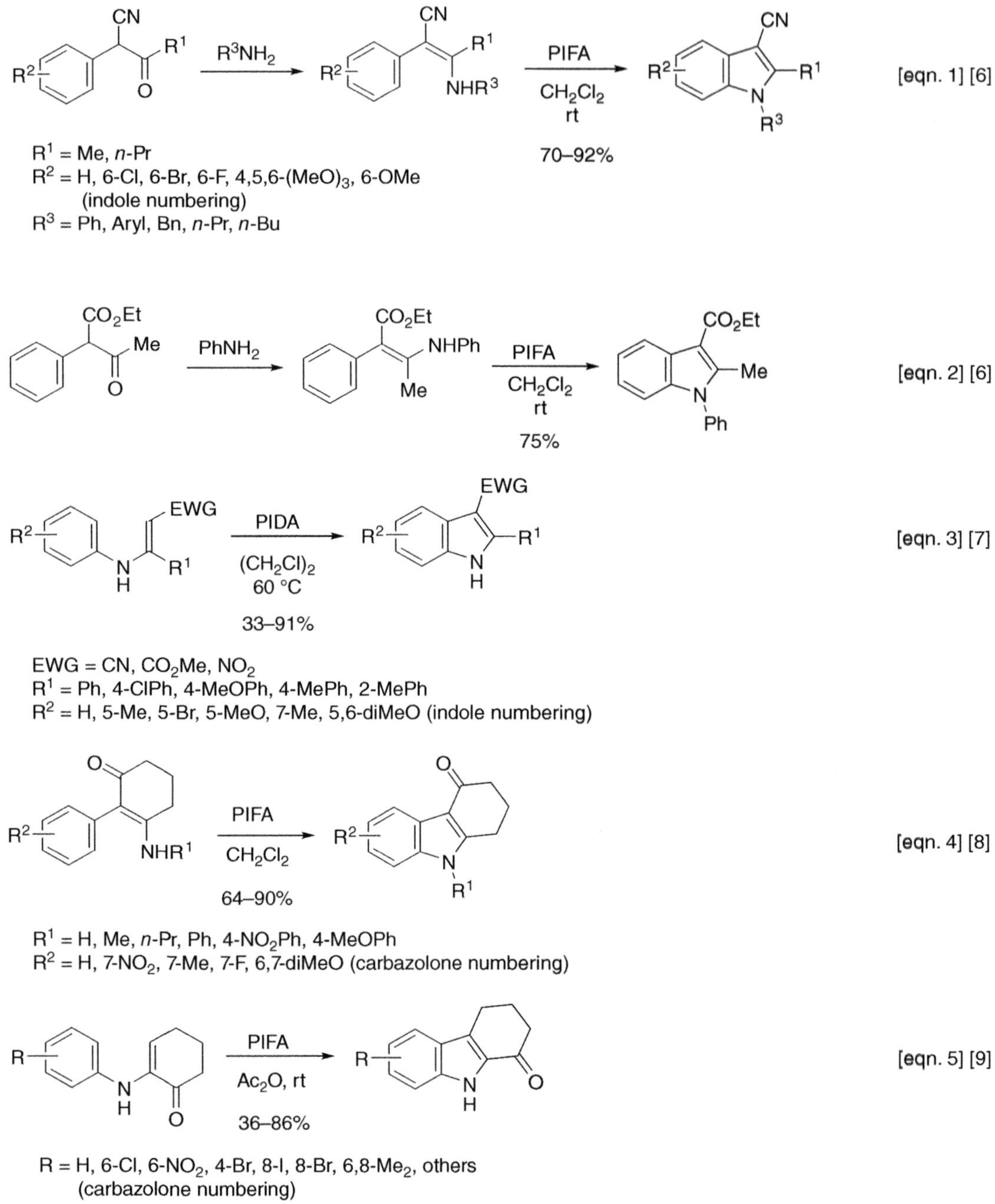

Scheme 3 *Zhao Synthesis of Indoles via PIFA/PIDA Oxidative Cyclization*

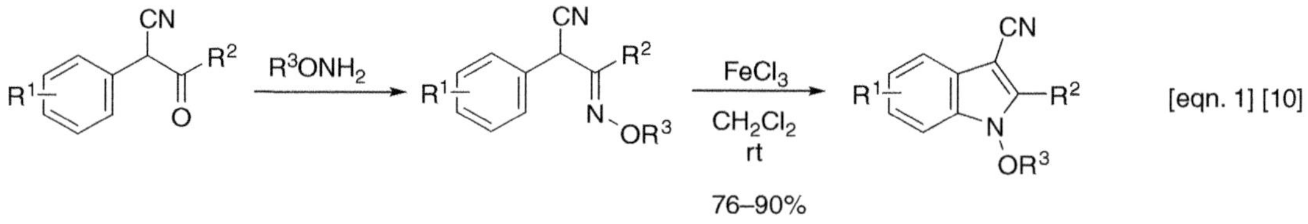

R^1 = H, 6-Me, 6-Cl, 4-I, 6-F, 5-MeO (indole numbering)
R^2 = H, Me, Ph, *n*-Pr, Bn
R^3 = Me, Bn, *n*-Bu

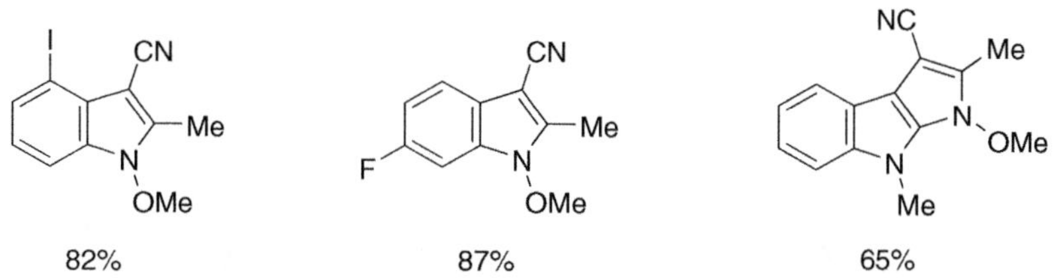

Scheme 4 *Zhao Indole Synthesis via $FeCl_3$-mediated Cyclization*

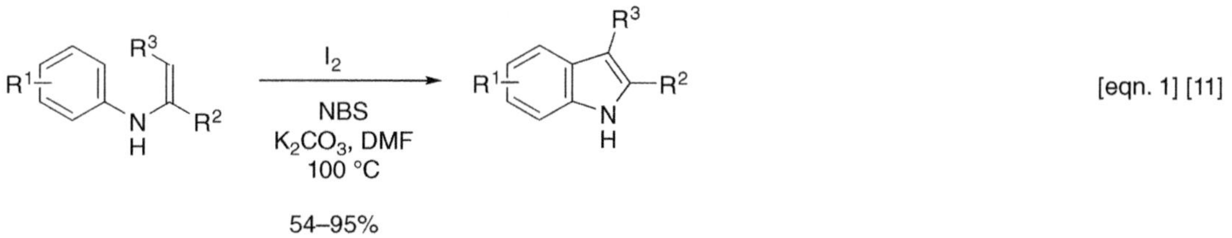

R^1 = H, 5-OMe, 5-Me, 5-Br, 5-I, 7-Me, 4-OMe, 4-Me (indole numbering)
R^2 = Ph, *i*-Pr
R^3 = CO_2Et, CO_2Me, CONHPh

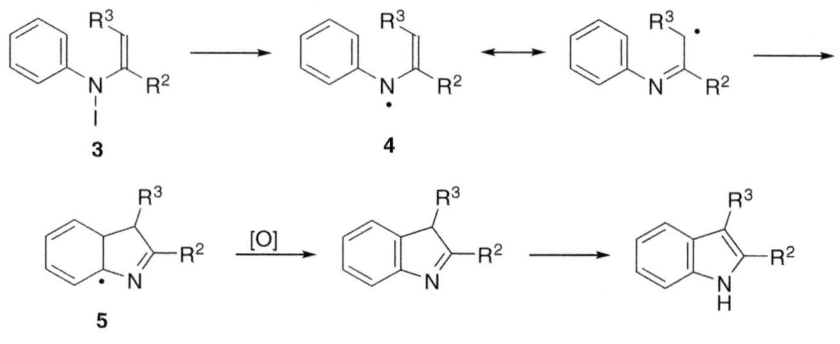

Scheme 5 *Li Synthesis of Indoles*

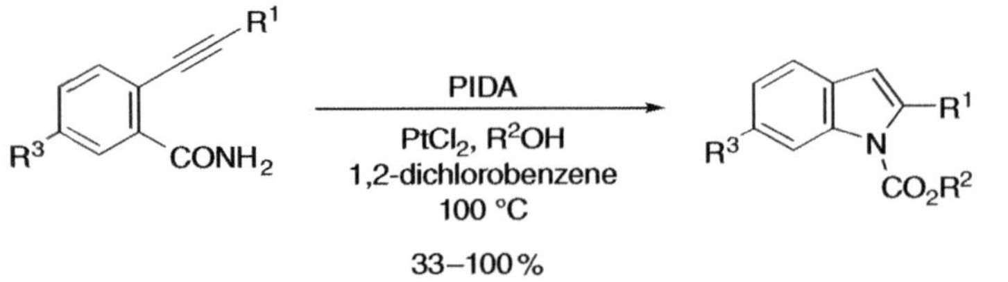

[eqn. 1] [12]

R^1 = *n*-Bu, Ph, *p*-Tol, H, TMS, $(CH_2)_3OTs$
R^2 = Et, Bn, *t*-Bu
R^3 = H, NO_2, OMe

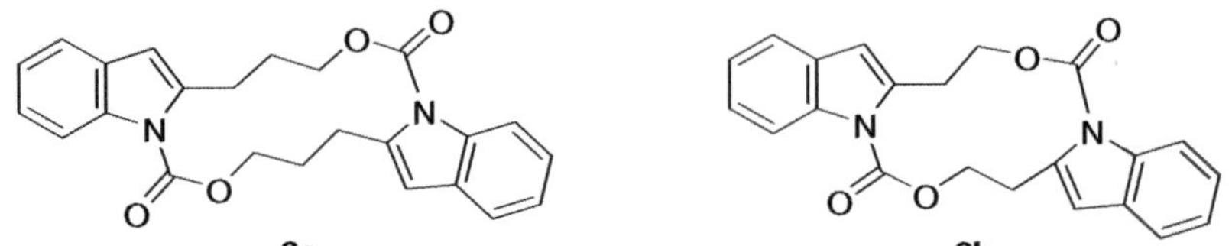

6a **6b**

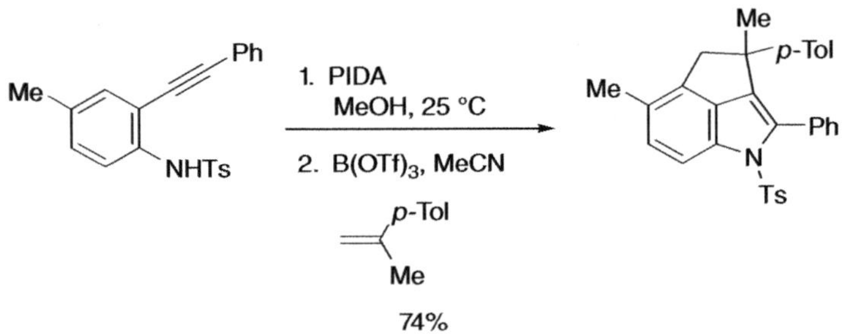

[eqn. 2] [13]

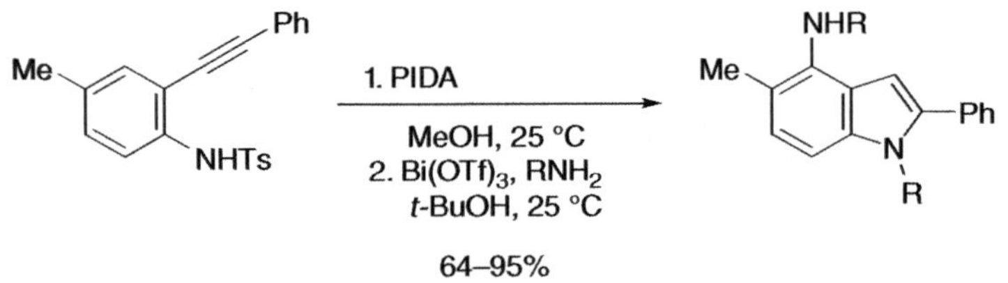

[eqn. 3] [14]

R = Ph, *p*-Tol, 4-BrPh, 4-ClPh, 4-FPh, 4-pyridyl

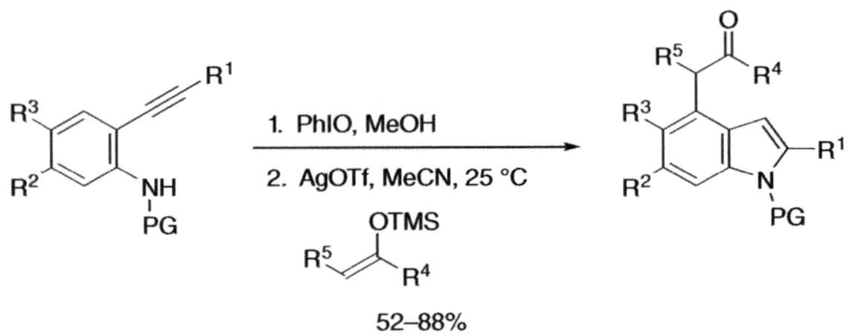

[eqn. 4] [15]

R^1 = H, Ph, cyclopropyl, aryl
R^2 = H, Me
R^3 = H, Me
R^4 = Aryl
R^5 = H, Me
PG = Ts, Bz

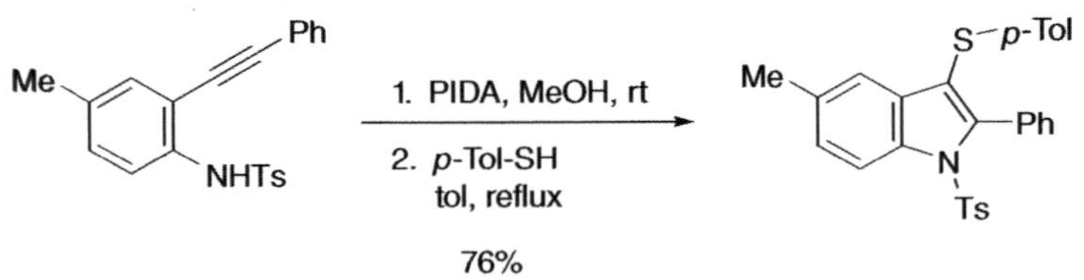

[eqn. 5] [16]

Scheme 6 *Oxidative Cyclization of 2-Alkynylanilines to Indoles*

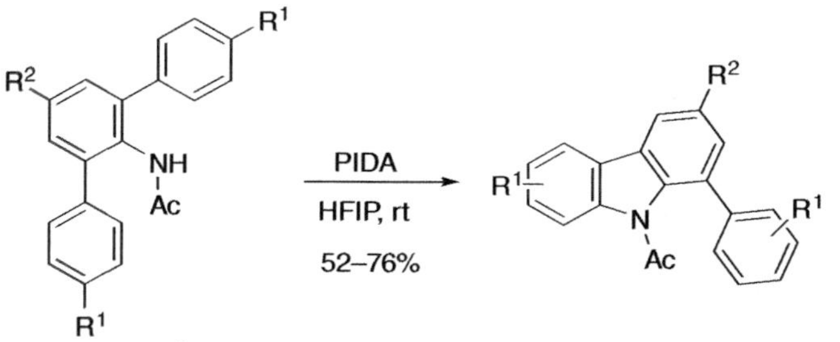

[eqn. 1] [17]

R^1 = H, F, Cl, *t*-Bu, Me
R^2 = H, Cl, *i*-Pr, *n*-Bu, *t*-Bu, CO_2Me

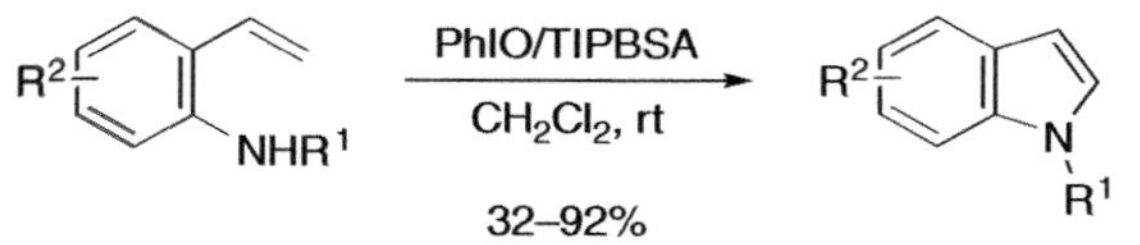

[eqn. 2] [18]

R^1 = Cbz, CO_2Me
R^2 = H, 5/6-Me, 5/6-F, 5/6-Cl, 5-Br, 5/6-CF_3, 4-CHO, 5-Ac, 5-CO_2Me, 5-CN, 6-OMe, 6-OCF_3, 4-Me, others (indole numbering)

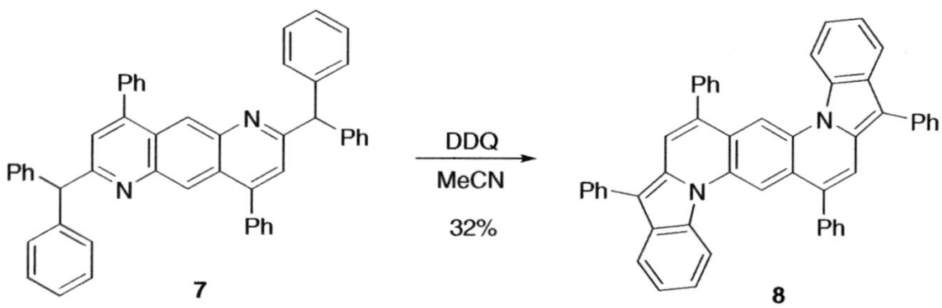

[eqn. 3] [19]

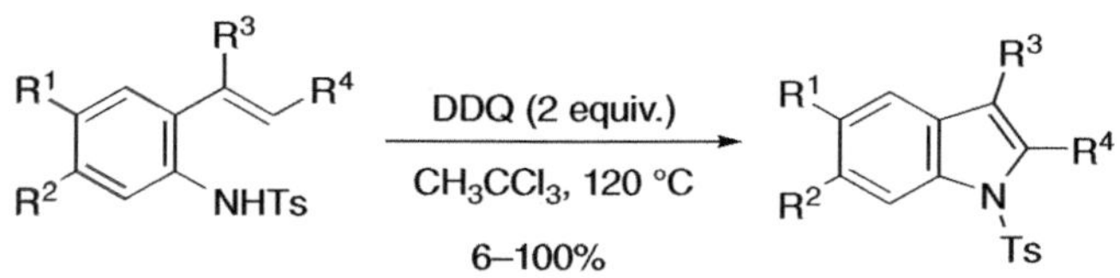

[eqn. 4] [20]

R^1 = H, OMe, Me, Cl, CF_3, NO_2
R^2 = H, OMe, Me, Cl, NO_2
R^3 = H, Ph
R^4 = Ph, Aryl, 1-naphthyl, 3-thienyl, H

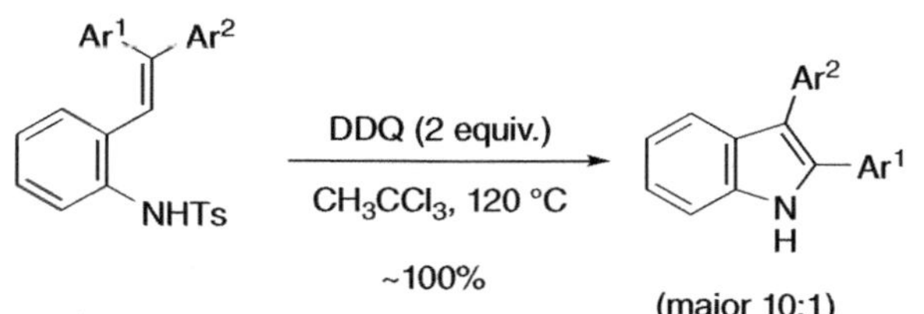

[eqn. 5] [20]

Ar^1 = Ph
Ar^2 = 4-MeOPh

Scheme 7 *Miscellaneous Oxidative Cyclization to Indoles*

References

[1] G.A. Swan, *Prog. Chem. Org. Nat. Prod.*, 1974, **31**, 521–582.
[2] M.-I. Lim and D.G. Patil, *Tetrahedron Lett.*, 1987, **28**, 3775–3778.
[3] J.F. Carpender, *J. Org. Chem.*, 1993, **58**, 1607–1609.
[4] K. Kang, S. Lee, R. Kim, *et al.*, *Angew. Chem. Int. Ed.*, 2012, **51**, 13101–13104.
[5] Y. Kita, H. Tohma, M. Inagaki, and K. Hatanaka, *Heterocycles*, 1992, **33**, 503–506.
[6] Y. Du, R. Liu, G. Linn, and K. Zhao, *Org. Lett.*, 2006, **8**, 5919–5922.
[7] W. Yu, Y. Du, and K. Zhao, *Org. Lett.*, 2009, **11**, 2417–2420.
[8] X. Ban, Y. Pan, Y. Lin, *et al.*, *Org. Biomol. Chem.*, 2012, **10**, 3606–3609.
[9] H. Shi, T. Guo, D. Zhang-Negrerie, *et al.*, *Tetrahedron*, 2014, **70**, 2753–2760.
[10] Y. Du, J. Chang, J. Reiner, and K. Zhao, *J. Org. Chem.*, 2008, **73**, 2007–2010.
[11] Z. He, W. Liu, and Z. Li, *Chem. Asian J.*, 2011, **6**, 1340–1343.
[12] N. Okamoto, Y. Miwa, H. Minami, *et al.*, *Angew. Chem. Int. Ed.*, 2009, **48**, 9693–9696.
[13] L. Wang and R. Fan, *Org. Lett.*, 2012, **14**, 3596–3599.
[14] L. Zhang, Z. Li, and R. Fan, *Org. Lett.*, 2012, **14**, 6076–6079.
[15] X. Feng, H. Wang, B. Yang, and R. Fan, *Org. Lett.*, 2014, **16**, 3600–3603.
[16] D. Han, Z. Li, and R. Fan, *Org. Lett.*, 2014, **16**, 6508–6511.
[17] R. Samanta, K. Kulikov, C. Strohmann, and A.P. Antonchick, *Synthesis*, 2012, **44**, 2325–2332.
[18] L. Fra, A. Millán, J.A. Souto, and K. Muñiz, *Angew. Chem. Int. Ed.*, 2014, **53**, 7349–7353.
[19] E. Ahmed, A.L. Briseno, Y. Xia, and S.A. Jenekhe, *J. Am. Chem. Soc.*, 2008, **130**, 1118–1119.
[20] Y.H. Jang and S.W. Youn, *Org. Lett.*, 2014, **16**, 3720–3723.

PART VI
Radical Cyclization

Unlike their polar species counterparts, carbocations and carbanions, radicals did not play a major role in organic synthesis until recent times. But now they feature prominently in all aspects of organic chemistry, including indole ring synthesis.

Indole Ring Synthesis: From Natural Products to Drug Discovery, First Edition. Gordon W. Gribble.

49

Fukuyama Indole Synthesis

Based on the early studies of radical generation from organotin hydrides and their addition to multiple bonds [1–3], particularly in cyclization [4–7], several groups applied this chemistry to the synthesis of indoles and indolines. For example, in 1975 Beckwith and colleagues found that *o*-(*N*-allyl-*N*-methylamino)iodobenzene was converted to 1,3-dimethylindoline (78% yield) upon treatment with tri-*n*-butylstannane (prepared by the reduction of tri-*n*-butyltin chloride with $LiAlH_4$) in the presence of AIBN (azobisisobutyronitrile) and benzene at 130 °C in a sealed tube [4].

However, guided by the observation by Saegusa and colleagues that tri-*n*-butyltin radical, generated from tri-*n*-butyltin hydride and di-*t*-butyl peroxide, adds to isocyanides [8], it was Fukuyama and coworkers who developed an extraordinarily useful tin-based radical cyclization of α-stannoimidoyl radicals **2** generated from isocyanides **1** to afford indoles (Scheme 1, equations 1–3) [9–17], summarized by Tokuyama and Fukuyama [18]. The general pathway is shown in equation 1, with examples in equation 2, including a Stille coupling of the 2-stannylindole (**3**) (equation 3) [9]. In rare cases, cyclization can occur in an *endo* fashion to give a tetrahydroquinoline.

A selection of additional indoles prepared by Fukuyama via the corresponding *o*-isocyanostyrene is shown in Scheme 2.

A powerful extension of this method is the Stille-iodination of the initially formed 2-stannylindoles with iodine or *N*-iodosuccinimide (NIS) and further Pd-catalyzed cross-coupling reactions (i.e., Sonogashira, Stille, Heck) (Scheme 3).

Fukuyama and colleagues employed this indole ring synthesis in total syntheses of (±)-vincadifformine [11], (–)-tabersonine [11], (–)-vindoline [13], (–)-aspidophytine [15, 16], and tryprostatins A and B [17]. For reviews of isocyanide-based indole synthesis and organic synthesis using isocyanides in general, see Campo and colleagues and Tobisu and colleagues [19, 20].

In 1999 Fukuyama's group described a new indole ring construction based on *o*-alkenylthioanilides and their radical addition, followed by indolization (Scheme 4, equations 1 and 2) [21]. Fukuyama applied this variant indole synthesis to the synthesis of (±)-catharanthine [22], (+)-vinblastine [23–25], (+)-vincristine [26], and (–)-strychnine [27]. The catharanthine synthesis required hypophosphorous acid and AIBN to achieve indole formation (equation 3).

In a third variation of a tin-mediated indole ring formation, Fukuyama used imidoyltellurides to afford 2,3-dialkylindoles (Scheme 5) [28].

The tin-mediated Fukuyama indolization from isocyanides has been used by Rainier [29, 30], Nakajima [31], Bénéteau [32], and Jeon [33] to craft various indoles. Notably, Rainier and coworkers effected a tin-mediated isocyanide-alkyne cyclization, but they attempted unsuccessfully to ambush the intermediate indolenine [29, 30], although this was successful under a sulfur-mediated radical cyclization (Chapter 52).

Indole Ring Synthesis: From Natural Products to Drug Discovery, First Edition. Gordon W. Gribble.

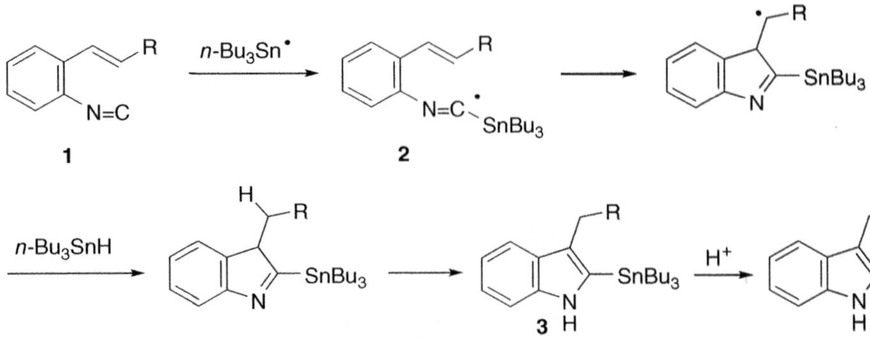

[eqn. 1] [9]

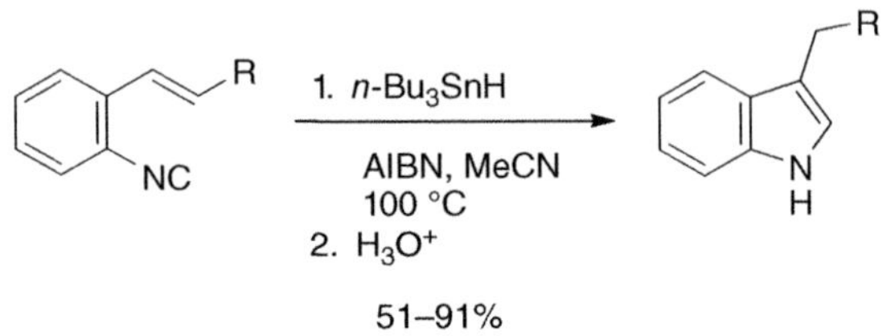

[eqn. 2] [9]

R = CO_2Me, CH_2OTHP, Ph, *n*-Bu, CH_2OBn

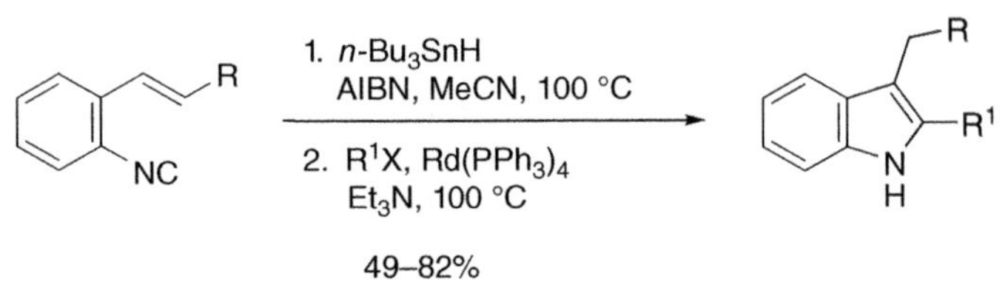

[eqn. 3] [9]

R = CO_2Me, *n*-Bu, CH_2OTHP
R^1 = Ph, 4-AcPh, Bn, CH=CH*n*-Bu
X = Br, I, OTf

Scheme 1 *Fukuyama Indole Synthesis*

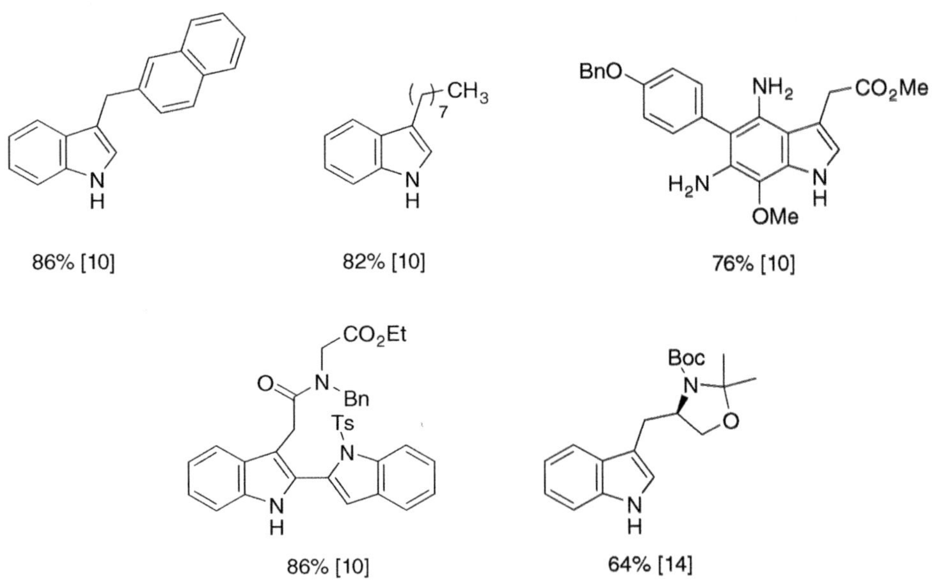

Scheme 2 *Representative Indoles Prepared by Fukuyama*

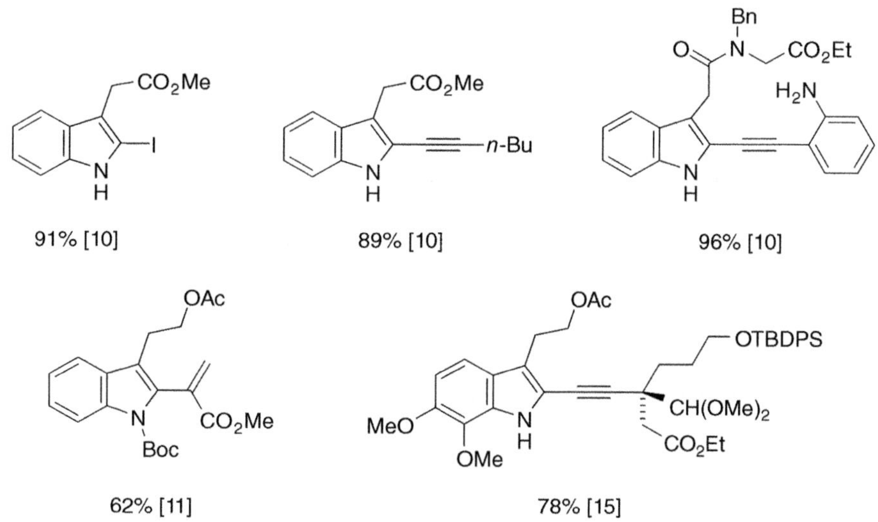

Scheme 3 *C-2 Substituted Indoles Prepared by Fukuyama*

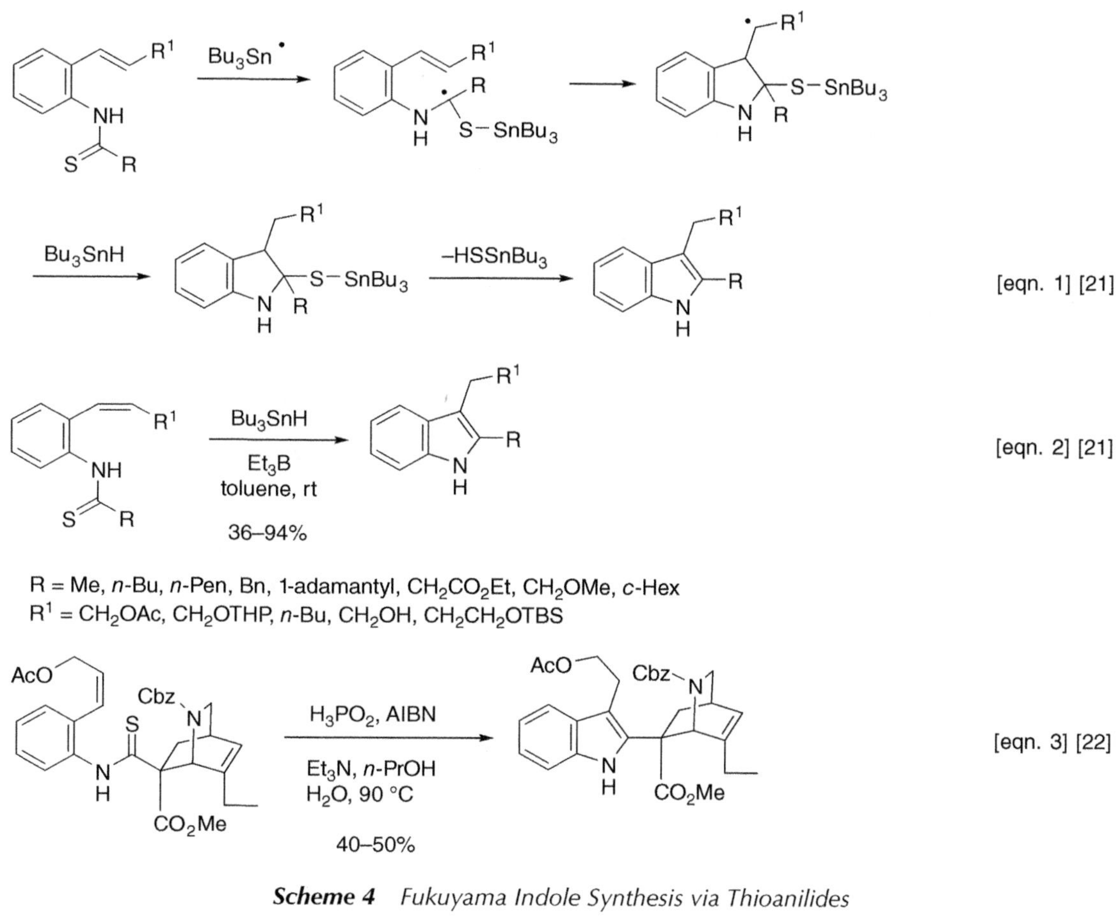

Scheme 4 *Fukuyama Indole Synthesis via Thioanilides*

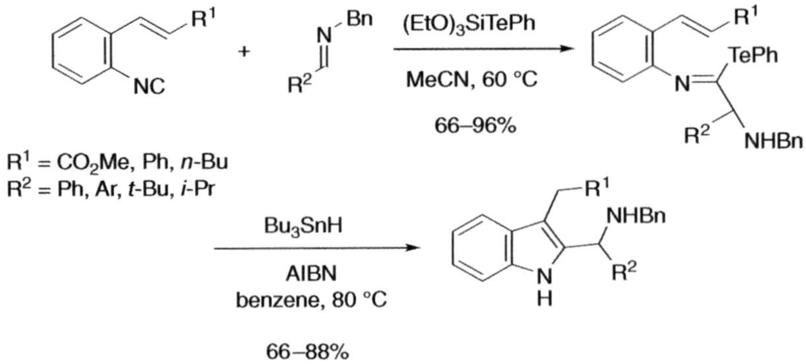

Scheme 5 *Fukuyama Indole Synthesis via Imidoyltellurides*

References

[1] W.P. Neumann, H. Niermann, and R. Sommer, *Angew. Chem.*, 1961, **73**, 768.
[2] H.G. Kuivila, W. Rahman, and R.H. Fish, *J. Am. Chem. Soc.*, 1965, **87**, 2835–2840.
[3] H.G. Kuivila, *Acc. Chem. Res.*, 1968, **1**, 299–305.
[4] A.L.J. Beckwith and W.B. Gara, *J. Chem. Soc., Perkin II*, 1975, 795–802.
[5] M. Julia, *Pure Appl. Chem.*, 1967, **15**, 167–183.
[6] M. Julia, *Acc. Chem. Res.*, 1971, **4**, 386–392.
[7] B.M. Loertscher and S.L. Castle (2014) Radical cyclizations and sequential radical reactions, in *Comprehensive Organic Synthesis II*, vol. **4** (eds P. Knochel and G.A. Molander), Elsevier Ltd, Oxford, UK, pp. 742–809.
[8] T. Saegusa, S. Kobayashi, Y. Ito, and N. Yasuda, *J. Am. Chem. Soc.*, 1968, **90**, 4182.
[9] T. Fukuyama, X. Chen, and G. Peng, *J. Am. Chem. Soc.*, 1994, **116**, 3127–3128.
[10] Y. Kobayashi and T. Fukuyama, *J. Heterocycl. Chem.*, 1998, **35**, 1043–1055.
[11] S. Kobayashi, G. Peng, and T. Fukuyama, *Tetrahedron Lett.*, 1999, **40**, 1519–1522.
[12] H. Tokuyama, Y. Kaburagi, X. Chen, and T. Fukuyama, *Synthesis*, 2000, 429–434.
[13] S. Kobayashi, T. Ueda, and T. Fukuyama, *Synlett*, 2000, 883–886.
[14] H. Tokuyama, M. Watanabe, Y. Hayashi, *et al.*, *Synlett*, 2001, 1403–1406.
[15] S. Sumi, K. Matsumoto, H. Tokuyama, and T. Fukuyama, *Org. Lett.*, 2003, **5**, 1891–1893.
[16] S. Sumi, K. Matsumoto, H. Tokuyama, and T. Fukuyama, *Tetrahedron*, 2003, **59**, 8571–8587.
[17] T. Yamakawa, E. Ideue, J. Shimokawa, and T. Fukuyama, *Angew. Chem. Int. Ed.*, 2010, **49**, 9262–9265.
[18] H. Tokuyama and T. Fukuyama, *Chem. Rec.*, 2002, **2**, 37–45.
[19] J. Campo, M. García-Valverde, S. Marcaccini, *et al.*, *Org. Biomol. Chem.*, 2006, **4**, 757–765.
[20] M. Tobisu and N. Chatani, *Chem. Lett.*, 2011, **40**, 330–340.
[21] H. Tokuyama, T. Yamashita, M.T. Reding, *et al.*, *J. Am. Chem. Soc.*, 1999, **121**, 3791–3792.
[22] M.T. Reding and T. Fukuyama, *Org. Lett.*, 1999, **1**, 973–976.
[23] S. Yokoshima, T. Ueda, S. Kobayashi, *et al.*, *J. Am. Chem. Soc.*, 2002, **124**, 2137–2138.
[24] S. Yokoshima, T. Ueda, S. Kobayashi, *et al.*, *Pure Appl. Chem.*, 2003, **75**, 29–38.
[25] T. Miyazaki, S. Yokoshima, S. Simizu, *et al.*, *Org. Lett.*, 2007, **9**, 4737–4740.
[26] T. Kuboyama, S. Yokoshima, H. Tokuyama, and T. Fukuyama, *Proc. Natl. Acad. Sci.*, 2004, **101**, 11966–11970.
[27] Y. Kaburagi, H. Tokuyama, and T. Fukuyama, *J. Am. Chem. Soc.*, 2004, **126**, 10246–10247.
[28] M. Kotani, S. Yamago, A. Satoh, *et al.*, *Synlett*, 2005, 1893–1896.
[29] J.D. Rainier, A.R. Kennedy, and E. Chase, *Tetrahedron Lett.*, 1999, **40**, 6325–6327.
[30] J.D. Rainier and A.R. Kennedy, *J. Org. Chem.*, 2000, **65**, 6213–6216.
[31] T. Shinada, M. Miyachi, Y. Itagaki, *et al.*, *Tetrahedron Lett.*, 1996, **37**, 7099–7102.
[32] N. Henry, J. Blu, V. Bénéteau, and J.-Y. Mérour, *Synthesis*, 2006, 3895–3901.
[33] H.J. Gim, H. Li, E. Lee, *et al.*, *Bioorg. Med. Chem. Lett.*, 2013, **23**, 513–517.

50

Other Tin-Mediated Indole Syntheses

The applicability of tin-mediated radical cyclizations of non-isocyanides to form indoles has been pursued by several other investigators. Uneyama and colleagues reported the tin-mediated deiodination of imidoyl iodides and subsequent indolization (Scheme 1, equation 1) [1]. This radical cyclization is also effected photochemically. Bowman and colleagues exploited the tin-promoted radical generation from imidoyl selanides and cyclization to form indoles [2, 3], including a synthesis of ellipticine [4] (Scheme 1, equations 2–3). The requisite precursors were prepared in two steps from the corresponding amides (1. $COCl_2$; 2. $PhSe^-$). Depending on substituents, a quinolone can form via a 6-*endo*-dig pathway.

Orito described a radical cyclization to the indolo[2,1-*a*]isoquinoline ring system rather than to the anticipated aporphine skeleton (Scheme 2, equation 1) [5]. Kim and

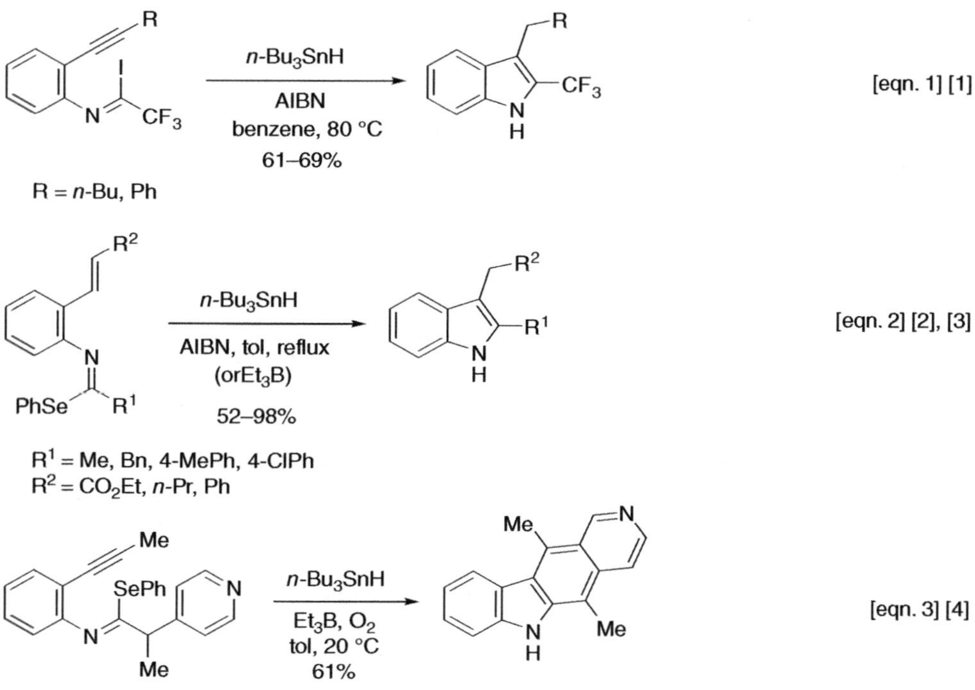

Scheme 1 *Tin-Mediated Radical Cyclization of Imidoyl Substrates*

Indole Ring Synthesis: From Natural Products to Drug Discovery, First Edition. Gordon W. Gribble.

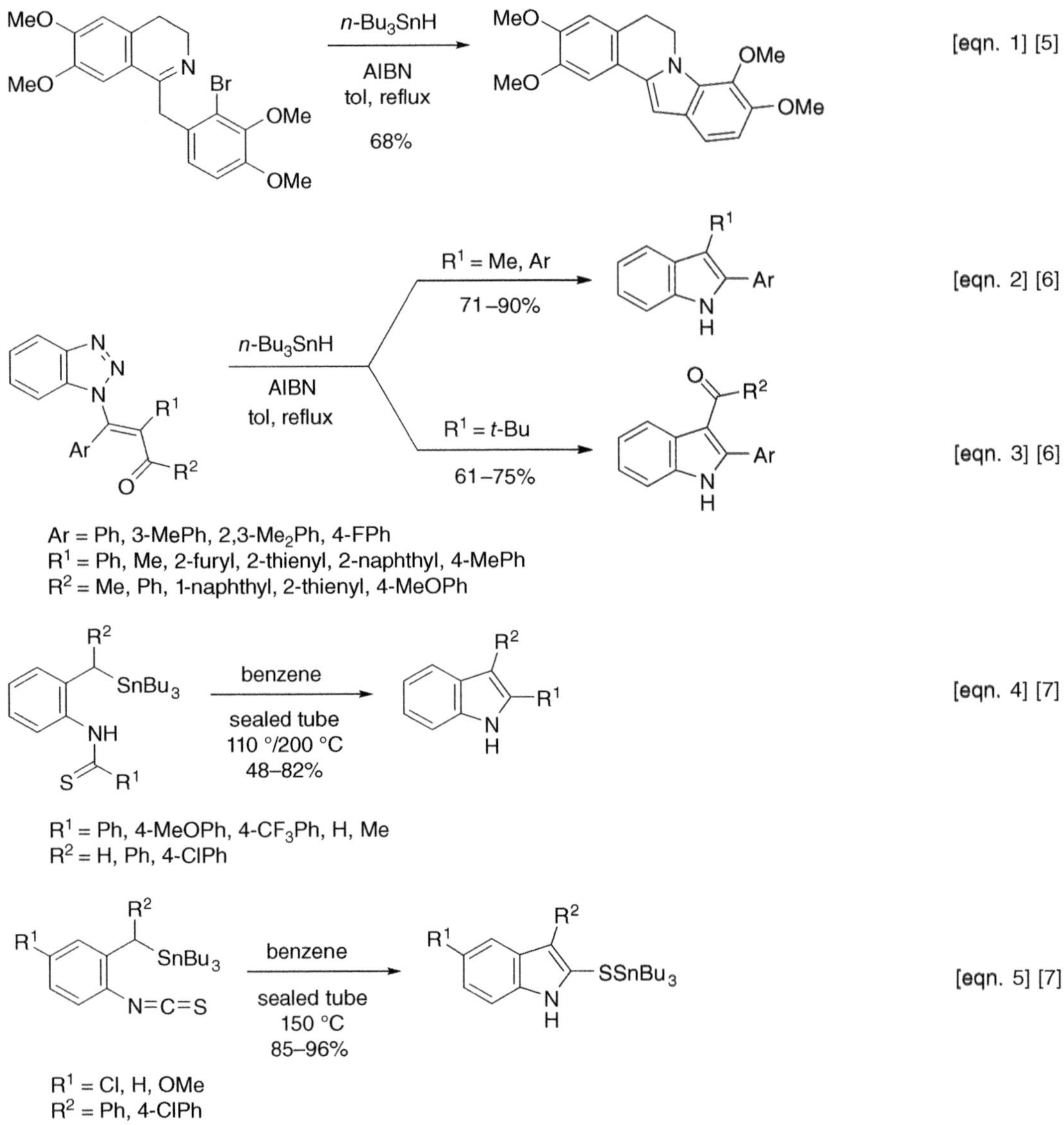

Scheme 2 *Other Tin-Mediated Radical Indolizations*

Kim demonstrated that the substituted benzotriazoles can afford either indoles or 3-acylindoles depending on the R^1 group (equations 2 and 3) [6]. As might be expected, phenanthridines accompany the 3-acylindole pathway (up to 30%). This reaction is initiated by formation of an *O*-stannyl ketyl radical and loss of N_2. Further cyclization and loss of either *t*-butyl radical or R^2CHO affords the indole products. Minakata, Komatsu, and coworkers generated 1,5-dipoles from either *O*-stannylmethylated thioanilides or isothiocyanates that cyclize to form indoles (equations 4 and 5) [7]. The instability of the stannylthio group (equation 5) was circumvented by transforming this product to the corresponding disulfide in air (e.g., MeI, TBAF; 88%), to 2-pyridylthio (2-bromopyridine, $Pd(PPh_3)_4$; 86%), and to acylthio (e.g., PhCOCl; 74%).

The several tin-promoted cyclizations that afford *indolines* (2,3-dihydroindoles) are not covered herein. Rather, if subsequent *indole* formation is reported, then that work is in the later chapter covering the conversion of indolines to indoles.

References

[1] Y. Dan-oh, H. Matta, J. Uemura, *et al.*, *Bull. Chem. Soc. Jpn.*, 1995, **68**, 1497–1507.
[2] W.R. Bowman, A.J. Fletcher, P.J. Lovell, and J.M. Pedersen, *Synlett*, 2004, 1905–1908.
[3] W.R. Bowman, A.J. Fletcher, J.M. Pedersen, *et al.*, *Tetrahedron*, 2007, **63**, 191–203.
[4] J.M. Pedersen, W.R. Bowman, M.R.J. Elsegood, *et al.*, *J. Org. Chem.*, 2005, **70**, 10615–10618.
[5] K. Orito, S. Uchiito, Y. Satoh, *et al.*, *Org. Lett.*, 2000, **2**, 307–310.
[6] T. Kim and K. Kim, *Tetrahedron Lett.*, 2010, **51**, 868–871.
[7] S. Minakata, Y. Kasano, H. Ota, *et al.*, *Org. Lett.*, 2006, **8**, 3693–3695.

51

The Murphy Indole Synthesis

Following the exploration of the synthetic utility of arenediazonium salts in radical formation and cyclization—"radical-polar crossover" [1–6]—Murphy described a new indole synthesis based on his methodology (Scheme 1, equations 1–4) [7–9]. The requisite precursor diazonium salts can be rapidly assembled from *o*-nitroaniline. In cases where the fully aromatic indole ring is not obtained, isomerization can be achieved with acid (equation 3) [8]. The Murphy radical indolization avoids the use of toxic tin compounds and is often superior in product yield. Murphy and coworkers presented tetrakis(dimethylamino)ethylene (TDAE) as a new reagent for the generation of aryl radicals from arenediazonium salts (equation 4) [9].

Indole Ring Synthesis: From Natural Products to Drug Discovery, First Edition. Gordon W. Gribble.

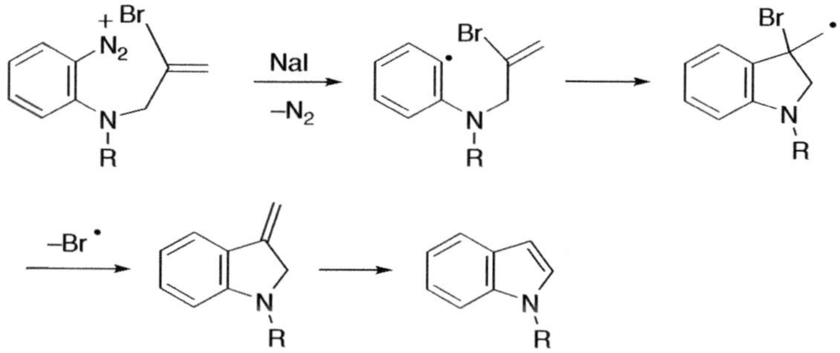

[eqn. 1] [7], [8]

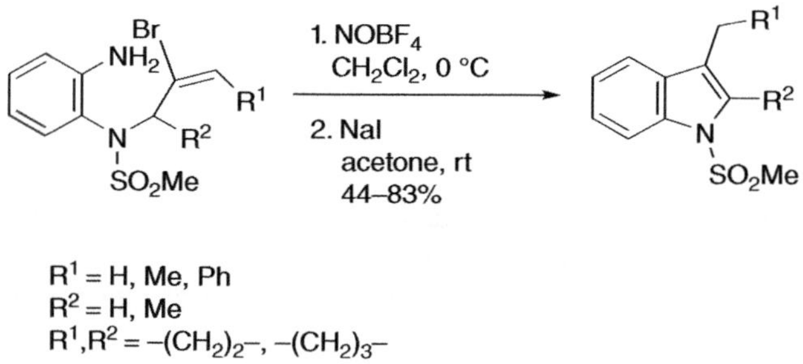

[eqn. 2] [7], [8]

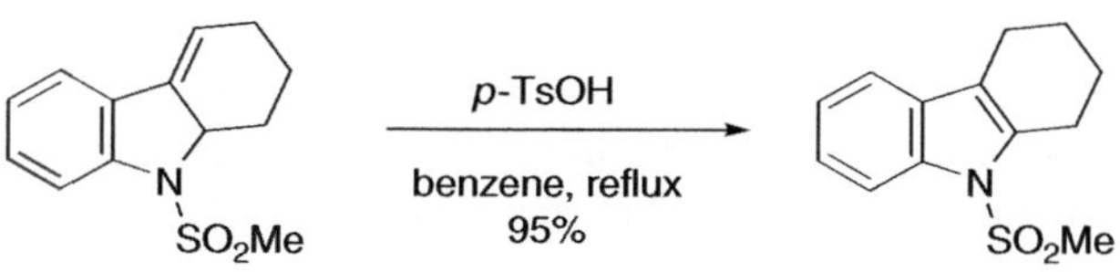

[eqn. 3] [8]

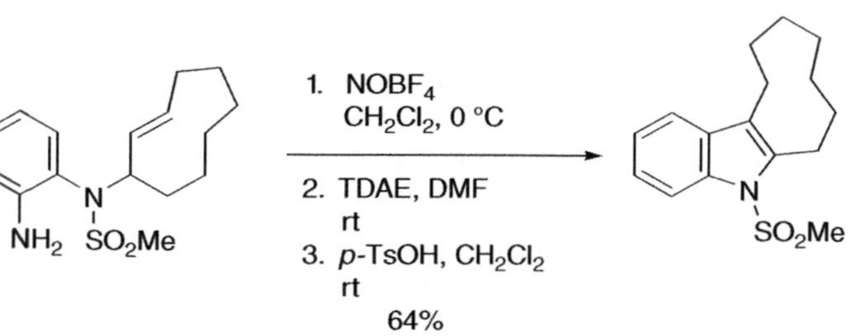

[eqn. 4] [9]

Scheme 1 *Murphy Indole Synthesis*

References

[1] C. Lampard, N. Lewis, and J.A. Murphy, *J. Chem. Soc., Chem. Commun.*, 1993, 295–297.
[2] C. Lampard, J.A. Murphy, F. Rasheed, *et al.*, *Tetrahedron Lett.*, 1994, **35**, 8675–8678.
[3] M. Kizil, C. Lampard, and J.A. Murphy, *Tetrahedron Lett.*, 1996, **37**, 2511–2514.
[4] J.A. Murphy, F. Rasheed, S. Gastaldi, *et al.*, *J. Chem. Soc., Perkin Trans. 1*, 1997, 1549–1558.
[5] R. Fletcher, M. Kizil, C. Lampard, *et al.*, *J. Chem. Soc., Perkin Trans. 1*, 1998, 2341–2351.
[6] B. Patro, M. Merrett, J.A. Murphy, *et al.*, *Tetrahedron Lett.*, 1999, **40**, 7857–7860.
[7] J.A. Murphy, K.A. Scott, R.S. Sinclair, and N. Lewis, *Tetrahedron Lett.*, 1997, **38**, 7295–7298.
[8] J.A. Murphy, K.A. Scott, R.S. Sinclair, *et al.*, *J. Chem. Soc., Perkin Trans. 1*, 2000, 2395-2408.
[9] M. Mahesh, J.A. Murphy, F. LeStrat, and H.P. Wessel, *Beilstein J. Org. Chem.*, 2009, **5**, 1–12.

52

Miscellaneous Radical-Promoted Indole Syntheses

The proclivity of radicals to add to multiple bonds has led numerous investigators to invent new indole ring construction via radical cyclization in addition to those methods previously discussed.

Inanaga and colleagues have found that the facile reduction of organic halides with samarium diiodide [1] can be extended to an indole synthesis (Scheme 1, equation 1) [2]. With allylamino substituents, the SmI_2 radical cyclization affords indolines (59%–99% yields). A control experiment indicated that the intermediate *exo*-cyclic double bond product is not a vinyl anion. A different SmI_2-induced indole ring formation involves reductive coupling of acylamido carbonyl compounds, as described by Zhang and Fan (equation 2) [3]. When R = alkyl (Me, Et) the yields are much lower (38%–43%), and the (often) major product is the corresponding 2,3-dihydroxyindoline, which is the major product when the reaction in equation 2 is carried out at lower temperatures (0 to –20 °C). A ketyl radical anion is proposed as the first intermediate. Further reduction to a dianion and cyclization leads to the indole. Montevecchi and coworkers reported the sulfanyl radical addition to *o*-alkynylphenyl azides to form indoles (equation 3) [4, 5] and the related sulfanyl radical cyclization to *o*-alkynylanilines (equation 4) [6]. Benzyl mercaptan can also be the radical trigger. The reaction in equation 4 can be effected using PhSSPh in the presence of AIBN at 150 °C.

Fukuyama and colleagues employed hypophosphorous acid to induce indolization of *o*-alkenylthioanilides (Scheme 2, equation 1), including a synthesis of (±)-catharanthine and a 2-adamantylindole [7, 8]. Rainier and Kennedy succeeded in capturing the intermediate indolenine from the radical cyclization of the isocyanide **1** to indole **2** (equation 2). Furthermore, **2** (R = Et) was smoothly converted to substitution products (e.g., **3**; equation 3) in a process believed to involve a phosphonium ylide (e.g., **4**, equation 4) [9].

Manganese (III) acetate has found utility in indole ring formation via a free radical cyclization. Thus, Chuang and colleagues used this oxidant to couple 2-aminonaphthoquinones with β-dicarbonyl compounds to form benzoindoloquinones (Scheme 3, equation 1) [10–12]. Cerium (IV) also effects this indolization [13]. Velu and colleagues used this oxidation in a synthesis of calothrixins A and B (key step: equation 2) [14]. Uneyama and coworkers generated α-trifluoroacetimidoyl radicals from the photolysis of phenyltellurotrifluoroacetimidates to give indoles (equation 3) [15]. Ogawa used a similar photochemical process with diphenyl ditelluride, isocyanides, and disulfides to prepare indoles (equation 4) [16, 17], *à la* Rainier's isocyanide method [9, 18–22]. This radical process is initiated by the photo dissociation of diphenyl ditelluride.

Studer and Zhang developed a radical trifluoromethylation of isocyanides to give 2-trifluoromethylated indoles using the Togni reagent (**5**) (Scheme 4, equation 1) [23]. The *E*-isomer is generally favored. Addition of the trifluoromethyl radical to the isocyanide carbon initiates the process, as we saw earlier. The vinyl double bond in the product can be reduced (H_2, Pd/C or $CeCl_3$, $NaBH_4$). An electrochemical radical initiation has been applied to indolization by several investigators. Boujlel's group employed a Bunnett $S_{RN}1$ process (equation 2) [24]. Fox and colleagues used a nickel catalyst in an electrochemical indole ring formation (equation 3) [25], and Grimshaw's team found that indolines are formed under

Indole Ring Synthesis: From Natural Products to Drug Discovery, First Edition. Gordon W. Gribble.

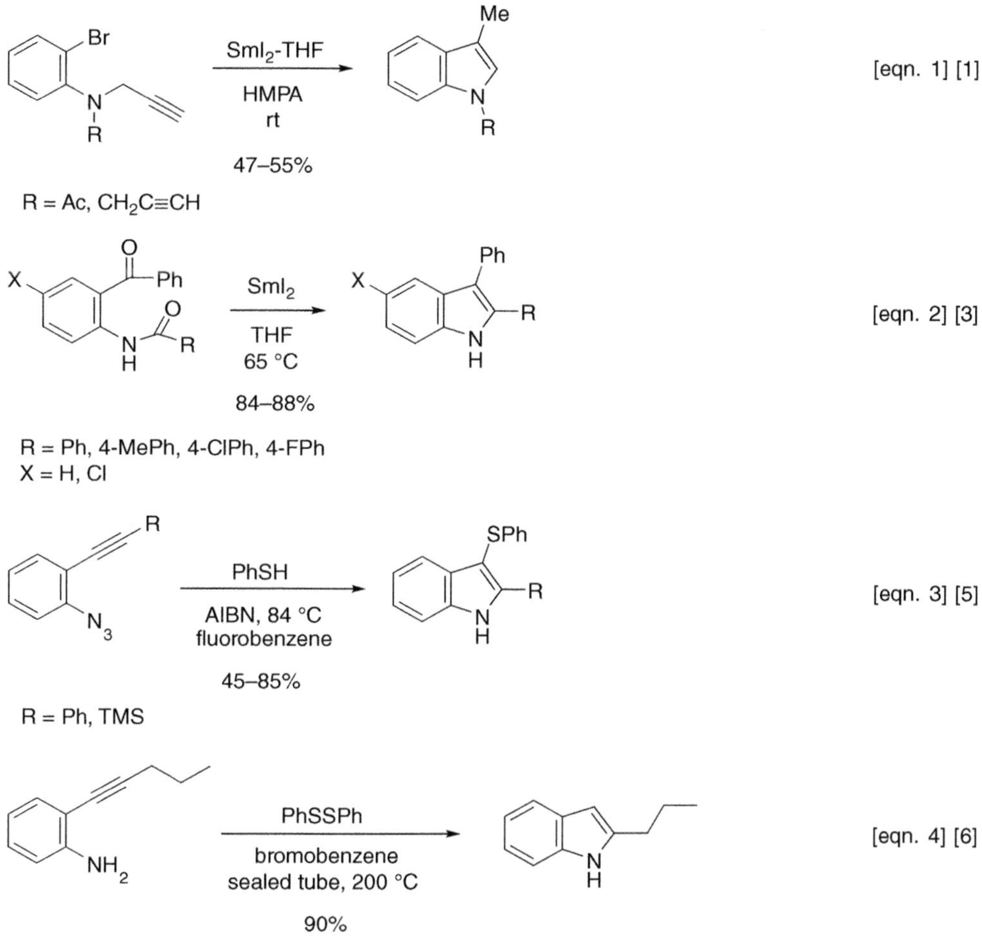

Scheme 1 *Miscellaneous Radical Cyclizations to Form Indoles — 1*

electrochemical conditions when the aniline substrate lacks a potential leaving group [26]. Bremner and Rezaie employed a presumed bromine atom indolization (equation 4) [27]. In addition to lactam **6**, two ring-brominated indole lactams were obtained (26%). An amidyl radical mechanism is suggested.

Peroxides are notorious for generating radicals, and this tactic has furnished some novel indole syntheses. Sortais and colleagues found that a subtle modification of their indoline synthesis can lead to indoles (Scheme 5, equation 1) including a synthesis of melatonin [28]. The C-5 methoxyl group may be necessary to assist in the ejection of $MeSO_3H$. The combination of a hypophosphite and oxygen has paved the way for an indole synthesis as discovered by Wilden and Gray (equation 2) [29]. The excellent leaving group (trichlorophenylsulfonate) ensures ultimate indolization. Simple pyrolysis of suitable substrates can yield radicals appropriate for cyclization to indoles. For example, Yonemitsu and colleagues showed that thermolysis of *N*-acylacetylphenylhydroxylamines leads to indoles (equation 3) [30]. A radical mechanism is indicated because AIBN increases the yield in the case R = Me (67%–82%), and the radical trap diphenylpicrylhydrazyl (DPPH) shuts down the reaction. Aiken and Murray found that flash vacuum pyrolysis (FVP) of various ylides led to fused indoles (equation 4) [31]. The reaction may proceed to the *o*-alkynylaminyl radical by loss of triphenylphosphine oxide and methanesulfonyl radical and then cyclization.

The cycloaddition of nitrosoarenes with alkynes as reported by Penoni and colleagues seems to involve a radical-mediated indole ring formation (Scheme 6, equations 1, 2) [32]. Other groups can be used to cap the labile *N*-hydroxyindoles (Boc, Bn, CO_2Et, Cbz, Ts, Bz). Li and colleagues have discovered a new synthesis of

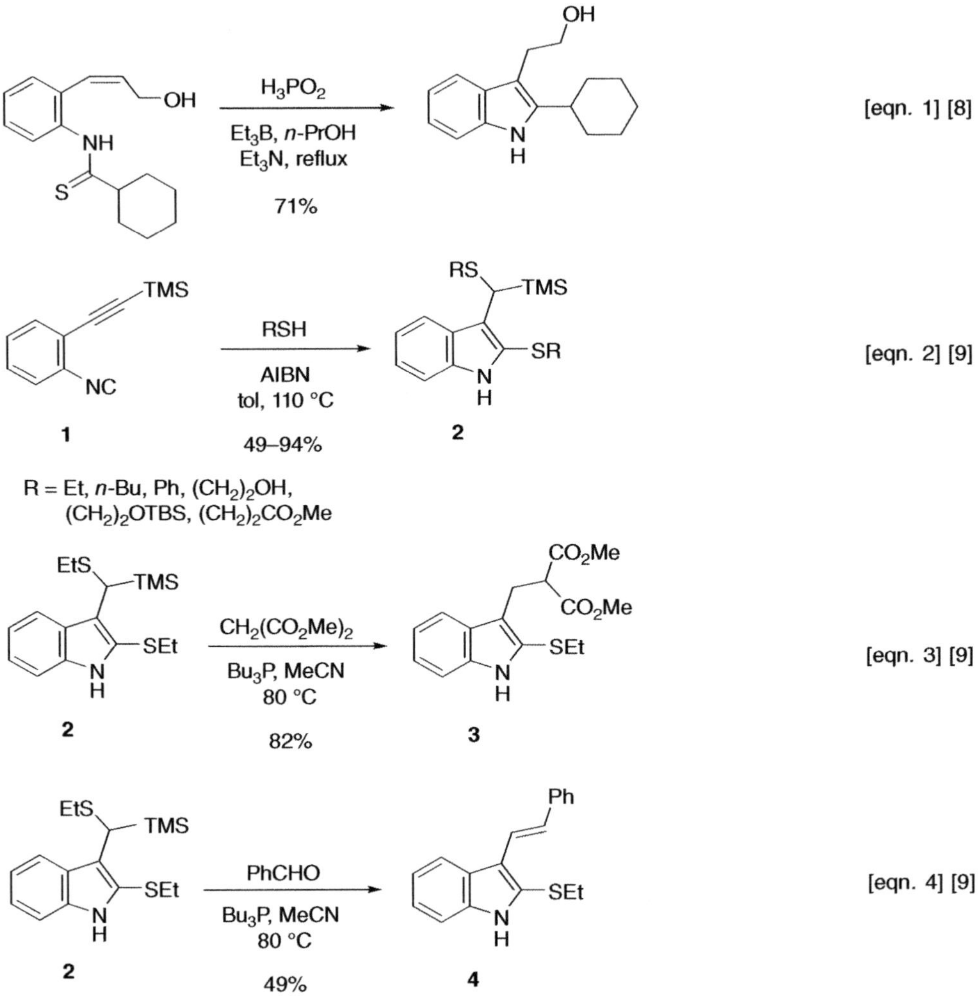

Scheme 2 *Miscellaneous Radical Cyclizations to Form Indoles — 2*

pyrrolo[4,3,2-*de*]quinolinones that involves *t*-butyl nitrite-promoted (nitrogen dioxide or nitric oxide) radical cyclization of *o*-alkynylamides (equations 3, 4) [33]. Many examples were synthesized by this group.

The separate groups of Molina, Alajarín, and Vidal [34–38] and of Wang [39–46] have made extensive use of a biradical cycloaromatization reaction to form both indoles and highly fused indoles. Only a partial reference list is cited here. For example, Vidal and colleagues reported the radical-mediated cyclization of ketenimine xanthates **7** to yield 2-alkylindoles (Scheme 7, equation 1) [36]. Substrates **7** were prepared from 2-azidobenzyl bromides and the xanthate salt, KSC(S)OEt, then reaction with triphenylphosphane and treatment of the resulting triphenylphosphazene with the appropriate ketene. Related chemistry by this group afforded quinindoline (**8**), the precursor of the alkaloid cryptotackieine (equation 2) [35]. The major product is 2-anilinoquinoline (40%).

Likewise, Wang and his coworkers have made extensive use of the biradicals from *o*-alkynylketenimines and carbodiimides in the synthesis of fused indoles (Scheme 8, equations 1, 2). When R = H (equation 1) the corresponding quinoline is formed (49%) [40]. Some of Wang's other successes are **9–12** with the yield of the cycloaromatization step shown. Pieters and coworkers employed Wang's method to prepare a series of neocryptolepine derivatives for biological screening. Indeed, 2-bromoneocryptolepine was very active in an antimalarial screen [47]. Benati's team described a similar protocol leading to thiochromeno[2,3-*b*]indoles from the reaction of aryl radicals (from aryl diazonium salts) and *o*-alkynyl isothiocyanates [48].

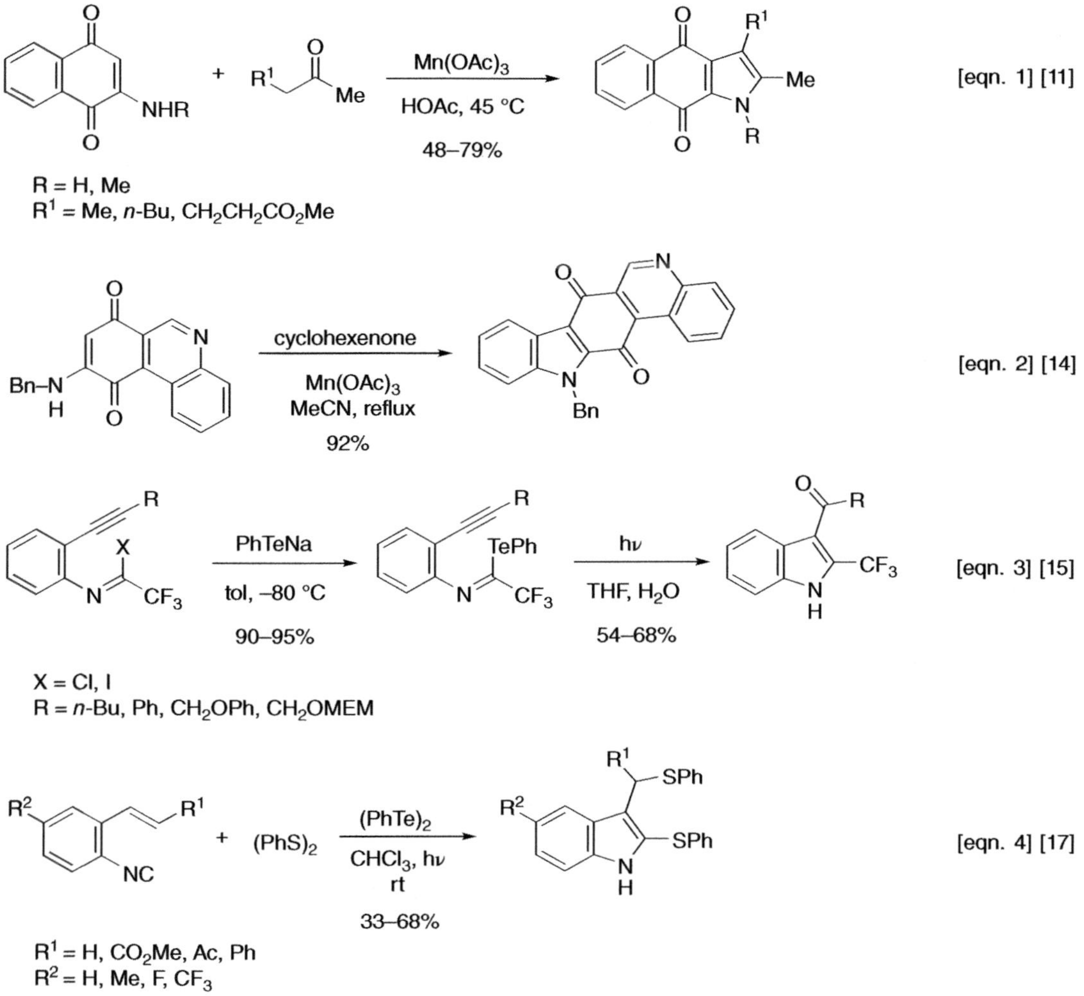

Scheme 3 *Miscellaneous Radical Cyclizations to Form Indoles — 3*

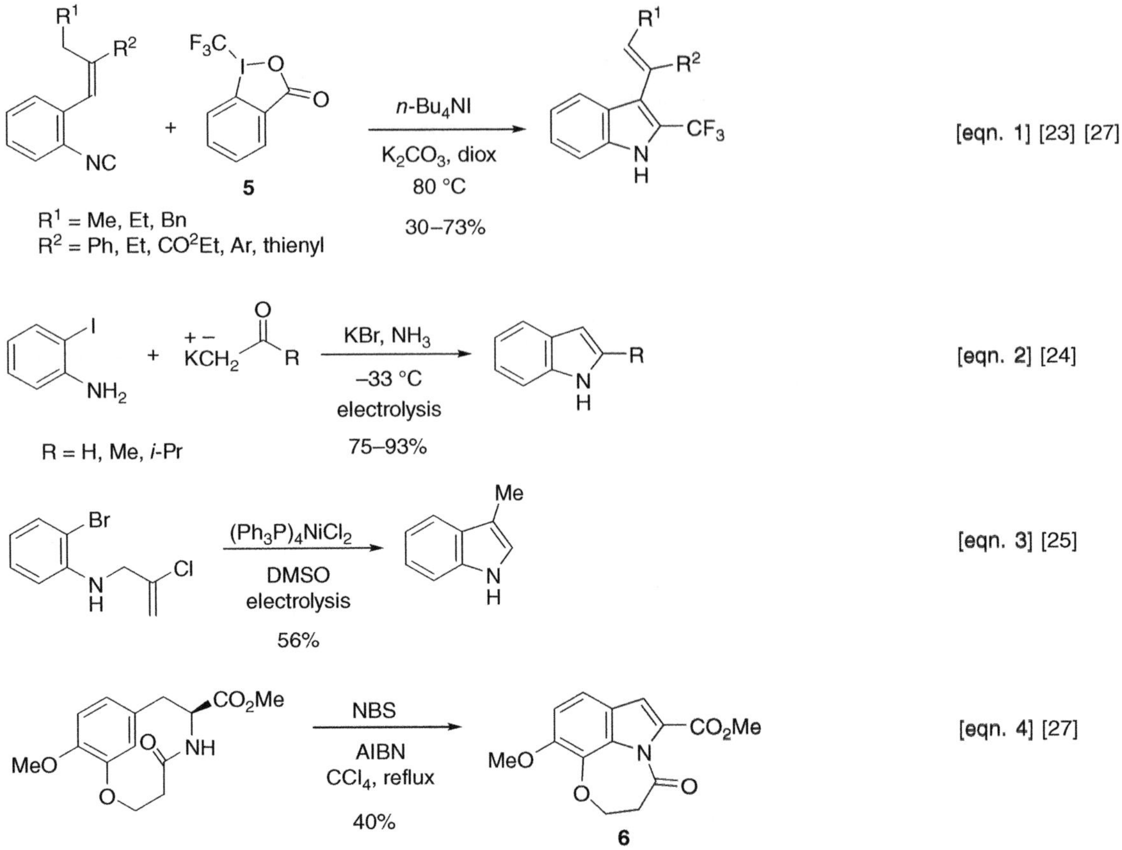

Scheme 4 *Miscellaneous Radical Cyclizations to Form Indoles — 4*

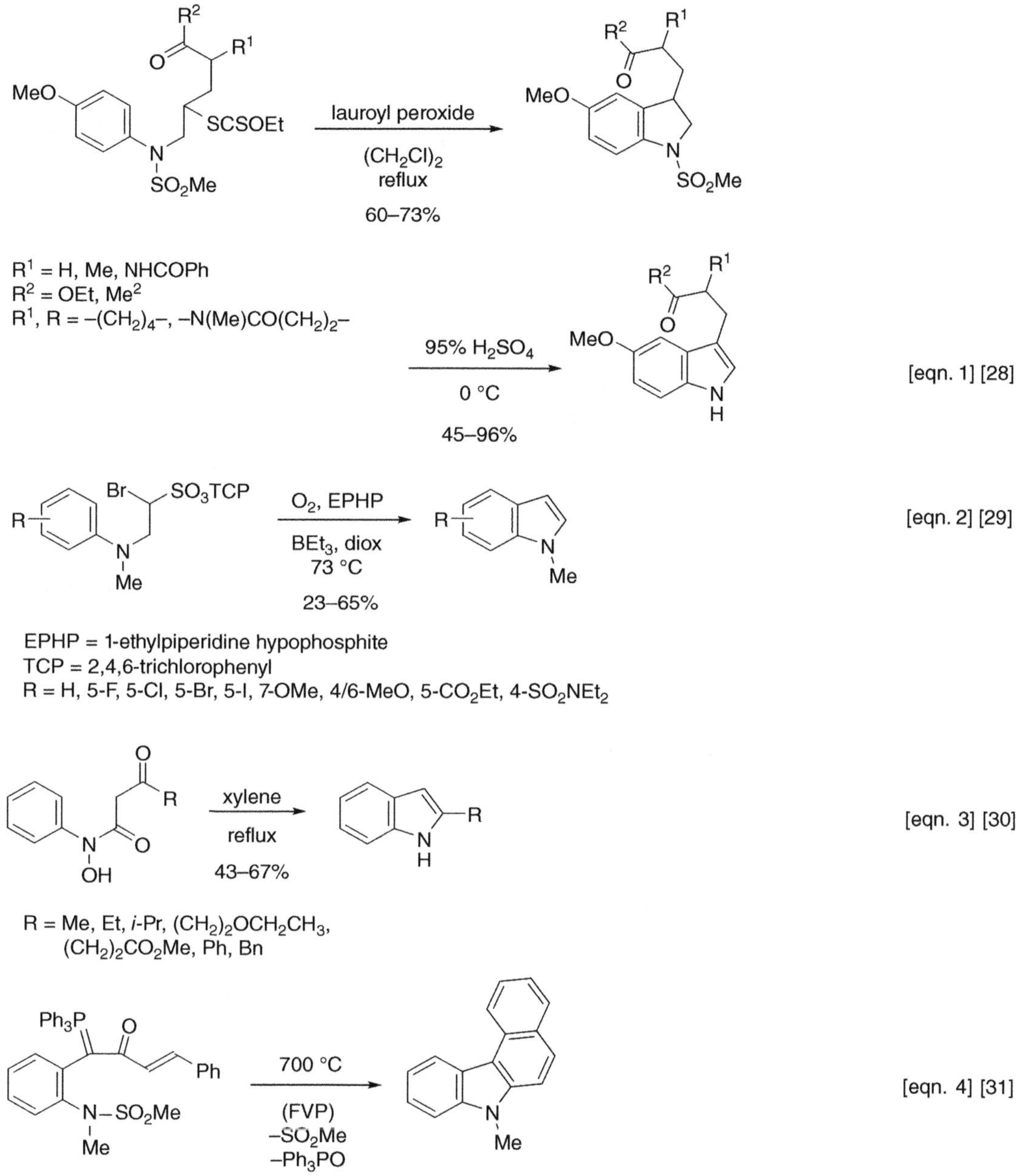

Scheme 5 *Miscellaneous Radical Cyclizations to Form Indoles — 5*

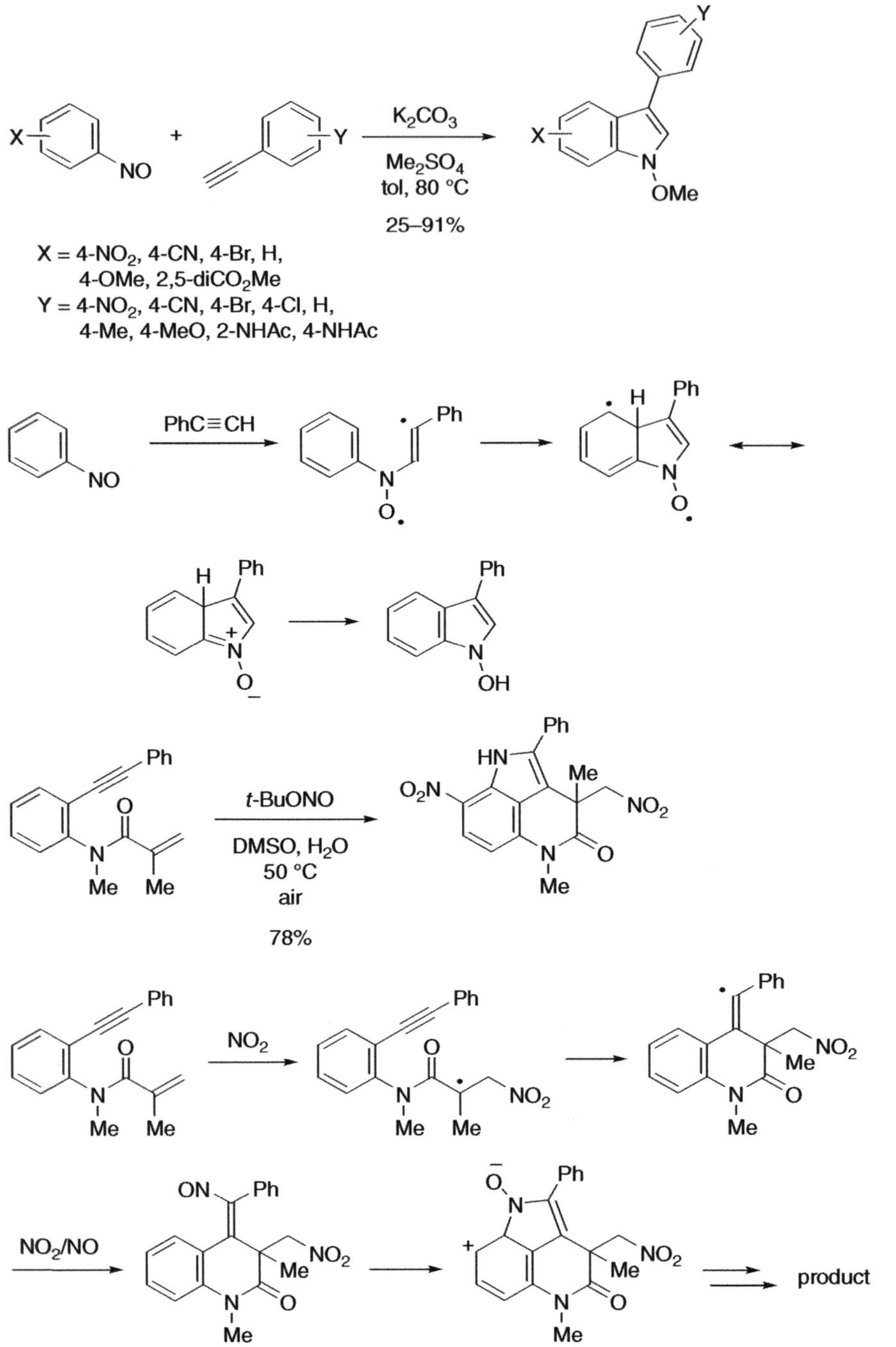

[eqn. 1] [32]

[eqn. 2] [32]

[eqn. 3] [33]

[eqn. 4] [33]

Scheme 6 *Miscellaneous Radical Cyclizations to Form Indoles — 6*

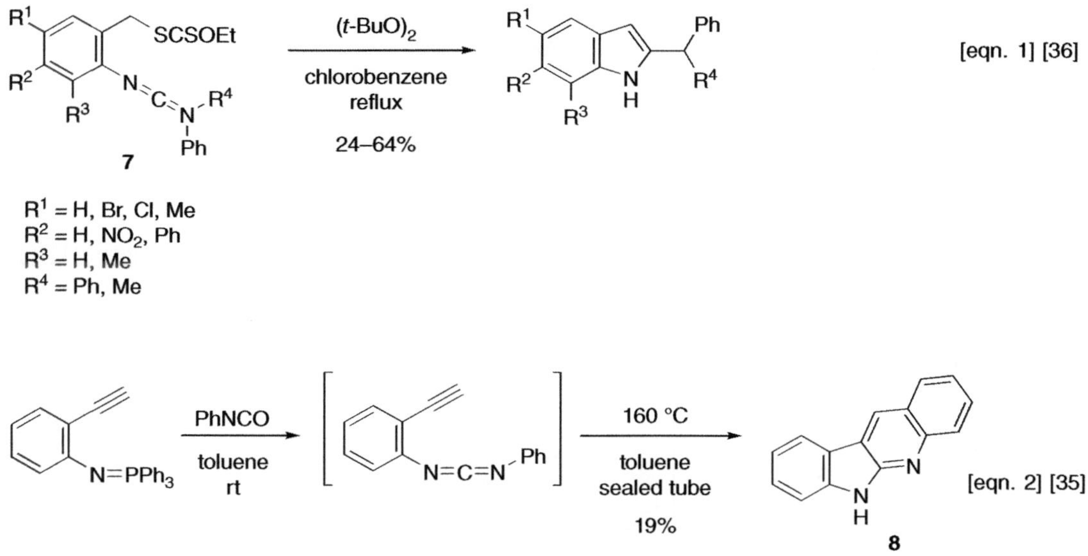

Scheme 7 *Molina, Vidal, and Alajarín Indole Ring Synthesis*

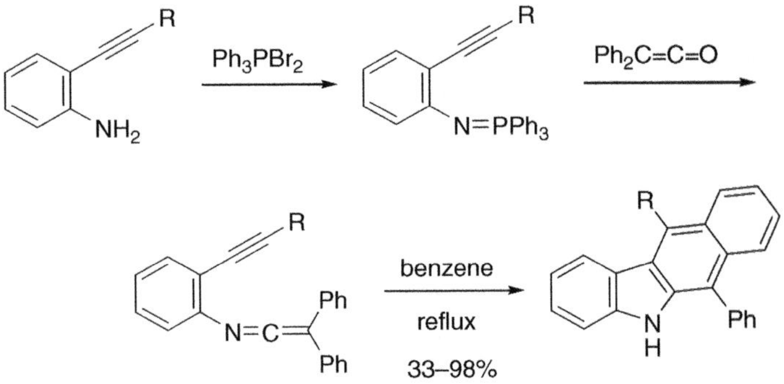

R = *t*-Bu, TMS, *n*-Pr, Ph

[eqn. 1] [40]

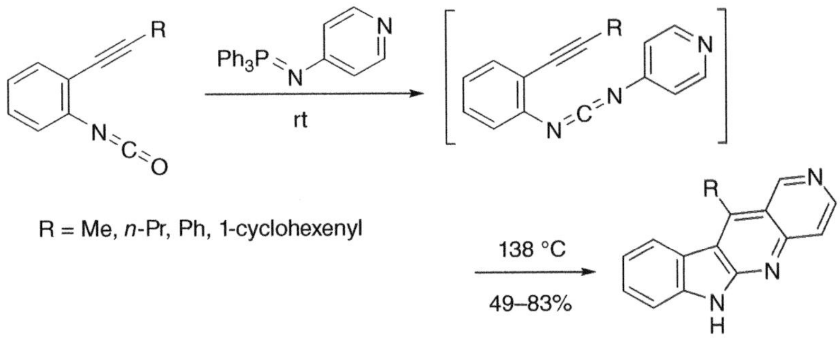

R = Me, *n*-Pr, Ph, 1-cyclohexenyl

[eqn. 2] [42]

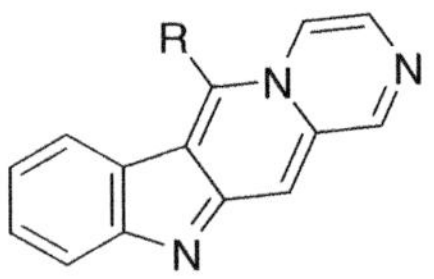

9 (R = *n*-Pr, Ph) [45]

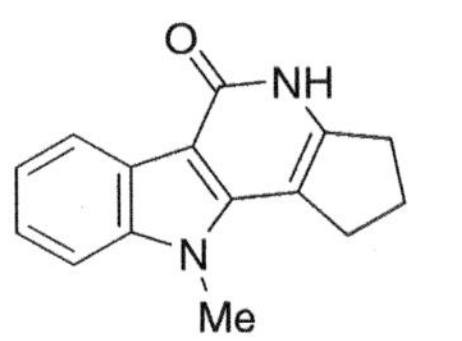

10 (75%) [46]

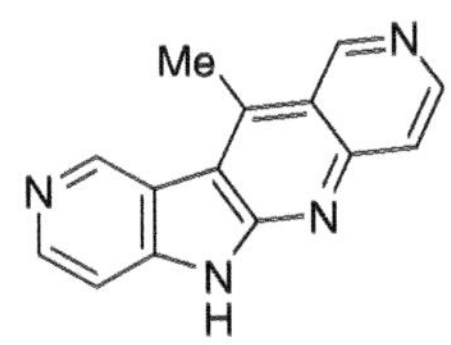

11 (64%) [43]

(47–56%)

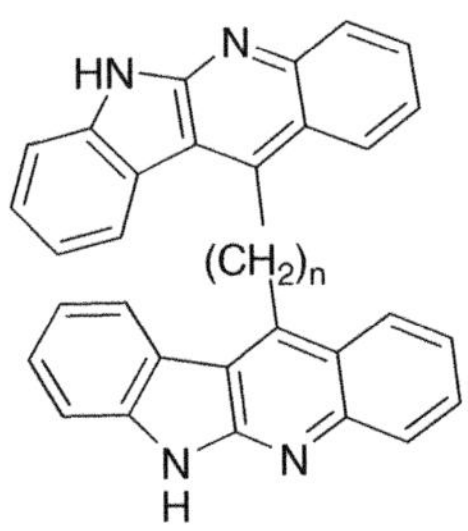

12 (n = 3, 5) (74–77%) [41]

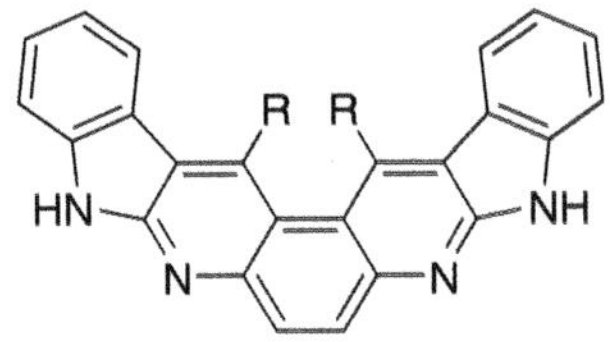

13 (R = *n*-C_8H_{17}) (66%) [41]

Scheme 8 *Wang Indole Ring Synthesis*

References

[1] J. Inanaga, M. Ishikawa, and M. Yamaguchi, *Chem. Lett.*, 1987, 1485–1486.
[2] J. Inanaga, O. Ujikawa, and M. Yamaguchi, *Tetrahedron Lett.*, 1991, **32**, 1737–1740.
[3] X. Fan and Y. Zhang, *Tetrahedron*, 2003, **59**, 1917–1923.
[4] P.C. Montevecchi, M.L. Navacchia, and P. Spagnolo, *Tetrahedron Lett.*, 1997, **38**, 7913–7916.
[5] P.C. Montevecchi, M.L. Navacchia, and P. Spagnolo, *Eur. J. Org. Chem.*, 1998, 1219–1226.
[6] P.C. Montevecchi and M.L. Navacchia, *Tetrahedron Lett.*, 1998, **39**, 9077–9080.
[7] M.T. Reding and T. Fukuyama, *Org. Lett.*, 1999, **1**, 973–976.
[8] M.T. Reding, Y. Kaburagi, H. Tokuyama, and T. Fukuyama, *Heterocycles*, 2002, **56**, 313–330.
[9] J.D. Rainier and A.R. Kennedy, *J. Org. Chem.*, 2000, **65**, 6213–6216.
[10] M.-C. Jiang and C.-P. Chuang, *J. Org. Chem.*, 2000, **65**, 5409–5412.
[11] Y.-L. Wu, C.-P. Chuang, and P.-Y. Lin, *Tetrahedron*, 2001, **57**, 5543–5549.
[12] C.-M. Tseng, Y.-L. Wu, and C.-P. Chuang, *Tetrahedron*, 2004, **60**, 12249–12260.
[13] C.-C. Tseng, Y.-L. Wu, and C.-P. Chuang, *Tetrahedron*, 2002, **58**, 7625–7633.
[14] S. Xu, T. Nguyen, I. Pomilio, *et al.*, *Tetrahedron*, 2014, **70**, 5928–5933.
[15] Y. Ueda, H. Watanabe, J. Uemura, and K. Uneyama, *Tetrahedron Lett.*, 1993, **34**, 7933–7934.
[16] T. Mitamura, Y. Tsuboi, K. Iwata, *et al.*, *Tetrahedron Lett.*, 2007, **48**, 5953–5957.
[17] T. Mitamura, K. Iwata, and A. Ogawa, *J. Org. Chem.*, 2011, **76**, 3880–3887.
[18] A.R. Kennedy, M.H. Taday, and J.D. Rainier, *Org. Lett.*, 2001, **3**, 2407–2409.
[19] A.V. Novikov, A.R. Kennedy, and J.D. Rainier, *J. Org. Chem.*, 2003, **68**, 993–996.
[20] A.M. Nyong and J.D. Rainier, *J. Org. Chem.*, 2005, **70**, 746–748.
[21] A.V. Novikov, A. Sabahi, A.M. Nyong, and J.D. Rainier, *Tetrahedron Asymm.*, 2003, **14**, 911–913.
[22] A. Sabahi and J.D. Rainier, *ARKIVOC*, 2010, 116–125.
[23] B. Zhang and A. Studer, *Org. Lett.*, 2014, **16**, 1216–1219.
[24] K. Boujlel, J. Simonet, G. Roussi, and R. Beugelmans, *Tetrahedron Lett.*, 1982, **23**, 173–176.
[25] M.A. Fox, D.A. Chandler, and C. Lee, *J. Org. Chem.*, 1991, **56**, 3246–3255.
[26] M. Dias, M. Gibson, J. Grimshaw, *et al.*, *Acta Chem. Scand.*, 1998, **52**, 549–554.
[27] R. Rezaie and J.B. Bremner, *Synlett*, 1996, 1061–1062.
[28] B. Quiclet-Sire, B. Sortais, and S.Z. Zard, *Chem. Commun.*, 2002, 1692–1693.
[29] V.J. Gray and J.D. Wilden, *Tetrahedron Lett.*, 2012, **53**, 41–44.
[30] K. Hirao, K. Mohri, O. Yonemitsu, *et al.*, *Tetrahedron Lett.*, 1992, **33**, 1459–1462.
[31] R.A. Aitken and L. Murray, *J. Org. Chem.*, 2008, **73**, 9781–9783.
[32] G. Ieronimo, A. Mondelli, F. Tibiletti, *et al.*, *Tetrahedron*, 2013, **69**, 10906–10920.
[33] Y. Liu, J.-L. Zhang, R.-J. Song, *et al.*, *Angew. Chem. Int. Ed.*, 2014, **53**, 9017–9020.
[34] P. Molina, M. Alajarín, and A. Vidal, *J. Chem. Soc., Chem. Commun.*, 1990, 1277–1279.
[35] M. Alajarín, P. Molina, and A. Vidal, *J. Nat. Prod.*, 1997, **60**, 747–748.
[36] M. Alajarín, A. Vidal, and M.-M. Ortin, *Tetrahedron Lett.*, 2003, **44**, 3027–3030.
[37] M. Alajarín, B. Bonillo, Á. Vidal, and D. Bautista, *J. Org. Chem.*, 2008, **73**, 291–294.
[38] M. Alajarín, B. Bonillo, M.-M. Ortin, *et al.*, *Org. Biomol. Chem.*, 2011, **9**, 6741–6749.
[39] K.K. Wang, *Chem. Rev.*, 1996, **96**, 207–222.
[40] C. Shi and K.K. Wang, *J. Org. Chem.*, 1998, **63**, 3517–3520.
[41] C. Shi, Q. Zhang, and K.K. Wang, *J. Org. Chem.*, 1999, **64**, 925–932.
[42] Q. Zhang, C.S. Shi, H.R. Zhang, and K.-K. Wang, *J. Org. Chem.*, 2000, **65**, 7977–7983.
[43] X. Lu, J.L. Petersen, and K.K. Wang, *J. Org. Chem.*, 2002, **67**, 5412–5415.
[44] X. Lu, J.L. Petersen, and K.K. Wang, *Org. Lett.*, 2003, **5**, 3277–3280.
[45] X. Lu, J.L. Petersen, and K.K. Wang, *J. Org. Chem.*, 2002, **67**, 7797–7801.
[46] H. Li, H. Yang, J.L. Petersen, and K.K. Wang, *J. Org. Chem.*, 2004, **69**, 4500–4508.
[47] T.H.M. Jonckers, S. van Miert, K. Cimanga, *et al.*, *J. Med. Chem.*, 2002, **45**, 3497–3508.
[48] L. Benati, G. Calestani, R. Leardini, *et al.*, *J. Org. Chem.*, 2003, **68**, 3454–3464.

53

The Graebe–Ullmann Carbazole-Carboline Synthesis

Unlike the previously discussed modern free radical–promoted indolizations (Chapters 49–52), the venerable Graebe–Ullmann indole ring synthesis enjoys the longevity of other classic indole synthesis name reactions (Fischer, Madelung, Bischler, Nenitzescu, Reissert). In 1896, Carl Graebe and Fritz Ullmann reported a novel synthesis of carbazole (Scheme 1, equation 1) [1]. Thus, diazotization of *o*-aminodiphenylamine (**1**) yielded 1-phenyl-1,2,3-benzotriazole (**2**), which, upon heating to 360 °C, lost nitrogen to form diradical **3** in equilibrium with **4**, and the latter cyclized to carbazole (**5**) in essentially quantitative yield. Ullmann pursued this carbazole synthesis for several years, describing the syntheses of 2- and 3-chlorocarbazole, 1- and 3-methylcarbazole, 1,3-dimethylcarbazole, 3-aminocarbazole, benzo[*c*]carbazole, and a few others [2–4]. Other workers were less successful in employing the Graebe–Ullmann method to the synthesis of substituted carbazoles [5, 6]. For brief reviews, see Li and Cook [7], Rachwal and Katritzky [8], and Campbell and Barclay [9].

Support for the diradical mechanism was shown by Wentrup [10] and Ohashi and colleagues [11], who established that a benzazirine intermediate *does not* form in the thermolysis of **2**. Thus, the pyrolysis of 5-methyl-1-phenylbenzotriazole (**6**) at 780–1100 °C gave only 3-methylcarbazole (**7**) but no 2-methylcarbazole (**9**), which would have formed had benzazirine (**8**) intervened (Scheme 2, equation 1) [10]. Likewise, only 3,6-dichlorocarbazole (**11**) is obtained from **10**, and not 2,6-dichlorocarbazole (**12**) (equation 2) [11]. Moreover, irradiation of the isomeric *N*-methylbenzotriazoles **13** and **15** affords phenylated products **14** and **16**, respectively (equations 3, 4), which excludes the intermediacy of benzazirine **17** [12]. In contrast, the photolysis of 3,3-disubstituted indazenes affords a diradical that cyclizes to the (isolable) benzocyclopropene ring system [13, 14].

Robinson and colleagues were apparently the first to extend the Graebe–Ullmann carbazole synthesis to *carbolines* (Scheme 3, equation 1) [15]. Thus, 1-α-pyridylbenzotriazole, prepared as shown, was converted to α-carboline (equation 1). "Quinindoline" (6*H*-indolo[2,3-*b*]quinolone) was similarly prepared from 2-chloroquinoline. Kermack and Smith found that the Graebe–Ullmann carbazole synthesis could be performed at lower temperature in "syrupy" phosphoric acid to give, for example, "2:3-benz-γ-carboline" (11*H*-indolo[3,2-*c*]quinolone) (equation 2) [16, 17]. A number of analogous benzocarbolines were described [17]. Several of the early carbazoles and carbolines that were crafted using the Graebe–Ullmann benzotriazole methodology under thermal conditions are shown in Scheme 3 [18–22]. The origin of the indole ring is shown in bold. In many cases, yields are not cited. The synthesis of 1-nitrocarbazole was carried out in the presence of copper bronze [22]. The benzotriazoles **18** and **19** failed to undergo the Graebe–Ullmann carbazole synthesis [22]. The Graebe–Ullmann modification using hot syrupy phosphoric acid or polyphosphoric acid was employed by Freak and Robinson to synthesize carbolines such as "9:10-benzo-3-carboline" (7*H*-benzo[*e*]pyrido[2,3-*b*]indole) [23], by Witkop to prepare α-carboline (50% yield) [24], and by Ashton and Suschitzky to forge 6-chloro-α-carboline (37%) and 6-methyl-α-carboline (46%) [25].

In 1968, Burgess and colleagues discovered the photochemical Graebe–Ullmann carbazole synthesis. Thus, irradiation of a solution of 1-phenylbenzotriazole in benzene (0.1 M) afforded a nearly quantitative yield of carbazole (Scheme 4, equation 1) [26]. Schmid and colleagues

Indole Ring Synthesis: From Natural Products to Drug Discovery, First Edition. Gordon W. Gribble.

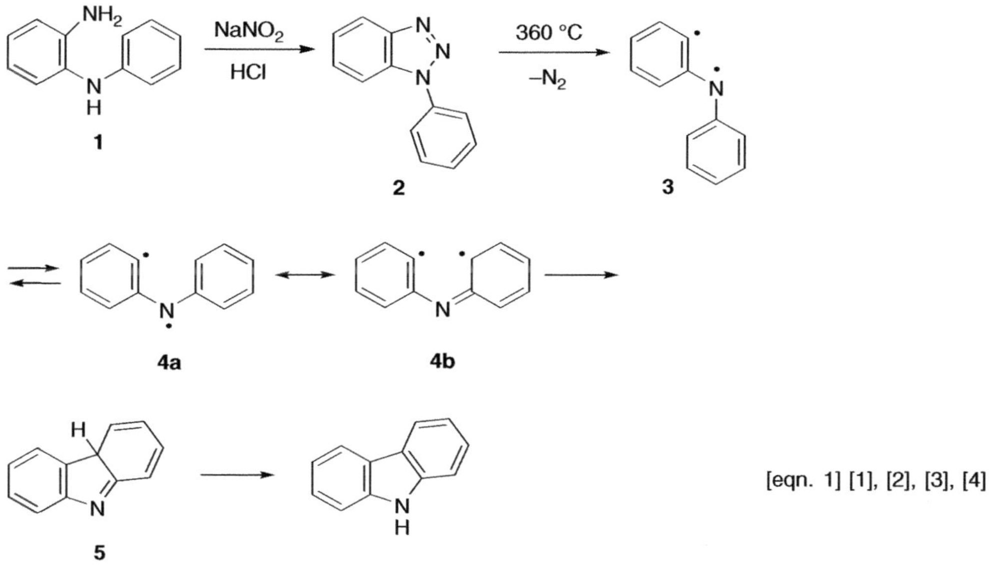

Scheme 1 *Graebe-Ullmann Carbazole Synthesis*

somewhat later encountered this new indole–carbazole synthesis [27, 28], as did Hubert who extended this benzotriazole photolysis to obtain, for example, pyrido[1,2-*a*] benzimidazole (equation 2), benzimidazo[2,1-*a*]isoquinoline (equation 3), and thiazolo[3,2-*a*]benzimidazole (equation 4) [29, 30]. In each case the yield from thermolysis is either lower or affords a different product (α-carboline) from the photochemical method. Strausz and colleagues studied the electron spin resonance spectrum of the diradical **3** ⇄ **4** (Scheme 1) generated photochemically (Vycor-filtered) at 77 K [31]. These diradicals exist as rotameric conformers, with **4** being thermally less stable than **3**.

The photochemical Graebe–Ullmann carbazole–carboline synthesis has furnished several compounds that would be difficultly accessible by other methods (Scheme 5). For a relevant review, see Claus, Doppler, Gakis, *et al.* [32]. For example, although photolysis was successful, under thermal conditions no azaellipticine **20** (equation 1) was obtained [33]. Photolysis of several 1-(1-naphthyl)benzotrizoles gave the novel cycloocta[*def*]carbazole ring system (e.g., equation 2) [34]. Space does not permit presentation of this extraordinarily rich mechanism and the varied photolysis products that other 1-(1-naphthyl)benzotriazoles afford. Similarly, Rees and his coworkers have photolyzed several 1-aryl-benzotriazoles [35] with *ortho*-blocked substituents to afford an array of products such as cyclopentaquinones (e.g., equation 3) [36]. Under thermal conditions, carbazoles are the major products, having undergone [1, 5]methyl shifts.

The adaptation of the Graebe–Ullmann carbazole synthesis to indoles was apparently first described by Schmid's group [27, 28], but Wender and Cooper parlayed the photochemistry of 1-alkenylbenzotriazoles into a viable indole synthesis (Scheme 6, equations 1, 2) [37]. The yield of cyclopent[*b*]indole was 44%. Nagawa and colleagues photolyzed naphthyl-1,2,3-triazoles to afford indoles, including the new indolo[6,7-*g*]indole ring system (equations 3, 4) [38, 39]. Irradiation of **20** for only 2 minutes allowed the isolation of the "mono-indole" intermediate [39]. Barker and Storr used flash vacuum pyrolysis (FVP) to prepare simple indoles from 1-alkenylbenzotriazoles (equations 5 and 6) [40]. Edstrom and Yuan synthesized pyrrolo[2,3-*d*]pyrimidines from the photolysis of triazolyluracils (e.g., equation 7) [41].

Rees and colleagues explored the mechanism of the pyrolysis of 1,2,3-triazoles and established in elegant fashion that 1*H*-azirines *are* intermediates (Scheme 7) [42]. Thus, pyrolysis of triazoles **21** and **22** yields the same two 2*H*-azirines **24** and **25** and subsequent products via the single intermediate 1*H*-azirine **23**. Pyrolysis of **24** affords *N*-vinylphthalimide (**26**) and benzonitrile, whereas azirine **25** gives indole **27**, with no crossover. Moreover, there is no equilibration of the triazoles **21** and **22**.

Parrick and colleagues used the photochemical Graebe–Ullmann method to synthesize a series of dialkoxycarbolines (Scheme 8, equation 1) [43]. Warburton and Stephenson prepared several α-carbolines via an ortho-phosphoric acid–mediated synthesis (150–200 °C) [44], and Nantka-Namirski summarized his work on the synthesis of numerous α-carbolines using phosphoric acid [45]. In 1993, Alvarez-Builla and coworkers reported the first use of microwave irradiation in a one-pot Graebe–Ullmann synthesis (equation 2) [46]. The intermediate *N*-pyridylbenzotriazole was not

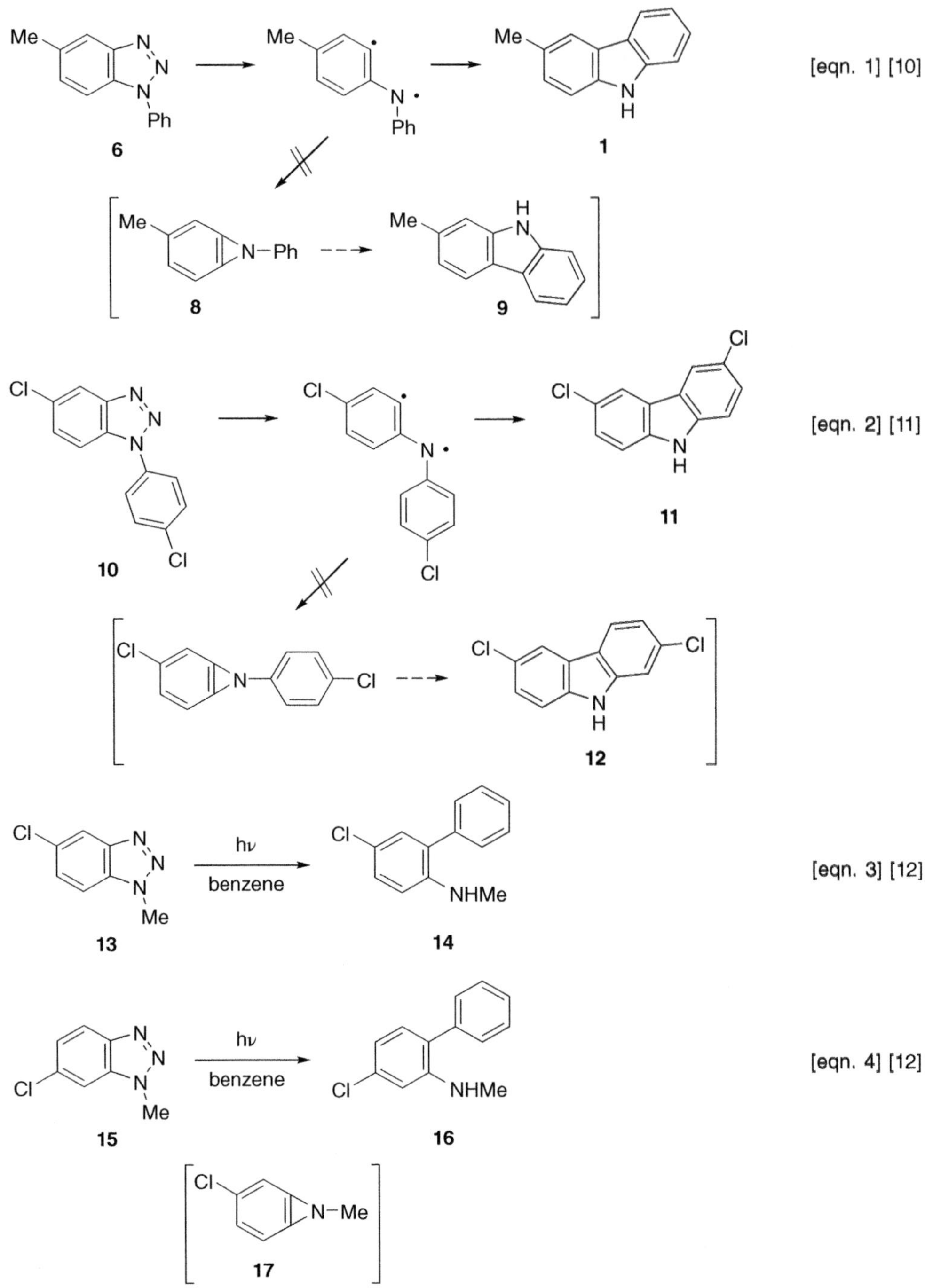

Scheme 2 *Thermolysis of 1-Phenylbenzotriazoles Does Not Involve a Benzazirine [10–12]*

isolated but rather was treated with pyrophosphoric acid and subjected to additional microwave irradiation. This research team extended this method to the synthesis of α-carbolines (equation 3) [47]. In this case the pyridylbenzotriazole was isolated. A selected group of carbolines synthesized by this method is shown in Scheme 8. Yurovskaya and colleagues also reported microwave heating in conjunction with polyphosphoric acid in the synthesis of isomeric aza-γ-carbolines [48]. Two examples from this work are shown.

The Graebe–Ullmann methodology has proved especially valuable in the synthesis of ellipticine and analogues. Using flash vacuum pyrolysis, Miller and Stowell

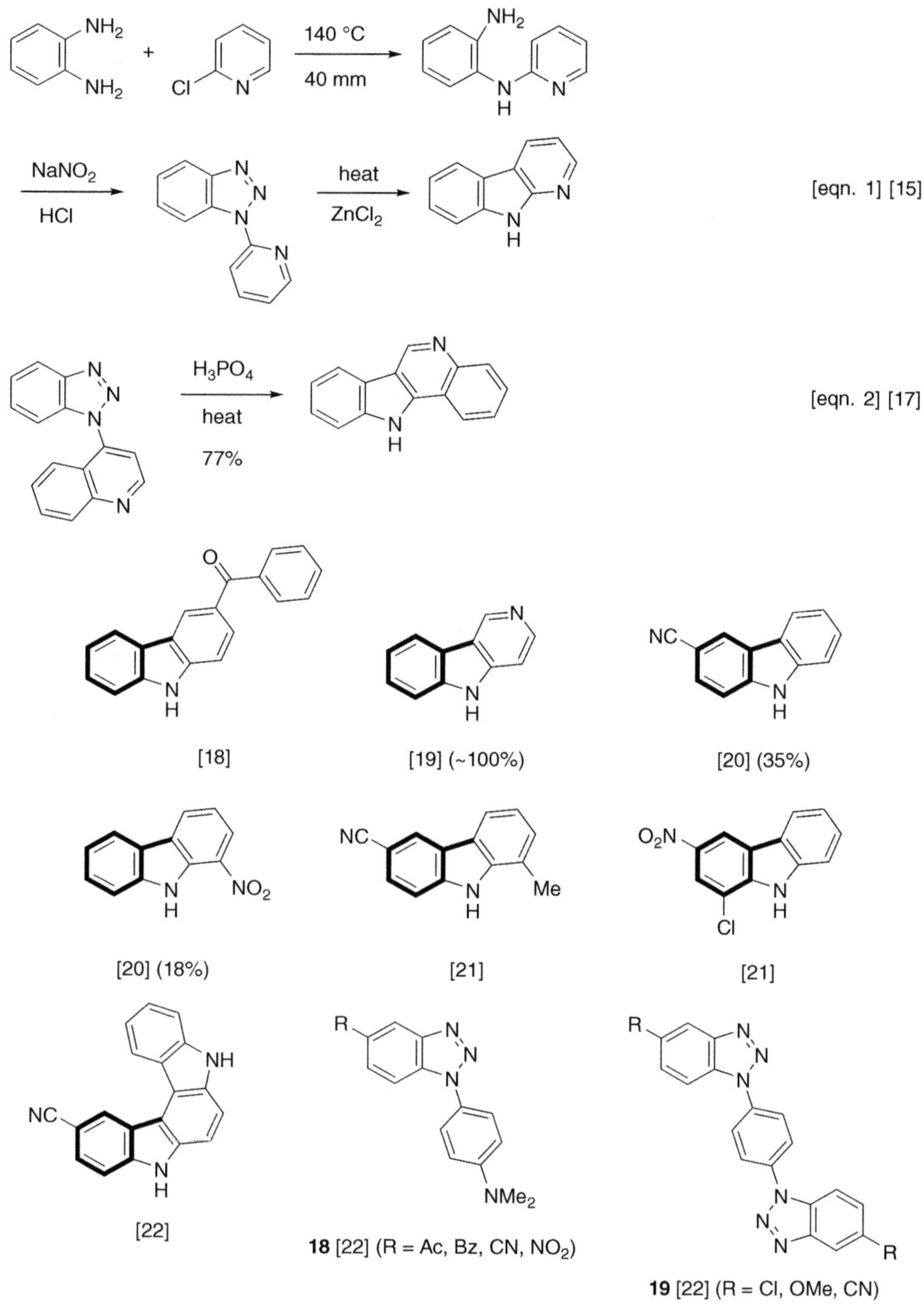

Scheme 3 *Early Applications of the Graebe-Ullmann Carbazole-Carboline Synthesis*

synthesized both ellipticine and 9-methoxyellipticine (Scheme 9, equation 1) [49]. Conventional heating in PPA (220 °C) and photolysis were less successful, 16% and 33%, respectively. Bisagni's group synthesized 2-azaellipticine (6,11-dimethyl-5*H*-pyrido[3′,4′:4,5]pyrrolo[2,3-*a*] isoquinoline) from the respective triazolopyridine at 320–340 °C in paraffin (41%) [50]. Using polyphosphoric acid at temperatures up to 180 °C, Peczyńska-Czoch and colleagues prepared a series of novel cytotoxic and DNA topoisomerase II inhibitors, 5*H*- and 6*H*-indolo[2,3-*b*]quinolones (e.g., equation 2) [51, 52]. The yields are 14% to 43%, and the starting benzotriazoles are readily prepared

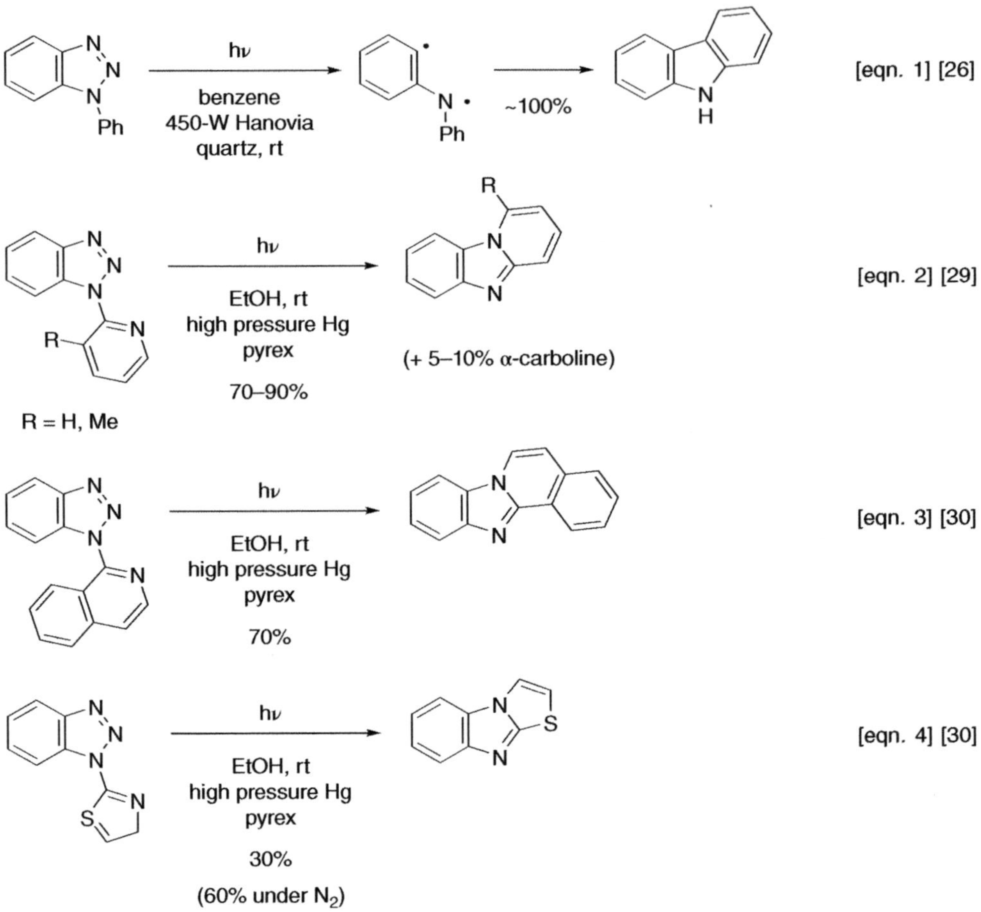

Scheme 4 *The Photochemical Graebe-Ullmann Carbazole Synthesis*

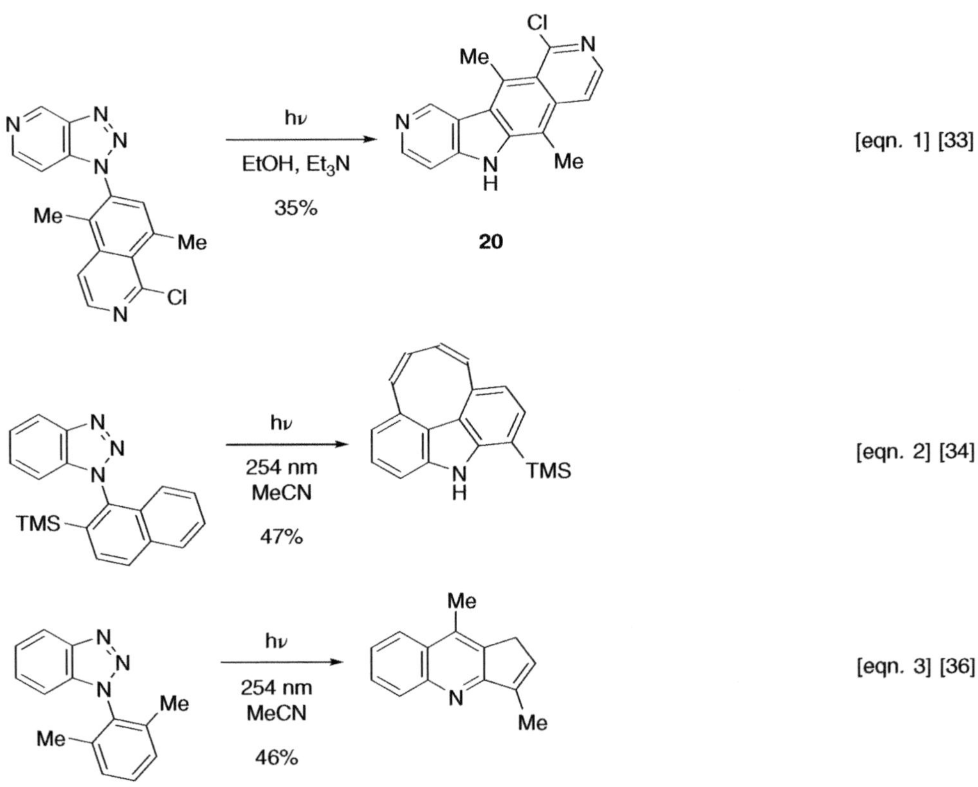

Scheme 5 *Applications of the Photochemical Graebe-Ullmann Carbazole-Carboline Synthesis*

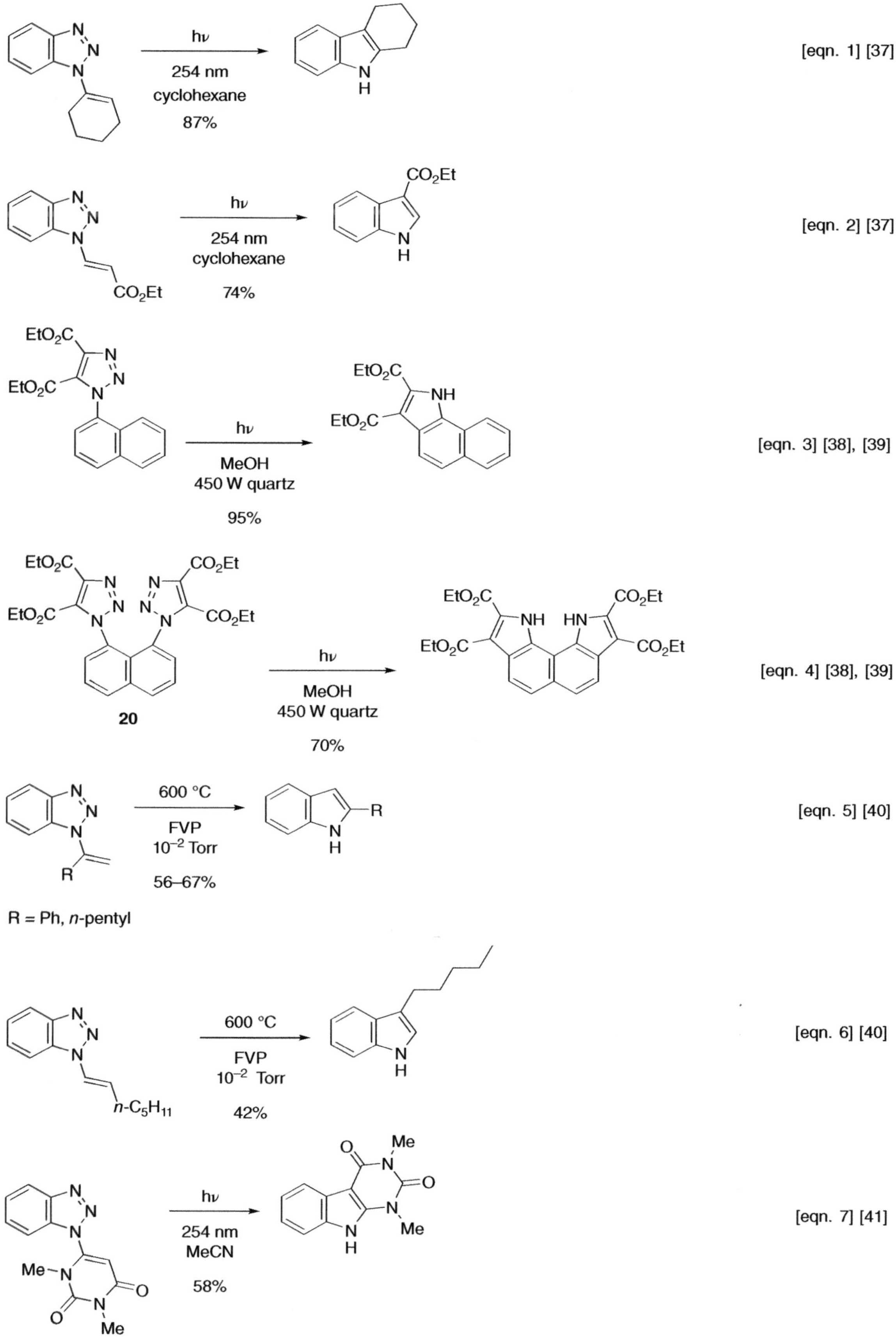

Scheme 6 *Application of the Graebe-Ullmann Carbazole Synthesis to Indoles*

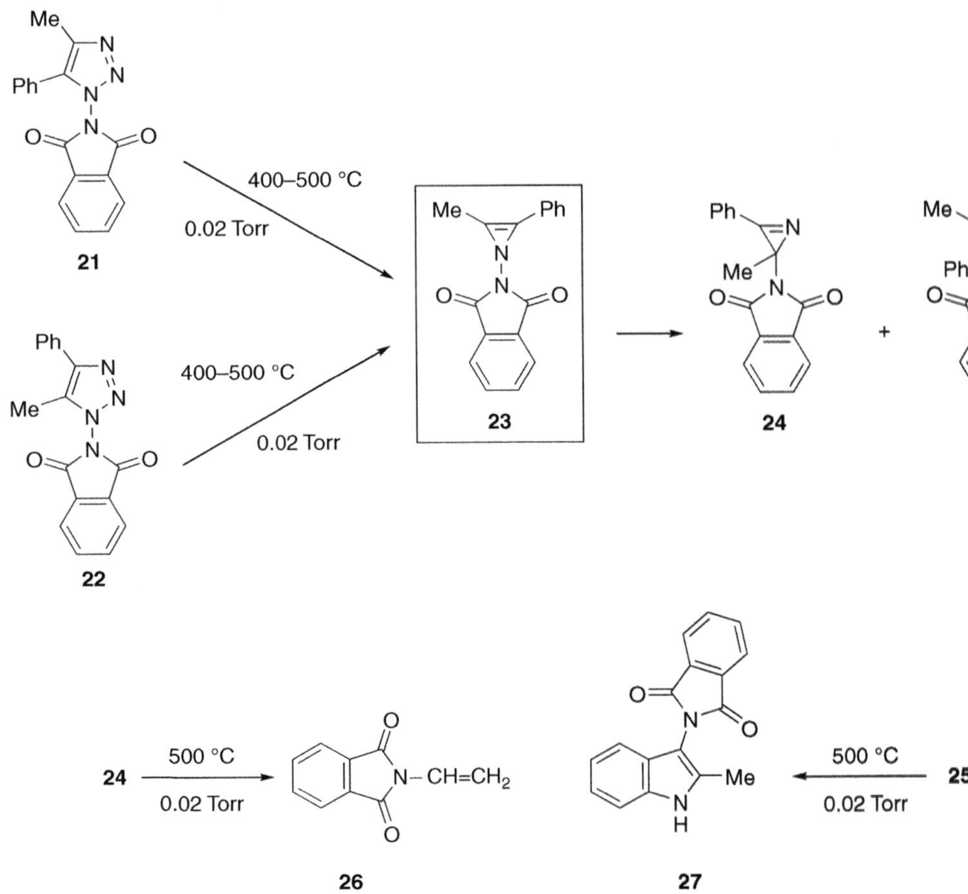

Scheme 7 *Intermediacy of 1*H*-Azirine in the Pyrolysis of 1,2,3-Triazoles [42]*

by condensing 2-chloroquinolines with benzotriazole or via the conventional diazotization route. Chen and colleagues reported a similar route to a series of 11-aminosubstituted 6*H*-indolo[2,3-*b*]quinolones using PPA, 140–150 °C [53]. Notably, several 11-anilino derivatives showed GI_{50} = 0.09–0.78 μM cytotoxicity against human cancer cell lines. Related Graebe–Ullmann chemistry by Augustyns and coworkers led to N^1,N^1-diethyl-N^4-(5-methyl-5*H*-indolo[2,3-*b*]quinolin-8-yl)pentane-1,4-diamine, which displayed excellent antiplasmodial activity, IC_{50} = 0.01 μM [54]. Murray and coworkers observed antimalarial activity in a series of synthetic isocryptolepines (5-methyl-11*H*-indolo[3,2-*c*]quinoline), one of which, 8-bromo-2-chloroisocryptolepine, exhibited the highest selectivity index, 106 (ratio of cytotoxicity [IC_{50} = 9,005 nM] to antimalarial activity [IC_{50} = 85 nM]) [55]. A PPA (130–140 °C) Graebe–Ullmann reaction was used to construct these compounds, and the requisite benzotriazoles were forged from 4-chloroquinolines and benzotriazole. Thiéry's team used a Graebe–Ullmann protocol to prepare novel thiazoloindolo[3,2-*c*]quinolines and 8-*N*-substituted-11*H*-indolo[3,2-*c*]quinolines from the requisite benzotriazoles [56], and Bergman and coworkers synthesized 11-chloro-6*H*-indolo[2,3-*b*]quinolone in low yield via a PPA Graebe–Ullmann reaction [57]. Moreau and colleagues synthesized the novel pyrrolopyridoindole and pyrrolopyridobenzimidazole ring systems via a thermal Graebe–Ullmann reaction (equation 3) [58]. Alvarez-Builla and coworkers employed pyrophosphoric acid with microwave heating to prepare 11*H*-indolo[3,2-*c*]quinolines (**28**), 5*H*-benzo[*f*]pyrido[4,3-*b*]indoles (**29**), and 13*H*-benz [5, 6] indolo[3,2-*c*]quinolines (**30**), some of which were derivatized to bis-intercalators [59].

As featured in reviews, the Graebe–Ullmann reaction has played an important role in the synthesis of

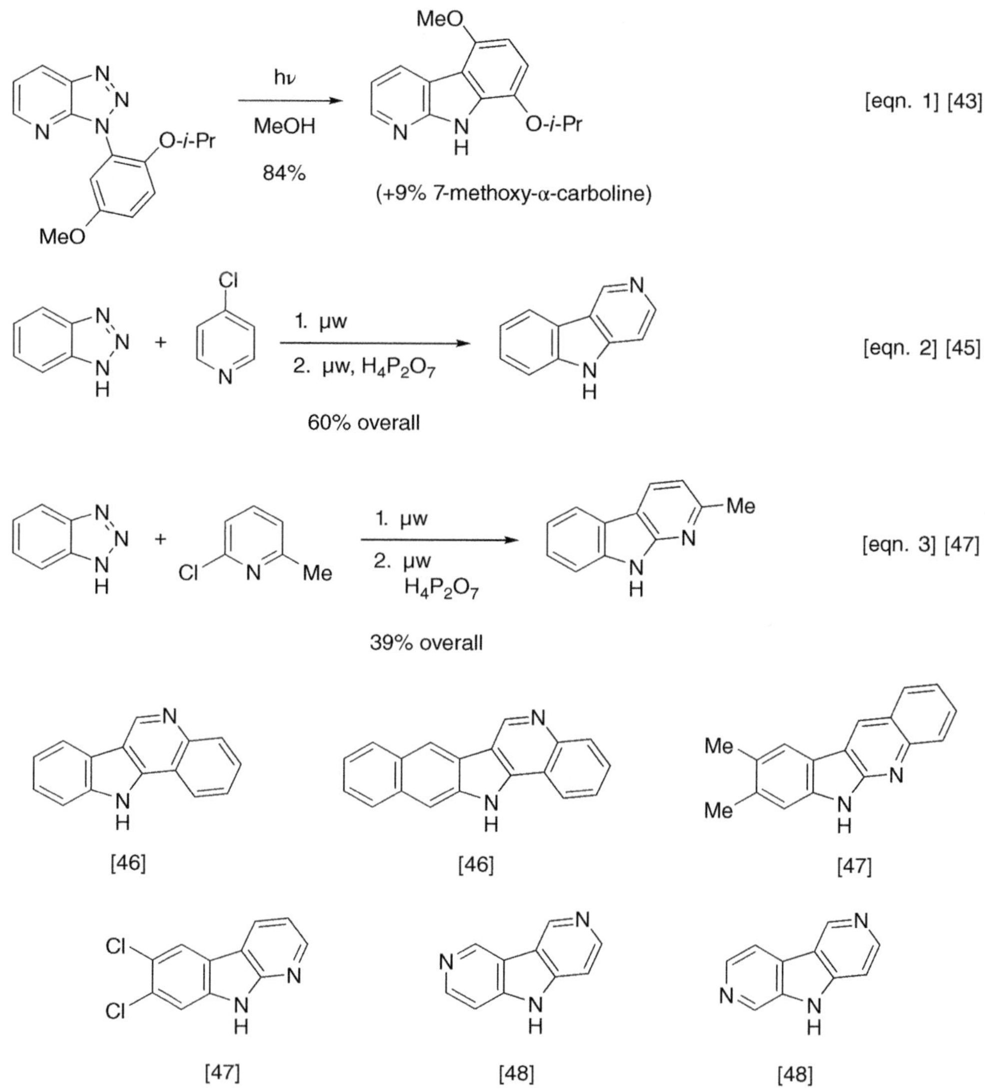

Scheme 8 *Graebe-Ullmann Carboline Synthesis with Microwave Irradiation*

carbolines (pyrido[*b*]indoles) [60, 61]. Goekjian and colleagues used a Graebe–Ullmann synthesis of α-carboline to access 6-substituted analogues, which have been rarely studied (Scheme 10, equation 1) [62]. The 6-nitro derivative was prepared by nitration of the *N*-Boc derivative. Le and colleagues found that benzotriazole **31** gave the novel azepino[1′,2′:1,2]imidazo[4,5-*b*]carbazole **32** in refluxing toluene in the presence of Hünig's base (equation 2) [63].

The Graebe–Ullmann reaction of *N*-arylbenzotriazoles is an extraordinarily versatile and powerful synthesis of carbazoles and carbolines. Despite the commonly encountered low yields, the ease of preparation of the requisite *N*-arylbenzotriazoles, either by the diazotization of 2-aminodiarylamines or the coupling of benzotriazole with the appropriate aromatic or heteroaromatic partner, elevates the Graebe–Ullmann reaction to one of distinction for the synthesis of carbazoles and carbolines.

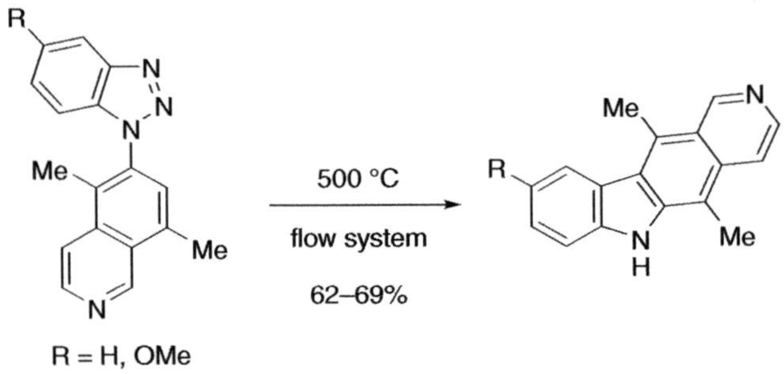

[eqn. 1] [49]

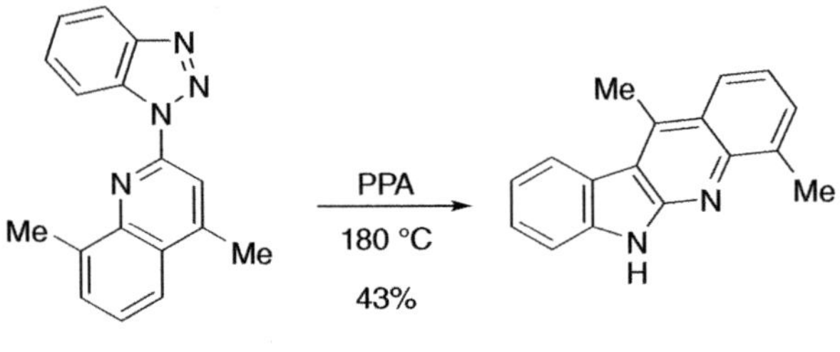

[eqn. 2] [51]

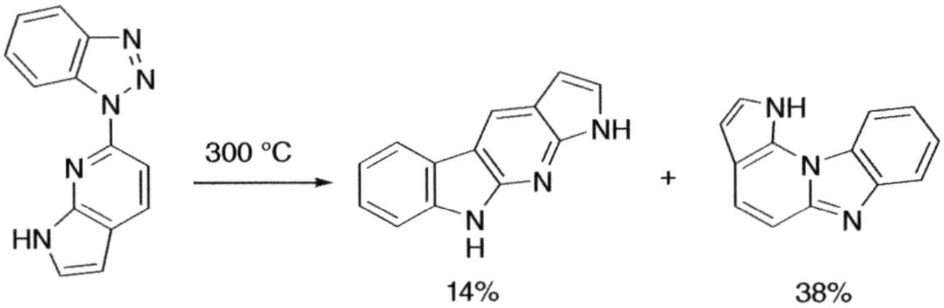

[eqn. 3] [58]

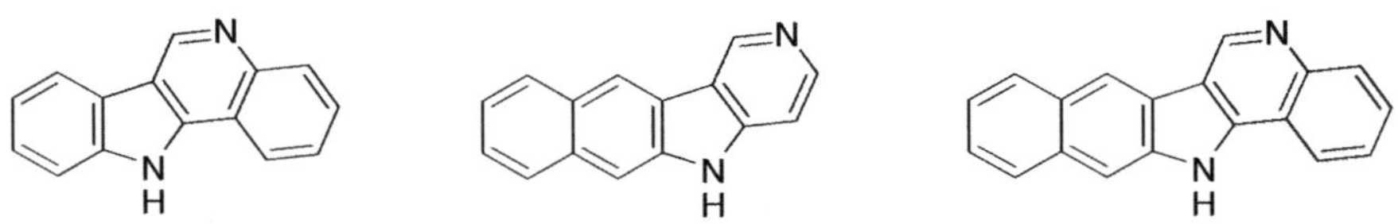

28 **29** **30**

Scheme 9 *Graebe-Ullmann Synthesis of Ellipticine and Fused Carbolines*

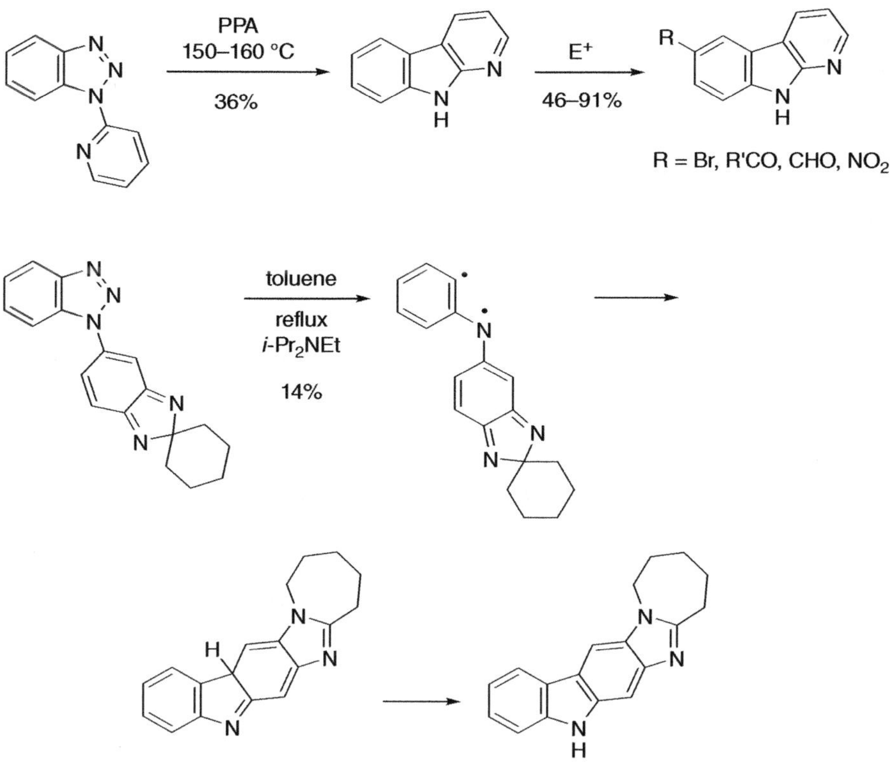

[eqn. 1] [62]

[eqn. 2] [63]

Scheme 10 *Miscellaneous Graebe-Ullmann Syntheses*

References

[1] C. Graebe and F. Ullmann, *Liebigs Ann. Chem.*, 1896, **291**, 16–17.
[2] F. Ullmann, *Chem. Ber.*, 1898, **31**, 1697–1698.
[3] E. Delétra and F. Ullmann, *Arch. Sci. Phys. Nat. Geneva*, 1904, [v] **17**, 78–92.
[4] F. Ullmann, *Liebigs Ann. Chem.*, 1904, **332**, 82–104.
[5] W. Borsche and M. Feise, *Chem. Ber.*, 1907, **40**, 378–386.
[6] A.V. Blom, *Helv. Chim. Acta*, 1921, **4**, 1036–1039.
[7] J. Li, and J.M. Cook (2005) Graebe–Ullmann carbazole synthesis, in *Name Reactions in Heterocyclic Chemistry*, (ed. J.J. Li), John Wiley & Sons, Hoboken, New Jersey, pp. 132–134.
[8] S. Rachwal and A.R. Katritzky (2008) 1,2,3-Triazoles, in *Comprehensive Heterocyclic Chemistry III*, (eds. A.R. Katritzky, C.A. Ramsden, E.F.V. Scriven, and R.J.K. Taylor), Elsevier, Oxford, pp. 1–158.
[9] N. Campbell and B.M. Barclay, *Chem. Rev.*, 1947, **40**, 359–380.
[10] C. Wentrup, *Helv. Chim. Acta*, 1972, **55**, 1613–1617.
[11] M. Ohashi, K. Tsujimoto, and T. Yonezawa, *Chem. Commun.*, 1970, 1089–1090.
[12] M. Märky, H. Schmid, and H.-J. Hansen, *Helv. Chim. Acta*, 1979, **62**, 2129–2153.
[13] R. Anet and F.A.L. Anet, *J. Am. Chem. Soc.*, 1964, **86**, 525–526.
[14] G.W. Gribble, unpublished results.
[15] W. Lawson, W.H. Perkin, Jr., and R. Robinson, *J. Chem. Soc.*, 1924, **125**, 626–657.
[16] W.O. Kermack and J.F. Smith, *J. Chem. Soc.*, 1930, 1999–2010.
[17] W.O. Kermack and N.E. Storey, *J. Chem. Soc.*, 1950, 607–612.
[18] W.H. Hunter and S.F. Darling, *J. Am. Chem. Soc.*, 1931, **54**, 4183–4186.
[19] O. Bremer, *Liebigs Ann. Chem.*, 1934, **514**, 279–291.
[20] R.W.G. Preston, S.H. Tucker, and J.M.L. Cameron, *J. Chem. Soc.*, 1942, 500–504.
[21] G.G. Coker, S.G.P. Plant, and P.B. Turner, *J. Chem. Soc.*, 1951, 110–115.
[22] P.V. Clifton and S.G.P. Plant, *J. Chem. Soc.*, 1951, 461–466.
[23] R.H. Freak and R. Robinson, *J. Chem. Soc.*, 1938, 2013–2015.
[24] B. Witkop, *J. Am. Chem. Soc.*, 1953, **75**, 3361–3370.

[25] B.W. Ashton and H. Suschitzky, *J. Chem. Soc.*, 1957, 4559–4562.
[26] E.M. Burgess, R. Carithers, and L. McCullagh, *J. Am. Chem. Soc.*, 1968, **90**, 1923–1924.
[27] M. Märky, Th. Doppler, H.-J. Hansen, and H. Schmid, *Chimia*, 1969, **23**, 230–231.
[28] M. Märky, H. Schmid, and H.-J. Hansen, *Helv. Chim. Acta*, 1979, **62**, 2129–2153.
[29] A.J. Hubert, *J. Chem. Soc. (C)*, 1969, 1334–1336.
[30] A.J. Hubert, *Chem. Commun.*, 1969, 328.
[31] H. Murai, M. Torres, and O.P. Strausz, *J. Am. Chem. Soc.*, 1980, **102**, 1421–1422.
[32] P. Claus, Th. Doppler, N. Gakis, *et al.*, *Pure Appl. Chem.*, 1973, **33**, 339–361.
[33] C. Rivalle, C. Ducrocq, J.-M. Lhoste, and E. Bisagni, *J. Org. Chem.*, 1980, **45**, 2176–2180.
[34] G. Mitchell and C.W. Rees, *J. Chem. Soc., Perkin Trans. 1*, 1987, 403–412.
[35] J.J. Kulagowski, C.J. Moody, and R.W. Rees, *J. Chem. Soc., Chem. Commun.*, 1982, 548–550.
[36] J.J. Kulagowski, C.J. Moody, and C.W. Rees, *J. Chem. Soc., Perkin Trans. 1*, 1985, 2725–2732.
[37] P.A. Wender and C.B. Cooper, *Tetrahedron*, 1986, **42**, 2985–2991.
[38] Y. Nagawa, K. Honda, and H. Nakanishi, *J. Chem. Soc., Chem. Commun.*, 1988, 989–990.
[39] Y. Nagawa, M. Goto, K. Honda, and H, Nakanishi, *Bull. Chem. Soc. Jpn.*, 1989, **62**, 3109–3113.
[40] S.J. Barker and R.C. Storr, *J. Chem. Soc., Perkin Trans. 1*, 1990, 485–488.
[41] E.D. Edstrom and W. Yuan, *Tetrahedron Lett.*, 1991, **32**, 323–326.
[42] T.L. Gilchrist, G.E. Gymer, and C.W. Rees, *J. Chem. Soc. D, Chem. Commun.*, 1971, 1519–1520.
[43] L.K. Mehta, J. Parrick, and F. Payne, *J. Chem. Soc., Perkin Trans. 1*, 1993, 1261–1267.
[44] L. Stephenson and W.K. Warburton, *J. Chem. Soc. (C)*, 1970, 1355–1364.
[45] P. Nantka-Namirski and J. Kalinowski, *Seria Chem.*, 1975, **18**, 259–271.
[46] A. Molina, J.J. Vaquero, J.L. García-Navio, and J. Alvarez-Builla, *Tetrahedron Lett.*, 1993, **34**, 2673–2676.
[47] P. Vera-Luque, R. Alajarín, J. Alvarez-Builla, and J.J. Vaquero, *Org. Lett.*, 2006, **8**, 415–418.
[48] R.S. Alekseev, A.V, Kurkin, and M.A. Yurovskaya, *Chem. Heterocycl. Compd.*, 2012, **48**, 1235–1250.
[49] R.B. Miller and J.G. Stowell, *J. Org. Chem.*, 1983, **48**, 886–888.
[50] C. Rivalle, C. Ducrocq, and E. Bisagni, *J. Chem. Soc., Perkin I*, 1979, 138–141.
[51] W. Peczynska-Czoch, F. Pognan, L. Kaczmarek, and J. Boratynski, *J. Med. Chem.*, 1994, **37**, 3503–3510.
[52] L. Kaczmarek, W. Peczyńska-Czoch, J. Osiadacz, *et al.*, *Bioorg. Med. Chem.*, 1999, **7**, 2457–2464.
[53] Y.-L. Chen, H.-M. Hung, C.-M. Lu, *et al.*, *Bioorg. Med. Chem.*, 2004, **12**, 6539–6546.
[54] I. El Sayed, P. Van der Veken, K. Steert, *et al.*, *J. Med. Chem.*, 2009, **52**, 2979–2988.
[55] L.R. Whittell, K.T. Batty, R.P.M. Wong, *et al.*, *Bioorg. Med. Chem.*, 2011, **19**, 7519–7525.
[56] A. Beauchard, H. Chabane, S. Sinbandhit, *et al.*, *Tetrahedron*, 2006, **62**, 1895–1903.
[57] J. Bergman, R. Engqvist, C. Stålhandske, and H. Wallberg, *Tetrahedron*, 2003, **59**, 1033–1048.
[58] F. Bouchikhi, M. Sassatelli, F. Anizon, *et al.*, *Synthesis*, 2009, 755–758.
[59] A. Molina, J.J. Vaquero, J.L. Garcia-Navio, *et al.*, *J. Org. Chem.*, 1996, **61**, 5587–5599.
[60] A.A. Semenov and V.V. Tolstikhina, *Chem. Heterocycl. Compd.*, 1984, **20**, 345–356.
[61] O.B. Smirnova, T.V. Golovko, and V.G. Granik, *Pharm. Chem. J.*, 2011, **44**, 654–678.
[62] C. Schneider, D. Gueyrard, F. Popowycz, *et al.*, *Synlett*, 2007, 2237–2241.
[63] H.P. Le, A. Kelbig, A. Lindauer, *et al.*, *J. Chem. Res.*, 2004, 453–456.

PART VII

Cycloaddition and Electrocyclization

Cycloaddition reactions and the related electrocyclization reactions rank at or near the top of the hierarchy of organic synthesis. Several clever adaptations of these reactions have been described for the synthesis of indoles. Methods that *originate* with an intact indole ring—for example, to give carbazoles—are not included in this chapter. Thus, the several elegant Diels–Alder cycloadditions of vinylindoles to give carbazoles are not covered.

Indole Ring Synthesis: From Natural Products to Drug Discovery, First Edition. Gordon W. Gribble.

54

Diels–Alder Cycloaddition

The prominence of the Diels–Alder reaction of quino-2,3-dimethanes (also known as *ortho*-quinodimethanes and *ortho*-xylylenes) has been established for nearly six decades since the seminal work by Cava [1, 2]. Several excellent reviews are available [3–7]. The logical extension of this chemistry to the synthesis and Diels–Alder reactions of indole-2,3-quinodimethanes (2,3-dimethylene-2,3-dihydro-1*H*-indoles) has been recognized as an efficient route to indoles, carbazoles, and related fused indoles [8].

The first example of an indole-2,3-quinodimethane (IQM) undergoing a Diels–Alder cycloaddition to furnish a carbazole was reported by Plieninger and coworkers in 1964 [9]. Thus, indole-3-acetic acid was readily converted to pyrano[3,4-*b*]indol-3-ones upon treatment with carboxylic acid anhydrides (Scheme 1, equation 1). These stable synthetic equivalents of IQMs undergo Diels–Alder reactions with electron-deficient dienophiles (*N*-phenylmaleimide, maleic anhydride, dimethyl acetylenedicarboxylate) (equation 2). Plieninger's discovery notwithstanding, it was Moody and coworkers who parlayed this chemistry into a powerful carbazole synthesis (equations 3, 4) [10–18].

In addition to improving the Plieninger conditions for preparing simple carbazoles, Moody greatly extended this methodology to encompass a range of carbazoles and related indole-fused heterocycles (Scheme 2). The Diels–Alder/loss of CO_2 reaction in equation 2 was used to synthesize the alkaloids carbazomycins A and B and, using a similar tactic, hyellazole [16, 18]. Depending on both steric and electronic factors, the reaction can be highly regioselective (equations 3, 4) [15].

Moody and coworkers synthesized and explored the Diels–Alder chemistry of the isomeric pyrano[4,3-*b*]indol-3-ones (Scheme 3, equations 1, 2) [19, 20]. The requisite pyranoindoles are prepared from ethyl indol-2-ylacetate and then electrophilic acylation at C-3, followed by acid anhydride cyclization. A comparison of the twin reactions of pyrano[3,4-*b*]indol-3-ones with pyrano[4,3-*b*]indol-3-ones reveals that the pyrano ring oxygen plays a directing role in the Diels–Alder reactions. In a major diversion of his method, Moody prepared the 1,4-dihydropyrano[3,4-*b*]indolone ring system and found, upon heating and loss of CO_2, that the resulting indole-2,3-quinodimethane could be trapped with dienophiles (equation 4) [21]. Other dienophiles were 1,4-benzoquinone and 1,4-naphthoquinone. Yields were much lower with methyl vinyl ketone, dimethyl fumarate, and dimethyl maleate. Moody employed a pyrano[4,3-*b*]indol-3-one to prepare the aglycone of staurosporine [22].

Several other investigators have used the Plieninger-Moody chemistry in various formats to synthesize carbazoles and indoles. Interestingly, before Moody's work, Rokach and colleagues in 1971 photolyzed pyrano[3,4-*b*]indol-3-ones in the presence of oxygen to give indole 2,3-dicarbonyl compounds (Scheme 4, equation 1) [23]. A peroxide intermediate **1** is implicated because $^{18}O_2$ is incorporated in both carbonyl oxygens. Independently from Moody, in 1985 Narasimhan and Gokhale described a new synthesis of pyridocarbazoles, later to include the alkaloid olivacine, that featured Diels–Alder reactions of pyrano[3,4-*b*]indol-3-ones (equation 2) [24, 25]. The presence of the chlorine leaving group in the dienophiles allowed the production of the fully aromatic carbazole. Olivacine was prepared from 3-acetyl-1-methylcarbazole in four steps [25]. Pindur and colleagues have also explored the Diels–Alder cycloadditions of pyrano[3,4-*b*]indol-3-ones with benzyne, *N*-phenylmaleimide, *p*-benzoquinone, and *p*-naphthoquinone to give the expected annelated

Indole Ring Synthesis: From Natural Products to Drug Discovery, First Edition. Gordon W. Gribble.

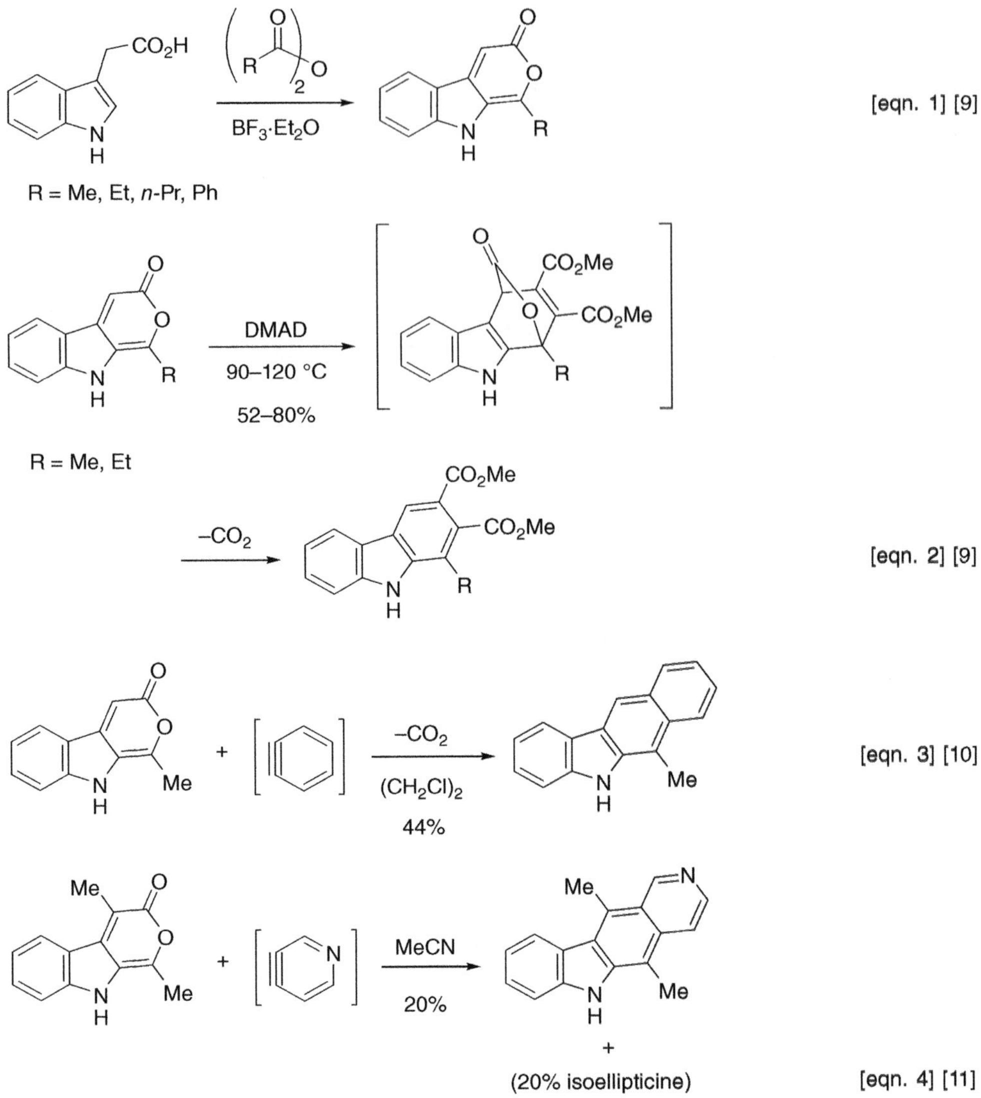

Scheme 1 *Plieninger-Moody Carbazole Synthesis*

carbazoles [26, 27] and with acyclic unsymmetrical dienophiles to give functionalized carbazoles [28]. Hoornaert and coworkers isolated dihydrocarbazoles from the Diels–Alder reaction of pyrano[3,4-*b*]indol-3-ones with olefinic dienophiles. In this study these workers identified a stepwise mechanism that in some cases gives 1,2-dihydrocarbazoles in excellent yield (equation 3) [29, 30]. Nandin de Carvalho and colleagues independently discovered this synthesis of 1,2-dihydrocarbazoles with an electron-withdrawing group at C-3 [31].

Nandin de Carvalho, Dmitrienko, and their colleagues effected decarboxylation of 1-methyl-1,4-dihydropyrano[3,4-*b*]indol-3-one (prepared by hydrogenation) and trapped the intermediate indole-2,3-quinodimethane with unsymmetrical dienophiles (Scheme 5, equation 1) [32, 33], a reaction that was described later by Moody (Scheme 3, equation 4) [21]. Guitián's team improved Moody's ellipticine synthesis (Scheme 1, equation 4) [11] by incorporating a chlorine at C-2 of 3,4-pyridyne (equation 2) [34]. No isoellipticine isomer was detected.

In 1977 Bergman and Carlsson described the first thermal generation and intramolecular trapping of an IQD intermediate **2** en route to a synthesis of ellipticine (Scheme 6, equation 1) [35] and later olivacine [36]. Also isolated was a dimer consistent with the generation of **2**. Similar chemistry that also targeted ellipticine was reported by Sainsbury [37] and Kano [38]. A major development in the application of IQD chemistry to natural product synthesis was described by Magnus and Gallagher in 1981. This elegant methodology involves the generation and

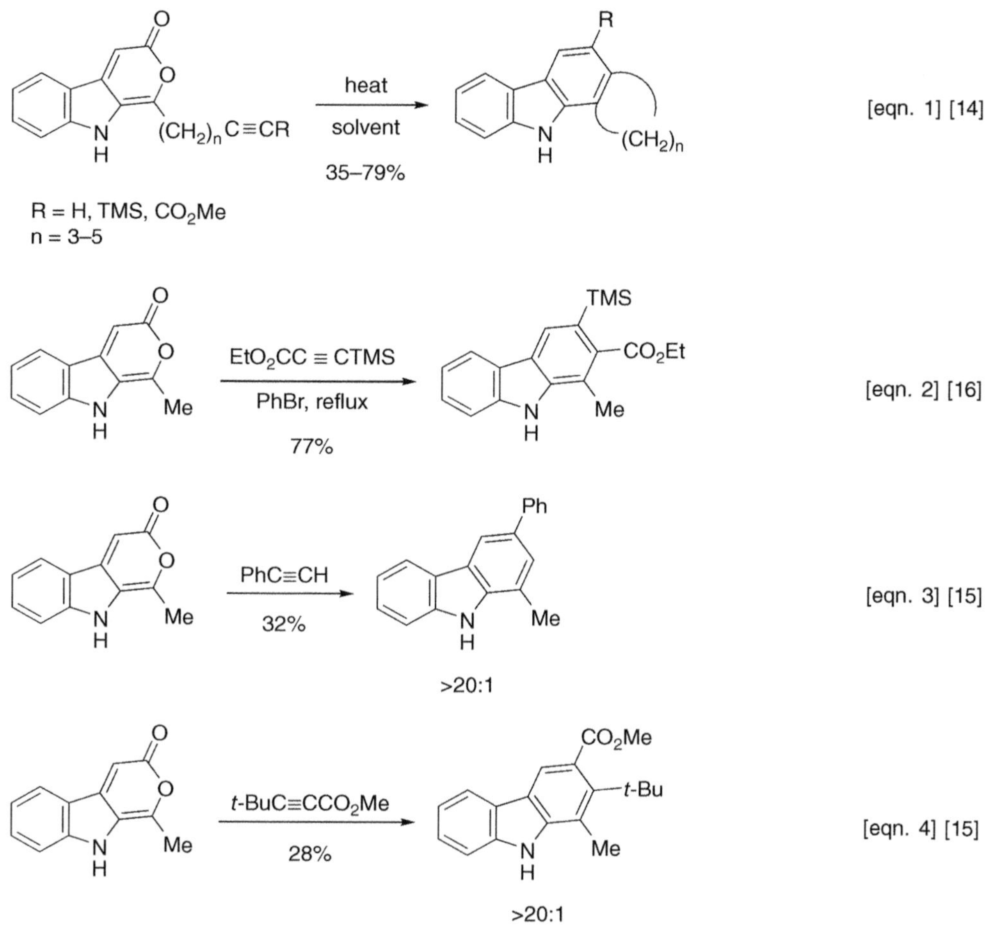

Scheme 2 *Moody Carbazole Synthesis from Pyrano[3,4-*b*]indol-3-ones*

intramolecular trapping of IQDs as a novel route to indole alkaloids (equation 2) [39–41]. For example, Magnus synthesized (±)-kopsanone [42], (±)-aspidospermidine [43], (±)-eburnamonine [44], staurosporinone analogues [45], (+)- and (–)-16-methoxytabersonine [46], *Aspidosperma* model alkaloids [47], and 1,4-dihydrocarbazoles (equation 3) [48]. Ciganek and Schubert adapted the Magnus protocol to a synthesis of various pyrrolo[3,4-*c*]- and pyrrolo[4,3-*c*] carbazoles (equation 4) [49].

The first generation and intermolecular capture of an unfettered indole-2,3-quinodimethane (**3**) was reported by Marinelli (Scheme 7, equation 1) [50]. IQD **3** also reacts in anticipated Diels–Alder fashion with methyl acrylate and with another molecule of **3**. A different means to generate an IQD was described by Srinivasan and Saroja (equation 2) [51], a method founded on the early work of Cava involving α,α′-dibromo-*o*-xylene [1, 2]. Thus, treatment of *N*-benzoyl-2,3-bis(bromomethyl)indole with sodium iodide (DMF, 50 °C) in the presence of dienophiles (*p*-benzoquinone, dimethyl acetylenedicarboxylate, *N*-phenylmaleimide) afforded the expected Diels–Alder adducts from the trapping of IQD **4**. Independently, Nandin de Carvalho and colleagues discovered this same generation and trapping of IQD **5** (equation 3) [32]. Sulfur dioxide also captured **5** (R=*t*-Boc) to give 1,3-dihydrothieno[3,4-*b*] indole-2,2-dioxide (75% yield), analogous to the familiar reaction of sulfur dioxide with 1,3-dienes to give 2,5-dihydrothiophene-1,1-dioxides. Upon heating (80–100 °C), this indole sulfone undergoes a *retro* Diels–Alder reaction to regenerate IQD **5**.

Pindur and Haber explored the chemistry shown in equation 2 and synthesized several new cycloadducts (Scheme 7, **6–12**) from IQD **4** [52]. The observed regiochemistry with unsymmetrical dienophiles (methyl acrylate, nitrosobenzene, 1,1-bis(phenylsulfonyl)ethane, methyl vinyl sulfone, and methyl vinyl ketone) was low to modest, with the major isomer as predicted by simple FMO theory. However, dienophiles acrolein, propynoates, and diethyl maleate were unreactive relative to dimerization of IQD **4**. Pindur and Meyer prepared several bis(tetrahydropyrrolo[3,4-*b*] carbazoles) by exposing IQDs (generated as shown in equation 2) to alkane-tethered bismaleimides [53]. In two

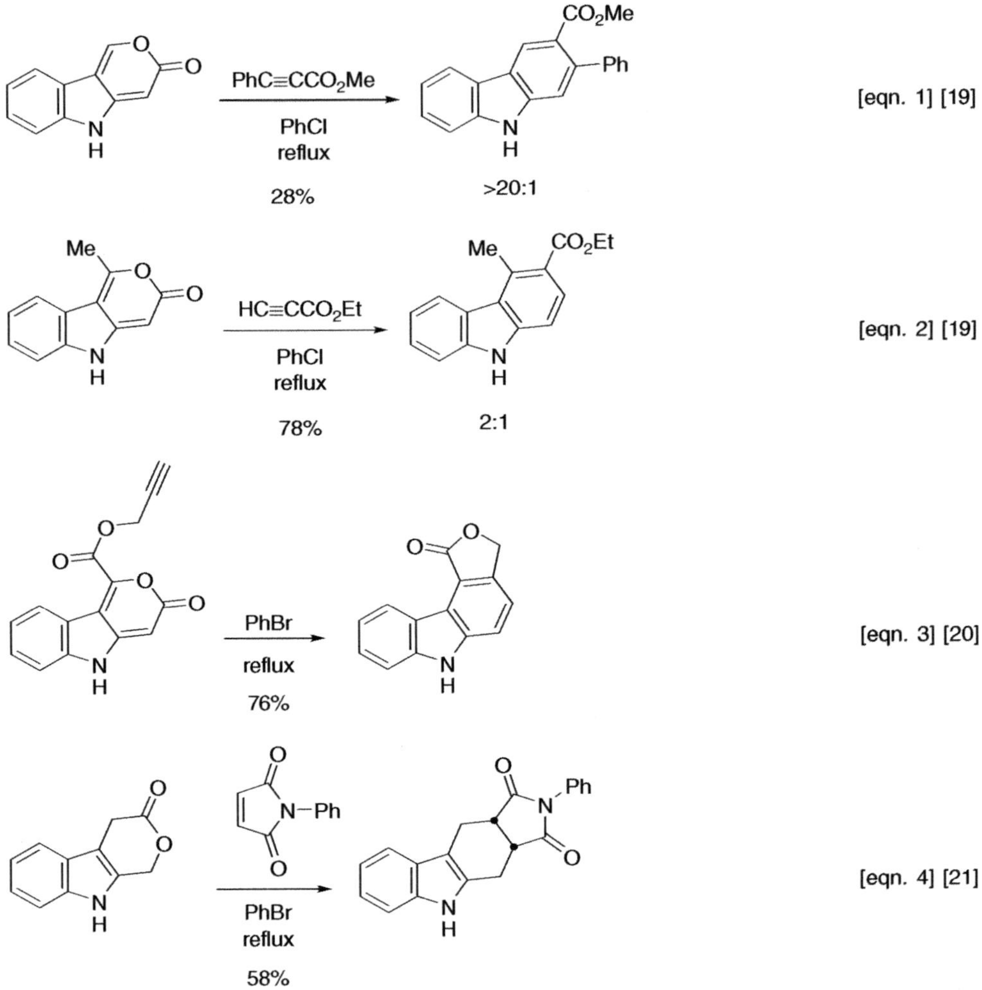

Scheme 3 *Moody Carbazole Synthesis from Pyrano[4,3-*b*]indol-3-ones*

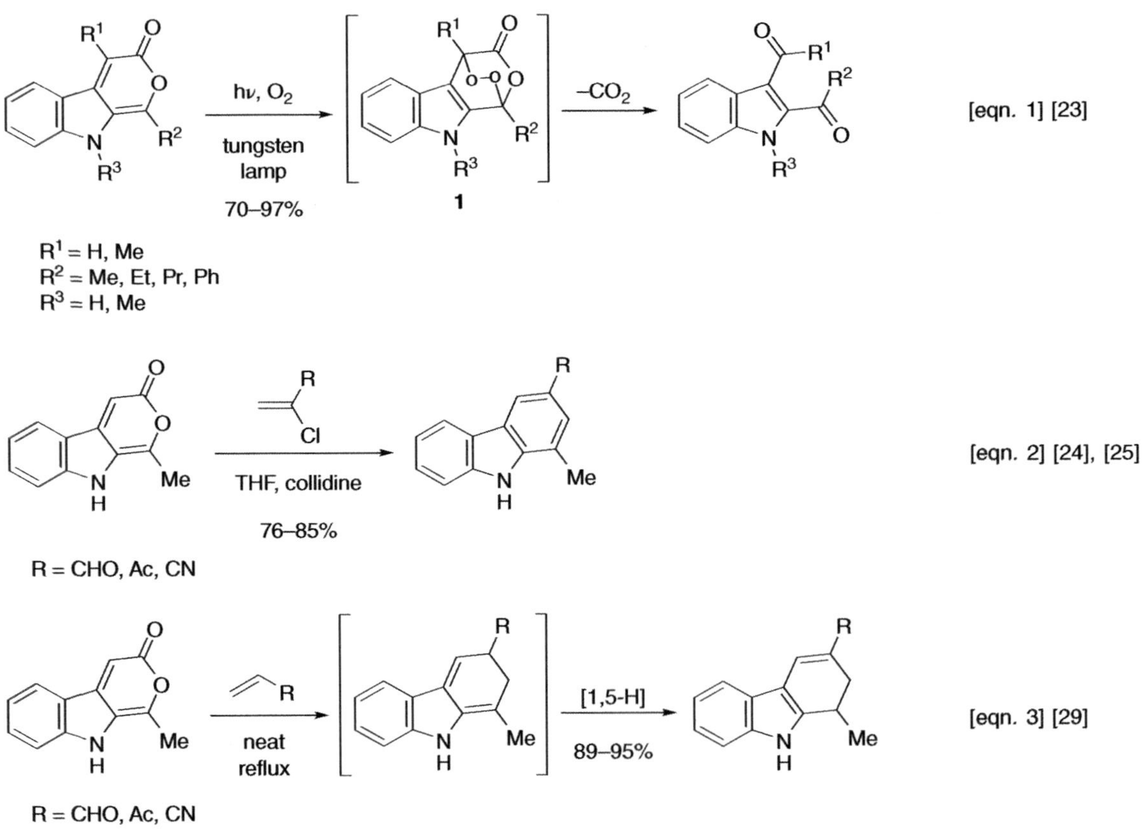

Scheme 4 *Variations on the Plieninger-Moody Carbazole Synthesis*

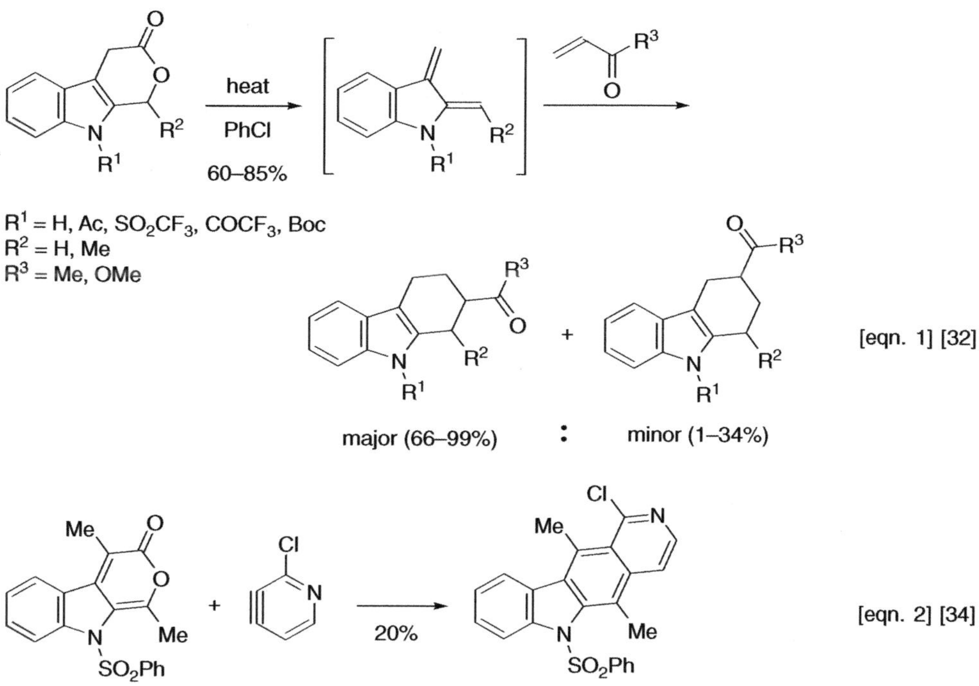

Scheme 5 *Further Variations of the Plieninger-Moody Carbazole Synthesis*

separate studies, Stephanidou-Stephanatou and colleagues synthesized a series of tetrahydrochromeno[2,3-*b*]carbazoles from IQDs (generated as shown in equation 2) and chromone-3-carboxaldehydes [54] and novel dispiroisoxazolines from IQDs and stable aryl nitrile oxides [55]. Thus, the simple generation of indole-2,3-quinodimethanes in Scheme 7 (equations 2, 3) has proved to be extraordinarily versatile in Diels–Alder and other cycloaddition reactions.

Following the pioneering work of Plieninger, Bergman, Moody, Marinelli, Narasimhan, Magnus, Pindur, Nandin de Carvalho, and Dmitrienko, the spark was lit for the invention of new ways to generate (even isolate?!) indole-2,3-quinodimethanes. For unsuccessful attempts to isolate IQDs, see Rinderspacher [56].

Chou and Ko prepared *N*-acetylindolo-3-sulfolene **13** (1,3-dihydrothieno[3,4-*b*]indole-2,2-dioxide) and found, as did Vice and colleagues earlier [32], that it readily generates IQD **14** on heating, which can be trapped with *N*-phenylmaleimide (Scheme 8, equation 1) [57, 58]. Sapi and coworkers cleverly parlayed a [1,5] hydrogen shift to afford IQDs, which were subsequently trapped by dienophiles (equation 2) [59]. The use of *N-tert*-butyl-3(2*H*)isothiazolone-1-oxide as dienophile with IQDs gave novel isothiazolo[4,5-*b*]carbazoles, some of which are active against GSK-3β kinase and human carbonic anhydrase I [60]. Mukai and colleagues employed an S_N2' displacement reaction to generate and capture an intermediate IQD (equation 3) [61, 62]. The usual complement of dienophiles was employed. In their second paper, Mukai's group used a palladium-catalyzed IQD **15** generation and Diels–Alder trapping protocol (equation 4) [62]. IQD **15** could actually exist as a palladium complex. Independently, Fuwa and colleagues described the palladium-catalyzed IQD formation from both allenamides and α-phosphono enecarbamates and from subsequent Diels–Alder reactions to give the usual suite of indoles [63, 64].

Tamura and colleagues generated an IQD (**16**) from 3-carboxy-1-methylindole-2-acetic anhydride by treatment with sodium hydride. Subsequent trapping with acetylenic dienophiles and a chloroquinone afforded the expected cycloadducts (Scheme 9, equation 1, and **17**, **18**) [65]. Ila, Junjappa, and coworkers also employed basic conditions to generate and trap a bright red IQD (**19**) from 1,2-dimethyl-3-carboxyaldehyde (equation 2) [66]. A large number of carbazoles (**20–22**) and 1,2-dihydrocarbazoles (**23–25**) were synthesized in very good yields. Either silica gel or pyridinium tosylate was required to dehydrate/aromatize the initially formed carbinol intermediate.

A formal [4+2] cycloaddition between arylidenoxindoles and ethyl allenoate afforded dihydropyran-fused indoles, as shown by Wang and colleagues (Scheme 10, equation 1) [67]. Density functional theory (DFT) implicates a stepwise mechanism. A not-unrelated synthesis of indole-annulated dihydropyrano[3,4-*c*]chromenes via a hetero-Diels–Alder reaction was discovered by Jha and colleagues (equation 2) [68]. The requisite precursors **26**

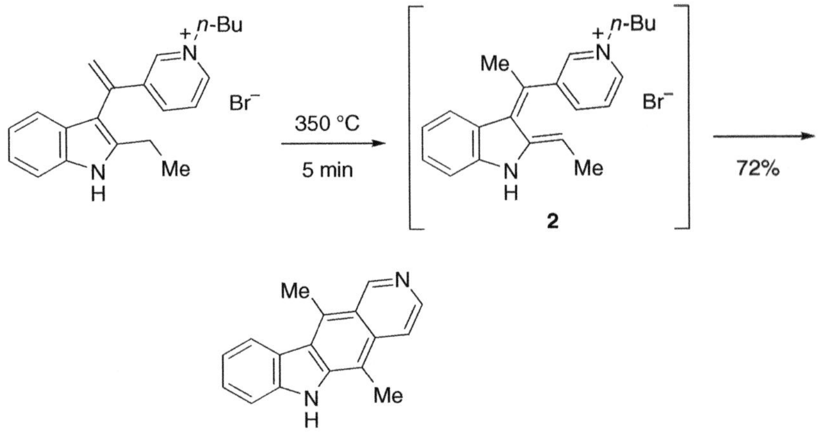

[eqn. 1] [35]

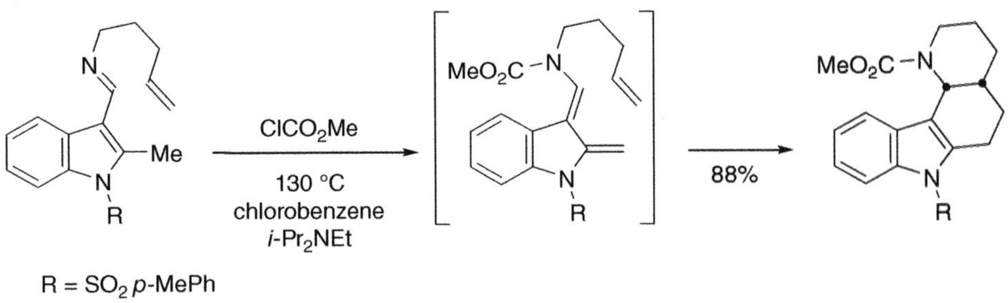

[eqn. 2] [39]

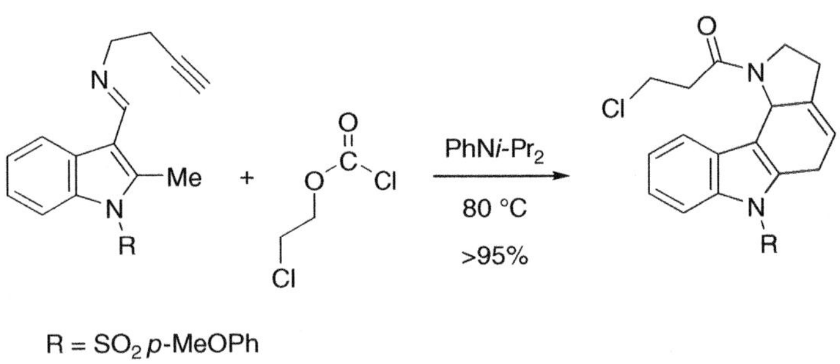

[eqn. 3] [48]

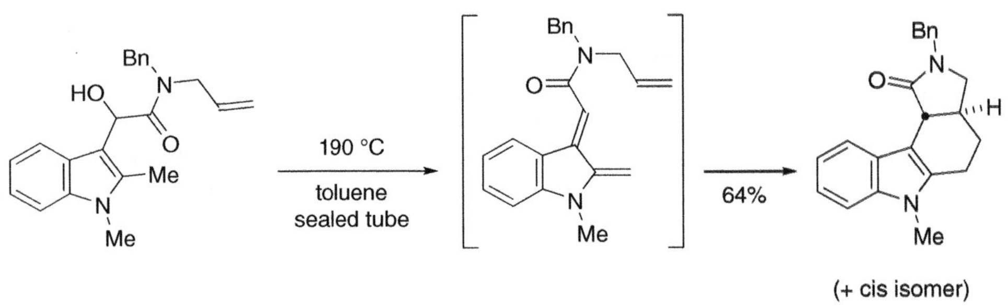

[eqn. 4] [49]

Scheme 6 *Bergman and Magnus Carbazole-Indole Syntheses*

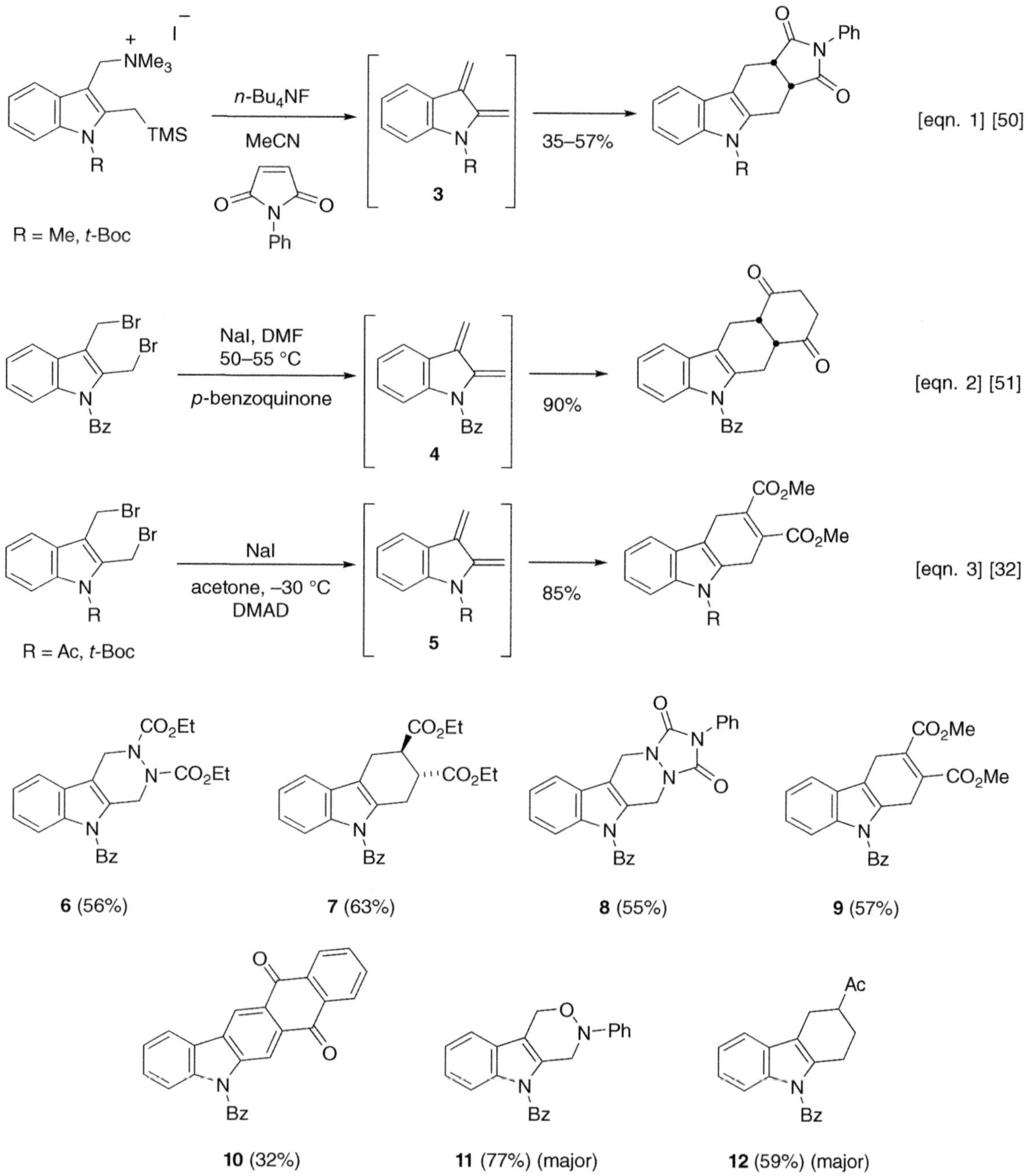

Scheme 7 *Generation and Intermolecular Trapping of Indole-2,3-quinodimethane*

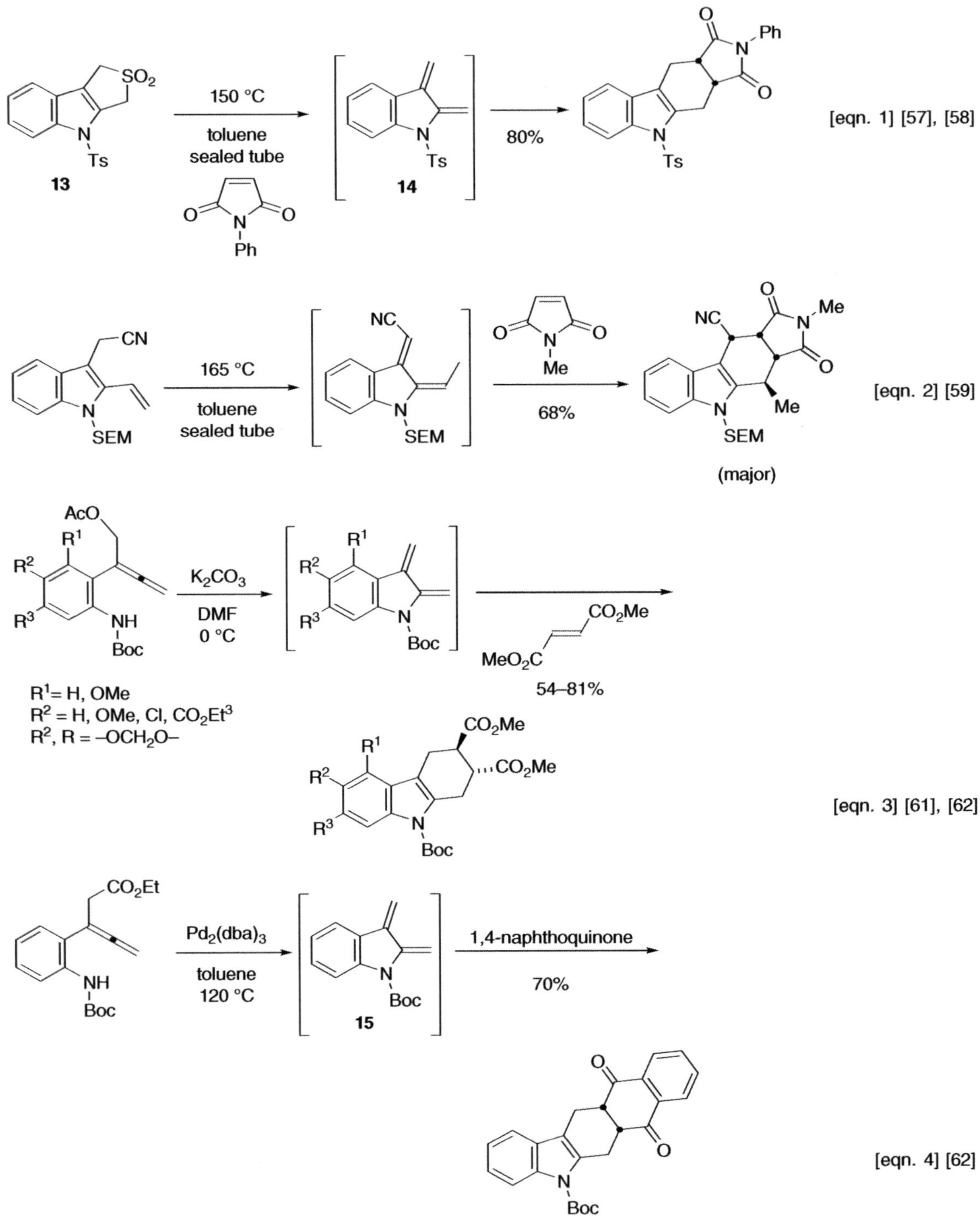

Scheme 8 *Novel Generation of Indole-2,3-quinodimethanes*

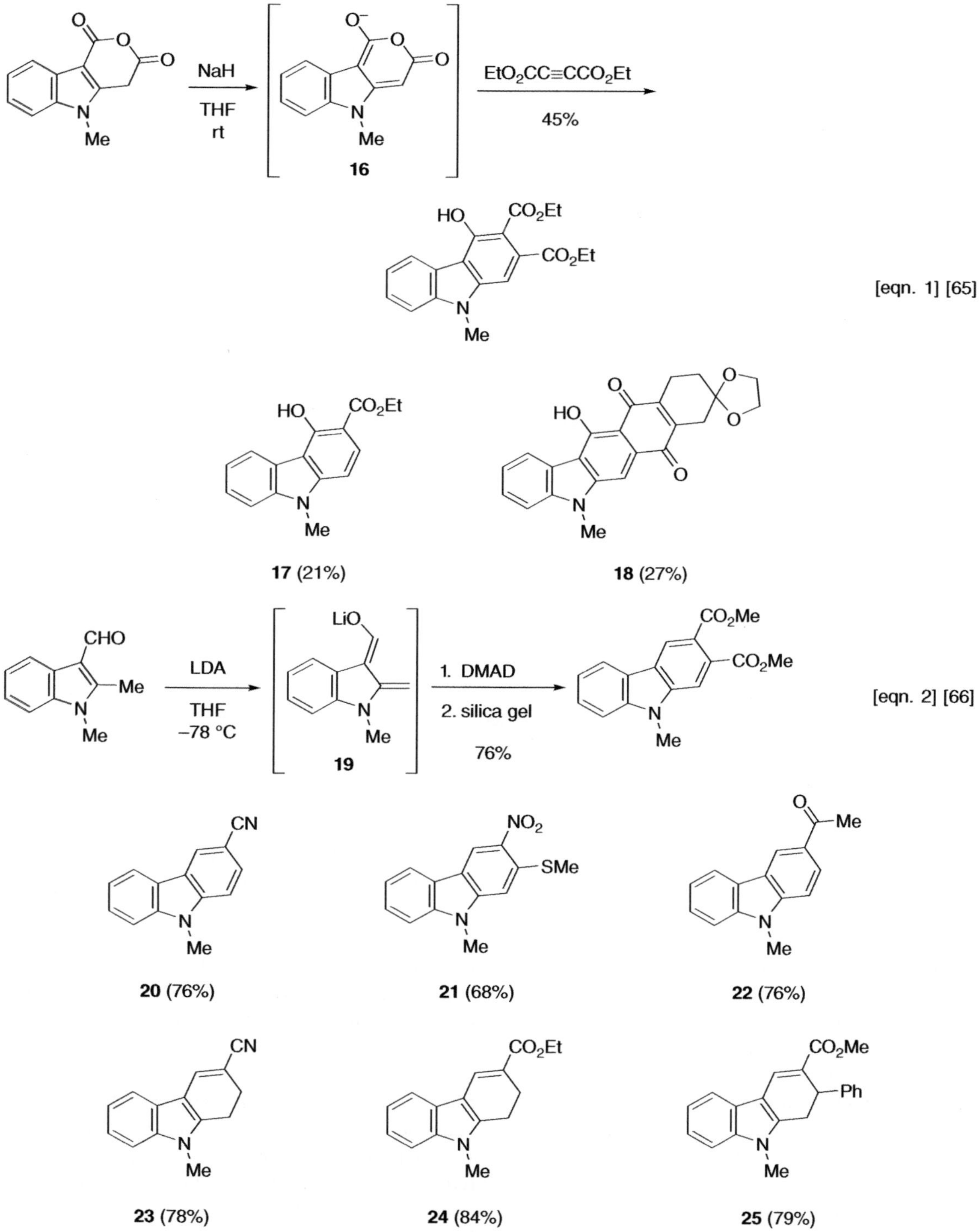

Scheme 9 *Base Generation and Trapping of Indole-2,3-quinodimethanes*

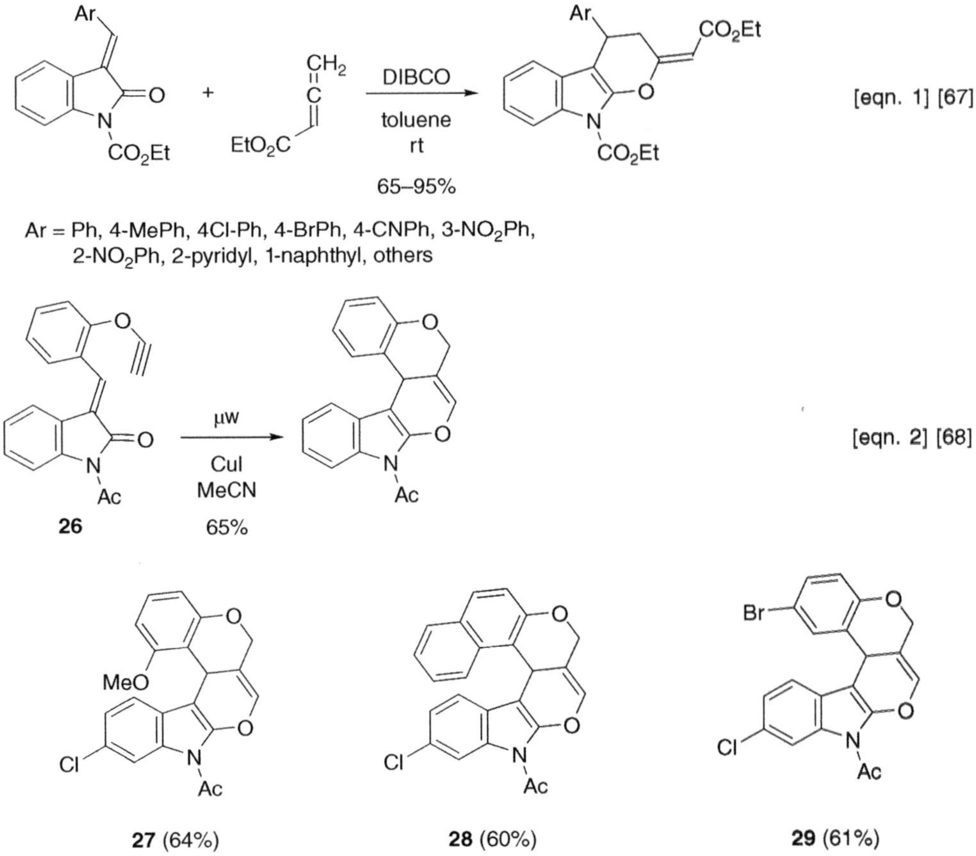

Scheme 10 *Hetero-Diels-Alder Syntheses of Indoles*

were prepared by a Knoevenagel condensation between oxindole and the *O*-propargylated salicyladehyde. Several novel indole-fused pyranochromenes were synthesized in this study (e.g., **27–29**). The cyclization may be facilitated by copper activation of the alkyne.

Melchione's team reported the asymmetric catalysis of Diels–Alder reactions of IQDs (Scheme 11, equation 1) [69, 70]. In addition to other nitro-substituted arylethenes, methyleneindolinones were employed as dienophiles. A limited selection of the compounds synthesized is shown in Scheme 11 (**30–32**). The third compound (**32**) is the result of a final cross-benzoin condensation. Chen and colleagues effected an asymmetric Diels–Alder reaction of IQDs (**33**) generated under mild acidic conditions from 2-methyl-3-indolemethanols and α,β-unsaturated aldehydes (equation 2) [71]. Three representative indoles that were prepared in this fashion are **34** to **36**. The IQD **33** is presumed to be in equilibrium with the 3-vinylindolenium species. A wide range of substituted *trans*-cinnamaldehydes was successfully employed. Although other acids (HOAc, TFA, $PhCO_2H$, silica gel) effected the reaction, Montmorillonite K10 clay was superior in terms of yield, enantioselectivity, and diastereoselectivity.

Kurihara and colleagues generated and trapped IQD **37** from the base-treatment of 2-cyano-3-indoleacetonitrile (Scheme 12, equations 1, 2) [72, 73], in chemistry reminiscent of that presented in Scheme 9. Presumed **37** is dark red (NaH/THF, –5 °C). Trapping of **37** with 3-pyridyne (3-chloropyridine, LDA, –78 °C, THF) gave the isomeric 5-amino-11-cyano-6-methylpyrido[4,3-*b*]carbazole (45%) (ellipticine type) and 11-amino-5-cyano-10-methylpyrido[3,4-*b*]carbazole (27%) (isoellipticine type). Moghaddam and Kiamehr synthesized a series of indole-annulated thiopyrano-chromenes via a domino Knoevenagel-hetero-Diels–Alder reaction of thiodiene **38** (equation 3) [74], similar to the approach used by Jha in Scheme 10 [68]. The use of ZnO as a heterogeneous catalyst is relatively rare.

In addition to the pyrano[3,4-*b*]indol-3-one ring system, another group of stable indole-2,3-quinodimethane synthetic equivalents are the fused indole ring systems, pyrrolo[3,4-*b*]indole, furo[3,4-*b*]indole, thieno[3,4-*b*] indole, and seleno[3,4-*b*]indole, each of which is known.

In 1976 Welch prepared the first example of a pyrrolo[3,4-*b*]indole, 2-benzyl-4-phenyl-2,4-dihydropyrrolo[3,4-*b*] indole (**39**) along with the fully reduced **40** (Scheme 13,

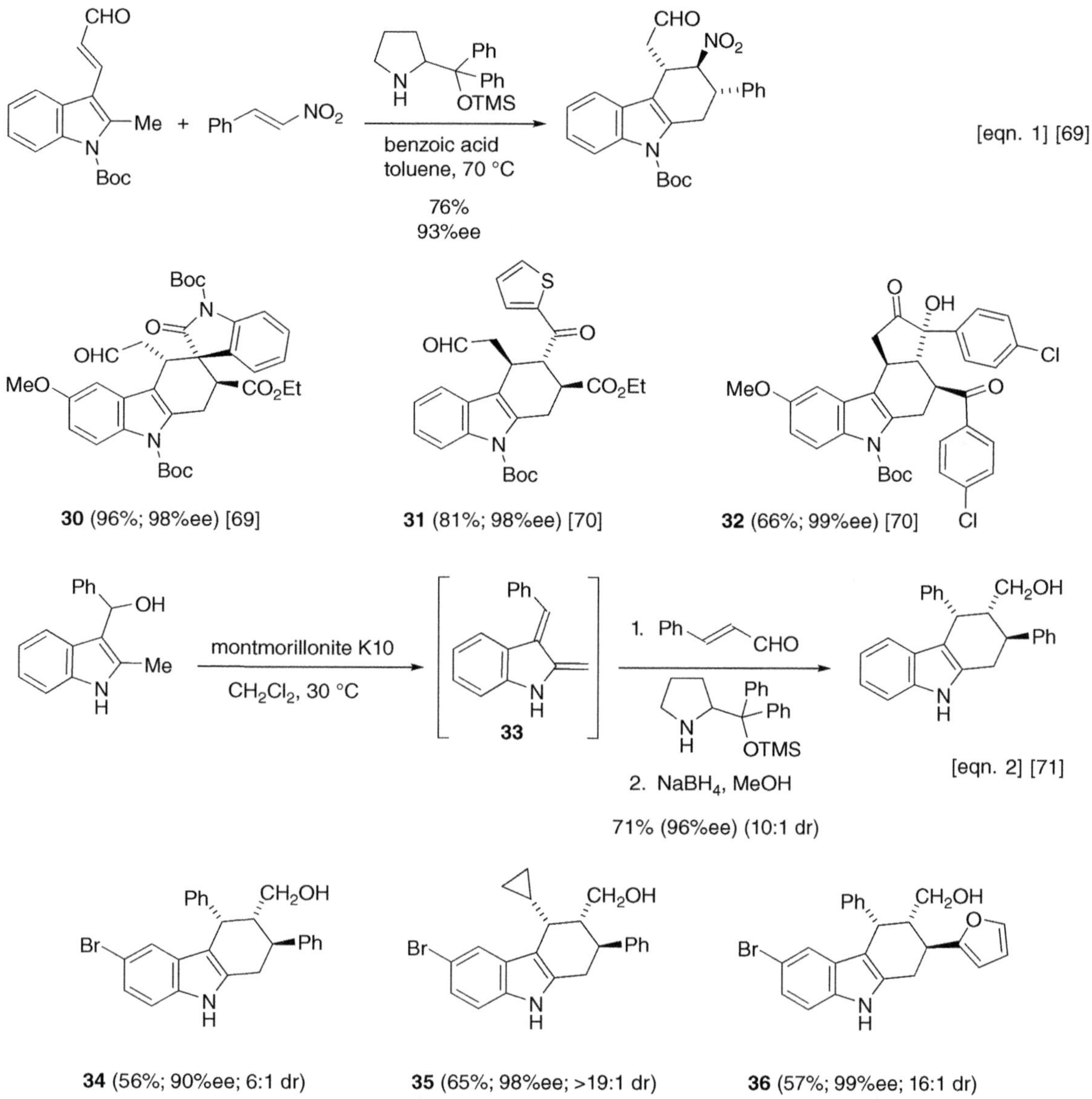

Scheme 11 *Asymmetric Syntheses of Indoles via Indole-2,3-quinodimethanes*

equation 1) [75] but reported no Diels–Alder chemistry. Sha and colleagues developed a different approach and reported Diels–Alder reactions of **41** (equation 2) [76, 77]. Whereas **41** failed to engage in Diels–Alder reactions, **42** reacted with benzyne (from benzenediazonium-2-carboxylate, THF, reflux) to give the expected adduct. Also undergoing Diels–Alder cycloaddition with **42** are *N*-phenylmaleimide (57% *endo* + 19% *exo*) and dimethyl acetylenedicarboxylate (52%). Sha subsequently adapted his chemistry to syntheses of ellipticine and isoellipticine and their amino analogues via the appropriate 1,3-dimethyl-2,4-dihydropyrrolo[3,4-*b*]indole [78], and he also examined some further reactions of these Diels–Alder adducts including a *retro*-Diels–Alder synthesis of an isoindole [79].

Kreher improved upon Welch's lactam reduction by using diisobutylaluminum hydride (DIBAL) to reduce either lactam (Scheme 14, equation 1) [80]. Srinivasan and Jeevanandam employed a route similar to Sha's [78] to prepare a series of 2,4-dihydropyrrolo[3,4-*b*]indoles **43** (equation 2) [81]. A Diels–Alder reaction of **43** (R = Me, R′ = Bn) with DMAD gave the cycloadduct in 70% yield, which on exposure to tosic acid yielded dimethyl 4-benzylamino-1-methyl-5-(phenylsulfonyl)carbazole-2,3-dicarboxylate (68% yield). Snyder and coworkers described a novel pyridazine reductive ring contraction, as explored by Boger

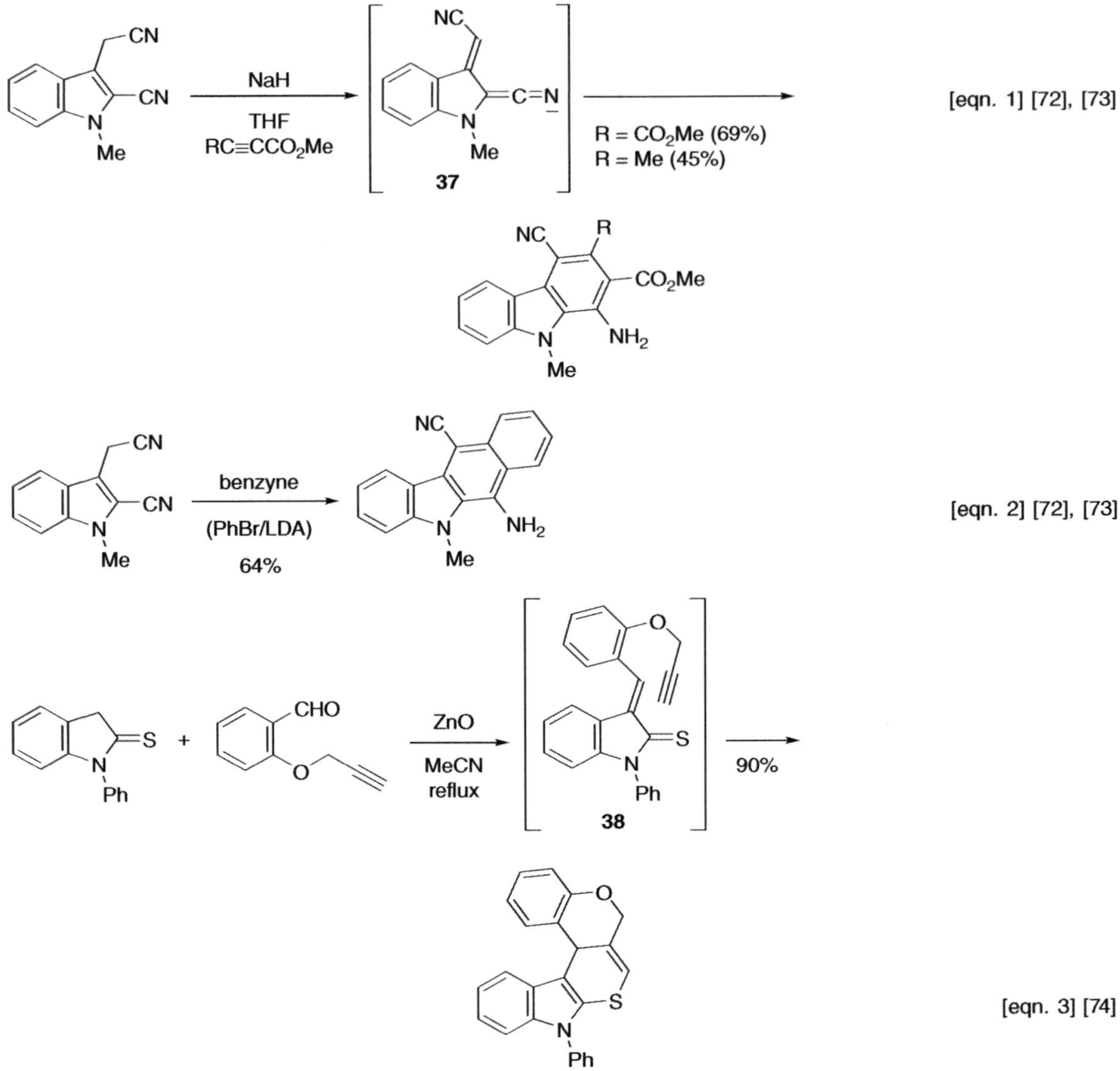

Scheme 12 *Generation and Trapping of Novel Indole-2,3-quinodimethanes*

[82], to access the pyrrolo[3,4-*b*] ring system (equation 3) [83]. The starting pyridazino[4,5-*b*]indoles were prepared from the *N*-protected indole and dimethyl 1,2,4,5-tetrazine-3,6-dicarboxylate in 70% to 95% yield.

Gribble and Pelkey adapted the Barton–Zard pyrrole synthesis [84] to 3-nitroindoles and ethyl isocyanoacetate (and tosylmethyl isocyanide) to give 2,4-dihydropyrrolo [3,4-*b*[indoles (Scheme 15, equation 1) [85, 86]. In complementary work, Gribble and coworkers synthesized pyrrolo[3,4-*b*]indoles from the 1,3-dipolar cycloaddition of 2- and 3-nitroindoles with münchnones (1,3-oxazolium-5 olates) (equations 2, 3) [87, 88]. The mesoionic münchnones were generated *in situ* from the appropriate *N*-benzyl-*N*-acylamino acids. Gribble and Kishbaugh prepared 1,2,3,4-tetrapyrrolo[3,4-*b*]indoles from the corresponding 2,3-bis(bromomethyl)indoles and primary amines and oxidize the former to the respective 2,4-dihydropyrrolo[3,4-*b*]indoles with DDQ (equation 4) [89]. In another route to these indole-2,3-quinodimethane analogues, Gribble and colleagues used a 1,3-dipolar cycloaddition between 3-nitroindoles and azomethine ylides, followed by denitration and oxidation to the pyrrolo[3,4-*b*] indole (equation 5) [90].

With the knowledge that furans are generally better partners in Diels–Alder reactions than pyrroles [91], Gribble and Saulnier in 1983 were the first to synthesize the 4*H*-furo-3,4-*b*]indole (4*H*-furano[3,4-*b*]indole) ring system (**44**) and evaluate its potential as an indole-2,3-quinodimethane

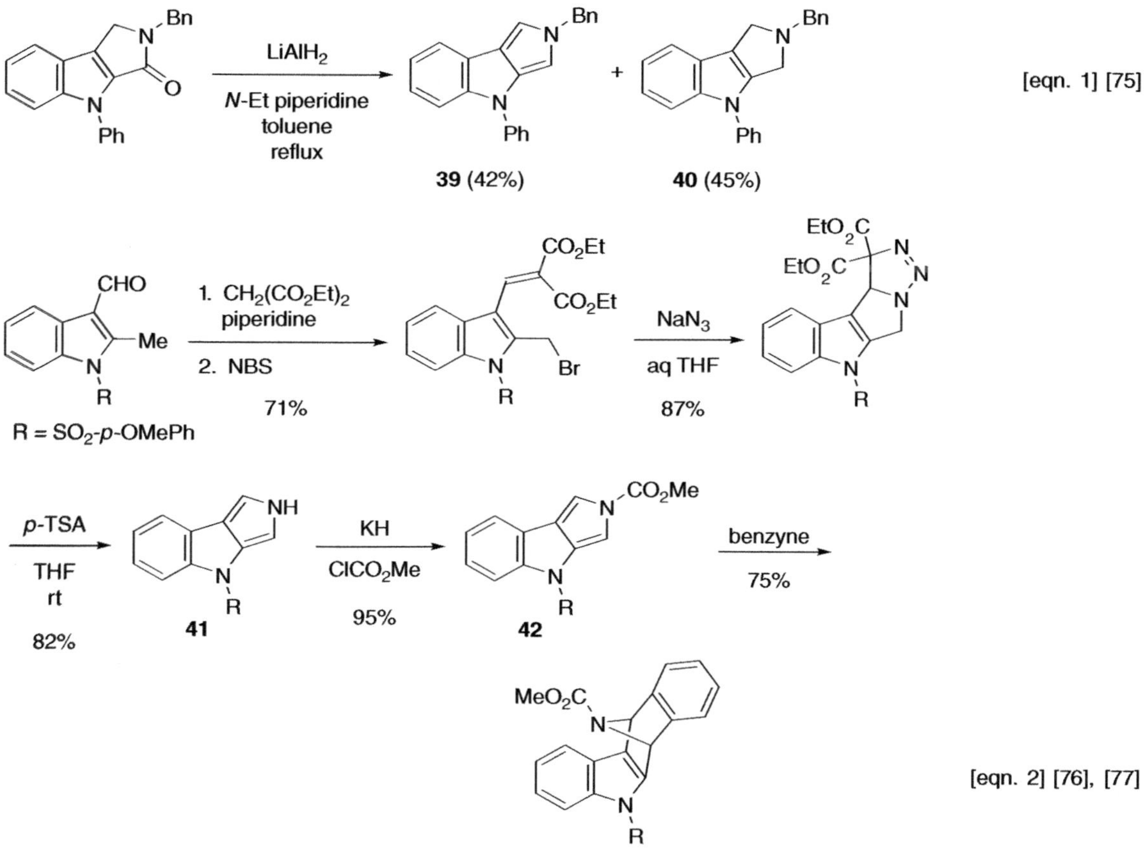

Scheme 13 *Welch and Sha Syntheses and Diels-Alder Reactions of 2,4-Dihydropyrrolo[3,4-*b*]indoles*

in Diels–Alder reactions. Reaction of **44** with benzyne (*o*-bromofluorobenzene, Mg) gave the expected adduct in 38% yield. Subsequent conversion to 5*H*-benzo[*b*]carbazole ($NaBH_4$/TFA; NaOH, MeOH) proceeded in 88% yield (Scheme 16, equation 1) [92]. The dimethylfuro[3,4-*b*]indole analogue was similarly prepared (equation 2), and this strategy was used to synthesize 1,10-bis(6-methyl-5*H*-benzo[*b*]carbazol-11-yl)decane (equation 3) [93]. The reaction of dimethylfuroindole **45** with benzyne gave the Diels–Alder adduct in 93% yield, much higher than the benzyne reaction with furoindole **44**, which perhaps reflects the presence of (acidic) protons adjacent to the furan oxygen in **44** [94].

A more efficient synthesis of 1,3-dimethyl-4*H*-furo[3,4-*b*]indole **45** from 3-ethylindole was described by Gribble and colleagues (Scheme 17, equation 1) [94]. Whereas furoindoles **44** and (especially) **45** are air-stable colorless solids, the desulfonylated furoindoles (NaOH) rapidly decompose. Diels–Alder reactions of **45** with DMAD and *N*-phenylmaleimide gave the expected adducts in quantitative yield. The reaction of **45** with 3,4-pyridyne gave the Diels–Alder adducts in 38% yield. Subsequent conversion to a mixture of ellipticine (23%) and isoellipticine (29%) was accomplished with $NaBH_4$ (NaOH, MeOH, reflux). This poor regioselectivity was circumvented by the use of a less-reactive dienophile than 3,4-pyridyne, as shown in equation 2 [95, 96]. The TMSOTf-acceleracted Diels–Alder reaction of **45** with 5,6-dihydropyridone **46** ultimately gave *only* ellipticine; no isoellipticine could be detected at levels above 1%. This unprecedented regiochemistry for an indole-2,3-quinodimethane equivalent was duplicated for the Diels–Alder reaction of **45** with ethyl acrylate to give only 3-carbethoxy-1,4-dimethyl-9-(phenylsulfonyl)carbazole (equation 3); *none* of the 2-carbethoxy isomer could be detected. Deprotection with Na(Hg) gave the known 3-carbethoxy-1,4-dimethylcarbazole. This regiochemistry was predicted by a consideration of FMO calculations [96]. In addition to **44** and **45**, Gribble and coworkers synthesized furo[3,4-*b*]indoles **47–52** [96,97], including greatly improved syntheses of the parent furo[3,4-*b*]indole **44** [97] and the dimethylfuro[3,4-*b*] indole **45** [98].

Gribble and coworkers employed an intramolecular Diels–Alder reaction of 4*H*-furo[3,4-*b*]indoles to access

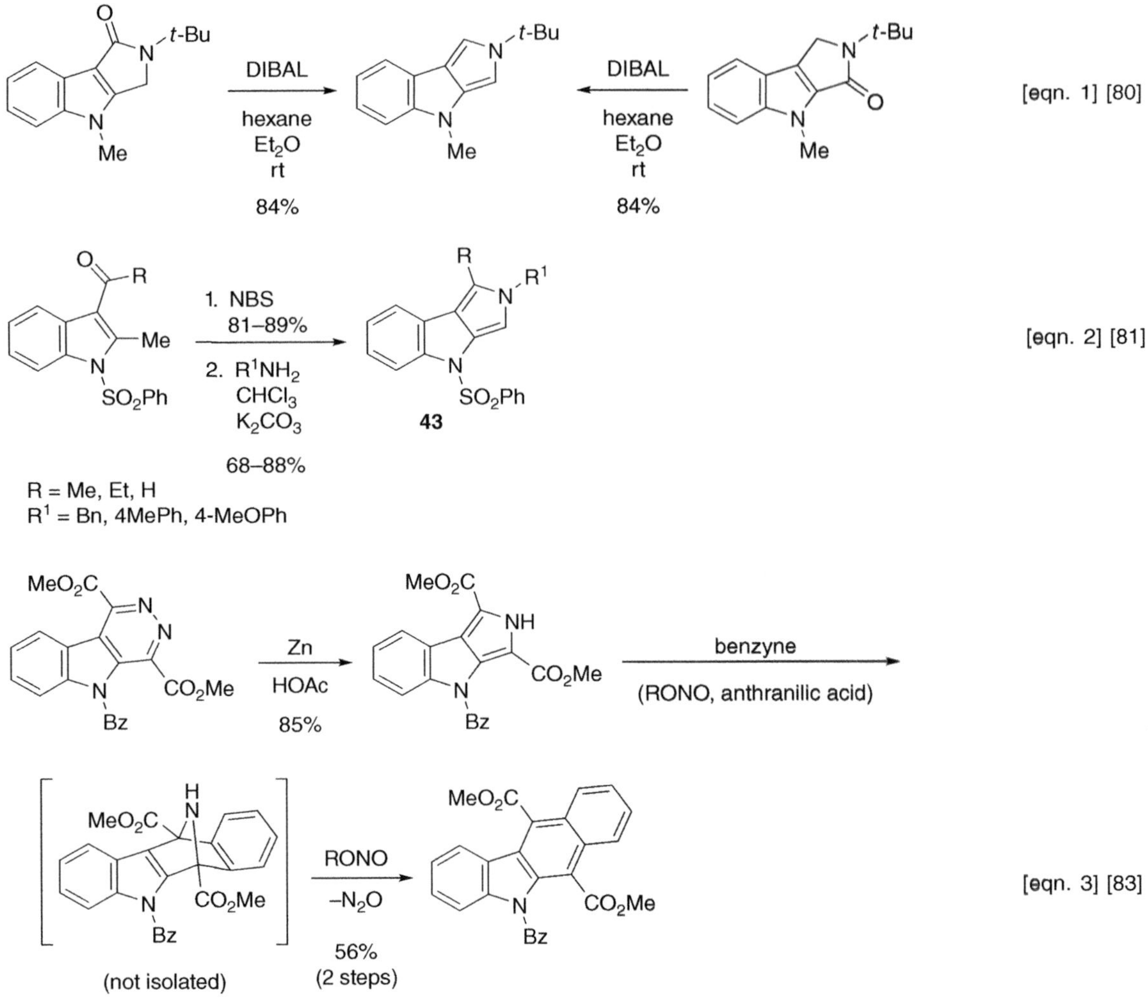

Scheme 14 *Kreher, Srinivasan, and Snyder Syntheses of 2,4-Dihydropyrrolo[3,4-*b*]indoles*

the benzo[*a*]carbazole and benzo[*c*]carbazole ring systems (Scheme 18) [99].

Following the syntheses and exploration of 4*H*-furo[3,4-*b*]indoles by Gribble, several investigators have picked up the baton and discovered new routes to these IQD-synthetic analogues. Friedrichsen and colleagues synthesized and trapped 4*H*-furo[3,4-*b*]indoles by the copper-catalyzed extrusion of nitrogen from diazoesters (Scheme 19, equations 1, 2) [100, 101]. The furoindoles were not isolated. Padwa used a similar generation and Diels–Alder reactions of furoindoles using Cu(II)- or Rh(II)-catalyzed carbenoid cyclization from α-diazo esters (equations 3, 4) [102]. In one case, a furoindole (**53**) was isolated by Padwa, but intramolecular Diels–Alder cyclization was not observed (equation 4).

Miki and Hachiken synthesized the alkaloid murrayaquinone A via a furoindole Diels–Alder approach (Scheme 20, equation 1) [103]. As Gribble earlier found (Scheme 17, equation 3) [95, 96], the reaction of furoindole **54** and methyl acrylate is highly regioselective to give a single carbazole. Conversion to murrayaquinone A was accomplished in a few steps (1. $LiAlH_4$, 2. Ac_2O, 3. $AlCl_3$, 4. $(KSO_3)_2NO$ [Fremy's salt]; 22%). Furoindole **49** also reacts with DMAD (56%) and *N*-phenylmaleimide (87%) to give the corresponding hydroxy carbazoles. Fang and coworkers synthesized several 4*H*-furo[3,4-*b*]indoles via the acid-catalyzed cyclization of *N*-methyl-3-acyl-2-hydroxyalkylindoles [104, 105], and Iwasaki, Kondo, and colleagues prepared the known furoindole **50** using a modification of the Gribble method [96] and obtained 4-hydroxycarbazoles upon reaction and ring-opening with DMAD (equation 2) [106]. Pujol and coworkers used an acid-resin to cyclize hydroxyacetals to the 4*H*-furo[3,4-*b*] indoles (e.g., **55**) (equation 3) [107], in an extension of the

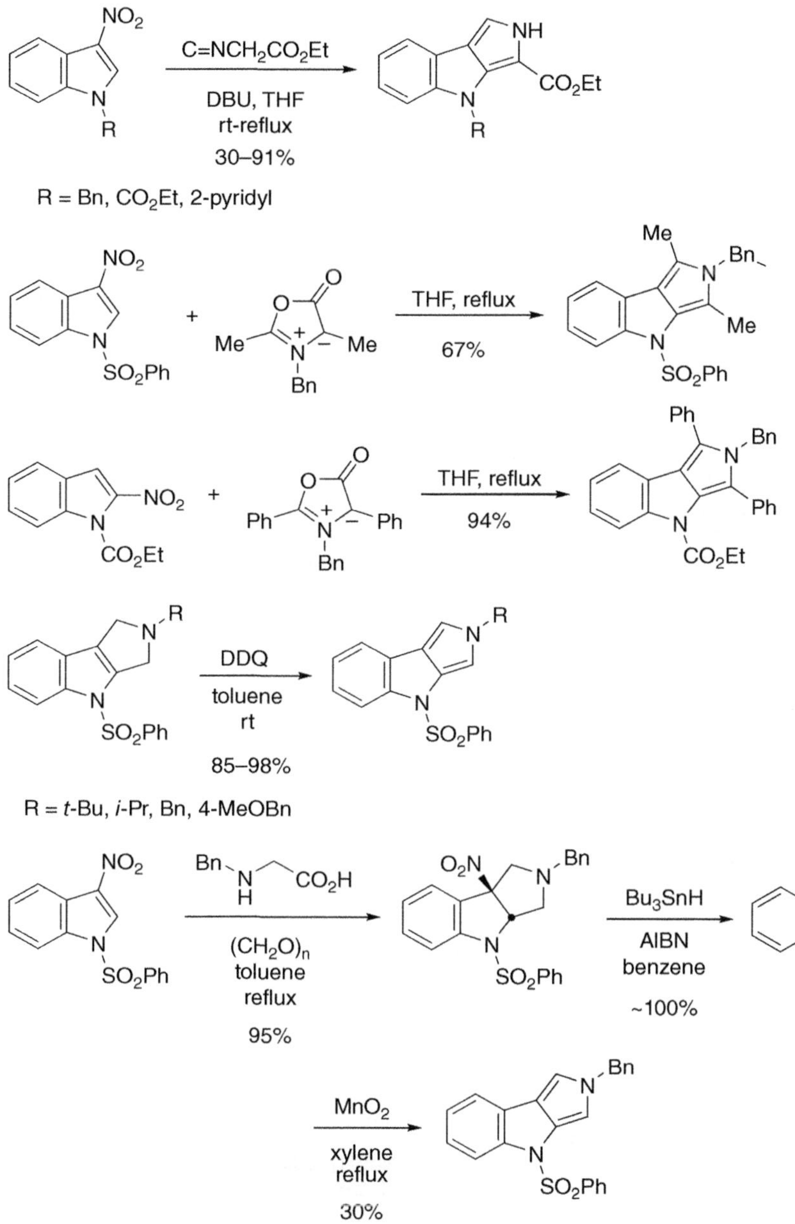

[eqn. 1] [85]

[eqn. 2] [88]

[eqn. 3] [88]

[eqn. 4] [89]

[eqn. 5] [90]

Scheme 15 *Gribble Syntheses of 2,4-Dihydropyrrole[3,4-*b*]indoles*

Gribble method [97], and used the 1,3-dimethyl analogue in a synthesis of ellipticine and isoellipticine. Mal's team used a base-generated furoindole **56** formation and trapping (equation 4) [108], similar to the chemistry depicted in equation 1 [103]. These authors exploited this chemistry to synthesize the alkaloids clausine E and mukonine and their 4-prenylated analogues. Guitián and colleagues employed furoindole **45** and 2-chloro-3, 4-pyridyne in an improved synthesis of ellipticine *vis-à-vis* isoellipticine [109].

Padwa and Kappe synthesized 4*H*-furo[3,4-*b*]indoles **57** and **58** via a novel Pummerer-induced cyclization and subsequent Diels–Alder reactions (Scheme 21) [110]. The reaction of **57** with DMAD proceeds directly to the ring-opened carbazole **59**, and furoindole **58** is significantly less reactive than **57** in Diels–Alder reactions with maleic anhydride and *N*-phenylmaleimide.

In contrast to the extensive chemistry of 1,3-dihydropyrrolo[3,4-*b*]indoles and 4*H*-furo[3,4-*b*]indoles, only one paper, by Shafiee and Sattari, describes the

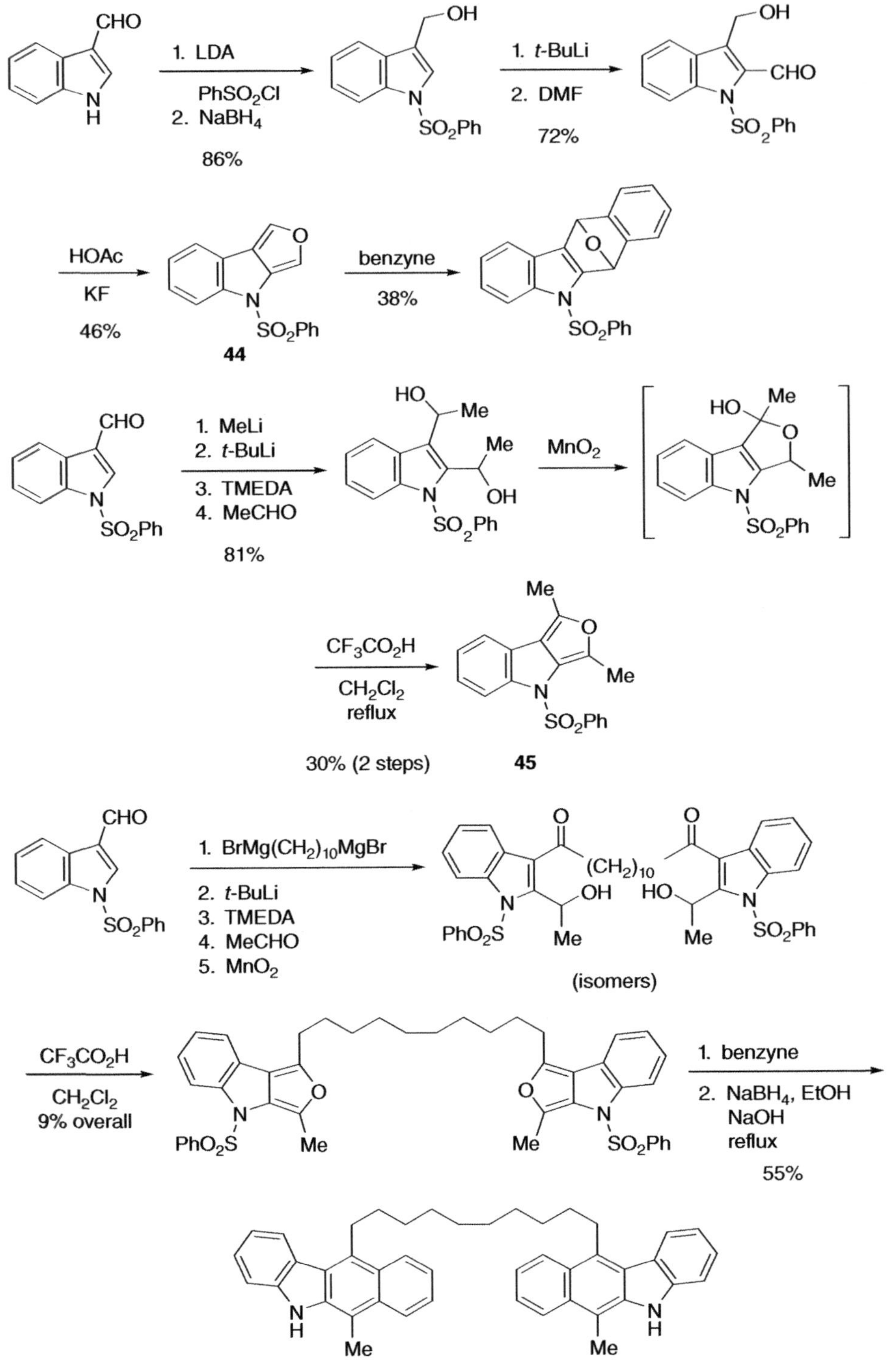

[eqn. 1] [92]

[eqn. 2] [92]

[eqn. 3] [93]

Scheme 16 *Gribble Synthesis and Diels-Alder Reactions of 4H-Furo[3,4-b]indoles*

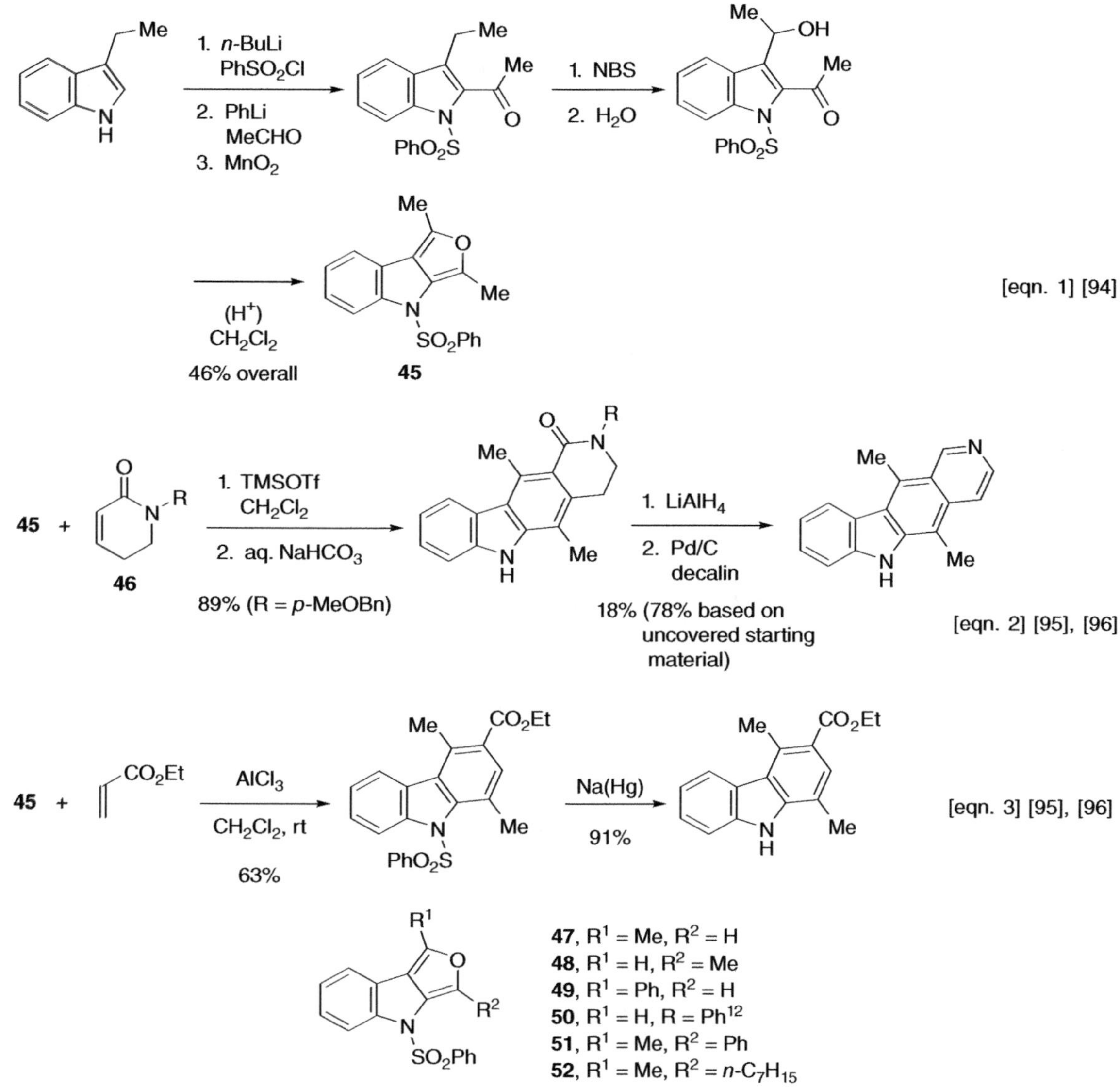

Scheme 17 *Gribble Improved Syntheses of 4H-Furo[3,4-b]indoles and Ellipticine*

synthesis and Diels–Alder reactions of the 4*H*-thieno[3,4-*b*]indole and 4*H*-seleno[3,4-*b*]indole ring systems (Scheme 22) [111]. As might be expected, the Diels–Alder reactions of the (more aromatic) thieno (**60**) and seleno (**61**) rings are lower yielding (with DMAD). Also prepared were several 1-aryl analogues of both ring systems. Further chemistry of these fused-indoles remains to be uncovered.

Beyond Diels–Alder reactions of indole-2,4-quinodimethanes to construct indoles, there is a wealth of indole ring synthesis involving Diels–Alder cycloadditions and electrocyclizations that *do not* begin with an indole or indolone. Indeed, a number of ingenious indole ring syntheses starting from scratch are known.

Ghosez and Differding effected an intramolecular Diels–Alder cycloaddition of acetylenic vinylketenimines to yield carbazoles (Scheme 23, equation 1) [112]. The starting anilides **63** were prepared in 71% to 80% yield from the appropriate β,γ-unsaturated carboxylic acid chloride and aniline. The reacting π bonds are in bold. The method also led to the pyridocarbazole alkaloid *N*-methyl-tetrahydroellipticine. In a series of papers, Molina and colleagues explored the intramolecular Diels–Alder reactions of conjugated carbodiimides and related substrates to form carbazoles, carbolines, and benzindoles (equations 2, 3) [113–116]. The diarylcarbodiimides (**65**) were prepared from the appropriate iminophosphorane and aromatic

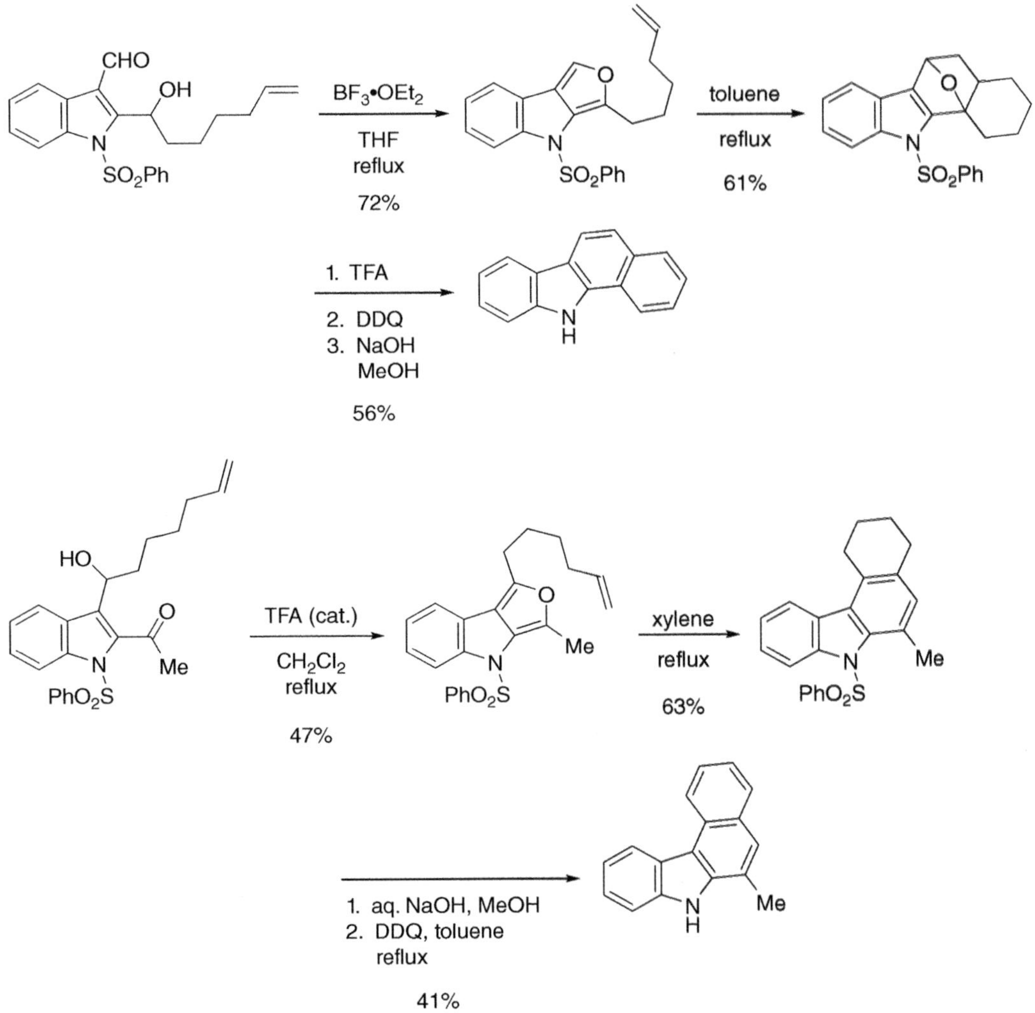

Scheme 18 *Intramolecular Diels-Alder Reactions of 4H-Furo[3,4-b]indoles*

isocyanate [113]. Ketenimines **66** were generated *in situ* by reaction of iminophosphoranes with diphenylketene [116]. Saito and colleagues independently described very similar intramolecular cycloadditions of conjugated carbodiimides to give α-carbolines [117].

Kanematsu and coworkers developed the intramolecular allene + diene Diels–Alder cycloaddition into a new indole synthesis (Scheme 24, equations 1, 2) [118–121]. The allene functionality in **67** was prepared by homologation (HCHO, CuBr, *i*-Pr_2NH). This cycloaddition strategy was successfully applied to syntheses of hippadine (equation 2) [119], and (±)- and (+)-*cis*-trikentrin B [120, 121].

Several groups have discovered that intramolecular Diels–Alder cycloaddition of suitably tethered ynamides gave indoles and/or indolines. Witulski and colleagues reported the simple example in Scheme 25 (equation 1) [122]. Interestingly, the final oxidation with DDQ halted at the indoline stage (**67**). Better results were obtained for this dienyne cyclization with other substrates using Rh(I) as a catalyst ($RhCl(PPh_3)_3/AgSbF_6$) (toluene, 20 °C). Saá and coworkers synthesized a plethora of complex carbazoles via the intramolecular dehydro Diels–Alder cycloaddition of ynamides (equations 2, 3) [123–125]. One of the lowest yielding cyclizations was that to give the demethylated

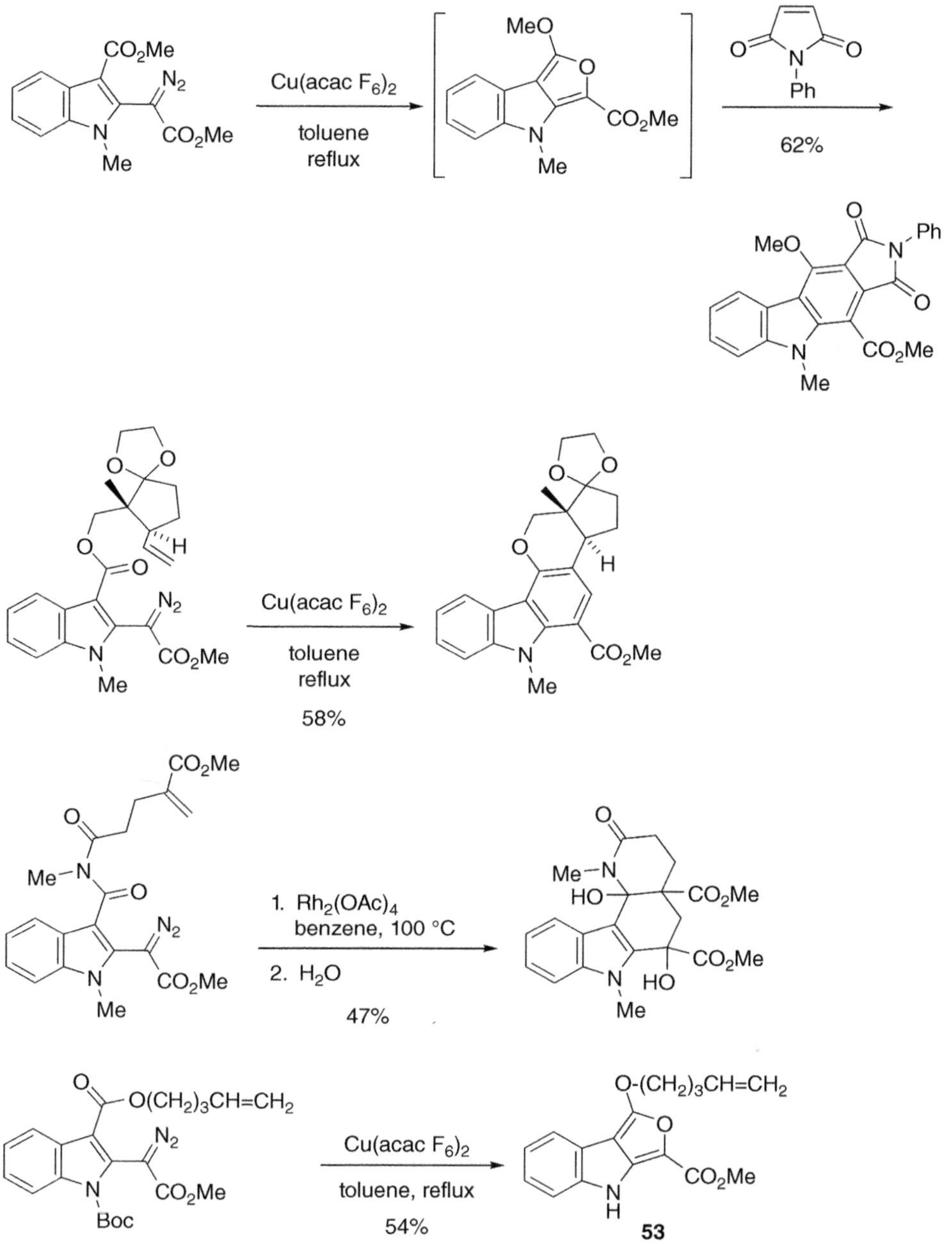

[eqn. 1] [100]

[eqn. 2] [101]

[eqn. 3] [102]

[eqn. 4] [102]

Scheme 19 *Generation and Trapping of 4H-Furo[3,4-b]indoles from α-Diazoesters*

ellipticine (equation 3) [125]. Other benzo[*b*]carbazoles were obtained in yields up to 95%. Danheiser and Dunetz described similar cycloadditions [126]. Wasserman and Blum reported the intermolecular Diels–Alder cyclization to lead ultimately to 3-hydroxyindoles (equation 4) [127].

Franck and Soll constructed the benz[*c,d*]indole framework from an isoquinoline network (Scheme 26, equation 1) [128]. This remarkable transformation involves a Diels–Alder reaction between the 1,4-carbons in the isoquinolinium ring and phenyl vinyl sulfide, followed by hydrolysis of the iminium ion to give an aldehyde that cyclizes to carbinol amine **69**. Dehydration provides the indole. Padwa and his coworkers developed a versatile synthesis of indolines (and indoles) that entailed the intramolecular Diels–Alder cycloaddition of 2-amidofurans with pendant dienophiles [129–136]. Although the initial

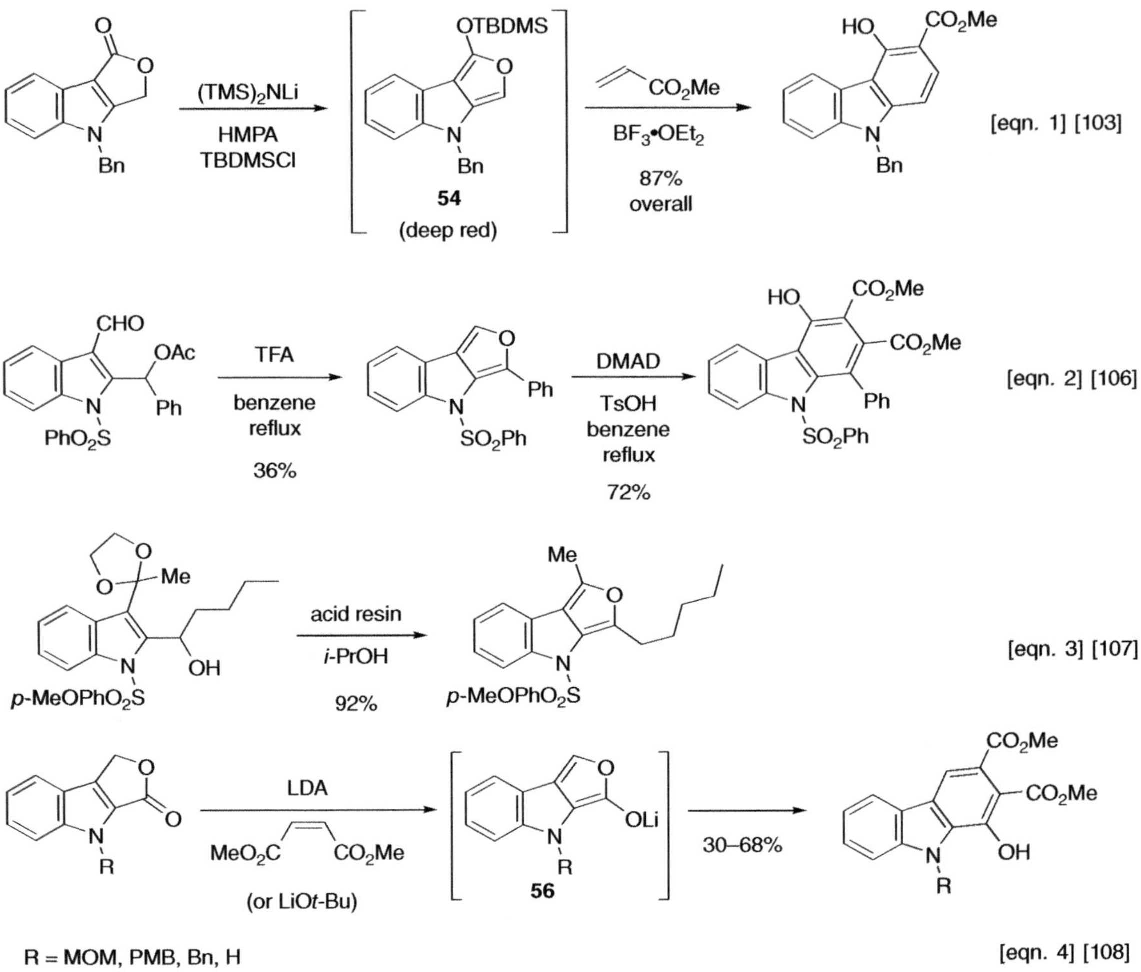

Scheme 20 *Generation and Trapping of 4H-Furo[3,4-b]indoles in Carbazole Synthesis*

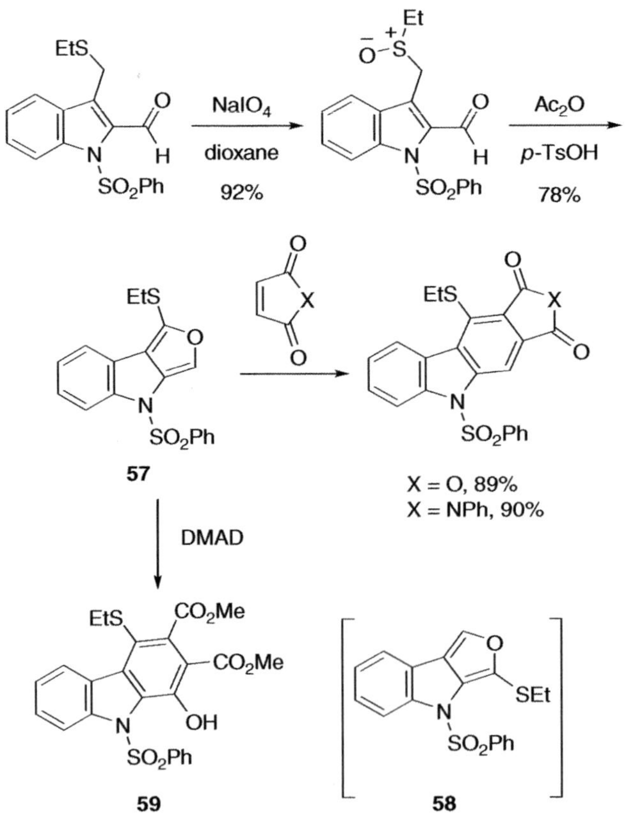

Scheme 21 *Padwa's Synthesis of 4H-Furo[3,4-b]indoles and Diels-Alder Reactions*

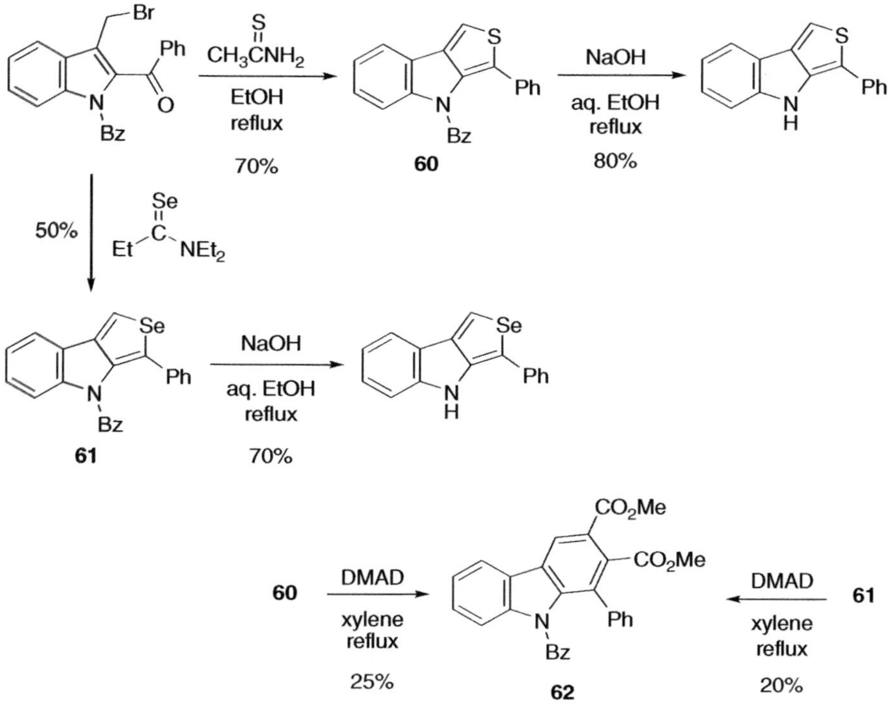

[eqn. 1]

Scheme 22 *Shafiee's Synthesis and Diels-Alder Reactions of 4H-Thieno[3,4-b]indole and 4H-Seleno[3,4-b]indole*

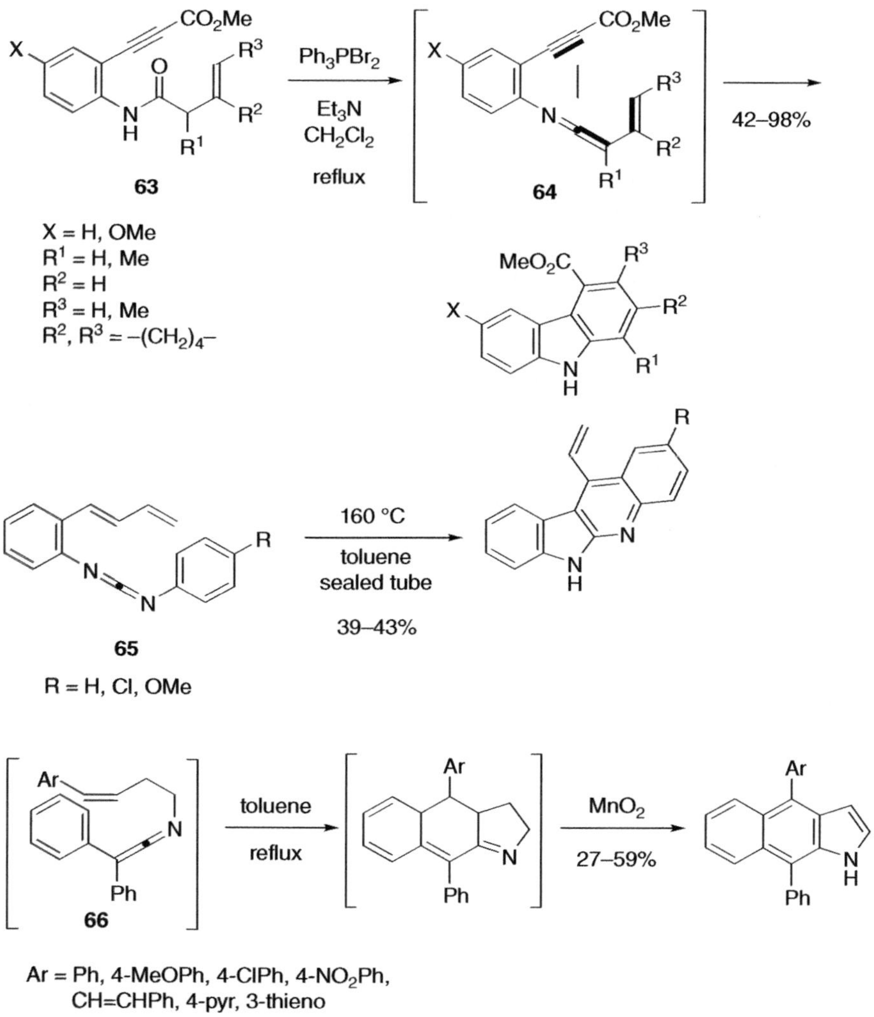

[eqn. 1] [112]

[eqn. 2] [113]

[eqn. 3] [116]

Scheme 23 *Ghosez and Molina Intramolecular Diels-Alder Cycloadditions*

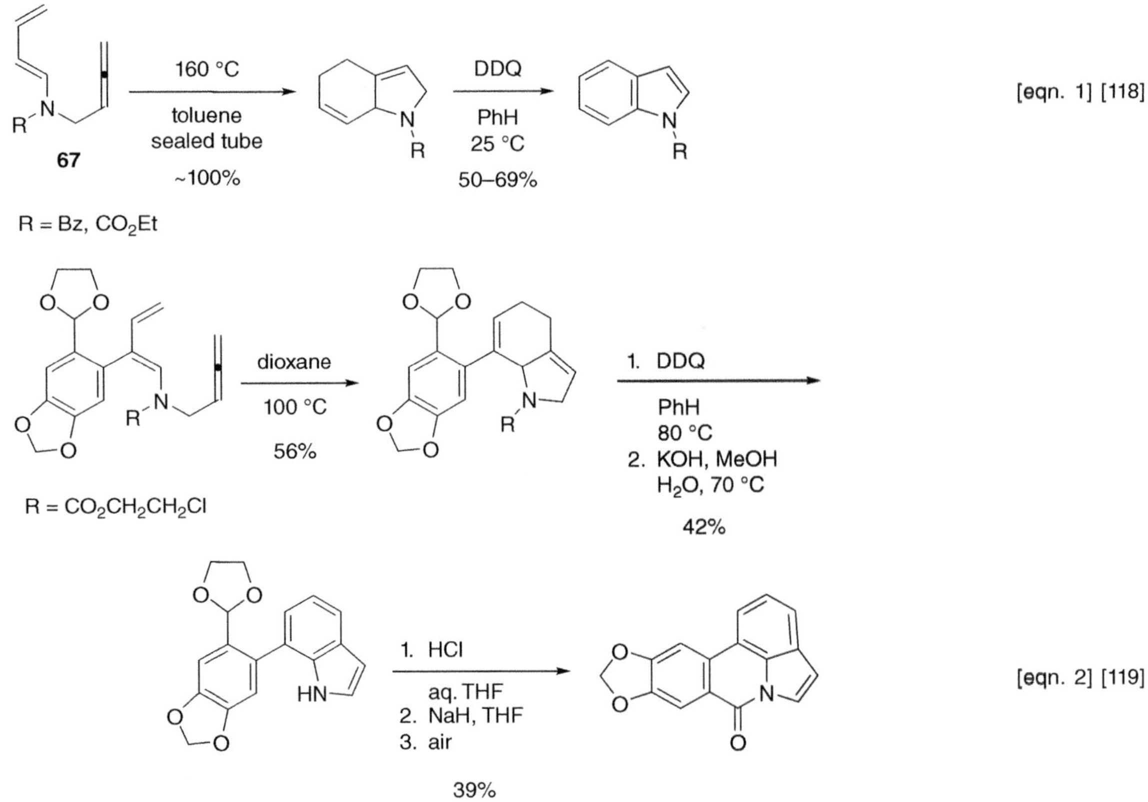

Scheme 24 *Kanematsu Indole Synthesis*

cycloadducts were typically indolines (equation 2), indoles can also be accessed by this sparkling methodology, such as the famous Uhle's ketone (**70**) (equation 3). Wipf and coworkers reported an intramolecular 2-amidofuran Diels–Alder cycloaddition leading, after thermal dehydration and deprotection, to 4-substituted indoles (equation 4) [137]. This novel indole synthesis was elegantly applied to a synthesis of the ergot alkaloid cycloclavine [138].

Hoornaert and colleagues synthesized α-carbolines and β-carbolinones via the intramolecular Diels–Alder cycloaddition of 2(1*H*)-pyrazinones (Scheme 27, equations 1, 2) [139]. The striking difference in product formation is a function of substrate structure and the mode of *retro*-Diels–Alder (loss of BnN=C=O vs. ClCN). A pioneer in azine Diels–Alder cycloadditions and *retro*-Diels–Alder reactions is Boger [140–144], and some examples are shown in equations 3 and 4. Both (±)-*cis*-(**71**) and (±)-*trans*-trikentrin A were synthesized in this fashion [142].

This chapter has presented an abundance of material on indole–carbazole–carboline syntheses that used the venerable Diels–Alder reaction in ways that would bring joy to Otto Diels and Kurt Alder.

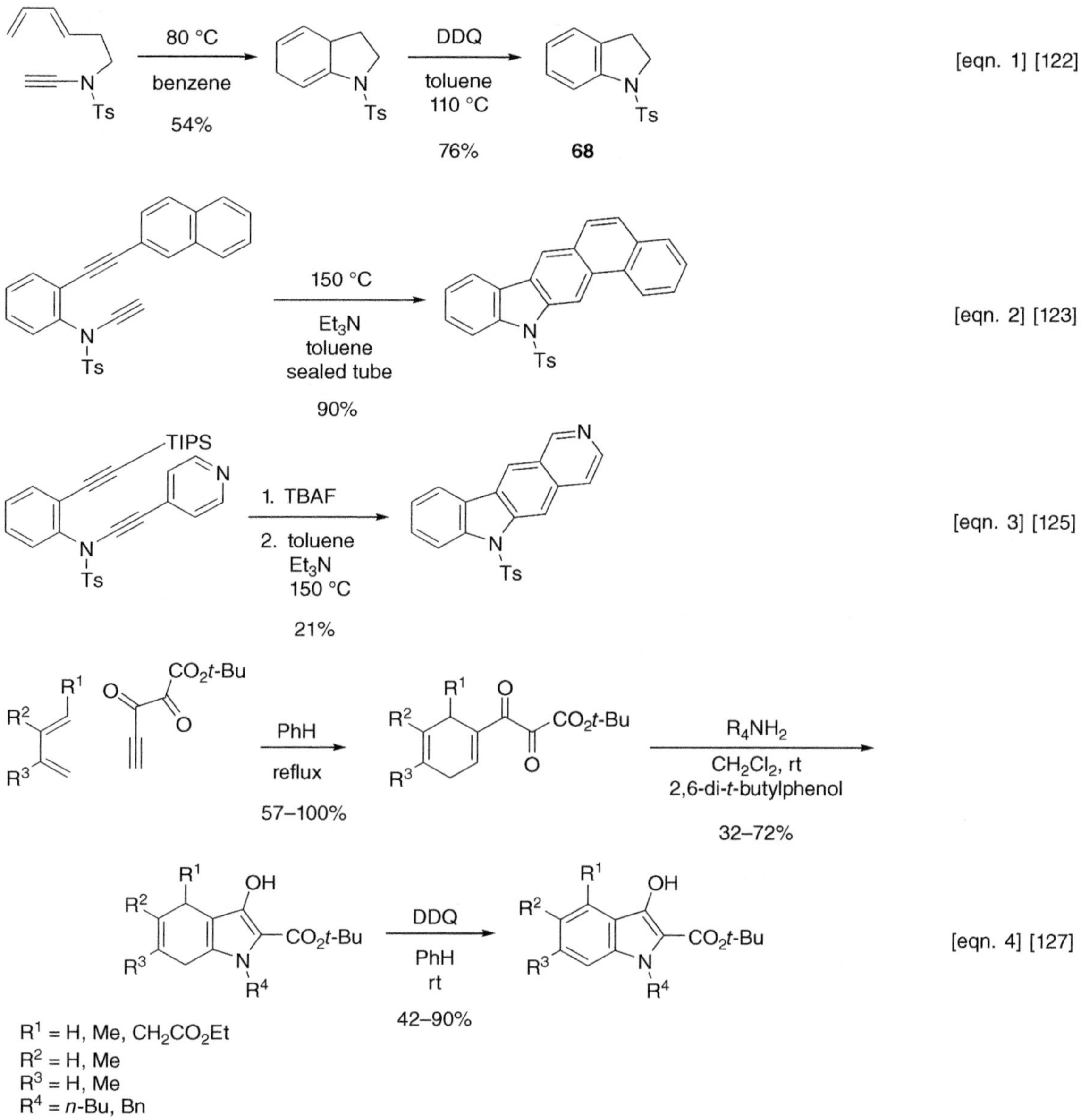

Scheme 25 *Witulski, Saá, and Wasserman Diels-Alder Cycloadditions*

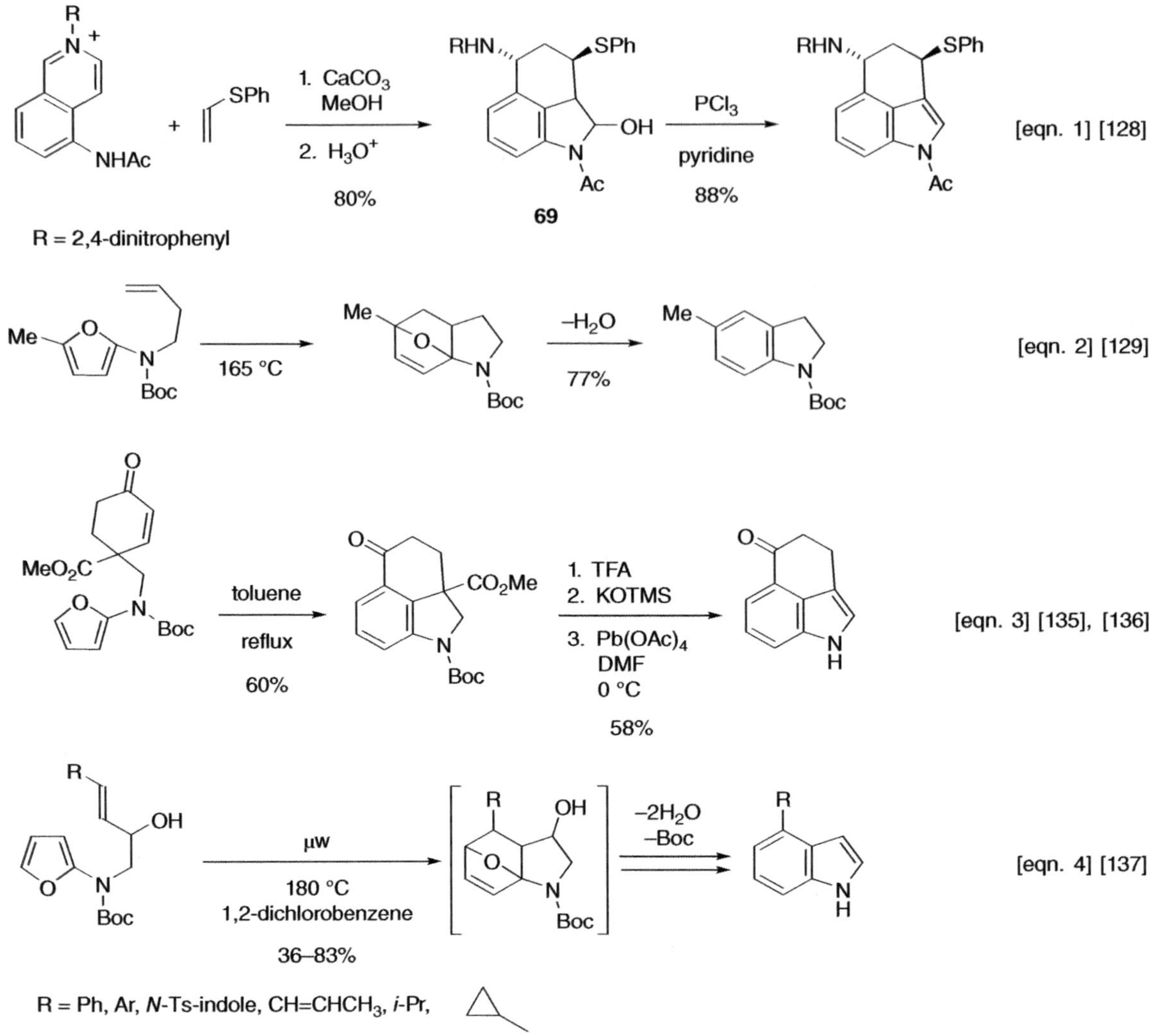

Scheme 26 *Franck, Padwa, and Wipf Diels-Alder Cycloadditions*

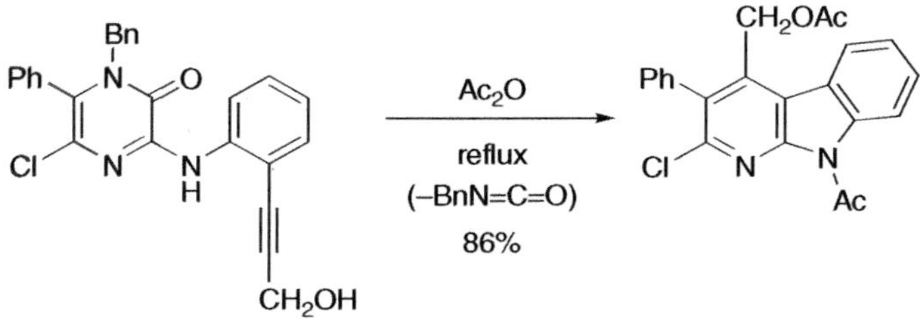

[eqn. 1] [139]

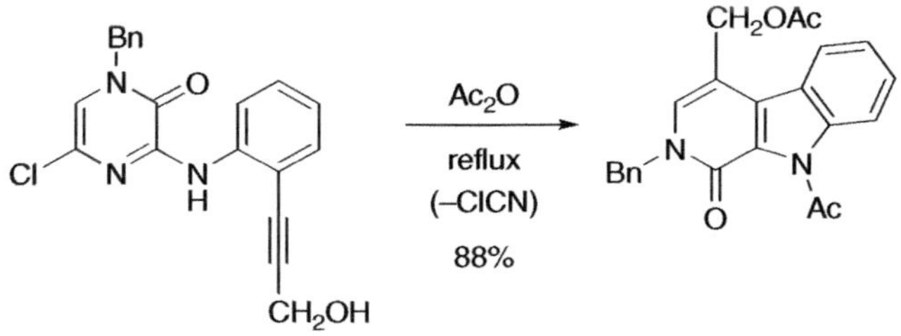

[eqn. 2] [139]

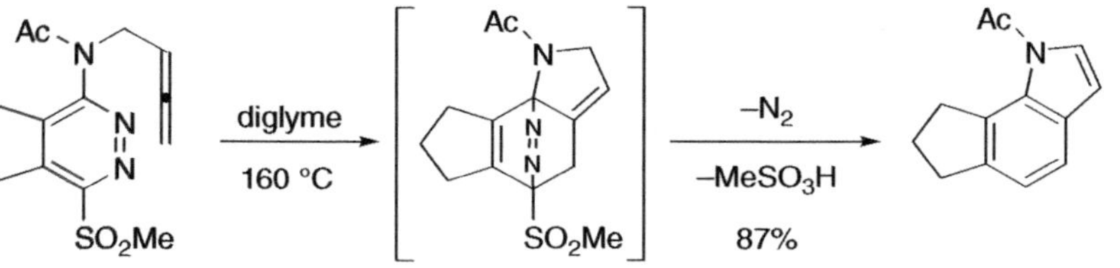

[eqn. 3] [141]

(TIPB = tri-*i*-Pr benzene)

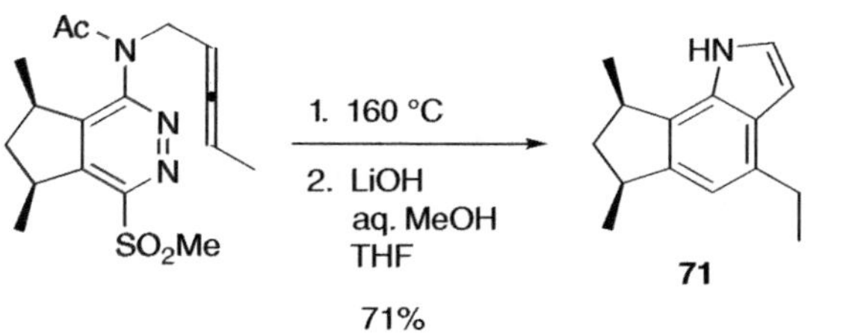

[eqn. 4] [142]

Scheme 27 *Hoornaert and Boger Intramolecular Diels-Alder Cycloadditions*

References

[1] M.P. Cava and D.R. Napier, *J. Am. Chem. Soc.*, 1957, **79**, 1701–1705.
[2] M.P. Cava, A.A. Deana, and K. Muth, *J. Am. Chem. Soc.*, 1959, **81**, 6458–6460.
[3] W. Oppolzer, *Synthesis*, 1978, 793–802.
[4] J.L. Charlton and M.M. Alauddin, *Tetrahedron*, 1987, **43**, 2873–2889.
[5] N. Martin, C. Seoane, and C. Hanack, *Org. Prep. Proc. Int.*, 1991, **23**, 237–272.
[6] J.L. Segura and N. Martín, *Chem. Rev.*, 1999, **99**, 3199–3246.
[7] T. Chou, *Rev. Heteroatom Chem.*, 1993, **8**, 65–104.
[8] U. Pindur and H. Erfanian-Abdoust, *Chem. Rev.*, 1989, **89**, 1681–1689.
[9] H. Plieninger, W. Müller, and K. Weinerth, *Chem. Ber.*, 1964, **97**, 667–681.
[10] C.J. Moody, *J. Chem. Soc., Chem. Commun.*, 1984, 925–926.
[11] C. May and C.J. Moody, *J. Chem. Soc., Chem. Commun.*, 1984, 926–927.
[12] C.J. Moody, *J. Chem. Soc., Perkin Trans. 1*, 1985, 2505–2508.
[13] C. May and C.J. Moody, *J. Chem. Soc., Perkin Trans. 1*, 1988, 247–250.
[14] C.J. Moody and P. Shah, *J. Chem. Soc., Perkin Trans. 1*, 1988, 3249–3254.
[15] C.J. Moody and P. Shah, *J. Chem. Soc., Perkin Trans. 1*, 1988, 1407–1415.
[16] C.J. Moody and P. Shah, *J. Chem. Soc., Perkin Trans. 1*, 1989, 2463–2471.
[17] C.J. Moody and P. Shah, *Tetrahedron Lett.*, 1988, **29**, 2693–2696.
[18] C.J. Moody and P. Shah, *J. Chem. Soc., Perkin Trans. 1*, 1989, 376–377.
[19] C.J. Moody and K.F. Rahimtoola, *J. Chem. Soc., Perkin Trans. 1*, 1990, 673–679.
[20] E.B. Fray, C.J. Moody, and P. Shah, *Tetrahedron*, 1993, **49**, 439–450.
[21] C.J. Moody and A.I. Morrell, *J. Indian Chem. Soc.*, 1994, **71**, 309–314.

[22] C.J. Moody, K.F. Rahimtoola, B. Porter, and B.C. Ross, *J. Org. Chem.*, 1992, **57**, 2105–2114.
[23] J. Rokach, D. McNeill, and C.S. Rooney, *Chem. Commun.*, 1971, 1085–1086.
[24] N.S. Narasimhan and S.M. Gokhale, *J. Chem. Soc., Chem. Commun.*, 1985, 86–87.
[25] N.S. Narasimhan and S.M. Gokhale, *J. Indian Inst. Sci.*, 2001, **81**, 135–138.
[26] U. Pindur and H. Erfanian-Abdoust, *Liebigs Ann. Chem.*, 1988, 803–805.
[27] U. Pindur, M. Haber, and H. Erfanian-Abdoust, *Heterocycles*, 1992, **34**, 781–790.
[28] U. Pindur and H. Erfanian-Abdoust, *Liebigs Ann. Chem.*, 1989, 227–230.
[29] P. Van Doren, D. Vanderzande, S. Toppet, and G. Hoornaert, *Tetrahedron*, 1989, **45**, 6761–6770.
[30] P. Van Doren, F. Compernolle, and G. Hoornaert, *Tetrahedron*, 1990, **46**, 4023–4030.
[31] H. Nandin de Carvalho, G.I. Dmitrienko, and K.E. Nielson, *Tetrahedron*, 1990, **46**, 5523–5532.
[32] S.F. Vice, H. Nandin de Carvalho, N.G. Taylor, and G.I. Dmitrienko, *Tetrahedron Lett.*, 1989, **30**, 7289–7292.
[33] O.M. Jakiwczyk, K.E. Nielsen, H. Nandin de Carvalho, and G.I. Dmitrienko, *Tetrahedron Lett.*, 1997, **38**, 6541–6544.
[34] M. Díaz, A. Cobas, E. Guitián, and L. Castedo, *Eur. J. Org. Chem.*, 2001, 4543–4549.
[35] J. Bergman and R. Carlsson, *Tetrahedron Lett.*, 1977, **18**, 4663–4666.
[36] J. Bergman and R. Carlsson, *Tetrahedron Lett.*, 1978, **19**, 4055–4058.
[37] M. Driver, I.T. Matthews, and M. Sainsbury, *J. Chem. Soc. Perkin Trans. 1*, 1979, 2506–2510.
[38] S. Kano, E. Sugino, S. Shibuya, and S. Hibino, *J. Org. Chem.*, 1981, **46**, 2979–2981.
[39] T. Gallagher and P. Magnus, *Tetrahedron*, 1981, **37**, 3889–3897.
[40] T. Gallagher and P. Magnus, *J. Am. Chem. Soc.*, 1982, **104**, 1140–1141.
[41] P. Magnus, T. Gallagher, P. Brown, and P. Pappalardo, *Acc. Chem. Res.*, 1984, **17**, 35–41.
[42] T. Gallagher and P. Magnus, *J. Am. Chem. Soc.*, 1983, **105**, 2086–2087.
[43] T. Gallagher, P. Magnus, and J.C. Huffman, *J. Am. Chem. Soc.*, 1983, **105**, 4750–4757.
[44] P. Magnus, P. Pappalardo, and I. Southwell, *Tetrahedron*, 1986, **42**, 3215–3222.
[45] P.D. Magnus, C. Exon, and N.L. Sear, *Tetrahedron*, 1983, **39**, 3725–3729.
[46] K. Cardwell, B. Hewitt, M. Ladlow, and P. Magnus, *J. Am. Chem. Soc.*, 1988, **110**, 2242–2248.
[47] P. Magnus and P.A. Pappalardo, *J. Am. Chem. Soc.*, 1986, **108**, 212–217.
[48] C. Exon, T. Gallagher, and P. Magnus, *J. Chem. Soc., Chem. Commun.*, 1982, 613–614.
[49] E. Ciganek and E.M. Schubert, *J. Org. Chem.*, 1995, **60**, 4629–4634.
[50] E.R. Marinelli, *Tetrahedron Lett.*, 1982, **23**, 2745–2748.
[51] B. Saroja and P.C. Srinivasan, *Tetrahedron Lett.*, 1984, **25**, 5429–5430.
[52] M. Haber and U. Pindur, *Tetrahedron*, 1991, **47**, 1925–1936.
[53] K.U. Meyer and U. Pindur, *J. Chem. Soc., Perkin Trans. 1*, 2001, 695–700.
[54] M. Terzidis, C.A. Tsoleridis, and J. Stephanidou-Stephanatou, *Tetrahedron Lett.*, 2005, **46**, 7239–7242.
[55] C.A. Tsoleridis, J. Dimtsas, D. Hatzimimikou, and J. Stephanidou-Stephanatou, *Tetrahedron*, 2006, **62**, 4232–4242.
[56] K.A. Rinderspacher, Ph.D. Thesis, Dartmouth College, Hanover, New Hampshire, 2006.
[57] C.-W. Ko and T. Chou, *Tetrahedron Lett.*, 1997, **38**, 5315–5318.
[58] C.-W. Ko and T. Chou, *J. Org. Chem.*, 1998, **63**, 4645–4653.
[59] M. Laronze and J. Sapi, *Tetrahedron Lett.*, 2002, **43**, 7925–7928.
[60] R.F. Miambo, M. Laronze-Cochard, A.-M. Lawson, *et al.*, *Tetrahedron*, 2014, **70**, 8286–8302.
[61] N. Kuroda, Y. Takahashi, K. Yoshinaga, and C. Mukai, *Org. Lett.*, 2006, **8**, 1843–1845.
[62] F. Inagaki, M. Mizutani, N. Kuroda, and C. Mukai, *J. Org. Chem.*, 2009, **74**, 6402–6405.
[63] H. Fuwa, T. Tako, M. Ebine, and M. Sasaki, *Chem. Lett.*, 2008, **37**, 904–905.
[64] H. Fuwa and M. Sasaki, *Chem. Comm.*, 2007, 2876–2878.
[65] Y. Tamura, S. Mohri, H. Maeda, *et al.*, *Tetrahedron Lett.*, 1984, **25**, 309–312.
[66] M.V. Basaveswara Rao, J. Satyanarayana, H. Ila, and H. Junjappa, *Tetrahedron Lett.*, 1995, **36**, 3385–3388.
[67] X.-Y. Chen, M.-W. Wen, S. Ye, and Z.-X. Wang, *Org. Lett.*, 2011, **13**, 1138–1141.
[68] M. Jha, S. Guy, and T.-Y. Chou, *Tetrahedron Lett.*, 2011, **52**, 4337–4341.
[69] Y. Liu, M. Nappi, E. Arceo, *et al.*, *J. Am. Chem. Soc.*, 2011, **133**, 15212–15218.
[70] Y. Liu, M. Nappi, E.C. Escudero-Adán, and P. Melchiorre, *Org. Lett.*, 2012, **14**, 1310–1313.
[71] Y.-C. Xiao, Q.-Q. Zhou, L. Dong, *et al.*, *Org. Lett.*, 2012, **14**, 5940–5943.
[72] T. Kurihara, M. Hanakawa, T. Wakita, and S. Harusawa, *Heterocycles*, 1985, **23**, 2221–2224.
[73] T. Kurihara, M. Hanakawa, S. Harusawa, and R. Yoneda, *Chem. Pharm. Bull.*, 1986, **34**, 4545–4553.
[74] M. Kiamehr and F.M. Moghaddam, *Tetrahedron Lett.*, 2009, **50**, 6723–6727.
[75] W.M. Welch, *J. Org. Chem.*, 1976, **41**, 2031–2032.
[76] C.-K. Sha, K.-S. Chuang, and J.-J. Young, *J. Chem. Soc., Chem. Commun.*, 1984, 1552–1554.
[77] C.-K. Sha, K.-S. Chuang, and S.-J. Wey, *J. Chem. Soc., Perkin Trans. 1*, 1987, 977–980.
[78] C.-K. Sha and J.-F. Yang, *Tetrahedron*, 1992, **48**, 10645–10654.
[79] C.-K. Sha, J.-F. Yang, and C.-J. Chang, *Tetrahedron Lett.*, 1996, **37**, 3487–3488.
[80] R.P. Kreher and G. Dyker, *Z. Naturforsch.*, 1987, **42b**, 473–477.
[81] A. Jeevanandam and P.C. Srinivasan, *J. Chem. Soc., Perkin Trans. 1*, 1995, 2663–2665.
[82] D.L. Boger, R.S. Coleman, J.S. Panek, and D. Yohannes, *J. Org. Chem.*, 1984, **49**, 4405–4409.
[83] K. Daly, R. Nomak, and J.K. Snyder, *Tetrahedron Lett.*, 1997, **38**, 8611–8614.
[84] D.H.R. Barton and S.Z. Zard, *J. Chem. Soc., Chem. Commun.*, 1985, 1098–1100.
[85] E.T. Pelkey and G.W. Gribble, *Chem. Commun.*, 1997, 1873–1874.

[86] E.T. Pelkey and G.W. Gribble, *Synthesis*, 1999, 1117–1122.
[87] G.W. Gribble, E.T. Pelkey, and F.L. Switzer, *Synlett*, 1998, 1061–1062.
[88] G.W. Gribble, E.T. Pelkey, W.M. Simon, and H.A. Trujillo, *Tetrahedron*, 2000, **56**, 10133–10140.
[89] T.L.S. Kishbaugh and G.W. Gribble, *Synth. Commun.*, 2002, **32**, 2003–2008.
[90] S. Roy, T.L.S. Kishbaugh, J.P. Jasinski, and G.W. Gribble, *Tetrahedron Lett.*, 2007, **48**, 1313–1316.
[91] L.A. Paquette, "Principles of Modern Heterocyclic Chemistry," W.A. Benjamin, New York, 1968, pp. 136–138.
[92] M.G. Saulnier and G.W. Gribble, *Tetrahedron Lett.*, 1983, **24**, 5435–5438.
[93] G.W. Gribble and M.G. Saulnier, *J. Chem. Soc., Chem. Commun.*, 1984, 168–169.
[94] G.W. Gribble, M.G. Saulnier, M.P. Sibi, and J.A. Obaza-Nutaitis, *J. Org. Chem.*, 1984, **49**, 4518–4523.
[95] D.A. Davis and G.W. Gribble, *Tetrahedron Lett.*, 1990, **31**, 1081–1084.
[96] G.W. Gribble, D.J. Keavy, D.A. Davis, *et al.*, *J. Org. Chem.*, 1992, **57**, 5878–5891.
[97] G.W. Gribble, J. Jiang, and Y. Liu, *J. Org. Chem.*, 2002, **67**, 1001–1003.
[98] J. Jiang and G.W. Gribble, *Org. Prep. Proc. Int.*, 2002, **34**, 533–535.
[99] G.W. Gribble, R.A. Silva, and M.G. Saulnier, *Synth. Commun.*, 1999, **29**, 729–747.
[100] J. Nagel, W. Friedrichsen, and T. Debaerdemaeker, *Z. Naturforsch.*, 1993, **48b**, 213–223.
[101] O. Peters and W. Friedrichsen, *Heterocycl. Commun.*, 1996, **2**, 203–213.
[102] A. Padwa, Y. Zou, B. Cheng, *et al.*, *J. Org. Chem.*, 2014, **79**, 3173–3184.
[103] Y. Miki and H. Hachiken, *Synlett*, 1993, 333–334.
[104] J.-S. Shiue and J.-M. Fang, *J. Chem. Soc., Chem. Commun.*, 1993, 1277–1278.
[105] S.-C. Lin, F.-D. Yang, J.-S. Shiue, *et al.*, *J. Org. Chem.*, 1998, **63**, 2909–2917.
[106] T. Kuroda, M. Takahashi, T. Ogiku, *et al.*, *J. Org. Chem.*, 1994, **59**, 7353–7357.
[107] J. Basset, M. Romero, T. Serra, and M.D. Pujol, *Tetrahedron*, 2012, **68**, 356–362.
[108] A.K. Jana, P. Pahari, and D. Mal, *Synlett*, 2012, **23**, 1769–1774.
[109] M.T. Díaz, A. Cobas, E. Guitián, and L. Castedo, *Synlett*, 1998, 157–158.
[110] C.O. Kappe and A. Padwa, *J. Org. Chem.*, 1996, **61**, 6166–6174.
[111] A. Shafiee and S. Sattari, *J. Heterocyclic Chem.*, 1982, **19**, 227–231.
[112] E. Differding and L. Ghosez, *Tetrahedron Lett.*, 1985, **26**, 1647–1650.
[113] P. Molina, M. Alajarín, and A. Vidal, *J. Chem. Soc., Chem. Commun.*, 1990, 1277–1279.
[114] P. Molina, A. Arques, A. Alías, and M.V. Vinader, *Tetrahedron Lett.*, 1991, **32**, 4401–4404.
[115] P. Molina, M. Alajarín, A. Vidal, and P. Sánchez-Andrada, *J. Org. Chem.*, 1992, **57**, 929–939.
[116] P. Molina and C. López-Leonardo, *Tetrahedron Lett.*, 1993, **34**, 2809–2812.
[117] T. Saito, H. Ohmori, E. Furuno, and S. Motoki, *J. Chem. Soc., Chem. Commun.*, 1992, 22–24.
[118] K. Hayakawa, T. Yasukouchi, and K. Kanematsu, *Tetrahedron Lett.*, 1986, **27**, 1837–1840.
[119] K. Hayakawa, T. Yasukouchi, and K. Kanematsu, *Tetrahedron Lett.*, 1987, **28**, 5895–5898.
[120] T. Yasukouchi and K. Kanematsu, *Tetrahedron Lett.*, 1989, **30**, 6559–6562.
[121] M. Lee, I. Ikeda, T. Kawabe, *et al.*, *J. Org. Chem.*, 1996, **61**, 3406–3416.
[122] B. Witulski, J. Lumtscher, and U. Bergsträßer, *Synlett*, 2003, 708–710.
[123] M.F. Martínez-Esperón, D. Rodríguez, L. Castedo, and C. Saá, *Org. Lett.*, 2005, **7**, 2213–2216.
[124] M.F. Martínez-Esperón, D. Rodríguez, L. Castedo, and C. Saá, *Tetrahedron*, 2006, **62**, 3843–3855.
[125] M.F. Martínez-Esperón, D. Rodríguez, L. Castedo, and C. Saá, *Tetrahedron*, 2008, **64**, 3674–3686.
[126] J.R. Dunetz and R.L. Danheiser, *J. Am. Chem. Soc.*, 2005, **127**, 5776–5777.
[127] H.H. Wasserman and C.A. Blum, *Tetrahedron Lett.*, 1994, **35**, 9787–9790.
[128] C.E. Soll and R.W. Franck, *Heterocycles*, 2006, **70**, 531–540.
[129] A. Padwa, M.A. Brodney, B. Liu, *et al.*, *J. Org. Chem.*, 1999, **64**, 3595–3607.
[130] For a review of his work in this area, see A. Padwa, *Chem. Commun.*, 1998, 1417–1424.
[131] S.K. Bur, S.M. Lynch, and A. Padwa, *Org. Lett.*, 2002, **4**, 473–476.
[132] P. Rashatasakhon and A. Padwa, *Org. Lett.*, 2003, **5**, 189–191.
[133] A. Padwa, P. Rashatasakhon, and M. Rose, *J. Org. Chem.*, 2003, **68**, 5139–5146.
[134] A. Padwa, W.S. Kissell, and C.K. Eidell, *Can. J. Chem.*, 2001, **79**, 1681–1693.
[135] S.K. Bur and A. Padwa, *Org. Lett.*, 2002, **4**, 4135–4137.
[136] A. Padwa, S.K. Bur, and H. Zhang, *J. Org. Chem.*, 2005, **70**, 6833–-6841.
[137] F. Petronijevic, C. Timmons, A. Cuzzupe, and P. Wipf, *Chem. Commun.*, 2009, 104–106.
[138] F.R. Petronijevic and P. Wipf, *J. Am. Chem. Soc.*, 2011, **133**, 7704–7707.
[139] A. Tahri, K.J. Buysens, E.V. Van der Eycken, *et al.*, *Tetrahedron*, 1998, **54**, 13211–13226.
[140] D.L. Boger, *Chem. Rev.*, 1986, **86**, 781–793.
[141] D.L. Boger and S.M. Sakya, *J. Org. Chem.*, 1988, **53**, 1415–1423.
[142] D.L. Boger and M. Zhang, *J. Am. Chem. Soc.*, 1991, **113**, 4230–4234.
[143] D.L. Boger, *Bull. Soc. Chim. Belg.*, 1990, **99**, 599–615.
[144] D.L. Boger, *Tetrahedron*, 1983, **39**, 2869–2939.

55

Plieninger Indole Synthesis

In 1956, Plieninger and Suhr revealed a novel synthesis of indoles that entailed ozone cleavage of 1-acetamido-5,8-dihydronaphthalene followed by rapid cyclization and indolization (Scheme 1, equations 1, 2, 3) [1]. This initial report was followed by several companion papers and improvements by Plieninger and coworkers on their methodology [2–5]. The starting 1-acetamidodihydronaphthalene was prepared by a Birch reduction of 1-aminonaphthalene followed by acetylation. Plieninger adapted this reaction to the synthesis of aldehyde derivatives such as 4-cyanomethylindole [1, 2] and Uhle's ketone [4]. The sequence shown in equation 3 was carried on to 4-(3-methyl-2-butenyl)tryptophan, a putative ergot alkaloid biosynthetic intermediate [5].

Kraus and Yue discovered a similar indole ring synthesis that featured a Diels–Alder synthesis of the requisite dihydronaphthalene **1**. Ozonolysis gave the twin cyclized product **2** in excellent yield (Scheme 2, equation 1) [6]. Using a similar strategy, Kraus and coworkers synthesized the left half of CC-1065 (**3**), a potent antitumor metabolite from *Streptomyces zelensis* (equation 2) [7]. Maehr and Smallheer used the original Plieninger indolization to synthesize the alga metabolites (±)-rivularins D_1 and D_3 via 5-bromo-7-methoxy-1-[(4-methylphenyl)sulfonyl]-1*H*-indole-4-acetaldehyde, which was prepared via a Birch reduction of 1-amino-2-methoxynaphthalene followed by ozonolysis [8].

Even though a Birch reduction to give the requisite dihydronaphthalene for the Plieninger indolization is obviously successful, it is not an attractive or general method for providing these dihydronaphthalenes. In 1995, Kerr combined a high-pressure Diels–Alder cycloaddition of quinone mono ketals with 1,3-butadienes to give cycloadducts that readily aromatize to 1,4-dihydronaphthalenes (Scheme 3, equation 1) [9–12]. A Plieninger indole ring synthesis using osmium tetroxide rather than ozone completes the process (equation 2) [13]. Kerr has summarized his elegant applications of his chemistry, which has clearly put the Plieninger indolization on the map [12]. Some of the indoles that Kerr has synthesized are shown in Scheme 3 (**4–6**).

Kerr and his coworkers parlayed the high-pressure Diels–Alder–Plieninger protocol into several syntheses of indole natural products: (±)-herbindole B [14], [15], (±)-*cis*-trikentrin B [14], (±)-herbindole A [15], (±)-*cis*-trikentrin A [15], eustifolines A–D [16], glycomaurrol [16], (±)-CC-1065 subunit [17], (±)-decursivine [18], clausamines A–C [19], and clausevatine D [19]. In addition, they described a route to the ergot alkaloid skeleton [20], mimicking Plieninger's original syntheses of the benzo[*cd*]indole ring system [3, 4] and the western half of the lolicines and lolitrems [21]. Scheme 4 illustrates Kerr's syntheses of 5-alkylindoles (equation 1) [22] and 5-triflyloxyindoles (equation 2) [23]. For both reactions, many examples were reported. The indolyl triflates were employed in several palladium-catalyzed cross-coupling reactions.

Boger and colleagues employed a Plieninger–Kraus–Kerr indolization sequence to synthesize (+)-yatakemycin and *ent*-(–)-yatakemycin (Scheme 5) [24, 25].

The marriage of Plieninger's 1956 indole synthesis from 1-amino-5,8-dihydronaphthalenes to Kerr's 1995 high-pressure Diels–Alder synthesis of these primed indole precursors has been highly successful, as illustrated by the applications presented in this chapter. The method is particularly attractive for the preparation of C-4 functionalized indoles.

Indole Ring Synthesis: From Natural Products to Drug Discovery, First Edition. Gordon W. Gribble.

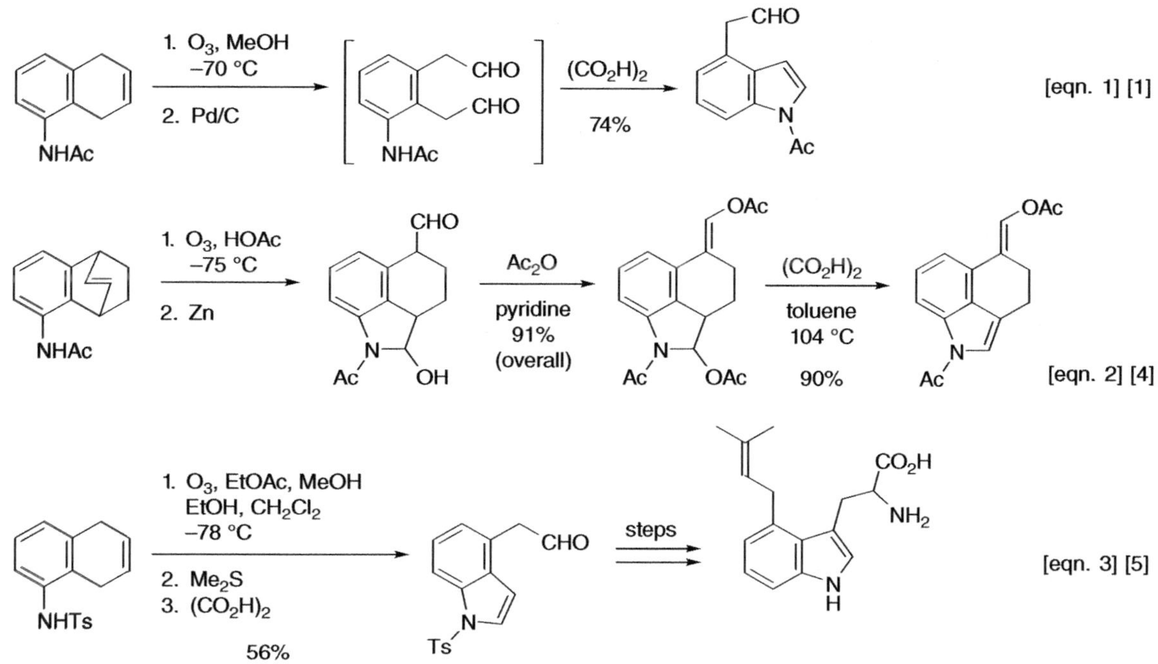

Scheme 1 *Plieninger Indole Synthesis*

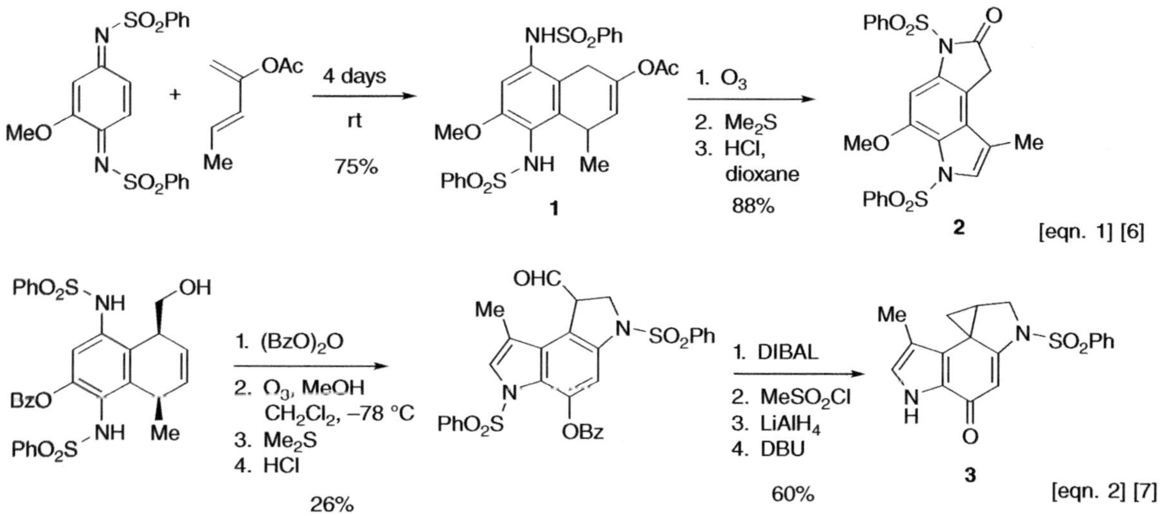

Scheme 2 *Kraus Indole Synthesis*

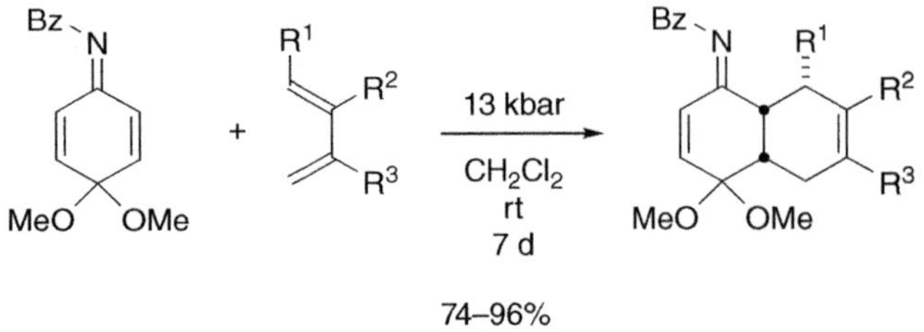

[eqn. 1] [9], [11]

R^1 = H, Me, OTBS, CH_2OTBS, OMe
R^2 = H, Me
R^3 = H, Me, OTBS

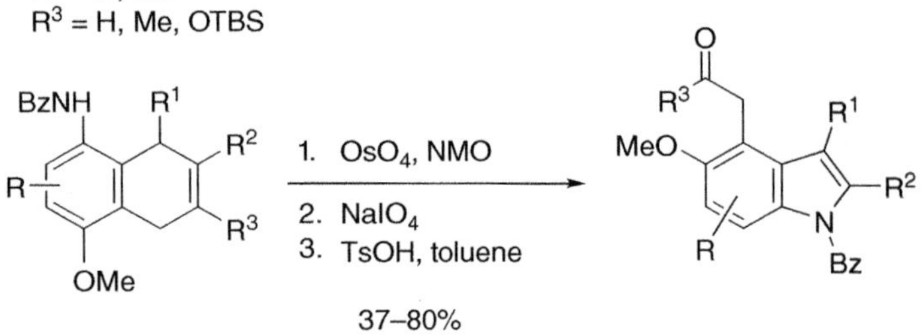

[eqn. 2] [13]

R^1= H, Me, CH_2OPiv
R^2 = H, Me
R^3 = H, Me
R = H, Cl, MeO

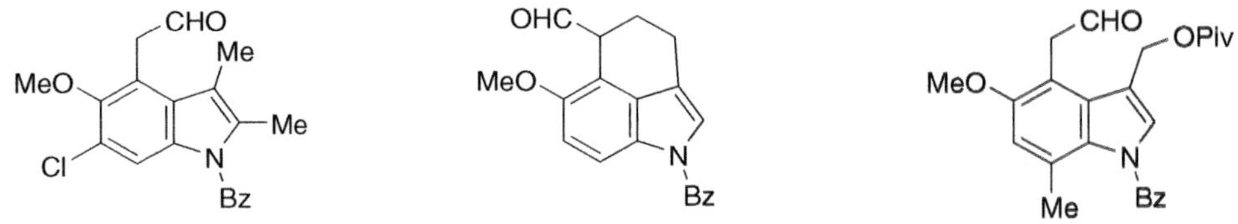

4 (70%) **5** (71%) **6** (46%)

Scheme 3 *Kerr Indole Synthesis*

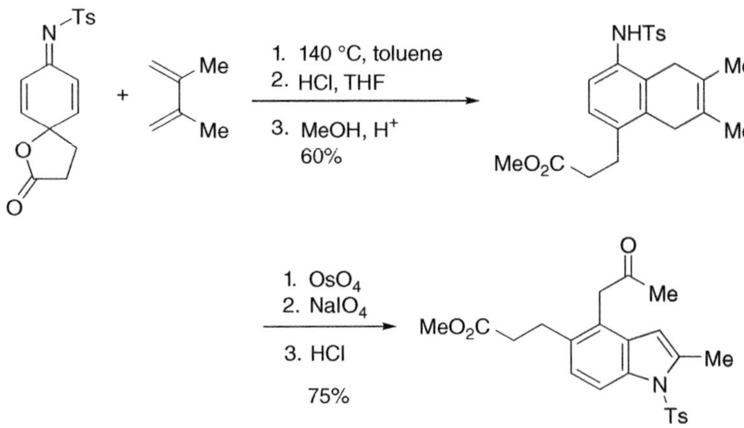

[eqn. 1] [22]

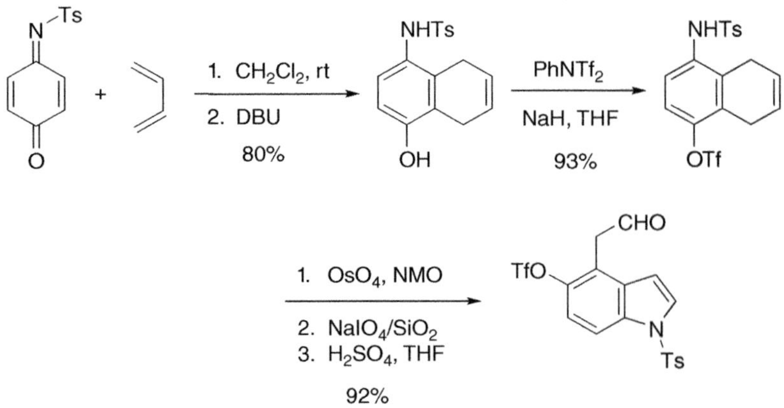

[eqn. 2] [23]

Scheme 4 *Kerr Indole Synthesis – 2*

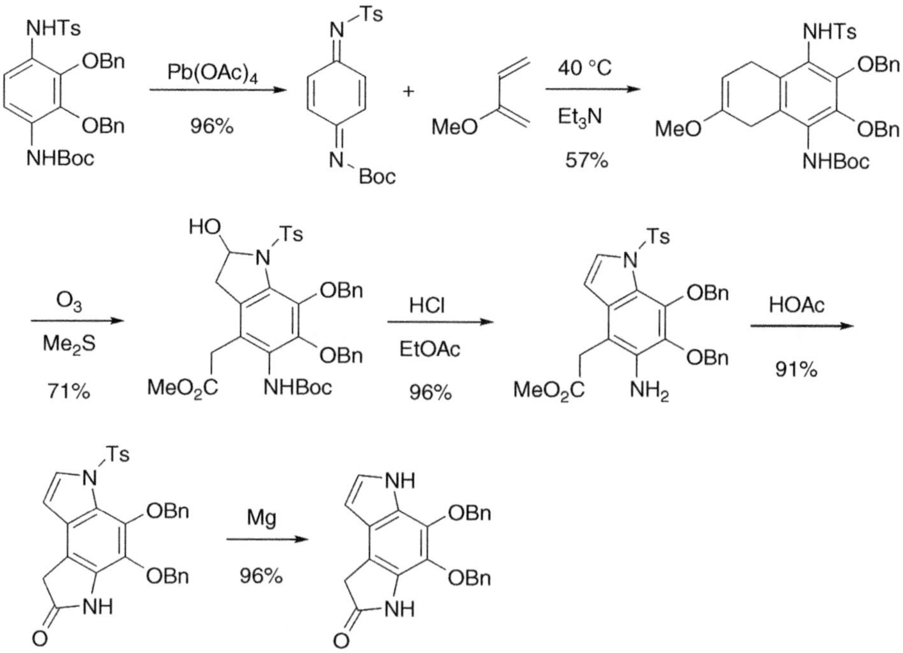

Scheme 5 *Boger's Synthetic Approach to Yatakemycin*

References

[1] H. Plieninger and K. Suhr, *Chem. Ber.*, 1956, **89**, 270–278.
[2] H. Plieninger and K. Suhr, *Chem. Ber.*, 1957, **90**, 1980–1984.
[3] H. Plieninger and W. Lehnert, *Chem. Ber.*, 1967, **100**, 2427–2434.
[4] H. Plieninger and A. Völkl, *Chem. Ber.*, 1976, **109**, 2121–2125.
[5] H. Plieninger, E. Meyer, F. Sharif-Nassirian, and E. Weidmann, *Liebigs Ann. Chem.*, 1976, 1475–1486.
[6] G.A. Kraus and S. Yue, *J. Chem. Soc., Chem. Commun.*, 1983, 1198–1199.
[7] G.A. Kraus, S. Yue, and J. Sy, *J. Org. Chem.*, 1985, **50**, 283–284.
[8] H. Maehr and J. Smallheer, *J. Am. Chem. Soc.*, 1985, **107**, 2943–2945.
[9] M.A. Kerr, *Synlett*, 1995, 1165–1167.
[10] E.R. Jarvo, S.R. Boothroyd, and M.A. Kerr, *Synlett*, 1996, 897–899.
[11] S.C. Banfield and M.A. Kerr, *Can. J. Chem.*, 2004, **82**, 131–138.
[12] K. Sapeta, T.P. Lebold, and M.A. Kerr, *Synlett*, 2011, 1495–1514.
[13] S.C. Banfield, D.B. England, and M.A. Kerr, *Org. Lett.*, 2001, **3**, 3325–3327.
[14] S.K. Jackson, S.C. Banfield, and M.A. Kerr, *Org. Lett.*, 2005, **7**, 1215–1218.
[15] S.K. Jackson and M.A. Kerr, *J. Org. Chem.*, 2007, **72**, 1405–1411.
[16] T.P. Lebold and M.A. Kerr, *Org. Lett.*, 2007, **9**, 1883–1886.
[17] M.D. Ganton and M.A. Kerr, *J. Org. Chem.*, 2007, **72**, 574–582.
[18] A.B. Leduc and M.A. Kerr, *Eur. J. Org. Chem.*, 2007, 237–240.
[19] T.P. Lebold and M.A. Kerr, *Org. Lett.*, 2008, **10**, 997–1000.
[20] S.C. Banfield and M.A. Kerr, *Synlett*, 2001, 436–438.
[21] D.B. England, J. Magolan, and M.A. Kerr, *Org. Lett.*, 2006, **8**, 2209–2212.
[22] P.V. Zawada, S.C. Banfield, and M.A. Kerr, *Synlett*, 2003, 971–974.
[23] D.B. England and M.A. Kerr, *J. Org. Chem.*, 2005, **70**, 6519–6522.
[24] M.S. Tichenor, D.B. Kastrinsky, and D.L. Boger, *J. Am. Chem. Soc.*, 2004, **126**, 8396–8398.
[25] M.S. Tichenor, J.D. Trzupek, D.B. Kastrinsky, *et al.*, *J. Am. Chem. Soc.*, 2006, **128**, 15683–15696.

56

Photochemical Synthesis of Indoles and Carbazoles

An early application of photochemistry to the synthesis of indoles and carbazoles was the photocyclization of diphenylamines to carbazoles, but many of the initial studies were solely mechanistic in nature. Several limited reviews are available [1–4]. The first report of the irradiation of diphenylamine to form carbazole (250–300 nm) was that of Parker and Barnes in 1957 [5]. This was followed in 1963 by the work of Bowen and Eland [6] and Grellmann and colleagues [1]. The former team observed a quantum yield of 0.1 in dilute solvent (MeOH, *i*-PrOH, or hexane at room temperature) for the photolysis of diphenylamine to form carbazole, and they also found that oxygen is detrimental to the reaction. Grellmann and colleagues irradiated *N*-methyldiphenylamine and triphenylamine to yield *N*-methyl- and *N*-phenylcarbazole (70% and 65% yield, respectively) and provided spectroscopic evidence for the pathway shown in Scheme 1 (equation 1) [7, 8]. Subsequently, several groups intensely studied the mechanism of this photocyclization: Kemp [8], Terry [9], Tanaka [10, 11], Wentrup [12], Grellmann [13–19], Troe [20], Fox [21], Johnston and Redmond [22], Moriwaki [23], Obi [24], Chattopadhyay [25, 26], Görner [27, 28], and Amano [29]. This wealth of photochemistry implicated a singlet to triplet state (**1**) interconversion, followed by cyclization to the triplet state (**2**) of 4a,4b-dihydrocarbazole, which decays to the ground state and then to carbazole by loss of dihydrogen. Oxygen quenches these two triplet states, but it also can react with 4a,4b-dihydrocarbazole to afford carbazole and hydrogen peroxide [27, 28]. In any event, the photocyclodehydrogenation of diphenylamines is a useful carbazole (and carboline) synthesis (equations 2–4) [2, 30–32]. As seen in Scheme 1, the carbazole yields vary widely, perhaps as a function of oxygen, the irradiation wavelength, and/or the substrate. For example, whereas 2-anilino-*p*-xylene afforded the carbazole in excellent yield (equation 5), photolysis of 6-anilino-5,8-dimethylisoquinoline under the same conditions (to give ellipticine) failed [33].

Some additional examples are depicted in Scheme 2, such as a convenient synthesis of 11*H*-benzo[*a*]carbazole (equation 1) [34], synthesis of *N*-substituted carbazoles (equations 2 and 3) [18, 35], and the use of *N*,*N*-diarylsulfonamides (equation 4) to construct ellipticine precursors [36, 37]. Interestingly, photolysis of 2-cyanodiphenylamine (254 nm, EtOH, MeCN) afforded carbazole in 87% yield, but a similar photochemical reaction of 2-aminophenylacetylene gave only a trace of indole (1%–2%) [38].

Clark and coworkers were the first to synthesize carbolines from the photolysis of aniline-pyridines (Scheme 3, equations 1, 2) [39]. The photo reaction of 3-anilinopyridine gave a mixture of β- (24%) and γ-carboline (46%), and diphenylamine afforded carbazole in 62% yield [39]. Kobayashi and colleagues found a similar distribution of β- and γ-carbolines (31% vs. 45%) for the photolysis of *N*-(4-methoxy-3,5-dimethylphenyl)pyridine-3-amine in a synthesis of eudistomin D analogues [40]. Mohan and coworkers employed photochemistry to synthesize the alkaloids cryptotackienine, cryptosanguinolentine, and cryptolepine as shown for two precursors (equations 3 and 4) [41, 42]. The latter reactions illustrate the use of a leaving group (chloride) to facilitate the aromatization. Similarly, Groundwater's group found that fluoride was lost in the photocyclization of fluorine-containing dimethyldiphenylamines to 1,4-dimethylcarbazoles [43]. Chakrabarty and colleagues prepared indolocarbazoles by irradiating several *N*,*N*′-diphenyl-phenylenediamines [44], and Collins and his group used a continuous flow UV

Indole Ring Synthesis: From Natural Products to Drug Discovery, First Edition. Gordon W. Gribble.

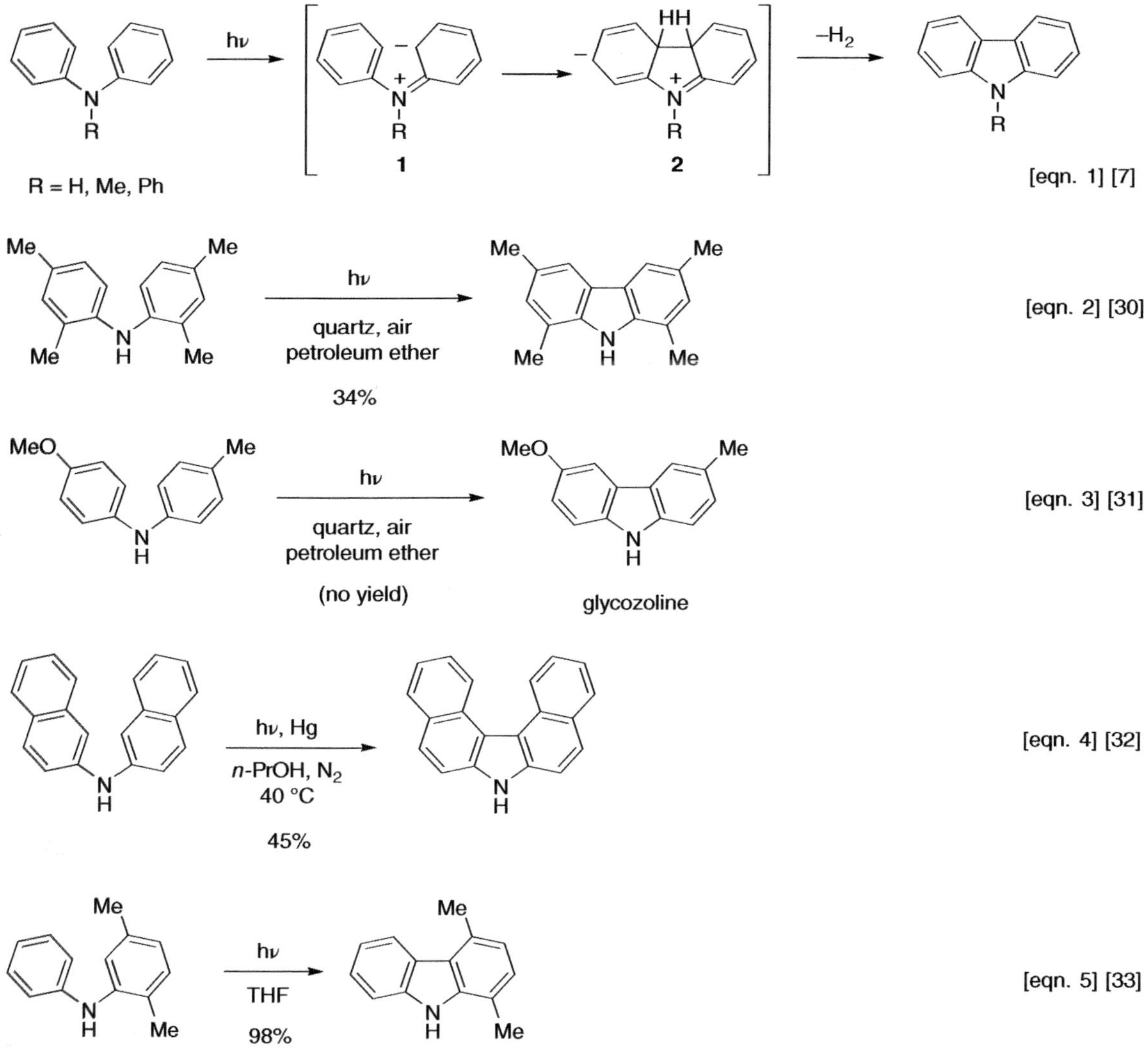

Scheme 1 *Synthesis of Carbazoles by Photocyclodehydrogenation of Diphenylamines*

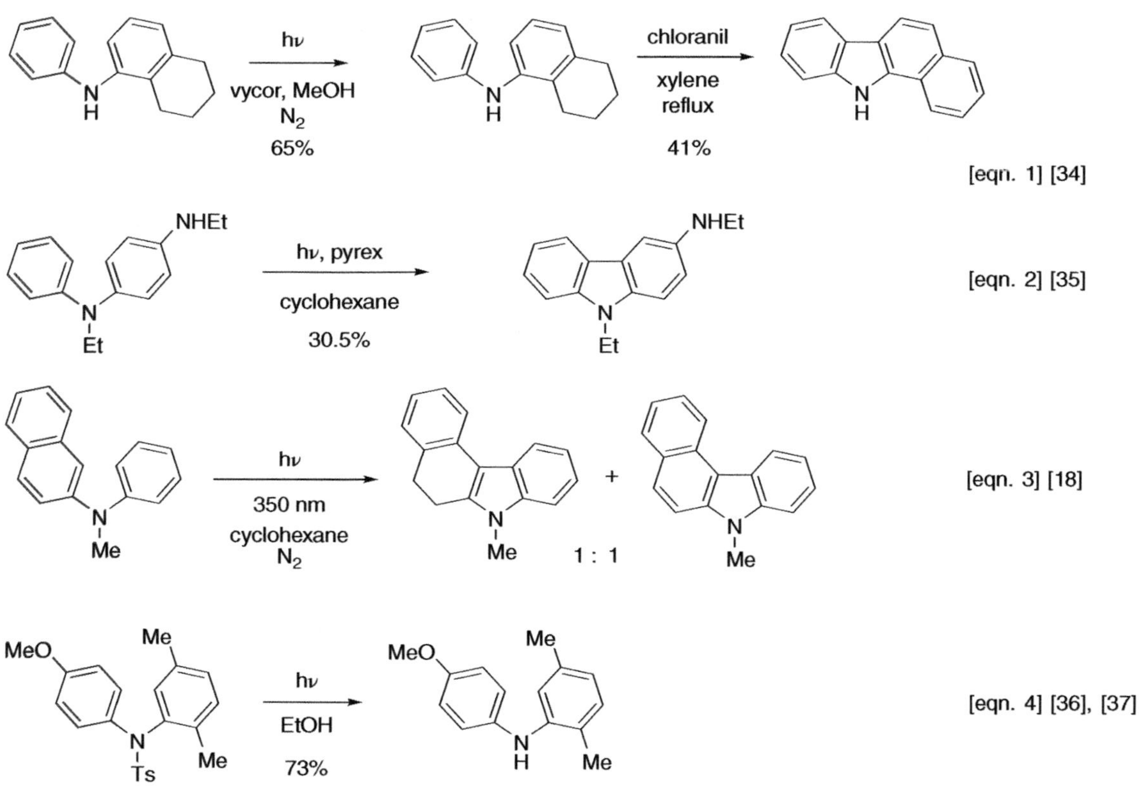

Scheme 2 *Synthesis of Carbazoles by Photocyclodehydrogenation of Diphenylamines – 2*

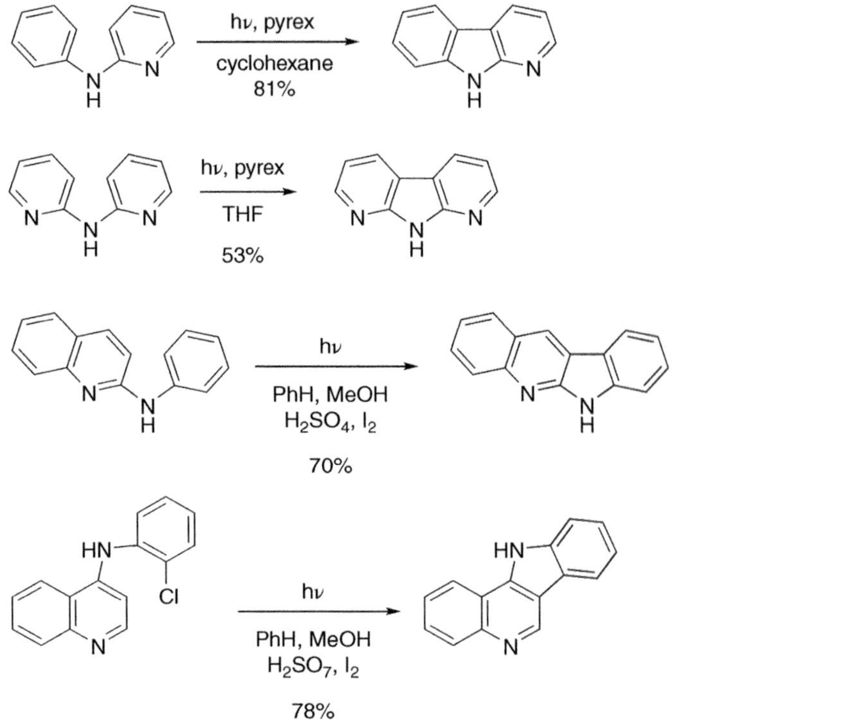

[eqn. 1] [39]

[eqn. 2] [39]

[eqn. 3] [41]

[eqn. 4] [41]

Scheme 3 *Synthesis of Carbolines by Photocyclodehydrogenation of Anilinopyridines and Anilinoquinolines*

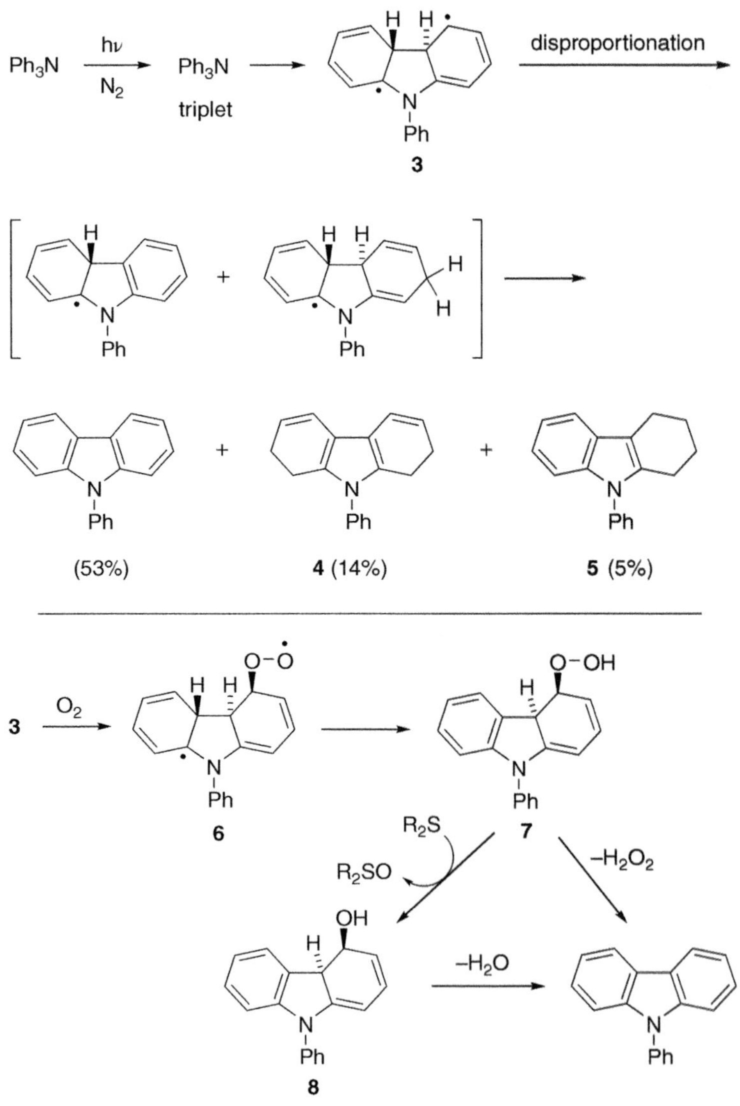

Scheme 4 *Albini's Proposed Mechanism for the Photocyclization of Triphenylamine [46]*

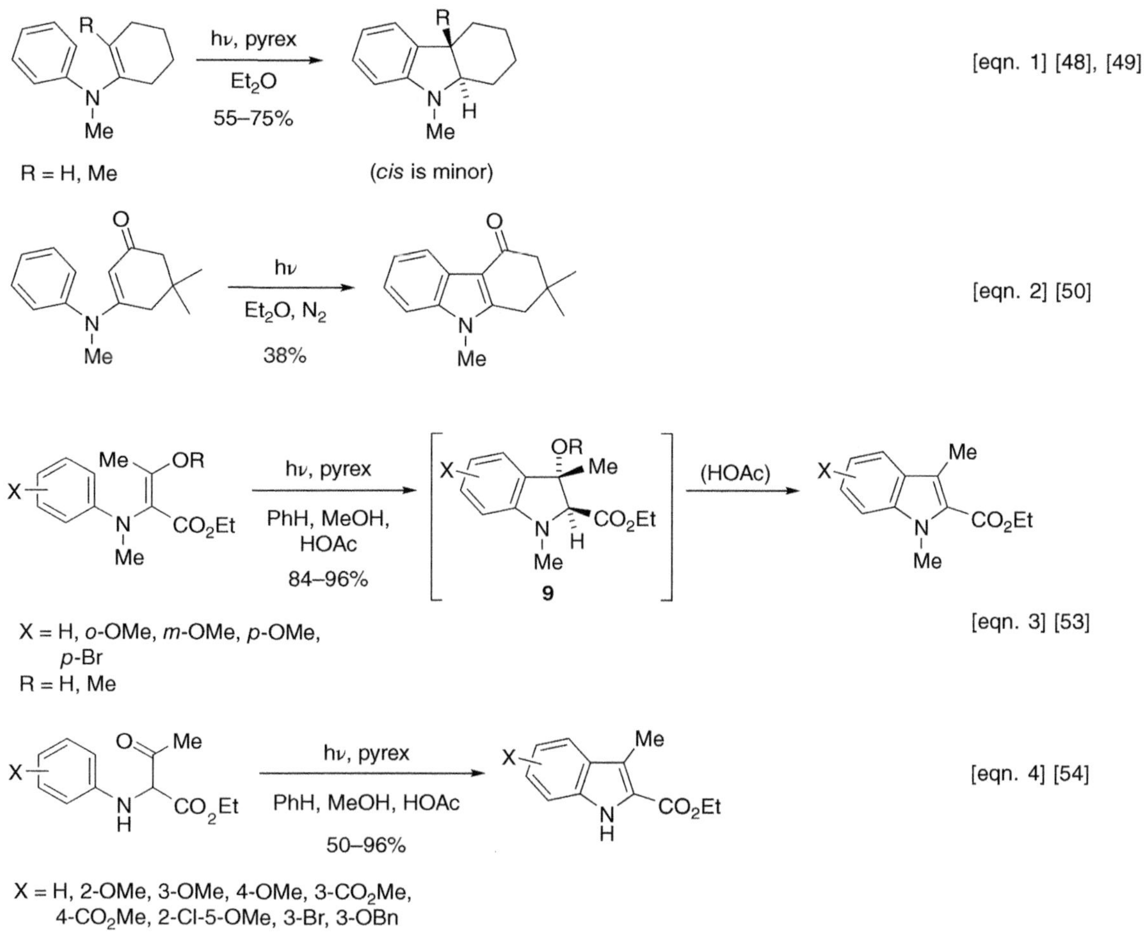

Scheme 5 *Chapman, Yamada, and Schultz Indoline-Indole Syntheses*

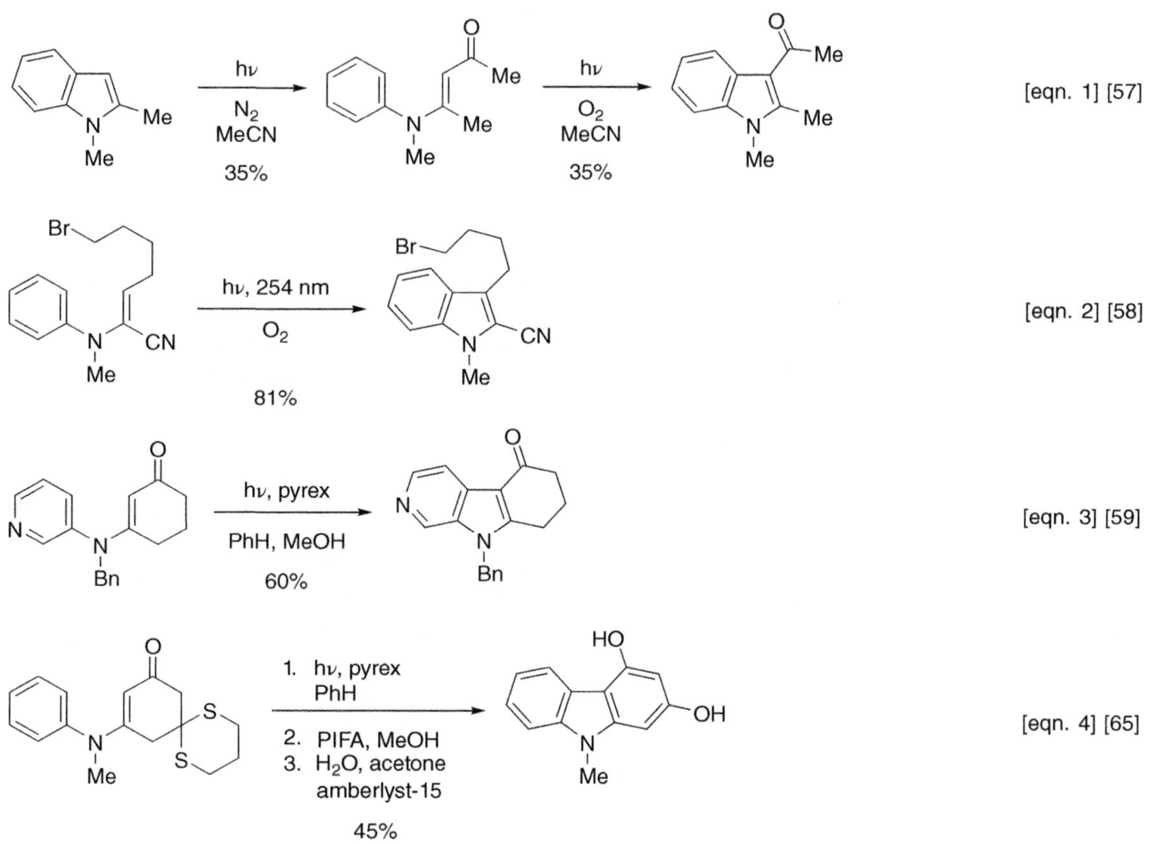

Scheme 6 *Photocyclization of Enamino Ketones and Nitriles to Indoles*

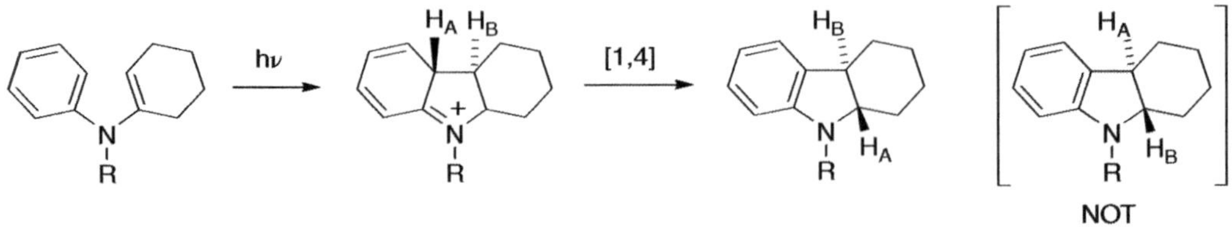

Scheme 7 *Grellmann Mechanism for Photocyclization of* N*-Aryl Enamines to Indolines [66]*

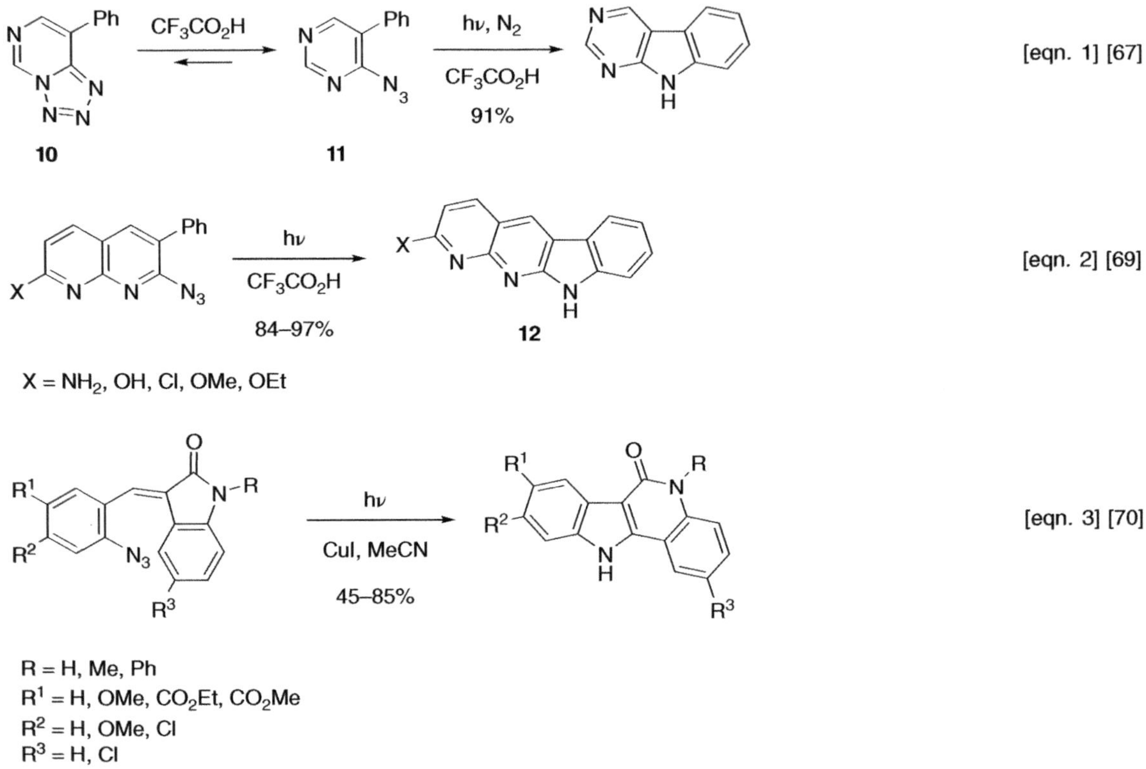

Scheme 8 *Photolysis of Aryl Azides to Form Indoles*

reactor to prepare derivatives of the antiinflammatory carbazole drug carprofen [45]. Under these conditions, triphenylamine was converted to *N*-phenylcarbazole in 93% yield.

Albini and colleagues proposed a mechanism for the photocyclization of triphenylamine (Scheme 4) under both oxygen-free and oxygen-present conditions [46]. Under the oxygen-free conditions, disproportionation products **4** and **5** were isolated. Not shown is reversion of triplet dihydrocarbazole **3** back to triphenylamine via dihydrocarbazole singlet. In the presence of oxygen, Albini's pathway proceeds to triplet diradical **6**, intramolecular hydrogen atom abstraction to give hydroperoxide **7**, oxygen capture by a sulfide to give alcohol **8**, and dehydration to afford *N*-phenylcarbazole. Direct loss of hydrogen peroxide can also give *N*-phenylcarbazole.

Equally important to the irradiation of diphenylamines to form carbazoles is the photolysis of *N*-aryl enamines and enamino ketones to indolines and/or indoles [1–4, 47], as first reported by Chapman in 1968 (Scheme 5, equations 1, 2) [48, 49]. Although indolines (2,3-dihydroindoles) were the primary products, indoles are obtained if the photolysis is performed in the presence of air. The mechanism is proposed to involve a Woodward–Hoffmann-allowed photochemical conrotatory electrocyclization, followed by electron demotion to a zwitterionic species (similar to **2**, Scheme 1) and a suprafacial [1,4] sigmatropic hydrogen shift to give the *trans* product (equation 1). However, deuterium labeling experiments by Chapman also implicated a sequence of two [1,2] sigmatropic hydrogen shifts [49]. Notwithstanding the beauty of this chemistry, the remainder of this section includes only examples of *indole*

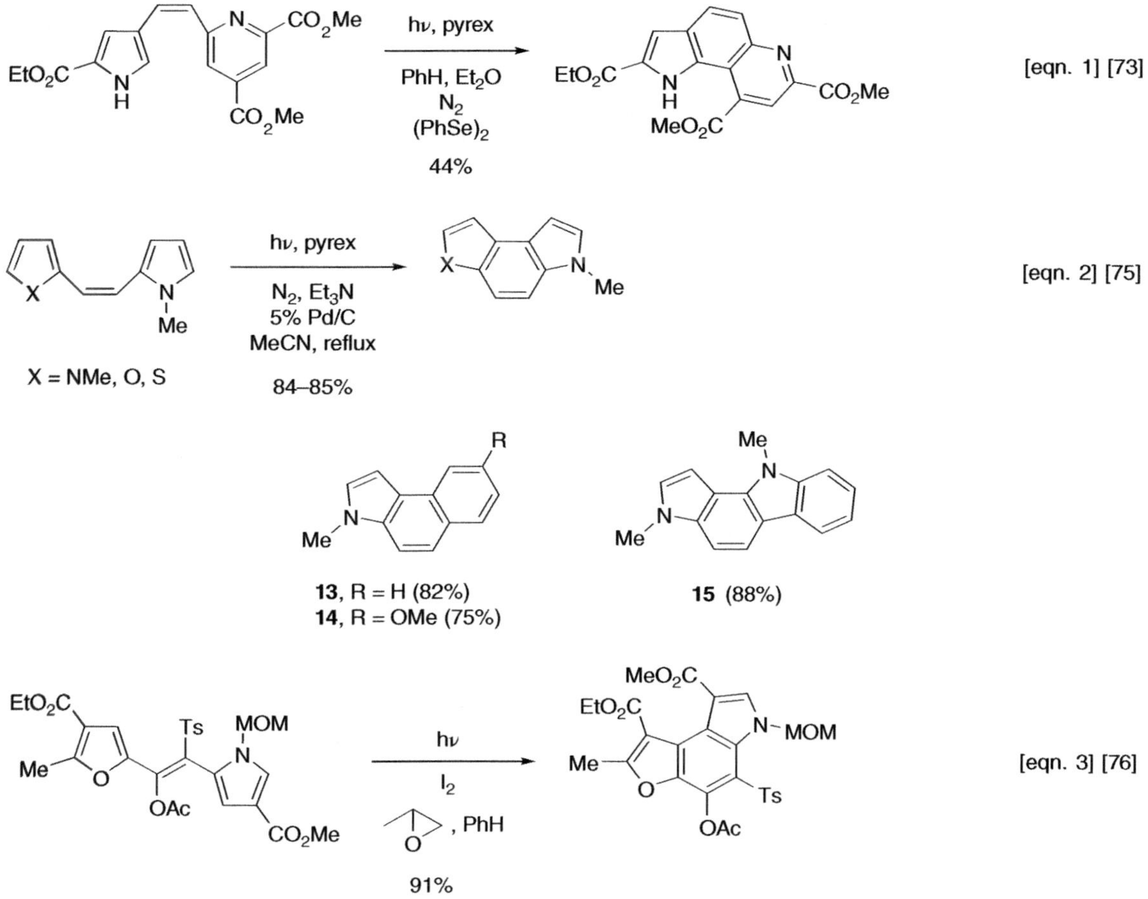

Scheme 9 *Photocyclization of Pyrrole Analogues of Stilbene*

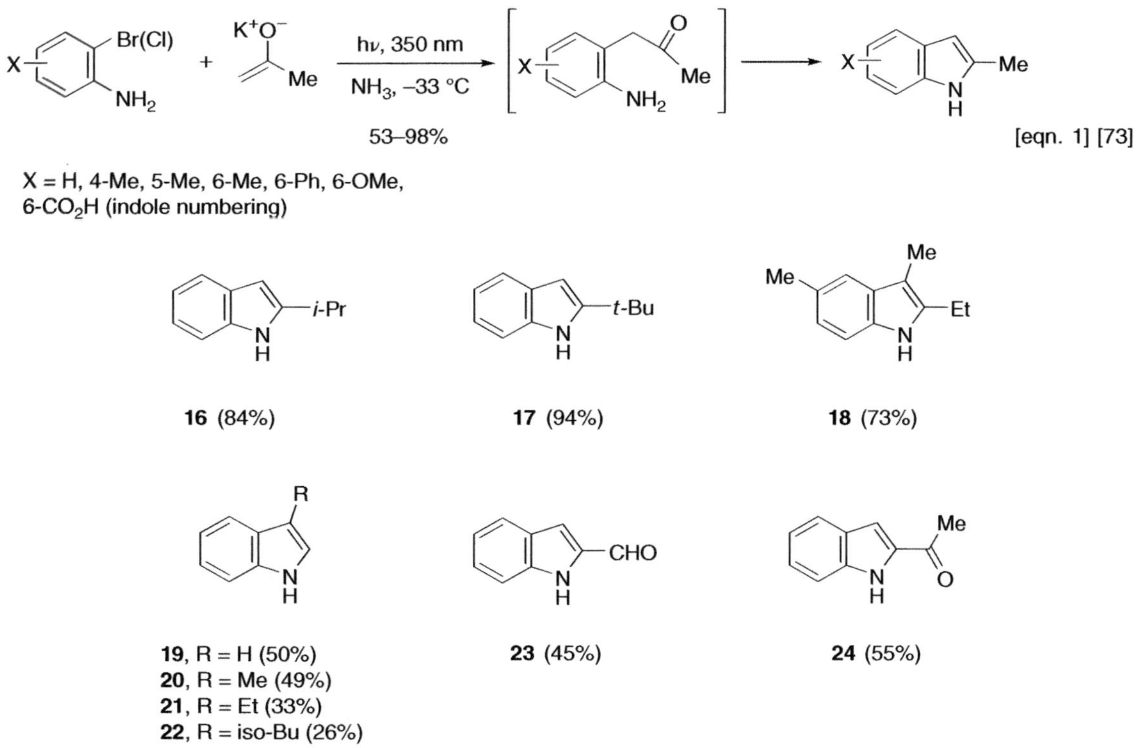

Scheme 10 *The Bunnett-Beugelmans Indole Synthesis*

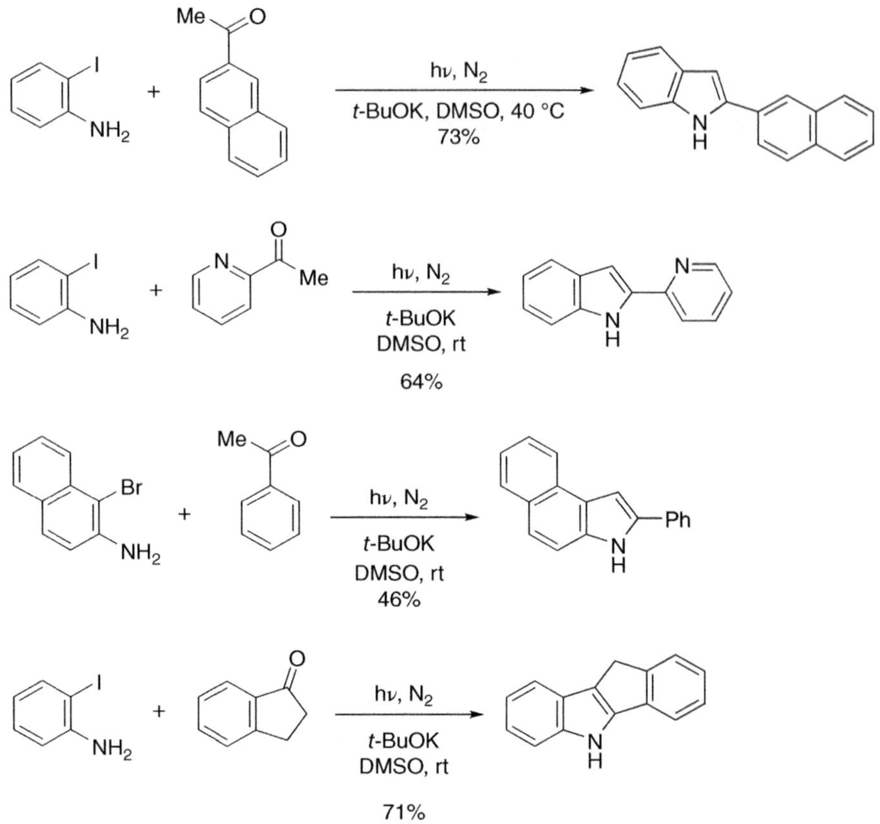

[eqn. 1] [84]

[eqn. 2] [85]

[eqn. 3] [85]

[eqn. 4] [85]

Scheme 11 *Rossi Synthesis of Indoles*

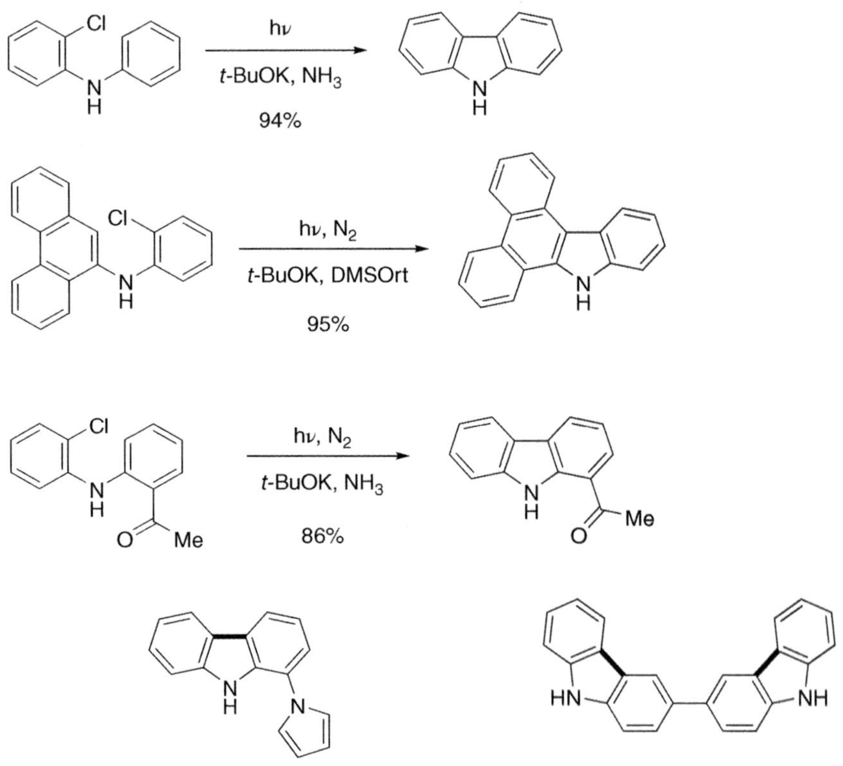

[eqn. 1] [87]

[eqn. 2] [87]

[eqn. 3] [88]

25 (91%) [87]

26 (67%) [87]

Scheme 12 *Rossi Synthesis of Carbazoles*

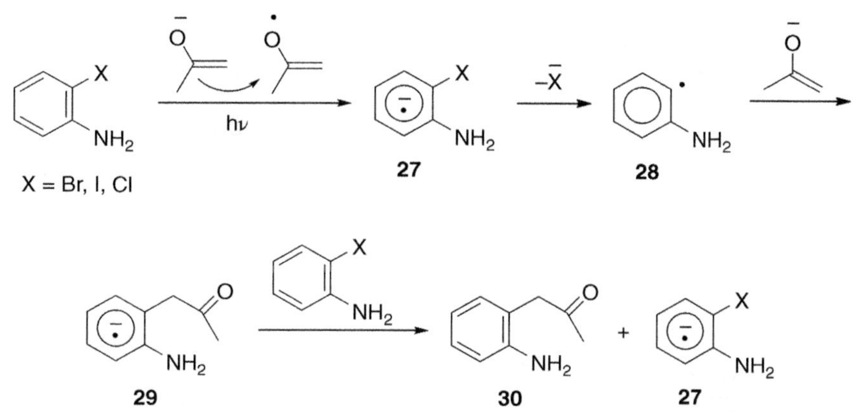

Scheme 13 *General Mechanism for the Bunnett-Beugelmans Indole Synthesis*

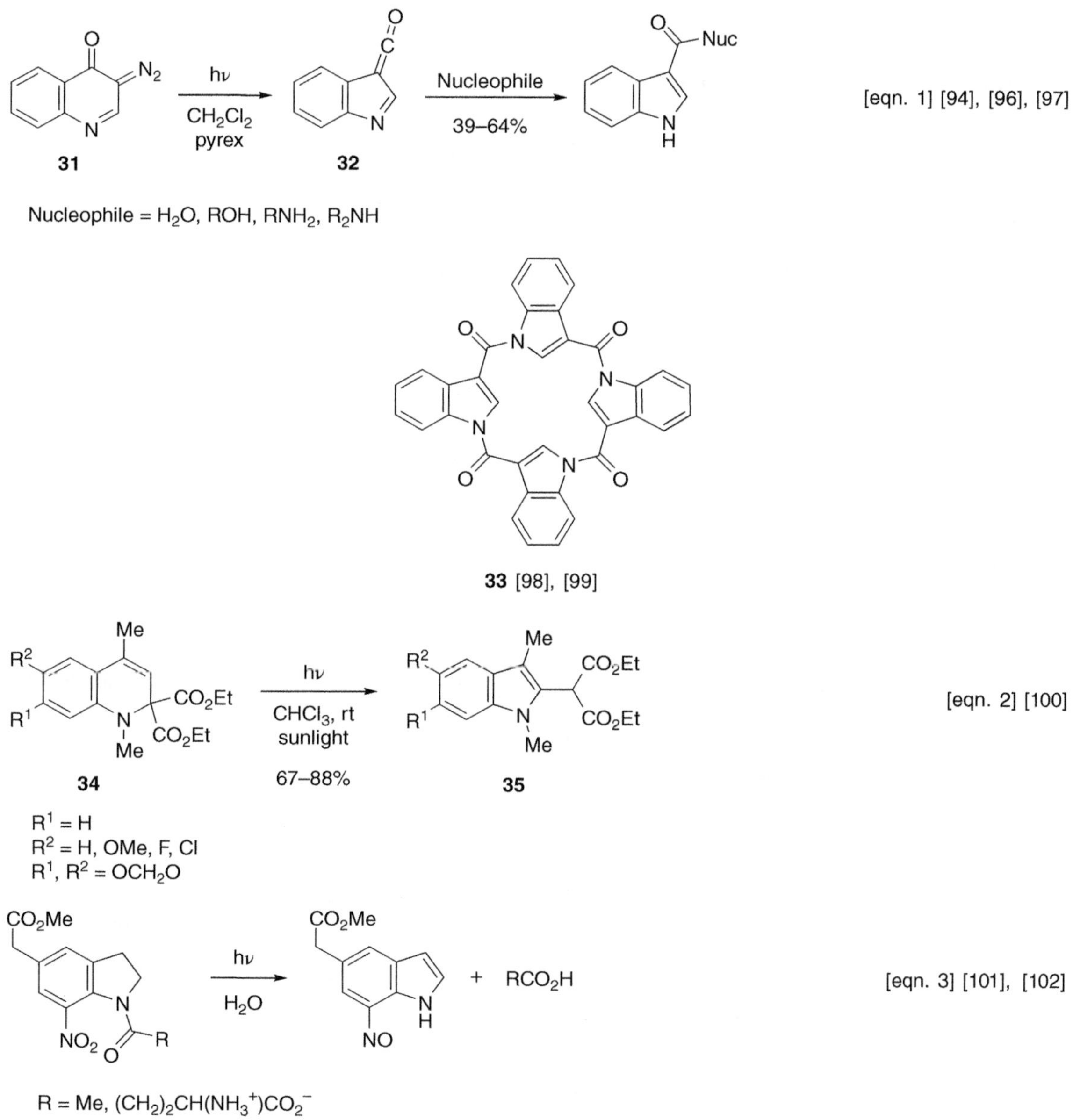

Scheme 14 *Photochemical Indole Syntheses – 1*

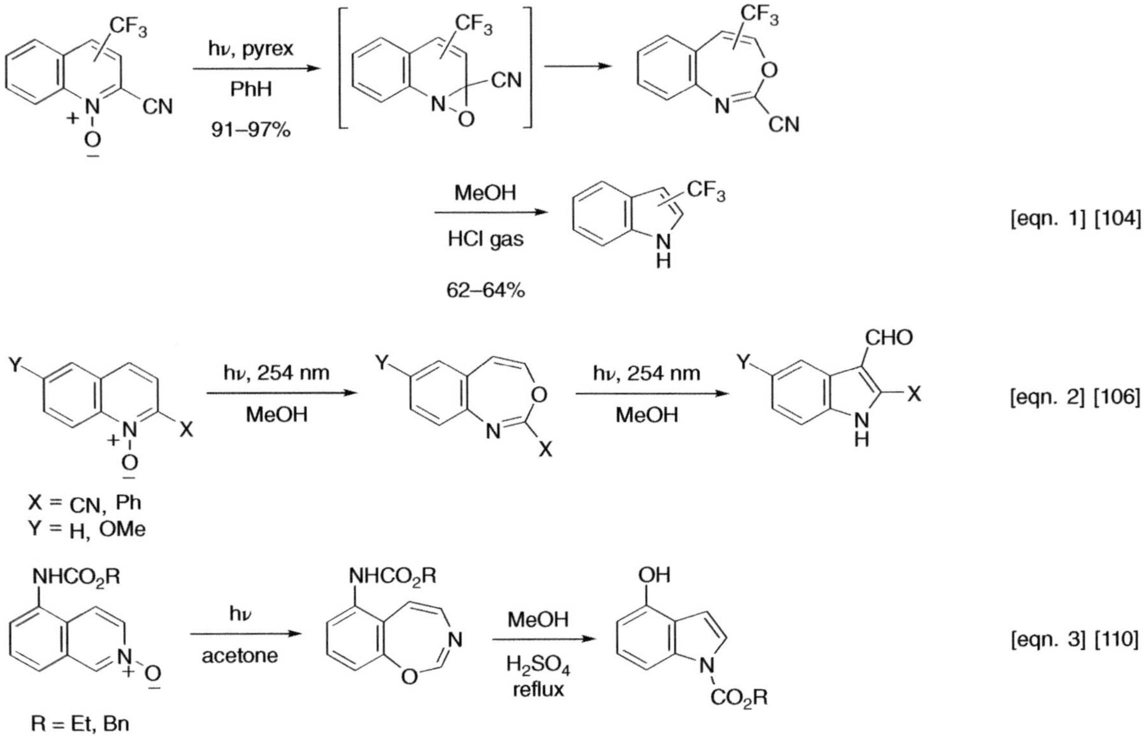

Scheme 15 *Photolysis of Quinoline and sIsoquinoline* N*-oxides*

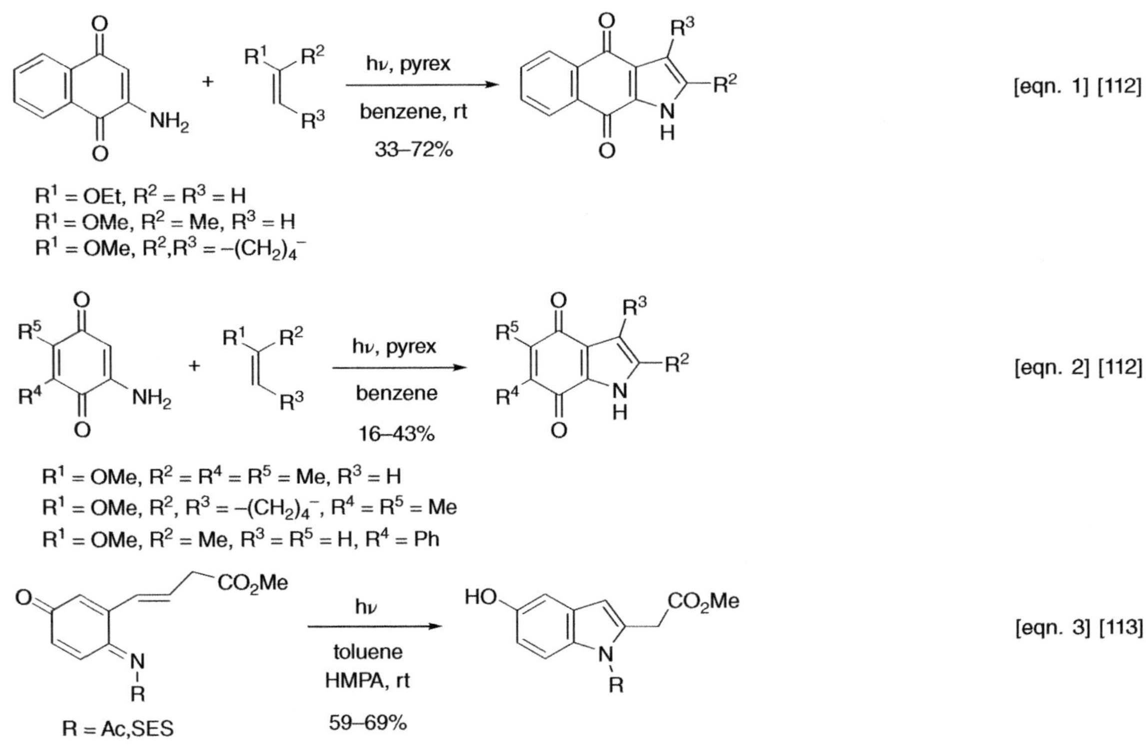

Scheme 16 *Photochemical Synthesis of Indoles from Quinones and Quinone Imides*

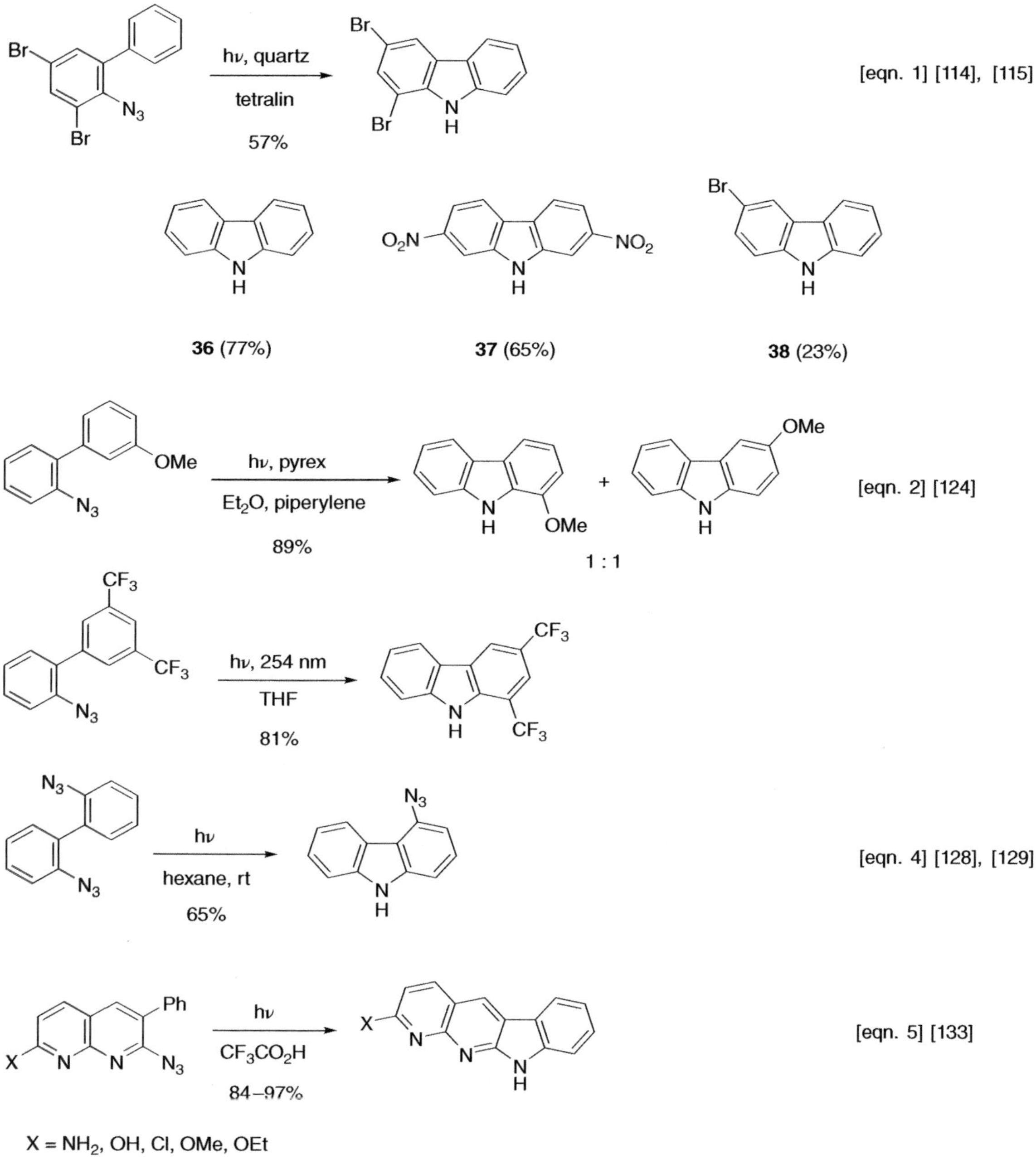

Scheme 17 *The Smith Synthesis of Carbazoles from 2-Azidobiphenyls*

ring synthesis. In contrast to Chapman's result (equation 1), Yamada and colleagues found that irradiation of enamino ketones afforded the ketoindole product directly, even under N_2 (equation 2) [50–52]. The carbonyl group may be promoting indolization via the enol tautomer. This seminal result has been observed by many groups in the succeeding years. Notably, Schultz adapted this photocyclization to a powerful indole synthesis (equations 3 and 4) [53–55] and, by extension, to a synthesis of an *Aspidosperma* alkaloid model compound [56]. Intermediate indoline **9** could be isolated (R=X=H) by conducting the photolysis in the presence of solid sodium carbonate. Subsequently, Schultz extended his method to the preparation of *N*-unsubstituted indoles (equation 4) [54] and to 3-carboethoxyoxindoles by treatment of 3-hydroxyindolines (i.e., **9**) with lead tetraacetate [55] (not shown).

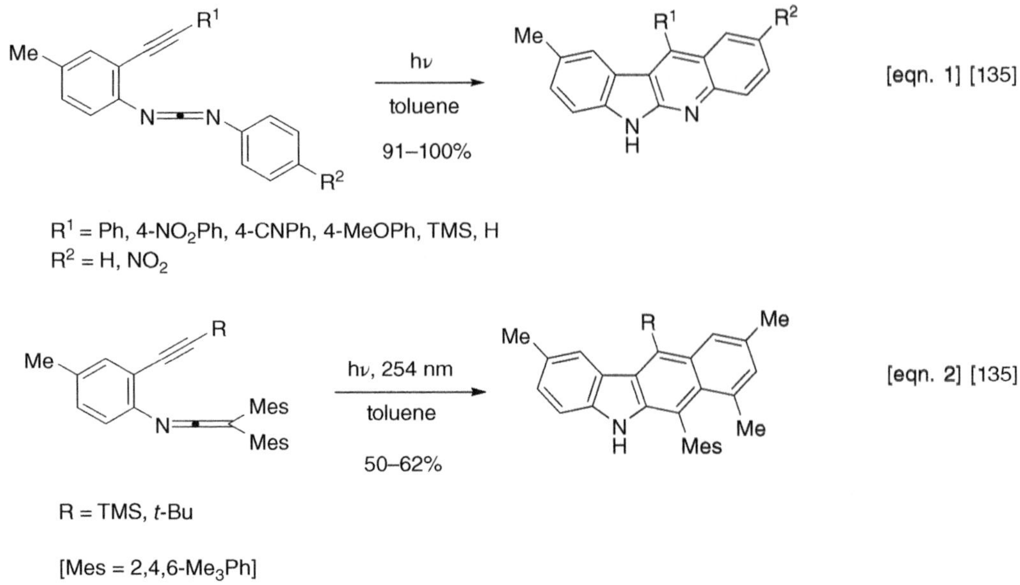

Scheme 18 *The Schmittel Photochemical Synthesis of Indoloquinolines and Benzo[b]carbazoles [135]*

Watson and Dillin observed that the irradiation of enamino ketone **10** afforded two products, depending on the reaction atmosphere (Scheme 6, equation 1) [57]. Loss of acetaldehyde to form 1,2-dimethylindole is novel, and the authors propose two consecutive 1,2-hydride shifts. Fang and colleagues synthesized a series of 2-cyanoindoles via the photocyclization of 2-anilino alkenenitriles (equation 2), en route to spiro-annelated indolines [58]. Teulade and coworkers prepared an azacarbolinone from the respective *N*-benzylenaminone (equation 3) [59], and this same research group described related photochemical cyclizations [60–64], including one to form a dihydroxycarbazole (equation 4) [65].

Grellmann and colleagues present kinetic evidence that the non-oxidative photocyclization of *N*-aryl enamines to indolines involves a tunneling contribution to the sigmatropic hydrogen shift. They conclude that *only* a [1,4] sigmatropic shift takes place, and "it is very unlikely that [1,2] hydrogen shifts play a substantial role in the reaction" (Scheme 7) [66].

Aryl azides undergo indolization upon photolysis, and several examples are known (Scheme 8) [67–70]. Thus, *p*-phenyl–substituted (Me, OMe, Cl) azidopyrimidines **11** were converted to the corresponding 9*H*-pyrimido[4,5-*b*]indole (equation 1) [67, 68]. The azide precursor is in equilibrium with the ring-opened 8-phenyltetrazolo[1,5-*c*]pyrimidine (**10**) (in trifluoroacetic acid). Similarly, Da Settimo and colleagues synthesized the new ring system 6*H*-indolo[2,3-*b*][1,8]naphthyridine **12** by photolysis in trifluoroacetic acid of the equilibrium mixture of 4-phenyltetrazolo[1,5-*a*][1,8]naphthyridines and its ring-opened azide (equation 2) [69]. Zhang and coworkers prepared several indolo[3,2-*c*]quinolin-6-ones (equation 3), pyrido[4,3-*b*]indol-1-ones, and related heterocycles by photolysis of azide precursors [70].

Given the enormous success of the stilbene photocyclization to phenanthrenes—the Mallory reaction [2, 71, 72]—it is no surprise that the heterocyclic counterpart has been extended to indole ring synthesis by the groups of Hendrickson [73], Cava [74, 75], and Tojo [76] (Scheme 9). Hendrickson and deVries were the first to use this heterocyclic stilbene photocyclization, employing it in a synthesis of methoxatin (equation 1). Targeting the antitumor compound CC-1065 and analogues, Cava's group effected this photocyclization to give the fused indoles (equation 2). Cava found that a catalytic amount of Pd/C increased the yields [75] over those in the original report [74]. Also prepared in this study were indoles **13** to **15**. Tojo had a similar route to analogues of CC-1065 but used iodine to facilitate the final aromatization, and propylene oxide destroyed the HI that was generated (equation 3) [76].

In 1970, Bunnett and Kim reported a new nucleophilic aromatic substitution reaction: the $S_{RN}1$ mechanism (substitution, radical-nucleophilic, first order) [77–80]. Nine years later, Bunnett [81] and Beugelmans [82, 83] independently discovered the application of the $S_{RN}1$ reaction to the synthesis of indoles (Scheme 10). In Bunnett's work (equation 1) the acetone enolate (and others) was generated from acetone and potassium *tert*-butoxide (*tert*-butyl alcohol) in the presence of potassium amide in liquid ammonia [81]. Indoles **16** to **18** were also prepared by Bard and Bunnett. Beugelmans and Roussi employed a similar procedure using *o*-iodoaniline, but they explored a wider range of enolates, including aldehyde enolates [82, 83]. A selection of Beugelmans's indoles is **19** through **24**, in addition to indoles that Bunnett also prepared. Indoles **23** and **24** were obtained after hydrolysis of the corresponding dimethyl acetals.

In subsequent years, the Bunnett–Beugelmans indole synthesis has been pursued by Rossi and his coworkers [84–88]. A selection of Rossi's indole syntheses is in Scheme 11. The photostimulation of *o*-iodoaniline can be run in DMSO (equation 1), avoiding the use of ammonia. It can also be accomplished with Fe(II) without irradiation. In all cases the dark reaction is sluggish and lower yielding.

Rossi and colleagues extended the Bunnett–Beugelmans indole synthesis to carbazoles (Scheme 12) [87, 88]. The reaction may be run in liquid ammonia, DMSO, or THF. For the process in equation 1, there is no reaction in the dark after 60 minutes, and an $S_{RN}1$ mechanism was implicated. The yields are generally very good to excellent, and the methodology was adapted to the synthesis of other benzo-fused heterocycles [88].

The mechanism of the Bunnett–Beugelmans indole synthesis is closely aligned with the earlier studies of Kornblum [89] and Russell [90]. A general mechanism is shown in Scheme 13 [79]. Single electron transfer from a radical initiator to *o*-haloaniline from the donor (enolate?) is facilitated by light to give the radical anion **27**. Loss of halide furnishes radical **28**, which adds the nucleophilic enolate to give a new radical anion **29**. Electron transfer from **29** to the *o*-haloaniline substrate gives the acyl product **30** and a fresh radical anion **27** that reenters the chain process, and amino carbonyl **30** forms the final indole product. For reviews on this reaction see Rossi, Pierini, and Santiogo [91]; Rossi [92]; and Rossi, Pierini, and Peñéñory [93].

The final section in this chapter illustrates relatively little explored—even rare—photochemical indole ring syntheses, but these reactions are deserving of exposure in the book.

The well-known 3-diazo-4-oxo-3,4-dihydroquinoline (**31**) undergoes photolysis (intramolecular Wolff rearrangement) to give 3-carbonyl-3*H*-indole (**32**), which can be trapped with various nucleophiles to give a carboxylic acid, esters, and amides (Scheme 14, equation 1) [94–97]. Ketene **32** undergoes tetramerization to form 1,3′-tetrakis(indolylmethanone) (**33**) [98, 99]. The novel ring contraction of dihydroquinolines **34** to indoles **35** by sunlight was observed by Gagosz and colleagues (equation 2) [100]. The authors propose electrocyclic ring opening, intramolecular nucleophilic attack by nitrogen on the β-carbon of the alkylidene malonate, and reorganization to the indole. Corrie and colleagues described a fascinating photochemical metathesis of 7-nitroindolines to 7-nitrosoindoles (equation 3) [101, 102] in a search for a photolytic generation of L-glutamate [103]. The authors propose oxygen transfer from the nitro group to the carbonyl, followed by loss of carboxylate.

Several investigators discovered the photochemical rearrangement of quinoline *N*-oxides to the corresponding benzoxazepines and then to indoles. For example, Kobayashi and colleagues synthesized for the first time 2- and 3-trifluoromethylindoles via this route (Scheme 15, equation 1) [104]. In a series of papers, Kaneko and coworkers described the photochemical synthesis of indoles from quinoline *N*-oxides via benzoxazepines, which were not isolated (equation 2) [105–109]. A similar photolysis of 5-(alkoxycarbonylamino)isoquinoline 2-oxides afforded 1-alkoxycarbonyl-4-hydroxyindoles (equation 3) [110]. Irvine's group reported similar photochemical syntheses of indoles from quinoline *N*-oxides albeit in low yields [111].

Suginome and coworkers discovered the novel [3+2] regioselective photoaddition of 2-amino-1,4-naphthoquinones with electron-rich alkenes to afford 1*H*-benz[*f*]indole-4,9-diones (Scheme 16, equation 1) [112]. Likewise, amino-1,4-benzoquinones with alkenes gave 1*H*-indole-4,7-diones (equation 2). If the initial photoproducts could not indolize by loss of alcohol, then the corresponding dihydro compounds were isolated. Parker and Mindt found that quinone monoimides were smoothly photocyclized to the respective 5-hydroxyindoles (equation 3) [113].

As we saw briefly in Chapter 53, the photochemical version of the Graebe–Ullmann carbazole–carboline synthesis can be an excellent route to these heterocycles. Similar to this benzotriazole reaction is the photolysis of aryl azides to give carbazoles and related fused indoles.

In 1951, Smith and Brown were the first to show that the photolysis of aryl azides led to carbazoles (Scheme 17, equation 1) [114], a reaction that also occurs upon heating. Other carbazoles so prepared are **36** to **39**. This reaction has been reviewed [115–119], and has been of more interest mechanistically than synthetically, by the groups of Reiser [120, 121], Swenton [122–124], Berry [125], Sundberg [126, 127], Yabe [128–130], Meth-Cohn [131], and Spagnolo [132]. Sundberg provides an excellent summary of the possible mechanisms involved in the photolysis of 2-azidobiphenyl to carbazole [126, 127], and his own work indicates that a triplet nitrene may not be "the sole or major carbazole precursor" [127]. In any event, the photochemical transformation of aryl azides is a useful synthesis of carbazoles. Some additional examples are shown in Scheme 17 [114, 115, 124, 128, 129, 133]. In addition, Sauer and Engels obtained, as expected, a nearly 1:1 ratio of 1- and 3-methylcarbazoles (85%) on photolysis of 2-azido-3'-methylbiphenyl [134].

A photochemical cyclization of enyne-heteroallenes was described by Schmittel and coworkers (Scheme 18) [135]. Thus, either direct or triplet-sensitized photolysis of enyne-carbodiimides afforded high yields of indoloquinolines (equation 1), and enyne-ketenimines afforded benzo[*b*]carbazoles (equation 2), in what might be viewed as the photochemical version of the Myers–Saito thermal cyclization of enyne-allenes [136, 137]. The authors provide evidence for a triplet biradical intermediate.

Although the techniques of photochemistry may be inconvenient, especially for large-scale synthesis, the remarkable indole syntheses shown in this chapter—many of which have no nonphotochemical counterpart—should not be ignored by the organic chemist seeking to prepare a target indole or carbazole.

References

[1] J. A. Joule, *Adv. Heterocycl. Chem.*, 1984, **35**, 181–183.
[2] F.B. Mallory and C.W. Mallory, *Org. React.*, 2004, **30**, 1–456.
[3] J.-S. Yang (2007), Spectroscopy, photophysics and photochemistry of anilines, in *The Chemistry of Anilines*, part 2 (ed. Z. Rappoport), John Wiley & Sons, Chichester, UK, pp. 799–809.
[4] D.C. Neckers, *Mechanistic Organic Photochemistry*, Reinhold, New York, pp. 239–240.
[5] C.A. Parker and W.J. Barnes, *Analyst*, 1957, **82**, 606–618.
[6] E.J. Bowen and J.H.D. Eland, *Proc. Chem. Soc.*, 1963, 202.
[7] K.-H. Grellmann, G.M. Sherman, and H. Linschitz, *J. Am. Chem. Soc.*, 1963, **85**, 1881–1882.
[8] T.J. Kemp, J.P. Roberts, G.A. Salmon, and G.F. Thompson, *J. Phys. Chem.*, 1968, **72**, 1464–1470.
[9] G.C. Terry, V.E. Uffindell, and F.W. Willets, *Nature*, 1969, **223**, 1050–1051.
[10] H. Shizuka, Y. Takayama, I. Tanaka, and T. Morita, *J. Am. Chem. Soc.*, 1970, **92**, 7270–7277.
[11] H. Shizuka, Y. Takayama, T. Morita, *et al.*, *J. Am. Chem. Soc.*, 1971, **93**, 5987–5992.
[12] C. Wentrup and M. Gaugaz, *Helv. Chim. Acta*, 1971, **54**, 2108–2111.
[13] E.W. Förster and K.H. Grellmann, *Chem. Phys. Lett.*, 1972, **14**, 536–538.
[14] E. W. Förster, K.H. Grellmann, and H. Linschitz, *J. Am. Chem. Soc.*, 1973, **95**, 3108–3115.
[15] G. Fischer, E. Fischer, K.H. Grellmann, *et al.*, *J. Am. Chem. Soc.*, 1974, **96**, 6267–6269.
[16] K.H. Grellmann, W. Kühnle, and Th. Wolff, *Z. Phys. Chem. Neue Fol.*, 1976, **101**, 295–306.
[17] K.-H. Grellmann, W. Kühnle, H. Weller, and T. Wolff, *J. Am. Chem. Soc.*, 1981, **103**, 6889–6893.
[18] K.H. Grellmann and U. Schmitt, *J. Am. Chem. Soc.*, 1982, **104**, 6267–6272.
[19] H. Weller and K.-H. Grellmann, *J. Am. Chem. Soc.*, 1983, **105**, 6268–6273.
[20] R. Rahn, J. Schroeder, J. Troe, and K.H. Grellmann, *J. Phys. Chem.*, 1989, **93**, 7841–7846.
[21] M.A. Fox, M.T. Dulay, and K. Krosley, *J. Am. Chem. Soc.*, 1994, **116**, 10992–10999.
[22] L.J. Johnston and R.W. Redmond, *J. Phys. Chem. A*, 1997, **101**, 4660–4665.
[23] K. Moriwaki, S. Yoshikawa, Y. Kotani, *et al.*, *J. Photopolymer Sci. Tech.*, 1999, **12**, 777–780.
[24] T. Suzuki, Y. Kajii, K. Shibuya, and K. Obi, *Bull. Chem. Soc. Jpn.*, 1992, **65**, 1084–1088.
[25] D. Sur, P. Purkayastha, S.C. Bera, and N. Chattopadhyay, *J. Mol. Liq.*, 2000, **89**, 175–188.
[26] N. Chattopadhyay, C. Serpa, L.G. Arnaut, and S.J. Formosinho, *Phys. Chem. Chem. Phys.*, 2001, **3**, 3690–3695.
[27] H. Görner, *J. Phys. Chem. A*, 2008, **112**, 1245–1250.
[28] H. Görner, *J. Photochem. Photobiol A: Chem.*, 2010, **211**, 1–6.
[29] K. Amano, T. Hinohara, and M. Hoshino, *Photochem. Photobiol. A: Chem.*, 1991, **59**, 43–54.
[30] W. Carruthers, *J. Chem. Soc. (C)*, 1968, 2244–2247.
[31] W. Carruthers, *J. Chem. Soc., Chem. Commun.*, 1966, 272.
[32] M. Zander and W.H. Franke, *Chem. Ber*, 1966, 2449–2453.
[33] R.B. Miller and T. Moock, *Tetrahedron Lett.*, 1980, **21**, 3319–3322.
[34] R.J. Olsen and O.W. Cummings, *J. Heterocycl. Chem.*, 1981, **18**, 439–440.
[35] J.-D. Cheng and H.J. Shine, *J. Org. Chem.*, 1974, **39**, 336–340.
[36] A.-M. Oliveira-Campos, M.J.R.P. Queiroz, and P.V.R. Shannon, *Chem. Ind.*, 1991, 352–353.
[37] R.J. Hall, J. Marchant, A.M.F. Oliveira-Campos, *et al.*, *J. Chem. Soc., Perkin Trans. 1*, 1992, 3439–3450.
[38] J.P. Ferris and F.R. Antonucci, *J. Am. Chem. Soc.*, 1974, **96**, 2010–2014.
[39] V.M. Clark, A. Cox, and E.J. Herbert, *J. Chem. Soc. (C)*, 1968, 831–833.
[40] H. Ishiyama, K. Ohshita, T. Abe, *et al.*, *Bioorg. Med. Chem.*, 2008, **16**, 3825–3830.
[41] T. Dhanabal, R. Sangeetha, and P.S. Mohan, *Tetrahedron*, 2006, **62**, 6258–6263.
[42] R.N. Kumar, T. Suresh, and P.S. Mohan, *Tetrahedron Lett.*, 2002, **43**, 3327–3328.
[43] P.W. Groundwater, D. Hughes, M.B. Hursthouse, and R. Lewis, *J. Chem. Soc., Perkin Trans. 1*, 1996, 669–673.
[44] M. Chakrabarty, A. Batabyal, and S. Khasnobis, *Synth. Commun.*, 2000, **30**, 3651–3668.
[45] A. Caron, A.C. Hernandez-Perez, and S.K. Collins, *Org. Process Res. Dev.*, 2014, **18**, 1571–1574.
[46] S.M. Bonesi, D. Dondi, S. Protti, *et al.*, *Tetrahedron Lett.*, 2014, **55**, 2932–2935.
[47] P.J. Aragon and Y. Blache, *Trends Heterocycl. Chem.*, 2003, **9**, 47–59.
[48] O.L. Chapman and G.L. Eian, *J. Am. Chem. Soc.*, 1968, **90**, 5329–5330.
[49] O.L. Chapman, G.L. Eian, A. Bloom, and J. Clardy, *J. Am. Chem. Soc.*, 1971, **93**, 2918–2928.
[50] K. Yamada, T. Konakahara, S. Ishihara, *et al.*, *Tetrahedron Lett.*, 1972, **13**, 2513–2516.

[51] K. Yamada, T. Konakahara, and H. Iida, *Bull. Chem. Soc. Jpn.*, 1973, **46**, 2504–2511.
[52] H. Iida, Y. Yuasa, and C. Kibayashi, *J. Org. Chem.*, 1979, **44**, 1236–1241.
[53] A.G. Schultz and W.K. Hagmann, *J. Chem. Soc., Chem. Commun.*, 1976, 726–727.
[54] A.G. Schultz and W.K. Hagmann, *J. Org. Chem.*, 1978, **43**, 3391–3393.
[55] A.G. Schultz and W.K. Hagmann, *J. Org. Chem.*, 1978, **43**, 4231–4233.
[56] A.G. Schultz and I-C. Chiu, *J. Chem. Soc., Chem. Commun.*, 1978, 29.
[57] D. Watson and D.R. Dillin, *Tetrahedron Lett.*, 1980, **21**, 3969–3970.
[58] C.-C. Yang, H.-T. Chang, and J.-M. Fang, *J. Org. Chem.*, 1993, **58**, 3100–3105.
[59] Y. Blache, O. Chavignon, M.E. Sinibaldi-Troin, *et al.*, *Heterocycles*, 1994, **38**, 1241–1246.
[60] Y. Blache, M.-E. Sinibaldi-Troin, A. Voldoire, *et al.*, *J. Org. Chem.*, 1997, **62**, 8553–8556.
[61] Y. Blache, M.-E. Sinibaldi-Troin, M. Hichour, *et al.*, *Tetrahedron*, 1999, **55**, 1959–1970.
[62] C. Tietcheu, C. Garcia, D. Gardette, *et al.*, *J. Heterocycl. Chem.*, 2002, **39**, 965–973.
[63] P.-J. Aragon, A.-D. Yapi, F. Pinguet, *et al.*, *Chem. Pharm. Bull.*, 2004, **52**, 659–663.
[64] P.-J. Aragon, A.-D. Yapi, F. Pinguet, *et al.*, *Chem. Pharm. Bull.*, 2007, **55**, 1349–1355.
[65] M. Ibrahim-Ouali, A. Missoum, M.-E. Sinibaldi, *et al.*, *Synth. Commun.*, 1996, **26**, 657–670.
[66] U. Baron, G. Bartelt, A. Eychmüller, *et al.*, *J. Photochem.*, 1985, **28**, 187–195.
[67] J.A. Hyatt and J.S. Swenton, *J. Heterocycl. Chem.*, 1972, **9**, 409–410.
[68] J.A. Hyatt and J.S. Swenton, *J. Org. Chem*, 1972, **37**, 3216–3220.
[69] A. Da Settimo, G. Primofiore, V. Santerini, *et al.*, *J. Org. Chem.*, 1977, **42**, 1725–1728.
[70] Z. Shi, Y. Ren, B. Li, *et al.*, *Chem. Commun.* 2010, **46**, 3973–3975.
[71] D.T. McQuade, A.G. O'Brien, M. Dörr, *et al.*, *Chem. Sci.*, 2013, **4**, 4067–4070.
[72] K.B. Jørgensen, *Molecules*, 2010, **15**, 4334–4358.
[73] J.B. Hendrickson and J.G. deVries, *J. Org. Chem.*, 1982, **47**, 1148–1150.
[74] V.H. Rawal and M.P. Cava, *J. Chem. Soc., Chem. Commun.*, 1984, 1526–1527.
[75] V.H. Rawal, R.J. Jones, and M.P. Cava, *Tetrahedron Lett.*, 1985, **26**, 2423–2426.
[76] J. Enjo, L. Castedo, and G. Tojo, *Org. Lett.*, 2001, **3**, 1343–1345.
[77] J.K. Kim and J.F. Bunnett, *J. Am. Chem. Soc.*, 1970, **92**, 7463–7464.
[78] J.K. Kim and J.F. Bunnett, *J. Am. Chem. Soc.*, 1970, **92**, 7464–7466.
[79] J.F. Bunnett, *Acc. Chem. Res.*, 1978, **11**, 413–420.
[80] J.F. Wolfe and D.R. Carver, *Org. Prep. Proc. Int.*, 1978, **10**, 225–253.
[81] R.R. Bard and J.F. Bunnett, *J. Org. Chem.*, 1980, **45**, 1546–1547.
[82] R. Beugelmans and G. Roussi, *J. Chem. Soc., Chem. Commun.*, 1979, 950–951.
[83] R. Beugelmans and G. Roussi, *Tetrahedron*, 1981, **37**, 393–397.
[84] M.T. Baumgartner, M.A. Nazareno, M.C. Murguia, *et al.*, *Synthesis*, 1999, 2053–2056.
[85] S.M. Barolo, A.E. Lukach, and R.A. Rossi, *J. Org. Chem.*, 2003, **68**, 2807–2811.
[86] S.M. Barolo, C. Rosales, J.E.A. Guío, and R.A. Rossi, *J. Heterocycl. Chem.*, 2006, **43**, 695–699.
[87] M.E. Budén, V.A. Vaillard, S.E. Martin, and R.A. Rossi, *J. Org. Chem.*, 2009, **74**, 4490–4498.
[88] J.F. Guastavino and T.A. Rossi, *J. Org. Chem.*, 2012, **77**, 460–472.
[89] N. Kornblum, R.E. Michel, and R.C. Kerber, *J. Am. Chem. Soc.*, 1966, **88**, 5662–5663.
[90] G.A. Russell and W.C. Danen, *J. Am. Chem. Soc.*, 1966, **88**, 5663–5665.
[91] R.A. Rossi, A.B. Pierini, and A.N. Santiago, *Org. React.*, 1999, **54**, 1–271.
[92] R.A. Rossi, *Mol. Supramol. Photochem.*, 2005, **12**, 495–527.
[93] R.A. Rossi, A.B. Pierini, and A.B. Peñéñory, *Chem. Rev.*, 2003, **103**, 71–167.
[94] O. Süs, M. Glos, K. Möller, and H.-D. Eberhardt, *Liebigs Ann. Chem.*, 1953, **583**, 150–160.
[95] O. Süs and K. Möller, *Liebigs Ann. Chem.*, 1955, **593**, 91–126.
[96] B. Stanovnik, M. Tišler, and J.T. Carlock, *Synthesis*, 1976, 754–755.
[97] J.T. Carlock, J.S. Bradshaw, B. Stanovnik, and M. Tišler, *J. Org. Chem.*, 1977, **42**, 1883–1885.
[98] G.G. Qiao and C. Wentrup, *Tetrahedron Lett.*, 1995, **36**, 3913–3916.
[99] G.G. Qiao, M.W. Wong, and C. Wentrup, *J. Org. Chem.*, 1996, **61**, 8125–8131.
[100] C. Gronnier, Y. Odabachian, and F. Gagosz, *Chem. Commun.*, 2011, **47**, 218–220.
[101] G. Papageorgiou, D.C. Ogden, A. Barth, and J.E.T. Corrie, *J. Am. Chem. Soc.*, 1999, **121**, 6503–6504.
[102] G. Papageorgiou and J.E.T. Corrie, *Tetrahedron*, 2000, **56**, 8197–8205.
[103] G. Papageorgiou and J.E.T. Corrie, *Tetrahedron*, 2007, **63**, 9668–9676.
[104] Y. Kobayashi, I. Kumadaki, Y. Hirose, and Y. Hanzawa, *J. Org. Chem.*, 1974, **39**, 1836–1838.
[105] C. Kaneko and Sa. Yamada, *Chem. Pharm. Bull.*, 1966, **14**, 555–557.
[106] C. Kaneko and R. Kitamura, *Heterocycles*, 1977, **6**, 111–115.
[107] C. Kaneko and R. Kitamura, *Heterocycles*, 1977, **6**, 117–121.

[108] R. Kitamura, H. Fujii, K. Hashiba, *et al.*, *Tetrahedron Lett.*, 1977, **18**, 2911–2914.
[109] C. Kaneko, A. Yamamoto, and M. Hashiba, *Chem. Pharm. Bull.*, 1979, **27**, 946–952.
[110] C. Kaneko, W. Okuda, Y. Karasawa, and M. Somei, *Chem. Lett.*, 1980, 547–550.
[111] R.W. Irvine, J.C. Summers, and W.C. Taylor, *Aust. J. Chem.*, 1983, **36**, 1419–1430.
[112] K. Kobayashi, H. Takeuchi, S. Seko, *et al.*, *Helv. Chim. Acta*, 1993, **76**, 2942–2950.
[113] K.A. Parker and T.L. Mindt, *Org. Lett.*, 2002, **4**, 4265–4268.
[114] P.A.S. Smith and B.B. Brown, *J. Am. Chem. Soc.*, 1951, **73**, 2435–2437.
[115] B.C.G. Soderberg, *Curr. Org. Chem.*, 2000, **4**, 727–764.
[116] R.A. Abramovitch and B.A. Davis, *Chem. Rev.*, 1964, **64**, 149–185.
[117] S.T. Reid, *Adv. Heterocycl. Chem.*, 1970, **11**, 1–121.
[118] J.A. Joule, *Adv. Heterocycl. Chem.*, 1984, **35**, 167–181.
[119] C. Wentrup, *Adv. Heterocycl. Chem.*, 1981, **28**, 231–361.
[120] A. Reiser, H. Wagner, and G. Bowes, *Tetrahedron Lett.*, 1966, **7**, 2635–2641.
[121] A. Reiser, F.W. Willets, G.C. Terry, *et al.*, *Trans. Faraday Soc.*, 1968, 3265–3275.
[122] J.S. Swenton, *Tetrahedron Lett.*, 1968, **9**, 3421–3424.
[123] J.S. Swenton, T.J. Ikeler, and B.H. Williams, *Chem. Comm.*, 1969, 1263–1264.
[124] J.S. Swenton, T.J. Ikeler, and B.H. Williams, *J. Am. Chem. Soc.*, 1970, **92**, 3103–3109.
[125] P.A. Lehman and R.S. Berry, *J. Am. Chem. Soc.*, 1973, **95**, 8614–8620.
[126] R.J. Sundberg and R.W. Heintzelman, *J. Org. Chem.*, 1974, **39**, 2546–2552.
[127] R.J. Sundberg, D.W. Gillespie, and B.A. DeGraff, *J. Am. Chem. Soc.*, 1975, **97**, 6193–6196.
[128] A. Yabe and K. Honda, *Tetrahedron Lett.*, 1975, **16**, 1079–1082.
[129] A. Yabe and K. Honda, *Bull. Chem. Soc. Jpn.*, 1976, **49**, 2495–2499.
[130] A. Yabe, *Bull. Chem. Soc. Jpn.*, 1980, **53**, 2933–2937.
[131] J.M. Lindley, I.M. McRobbie, O. Meth-Cohn, and H. Suschitzky, *J. Chem. Soc., Perkin Trans.* **1**, 1977, 2194–2204.
[132] P. Spagnolo, A. Tundo, and P. Zanirato, *J. Org. Chem.*, 1977, **42**, 292–296.
[133] A. Da Settimo, G. Primofiore, V. Santerini, *et al.*, *J. Org. Chem.*, 1977, **42**, 1725–1728.
[134] J. Sauer and J. Engels, *Tetrahedron Lett.*, 1969, **10**, 5175–5178.
[135] M. Schmittel, D. Rodríguez, and J.-P. Steffen, *Angew. Chem. Int. Ed.*, 2000, **39**, 2152–2155.
[136] A.G. Myers, E.Y. Kuo, and N.S. Finney, *J. Am. Chem. Soc.*, 1989, **111**, 8057–8059.
[137] R. Nagata, H. Yamanaka, E. Okazaki, and I. Saito, *Tetrahedron Lett.*, 1989, **30**, 4995–4998.

57

Dipolar Cycloaddition, Anionic, and Electrocyclization Reactions

The most common dipolar cycloaddition reaction is the 1,3-dipolar cycloaddition (1,3-DCA) [1–4], and several investigators have employed this reaction in the synthesis of indoles. Kobayashi and colleagues found that 2,3-bis (trifluoromethyl)indole was assembled from the 1,3-DCA reaction between nitrone **1** and hexafluorobut-2-yne (Scheme 1, equation 1) [5]. Intermediates **2** through **5** were each isolated and characterized, and each individually was converted to the indole product by chromatography or acid treatment. No yields were reported. The isoxazoline-to-oxazoline (**2**) rearrangement was proposed to involve an aziridine. Cambon and colleagues described a similar synthesis of fluorinated indoles (equation 2) [6].

Kozikowski and colleagues employed a [3+2] DCA between a nitrile oxide and *N*-ethoxycarbonyl-2-allylpyrrole to give indole *C*-glycoside **6**, after hydrogenation of the isoxazoline cycloadduct, cyclization of the β-ketol, and deprotection of the glycoside (Scheme 2, equation 1) [7]. The nitrile oxide was prepared from the corresponding nitro compound and phenylisocyanate. An intramolecular variation of this 1,3-DCA provided Kozikowski and Mugrage a route to mitomycin analogue **7** (equation 2) [8]. The adduct was deprotected and the N–O bond was cleaved with Na/NH_3 and cyclized to **7** upon silica gel chromatography.

An indoloquinone was synthesized by Vedejs and Monahan in a reaction sequence that featured an intramolecular 1,3-DCA of azomethine ylide **9**, which was generated in situ from oxazolium salt **8** (Scheme 3, equation 1) [9]. A final oxidative workup with DDQ furnished the indoloquinone. For Vedejs's related synthesis of an aziridinomitosene A analogue, see Bobeck, Warner, and Vedejs [10]. Kerr and colleagues have employed a 1,3-DCA reaction between nitrones and 1,1-cyclopropanediesters to access tetrahydro-1,2-oxazines, which can be transformed into pyrrolo[1,2-*a*]indoles (equation 2) [10–12]. This tricycle is commonly found in natural products such as the well-known mitomycins.

In an approach to the mitosene skeleton, Iwasawa's group effected a 1,3-DCA between azomethine ylides and vinyl ethers, as catalyzed by third-row transition metal complexes ($PtCl_2$, $AuBr_3$, $Re(CO)_5Cl$, $W(CO)_6$, and others), with $PtCl_2$ being the most efficient (Scheme 4, equation 1) [13]. Formation of the metal-containing azomethine ylide **10** was followed by the [3+2] cycloaddition with a vinyl ether to give a carbene complex. This underwent a 1,2-migration of the R group from C-2 to C-3 to yield the indole product. Typically, the *cis*-isomer was slightly favored (e.g., 73:27). Shi and colleagues found that a [3+2] cyclization took place under gentle conditions between aniline-tethered alkylidenecyclopropanes and aldehydes to furnish pyrrolo[1,2-*a*]indoles in excellent yield (equation 2) [14]. A control experiment revealed that the reaction proceeded initially to form the imine, which then underwent a [3+2] thermal cyclization between the imine group and the cyclopropane ring. Feldman and coworkers adapted an azide–allene intramolecular 1,3-DCA to a preparation of either cyclopent[*b*]indole **11** or cyclopent[*a*]indole **12**, depending on the reaction conditions (equation 3) [15–17].

Vidal and coworkers found that an intramolecular 1,3-DCA occurred with azido-ketenimines to give indolo[1,2-*a*]quinazolines **13** or, when R^3=Me, indolo[2,1-*b*]quinazolines **14** (Scheme 5, equation 1) [18].

Foucaud and colleagues described the [1+4] cycloaddition reaction between β-nitrostyrenes and isonitriles to give *N*-hydroxyindoles (Scheme 6 equation 1) [19, 20]. The authors propose the pathway shown where the nitro

Indole Ring Synthesis: From Natural Products to Drug Discovery, First Edition. Gordon W. Gribble.

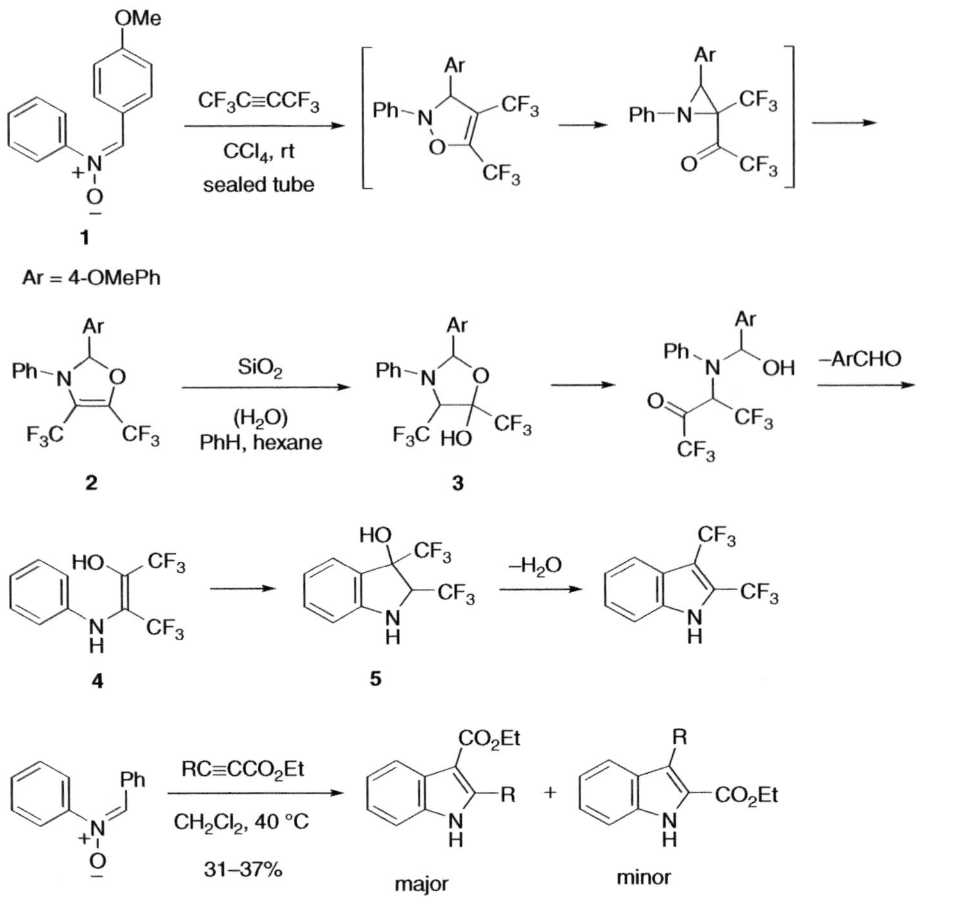

[eqn. 1] [5]

[eqn. 2] [6]

R = CF_3, n-C_3F_7, n-C_5H_{11}, n-C_3H_{13}, n-C_7H_{15}

Scheme 1 *Synthesis of Fluorinated Indoles from Nitrones and Alkynes*

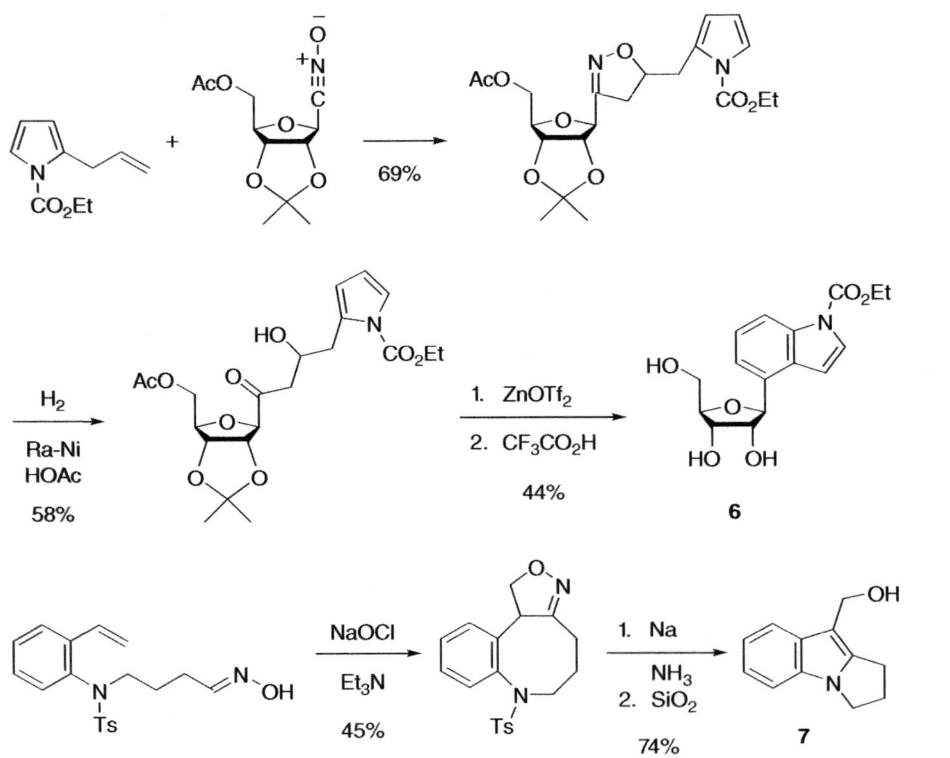

[eqn. 1] [7]

[eqn. 2] [8]

Scheme 2 *Kozikowski Nitrile Oxide 1,3-Dipolar Cycloaddition Indole Synthesis*

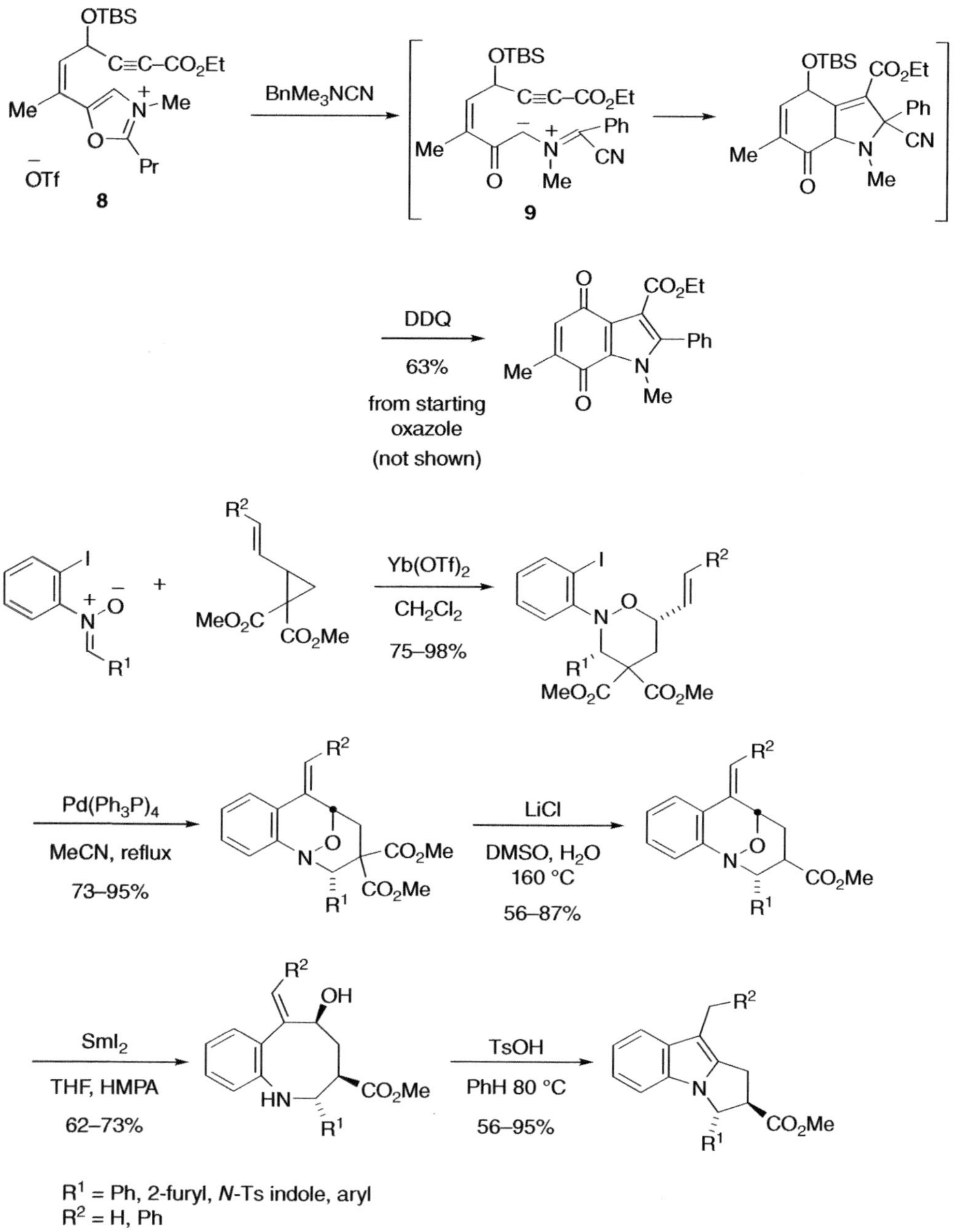

[eqn. 1] [9]

[eqn. 2] [11]

Scheme 3 *Vedejs and Kerr 1,3-Dipolar Cycloaddition Indole Syntheses*

group attacks the electron-deficient isocyanide carbon, which then cyclizes back onto the β-carbon of the nitrostyrene to give intermediate **15**. Fragmentation of **15** and subsequent indolization completes the process.

In a series of papers, Blechert and coworkers reported several closely related indole syntheses involving, for example, the reaction of *N*-phenylhydroxyamines with electron-withdrawing allenes to give *o*-alkylated aniline intermediates, which cyclized to 1,2-disubstituted indoles in an oxy-Cope rearrangement (Scheme 7, equation 1) [21–27]. In a related process that involves a [3+2] DCA, Blechert and Wirth found that *N*-phenylnitrones react with functionalized cycloallenes to give 2,3-disubstituted indoles in a one-pot reaction (equation 2) [26]. Following the initial dipolar cycloaddition to yield isoxazoline **16**, a hetero-Cope rearrangement ensued to give **17**. This suffered a *retro*-Michael reaction to **18**, which underwent a final indolization. Liu and colleagues reported the 1,3-DCA reaction of *C*-aryl-*N*-phenylnitrones with fluorinated alkenes to give 2-aryl-3-fluorinated-alkylindoles [28].

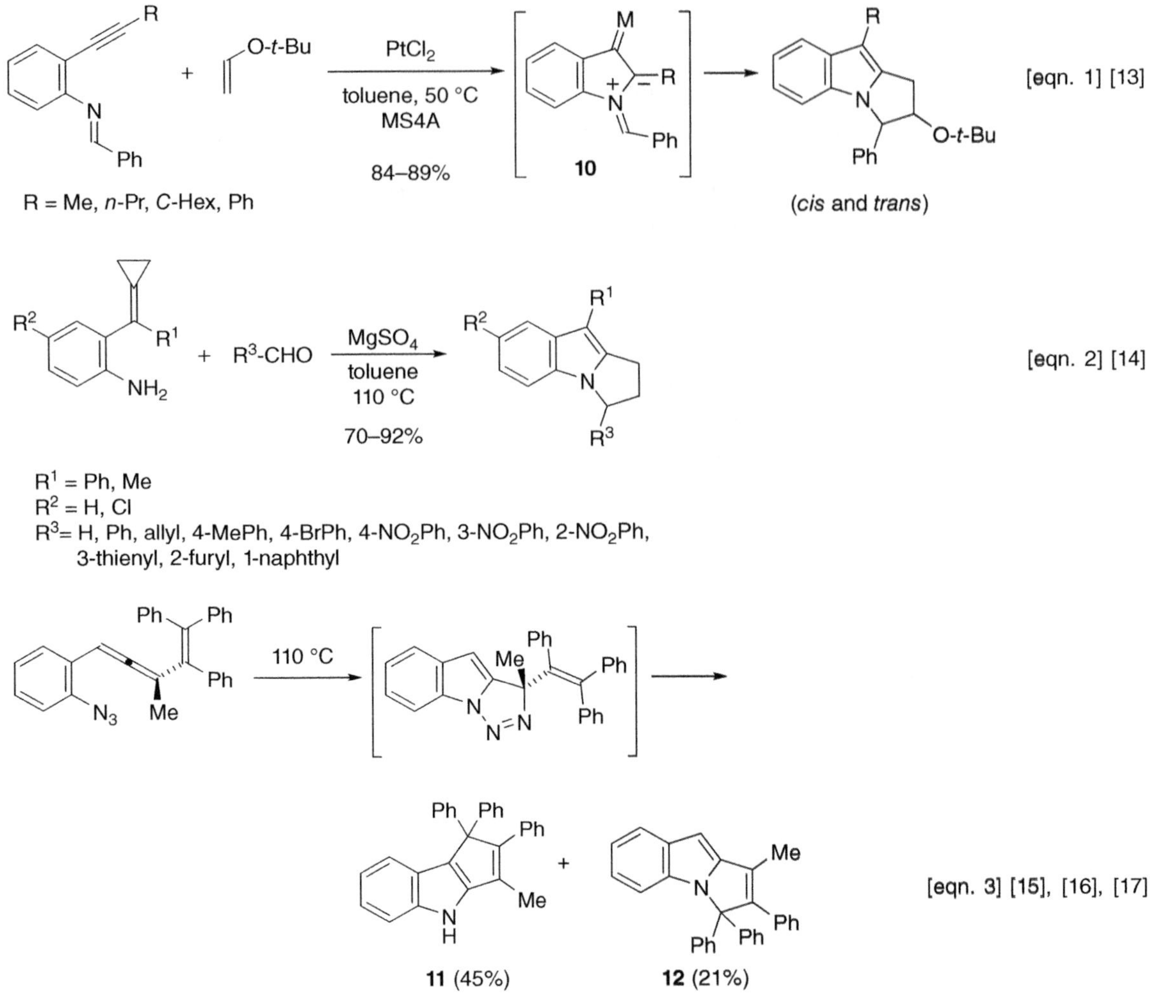

Scheme 4 Iwasawa, Shi, and Feldman Syntheses of Pyrrolo[1,2-a]indoles via [3 + 2] Cycloadditions

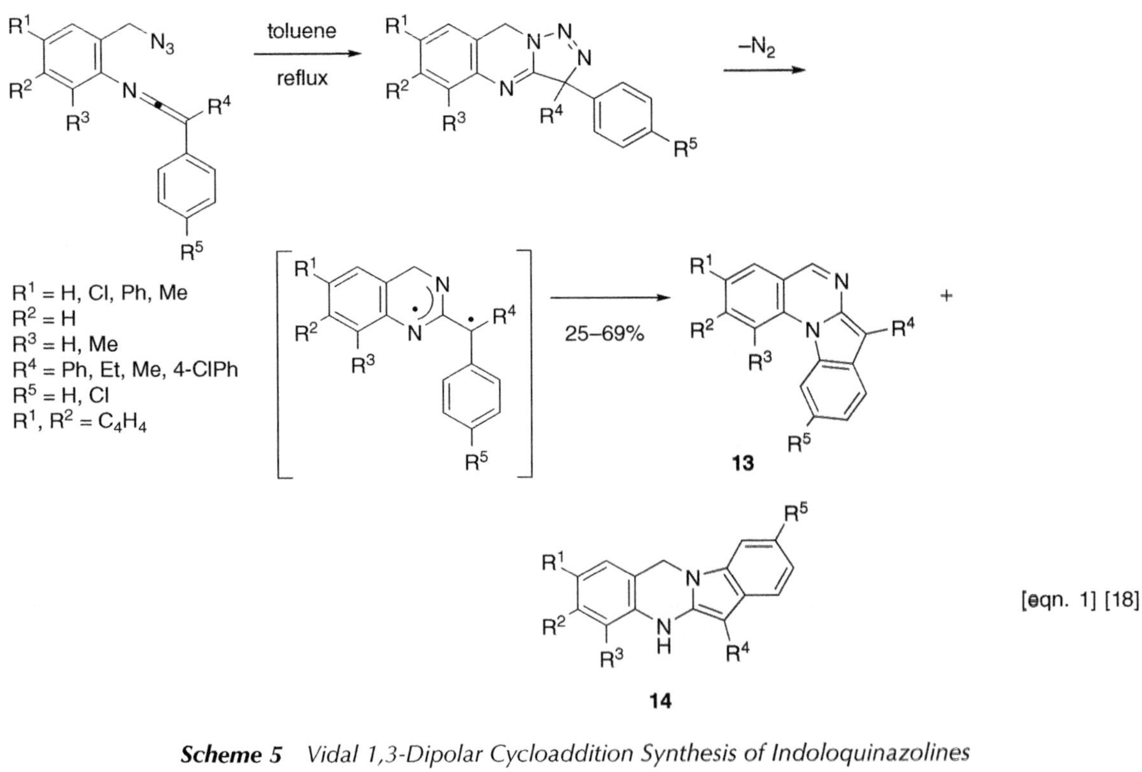

Scheme 5 Vidal 1,3-Dipolar Cycloaddition Synthesis of Indoloquinazolines

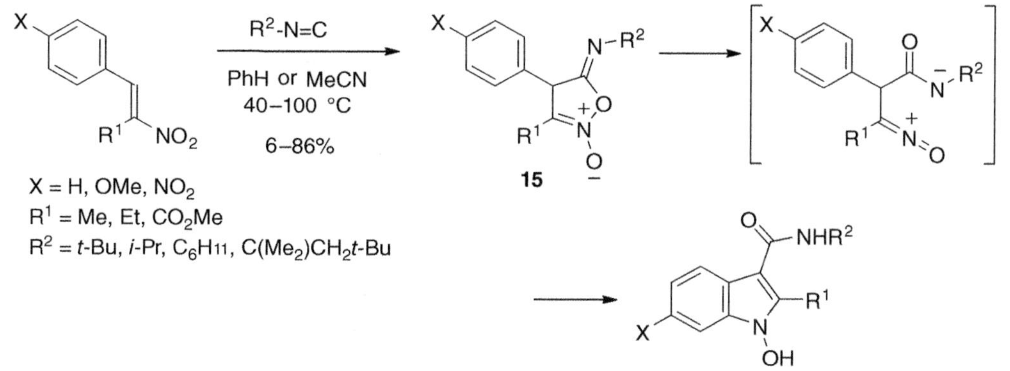

Scheme 6 *Foucaud Synthesis of* N *Hydroxyindoles*

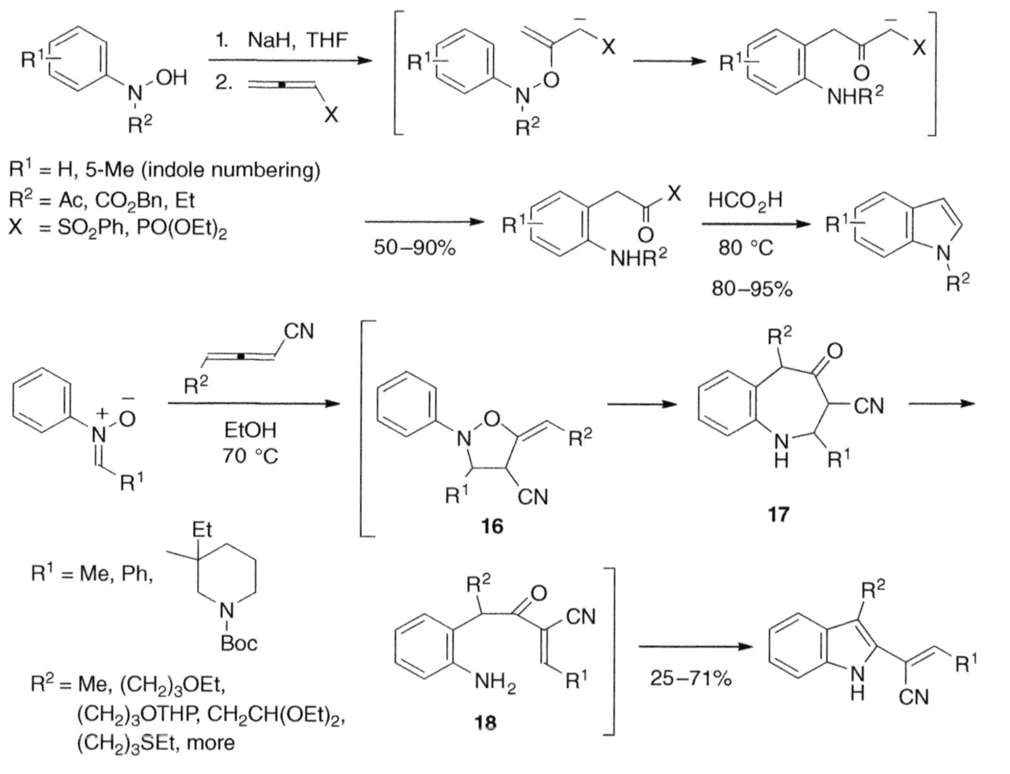

Scheme 7 *Blechert Indole Synthesis*

Although we have encountered sigmatropic rearrangements earlier in the guise of specific name reactions (Chapters 2–7), several related newer indole syntheses involve similar pathways and are presented in this chapter.

Cheng and Cheng found that β-lactam carbenes react in a [2+2] fashion with aryl isonitriles to give 2-azetidinonylidene indoles **19**, which underwent an acid-catalyzed rearrangement to give δ-carbolin-2,4-dione **20** (Scheme 8, equation 1) [29]. Many examples are described, and a different carbolin-2,4-dione was obtained upon treatment of **19** with hydrochloric acid. Xiao and colleagues discovered a simple cascade reaction of sulfur ylides and *N*-(*o*-chloromethyl)aryl amides leading to indoles (equation 2) [30]. The authors suggest a pathway involving the generation of aza-*o*-quinodimethane **21**, followed by conjugate addition of the sulfur ylide, expulsion of dimethylsulfide to give **22**, base elimination of *p*-tolylsulfonimide, and final indolization.

Penoni, Nicholas, HJouk, and their coworkers reported the synthesis of indoles via the one-pot reaction between nitrosoarenes and alkynes [31–33], in contrast to the reaction of alkynes with *nitro*arenes to give *N*-hydroxyindoles

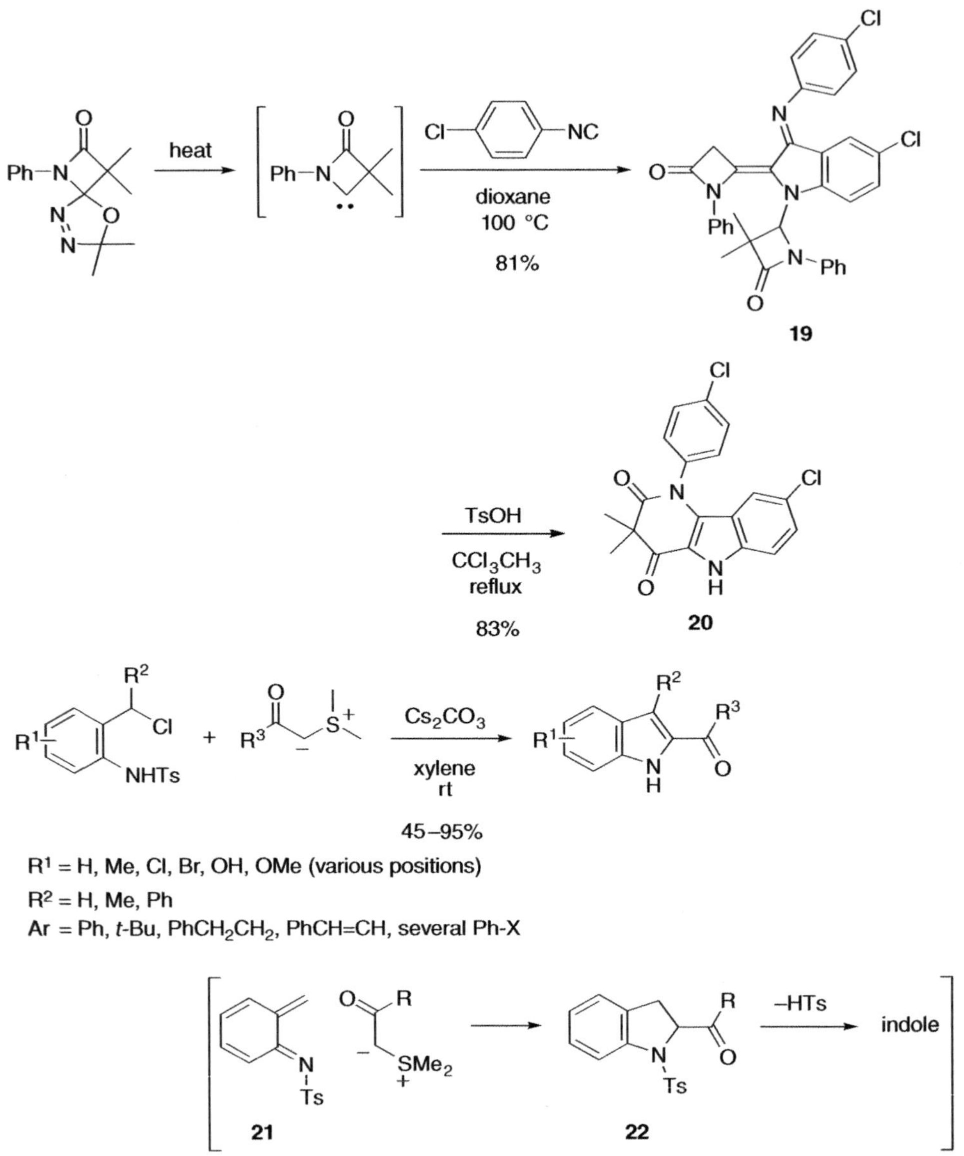

[eqn. 1] [29]

[eqn. 2] [30]

Scheme 8 *Cheng and Xiao Indole Syntheses*

as catalyzed by Fe and Ru [34]. This methodology was put to good use in the synthesis of the meridianin marine alkaloids and analogues (Scheme 9, equation 1). Computational studies (density functional theory) implicated a stepwise diradical path for this reaction [31]. Danishefsky and Phillips described a new route to 4-substituted indoles that used an aza Claisen rearrangement (equation 2) [35]. The unstable aldehyde was converted to the ethyl ester (1. Ag_2O, aq. MeOH; 2. EtOH, HCl). Substituents other than cyano (CO_2Me, NO_2, F, CF_3) were less favorable for providing the 1,2,3-substituted product. Using a 6π-electrocyclic ring closure as the key step, Funk and coworkers developed a novel indole ring construction (equation 3) [36] and used it in syntheses toward the alkaloids welwistatin and dragmacidin E [37, 38]. Also prepared were indoles **23** to **26** [36, 37].

In a cycloaddition reminiscent of the Moody indole synthesis (Chapter 54), Kočevar and Kranjc effected the high-pressure Diels–Alder cycloaddition of 2*H*-pyran-2-ones with (*Z*)-1-methoxy-1-en-3-yne to afford indoles after acid-cyclization (Scheme 10, equation 1) [39]. In some cases the 2,2-dimethoxyethyl intermediates were formed, but these were transformed to indoles under the same acidic conditions. An intramolecular [4+2] cycloaddition between ynamides and conjugated enynes to indolines and indoles was featured by Danheiser and Dunetz (equation 2)

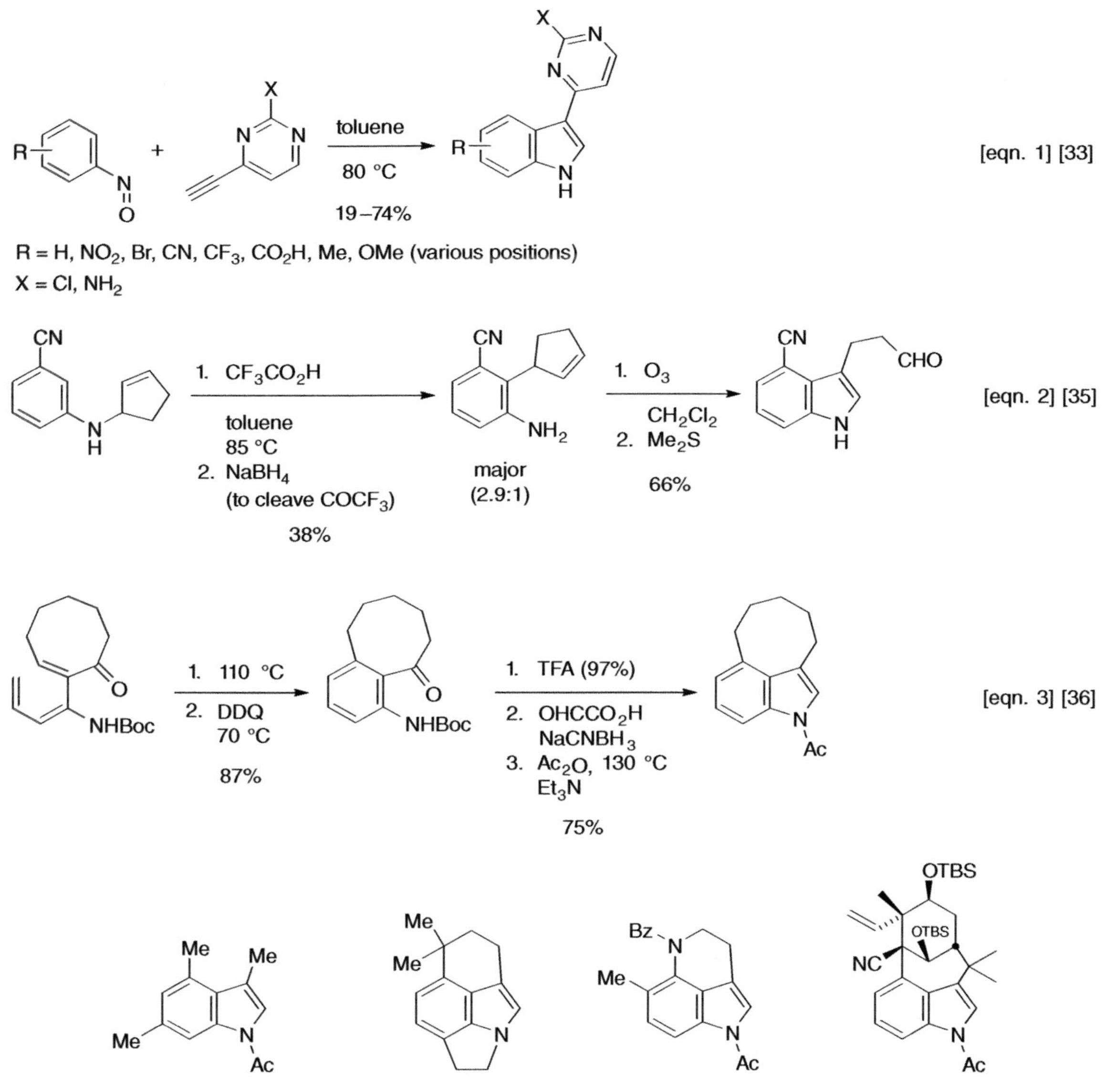

Scheme 9 *Penoni, Danishefsky, and Funk Indole Syntheses*

[40]. An interesting indolization was discovered by Meslin and colleagues when they treated benzyl imine **27** with phosphorus pentasulfide in refluxing toluene or xylene, conditions that yielded an indole (equation 3) [41]. Without the methoxyl group, the reaction took a different course and 2*H*-benzo-1,4-thiazines and benzothiazoles were the products.

Wu and colleagues found that an intramolecular anionic cycloaromatization of (3(Z)-hexen-1,5-diynl)anilines afforded carbazoles (Scheme 11, equation 1) [42]. The corresponding indoles were formed first and could be isolated in yields up to 40%, but carbazoles were always the major product. Würthwein and coworkers discovered a similar base-induced ring closure reaction to form 3-aminoindoles and δ-carbolines (equations 2, 3) [43]. The initially formed (unstable) 3-aminoindoles were capped as *tert*-butyl amides.

Langer and colleagues explored the generation and cyclization of cyano-stabilized anions to give indoles and carbolines [44–46], For example, δ-carbolines were accessible via 2-alkylidene-3-iminoindoles (Scheme 12, equations 1, 2) [45]. These workers uncovered the remarkable double domino cyclization/electrocyclization/elimination reaction (equation 3) [46], a fitting reaction to conclude this chapter!

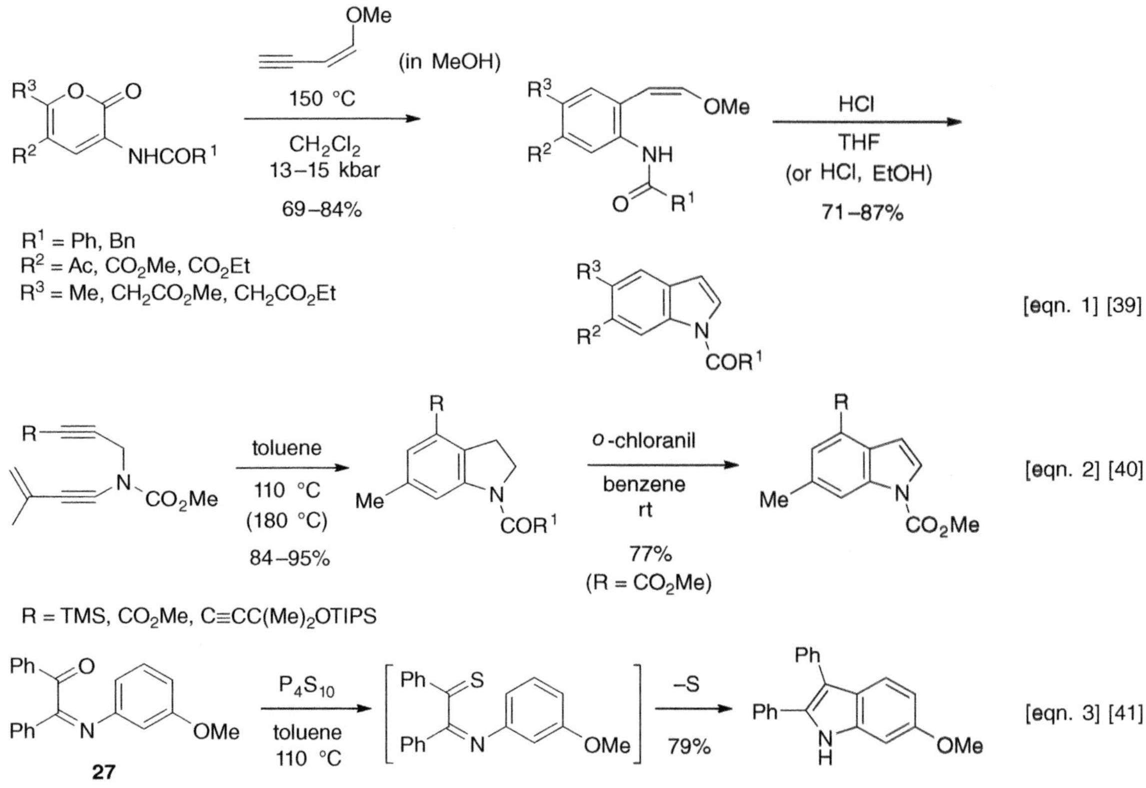

Scheme 10 *Kocevar, Danheiser, and Meslin Indole Syntheses*

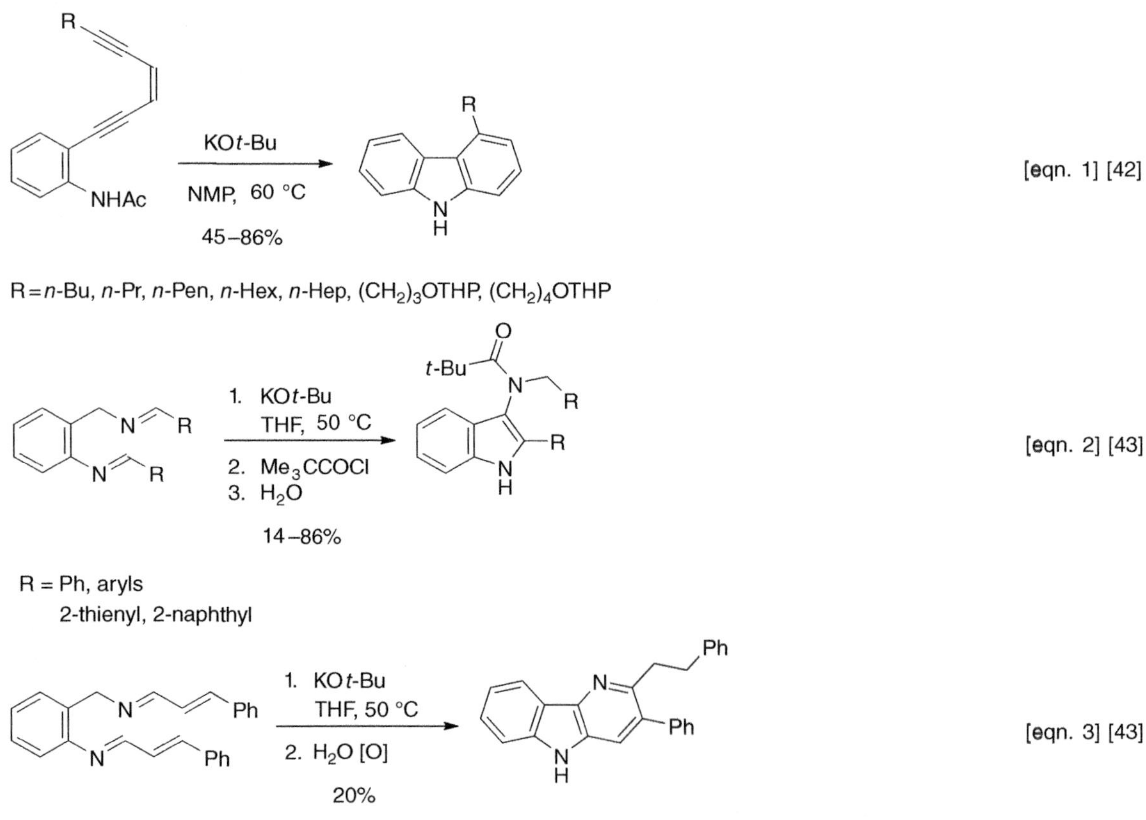

Scheme 11 *Wu and Würthwein Indole Syntheses*

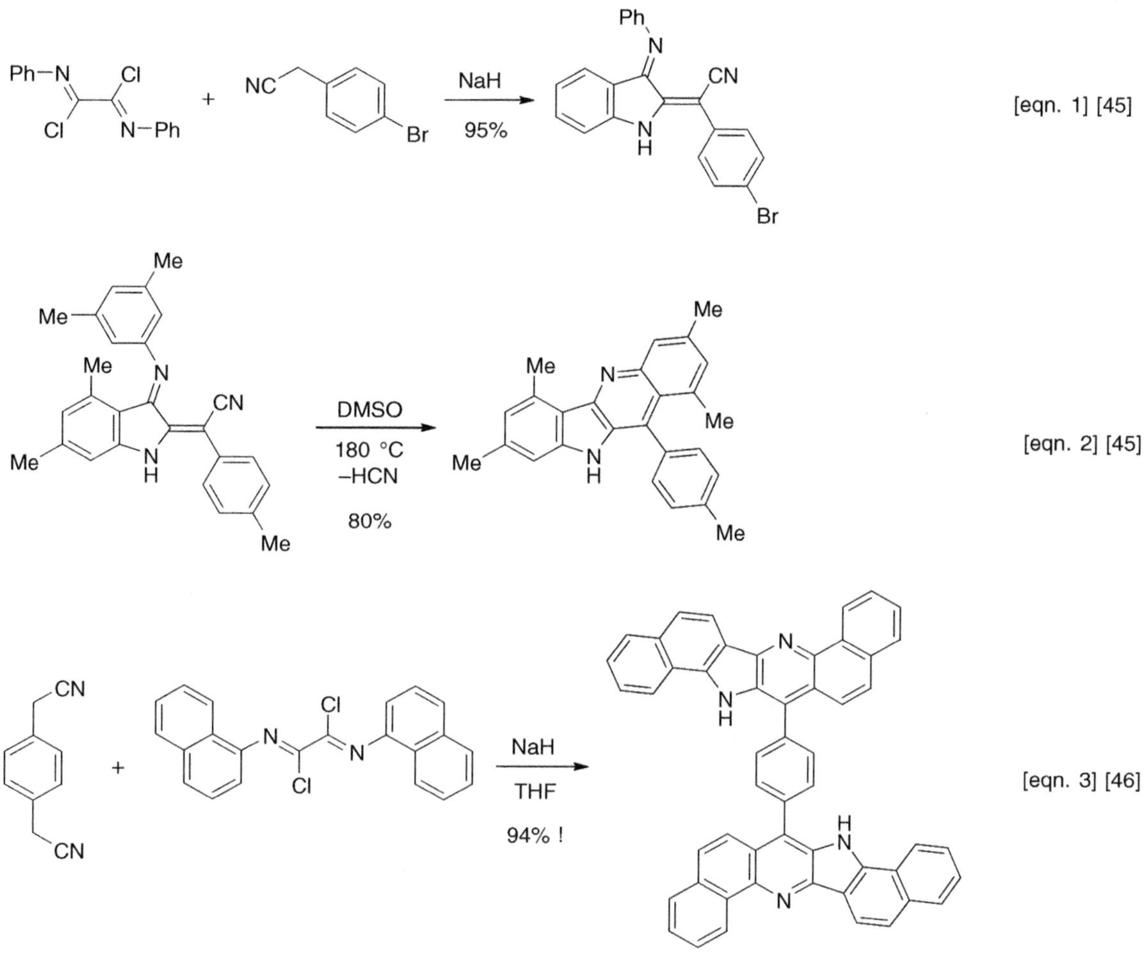

Scheme 12 Langer Indole Synthesis

References

[1] A. Padwa (2002) Synthetic applications of 1,3-dipolar cycloaddition chemistry toward heterocycles and natural products, in *The Chemistry of Heterocyclic Compounds*, vol. **59** (eds A. Padwa and W.H. Pearson), John Wiley & Sons, Chichester, UK.
[2] I. Coldham and R. Hufton, *Chem. Rev.*, 2005, **105**, 2765–2809.
[3] L.-J. Wang and Y. Tang (2014) Intermolecular 1,3-dipolar cycloadditions of alkenes, alkynes, and allenes, in *Comprehensive Organic Synthesis*, 2nd ed., vol. **4** (eds P. Knochel and G.A. Molander), John Wiley & Sons, Chichester, UK, pp. 1342–1383.
[4] R.S. Menon and V. Nair (2014) Intramolecular 1,3-dipolar cycloadditions of alkenes, alkynes, and allenes, in *Comprehensive Organic Synthesis*, 2nd ed., vol. **4** (eds P. Knochel and G.A. Molander), Elsevier, Amsterdam, pp. 1281–1341.
[5] Y. Kobayashi, I. Kumadaki, and T. Yoshida, *Heterocycles*, 1977, **8**, 387–390.
[6] J. Fayn, A. Nezis, and A. Cambon, *J. Fluorine Chem.*, 1987, **36**, 479–482.
[7] A.P. Kozikowski and X.-M. Cheng, *J. Chem. Soc., Chem. Commun.*, 1987, 680–683.
[8] A.P. Kozikowski and B.B. Mugrage, *J. Chem. Soc., Chem. Commun.*, 1988, 198–200.
[9] E. Vedejs and S.D. Monahan, *J. Org. Chem.*, 1997, **62**, 4763–4769.
[10] D.R. Bobeck, D.L. Warner, and E. Vedejs, *J. Org. Chem.*, 2007, **72**, 8506–8518.
[11] M.B. Johansen and M.A. Kerr, *Org. Lett.*, 2008, **10**, 3497–3500.
[12] I.S. Young and M.A. Kerr, *Org. Lett.*, 2004, **6**, 139–141.
[13] J. Takaya, Y. Miyashita, H. Kusama, and N. Iwasawa, *Tetrahedron*, 2011, **67**, 4455–4466.
[14] K. Chen, Z. Zhang, Y. Wei, and M. Shi, *Chem. Commun.*, **48**, 7696–7698.
[15] K.S. Feldman, M.R. Iyer, and D.K. Hester II, *Org. Lett.*, 2006, **8**, 3113–3116.
[16] K.S. Feldman, D.K. Hester II, C.S. López, and O.N. Faza, *Org. Lett.*, 2008, **10**, 1665–1668.
[17] K.S. Feldman, D.K. Hester II, M.R. Iyer, *et al.*, *J. Org. Chem.*, 2009, **74**, 4958–4974.

[18] M. Alajarin, B. Bonillo, M.-M. Ortin, *et al.*, *Org. Biomol. Chem.*, 2011, **9**, 6741–6749.
[19] H. Person, M. Del Aguila Pardo, and A. Foucaud, *Tetrahedron Lett.*, 1980, **21**, 281–284.
[20] A. Foucaud, C. Razorilalana-Rabearivony, E. Loukakou, and H. Person, *J. Org. Chem.*, 1983, **48**, 3639–3644.
[21] S. Blechert, *Tetrahedron Lett.*, 1984, **25**, 1547–1550.
[22] S. Blechert, *Liebigs Ann. Chem.*, 1985, 673–682.
[23] S. Blechert, *Helv. Chim. Acta*, 1985, **68**, 1835–1843.
[24] R. Höfelmeier and S. Blechert, *Tetrahedron Lett.*, 1985, **26**, 5281–5284.
[25] J. Wilkens, A. Kühling, and S. Blechert, *Tetrahedron*, 1987, **43**, 3237–3246.
[26] T. Wirth and S. Blechert, *Synlett*, 1994, 717–718.
[27] S. Blechert, R. Knier, H. Schroers, and T. Wirth, *Synthesis*, 1995, 592–604.
[28] J.-T. Liu and H.-J. Lu, *Chin, J. Chem.*, 2002, **20**, 1330–1333.
[29] Y. Cheng and L.-Q. Cheng, *J. Org. Chem.*, 2007, **72**, 2625–2630.
[30] Q.-Q. Yang, C. Xiao, L.-Q. Lu, *et al.*, *Angew. Chem. Int. Ed.*, 2012, **51**, 9137–9140.
[31] A. Penoni, J. Volkmann, and K.M. Nicholas, *Org. Lett.*, 2002, **4**, 699–701.
[32] A. Penoni, G. Palmisano, Y.-L. Zhao, *et al.*, *J. Am. Chem. Soc.*, 2009, **131**, 653–661.
[33] F. Tibiletti, M. Simonetti, K.M. Nicholas, *et al.*, *Tetrahedron*, 2010, **66**, 1280–1288.
[34] A. Penoni and K.M. Nicholas, *Chem. Commun.*, 2002, 484–485.
[35] S.J. Danishefsky and G.B. Phillips, *Tetrahedron Lett.*, 1984, **25**, 3159–3162.
[36] T.J. Greshock and R.L. Funk, *J. Am. Chem. Soc.*, 2006, **128**, 4946–4947.
[37] T.J. Greshock and R.L. Funk, *Org. Lett.*, 2006, **8**, 2643–2645.
[38] R.J. Huntley and R.L. Funk, *Org. Lett.*, 2006, **8**, 4775–4778.
[39] K. Kranjc and M. Kočevar, *Tetrahedron*, 2008, **64**, 45–52.
[40] J.R. Dunetz and R.L. Danheiser, *J. Am. Chem. Soc.*, 2005, **127**, 5776–5777.
[41] J.-D. Charrier, C. Landreau, D. Deniaud, *et al.*, *Tetrahedron*, 2001, **57**, 4195–4202.
[42] C.-Y. Lee, C.-F. Lin, J.-L. Lee, *et al.*, *J. Org. Chem.*, 2004, **69**, 2106–2110.
[43] B. Neue, R. Reiermann, K. Gerdes, *et al.*, *J. Org. Chem.*, 2011, **76**, 8794–8806.
[44] P. Langer and J.T. Anders, *Eur. J. Org. Chem.*, 2002, 686–691.
[45] P. Langer, J.T. Anders, K. Weisz, and J. Jähnchen, *Chem. Eur. J.*, 2003, **9**, 3951–3964.
[46] J.T. Anders and P. Langer, *Eur. J. Org. Chem.*, 2004, 5020–5026.

PART VIII
Indoles from Pyrroles

An obvious approach to the indole ring is to begin with a pyrrole ring. Accordingly, numerous indole ring syntheses begin with a pyrrole. The next few chapters cover such indole ring syntheses.

Indole Ring Synthesis: From Natural Products to Drug Discovery, First Edition. Gordon W. Gribble.

58

Electrophilic Cyclization of Pyrrole

In a series of papers spanning 20 years, Natsume, Muratake, and coworkers reported a versatile synthesis of indoles exploiting the reactivity of the pyrrole ring toward intramolecular electrophilic attack at the C-2 or C-3 position [1–16]. In their seminal paper, Natsume and Muratake reported a simple synthesis of 4-alkylpyrroles (Scheme 1, equation 1) [1] that derived from their earlier work involving the formation and reaction of heterocyclic endoperoxides with nucleophiles [2]. Using this strategy, these workers prepared the ergot biogenetic precursor

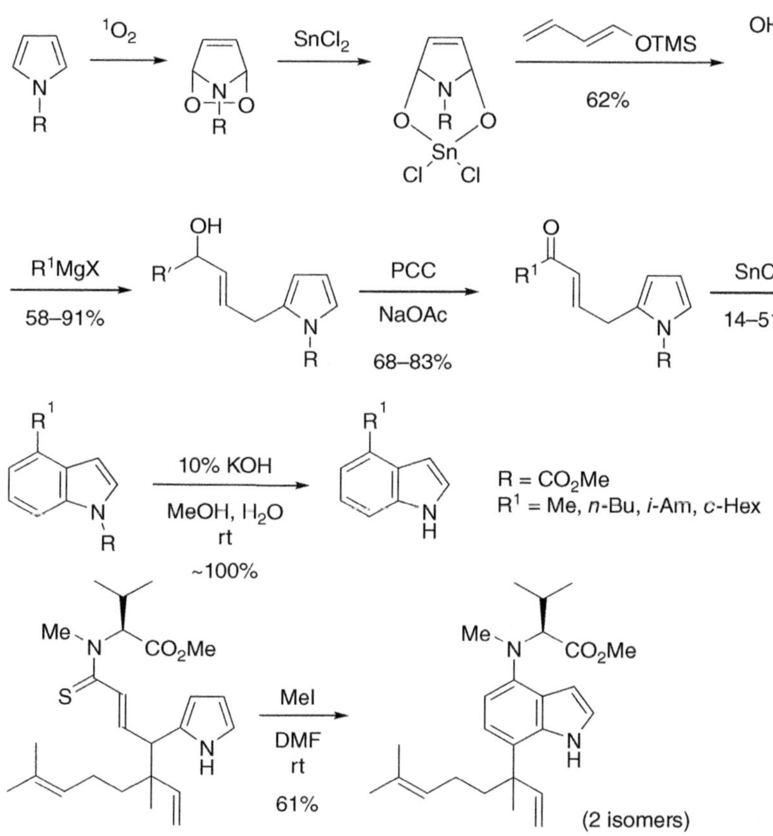

[eqn. 1] [1]

[eqn. 2] [3]

Scheme 1 *Natsume and Muratake Indole Synthesis*

Indole Ring Synthesis: From Natural Products to Drug Discovery, First Edition. Gordon W. Gribble.

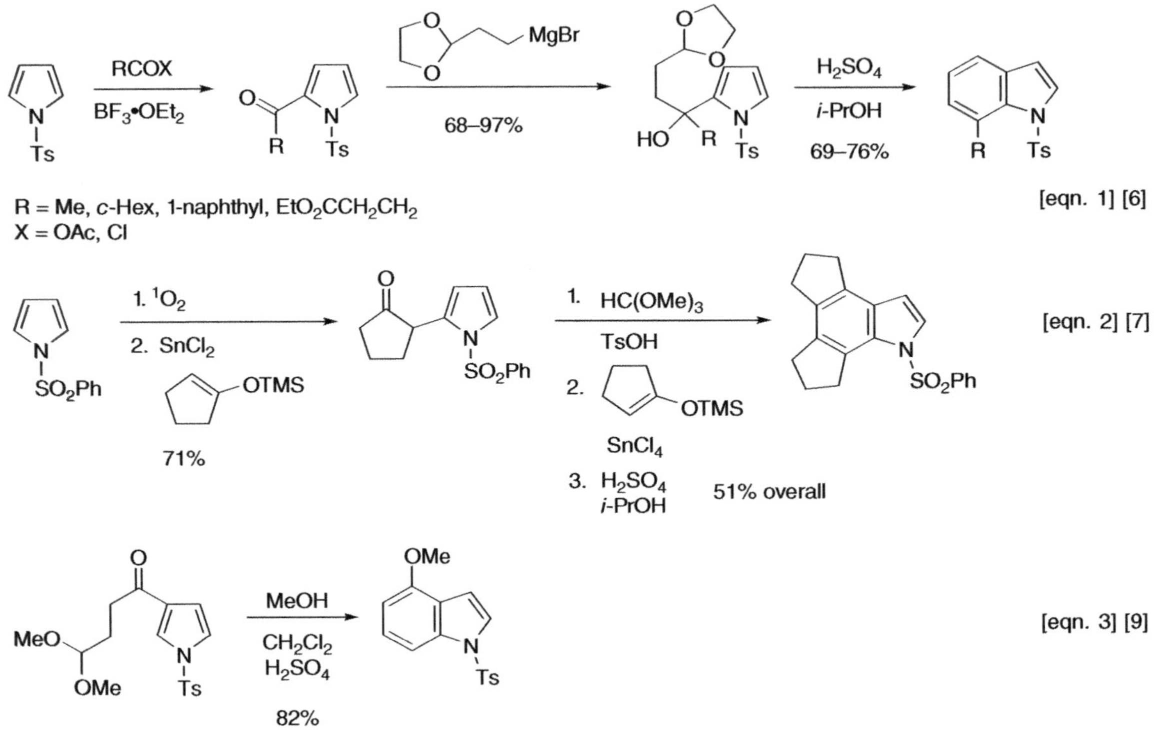

Scheme 2 *Natsume and Muratake Indole Synthesis – 2*

4-(3-methyl-2-butenyl)indole, and, using direct C-2 intermolecular electrophilic attack on *N*-tosylpyrrole followed by intramolecular C-3 electrophilic cyclization, Muratake and Natsume synthesized the tumor promoters lyngbyatoxin A (= teleocidin A-1) and teleocidin A-2; the key indole-forming step is shown in equation 2 [3]. In addition to the two isomeric products, the C-4 *S*-methyl compound was also obtained (30%).

Natsume and Muratake soon molded their original method to an excellent indole synthesis (Scheme 2, equations 1 and 2) [4–8]. The tosyl group may be removed with Mg, MeOH or 10% KOH/DME-MeOH-H_2O. Interestingly, the use of $AlCl_3$ induced cyclization to C-3, leading ultimately to 4-substituted indoles.

In a *tour de force* over the next ten years, Natsume, Muratake, and colleagues applied their pyrrole-to-indole methodology to the syntheses of the trikentrins [10–13], pendolmycin [14], (±)- and (*S*)-pindolol [15], the herbindoles [13, 16], hapalindole O [17], mitosene analogues [18], and the duocarmycins [19, 20]. Several related companion papers described model studies on these natural products [21–23].

Several other investigators have described related indole ring formations via initial electrophilic substitution on a pyrrole. Thus, Murakami and colleagues devised a new synthesis of 7-substituted indoles similar to the Natsume–Muratake approach (Scheme 3, equation 1) [24]. Kozikowski and coworkers reported a clever approach to C-4 and C-7 substituted indoles that involved the addition of a nitrile oxide to an allylpyrrole, culminating with a metal triflate cyclization to an indole (equation 2) [25]. These workers applied their methodology to syntheses of lyngbyatoxin A analogues [26–29]. An observed minor product was the corresponding C-7 substituted indole that arose via rearrangement of hydroxyl ketone **1**. One of Kozikowski's lyngbyatoxin A analogues is shown in equation 3 [29].

Trost and coworkers employed a thionium ion cyclization in a novel synthesis of 4-substituted indoles from *N*-methylpyrrole (Scheme 4, equation 1) [30]. Ishibashi's

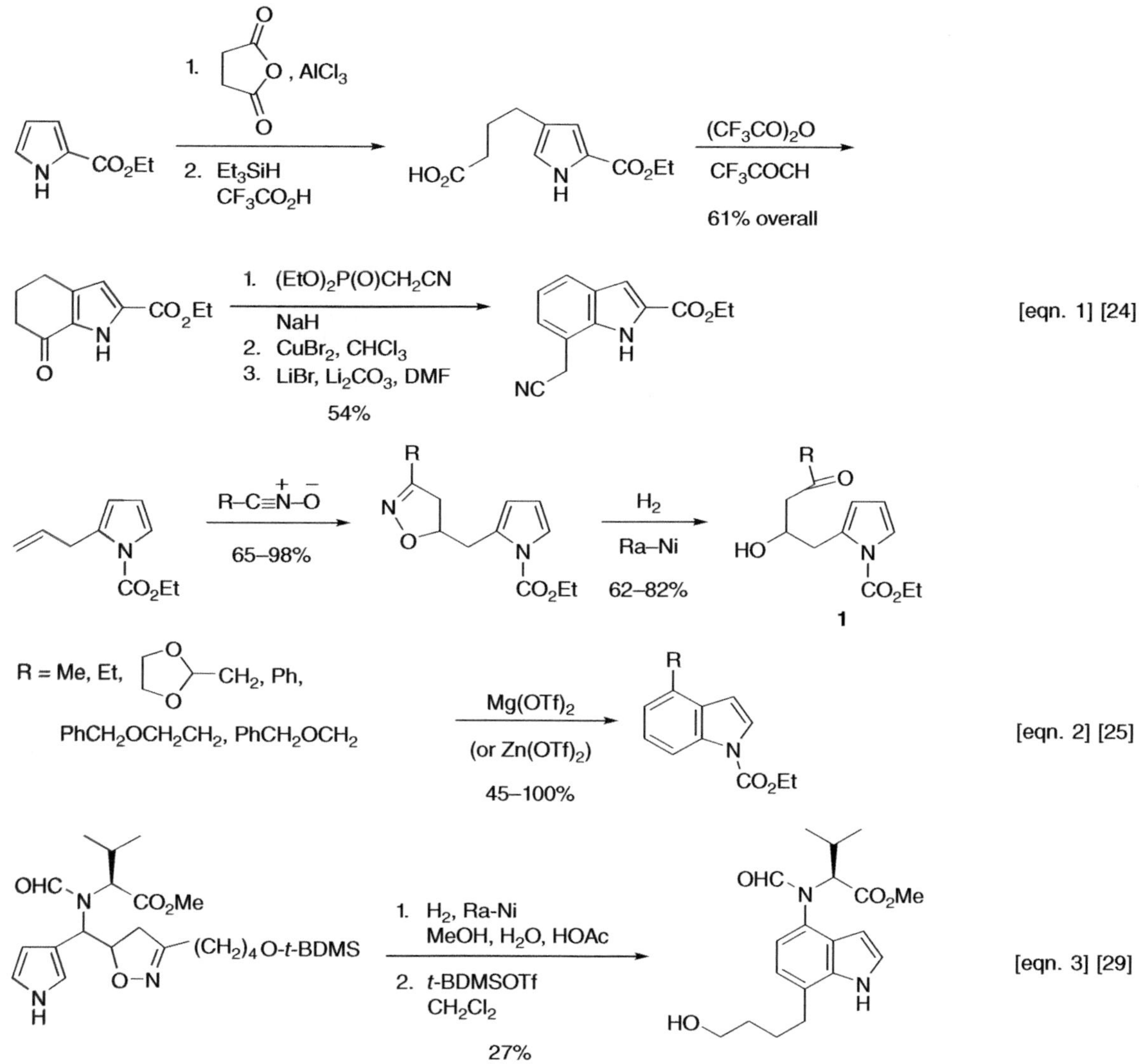

Scheme 3 *Murakami and Kozikowski Indole Syntheses*

team devised a related approach to 4-substituted indoles that featured syntheses of (*S*)-(–)-pindolol (**2**) and (±)-chuangxinmycin (**3**) (equation 2) [31, 32].

Demopoulos and colleagues created a synthesis of a potent dopamine-receptor agonist (**4**) starting from a 3-substituted pyrrole (Scheme 5, equation 1) [33]. The key indolization involved a Natsume–Muratake cyclization. Likewise, Gelb's group synthesized a potent inhibitor against human group X secreted phospholipase A_2 via the electrophilic cyclization of a keto pyrrole as the indole-forming step (equation 2) [34]. Latos-Grażyński and coworkers discovered a fascinating preparation of 4,7-diarylindoles by the acid-catalyzed rearrangement of 1,4-diaryl-1,4-di(pyrrol-2-yl)but-2-ynes (equation 3) [35]. The venerable acid-catalyzed indole dimerization occurred with these 4,7-diarylindoles on further treatment with trifluoroacetic acid.

Using a [3+2] annulation of 2-(benzotriazol-1-ylmethyl) pyrroles with enones and enals, Katritzky and colleagues fashioned a novel synthesis of polysubstituted indoles (Scheme 6, equations 1 and 2) [36, 37]. The requisite pyrroles were fashioned in three steps from propargylbenzotriazole as shown. In a modified C-3 benzotriazole-mediated reaction, Katritzky and coworkers developed a different

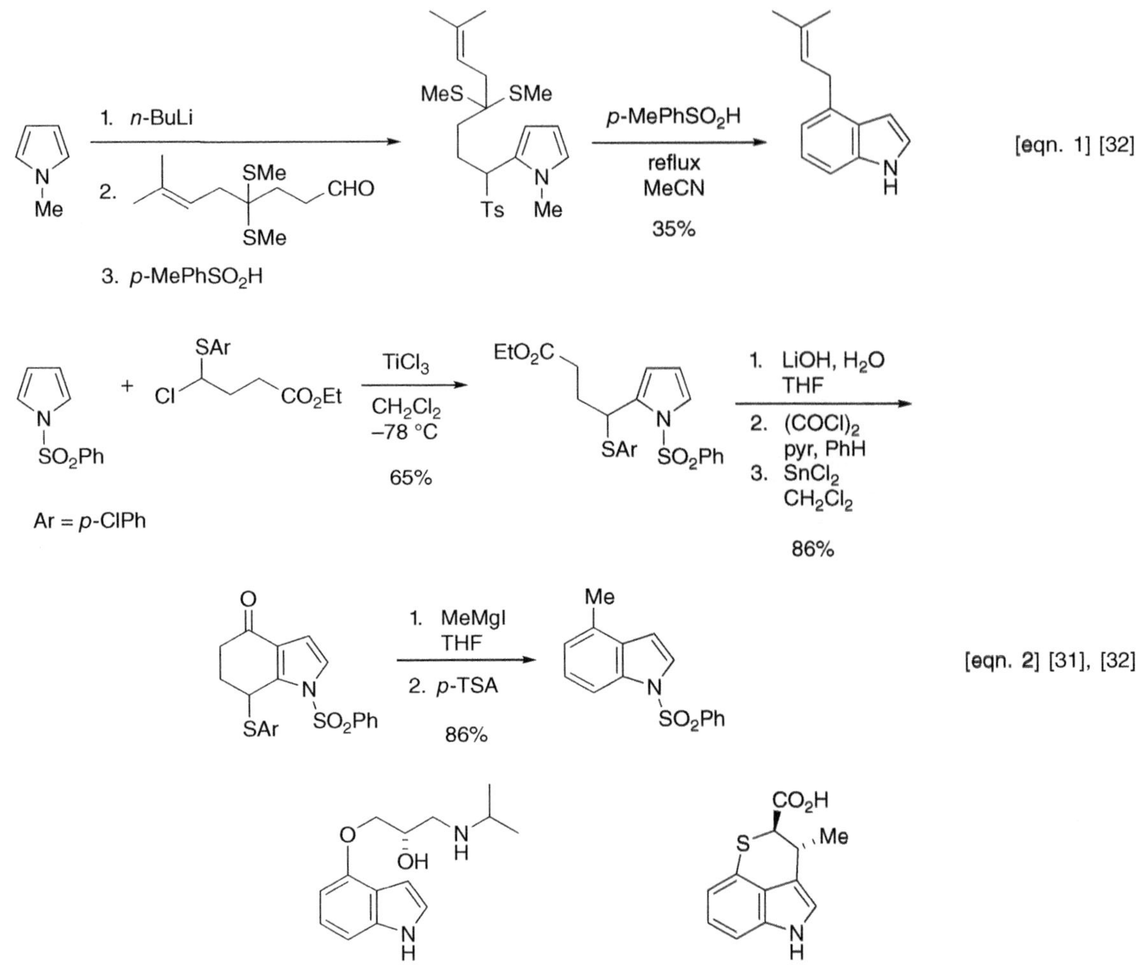

2 **3**

Scheme 4 *Trost and Ishibashi Indole Syntheses*

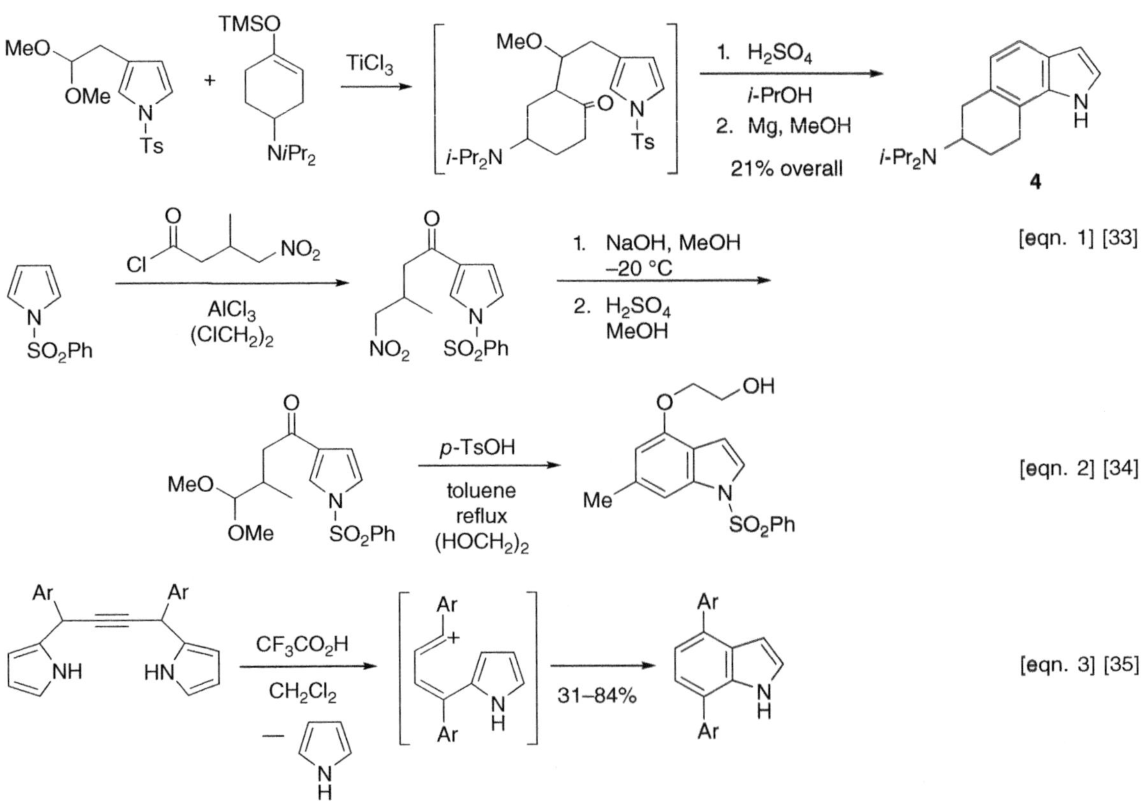

Ar = Ph, *p*-Tol, *p*-MeOPh, 2-Thienyl

Scheme 5 *Demopoulos, Gelb, and Latos-Grazynski Indole Syntheses*

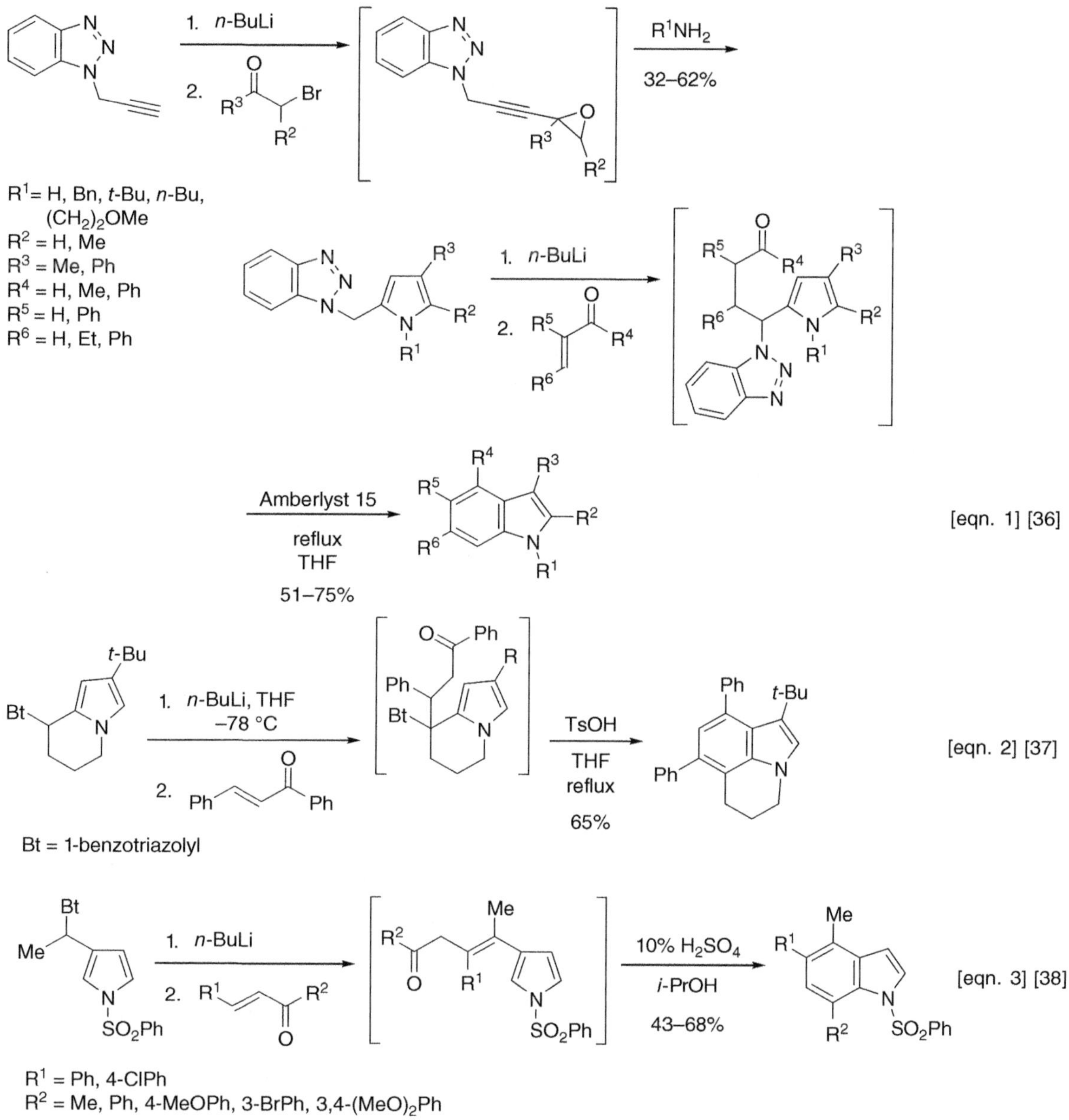

Scheme 6 *Katritzky Indole Syntheses*

synthesis of indoles (equation 3) [38]. In some cases the benzotriazole adduct was isolated from the conjugate addition to the enone.

Rather than using benzotriazole as an anion-stabilizing group, Ila, Junjappa, and their coworkers employed a cyano group to effect an indole ring synthesis (Scheme 7, equation 1) [39]. Attempts to isolate adduct **5** were unsuccessful as it slowly underwent indolization upon silica gel chromatography. Raney nickel desulfurization also converted the indole products **6** to the corresponding 7-methylindoles (85%–94% yield). Several novel condensed indoles were crafted by these investigators (**7–9**). France and colleagues employed a cyclopropene ring to furnish in elegant style indoles from pyrroles (equation 2) [40]. The pyrrole cyclopropene precursor was synthesized from the corresponding diazo-β-ester and an alkyne in the presence of Rh_2esp_2 (dirhodium α,α,α′, α′-tetramethyl-1,3-benzene dipropanoate.

Not surprisingly, the well-known Paal-Knorr pyrrole synthesis [41] can be pushed to yield indoles and even carbazoles, as shown by Kashima [42] and Török [43, 44]

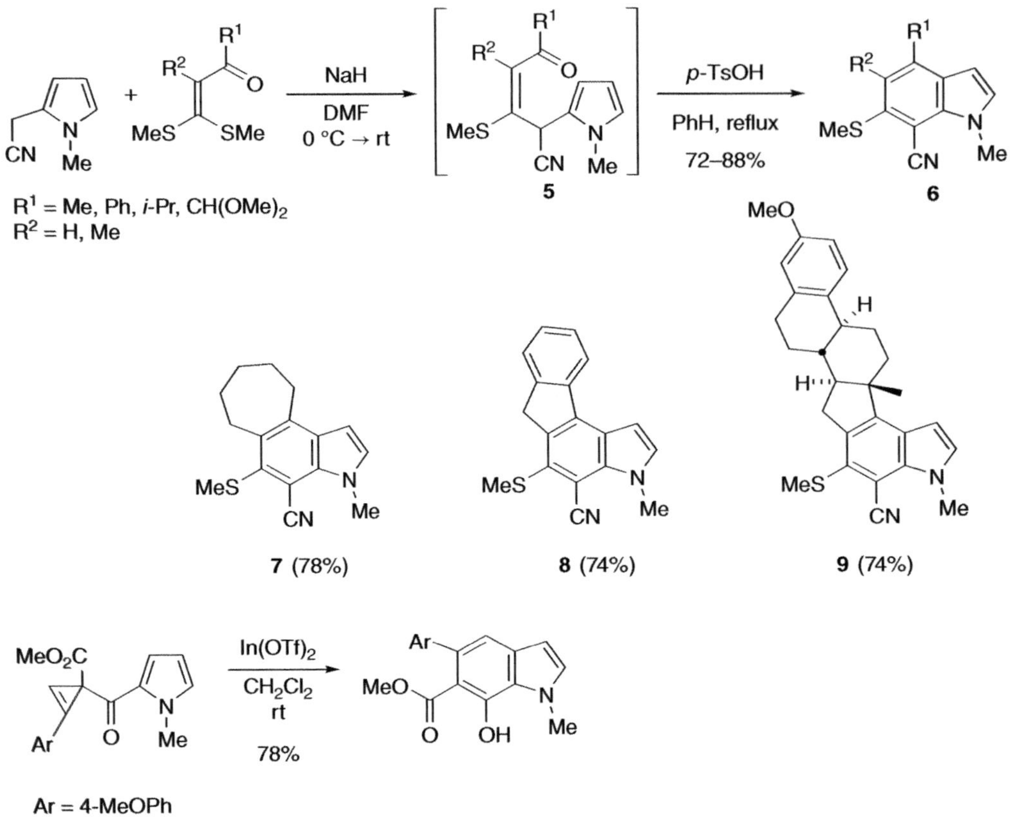

[eqn. 1] [39]

[eqn. 2] [40]

Scheme 7 *Ila-Junjappa and France Indole Syntheses*

(Scheme 8). Kashima's team showed that, as expected, the corresponding pyrrole and, subsequently, the indole precede the final *N*-substituted carbazole (equation 1). Török and colleagues described a similar sequence as Kashima but were able to isolate pyrroles, indoles, or carbazoles depending on the triflic acid concentration (equations 2 and 3) [43]. Using 0.05 equivalents of triflic acid, Török obtained *N*-arylsulfonyl pyrroles in 80% to 92% yield. This group also synthesized *N*-acylindoles from 2,5-dimethoxytetrahydrofuran and aryl amides under the influence of montmorillonite K-10 in yields of 60% to 85% [44]. The corresponding pyrrole is the presumed intermediate (not isolated), and only traces of carbazoles (1%–2%) were isolated, even with excess 2,5-dimethoxytetrahydrofuran, higher temperatures, and longer reaction times. Barluenga and colleagues effected a shrewd indole synthesis from the annulation of 3-alkynylpyrrolo-2-carboxyaldehydes with alkenes (equation 4) [45, 46]. The sequence involved formation of pyrylium cation **10** that underwent nucleophilic addition by the alkene and fragmentation to the 4-acylindole. A selection of indoles prepared by Barluenga is shown (**11–15**). Asao and Aikawa independently discovered a similar indole synthesis from 3-alkynylpyrroles and aldehydes using copper triflate [47].

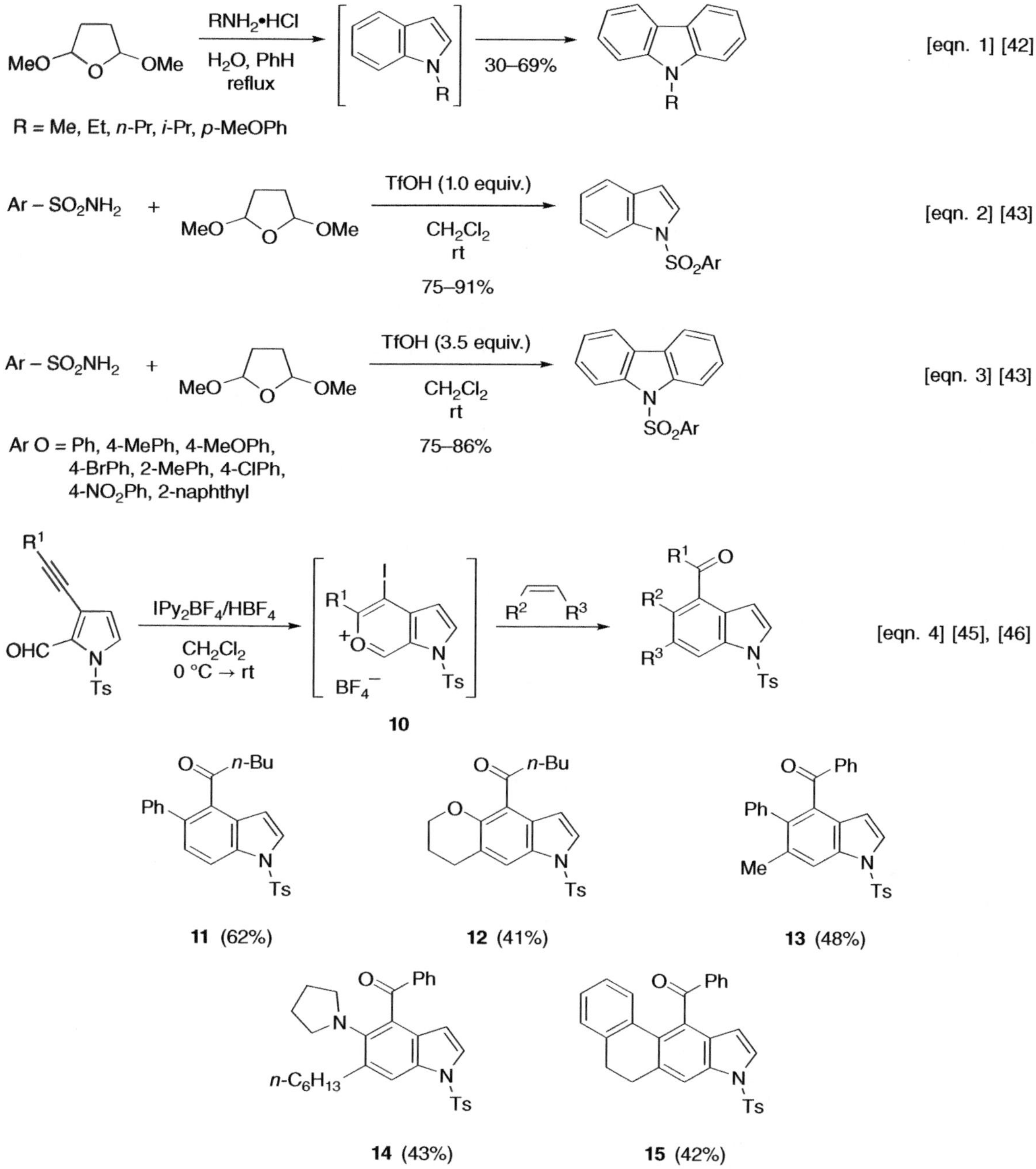

Scheme 8 *Kashima, Török, and Barluenga Indole Syntheses*

References

[1] M. Natsume and H. Muratake, *Tetrahedron Lett.*, 1979, **20**, 3477–3480.
[2] M. Natsume, Y. Sekine, M. Ogawa, *et al.*, *Tetrahedron Lett.*, 1979, **20**, 3473–3476.
[3] H. Muratake and M. Natsume, *Tetrahedron Lett.*, 1987, **28**, 2265–2268.
[4] H. Muratake and M. Natsume, *Heterocycles*, 1989, **29**, 771–782.
[5] H. Muratake and M. Natsume, *Heterocycles*, 1989, **29**, 783–794.
[6] H. Muratake and M. Natsume, *Heterocycles*, 1990, **31**, 683–690.
[7] H. Muratake and M. Natsume, *Heterocycles*, 1990, **31**, 691–700.
[8] M. Fuji, H. Muratake, and M. Natsume, *Chem. Pharm. Bull.*, 1992, **40**, 2338–2343.
[9] M. Fuji, H. Muratake, and M. Natsume, *Chem. Pharm. Bull.*, 1992, **40**, 2344–2352.
[10] H. Muratake and M. Natsume, *Tetrahedron Lett.*, 1989, **30**, 5771–5772.
[11] H. Muratake, M. Watanabe, K. Goto, and M. Natsume, *Tetrahedron*, 1990, **46**, 4179–4192.
[12] H. Muratake, T. Seino, and M. Natsume, *Tetrahedron Lett.*, 1993, **34**, 4815–4818.
[13] H. Muratake, A. Mikawa, T. Seino, and M. Natsume, *Chem. Pharm. Bull.*, 1994, **42**, 854–864.
[14] K. Okabe, H. Muratake, and M. Natsume, *Tetrahedron*, 1990, **46**, 5113–5120.
[15] M. Fuji, H. Muratake, M. Akiyama, and M. Natsume, *Chem. Pharm. Bull.*, 1992, **40**, 2353–2357.
[16] H. Muratake, A. Mikawa, and M. Natsume, *Tetrahedron Lett.*, 1992, **33**, 4595–4598.
[17] M. Sakagami, H. Muratake, and M. Natsume, *Chem. Pharm. Bull.*, 1994, **42**, 1393–1398.
[18] I. Utsunomiya, H. Muratake, and M. Natsume, *Chem. Pharm. Bull.*, 1995, **43**, 37–48.
[19] H. Muratake, I. Abe, and M. Natsume, *Chem. Pharm. Bull.*, 1996, **44**, 67–79.
[20] H. Muratake, N. Matsumura, and M. Natsume, *Chem. Pharm. Bull.*, 1998, **46**, 559–571.
[21] I. Utsunomiya, H. Muratake, and M. Natsume, *Chem. Pharm. Bull.*, 1992, **40**, 2358–2361.
[22] I. Utsunomiya, M. Fuji, T. Sato, and M. Natsume, *Chem. Pharm. Bull.*, 1993, **41**, 854–860.
[23] H. Muratake, A. Mikawa, T. Seino, and M. Natsume, *Chem. Pharm. Bull.*, 1994, **42**, 846–851.
[24] Y. Murakami, M. Tani, T. Ariyasu, *et al.*, *Heterocycles*, 1988, **27**, 1855–1860.
[25] A.P. Kozikowski, X.-M. Cheng, C.-S. Li, and J.G. Scripko, *Isr. J. Chem.*, 1986, **27**, 61–65.
[26] A.P. Kozikowski and X.-M. Cheng, *Tetrahedron Lett.*, 1985, **26**, 4047–4050.
[27] A.P. Kozikowski and X.-M. Cheng, *Tetrahedron Lett.*, 1987, **28**, 3189–3192.
[28] A.P. Kozikowski, K. Sato, A. Basu, and J.S. Lazo, *J. Am. Chem. Soc.*, 1989, **111**, 6228–6234.
[29] A.P. Kozikowski, P.W. Shum, A. Basu, and J.S. Lazo, *J. Med. Chem.*, 1991, **34**, 2420–2430.
[30] B.M. Trost, M. Reiffen, and M. Crimmin, *J. Am. Chem. Soc.*, 1979, **101**, 257–259.
[31] H. Ishibashi, T. Tabata, K. Hanaoka, *et al.*, *Tetrahedron Lett.*, 1993, **34**, 489–492.
[32] H. Ishibashi, S. Akamatsu, H. Iriyama, *et al.*, *Chem. Pharm. Bull.*, 1994, **42**, 271–276.
[33] V.J. Demopoulos, A. Gavalas, G. Rekatas, and E. Tani, *J. Heterocycl. Chem.*, 1995, **32**, 1145–1148.
[34] B.P. Smart, R.C. Oslund, L.A. Walsh, and M.H. Gelb, *J. Med. Chem.*, 2006, **49**, 2858–2860.
[35] E. Nojman, L. Latos-Grażyński, and L. Szterenberg, *Eur. J. Org. Chem.*, 2012, 4115–4122.
[36] A.R. Katritzky, J.R. Levell, and J. Li, *Tetrahedron Lett.*, 1996, **37**, 5641–5644.
[37] A.R. Katritzky, C.N. Fali, and J. Li, *J. Org. Chem.*, 1997, **62**, 4148–4154.
[38] A.R. Katritzky, S. Ledoux, and S.K. Nair, *J. Org. Chem.*, 2003, **68**, 5728–5730.
[39] J.R. Suresh, P.K. Patra, H. Ila, and H. Junjappa, *Tetrahedron*, 1997, **53**, 14737–14748.
[40] L.H. Phun, J. Aponte-Guzman, and S. France, *Angew. Chem. Int. Ed.*, 2012, **51**, 3198–3202.
[41] G.W. Gribble, (2004) Knorr and Paal-Knorr pyrrole syntheses, in *Name Reactions in Heterocyclic Chemistry* (ed. J.J. Li), John Wiley & Sons, Hoboken, New Jersey, pp. 79–88.
[42] C. Kashima, S. Hibi, T. Maruyama, and Y. Omote, *Tetrahedron Lett.*, 1986, **27**, 2131–2134.
[43] M. Abid, L. Teixeira, and B. Török, *Tetrahedron Lett.*, 2007, **48**, 4047–4050.
[44] M. Abid, O. De Paolis, and B. Török, *Synlett*, 2008, 410–412.
[45] J. Barluenga, H. Vázquez-Villa, A. Ballesteros, and J.M. González, *Adv. Synth. Catal.*, 2005, **347**, 526–530.
[46] J. Barluenga, H. Vázquez-Villa, I. Merino, *et al.*, *Chem. Eur. J.*, 2006, **12**, 5790–5805.
[47] N. Asao and H. Aikawa, *J. Org. Chem.*, 2006, **71**, 5249–5253.

59

Palladium-Catalyzed Cyclization of Pyrroles

Although most of the palladium-catalyzed syntheses of indoles are discussed in later chapters, it is appropriate to present now those few examples that originate from pyrroles.

Murakami and coworkers reported the synthesis of indoles via the Pd-catalyzed cyclization of the 3-acylpyrrole **1** and the acetoxy derivative **2** (Scheme 1, equations 1 and 2) [1]. Once again, we encounter the selective C-3 acylation of a pyrrole with $AlCl_3$ as the Lewis acid. A complementary synthesis of 4-acetoxyindoles **3** was reported by Hidai and colleagues (equation 3) [2]. A dimer byproduct was also formed in lesser yield (up to 28%) when $R = CH_2OMe$. Muratake and Natsume described the use of a Pd-catalyzed indolization en route to their synthesis of (±)-duocarmycin 3A (equation 4) [3].

Lautens and Hulcoop developed a Pd-catalyzed annulation of 3-aryl substituted pyrroles to yield indoles, following retro-Diels–Alder loss of cyclopentadiene (Scheme 2, equations 1 and 2) [4]. Miura, Satoh, and coworkers found that indoles were formed from *N*-methylpyrrole-2 carboxylic acid and alkynes as catalyzed by palladium (equation 3) [5, 6]. The indoles themselves undergo this oxidative coupling with alkynes to form octa-substituted carbazoles, e.g. **4** [6].

Indole Ring Synthesis: From Natural Products to Drug Discovery, First Edition. Gordon W. Gribble.

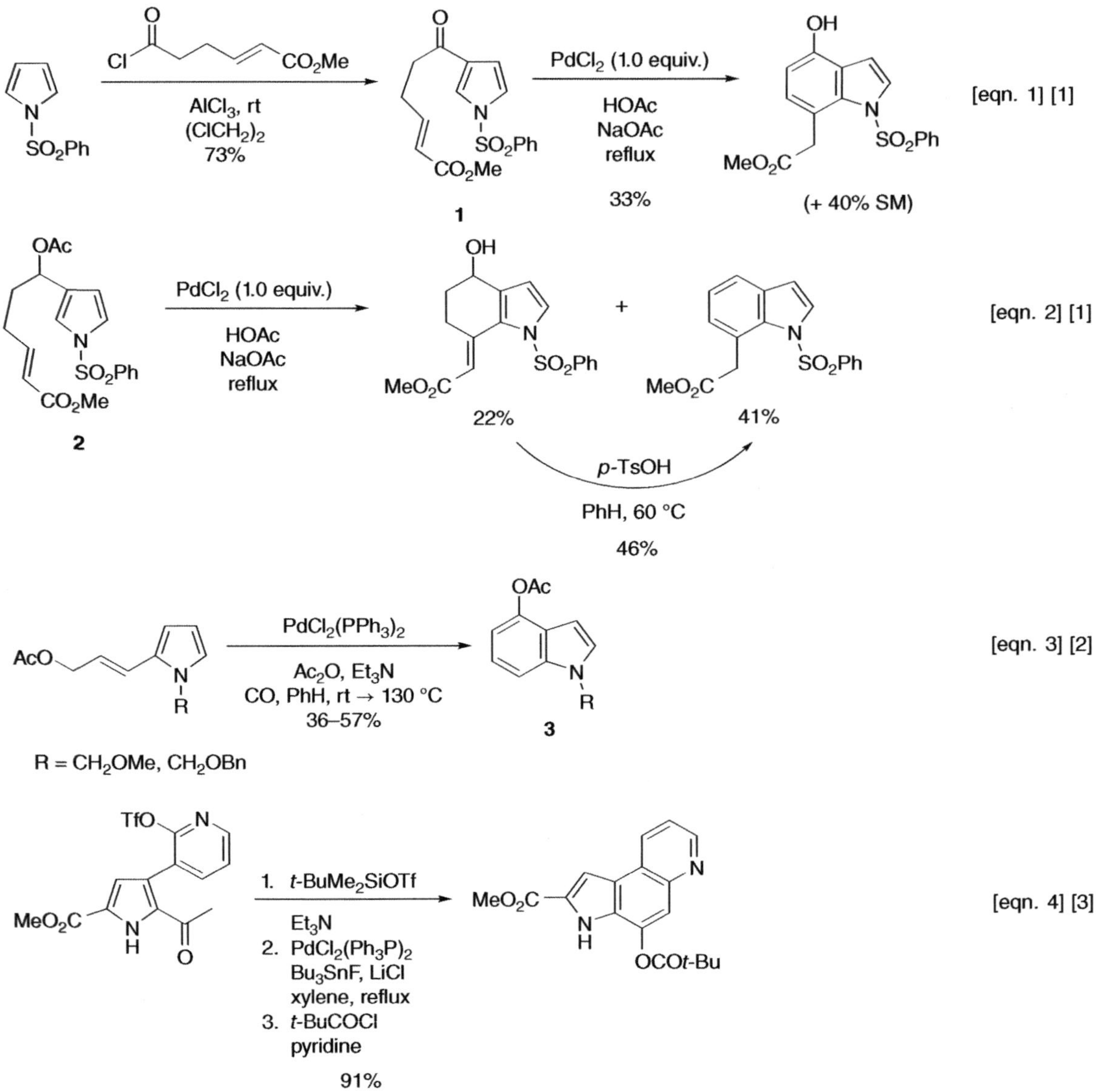

Scheme 1 *Murakami, Hidai, and Muratake-Natsume Indole Syntheses*

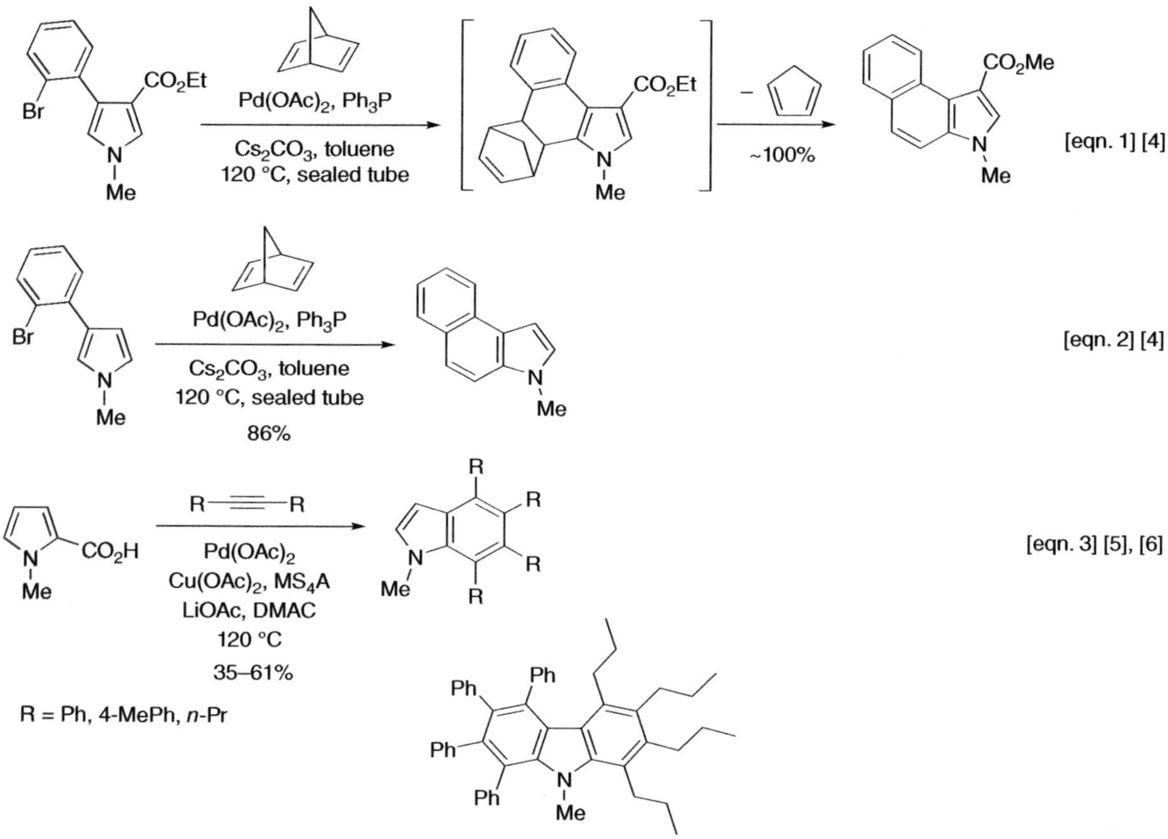

4 (77%)

Scheme 2 *Lautens and Miura-Satoh Indole Syntheses*

References

[1] Y. Yokoyama, H. Suzuki, S. Matsumoto, *et al.*, *Chem. Pharm. Bull.*, 1991, **39**, 2830–2836.
[2] M. Iwasaki, Y. Kobayashi, J.-P. Li, *et al.*, *J. Org. Chem.*, 1991, **56**, 1922–1927.
[3] H. Muratake, M. Tonegawa, and M. Natsume, *Chem. Pharm. Bull.*, 1998, **46**, 400–412.
[4] D.G. Hulcoop and M. Lautens, *Org. Lett.*, 2007, **9**, 1761–1764.
[5] M. Yamashita, K. Hirano, T. Satoh, and M. Miura, *Org. Lett.*, 2009, **11**, 2337–2340.
[6] M. Yamashita, H. Horiguchi, K. Hirano, *et al.*, *J. Org. Chem.*, 2009, **74**, 7481–7488.

60

Cycloaddition Syntheses from Vinyl Pyrroles

Like the Diels–Alder cycloaddition reaction of 2- and 3-vinylindoles to give carbazoles, the cycloaddition of 2- and 3-vinylpyrroles to give indoles is a well-established methodology [1].

An early study was reported by Hiremath and Schneller (Scheme 1, equations 1 and 2) [2], where loss of nitrous acid was the driving force for aromatization. Acheson and Woollard described the reaction between acetylacetylene and pyrrole to form 2-vinylpyrrole **1**, indole **2**, and 1,3,5-triacetylbenzene (equation 3) [3].

Another pioneer in the Diels–Alder reactions of vinylpyrroles was Noland, who also developed the reactions of vinylindoles to yield carbazoles. Some examples of the former are shown in Scheme 2 (equations 1 and 2) [4–7]. Jones and his colleagues were equally active in this cycloaddition chemistry of vinylpyrroles (equations 3 and 4) [8–12]. These workers measured the rates of the reaction between 1-methyl-2-vinylpyrrole and seven dienophiles, with maleic anhydride being 4800 to 50,000 times more reactive than the other dienophiles (DMAD, maleonitrile, fumaronitrile, dimethyl maleate, methyl acrylate, and acrylonitrile) [8]. In a clever tactic to thwart the formation of dihydroindoles, Jones used an excess of methyl propiolate to convert the initial adduct to a second Diels–Alder cycloadduct that subsequently loses ethene by a *retro*-Diels–Alder reaction to afford the dimethyl 1-methyl (phenyl)-4,7-dicarboxylates (equation 4). The reactions are concerted and were consistent with FMO calculations (HOMO[vinylpyrrole]-LUMO[alkene]). The yields are 54% to 81%, but attempts to dehydrogenate the tetrahydroindole products to indoles were unsuccessful. 2-Vinylpyrrole itself undergoes Michael additions and polymerization with these dienophiles. Domingo, Jones, and coworkers subsequently described more *ab initio* calculations on the reaction between 1-methyl-2-vinylpyrroles and DMAD, and they concluded that both a concerted and stepwise mechanism can exist concurrently [11, 12].

An excellent 4-acylindole synthesis using the Diels–Alder cycloaddition of a 2-(2-phenylsulfinylvinyl)pyrrole was described by Muchowski and Scheller (Scheme 3, equation 1) [13]. Desilylation of the TIPS-protected indoles occurred readily (n-Bu_4NF, THF, 0 °C). The 2-vinylpyrrole was prepared from pyrrole in four steps (50% yield). This method provided an excellent synthesis of 4-formylindole ($R^1 = R^2 = H$). Seitz and colleagues reported a synthesis of 4-bromoindole **3** via the cycloaddition of *N*-tosyl-2-vinylpyrrole and tetrabromocyclopropene (equation 2) [14]. Base treatment of **3** afforded 4-bromo-5-formylindole (30%). Eguchi and coworkers used *N*-ethoxycarbonyl-2-[(trimethylsilyl)vinyl]pyrrole as the diene in cycloaddition reactions with dienophiles to give indoles or (usually) tetrahydroindoles (equation 3) [15, 16]. The yield of indole **4** was increased to 28% if the reaction was run under oxygen. Ketcha and Xiao described the Diels–Alder reactions of both 2- and 3-vinyl-1-(phenylsulfonyl)pyrroles with electron-deficient vinylic dienophiles to give tetrahydroindoles [17]. The vinylindoles were crafted in two steps from the *N*-phenylsulfonyl 2- and 3-acetylpyrroles.

Murase and coworkers adopted 3-thioacetylpyrrole **5** as the source of 3-vinylpyrroles in cycloaddition reactions (Scheme 4, equations 1 and 2) [18, 19]. To aromatize the intermediate hydroindoles, if necessary, the authors found that DDQ was effective. The thioacetylpyrrole **5** was prepared from 3-acetyl-1-methylpyrrole with Lawesson's reagent (87%). This novel generation and trapping of the vinylmercaptans was employed by Murase's group in

Indole Ring Synthesis: From Natural Products to Drug Discovery, First Edition. Gordon W. Gribble.

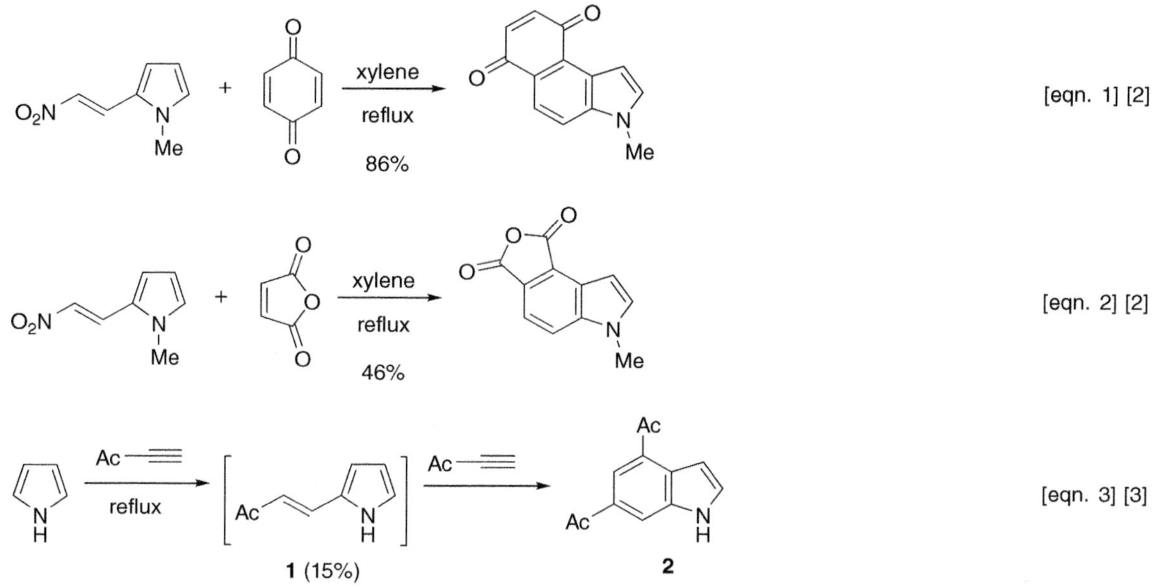

Scheme 1 *Hiremath-Schneller and Acheson Indole Syntheses*

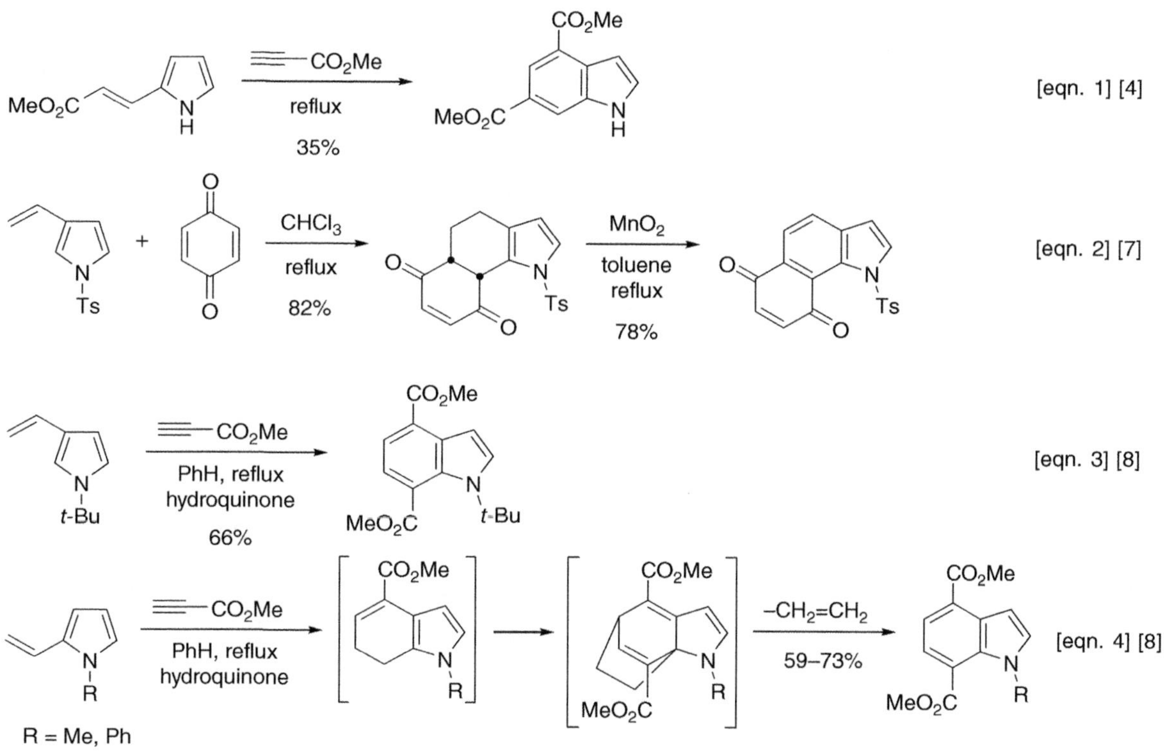

Scheme 2 *Noland and Jones Indole Syntheses*

the synthesis of chuangxinmycin derivatives [19]. Tao and coworkers prepared several pyrroloindoles via a 2-vinylpyrrole Diels–Alder strategy (equation 3) [20].

Yamashita and colleagues reported an indole synthesis from pyrrole–carbene chromium complexes reacting with alkynes (Scheme 5, equation 1) [21]. The acetylation occurred after the 1,6-cycloaddition step. Indoles **6** to **8** were prepared in this fashion, with the latter two in the absence of Ac_2O. The pyrrole carbene complex was prepared from 2-lithio-1-methylpyrrole and chromium

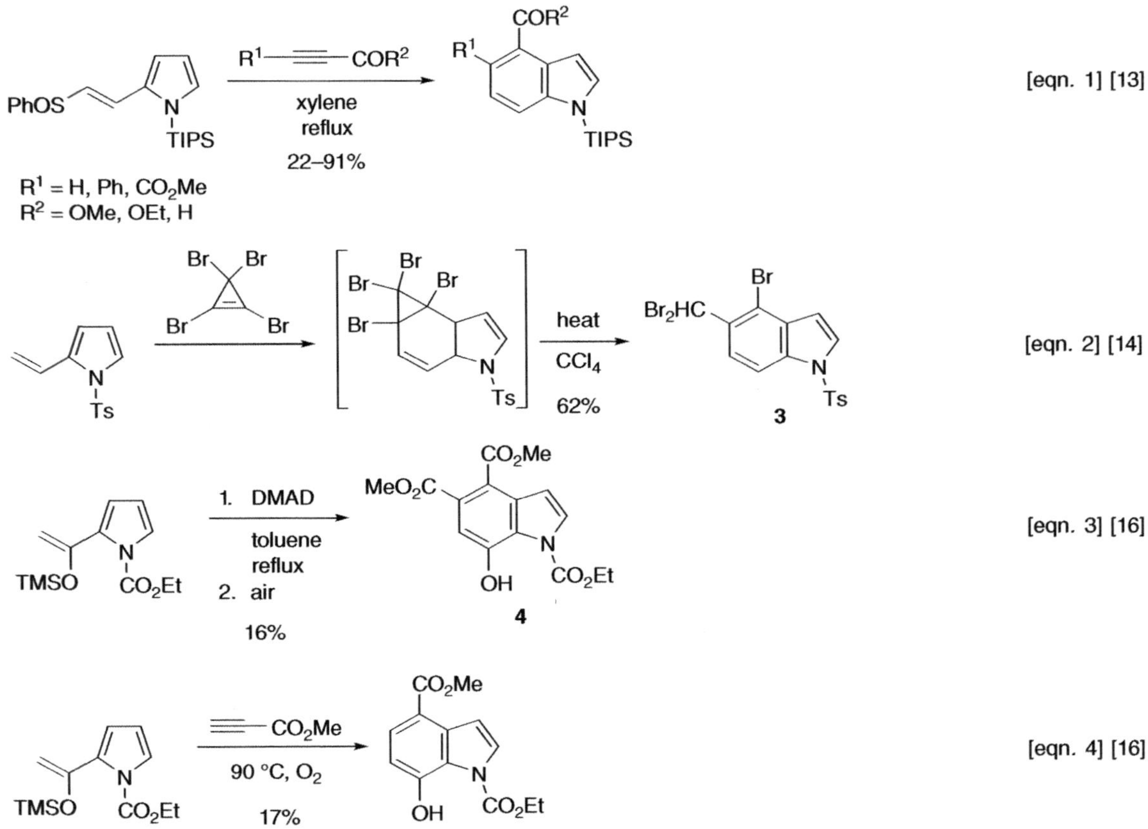

Scheme 3 *Muchowski, Seitz, and Eguchi Indole Syntheses*

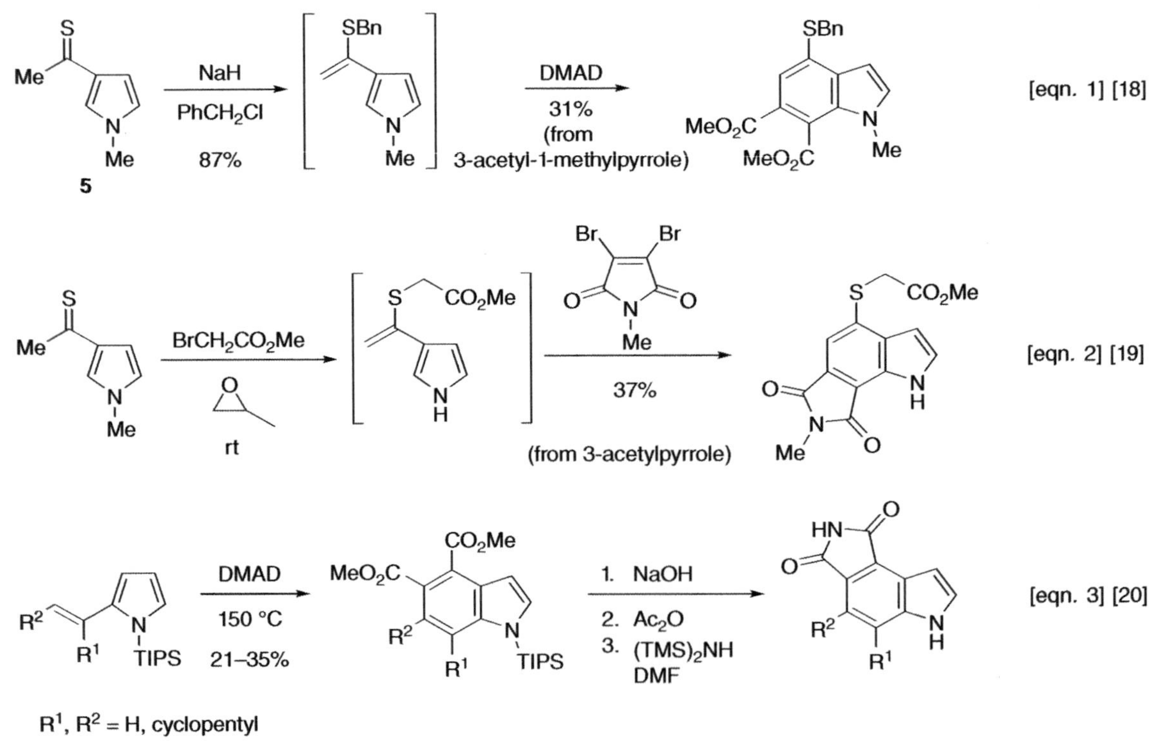

Scheme 4 *Murase and Tao Indole Syntheses*

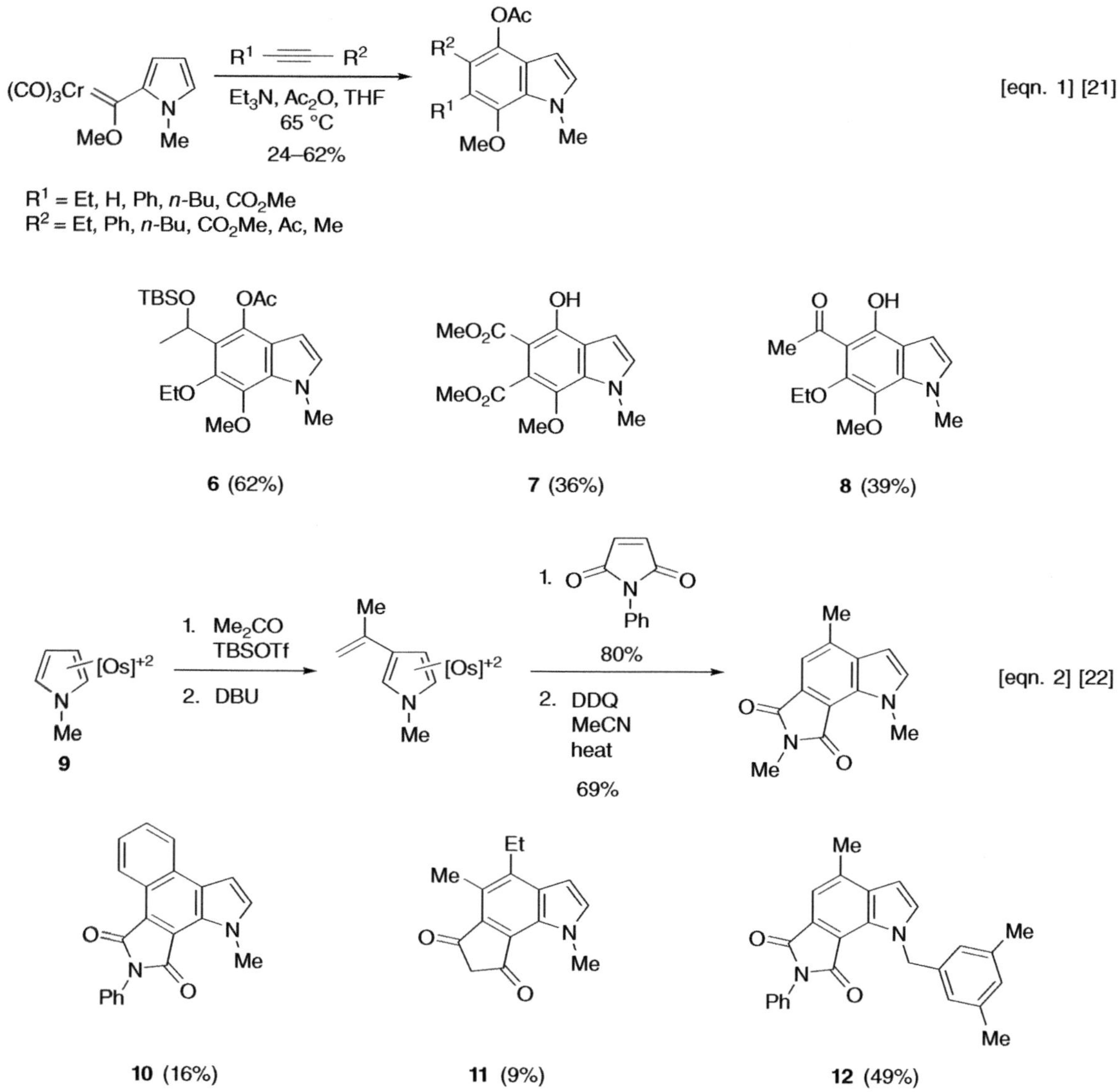

Scheme 5 *Yamashita and Harman Indole Syntheses*

hexacarbonyl followed by methylation with trimethyloxonium tetrafluoroborate (64% yield). Harman and coworkers employed a 2-vinylpyrrole osmium complex in a new indole synthesis (equation 2) [22, 23]. The osmium complex of the cycloaddition adduct was obtained in 80% yield. The starting complex was prepared in 90% to 95% yield from *N*-methylpyrrole and $Os(NH_3)_5(OTf)_3$ in DMAC. Some of the indoles that were prepared in this study are shown, **10** to **12** (overall yields from the pyrrole) [23].

One of the few high-yielding indole syntheses featuring a vinylpyrrole Diels–Alder reaction was reported by Baell and colleagues (Scheme 6, equation 1) [24]. In nifty fashion, using the Jones *retro*-Diels–Alder tactic [8], indole **13** was converted to an indole peptidomimetic (**14**) that inhibits the Kv1.3 potassium channel in T-lymphocytes. The starting 2-vinylindole was prepared from 2-formylindole in two steps (*N*-alkylation; Wittig; 71%). Yoshida and Yanagisawa were the first to use a ring-closing olefin metathesis (RCM) to construct indoles from 3-vinylpyrroles (equations 2, 3, and 4) [25]. The requisite 3-vinylpyrroles were prepared by a sequence of C-3 Pd-catalyzed cross-coupling and functionalization of C-2 (usually an aldehyde), and the Grubbs second-generation ruthenium catalyst was used.

A review on the syntheses of vinylpyrroles is available [26].

Thus, although the yields of indoles from vinylpyrroles can hardly be touted as generally good to excellent, the method does provide rapid access to indoles that might be difficult to obtain other ways. Notably, the RCM approach developed by Yoshida and Yanagisawa is a very promising avenue for future exploration.

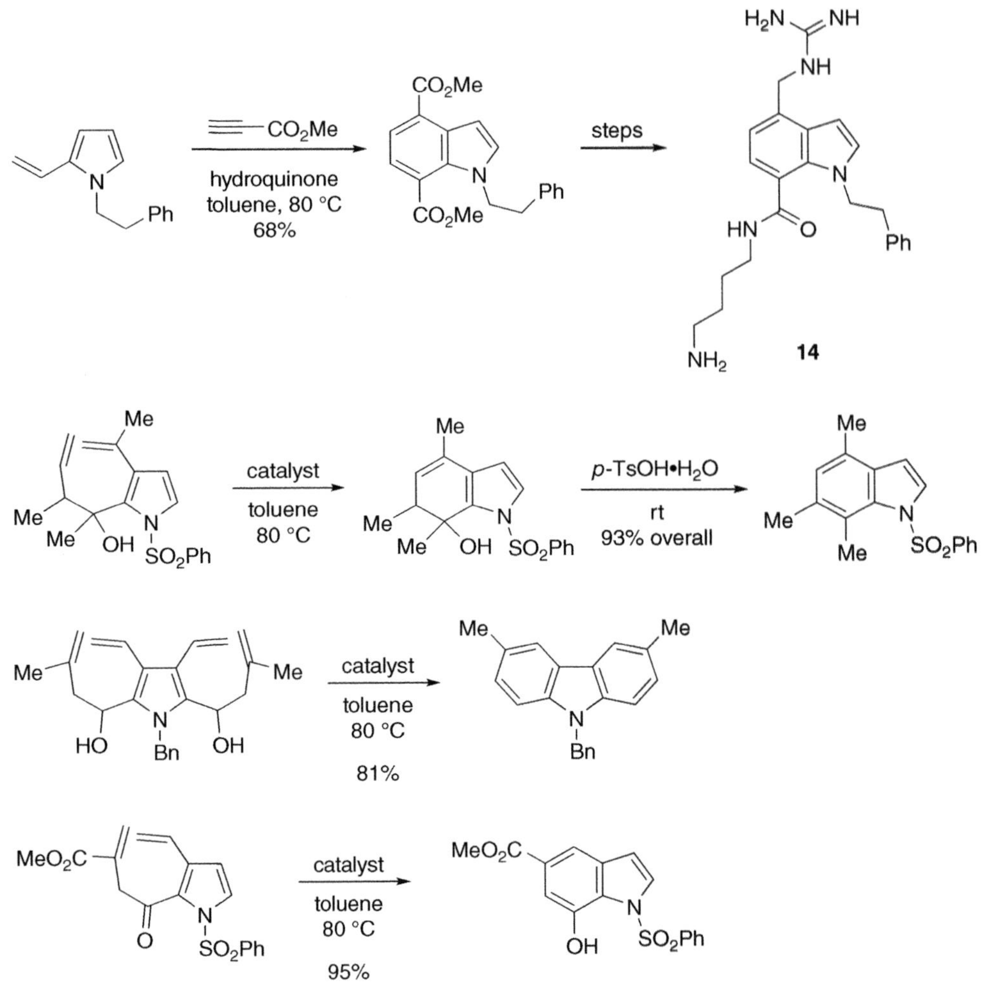

[eqn. 1] [24]

[eqn. 2] [25]

[eqn. 3] [25]

[eqn. 4] [25]

Scheme 6 *Baell and Yoshida-Yanagisawa Indole Syntheses*

References

[1] B.A. Trofimov, L.N. Sobenina, A.P. Demenev, and A.I. Mikhaleva, *Chem. Rev.*, 2004, **104**, 2481–2506.
[2] R.S. Hosmane, S.P. Hiremath, and S.W. Schneller, *J. Chem. Soc., Perkin Trans. 1*, 1973, 2450–2453.
[3] R.M. Acheson and J. Woollard, *J. Chem. Soc., Perkin Trans. 1*, 1975, 446–451.
[4] W.E. Noland, C.K. Lee, S.K. Bae *et al.*, *J. Org. Chem.*, 1983, **48**, 2488–2491.
[5] W.E. Noland and G. Pardi, *J. Heterocycl. Chem.*, 2005, **42**, 1149–1154.
[6] W.E. Noland, N.P. Lanzatella, E.P. Sizova, *et al.*, *J. Heterocycl. Chem.*, 2009, **46**, 503–534.
[7] W.E. Noland and N.P. Lanzatella, *J. Heterocycl. Chem.*, 2009, **46**, 1285–1295.
[8] R.A. Jones, M.T.P. Marriott, W.P. Rosenthal, and J.S. Arques, *J. Org. Chem.*, 1980, **45**, 4515–4519.
[9] R.A. Jones and J.S. Arques, *Tetrahedron*, 1981, **37**, 1597–1599.
[10] R.A. Jones, T.A. Saliente, and J.S. Arques, *J. Chem. Soc., Perkin Trans. 1*, 1984, 2541–2543.
[11] L.R. Domingo, R.A. Jones, M.T. Picher, and J. Sepúlveda-Arques, *Tetrahedron*, 1995, **51**, 8739–8748.
[12] L.R. Domingo, M.T. Picher, J. Andrés, *et al.*, *Tetrahedron*, 1996, **52**, 10693–10704.
[13] J.M. Muchowski and M.E. Scheller, *Tetrahedron Lett.*, 1987, **28**, 3453–3456.
[14] J.-M. Keil, T. Kämpchen, and G. Seitz, *Tetrahedron Lett.*, 1990, **31**, 4581–4584.
[15] M. Ohno, S. Shimizu, and S. Eguchi, *Tetrahedron Lett.*, 1990, **31**, 4613–4616.
[16] M. Ohno, S. Shimizu, and S. Eguchi, *Heterocycles*, 1991, **32**, 1199–1202.
[17] D. Xiao and D.M. Ketcha, *J. Heterocycl. Chem.*, 1995, **32**, 499–503.
[18] M. Murase, S. Yoshida, T. Hosaka, and S. Tobinaga, *Chem. Pharm. Bull.*, 1991, **39**, 489–492.
[19] T. Yoshida, A. Ito, K. Ibusuki, *et al.*, *Chem. Pharm. Bull.*, 2001, **49**, 1198–1202.
[20] M. Tao, C.H. Park, R. Bihovsky, *et al.*, *Bioorg. Med. Chem. Lett.*, 2006, **16**, 938–942.

[21] A. Yamashita, T.A. Scahill, and A. Toy, *Tetrahedron Lett.*, 1985, **26**, 2969–2972.
[22] L.M. Hodges, M.W. Moody, and W.D. Harman, *J. Am. Chem. Soc.*, 1994, **116**, 7931–7932.
[23] L.M. Hodges, M.L. Spera, M.W. Moody, and W.D. Harman, *J. Am. Chem. Soc.*, 1996, **118**, 7117–7127.
[24] A.J. Harvey, R.W. Gable, and J.B. Baell, *Bioorg. Med. Chem. Lett.*, 2005, **15**, 3193–3196.
[25] K. Yoshida, K. Hayashi, and A. Yanagisawa, *Org. Lett.*, 2011, **13**, 4762–4765.
[26] L.N. Sobenina, A.P. Demenev, A.I. Mikhaleva, and B.A. Trofimov, *Russ. Chem. Rev.*, 2002, **71**, 563–591.

61

Electrocyclization of Pyrroles

A variation of indole synthesis involving *inter*molecular Diels–Alder reactions of vinylpyrroles (Chapter 60) is *intra*molecular electrocyclization reactions of vinyl- and divinylpyrroles to give indoles.

Although it is not an indole synthesis *per se*, Weber and Rosen were the first investigators to describe the gas phase vacuum pyrolysis of 1-(2′-*N*-methylpyrryl)-1, 3-butadiene to give 4,5-dihydro-*N*-methylindole (Scheme 1, equation 1) [1]. Similarly, 4,5-dihydro-4,*N*-dimethylindole was prepared in 56% yield. van Leusen and his colleagues extensively developed an indole synthesis that featured the thermal or photochemical cyclization of 2,3-divinylpyrroles to indoles (equations 2–4) [2–5]. The requisite pyrroles were assembled by the base-induced condensation of unsaturated ketones and esters with 1-tosylalk-1-enyl isocyanides. The overall yields were as high as 98%. The dehydrogenation of the dihydroindole product formed in equation 3 is effected by the nitrobenzene solvent (similar to the Skraup quinoline synthesis) [4, 5]. Similarly, nitroindoles **1–3** were synthesized by van Leusen [5].

As cited in Chapter 56, Hendrickson and DeVries were the first to photocyclize a pyrrole-stilbene to form an indole en route to the coenzyme methoxatin [6]. A similar photocyclization of dithienylpyrroles leading to phenyldithienoindoles was described by Kagan and coworkers (Scheme 2, equation 1) [6]. Also prepared were the three other analogues with isomeric thieno rings (28%–47% yields). As illustrated in Chapter 56, Tojo and colleagues extended the photocyclization of pyrrole-stilbenes to give CC-1065 analogues [8]. The groups of Oda [9] and Ho [10] reported similar photocyclizations of styrylpyrroles to form indoles. In particular, Oda, Machida and coworkers converted 2-phenyl-3-furanylpyrrole to fused indole **4** (equation 2) [9]. Shannon and Chunchatprasert employed classical Vilsmeier chemistry to construct pyrrolo[3,2-*f*] indoles and novel pyrrolo[3,2-*f*;4,5-*f*′]diindoles; an example of the former is shown (equation 3) [11]. Vedejs and Kim found that vinylpyrrole **5** is transformed to indole **7** along with an isomeric indolizine (equation 4) [12]. The reaction proceeds via vinyl ketene **6** that can cyclize two different ways.

Following their earlier work on the electrocyclization of indole-based allenes to give carbazoles [13, 14], Hibino and colleagues stretched this methodology to the synthesis of indoles from allenylic pyrroles (Scheme 3, equations 1 and 2) [15, 16]. Both routes were used to prepare indole-4,7-quinones, some of which showed antitumor activity against the human cell lines NCI-H460 (lung) and MDA-MB-231 (breast). For example, *N*-phenylsulfonyl-5-methylindole-4, 7-quinone was more active than cisplatin toward MDA-MB-231 [16]. Townsend and Outlaw synthesized a series of 7-aminoindoles from pyrrole-3-carboxyaldehydes via the intermediate Wittig vinylpyrroles (**8**) and a Lewis acid–catalyzed cyclization (equation 3) [17]. Functionalization of **8** (LDA, RX) provided the synthesis of 6-substituted analogues of indole **9** (Me, Et, benzyl, allyl, $CH_2C{\equiv}CH$, CH_2CO_2Et). Langer and Toguem converted 2,4,5-tribromo-*N*-methylpyrrole to 2,5,6-trisubstituted *N*-methylindoles via a sequence of Suzuki, Heck, and electrocyclization reactions (equation 4) [18]. The regioselectivity of the Suzuki reaction is noteworthy, and the conversion of the 2,3-divinylpyrroles to indoles (6π-electrocyclization, dehydrogenation) is in one pot.

Funk and coworkers also employed 6π-electrocyclizations to fashion indoles (Scheme 4, equation 1) [19, 20]. Thus, in total syntheses of (±)-*cis*-trikentrin A and B,

Indole Ring Synthesis: From Natural Products to Drug Discovery, First Edition. Gordon W. Gribble.

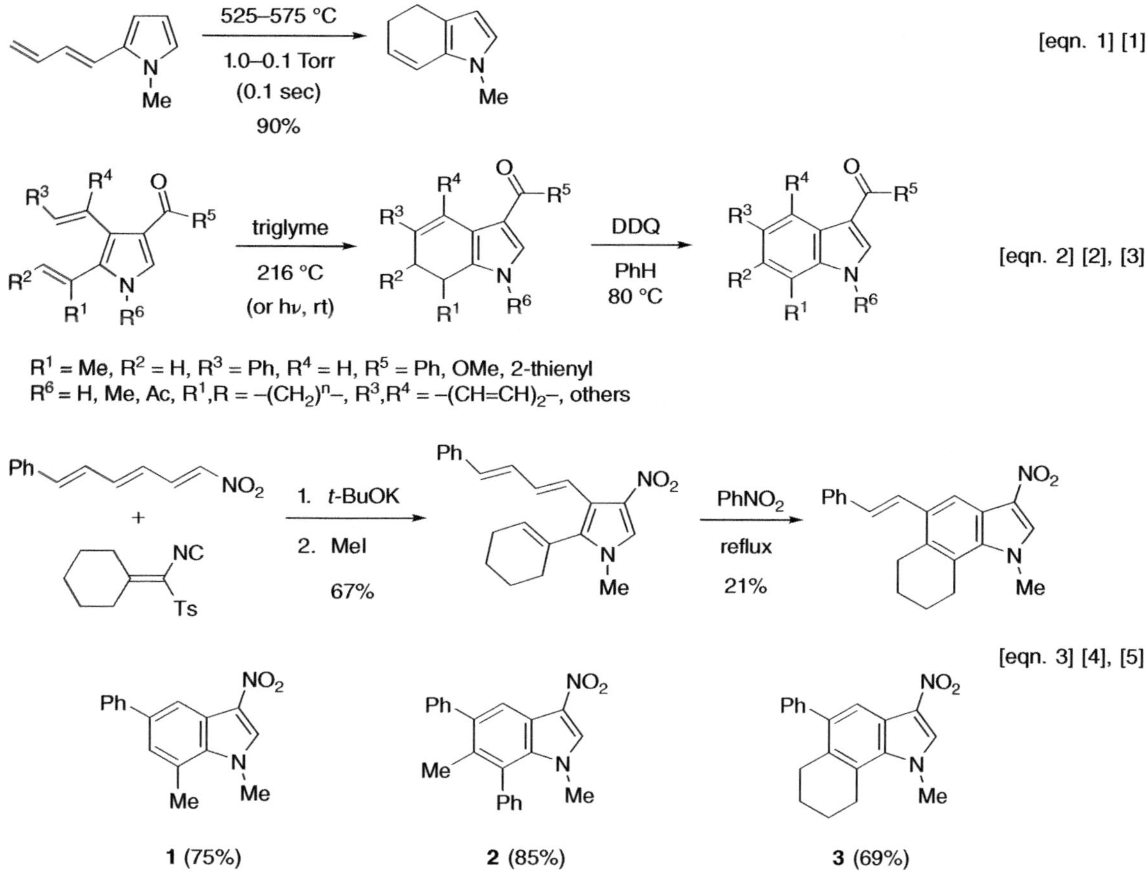

Scheme 1 *Weber and van Leusen Indole Syntheses*

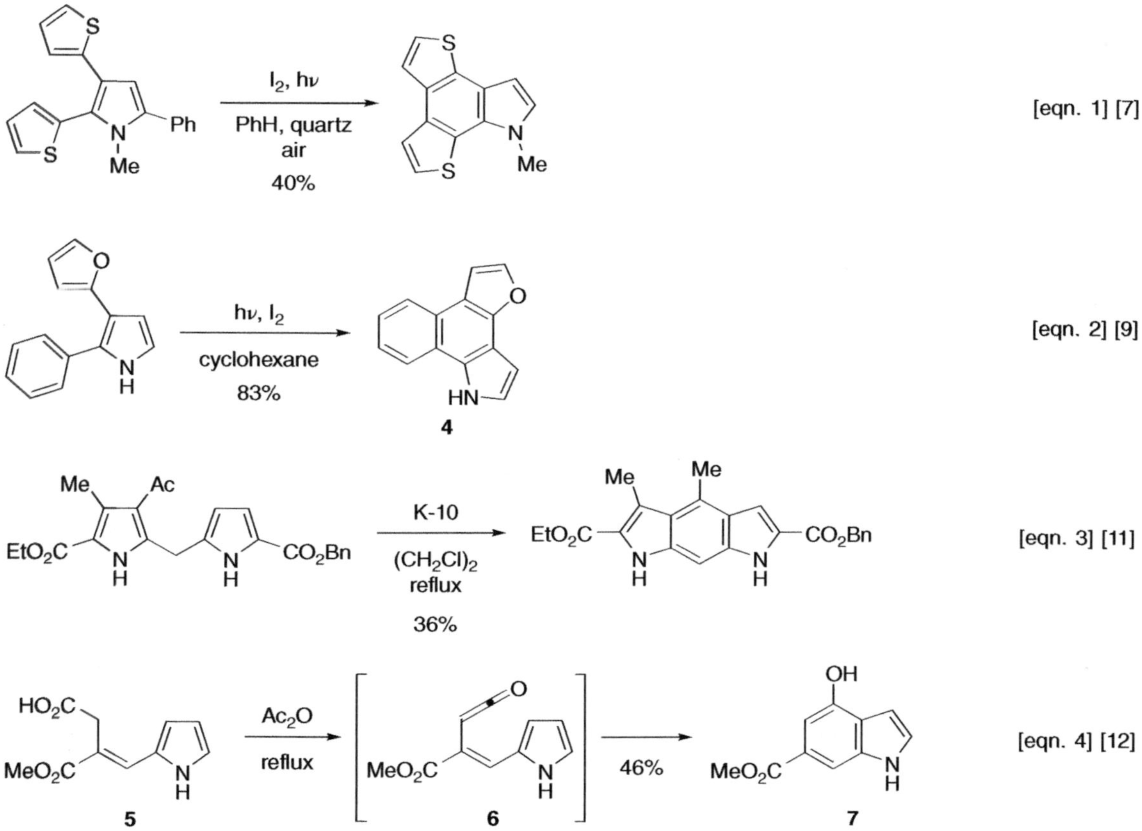

Scheme 2 *Kagan, Oda-Machida, Shannon, and Vedejs Indole Syntheses*

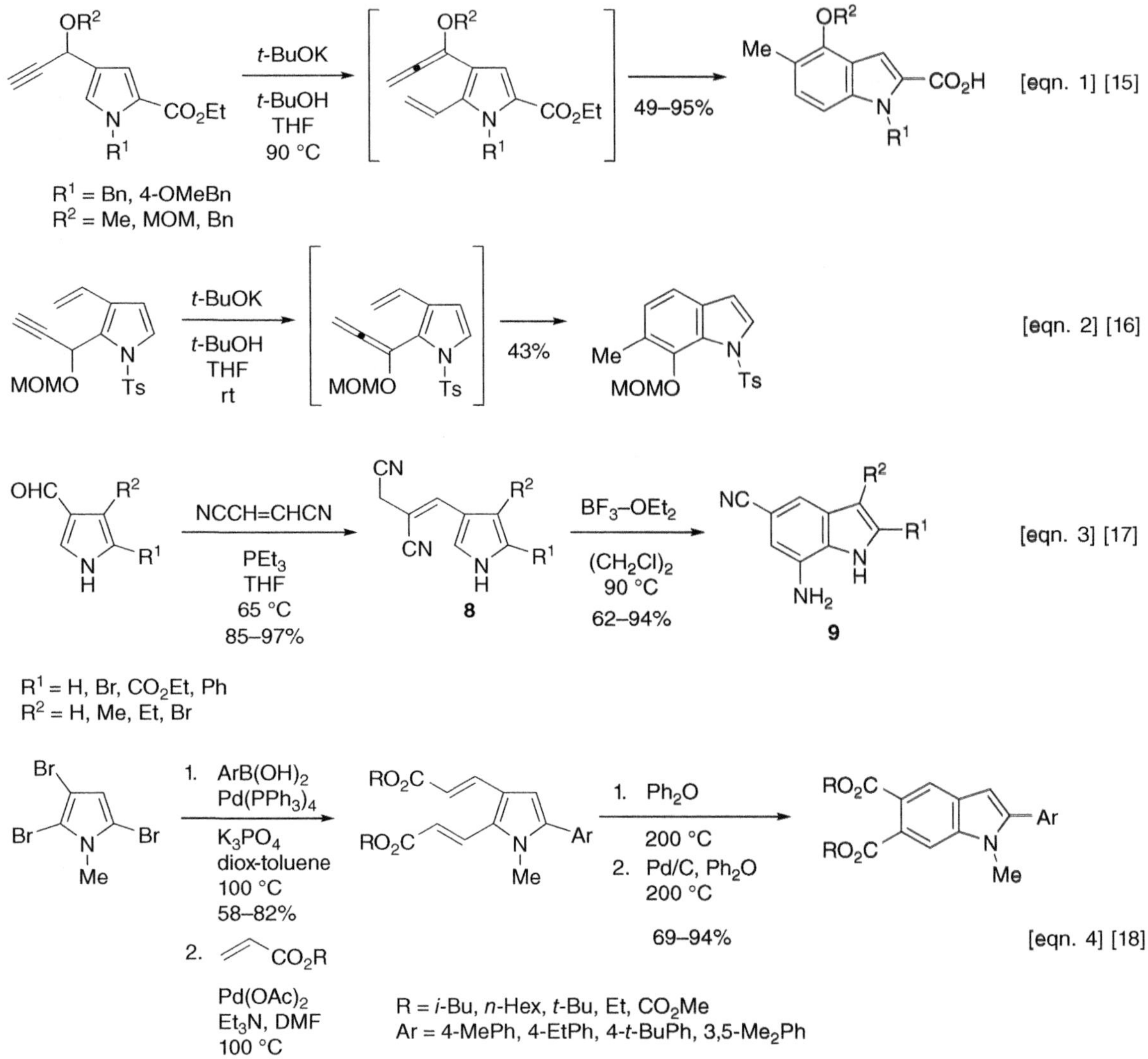

Scheme 3 Hibino, Townsend, and Langer Indole Syntheses

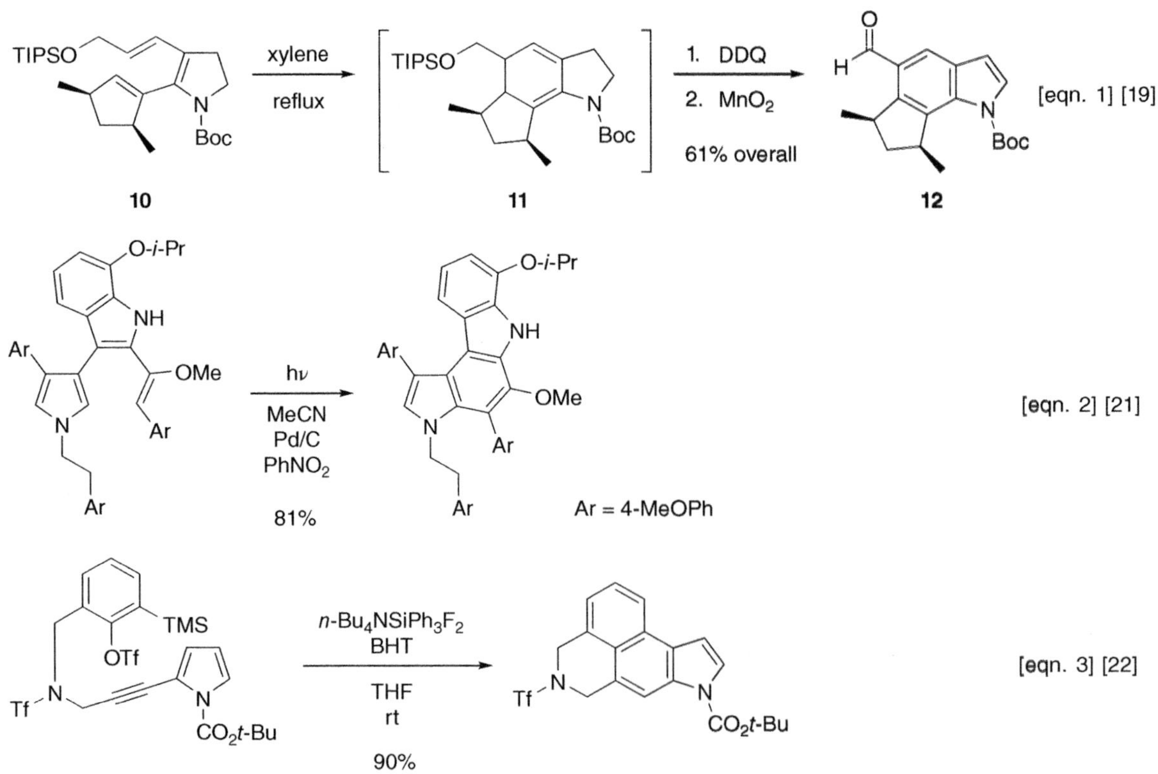

Scheme 4 Funk, Fürstner, and Danheiser Indole Syntheses

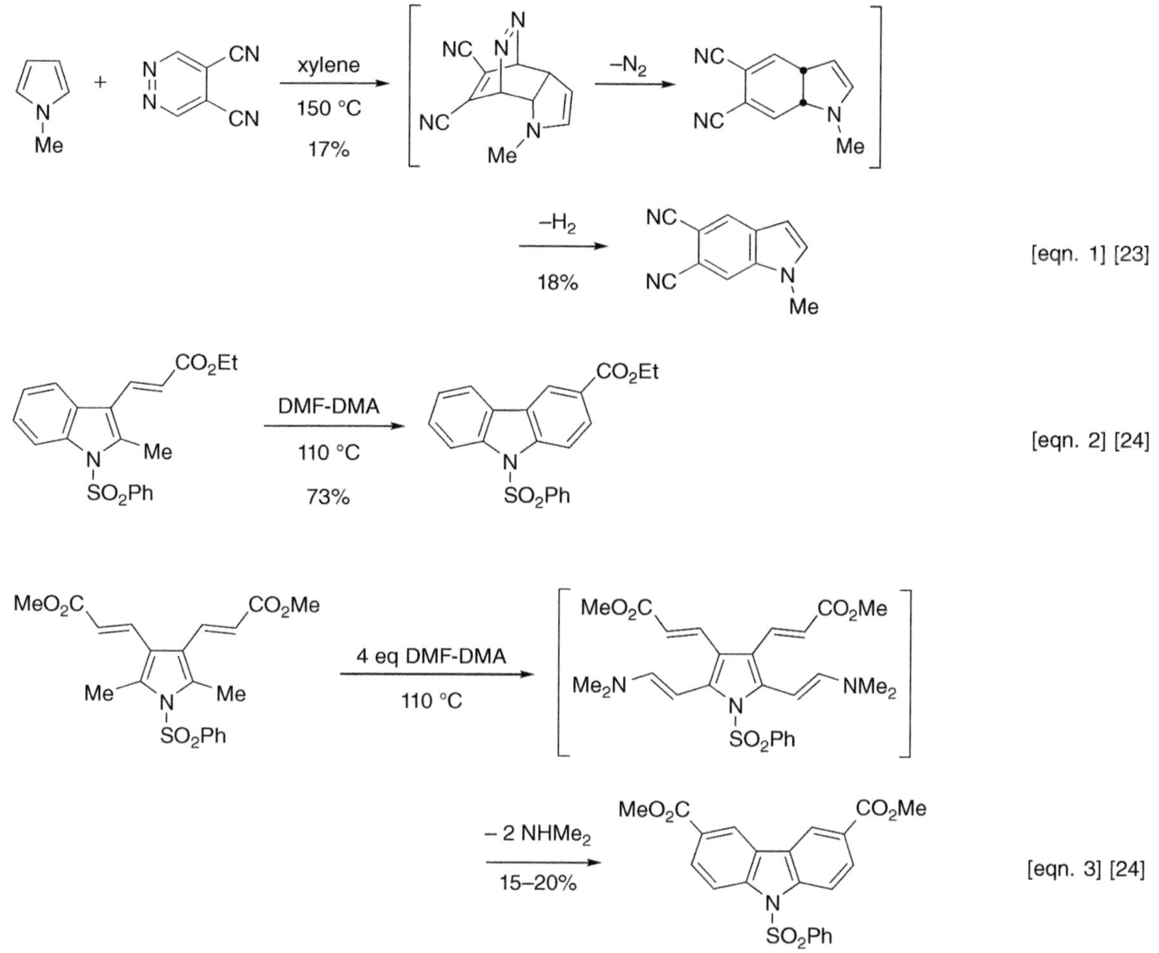

Scheme 5 *Giomi and Mohanakrishnan Indole and Carbazole Syntheses*

Funk and Huntley converted divinylpyrrolidine **10** to diene **11**, and thence to an indole precursor **12** of (±)-*cis*-trikentrin B. A similar sequence was used to synthesize (±)-*cis*-trikentrin B. Dihydropyrrole **10** was assembled in five steps from *N*-Boc-2-pyrrolidinone. Fürstner and coworkers synthesized several pyrrolo[2,3-*c*]carbazole alkaloids, the dictyodendrins B, C, and E, that featured a photochemical 6π-electrocyclization and *in situ* dehydrogenation (equation 2) [21]. Danheiser and colleagues effected an intramolecular benzyne generation and cycloaddition onto an alkynylpyrrole to afford a polycyclic indole (equation 3) [22].

A different mode of a [4+2] cycloaddition with *N*-methylpyrrole was employed by Giomi and Cecchi to prepare 5,6-dicyano-*N*-methylindole albeit in low yield (Scheme 5, equation 1) [23]. Attempts to increase the yield with CuI led only to pyrrole electrophilic addition products with the 4,5-dicyanopyrazine. This pyrazine reacted with both indole and *N*-methylindole to give the corresponding 2,3-dicyanocarbazoles (57% and 62%, respectively). Attempts by Mohanakrishnan and Balamurugan to expand their 2,3-divinylindole-to-carbazole synthesis (equation 2) to similar twin reactions of *in situ*-generated 2,3-divinylpyrroles were only marginally successful (equation 3) [24].

References

[1] B.I. Rosen and W.P. Weber, *Tetrahedron Lett.*, 1977, **18**, 151–154.
[2] J. Moskal, R. van Stralen, D. Postma, and A.M. van Leusen, *Tetrahedron Lett.*, 1986, **27**, 2173–2176.
[3] J. Moskal and A.M. van Leusen, *J. Org. Chem.*, 1986, **51**, 4131–4139.
[4] F.R. Leusink, R. ten Have, K.J. van den Berg, and A.M. van Leusen, *J. Chem. Soc., Chem. Commun.*, 1992, 1401–1402.
[5] R. ten Have and A.M. van Leusen, *Tetrahedron*, 1998, **54**, 1913–1920.

[6] J.B. Hendrickson and J.G. DeVries, *J. Org. Chem.*, 1982, **47**, 1148–1150.
[7] D.M. Perrine, J. Kagan, D.-B. Huang, *et al.*, *J. Org. Chem.*, 1987, **52**, 2213–2216.
[8] A.G. Neo, A. Pérez, C. López, *et al.*, *J. Org. Chem.*, 2009, **74**, 3203–3206.
[9] K. Oda, H. Tsujita, M. Sakai, and M. Machida, *Chem. Pharm. Bull.*, 1998, **46**, 1522–1526.
[10] J.-Y. Wu, J.-H. Ho, S.-M. Shih, *et al.*, *Org. Lett.*, 1999, **1**, 1039–1041.
[11] L. Chunchatprasert and P.V.R. Shannon, *J. Chem. Soc., Perkin Trans. 1*, 1996, 1787–1795.
[12] M. Kim and E. Vedejs, *J. Org. Chem.*, 2004, **69**, 6945–6948.
[13] T. Choshi, T. Sada, H. Fujimoto, *et al.*, *Tetrahedron Lett.*, 1996, **37**, 2593–2596.
[14] H. Hagiwara, T. Choshi, J. Nobuhiro, *et al.*, *Chem. Pharm. Bull.*, 2001, **49**, 881–886.
[15] M. Hirayama, T. Choshi, T. Kumemura, *et al.*, *Heterocycles*, 2004, **63**, 1765–1770.
[16] J. Nobuhiro, M. Hirayama, T. Choshi, *et al.*, *Heterocycles*, 2006, **70**, 491–499.
[17] V.K. Outlaw and C.A. Townsend, *Org. Lett.*, 2014, **16**, 6334–6337.
[18] S.-M.T. Toguem and P. Langer, *Synlett*, 2011, 513–516.
[19] R.J. Huntley and R.L. Funk, *Org. Lett.*, 2006, **8**, 3403–3406.
[20] T.J. Greshock and R.L. Funk, *J. Am. Chem. Soc.*, 2006, **128**, 4946–4947.
[21] A. Fürstner, M.M. Domostoj, and B. Scheiper, *J. Am. Chem. Soc.*, 2006, **128**, 8087–8094.
[22] M.E. Hayes, H. Shinokubo, and R.L. Danheiser, *Org. Lett.*, 2005, **7**, 3917–3920.
[23] D. Giomi and M. Cecchi, *Tetrahedron*, 2002, **58**, 8067–8071.
[24] A.K. Mohanakrishnan and R. Balamurugan, *Tetrahedron Lett.*, 2005, **46**, 4045–4048.

62

Indoles from Pyrrolo-2,3-Quinodimethanes

The extraordinary versatility of indole-2,3-quinodimethanes in the Diels–Alder cycloaddition to give carbazoles is presented in Chapter 54. Not surprisingly, several investigators have adapted this strategy to the generation and Diels–Alder trapping of pyrrole-2,3-quinodimethanes (2,3-dimethylene pyrroles) and fused analogues to a synthesis of indoles. Although these Diels–Alder reactions could have been included in Chapter 60, it was deemed appropriate that they occupy a separate presentation.

Sha and colleagues synthesized several examples of the novel (isolable) 2,4-dihydropyrrolo[3,4-*b*]pyrrole ring system and found that the Diels–Alder reaction with DMAD occurred smoothly (Scheme 1, equation 1) [1]. Subsequent deamination using the Gribble and Allen method [2]

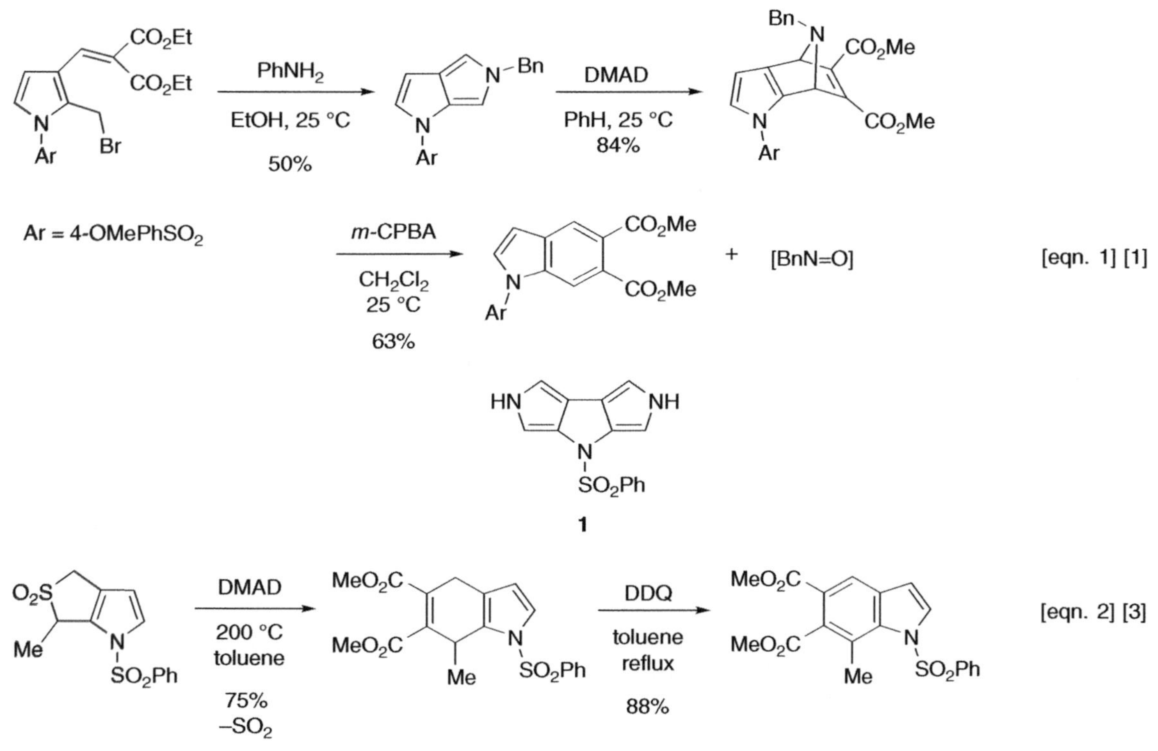

Scheme 1 *Sha and Chou Indole Syntheses*

Indole Ring Synthesis: From Natural Products to Drug Discovery, First Edition. Gordon W. Gribble.

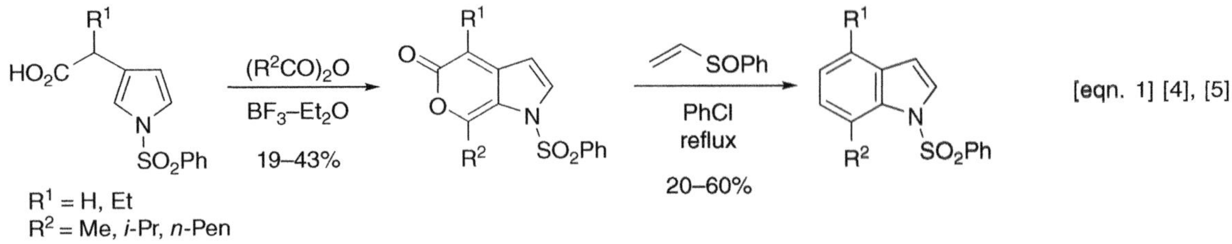

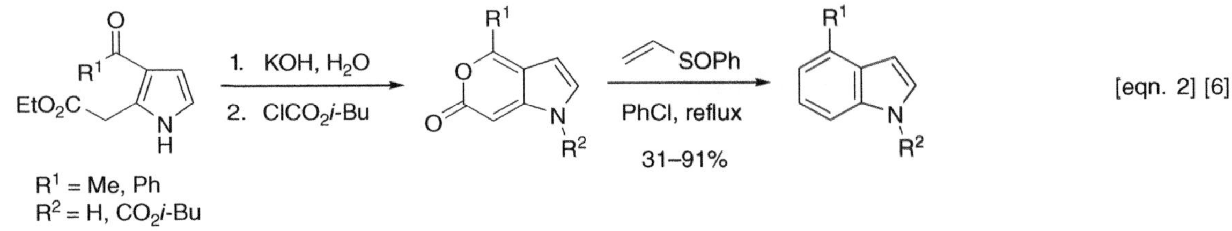

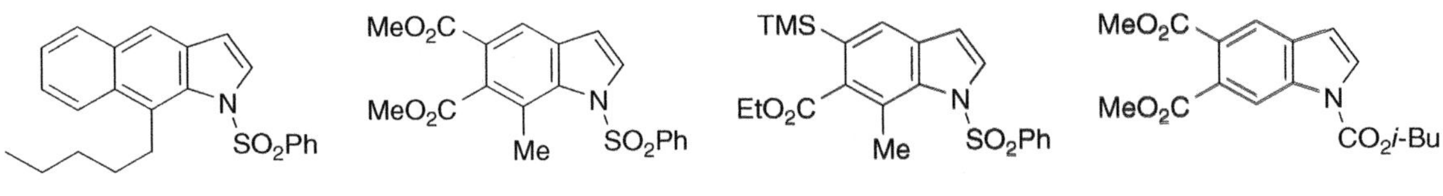

3 (77%) [4] **4** (70%) [4] **5** (53%) [4] **6** (99%) [5]

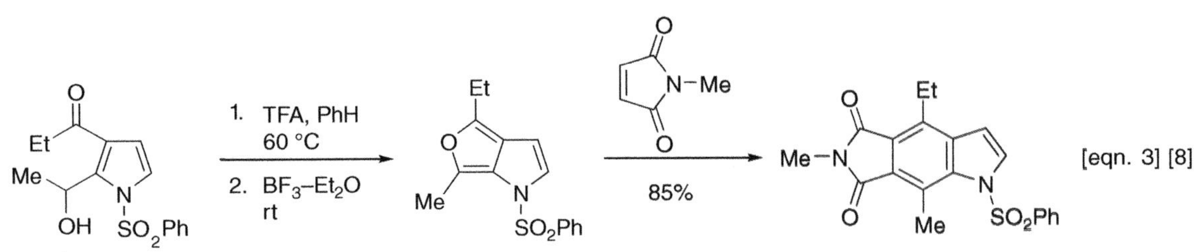

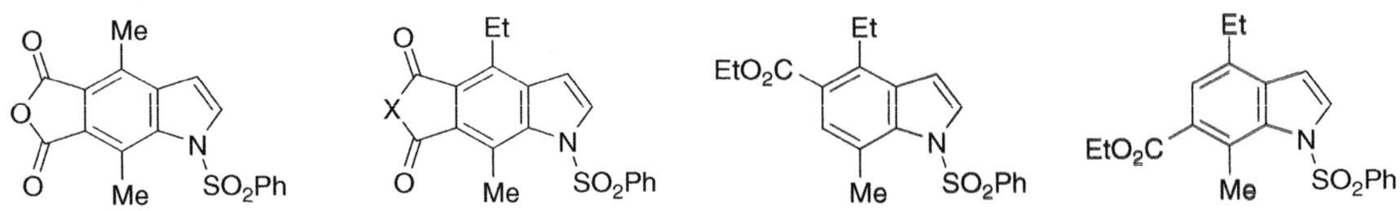

8 (79%) **9**, X = NPh (71%) **10**, X = O (70%) **11** **12**

Scheme 2 *Moody and Gribble Indole Syntheses*

yielded the 5,6-dicarbomethoxyindoles. The starting 3-vinylpyrrole was prepared in a few steps from a known pyrrole-3-carboxaldehyde. These workers also prepared the stable 4,6-dihydro-2*H*-dipyrrolo[3,4-*b*:3′,4″-*d*]pyrrole (**1**), but further chemistry of this novel ring system was not disclosed. Chou and Chang reported the generation and trapping of a pyrrolo-3-sulfolene to give, after a final oxidation, indoles (equation 2) [3].

In a series of papers, Moody and coworkers described the synthesis and Diels–Alder reactions of 1,5-dihydropyrano[3,4-*b*]pyrrol-5-ones and the isomeric 1,6-dihydropyrano[4,3-*b*]pyrrol-6-ones to give indoles (Scheme 2, equations 1 and 2) [4–7]. Other indoles prepared by Moody in these reactions are **3–6**. Unsymmetrical alkynes often gave mixtures of adducts (exception: **5**). The isobutyl carbamate protecting group was removed with NH_3/pyridine. Several cyclopent[*b*]indoles were also synthesized by Moody and colleagues [7]. Gribble and Moskalev synthesized the furo[3,4-*b*]indole ring system and effected Diels–Alder reactions to give indoles after spontaneous dehydration of the adduct (equation 3) [8]. Although somewhat unstable, furopyrrole **7** could be

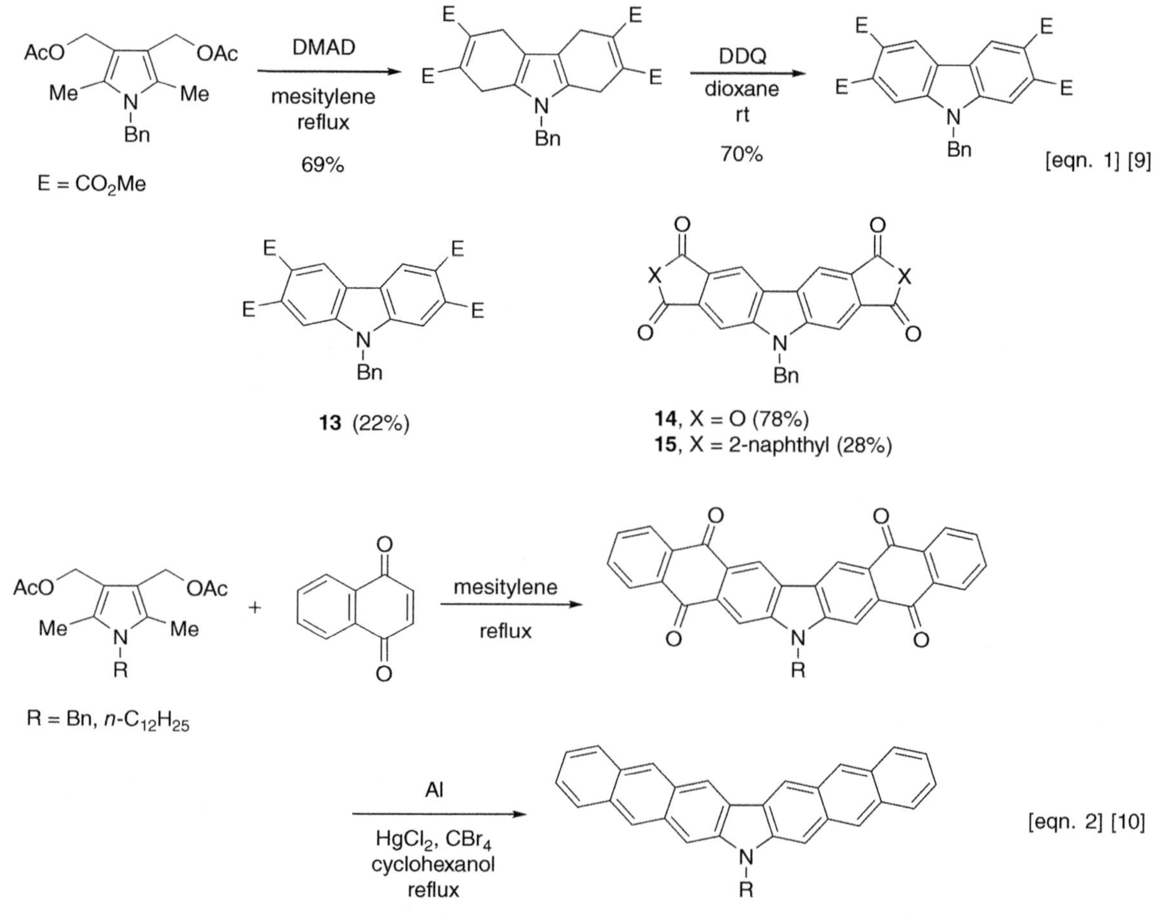

Scheme 3 *Petillo Carbazole Synthesis*

isolated in 71% yield. Indoles **8–12** were also prepared in this study. Ethyl acrylate afforded indoles **11** and **12** in a 1.5:1 ratio (50% yield).

Petillo and colleagues generated a formal bis-pyrrole-2,3-quinodimethane that underwent two subsequent Diels–Alder cycloaddition reactions to form carbazoles after DDQ oxidation (Scheme 3, equation 1) [9]. Carbazoles **13–15** were prepared similarly, among others. The olefinic dienophiles required more extensive dehydrogenation of the adducts with DDQ. Wudl and colleagues employed the Petillo methodology to craft novel azaheptacenes (equation 2) [10].

References

[1] C.-K. Sha, J.-M. Liu, R.-K. Chiang, and S.-L. Wang, *Heterocycles*, 1990, **31**, 603–609.
[2] G.W. Gribble, R.W. Allen, P.A. Anderson, *et al., Tetrahedron Lett.*, 1976, **17**, 3673-3676.
[3] T. Chou and R.-C. Chang, *J. Chem. Soc., Chem. Commun.*, 1992, 549–551.
[4] P.M. Jackson and C.J. Moody, *J. Chem. Soc., Perkin Trans. 1*, 1990, 2156–2158.
[5] P.M. Jackson and C.J. Moody, *Tetrahedron*, 1992, **48**, 7447–7466.
[6] J.F.P. Andrews, P.M. Jackson, and C.J. Moody, *Tetrahedron*, 1993, **49**, 7353–7372.
[7] C.-A. Harrison, P.M. Jackson, C.J. Moody, and J.M.J. Williams, *J. Chem. Soc., Perkin Trans. 1*, 1995, 1131–1136.
[8] N.V. Moskalev and G.W. Gribble, *Tetrahedron Lett.*, 2002, **43**, 197–201.
[9] J.T. Vessels, S.Z. Janicki, and P.A. Petillo, *Org. Lett.*, 2000, **2**, 73–76.
[10] T.V. Pho, J.D. Yuen, J.A. Kurzman, *et al., J. Am. Chem. Soc.*, 2012, **134**, 18185–18188.

63

Indoles via Dehydrogenation of Pyrroles

Whereas the dehydrogenation of pyrroles to indoles has been described sporadically in preceding chapters, the present chapter covers those examples from tetrahydroindoles, such as the often encountered 4-keto-4,5,6,7-tetrahydroindole. The dehydrogenation of indolines (2,3-dihydroindoles) to indoles is covered in a later chapter.

Remers and Weiss reported a versatile indole synthesis via 4-keto-4,5,6,7-tetrahydroindoles (Scheme 1, equations 1 and 2) [1–3], which were prepared by an early method involving 1,3-cyclohexanediones, α-haloketones, and ammonia or primary amines (the Stetter synthesis) [4]. Thus, following elaboration of the ketotetrahydroindole, dehydrogenation was accomplished with Pd/C or DDQ. Repke and colleagues used 10% Pd/C in refluxing *p*-cymene to dehydrogenate 4-keto-4,5,6,7-tetrahydroindole to 4-hydroxyindole in 77% yield [5], and Torii and coworkers obtained a 69% yield of 4-hydroxyindole using the same conditions [6]. Baker and Lebar obtained a 49% yield of 4-hydroxyindole using 10% Pd/C in refluxing diisobutyl ketone (190 °C) [7]. In contrast, Edstrom employed DDQ in refluxing ethyl acetate or benzene to dehydrogenate ketopyrroles **1** and **2** to their respective 4-hydroxyindoles in 72% and 82% yields (Scheme 1) [8, 9]. Likewise, Nikolaropoulos and colleagues found DDQ to be the oxidant of choice for the conversion of dihydroindoles **3** to the corresponding indoles [10] in work patterned after Remers [2]. Mori and Doi converted *N*-tosyl-4-keto-4,5,6,7-tetrahydroindole to an enol triflate **4** and subsequently to the 4-carbomethoxyindole (**5**) using 10% Pd/C (equation 3) along with other 4-substituted indoles featuring Stille and Suzuki reactions [11]. Taddei's group reported a microwave-assisted Stetter synthesis of 4-ketotetrahydroindoles [12].

Davies and Manning engineered a unique synthesis of 4-substituted indoles that involved the rhodium(II)-catalyzed C-H activation of 4-acetoxy-6,7-dihydroindole **6**, followed by a Cope rearrangement and elimination sequence (Scheme 2, equation 1) [13]. This notion of employing a leaving group to aromatize the dihydroindole was achieved by other investigators. Thus, Matsumoto and colleagues used a halogenation tactic to prepare 4-hydroxyindoles from 4-keto-4,5,6,7-tetrahydroindoles (equation 2) [14]. Matsumoto and Hatanaka prepared the halo-substituted cyanohydrin silyl ethers **7**, which could be dehydrohalogenated to afford 4-cyanoindoles (equation 3) [15]. The cyanohydrin silyl ether of 4-keto-4,5,6,7-tetrahydroindole was also converted to 4-cyanoindole by acid-catalyzed elimination of TMSOH (84%) and then either DDQ dehydrogenation (93%) or chlorination and dehydrochlorination (53%) [15].

Murakami and coworkers invoked a similar halogenation/dehydrohalogenation tactic to prepare 7-hydroxyindoles from 7-keto-4,5,6,7-tetrahydroindole **8** (Scheme 3, equation 1) [16]. An appropriate Wittig reaction on **8** followed by bromination/elimination gave, for example, ethyl 7-cyanomethylindole-2-carboxylate (63% overall). Tobinaga and colleagues developed a simple synthesis of 7-alkylthioindoles that featured a final DDQ dehydrogenation (equation 2) [17]. This method was used in a formal synthesis of chuangxinmycin. Saito's team developed a 4-acetoxyindole synthesis involving air oxidation (equation 3) [18]. A new synthesis of *N*-substituted 7-keto-4,5,6,7-tetrahydroindoles was described by Montalban and coworkers, and these compounds along with the 4-keto analogues were aromatized to the 7- and 4-hydroxyindoles via a halogenation/dehydrohalogenation protocol (equation 4) [19].

Indole Ring Synthesis: From Natural Products to Drug Discovery, First Edition. Gordon W. Gribble.

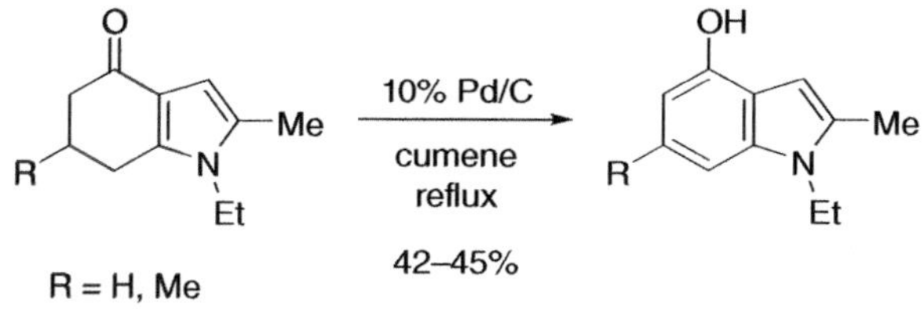

[eqn. 1] [1]

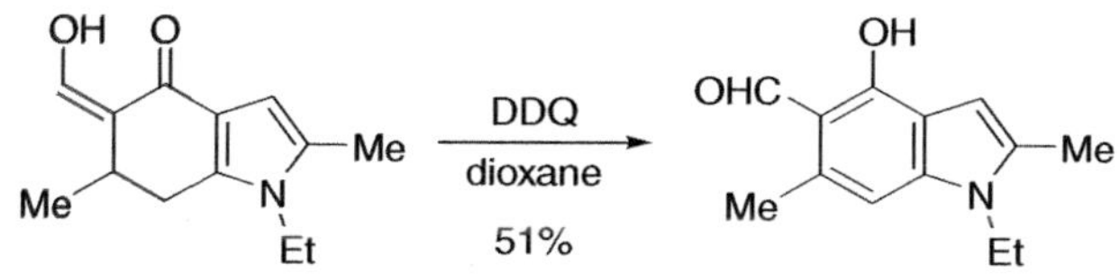

[eqn. 2] [1]

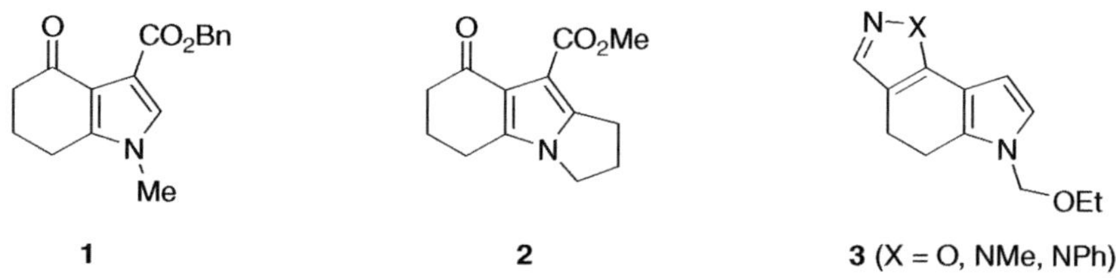

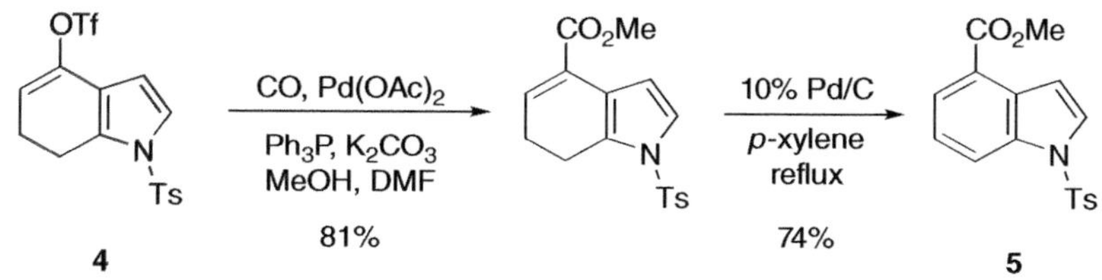

[eqn. 3] [11]

Scheme 1 *Remers Indole Synthesis*

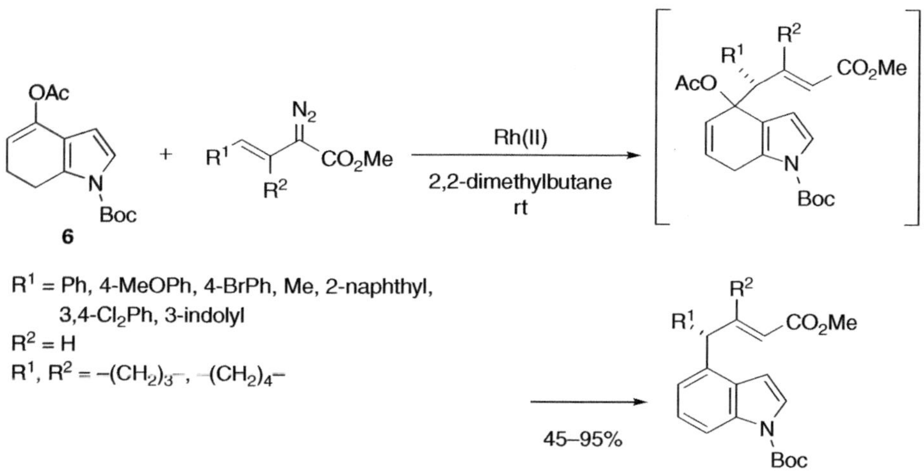

[eqn. 1] [13]

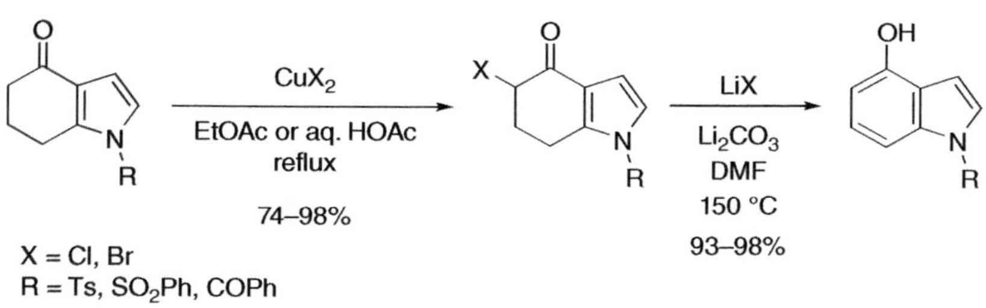

[eqn. 2] [14]

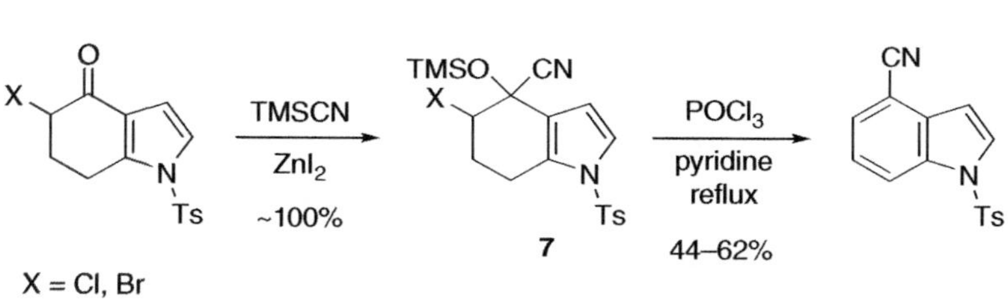

[eqn. 3] [15]

Scheme 2 *Davies and Matsumoto Indole Syntheses*

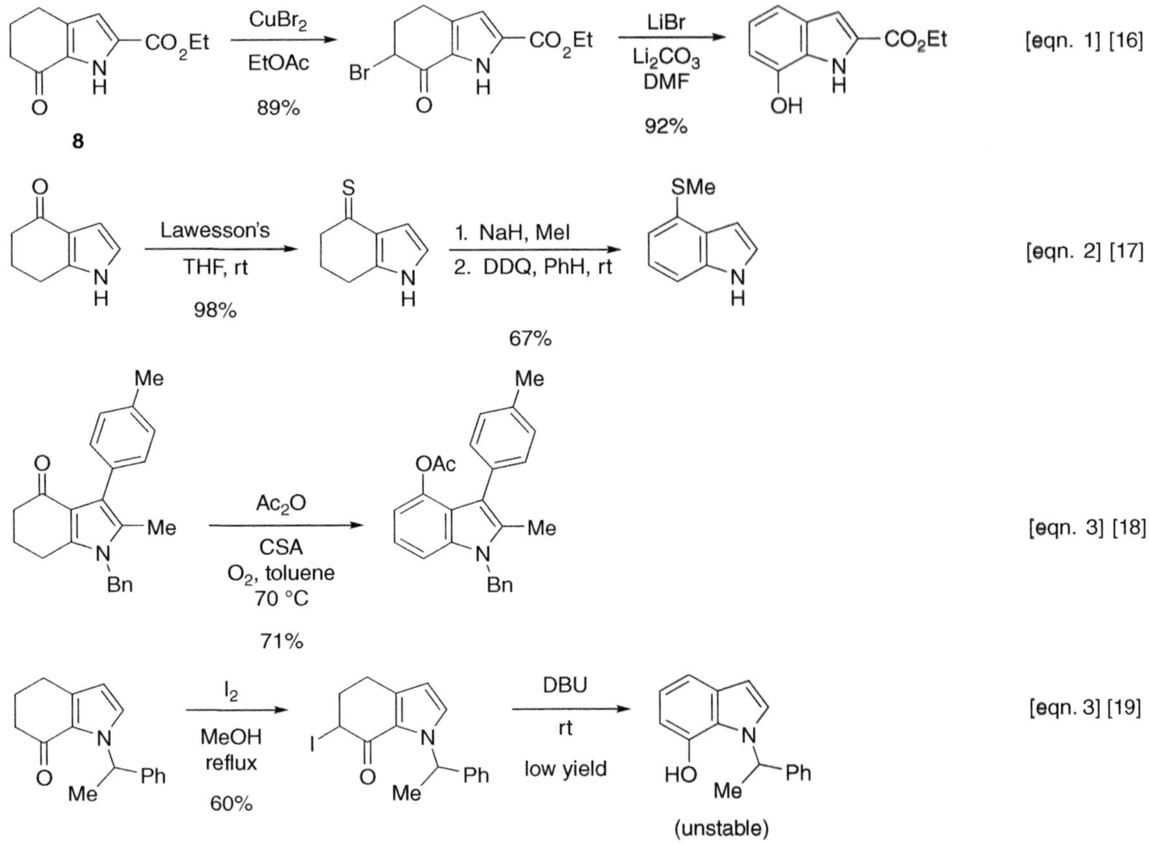

Scheme 3 *Murakami, Tobinaga, and Saito Indole Syntheses*

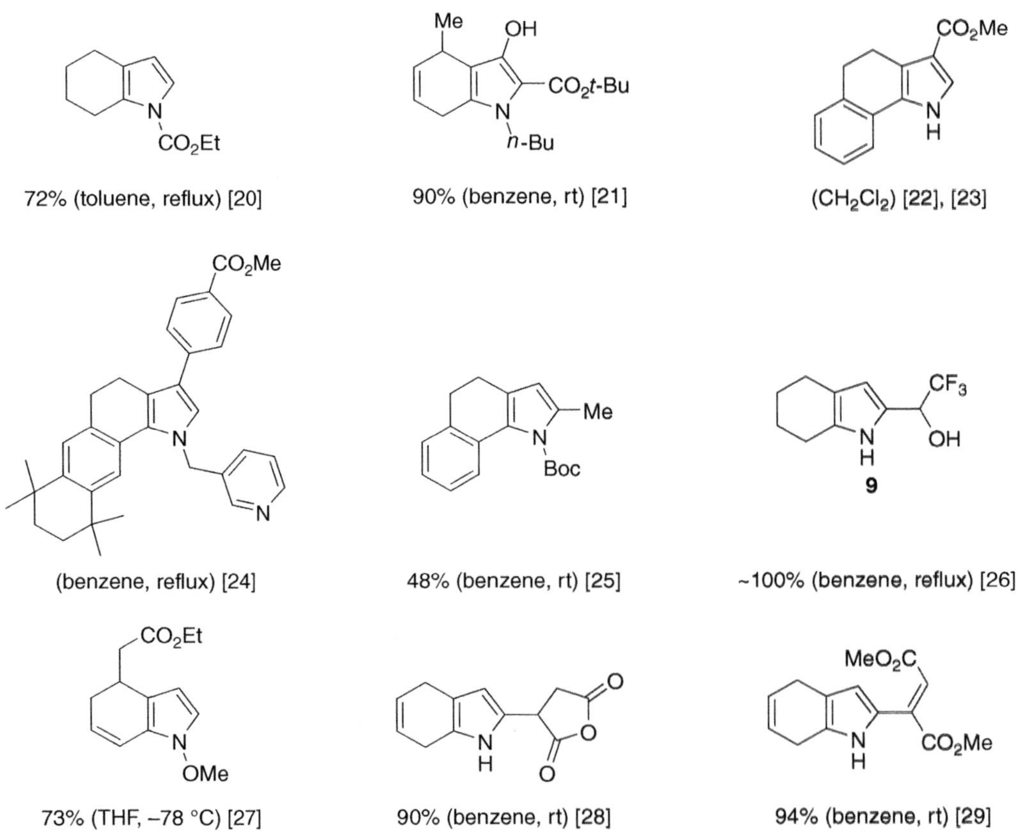

Scheme 4 *DDQ Dehydrogenation of 4,5,6,7-Tetrahydroindoles*

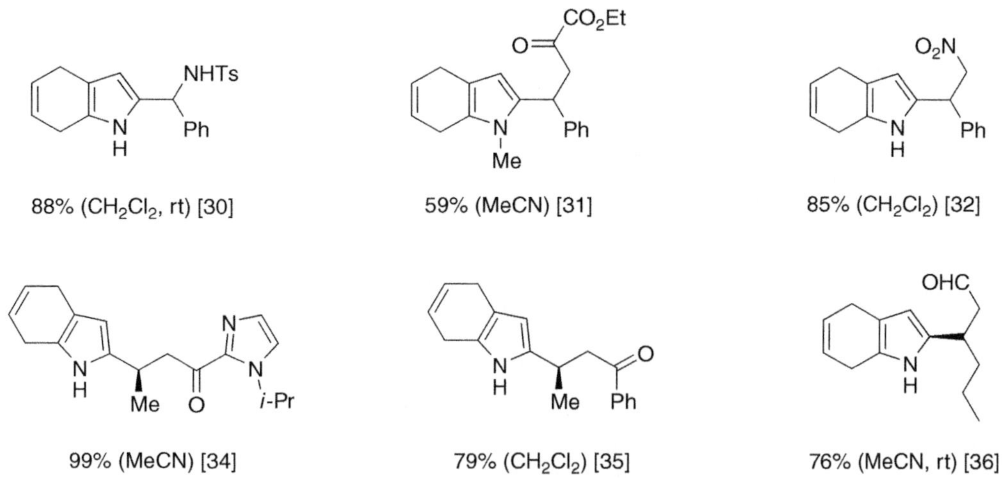

Scheme 5 *p-Benzoquinone Dehydrogenation of 4,7-Dihydroindoles*

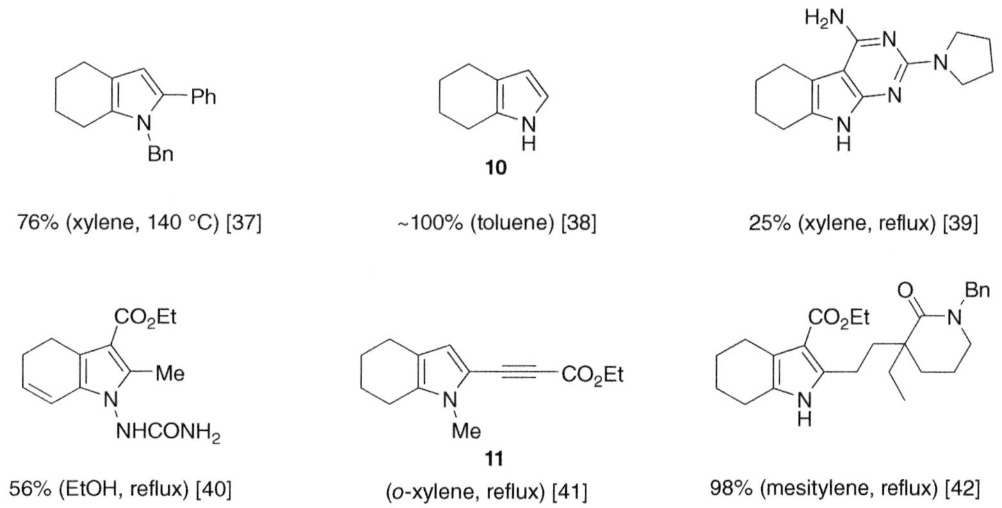

Scheme 6 *Palladium/Charcoal Dehydrogenation of 4,5,6,7-Tetrahydroindoles*

As we have seen, a large number of pyrrole → indole dehydrogenations involved the use of a benzoquinone, usually DDQ, and a selection of the substrates that were converted to indoles is summarized in Scheme 4 [20–29]. In each case the compound shown afforded the fully aromatic indole; in some cases the yields were not reported. The oxidation of trifluoro alcohol **9** afforded the keto indole. Space does not permit the illustration of the syntheses of the substrates shown in Scheme 4.

The less-reactive *p*-benzoquinone has also been employed to dehydrogenate 4,7-dihydroindoles, such as those obtained via Friedel–Crafts reactions onto the pyrrole C-2 position. A selection of these substrates so oxidized is shown in Scheme 5 [30–32, 34–36]. In each case *p*-benzoquinone was the oxidant used. You and coworkers have been particularly active in this arena [30–33].

As we have already seen in this chapter, the classical dehydrogenation conditions of palladium-on-charcoal in an inert solvent at elevated temperatures remain a popular method to dehydrogenate perhydroindoles to indoles. A selection of such substrates is shown in Scheme 6 [37–42]. The synthesis of indole from tetrahydroindole **10** involved an extensive survey of Pd and other catalysts to give indole in quantitative yield [38]. The indole product from **11** had the alkyne reduced to a mixture of alkene and alkane [41].

References

[1] W.A. Remers and M.J. Weiss, *J. Am. Chem. Soc.*, 1965, **87**, 5262–5264.
[2] W.A. Remers, R.H. Roth, G.J. Gibs, and M.J. Weiss, *J. Org. Chem.*, 1971, **36**, 1232–1240.
[3] W.A. Remers and M.J. Weiss, *J. Org. Chem.*, 1971, **36**, 1241–1247.
[4] H. Stetter and R. Lauterbach, *Ann.*, 1962, **655**, 20–26.
[5] D.B. Repke, W.J. Ferguson, and D.K. Bates, *J. Heterocycl. Chem.*, 1977, **14**, 71–74.
[6] S. Torii, K. Uneyama, T. Onishi, *et al.*, *Chem. Lett.*, **1980**, 1603–1604.
[7] M.D. Lebar and B.J. Baker, *Aust. J. Chem.*, 2010, **63**, 862–866.
[8] E.D. Edstrom, *Synlett*, 1995, 49–50.
[9] E.D. Edstrom, T. Yu, and Z. Jones, *Tetrahedron Lett.*, 1995, **36**, 7035–7038.
[10] K. Spyridonidou, M. Fousteris, M. Antonia, *et al.*, *Bioorg. Med. Chem. Lett.*, 2009, **19**, 4810–4813.
[11] K. Doi and M. Mori, *Heterocycles*, 1996, **42**, 113–116.
[12] L. Piras, C. Ghiron, G. Minetto, and M. Taddei, *Tetrahedron Lett.*, 2008, **49**, 459–462.
[13] H.M.L. Davies and J.R. Manning, *J. Am. Chem. Soc.*, 2006, **128**, 1060–1061.
[14] M. Matsumoto, Y. Ishida, and N. Watanabe, *Heterocycles*, 1985, **23**, 165–170.
[15] N. Hatanaka and M. Matsumoto, *Heterocycles*, 1986, **24**, 1963–1971.
[16] M. Tani, T. Ariyasu, M. Ohtsuka, *et al.*, *Chem. Pharm. Bull.*, 1996, **44**, 55–61.
[17] M. Murase, T. Hosaka, and S. Tobinaga, *Heterocycles*, 1990, **30**, 905–908.
[18] M. Arai, Y. Miyauchi, T. Miyahara, *et al.*, *Synlett*, 2009, 122–126.
[19] A.G. Montalban, S.M. Baum, J. Cowell, and A. McKillop, *Tetrahedron Lett.*, 2012, **53**, 4276–4279.
[20] E.M.M. van den Berg, F.J.H.M. Jansen, A.T.J.W. de Goede, *et al.*, *Recl. Trav. Chim. Pays-Bas*, 1990, **109**, 287–297.
[21] H.H. Wasserman and C.A. Blum, *Tetrahedron Lett.*, 1994, **35**, 9787–9790.
[22] G.A. Pinna, M.A. Pirisi, and G. Paglietti, *J. Chem. Res. (S)*, 1990, 360–361.
[23] G. Murineddu, G. Cignarella, G. Chelucci, *et al.*, *Chem. Pharm. Bull.*, 2002, **50**, 754–759.
[24] H. Yoshimura, M. Nagai, S. Hibi, *et al.*, *J. Med. Chem.*, 1995, **38**, 3163–3173.
[25] P. Nagafuji and M. Cushman, *J. Org. Chem.*, 1996, **61**, 4999–5003.
[26] A.L. Sigan, D.V. Gusev, N.D. Chkanikov, *et al.*, *Tetrahedron Lett.*, 2011, **52**, 5025–5028.
[27] E.L. Fisher, S.M. Wilkerson-Hill, and R. Sarpong, *J. Am. Chem. Soc.*, 2012, **134**, 9946–9949.
[28] H. Çavdar and N. Saraçuğlu, *Tetrahedron*, 2005, **61**, 2401–2405.
[29] H. Çavdar and N. Saraçuğlu, *J. Org. Chem.*, 2006, **71**, 7793–7799.
[30] Q. Kang, X.-J. Zheng, and S.-L. You, *Chem. Eur. J.*, 2008, **14**, 3539–3542.
[31] M. Zeng, Q. Kang, Q.-L. He, and S.-L. You, *Adv. Synth. Catal.*, 2008, **350**, 2169–2173.
[32] Y.-F. Sheng, G.-Q. Li, Q. Kang, *et al.*, *Chem. Eur. J.*, 2009, **15**, 3351–3354.
[33] M. Zeng and S.-L. You, *Synlett*, 2010, 1289–1301.
[34] D.A. Evans and K.R. Fandrick, *Org. Lett.*, 2006, **8**, 2249–2252.
[35] G. Blay, I. Fernández, J.R. Pedro, and C. Vila, *Tetrahedron Lett.*, 2007, **48**, 6731–6734.
[36] L. Hong, C. Liu, W. Sun, *et al.*, *Org. Lett.*, 2009, **11**, 2177–2180.
[37] S. Lim, I. Jabin, and G. Revial, *Tetrahedron Lett.*, 1999, **40**, 4177–4180.
[38] M.A. Ryashentseva, *Chem. Heterocycl. Comp.*, 2006, **42**, 1018–1020.
[39] B. Dotzauer, R. Grünert, P.J. Bednarski, *et al.*, *Bioorg. Med. Chem.*, 2006, **14**, 7282–7292.
[40] O.A. Attanasi, G. Favi, P. Filippone, *et al.*, *Adv. Synth. Catal.*, 2007, **349**, 907–915.
[41] B.A. Trofimov, L.N. Sobenina, Z.V. Stepanova, *et al.*, *Tetrahedron Lett.*, 2008, **49**, 3946–3949.
[42] C.L. Morales and B.L. Pagenkopf, *Org. Lett.*, 2008, **10**, 157–159.

64

Miscellaneous Indole Syntheses from Pyrroles

This short chapter presents several new indole syntheses from pyrroles that do not logically fit into the previous pyrrole chapters but that nevertheless deserve to be revealed as being clever and promising for future applications.

Quéguiner and coworkers found a simple benzannulation of methylpyrrole-2,3-dicarboxyaldehyde to give 5,6-disubstituted indoles (Scheme 1, equations 1 and 2) [1]. Reaction of this dialdehyde with sulfones afforded thiepino[4,5-*b*]pyrrole 6,6-dioxides. In contrast, *N*-methylpyrrole-3,4-dicarboxyaldehyde (not shown) yielded solely the thiepino[4,5-*c*]pyrroles or the 6,6-dioxides. Mancini and colleagues effected a Diels–Alder reaction between 2- and 3-nitropyrroles and activated dienes to form indoles in modest yields (equations 3, 4) [2–4].

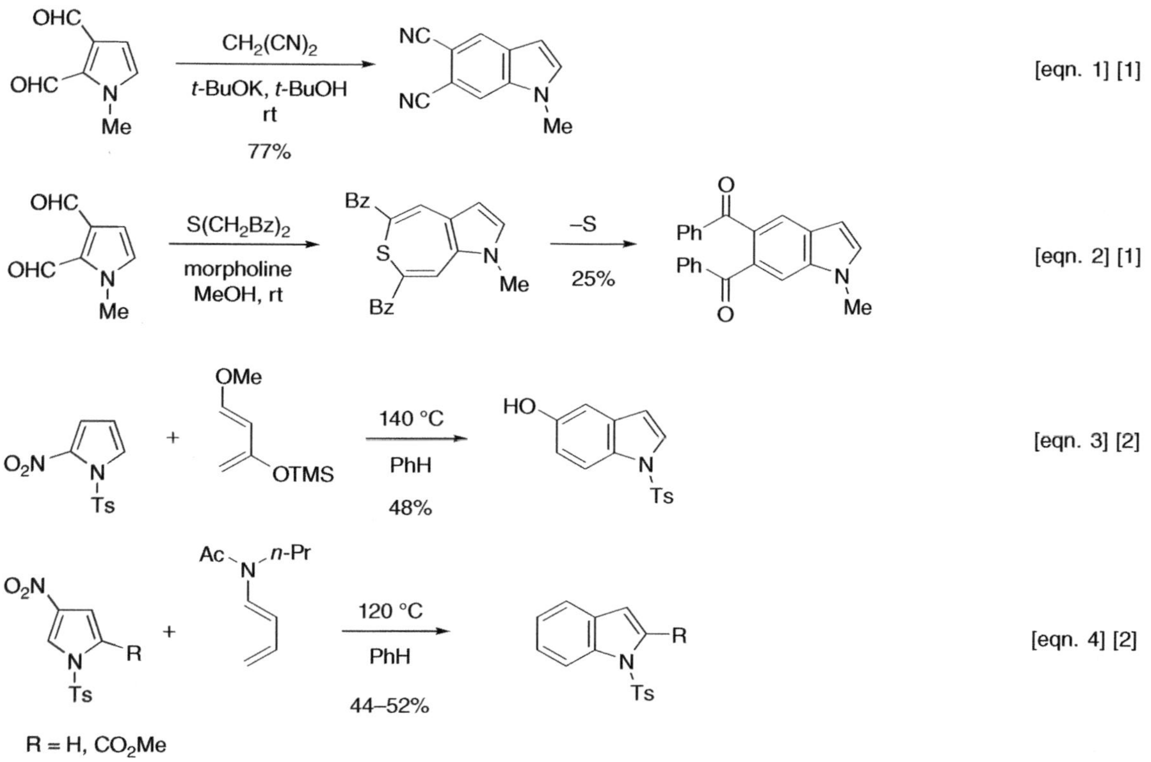

Scheme 1 *Quéguiner and Mancini Indole Syntheses*

Indole Ring Synthesis: From Natural Products to Drug Discovery, First Edition. Gordon W. Gribble.

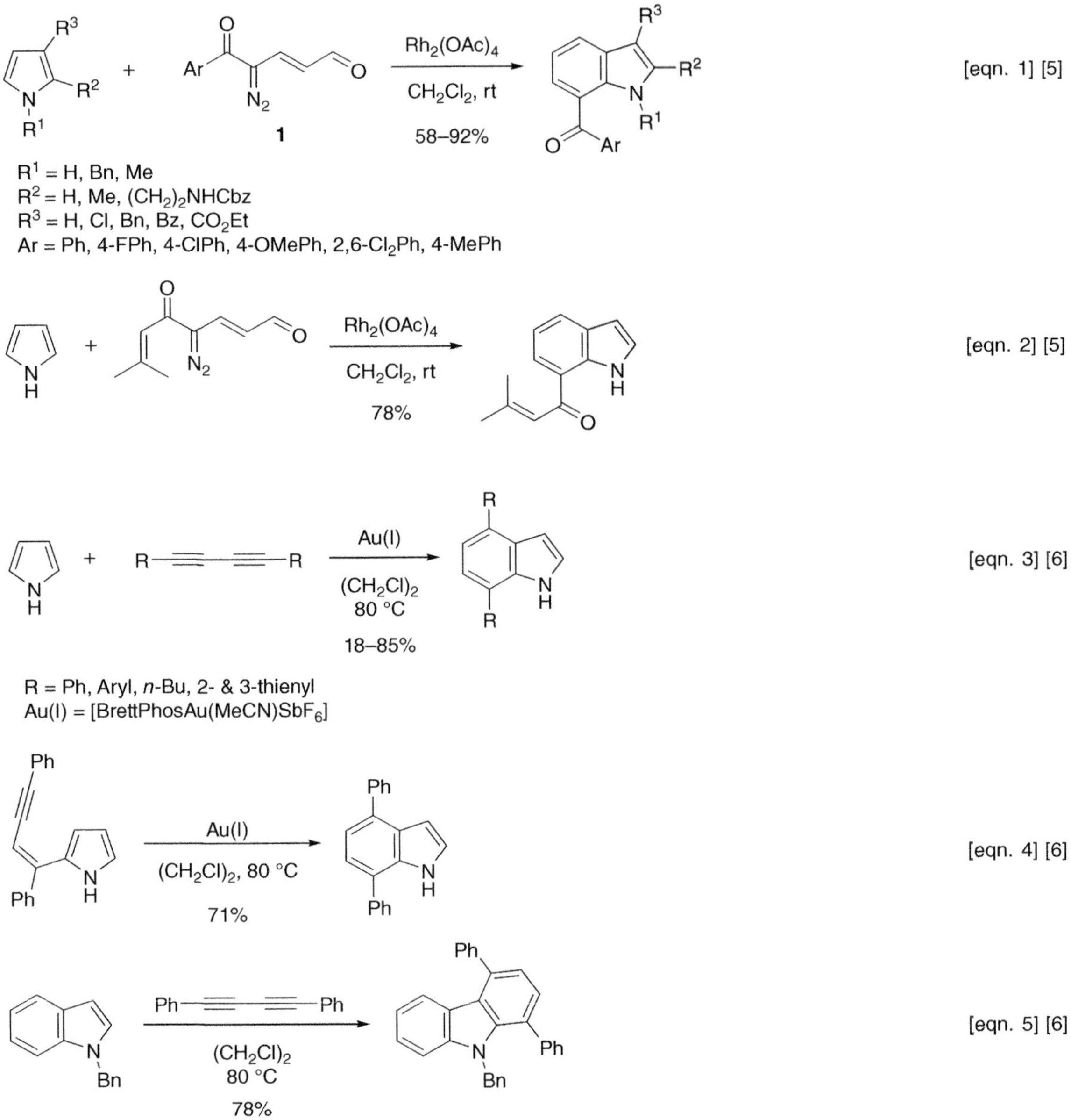

Scheme 2 *Katukojvala and Fujii-Ohno Indole Syntheses*

A rhodium(II)-catalyzed [4+2] benzannulation of pyrroles was reported by Katukojvala and colleagues to give 7-acylindoles (Scheme 2, equation 1) [5]. The simple natural product leicarpone was also synthesized (equation 2). The requisite enaldiazo compounds **1** were prepared from the corresponding aryl enones in three steps. The initial step in the process involves Rh(II)-catalyzed generation of an enalcarbenoid and functionalization at C-2 of the pyrrole. Subsequent cyclization eventually affords the 7-acylindole. Fujii and Ohno employed gold in a catalyzed indole synthesis that also involved a formal [4+2] reaction between pyrroles and 1,3-diynes (equation 3) [6]. The mechanism is thought to involve electrophilic attack at C-2 of pyrrole by a Ag(I)-diyne complex followed by an intramolecular 6-*endo-dig* hydroarylation. The initial C-2 adduct can be isolated and further converted to the indole (equation 4). Not unexpectedly, carbazoles were formed from indoles under these conditions (equation 5).

References

[1] J. Duflos, G. Dupas, and G. Quéguiner, *J. Heterocycl. Chem.*, 1983, **20**, 1191–1193.
[2] C. Della Rosa, M. Kneeteman, and P. Mancini, *Tetrahedron Lett.*, 2007, **48**, 1435–1438.
[3] C. Della Rosa, C. Ormachea, M.N. Kneeteman, *et al.*, *Tetrahedron Lett.*, 2011, **52**, 6754–6757.
[4] P.M.E. Mancini, C.M. Ormachea, C.D. Della Rosa, *et al.*, *Tetrahedron Lett.*, 2012, **53**, 6508–6511.
[5] S.G. Dawande, V. Kanchupalli, J. Kalepu, *et al.*, *Angew. Chem. Int. Ed.*, 2014, **53**, 4076–4080.
[6] Y. Matsuda, S. Naoe, S. Oishi, *et. al.*, *Chem. Eur. J.*, 2015, **21**, 1463–1467.

65

Indoles via Arynes

The utility of benzyne and other arynes in synthesis has exploded in the past decade. No longer just a fleeting curiosity, benzyne has become a bona fide chemical reagent with enormous synthetic versatility. Accordingly, arynes have found employment in the synthesis of heterocycles, including indoles [1, 2]. The two main categories of arynes in indole and carbazole synthesis are Diels–Alder reactions and nucleophilic addition reactions. The coverage herein is roughly chronological.

The notion of generating and trapping intramolecularly an aryne as a general principle of synthesis was first advanced and executed by Bunnett and Hrutfiord in 1961 [3]. In addition to developing new syntheses of benzothiazole, benzoxazole, and phenothiazine, 2-methylindole was obtained via a benzyne intermediate (Scheme 1, equation 1). Another possible pathway involves addition of amide to the carbonyl group first and then aryne formation. Despite this seminal investigation, the use of arynes in synthesis got off to a slow start. Miwa and colleagues uncovered an unusual reaction of benzyne with its precursor benzenediazonium *o*-carboxylate (equation 2) [4]. Benzyne addition product **1** reacted with a second benzyne to give **2**. Benzoic acid and biphenylene were also formed. Bauer and colleagues reported a similar decomposition of *o*-carboxybenzenediazonium halides to afford *N*-(carbazolyl)anilines in minuscule yields (equation 3) [5]. Pascal and coworkers described similar chemistry leading to carbazole formation from the diazotization of anthranilic acids [6]. Nair and Kim described a modest yield of indoles from the reaction of 1-azirines with benzyne, generated from benzenediazonium-*o*-carboxylate (equation 4) [7]. The reaction of benzyne with 2,3-diphenyl-1-azirine also gave 14% of 1,2,3-triphenylindole.

The first practical synthesis of indoles using aryne technology was that of Caubère and colleagues [8–12]. A selection of these anionic intramolecular indolizations is shown in Scheme 2, all of which utilize the complex superbase $NaNH_2$-*t*-BuONa [13]. One can alkylate the intermediate cyclized anion on nitrogen (equation 3).

Brown, Eastwood, and colleagues pyrolyzed quinolone-3,4-dicarboxylic anhydrides to give fused indoles (Scheme 3, equations 1 and 2) [14]. The unstable indeno[1,2-*b*]indole (**1**) reacted with various nucleophiles in low yield (17%–29%) (equation 3).

Guitián and colleagues employed aryne generation and intramolecular cyclization to access ergot, lycorine, and related ring systems [15–18], an example of which is shown in Scheme 4 [18]. Oxidation of lactam indoline **3** with DDQ gave the corresponding indole (99%). Iwao, Watanabe, and coworkers reported a simple and high-yielding carbazole synthesis via aryne anionic cyclization (equation 2) [19]. The chlorine can also be at the *meta* position, but the yield is slightly lower (70%). The starting *N*-Boc anilines were prepared using a Stille reaction. The carbazole alkaloids glycozolinine and glycozolidine were prepared in the same fashion. Bailey and Carson reported a 3,4-disubstituted indoline synthesis via the lithiation of 2- or 3-fluoro-*N*,*N*-diallylamines and subsequent anionic cyclization and protonation [20]. This strategy was extended by Barluenga and coworkers to the synthesis of indoles (equation 3) [21–23]. The method was adapted to the syntheses of NH indoles, serotonin analogues, and tetrahydrocarbazoles [23].

In an approach similar to that of Caubère (Scheme 2), but employing LDA, Kudzma synthesized a series of indoles and tetrahydrocarbazoles via the lithiation of

Indole Ring Synthesis: From Natural Products to Drug Discovery, First Edition. Gordon W. Gribble.

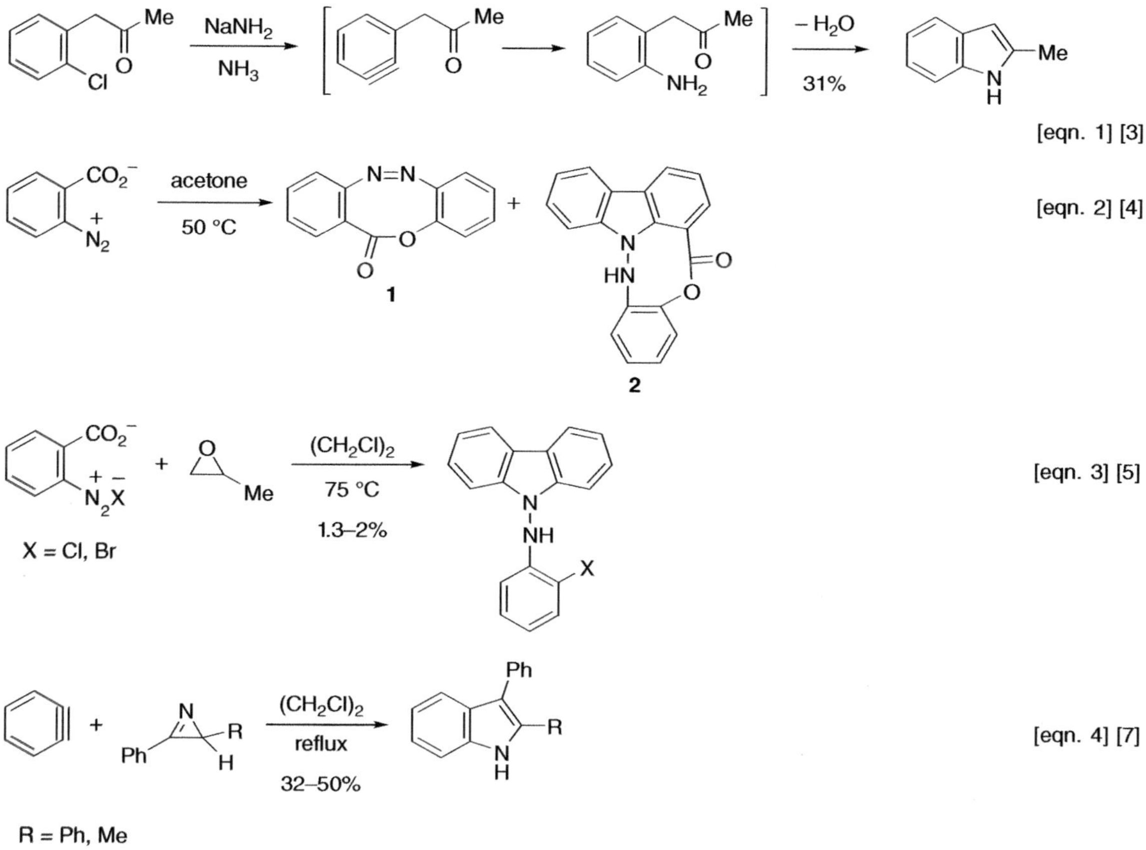

Scheme 1 *Bunnett and Other Early Indole Syntheses via Arynes*

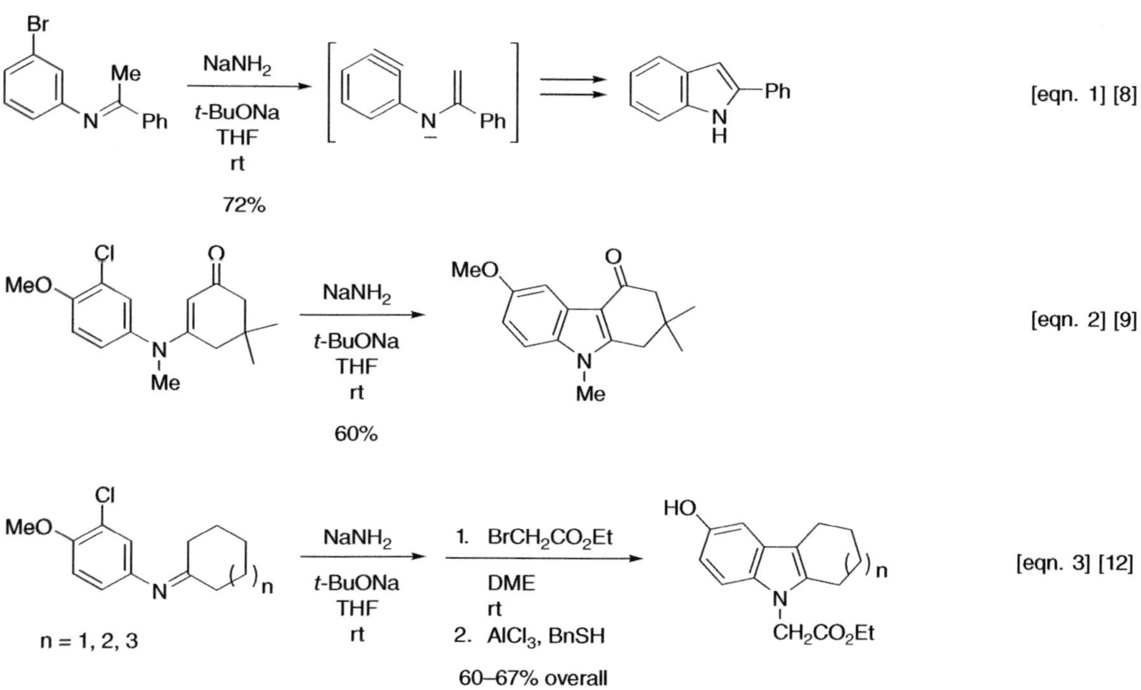

Scheme 2 *Caubère Indole Synthesis*

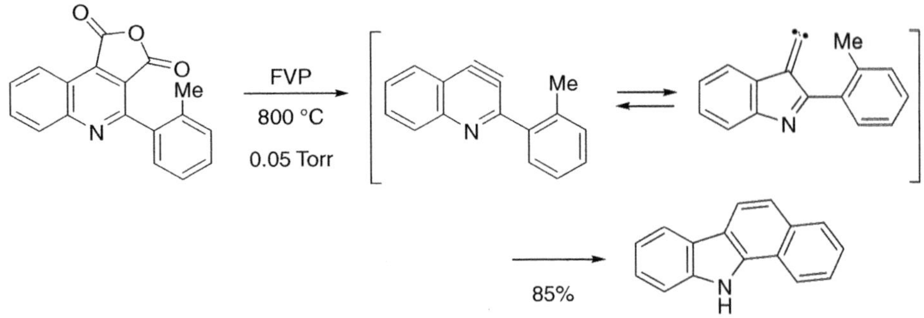

[eqn. 1] [14]

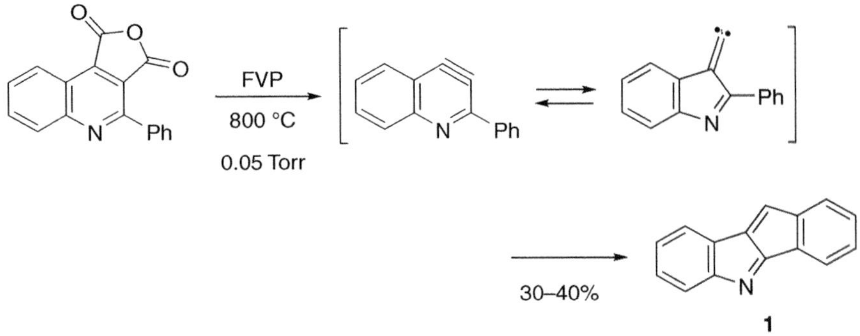

[eqn. 2] [14]

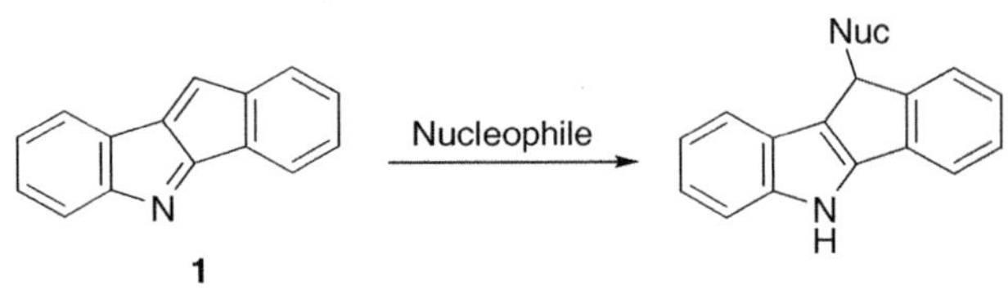

[eqn. 3] [14]

Nuc = H, Et_2N, Ph, OMe, Me, $CH(CO_2Me)_2$, $CMeNO_2$

Scheme 3 *Brown-Eastwood Indole Synthesis*

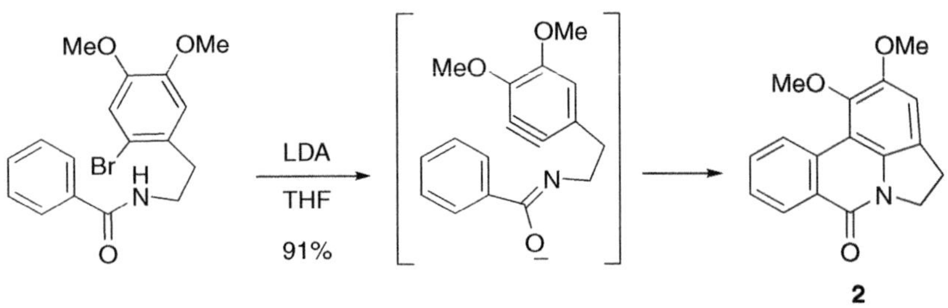

[eqn. 1] [18]

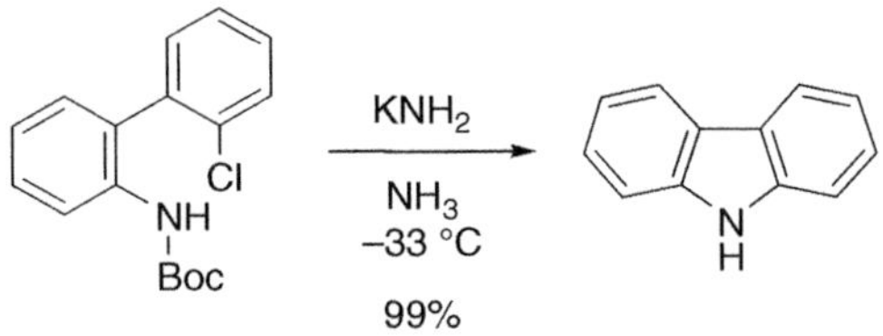

[eqn. 2] [19]

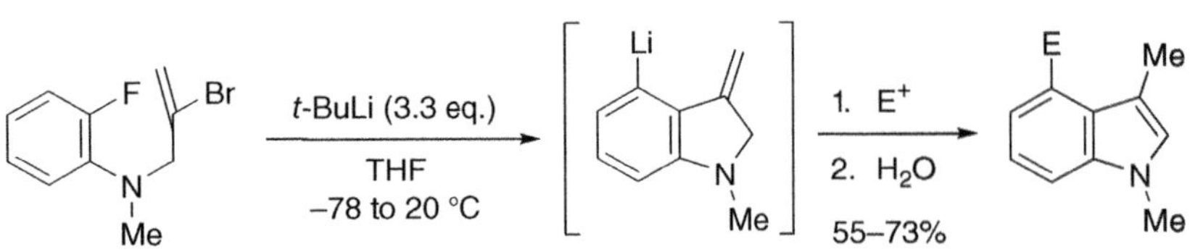

[eqn. 3] [21]

E = D, $SnBu_3$, PhCHOH, CO_2Et, Me_2COH, 4ClPhCO, PhCHNHPh

Scheme 4 *Guitián, Iwao-Watanabe, and Barluenga Indole Syntheses*

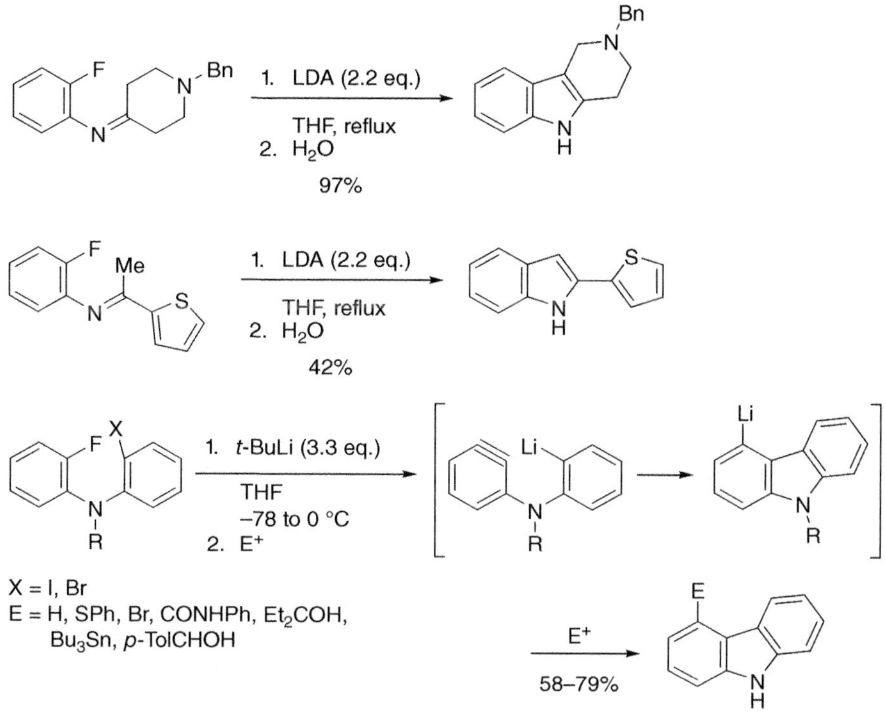

[eqn. 1] [24]

[eqn. 2] [24]

[eqn. 3] [25]

Scheme 5 *Kudzma and Sanz Indole and Carbazole Syntheses*

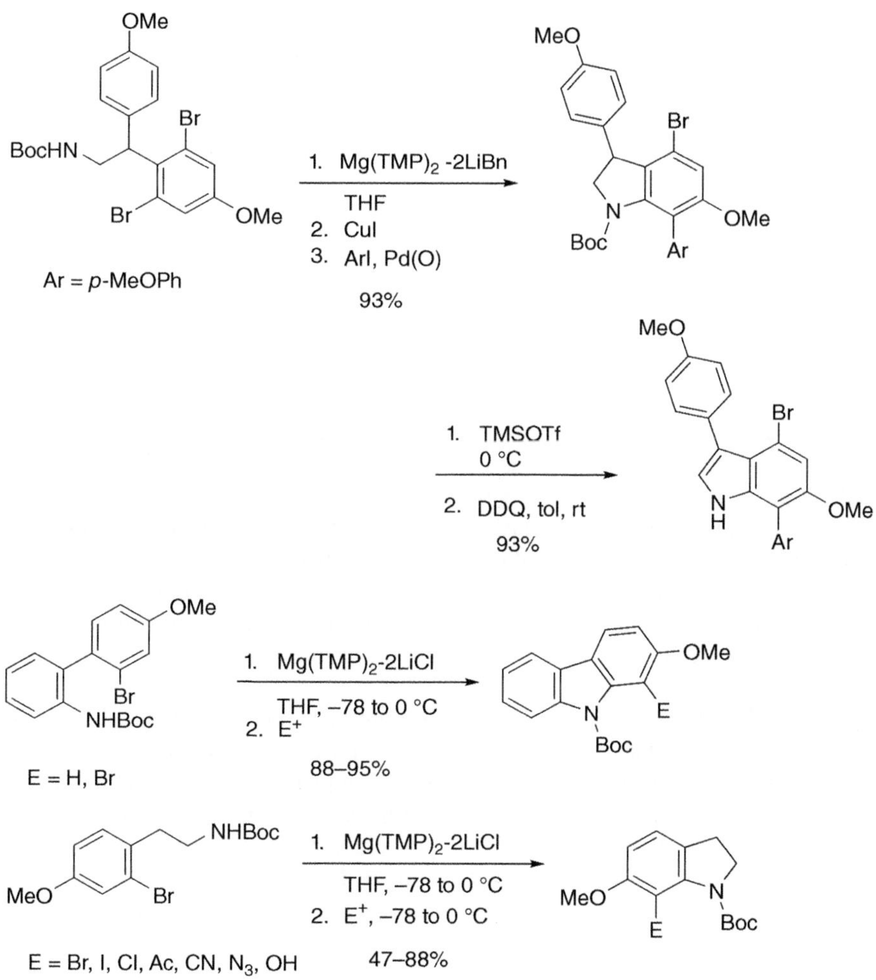

[eqn. 1] [26]

[eqn. 2] [27]

[eqn. 3] [27]

Scheme 6 *Tokuyama Indoline – Carbazole Synthesis*

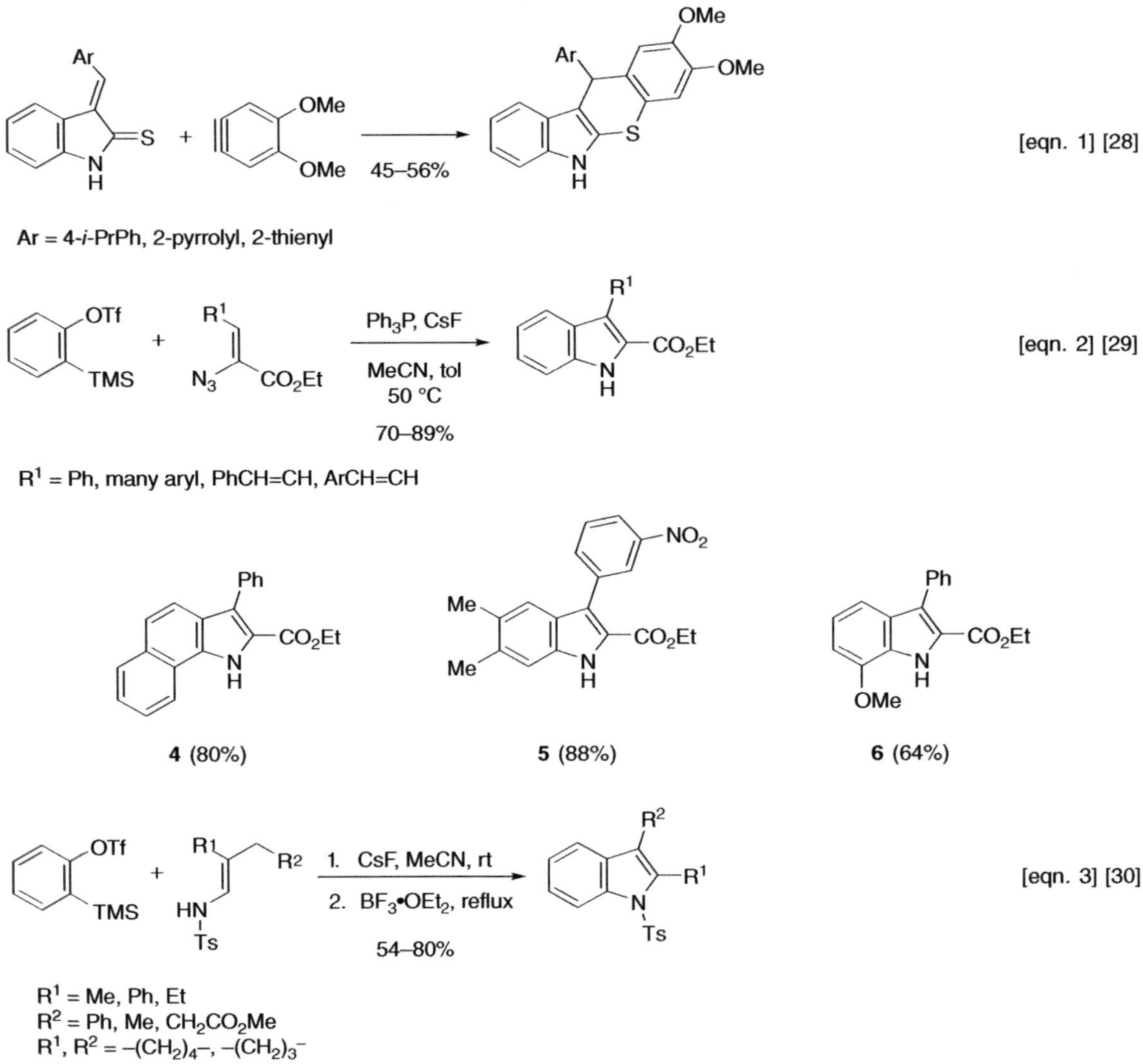

Scheme 7 *Biehl, Lin-Wang, and Greaney Indole Syntheses*

2-fluorophenyl imines, aryne formation, and cyclization (Scheme 5, equations 1 and 2) [24]. The enaminone from 1,3-cyclohexanone and 2-fluoroaniline gave 1,2,3,9-tetrahydro-4*H*-carbazole-4-one (33% yield). Deuteration experiments implicated aryne formation in this process. In an extension of previous aryne cyclization studies, Sanz's team synthesized functionalized carbazoles through cyclization of aryne-linked aryllithiums (equation 3) [25]. The requisite diaryl amines were prepared by a Pd-catalyzed amination protocol between an aniline and a halobenzene.

Tokuyama and colleagues employed a magnesium base to generate and cyclize an aryne intermediate leading to carbazoles and indolines (Scheme 6, equations 1 and 2) [26, 27]. Originally discovered in their syntheses of the dictyodendrins A–E (equation 1) [26], the method was subsequently fleshed out to become an excellent carbazole synthesis (equation 2), as well as an indoline synthesis (equation 3) [27]. The carbazole alkaloid heptaphylline was synthesized by this method (five operations from bromoanisaldehyde). Of the several magnesium amide bases studied, $Mg(TMP)_2$-2LiCl was best.

In addition to the aforementioned *intra*molecular indolizations involving aryne cyclizations, *inter*molecular reactions of arynes leading to indoles have been described.

Biehl and Kamila synthesized benzothiopyrano[2,3-*b*] indoles from the reaction between arynes and indole-2-thiones (Scheme 7, equation 1) [28]. 4,5-Dimethoxybenzyne was generated from 2-diazonio 4,5-dimethoxybenzenecarboxylate hydrochloride. Other aryne sources were less efficient or failed. Lin, Wang, and colleagues discovered that benzynes react with 2-azidoacrylates in the presence

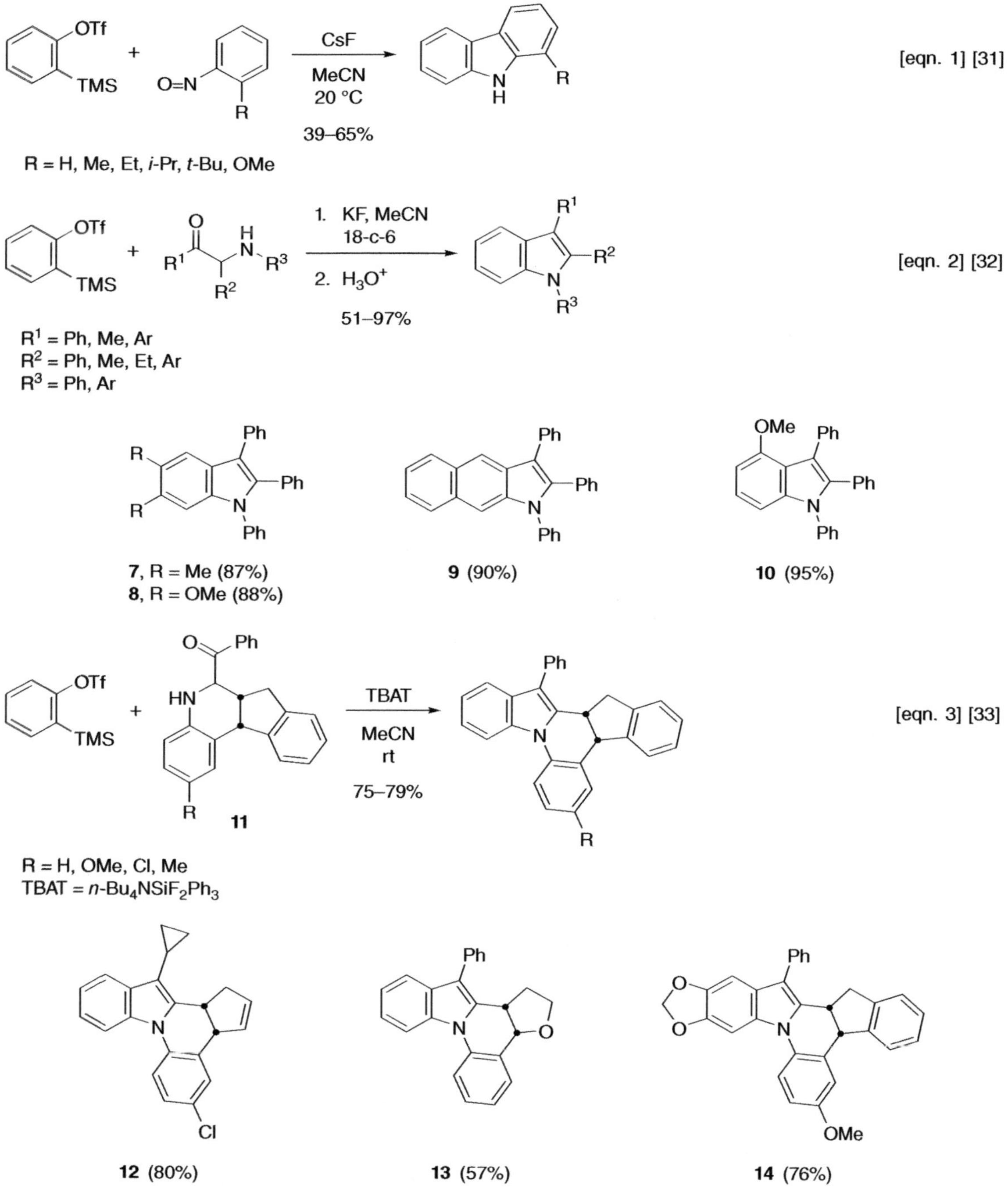

Scheme 8 *Studer, He, and Zhu Indole Syntheses*

of Ph_3P/CsF to give indoles (equation 2) [29]. Several substituted benzynes were explored and led to, for example, indoles **4–6**. The authors proposed an initial vinyl iminophosphorane intermediate (a Staudinger–Meyer reaction) that reacts with benzyne to give an adduct that is hydrolyzed to an indoline that undergoes arial oxidation to the indole. The reactive adduct was detected by mass spectrometry and ^{31}P NMR. Greaney and coworkers described a benzyne Fischer indole synthesis as presented in equation 3 [30]. The reaction was shown to involve *N*-arylation of the hydrazones with benzyne, and a subsequent standard Fischer indolization. Some substituted benzynes were successfully used in this one-pot reaction (e.g., 2-naphthalyne, and 3- and 4-methylbenzyne).

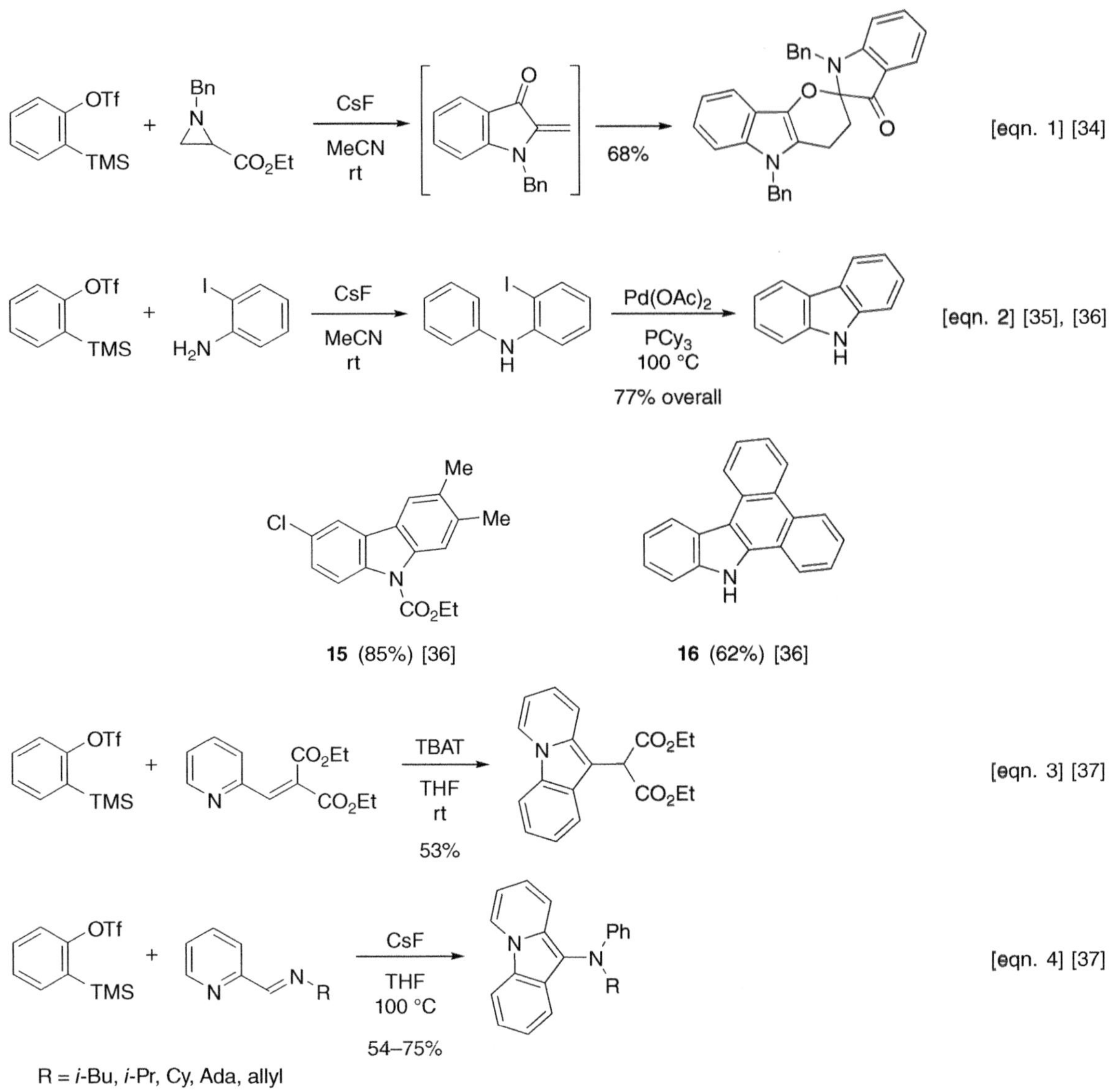

Scheme 9 *Wu-Sha and Larock Indole Syntheses*

Studer and colleagues discovered a novel carbazole synthesis involving benzynes reacting with nitrosoarenes to give carbazoles (Scheme 8, equation 1) [31]. This reaction may involve a [2+2] cycloaddition to give a 1,2-oxazetidine that opens to a quinone that eventually yields a carbazole by an unclear path. Interestingly, the use of Bu_4N/Ph_3SiF_2 with 2.2 equivalents of nitrosoarene yields the *N*-arylcarbazole. Several substituted benzynes were also generated and formed carbazoles upon reaction with nitrosoarenes. He's group effected the coupling of arynes and α-amino ketones to produce *N*-arylindoles (equation 2) [32]. This simple condensation is thought to give a 3-hydroxyindole, which suffers dehydration on exposure to acid. Substituted benzynes led to indoles **7–10**. A heteroannulation of benzynes with tetrahydroisoquinolines **7** to afford 5,6-dihydroindolo[1,2-*a*]quinolines was developed by Zhu and colleagues (equation 3) [33]. A Povarov reaction between α-oxo aldehydes, anilines, and dienes provides the requisite precursor **7**. A range of substituted benzynes and reaction partners led to indoles **8–10**, among others.

Wu, Sha, and coworkers found that aziridines react with benzyne to give novel spiroindoles (Scheme 9, equation 1) [34]. The products are [4+2] dimers of 1-benzyl-2-methyleneindolin-3-one. Similar compounds were obtained using *N*-substituted aziridines and substituted benzynes. This reaction did not occur with KF or TBAF; instead, α-fluoro-β-amino acid derivatives were the products. When water was present, the intermediate aziridinium ion-aryl

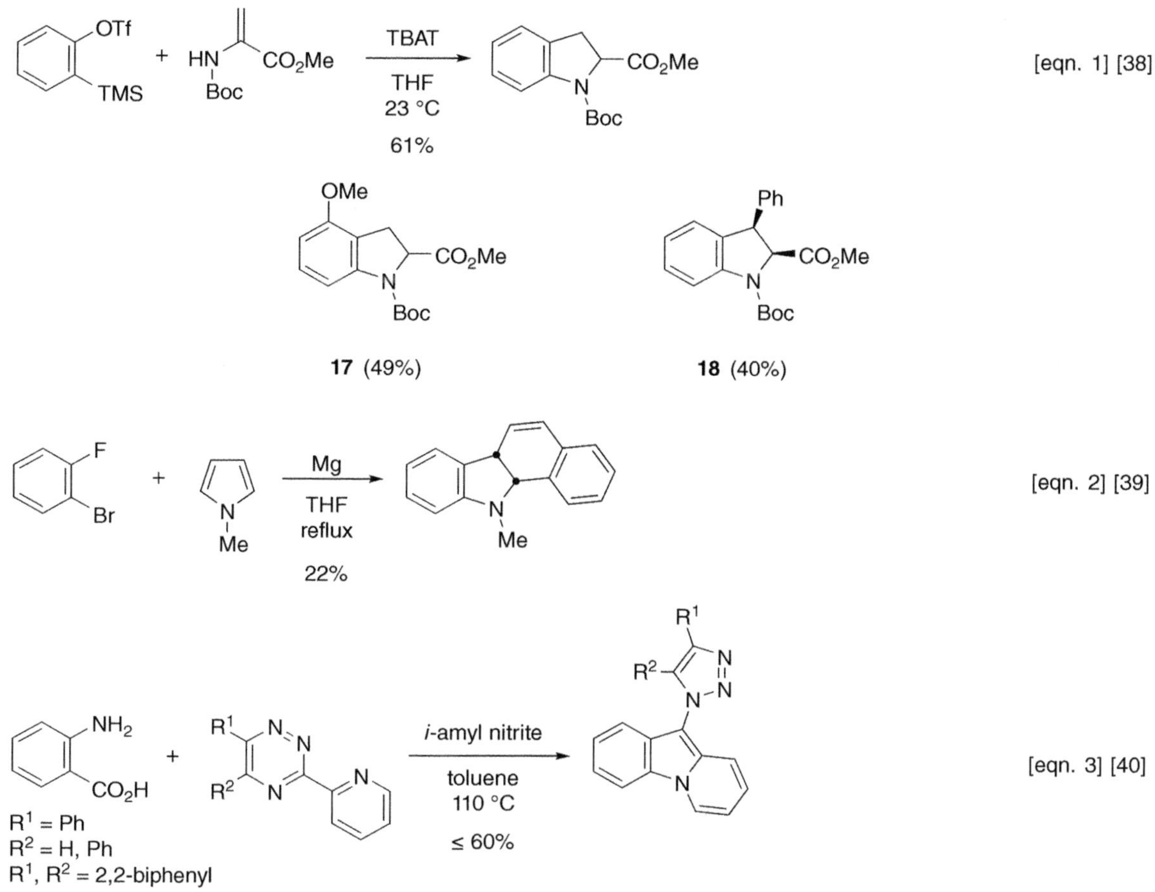

Scheme 10 *Stoltz, Greaney, and Zyryanov Indoline-Indole Syntheses*

anion was protonated, preventing ring closure to the methyleneindolin-3-one. Larock and Liu reported a carbazole synthesis via the reaction of arynes with *o*-iodoanilines followed by Pd-catalyzed cyclization (equation 2) [35, 36]. The carbazole alkaloid mukonine was prepared 76% yield (three steps), and indoles **15** and **16** were constructed using substituted arynes. Larock's team developed a synthesis of unique pyrido[1,2-*a*]indole malonates and amines via aryne annulation (equation 3) [37]. Both sequences resulted in the production of many analogues.

Stoltz and his group employed benzyne to prepare indolines by reaction with ene carbamates (Scheme 10, equation 1) [38]. Indolines **17** and **18** also emerged from this study. In a study of the benzyne aza-Claisen rearrangement, Greaney and colleagues reexamined the original Wittig trapping of benzyne with *N*-methylpyrrole and obtained a 22% yield of the Wittig indoline (compared to Wittig's 12%) (equation 2) [39]. Interestingly, a modern benzyne generation method (*o*-trimethylsilyl phenyltriflate, CsF, MeCN) gave only an α-naphthylamine product and no Diels–Alder adduct. Zyryanov and coworkers discovered an unusual pyrido[1,2-*a*]indole synthesis stemming from the benzyne reaction with aryl-substituted 3-(2-pyridyl)-1,2,4-triazines (equation 3) instead of the expected pyridyl-isoquinolines that would have arisen via a Diels–Alder cycloaddition with the pyridazine ring [40].

References

[1] D. Peña, D. Pérez, and E. Guitián, *Heterocycles*, 2007, **74**, 89–100.
[2] R. Sanz, *Org. Prep. Proc. Int.*, 2008, **40**, 215–291.
[3] J.F. Bunnett and J.F. Hrutfiord, *J. Am. Chem. Soc.*, 1961, **83**, 1691–1697.
[4] T. Miwa, M. Kato, and T. Tamano, *Tetrahedron Lett.*, 1968, **9**, 2743–2745.
[5] R. Ghosh, E.B. Sheinin, C.L. Bell, and L. Bauer, *J. Heterocycl. Chem.*, 1975, **12**, 203–206.
[6] X. Qiao, D.M. Ho, and R.A. Pascal, Jr. *J. Org. Chem.*, 1996, **61**, 6748–6750.

[7] V. Nair and K.H. Kim, *J. Org. Chem.*, 1975, **40**, 3784–3786.
[8] L. Lalloz and P. Caubère, *J. Chem. Soc., Chem. Commun.*, 1975, 745.
[9] C. Caubère, P. Caubère, P. Renard, *et al.*, *Tetrahedron Lett.*, 1993, **34**, 6889–6892.
[10] C. Kuehm-Caubère, I. Rodriguez, B. Pfeiffer, *et al.*, *J. Chem. Soc., Perkin Trans. 1*, 1997, 2857–2862.
[11] S. Christophe, C. Kuehm-Caubère, P. Renard, *et al.*, *Tetrahedron Lett.*, 1998, **39**, 9431–9434.
[12] C. Kuehm-Caubère, P. Caubère, B. Jamart-Grégoire, *et al.*, *Eur. J. Med. Chem.*, 1999, **34**, 51–61.
[13] P. Caubère, *Chem. Rev.*, 1993, **93**, 2317–2334.
[14] R.F.C. Brown, K.J. Coulston, F.W. Eastwood, and M.R. Moffat, *Tetrahedron*, 1992, **48**, 7763–7773.
[15] B. Gómez, E. Guitián, and L. Castedo, *Synlett*, 1992, 903–904.
[16] C. González, D. Pérez, E. Guitián, and L. Castedo, *J. Org. Chem.*, 1995, **60**, 6318–6326.
[17] C. González, E. Guitián, and L. Castedo, *Tetrahedron Lett.*, 1996, **37**, 405–406.
[18] C. González, E. Guitián, and L. Castedo, *Tetrahedron*, 1999, **55**, 5195–5206.
[19] M. Iwao, H. Takehara, S. Furukawa, and M. Watanabe, *Heterocycles*, 1993, **36**, 1483–1488.
[20] W.F. Bailey and M.W. Carson, *Tetrahedron Lett.*, 1997, **38**, 1329–1332.
[21] J. Barluenga, F.J. Fañanás, R. Sanz, and Y. Fernández, *Tetrahedron Lett.*, 1999, **40**, 1049–1052.
[22] J. Barluenga, R. Sanz, and F. J. Fañanás, *Tetrahedron Lett.*, 1997, **38**, 2763–2766.
[23] J. Barluenga, F.J. Fañanás, R. Sanz, and Y. Fernández, *Chem. Eur. J.*, 2002, **8**, 2034–2046.
[24] L.V. Kudzma, *Synthesis*, 2003, 1661–1666.
[25] R. Sanz, Y. Fernández, M.P. Castroviejo, *et al.*, *J. Org. Chem.*, 2006, **71**, 6291–6294.
[26] H. Tokuyama, K. Okano, H. Fujiwara, *et al.*, *Chem. Asian J.*, 2011, **6**, 560–572.
[27] T. Noji, H. Fujiwara, K. Okano, and H. Tokuyama, *Org. Lett.*, 2013, **15**, 1946–1949.
[28] S. Kamila and E. Biehl, *Heterocycles*, 2004, **63**, 2785–2795.
[29] D. Hong, Z. Chen, X. Lin, and Y. Wang, *Org. Lett.*, 2010, **12**, 4608–4611.
[30] D. McAusland, S. Seo, D.G. Pintori, *et al.*, *Org. Lett.*, 2011, **13**, 3667–3669.
[31] S. Chakrabarty, I. Chatterjee, L. Tebben, and A. Studer, *Angew. Chem. Int. Ed.*, 2013, **52**, 2968–2971.
[32] L. He, J.-X. Pian, J.-F. Shi, *et al.*, *Tetrahedron*, 2014, **70**, 2400–2405.
[33] A. Bunescu, Q. Wang, and J. Zhu, *Org. Lett.*, 2014, **16**, 1756–1759.
[34] C.-Y. Tang, G. Wang, X.-Y. Yang, *et al.*, *Tetrahedron Lett.*, 2014, **55**, 6447–6450.
[35] Z. Liu and R.C. Larock, *Org. Lett.*, 2004, **6**, 3739–3741.
[36] Z. Liu and R.C. Larock, *Tetrahedron*, 2007, **63**, 347–355.
[37] D.C. Rogness, N.A. Markina, J.P. Waldo, and R.C. Larock, *J. Org. Chem.*, 2012, **77**, 2743–2755.
[38] C.D. Gilmore, K.M. Allan, and B.M. Stoltz, *J. Am. Chem. Soc.*, 2008, **130**, 1558–1559.
[39] A.A. Cant, G.H.V. Bertrand, J.L. Henderson, *et al.*, *Angew. Chem. Int. Ed.*, 2009, **48**, 5199–5202.
[40] I.L. Nikonov, D.S. Kopchuk, I.S. Kovalev, *et al.*, *Tetrahedron Lett.*, 2013, **54**, 6427–6429.

PART IX

Indoles from Indolines

The fact that indoline – 2,3-dihydroindole – is only a simple dehydrogenation step to indole, this conversion represents an obvious and important indole synthesis.

Indole Ring Synthesis: From Natural Products to Drug Discovery, First Edition. Gordon W. Gribble.

66

Indoline Dehydrogenation

An extremely important and versatile indole synthesis is the dehydrogenation (oxidation) of indolines (2,3-dihydroindoles), as we have already seen in several previous chapters. This method of indole ring generation is presented here in full. For an excellent summary of the early methods, see Sundberg [1].

The venerable method of dehydrogenation using palladium on charcoal has been employed extensively to dehydrogenate indolines to indoles, and some early examples are shown in Scheme 1 (equations 1–5) [2–6]. In addition to methoxy- and cyanoindolines, both indoline-5- and indoline-7-carboxylic acids were smoothly dehydrogenated to the corresponding indole carboxylic acids using 10% Pd/C (refluxing xylene; 80%–81%) [3].

Kuehne and Hall explored several conditions to dehydrogenate indoline using palladium and found that palladium dichloride in methanol/triethylamine afforded the highest yield of indole (83%) [7]. Naito and coworkers reported a one-pot synthesis of indoles from *N*-benzylindolines (Scheme 2, equation 1), a reaction that proceeds by initial debenzylation [8]. Gribble and Pelcman employed Pd/C indoline dehydrogenation in the first total synthesis of the marine alkaloid fascaplysin (equations 2 and 3) [9, 10]. In this study, a mixture of indolines **1** and **2** afforded **3** in refluxing ethyl acetate (equation 2), but the fully aromatic **4** was obtained in refluxing 2-ethoxy ether (180–190 °C) (equation 3). The dehydrogenation of a hexahydrocarbazole was featured in Clive's synthesis of carbazomycin B (equation 4) [11]. An example of the Pd/C dehydrogenation of a tetrahydrocarbazole to a carbazole was described by Brunton and colleagues [12], and Ishii's team reported the synthesis of *N*-butyl-2-methyl-3-propylindole in greater than 99% yield from the corresponding 4,5,6,7-tetrahydroindole (5% Pd/C, decalin, reflux) [13].

A selective compilation of indoles that were prepared by Pd/C dehydrogenation of the corresponding indolines is shown in Scheme 3 [14–21]. The newly introduced bond is shown in bold. In most cases 10% Pd/C was used.

Kaneda and colleagues found that a reusable hydroxyapatite-bound palladium catalyst was effective for the dehydrogenation of simple indolines to indoles in generally excellent yields (>92%) (toluene, 100 °C) [22]. The authors propose an indoline nitrogen-Pd(0) species, followed by oxidative addition at the adjacent C–H bond to give a σ-alkyl Pd species. Subsequent β-hydride elimination affords the indole. The catalyst was prepared from calcium hydroxyapatite and $PdCl_2(PhCN)_2$ in acetone. Crabtree and colleagues studied the indoline dehydrogenation with various heterogeneous catalysts, both computationally and experimentally, and found that indoline is fully dehydrogenated in 30 minutes with Pd/C or Rh/C at 110 °C in toluene [23].

Another extremely useful and versatile indoline dehydrogenation reagent is 2,3-dichloro-5,6-dicyano-1,4-benzoquinone (DDQ) and related benzoquinones such as tetrachloro-1,4-benzoquinone (chloranil). Braude and colleagues were the first to recognize the general oxidative power of these reagents [24]. These quinones, particularly DDQ and chloranil, are the two most widely used indoline–to–indole dehydrogenation reagents. An early example was that of Terent'ev and colleagues who employed chloranil in the synthesis of 5-acylindoles (Scheme 4,

Indole Ring Synthesis: From Natural Products to Drug Discovery, First Edition. Gordon W. Gribble.

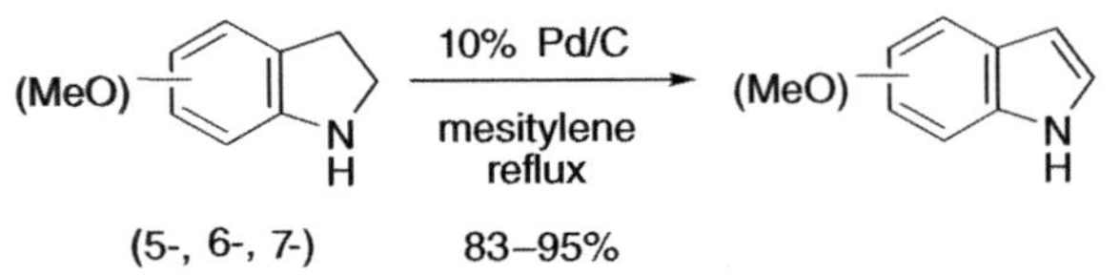

[eqn. 1] [2]

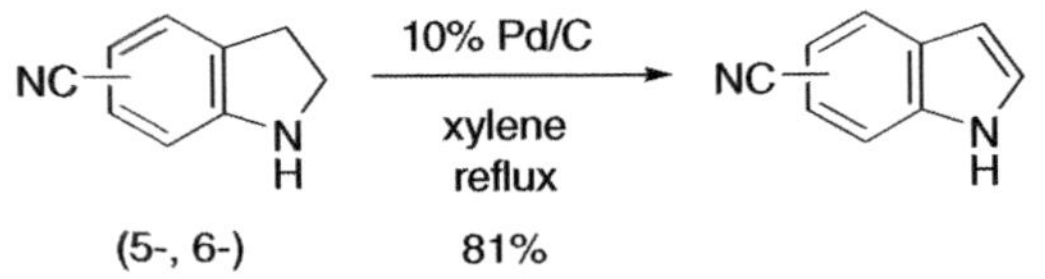

[eqn. 2] [3]

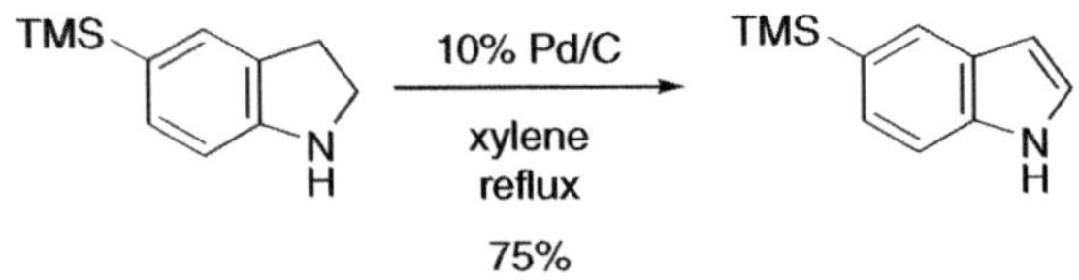

[eqn. 3] [4]

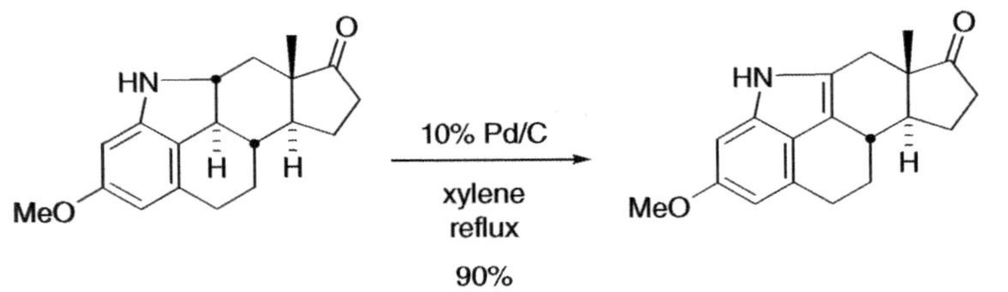

[eqn. 4] [5]

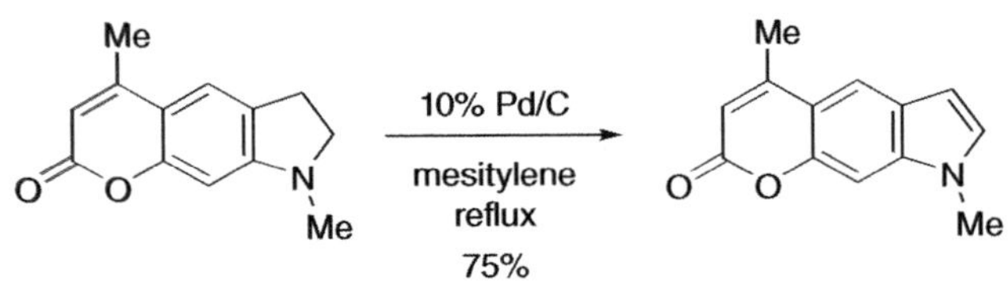

[eqn. 5] [6]

Scheme 1 *Dehydrogenation of Indolines*

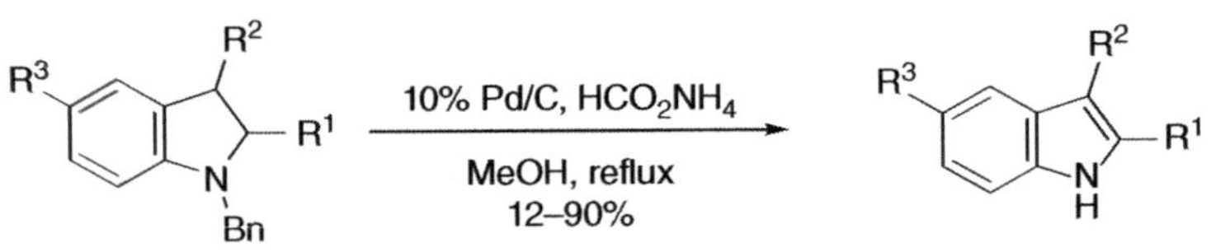

[eqn. 1] [8]

R^1 = H, Me
R^2 = H, Me, CH_2NHBz
R^3 = H, OMe
R^1, R^2 = $-(CH_2)_4-$

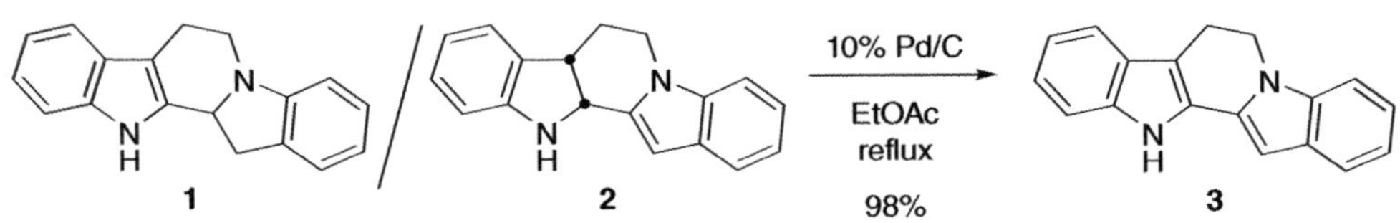

[eqn. 2] [9], [10]

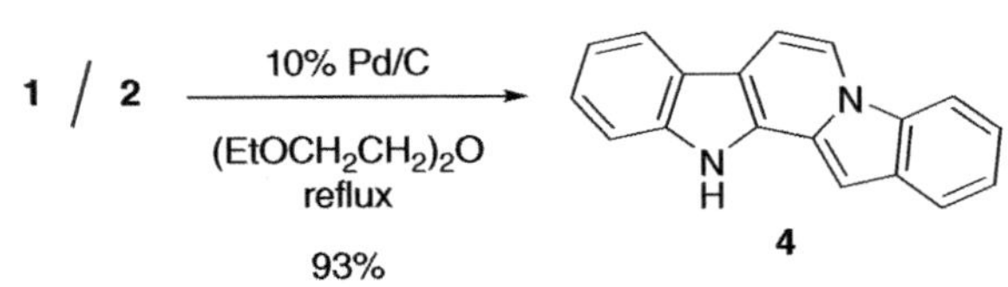

[eqn. 3] [9], [10]

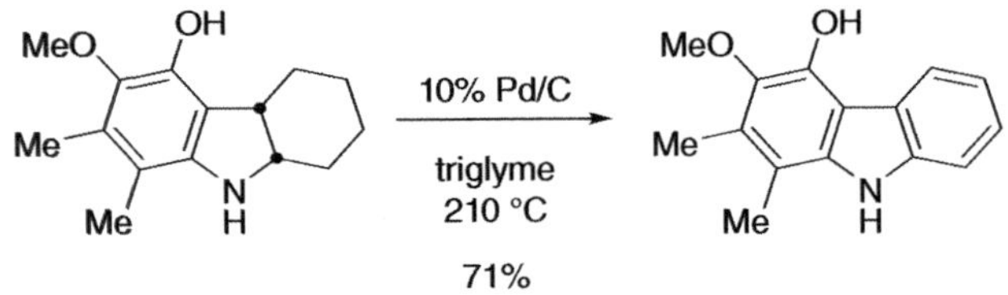

[eqn. 4] [11]

Scheme 2 *Naito, Gribble, and Clive Indoline Dehydrogenations*

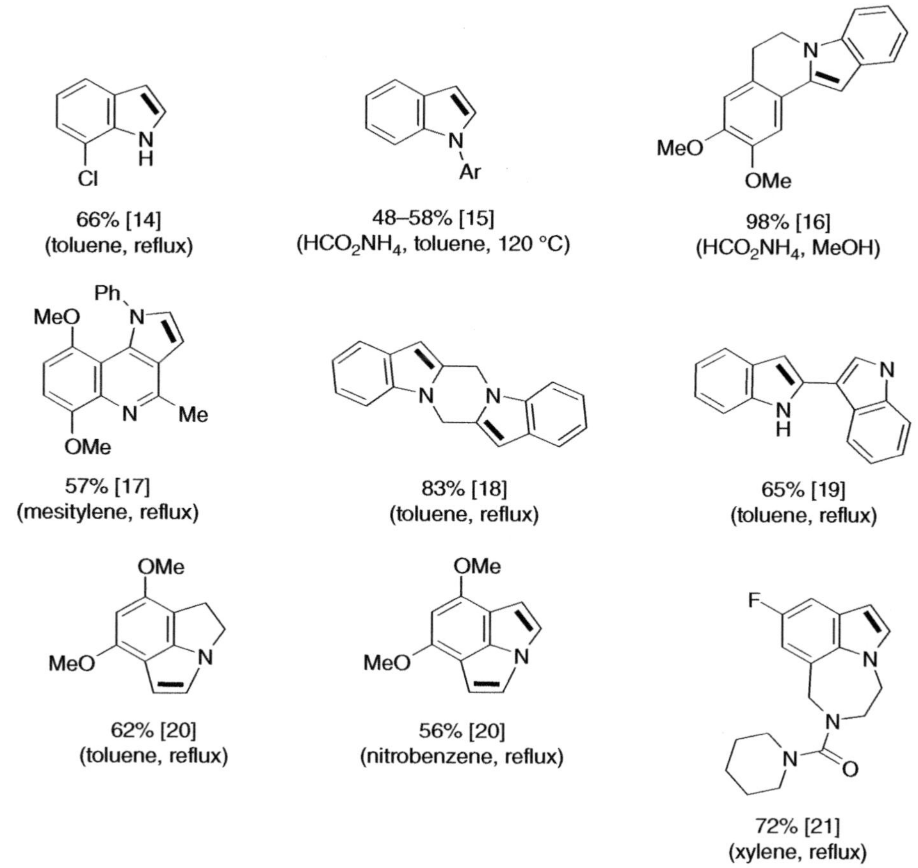

Scheme 3 *Indoles from Pd/C Dehydrogenation of Indolines*

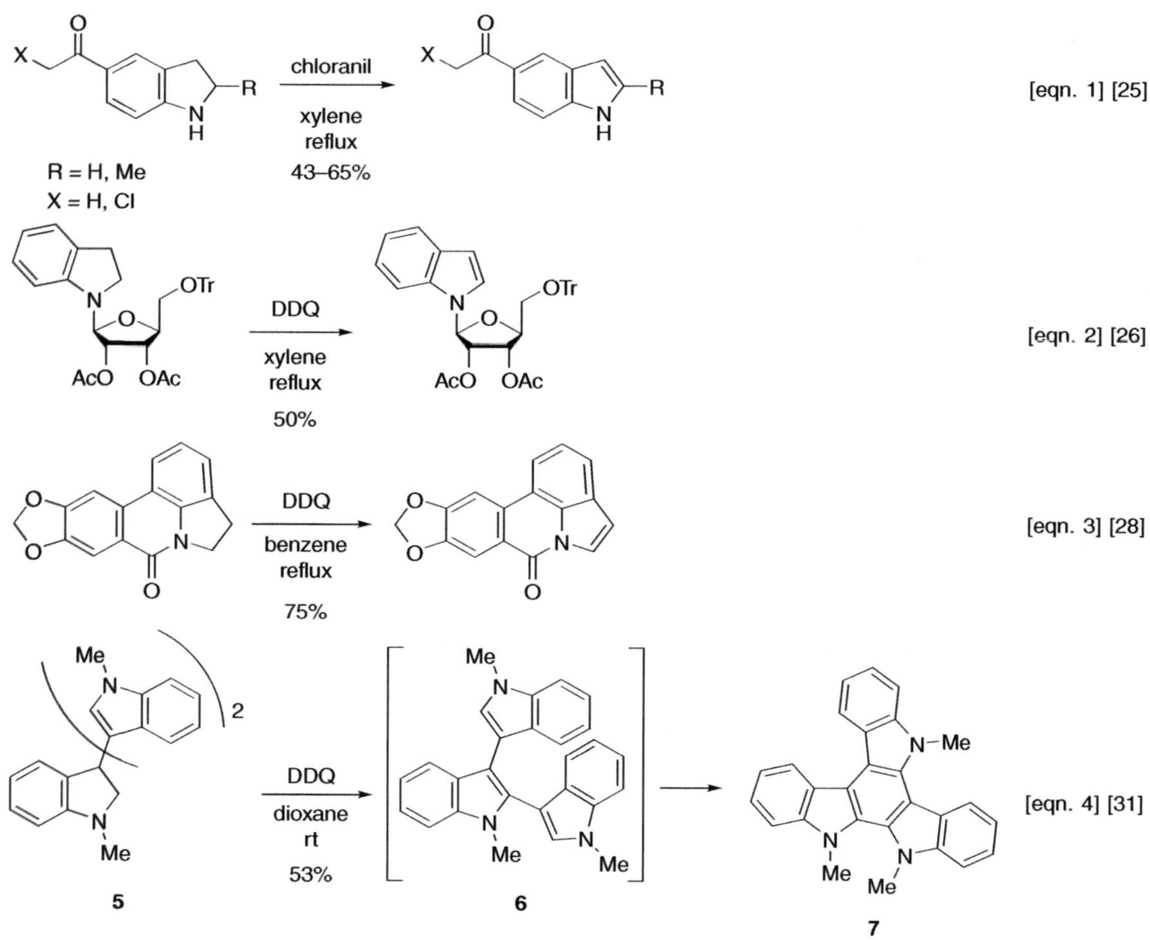

Scheme 4 *Indoles from Chloranil and DDQ Dehydrogenation of Indolines*

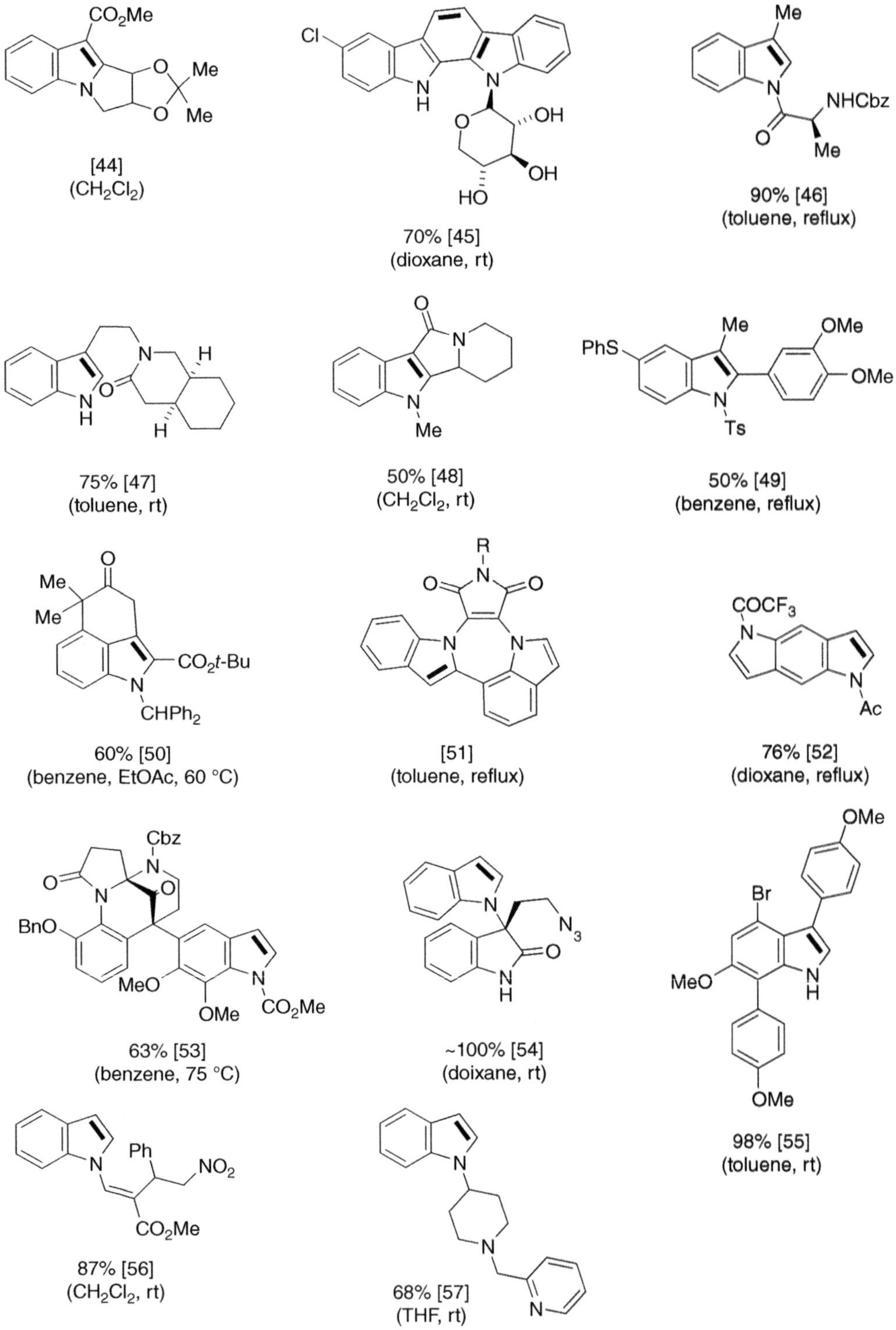

Scheme 5 *Indoles from DDQ Dehydrogenation of Indolines*

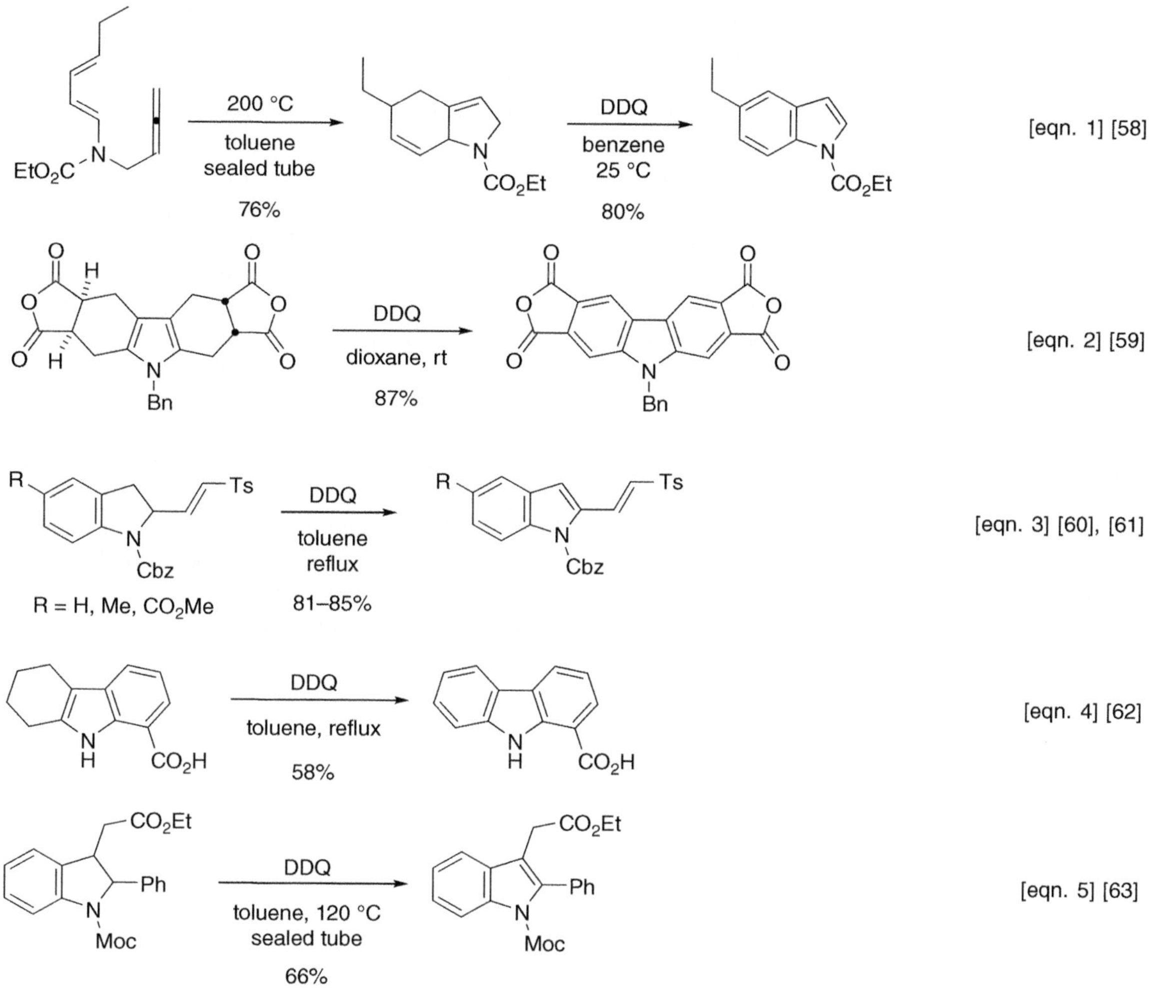

Scheme 6 *Miscellaneous Indoline to Indole DDQ Dehydrogenations*

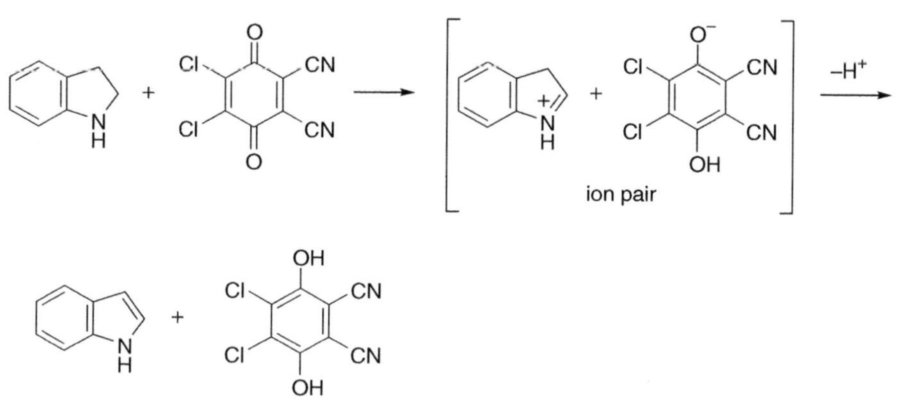

Scheme 7 *A Mechanism of DDQ-Induced Dehydrogenation of Indoline [75]*

equation 1) [25]. An early use of DDQ was reported by Preobrazhenskaya and colleagues in the synthesis of a 1-(D-β-ribofuranosyl)indole (equation 2) [26]. Walton and colleagues described similar DDQ dehydrogenations of several indoline nucleosides, and they also noted the superiority of DDQ over chloranil (90% vs. 30%) [27]. Ghosal's team employed DDQ dehydrogenation as the final step in a synthesis of hippadine (equation 3) [28]. Black and coworkers synthesized several pyrrolophenanthridone alkaloids and analogues in similar fashion [29, 30]. Bergman and Eklund reported some surprising DDQ-induced oxidative cyclizations (**5**→**6**) with tris-indolobenzenes (equation 4) [31].

Dehydrogenation of indolines using DDQ has been used to prepare 6-nitroindole (benzene, reflux, 86%) [32], 7-bromo-5-nitroindole (98%) [33], 6-azido-4,5,7-trifluoroindole (benzene, 88%) [34], 4,5,6,7-tetrahydroindole (benzene, reflux, 68%) [35], methyl 1-acetyl-2-cyclopropyl-5-methoxyindole-3-carboxylate (toluene, reflux, 87%) [36], 5,6-dicyano-1-methylindole (benzene, reflux, 97%) [37], 5,6-dicyano-1-methyl-4-(methylthio)indole (benzene, reflux, 97%) [37], 1-(*N*,*N*-dimethylsulfamoyl)indole-5-(*N*-phenyl)carboxamide (toluene, reflux, 76%) [38], 1-benzylindole-5-carbaldehyde (CH_2Cl_2, 0 °C, 75%) [39], 1-acryloylindole (toluene, reflux, 81%) [40], and several glycosyl-indoles (dioxane, rt to 50 °C) [41–43]. A compilation of interesting indoles and indole natural products that were synthesized via a DDQ oxidation of the corresponding indoline is shown in Scheme 5 [44–57]. The bond so formed is depicted in bold. Clearly, an extraordinarily rich collection of indoles is available using DDQ dehydrogenation of the appropriate indolines.

The use of DDQ in more-extended dehydrogenations has been described by several investigators (Scheme 6, equations 1–5) [58–63]. The examples in equations 1 and 2 are representative of a large group of indole syntheses. Reiser and coworkers found that DDQ in toluene (sealed tube, microwave, 120 °C) was superior both to other conditions with DDQ and to MnO_2, NBS, and CAN in effecting the indolization in equation 5 [63].

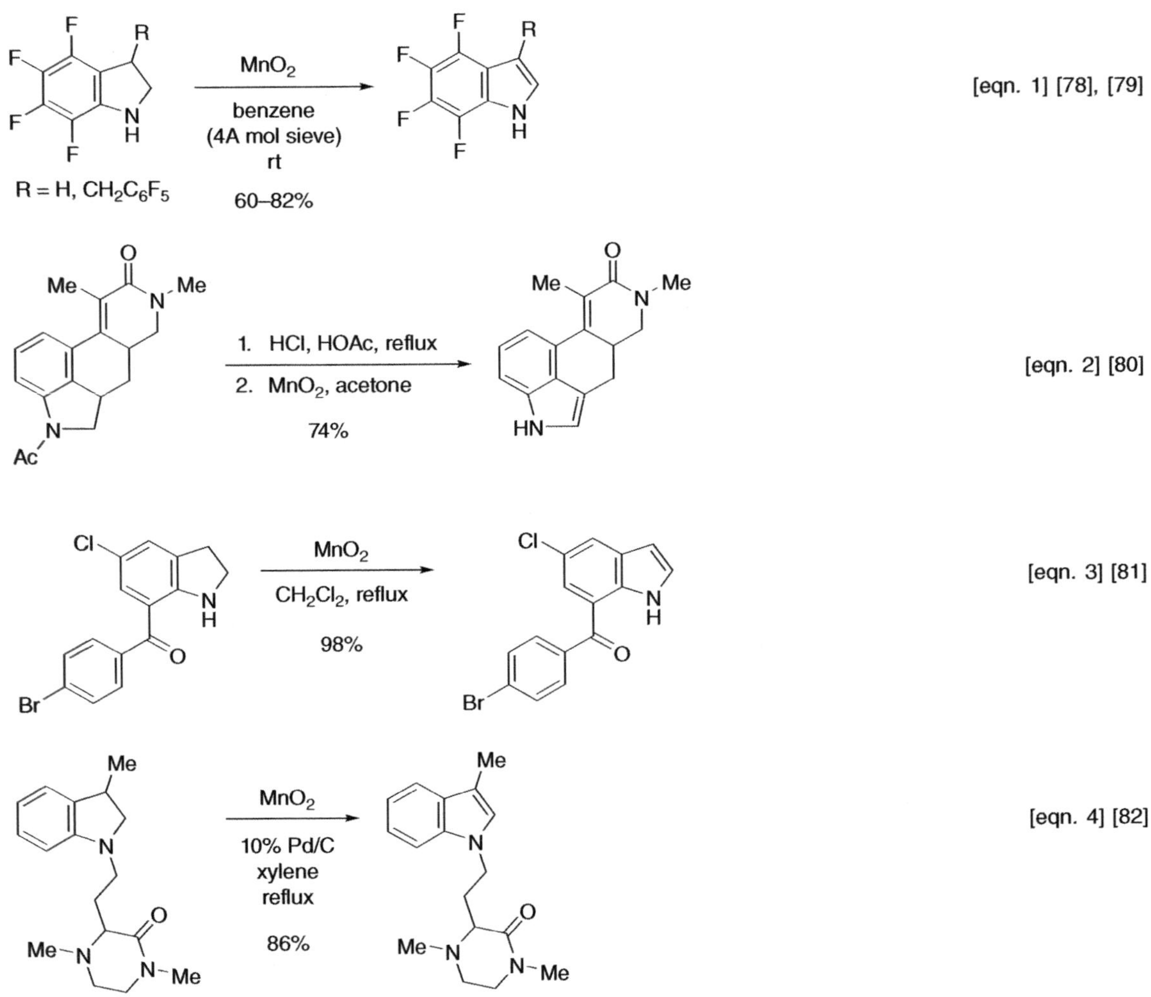

Scheme 8 *Indoles from MnO_2 Dehydrogenation of Indolines*

Dehydrogenation of indolines with DDQ to form indoles has been used to access the plant alkaloid 9-methoxycarbazole-3-carbaldehyde (benzene, 29%) [64] and in syntheses of a novel diazaindacene BODIPY dye (CH_2Cl_2, rt, 93%) [65], *N*-glycosylated indoxyls (dioxane, 20 °C, 80%) [66], 3a,9b-dihydrobenzo[g]indoles (Et_2O, rt, 79%) [67], new *N*-aryl indole-derived axially chiral phosphine ligands (xylene, reflux, 51%) [68], enantioenriched *N*-allylindoles (THF, 50 °C, 76%–92%) [69], diazepinoindoles as glycogen synthase kinase-3 inhibitors (Et_2O, toluene, rt, 50%) [70], and (the final step) in a synthesis of rizatriptan (toluene, reflux, 83.4%) [71]. The conditions and yields shown refer to the indolization step. *o*-Chloranil was used by Bailey and Jiang to dehydrogenate *N*-allyl-3-indoline to its indole (benzene, rt, 82%) [72] and by Danheiser and Dunetz to prepare a 4-carbomethoxyindole analogue (benzene, rt, 88%) [73]. Ikan and Rapaport synthesized 5-cyanoindole and 5-bromoindole from the respective indolines using chloranil (xylene, reflux) in 45% and 42% yields, respectively [3]. Saracoglu and coworkers employed *p*-benzoquinone to dehydrogenate several 4,7-dihydroindoles to indoles [74]. Based on calculations by Crabtree and colleagues, a simple mechanism for the DDQ indoline dehydrogenation can be formulated (Scheme 7) [75].

Another extremely attractive and popular indoline to indole dehydrogenation reagent is manganese dioxide. In contrast to the commercially available DDQ (and chloranil), MnO_2 often needs to be freshly prepared for maximum activity.

Pratt and McGovern first described the dehydrogenation of indoline to indole in 59% yield using manganese dioxide (benzene, reflux, water removal), although the authors state that the "yield could probably be raised by stopping reaction when yield of water reaches 100%" [76]. Independently, Jansen and coworkers reported an indole yield of 56% when a mixture of indoline, MnO_2, and benzene "was shaken overnight" [77] A selection of early applications of the MnO_2 indoline dehydrogenation is shown in Scheme 8 (equations 1–4) [78–82]. Filler and colleagues employed molecular sieves (Linde Type 4A) to dehydrogenate 4,5,6,7-tetrafluoroindoline (equation 1) [78] and used the activated MnO_2 procedure of Attenburrow [83]. In their indole synthesis (equation 2), Hunter and colleagues used the Carpino method to activate MnO_2 [84]. Walsh's group achieved excellent yields using commercial activated MnO_2 to prepare a series of related 7-benzoylindoles (equation 3) [81], and Jirkovsky and coworkers employed a mixture of MnO_2 and Pd/C to effect dehydrogenation of indolines in a synthesis of pyrazino-pyrido[1,2-*a*]indole antihypertensive agents (equation 4) [82].

Ketcha discovered that $Mn(OAc)_3$ glacial acetic acid is an excellent method to dehydrogenate *N*-protected

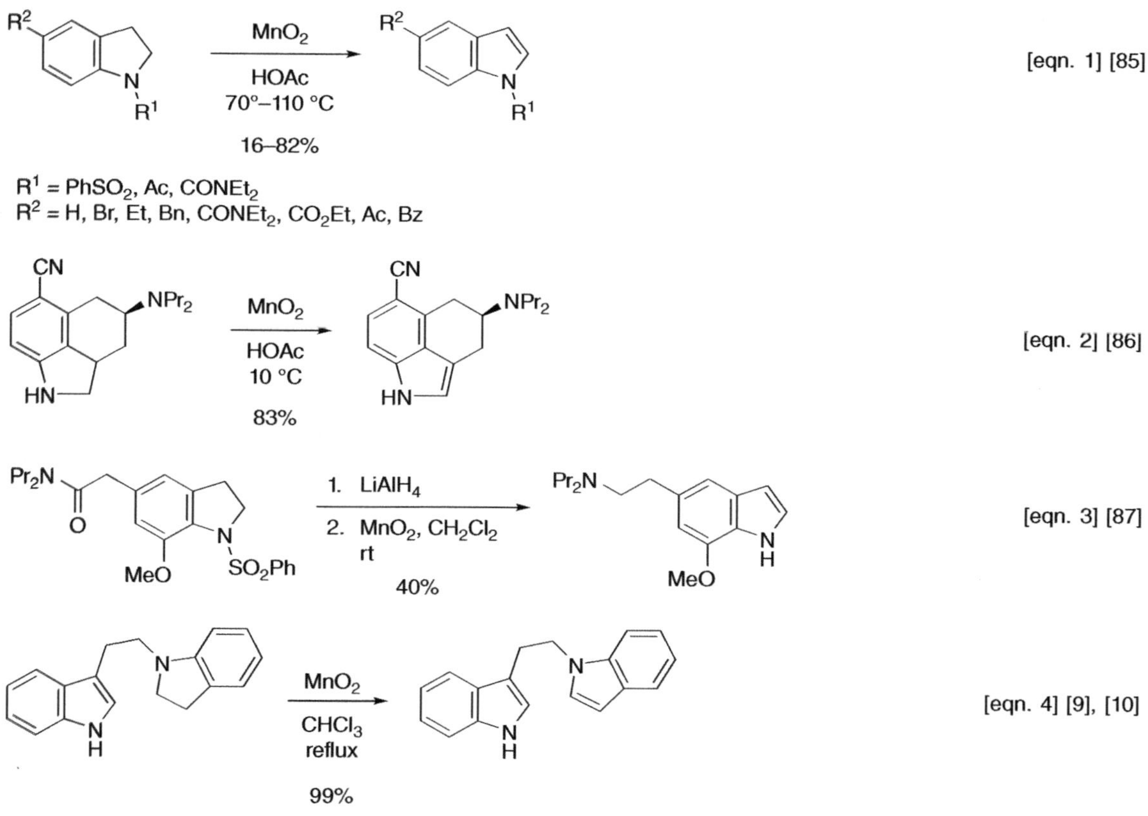

Scheme 9 *Indoles from MnO_2 Dehydrogenation of Indolines – 2*

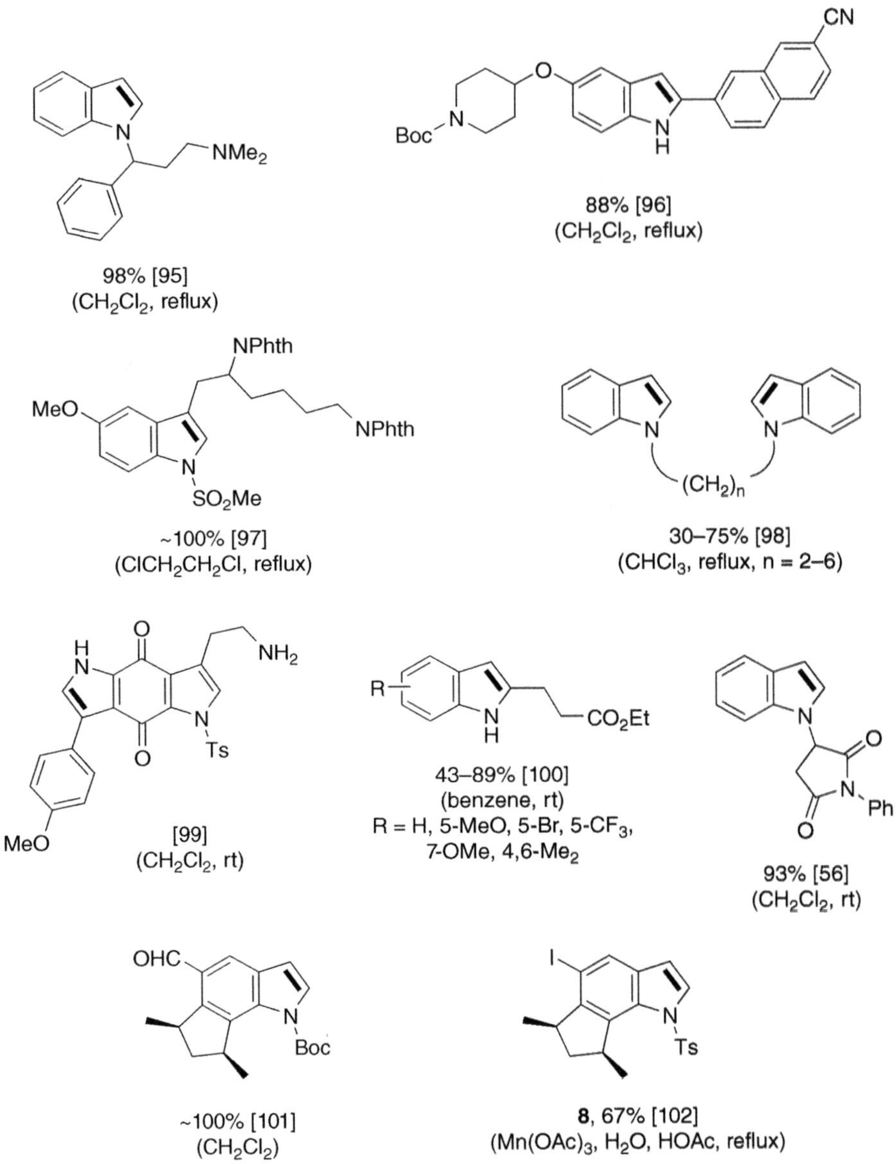

Scheme 10 *Indoles from MnO_2 Dehydrogenation of Indolines – 3*

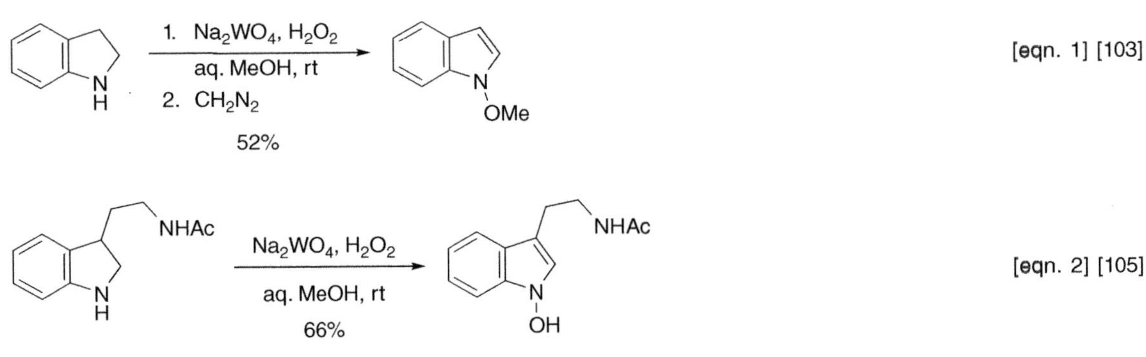

Scheme 11 *Somei N-Hydroxyindole Synthesis*

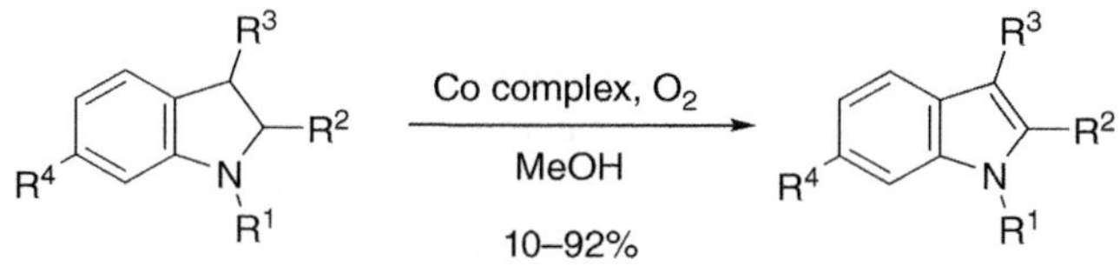

[eqn. 1] [110]

R^1 = H, Me
R^2 = H, Me
R^3 = H, Me, $(CH_2)_3CO_2Et$, $CH_2CH(NHAc)CO_2Et$
R^4 = H, NO_2

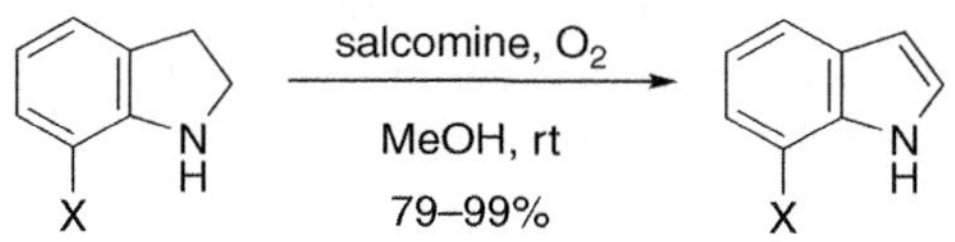

[eqn. 2] [112]

X = Cl, Br, I

Scheme 12 *Inada Indole Syntheses*

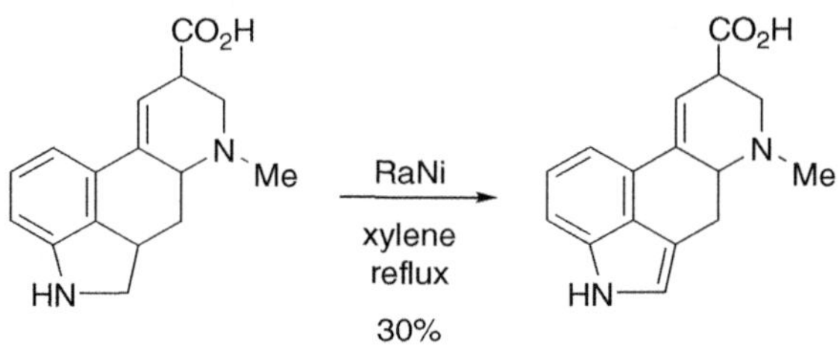

[eqn. 1] [113]

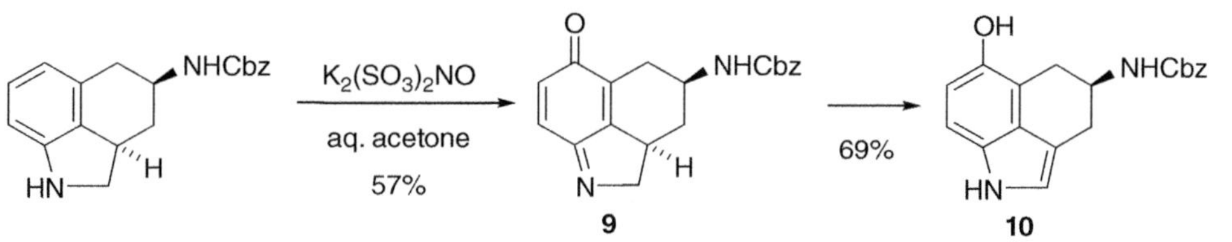

[eqn. 2] [114]

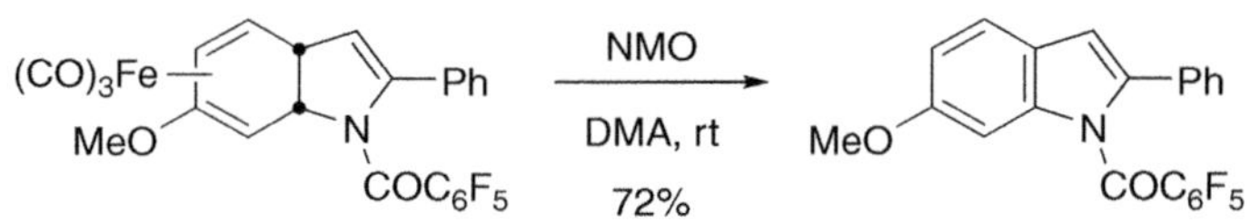

[eqn. 3] [115]

NMO = *N*-methylmorpholine *N*-oxide

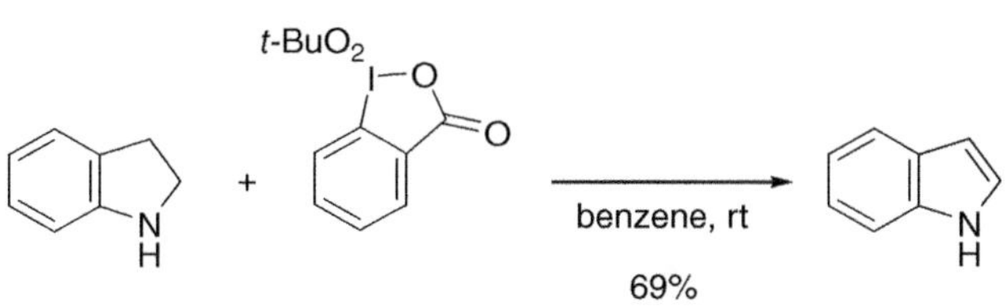

[eqn. 4] [117]

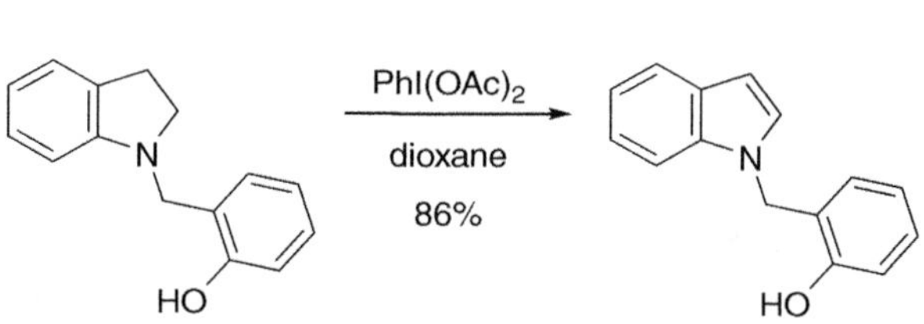

[eqn. 5] [118]

Scheme 13 *Miscellaneous Indoline to Indole Dehydrogenations*

indolines (Scheme 9, equation 1) [85]. This would appear to be an attractive route to *N*-protected indoles provided that the benzene ring is not strongly deactivated. Several additional indole syntheses using MnO_2 are illustrated (equations 2–4) [9, 10, 86, 87]. Like Ketcha, Martinelli's group noted the strong activating effect of acetic acid, relative to methylene chloride, on the rate of dehydrogenation (equation 2) [86].

The MnO_2 indoline dehydrogenation was used to synthesize 4,5,6,7-tetrafluoroindole (benzene, rt, 68%) [35], α-indoline nucleosides (benzene, 40–50 °C; or CH_2Cl_2, 30–40 °C; molecular sieves, 86%–95%) [88], *N*-reverse prenylated indoles (toluene, rt, 86%) [89, 90], 7-hydroxyindole (benzene, reflux, 81%) [91], 7-prenylindole (CH_2Cl_2, reflux, 88%) [92], 3-(1*H*-indol-1-yl)cyclohexanone (CH_2Cl_2, rt, 90%) [93], and methyl 5-bromoindole-7-carboxylate (THF, rt) [94]. A selection of indoles synthesized from the corresponding indolines using MnO_2 is shown in Scheme 10 [56, 95–102]. Like DDQ, MnO_2 is capable of dehydrogenating a wide range of indolines. The bond so formed is in bold. The last example, indole **8**, features the $Mn(OAc)_3$, HOAc Ketcha method [102].

Given the striking synthetic power of Pd/C, DDQ, and MnO_2 to dehydrogenate indolines to indoles, the synthetic chemist has no excuse for being unable to convert an indoline to an indole!

Nevertheless, there are several other oxidation (dehydrogenation) routes to indoles from indolines. In a series of papers, Somei and colleagues used sodium tungstate to synthesize *N*-hydroxyindole derivatives from the corresponding indolines (Scheme 11, equations 1 and 2) [64, 103–107]. The simple 1-hydroxy-6-nitroindole [106] and 4-, 6-, and 7-ethoxy-1-methoxyindoles were synthesized in similar fashion [107]. Pedras and coworkers employed the Somei method to prepare the natural phytoalexin methyl 1-methoxyindole-3-carboxylate [108], and McNab and

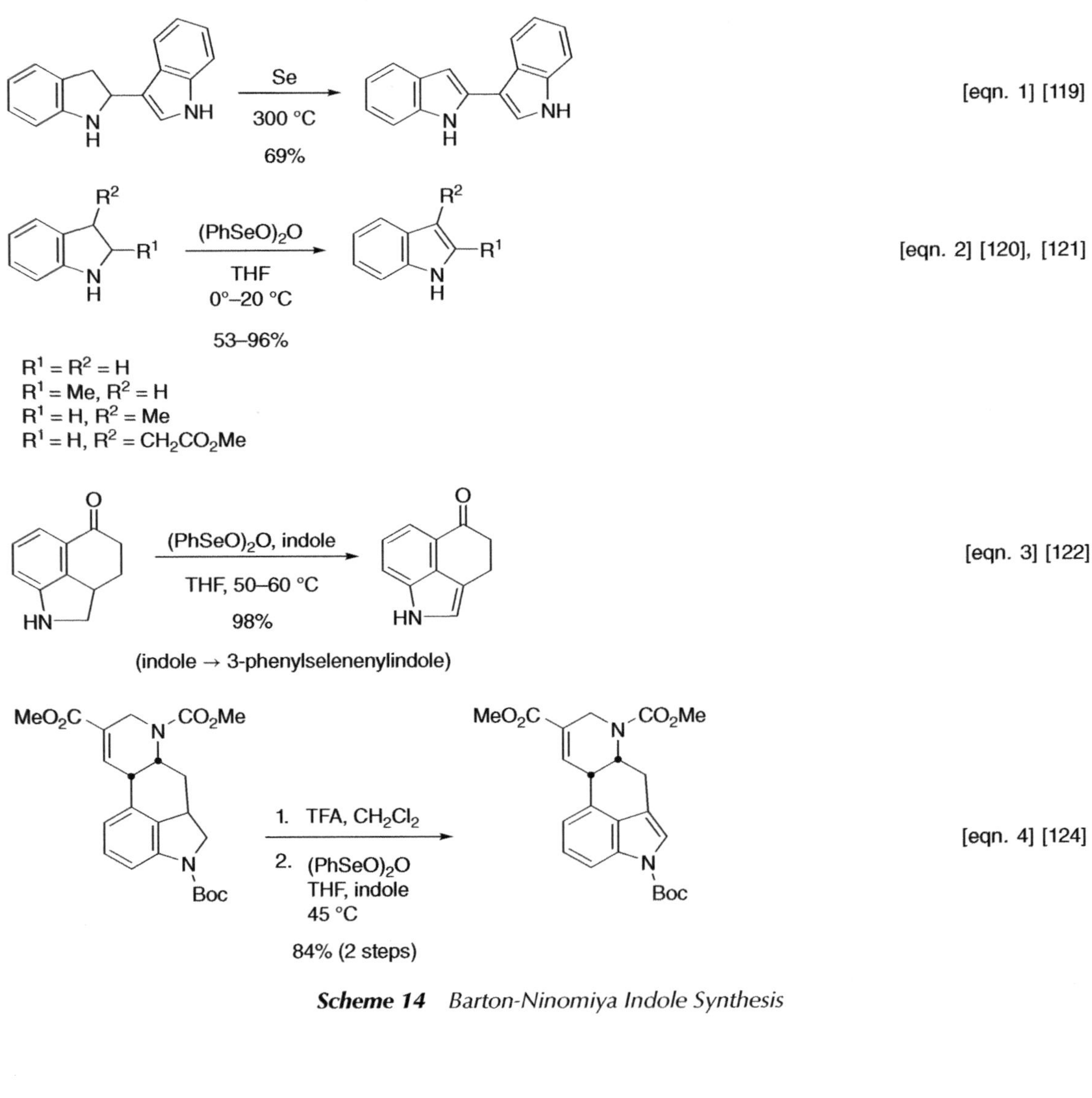

Scheme 14 *Barton-Ninomiya Indole Synthesis*

colleagues employed a tungsten trioxide catalyst (FVP, 525 °C) to synthesize pyrrolo[3,2,1-*ij*]quinolin-6-one from the corresponding indoline [109].

Inada and colleagues found that the combination of oxygen and the cobalt Schiff bases bis(salicylidene) ethylenediaminatocobalt(II) [Co(salen)] or bis(3-methoxysalicylidene)ethylenediaminatocobalt(II) [Co(MeO-salen)] effects the dehydrogenation of indolines to indoles (Scheme 12, equation 1) [110], in some cases superior to $PdCl_2$, MnO_2, and Raney nickel. Somei prepared 7-haloindoles using the Inada method (70%–77%) [111], and De Kimpe synthesized 7-chloroindole (65%) in similar fashion [14]. Yamada and colleagues obtained 7-haloindoles in excellent yields under these conditions (equation 2) [112].

A number of other indoline-to-indole dehydrogenations are known, although they have received lesser attention from the synthetic community compared to the previous methods.

The noteworthy synthesis of lysergic acid by Woodward and colleagues featured a Raney nickel final oxidation (Scheme 13 [113–115, 117, 118], equation 1 [113]). Fremy's salt was employed by Schaus and Giethlen in a similar transformation (equation 2) [114]. The conversion of quinone imine **9** to indole **10** occurs upon standing (more stable tautomer). Ochiai, Nagao, Moriarty, and coworkers found that iodosobenzene and *tert*-butylperoxyiodane dehydrogenate indoline to indole (equation 4) [116, 117]. The latter reagent gave the highest yield (69% vs. 38%). Sun, Pan, and colleagues used this dehydrogenation to synthesize indole **11** (equation 5) [118].

Selenium has been found by a few investigators to dehydrogenate indolines (Scheme 14). Bergman converted 2-(3-indolyl)indoline (indole dimer) to 2,3′-biindolyl (equation 1) [119]. Barton, Ninomiya, and colleagues showed that phenylseleninic anhydride converts indolines to indoles (equations 2, 3) 120–122]. If the indoline β-position is unsubstituted, then the β-phenylselenoindole is formed (96%–98%), which can be reduced by nickel boride to the indole product. The method was greatly improved by adding a sacrificial enamine to scavenge the PhSeOH that would otherwise react with indole product. Thus, indole (or dihydropyran) served this function (equation 3) [122]. The yield of the scavenged 3-phenylselenenylindole was approximately 100%. This indolization was adapted by Ninomiya, Barton, and coworkers into syntheses of the ergot alkaloids (±)-lysergol, (±)-isolysergol, and (±)-elymoclavine [123]. Fukuyama's team employed this phenylseleninic anhydride dehydrogenation in a total synthesis of (+)-lysergic acid (equation 4) [124]. Nicolaou and colleagues used a polymer-bound selenenyl bromide resin to

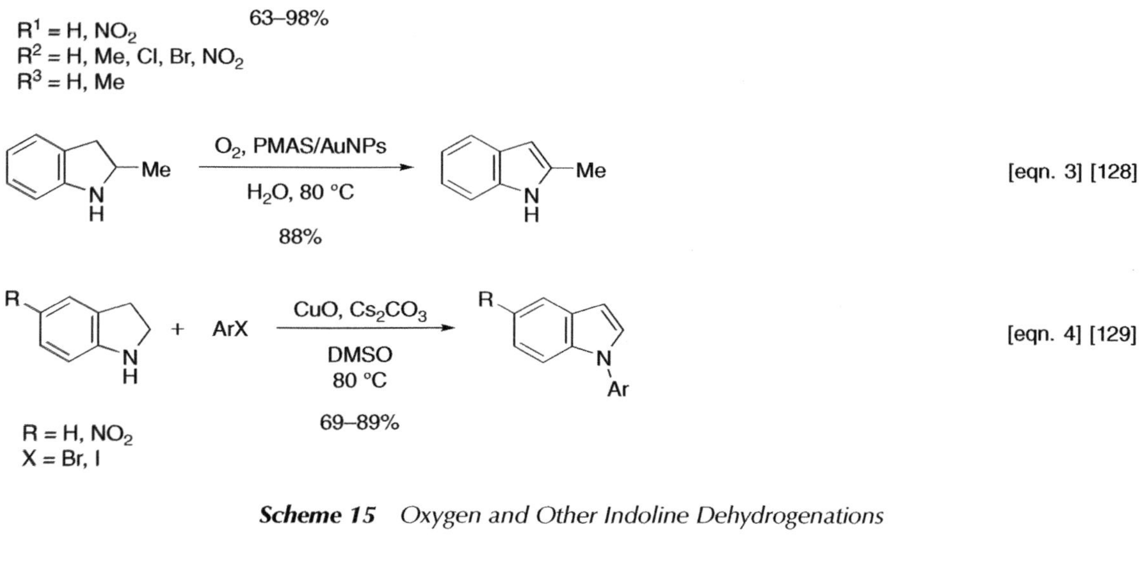

Scheme 15 *Oxygen and Other Indoline Dehydrogenations*

develop a series of indoline and indole platforms, leading to libraries of *N*-functionalized 2-methylindoles [125].

Given that indolines are prone to undergo aeration oxidation [126] (Scheme 15, equation 1) [60], it is not surprising that several metal-catalyzed aeration indoline-to-indole conversions are known (equations 2–4) [127–129]. The catalyst in equation 3, PMAS/Au NPs, is poly(2-methoxyaniline-5-sulfonic acid) on gold nanoparticles [128]. The novel copper-catalyzed indoline dehydrogenation followed by *N*-arylation was discovered by Nageswar and coworkers (equation 4) [129]. A large number of aryl halides (usually iodides) were employed. Indoline-2-carboxylic acids are likewise converted to the *N*-arylindoles. Indoline was oxidized to indole in 73% yield by catalytic tetra-*n*-propylammonium perruthenate and *N*-methylmorpholine *N*-oxide [130].

References

[1] R.J. Sundberg (1970) *The Chemistry of Indoles*, Academic Press, New York, pp. 132–135.
[2] R.R. Hunt and R.L. Rickard, *J. Chem. Soc. (C)*, 1966, 344–345.
[3] R. Ikan and E. Rapaport, *Tetrahedron*, 1967, **23**, 3823–3827.
[4] I. Belsky, D. Gertner, and A. Zilkha, *J. Org. Chem.*, 1968, **33**, 1348–1350.
[5] E.W. Cantrall, R.B. Conrow, and S. Bernstein, *J. Org. Chem.*, 1967, **32**, 3445–3452.
[6] E. Quanten, P. Adriaens, F.C. De Schryver, *et al.*, *Photochem. Photobiol.*, 1986, **43**, 485–492.
[7] M.E. Kuehne and T.C. Hall, *J. Org. Chem.*, 1976, **41**, 2742–2746.
[8] T. Kiguchi, N. Kuninobu Y, Takahashi, *et al.*, *Synthesis*, 1989, 778–781.
[9] B. Pelcman and G.W. Gribble, *Tetrahedron Lett.*, 1990, **31**, 2381–2384.
[10] G.W. Gribble and B. Pelcman, *J. Org. Chem.*, 1992, **57**, 3636–3642.
[11] D.L.J. Clive, N. Etkin, T. Joseph, and J.W. Lown, *J. Org. Chem.*, 1993, **58**, 2442–2445.
[12] R.J. Brunton, F.K. Drayson, S.G.P. Plant, and M.L. Tomlinson, *J. Chem. Soc.*, 1956, 4783–4785.
[13] H. Shiraishi, T. Nishitani, S. Sakaguchi, and Y. Ishii, *J. Org. Chem.*, 1998, **63**, 6234–6238.
[14] N. De Kimpe and M. Keppens, *Tetrahedron*, 1996, **52**, 3705–3718.
[15] M. Beller, C. Breindl, T.H. Riermeier, and A. Tillack, *J. Org. Chem.*, 2001, **66**, 1403–1412.
[16] R. Omar-Amrani, A. Thomas, E. Brenner, *et al.*, *Org. Lett.*, 2003, **5**, 2311–2314.
[17] R.A. Tapia, C. López, A. Morello, *et al.*, *Synthesis*, 2005, 903–906.
[18] D.P. Zlotos, C. Tränkle, A. Abdelrahman, *et al.*, *Bioorg. Med. Chem. Lett.*, 2006, **16**, 1481–1485.
[19] N. Wahlström, J. Slätt, B. Stensland, *et al.*, *J. Org. Chem.*, 2007, **72**, 5886–5889.
[20] Jumina, N. Kumar, and D.St.C. Black, *Tetrahedron*, 2009, **65**, 2591–2598.
[21] N.A. Magnus, C.P. Ley, P.M. Pollock, and J.P. Wepsiec, *Org. Lett.*, 2010, **12**, 3700–3703.
[22] T. Hara, K. Mori, T. Mizugaki, *et al.*, *Tetrahedron Lett.*, 2003, **44**, 6207–6210.
[23] A. Moores, M. Poyatos, Y. Luo, and R.H. Crabtree, *New J. Chem.*, 2006, **30**, 1675–1678.
[24] E.A. Braude, A.G. Brook, and R.P. Linstead, *J. Chem. Soc.*, 1954, 3569–3574.
[25] A.P. Terent'ev, M.N. Preobrazhenskaya, and G.M. Sorokina, *J. Gen. Chem. USSR Engl. Trans.*, 1959, **29**, 2835–2841.
[26] M.N. Preobrazhenskaya, M.M. Vigdorchik, and N.N. Suvorov, *Tetrahedron*, 1967, **23**, 4653–4660.
[27] E. Walton, F.W. Holly, and S.R. Jenkins, *J. Org. Chem.*, 1968, **33**, 192–197.
[28] S. Ghosal, P.H. Rao, D.K. Jaiswal, *et al.*, *Phytochemistry*, 1981, **20**, 2003–2007.
[29] D.St.C. Black, P.A. Keller, and N. Kumar, *Tetrahedron Lett.*, 1989, **30**, 5807–5808.
[30] D.St.C. Black, P.A. Keller, and N. Kumar, *Tetrahedron*, 1993, **49**, 151–164.
[31] J. Bergman and N. Eklund, *Tetrahedron*, 1980, **36**, 1445–1450.
[32] L.L. Melhado and N.J. Leonard, *J. Org. Chem.*, 1983, **48**, 5130–5133.
[33] L.L. Melhado, C.J. Pearce, M. D'Alarcao, and N.J. Leonard, *Phytochemistry*, 1982, **21**, 2879–2885.
[34] M.W. Shaffer and M.S. Platz, *Tetrahedron Lett.*, 1989, **30**, 6465–6468.
[35] R. Filler, W. Chen, and S.M. Woods, *J. Fluorine Chem.*, 1995, **73**, 95–100.
[36] M.A. Naylor, M. Jaffar, J. Nolan, *et al.*, *J. Med. Chem.*, 1997, **40**, 2335–2346.
[37] D. Giomi and M. Cecchi, *J. Org. Chem.*, 2003, **68**, 3340–3343.
[38] J.-F. Rousseau and R.H. Dodd, *Heterocycles*, 2001, **55**, 2289–2304.
[39] L. Poszávácz, G. Simig, J. Fetter, and F. Bertha, *Heterocycles*, 2006, **68**, 713–719.
[40] J. Magolan, C.A. Carson, and M.A. Kerr, *Org. Lett.*, 2008, **10**, 1437–1440.
[41] M. Sassatelli, E. Saab, F. Anizon, *et al.*, *Tetrahedron Lett.*, 2004, **45**, 4827–4830.
[42] N. Ding, X. Du, W. Zhang, *et al.*, *Bioorg. Med. Chem. Lett.*, 2011, **21**, 3531–3535.
[43] S. Libnow, M. Hein, and P. Langer, *Tetrahedron Lett.*, 2008, **49**, 289–291.
[44] H.-J. Lim and G.A. Sulikowski, *J. Org. Chem.*, 1995, **60**, 2326–2327.
[45] E.J. Gilbert, J.W. Ziller, and D.L. Van Vranken, *Tetrahedron*, 1997, **53**, 16553–16564.
[46] F. He, B.M. Foxman, and B.B. Snider, *J. Am. Chem. Soc.*, 1998, **120**, 6417–6418.
[47] S.M. Sparks and K.J. Shea, *Tetrahedron Lett.*, 2000, **41**, 6721–6724.
[48] G.W. Gribble, H.L. Fraser, and J.C. Badenock, *Chem. Commun.*, 2001, 805–806.
[49] S. Akai, N. Kawashita, N. Morita, *et al.*, *Heterocycles*, 2002, **58**, 75–78.

[50] A. Chandra, R. Viswanathan, and J.N. Johnston, *Org. Lett.*, 2007, **9**, 5027–5029.
[51] V.N. Danilenko, A.Y. Simonov, S.A. Lakatosh, *et al.*, *J. Med. Chem.*, 2008, **51**, 7731–7736.
[52] G.K.B. Prasad, A. Burchat, G. Weeratunga, *et al.*, *Tetrahedron Lett.*, 1991, **32**, 5035–5038.
[53] K.C. Nicolaou, S.M. Dalby, S. Li, *et al.*, *Angew. Chem. Int. Ed.*, 2009, **48**, 7616–7620.
[54] N. Takahashi, T. Ito, Y. Matsuda, *et al.*, *Chem. Commun.*, 2010, **46**, 2501–2503.
[55] K. Okano, H. Fujiwara, T. Noji, *et al.*, *Angew. Chem. Int. Ed.*, 2010, **49**, 5925–5929.
[56] H. Kilic, S. Bayindir, E. Erdogan, and N. Saracoglu, *Tetrahedron*, 2012, **68**, 5619–5630.
[57] M. Wang, L. Xu, M. Gao, *et al.*, *Bioorg. Med. Chem. Lett.*, 2011, **21**, 1649–1653.
[58] K. Hayakawa, T. Yasukouchi, and K. Kanematsu, *Tetrahedron Lett.*, 1986, **27**, 1837–1840.
[59] J.T. Vessels, S.Z. Janicki, and P.A. Petillo, *Org. Lett.*, 2000, **2**, 73–76.
[60] T.B. Back, R.J. Bethell, M. Parvez, and J.A. Taylor, *J. Org. Chem.*, 2001, **66**, 8599–8605.
[61] T.G. Back, A. Pandyra, and J.E. Wulff, *J. Org. Chem.*, 2003, **68**, 3299–3302.
[62] J. Kamata, T. Okada, Y. Kotake, *et al.*, *Chem. Pharm. Bull.*, 2004, **52**, 1071–1081.
[63] I. Prediger, T. Weiss, and O. Reiser, *Synthesis*, 2008, 2191–2198.
[64] T. Kawasaki and M. Somei, *Heterocycles*, 1990, **31**, 1605–1608.
[65] J. Chen, A. Burghart, A. Derecskei-Kovacs, and K. Burgess, *J. Org. Chem.*, 2000, **65**, 2900–2906.
[66] S. Libnow, M. Hein, and P. Langer, *Synlett*, 2009, 221–224.
[67] A. Tenaglia and S. Marc, *J. Org. Chem.*, 2008, **73**, 1397–1402.
[68] T. Mino, S. Komatsu, K. Wakui, *et al.*, *Tetrahedron: Asymm.*, 2010, **21**, 711–718.
[69] W.-B. Liu, X. Zhang, L.-X. Dai, and S.-L. You, *Angew. Chem. Int. Ed.*, 2012, **51**, 5183–5187.
[70] H. Gunosewoyo, A. Midzak, I.N. Gaisina, *et al.*, *J. Med. Chem.*, 2013, **56**, 5115–5129.
[71] Y. He, X. Li, J. Li, *et al.*, *Tetrahedron Lett.*, 2014, **55**, 3938–3941.
[72] W.F. Bailey and X.-L. Jiang, *J. Org. Chem.*, 1996, **61**, 2596–2597.
[73] J.R. Dunetz and R.L. Danheiser, *J. Am. Chem. Soc.*, 2005, **127**, 5776–5777.
[74] H. Kilic, S. Bayindir, and N. Saracoglu, *Curr. Org. Synth.*, 2014, **11**, 167–181.
[75] O.R. Luca, T. Wang, S.J. Konezny, *et al.*, *New J. Chem.*, 2011, **35**, 998–999.
[76] E.F. Pratt and T.P. McGovern, *J. Org. Chem.*, 1964, **29**, 1540–1543.
[77] A.B.A. Jansen, J.M. Johnson, and J.R. Surtees, *J. Chem. Soc.*, 1964, 5573–5577.
[78] R. Filler, S.M. Woods, and A.F. Freudenthal, *J. Org. Chem.*, 1973, **38**, 811–812.
[79] R. Filler, S.M. Woods, and W.L. White, *Can. J. Chem.*, 1989, **67**, 1837–1841.
[80] D.C. Horwell, D.E. Tupper, and W.H. Hunter, *J. Chem. Soc., Perkin Trans. 1*, 1983, 1545–1552.
[81] D.A. Walsh, H.W. Moran, D.A. Shamblee, *et al.*, *J. Med. Chem.*, 1984, **27**, 1379–1388.
[82] I. Jirkovsky, G. Santroch, R. Baudy, and G. Oshiro, *J. Med. Chem.*, 1987, **30**, 388–394.
[83] J. Attenburrow, A.F.B. Cameron, J.H. Chapman *et al.*, *J. Chem. Soc.*, 1952, 1094–1111.
[84] L.A. Carpino, *J. Org. Chem.*, 1970, **35**, 3971–3972.
[85] D.M. Ketcha, *Tetrahedron Lett.*, 1988, **29**, 2151–2154.
[86] M.J. Martinelli, M.R. Leanna, D.L. Varie, *et al.*, *Tetrahedron Lett.*, 1990, **31**, 7579–7582.
[87] C.G. Gourdoupis and I.K. Stamos, *Synth. Commun.*, 1993, **23**, 2241–2249.
[88] T. Chandra, S. Zou, and K.L. Brown, *Tetrahedron Lett.*, 2004, **45**, 7783–7786.
[89] H. Sugiyama, F. Yokokawa, T. Aoyama, and T. Shioiri, *Tetrahedron Lett.*, 2001, **42**, 7277–7280.
[90] F. Yokokawa, H. Sugiyama, T. Aoyama, and T. Shioiri, *Synthesis*, 2004, 1476–1480.
[91] K. Ishiyama and Y. Yamada, *Tetrahedron Lett.*, 2005, **46**, 1021–1022.
[92] X. Xiong and M.C. Pirrung, *J. Org. Chem.*, 2007, **72**, 5832–5834.
[93] S. Bayindir, E. Erdogan, H. Kilic, and N. Saracoglu, *Synlett*, 2010, 1455–1458.
[94] D.D. Miller, P. Bamborough, J.A. Christopher, *et al.*, *Bioorg. Med. Chem. Lett.*, 2011, **21**, 2255–2258.
[95] P.E. Mahaney, A.T. Vu, C.C. McComas, *et al.*, *Bioorg. Med. Chem.*, 2006, **14**, 8455–8466.
[96] T. Noguchi, N. Tanaka, T. Nishimata, *et al.*, *Chem. Pharm. Bull.*, 2006, **54**, 163–174.
[97] B. Quiclet-Sire, G. Revol, and S.Z. Zard, *Org. Lett.*, 2009, **11**, 3554–3557.
[98] C. Aubry, A.J. Wilson, D. Emmerson, *et al.*, *Bioorg. Med. Chem.*, 2009, **17**, 6073–6084.
[99] A. Rives, T. Delaine, L. Legentil, and E. Delfourne, *Tetrahedron Lett.*, 2009, **50**, 1128–1130.
[100] F. Brucelle and P. Renaud, *Org. Lett.*, 2012, **14**, 3048–3051.
[101] R.J. Huntley and R.L. Funk, *Org. Chem.*, 2006, **8**, 3403–3406.
[102] W. Liu, H.J. Lim, and T.V. RajaBabu, *J. Am. Chem. Soc.*, 2012, **134**, 5496–5499.
[103] M. Somei and T. Kawasaki, *Heterocycles*, 1989, **29**, 1251–1254.
[104] T. Kawasaki, A. Kodama, T. Nishida, *et al.*, *Heterocycles*, 1991, **32**, 221–227.
[105] M. Somei, Y. Fukui, M. Hasegawa, *et al.*, *Heterocycles*, 2000, **53**, 1725–1736.
[106] K. Yamada, T. Kawasaki, T. Fujita, and M. Somei, *Heterocycles*, 2001, **55**, 1151–1159.

[107] K. Yamada and M. Somei, *Heterocycles*, 2012, **84**, 785–799.
[108] M.S.C. Padras, F.I. Okanga, I.L. Zaharia, and A.Q. Khan, *Phytochemistry*, 2000, **53**, 161–176.
[109] H. McNab, D.J. Nelson, and E.J. Rozgowska, *Synthesis*, 2009, 2171–2174.
[110] A. Inada, Y. Nakamura, and Y. Morita, *Chem. Lett.*, 1980, 1287–1290.
[111] M. Somei, Y. Saida, T. Funamoto, and T. Ohta, *Chem. Pharm. Bull.*, 1987, **35**, 3146–3154.
[112] Y. Yamada, S. Arima, C. Okada, *et al.*, *Chem. Pharm. Bull.*, 2006, **54**, 788–794.
[113] E.C. Kornfeld, E.J. Fornefeld, G.B. Kline, *et al.*, *J. Am. Chem. Soc.*, 1956, **78**, 3087–3114.
[114] B. Giethlen and J.M. Schaus, *Tetrahedron Lett.*, 1997, **38**, 8483–8486.
[115] K. Tanaka, N. Yukimura, and K. Narasaka, *Bull. Chem. Soc. Jpn.*, 2004, **77**, 575–584.
[116] M. Ochiai, M. Inenaga, Y. Nagao, *et al.*, *Tetrahedron Lett.*, 1988, **29**, 6917–6920.
[117] M. Ochiai, D. Kajishima, and T. Sueda, *Heterocycles*, 1997, **46**, 71–76.
[118] H. Mao, R. Xu, J. Wan, *et al.*, *Chem. Eur. J.*, 2010, **16**, 13352–13355.
[119] J. Bergman, *J. Heterocycl. Chem.*, 1973, **10**, 121–122.
[120] D.H.R. Barton, X. Lusinchi, and P. Milliet, *Tetrahedron Lett.*, 1982, **23**, 4949–4952.
[121] D.H.R. Barton, X. Lusinchi, and P. Milliet, *Tetrahedron*, 1985, **41**, 4727–4738.
[122] I. Ninomiya, T. Kiguchi, C. Hashimoto, *et al.*, *Tetrahedron Lett.*, 1985, **26**, 4183–4186.
[123] I. Ninomiya, C. Hashimoto, T. Kiguchi, *et al.*, *J. Chem. Soc., Perkin Trans. 1*, 1990, 707–713.
[124] T. Kurokawa, M. Isomura, H. Tokuyama, and T. Fukuyama, *Synlett*, 2009, 775–778.
[125] K.C. Nicolaou, A.J. Roecker, R. Hughes, *et al.*, *Bioorg. Med. Chem.*, 2003, **11**, 465–476.
[126] G.W. Gribble, unpublished observations.
[127] Y. Nomura, Y. Kawashita, and M. Hayashi, *Heterocycles*, 2007, **74**, 629–635.
[128] T. Amaya, T. Ito, Y. Inada, *et al.*, *Tetrahedron Lett.*, 2012, **53**, 6144–6147.
[129] K.H.V. Reddy, G. Satish, K. Ramesh, *et al.*, *Tetrahedron Lett.*, 2012, **53**, 3061–3065.
[130] A. Goti and M. Romani, *Tetrahedron Lett.*, 1994, **35**, 6567–6570.

67

Indolines to Indoles by Functionalized Elimination

Whereas the wealth of indoline-to-indole transformations involves dehydrogenation (Chapter 66), the functionalization of an indoline and subsequent elimination to afford an indole is a method that has become increasingly popular.

A simple example of this protocol is the synthesis of 4-cyanoindole by Plieninger and Klinga (Scheme 1, equation 1) [1]. The 4-cyanoindoline was prepared in a few steps from 4-keto-4,5,6,7-tetrahydroindole. The use of an indoline as a relay compound between two indoles was first discovered by Thesing and colleagues (equation 2) [2]. Kikugawa and coworkers employed azasulfonium salts to convert indolines to indoles (equation 3) [3–5]. This clever indole synthesis was applied to syntheses of 6-bromoindoles (and 6-methoxyindoles) [4]. Kikugawa subsequently found that *N*-chloroindolines also could be employed in a similar indole synthesis in an elimination process (equation 4) [5].

Similar to Thesing (Scheme 1, equation 1), Chou and Ko prepared a thieno[3,4-*b*]indole via NBS benzylic bromination of the corresponding indoline followed by elimination of HBr [6]. Mukaiyama's group [7] and Tilstam [8] discovered that the indoline-to-indole transformation was possible by *N*-functionalization of indoline using *N*-*tert*-butylphenylsulfinimidoyl chloride and trichloroisocyanuric acid, respectively, followed by DBU-induced elimination. In the latter case, indole yields were 72% to 89%, which included 5- and 6-nitroindoles [8]. Overton and Keirs described Swern oxidation conditions to convert indolines to indoles (Scheme 2, equation 1) [9]. Faul and colleagues employed similar Swern conditions (TFAA, DMSO) to synthesize ethyl (1*H*-indol-7-yl)-α-oxoacetate from ethyl 7-indolineglyoxylate (99%) [10]. Dalla Croce and colleagues used a sulfur ylide approach to indolines and indoles (equations 2 and 3) [11, 12]. Bromination (NBS, Et_3N) of indoline **1** ($R=R^1=Ph$, $R^2=H$) gave 2-benzoyl-3-phenyl-1-phenylsulfonylindole (90%).

The dehydration of 3-hydroxyindolines (cf. **2** in Scheme 2) was the tactic for indolization in other indole syntheses. For example, Bonnet-Delpon's team prepared 3-trifluoromethylindoles from the corresponding 3-hydroxyindolines ($SOCl_2$, pyridine, 80%–87%), which were obtained from the reaction of aromatic amines and trifluoromethyl epoxy ethers [13], and Swenton and coworkers prepared 2,3-dimethyl-5-methoxyindole by acid-catalyzed dehydration of the 3-hydroxyindoline, and they also effected loss of methanol from C-5 of quinone imine ketals to give, for example, 5,6-dimethoxyindole (73%) [14]. A more common indolization of indolines is the elimination of sulfinate from *N*-substituted-sulfonylindoles, and both base- and acid-catalyzed examples are known (Scheme 3, equations 1–3) [15–17].

Several examples of remote elimination to give indoles are known (Scheme 4, equations 1 and 2) [18, 19], as are other indolizations featuring Lewis-acid catalysis (equation 3) [20] (where the range of C-2 heterocycles includes indoles, pyrroles, and furans), and double-bond isomerizations (equation 4) [21], or S_N2'-type reactions (equations 5 and 6) [22, 23].

Examples of the conversion of indoline-2-carboxylic acid to indoles are known (Scheme 5, equations 1–3) [24–26]. The novel transformation in equation 1 is an anomalous Dakin–West reaction that proceeds through a mesoionic 1,3-oxazolium-5-olate **3** (a münchnone) [24],

Indole Ring Synthesis: From Natural Products to Drug Discovery, First Edition. Gordon W. Gribble.

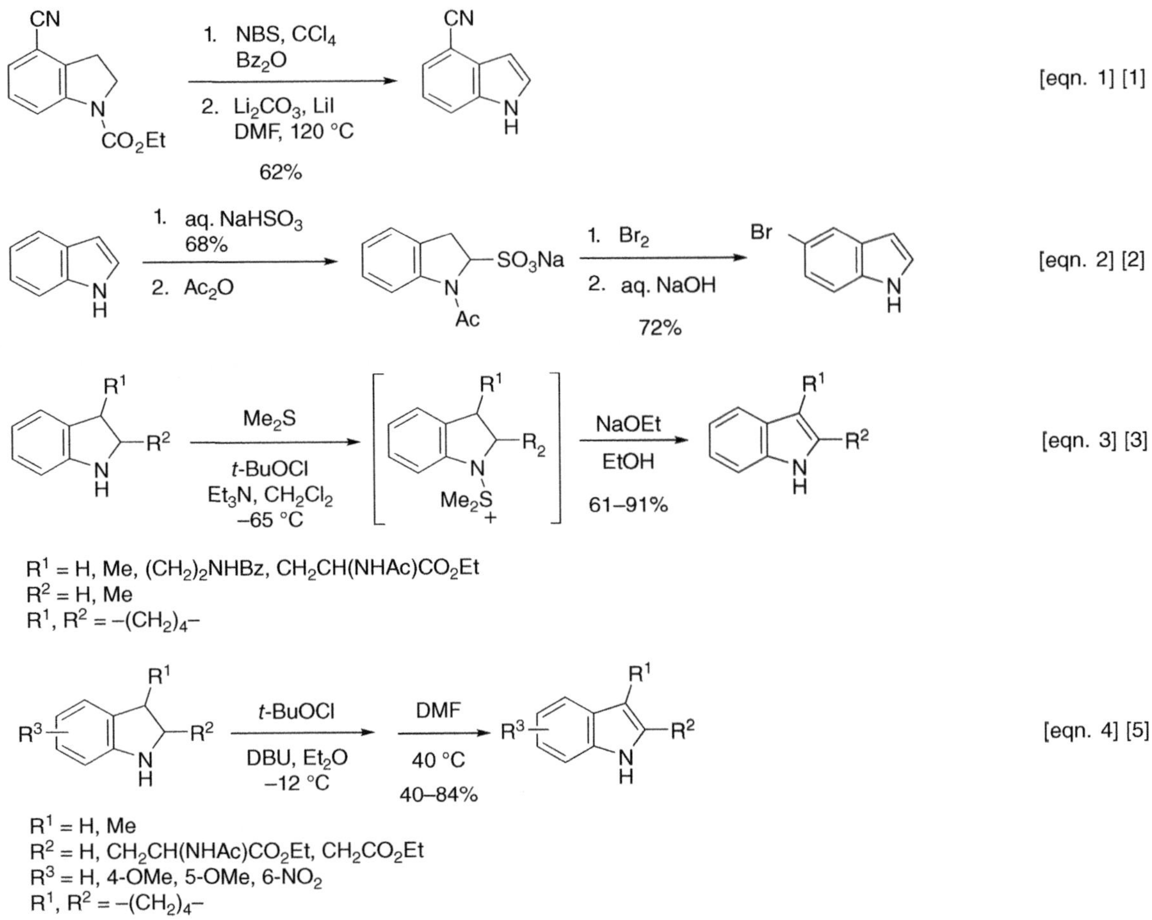

Scheme 1 *Plieninger, Thesing, and Kikugawa Indole Syntheses*

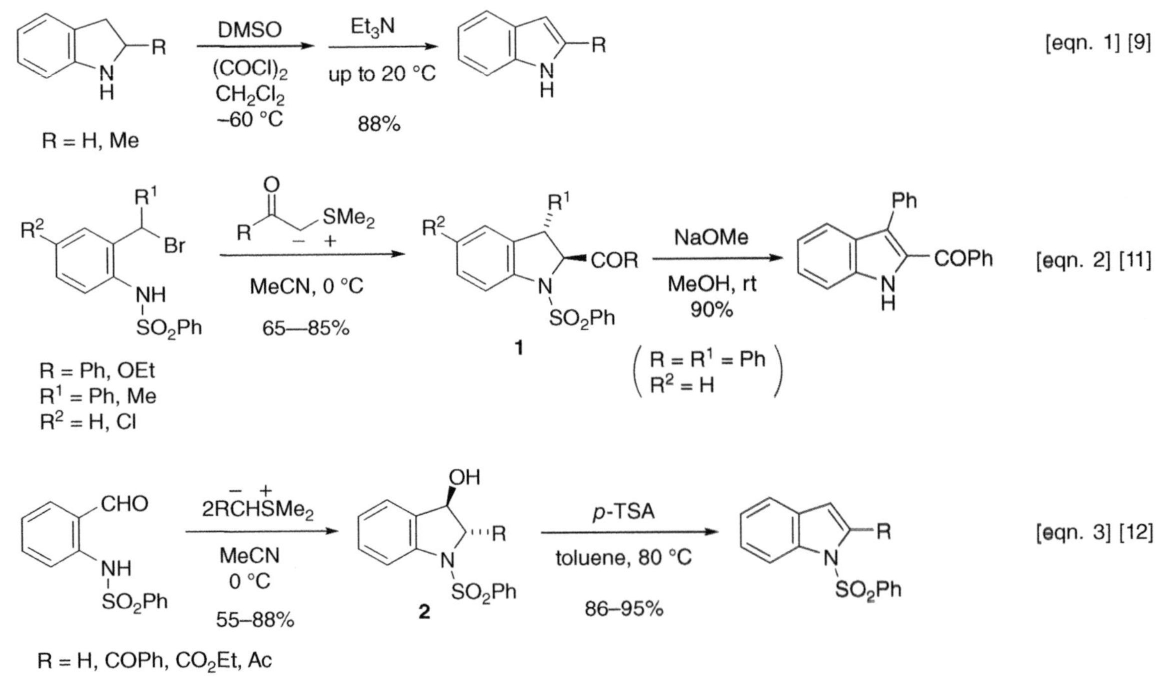

Scheme 2 *Overton and Dalla Croce Indole Syntheses*

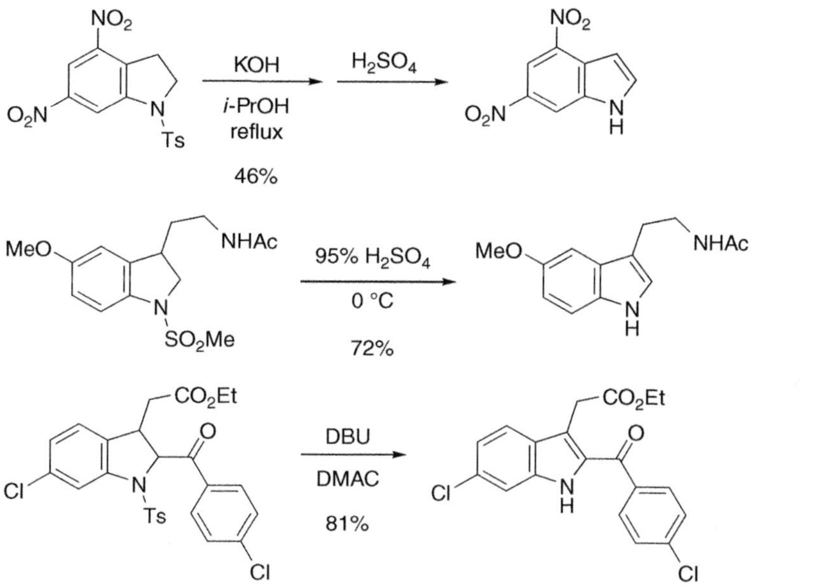

Scheme 3 *Indoles via Elimination of N-Sufinate*

and the authors propose a "decarboxylative redox amination" pathway for equation 2 [25]. A similar redox reaction from indolines to indoles was discovered by Seidel and coworkers (equation 3), where the authors supported the intermediacy of an azomethine ylide by trapping experiments [26]. Independently, Pan, Sun, and colleagues described the same chemistry leading to *N*-alkylindoles [27].

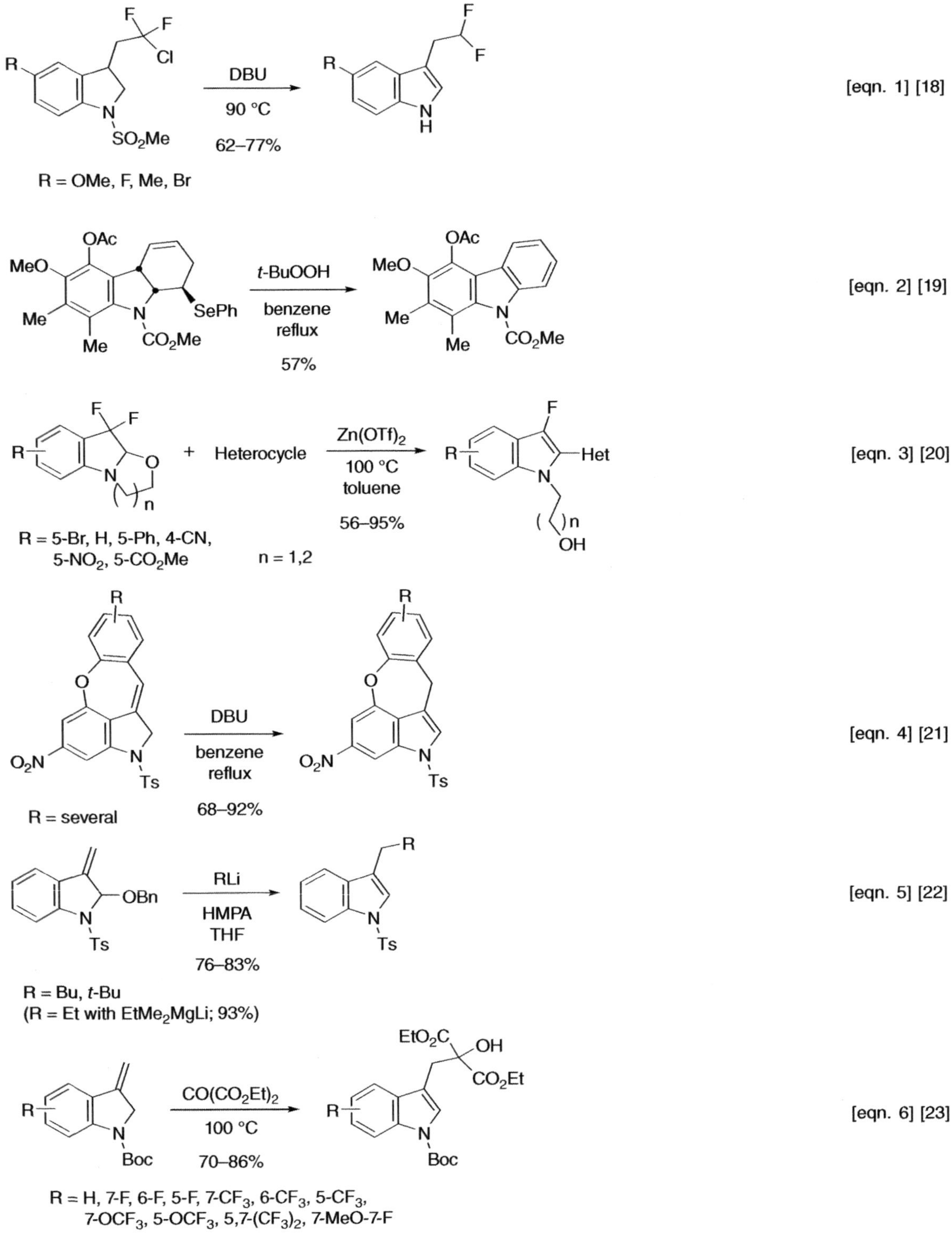

Scheme 4 *Miscellaneous Indoline to Indole Conversions*

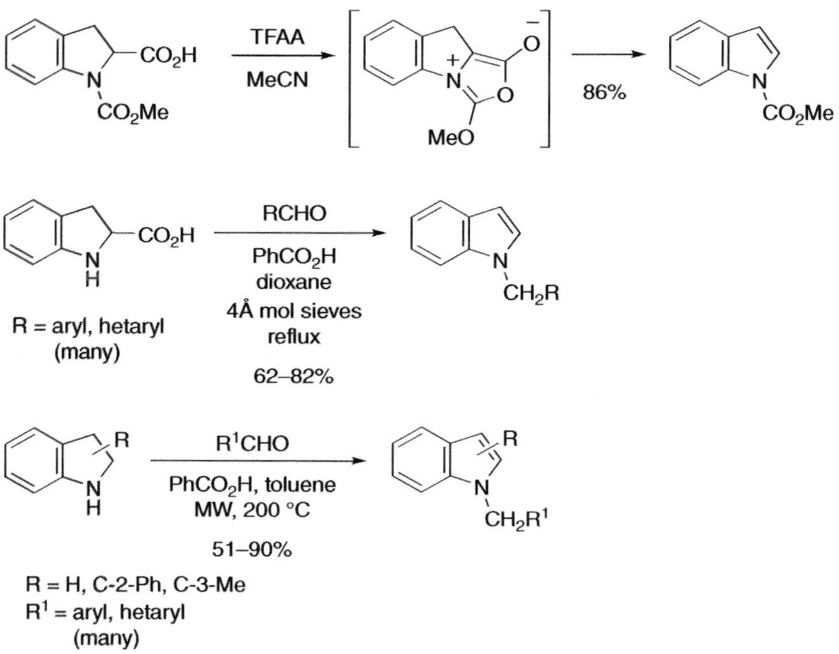

[eqn. 1] [24]

[eqn. 2] [25]

[eqn. 3] [26]

Scheme 5 *Miscellaneous Indoline to Indole Conversions – 2*

References

[1] H. Plieninger and K. Klinga, *Chem. Ber.*, 1968, **101**, 2605–2607.
[2] J. Thesing, G. Semler, and G. Mohr, *Chem. Ber.*, 1962, **95**, 2205–2211.
[3] Y. Kikugawa and M. Kawase, *Chem. Lett.*, 1981, 445–446.
[4] Y. Miyake and Y. Kikugawa, *J. Heterocycl. Chem.*, 1983, **20**, 349–352.
[5] M. Kawase, Y. Miyake, and Y. Kikugawa, *J. Chem. Soc., Perkin Trans. 1*, 1984, 1401–1404.
[6] C.-W. Ko and T. Chou, *J. Org. Chem.*, 1998, **63**, 4645–4653.
[7] T. Mukaiyama, A. Kawana, Y. Fukuda, and J. Matsuo, *Chem. Lett.*, 2001, 390–391.
[8] U. Tilstam, M. Harre, T. Heckrodt, and H. Weinmann, *Tetrahedron Lett.*, 2001, **42**, 5385–5387.
[9] D. Keirs and K. Overton, *J. Chem. Soc., Chem. Commun.*, 1987, 1660–1661.
[10] M.M. Faul, T.A. Engler, K.A. Sullivan, *et al.*, *J. Org. Chem.*, 2004, **69**, 2967–2975.
[11] G. Cremonesi, P. Dalla Croce, and C. La Rosa, *Heterocycles*, 2005, **66**, 557–562.
[12] G. Cremonesi, P. Dalla Croce, F. Fontana, and C. La Rosa, *Heterocycles*, 2007, **73**, 873–876.
[13] I. Rodrigues, D. Bonnet-Delpon, and J.-P. Bégué, *J. Org. Chem.*, 2001, **66**, 2098–2103.
[14] J.S. Swenton, C. Shih, C.-P. Chen, and C.-T. Chou, *J. Org. Chem.*, 1990, **55**, 2019–2026.
[15] A.V. Samet, E.P. Zakharov, V.V. Semenov, *et al.*, *Synth. Commun.*, 2001, **31**, 1441–1445.
[16] B. Quiclet-Sire, B. Sortais, and S.Z. Zard, *Chem. Commun.*, 2002, 1692–1693.
[17] S. Caron, E. Vazquez, R.W. Stevens, *et al.*, *J. Org. Chem.*, 2003, **68**, 4104–4107.
[18] P. Salomon and S.Z. Zard, *Org Lett.*, 2014, **16**, 2926–2929.
[19] D. Crich and S. Rumthao, *Tetrahedron*, 2004, **60**, 1513–1516.
[20] J.Z.M. Fong, S.S.S. Choo, J.-A. Richard, *et al.*, *Eur. J. Org. Chem.*, 2015, 995–1006.
[21] A.V. Samet, A.N. Yamskov, Y.A. Strelenko, and V.V. Semenov, *Tetrahedron*, 2009, **65**, 6868–6872.
[22] K. Inamoto, A. Yamamoto, K. Ohsawa, *et al.*, *Chem. Pharm. Bull.*, 2005, **53**, 1502–1507.
[23] F. Bellezza. A. Cipiciani, R. Ruzziconi, and S. Spizzichino, *J. Fluorine Chem.*, 2008, **129**, 97–107.
[24] M. Kawase, M. Hirabayashi, H. Koiwai, *et al.*, *Chem. Commun.*, 1998, 641–642.
[25] H. Mao, S. Wang, P. Yu, *et al.*, *J. Org. Chem.*, 2011, **76**, 1167–1169.
[26] I. Deb, D. Das, and D. Seidel, *Org. Lett.*, 2011, **13**, 812–815.
[27] H. Mao, R. Xu, J. Wan, *et al.*, *Chem. Eur. J.*, 2010, **16**, 13352–13355.

68

Indolines from Oxindoles, Isatins, and Indoxyls

An important route to indoles is the conversion—usually by reduction or the equivalent manipulation—of oxindoles, isatins, and indoxyls, the most common involving oxindoles [1, 2].

The reduction of oxindoles to indoles rarely if ever involves the isolation of the 2-hydroxyindoline intermediate. Julian's early studies on indoles, which included the toxic alkaloid physostigmine (the drug of "divine justice"), uncovered the Na/EtOH reduction of oxindole **1** and the subsequent cyclization to desoxynoreseroline, presumably via the indolenine (Scheme 1, equation 1) [3]. Dolby and Lord employed $LiAlH_4$ (LAH) to synthesize 1,3-diphenylindole in good yield from the oxindole (equation 2) [4]. Interestingly, Hudson and Robertson found that LAH reductively cleaved 3-(1-pyrrolyl)oxindole to indole (29% yield) [5]. The corresponding *N*-methyloxindole gave 1-methyl-3-(1-pyrrolyl)indole (70%) and 1-methyl-3-(1-pyrrolyl)indoline (18%). In a synthesis of the *Aristotelia* alkaloids, Lévy and colleagues used LAH to indolize oxindole **2** (equation 3) [6]. Joule, Alvarez, and colleagues selected LAH to reduce oxindole **3** to the desired pyrrolo[4,3,2-*de*]quinoline in excellent yield (equation 4) [7]. This important ring system is embodied in several marine alkaloids. Stauss found that 3-(benzylidien)indolin-2-one is reduced to 3-benzylindole with LAH [8], and Black obtained 4,6-dimethoxyindole (24%) and 4,6-dimethoxy-3-methylindole (50%) from 3-carbethoxy-4,6-dimethoxy-3-hydroxyoxindole from LAH treatment [9]. The reduction of oxindoles with LAH often affords *indolines*, rather than *indoles*, and, of course, this is *fait accompli* with 3,3-disubstituted oxindoles.

After a less-than-auspicious start [10, 11], researchers soon realized that boron hydrides do reduce oxindoles to indoles (Scheme 2, equations 1–4) [11–14]. Garden and coworkers employed the commercial borane–tetrahydrofuran complex to synthesize 3-fluoroindoles from the 3,3-difluorooxindoles (equation 2) [12]. These relatively labile 3-fluoroindoles were capped as the *N*-tosyl derivatives. Williams and colleagues found the combination of $NaBH_4$/BF_3-OEt_2 to be effective in reducing oxindole **4** (equation 3) en route to a total synthesis of (+)-paraherquamide B [13]. A two-step borane reduction/thionyl chloride–pyridine elimination achieved the conversion of 3-hydroxyoxindoles to the 3-trifluoromethylindoles (equation 4) [14]. The oxindoles were prepared from the corresponding isatins via C-3 addition of (trifluoromethyl)trimethylsilane ($TMSCF_3$).

Wierenga discovered that borane–dimethyl sulfide reduced 3-(alkylthio)oxindole **5** to the desired indole (Scheme 3, equation 1), because LAH was "sluggish or completely ineffective" [15]. Subsequently, Wierenga generalized this smooth reduction of 3-(alkylthio)oxindoles (equation 2) [16]. Also reduced to 3-alkylindoles with BH_3-SMe_2 were 3-hydroxyoxindoles (88–93%), which were prepared from isatins by selective C-3 addition of Grignards. Sano and coworkers used aluminum hydride to reduce 3-(phenylthio)oxindole **6** to the desired indole (equation 3) [17]. Some indoline (14%) was also produced. Wróbel employed diisobutylaluminum hydride (DIBAL) in an oxindole reduction to a pyrrolo[4,3,2-*de*]quinoline (equation 4) [18]. Katayama's group found DIBAL to be superior to LAH in reducing 2-oxo-1,2,4,5,6,7-hexahydroazepino[3,2,1-*hi*]indole to 4,5,6,7-tetrahydroazepino[3,2,1-*hi*]indole (66% vs. 18%) [19]. Nishio and colleagues also reported the superiority of DIBAL vs. LAH in the reduction of a large number of oxindoles to indoles (equation 5) [20], in large part because LAH forms more indolines than DIBAL.

Indole Ring Synthesis: From Natural Products to Drug Discovery, First Edition. Gordon W. Gribble.

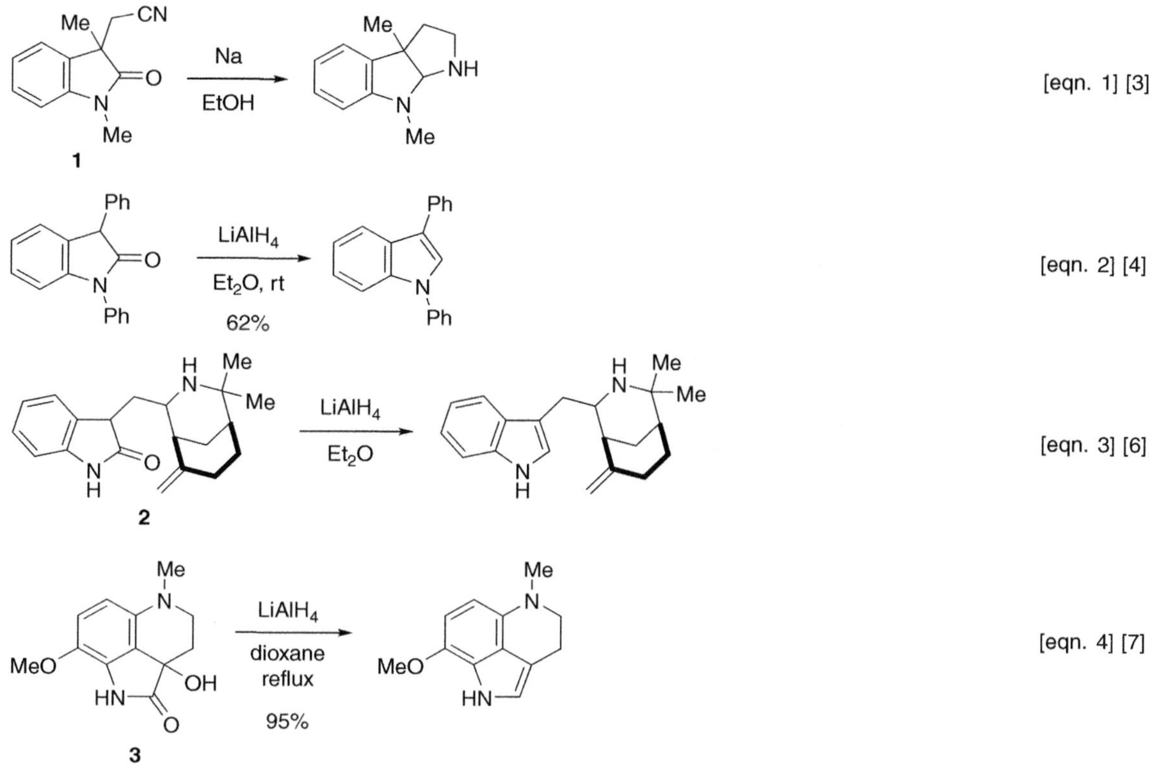

Scheme 1 *Reduction of Oxindoles with Sodium and Lithium Aluminum Hydride*

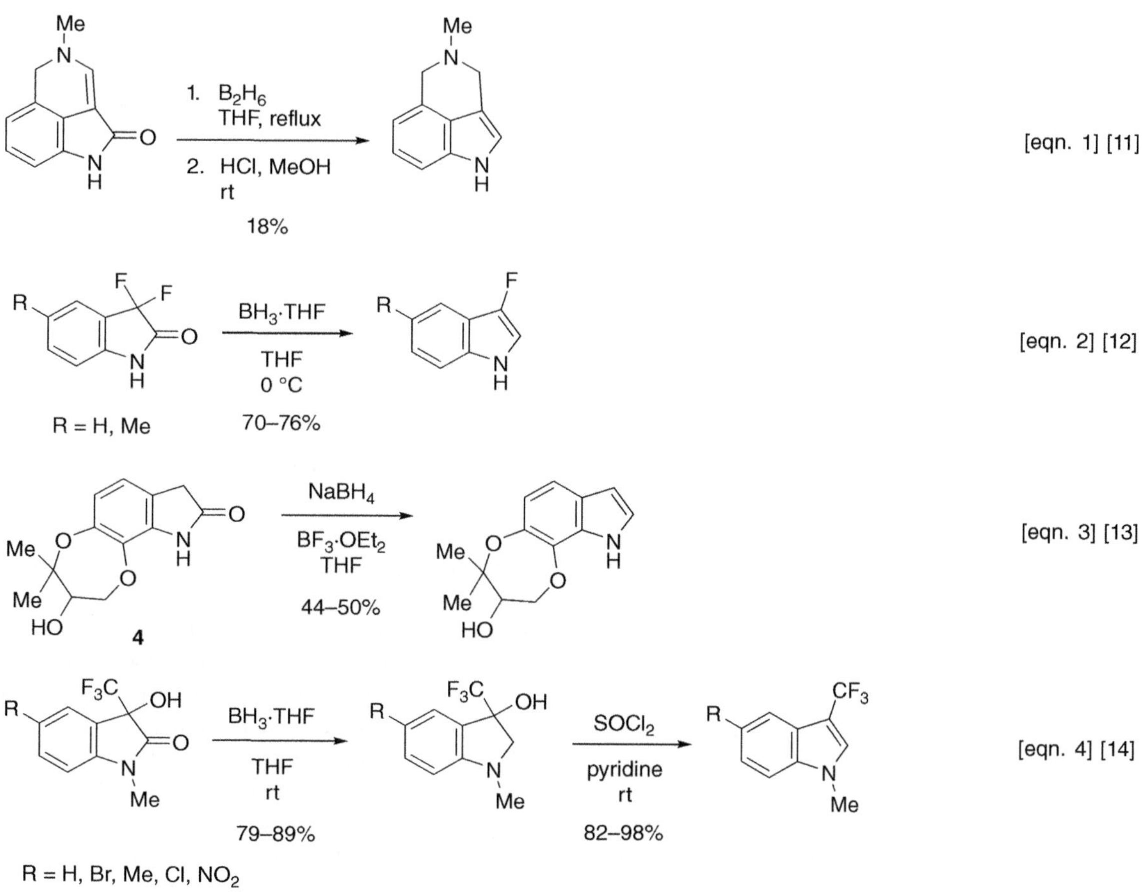

Scheme 2 *Borane Reduction of Oxindoles to Indoles*

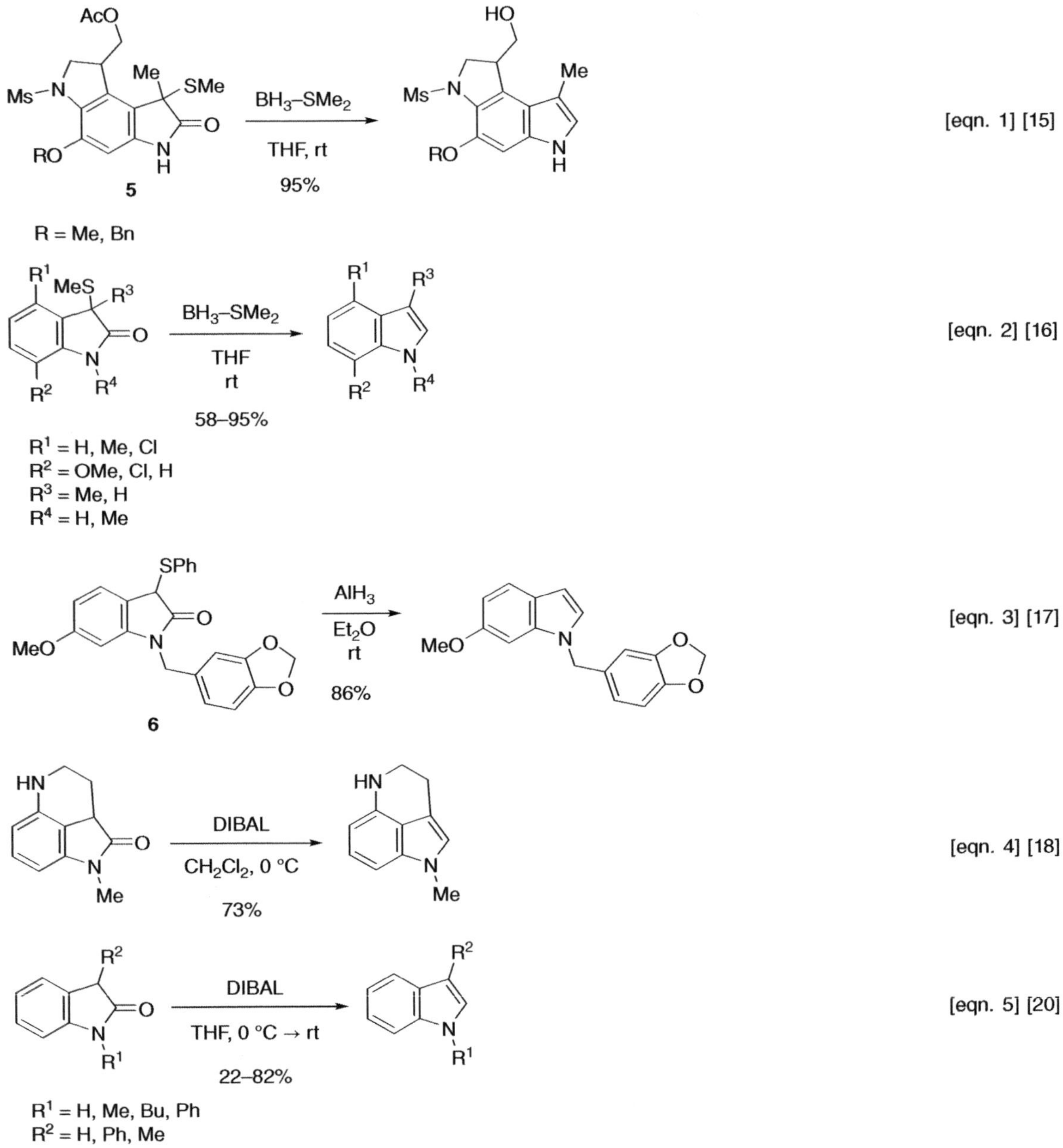

Scheme 3 *Other Hydride Reductions of Oxindoles to Indoles*

A second important and widely used reaction of oxindoles is activation of the carbonyl group with $POCl_3$ ($POBr_3$) or acid chlorides, followed by subsequent chemistry at C-2. There is a vast literature on this reaction!

One of the first examples of the activation of oxindole with $POCl_3$ was that of Powers, who described the synthesis of (unstable) 2-chloroindole via the Vilsmeier salt **7** (Scheme 4, equation 1) [21]. 1-Benzyl-2-chloroindole was similarly prepared from 1-benzyloxindole. Independently from Powers, Schulte's team used $POCl_3$/DMF to synthesize 2-chloroindole-3-carbaldehydes **8** (equation 2) and effect subsequent nucleophilic reactions (equation 3) [22, 23]. Kubo and Nakai extended Powers's 2-chloroindole synthesis to the preparation of 3-alkylindoles (equation 4) [24]. Erickson and coworkers synthesized 2-bromoindole ($POBr_3$, 15%) and 2-chloroindole ($POCl_3$, 26%) in a program to prepare the naturally occurring 2,3-dihaloindoles [25]. Both compounds were extremely labile.

Schulte's 2-chloroindole-3-carbaldehyde (**8**, R=H) has been the figurehead in many indole syntheses, such as moroidin analogues (Moody [26]), the carbazole alkaloid furostifoline (Knölker [27]), thiopyrano[2,3-*b*]indoles

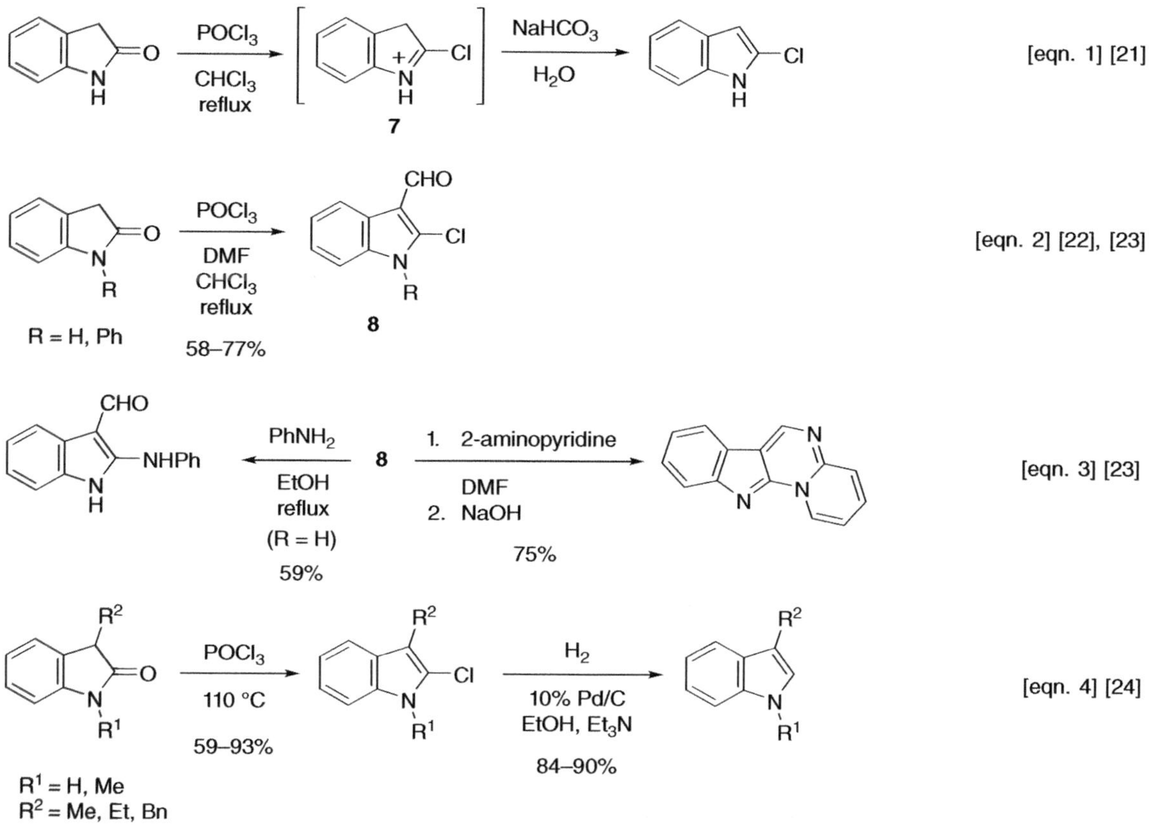

Scheme 4 *Powers, Schulte, and Kubo Indole Syntheses from Oxindoles*

(Bhuyan [28]), α-carbolines (Bhuyan [29]), and 2,2′′-diselenobis(indoles) as protein kinase inhibitors (Showalter [30]). The latter study described an improved synthesis of Schulte's **8** (R=H), and Somei used 2-bromoindole-3-carbaldehyde (**9**) in a synthesis of the alkaloid borrerine via Stille coupling (Scheme 5, equation 1) [31]. Belmont and Tiano improved the preparation of **9** ($POBr_3$, DMF, CH_2Cl_2, reflux, 98% yield) and performed Sonogashira coupling reactions to access a collection of novel carbazoles [32]. Zhang and colleagues used 2-bromo-7-fluoro-5-methoxyindole-3-carbaldehyde for the preparation of antitumor 5-ureidobenzofuranone indoles [33]. Erba and coworkers synthesized **8** (R=H) in a different fashion and employed it to make pyrimido[4,5-*b*]indoles via a C-2 azide (equation 2) [34]. The isomeric 3-chloroindole-2-carbaldehyde was prepared by Su's group (equation 3) [35].

Despite the inherent instability of 2-chloroindole, derivatives have found many synthetic applications. Eissenstat and colleagues prepared more than 140 cannabinoid mimetics **10** (Scheme 6, equation 1) [36]. Interestingly, longer reaction times for this chlorination led to trimer **11** (72%). Similarly, Hiyoshi synthesized the bromo trimer **12** [37], tetramer **13** [38], and alkenyl derivatives via Suzuki coupling as novel donor-π-acceptor molecules. Fang [39] and Lai [40] synthesized related compounds via the oxindole → 2-haloindole transformation.

Townsend improved the syntheses of 2-chloro- and 2-bromoindoles from 5,6-dichlorooxindole by incorporating the weak base imidazole, as an acid neutralizer, to give 2,5,6-trichloroindole (80%) and 2-bromo-5,6-dichloroindole (POX_3, $ClCH_2CH_2Cl$, imidazole) [41, 42], which were converted to novel indole *N*-nucleosides. Magnus and colleagues used triphenylphosphine and carbon tetrachloride to convert 4-(2-methoxyphenyl)oxindole to the corresponding 2-chloroindole (83%), a reaction that failed using $POCl_3$/CH_2Cl_2 [43]. Bergman activated several oxindoles to the corresponding 2-chloroindole-3-carbonyl chlorides with phosgene, and 2-chloroindole-3-glyoxylyl chlorides with oxalyl chloride, for the preparation of several indole-fused heterocycles [44]. Using the Bergman method, Sahagún synthesized several indolopyrrolocarbazoles from 2-chloroindole-3-glyoxylyl chlorides [45]. Estévez and colleagues employed $POBr_3$/anisole to prepare 2-bromoindoles from oxindoles en route to indolo[2,3-*b*]naphthalene-6, 11-diones [46]. Melatonin analogues were crafted by Lozinskaya's group from the oxindole-derived (2-chloro-5-methoxyindole)acetonitrile ($POCl_3$, reflux, 82%) [47]. Yamato used Suzuki coupling to synthesize 2,5-diarylindoles from

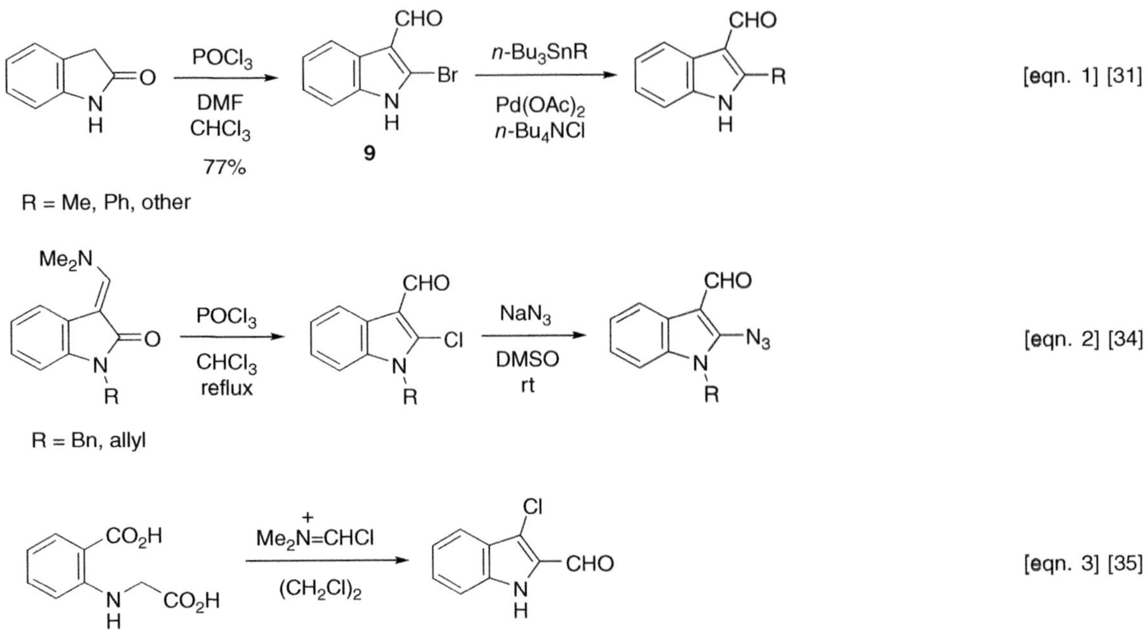

Scheme 5 *Synthetic Applications of 2-Haloindole-3-carbaldehydes*

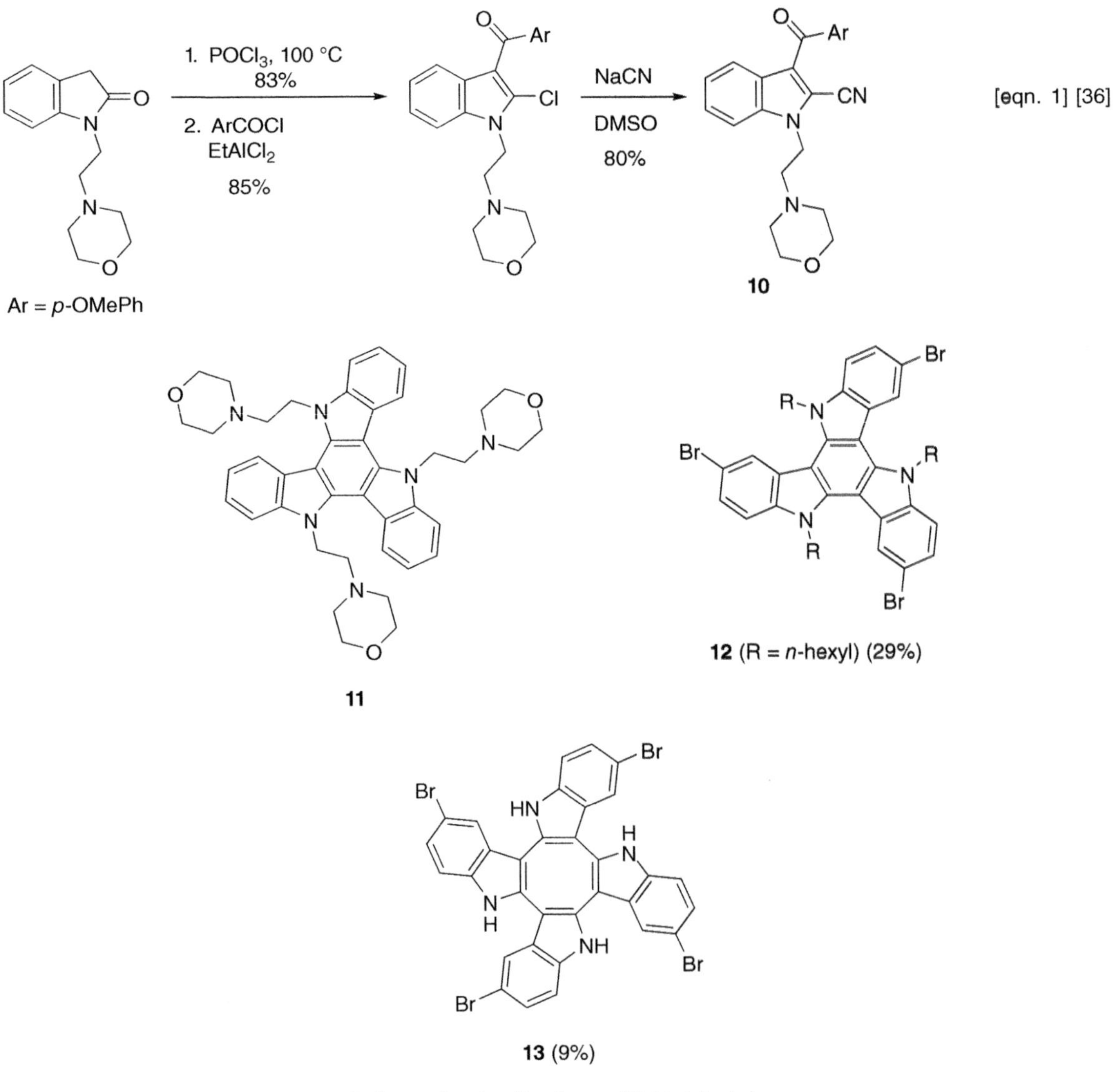

Scheme 6 *Applications of 2-Haloindoles*

5-bromo-2-chloro-1-methylindole, which was obtained from the corresponding oxindole ($POCl_3$, chlorobenzene, 100 °C, 48%) [48]. Illustrative of the applicability of the original Powers 2-chloroindole synthesis are the examples in Scheme 7. Black prepared several biindolyls via an activated oxindole and its nucleophilic coupling of the resulting Vilsmeier salt with a second indole unit (equation 1) [49]. Owa's synthesis of 7-amino-2-chloroindole (equation 2) was followed by conversion to potential cell cycle inhibitors, *N*-(7-indolyl)benzenesulfonamides [50]. Boger employed anisole to prevent ring bromination and imidazole to preclude debenzylation and deprotection in his synthesis of the CC-1065 subunit **14** (equation 3) [51]. Pfau and coworkers discovered an interesting $POCl_3$-induced indolization (equation 4) [52].

Oxindoles can also be activated at the carbonyl position by, for example, triflate formation. Gribble and Conway synthesized 1-(trifluoromethanesulfonyl)indol-2-yl trifluoromethanesulfonate from oxindole (triflic anhydride, CH_2Cl_2, 2,6-di-*tert*-butylpyridine, 72%) [53], and Prandi and colleagues prepared the corresponding *N*-methyl analogue and subjected it to Pd-catalyzed coupling (Scheme 8, equation 1) [54, 55]. Shea reported a similar activation of an oxindole in the context of a welwitindolinone alkaloid core synthesis [56]. Ghandi and coworkers activated oxindole with triflic anhydride and coupled the triflate with 2-naphthol (equation 2) [57]. Möhrle and Dwuletzki converted *N*-methyloxindole to the corresponding 2-alkoxyindoles with trialkyloxonium tetrafluoroborates, which underwent a Diels–Alder cycloaddition with a triazine to afford a γ-carboline [58]. Jha and colleagues prepared a series of polyacetylated indoles from oxindoles with acetic anhydride [59], and Yang, Xie, and colleague converted 1-acetyl-5-chlorooxindole to an azachromeno[2,3-*b*]indol-11(6*H*)-one (equation 3) [60].

Myriad examples are known of oxindoles reacting to form indoles without prior carbonyl activation. Given the vast scope of this subject, only a few early examples will be cited in addition to recent studies. Several groups have annulated oxindoles to access fused-indole heterocycles. Ila, Junjappa, and coworkers prepared pyrido[2,3-*b*]indoles [61, 62] and 6*H*-indolo[2,3-*b*]quinolines [63] using this approach (Scheme 9, equation 1). Adib and colleagues reported a simple synthesis of 9*H*-pyrimido[4,5-*b*]indoles from oxindoles (equation 2) [64], and Bazgir and coworkers described a four-component one-pot α-carboline synthesis

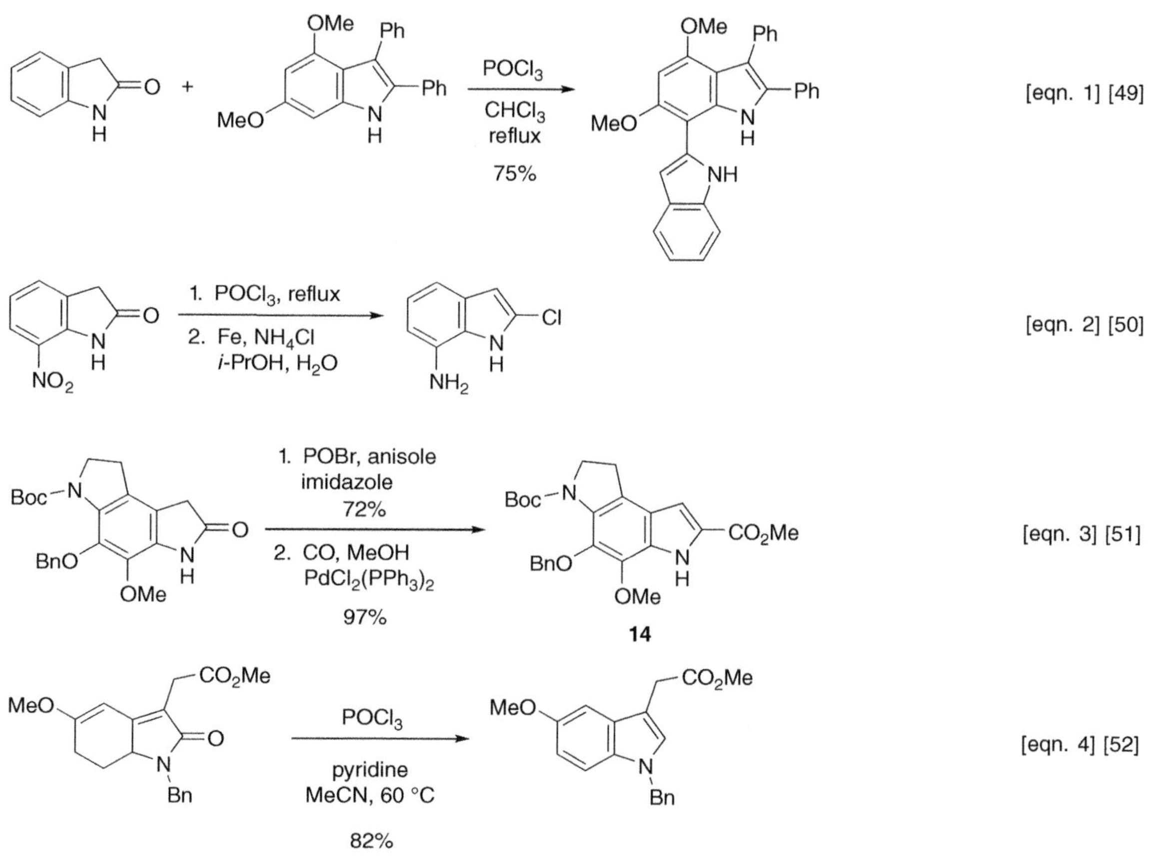

Scheme 7 *Applications of 2-Haloindoles – 2*

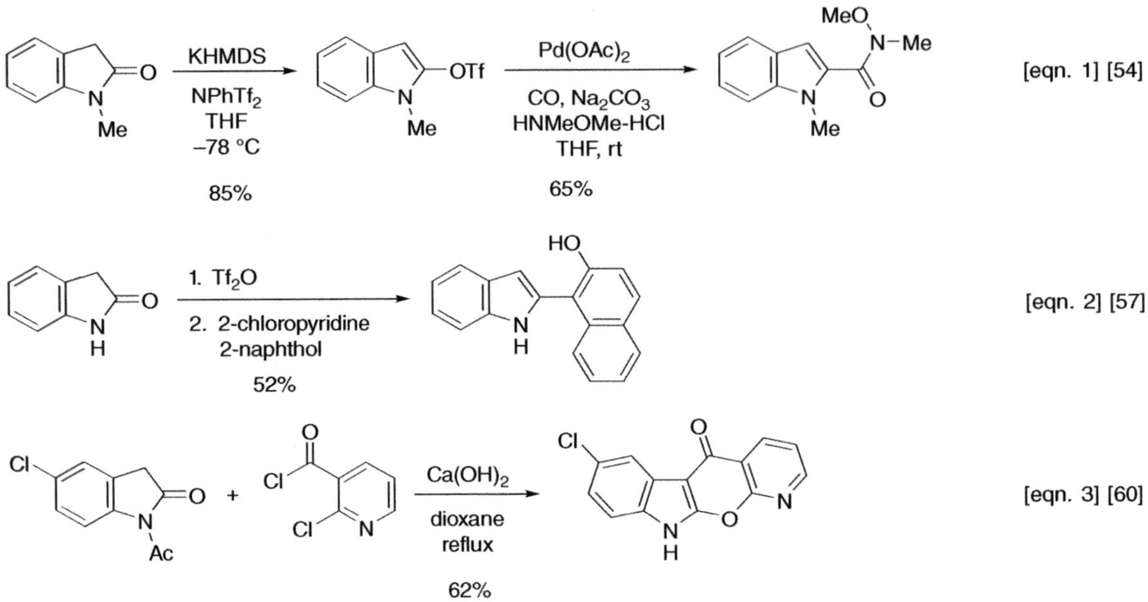

Scheme 8 Applications of 2-Oxygenated Indoles

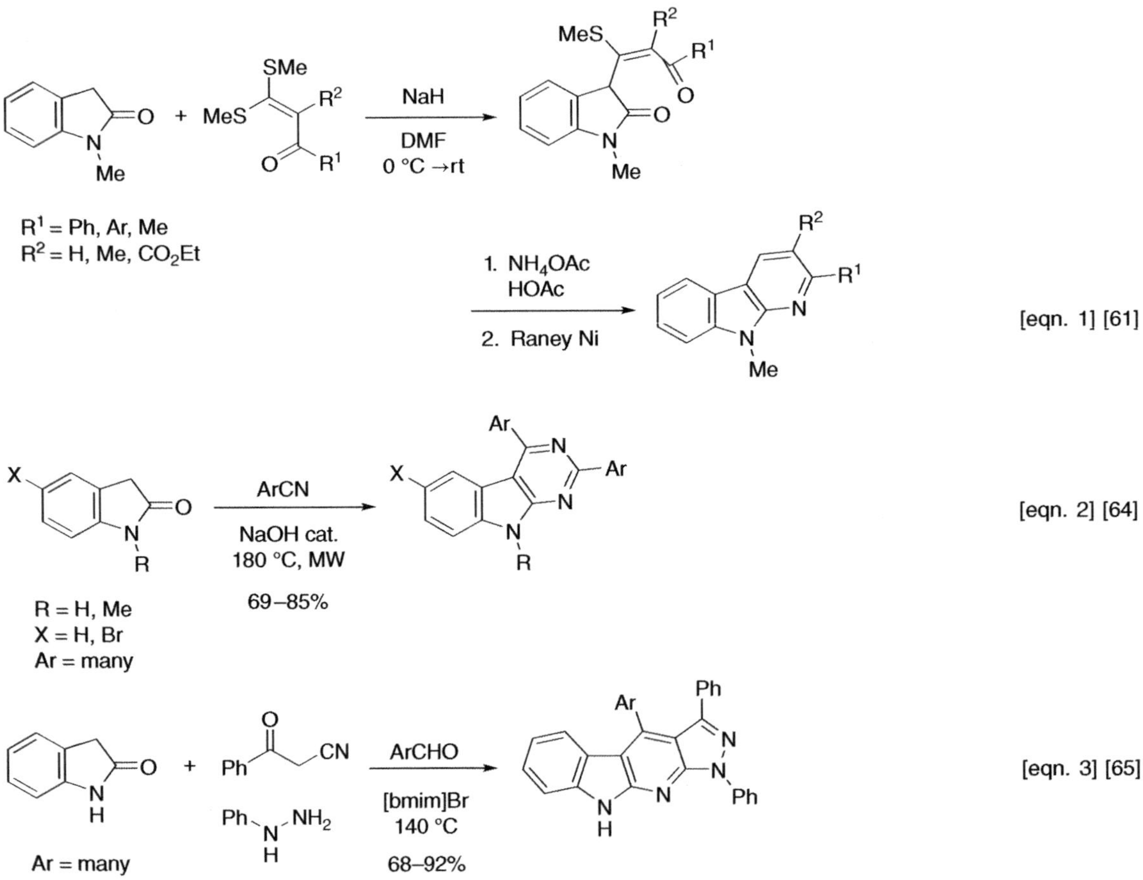

Scheme 9 Annulation of Oxindoles to Give Fused-Indoles

from oxindole (equation 3) [65]. Procter and coworkers prepared indoloquinolines from oxindoles [66], and Zhong [67] and Cheng [68] independently synthesized pyrano [2,3-*b*]indoles by the annulation of oxindoles.

Thiooxindoles (2-indolinethiones) are easily prepared from oxindoles with phosphorus pentasulfide [69] or Lawesson's reagent [70], and they display a wealth of interesting chemistry. Wenkert's group discovered a simple 2-alkyl (aryl) indole synthesis via thiooxindole (Scheme 10, equation 1) [70]. Pedras and Zaharia prepared the cruciferous phytoalexins, sinalexin and brassilexin, from thiooxindole (equation 2) [71, 72]. Three groups simultaneously reported a (different) synthesis of the naturally occurring thieno[2,3-*b*]indole ring system from indoline-2-thiones [73–75] (equation 3) [74]. Moghaddam and coworkers achieved cycloaddition reactions between indoline-2-thiones and isoquinolinium and quinolinium salts (equation 4) [76, 77].

Other synthetic applications of indoline-2-thiones include the synthesis of thiopyrano[2,3-*b*]indoles and thieno[2,3-*b*]indoles [78–80], *S*-alkylated indoles [81, 82], thiopyrano[2,3-*b*:6,5-*b*′]diindoles [83], and thiacarbazol-2(9*H*)-ones [84]. Bergman and Janosik oxidized thiooxindole with *p*-toluenesulfonyl azide to produce a trimer that apparently forms via a thiooxindole carbine [85]. Thiooxindoles are usually desulfurized to indolines and/or indoles with Raney nickel [20].

Although isatins having adjacent carbonyls of very different reactivity—amide vs. ketone—undergo a variety of chemical reactions [86, 87], the focus here is on synthetic transformations to indoles. Accordingly, isatins can be reduced directly to indoles (Scheme 11, equations 1–7) [16, 89–93]. Given the availability of isatins from aromatic amines (e.g., Sandmeyer synthesis [86], Gassman synthesis [88]), the reduction of isatins to indoles can be an important alternative to other methods.

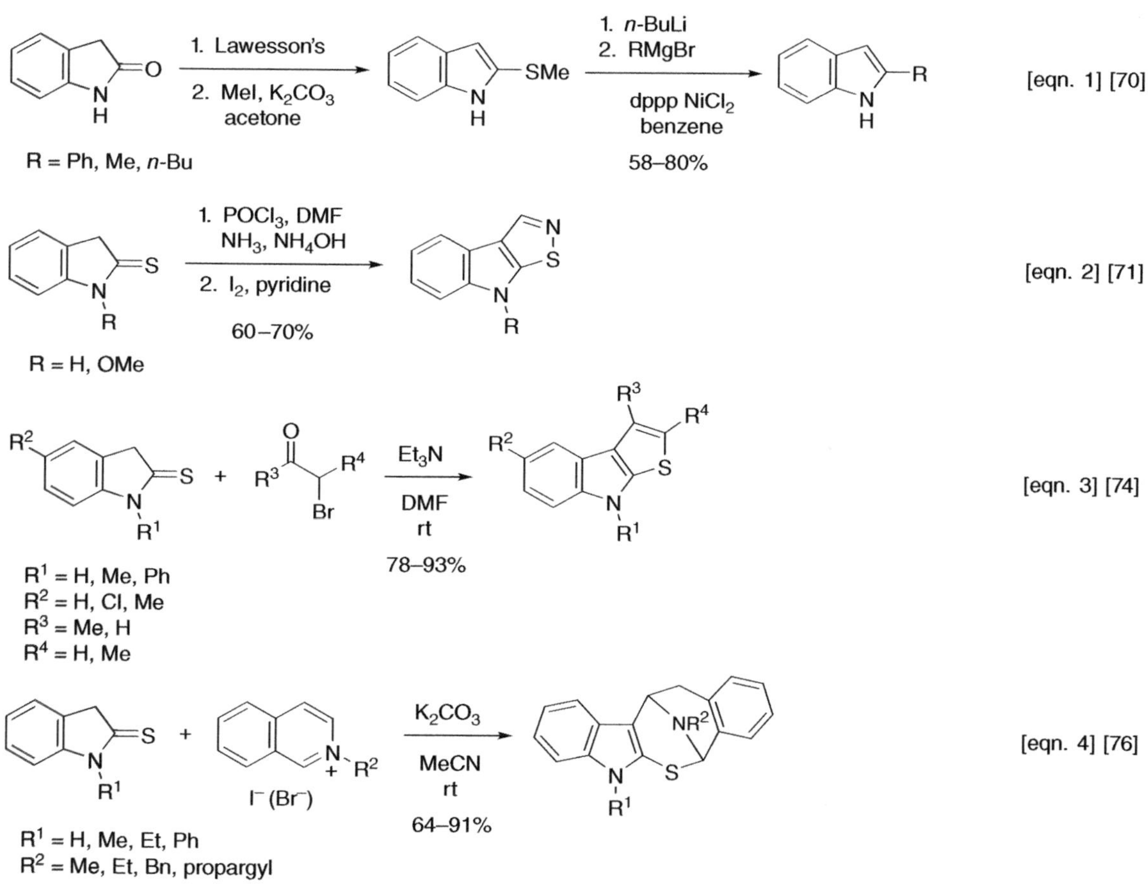

Scheme 10 *Applications of Indoline-2-thiones in Synthesis*

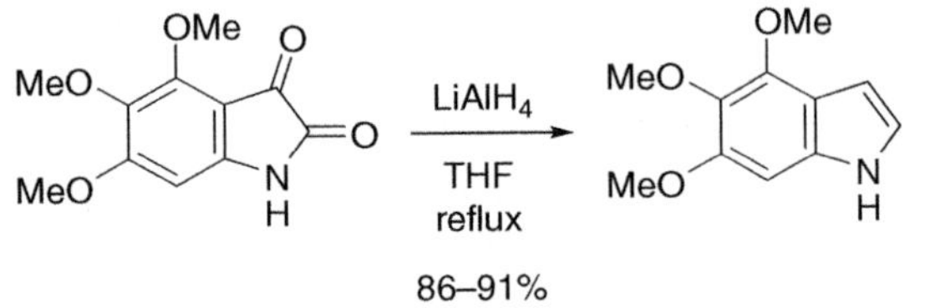

[eqn. 1] [89]

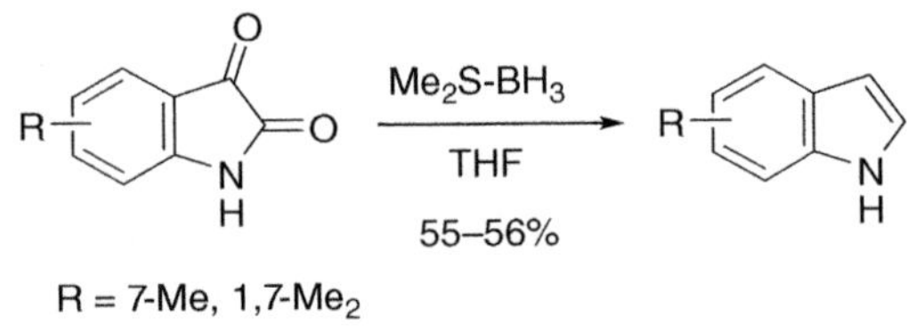

[eqn. 2] [16]

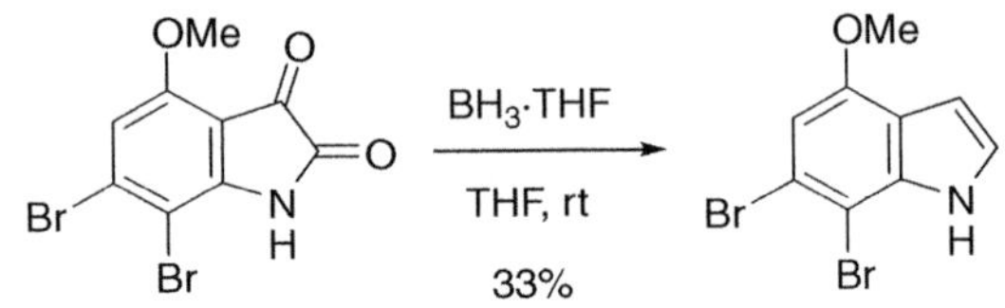

[eqn. 3] [90]

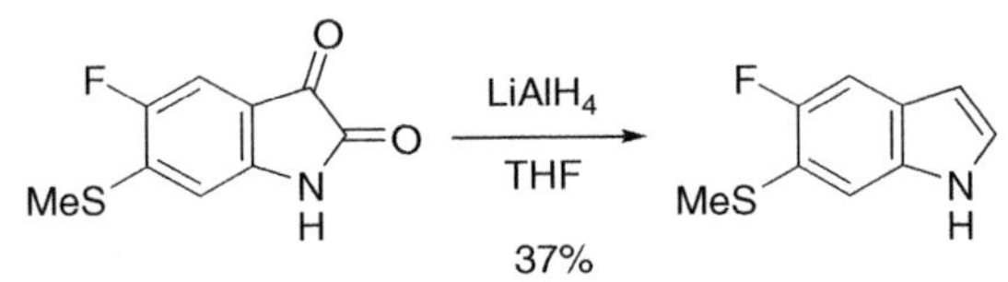

[eqn. 4] [91]

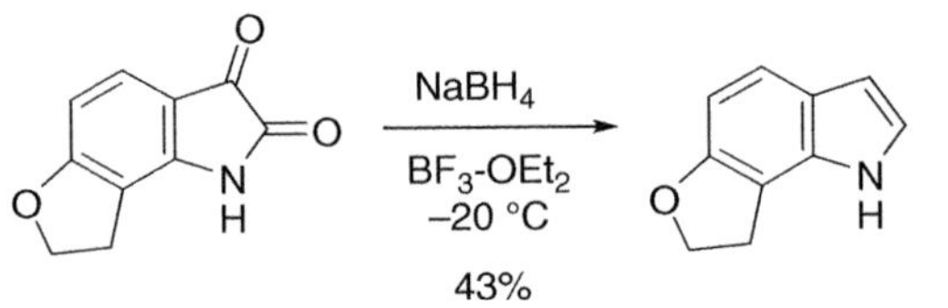

[eqn. 5] [91]

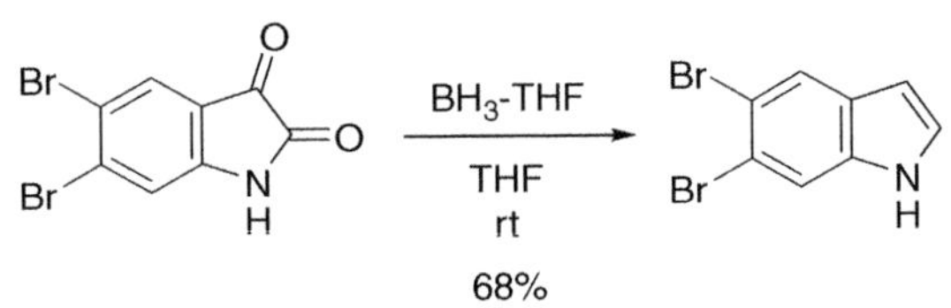

[eqn. 6] [92]

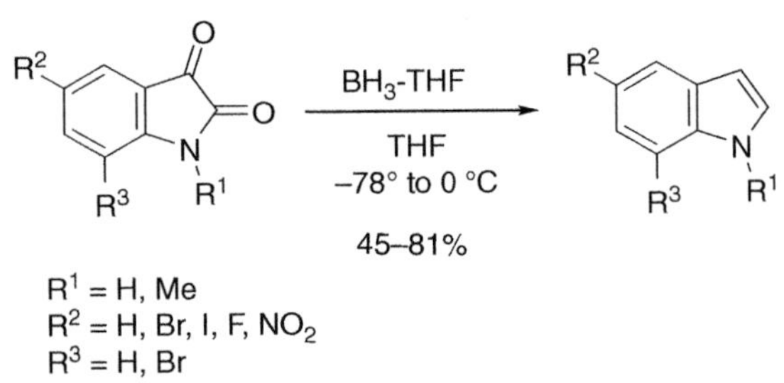

[eqn. 7] [93]

Scheme 11 *Reduction of Isatins to Indoles*

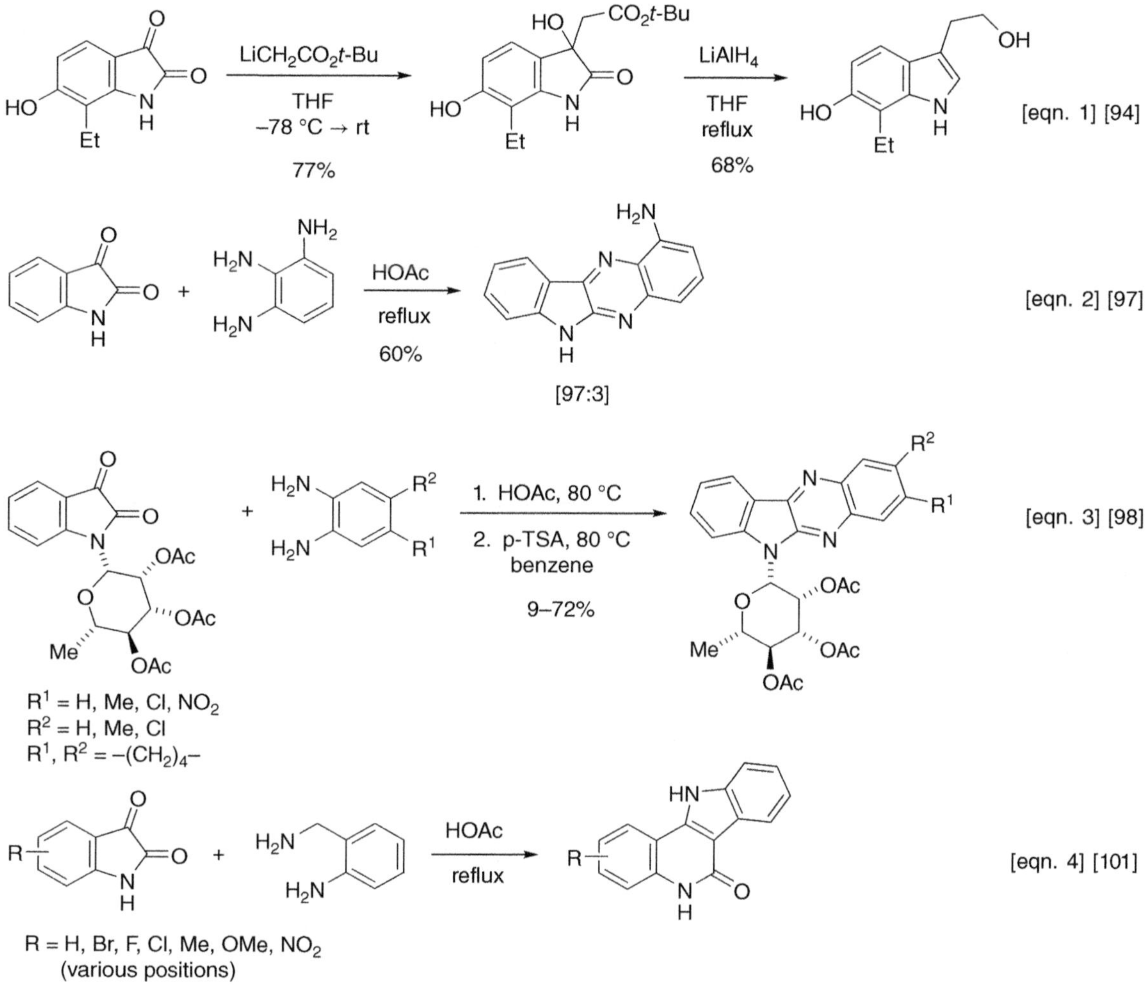

Scheme 12 *Indole-Forming Reactions of Isatins*

The inherent reactivity difference between the two isatin carbonyl groups has been exploited in synthesis by many investigators. Soll [93], Humber [94], and Demerson [95, 96] reported an efficient tryptophol synthesis from isatins (Scheme 12, equation 1) [93]. The venerable condensation of isatins with diamines affords an array of heterocyclic ring systems, and this classical reaction has been known for a long time. Some recent examples are shown in equations 2 to 4. The more nucleophilic 2-amino group in 1,2,3-triaminobenzene reacts with the more electrophilic C-3 carbonyl (keto) of isatin (equation 2) [97]. Langer's team has synthesized cytotoxic 6*H*-indolo[2,3-*b*]quinoxaline-*N*-glycosides (equation 3) [98]. Renault and coworkers reported related synthetic studies of these DNA intercalators [99]. A well-known isatin condensation is that with 2-aminobenzylamines to give indolo[3,2-*c*]quinolin-6-ones, as first described by Bergman [100]. A later example was provided by Inokuchi (equation 4) [101], in syntheses of potential new antimalarials.

Indoxyl, the major tautomer of 3-hydroxyindole, is an unstable compound that is readily oxidized to the natural pigment indigo [102] (see also Joule and Mills [87], pp. 348–349, 367). *N*-protected indoxyls are stable, and an early example is that of Vorlander (Scheme 13, equation 1) [103]. Suitably *N*-protected, indoxyls undergo many reactions typical of ketone carbonyls. For example, Pretka and Lindwall effected Grignard and Reformatsky reactions on these *N*-acetylindoxyls (equation 2) [104]. Indolization is facile (equation 3) [105], as are Wittig reactions (equation 4) [106]. In this latter study, the intermediate *exo*-cyclic alkene was not observed, and only in one case did deacetylation occur. Mérour and coworkers reported similar Wittig and Hormer–Emmons reactions of *N*-Ac- and *N*-(phenylsulfonyl)indoxyls, and in several instances the

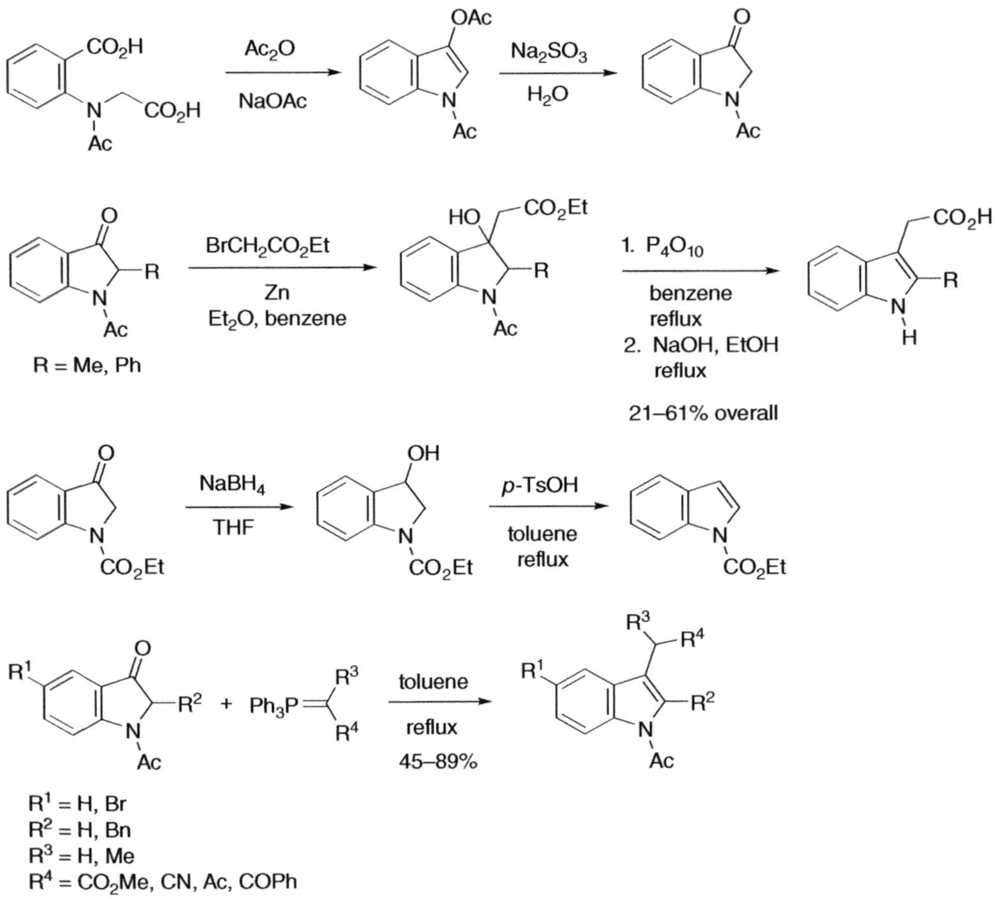

Scheme 13 *Ketone Reactions of* N-*Protected Indoxyls*

exo-alkene was isolated [107]. Kawasaki and colleagues employed Wittig olefination of 1-acetoxy-5-bromo-2-methoxyindoxyl to synthesize the marine alkaloids, hamacanthins [108].

Reactions at C-2 of *N*-protected indoxyls provide entry to novel indoles that are difficult to access other ways. Sakamoto and colleagues discovered a tandem Wittig reaction and Cope (or Claisen) rearrangement to provide 3-substituted indoles (Scheme 14, equation 1) [109–111]. The method is exemplified by a synthesis of the marine alkaloid flustramine C (**15**) [112]. Mérour and colleagues employed an intramolecular hetero-Diels–Alder cycloaddition to prepare pyrano[3,2-*b*]indoles (equation 2) [113]. Ila and Junjappa prepared a series of benzo- and heterocyclo-fused carbazoles and indoles via reactions of 2-bis(methylthio)methylene-1-methylindoxyl [114].

The Friedlander quinoline synthesis was applied in a one-step synthesis of 6*H*-indolo[3,2-*b*]naphthyridines (azaellipticines) by Quéguiner's team by condensing *N*-acetylindoxyl with aminofornyl pyridines (Scheme 15, equation 1) [115]. In addition to 6*H*-indolo[3,2-*b*]naphthyridine-1,7 (**16**), the -1,5 (**17**) and -1,8 (**18**) isomers were also prepared from the appropriate pyridines. Merlic and Quinn also engaged the Friedlander quinoline synthesis in a preparation of the novel RNA-binding fluorochrome, Fluoro Nissl Green (**19**) (equation 2) [116].

Gribble and Conway described the synthesis of a 3-indolyl triflate **20** (1-(phenylsulfonyl)indol-3-yl trifluoromethanesulfonate) from the corresponding *N*-protected indoxyl, and its use in Stille and Sonogashira couplings (Scheme 16, equations 1 and 2) [117, 118]. Thus, indole triflate **20** can serve as an alternative to 3-haloindoles as a substrate for palladium-catalyzed cross-coupling reactions. Mérour and Malapel-Andrieu reported a similar use of *N*-acetyl 3-indolyl triflate **21** and Pd-catalyzed reaction with 1,2-diamines to give indole-fused heterocycles such as pyrazino[2,3-*b*]indole (equation 3) [119]. The yield was much lower (31%) and the reaction slower without palladium. Also synthesized using this reaction from **21** were heterocycles **22** and a 1:1 mixture of **23** and **24**.

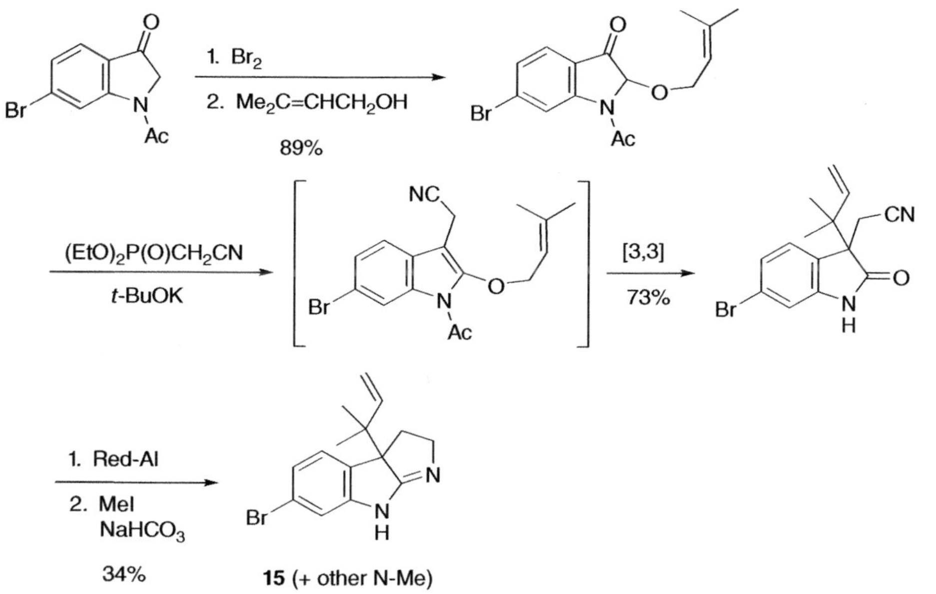

[eqn. 1] [110]

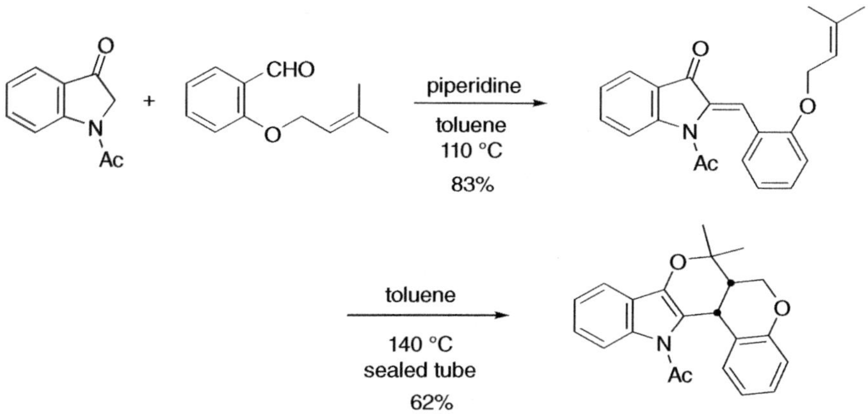

[eqn. 2] [113]

Scheme 14 *C-2 Reactions of* N-*Protected Indoxyls*

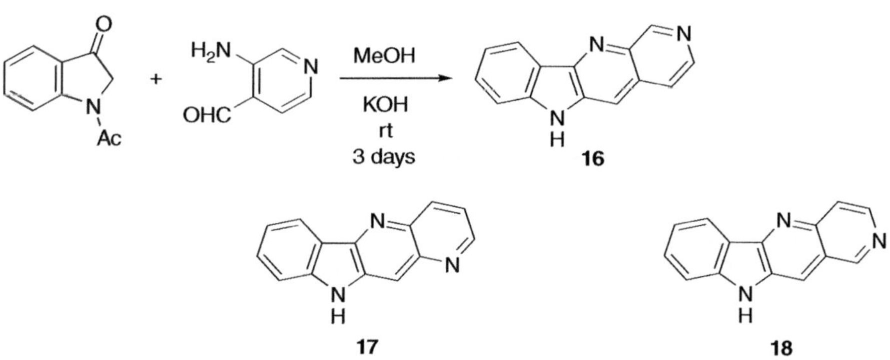

[eqn. 1] [115]

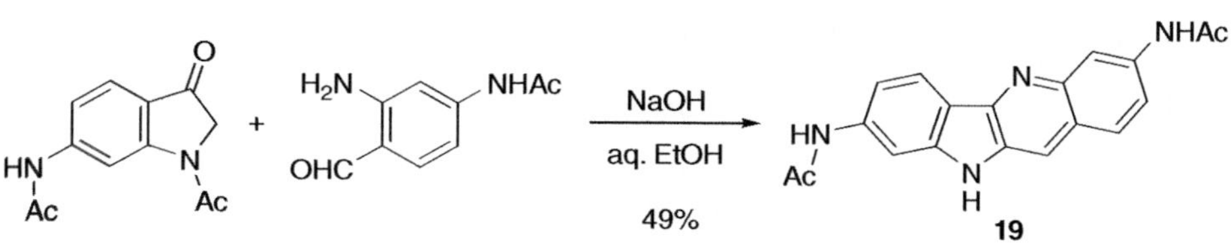

[eqn. 2] [116]

Scheme 15 *Applications of the Friedlander Synthesis to Indole-Fused Heterocycles*

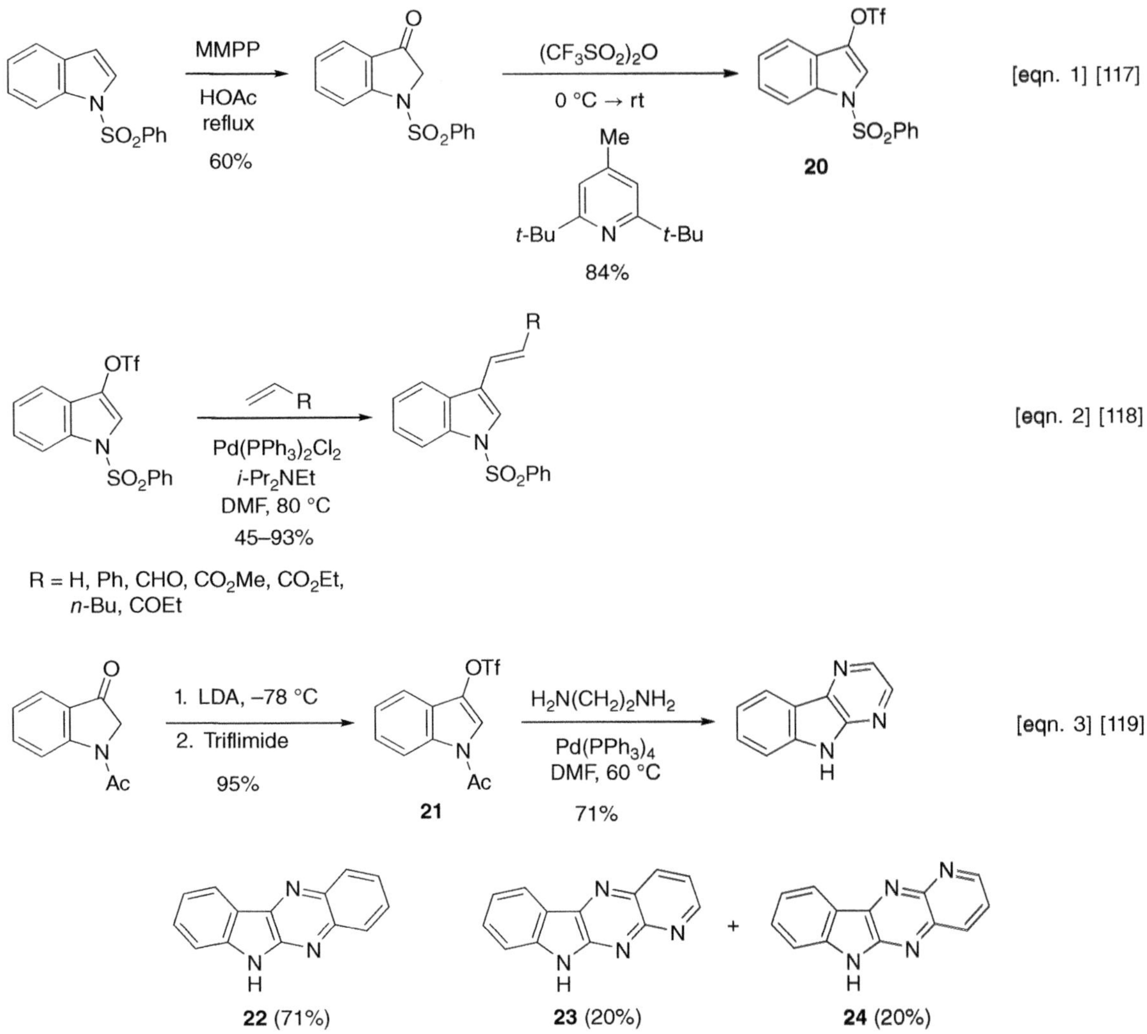

Scheme 16 *Applications of 3-Indolyl Triflates in Indole Synthesis*

References

[1] W.C. Sumpter, *Chem. Rev.*, 1945, **37**, 443–479.
[2] G.M. Karp, *Org. Prep. Proc. Int.*, 1993, **25**, 481–513.
[3] P.L. Julian, J. Pikl, and D. Boggess, *J. Am. Chem. Soc.*, 1934, **56**, 1797–1801.
[4] L.J. Dolby and P.D. Lord, *J. Org. Chem.*, 1969, **34**, 2988–2993.
[5] C.B. Hudson and A.V. Robertson, *Aust. J. Chem.*, 1967, **20**, 1699–1704.
[6] C. Mirand, G. Massiot, and J. Lévy, *J. Org. Chem.*, 1982, **47**, 4169–4170.
[7] P. Balczewski, J.A. Joule, C. Estévez, and M. Alvarez, *J. Org. Chem.*, 1994, **59**, 4571–4575.
[8] U. Stauss, H.P. Härter, M. Neuenschwander, and O. Schindler, *Helv. Chim. Acta*, 1972, **55**, 771–780.
[9] D.St.C. Black, N.E. Rothnie, and L.C.H. Wong, *Aust. J. Chem.*, 1983, **36**, 2407–2412.
[10] F.J. McEvoy and G.R. Allen, Jr., *J. Org. Chem.*, 1973, **38**, 3350–3352.
[11] M. Somei, K. Hashiba, F. Yamada, *et al.*, *Chem. Lett.*, 1978, 1245–1248.
[12] J.C. Torres, S.J. Garden, A.C. Pinto, *et al.*, *Tetrahedron*, 1999, **55**, 1881–1892.
[13] T.D. Cushing, J.F. Sanz-Cervera, and R.M. Williams, *J. Am. Chem. Soc.*, 1996, **118**, 557–579.
[14] M.M. Bastos, L.M.U. Mayer, E.C.S. Figueira, *et al.*, *J. Heterocycl. Chem.*, 2008, **45**, 969–973.
[15] W. Wierenga, *J. Am. Chem. Soc.*, 1981, **103**, 5621–5623.
[16] W. Wierenga, J. Griffin, and M.A. Warpehoski, *Tetrahedron Lett.*, 1983, **24**, 2437–2440.
[17] Y. Horiguchi, A. Sonobe, T. Saitoh, *et al.*, *Chem. Pharm. Bull.*, 2001, **49**, 1132–1137.
[18] Z. Wróbel, *Tetrahedron Lett.*, 2001, **42**, 5537–5539.

[19] H. Katayama, K. Maeda, and K. Kaneko, *J. Heterocycl. Chem.*, 1988, **25**, 937–942.
[20] T. Nishio, N. Okuda, and C. Kashima, *Helv. Chim. Acta*, 1990, **73**, 1719–1723.
[21] J.C. Powers, *J. Org. Chem.*, 1966, **31**, 2627–2631.
[22] K.E. Schulte, J. Reisch, and U. Stoess, *Angew. Chem. Int. Ed.*, 1965, **4**, 1081–1082.
[23] K.E. Schulte, J. Reisch, and U. Stoess, *Arch. Pharmz.*, 1972, **305**, 523–533.
[24] A. Kubo and T. Nakai, *Synthesis*, 1980, 365–366.
[25] M.R. Brennan, K.L. Erickson, F.S. Szmalc, *et al.*, *Heterocycles*, 1986, **24**, 2879–2885.
[26] M.F. Comber and C.J. Moody, *Synthesis*, 1992, 731–733.
[27] W. Fröhner, M.P. Krahl, K.R. Reddy, and H.-J. Knölker, *Heterocycles*, 2004, **63**, 2393–2407.
[28] S. Majumder and P.J. Bhuyan, *Tetrahedron Lett.*, 2012, **53**, 762–764.
[29] S. Majumder and P.J. Bhuyan, *Synlett*, 2011, 173–176.
[30] H.D.H. Showalter, A.D. Sercel, B.M. Leja, *et al.*, *J. Med. Chem.*, 1997, **40**, 413–426.
[31] M. Somei, S. Sayama, K. Naka, and F. Yamada, *Heterocycles*, 1988, **27**, 1585–1587.
[32] M. Tiano and P. Belmont, *J. Org. Chem.*, 2008, **73**, 4101–4109.
[33] N. Zhang, S. Ayral-Kaloustian, J.T. Anderson, *et al.*, *Bioorg. Med. Chem. Lett.*, 2010, **20**, 3526–3529.
[34] E. Erba, D. Pocar, and M. Valle, *J. Chem. Soc., Perkin Trans. 1*, 1999, 421–425.
[35] Z.H. Li, Z.R. Hu, R.E. Chen, and W.K. Su, *Org. Prep. Proc. Internat.*, 2010, **42**, 479–484.
[36] M.A. Eissenstat, M.R. Bell, T.E. D'Ambra, *et al.*, *J. Med. Chem.*, 1995, **38**, 3094–3105.
[37] I. Hiyoshi, H. Kumagai, H. Ooi, *et al.*, *Heterocycles*, 2007, **72**, 231–238.
[38] H. Hiyoshi, T. Sonoda, and S. Mataka, *Heterocycles*, 2006, **68**, 763–769.
[39] L. Ji, Q. Fang, M. Yuan, *et al.*, *Org. Lett.*, 2010, **12**, 5192–5195.
[40] F. Wang, X.-C. Li, W.-Y. Lai, *et al.*, *Org. Lett.*, 2014, **16**, 2942–2945.
[41] J.J. Chen, Y. Wei, J.C. Drach, and L.B. Townsend, *J. Med. Chem.*, 2000, **43**, 2449–2456.
[42] J.D. Williams, J.J. Chen, J.C. Drach, and L.B. Townsend, *J. Med. Chem.*, 2004, **47**, 5753–5765.
[43] F. Chan, P. Magnus, and E.G. McIver, *Tetrahedron Lett.*, 2000, **41**, 835–838.
[44] J. Bergman, R. Carlsson, and B. Sjöberg, *J. Heterocycl. Chem.*, 1977, **14**, 1123–1134.
[45] W. Fröhner, B. Monse, T.M. Braxmeier, *et al.*, *Org. Lett.*, 2005, **7**, 4573–4576.
[46] M. Fernandez, C. Barcia, J.C. Estévez, *et al.*, *Synlett*, 2004, 267–270.
[47] N.A. Lozinskaya, S.E. Sosonyuk, M.S. Volkova, *et al.*, *Synthesis*, 2011, 273–276.
[48] H. Hiyoshi, J.-H. Do, X. Feng, *et al.*, *Heterocycles*, 2011, **83**, 1017–1027.
[49] D.StC. Black, A.J. Ivory, and N. Kumar, *Tetrahedron*, 1996, **52**, 4697–4708.
[50] T. Owa, T. Okauchi, K. Yoshimatsu, *et al.*, *Bioorg. Med. Chem. Lett.*, 2000, **10**, 1223–1226.
[51] M.S. Tichenor, J.D. Trzupek, D.B. Kastrinsky, *et al.*, *J. Am. Chem. Soc.*, 2006, **128**, 15683–15696.
[52] G. Revial, I. Jabin, S. Lim, and M. Pfau, *J. Org. Chem.*, 2002, **67**, 2252–2256.
[53] S.C. Conway and G.W. Gribble, *Synth. Commun.*, 1992, **22**, 2987–2995.
[54] C. Bhattacharya, P. Bonfante, A. Deagostino, *et al.*, *Org. Biomol. Chem.*, 2009, **7**, 3413–3420.
[55] C. Prandi, E.G. Occhiato, S. Tabasso, *et al.*, *Eur. J. Org. Chem.*, 2011, 3781–3793.
[56] J.A. Brailsford, R. Lauchli, and K.J. Shea, *Org. Lett.*, 2009, **11**, 5330–5333.
[57] M. Ghandi, S. Salahi, and M. Hasani, *Tetrahedron Lett.*, 2011, **52**, 270–273.
[58] H. Möhrle and H. Dwuletzki, *Z. Naturforsch.*, 1987, **42b**, 1032–1034.
[59] M. Jha, T.-Y. Chou, and B. Blunt, *Tetrahedron*, 2011, **67**, 982–989.
[60] Y. Chen, C. Yang, and Y. Xie, *Heterocycles*, 2010, **80**, 251–258.
[61] O. Barun, P.K. Patra, H. Ila, and H. Junjappa, *Tetrahedron Lett.*, 1999, **40**, 3797–3800.
[62] J.R. Suresh, U.K. Syam Kumar, H. Ila, and H. Junjappa, *Tetrahedron*, 2001, **57**, 781–789.
[63] G.S.M. Sundaram, C. Venkatesh, U.K. Syam Kumar, *et al.*, *J. Org. Chem.*, 2004, **69**, 5760–5762.
[64] M. Adid, B. Mohammadi, and H. Reza Bijanzadeh, *Synlett*, 2008, 177–180.
[65] R. Ghahremanzadeh, S. Ahadi, and A. Bazgir, *Tetrahedron Lett.*, 2009, **50**, 7379–7381.
[66] K.M. James, N. Willetts, and D.J. Procter, *Org. Lett.*, 2008, **10**, 1203–1206.
[67] L. Yang, F. Wang, P.J. Chua, *et al.*, *Org. Lett.*, 2012, **14**, 2894–2897.
[68] Z. Mao, W. Li, Y. Shi, *et al.*, *Chem. Eur. J.*, 2013, **19**, 9754–9759.
[69] T. Hino, K. Tsuneoka, M. Nakagawa, and S. Akaboshi, *Chem. Pharm. Bull.*, 1969, **17**, 550–558.
[70] E. Wenkert, J.M. Hana, Jr., M.H. Leftin, *et al.*, *J. Org. Chem.*, 1985, **50**, 1125–1126.
[71] M.S.C. Pedras and I.L. Zaharia, *Org. Lett.*, 2001, **3**, 1213–1216.
[72] M.S.C. Pedras and M. Jha, *J. Org. Chem.*, 2005, **70**, 1828–1834.
[73] F. Matloubi Moghaddam, H. Saeidian, Z. Mirjafary, *et al.*, *Synlett*, 2009, 1047–1050.
[74] H. Zali Boeini, *Helv. Chim. Acta*, 2009, **92**, 1268–1272.
[75] T. Otani, S. Kunimatsu, T. Takahashi, *et al.*, *Tetrahedron Lett.*, 2009, **50**, 3853–3856.
[76] F. Matloubi Moghaddam, S. Taheri, Z. Mirjafary, and H. Saeidian, *Synlett*, 2010, 123–127.
[77] F. Matloubi Moghaddam, Z. Mirjafary, H. Saeidian, *et al.*, *Tetrahedron*, 2010, **66**, 134–138.
[78] K.C. Majumdar, P. Debnath, S. Alam, and P.K. Maji, *Tetrahedron Lett.*, 2007, **48**, 7031–7033.
[79] K.C. Majumdar, A. Taher, and K. Ray, *Tetrahedron Lett.*, 2009, **50**, 3889–3891.
[80] K.C. Majumdar, A. Taher, and S. Ponra, *Tetrahedron Lett.*, 2010, **51**, 147–150.
[81] M. Jha, O. Enaohwo, and A. Marcellus, *Tetrahedron Lett.*, 2009, **50**, 7184–7187.

[82] M. Jha, O. Enaohwo, and S. Guy, *Tetrahedron Lett.*, 2011, **52**, 684–687.
[83] M. Jha, M. Edmunds, K. Lund, and A. Ryan, *Tetrahedron Lett.*, 2014, **55**, 5691–5694.
[84] S. Chen, J. Pan, Y. Wang, and Z. Zhou, *Eur. J. Org. Chem.*, 2014, 7940–7947.
[85] T. Janosik and J. Bergman, *Heterocycles*, 2002, **57**, 1273–1278.
[86] W.C. Sumpter, *Chem. Rev.*, 1944, **34**, 393–434.
[87] J.A. Joule and K. Mills (2010) *Heterocyclic Chemistry*, 5th edn, Blackwell, London, p. 399.
[88] P.G. Gassman, B.W. Cue, Jr., and T.-Y. Luh, *J. Org. Chem.*, 1977, **42**, 1344–1348.
[89] M.J.E. Hewlins, A.H. Jackson, A.-M. Oliveira-Campos, and P.V.R. Shannon, *J. Chem. Soc., Perkin Trans. 1*, 1981, 2906–2911.
[90] B. Jiang, J.M. Smallheer, C. Amaral-Ly, and M.A. Wuonola, *J. Org. Chem.*, 1994, **59**, 6823–6827.
[91] J.M. Bentley, D.R. Adams, D. Bebbington, *et al.*, *Bioorg. Med. Chem. Lett.*, 2004, **14**, 2367–2370.
[92] A. Mollica, A. Stefanucci, F. Feliciani, *et al.*, *Tetrahedron Lett.*, 2011, **52**, 2583–2585.
[93] H. Sirowej, S. Ahmad Khan, and H. Plieninger, *Synthesis*, 1972, 84.
[94] R.M. Soll, C. Guinosso, and A. Asselin, *J. Org. Chem.*, 1988, **53**, 2844–2847.
[95] C.A. Demerson, L.G. Humber, A.H. Philipp, and R.R. Martel, *J. Med. Chem.*, 1976, **19**, 391–395.
[96] A.H. Katz, C.A. Demerson, C.-C. Shaw, *et al.*, *J. Med. Chem.*, 1988, **31**, 1244–1250.
[97] A. Jaouen, P. Helissey, S. Desbène-Finck, and S. Giorgi-Renault, *Heterocycles*, 2008, **75**, 2745–2759.
[98] K.M. Driller, S. Libnow, M. Hein, *et al.*, *Org. Biomol. Chem.*, 2008, **6**, 4218–4223.
[99] P. Helissey, S. Desbène-Finck, and S. Giorgi-Renault, *Eur. J. Org. Chem.*, 2005, 410–415.
[100] J. Bergman, C. Damberg, and H. Vallberg, *Recl. Trav. Chim. Pays-Bas*, 1996, **115**, 31–36.
[101] N. Wang, K.J. Wicht, K. Imai, *et al.*, *Bioorg. Med. Chem.*, 2014, **22**, 2629–2642.
[102] R.M. Acheson (1962) *An Introduction to the Chemistry of Heterocyclic Compounds*, Interscience, New York, pp. 136–137.
[103] D. Vorlander, *Ber.*, 1902, **35**, 1683–1698.
[104] J.E. Pretka and H.G. Lindwall, *J. Org. Chem.*, 1954, **19**, 1080–1088.
[105] J.L. Garcia Ruano, C. Pedregal, and J.H. Rodriguez, *Tetrahedron*, 1989, **45**, 203–214.
[106] T. Kawasaki, Y. Nonaka, M. Uemura, and M. Sakamoto, *Synthesis*, 1991, 701–702.
[107] J.-Y. Mérour, P. Gadonneix, B. Malapel-Andrieu, and E. Desarbre, *Tetrahedron*, 2001, **57**, 1995–2002.
[108] T. Kouko, K. Matsumura, and T. Kawasaki, *Tetrahedron*, 2005, **61**, 2309–2318.
[109] T. Kawasaki, K. Watanabe, K. Masuda, and M. Sakamoto, *J. Chem. Soc., Chem. Commun.*, 1995, 381–382.
[110] T. Kawasaki, R. Terashima, K. Sakaguchi, *et al.*, *Tetrahedron Lett.*, 1996, **37**, 7525–7528.
[111] T. Kawasaki, Y. Nonaka, K. Watanabe, *et al.*, *J. Org. Chem.*, 2001, **66**, 1200–1204.
[112] J.S. Carle and C. Christophersen, *J. Org. Chem.s*, 1981, **46**, 3440–3443.
[113] Y. Davion, B. Joseph, V. Bénéteau, *et al.*, *Helv. Chim. Acta*, 2003, **86**, 2687–2697.
[114] M.V. Basaveswara Rao, U.K. Syam Kumar, H. Ila, and H. Junjappa, *Tetrahedron*, 1999, **55**, 11563–11578.
[115] F. Nivoliers, A. Decormeille, A. Godard, and G. Quéguiner, *Tetrahedron Lett.*, 1980, **21**, 4485–4486.
[116] C.A. Merlic, S. Motamed, and B. Quinn, *J. Org. Chem.*, 1995, **60**, 3365–3369.
[117] S.C. Conway and G.W. Gribble, *Heterocycles*, 1990, **30**, 627–633.
[118] G.W. Gribble and S.C. Conway, *Synth. Commun.*, 1992, **22**, 2129–2141.
[119] B. Malapel-Andrieu and J.-Y. Mérour, *Tetrahedron*, 1998, **54**, 11095–11110.

PART X

Metal-Catalyzed Indole Synthesis

The use of metals as catalysts to construct the indole ring has revolutionized this area of organic synthesis. In the following chapters we present the role of copper, palladium, rhodium, ruthenium, titanium, zirconium, gold, and other metals to prepare indoles.

Indole Ring Synthesis: From Natural Products to Drug Discovery, First Edition. Gordon W. Gribble.

69

Copper-Catalyzed Indole Synthesis

The relative inexpensive copper and its variable oxidation states have elevated it to prominence for indole synthesis; numerous recent reviews are available [1–8].

The first reported indole ring synthesis using copper was that of Castro and Stephens [9, 10], and subsequently, Castro and other colleagues [11, 12] (Scheme 1, equation 1). For an excellent review of the Castro–Stephens reaction, see Gray [13]. The synthesis may be carried out in one step (DMF, 120 °C) from a mixture of 2-iodoaniline and cuprous phenylacetylide (prepared in quantitative yield from phenylacetylene, cuprous iodide, and NH_4OH) or in two steps by isolating the intermediate 2-aminotolane (2-aminodiphenylacetylene). Indoles **1–4** were also synthesized by Castro [11]. Independently, Reisch described a synthesis of indole (and 7-azaindole) using similar conditions to those of Castro and Stephens (equation 2) [14]. Castro and coworkers studied the mechanism and kinetics of this reaction and proposed a four-center process for the aryl substitution [12] (see Gray [13]).

The Castro–Stephens reaction has found utility in indole synthesis *per se* and more particularly as the final copper-induced indolization step in the palladium-catalyzed Sonogashira reaction, which is presented in Chapter 70. As Gray noted regarding the comparison of the Castro–Stephens coupling with the Sonogashira coupling: “Advances in both methodologies, particularly the *in situ* formation of copper acetylides for Castro–Stephens reactions, have blurred the lines between these two methodologies, and under some conditions, they share key mechanistic features” [13]. A few of the examples that follow do involve a Sonogashira synthesis of the precursor aryl acetylene.

Applications of the original Castro–Stephens indole synthesis were described by Yamamoto (*N*-benzylindole, 73%) [15], Katritzky (2-(benzotriazol-1-ylmethyl)indole, 60%) [16], and Vasilevsky (naphtho[2,3-*g*]indole-6,11-diones, 58%–98%) [17]. Several modifications and extensions of the Castro–Stephens indolization have been discovered. Cacchi and coworkers effected an indole synthesis via a “domino copper-catalyzed coupling–cyclization process” (Scheme 2, equation 1) [18]. Lower yields were observed for 1-hexyne and for conditions that employed CuI/PPh_3 in dioxane (110 °C) (60%–85%). Independently, Venkataraman and colleagues described a very similar 2-arylindole synthesis from 2-iodoaniline using Cs_2CO_3 and *t*-BuONa as bases [19]. They also prepared 2-arylindoles from *N*-(2-ethynylphenyl)acetamide and aryliodides (equation 2). Ma and Liu employed a CuI/L-proline catalyst (DMF, K_2CO_3, 80 °C) to access a variety of 2-aryl- and 2-alkylindoles from 2-bromotrifluoroacetanilides and terminal alkynes [20]. Likewise, Zhao’s group described similar chemistry to prepare *o*-alkynylacetanilides (CuI/*N*-methylpyrrolidine-2-carboxamide) (not carried onto indoles) [21]. An aqueous recyclable sulfonato-Cu(II)(salen) catalyst was employed by Zhou to prepare 2-arylindoles from 2-iodoanilines and aryl acetylenes (92%–96%) [22]. A related aqueous recyclable copper catalyst was invented by Wan to craft both 2-aryl- and 2-alkylindoles from *N*-(2-ethynylphenyl)-sulfonamides (CuI/polystyrene-supported pyrrole-2-carbohydrazide/tetra-*n*-butylammonium bromide) [23]. Lu and Shen reported a 3-haloindole synthesis from a halocyclization of 2-alkynylanilines with cupric halides (equation 3) [24]. The proposed pathway involves a copper-coordinated alkyne complex that cyclizes to an indole C-3 copper species; reductive elimination then produces the 3-haloindole and Cu(0). Oxidation of Cu(0) by CuX_2 provides CuX. Pyne

Indole Ring Synthesis: From Natural Products to Drug Discovery, First Edition. Gordon W. Gribble.

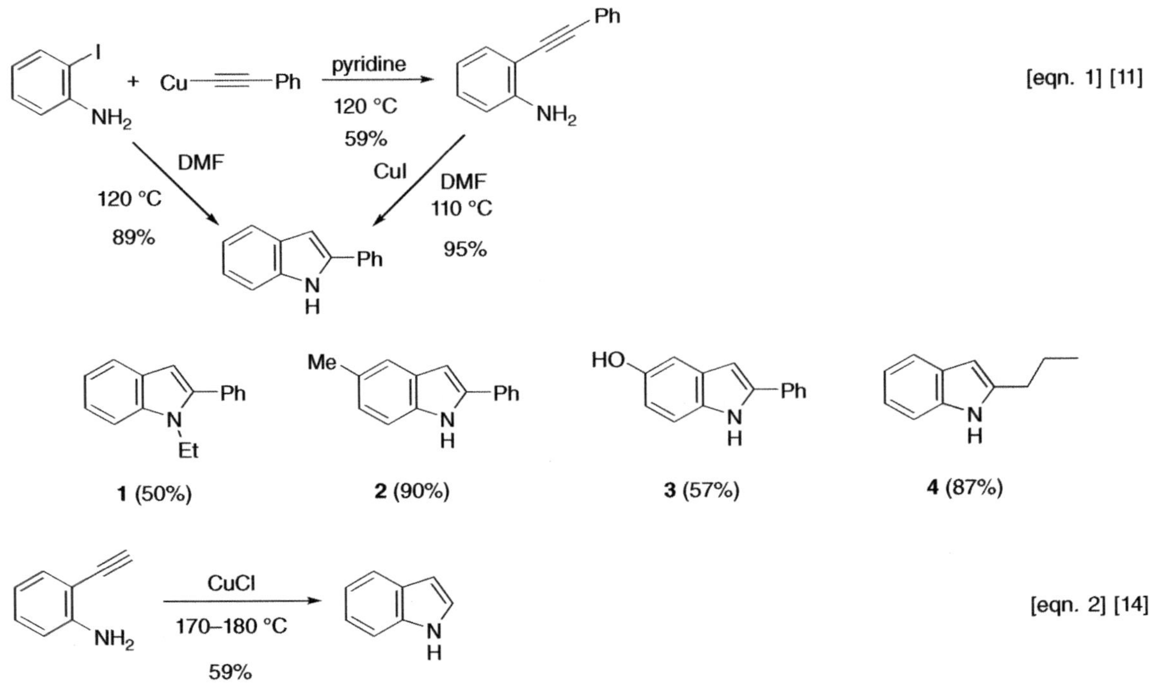

Scheme 1 The Castro-Stephens Indole Synthesis

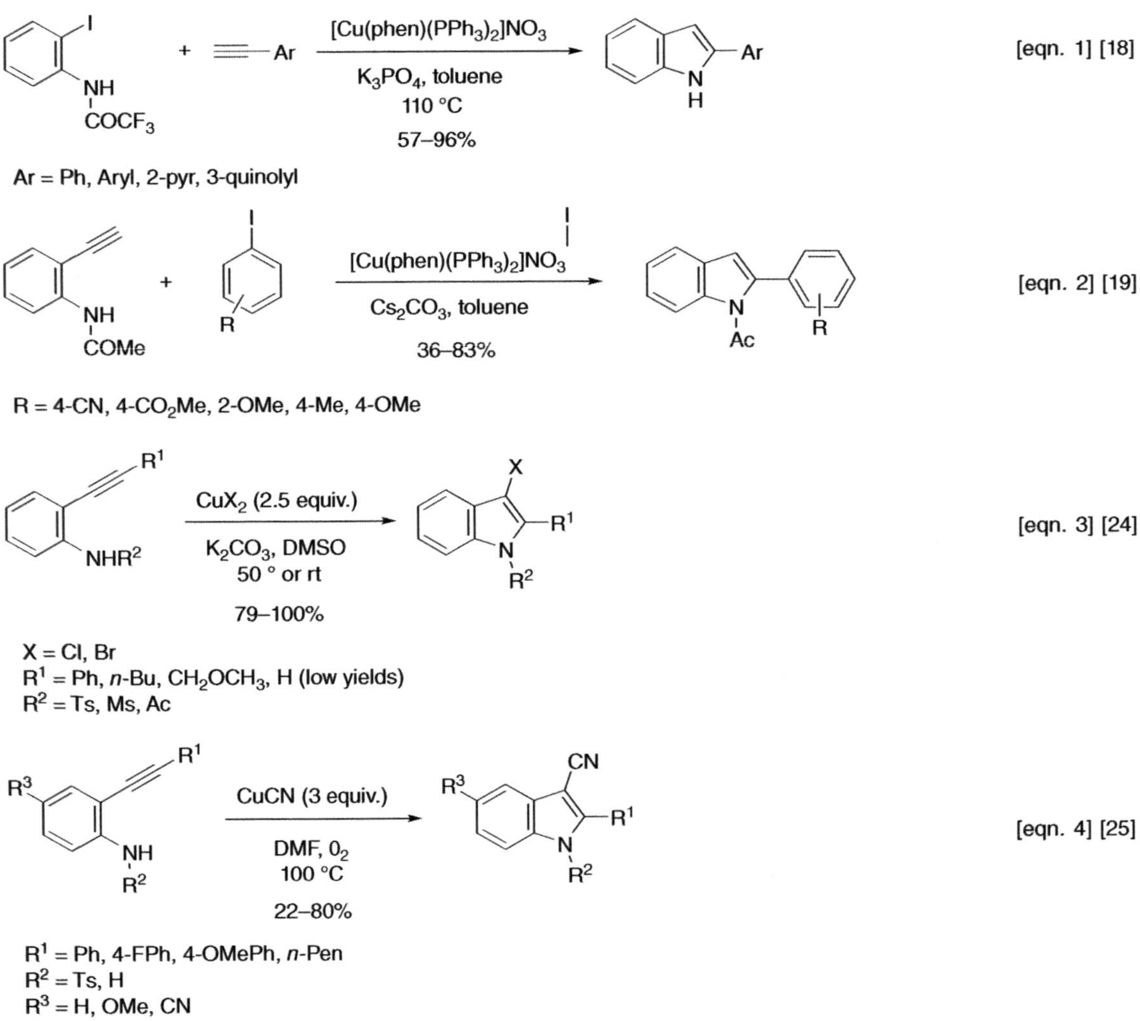

Scheme 2 Copper-Mediated Indole Syntheses from 2-Alkynylanilines – 1

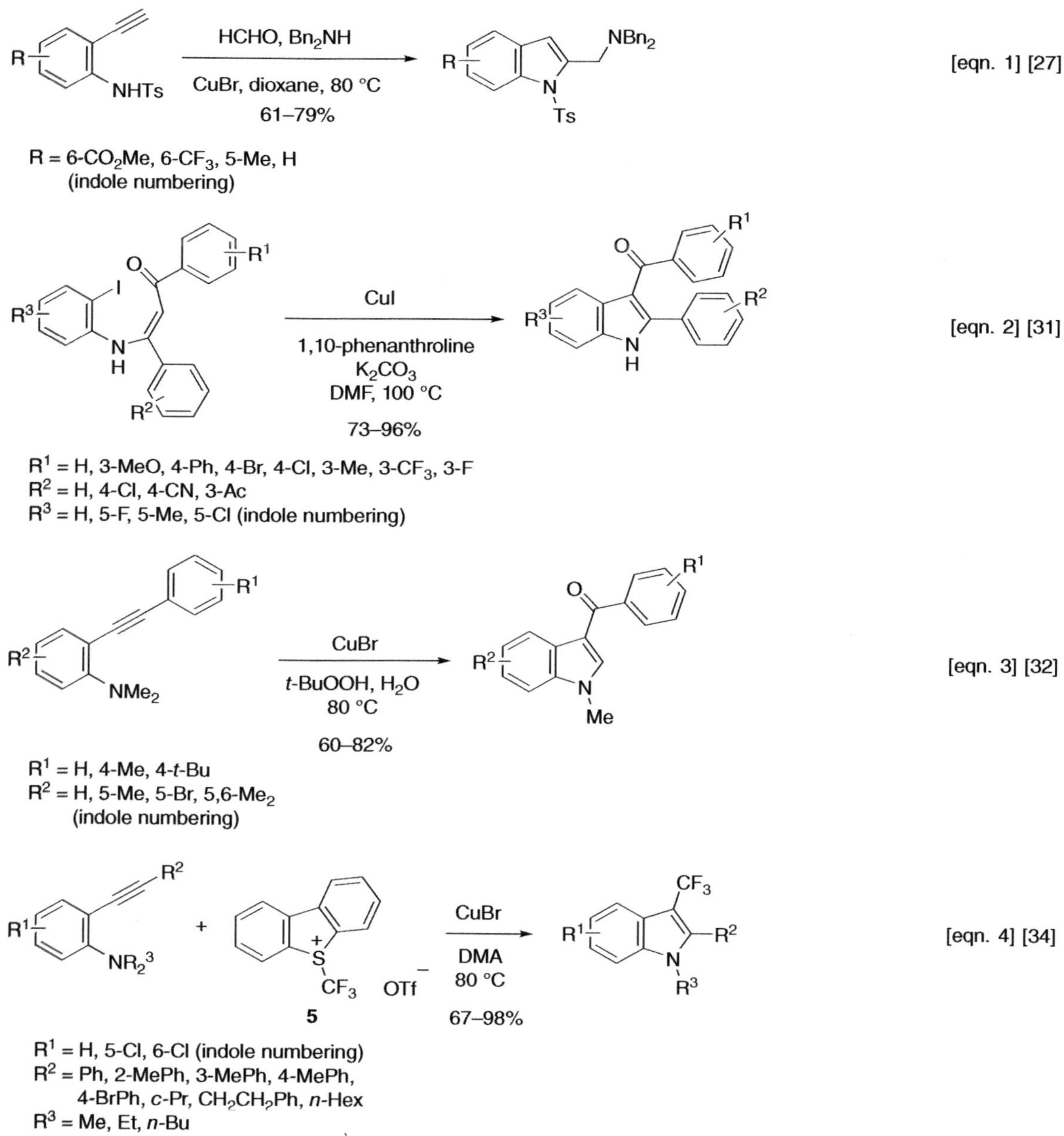

Scheme 3 *Copper-Mediated Indole Syntheses from 2-Alkynylanilines – 2*

and colleagues described a copper-mediated 3-cyanoindole synthesis (equation 4) [25].

Several investigators have uncovered additional copper-catalyzed indole syntheses from 2-ethynylanilines accompanied by *in situ* indole functionalization. Yamamoto and colleagues reported the synthesis of *N*-(alkoxybenzyl) indoles from the copper-catalyzed tandem reaction between 2-alkynyl-*N*-arylideneanilines and alcohols (CuI, toluene, 100 °C, 55%–83%) [26]. Ohno, Fujii, and coworkers employed a domino three-component copper-catalyzed process to prepare 2-(aminomethyl)indoles [27, 28] and β-carbolines [29] (Scheme 3, equation 1). A variety of secondary amines were employed, and the method was adapted to the preparation of pyrrolo[3,2-*f*]indoles and dipyrrolo[3,2-*b*:2′,3′-*e*]pyridines [30]. Cacchi and colleagues synthesized 3-aroylindoles from the copper-catalyzed cyclization of *N*-(2-iodoaryl)enaminones (equation 2) [31]. The latter were prepared from 2-haloanilines and α,β-ynones, which can react directly with 2-iodoaniline under the copper protocol to give 3-benzoyl-2-phenylindole (76%), thus bypassing the enaminone. Patel described a copper-catalyzed 3-aroylindole synthesis from 2-alkynylated

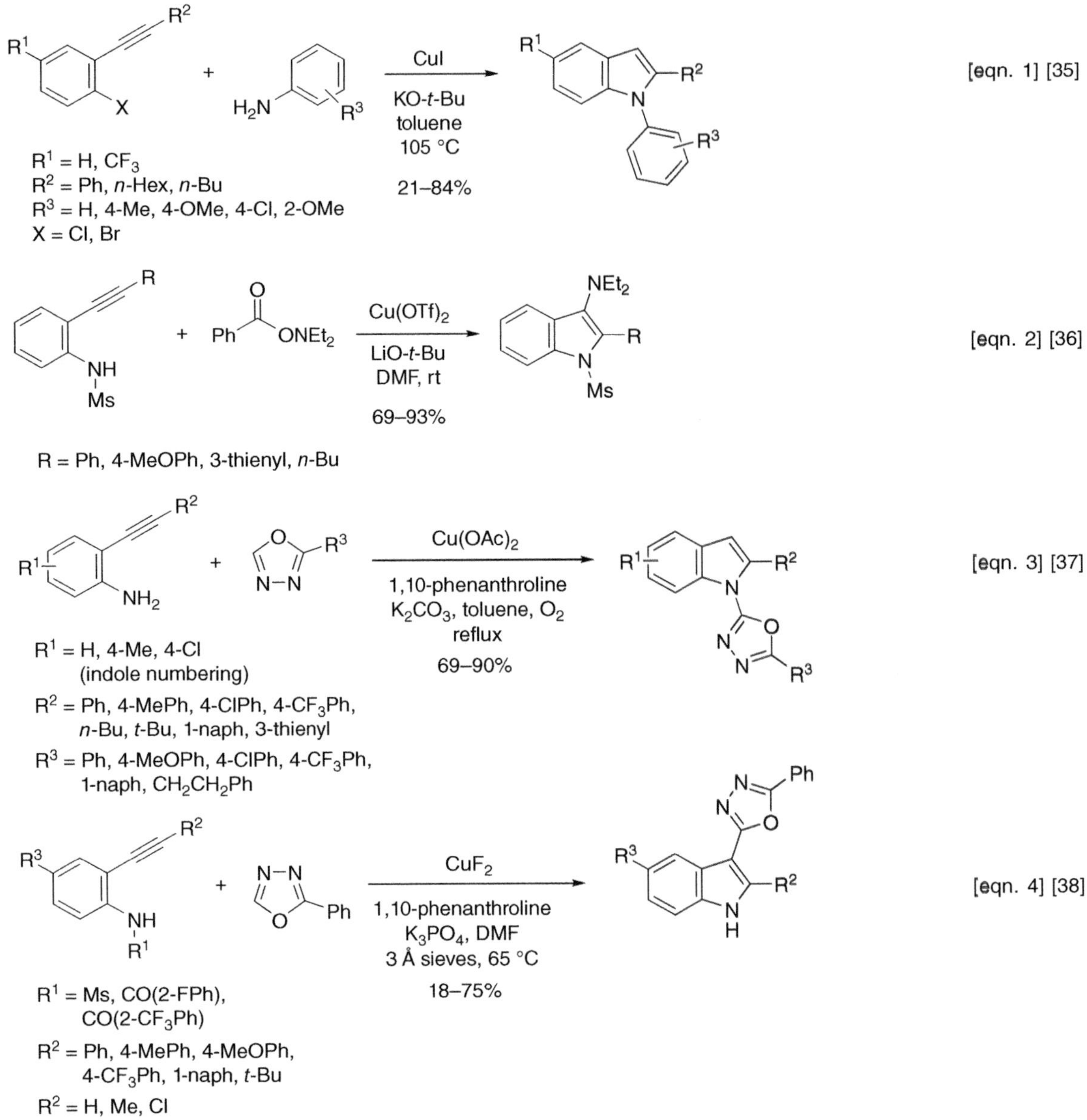

Scheme 4 *Copper-Mediated Indole Syntheses from 2-Alkynylanilines – 3*

N,N-dimethylanilines (equation 3) [32]. Independently, Liang reported a similar synthesis of 3-acylindoles [33]. Hou and colleagues uncovered a novel 3-trifluoromethylindole synthesis that used a CuBr-promoted reaction of 2-alkynylaniline in the presence of Umemoto's reagent (**5**) (equation 4) [34].

Ackermann engineered a copper-catalyzed process to achieve the *N*-arylation/hydroamination synthesis of indoles (Scheme 4, equation 1) [35]. In similar fashion, *N*-acylindoles and NH-indoles were made available using *N*-unsubstituted amides and carbamates in place of anilines. In these latter reactions, a vicinal diamine ligand improved the cyclizations. Miura and Hirano reported a 3-aminoindole synthesis using cupric triflate as catalyst and electrophilic amination of a presumed 3-indolyl copper species (equation 2) [36]. This research team also described the preparation of *N*-azolylindoles (equation 3) [37] and C-3 azolylindoles (equation 4) [38] by copper-mediated annulative coupling of 2-alkynylanilines.

Srivastava and colleagues described a direct synthesis of 3-arylindoles from their respective nitrosoarenes and arylacetylenes (Scheme 5, equation 1) [39], and Muthusubramanian and Ponpandian developed a domino decarboxylative synthesis of 2-arylindoles catalyzed by copper (equation 2) [40]. Jeong and coworkers synthesized a 2,2-biindolyl en route to a new class of anion receptors

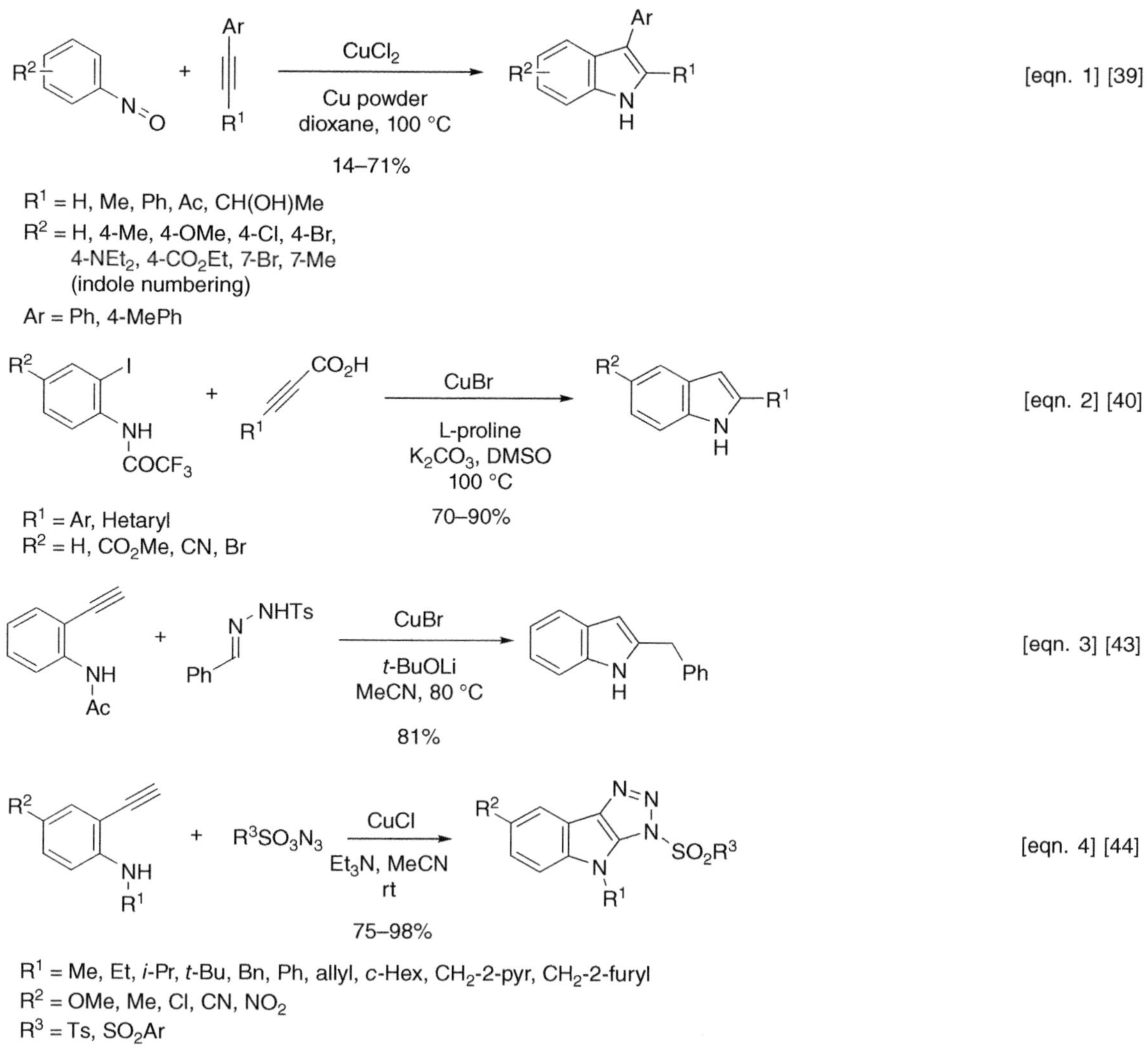

Scheme 5 *Copper-Mediated Indole Syntheses from 2-Ethynylanilines*

via oxidative dimerization (Glaser coupling) (Cu(OAc)$_2$, pyridine, rt, 95%) of 2-ethynyl-6-iodoaniline followed by twin indolization (CuI, DMF, 100–110 °C, 81%) to give 7,7′-diiodo-2,2′-biindole [41]. Hiroya and coworkers reported that aqueous methanol conditions with Cu(OCOCF$_3$)$_2$/1-ethylpiperidine effect indolization of 2-ethynylanilines to 2-substituted indoles [42]. Zhou and colleagues described the use of *N*-tosylhydrazones in a CuBr-catalyzed coupling-allenylation-cyclization path to yield indoles (equation 3) [43]. Ketone *N*-tosylhydrazones also led to indoles. Wang, Lu, and their colleagues prepared triazoindoles via a copper catalysis process (equation 4) [44]. These novel [1,2,3]triazolo[4,5-*b*]indoles served as α-imino rhodium carbene precursors for the construction of a variety of fused indoles and biindoles.

Although they differ in subtleties, the aforementioned copper-induced cyclizations invariably involve a cuprous acetylide, but several other indole syntheses not involving cuprous acetylides are known. One well-explored example is copper-mediated intramolecular amination Buchwald coupling [45, 46] of a suitable aryl halide. Scheme 6 (equations 1–5) presents some simple indole syntheses that used this ring formation. Fukuyama employed this amination (equation 1) in a synthesis of the duocarmycins [47] and also in the synthesis of (+)-yatakemycin [48]. The method of Karchava shown in equation 2 was also extended to the preparation of 1-aminoindoles [40] and *N*-substituted indole-3-carboxylic acid derivatives [50]. Nenajdenko's team reported a copper-promoted one-pot synthesis of 2-trifluoromethylindoles from β-halo-β-trifluoromethyl-*o*-halostyrenes and primary amines (equation 3) [51]. The copper-catalyzed double amination of 2-(2-bromoalkenyl) bromobenzene (equation 4) could also employ CuOAc and CuTC (copper(I) thiophene-2-carboxylate) and other

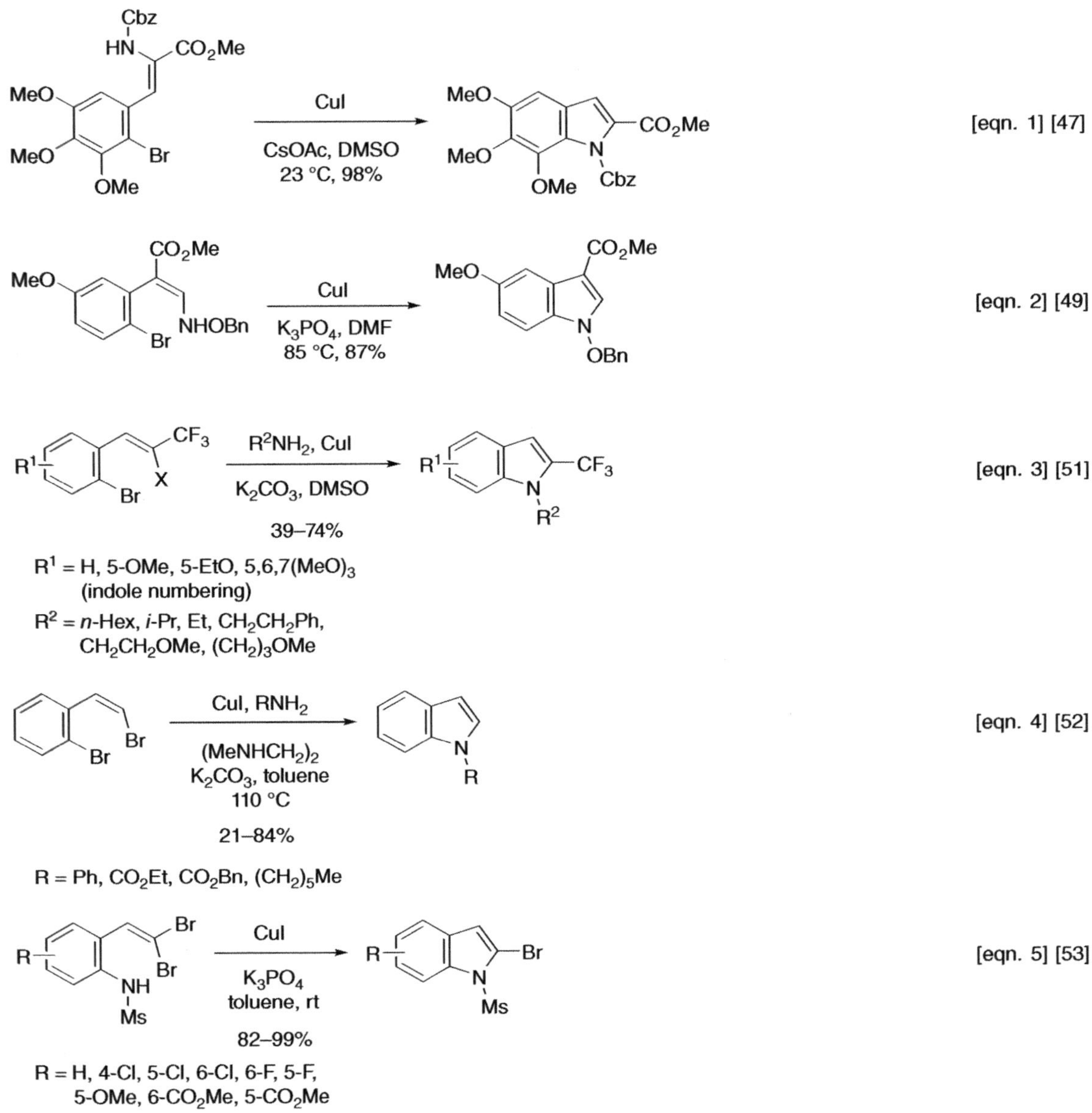

Scheme 6 *Copper-Catalyzed Indole Synthesis via Amination*

ring-substituted 2-(2-haloalkenyl)aryl halides [52]. Zhang described an excellent route to 2-bromoindoles from *gem*-dibromovinylanilines (equation 5) [53], which were prepared in a few steps from the requisite 2-nitrobenzaldehyde and CBr_4. Lv, Wang, and coworkers used the intramolecular amination of *o*-*gem*-dibromovinyl substrates to synthesize oxazino[3,2-*a*]indoles, thiazino[3,2-*a*]indoles, and indolo[2,1-*b*]quinazolines, in which the second bromine undergoes copper-catalyzed cyclization with an oxygen, sulfur, or nitrogen nucleophile to give the aforementioned ring systems [54]. The groups of Zeng-Chen and Peng separately reported copper-catalyzed syntheses of indole-2-carboxylic acid esters, respectively, from 3-(2-aminophenyl)-2-bromoacrylates [55] and 2-amino-3-(2′-bromophenyl)acrylates [56]. Koenig described a similar indole-2-carboxylate preparation that involved condensation between 2-halobenzaldehydes and glycine amidoesters followed by copper-catalyzed indolization [57].

The intramolecular amine–halogen copper-promoted indolization was applied to the synthesis of more-complex indoles (Scheme 7, equations 1–4) [58–61]. Other examples in this category of copper-catalyzed amination are syntheses of the carbazole alkaloids murrayaquinone-A and (±)-bismurrayaquinone-A [62], 3-aryl β-carbolin-1-ones [63, 64], carbazoles from double C–N cyclization [65], pyrrolo[2,3-*c*]

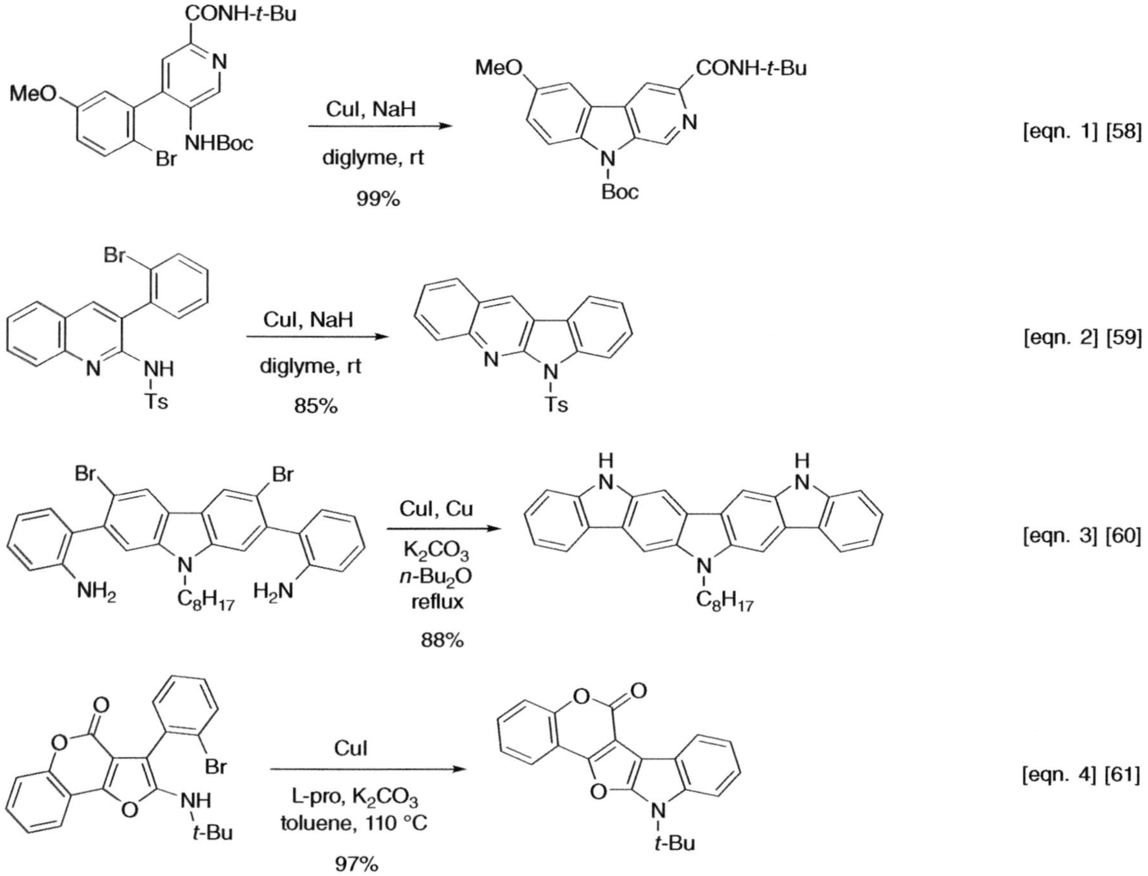

Scheme 7 *Copper-Catalyzed Synthesis of Complex Indoles via Amination*

pyridines and functionalized indoles [66], benzofuro[3,2-*b*]indoles [67], PDE-II [68], canthin-6-ones [69], benzoindoloquinolizines [70], and extended π-conjugated heteroacenes [71]. May's group described a simple *N*-substituted indole synthesis from the reaction of primary amines with α-(*o*-haloaryl)ketone (or aldehyde) followed by copper-catalyzed intramolecular amination [72]. Tertiary and aromatic primary amines fared very well in this indole synthesis. Cai and colleagues described an indole-2-carboxylic acid ester preparation that entailed the reaction of 2-halo aryl aldehydes/ketones with ethyl isocyanoacetate, as a nitrogen partner in this novel copper-catalyzed cyclization-deformylation process [73].

Another copper-mediated indole formation is the cyclization of *N*-2-haloaryl enaminones (Heck reaction), and related protocols and numerous examples exist (Scheme 8, equations 1–4). An early report was that of Suzuki's group giving the product indoles in excellent yield, except when X=Cl (4%–6%) (equation 1) [74]. The authors suggested a cuprous enolate intermediate, and they subsequently developed a one-pot synthesis of simple indoles using this strategy (equation 2) [75]. d'Angelo and Desmaele employed the Suzuki conditions to prepare a (*R*)-carbazolone intermediate for the synthesis of *Aspidosperma* alkaloids [76], and Jiang similarly synthesized a series of carbazol-4-ones and 3,4-dihydrocyclopentylindol-1-ones from *N*-2-iodoaryl enaminones using CuI/proline/K_2CO_3/DMSO [77]. Punniyamurthy and Ali established that the pathway in their reaction conditions between 2-iodoanilines and 1,3-dicarbonyl compounds involved amine–carbonyl condensation followed by copper-catalyzed ring formation (equation 3) [78]. A suggested mechanism is shown starting from imine **6** (equation 4). Zhu and colleagues reported a similar 2,3-disubstituted indole synthesis (and mechanism) involving the copper-catalyzed coupling of 2-aminophenylboronic acid pinacol esters with β-keto esters [79]. Both studies demonstrated that iodobenzene itself does not undergo C–C bond formation under the reaction conditions [78, 79].

Huang [80] and Liu [81] independently described a copper-catalyzed indole synthesis involving 2-iodoaniline and allenes (e.g., ethyl buta-2,3-dienoate) in a process that involved *in situ* formation of an aza-Michael addition adduct (vinylogous carbamate) and intramolecular Heck-like cyclization. Both groups used similar reaction

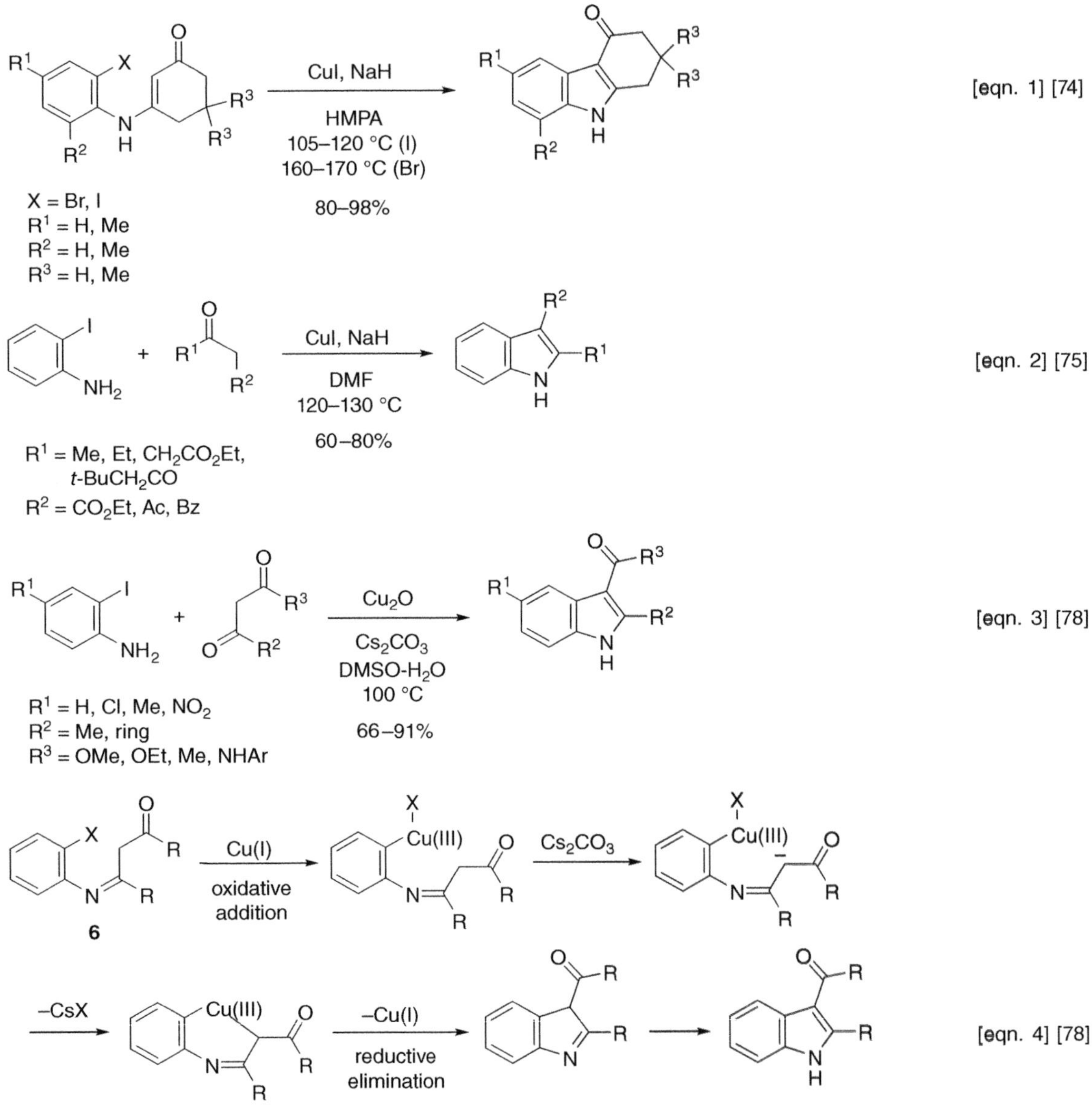

Scheme 8 *Copper-Catalyzed Indole Syntheses from 2-Haloaryl Enaminones*

conditions (CuI, K_2CO_3, dioxane, 120 °C) to obtain a collection of C-2- and ring-substituted indole-3-carboxylates. Cao and colleagues employed an aza-Michael addition between 2-haloanilines and methyl perfluoroalk-2-ynoates to give the *N*-(2-haloaryl)enamines (vinylogous carbamates), which subsequently undergo CuBr-catalyzed cyclization to give 2-(perfluoroalkyl)indole-3-carboxylates (CuBr, K_2CO_3, DMSO, 60 °C, 55%–93%) [82]. Interestingly, other reaction substrates and conditions were found to involve initial copper-catalyzed C–C bond formation, followed by amine-carbonyl condensation. Miura's synthesis of 3-acetyl-2-methylindole implicates copper-catalyzed carbon–carbon coupling of enolates with 2-iodoaniline, followed by indolization (Scheme 9, equation 1) [83]. Thus, CuI-catalyzed condensation of iodobenzene with 2,4-pentanedione gave 3-phenylpentane-2,4-dione (65%) [83]. Ma and colleagues reported similar indole syntheses (equation 2) [84, 85], where the formation of 2-trifluoromethylindoles entails deacylation of COR^3 and retention of the trifluoromethyl group. Fu, Oiao, and coworkers [86] and Kobayashi [87] independently arrived at the same 2-aminoindole synthesis involving the copper-catalyzed reaction between 2-haloanilines and cyanoacetates or malononitriles (equations 3 and 4).

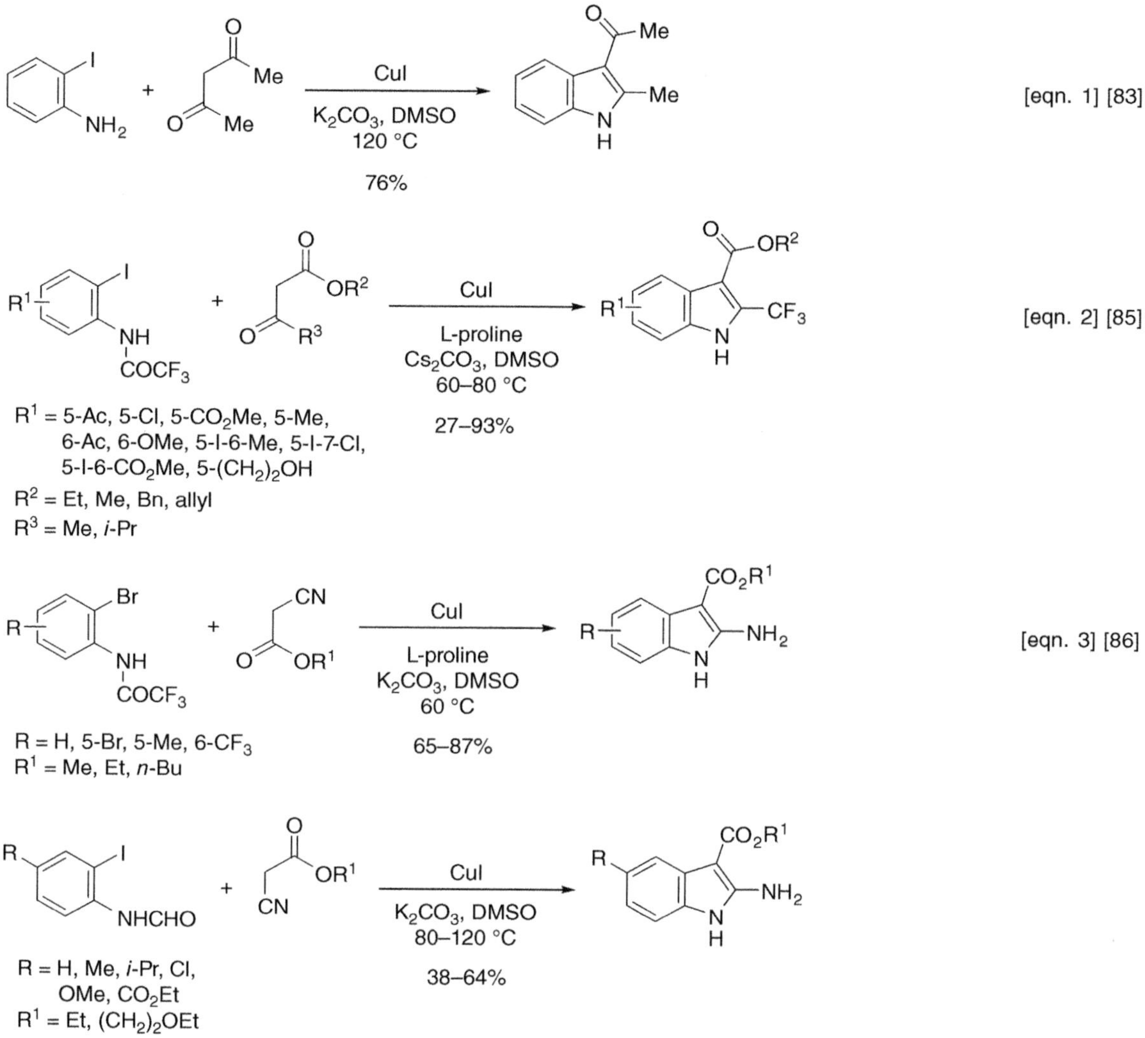

Scheme 9 *Copper-Catalyzed Indole Synthesis from 2-Haloanilines and β-Dicarbonyl Compounds*

The 2-amino-3-cyanoindoles were obtained in 72% to 93% yield with malononitrile [86].

The copper(I)-catalyzed reaction between 2-haloanilines and β-dicarbonyls (or equivalent compounds), irrespective of the actual pathway, is an extremely versatile and efficient indole synthesis, and many applications are known.

Sellitto and colleagues synthesized a series of novel ethyl indole-3-carboxylates with promising anti–hepatitis C activity [88], and Zhao's group fashioned the aforementioned 2-aminoindole synthesis to a one-pot domino synthesis of 5,12-dihydroindolo[2,1-*b*]quinazolines, wherein the final step is a copper-catalyzed amination between the C-2 amino group and a pendant arylbromide [89]. Filosa and her coworkers prepared a series of indole-3-carboxylate arbidol analogues as new inhibitors of chikungunya virus replication [90]. González-Muñiz prepared a series of 2,3-disubstituted indoles using amino acid β-keto esters and 2-iodoaniline in a copper-catalyzed union [91]. Sulikowski and colleagues employed copper-catalyzed C-H insertion of α-diazo carbonyl compounds to access mitosenes, following oxidation by DDQ or chloranil [92–94]. A similar copper-catalyzed decomposition of alkyl diazoacetates in the presence of enaminones afforded indoles, following oxidation of the 2,3,5,6-tetrahydroindoles [95]. Reddy employed $Cu(OTf)_2$ in the synthesis of 2,3-disubstituted indoles via α-diazoketones and 2-aminoaryl ketones (Scheme 10, equation 1) [96]. The method was extended to a synthesis of homofascaplysin C. Duan and colleagues found that aqueous ammonia and 2-bromo arylacetylenes provides a simple and direct copper(II)-catalyzed indole synthesis (equation 2) [97]. A new synthesis of 3-aminoindoles (and indolines) was developed by Gevorgyan from 2-aminobenzaldehyde, a secondary

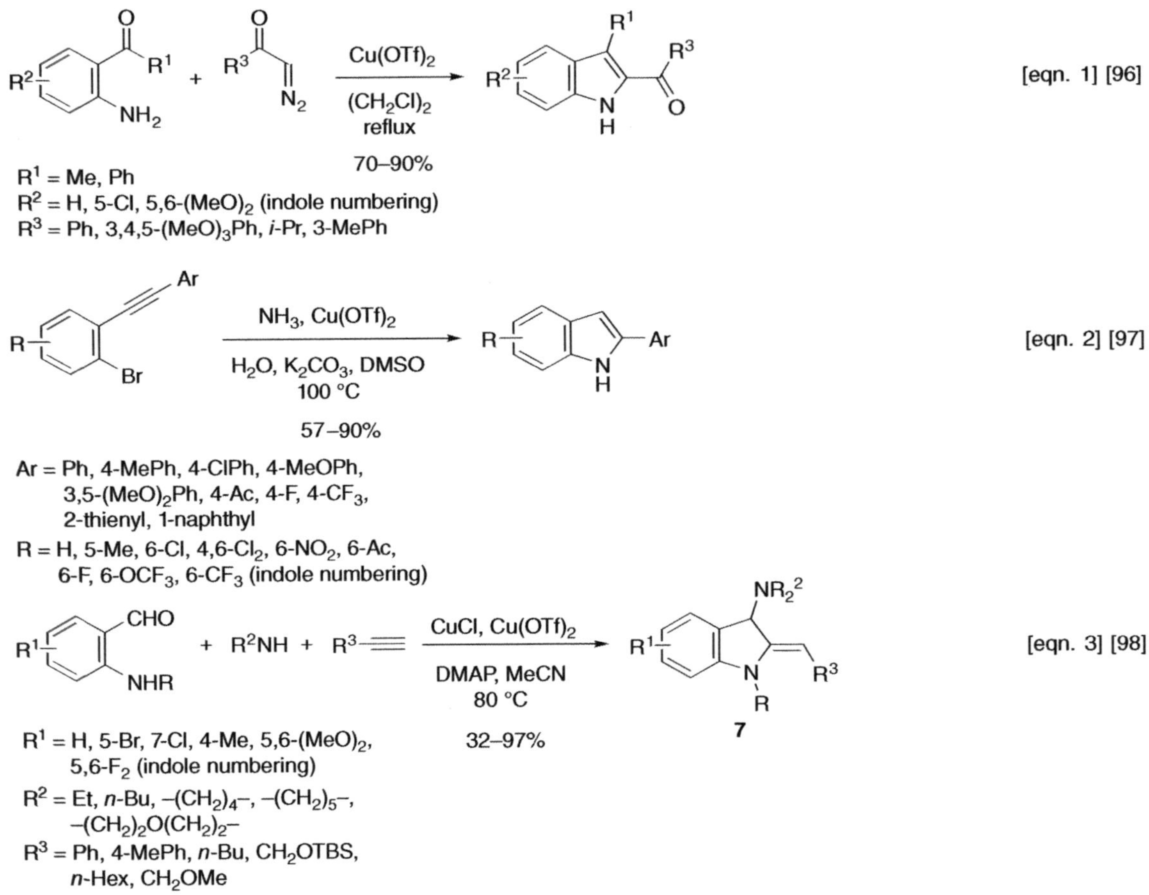

Scheme 10 *Miscellaneous Copper-Catalyzed Indole Syntheses – 1*

amine, and an alkyne (equation 3) [98]. The initially formed (isolable) 3-aminoindolines were easily isomerized to the 3-aminoindoles (Cs_2CO_3, THF, MeOH, 65%–87%).

Cacchi reported a copper-catalyzed C-H activation synthesis of 2-aryl-3-aroylindoles from *N*-aryl enaminones (Scheme 11, equation 1) [99], and Chang and coworkers uncovered a copper-catalyzed carbazole synthesis that does not require a haloaryl substrate and is quite general (equation 2) [100]. Miura, Hirano, and coworkers described a similar carbazole synthesis involving copper-catalyzed intramolecular C-H amination, the key feature of which was the presence of the *N*-picolinamide directing group [101]. Vyskočil and colleagues discovered the copper(II)-mediated oxidative coupling of 2-aminonaphthalene homologues to give symmetrical carbazoles [102]. For example, 9-aminophenanthrene gave 9*H*-tetrabenzo[*a*,*c*,*g*,*i*]carbazole in 78% yield ($CuCl_2$, MeOH). Knochel reported the oxidative amination of zinc organometallics, as mediated by CuCl, to give fused indoles (equation 3) [103]. The use of CuCN in a similar indole synthesis from the double lithiation of 2-bromo-*N*-(2-bromoallyl)anilines to generate, for example, 3-lithiomethylindoles was developed by Barluenga [104]. A clever synthesis of 2-silyl- and 2-borylindoles was reported by Chatani and colleagues that entailed the copper-catalyzed cyclization of 2-alkenylaryl isocyanides (equation 4) [105]. Correia and Taylor found that (*Z*)-ethyl-3-(4-methoxyphenyl)-3-(2-nitrophenyl)acrylate was converted to ethyl 3-(4-methoxyphenyl)indole-2-carboxylate (58% yield) upon treatment with $Cu(OAc)_2$ in the presence of *R*-JOSIPHOS [106]. Kanai's team synthesized 2-(2-hydroxyethyl) indoles that involved a copper-catalyzed intramolecular amido-cupration of an allene to generate an allyl copper species. This was followed by reaction with aldehydes and ketones to form enantioenriched (77%–96% *ee*) indole products [107].

The remarkable synthetic versatility and robustness of copper(I) and copper(II) in constructing the indole ring is amply clear from this chapter and is a credit to the ingenuity of the cited chemists.

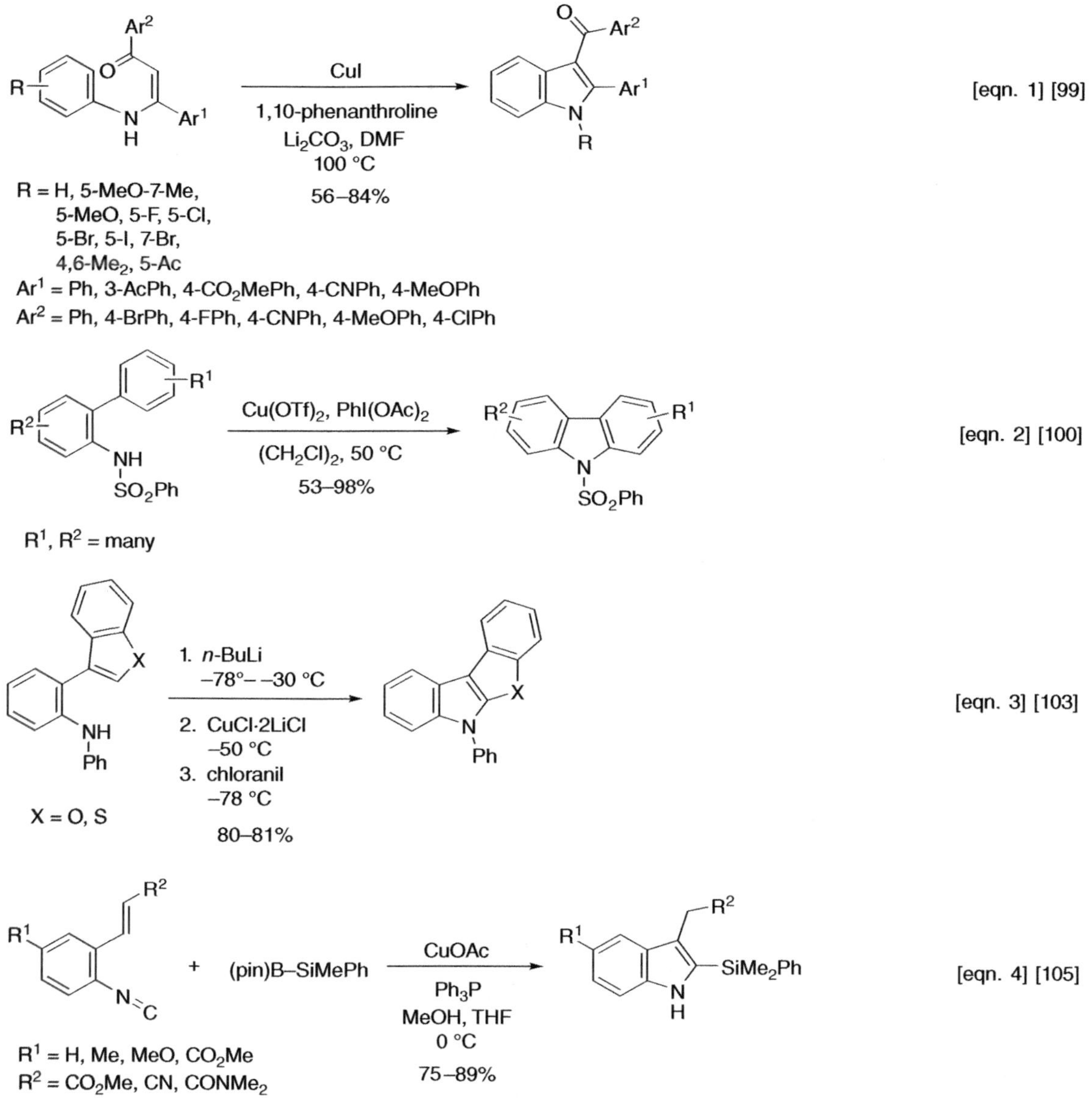

Scheme 11 *Miscellaneous Copper-Catalyzed Indole Syntheses – 2*

References

[1] S.R. Chemler and P.H. Fuller, *Chem. Soc. Rev.*, 2007, **36**, 1153–1160.
[2] D. Ma and Q. Cai, *Acc. Chem. Res.*, 2008, **41**, 1450–1460.
[3] S. Cacchi, G. Fabrizi, and A. Goggiamani, *Org. Biomol. Chem.*, 2011, **9**, 641–652.
[4] J.E.R. Sadig and M.C. Willis, *Synthesis*, 2011, 1–22.
[5] Y. Liu and J.-P. Wan, *Org. Biomol. Chem.*, 2011, **9**, 6873–6894.
[6] K. Hirano and M. Miura, *Chem. Commun.*, 2012, **48**, 10704–10714.
[7] T. Liu and H. Fu, *Synthesis*, 2012, **44**, 2805–2824.
[8] T. Guo, F. Huang, L. Yu, and Z. Yu, *Tetrahedron Lett.*, 2015, **56**, 296–302.
[9] C.E. Castro and R.D. Stephens, *J. Org. Chem.*, 1963, **28**, 2163.
[10] R.D. Stephens and C.E. Castro, *J. Org. Chem.*, 1963, **28**, 3313–3315.
[11] C.E. Castro, E.J. Gaughan, and D.C. Owsley, *J. Org. Chem.*, 1966, **31**, 4071–4078.
[12] C.E. Castro, R. Havlin, V.K. Honwad, *et al.*, *J. Am. Chem. Soc.*, 1969, **91**, 6464–6470.

[13] D.L. Gray (2009) in *Name Reactions for Homologations – I* (ed. J.J. Li), Wiley, Hoboken, New Jersey, pp. 212–235.
[14] J. Reisch, *Chem. Ber.*, 1964, **97**, 2717–2718.
[15] J. Fujiwara, Y. Fukutani, H. Sano, *et al.*, *J. Am. Chem. Soc.*, 1983, **105**, 7177–7179.
[16] A.R. Katritzky, J. Li, and C.V. Stevens, *J. Org. Chem.*, 1995, **60**, 3401–3404.
[17] S.F. Vasilevsky, L.M. Gornostaev, A.A. Stepanov, *et al.*, *Tetrahedron Lett.*, 2007, **48**, 1867–1870.
[18] S. Cacchi, G. Fabrizi, and L.M. Parisi, *Org. Lett.*, 2003, **5**, 3843–3846.
[19] P. Saejueng, C.G. Bates, and D. Venkataraman, *Synthesis*, 2005, 1706–1712.
[20] F. Liu and D. Ma, *J. Org. Chem.*, 2007, **72**, 4844–4850.
[21] H. Jiang, H. Fu, R. Qiao, *et al.*, *Synthesis*, 2008, 2417–2426.
[22] L. Yu, X. Jiang, L. Wang, *et al.*, *Eur. J. Org. Chem.*, 2010, 5560–5562.
[23] S. Song, M. Huang, W. Li, *et al.*, *Tetrahedron*, 2015, **71**, 451–456.
[24] Z. Shen and X. Lu, *Adv. Synth. Catal.*, 2009, **351**, 3107–3112.
[25] N. Kumara Swamy, A. Yazici, and S.G. Pyne, *J. Org. Chem.*, 2010, **75**, 3412–3419.
[26] S. Kamijo, Y. Sasaki, and Y. Yamamoto, *Tetrahedron Lett.*, 2004, **45**, 35–38.
[27] H. Ohno, Y. Ohta, S. Oishi, and N. Fujii, *Angew. Chem. Int. Ed.*, 2007, **46**, 2295–2298.
[28] Y. Ohta, H. Chiba, S. Oishi, *et al.*, *J. Org. Chem.*, 2009, **74**, 7052–7058.
[29] Y. Ohta, S. Oishi, N. Fujii, and H. Ohno, *Org. Lett.*, 2009, **11**, 1979–1982.
[30] Y. Suzuki, Y. Ohta, S. Oishi, *et al.*, *J. Org. Chem.*, 2009, **74**, 4246–4251.
[31] R. Bernini, S. Cacchi, G. Fabrizi, *et al.*, *Synlett*, 2009, 1480–1484.
[32] A. Gogoi, S. Guin, S. Kumar Rout, and B.K. Patel, *Org. Lett.*, 2013, **15**, 1802–1805.
[33] X.-F. Xia, L.-L. Zhang, X.-R. Song, *et al.*, *Chem. Commun.*, 2013, **49**, 1410–1412.
[34] G. Ge, X. Huang, C. Ding, *et al.*, *Chin. J. Chem.* 2014, **32**, 727–733.
[35] L. Ackermann, S. Barfüßer, and H.K. Potukuchi, *Adv. Synth. Catal.*, 2009, **351**, 1064–1072.
[36] N. Matsuda, K. Hirano, T. Satoh, and M. Miura, *Synthesis*, 2012, **44**, 1792–1797.
[37] Y. Oda, K. Hirano, T. Satoh, and M. Miura, *Org. Lett.*, 2012, **14**, 664–667.
[38] Y. Oda, N. Matsuyama, K. Hirano, *et al.*, *Synthesis*, 2012, **44**, 1515–1520.
[39] S. Murru, A.A. Gallo, and R.S. Srivastava, *Eur. J. Org. Chem.*, 2011, 2035–2038.
[40] T. Ponpandian and S. Muthusubramanian, *Tetrahedron Lett.*, 2012, **53**, 4248–4252.
[41] K.-J. Chang, D. Moon, M.S. Lah, and K.-S. Jeong, *Angew. Chem. Int. Ed.*, 2005, **44**, 7926–7929.
[42] K. Hiroya, S. Itoh, and T. Sakamoto, *Tetrahedron*, 2005, **61**, 10958–10964.
[43] T. Xiao, X. Dong, and L. Zhou, *Org. Biomol. Chem.*, 2013, **11**, 1490–1497.
[44] Y. Xing, G. Sheng, J. Wang, *et al.*, *Org. Lett.*, 2014, **16**, 1244–1247.
[45] A. Klapars, X. Huang, and S.L. Buchwald, *J. Am. Chem. Soc.*, 2002, **124**, 7421–7428.
[46] F.Y. Kwong, A. Klapars, and S.L. Buchwald, *Org. Lett.*, 2002, **4**, 581–584.
[47] K. Yamada, T. Kurokawa, H. Tokuyama, and T. Fukuyama, *J. Am. Chem. Soc.*, 2003, **125**, 6630–6631.
[48] K. Okano, H. Tokuyama, and T. Fukuyama, *J. Am. Chem. Soc.*, 2006, **128**, 7136–7137.
[49] F. Melkonyan, A. Topolyan, M. Yurovskaya, and A. Karchava, *Eur. J. Org. Chem.*, 2008, 5952–5956.
[50] F.S. Melkonyan, A.V. Karchava, and M.A. Yurovskaya, *J. Org. Chem.*, 2008, **73**, 4275–4278.
[51] M.G. Mokrushin, A.V. Shastin, V.M. Muzalevskiy, *et al.*, *Mendeleev Commun.*, 2008, **18**, 327–328.
[52] R.C. Hodgkinson, J. Schulz, and M.C. Willis, *Org. Biomol. Chem.*, 2009, **7**, 432–434.
[53] B. Jiang, K. Tao, W. Shen, and J. Zhang, *Tetrahedron Lett.*, 2010, **51**, 6342–6344.
[54] Z. Xia, K. Wang, J. Zheng, *et al.*, *Org. Biomol. Chem.*, 2012, **10**, 1602–1611.
[55] X. Xiao, T.-Q. Chen, J. Ren, *et al.*, *Tetrahedron Lett.*, 2014, **55**, 2056–2060.
[56] Z. Zhu, J. Yuan, Y. Zhou, *et al.*, *Eur. J. Org. Chem.*, 2014, 511–514.
[57] S.G. Koenig, J.W. Dankwardt, Y. Liu, *et al.*, *Tetrahedron Lett.*, 2010, **51**, 6549–6551.
[58] W. Schlecker, A. Huth, E. Ottow, and J. Mulzer, *Tetrahedron*, 1995, **51**, 9531–9542.
[59] P. Molina, P.M. Fresneda, and S. Delgado, *Synthesis*, 1999, 326–329.
[60] S. Wakim, J. Bouchard, N. Blouin, *et al.*, *Org. Lett.*, 2004, **6**, 3413–3416.
[61] X. Zhu, X.-P. Xu, C. Sun, *et al.*, *Tetrahedron*, 2011, **67**, 6375–6381.
[62] W.S. Murphy and M. Bertrand, *J. Chem. Soc., Perkin Trans. 1*, 1998, 4115–4119.
[63] S. Wang, J. Sun, G. Yu, *et al.*, *Org. Biomol. Chem.*, 2004, **2**, 1573–1574.
[64] S. Wang, Y. Dong, X. Wang, *et al.*, *Org. Biomol. Chem.*, 2005, **3**, 911–916.
[65] E. Li, X. Xu, H. Li, *et al.*, *Tetrahedron*, 2009, **65**, 8961–8968.
[66] C. Barberis, T.D. Gordon, C. Thomas, *et al.*, *Tetrahedron Lett.*, 2005, **46**, 8877–8880.
[67] M. Carril, R. SanMartin, E. Domínguez, and I. Tellitu, *Green Chem.*, 2007, **9**, 219–220.
[68] K. Okano, N. Mitsuhashi, and H. Tokuyama, *Chem. Commun.*, 2010, **46**, 2641–2643.
[69] H.A. Ioannidou, A. Martin, A. Gollner, and P.A. Koutentis, *J. Org. Chem.*, 2011, **76**, 5113–5122.
[70] R. Worayuthakarn, P. Nealmongkol, S. Ruchirawat, and N. Thasana, *Tetrahedron*, 2012, **68**, 2864–2875.
[71] Y. Huang, D. Wu, J. Huang, *et al.*, *Angew. Chem. Int. Ed.*, 2014, **53**, 12158–12162.
[72] R. Besandre, M. Jaimes, and J.A. May, *Org. Lett.*, 2013, **15**, 1666–1669.
[73] Q. Cai, Z. Li, J. Wei, *et al.*, *Chem. Commun.*, 2009, 7581–7583.
[74] A. Osuka, Y. Mori, and H. Suzuki, *Chem. Lett.*, 1982, 2031–2034.
[75] H. Suzuki, S.V. Thiruvikraman, and A. Osuka, *Synthesis*, 1984, 616–617.

[76] J. d'Angelo and D. Desmaele, *Tetrahedron Lett.*, 1990, **31**, 879–882.
[77] S. Yan, H. Wu, N. Wu, and Y. Jiang, *Synlett*, 2007, 2699–2702.
[78] M.A. Ali and T. Punniyamurthy, *Synlett*, 2011, 623–626.
[79] A. Bunescu, Q. Wang, and J. Zhu, *Synthesis*, 2012, **44**, 3811–3814.
[80] X. Wang, J. Liu, H. Guo, *et al.*, *Synthesis*, 2012 **44**, 1037–1042.
[81] B. Huang, D. Hu, J. Wang, *et al.*, *Tetrahedron Lett.*, 2015, **56**, 2551–2554.
[82] L. Cao, D. Shen, J. Wei, *et al.*, *Eur. J. Org. Chem.*, 2014, 2460–2467.
[83] K. Okuro, M. Furuune, M. Miura, and M. Nomura, *J. Org. Chem.*, 1993, **58**, 7606–7607.
[84] Y. Chen, X. Xie, and D. Ma, *J. Org. Chem.*, 2007, **72**, 9329–9334.
[85] Y. Chen, Y. Wang, Z. Sun, and D. Ma, *Org. Lett.*, 2008, **10**, 625–628.
[86] X. Yang, H. Fu, R. Qiao, *et al.*, *Adv. Synth. Catal.*, 2010, **352**, 1033–1038.
[87] K. Kobayashi, T. Komatsu, Y. Yokoi, and H. Konishi, *Synthesis*, 2011, 764–768.
[88] G. Sellitto, A. Faruolo, P. de Caprariis, *et al.*, *Bioorg. Med. Chem.*, 2010, **18**, 6143–6148.
[89] M. Jiang, J. Li, F. Wang, *et al.*, *Org. Lett.*, 2012, **14**, 1420–1423.
[90] A. Di Mola, A. Peduto, A. La Gatta, *et al.*, *Bioorg. Med. Chem.*, 2014, **22**, 6014–6025.
[91] M.I. García-Aranda, M.T. García-López, M.J. Pérez de Vega, and R. González-Muñiz, *Tetrahedron Lett.*, 2014, **55**, 2142–2145.
[92] H.-J. Lim and G.A. Sulikowski, *J. Org. Chem.*, 1995, **60**, 2326–2327.
[93] S. Lee, H.-J. Lim, K.L. Cha, and G.A. Sulikowski, *Tetrahedron*, 1997, **53**, 16521–16532.
[94] S. Lee, W.-M. Lee, and G.A. Sulikowski, *J. Org. Chem.*, 1999, **64**, 4224–4225.
[95] A. Müller, A. Maier, R. Neumann, and G. Maas, *Eur. J. Org. Chem.*, 1998, 1177–1187.
[96] B.V.S. Reddy, M.R. Reddy, Y.G. Rao, *et al.*, *Org. Lett.*, 2013, **15**, 464–467.
[97] H. Wang, Y. Li, L. Jiang, *et al.*, *Org. Biomol. Chem.*, 2011, **9**, 4983–4986.
[98] D. Chernyak, N. Chernyak, and V. Gevorgyan, *Adv. Synth. Catal.*, 2010, **352**, 961–966.
[99] R. Bernini, G. Fabrizi, A. Sferrazza, and S. Cacchi, *Angew. Chem. Int. Ed.*, 2009, **48**, 8078–8081.
[100] S.H. Cho, J. Yoon, and S. Chang, *J. Am. Chem. Soc.*, 2011, **133**, 5996–6005.
[101] K. Takamatsu, K. Hirano, T. Satoh, and M. Miura, *Org. Lett*, **16**, 2892–2895.
[102] S. Vyskočil, M. Smrčina, M. Lorenc, *et al.*, *J. Org. Chem.*, 2001, **66**, 1359–1365.
[103] M. Kienle, A.J. Wagner, C. Dunst, and P. Knochel, *Chem. Asian J.*, 2011, **6**, 517–523.
[104] J. Barluenga, R. Sanz, A. Granados, and F.J. Fañanás, *J. Am. Chem. Soc.*, 1998, **120**, 4865–4866.
[105] M. Tobisu, H. Fujihara, K. Koh, and N. Chatani, *J. Org. Chem.*, 2010, **75**, 4841–4847.
[106] J.G. Taylor and C.R.D. Correia, *J. Org. Chem.*, 2011, **76**, 857–869.
[107] P. Kumara Chikkade, Y. Shimizu, and M. Kanai, *Chem. Sci.*, 2014, **5**, 1585–1590.

70

Palladium-Catalyzed Indole Ring Synthesis: Hegedus

Given the truly voluminous literature on palladium-catalyzed indole ring synthesis and the large number of recent reviews on this topic [1–11], coverage in this chapter and in succeeding ones is strictly limited. Additional reviews are cited in Chapter 1.

One of the first applications of palladium to indole synthesis was due to Hegedus [12–17], who adapted the known Pd-mediated amination of alkenes [18] to the intramoleclar amination of *o*-vinyl- or *o*-allylanilines (Scheme 1, equations 1 and 2). This Wacker-like process involves Pd(II) and can be stoichiometric or catalytic. For an excellent review of the Hegedus indole synthesis, see Johnston [19]. Although it is rarely referred to as such, the Hegedus reaction has found several applications and extensions in indole synthesis (equations 3–5) [20, 21, 23]. The preparation of ethyl indole-2-carboxylates (equation 4) [21] has been improved by McNulty and Keskar [22], and Danheiser synthesized 6-*n*-butyl-1,2-dimethyl-4-hydroxyindole via a benzannulation of a cyclobutenone and a ynamide, and subsequent Pd-catalyzed cyclization (equation 5) [23]. The synthesis of both paspalinine (equation 6) [24] and (±)-lecanindole-D [25] featured a Hegedus indolization.

Li, Deng, and coworkers employed allylic esters with *o*-iodoanilines and $PdCl_2$ (n-Bu_4NCl, K_2CO_3, DMA, 100 °C) to give indoles in yields up to 88% [26], and Ghorai and colleagues adapted the Hegedus protocol to access a series of 3-substituted 2-benzylindoles (Scheme 2, equation 1) [27]. The *o*-allylanilines were prepared from the corresponding *o*-bromoanilines and allyl alcohols via nickel coupling. Buchwald and Tsvelikhovsky synthesized several *N*-arylindoles (equation 2) or 2-vinylcarbazoles (equation 3) (aryl-Heck), depending on the precise reaction conditions [28]. Mukai's group used a Hegedus procedure with *o*-allenylanilines to synthesize 2-vinylindoles (equation 4) [29]. A Stille coupling afforded the starting allenylanilines.

Indole Ring Synthesis: From Natural Products to Drug Discovery, First Edition. Gordon W. Gribble.

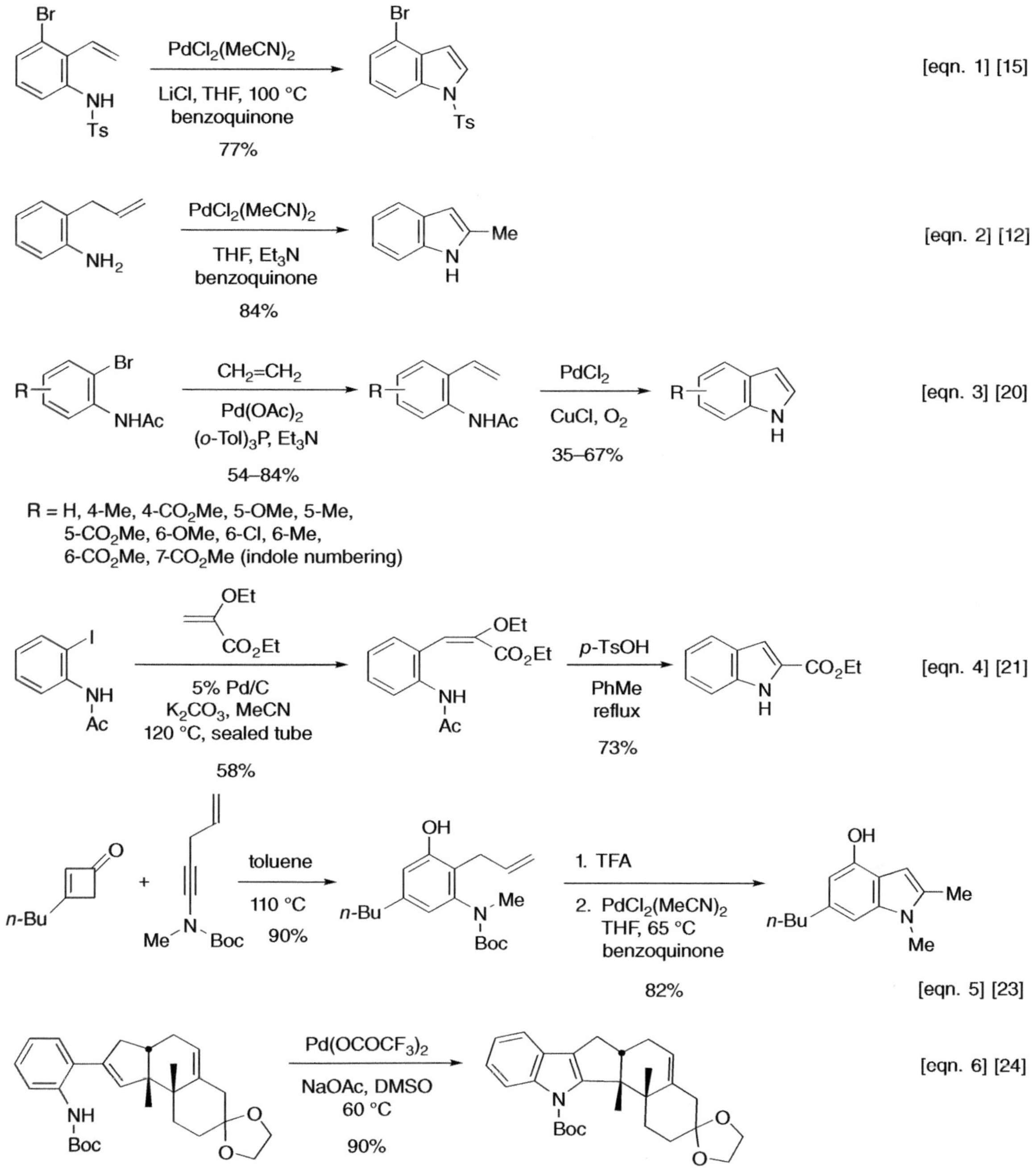

Scheme 1 *Hegedus Indole Synthesis*

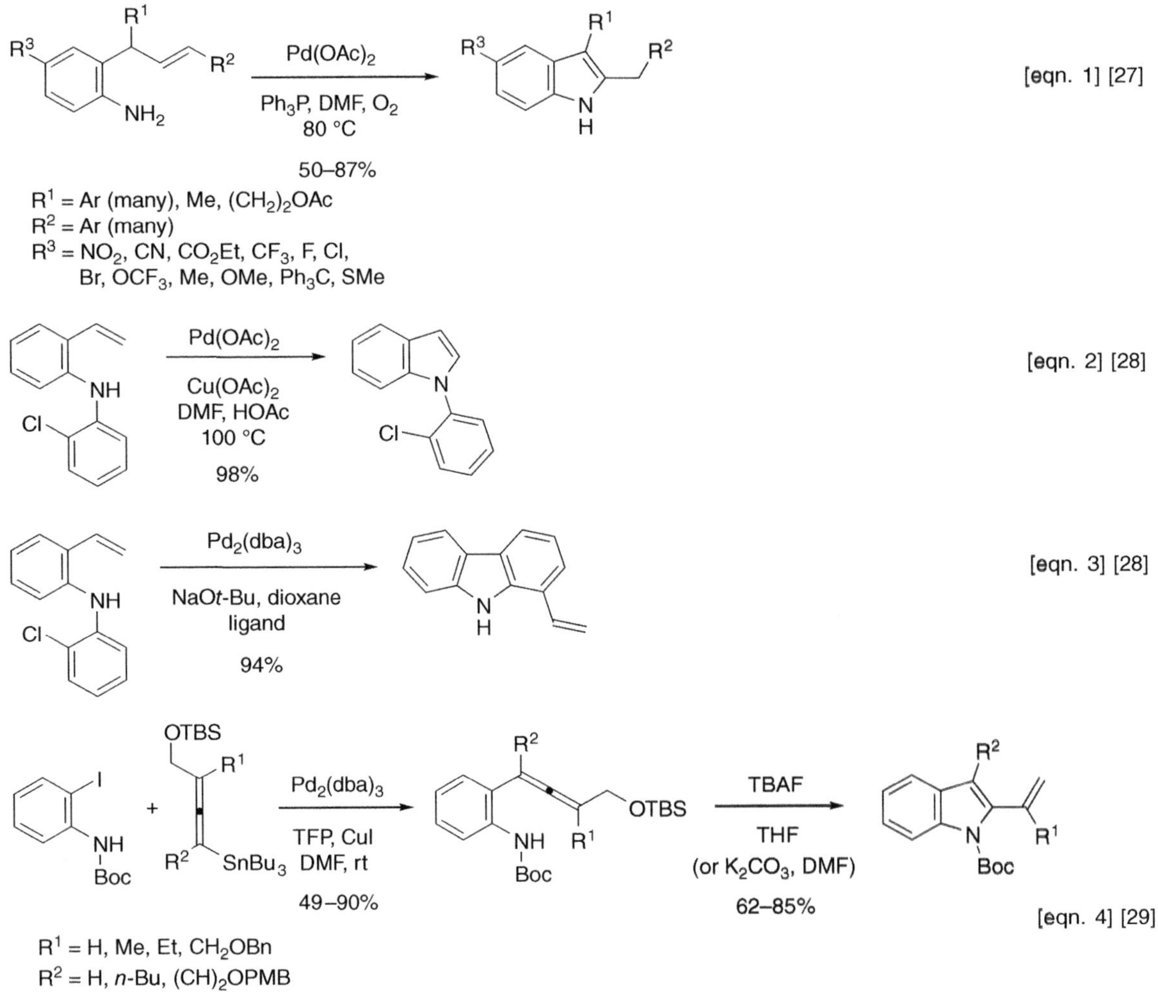

Scheme 2 *Applications of the Hegedus Indole Synthesis*

References

[1] L.S. Hegedus, *Angew. Chem. Int. Ed.*, 1988, **27**, 1113–1226.
[2] J.-P. Corbet and G. Mignani, *Chem. Rev.*, 2006, **106**, 2651–2710.
[3] J.J. Song, J.T. Reeves, D.R. Fandrick, *et al.*, *ARKIVOC*, 2010, 390–449.
[4] S. Cacchi and G. Fabrizi, *Chem. Rev.*, 2011, **111**, PR215–PR283.
[5] M. Platon, R. Amardeil, L. Djakovitch, and J.-C. Hierso, *Chem. Soc. Rev.*, 2012, **41**, 3929–3968.
[6] K. Inamoto, *Chem. Pharm. Bull.*, 2013, **61**, 987–996.
[7] N. Yoshikai and Y. Wei, *Asian J. Org. Chem.*, 2013, **2**, 466–478.
[8] J. Roy, A.K. Jana, and D. Mal, *Tetrahedron*, 2012, **68**, 6099–6121.
[9] T. Guo, F. Huang, L. Yu, and Z. Yu, *Tetrahedron Lett.*, 2015, **56**, 296–302.
[10] G.W. Gribble (2007) *Palladium in Heterocyclic Chemistry*, 2nd edn (eds J.J. Li and G.W. Gribble), Elsevier, Amsterdam, pp. 81–188.
[11] J.J. Lie and G.W. Gribble, *Top. Heterocycl. Chem.*, 2010, **26**, 193–234.
[12] L.S. Hegedus, G.F. Allen, and E.L. Waterman, *J. Am. Chem. Soc.*, 1976, **98**, 2674–2676.
[13] L.S. Hegedus, G.F. Allen, J.J. Bozell, and E.L. Waterman, *J. Am. Chem. Soc.*, 1978, **100**, 5800–5807.
[14] L.S. Hegedus, G.F. Allen, and D.J. Olsen, *J. Am. Chem. Soc.*, 1980, **102**, 3583–3587.
[15] P.J. Harrington and L.S. Hegedus, *J. Org. Chem.*, 1984, **49**, 2657–2662.
[16] L.S. Hegedus, P.R. Weider, T.A. Mulhern, *et al.*, *Gazz. Chim. Ital.*, 1986, **116**, 213–219.
[17] P.J. Harrington, L.S. Hegedus, and K.F. McDaniel, *J. Am. Chem. Soc.*, 1987, **109**, 4335–4338.
[18] B. Åkermark, J.E. Bäckvall, L.S. Hegedus, *et al.*, *J. Organometal. Chem.*, 1974, **72**, 127–138.
[19] J.N. Johnston (2005) In: *Name Reactions in Heterocyclic Chemistry II*, (eds. J.J. Li and E.J. Corey), Wiley, Hoboken, NJ, pp. 135–139.

[20] A. Kasahara, T. Izumi, S. Murakami, *et al.*, *J. Heterocycl. Chem.*, 1989, **26**, 1405–1413.
[21] T. Sakamoto, Y. Kondo, and H. Yamanaka, *Heterocycles*, 1988, **27**, 453–456.
[22] J. McNulty and K. Keskar, *Eur. J. Org. Chem.*, 2011, 6902–6908.
[23] T.Y. Lam, Y.-P. Wang, and R.L. Danheiser, *J. Org. Chem.*, 2013, **78**, 9396–9414.
[24] M. Enomoto, A. Morita, and S. Kuwahara, *Angew. Chem. Int. Ed.*, 2012, **51**, 12833–12836.
[25] A. Asanuma, M. Enomoto, T. Nagasawa, and S. Kuwahara, *Tetrahedron Lett.*, 2013, **54**, 4561–4563.
[26] Y. Liu, B. Yao, C.-L. Deng, *et al.*, *Org. Lett.*, 2011, **13**, 1126–1129.
[27] R. Nallagonda, M. Rehan, and P. Ghorai, *Org. Lett.*, 2014, **16**, 4786–4789.
[28] D. Tsvelikhovsky and S.L. Buchwald, *J. Am. Chem. Soc.*, 2010, **132**, 14048–14051.
[29] Y.A.M. Mohamed, F. Inagaki, R. Takahashi, and C. Mukai, *Tetrahedron*, 2011, **67**, 5133–5141.

71

Palladium-Catalyzed Indole Ring Synthesis: Mori–Ban–Heck

The Mori–Ban indole synthesis is a Pd-catalyzed cyclization of *N*-allyl-*o*-haloanilines in what is an intramolecular Heck reaction [1] (Scheme 1, equation 1) [2, 3]. This indolization also transpires with nickel catalysts [4]. Independently, Heck and Terpko discovered the same reaction, but to give oxindoles [5]; somewhat later Hegedus described improved cyclization yields (equation 2) [6]. Larock and Babu improved the Mori–Ban indole synthesis (equation 3) [7] by adopting the conditions reported by Jeffery in his careful studies of the general Heck reaction [8–11]. Several groups extended the Mori–Ban–Heck indolization to *o*-halo enaminones as a route to 3-acylindoles and 4-oxo-β-carbolines (equation 4) [12–14]. A major convenience of this reaction is that the substrates are readily fashioned from *o*-haloanilines and 1,3-dicarbonyls (47%–88% yield) [13]. As seen from the work of Nishida, the conjugated double bond (vinylogous carbamate) can isomerize as necessary to set up the Heck cyclization (equation 5) [15]. A Mori–Ban indole synthesis has featured prominently in the syntheses of (±)-aspidophytine [16], stephacidin A [17], 1,1′-bisindoles [18], a potent EP_3 receptor antagonist (DG-041) [19], kinesin spindle protein inhibitors [20], hepatitis C protein inhibitors [21], and $S1P_1$ agonists [22].

In addition to the numerous applications of the Mori–Ban indole synthesis, there have been developed several new methods that extend and improve the original procedure. A Pd-catalyzed one-pot *N*-alkylation/Heck cyclization was published by both Jørgensen [23] and Beck [24] for the Mori–Ban indolization, and Kurth discovered a one-pot, three-component assembly to afford indoles (Scheme 2, equation 1) [25]. Kaim and Grimaud reported a one-pot, four-component Ugi–Smiles process to synthesize indoles (equation 2) [26]. Sun and colleagues synthesized 2-arylindoles via a Pd-catalyzed sequential Heck dealkylation protocol (equation 3) [27]. An iron-catalyzed domino isomerization/cyclodehydration sequence led to benzo[*b*]carbazoles following a Heck–Suzuki cyclization, as discovered by Jana and colleagues (equation 4) [28].

In addition to the aforementioned remarkable indole syntheses, the Mori–Ban–Heck procedures have been extremely valuable for the preparation of pyrrolo[3,2-*c*]quinolines [29], pyrido[1,2-*a*]indoles [30], pyrido[4,3-*b*]indoles [31], 6-arylindoles [32], and 1,2-bis(3-indolyl)ethanes as novel MRSA (methicillin-resistant *Staphylococcus aureus*) inhibitors [33]. Microwave heating facilitated the Mori–Ban–Heck indole synthesis [34–36]; the latter example provided a new route to 3-nitroindoles. Because *N*-vinyl-*o*-haloanilines are inherently labile (e.g., hydrolysis), only suitably deactivated analogues are feasible for a Heck indolization. However, Barluenga and coworkers effected a novel one-pot indole synthesis from *o*-haloanilines and alkenyl halides (Scheme 3, equation 1) [37]. This avoided the isolation of the intermediate enamines. Similarly, 2-substituted indoles were prepared employing 1-substituted alkenyl bromides. Fuwa and Sasaki utilized *N*-(*o*-halophenyl)enecarbamates in a Suzuki coupling/cyclization protocol to give 2-arylindoles (equation 2) [38]. This elegant chemistry was also executed as a one-pot reaction. Correia and colleagues parlayed a Heck–Matsuda reaction to an indole synthesis (equation 3) [39]. Related Pd-catalyzed indole syntheses are the α-arylation of β-(2-iodoaniline)carboxamides (Solé and Serrano) [40], a Heck cyclization of Baylis–Hillman adducts from

Indole Ring Synthesis: From Natural Products to Drug Discovery, First Edition. Gordon W. Gribble.

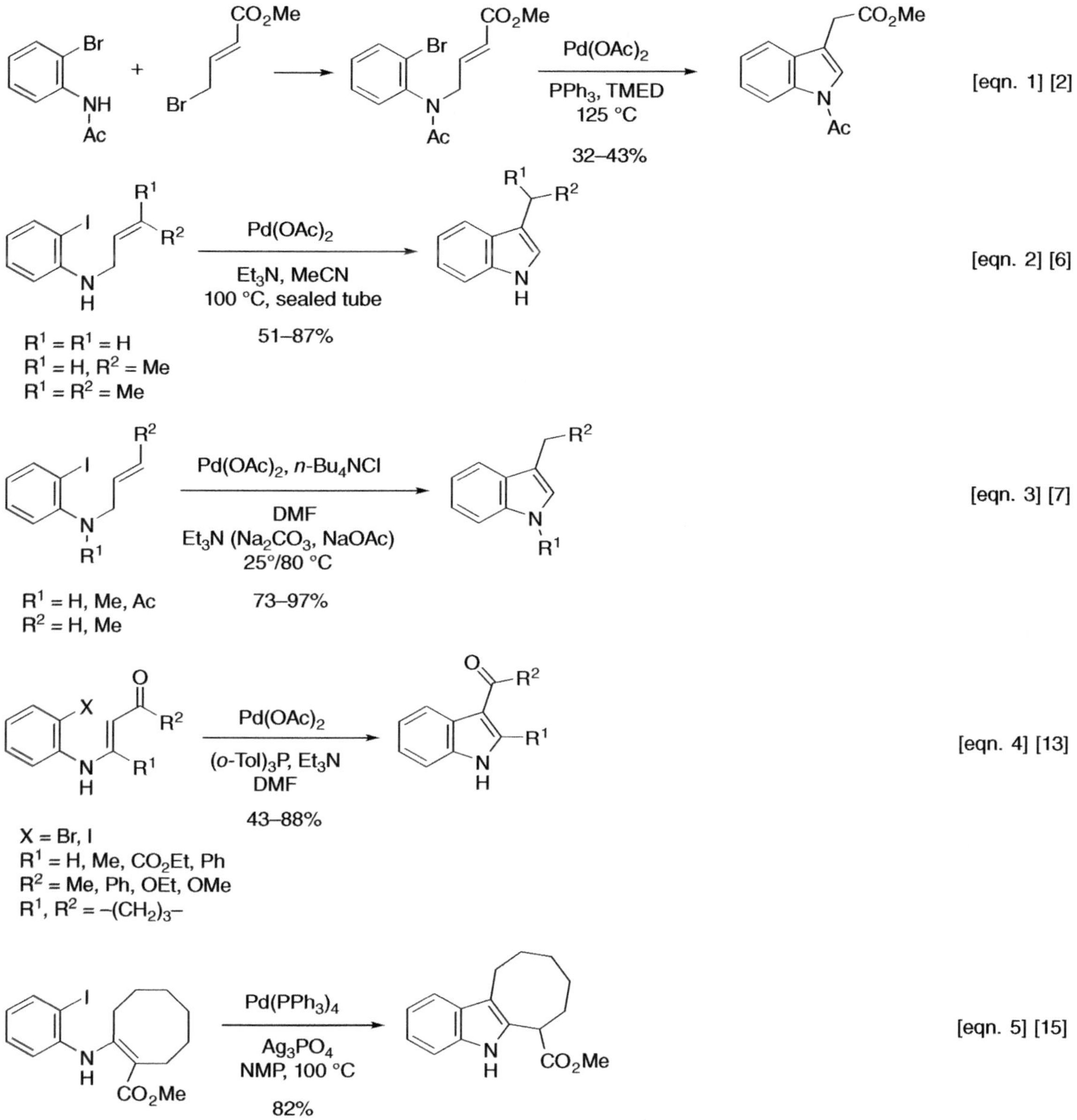

Scheme 1 *Mori-Ban Indole Synthesis*

o-bromoaniline (Kim) [41], the α-arylation of enones to give 2-alkenylindoles and carbazoles (Kapur) [42], and the coupling of α-iodo enaminones to give carbazolones (Du and Zhao) [43].

An important variation of the Mori–Ban–Heck indole synthesis is Pd-catalyzed oxidative cyclization of *N*-aryl enaminones and enamines, as largely developed by Glorius (Scheme 4, equations 1 and 2) [44–46]. The C-2 substituent was easily varied, and the method was extended to imine cyclization (equation 2). Related Pd-catalyzed oxidative cyclizations were described by Li (carbazolones) [47], Guan (1,3-disubstituted indoles) [48], Rueping (photoredox catalysis) [49], and Jiao (*N*-alkynylaniline hydroamination) [50]. Before Glorius, Lee and colleagues employed Pd-catalyzed oxidative cyclization in a synthesis of duocarmycin SA analogues [51]. Some additional examples of this oxidative cyclization to give indoles are in later chapters.

In summary, the original discovery by Mori and Ban of a new Pd-catalyzed indole synthesis involving an intramolecular Heck cyclization has been refined by many investigators to become a powerful method for the synthesis of diverse indoles.

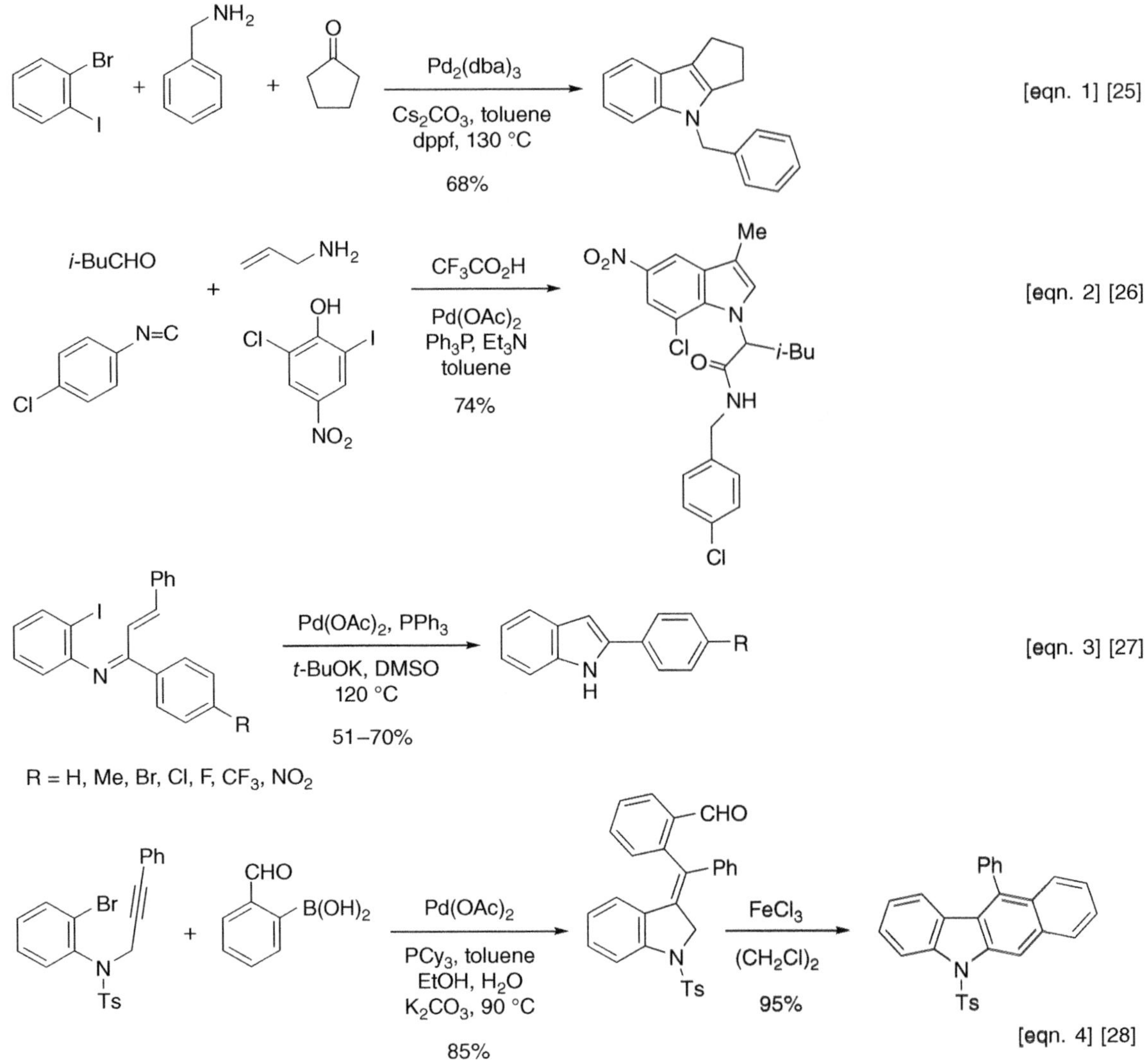

Scheme 2 *Applications of the Mori-Ban-Heck Indole Synthesis*

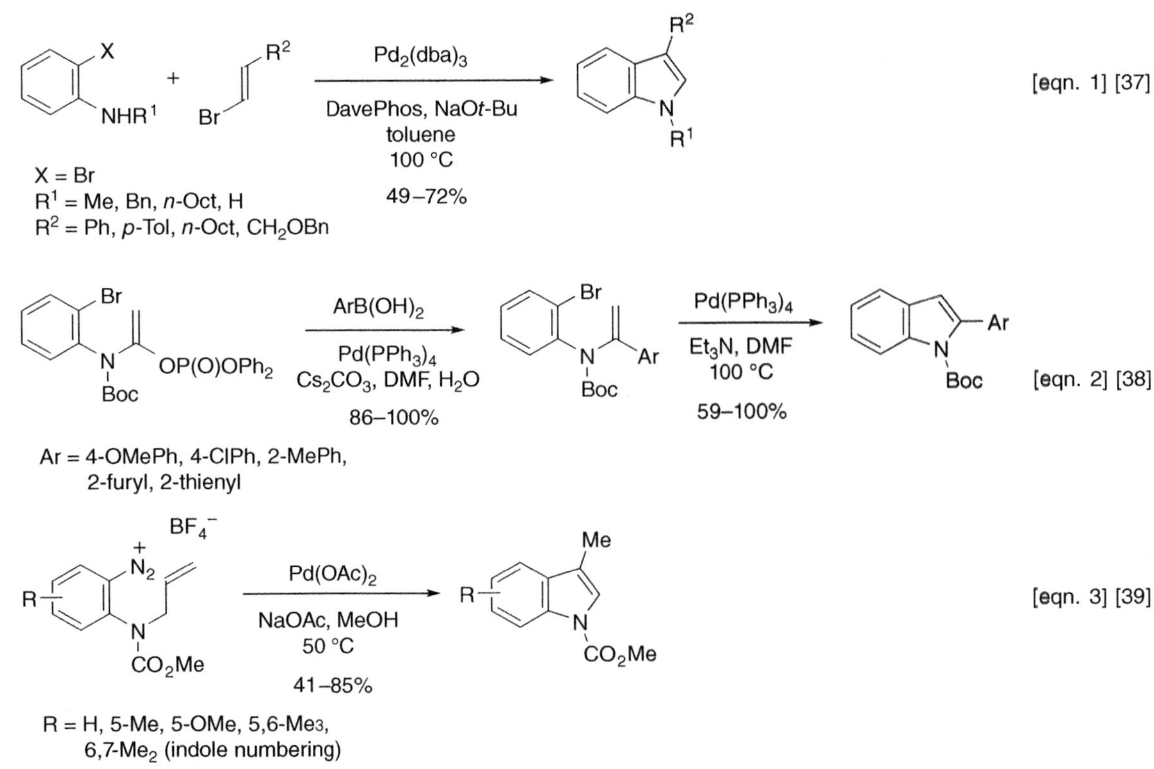

Scheme 3 *Barluenga, Fuwa-Sasaki, and Correia Indole Syntheses*

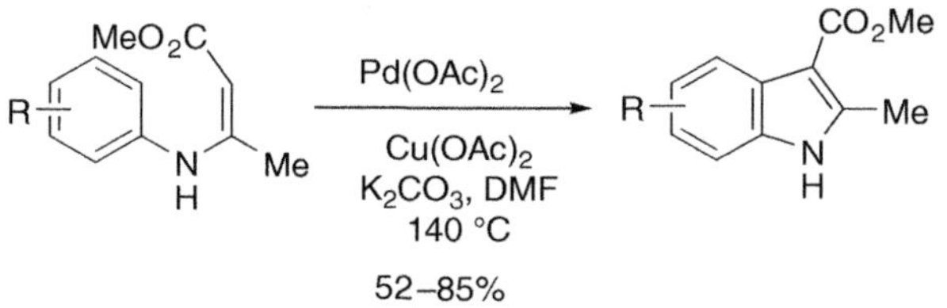

[eqn. 1] [44]

R = H, 7-Me, 5-Me, 7-OMe, 7-F, 5-CN, 5-Ac, 6-Ac, 5-F, 5-Cl, 5-CONEt$_2$, etc. (indole numbering)

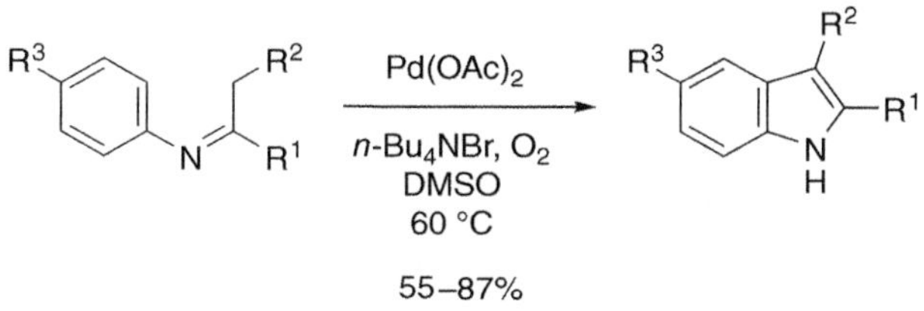

[eqn. 2] [46]

R^1 = Ph, *c*-propyl, 4-pyridyl, Me
R^2 = H, Ph, CN
R^3 = H, NO_2, OMe

Scheme 4 *Glorius Indole Synthesis*

References

[1] R.F. Heck and J.P. Nolley, Jr., *J. Org. Chem.*, 1972, **37**, 2320–2322.
[2] M. Mori, K. Chiba, and Y. Ban, *Tetrahedron Lett.*, 1977, **18**, 1037–1040.
[3] Y. Ban, T. Wakamatsu, and M. Mori, *Heterocycles*, 1977, **6**, 1711–1715.
[4] M. Mori and Y. Ban, *Tetrahedron Lett.*, 1976, **17**, 1803–1806.
[5] M.O. Terpko and R.F. Heck, *J. Am. Chem. Soc.*, 1979, **101**, 5281–5283.
[6] R. Odle, B. Blevins, M. Ratcliff, and L.S. Hegedus, *J. Org. Chem.*, 1980, **45**, 2709–2710.
[7] R.C. Larock and S. Babu, *Tetrahedron Lett.*, 1987, **28**, 5291–5294.
[8] T. Jeffery, *J. Chem. Soc., Chem. Commun.*, 1984, 1287–1289.
[9] T. Jeffery and J.-C. Galland, *Tetrahedron Lett.*, 1994, **35**, 4103–4106.
[10] T. Jeffery, *Tetrahedron*, 1996, **52**, 10113–10130.
[11] T. Jeffery, *Tetrahedron Lett.*, 1999, **40**, 1673–1676.
[12] H. Iida, Y. Yuasa, and C. Kibayashi, *J. Org. Chem.*, 1980, **45**, 2938–2942.
[13] T. Sakamoto, T. Nagano, Y. Kondo, and H. Yamanaka, *Synthesis*, 1990, 215–218.
[14] L.-C. Chen, S.-C. Yang, and H.-M. Wang, *Synthesis*, 1995, 385–386.
[15] T. Watanabe, S. Arai, and A. Nishida, *Synlett*, 2004, 907–909.
[16] J.M. Mejía-Oneto and A. Padwa, *Org. Lett.*, 2006, **8**, 3275–3278.
[17] P.S. Baran, C.A. Guerrero, N.B. Ambhaikar, and B.D. Hafensteiner, *Angew. Chem. Int. Ed.*, 2005, **44**, 606–609.
[18] C. Wang and J. Sperry, *Chem. Commun.*, 2013, **49**, 4349–4351.
[19] S. Zegar, C. Tokar, L.A. Enache, *et al.*, *Org. Process Res. Dev.*, 2007, **11**, 747–753.
[20] S. Oishi, T. Watanabe, J. Sawada, *et al.*, *J. Med. Chem.*, 2010, **53**, 5054–5058.
[21] S. Venkatraman, F. Velazquez, S. Gavalas, *et al.*, *Bioorg. Med. Chem.*, 2013, **21**, 2007–2017.
[22] D.J. Buzard, S. Han, L. Lopez, *et al.*, *Bioorg. Med. Chem. Lett.*, 2012, **22**, 4404–4409.
[23] T. Jensen, H. Pedersen, B. Band-Andersen, *et al.*, *Angew. Chem. Int. Ed.*, 2008, **47**, 888–890.
[24] M.L. Weinrich and H.P. Beck, *Tetrahedron Lett.*, 2009, **50**, 6968–6972.
[25] J.M. Knapp, J.S. Zhu, D.J. Tantillo, and M.J. Kurth, *Angew. Chem. Int. Ed.*, 2012, **51**, 10588–10591.
[26] L. El Kaim, M. Gizzi, and L. Grimaud, *Org. Lett.*, 2008, **10**, 3417–3419.
[27] H. Mao, J.-P. Wan, Y. Pan, and C. Sun, *Tetrahedron Lett.*, 2010, **51**, 1844–1846.
[28] K. Paul, K. Bera, S. Jalal, *et al.*, *Org. Lett.*, 2014, **16**, 2166–2169.
[29] M.J. Mphahlele, L.G. Lesenyeho, and H.R. Makelane, *Tetrahedron*, 2010, **66**, 6040–6046.
[30] N. Koay, D.L. Tonelli, and V.L. Truong, *Tetrahedron Lett.*, 2011, **52**, 122–124.
[31] M. Madaiah, M.K. Prashanth, and H.D. Revanasiddappa, *Tetrahedron Lett.*, 2013, **54**, 1424–1427.
[32] S.G. Hammer and M.R. Heinrich, *Tetrahedron*, 2014, **70**, 8114–8121.
[33] R. Zoraghi, S. Campbell, C. Kim, *et al.*, *Bioorg. Med. Chem. Lett.*, 2014, **24**, 5059–5062.

[34] R.P. Karuvalam, K.R. Haridas, A.M. Sajith, and A. Muralidharan, *Tetrahedron Lett.*, 2013, **54**, 5126–5129.
[35] W.-I. Lee, J.-W. Jung, J. Sim, *et al.*, *Tetrahedron*, 2013, **69**, 7211–7219.
[36] H.H. Nguyen and M.J. Kurth, *Org. Lett.*, 2013, **15**, 362–365.
[37] J. Barluenga, M.A. Fernández, F. Aznar, and C. Valdés, *Chem. Eur. J.*, 2005, **11**, 2276–2283.
[38] H. Fuwa and M. Sasaki, *J. Org. Chem.*, 2009, **74**, 212–221.
[39] F.A. Siqueira, J.G. Taylor, and C.R.D. Correia, *Tetrahedron Lett.*, 2010, **51**, 2102–2105.
[40] D. Solé and O. Serrano, *J. Org. Chem.*, 2008, **73**, 9372–9378.
[41] H.S. Kim, H.S. Lee, S.H. Kim, and J.N. Kim, *Tetrahedron Lett.*, 2009, **50**, 3154–3157.
[42] A.P. Kale, G.S. Kumar, A.R.K. Mangadan, and M. Kapur, *Org. Lett.*, 2015, **17**, 1324–1327.
[43] X.-L. Yun, W.-Y. Bi, J.-H, Huang, *et al.*, *Tetrahedron Lett.*, 2012, **53**, 5076–5080.
[44] S. Würtz, S. Rakshit, J.J. Newmann, *et al.*, *Angew. Chem. Int. Ed.*, 2008, **47**, 7230–7233.
[45] J.J. Neumann, S. Rakshit, T. Dröge, *et al.*, *Chem. Eur. J.*, 2011, **17**, 7298–7303.
[46] Z. Shi and F. Glorius, *Angew. Chem. Int. Ed.*, 2012, **51**, 9220–9222.
[47] B. Weng, R. Liu, and J.-H. Li, *Synthesis*, 2010, 2926–2930.
[48] X.-L. Lian, Z.-H. Ren, Y.-Y. Wang, and Z.-H. Guan, *Org. Lett.*, 2014, **16**, 3360–3363.
[49] J. Zoller, D.C. Fabry, M.A. Ronge, and M. Rueping, *Angew. Chem. Int. Ed.*, 2014, **53**, 13264–13268.
[50] L. Ren, Z. Shi, and N. Jiao, *Tetrahedron*, 2013, **69**, 4408–4414.
[51] K. Daniell, M. Stewart, E. Madsen, *et al.*, *Bioorg. Med. Chem. Lett.*, 2005, **15**, 177–180.

72

Palladium-Catalyzed Indole Ring Synthesis: Aryl-Heck

The aromatic ring version of the Mori–Ban–Heck indole synthesis can be called an aryl-Heck reaction. This palladium-catalyzed cyclization of *o*-halo diaryl amines has seen many applications in carbazole and carboline synthesis.

An early example is that of Ames and Bull, who synthesized indolo[3,2-*c*]cinnoline (Scheme 1, equation 1) [1]. Padwa discovered a new β-carbolinone synthesis using an aryl-Heck process (equation 2) [2], and Maes has exploited this strategy to prepare a number of nitrogen heterocycles: indolo[2,3-*c*]quinolines [3], indolo[3,2-*c*]isoquinolines [4], indolo[2,3-*c*]isoquinolines [4], isocryptolepine [5], α-carbolines [6], and neocryptolepines [7] (equation 3). Queiroz and coworkers employed similar tactics to synthesize thieno[3,2-*c*]carbazoles, thieno[2,3-*b*]carbazoles, and indolo[3,2-*b*]benzo[*b*]thiophenes [8, 9]. Fagnou and colleagues prepared a range of indoles and carbazoles via direct arylation of aryl chlorides, bromides, and iodides [10, 11]. A synthesis of the carbazole alkaloid mukonine is representative of the method (equation 4). A Buchwald–Hartwig amination of the appropriate triflate with *o*-chloroaniline gave the starting diaryl amine. Bedford and colleagues employed a similar consecutive amination/aryl-Heck cyclization to access a variety of carbazoles [12], including the rare fluorinated carbazoles [13], and murrayaquinone and analogues [14]. Likewise, Tamariz and Bernal synthesized murrayanine via a Pd-catalyzed aryl-Heck cyclization [15]. Other natural product syntheses that used an aryl-Heck reaction include carbazomycin A (Catellani) [16], dictyodendrins B and E (Jia) [17], and a novel ellipticine–makaluvamine pyrido[3,2-*b*]carbazole hybrid (Guillard) [18]. Cook and coworkers developed a two-step route to biologically active β-carbolines that entailed an amination/aryl-Heck cyclization sequence [19]. Bogányi and Kámán also adopted consecutive Pd-catalyzed reactions (Buchwald–Hartwig/aryl-Heck) to craft indoloquinoline alkaloids [20], as did Mineno's group to prepare a collection of α-carbolines [21]. In a similar vein, Cuny and coworkers developed a one-pot sequential amination/arylation procedure to synthesize α-carbolines [22].

Ackermann employed a different reaction order to meld 1,2-dihaloaryls together with anilines to access a series of carbazoles, including murrayafoline A (Scheme 2, equations 1 and 2) [23]. Jean and colleagues reported a tandem Suzuki/aryl-Heck protocol to prepare functionalized carbazoles and glycosinine (equation 3) [24]. Larock and Liu synthesized a series of carbazoles and the alkaloid mukonine via an aryl-Heck reaction, which featured an aryne mediated preparation of the requisite diaryl amines [25]. Willis and coworkers reported an indole synthesis involving Heck arylation followed by double bond isomerization (equation 4) [26]. Urabe and colleagues described a similar aryl-Heck indolization from β-bromo-*N*-aryl enamines giving 2-alkylindoles [27].

Buchwald reported a Pd-catalyzed aryl-Heck asymmetric dearomatization of diaryl amines (Scheme 3, equation 1) [28]. Chianese described an interesting dealkylation en route to a carbazole synthesis (equation 2) [29]. Although the corresponding 2,6-diethyl analogue does not undergo deethylation, the labile cyclized intermediate cannot be isolated, and it undergoes dimerization (60%).

Indole Ring Synthesis: From Natural Products to Drug Discovery, First Edition. Gordon W. Gribble.

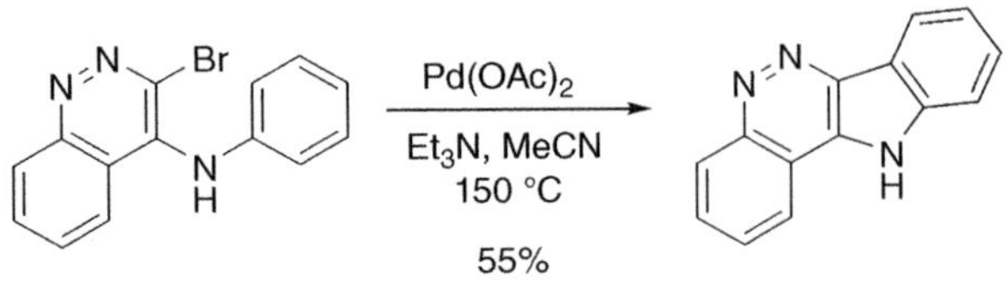

[eqn. 1] [1]

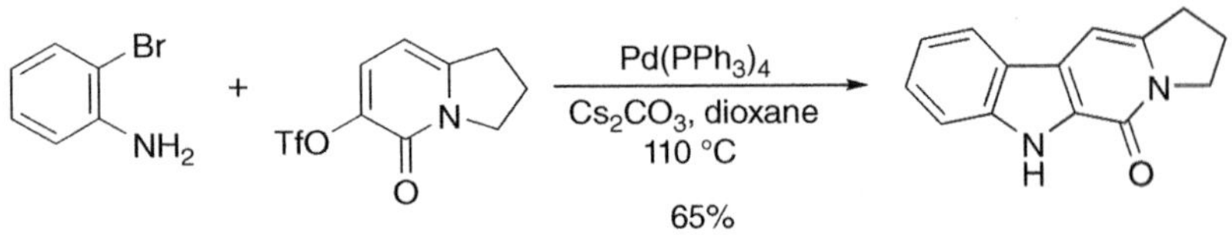

[eqn. 2] [2]

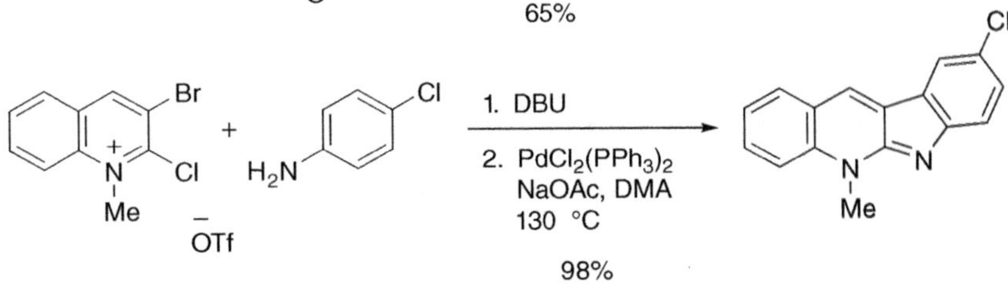

[eqn. 3] [7]

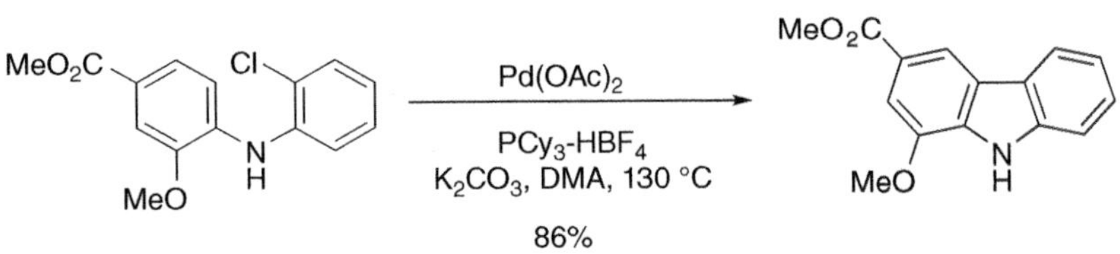

[eqn. 4] [11]

Scheme 1 *Aryl-Heck Carbazole Synthesis*

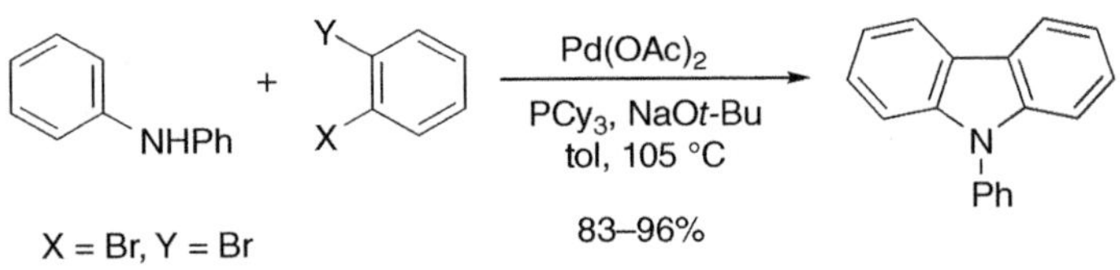

[eqn. 1] [23]

X = Br, Y = Br
X = I, Y = Cl
X = Cl, Y = Br

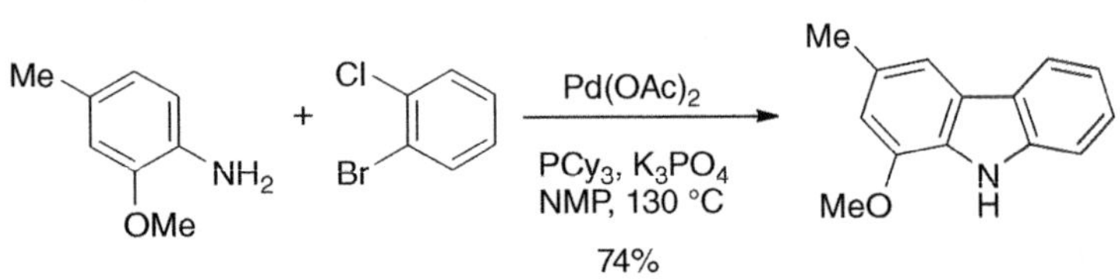

[eqn. 2] [23]

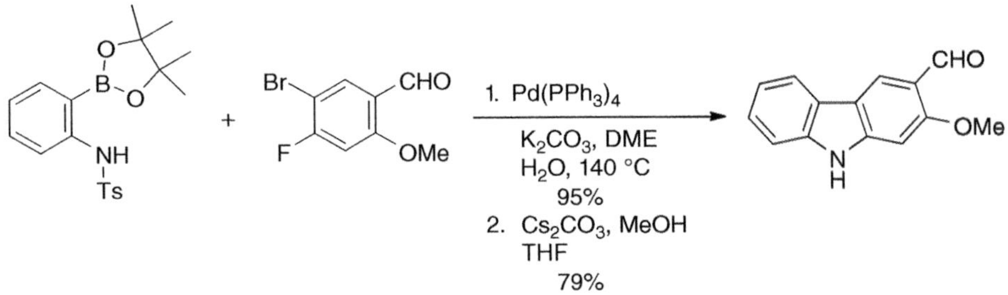

[eqn. 3] [24]

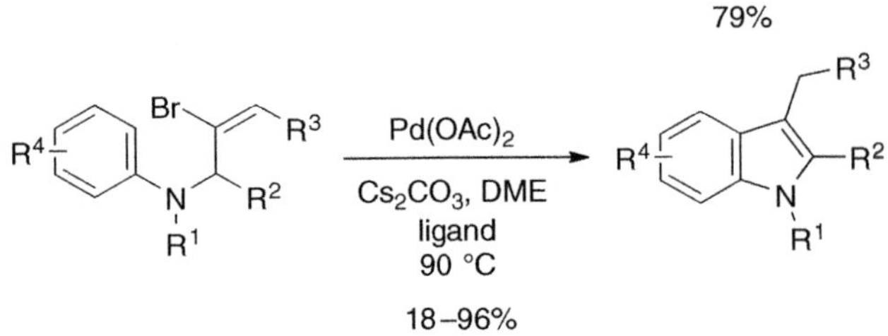

[eqn. 4] [26]

R^1 = Me, *i*-Pr, Bn, Ts
R^2 = Me
R^2, R^3 = $-(CH_2)_3-$, $-(CH_2)_3-$
R^4 = H, 5-OMe, 4-OMe, 5-Cl, 6-Me, 5-F,
5-CO_2Me (indole numbering)

Scheme 2 *Aryl-Heck Carbazole and Indole Syntheses*

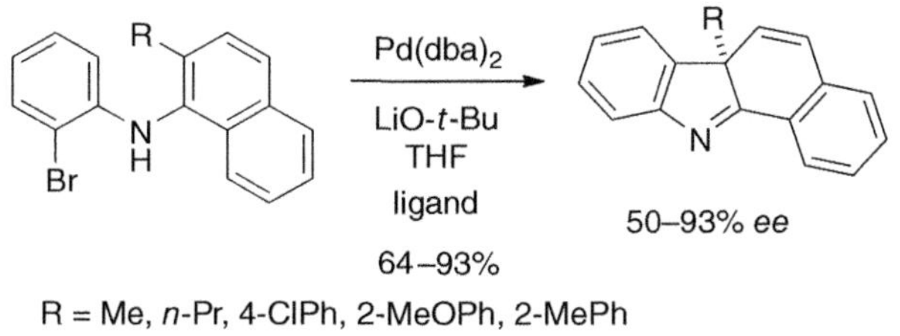

[eqn. 1] [28]

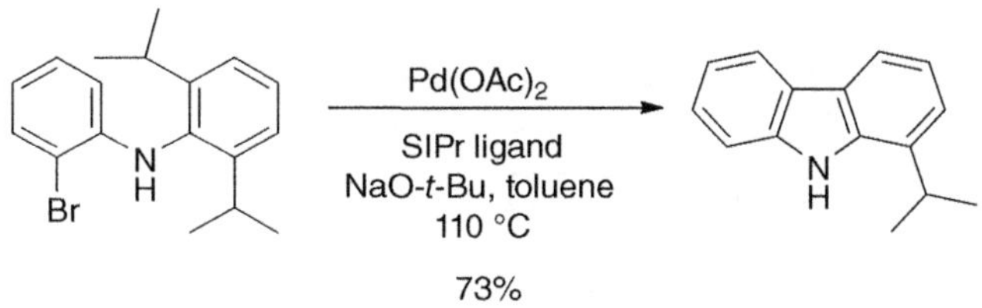

[eqn. 2] [29]

Scheme 3 *Unusual Aryl-Heck Carbazole Syntheses*

References

[1] D.E. Ames and D. Bull, *Tetrahedron*, 1982, **38**, 383–387.
[2] J.M. Harris and A. Padwa, *Org. Lett.*, 2003, **5**, 4195–4197.
[3] S. Hostyn, B.U.W. Maes, G. Van Baelen, *et al.*, *Tetrahedron*, 2006, **62**, 4676–4684.
[4] G. Van Baelen, C. Meyers, G.L.F. Lemière, *et al.*, *Tetrahedron*, 2008, **64**, 11802–11809.
[5] T.H.M. Jonckers, B.U.W. Maes, G.L.F. Lemière, *et al.*, *Synlett*, 2003, 615–618.
[6] S. Hostyn, G. Van Baelen, G.L.F. Lemière, and B.U.W. Maes, *Adv. Synth. Catal.*, 2008, **350**, 2653–2660.
[7] S. Hostyn, K.A. Tehrani, F. Lemière, *et al.*, *Tetrahedron*, 2011, **67**, 655–659.
[8] I.C.F.R. Ferreira, M.-J.R.P. Queiroz, and G. Kirsch, *Tetrahedron*, 2003, **59**, 3737–3743.
[9] M.-J.R.P. Queiroz, I.C.F.R. Ferreira, Y.De Gaetano, *et al.*, *Bioorg. Med. Chem.*, 2006, **14**, 6827–6831.
[10] L.-C. Campeau, P. Thansandote, and K. Fagnou, *Org. Lett.*, 2005, **7**, 1857–1860.
[11] L.-C. Campeau, M. Parisien, A. Jean, and K. Fagnou, *J. Am. Chem. Soc.*, 2006, **128**, 581–590.
[12] R.B. Bedford and M. Betham, *J. Org. Chem.*, 2006, **71**, 9403–9410.
[13] R.B. Bedford, M. Betham, J.P.H. Charmant, and A.L. Weeks, *Tetrahedron*, 2008, **64**, 6038–6050.
[14] R.B. Bedford, J.G. Bowen, and A.L. Weeks, *Tetrahedron*, 2013, **69**, 4389–4394.
[15] P. Bernal and J. Tamariz, *Helv. Chim. Acta*, 2007, **90**, 1449–1454.
[16] N. Della Ca', G. Sassi, and M. Catellani, *Adv. Synth. Catal.*, 2008, **350**, 2179–2182.
[17] J. Liang, W. Hu, P. Tao, and Y. Jia, *J. Org. Chem.*, 2013, **78**, 5810–5815.
[18] S. Boucl‌é and J. Guillard, *Synthesis*, 2011, 1616–1620.
[19] O.A. Namjoshi, A. Gryboski, G.O. Fonseca, *et al.*, *J. Org. Chem.*, 2011, **76**, 4721–4727.
[20] B. Bogányi and J. Kámán, *Tetrahedron*, 2013, **69**, 9512–9519.
[21] M. Mineno, M. Sera, T. Ueda, *et al.*, *Tetrahedron*, 2014, **70**, 5550–5557.
[22] J.K. Laha, P. Petrou, and G.D. Cuny, *J. Org. Chem.*, 2009, **74**, 3152–3155.
[23] L. Ackermann, A. Althammer, and P. Mayer, *Synthesis*, 2009, 3493–3503.
[24] D.J. St. Jean, Jr., S.F. Poon, and J.L. Schwarzbach, *Org. Lett.*, 2007, **9**, 4893–4896.
[25] Z. Liu and R.C. Larock, *Tetrahedron*, 2007, **63**, 347–355.
[26] M. Yagoubi, A.C.F. Cruz, P.L. Nichols, *et al.*, *Angew. Chem. Int. Ed.*, 2010, **49**, 7958–7962.
[27] M. Yamagishi, K. Nishigai, A. Ishii, *et al.*, *Angew. Chem. Int. Ed.*, 2012, **51**, 6471–6474.
[28] J. García-Fortanet, F. Kessler, and S.L. Buchwald, *J. Am. Chem. Soc.*, 2009, **131**, 6676–6677.
[29] A.R. Chianese, S.L. Rogers, and H. Al-Gattas, *Tetrahedron Lett.*, 2010, **51**, 2241–2243.

73

Palladium-Catalyzed Indole Ring Synthesis: Oxidative Cyclization

Prior to the discovery of the aryl-Heck reaction (Chapter 72), the direct Pd-promoted oxidative cyclization of diaryl amines to carbazoles was well known. In 1975 Åkermark reported this reaction (Scheme 1, equation 1) [1]. In addition, *N*-phenylanthranilic acid gave carbazole-1-carboxylic acid (60%). Miller and Moock used $Pd(OAc)_2$ to cyclize 6-anilino-5,8-dimethylisoquinoline to ellipticine in low yield [2]. The second advance in this chemistry was reported independently by Bittner [3] and Furukawa [4], who described the Pd-mediated (stoichiometric) oxidative conversion of 2-anilino-1,4-benzoquinones and 2-anilino-1,4-naphthoquinones to the corresponding carbazole-1,4-diones and benzo[*b*]carbazole-1,6-diones (equations 2, 3). Furukawa's studies included syntheses of several carbazolequinone alkaloids. In 1995 Åkermark and colleagues developed catalytic versions (i.e., using *tert*-butyl hydrogen peroxide [TBHP] or oxygen) of this cyclization (equation 3) [5, 6], which elevated the importance of this palladium oxidative cyclization, mainly because of the expense of $Pd(OAc)_2$. Somewhat earlier, Knölker used cupric acetate as a reoxidant in a synthesis of carbazole-1,4-quinones [7].

In a multitude of publications, Knölker and coworkers have parlayed the twin Pd-catalyzed oxidative cyclizations to give carbazoles or carbazole-1,4-quinones into an impressive collection of natural-product syntheses [8–12]; some recent examples are illustrated in Scheme 2 and include syntheses of glycozoline (equation 1) [13], isokoeniginequinone A (equation 2) [14], and mukonine (equation 3) [15]. Knölker and colleagues were able to isolate and crystallize the palladium complex **2** from the Pd-catalyzed reaction of diaryl amine **1** to give four isomeric carbazoles, from which an x-ray crystal structure was obtained [16].

Scheme 3 shows some recent carbazoles and carbazolequinones that were synthesized via the Pd-catalyzed oxidative cyclization of diaryl amines [17–21]. The so-formed bond is in bold, and the conditions are the typical $Pd(OAc)_2$ in refluxing acetic acid.

Fujii and Ohno developed a one-pot sequential *N*-arylation/oxidative cyclization procedure (Scheme 4, equation 1) [22, 23]. Several groups have extended this palladium-catalyzed oxidative cyclization to indole synthesis. Some examples are included in Chapter 71. Keller and Gamble applied their reaction to the synthesis of a 6,6′-biindole (equation 2) [24], and Du, Huang, and colleagues synthesized a large number of carbazolones from the appropriate 3-(arylamino)cyclohex-2-enones (equation 3) [25]. Interestingly, in the latter case, oxygen was better than $Cu(OAc)_2$ as an oxidant. Independently of Keller and Gamble, Jiao described a similar indole synthesis in one pot from anilines and dimethylacetylene dicarboxylate (and other acetylenic substrates) ($Pd(OAc)_2$, O_2, DMA, pivalic acid, 120 °C) [26]. Yoshikai reported the oxidative cyclization of *N*-arylimines to indoles ($Pd(OAc)_2$, O_2, *n*-Bu_4NBr, DMSO, 60 °C), particularly 2-arylindoles in generally excellent yields [27]. Li's group synthesized a series of *N*-(2-pyridyl)indoles from *N*-aryl-2-aminopyridines ($Pd(MeCN)_2Cl_2$, $CuCl_2$, DMF, 105 °C) [28]. For a concise review of C-H bond activation in the context of heterocycle synthesis, see Thansandote and Lautens [29]. Matsubara reported that these C-H activation processes were facilitated by a palladium (or platinum) catalyst in water [30]. Both diphenyl amine and 2-aminobiphenyl gave carbazole with either catalyst, but Pt/C was superior to Pd/C.

Indole Ring Synthesis: From Natural Products to Drug Discovery, First Edition. Gordon W. Gribble.

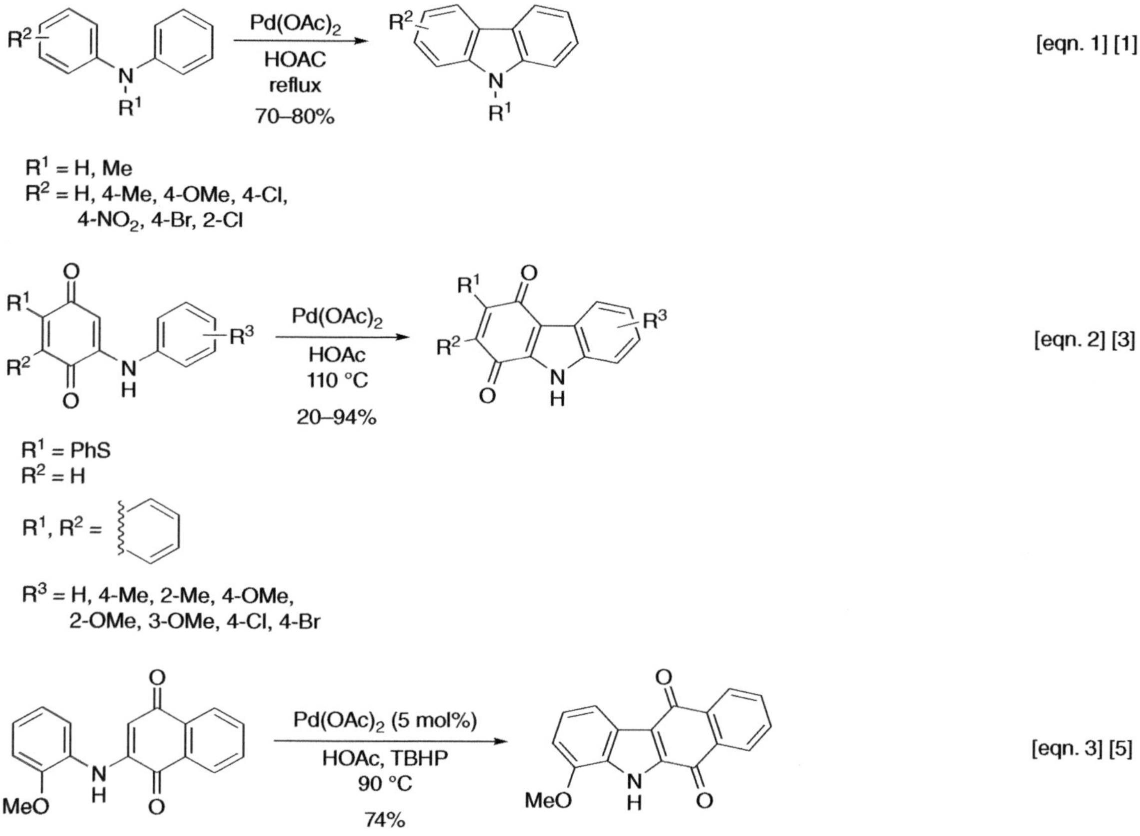

Scheme 1 *Åkermark Carbazole Synthesis*

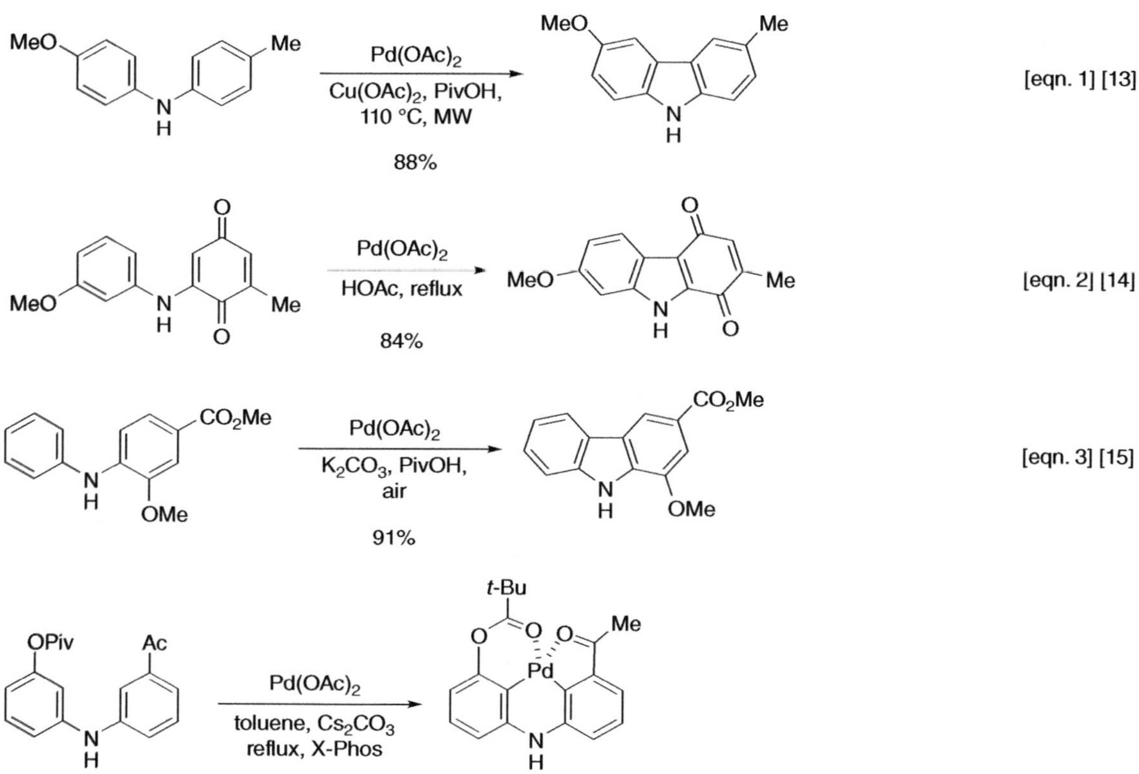

1 **2**

Scheme 2 *Knölker Palladium-Catalyzed Carbazole Synthesis*

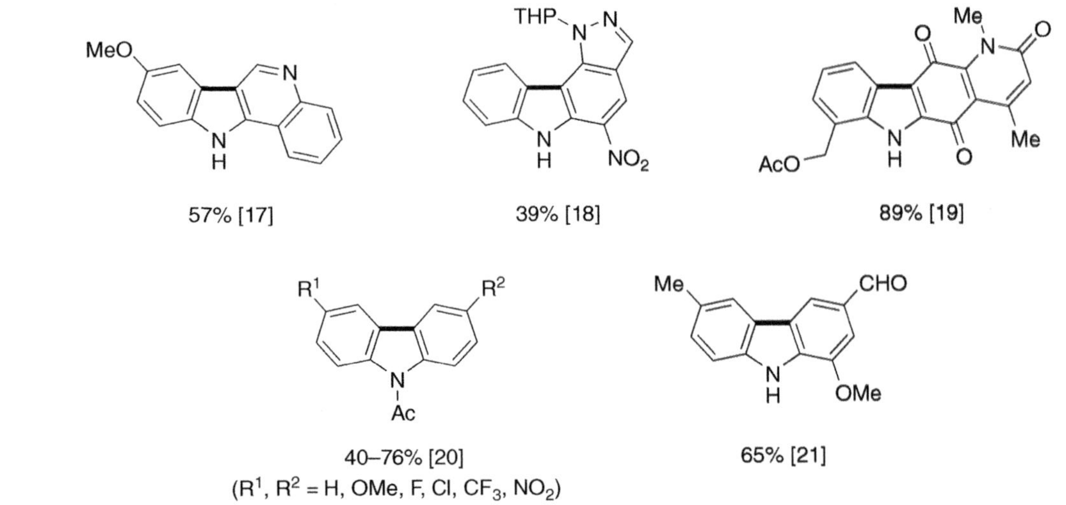

Scheme 3 *Carbazoles Synthesized via Palladium-Catalyzed Cyclization of Diaryl Amines*

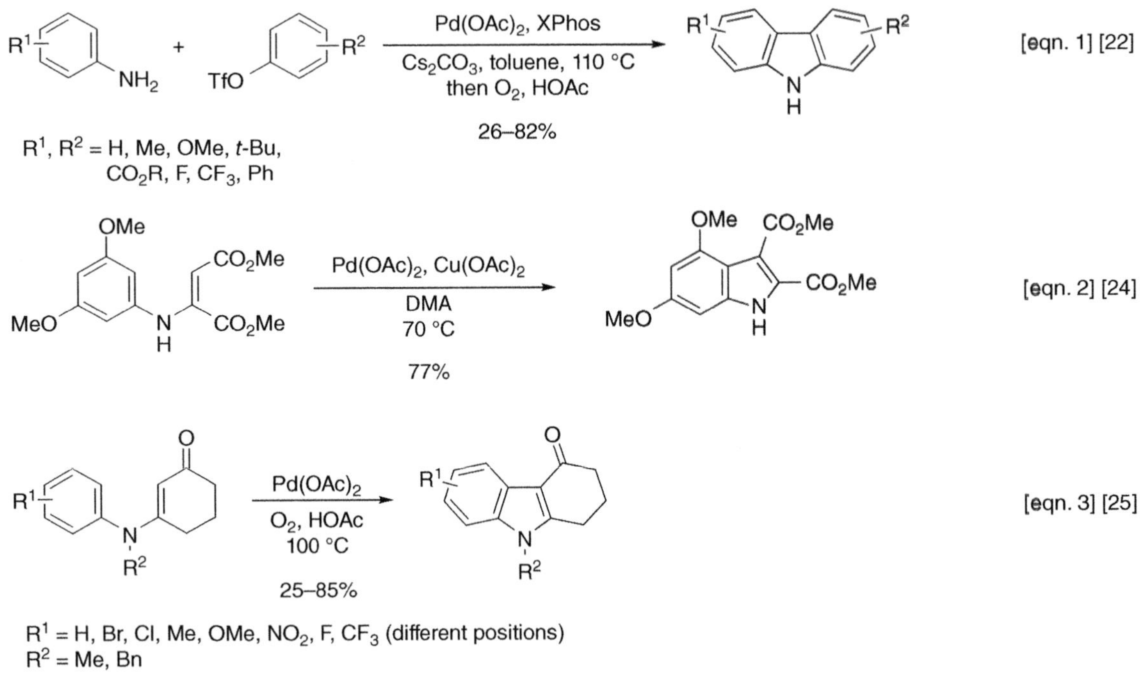

Scheme 4 *Palladium-Catalyzed Oxidative Cyclization*

References

[1] B. Åkermark, L. Eberson, E. Jonsson, and E. Pettersson, *J. Org. Chem.*, 1975, **40**, 1365–1367.
[2] R.B. Miller and T. Moock, *Tetrahedron Lett.*, 1980, **21**, 3319–3322.
[3] S. Bittner, P. Krief, and T. Massil, *Synthesis*, 1991, 215–216.
[4] M. Yogo, C. Ito, and H. Furukawa, *Chem. Pharm. Bull.*, 1991, **39**, 328–334.
[5] B. Åkermark, J.D. Oslob, and U. Heuschert, *Tetrahedron Lett.*, 1995, **36**, 1325–1326.
[6] H. Hagelin, J.D. Oslob, and B. Åkermark, *Chem. Eur. J.*, 1999, **5**, 2413–2416.
[7] H.-J. Knölker and N. O'Sullivan, *Tetrahedron*, 1994, **50**, 10893–10908.
[8] H.-J. Knölker, *Curr. Org. Synth.*, 2004, **1**, 309–331.
[9] H.-J. Knölker, *Top. Curr. Chem.*, 2005, **244**, 115–148.

[10] S. Agarwal, S. Cämmerer, S. Filali, *et al.*, *Curr. Org. Chem.*, 2005, **9**, 1601–1614.
[11] T.A. Choi, R. Czerwonka, R. Forke, *et al.*, *Med. Chem. Res.*, 2008, **17**, 374–385.
[12] H.-J. Knölker, *Chem. Lett.*, 2009, **38**, 8–13.
[13] R. Hesse, A.W. Schmidt, and H.-J. Knölker, *Tetrahedron*, 2015, **71**, 3485–3490.
[14] H.-J. Knölker and K.R. Reddy, *Heterocycles*, 2003, **60**, 1049–1052.
[15] C. Börger, A.W. Schmidt, and H.-J. Knölker, *Org. Biomol. Chem.*, 2014, **12**, 3831–3835.
[16] T. Gensch, M. Rönnefahrt, R. Czerwonka, *et al.*, *Chem. Eur. J.*, 2012, **18**, 770–776.
[17] L. He, H.-X. Chang, T.-C. Chou, *et al.*, *Eur. J. Med. Chem.*, 2003, **38**, 101–107.
[18] V. Suchaud, L. Gavara, E. Saugues, *et al.*, *Bioorg. Med. Chem.*, 2013, **21**, 4102–4111.
[19] J.D. Sánchez, C. Avendaño, and J.C. Menéndez, *Synlett*, 2008, 1371–1375.
[20] S. Wang, H. Mao, Z. Ni, and Y. Pan, *Tetrahedron Lett.*, 2012, **53**, 505–508.
[21] R. Bautista, P. Bernal, L.E. Montiel, *et al.*, *Synthesis*, 2011, 929–933.
[22] T. Watanabe, S. Oishi, N. Fujii, and H. Ohno, *J. Org. Chem.*, 2009, **74**, 4720–4726.
[23] S. Oishi, T. Watanabe, J. Sawada, *et al.*, *J. Med. Chem.*, 2010, **53**, 5054–5058.
[24] A.B. Gamble and P.A. Keller, *Chem. Commun.*, 2010, **46**, 4076–4078.
[25] W. Bi, X. Yun, Y. Fan, *et al.*, *Synlett*, 2010, 2899–2904.
[26] Z. Shi, C. Zhang, S. Li, *et al.*, *Angew. Chem. Int. Ed.*, 2009, **48**, 4572–4576.
[27] Y. Wei, I. Deb, and N. Yoshikai, *J. Am. Chem. Soc.*, 2012, **134**, 9098–9101.
[28] J. Chen, Q. Pang, Y. Sun, and X. Li, *J. Org. Chem.*, 2011, **76**, 3523–3526.
[29] P. Thansandote and M. Lautens, *Chem. Eur. J.*, 2009, **15**, 5874–5883.
[30] S. Matsubara, K. Asano, Y. Kajita, and M. Yamamoto, *Synthesis*, 2007, 2055–2059.

74

Palladium-Catalyzed Indole Ring Synthesis: Watanabe–Cenini–Söderberg

Although first discovered by Watanabe (Scheme 1, equation 1) [1, 2], the reductive *N*-heterocyclization of *o*-nitrostyrenes to give indoles was subsequently explored by the group of Cenini, Tollari, and Ragaini [3–7] (equation 2) and by Söderberg's team, who greatly amplified this methodology into a powerful synthetic method [8–14]. For reviews of this chemistry see Söderberg [15] and Ragaini *et al* [16].

A team at Merck has made use of this reductive *N*-heterocyclization to synthesize a number of complex carbazoles and indoles (Scheme 2, equation 1) [17–19]. Dong and Hsieh prepared a series of 3-arylindoles using a version of this methodology (i.e., with β-nitrostyrenes) (equation 2) [20], and Nishiyama and colleagues adapted this technology, in the absence of palladium, to an excellent synthesis of 2-arylindoles (equation 3) [21]. The authors proposed that elemental sulfur reacts with carbon monoxide to form carbonyl sulfide, which deoxygenates the nitro group, leading ultimately to a nitrene. Cyclization then affords an indole.

Indole Ring Synthesis: From Natural Products to Drug Discovery, First Edition. Gordon W. Gribble.

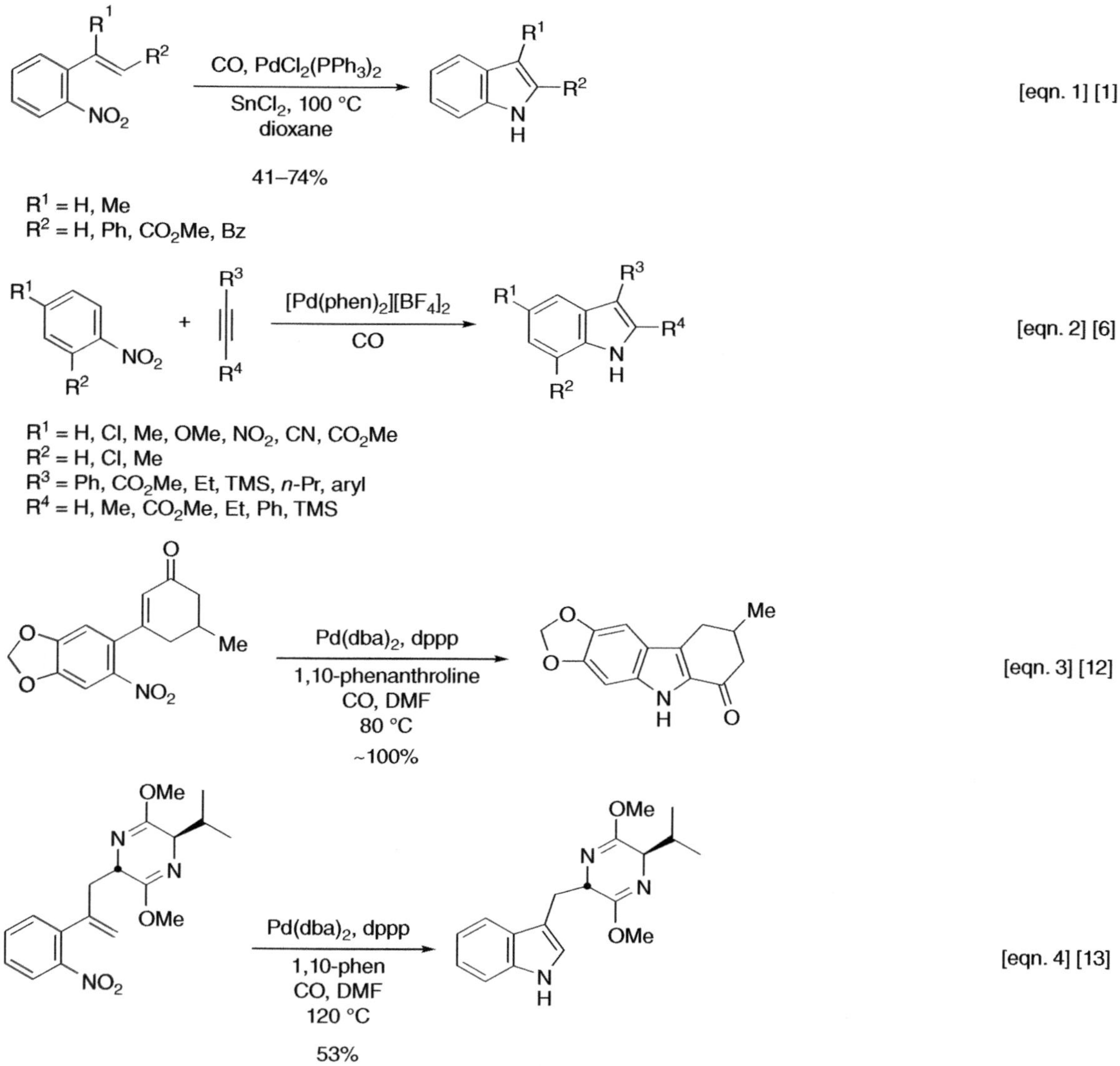

Scheme 1 *Watanabe-Cenini-Söderberg Indole Synthesis*

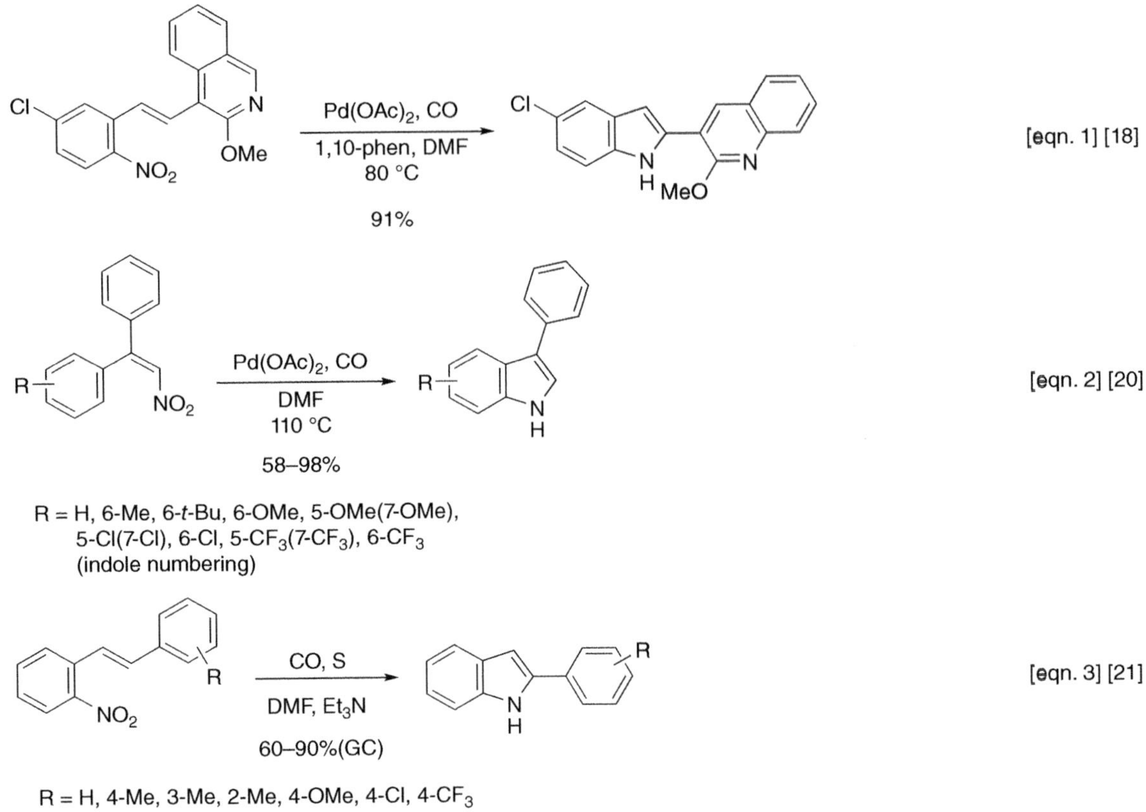

Scheme 2 *Applications of the Watanabe-Cenini-Söderberg Indole Synthesis*

References

[1] M. Akazome, T. Kondo, and Y. Watanabe, *Chem. Lett.*, 1992, 769–772.
[2] M. Akazome, T. Kondo, and Y. Watanabe, *J. Org. Chem.*, 1994, **59**, 3375–3380.
[3] S. Tollari, S. Cenini, A. Rossi, and G. Palmisano, *J. Mol. Catal. A: Chemical*, 1998, **135**, 241–248.
[4] F. Ragaini, P. Sportiello, and S. Cenini, *J. Organomet. Chem.*, 1999, **577**, 283–291.
[5] S. Tollari, A. Penoni, and S. Cenini, *J. Mol. Catal. A: Chemical*, 2000, **152**, 47–54.
[6] F. Ragaini, A. Rapetti, E. Visentin, *et al.*, *J. Org. Chem.*, 2006, **71**, 3748–3753.
[7] F. Ragaini, F. Ventriglia, M. Hagar, *et al.*, *Eur. J. Org. Chem.*, 2009, 2185–2189.
[8] B.C. Söderberg and J.A. Shriver, *J. Org. Chem.*, 1997, **62**, 5838–5845.
[9] T.L. Scott and B.C.G. Söderberg, *Tetrahedron Lett.*, 2002, **43**, 1621–1624.
[10] S.W. Dantale and B.C.G. Söderberg, *Tetrahedron*, 2003, **59**, 5507–5514.
[11] B.C.G. Söderberg, J.W. Hubbard, S.R. Rector, and S.N. O'Neil, *Tetrahedron*, 2005, **61**, 3637–3649.
[12] T.L. Scott, X. Yu, S.P. Gorugantula, *et al.*, *Tetrahedron*, 2006, **62**, 10835–10842.
[13] C.A. Dacko, N.G. Akhmedov, and B.C.G. Söderberg, *Tetrahedron: Asymmetry*, 2008, **19**, 2775–2783.
[14] S.R. Banini, M.R. Turner, M.M. Cummings, and B.C.G. Söderberg, *Tetrahedron*, 2011, **67**, 3603–3611.
[15] B.C.G. Söderberg, *Curr. Org. Chem.*, 2000, **4**, 727–764.
[16] F. Ragaini, S. Cenini, E. Gallo, *et al.*, *Curr. Org. Chem.*, 2006, **10**, 1479–1510.
[17] J.H. Smitrovich and I.W. Davies, *Org. Lett.*, 2004, **6**, 533–535.
[18] I.W. Davies, J.H. Smitrovich, R. Sidler, *et al.*, *Tetrahedron*, 2005, **61**, 6425–6437.
[19] J.T. Kuethe and I.W. Davies, *Tetrahedron*, 2006, **62**, 11381–11390.
[20] T.H.H. Hsieh and V.M. Dong, *Tetrahedron*, 2009, **65**, 3062–3068.
[21] R. Umeda, Y. Nishimoto, T. Mashino, and Y. Nishiyama, *Heterocycles*, 2013, **87**, 1241–1247.

75

Palladium-Catalyzed Indole Ring Synthesis: Yamanaka–Sakamoto–Sonogashira

In the first of three widely used indole syntheses involving Pd-catalyzed alkyne coupling with *o*-haloanilines, this chapter covers the Yamanaka–Sakamoto–Sonogashira reaction. This two-step process features an initial Sonogashira reaction with a terminal alkyne [1] followed by a facile indolization, which was first uncovered by Yamanaka, Sakamoto, and coworkers (Scheme 1, equations 1–4) [2–9]. The intermediate C-3 palladium species can undergo carbonylation with CO to give indole-3-carboxylates (equation 4) [7, 9]. For reviews, see Sakamoto, Kondo, and Yamanaka [10] and Heravi and Sadjadi [11].

The Yamanaka–Sakamoto–Sonogashira indole synthesis (now often truncated to "Sonogashira indole synthesis") has found myriad devotees who polished and modified the original reaction. The important work of Cacchi is presented separately in Chapter 77. Stille and Rudisill employed the coupling of alkynylstannanes with 2-bromo-(2-trifloxy)anilines followed by $PdCl_2$ indolization [12], and Srinivasan described a one-pot ligand-, copper-, and amine-free Sonogashira coupling and indolization to give 2-substituted indoles [13]. Sivakumar and colleagues used *tert*-butyl sulfinamide as an ammonia surrogate (Scheme 2, equation 1) [14], whereas Stradiotto found conditions that allowed ammonia itself (0.5 M in 1,4-dioxane) to function in the synthesis of 2-arylindoles from 2-alkynylbromoarenes [15]. The coupling partners methylamine and hydrazine were equally successful in this chemistry. Halland and colleagues also used hydrazines to access a variety of *N*-aminoindoles via a Sonogashira coupling/cyclization [16]. Ray's team reported that a Pd/C-$ZnCl_2$ catalyst was excellent for the direct indolization of *N*-tosyl-2-iodoanilines (equation 2) [17]. Mulcahy and Varelas discovered a novel synthesis of a pyrrolo-β-carboline via a Sonogashira coupling followed by an intramolecular [2+2+2] cyclization (equation 3) [18].

The diversity of indoles that can be synthesized by the Yamanaka–Sakamoto–Sonogashira method is illustrated in Scheme 3. The yields reflect the indolization step or the two-step coupling/cyclization. Indole **1** was synthesized on the solid phase in an indole library program [19], and pyrrolo[3.2.1-*ij*]quinoline **2** served in a pilot-scale synthesis of a herpes virus inhibitor [20]. Indole quinol **3** shows thioredoxin inhibitory properties [21], and the biindole–diazo conjugate **4** is a colorimetric anion receptor [22]. Indole **5** is the precursor to several iboga alkaloid analogues [23], and indole nucleoside **6** is a member of a growing number of novel nucleosides as biological agents [24]. The Sonogashira coupling is a new route to the important indolequinones such as **7** [25], and indole triazole **8** is an impurity in the antimigraine drug rizatriptan [26]. Epoxyindole **9** and isomers were prepared in a study of the stereochemistry of the alkaloid sespendole [27]. The Sonogashira reaction is an efficient path to pyrrolo[2,3-*c*]isoquinolines such as **10** [28] and also to C-2 substituted polyfluoroindoles (**11**) [29]. Indole **12** displays antiproliferative activity toward several human cancer cell lines [30].

Newer conditions to construct indoles from the Sonogashira 2-alkynylanilines are an electrochemical cyclization that avoids the use of metal catalysts and acid or base [31] and the employment of a recyclable silica-supported palladium catalyst [32]. In both cases the yields of indoles are excellent. Sanz reported a clever synthesis of 4-haloindoles, which can be further functionalized, from 2,3-dihalophenols via a Smiles rearrangement and

Indole Ring Synthesis: From Natural Products to Drug Discovery, First Edition. Gordon W. Gribble.

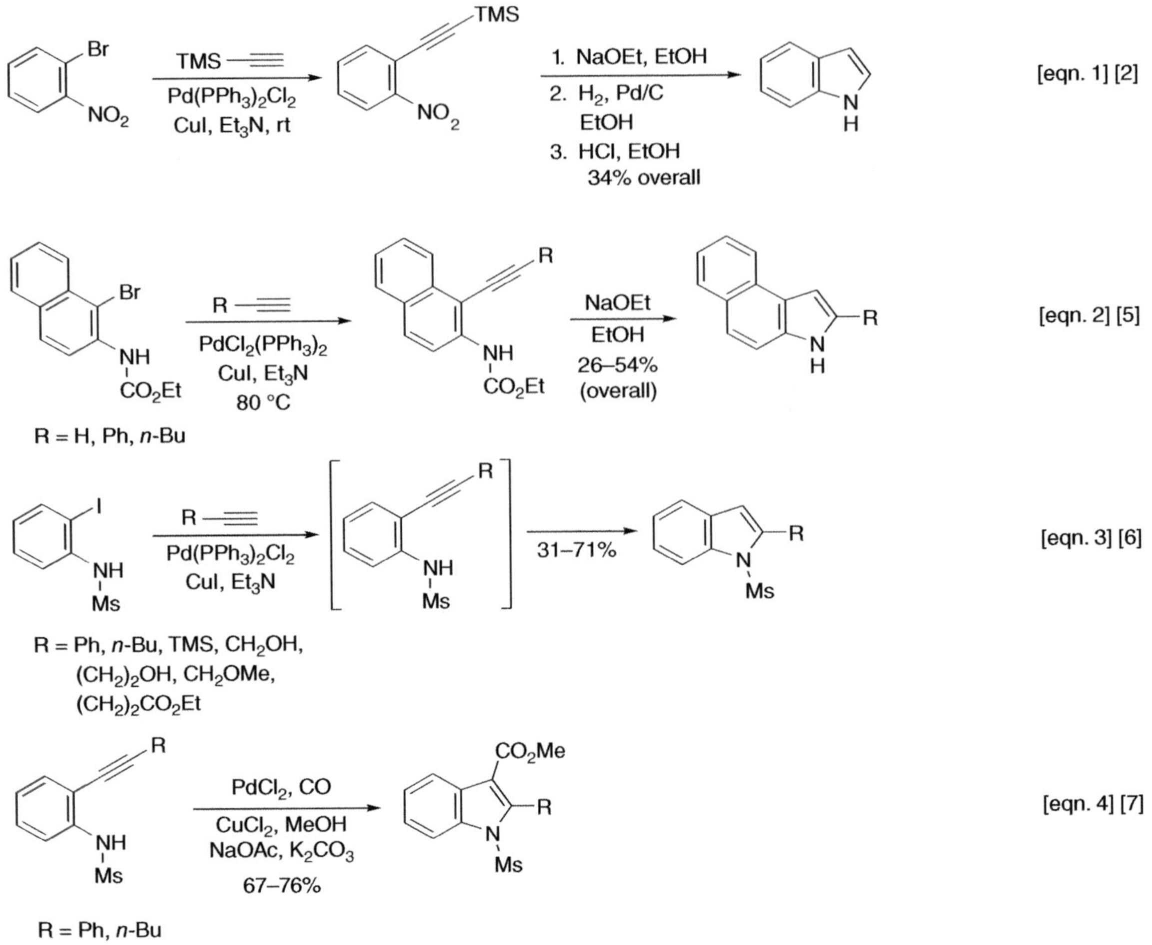

Scheme 1 *Yamanaka-Sakamoto-Sonogashira Indole Synthesis*

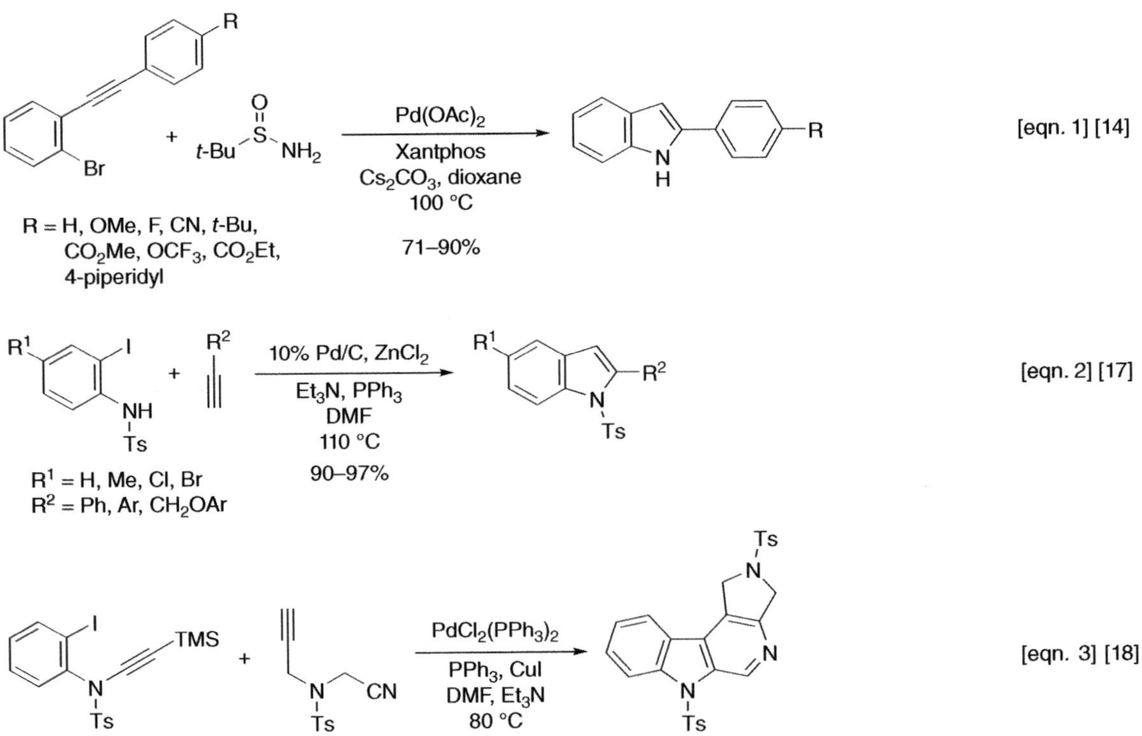

Scheme 2 *Applications of the Yamanaka-Sakamoto-Sonogashira Indole Synthesis*

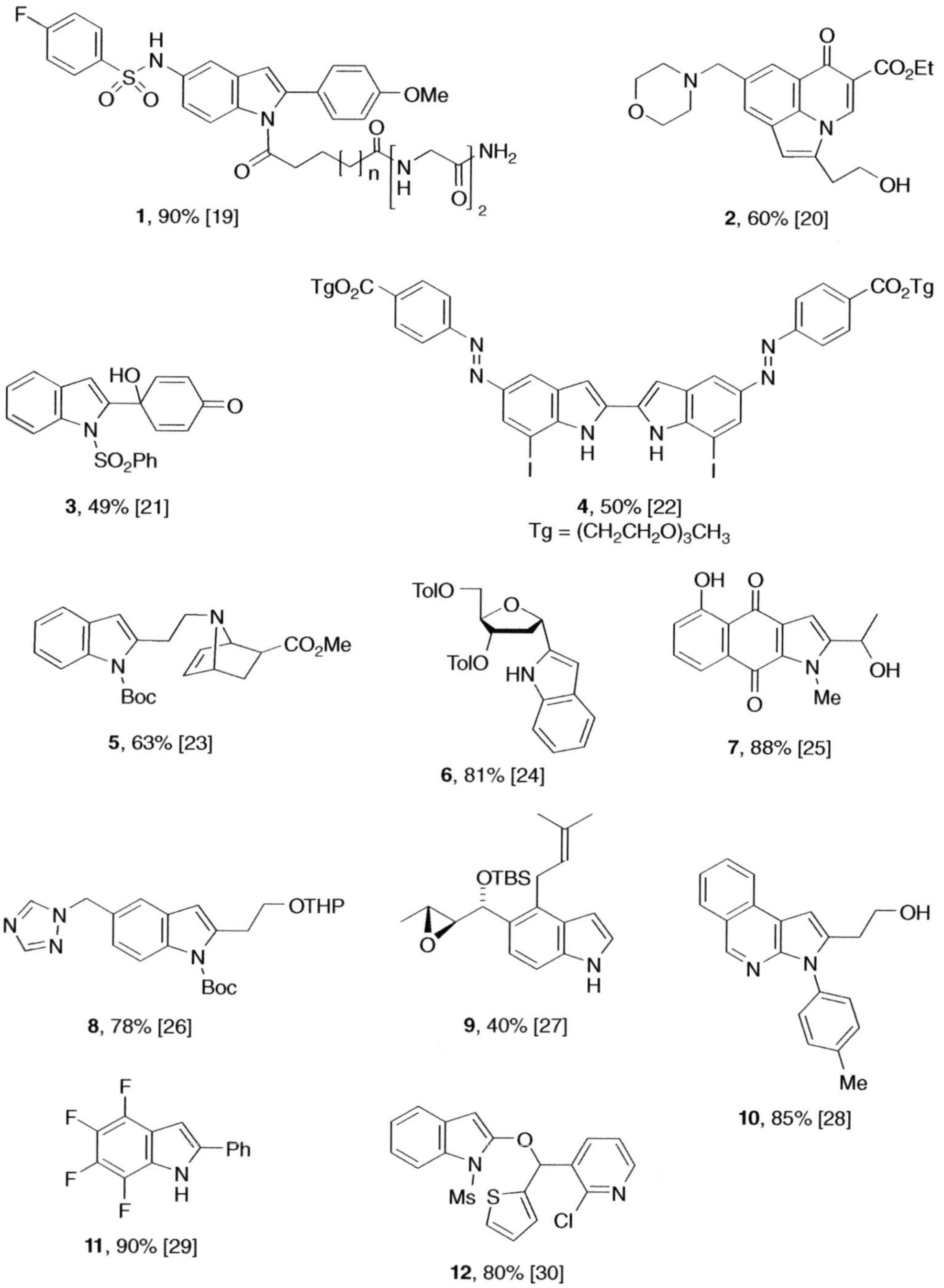

Scheme 3 *Indoles Synthesized by the Yamanaka-Sakamoto-Sonogashira Indole Synthesis*

Sonogashira coupling/cyclization [33]. Manabe and Yamaguchi described the synthesis and functionalization of 4-chloroindoles from 2,3-dichloroanilines. The catalyst-ligand system of Pd-dihydroxyterphenylphosphine allowed for selective *ortho*-Sonogashira coupling and subsequent indolization to give an assortment of 2,4-disubstituted indoles [34]. Li, Fan and colleagues have adapted the Sonogashira coupling to 2-alkynyl cyclohexadienimines to obtain *N*-heteroarylated indoles and two classes of azepinoindoles [35–37].

References

[1] K. Sonogashira, Y. Tohda, and N. Hagihara, *Tetrahedron Lett.*, 1975, **16**, 4467–4470.
[2] T. Sakamoto, Y. Kondo, and H. Yamanaka, *Heterocycles*, 1984, **22**, 1347–1350.
[3] T. Sakamoto, Y. Kondo, and H. Yamanaka, *Chem. Pharm. Bull.*, 1986, **34**, 2362–2368.
[4] T. Sakamoto, Y. Kondo, and H. Yamanaka, *Heterocycles*, 1986, **24**, 31–32.
[5] T. Sakamoto, Y. Kondo, and H. Yamanaka, *Heterocycles*, 1986, **24**, 1845–1847.
[6] T. Sakamoto, Y. Kondo, S. Iwashita, *et al.*, *Chem. Pharm. Bull.*, 1988, **36**, 1305–1308.
[7] Y. Kondo, T. Sakamoto, and H. Yamanaka, *Heterocycles*, 1989, **29**, 1013–1016.
[8] N. Suzuki, S. Yasaki, A. Yasuhara, and T. Sakamoto, *Chem. Pharm. Bull.*, 2003, **51**, 1170–1173.
[9] Y. Kondo, F. Shiga, N. Murata, *et al.*, *Tetrahedron*, 1994, **50**, 11803–11812.
[10] T. Sakamoto, Y. Kondo, and H. Yamanaka, *Heterocycles*, 1988, **27**, 2225–2249.
[11] M.M. Heravi and S. Sadjadi, *Tetrahedron*, 2009, **65**, 7761–7775.
[12] D.E. Rudisill and J.K. Stille, *J. Org. Chem.*, 1989, **54**, 5856–5866.
[13] S.S. Palimkar, P. Harish, Kumar, R.J. Lahoti, and K.V. Srinivasan, *Tetrahedron*, 2006, **62**, 5109–5115.
[14] A. Prakash, M. Dibakar, K. Selvakumar, *et al.*, *Tetrahedron Lett.*, 2011, **52**, 5625–5628.
[15] P.G. Alsabeh, R.J. Lundgren, L.E. Longobardi, and M. Stradiotto, *Chem. Commun.*, 2011, **47**, 6936–6938.
[16] N. Halland, M. Nazaré, J. Alonso, *et al.*, *Chem. Commun.*, 2011, **47**, 1042–1044.
[17] A. Ahmed, M. Ghosh, P. Sarkar, and J.K. Ray, *Tetrahedron Lett.*, 2013, **54**, 6691–6694.
[18] S.P. Mulcahy and J.G. Varelas, *Tetrahedron Lett.*, 2013, **54**, 6599–6601.
[19] L.-P. Sun and W.-M. Dai, *Angew. Chem. Int. Ed.*, 2006, **45**, 7255–7258.
[20] R.L. Dorow, P.M. Herrinton, R.A. Hohler, *et al.*, *Org. Proc. Res. Dev.*, 2006, **10**, 493–499.
[21] A.J. McCarroll, T.D. Bradshaw, A.D. Westwell, *et al.*, *J. Med. Chem.*, 2007, **50**, 1707–1710.
[22] G.W. Lee, N.-K. Kim, and K.-S. Jeong, *Org. Lett.*, 2010, **12**, 2634–2637.
[23] G.Kumar Jana and S. Sinha, *Tetrahedron Lett.*, 2010, **51**, 1994–1996.
[24] D. Nečas, D. Hidasová, M. Hocek, and M. Kotora, *Org. Biomol. Chem.*, 2011, **9**, 5934–5937.
[25] M. Yamashita, K. Ueda, K. Sakaguchi, *et al.*, *Chem. Pharm. Bull.*, 2011, **59**, 1289–1293.
[26] C. Pramanik, R. Bhumkar, G. Karhade, *et al.*, *Org. Process Res. Dev.*, 2012, 16, 507–511.
[27] M. Adachi, K. Higuchi, N. Thasana, *et al.*, *Org. Lett.*, 2012, **14**, 114–117.
[28] B. Dixit, J. Balog, Z. Riedl, *et al.*, *Tetrahedron*, 2012, **68**, 3560–3565.
[29] L.V. Politanskaya, I.P. Chuikov, and V.D. Shteingarts, *Tetrahedron*, 2013, **69**, 8477–8486.
[30] B. Dulla, E. Sailaja, U. Reddy CH, *et al.*, *Tetrahedron Lett.*, 2014, **55**, 921–926.
[31] A. Arcadi, G. Bianchi, A. Inesi, *et al.*, *Eur. J. Org. Chem.*, 2008, 783–787.
[32] E. Tyrrell, L. Whiteman, and N. Williams, *Synthesis*, 2009, 829–835.
[33] R. Sanz, V. Guilarte, and N. García, *Org. Biomol. Chem.*, 2010, **8**, 3860–3864.
[34] M. Yamaguchi and K. Manabe, *Org. Lett.*, 2014, **16**, 2386–2389.
[35] M. Yang, J. Tang, and R. Fan, *Chem. Commun.*, 2012, **48**, 11775–11777.
[36] L. Zhang, Z. Li, and R. Fan, *Org. Lett.*, 2012, **14**, 6076–6079.
[37] C. Zheng, J.J. Chen, and R. Fan, *Org. Lett.*, 2014, **16**, 816–819.

76

Palladium-Catalyzed Indole Ring Synthesis: Larock

The Larock indole synthesis is the internal alkyne version of the Yamanaka–Sakamoto–Sonogashira indole synthesis presented in Chapter 75. Also known as the Larock heteroannulation, it affords 2,3-disubstituted indoles, in contrast to 2-substituted indoles from the Sonogashira reaction (Scheme 1, equations 1–3) [1–12]. Like the Sonogashira, the Larock reaction has been of enormous utility in indole synthesis. With unsymmetrical alkynes, the regiochemistry is such that the more bulky group (e.g., *t*-butyl relative to methyl) is typically bonded to C-2 in the product (equation 2). One of several extensions discovered by Larock is the annulation of internal alkynes to afford isoindolo[2,1-*a*] indoles (equation 3) [6].

The Larock indole synthesis has been featured in syntheses of psilocin [13], (±)-lotthanongine [14], gardnerine [15], gardnutine [15], several vincamajine-related alkaloids [16], (–)-indolmycin [17], (–)-5-methoxyindolmycin [17], (±)-psychotrimine (Scheme 2, equation 1) [18],

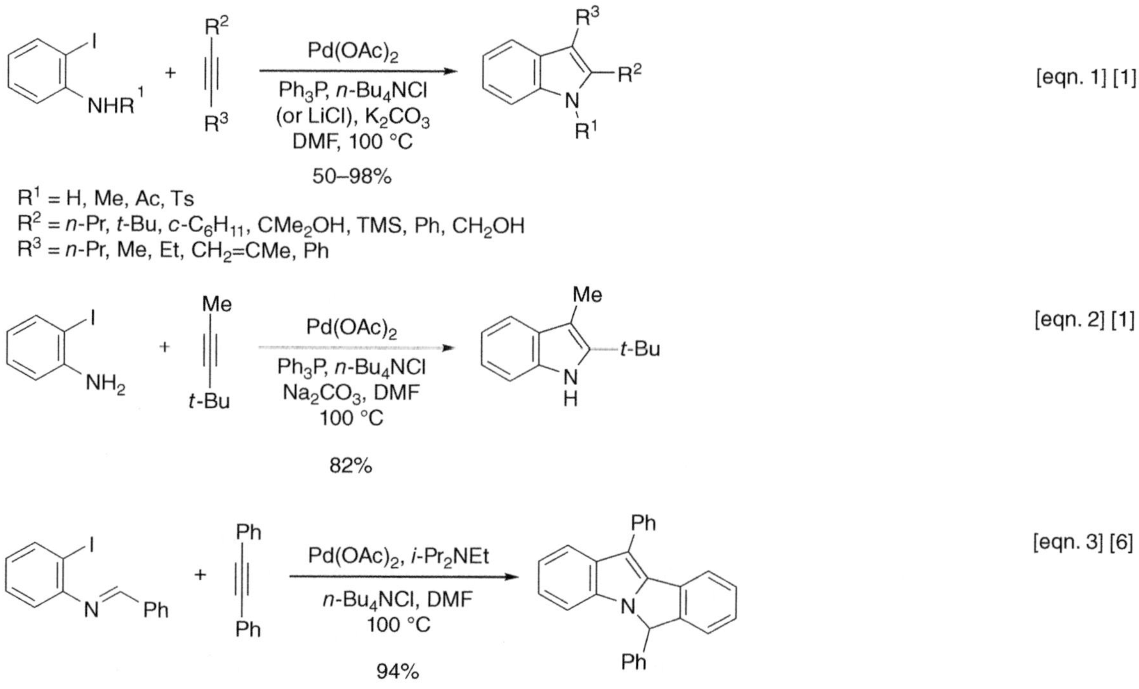

Scheme 1 *Larock Indole Synthesis*

Indole Ring Synthesis: From Natural Products to Drug Discovery, First Edition. Gordon W. Gribble.

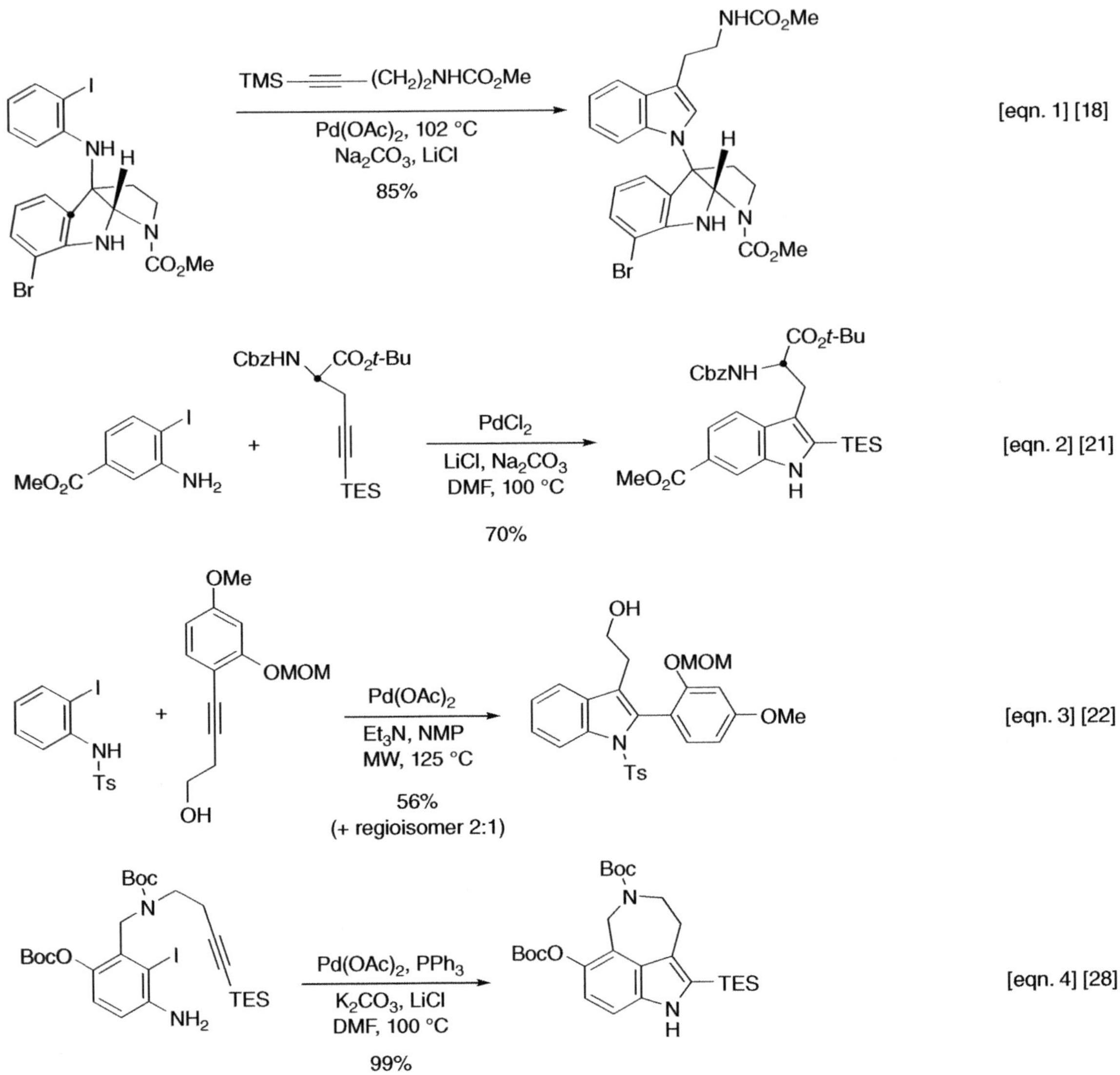

Scheme 2 *Applications of the Larock Indole Synthesis*

chloropeptins I and II [19, 20], celogentin C (equation 2) [21], phalarine (equation 3) [22], the iboga alkaloid scaffold [23], (+)-terreusinone [24], indomethacin [25], D-abrines [26], oxopropaline G [27], fargesine (equation 4) [28, 29], (±)-aspidospermidine [30], and the dictyodendrins [31]. Remarkable versatility!

The Larock indolization was applied to solid-phase indole synthesis (Smith) [32], to phosphine-free (Guo [33], Djakovitch [34]) and salt-free (Djakovitch [34]) conditions, to 2-chloroanilines (Senanayake [35], Ackermann [36]) (Scheme 3, equation 1), to the use of Pd/C (Sajiki) [37], to norbornadiene as an acetylene synthon (Lautens) [38], to indolization of *N*-aroylbenzotriazoles (Nakamura) (equation 2) [39], to sequential heteroannulation/silicon-based cross coupling (Denmark) [40], to a synthesis of 2-indolylphosphines (Yorimitsu, Oshima) [41], to a one-pot synthesis of 2-arylindoles from aryl iodides, TMS-acetylene, and 2-iodoanilines (Pal) [42], to a synthesis of δ-carbolines (Cao) (equation 3) [43], and to a one-pot multicomponent indole synthesis from 2-iodobenzoic acid [44]. Boger studied the macrocyclic Larock indole synthesis for use in the construction of chloropeptins I and II and analogues, an example of which is shown in equation 4 [45].

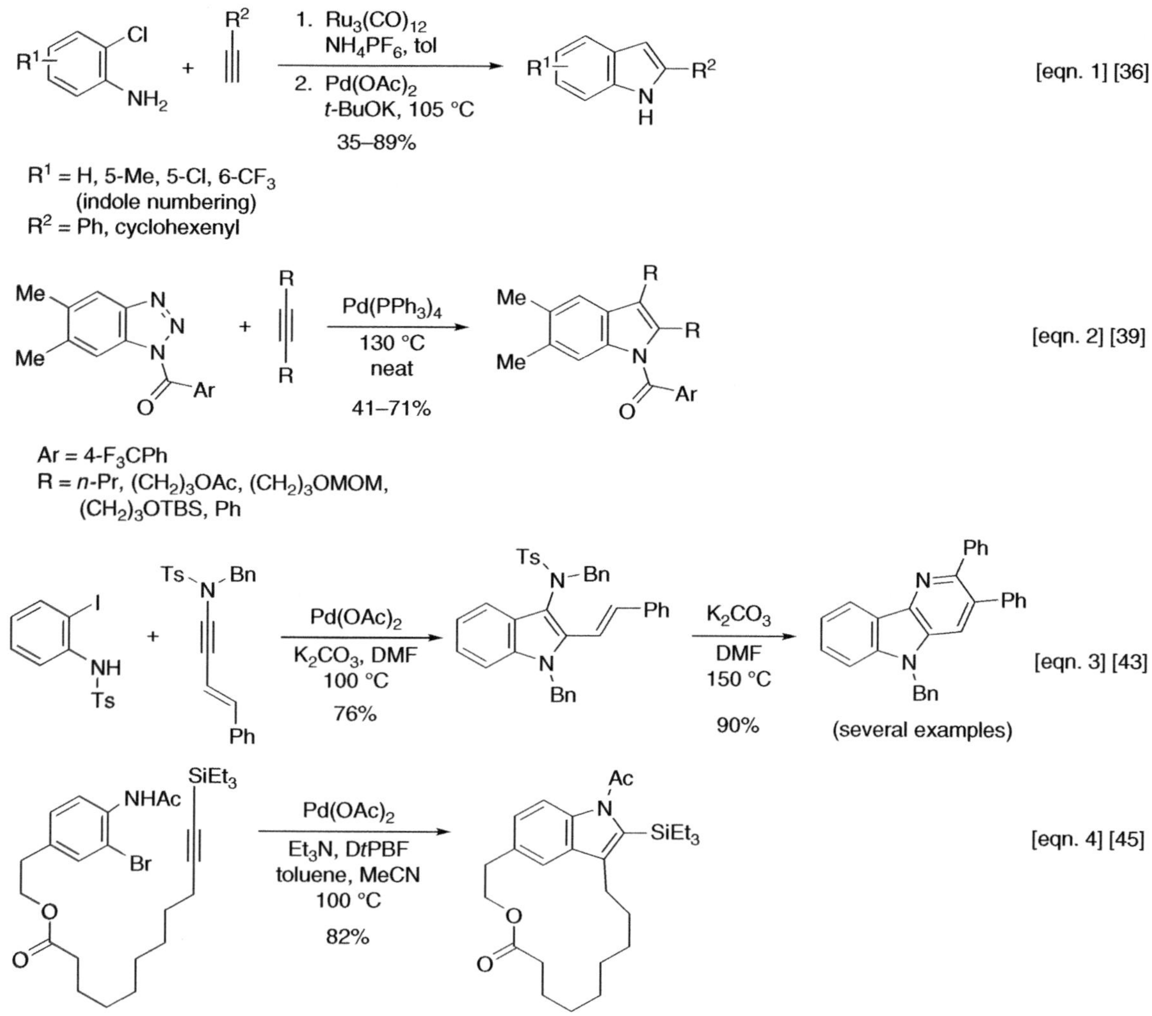

Scheme 3 *More Applications of the Larock Indole Synthesis*

References

[1] R.C. Larock and E.K. Yum, *J. Am. Chem. Soc.*, 1991, **113**, 6689–6690.
[2] R.C. Larock, E.K. Yum, and M.D. Refvik, *J. Org. Chem.*, 1998, **63**, 7652–7662.
[3] R.C. Larock, *J. Organometal. Chem.*, 1999, **576**, 111–124.
[4] K.R. Roesch and R.C. Larock, *Org. Lett.*, 1999, **1**, 1551–1553.
[5] R.C. Larock, *Pure Appl. Chem.*, 1999, **71**, 1435–1442.
[6] K.R. Roesch and R.C. Larock, *J. Org. Chem.*, 2001, **66**, 412–420.
[7] K.R. Roesch, H. Zhang, and R.C. Larock, *J. Org. Chem.*, 2001, **66**, 8042–8051.
[8] D. Yue and R.C. Larock, *J. Org. Chem.*, 2002, **67**, 1905–1909.
[9] J.J. Li and G.W. Gribble, *Top. Heterocycl. Chem.*, 2010, **26**, 196–199.
[10] R.C. Larock, *Top. Organometal. Chem.*, 2005, **14**, 147–182.
[11] G.W. Gribble (2007) *Palladium in Heterocyclic Chemistry*, 2nd edn (eds J.J. Li and G.W. Gribble), Elsevier, Amsteram, pp. 147–150.
[12] J.X. Qiao (2011) "Larock Indole Synthesis," in *Name Reactions in Heterocyclic Chemistry – II* (eds J.J. Li and E.J. Corey), Wiley, Hoboken, New Jersey, pp. 143–166.
[13] N. Gathergood and P.J. Scammells, *Org. Lett.*, 2003, **5**, 921–923.
[14] K. Hatakeyama, K. Ohmori, and K. Suzuki, *Synlett*, 2005, 1311–1315.
[15] H. Zhou, D. Han, X. Liao, and J.M. Cook, *Tetrahedrorn Lett.*, 2005, **46**, 4219–4224.
[16] J. Yu, X.Z. Wearing, and J.M. Cook, *J. Org. Chem.*, 2005, **70**, 3963–3979.
[17] N. Sutou, K. Kato, and H. Akita, *Tetrahedron: Asymmetry*, 2008, **19**, 1833–1838.
[18] T. Newhouse and P.S. Baran, *J. Am. Chem. Soc.*, 2008, **130**, 10886–10887.

[19] J. Garfunkle, F.S. Kimball, J.D. Trzupek, *et al.*, *J. Am. Chem. Soc.*, 2009, **131**, 16036–16038.
[20] H. Shimamura, S.P. Breazzano, J. Garfunkle, *et al.*, *J. Am. Chem. Soc.*, 2010, **132**, 7776–7783.
[21] B. Ma, B. Banerjee, D.N. Litvinov, *et al.*, *J. Am. Chem. Soc.*, 2010, **132**, 1159–1171.
[22] J.D. Trzupek, D. Lee, B.M. Crowley, *et al.*, *J. Am. Chem. Soc.*, 2010, **132**, 8506–8512.
[23] G. Kumar Jana and S. Sinha, *Tetrahedron Lett.*, 2012, **53**, 1671–1674.
[24] C. Wang and J. Sperry, *Tetrahedron*, 2013, **69**, 4563–4577.
[25] C. Zhu and S. Ma, *Org. Lett.*, 2013, **15**, 2782–2785.
[26] P. Danner, M. Morkunas, and M.E. Maier, *Org. Lett.*, 2013, **15**, 2474–2477.
[27] H. Song, Y. Liu, and Q. Wang, *Org. Lett.*, 2013, **15**, 3274–3277.
[28] D. Shan, Y. Gao, and Y. Jia, *Angew. Chem. Int. Ed.*, 2013, **52**, 4902–4905.
[29] Y. Gao, D. Shan, and Y. Jia, *Tetrahedron*, 2014, **70**, 5136–5141.
[30] G. Xia, X. Han, and X. Lu, *Org. Lett.*, 2014, **16**, 2058–2061.
[31] P. Tao, J. Liang, and Y. Jia, *Eur. J. Org. Chem.*, 2014, 5735–5748.
[32] A.L. Smith, G.I. Stevenson, C.J. Swain, and J.L. Castro, *Tetrahedron Lett.*, 1998, **39**, 8317–8320.
[33] X. Cui, J. Li, Y. Fu, *et al.*, *Tetrahedron Lett.*, 2008, **49**, 3458–3462.
[34] N. Batail, V. Dufaud, and L. Djakovitch, *Tetrahedron Lett.*, 2011, **52**, 1916–1918.
[35] M. Shen, G. Li, B.Z. Lu, *et al.*, *Org. Lett.*, 2004, **6**, 4129–4132.
[36] L. Ackermann and A. Althammer, *Synlett*, 2006, 3125–3129.
[37] Y. Monguchi, S. Mori, S. Aoyagi, *et al.*, *Org. Biomol. Chem.*, 2010, **8**, 3338–3342.
[38] P. Thansandote, D.G. Hulcoop, M. Langer, and M. Lautens, *J. Org. Chem.*, 2009, **74**, 1673–1678.
[39] I. Nakamura, T. Nemoto, N. Shiraiwa, and M. Terada, *Org. Lett.*, 2009, **11**, 1055–1058.
[40] S.E. Denmark and J.D. Baird, *Tetrahedron*, 2009, **65**, 3120–3129.
[41] A. Kondoh, H. Yorimitsu, and K. Oshima, *Org. Lett.*, 2010, **12**, 1476–1479.
[42] R.M. Rao, U. Reddy CH, Alinakhi, *et al.*, *Org. Biomol. Chem.*, 2011, **9**, 3808–3816.
[43] J. Cao, Y. Xu, Y. Kong, *et al.*, *Org. Lett.*, 2012, **14**, 38–41.
[44] O. Leogane and H. Lebel, *Angew. Chem. Int. Ed.*, 2008, **47**, 350–352.
[45] S.P. Breazzano, Y.B. Poudel, and D.L. Boger, *J. Am. Chem. Soc.*, 2013, **135**, 1600–1606.

77

Palladium-Catalyzed Indole Ring Synthesis: Cacchi

The third in the trilogy of palladium-catalyzed indole syntheses involving *o*-alkynylanilines as substrates is the Cacchi synthesis, which is an extension of the Yamanaka–Sakamoto–Sonogashira (Chapter 75) and the Larock (Chapter 76) indole syntheses so as to provide C-3 functionalized 2-substituted indoles [1, 2]. Although this reaction may have been first reported by Utimoto [3], it was Cacchi and his colleagues who molded it into a remarkably powerful indole synthesis. Utimoto succeeded in capturing the 3-indolylpalladium species with allyl chlorides (Scheme 1, equation 1) [3]. Shortly thereafter, Cacchi's group reported a similar ambush of the 3-indolylpalladium intermediate to yield a variety of 2,3-disubstituted indoles (equation 2), including indoles **1–3** [4].

During the following twenty years, Cacchi has improved upon and expanded this indole ring synthesis. Some selected examples are shown in Scheme 2 (equations 1–4) [5–8].

The Cacchi synthesis and closely related indole syntheses involving the Pd-catalyzed cyclization/trapping of 2-alkynylanilines have filled the organic literature, and three recent examples are shown in Scheme 3 (equations 1–3) [9–11]. Moreover, this chemistry has been applied to solid-phase synthesis (Ellingboe [12]), 3-iodoindoles (Barluenga [13], Knight [14], Larock [15], Wang [16], Flynn [17], Li [18]), 3-sulfenylindoles (Zhang [19], Larock [20]), 3-selenylindoles (Larock [20]), one-pot syntheses of 2,3-disubstituted indoles (Lu [21], Larock [22], Marinelli [23]), indoles with *N*-bulky alkyl groups (Ackermann [24]), 2-amidoindoles (Skrydstrup [25]), 3-hydroxymethylindoles (Lu [26]), indole-3-carboxylates (Gabriele [27]), indole-3-boronic esters (Harrity [28]), indole-3-carboxamides (Zhu [29]), polycyclic indoles (Liang [30]), indolylglycines (Sinha [31]), indolo[2,3-*b*]quinolines (Takemoto [32]), indolo[3,2-*c*]isoquinolinones (Zhu [33]), pyrrolo[3,2-*e*]indazoles (Chakrabarty [34]), benzo[*a*]carbazoles (Wu [35]), dibenzo[*a,c*]carbazoles (Goggiamani [36]), indole-3-carbaldehydes (Liu [37]), 3-acylindoles (Liang [38]), and murrapanine alkaloid analogues (Sinha [39]). Not to spoil the party, but Inamoto and colleagues reported the transition-metal-*free* cyclization of 2-ethynylanilines followed by CO_2 fixation to give indole-3-carboxylic acids [40].

The Cacchi synthesis of indoles via the Pd-catalyzed cyclization of 2-alkynylanilines and subsequent functionalization has extended enormously the original Yamanaka–Sakamoto–Sonogashira and Larock reactions. For a review of these methods see Beller [41].

Indole Ring Synthesis: From Natural Products to Drug Discovery, First Edition. Gordon W. Gribble.

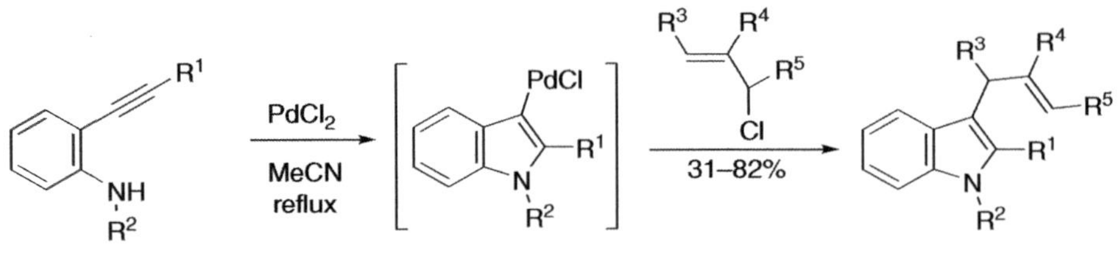

[eqn. 1] [3]

R^1 = *n*-Bu, *t*-Bu, Ph, CH=CHMe
R^2 = H, Ac, CO_2Me
R^3 = H, Me, Cl, CH_2Cl
R^4 = H, Me, Cl, CH_2Cl
R^5 = H, Me, Et, CH_2Cl, CH=CH_2

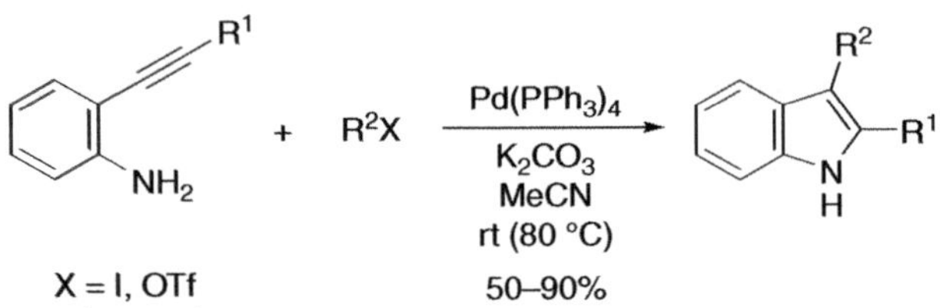

[eqn. 2] [4]

X = I, OTf
R^1 and R^1 = many

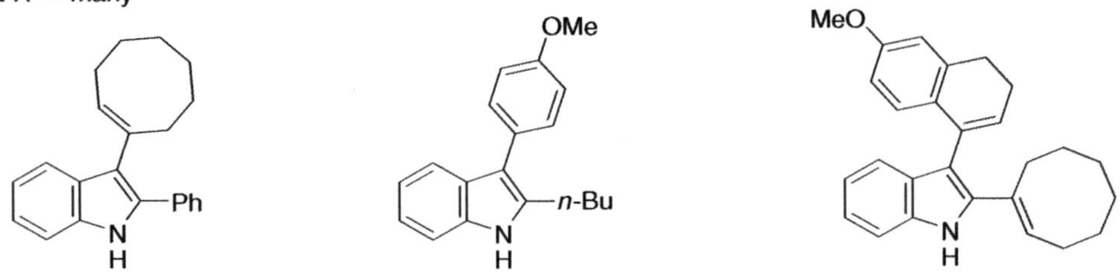

1 (74%) **2** (82%) **3** (89%)

Scheme 1 *Utimoto and Cacchi Indole Syntheses*

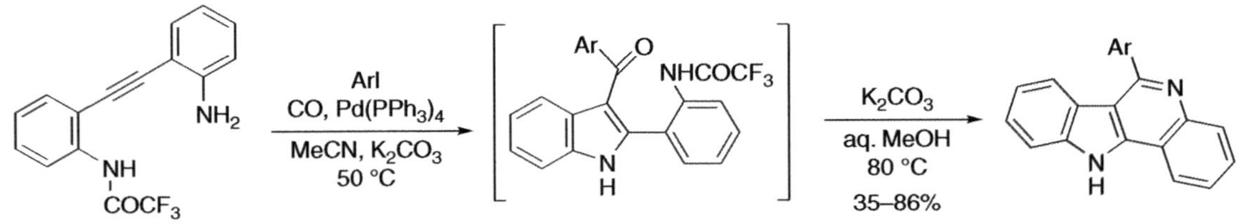

[eqn. 1] [5]

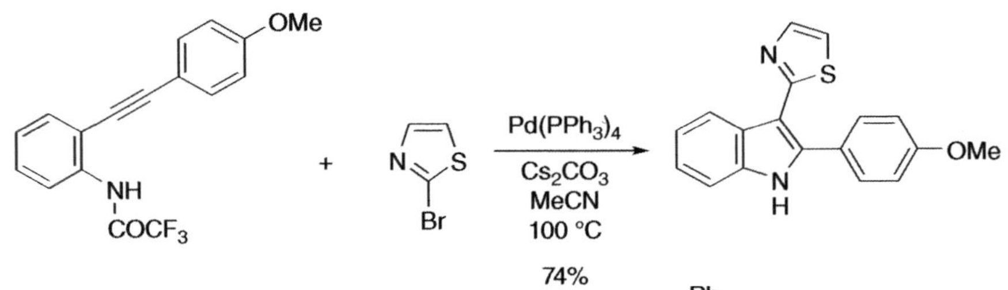

[eqn. 2] [6]

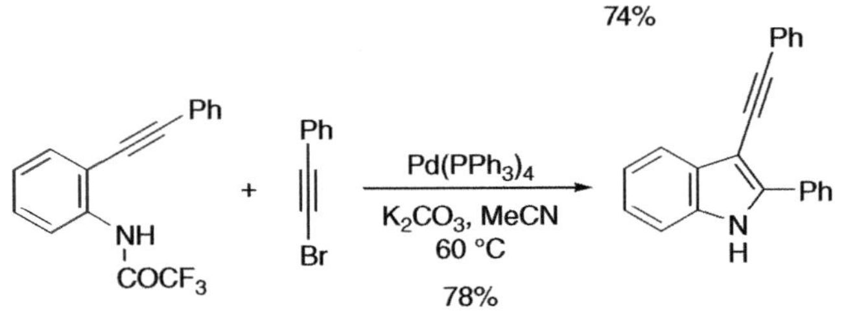

[eqn. 3] [7]

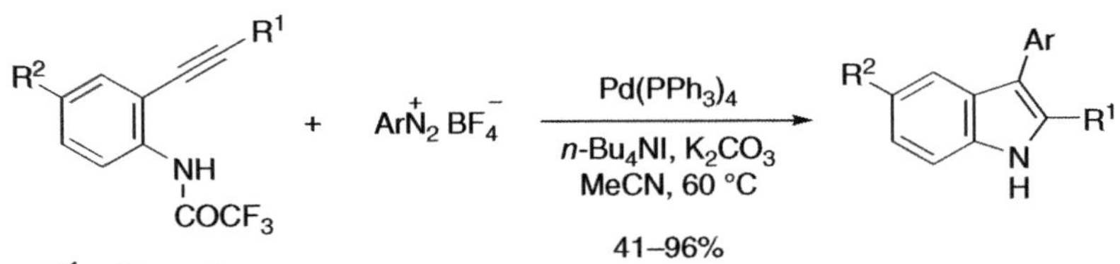

[eqn. 3] [8]

R^1 = Ph, aryl
R^2 = H, F, Me, OMe, Cl

Scheme 2 *Cacchi Indole Synthesis*

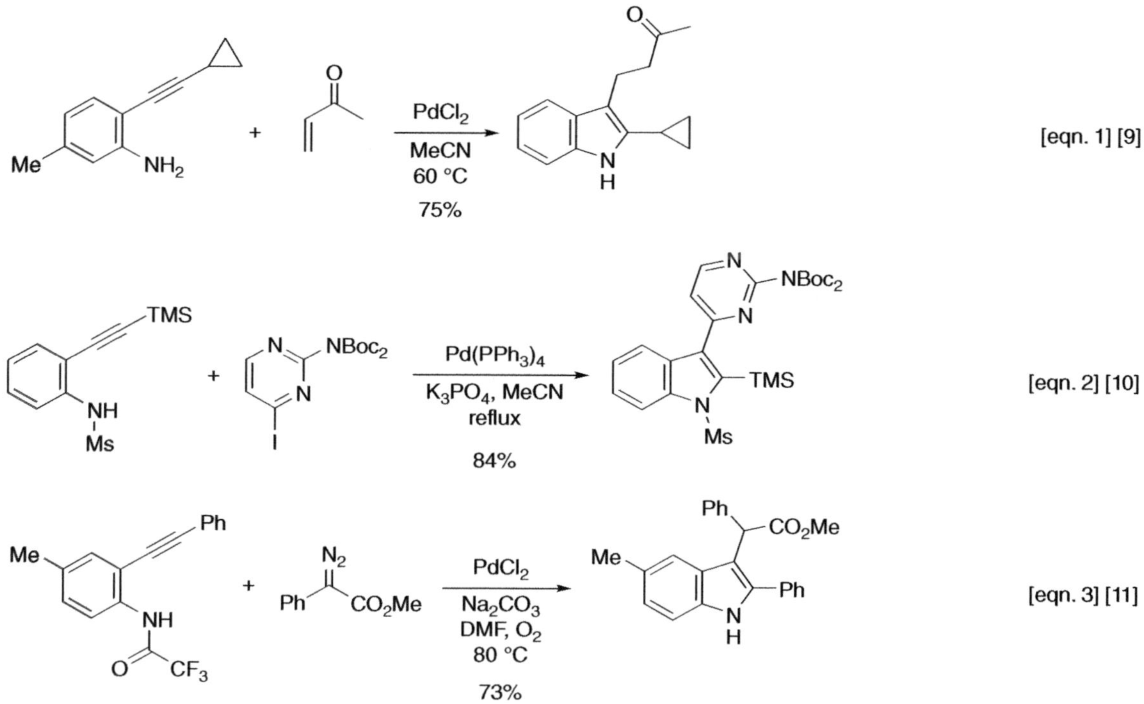

Scheme 3 *Applications of the Cacchi Indole Synthesis*

References

[1] S. Cacchi and G. Fabrizi, *Chem. Rev.*, 2005, **105**, 2873–2920.
[2] S. Cacchi and G. Fabrizi, *Chem. Rev.*, 2011, **111**, PR215–PR283.
[3] K. Iritani, S. Matsubara and K. Utimoto, *Tetrahedron Lett.*, 1988, **29**, 1799–1802.
[4] A. Arcadi, S. Cacchi, and F. Marinelli, *Tetrahedron Lett.*, 1992, **33**, 3915–3918.
[5] S. Cacchi, G. Fabrizi, P. Pace, and F. Marinelli, *Synlett*, 1999, 620–622.
[6] G. Battistuzzi, S. Cacchi, and G. Fabrizi, *Eur. J. Org. Chem.*, 2002, 2671–2681.
[7] A. Arcadi, S. Cacchi, G. Fabrizi, *et al.*, *J. Org. Chem.*, 2005, **70**, 6213–6217.
[8] S. Cacchi, G. Fabrizi, A. Goggiamani, *et al.*, *Org. Lett.*, 2010, **12**, 3279–3281.
[9] D. Janreddy, V. Kavala, C.-W. Kuo, *et al.*, *Tetrahedron*, 2013, **69**, 3323–3330.
[10] S.R. Walker, M.L. Czyz, and J.C. Morris, *Org. Lett.*, 2014, **16**, 708–711.
[11] Z. Hu, S. Luo, and Q. Zhu, *Adv. Synth. Catal.*, 2015, **357**, 1060–1064.
[12] M.D. Collini and J.W. Ellingboe, *Tetrahedron Lett.*, 1997, **38**, 7963–7966.
[13] J. Barluenga, M. Trincado, E. Rubio, and J.M. González, *Angew. Chem. Int. Ed.*, 2003, **42**, 2406–2409.
[14] M. Amjad and D.W. Knight, *Tetrahedron Lett.*, 2004, **45**, 539–541.
[15] D. Yue, T. Yao, and R.C. Larock, *J. Org. Chem.*, 2006, **71**, 62–69.
[16] S. Tang, Y.-X. Xie, J.-H. Li, and N.-X. Wang, *Synthesis*, 2007, 1841–1847.
[17] R. Halim, P.J. Scammells, and B.L. Flynn, *Org. Lett.*, 2008, **10**, 1967–1970.
[18] H.-P. Zhang, S.-C. Yu, Y. Liang, *et al.*, *Synlett*, 2011, 982–988.
[19] Y.-J. Guo, R.-Y. Tang, J.-H. Li, *et al.*, *Adv. Synth. Catal.*, 2009, **351**, 2615–2618.
[20] Y. Chen, C.-H. Cho, F. Shi, and R.C. Larock, *J. Org. Chem.*, 2009, **74**, 6802–6811.
[21] B.Z. Lu, H.-X. Wei, Y. Zhang, *et al.*, *J. Org. Chem.*, 2013, **78**, 4558–4562.
[22] Y. Chen, N.A. Markina, and R.C. Larock, *Tetrahedron*, 2009, **65**, 8908–8915.
[23] A. Arcadi, M. Chiarini, F. Marinelli, and S. Picchini, *Synthesis*, 2011, 4084–4090.
[24] L. Ackermann, R. Sandmann, M. Schinkel, and M.V. Kondrashov, *Tetrahedron*, 2009, **65**, 8930–8939.
[25] K. Dooleweerdt, T. Ruhland, and T. Skrydstrup, *Org. Lett.*, 2009, **11**, 221–224.
[26] X. Han and X. Lu, *Org. Lett.*, 2010, **12**, 3336–3339.
[27] B. Gabriele, L. Veltri, R. Mancuso, *et al.*, *Eur. J. Org. Chem.*, 2012, 2549–2559.
[28] J. Huang, S.J.F. Macdonald, and J.P.A. Harrity, *Chem. Commun.*, 2010, **46**, 8770–8772.
[29] Z. Hu, D. Liang, J. Zhao, *et al.*, *Chem. Commun.*, 2012, **48**, 7371–7373.

[30] X.-F. Xia, N. Wang, L.-L. Zhang, *et al.*, *J. Org. Chem.*, 2012, **77**, 9163–9170.
[31] K. Goswami, I. Duttagupta, and S. Sinha, *J. Org. Chem.*, 2012, **77**, 7081–7085.
[32] H. Takeda, T. Ishida, and Y. Takemoto, *Chem. Lett.*, 2009, **38**, 772–773.
[33] B. Yao, Q. Wang, and J. Zhu, *Angew. Chem. Int. Ed*, 2012, **51**, 5170–5174.
[34] S.K. Barik, M. Rakshit, G.K. Kar, and M. Chakrabarty, *ARKIVOC*, 2014, **v**, 1–15.
[35] C.-C. Chen, L.-Y. Chin, S.-C. Yang, and M.-J. Wu, *Org. Lett.*, 2010, **12**, 5652–5655.
[36] S. Cacchi, G. Fabrizi, A. Goggiamani, and A. Iazzetti, *Org. Biomol. Chem.*, 2012, **10**, 9142–9147.
[37] F. Zhao, D. Zhang, Y. Nian, *et al.*, *Org. Lett.*, 2014, **16**, 5124–5127.
[38] X.-F. Xia, L.-L. Zhang, X.-R. Song, *et al.*, *Chem. Commun.*, 2013, **49**, 1410–1412.
[39] A. Chakraborty and S. Sinha, *Tetrahedron Lett.*, 2011, **52**, 6635–6638.
[40] K. Inamoto, N. Asano, Y. Nakamura, *et al.*, *Org. Lett.*, 2012, **14**, 2622–2625.
[41] K. Krüger, A. Tillack, and M. Beller, *Adv. Synth. Catal.*, 2008, **350**, 2153–2167.

78

Palladium-Catalyzed Indole Ring Synthesis: Buchwald–Hartwig

The powerful Buchwald–Hartwig amination of *N*-aryl halides [1–5] has been employed to craft indolines [6] but only rarely to the (direct) synthesis of indoles. A notable exception is the work of Willis and coworkers (Scheme 1, equations 1–3) [7–9], which featured the preparation of indoles containing sterically demanding *N*-alkyl groups and a synthesis of demethylasterriquinone A1 [9, 10]. Xi and colleagues employed a Willis synthesis to prepare benzosilolo[2,3-*b*]indoles [11], and Li and colleagues similarly synthesized 2-trifluoromethylindoles following a Willis strategy [12].

The synthesis of carbazoles from suitably substituted 2,2′-biphenyls has been quite popular (Scheme 2, equations 1 and 2) [13–15]. Nozaki extended his carbazole synthesis (equation 1) to a synthesis of mukonine [13]. Chida and colleagues used a Nozaki carbazole synthesis to prepare (±)-murrayazoline [16], and Konakahara employed the Nozaki method to prepare the carbazole aldehyde that was a key intermediate for the synthesis of ellipticine (equation 3) [17]. Detert and Letessier synthesized *N*-δ-carbolines, including quindoline, via the reaction of benzopyridoiodolium salts with primary amines [18], and Langer synthesized benzothieno[3,2-*b*]indoles and thieno[3,2-*b*:4,5-*b*'] diindoles from the requisite bromothiophenes and primary amines using Buchwald–Hartwig amination [19]. Similarly, a double amination was reported by Wong's group to construct indolo[2,3-*b*]carbazoles from the appropriate diamino dichloro *m*-terphenyls [20]. Kan and coworkers described a one-pot Pd-catalyzed carbazole synthesis from 2-bromophenylboronic acid and 2-iodoaniline (71%) [21].

Hartwig and Tan reported a simple indole synthesis from an oxime acetate involving C-H functionalization (Scheme 3, equation 1) [22]. Buchwald described an excellent versatile carbazole synthesis that also involved C-H activation (equation 2) [23], as did Youn [24], Antonchick [25], Gault [26], and Shi [27], who synthesized 4-deoxycarbazomycin B (equation 3).

The Buchwald–Hartwig Pd-catalyzed amination methodology was used to synthesize 2-amidoindoles (Zhao [28]), 1,2-disubstituted indoles in one pot from *o*-chloro-substituted 1-phenyl-2-(aminoalkyl)alkynes (Doye [29]), staurosporine analogues (Snyder [30]), lavendamycin (Boger [31]), 3-arylindole-2-carboxylates (Queiroz [32]), functionalized indolines (Lautens [33]), 3-methylindoles (Baxter [34]), tetrahydro-β-carboline alkaloids (Orito [35]), one-pot 2-substituted indole synthesis from *o*-halochlorobenzene (Stradiotto [36]), 2-methylindoles from *o*-bromo(propa-1,2-dien-1-yl)arenes (Bräse [37]), and the use of 2*H*-azirines with aryl iodides to give indoles (Lautens [38]). Unfortunately, space does permit a full presentation of these elegant Pd-catalyzed indole ring syntheses.

Indole Ring Synthesis: From Natural Products to Drug Discovery, First Edition. Gordon W. Gribble.

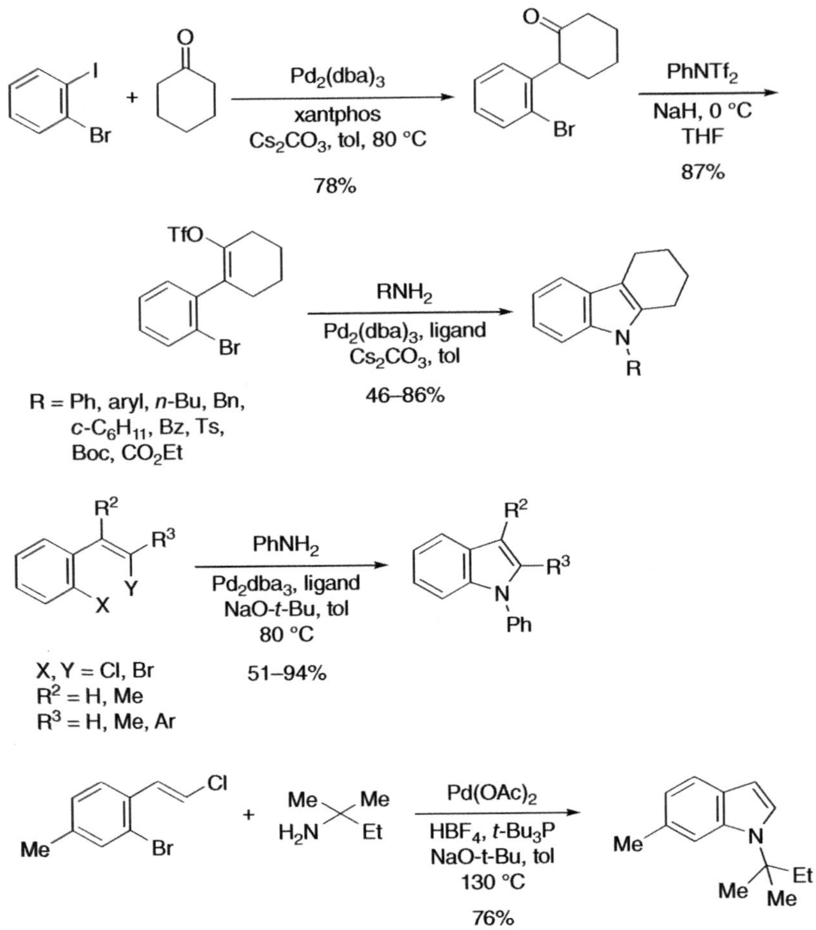

[eqn. 1] [7]

[eqn. 2] [8]

[eqn. 3] [9]

Scheme 1 *Willis Indole Synthesis*

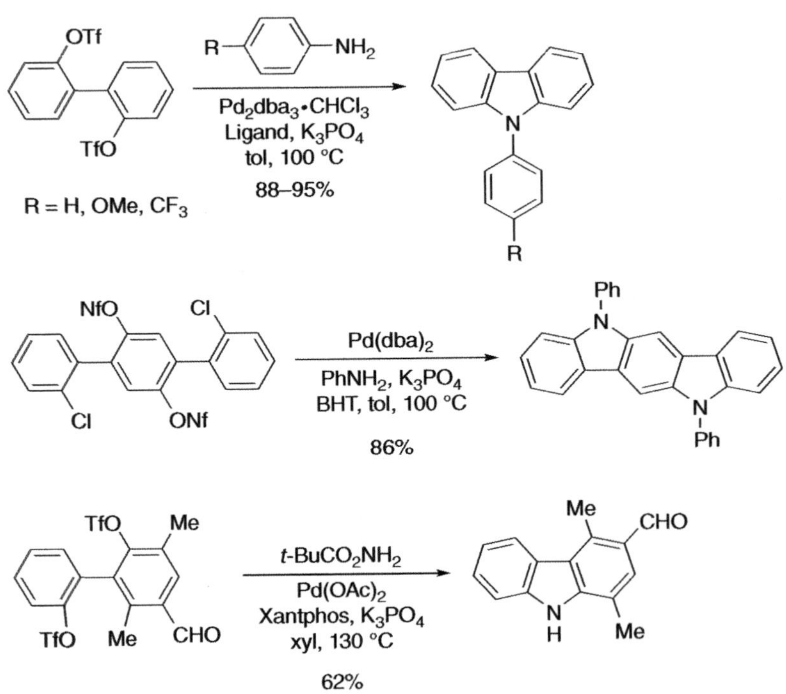

[eqn. 1] [13]

[eqn. 2] [14]

[eqn. 3] [17]

Scheme 2 *Nozaki Synthesis of Carbazoles*

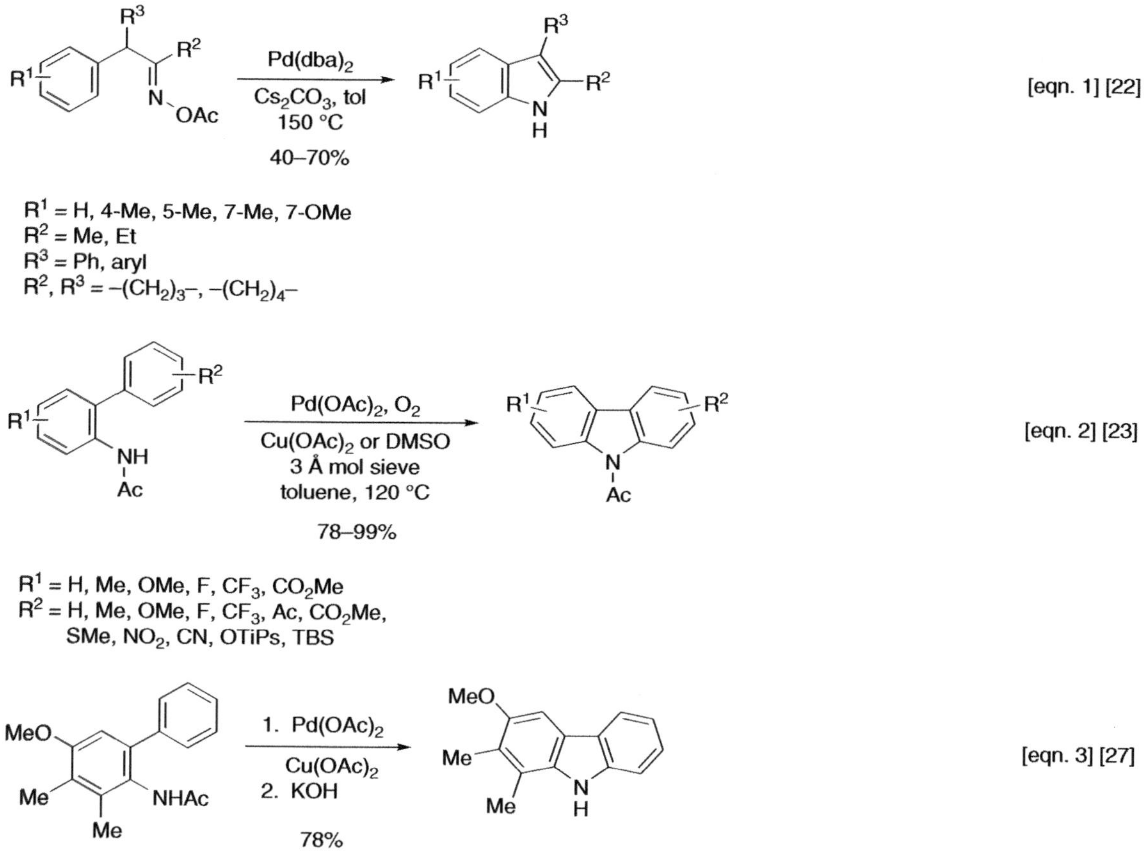

Scheme 3 *Applications of the Buchwald-Hartwig Amination Involving C-H Activation*

References

[1] J.P. Wolfe, S. Wagaw, J.-F. Marcoux, and S.L. Buchwald, *Acc. Chem. Res.*, 1998, **31**, 805–818.
[2] J.F. Hartwig, *Synlett*, 1997, 329–340.
[3] A.R. Muci and S.L. Buchwald, *Top. Curr. Chem.*, 2002, **219**, 131–209.
[4] J.F. Hartwig, *Acc. Chem. Res.*, 2008, **41**, 1534–1544.
[5] D.S. Surry and S.L. Buchwald, *Chem. Sci.*, 2011, **2**, 27–50.
[6] G.W. Gribble (2007) *Palladium in Heterocyclic Chemistry* (eds J.J. Li and G.W. Gribble), Elsevier, Amsterdam, pp. 161–163.
[7] M.C. Willis, G.N. Brace, and I.P. Holmes, *Angew. Chem. Int. Ed.*, 2005, **44**, 403–406.
[8] M.C. Willis, G.N. Brace, T.J.K. Findlay, and I.P. Holmes, *Adv. Synth. Catal.*, 2006, **348**, 851–856.
[9] A.J. Fletcher, M.N. Bax, and M.C. Willis, *Chem. Comm.*, 2007, 4764–4766.
[10] L.C. Henderson, M.J. Lindon, and M.C. Willis, *Tetrahedron*, 2010, **66**, 6632–6638.
[11] Y. Liang, S. Zhang, and Z. Xi, *J. Am. Chem. Soc.*, 2011 **133**, 9204–9207.
[12] S.-X. Dong, X.-G. Zhang, Q. Liu, *et al.*, *Synthesis*, 2010, 1521–1525.
[13] A. Kuwahara, K. Nakano, and K. Nozaki, *J. Org. Chem.*, 2005, **70**, 413–419.
[14] K. Kawaguchi, K. Nakano, and K. Nozaki, *J. Org. Chem.*, 2007, **72**, 5119–5128.
[15] K. Kawaguchi, K. Nakano, and K. Nozaki, *Org. Lett.*, 2008, **10**, 1199–1202.
[16] A. Ueno, T. Kitawaki, and N. Chida, *Org. Lett.*, 2008, **10**, 1999–2002.
[17] T. Konakahara, Y.B. Kiran, Y. Okuno, *et al.*, *Tetrahedron Lett.*, 2010, **51**, 2335–2338.
[18] J. Letessier and H. Detert, *Synthesis*, 2012, **44**, 290–296.
[19] T.Q. Hung, T.T. Dang, A. Villinger, *et al.*, *Org. Biomol. Chem.*, 2012, **10**, 9041–9044.
[20] J.-Y. Su, C.-Y. Lo, C.-H. Tsai, *et al.*, *Org. Lett.*, 2014, **16**, 3176–3179.
[21] Y. Kitamura, S. Yoshikawa, T. Furuta, and T. Kan, *Synlett*, 2008, 377–380.
[22] Y. Tan and J.F. Hartwig, *J. Am. Chem. Soc.*, 2010, **132**, 3676–3677.
[23] W.C.P. Tsang, R.H. Munday, G. Brasche, *et al.*, *J. Org. Chem.*, 2008, **73**, 7603–7610.
[24] S.W. Youn, J.H. Bihn, and B.S. Kim, *Org. Lett.*, 2011, **13**, 3738–3741.

[25] A.P. Antonchick, R. Samanta, K. Kulikov, and J. Lategahn, *Angew. Chem. Int. Ed.*, 2011, **50**, 8605–8608.
[26] J.A. Jordan-Hore, C.C.C. Johansson, M. Gulias, *et al.*, *J. Am. Chem. Soc.*, 2008, **130**, 16184–16186.
[27] B.-J. Li, S.-L. Tian, Z. Fang, and Z.-J. Shi, *Angew. Chem. Int. Ed.*, 2008, **47**, 1115–1118.
[28] P.-Y. Yao, Y. Zhang, R.P. Hsung, and K. Zhao, *Org. Lett.*, 2008, **10**, 4275–4278.
[29] H. Siebeneicher, I. Bytschkov, and S. Doye, *Angew. Chem. Int. Ed.*, 2003, **42**, 3042–3044.
[30] R. Nomak and J.K. Snyder, *Tetrahedron Lett.*, 2001, **42**, 7929–7933.
[31] D.L. Boger, S.R. Duff, J.S. Panek, and M. Yasuda, *J. Org. Chem.*, 1985, **50**, 5782–5789.
[32] M.-J.R.P. Queiroz, A.S. Abreu, E.M.S. Castanheira, and P.M.T. Ferreira, *Tetrahedron*, 2007, **63**, 2215–2222.
[33] P. Thansandote, M. Raemy, A. Rudolph, and M. Lautens, *Org. Lett.*, 2007, **9**, 5255–5258.
[34] C.A. Baxter, E. Cleator, M. Alam, *et al.*, *Org. Lett.*, 2010, **12**, 668–671.
[35] R. Harada, N. Nishida, S. Uchiito, *et al.*, *Eur. J. Org. Chem.*, 2012, 366–379.
[36] N.L. Rotta-Loria, A. Borzenko, P.G. Alsabeh, *et al.*, *Adv. Synth. Catal.*, 2015, **357**, 100–106.
[37] K.-S. Masters, M. Wallesch, and S. Bräse, *J. Org. Chem.*, 2011, **76**, 9060–9067.
[38] D.A. Candito and M. Lautens, *Org. Lett.*, 2010, **12**, 3312–3315.

79

Palladium-Catalyzed Indole Ring Synthesis: Miscellaneous

This final palladium chapter is a collection of relatively new approaches to indoles and carbazoles that do not precisely fit into the previous chapters, and in some instances the chemistry presented here was omitted from earlier coverage.

Ackermann and Althammer reported a carbazole/α-carboline synthesis from the domino amination/aryl-Heck reaction between anilines and 1,2-dihalobenzenes and a 2,3-dichloropyridine (Scheme 1, equation 1) [1]. *N*-unsubstituted and ring-substituted carbazoles were also prepared as was *N*-phenylcyclopent[*b*]indole and the carbazole alkaloid murrayafoline A. Bisseret [2] and Lautens [3] independently discovered a Pd-catalyzed tandem amination/Suzuki coupling of *o*-*gem*-dihalovinylanilines (equation 2) [3]. Lautens made extensive use of this *gem*-dibromovinyl functionality to prepare imidazoindolones [4], 2-vinylindoles [5], 2-alkynylindoles [6], an indol-2-yl quinolin-2-one and KDR kinase inhibitor [7], polycyclic heteroaromatics [8], 2-bromoindoles [9], and highly functionalized indoles [10]. Other investigators adopted this strategy for the synthesis of indole-2-carboxylates (Alper [11]), 2-aroylindoles (Florent and Pontikis [12]), indole-2-azoles (Wang [13] and Lan [14]), 2,2′-biindolyls (Bao [15]), heteroaryl-fused indoles (Bao [16, 17]), 2-alkynylindoles (Xi [18]), and the alkaloids cladoniamide G and F (Koert [19]). Hultin and Geary reported a one-pot Suzuki/aryl-Heck process to afford 2-substituted indoles (equation 3) [20]. Witulski and colleagues described a new Pd-catalyzed 2-aminoindole synthesis involving 2-alkynyl-2-haloanilides reacting with primary or secondary amines (equation 4) [21]. Several different amines were used (e.g., cyclopropyl amine, *N*-methylpiperazine, allyl amine, anilines, diethyl amine). Watanabe and colleagues discovered a synthesis of 1-aminoindoles via a Pd-catalyzed amination of arylchloride hydrazones (equation 5) [22]. Barluenga's group effected an indole synthesis by the Pd-catalyzed coupling of *o*-dihaloarenes or *o*-halofluorosulfonates and imines (equation 6) [23, 24]. A very large number of ring-substituted indoles were synthesized in this powerful C-C/C-N bond-forming sequence.

Mukai and coworkers described a 2,3-disubstituted indole synthesis via a Stille coupling between *N*-acyl-2-iodoanilines and 1-(tributylstannyl)-1-substituted allenes (Scheme 2, equation 1) [25]. The method was used to generate indole-2,3-quinodimethanes [26] and to synthesize the indole alkaloid (−)-goniomitine and its unnatural antipode [27]. Fuwa and Sasaki published similar chemistry (equation 2) [28] including the generation of indole-2,3-quinodimethanes [29]. Hiroi and colleagues reported the intramolecular carbopalladation of allenes followed by amination to give indoles [30], and Deagostino's team prepared 3-vinylindoles from *o*-iodoanilines and allenes (equation 3) [31]. Xu and colleagues used the combination of *o*-haloarylallenes and primary amines to access a range of functionalized indoles (equation 4) [32].

Previous chapters presented examples of Pd-catalyzed carbon–hydrogen activation as a route to indole ring formation, and several new examples are presented here. Lu and coworkers effected C-H activation of *N*-arylamides and subsequent coupling with alkynes to give 2,3-disubstituted indoles (Scheme 3, equation 1) [33]. Similar indoles were synthesized by Cabrera and colleagues from a Pd-catalyzed direct combination of anilines and α-diketones (equation 2) [34]. With 2,3-hexanedione, only 2-*n*-Pr-3-methylindole was obtained (95%), and 4-methoxyaniline afforded 2,3-diphenyl-5-methoxyindole in lower yield

Indole Ring Synthesis: From Natural Products to Drug Discovery, First Edition. Gordon W. Gribble.

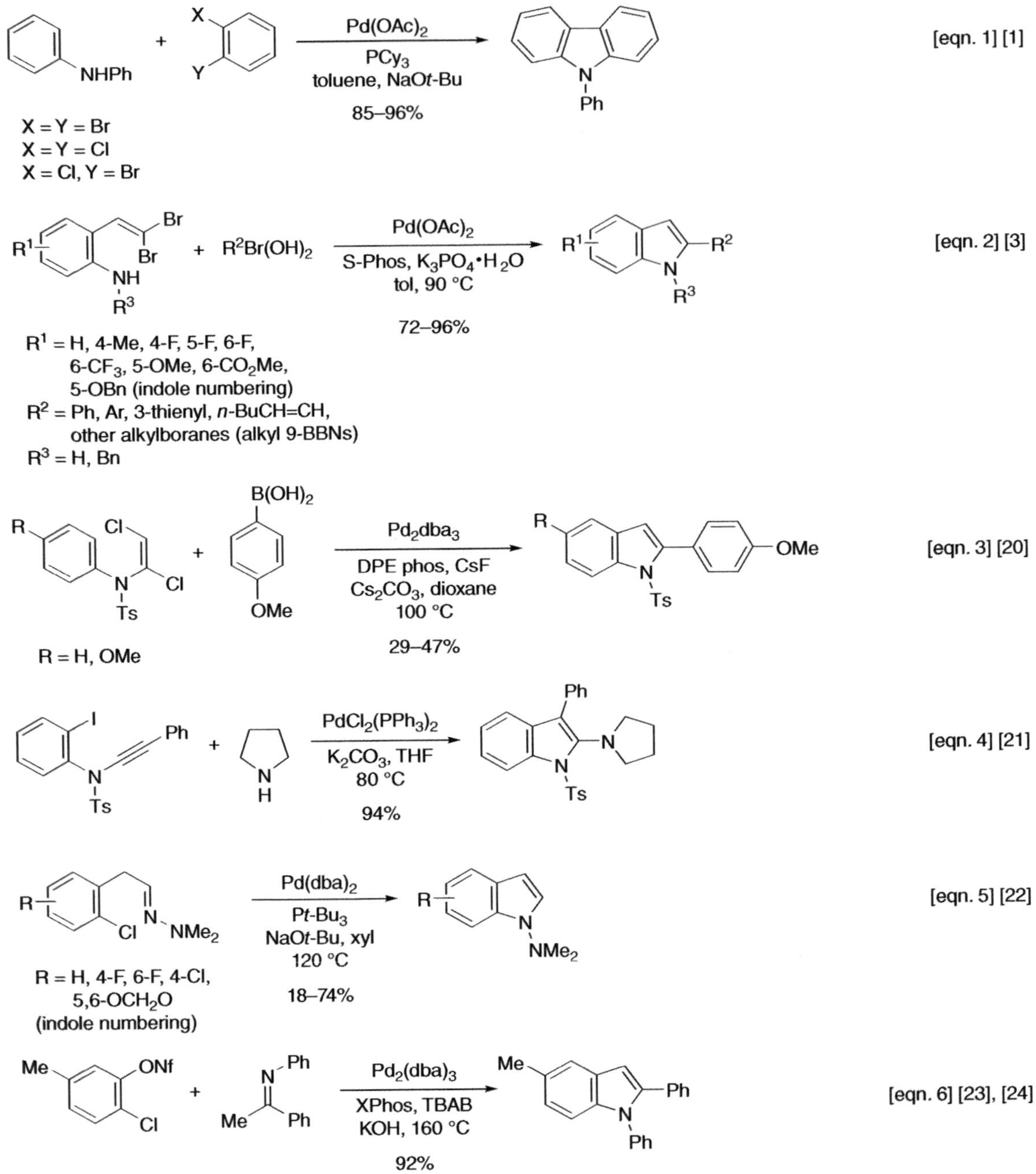

Scheme 1 *Miscellaneous Palladium-Catalyzed Indole Syntheses – 1*

(40%). Inamoto and Doi used C-H activation/cyclization of *N*-tosylenamines to prepare 3-arylindoles (equation 3) [35]. The substrates were prepared from symmetrically substituted benzophenones. Chiba's team employed *O*-acyloximes to effect a synthesis of indole-3-carboxamides (equation 4) [36]. The groups of Queiroz [37] and Thasana [38] employed C-H activation/C-C bond formation to prepare complex indoles (equations 5 and 6).

The direct coupling of diarylamines and alkenes (e.g., styrenes) with $Pd(OAc)_2$ was described by Maiti and colleagues (Scheme 4, equation 1) [39], and Wu and coworkers reported a one-pot sequential Beckmann rearrangement, cyclization, and indole C-3 chlorination to afford *N*-acylindoles (equation 2) [40]. Gong, Wu, and colleagues synthesized 2-trifluoromethyl-3-vinylindoles via a domino carbopalladation C-H activation protocol

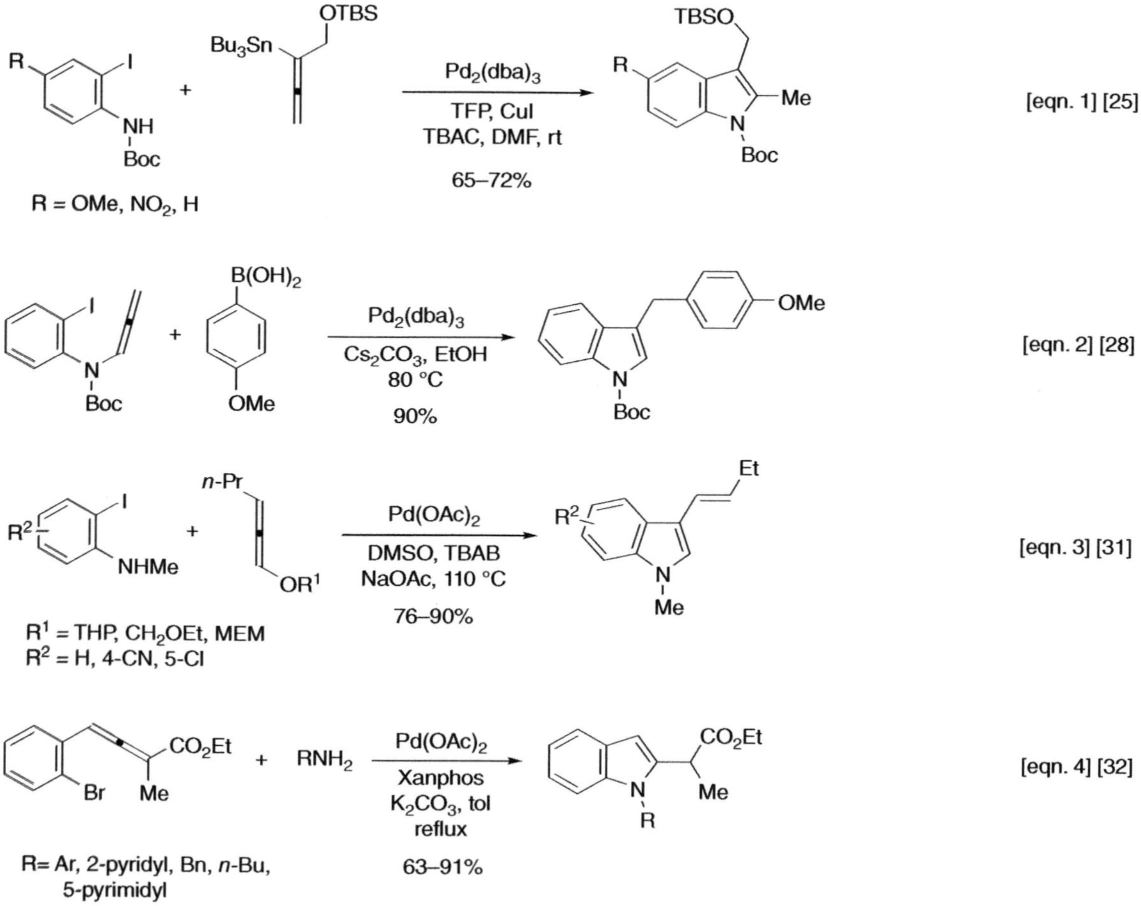

Scheme 2 *Palladium-Catalyzed Allene Heteroannulation*

(equation 3) [41]. The major thrust of this nice chemistry was to prepare 3-methylene-3*H*-indoles using aryl iodides in place of *n*-butyl iodide. These workers earlier used *o*-haloarylalkynylimines to achieve the synthesis of 2-trimethyl-3-aroylindoles with Pd(0) [42]. Yamamoto and coworkers described three new indole syntheses involving the Pd-catalyzed reactions of 2-(1-alkynyl)-*N*-alkylideneanilines (equation 4) [43], *o*-alkynylisocyanoarenes (equation 5) [44], and *o*-alkynylphenylisocyanates [45] to give indoles. Both Takemoto and Takahashi synthesized indoles via Pd-catalyzed coupling between halides and *o*-alkynylaryl isocyanides [46, 47] or *o*-alkenylaryl isocyanides [48].

A series of indole-2-acetates was synthesized by Gabriele and colleagues from the Pd-catalyzed carbonylation of 1-(2-aminoaryl)-2-yl-1-ols (Scheme 5, equation 1) [49]. The substrates were prepared by Grignard addition to *o*-aminoacetophenone. Both Chowdhury [50] and Reddy [51] described similar chemistry leading to 2-substituted indoles. Wang's group reported a new synthesis of 3-vinylindoles from a Pd-catalyzed coupling/cyclization of *N*-acetyl-*N*-(3-phenylprop-2-ynyl)-2-iodoaniline and *N*-tosylhydrazones (equation 2) [52]. The reaction can also be executed in one pot from arylhalide, aldehyde, and tosyl hydrazine. A novel 2-arylindole synthesis that was discovered by Rashinkar entails the Pd-catalyzed reaction of *o*-nitrobenzyl cyanides with arylboronic acids (equation 3) [53]. Tobisu and Chatani accomplished the direct Pd-catalyzed alkynylation of anilides leading to indoles (equation 4) [54]. Lee and colleagues induced nitriles into becoming indole-3-carboxylates through the Blaise reaction (equation 5) [55].

Ishar and coworkers discovered a spectacular indole synthesis involving a Diels–Alder reaction between azadienes and a silyl enol ether, followed by an aryl-Heck cyclization (Scheme 6, equation 1) [56]. Jiao's team developed a Pd-catalyzed intramolecular amination and C-H functionalization to give 2,3-dihydropyrrolo[1,2-*a*]indoles (equation 2) [57].

The final section of Pd-catalyzed indole synthesis is the simple annulation between *o*-haloanilines and aldehydes or ketones, which leads to the formation of enamines and thence to an aryl-Heck cyclization. A pioneering study was

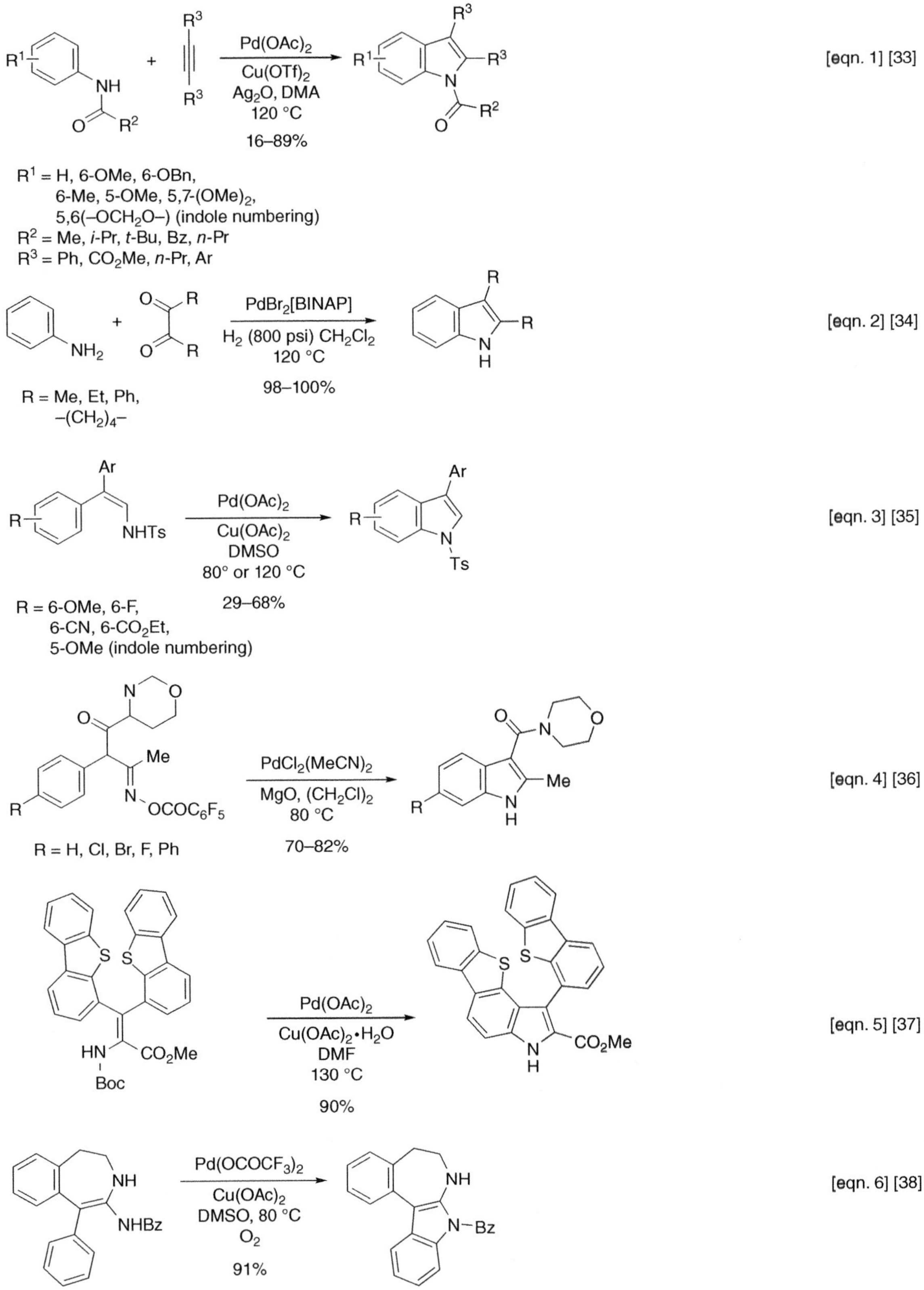

Scheme 3 *Palladium-Catalyzed Indole Syntheses via Carbon-Hydrogen Activation*

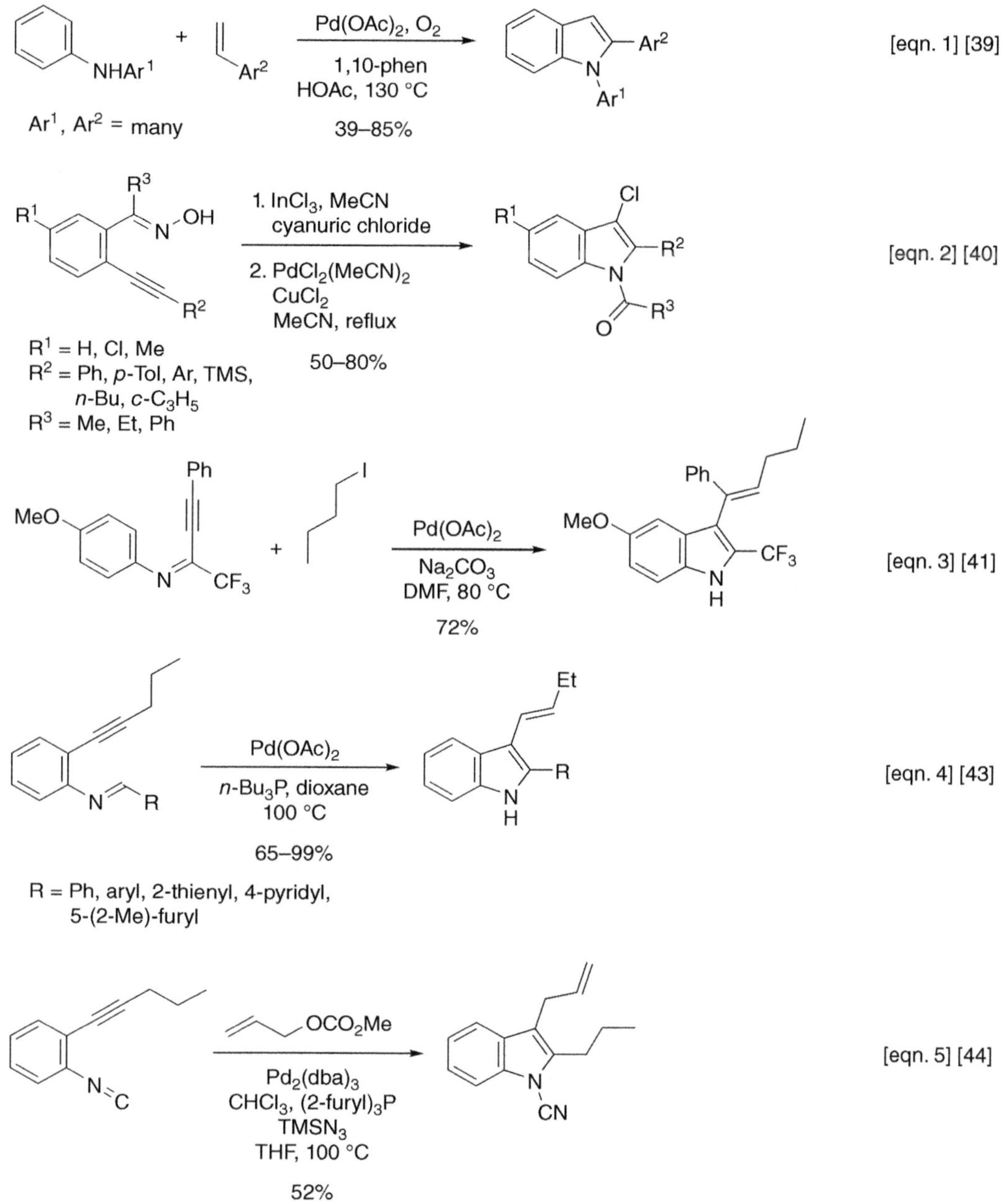

Scheme 4 *Palladium-Catalyzed Indole Syntheses via Carbon-Hydrogen Activation – 2*

that of Chen and colleagues (Scheme 7, equation 1) [58]. Many ketones underwent this simple indolization. Cho and Shim found that aldehydes and *o*-iodoaniline form indoles in modest yield (equation 2) [59]. Of the aldehydes examined, only 3,3-dimethylbutyraldehyde failed to react. Zhu and Jia described similar reactions with aldehydes to afford 3-substituted indoles [60–62], especially for the synthesis of tryptamine analogues (equation 3) [62]. Kurth's group crafted the reaction of *o*-bromoiodobenzene, benzyl amine, and cyclopentanone into a one-pot Pd-catalyzed synthesis of cyclopent[*b*]indoles, among others [63].

The true value of this simple enamine aryl-Heck indolization method is witnessed by the syntheses of prostaglandin D2 receptor antagonists (Shafiee [64] and Sturino [65]), stephacidin A (Baran [66]), avrainvillamide and the stephacidins (Baran [67]), the DEFG ring of complestatin and chloropeptin I (Zhu [68]), (+)-suaveolindole (Danishefsky [69]), celogentin C residue (Campagne and Michaux [70]), (–)-*cis*-clavicipitic acid (Jian [71]), duocarmycin SA analogues (Boger [72]), chloptosin (Yao [73]), (–)-indolactam V (Jia [74]), alstoscholarin (Neuville and Zhu [75]), and several maremycins (Jia [76]).

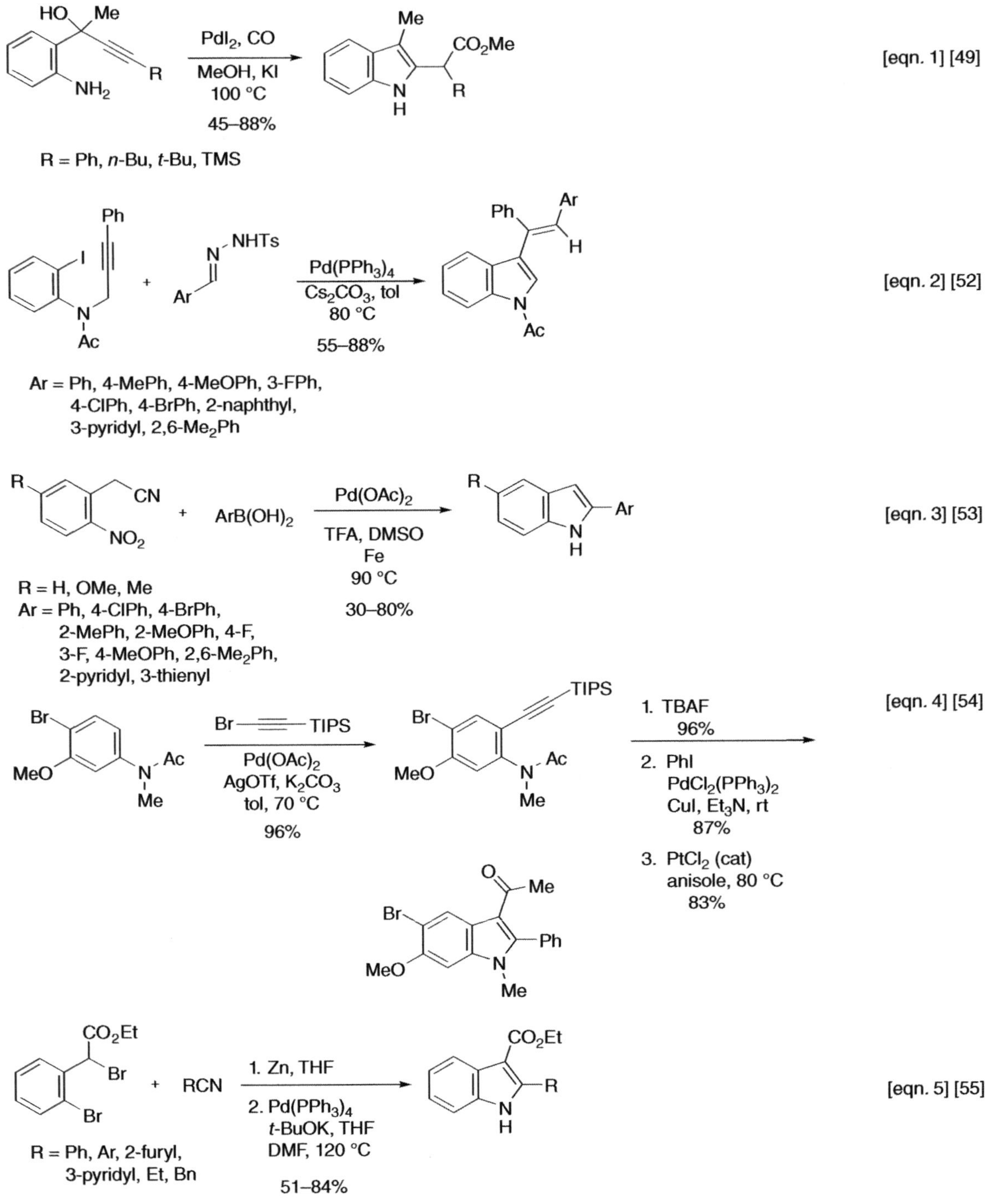

Scheme 5 *Miscellaneous Palladium-Catalyzed Indole Synthesis – 2*

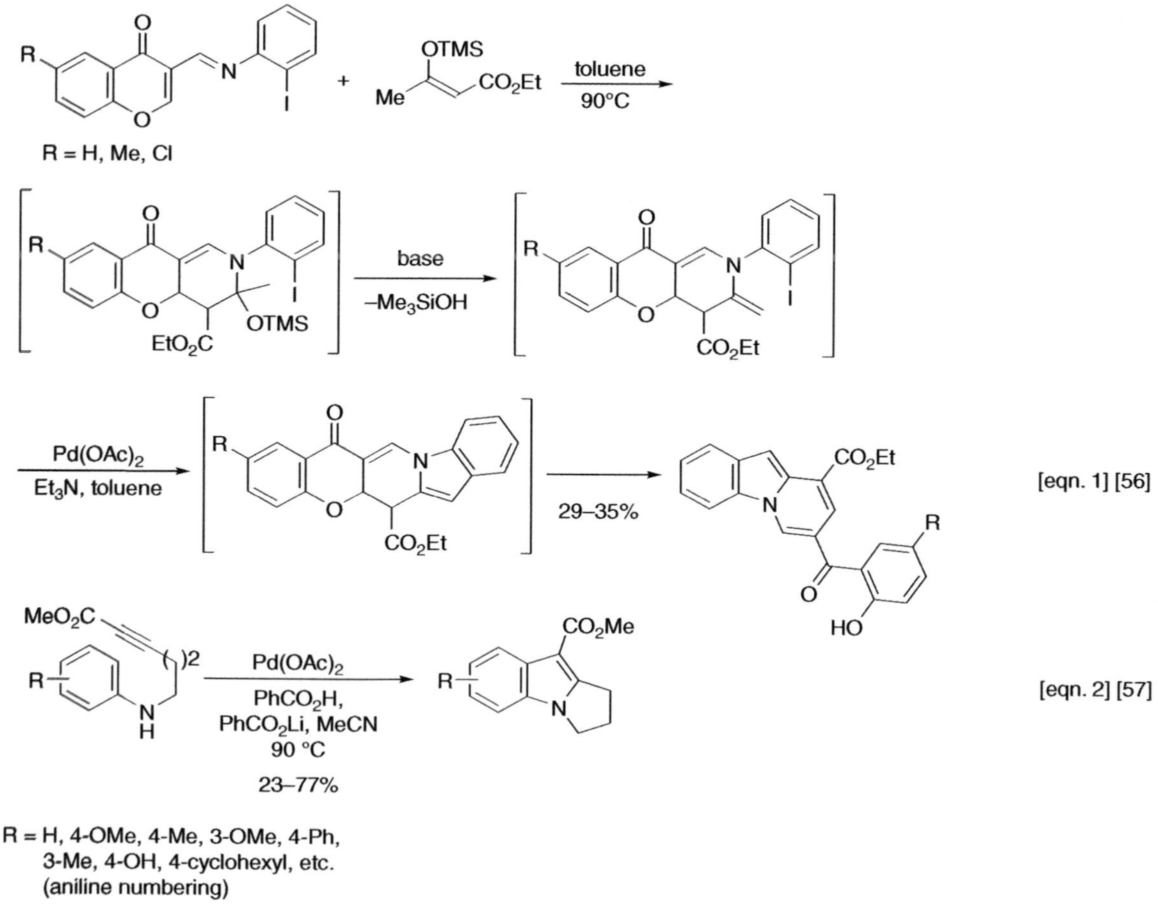

Scheme 6 *Miscellaneous Palladium-Catalyzed Indole Synthesis – 3*

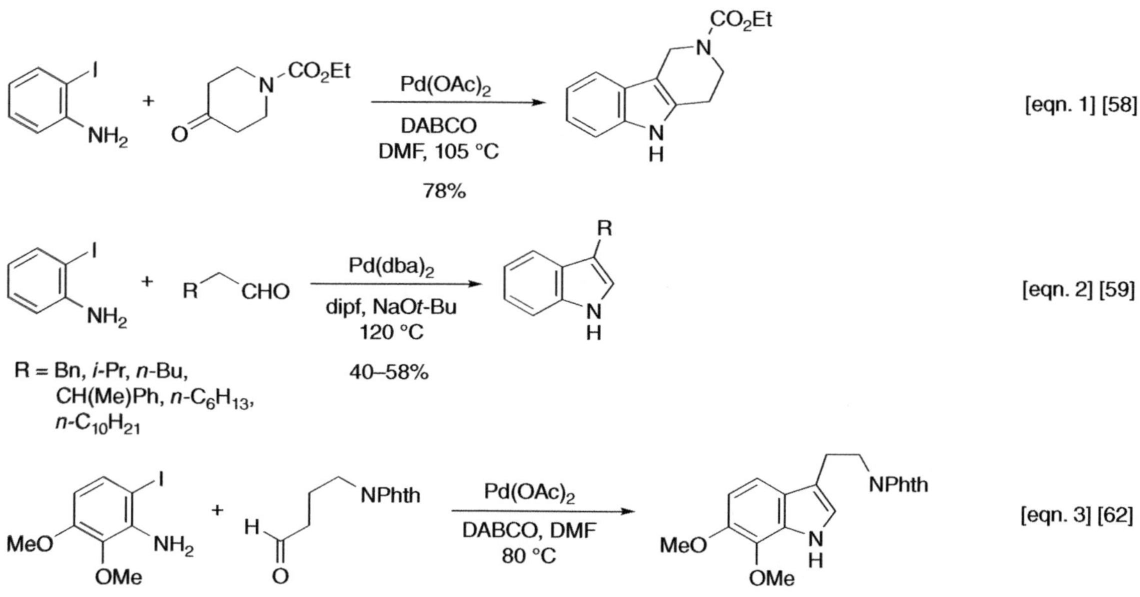

Scheme 7 *Palladium-Catalyzed Aryl-Heck Indole Syntheses*

References

[1] L. Ackermann and A. Althammer, *Angew. Chem. Int. Ed.*, 2007, **46**, 1627–1629.
[2] S. Thielges, E. Meddah, P. Bisseret, and J. Eustache, *Tetrahedron Lett.*, 2004, **45**, 907–910.
[3] Y.-Q. Fang and M. Lautens, *Org. Lett.*, 2005, **7**, 3549–3552.
[4] J. Yuen, Y.-Q. Fang, and M. Lautens, *Org. Lett.*, 2006, **8**, 653–656.
[5] A. Fayol, Y.-Q. Fang, and M. Lautens, *Org. Lett.*, 2006, **8**, 4203–4206.
[6] M. Nagamochi, Y.-Q. Fang, and M. Lautens, *Org. Lett.*, 2007, **9**, 2955–2958.
[7] Y.-Q. Fang, R. Karisch, and M. Lautens, *J. Org. Chem.*, 2007, **72**, 1341–1346.
[8] C.S. Bryan and M. Lautens, *Org. Lett.*, 2008, **10**, 4633–4636.
[9] S.G. Newman and M. Lautens, *J Am. Chem. Soc.*, 2010, **132**, 11416–11417.
[10] Y.-Q. Fang and M. Lautens, *J. Org. Chem.*, 2008, **73**, 538–549.
[11] T.O. Vieira, L.A. Meaney, Y.-L. Shi, and H. Alper, *Org. Lett.*, 2008, **10**, 4899–4901.
[12] M. Arthuis, R. Pontikis, and J.-C. Florent, *Org. Lett.*, 2009, **11**, 4608–4611.
[13] W. Chen, M. Wang, P. Li, and L. Wang, *Tetrahedron*, 2011, **67**, 5913–5919.
[14] X. Qin, X. Cong, D. Zhao, *et al.*, *Chem. Commun.*, 2011, **47**, 5611–5613.
[15] Z.-J. Wang, F. Yang, X. Lv, and W. Bao, *J. Org. Chem.*, 2011, **76**, 967–970.
[16] Z.-J. Wang, J.-G. Yang, F. Yang, and W. Bao, *Org. Lett.*, 2010, **12**, 3034–3037.
[17] H.-F. He, S. Dong, Y. Chen, *et al.*, *Tetrahedron*, 2012, **68**, 3112–3116.
[18] Y. Liang, T. Meng, H.-J. Zhang, and Z. Xi, *Synlett*, 2011, 911–914.
[19] J. Schütte, F. Kilgenstein, M. Fischer, and U. Koert, *Eur. J. Org. Chem.*, 2014, 5302–5311.
[20] L.M. Geary and P.G. Hultin, *Org. Lett.*, 2009, **11**, 5478–5481.
[21] B. Witulski, C. Alayrac, and L. Tevzadze-Saeftel, *Angew. Chem. Int. Ed.*, 2003, **42**, 4257–4260.
[22] M. Watanabe, T. Yamamoto, and M. Nishiyama, *Angew. Chem. Int. Ed.*, 2000, **39**, 2501–2504.
[23] J. Barluenga, A. Jiménez-Aquino, F. Aznar, and C. Valdés, *J. Am. Chem. Soc.*, 2009, **131**, 4031–4041.
[24] J. Barluenga, A. Jiménez-Aquino, F. Aznar, and C. Valdés, *Chem. Eur. J.*, 2010, **16**, 11707–11711.
[25] C. Mukai and Y. Takahashi, *Org. Lett.*, 2005, **7**, 5793–5796.
[26] N. Kuroda, Y. Takahashi, K. Yoshinaga, and C. Mukai, *Org. Lett.*, 2006, **8**, 1843–1845.
[27] M. Mizutani, F. Inagaki, T. Nakanishi, *et al.*, *Org. Lett.*, 2011, **13**, 1796–1799.
[28] H. Fuwa and M. Sasaki, *Org. Biomol. Chem.*, 2007, **5**, 2214–2218.
[29] H. Fuwa, T. Tako, M. Ebine, and M. Sasaki, *Chem. Lett.*, 2008, **37**, 904–905.
[30] K. Hiroi, Y. Hiratsuka, K. Watanabe, *et al.*, *Synlett*, 2001, 263–265.
[31] T. Boi, A. Deagostino, C. Prandi, *et al.*, *Org. Biomol. Chem.*, 2010, **8**, 2020–2027.
[32] B. Liu, X. Hong, D. Yan, *et al.*, *Org. Lett.*, 2012, **14**, 4398–4401.
[33] F. Zhou, X. Han, and X. Lu, *Tetrahedron Lett.*, 2011, **52**, 4681–4685.
[34] A. Cabrera, P. Sharma, M. Ayala, *et al.*, *Tetrahedron Lett.*, 2011, **52**, 6758–6762.
[35] K. Inamoto, T. Saito, K. Hiroya, and T. Doi, *Synlett*, 2008, 3157–3162.
[36] S. Chiba, L. Zhang, S. Sanjaya, and G.Y. Ang, *Tetrahedron*, 2010, **66**, 5692–5700.
[37] M.-J.R.P. Queiroz, E.M.S. Castanheira, M.S.D. Carvalho, *et al.*, *Tetrahedron*, 2008, **64**, 382–391.
[38] S. Boonya-udtayan, M. Eno, S. Ruchirawat, *et al.*, *Tetrahedron*, 2012, **68**, 10293–10301.
[39] U. Sharma, R. Kancherla, T. Naveen, *et al.*, *Angew. Chem. Int. Ed.*, 2014, **53**, 11895–11899.
[40] G. Qui, Q. Ding, H. Ren, *et al.*, *Org. Lett.*, 2010, **12**, 3975–3977.
[41] Z. Chen, J. Zhu, H. Xie, *et al.*, *Adv. Synth. Catal.*, 2011, **353**, 325–330.
[42] Z. Chen, J. Zhu, H. Xie, *et al.*, *Synlett*, 2010, 1418–1420.
[43] A. Tanaka, S. Kamijo, and Y. Yamamoto, *J. Am. Chem. Soc.*, 2000, **122**, 5662–5663.
[44] S. Kamijo and Y. Yamamoto, *J. Am. Chem. Soc.*, 2002, **124**, 11940–11945.
[45] S. Kamijo and Y. Yamamoto, *Angew. Chem. Int. Ed.*, 2002, **41**, 3230–3233.
[46] T. Nanjo, C. Tsukano, and Y. Takemoto, *Org. Lett.*, 2012, **14**, 4270–4273.
[47] T. Nanjo, S. Yamamoto, C. Tsukano, and Y. Takemoto, *Org. Lett.*, 2013, **15**, 3754–3757.
[48] K. Onitsuka, S. Suzuki, and S. Takahashi, *Tetrahedron Lett.*, 2002, **43**, 6197–6199.
[49] B. Gabriele, R. Mancuso, G. Salerno, *et al.*, *J. Org. Chem.*, 2008, **73**, 4971–4977.
[50] C. Chowdhury, B. Das, S. Mukherjee, and B. Achari, *J. Org. Chem.*, 2012, **77**, 5108–5119.
[51] N. Thirupathi, M.H. Babu, V. Dwivedi, *et al.*, *Org. Lett.*, 2014, **16**, 2908–2911.
[52] Z. Liu, Y. Xia, S. Zhou, *et al.*, *Org. Lett.*, 2013, **15**, 5032–5035.
[53] J. Jadhav, V. Gaikwad, R. Kurane, *et al.*, *Synlett*, 2012, 2511–2515.
[54] M. Tobisu, Y. Ano, and N. Chatani, *Org. Lett.*, 2009, **11**, 3250–3252.
[55] J.H. Kim, Y.S. Chun, H. Shin, and S. Lee, *Synthesis*, 2012, **44**, 1809–1817.
[56] A. Kapur, K. Kumar, L. Singh, *et al.*, *Tetrahedron*, 2009, **65**, 4593–4603.
[57] L. Ren, Z. Shi, and N. Jiao, *Tetrahedron*, 2013, **69**, 4408–4414.
[58] C. Chen, D.R. Lieberman, R.D. Larsen, *et al.*, *J. Org. Chem.*, 1997, **62**, 2676–2677.
[59] C.S. Cho, H.S. Shim, H.-J. Choi, *et al.*, *Bull. Korean Chem. Soc.*, 2004, **25**, 441–442.
[60] Y. Jia and J. Zhu, *J. Org. Chem.*, 2006, **71**, 7826–7834.
[61] Z. Xu, W. Hu, F. Zhang, *et al.*, *Synthesis,* 2008, 3981–3987.

[62] C. Hu, H. Qin, Y. Ciu, and Y. Jia, *Tetrahedron*, 2009, **65**, 9075–9080.
[63] J.M. Knapp, J.S. Zhu, D.J. Tantillo, and M.J. Kurth, *Angew. Chem. Int. Ed.*, 2012, **51**, 10588–10591.
[64] A. Shafiee, V. Upadhyay, E.J. Corley, *et al.*, *Tetrahedron Asymmetry*, 2005, **16**, 3094–3098.
[65] C.F. Sturino, G. O'Neill, N. Lachance, *et al.*, *J. Med. Chem.*, 2007, **50**, 794–806.
[66] P.S. Baran, C.A. Guerrero, N.B. Ambhaikar, and B.D. Hafensteiner, *Angew. Chem. Int. Ed.*, 2005, **44**, 606–609.
[67] P.S. Baran, B.D. Hafensteiner, N.B. Ambhaikar, *et al.*, *J. Am. Chem. Soc.*, 2006, **128**, 8678–8693.
[68] Y. Jia, M. Bois-Choussy, and J. Zhu, *Org. Lett.*, 2007, **9**, 2401–2404.
[69] E.J. Velthuisen and S.J. Danishefsky, *J. Am. Chem. Soc.*, 2007, **129**, 10640–10641.
[70] J. Michaux, P. Retailleau, and J.-M. Campagne, *Synlett*, 2008, 1532–1536.
[71] Z. Xu, Q. Li, L. Zhang, and Y. Jia, *J. Org. Chem.*, 2009, **74**, 6859–6862.
[72] W.M. Robertson, D.B. Kastrinsky, I. Hwang, and D.L. Boger, *Bioorg. Med. Chem. Lett.*, 2010, **20**, 2722–2725.
[73] S.-M. Yu, W.-X. Hong, Y. Wu, *et al.*, *Org. Lett.*, 2010, **12**, 1124–1127.
[74] Z. Xu, F. Zhang, L. Zhang, and Y. Jia, *Org. Biomol. Chem.*, 2011, **9**, 2512–2517.
[75] T. Gerfaud, C. Xie, L. Neuville, and J. Zhu, *Angew. Chem. Int. Ed.*, 2011, **50**, 3954–3957.
[76] Y. Liu, L. Zhang, and Y. Jia, *Tetrahedron Lett.*, 2012, **53**, 684–687.

80

Rhodium-Catalyzed Indole Ring Synthesis

After copper and palladium, rhodium is the third most important transition metal for the synthesis of the indole ring. For a 2007 review on this reaction, see Patil and Patil [1]. Some early examples (Scheme 1) are Alper's rhodium reaction of 2-aryl-2*H*-azirines to give 2-styrylindoles (equation 1) [2], Watanabe's Rh-catalyzed Fischer indole synthesis (equation 2) [3], Ucciani's 3-methylindole synthesis via the hydroformylation of *o*-nitrostyrene (equation 3) [4], and Durst's preparation of 3-acetyl-2-hydroxyindoles from the Rh-catalyzed decomposition and carbenoid aromatic C-H bond insertion (equation 4) [5]. Narasaka extended Alper's 2-aryl-2*H*-azirine reaction to a Rh(II)-catalyzed synthesis of 2,3-disubstituted indoles [6], and both Cenini [7] and Alper [8] stretched the deoxygenation of *o*-nitrostyrenes to give indoles. Durst's Rh-catalyzed decomposition of α-diazo carbonyl compounds was used by Dauban [9] and Jha [10] in the synthesis of substituted oxindoles.

Witulski and colleagues discovered a Rh-catalyzed alkyne cyclotrimerization leading to carbazoles and carbolines (Scheme 2, equation 1) [11–14]. Illustrative of this chemistry are syntheses of hyellazole [11], antiostatin A_1 [12], perlolyrine [13], and eudistomin U [14]. Ruthenium catalysts also function in this reaction. Saito and colleagues found that rhodium catalyzed an intramolecular Pauson–Khand-like reaction to give pyrrolo[2,3-b]indoles [15] and discovered a rearrangement of *N*-propargylanilines to give indoles (equation 2) [16]. Trost and McClory reported the Rh-catalyzed cycloisomerization of *o*-alkynylanilines to give indoles (equation 3) [17], and Fagnou's group described the Rh-catalyzed union of acetanilides with alkynes to give a wide range of indoles [18–20] (equation 4). Saito and Sato accomplished a synthesis of the herbindoles A, B, and C via a Rh-catalyzed intramolecular [2+2+2] cyclization between ynamide and diyne moieties (equation 5) [21] in a reaction reminiscent of Witulski's chemistry. Ruthenium, cobalt, nickel, and palladium catalysts were less effective.

In an extension of the Trost reaction [17], Lautens and coworkers developed a Rh-catalyzed cyclization of *o*-alkynylanilines followed by intermolecular conjugate addition to give 2,3-disubstituted indoles (Scheme 3, equation 1) [22, 23]. Huang and colleagues employed a Rh-catalyzed triazene-directed C-H activation/annulation of arenes with alkynes to form indoles (Scheme 3, equation 2) [24]. In similar fashion, Huang and Wang used the *N*-nitroso functionality to meld alkynes together with *N*-nitrosoanilines to form indoles (equation 3) [25]. Many indoles were synthesized in this manner. Glorius and colleagues developed a Rh-catalyzed hydrazine-directed C-H activation/annulation with alkynes to form indoles (equation 4) [26]. Like Huang's work, the overall process is redox-neutral and traceless. Unsymmetrical alkynes gave only one regioisomer, and several ring-substituted indoles were synthesized. Glorius reported α-halo- and α-tosylketones as alkyne equivalents in these Rh-catalyzed C-H activation processes [27]. Eilbracht described a Rh-catalyzed tandem hydroformulation/Fischer indole synthesis arising from arylhydrazines and alkenes [28]. Zhu and colleagues independently discovered the Rh-catalyzed annulation of *N*-nitrosoanilines with alkynes [29], and Matsuda and Tomaru employed alkylidene-hydrazines in a Rh-catalyzed coupling with internal alkynes to give 2,3-disubstituted indoles [30], a strategy similar to that of Glorius [26]. Nicholls

Indole Ring Synthesis: From Natural Products to Drug Discovery, First Edition. Gordon W. Gribble.

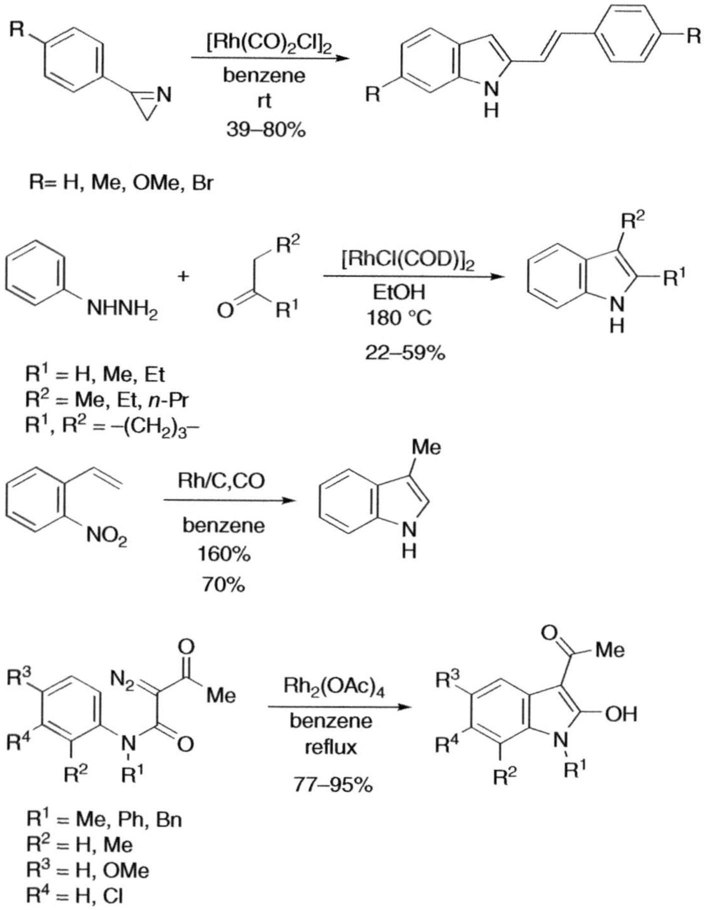

[eqn.1] [2]

[eqn. 2] [3]

[eqn.3] [4]

[eqn. 4] [5]

Scheme 1 *Early Rhodium-Catalyzed Indole Syntheses*

and Kathiravan used *N*-arylureas with internal alkynes and Rh-catalysis to forge indoles [31], Jin and colleagues were able to functionalize and indolize naphthylcarbamates with alkynes to form benzindoles in a Rh-catalytic reaction [32], and Li's group annulated phenylhydrazines with alkynes using 1,3-dinitrobenzene as an oxidant in a Rh-catalyzed process that led to 1-aminoindoles [33].

Driver and coworkers explored in depth the decomposition of azides with rhodium en route to indole ring formation (Scheme 4) [34–43]. The interesting nitro group migration (equation 4) is a function of the ring substitution and was also observed with other electron-withdrawing groups [39]. Yan and colleagues effected the direct C-H activation and ring closure of 2-aminobiphenyls to carbazoles with rhodium (RhCp*(OTf)$_2$, Cu(OAc)$_2$, B$_2$pin$_2$; 71%–93%) [44]. Field and colleagues used a Rh-catalyst to prepare benzo(dipyrroles) from the double cyclization of bis(*o*-alkynylanilines) [45].

Several investigators used a Rh-catalyzed intermolecular annulation of anilines with alkynes to prepare indoles. For example, Li and colleagues coupled *N*-aryl-2-aminopyridines with internal alkynes to give a swath of *N*-(2-pyridyl)indoles [46], and they coupled acrylamides with internal alkynes to give *N*-acryloylindoles [47]. Satoh, Miura, and coworkers fashioned carbazoles and pyranoindoles from the condensation of benzoic and anthranilic acids with internal alkynes and a rhodium catalyst [48]. Fu and colleagues used rhodium catalysis to prepare indoles via a C-H cross-coupling between acetanilides and allyl carbonates [49]. Saá independently described this same reaction [50]. Zhou, Li, and coworkers prepared a series of 3,4-fused indoles from the Rh-catalyzed intramolecular annulation of tethered alkynes to an *N*-nitrosoaniline ring [51], in what is an intramolecular analogue of Huang's chemistry (Scheme 3, equation 3) [25]. Tang's group discovered a Rh-catalyzed tandem annulation and [5+1] cycloaddition to give carbazoles (Scheme 5, equation 1) [52]. Wang, Wan, and coworkers observed that the Rh-catalyzed annulation of nitrones with unsymmetrical diaryl alkynes gave 2,3-diaryl indoles (equation 2) [53].

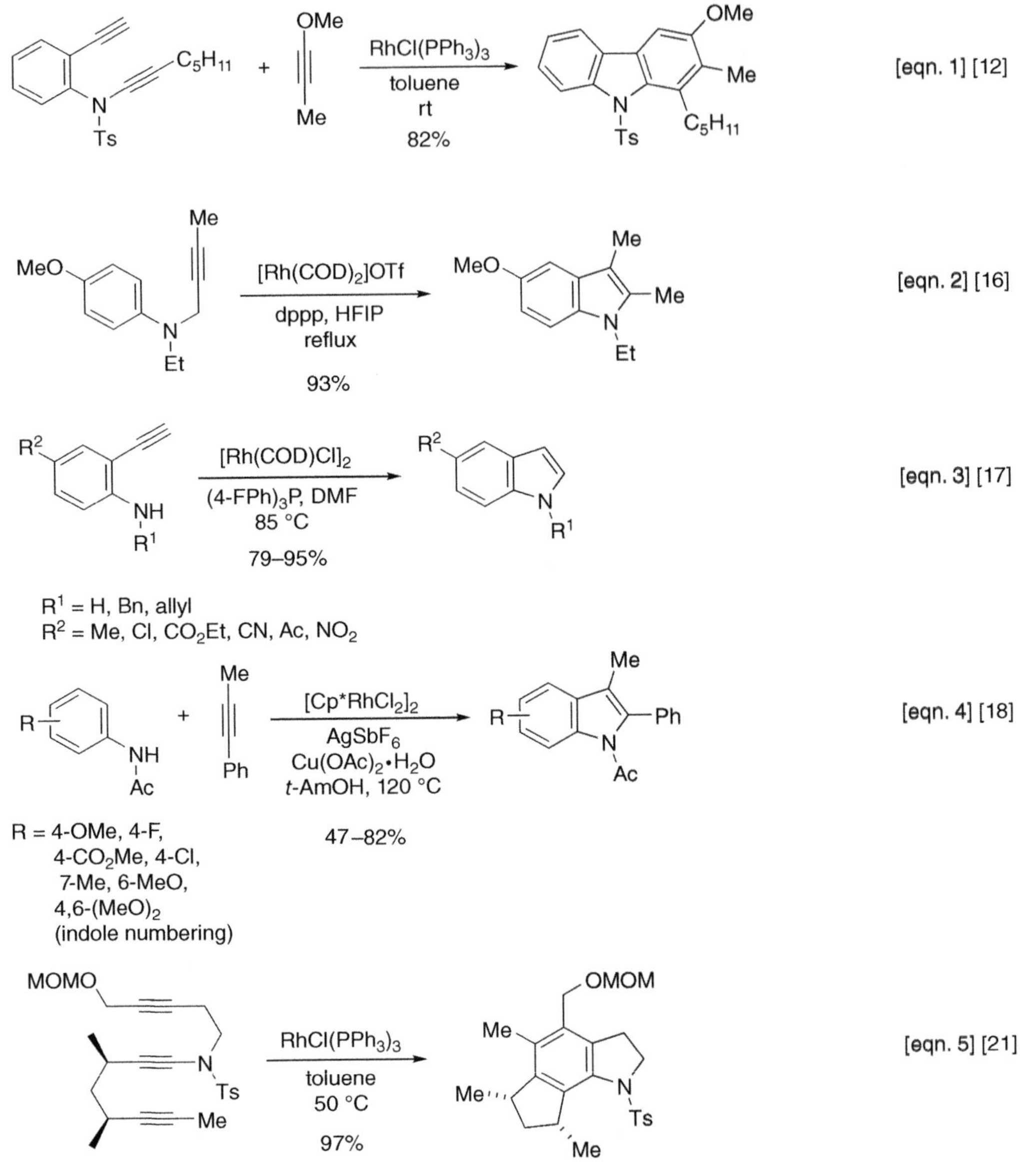

Scheme 2 *Witulski, Saito, Trost, and Fagnou Indole Syntheses*

The Rh-catalyzed annulation of triazoles has been explored by several investigators as a novel indole synthesis. Fokin's team generated azavinyl carbenes with rhodium and parlayed this reaction into an indole synthesis (equation 3) [54]. Davies and colleagues described a one-pot indole synthesis from tosyl azide and cyclohexenyl derivatives involving an *in situ* triazole formation (equation 4) [55], and Lin reported a synthesis of 3-indolylimines using a similar approach (equation 5) [56]. Miura and coworkers effected the Rh-catalyzed intramolecular synthesis of 3,4-fused indoles (equation 6) [57]. Starting from 5-phenylpentyne and tosyl azide, the yield of the fused indole was 70%, and the method was applicable to a preparation of Uhle's ketone.

Nakaura and Ukita devised a simple indole synthesis from the Rh-catalyzed condensation of α-diazophosphonates and *o*-acylanilines (Scheme 6, equation 1) [58]. Moody and colleagues combined diethyl diazomalonate and *N*-alkylanilines in a Rh-catalyzed synthesis of indoxylic acid esters (equation 2) [59]. This chemistry represents a simple synthesis of these rare indoxyl-2-carboxylates.

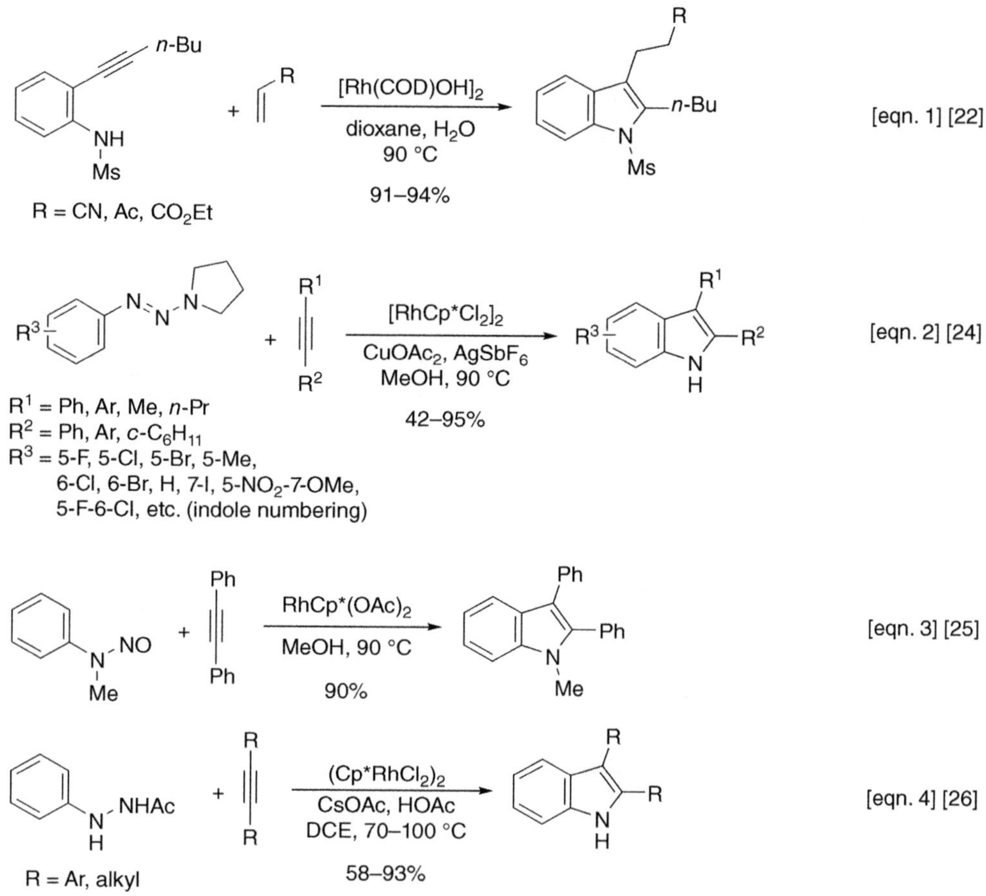

Scheme 3 *Lautens, Huang, and Glorius Indole Syntheses*

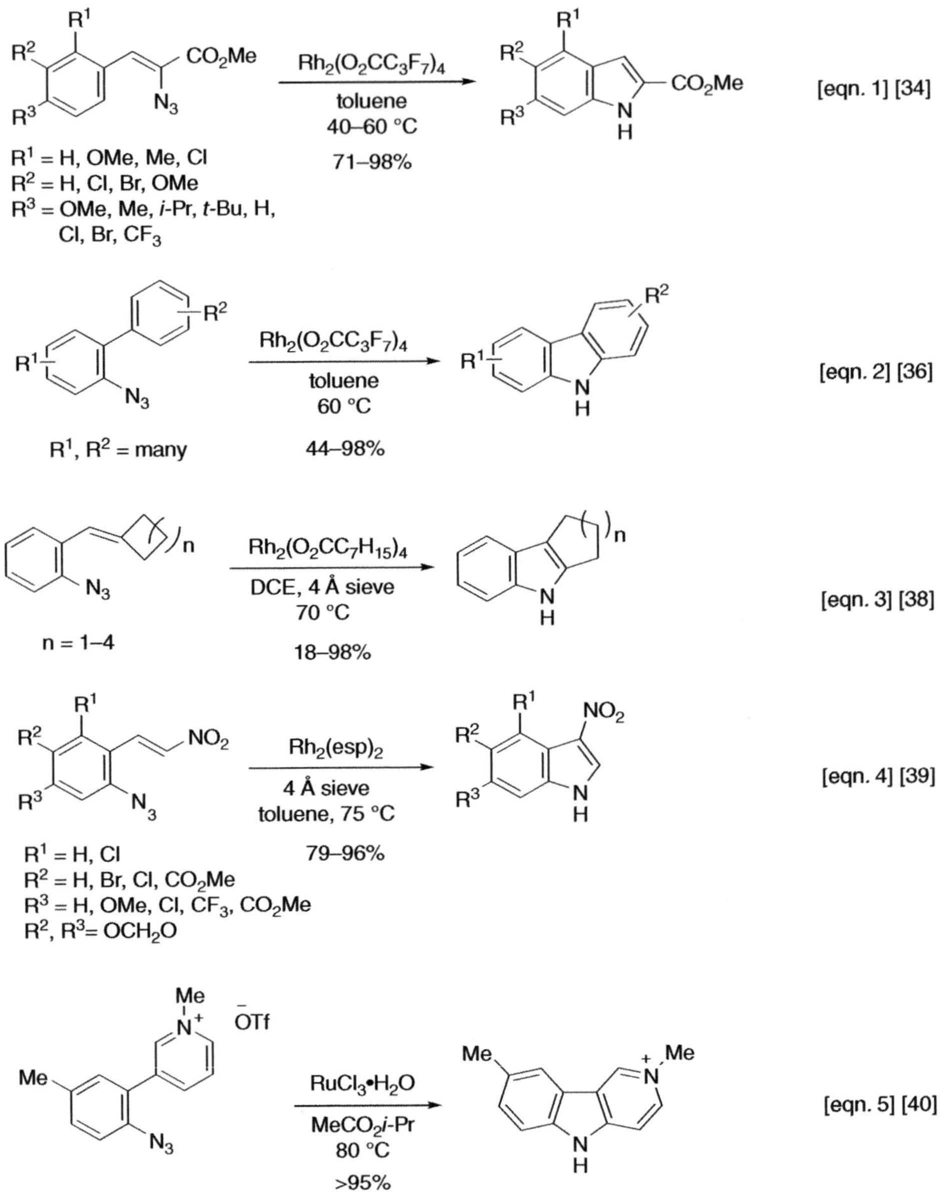

Scheme 4 *Driver Indole Synthesis*

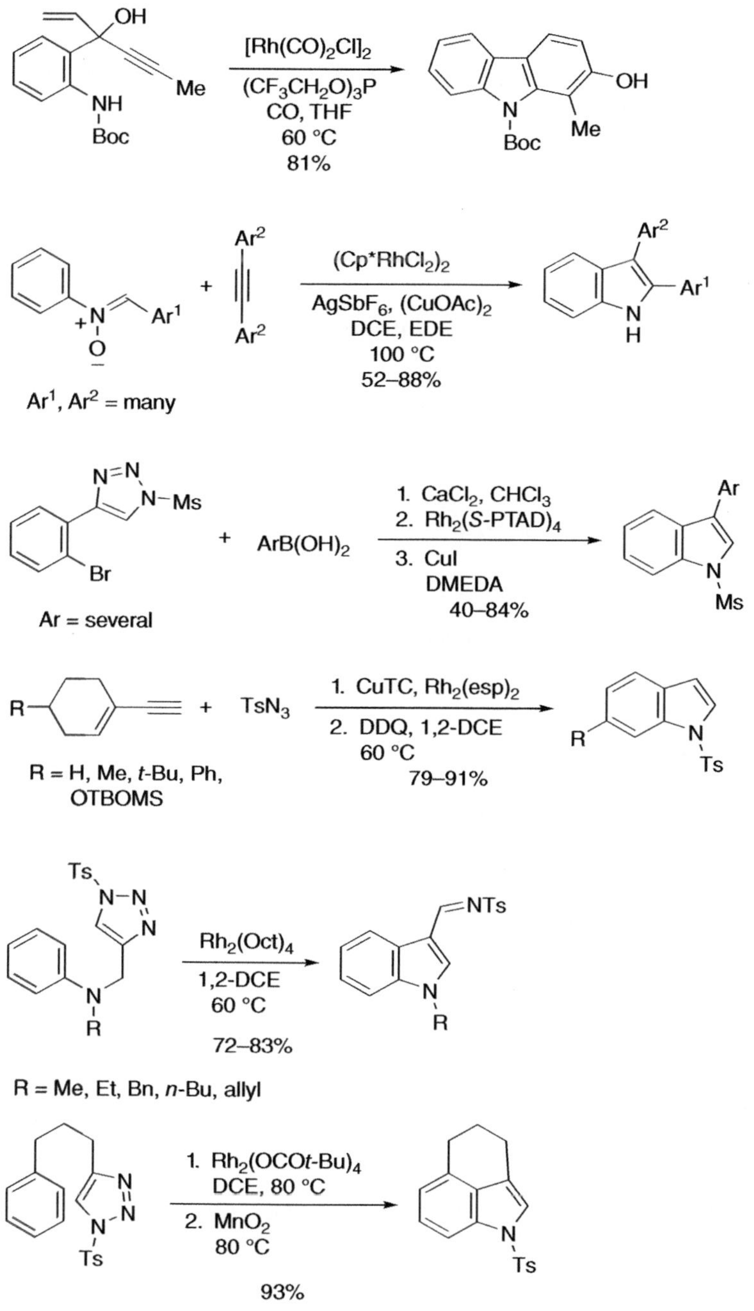

[eqn. 1] [52]

[eqn. 2] [53]

[eqn. 3] [54]

[eqn. 4] [55]

[eqn. 5] [56]

[eqn. 6] [57]

Scheme 5 *Miscellaneous Rhodium-Catalyzed Indole Syntheses*

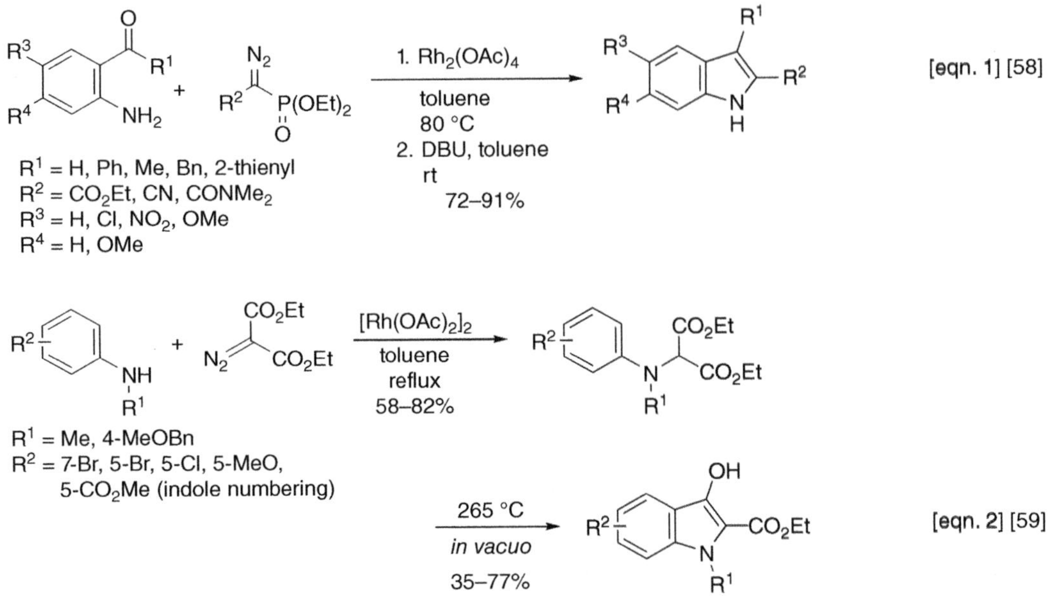

Scheme 6 *Nakamura and Moody Indole Syntheses*

References

[1] S. Patil and R. Patil, *Curr. Org. Synth.*, 2007, **4**, 201–222.
[2] H. Alper and J.E. Prickett, *J. Chem. Soc., Chem. Comm.*, 1976, 483.
[3] Y. Watanabe, M. Yamamoto, S.C. Shim, *et al.*, *Chem. Lett.*, 1980, 603–604.
[4] E. Ucciani and A. Bonfand, *J. Chem. Soc., Chem. Commun.*, 1981, 82–83.
[5] N. Etkin, S.D. Babu, C.J. Fooks, and T. Durst, *J. Org. Chem.*, 1990, **55**, 1093–1096.
[6] S. Chiba, G. Hattori, and K. Narasaka, *Chem. Lett.*, 2007, **36**, 52–53.
[7] C. Crotti, S. Cenini, B. Rindone, *et al.*, *J. Chem. Soc., Chem. Commun.*, 1986, 784–786.
[8] K. Okuro, J. Gurnham, and H. Alper, *J. Org. Chem.*, 2011, **76**, 4715–4720.
[9] D. Gauthier, R.H. Dodd, and P. Dauban, *Tetrahedron*, 2009, **65**, 8542–8555.
[10] M. Jha and B. Blunt, *Tetrahedron Lett.*, 2009, **50**, 6044–6047.
[11] B. Witulski and C. Alayrac, *Angew. Chem. Int. Ed.*, 2002, **41**, 3281–3284.
[12] C. Alayrac, D. Schollmeyer, and B. Witulski, *Chem. Commun.*, 2009, 1464–1466.
[13] B. Dassonneville, B. Witulski, and H. Detert, *Eur. J. Org. Chem.*, 2011, 2836–2844.
[14] F. Nissen, V. Richard, C. Alayrac, and B. Witulski, *Chem. Commun.*, 2011, **47**, 6656–6658.
[15] T. Saito, K. Sugizaki, T. Otani, and T. Suyama, *Org. Lett.*, 2007, **9**, 1239–1241.
[16] A. Saito, S. Oda, H. Fukaya, and Y. Hanzawa, *J. Org. Chem.*, 2009, **74**, 1517–1524.
[17] B.M. Trost and A. McClory, *Angew. Chem. Int. Ed.*, 2007, **46**, 2074–2077.
[18] D.R. Stuart, M. Bertrand-Laperle, K.M.N. Burgess, and K. Fagnou, *J. Am. Chem. Soc.*, 2008, **130**, 16474–16475.
[19] D.R. Stuart, P. Alsabeh, M. Kuhn, and K. Fagnou, *J. Am. Chem. Soc.*, 2010, **132**, 18326–18339.
[20] M.P. Huestis, L. Chan, D.R. Stuart, and K. Fagnou, *Angew. Chem. Int. Ed.*, 2011, **50**, 1338–1341.
[21] N. Saito, T. Ichimaru, and Y. Sato, *Org. Lett.*, 2012, **14**, 1914–1917.
[22] N. Isono and M. Lautens, *Org. Lett.*, 2009, **11**, 1329–1331.
[23] A. Boyer, N. Isono, S. Lackner, and M. Lautens, *Tetrahedron*, 2010, **66**, 6468–6482.
[24] C. Wang, H. Sun, Y. Fang, and Y. Huang, *Angew. Chem. Int. Ed.*, 2013, **52**, 5795–5798.
[25] C. Wang and Y. Huang, *Org. Lett.*, 2013, **15**, 5294–5297.
[26] D. Zhao, Z. Shi, and F. Glorius, *Angew. Chem. Int. Ed.*, 2013, **52**, 12426–12429.
[27] D.-G. Yu, F. de Azambuja, and F. Glorius, *Angew. Chem. Int. Ed.*, 2014, **53**, 2754–2758.
[28] P. Köhling, A.M. Schmidt, and P. Eilbracht, *Org. Lett.*, 2003, **5**, 3213–3216.
[29] B. Liu, C. Song, C. Sun, *et al.*, *J. Am. Chem. Soc.*, 2013, **135**, 16625–16631.
[30] T. Matsuda and Y. Tomaru, *Tetrahedron Lett.*, 2014, **55**, 3302–3304.
[31] S. Kathiravan and I.A. Nicholls, *Chem. Commun.*, 2014, **50**, 14964–14967.
[32] X. Zhang, W. Si, M. Bo, *et al.*, *Org. Lett.*, 2014, **16**, 4830–4833.
[33] D.Y. Li, H.J. Chen, and P.N. Liu, *Org. Lett.*, 2014, **16**, 6176–6179.
[34] B.J. Stokes, H. Dong, B.E. Leslie, *et al.*, *J. Am. Chem. Soc.*, 2007, **129**, 7500–7501.
[35] M. Shen, B.E. Leslie, and T.G. Driver, *Angew. Chem. Int. Ed.*, 2008, **47**, 5056–5059.

[36] B.J. Stokes, B. Jovanovic, H. Dong, *et al.*, *J. Org. Chem.*, 2009, **74**, 3225–3228.
[37] B.J. Stokes, K.J. Richert, and T.G. Driver, *J. Org. Chem.*, 2009, **74**, 6442–6451.
[38] K. Sun, S. Liu, P.M. Bec, and T.G. Driver, *Angew. Chem. Int. Ed.*, 2011, **50**, 1702–1706.
[39] B.J. Stokes, S. Liu, and T.G. Driver, *J. Am. Chem. Soc.*, 2011, **133**, 4702–4705.
[40] A.L. Pumphrey, H. Dong, and T.G. Driver, *Angew. Chem. Int. Ed.*, 2012, **51**, 5920–5923.
[41] Q. Nguyen, K. Sun, and T.G. Driver, *J. Am. Chem. Soc.*, 2012, **134**, 7262–7265.
[42] C. Kong, N. Jana, and T.G. Driver, *Org. Lett.*, 2013, **15**, 824–827.
[43] C. Jones, Q. Nguyen, and T.G. Driver, *Angew. Chem. Int. Ed.*, 2014, **53**, 785–788.
[44] Q. Jiang, D. Duan-Mu, W. Zhong, *et al.*, *Chem. Eur. J.*, 2013, **19**, 1903–1907.
[45] G.K.B. Clentsmith, L.D. Field, B.A. Messerle, *et al.*, *Tetrahedron Lett.*, 2009, **50**, 1469–1471.
[46] J. Chen, G. Song, C.-L. Pan, and X. Li, *Org. Lett.*, 2010, **12**, 5426–5429.
[47] Y. Su, M. Zhao, K. Han, *et al.*, *Org. Lett.*, 2010, **12**, 5462–5465.
[48] M. Shimizu, K. Hirano, T. Satoh, and M. Miura, *J. Org. Chem.*, 2009, **74**, 3478–3483.
[49] T.-J. Gong, W.-M. Cheng, W. Su, *et al.*, *Tetrahedron Lett.*, 2014, **55**, 1859–1862.
[50] A. Cajaraville, S. López, J.A. Varela, and C. Saá, *Org. Lett.*, 2013, **15**, 4576–4579.
[51] B. Zhou, Y. Yang, H. Tang, *et al.*, *Org. Lett.*, 2014, **16**, 3900–3903.
[52] X. Li, W. Song, and W. Tang, *J. Am. Chem. Soc.*, 2013, **135**, 16797–16800.
[53] H. Yan, H. Wang, X. Li, *et al.*, *Angew. Chem. Int. Ed.*, 2015, **54**, 10613–10617.
[54] N. Selander, B.T. Worrell, S. Chuprakov, *et al.*, *J. Am. Chem. Soc.*, 2012, **134**, 14670–14673.
[55] J.S. Alford, J.E. Spangler, and H.M.L. Davies, *J. Am. Chem. Soc.*, 2013, **135**, 11712–11715.
[56] B. Rajagopal, C.-H Chou, C.-C. Chung, and P.-C. Lin, *Org. Lett.*, 2014, **16**, 3752–3755.
[57] T. Miura, Y. Funakoshi, and M. Murakami, *J. Am. Chem. Soc.*, 2014, **136**, 2272–2275.
[58] Y. Nakamura and T. Ukita, *Org. Lett.*, 2002, **4**, 2317–2320.
[59] M.A. Honey and C.J. Moody, *Aust. J. Chem.*, 2014, **67**, 1211–1216.

81

Gold-Catalyzed Indole Ring Synthesis

In a relatively short time span, gold-catalyzed reactions have blossomed into a powerful new methodology in organic synthesis. Given several reviews in the period 2011 through 2014 on gold-catalyzed indole synthesis [1–4], this chapter focuses on a few seminal papers and then very recent work.

The first reported Au-catalyzed indole synthesis was that of Utimoto and coworkers (Scheme 1, equation 1) [5]. Like copper, palladium, and rhodium, gold effects the cyclization of *o*-alkynylanilines to indoles with ease. This basic electrophilic cyclization was improved upon by Marinelli [6, 7], was adapted to the synthesis of indole libraries [8, 9], was used with supported gold nanoparticles and *o*-alkynylnitroarenes under hydrogenation conditions in a one-pot indole synthesis [10], and was employed in total syntheses of the alkaloids (–)-mersicarpine [11] and voacangalactone [12]. There have been an abundance of extensions to the Au-catalyzed cyclization of *o*-alkynylanilines to afford 2,3-disubstituted indoles. Arcadi and colleagues introduced both α,β-enones [13] and 1,3-dicarbonyl compounds [14] into this Au-catalyzed reaction (equation 2) [14]. Perumal developed a domino Au-catalyzed synthesis of bis(indolyl)methanes and other C-3 functionalized indoles [15], and Majumdar described the syntheses of pyrrolocoumarins and pyrroloquinolines via a Au-catalyzed route [16]. The direct Au-catalyzed synthesis of 2-arylindoles from aryl alkynes and *o*-iodoanilines was reported by Wang (equation 3) [17]. The Au-catalyzed double hydroamination of *o*-alkynylanilines and terminal alkynes was found by Li to give *N*-vinylindoles [18]. Patil used a Au-catalyzed one-pot reaction between *o*-alkynylanilines and alkynols to give various 3-substituted indoles [19]. Zhang found that both *N*-arylhydroxylamines [20] and *o*-azidoarylalkynes [21] undergo indole formation under the influence of Au-catalysis (equation 4) [21]. Gagosz and Wetzel independently discovered this same 3-substituted indole synthesis from 2-alkynylarylazides [22].

In a series of papers, Ohno and coworkers described Au-catalyzed intramolecular cyclizations of diynes leading to fused indoles, annulated carbazoles, and indoloquinolines (Scheme 2, equation 1) [23–26]. The groups of Wang [27] and Arcadi [28] assembled 11*H*-indolo[3,2-*c*]quinolines by means of a Au-catalyzed two-step indolization of the appropriate *o*-amino diarylacetylene (e.g., *N*-[2-[(2-aminophenyl)ethynyl]phenyl]amide) followed by C-ring formation using the Hendrickson reagent (Tf_2O/Ph_3PO). Chan and colleagues discovered a Au-catalyzed tandem cycloisomerization/Friedel–Crafts alkylation of *o*-tosylaminophenylprop-1-yl-3-ols to give indoles (equation 2) [29]. By performing this chemistry in the presence of *N*-iodosuccinimide, Chan synthesized indole-2-carbaldehydes [30]. Zhu and colleagues forged a new synthesis of indolequinones from the Au-catalyzed reaction between 2-bromo-1,4-naphthoquinone and β-carbonyl enamines [31], and Helaja's team adapted the Au-catalyzed cycloisomerization of *o*-alkynylanilines using carbon-supported gold nanoparticles to a preparation of 3,3′-biindoles [32]. Blanc, Pale, and colleagues disclosed a novel Au-catalyzed rearrangement of *N*-aryl *o*-alkynylazetidines to pyrrolo[1,2-*a*]indoles (equation 3) [33]. Willis disclosed a synthesis of pyrimido[1,6-*a*]indolones via consecutive Ag-catalyzed cyclizations (equation 4) [34]. The gold catalyst was Echavarren's Au(I) catalyst. Reddy's group employed a gold catalyst to access spiroindolones and indolo[3,2-*c*]quinolones from 2-[(2-aminophenyl)ethynyl]phenylamines [35], and Arcadi and Michelet synthesized several

Indole Ring Synthesis: From Natural Products to Drug Discovery, First Edition. Gordon W. Gribble.

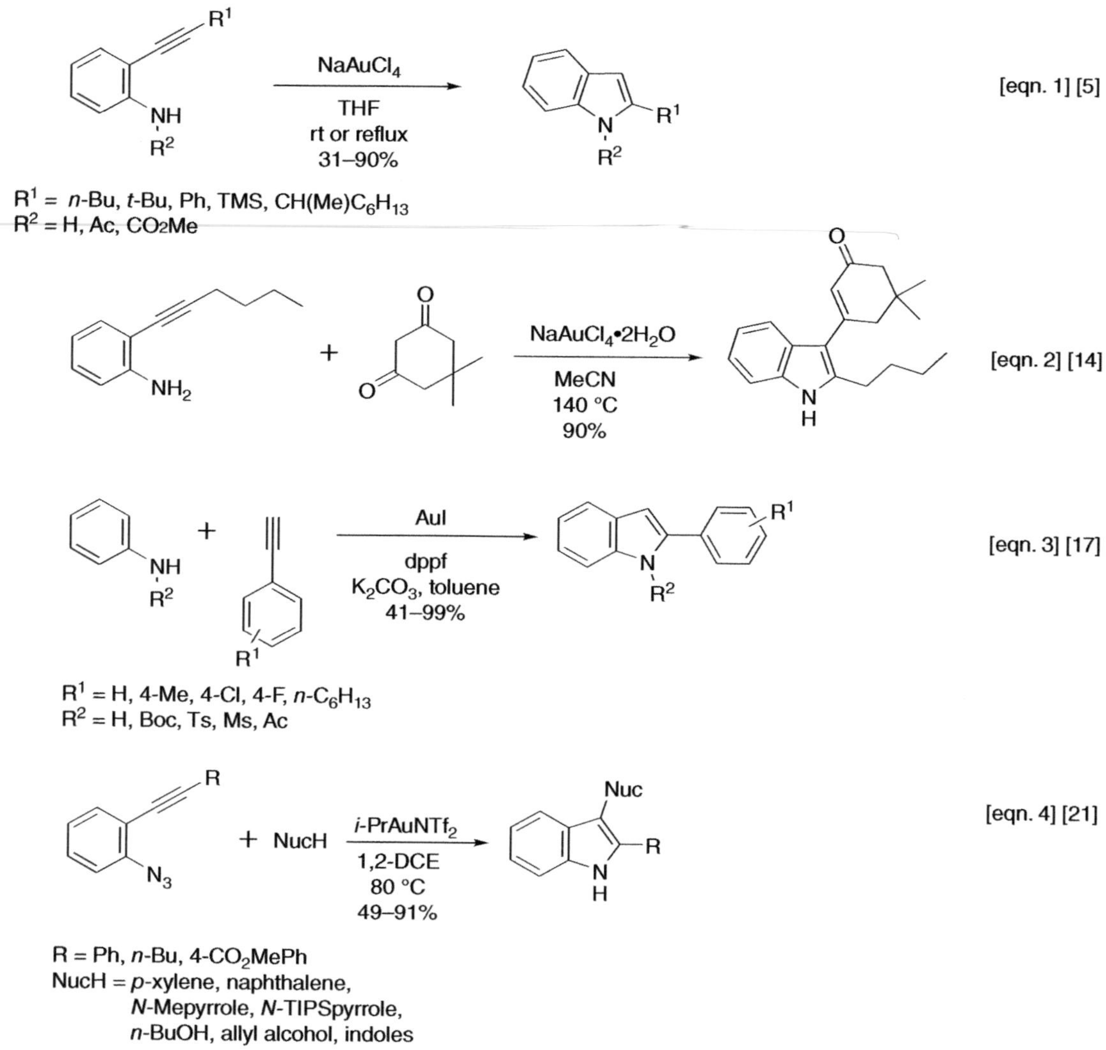

Scheme 1 *Gold-Catalyzed Indole Syntheses – 1*

3,3-difluoro-2-substituted indolenines and 2-aryl-3-fluoroindoles via a one-pot Au-catalyzed aminofluorination of *o*-alkynylanilines with Selectfluor [36].

Toste and colleagues developed a tandem cycloisomerization/Suzuki cross-coupling process with *o*-alkynylaniline boronates leading to 2-aryl indoles (Scheme 3, equation 1) [37]. Bandini and coworkers reported several Au-catalyzed cascade reactions leading to polycyclic indoles [38–40]. In addition to preparing azepino[1,2-*a*] indoles (equation 2) [39], Bandini achieved an asymmetric synthesis of oxazino-indoles via Au-catalysis (equation 3) [40]. Wang and colleagues found that visible light stimulated indolization of a Au-catalyzed coupling of anilines with alkynes (equation 4) [41]. The authors proposed a mechanism involving radical cation formation from the aniline–alkyne adduct, followed by radical cyclization and eventual aromatization to indole.

The gold-catalyzed chemistry presented in this chapter utilizes a variety of gold catalysts, but all are either gold(I) or gold(III) species. A common feature is that gold salts have an enormous affinity for multiple bonds and they behave as soft carbophilic Lewis acids, thus triggering subsequent chemistry. For excellent discussions of these mechanisms, see references [42–48].

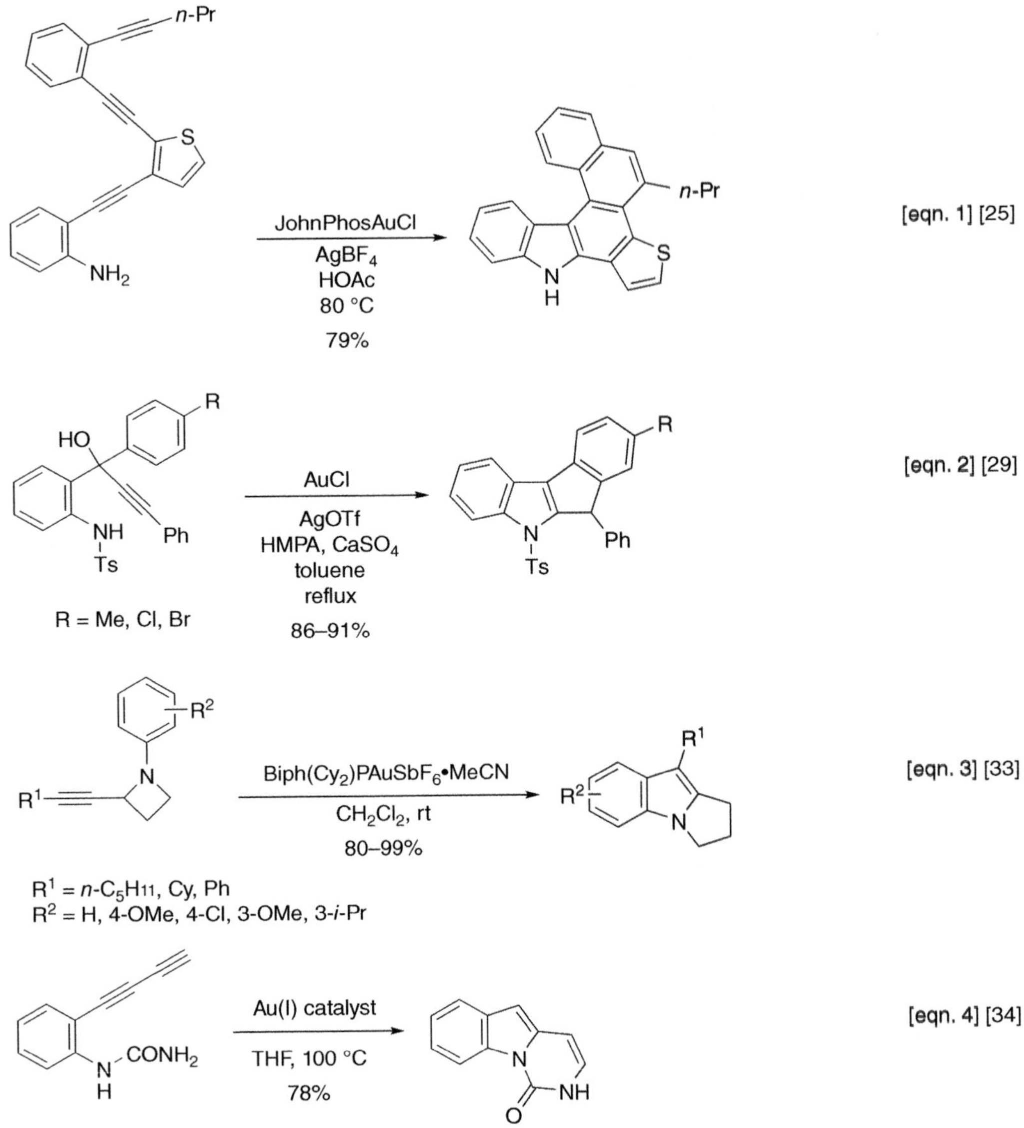

Scheme 2 *Gold-Catalyzed Indole Syntheses – 2*

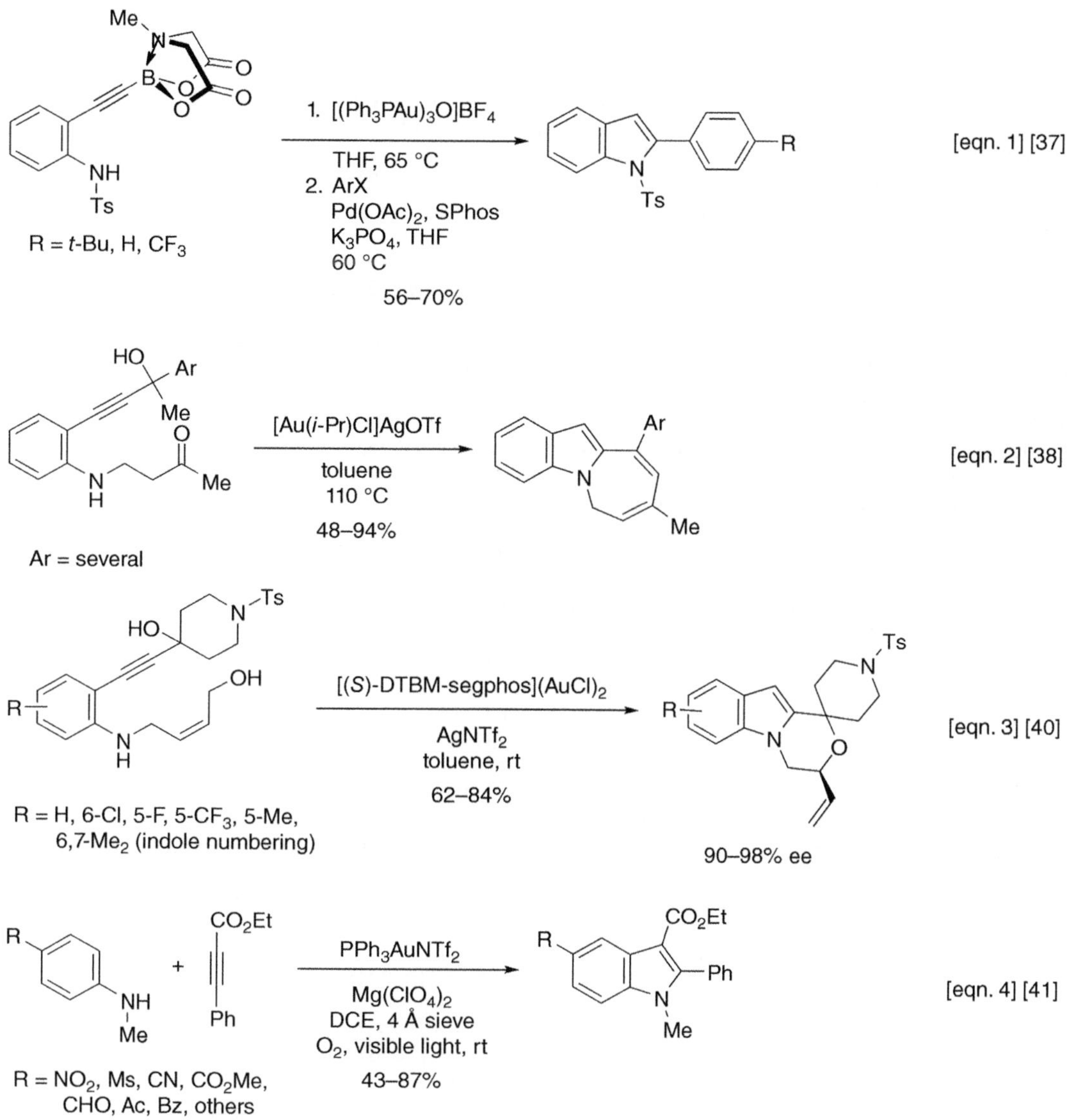

Scheme 3 *Gold-Catalyzed Indole Syntheses – 3*

References

[1] M. Rudolph and A.S.K. Hashmi, *Chem. Commun.*, 2011, **47**, 6536–6544.
[2] G. Abbiati, F. Marinelli, E. Rossi, and A. Arcadi, *Isr. J. Chem.*, 2013, **53**, 856–868.
[3] P.M. Barbour, L.J. Marholz, L. Chang, *et al.*, *Chem. Lett.*, 2014, **43**, 572–578.
[4] D.J. Gorin and F.D. Toste, *Nature*, 2007, **446**, 395–398.
[5] K. Iritani, S. Matsubara, and K. Utimoto, *Tetrahedron Lett.*, 1988, **29**, 1799–1802.
[6] A. Arcadi, G. Bianchi, and F. Marinelli, *Synthesis*, 2004, 610–618.
[7] I. Ambrogio, A. Arcadi, S. Cacchi, *et al.*, *Synlett*, 2007, 1775–1779.
[8] Y. Miyazaki and S. Kobayashi, *J. Comb. Chem.*, 2008, **10**, 355–357.
[9] A. La-Venia, S.A. Testero, M.P. Mischne, and E.G. Mata, *Org. Biomol. Chem.*, 2012, **10**, 2514–2517.
[10] Y. Yamane, X. Liu, A. Hamasaki, *et al.*, *Org. Lett.*, 2009, **11**, 5162–5165.
[11] R. Nakajima, T. Ogino, S. Yokoshima, and T. Fukuyama, *J. Am. Chem. Soc.*, 2010, **132**, 1236–1237.
[12] M. Harada, K.N. Asaba, M. Iwai, *et al.*, *Org. Lett.*, 2012, **14**, 5800–5803.
[13] M. Alfonsi, A. Arcadi, M. Aschi, *et al.*, *J. Org. Chem.*, 2005, **70**, 2265–2273.
[14] A. Arcadi, M. Alfonsi, G. Bianchi, *et al.*, *Adv. Synth. Catal.*, 2006, **348**, 331–338.
[15] C. Praveen, K. Karthikeyan, and P.T. Perumal, *Tetrahedron*, 2009, **65**, 9244–9255.
[16] K.C. Majumdar, B. Chattopadhyay, and S. Samanta, *Synthesis*, 2009, 311–317.
[17] P. Li, L. Wang, M. Wang, and F. You, *Eur. J. Org. Chem.*, 2008, 5946–5951.
[18] Y. Zhang, J.P. Donahue, and C.-J. Li, *Org. Lett.*, 2007, **9**, 627–630.
[19] N.T. Patil, V. Singh, A. Konala, and A.K. Mutyala, *Tetrahedron Lett.*, 2010, **51**, 1493–1496.
[20] Y. Wang, L. Ye, and L. Zhang, *Chem. Commun.*, 2011, **47**, 7815–7817.
[21] B. Lu, Y. Luo, L. Liu, *et al.*, *Angew. Chem. Int. Ed.*, 2011, **50**, 8358–8362.
[22] A. Wetzel and F. Gagosz, *Angew. Chem. Int. Ed.*, 2011, **50**, 7354–7358.
[23] K. Hirano, Y. Inaba, T. Watanabe, *et al.*, *Adv. Synth. Catal.*, 2010, **352**, 368–372.
[24] K. Hirano, Y. Inaba, N. Takahashi, *et al.*, *J. Org. Chem.*, 2011, **76**, 1212–1227.
[25] K. Hirano, Y. Inaba, K. Takasu, *et al.*, *J. Org. Chem.*, 2011, **76**, 9068–9080.
[26] Y. Tokimizu, S. Oishi, N. Fujii, and H. Ohno, *Org. Lett.*, 2014, **16**, 3138–3141.
[27] M. Xu, Q. Hou, S. Wang, *et al.*, *Synthesis*, 2011, 626–634.
[28] G. Abbiati, A. Arcadi, M. Chiarini, *et al.*, *Org. Biomol. Chem.*, 2012, **10**, 7801–7808.
[29] P. Kothandaraman, W. Rao, S.J. Foo, and P.W.H. Chan, *Angew. Chem. Int. Ed.*, 2010, **49**, 4619–4623.
[30] P. Kothandaraman, S.M. Mothe, S.S.M. Toh, and P.W.H. Chan, *J. Org. Chem.*, 2011, **76**, 7633–7640.
[31] A. Abdukader, Q. Xue, A. Lin, *et al.*, *Tetrahedron Lett.*, 2013, **54**, 5898–5900.
[32] J.E. Perea-Buceta, T. Wirtanen, O.-V. Laukkanen, *et al.*, *Angew. Chem. Int. Ed.*, 2013, **52**, 11835–11839.
[33] N. Kern, M. Hoffmann, A. Blanc, *et al.*, *Org. Lett.*, 2013, **15**, 836–839.
[34] P.P. Sharp, M.G. Banwell, J. Renner, *et al.*, *Org. Lett.*, 2013, **15**, 2616–2619.
[35] B.V.S. Reddy, M. Swain, S.M. Reddy, *et al.*, *Eur. J. Org. Chem.*, 2014, 3313–3318.
[36] A. Arcadi, E. Pietropaolo, A. Alvino, and V. Michelet, *Org. Lett.*, 2013, **15**, 2766–2769.
[37] J.M.W. Chan, G.W. Amarante, and F.D. Toste, *Tetrahedron*, 2011, **67**, 4306–4312.
[38] G. Cera, S. Piscitelli, M. Chiarucci, *et al.*, *Angew. Chem. Int. Ed.*, 2012, **51**, 9891–9895.
[39] M. Chiarucci, E. Matteucci, G. Cera, *et al.*, *Chem. Asian J.*, 2013, **8**, 1776–1779.
[40] M. Chiarucci, R. Mocci, L.-D. Syntrivanis, *et al.*, *Angew. Chem. Int. Ed.*, 2013, **52**, 10850–10853.
[41] S. Cai, K. Yang, and D.Z. Wang, *Org. Lett.*, 2014, **16**, 2606–2609.
[42] A.S.K. Hashmi and G.J. Hutchings, *Angew. Chem. Int. Ed.*, 2006, **45**, 7896–7936.
[43] E. Jiménez-Núñez and A.M. Echavarren, *Chem. Rev.*, 2008, **108**, 3326–3350.
[44] A. Fürstner and P.W. Davies, *Angew. Chem. Int. Ed.*, 2007, **46**, 3410–3449.
[45] A.S.K. Hashmi, *Chem. Rev.*, 2007, **107**, 3180–3211.
[46] N.D. Shapiro and F.D. Toste, *Synlett*, 2010, 675–691.
[47] G. Abbiati, F. Marinelli, E. Rossi, and A. Arcadi, *Isr. J. Chem.*, 2013, **53**, 856–868.
[48] H. Ohno, *Isr. J. Chem.*, 2013, **53**, 869–882.

82

Ruthenium-Catalyzed Indole Ring Synthesis

Several of the rhodium-catalyzed indolizations previously presented (Chapter 81) are equally accomplished with ruthenium. An example is the deoxygenation of 2-nitrostilbene to 2-phenylindole with $Ru_3(CO)_{12}$ [1]. The use of ruthenium to construct indoles may have been first reported by Watanabe and colleagues (Scheme 1, equation 1) [2, 3], and followed up by Izumi and Yokota (equation 2) [4]. In the latter study the Ru catalyst was superior to $Pd(OAc)_2$.

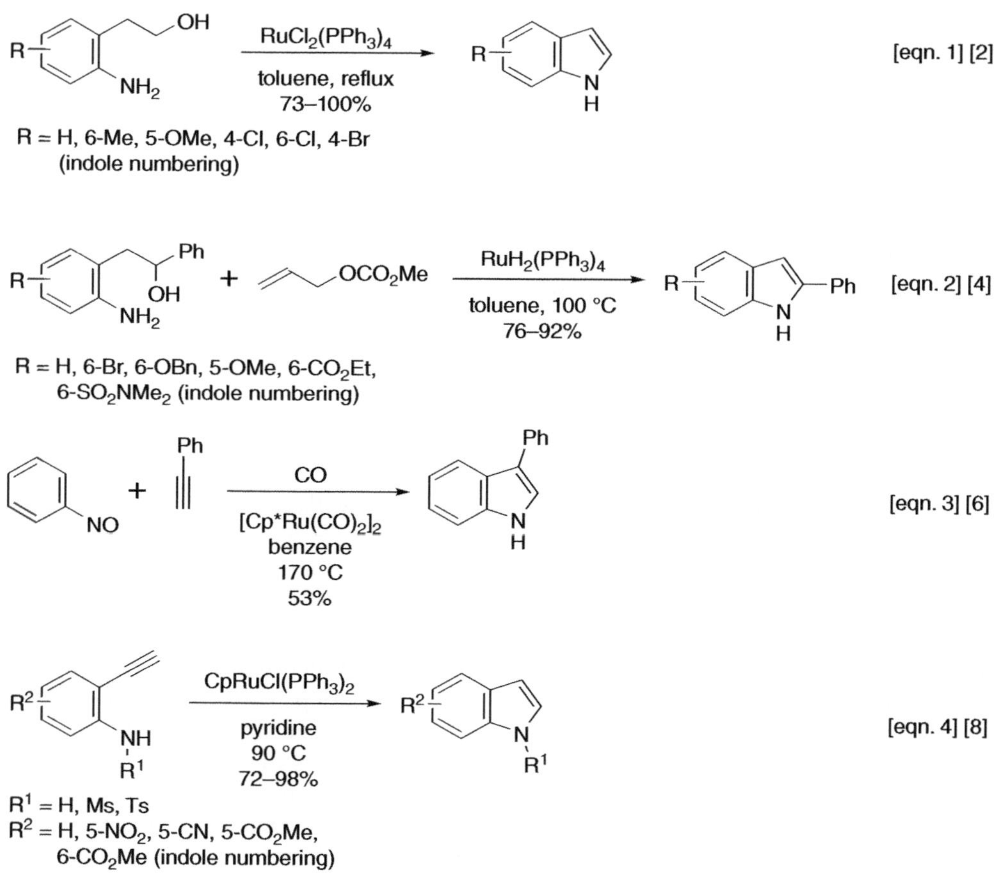

Scheme 1 *Ruthenium-Catalyzed Indole Syntheses – 1*

Indole Ring Synthesis: From Natural Products to Drug Discovery, First Edition. Gordon W. Gribble.

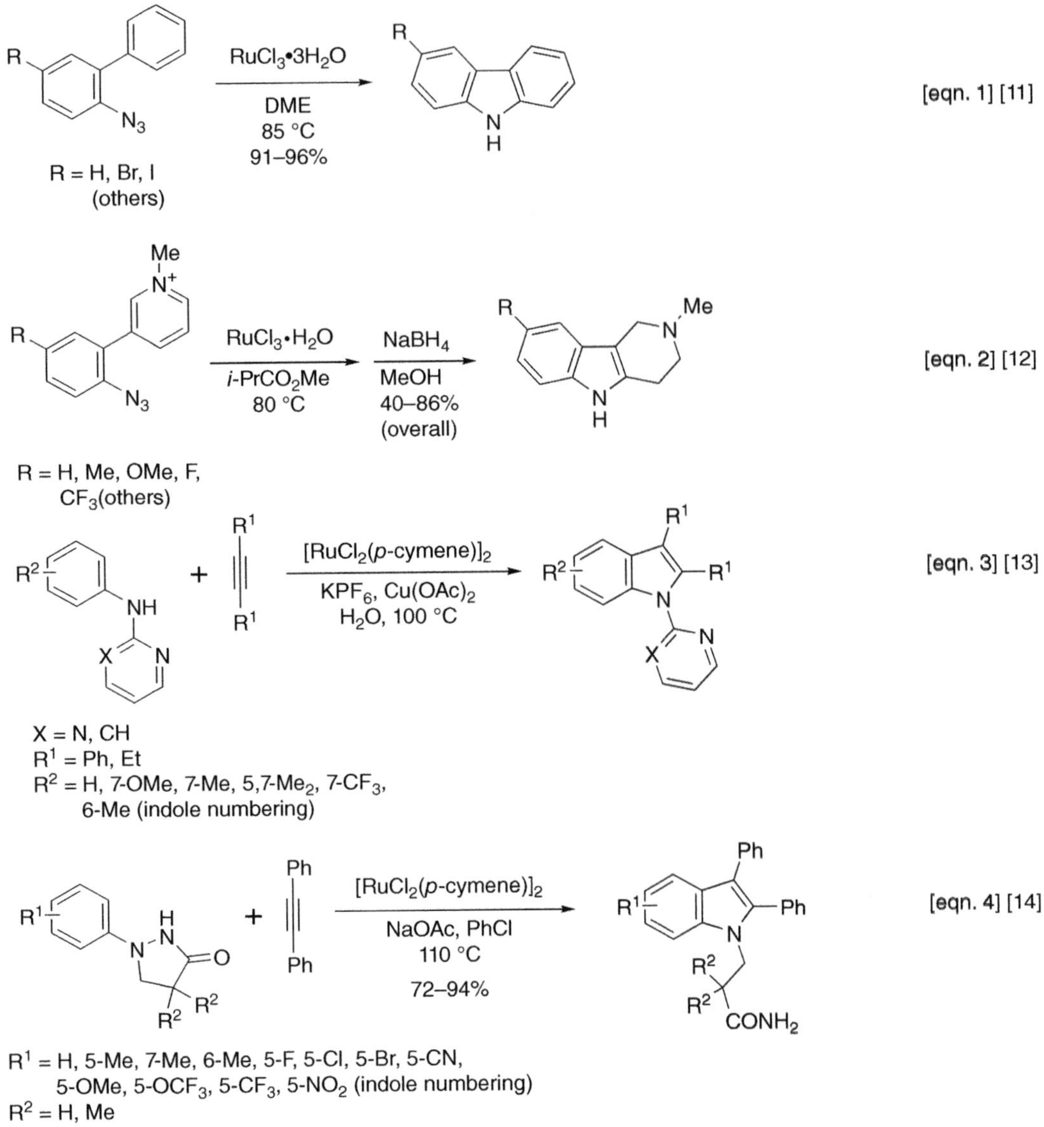

Scheme 2 *Ruthenium-Catalyzed Indole Syntheses – 2*

Tokunaga and Wakatsuki reported a one-pot indole synthesis from anilines and propargyl alcohols using $Ru_3(CO)_{12}$ [5], and Nicholas and colleagues reported a Ru-catalyzed indole synthesis via the reductive annulation of nitrosoarenes with alkynes (equation 3) [6, 7]. Saá and coworkers described the Ru-catalyzed cycloisomerization of *o*-alkynylanilines (equation 4) [8, 9]. Nissen and Detert reported a total synthesis of lavendamycin that featured a Ru-catalyzed [2+2+2] cycloaddition of an *o*-alkynylynamide, a method that was superior to rhodium catalysis both in terms of efficiency and regiochemistry [10].

Jia [11] and Driver [12] both discovered a Ru-catalyzed intramolecular amination of aryl azides to give indoles, carbazoles, and carbolines (Scheme 2, equations 1 and 2), including a synthesis of the γ-carboline alkaloid dimebolin [12]. Both groups surveyed several ruthenium catalysts. Ackermann and Lygin achieved an oxidative C-H activation in an indole synthesis from anilines and internal alkynes in water (equation 3) [13]. The *N*-pyrimidine protecting group was readily removed (NaOEt, DMSO, 120 °C), and regioselectivity was observed with unsymmetrical alkynes. Huang and colleagues discovered a redox-neutral C-H functionalization in their synthesis of 3-(indol-1-yl)propanamides (equation 4) [14]. Li's group uncovered a formyl group translocation in their Ru-catalyzed *o*-alkynylformamide cyclization giving indole-3-carbaldehydes [15].

An obvious and effective use of ruthenium was ring-closing metathesis with the two ruthenium-containing Grubbs catalysts. The groups of Arisawa and Nishida [16],

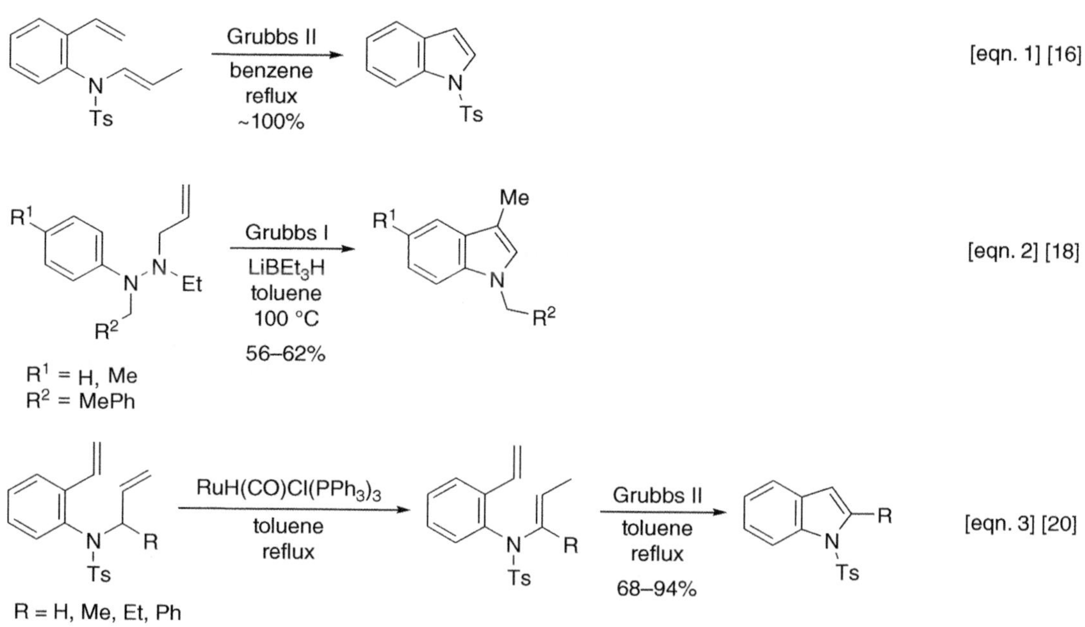

Scheme 3 *Ring-Closing Metathesis Indole Syntheses*

Mori and Sato [17], Rasmussen [18], and Shuto and Arisawa [19, 20] have each exploited the synthetic power of these ruthenium catalysts in indole synthesis (Scheme 3, equations 1–3). The Ru-catalyzed isomerization of alkylhydrazines is a fascinating application of the Grubbs I catalyst (equation 2) [18].

References

[1] C. Crotti, S. Cenini, R. Todeschini, and S. Tollari, *J. Chem. Soc., Faraday Trans.*, 1991, **87**, 2811–2820.
[2] Y. Tsuji, K.T. Huh, Y. Yokoyama, and Y. Watanabe, *J. Chem. Soc., Chem. Commun.*, 1986, 1575–1576.
[3] Y. Tsuji, S. Kotachi, K.T. Huh, and Y. Watanabe, *J. Org. Chem.*, 1990, **55**, 580–584.
[4] T. Izumi and T. Yokota, *J. Heterocycl. Chem.*, 1992, **29**, 1085–1090.
[5] M. Tokunaga, M. Ota, M. Haga, and Y. Wakatsuki, *Tetrahedron Lett.*, 2001, **42**, 3865–3868.
[6] A. Penoni, J. Volkmann, and K.M. Nicholas, *Org. Lett.*, 2002, **4**, 699–701.
[7] A. Penoni, G. Palmisano, G. Broggini, *et al.*, *J. Org. Chem.*, 2006, **71**, 823–825.
[8] A. Varela-Fernández, J.A. Varela, and C Saá, *Adv. Synth. Catal.*, 2011, **353**, 1933–1937.
[9] A. Varela-Fernández, J.A. Varela, and C. Saá, *Synthesis*, 2012, **44**, 3285–3295.
[10] F. Nissen and H. Detert, *Eur. J. Org. Chem.*, 2011, 2845–2853.
[11] W.G. Shou, J. Li, T. Guo, *et al.*, *Organometallics*, 2009, **28**, 6847–6854.
[12] H. Dong, R.T. Latka, and T.G. Driver, *Org. Lett.*, 2011, **13**, 2726–2729.
[13] L. Ackermann and A.V. Lygin, *Org. Lett.*, 2012, **14**, 764–767.
[14] Z. Zhang, H. Jiang, and Y. Huang, *Org. Lett.*, 2014, **16**, 5976–5979.
[15] C.-Y. Wu, M. Hu, Y. Liu, *et al.*, *Chem. Commun.*, 2012, **48**, 3197–3199.
[16] M. Arisawa, Y. Terada, K. Takahashi, *et al.*, *J. Org. Chem.*, 2006, **71**, 4255–4261.
[17] M. Mori, H. Wakamatsu, N. Saito, *et al.*, *Tetrahedron*, 2006, **62**, 3872–3881.
[18] S.D. Nielsen, T. Ruhland, and L.K. Rasmussen, *Synlett*, 2007, 443–446.
[19] T. Ogawa, T. Nakamura, T. Araki, *et al.*, *Eur. J. Org. Chem.*, 2012, 3084–3087.
[20] T. Kobayashi, M. Arisawa, and S. Shuto, *Org. Biomol. Chem.*, 2011, **9**, 1219–1224.

83

Platinum-Catalyzed Indole Ring Synthesis

Compared to other transition metals (Cu, Pd, Rh, Au), platinum has not received much attention with regard to indole ring construction. Nevertheless, several interesting applications have been reported with this metal.

Fürstner and Mamane described a synthesis of benzo[*g*]indoles using $PtCl_2$ (Scheme 1, equation 1) [1], and Fürstner and Davies subsequently generalized this indolization (equation 2) [2]. The atmosphere of carbon monoxide accelerates

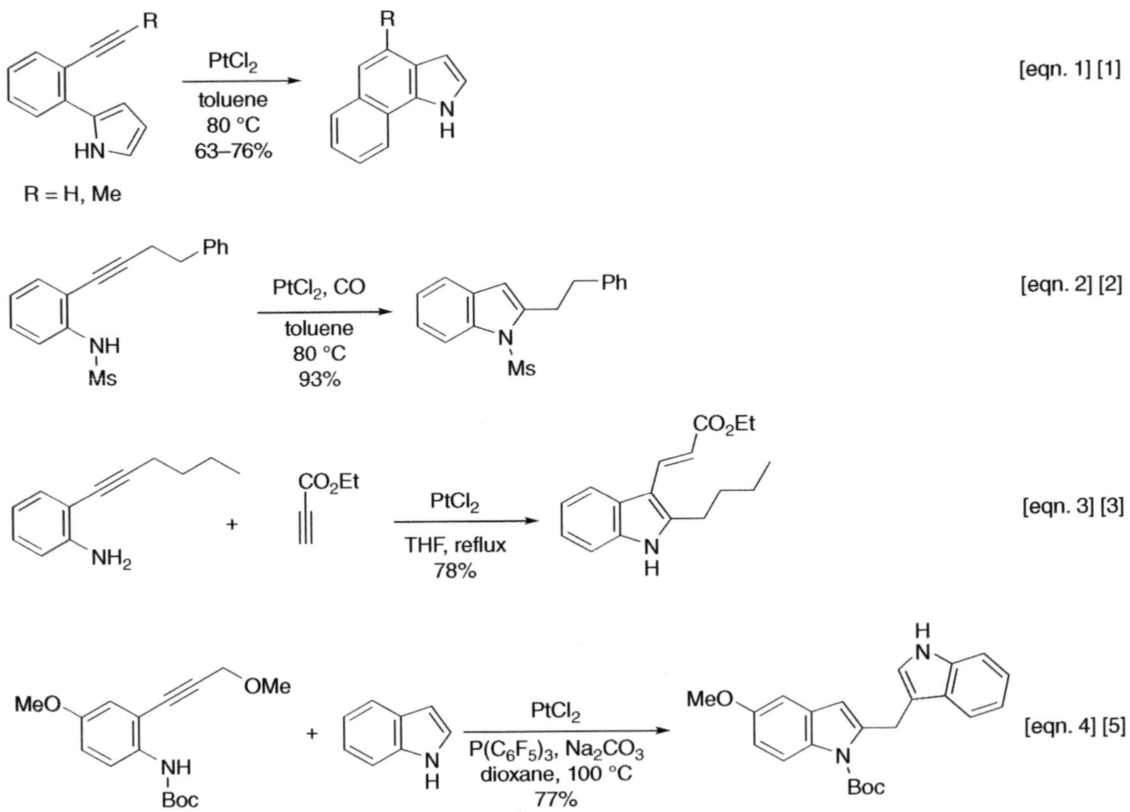

Scheme 1 *Platinum-Catalyzed Indole Syntheses – 1*

Indole Ring Synthesis: From Natural Products to Drug Discovery, First Edition. Gordon W. Gribble.

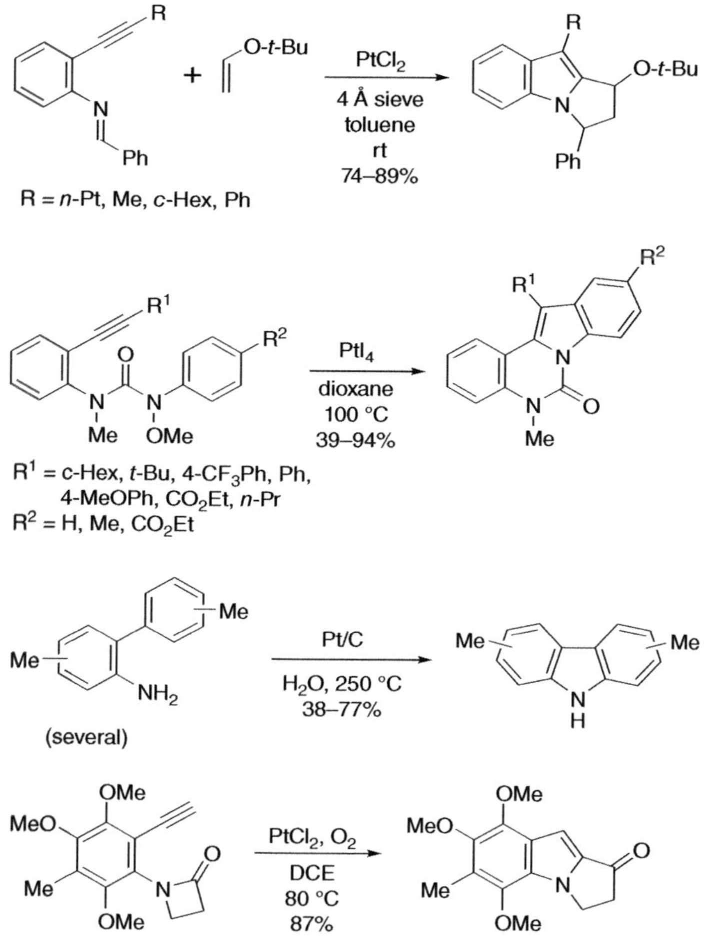

[eqn. 1] [6]

[eqn. 2] [7]

[eqn. 3] [9]

[eqn. 4] [11]

Scheme 2 *Platinum-Catalyzed Indole Syntheses – 2*

the reaction. Yu and colleagues extended the Pt-catalyzed cyclization of *o*-alkynylanilines in the presence of ethyl propiolate and dimethyl acetylenedicarboxylate to give 2,3-disubstituted indoles (equation 3) [3]. A similar Pt-catalyzed domino annulation reaction was found by Tang and coworkers to afford cyclohepta[*b*]indoles using diene traps [4] and, upon reaction with indoles, to give diindolylmethanes and indolo[3,2-*b*]carbazoles (equation 4) [5].

Iwasima and colleagues engineered the Pt-catalyzed [3+2] cycloaddition of *N*-(*o*-alkynylphenyl)imines with activated alkenes into a synthesis of the mitosene skeleton (Scheme 2, equation 1) [6]. The gold catalyst, $AuBr_3$, is equally effective. Nakamura's group described a novel Pt-catalyzed dehydroalkoxylation/cyclization cascade process to afford a unique synthesis of the indoloisoquinolinone and indoloquinazolinone ring systems (equation 2) [7]. This research group also reported a similar Pt-catalyzed synthesis of indole-3-carbamides and indole-3-carboxylates from *o*-alkynylphenylureas and *o*-alkynylphenyl carbamates, respectively [8]. The Pt-catalyzed C-H functionalization of 2-aminobiphenyls, in water as the reoxidant, gave carbazoles as discovered by Matsubara and Yamamoto (equation 3) [9]. Zhang and colleagues explored the use of platinum to forge cyclic ketone-fused indoles, including a formal synthesis of 7-methoxymitosene (equation 4) [10, 11]. Other investigators who developed Pt-catalyzed indole syntheses are Malacria and Fensterbank [12], Yanada [13], and Iwasawa [14].

References

[1] A. Fürstener and V. Mamane, *J. Org. Chem.*, 2002, **67**, 6264–6267.
[2] A. Fürstner and P.W. Davies, *J. Am. Chem. Soc.*, 2005, **127**, 15024–15025.
[3] X. Li, Jp.-Y. Wang, W. Yu, and L.-M. Wu, *Tetrahedron*, 2009, **65**, 1140–1146.

[4] D. Shu, W. Song, X. Li, and W. Tang, *Angew. Chem. Int. Ed.*, 2013, **52**, 3237–3240.
[5] D. Shu, G.N. Winston-McPherson, W. Song, and W. Tang, *Org. Lett.*, 2013, **15**, 4162–4165.
[6] H. Kusama, Y. Miyashita, J. Takaya, and N. Iwasawa, *Org. Lett.*, 2006, **8**, 289–292.
[7] I. Nakamura, Y. Sato, and M. Terada, *J. Am. Chem. Soc.*, 2009, **131**, 4198–4199.
[8] I. Nakamura, Y. Sato, S. Konta, and M. Terada, *Tetrahedron Lett.*, 2009, **50**, 2075–2077.
[9] M. Yamamoto and S. Matsubara, *Chem. Lett.*, 2007, **36**, 172–173.
[10] G. Li, X. Huang, and L. Zhang, *Angew. Chem. Int. Ed.*, 2008, **47**, 346–349.
[11] L. Liu, Y. Wang, and L. Zhang, *Org. Lett.*, 2012, **14**, 3736–3739.
[12] K. Cariou, B. Ronan, S. Mignani, *et al.*, *Angew. Chem. Int. Ed.*, 2007, **46**, 1881–1884.
[13] N. Okamoto, K. Takeda, and R. Yanada, *J. Org. Chem.*, 2010, **75**, 7615–7625.
[14] K. Saito, H. Sogou, T. Suga, *et al.*, *J. Am. Chem. Soc.*, 2011, **133**, 689–691.

84

Silver- and Zinc-Catalyzed Indole Ring Synthesis

Silver salts display similar electrophilic behavior as gold salts, and several indole ring syntheses have been described with silver catalysis. Ding and coworkers found that $AgNO_3$ outperforms Pd and Cu in the cyclization of (*o*-alkynylphenyl)guanidines generated *in situ* from the appropriate (*o*-alkynylphenyl)iminophosphoranes to give *N*-carboximidamides (Scheme 1, equation 1) [1]. *N*-carboximidoates were also prepared by substituting alcohols for secondary amines. Likewise, AgOTf was found by Oh and colleagues to be far superior to gold and platinum salts in the cyclization of *o*-alkynyl formimidates in the presence of dimethyl malonate to give indoles following a 1,3-alkenyl shift (equation 2) [2]. Waldmann, Kumar, and coworkers developed a route to the fascaplysin marine alkaloids from a Ag-catalyzed cyclization of a 3-ethynylindole-2-carbaldehyde (equation 3) [3]. This method was applied to the preparation of indolo[2,1-*a*]isoquinolines from 2-(phenylethynyl)benzaldehydes and di-*tert*-butyl-2-(2-aminophenyl)malonates. Chan's group devised a Ag-catalyzed domino heterocyclization/alkynylation synthesis of 2-alkynylindoles from the appropriate but-2-yne-1,4-diols (equation 4) [4].

A simple one-pot route to 3-fluoroindoles was discovered by You, Song, and coworkers (Scheme 2, equation 1) [5]. This strategy also provided 3,3-difluoroindolenines and 2-hydroxy-3,3-difluoroindolines. Bi and colleagues reported a Ag-catalyzed heteroannulation of propargylic alcohols with *p*-toluenesulfonylmethyl isocyanide (TosMIC) (equation 2) [6]. McNulty and Keskar invented a "highly efficient, chemically stable, and well-defined homogeneous silver(I) catalyst," **1** [7]. This catalyst is superior to Ag_2SO_4, Ag_2CO_3, AgOTf, $AgNO_3$, Ag_2O, AgOTs, and AgOAc in catalyzing the intramolecular hydroamination (cycloisomerization) of *o*-ethynylaniline to indoles.

Given the important biological role of zinc as a Lewis acid, it is not surprising that zinc salts can function in indole ring synthesis. Liu and Kumar described the $Zn(OTf)_2$-catalyzed annulation of anilines with propargyl alcohols to give substituted indoles (Scheme 3, equation 1) [8]. When R = ethyl, some of the isomeric 3-ethyl-2-methylindole is obtained. Zinc triflate (and $ZnCl_2$) also promoted the hydrohydrazination of terminal alkynes to give indoles, as discovered by Beller and colleagues (equation 2) [9]. Other catalysts ($FeCl_3$, $HAuCl_4$, H_2PtCl_6, $IrCl_3$, $Sc(OTf)_3$, $Yb(OTf)_3$) were much less effective or failed altogether in this novel Fischer indolization. Doyle and Zhou employed $Zn(OTf)_2$ to prepare indoles from *o*-imino methyl phenyldiazoacetates (equation 3) [10]. The diazo compounds were synthesized from the corresponding methyl 2-arylmethyleneaminophenylacetates (*p*-nitrobenzenesulfonylazide, DBU, MeCN, 60%–80%). Boron trifluoride etherate was equally effective in this indolization.

Okuma and colleagues found that $ZnBr_2$ and ZnI_2 catalyze the hydroamination of *o*-alkynylanilines to afford indoles (Scheme 4, equation 1) [11]. Indoles were also obtained from the corresponding nitrobenzenes under similar conditions (equation 2). Wang and colleagues extended this $ZnBr_2$-mediated hydroamination to a synthesis of several indolo[1,2-*c*]quinazolines (equation 3) [12]. Zinc bromide is far superior to $FeCl_3$, $AlCl_3$, CuI, and $PdCl_2$ and slightly better than $InBr_3$. Zhao and coworkers used diethylzinc to effect hydroamination of *o*-alkynylphenylsulfonamides to indoles (equation 4) [13]. When the cyclization intermediate was treated with an acid chloride (or *N*-bromosuccinimide), the corresponding 3-acylindoles (or 3-bromoindole) were obtained (equation 5).

Indole Ring Synthesis: From Natural Products to Drug Discovery, First Edition. Gordon W. Gribble.

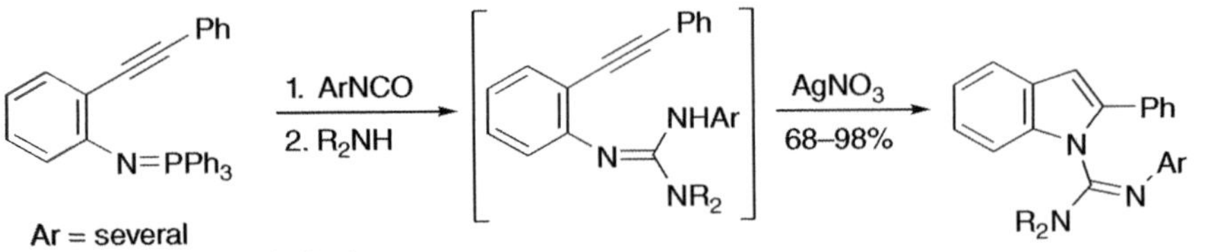

[eqn. 1] [1]

Ar = several
R = Et, *n*-Pr, morpholinyl, pyrrolidinyl

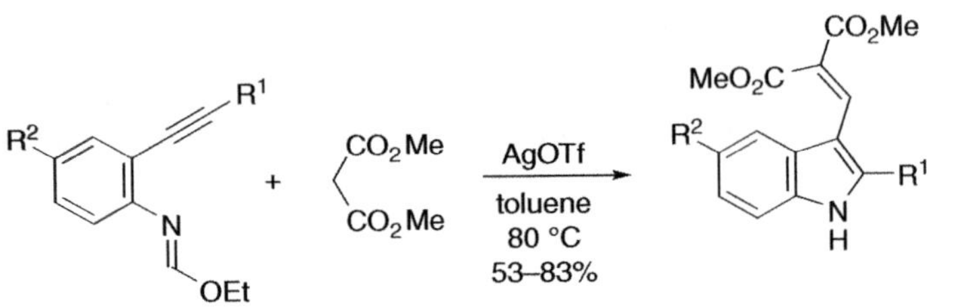

[eqn. 2] [2]

R^1 = *c*-C_6H_{11}, *n*-Bu, *t*-Bu, Ph, 4-MeOPh, 1-naphthyl, $(CH_2)_2OTBS$, $(CH_2)_4OBz$
R^2 = H, NO_2, OMe

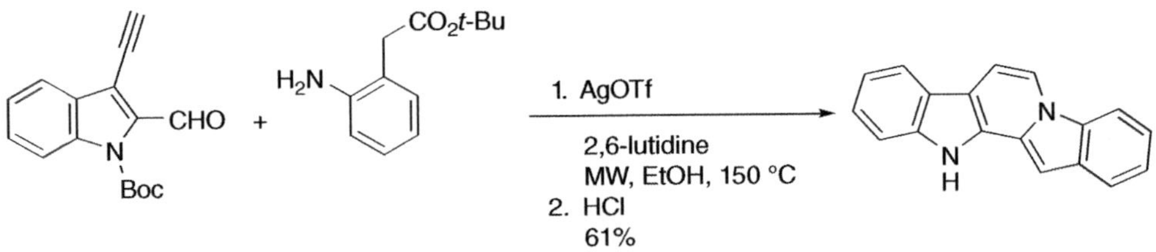

[eqn. 3] [3]

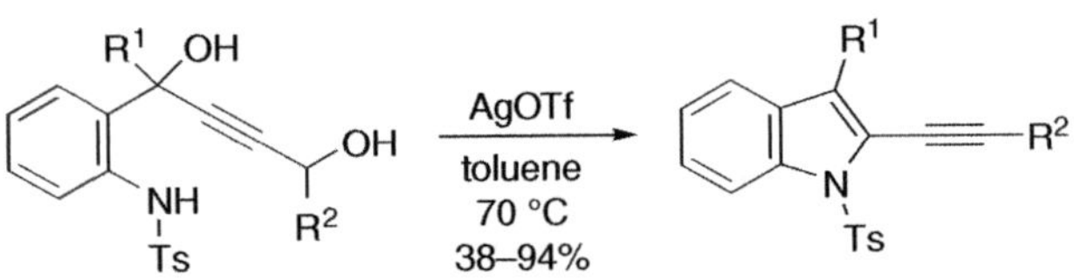

[eqn. 4] [4]

R^1 = Ph, Ar, H, 2-thienyl, ≡—Ph
R^2 = Ph, Ar, *c*-C_3H_5, *n*-Pen, H

Scheme 1 *Silver-Catalyzed Indole Syntheses – 1*

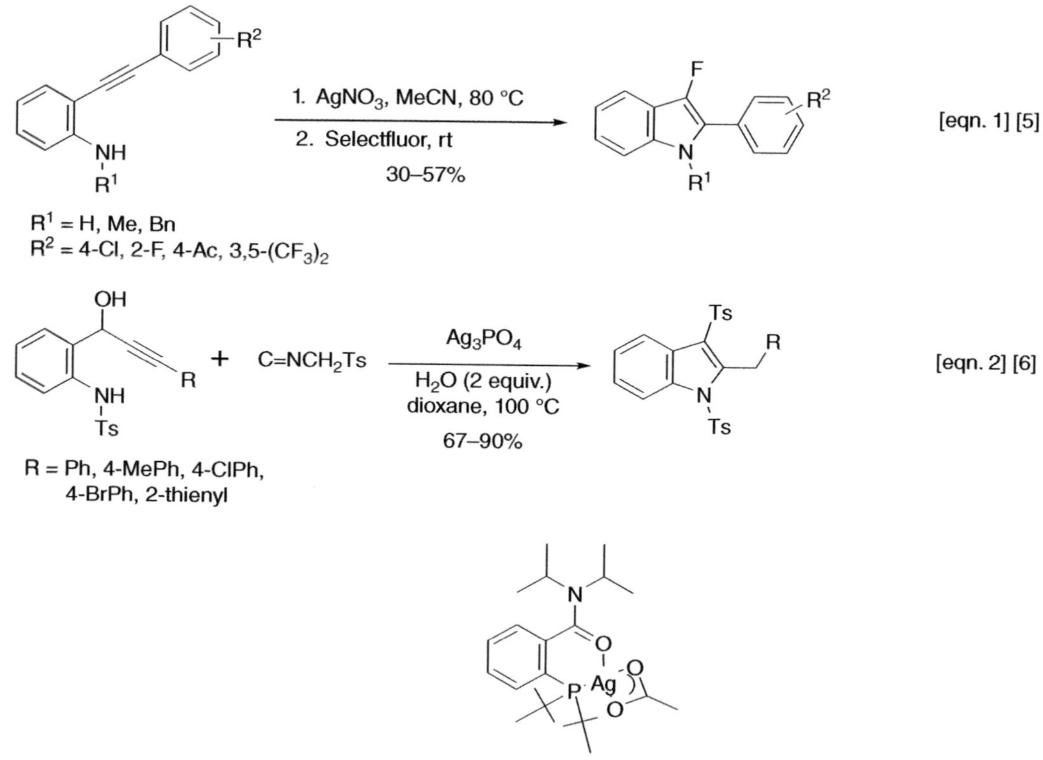

1 [7]

Scheme 2 *Silver-Catalyzed Indole Syntheses – 2*

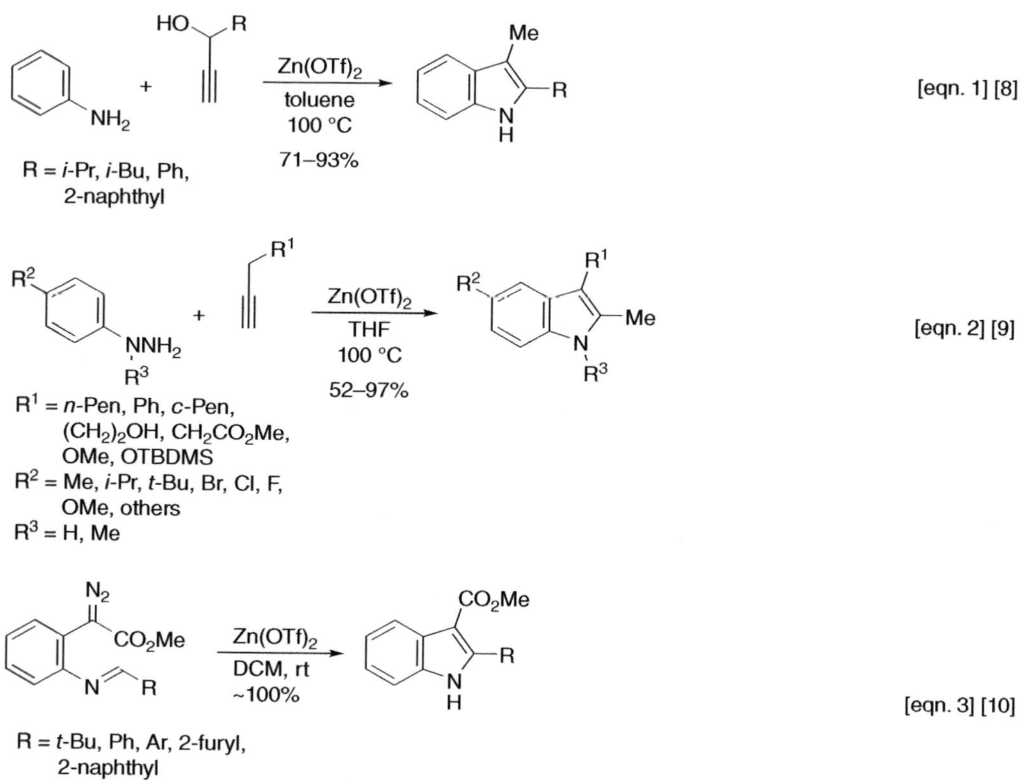

Scheme 3 *Zinc-Catalyzed Indole Syntheses – 1*

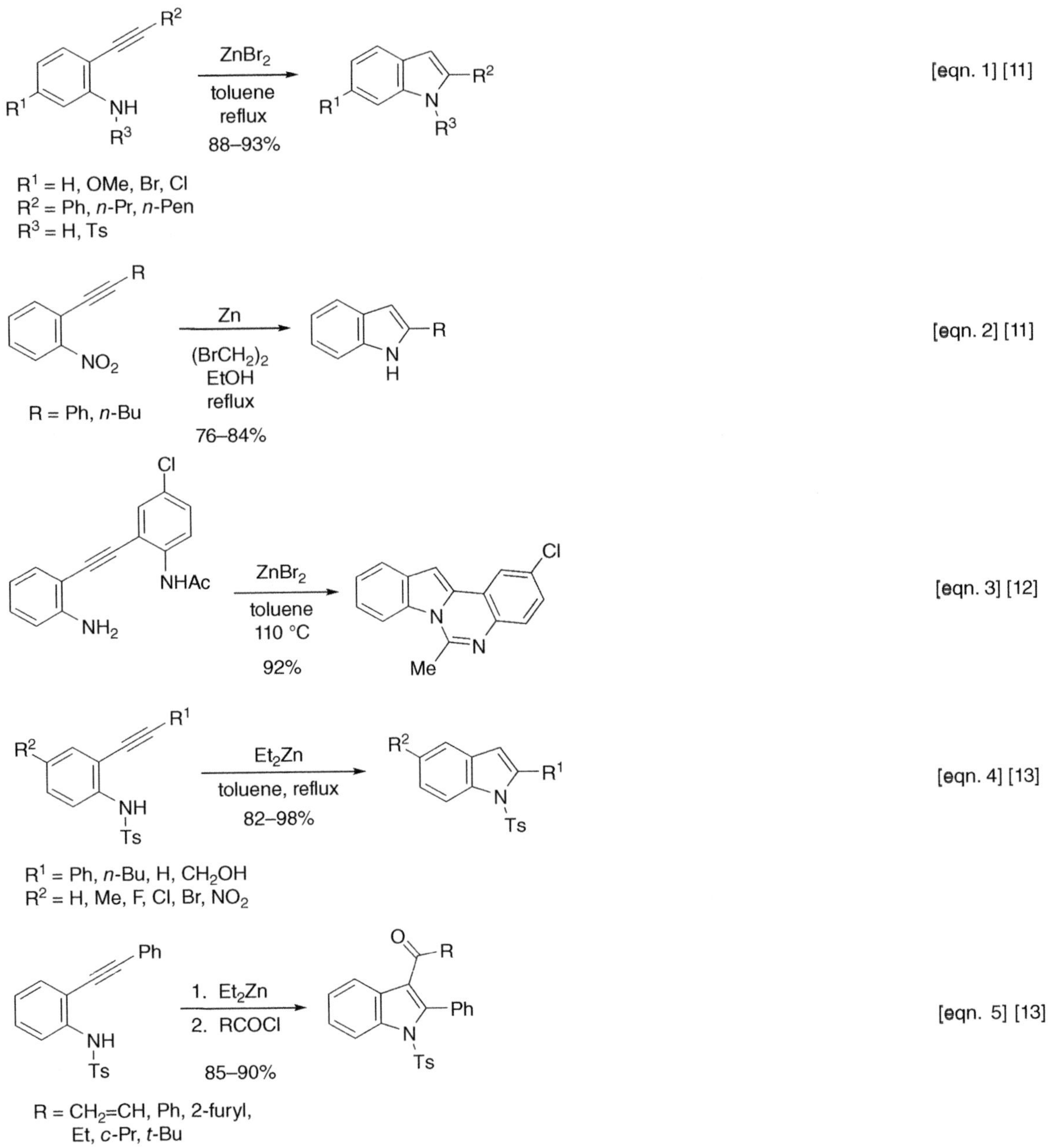

Scheme 4 *Zinc-Catalyzed Indole Syntheses – 2*

References

[1] N.-Y. Huang, M.-G. Liu, and M.-W. Ding, *J. Org. Chem.*, 2009, **74**, 6874–6877.
[2] C.H. Oh, S. Karmakar, HS. Park, *et al.*, *J. Am. Chem. Soc.*, 2010, **132**, 1792–1793.
[3] H. Waldmann, L. Eberhardt, K. Wittstein, and K. Kumar, *Chem. Commun.* 2010, **46**, 4622–4624.
[4] S.R. Mothe, P. Kothandaraman, S.J.L. Lauw, *et al.*, *Chem. Eur. J.*, 2012, **18**, 6133–6137.
[5] L. Yang, Y. Ma, F. Song, and J. You, *Chem. Commun.*, 2014, **50**, 3024–3026.
[6] J. Liu, Z. Liu, P. Liao, and X. Bi, *Org. Lett.*, 2014, **16**, 6204–6207.
[7] J. McNulty and K. Keskar, *Eur. J. Org. Chem.*, 2014, 1622–1629.
[8] M.P. Kumar and R.-S. Liu, *J. Org. Chem.*, 2006, **71**, 4951–4955.
[9] K. Alex, A. Tillack, N. Schwarz, and M. Beller, *Angew. Chem. Int. Ed.*, 2008, **47**, 2304–2307.
[10] L. Zhou and M.P. Doyle, *J. Org. Chem.*, 2009, **74**, 9222–9224.
[11] K. Okuma, J. Seto, K. Sakaguchi, *et al.*, *Tetrahedron Lett.*, 2009, **50**, 2943–2945.
[12] M. Xu, K. Xu, S. Wang, and Z.-J. Yao, *Tetrahedron Lett.*, 2013, **54**, 4675–4678.
[13] Y. Yin, Z. Chai, W.-Y. Ma, and G. Zhao, *Synthesis*, 2008, 4036–4040.

85

Iron-, Iridium-, and Indium-Catalyzed Indole Ring Syntheses

Like zinc, iron plays an important role in biochemistry, and this transition metal has found a place in indole ring synthesis. Nicholas and Lamar applied iron(II) phthalocyanine [Fe(Pc)] to an indole synthesis from aryl hydroxyamines and alkynes (Scheme 1, equation 1) [1]. A direct oxidative Fe-catalyzed intramolecular annulation of methyl 3-(phenylamino)but-2-enoates to form indole-3-carboxylates was uncovered by Liang and colleagues (equation 2) [2]. Zheng's team described the $FeCl_2$-catalyzed ring opening of 2*H*-azirines and subsequent indolization (equation 3) [3]. Interestingly, other possible catalysts ($Fe(OAc)_2$, $FeCl_3$, $AlCl_3$, BF_3•Et_2O, HCl) failed altogether, and copper was also ineffective (CuCl, 27%; $CuCl_2$, trace).

The iron-catalyzed version of the Hemetsberger indole synthesis was independently reported by Che [4] and Bolm [5]. Whereas Che employed a *meso*-tetrakis(pentafluorophenyl)porphyrin iron catalyst, Bolm used $Fe(OTf)_2$ (Scheme 2, equation 1) [5]. Both CuOTf and $Cu(OTf)_2$ failed to react. Driver and colleagues used $FeBr_2$ in an intramolecular C-H amination and 1,2-shift process with aryl azides to afford 2,3-disubstituted indoles (equations 2 and 3) [6]. Zhang-Negrerie and colleagues prepared a series of *N*-amino-3-cyanoindoles via an oxidative C-N bond formation using $FeBr_3$ in what may be a single electron transfer (SET) mechanism (equation 4) [7]. Sekar and Karthikeyan discovered a simple synthesis of pyrido[1,2-*a*]indoles that involved $FeCl_2$-catalyzed C-H bond activation of 2-benzhydrylpyridines (equation 5) [8]. A typical experiment is shown.

One of the new players in the indole ring synthesis game is iridium, in both Ir(I) and Ir(III) oxidation states. The familiar intramolecular alkyne hydroamination was reported by Crabtree's group to be catalyzed by an iridium hydride catalyst (**1**) (Scheme 3, equation 1) [9]. Similar alkyne hydroaminations using iridium catalysis were described by Liu ($[Ir(COD)Cl]_2$) [10], Shibata ($[Ir(COD)_2]OTf$) [11], and Leong ($[Cp^*IrCl_2]_2$) [12, 13]. Leong adapted their method to the synthesis of 2,2′-biindoles from *o*-ethynylanilines [13]. Driver and colleagues used $[(COD)Ir(OMe)]_2$ in toluene at room temperature to access indolines from *o*-homobenzylic aryl azides [14]. Shibata used $[Ir(COD)_2]BARF$ (BARF=tetrakis[3,5-bis(trifluoromethyl)phenyl]borate) to prepare 4-acetylindoles from α-arylamino ketones (equation 2) [15], Cossy and coworkers reported that $[IrCp^*Cl_2]_2$ effects a hydrogen transfer from benzylic alcohols leading ultimately to indoles (equation 3) [16].

Like iridium, indium is a new entry to the field of indole ring synthesis. In the classic cyclization (hydroamination) of *o*-ethynylanilines to indoles, Sakai and colleagues found that $InBr_3$ was superior to other Lewis acids (BF_3, $AlCl_3$, $Cu(OTf)_2$, $PdCl_2(PPh_3)_2$, $B(C_6F_5)_3$) (Scheme 4, equation 1) [17]. *N*-substituted indoles (*N*-Bn, *N*-Ac, *N*-$EtCO_2$) were also prepared. $InCl_3$ was slightly lower yielding, and InI_3 was very poor. Kim and colleagues found that $InCl_3$ reductively cyclized 1-(2-arylethynyl)-2-nitroarenes to 2-arylindoles (equation 2) [18]. Fujioka and coworkers used $InBr_3$ to forge indoles from *o*-alkynylanilines and β-keto esters (equation 3) [19]. Several β-keto esters were subjected to this In-catalyzed reaction as were simple ketones to give *N*-vinylindoles.

Indole Ring Synthesis: From Natural Products to Drug Discovery, First Edition. Gordon W. Gribble.

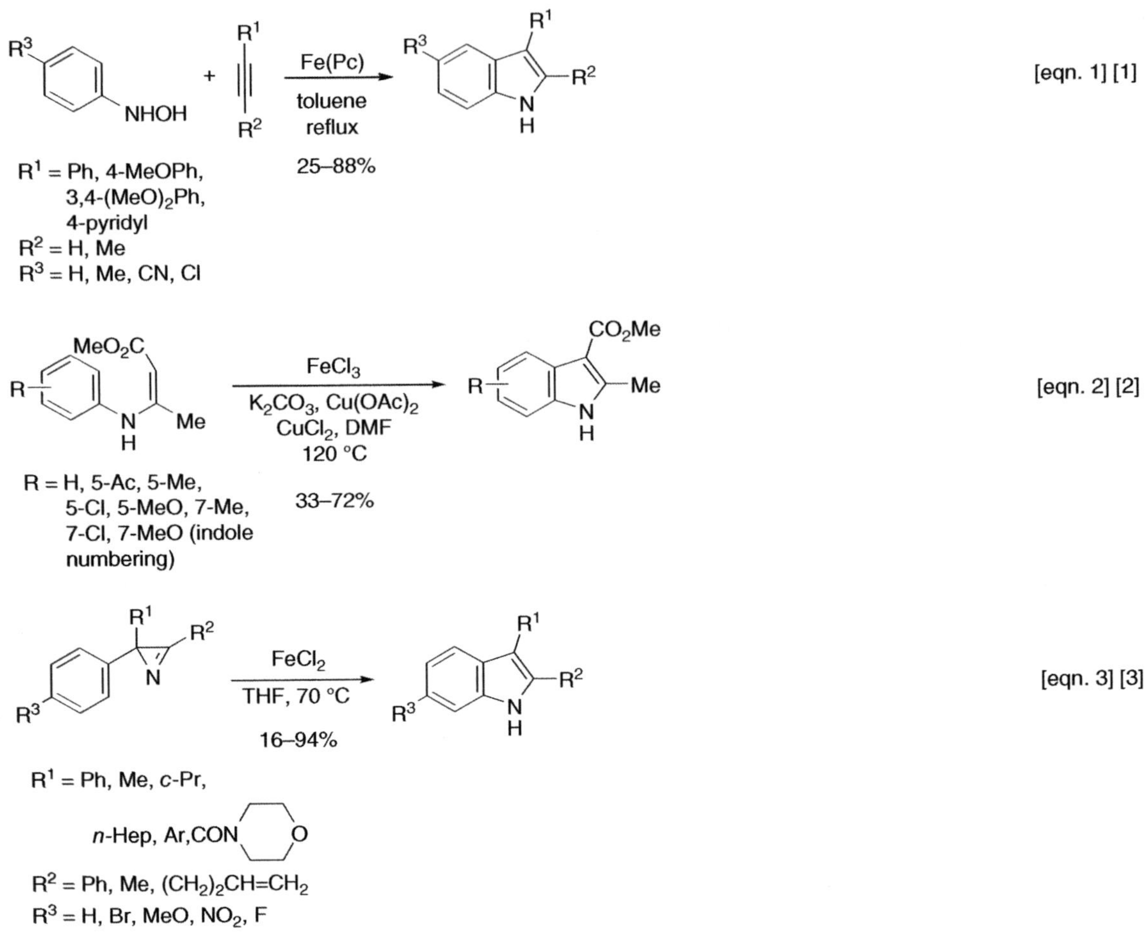

Scheme 1 *Iron-Catalyzed Indole Syntheses – 1*

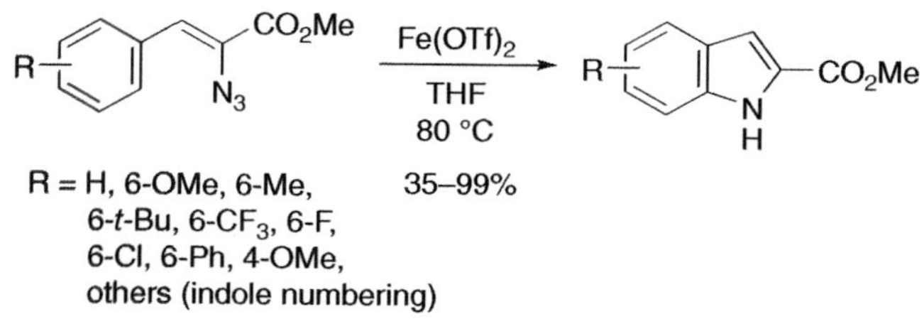

[eqn. 1] [5]

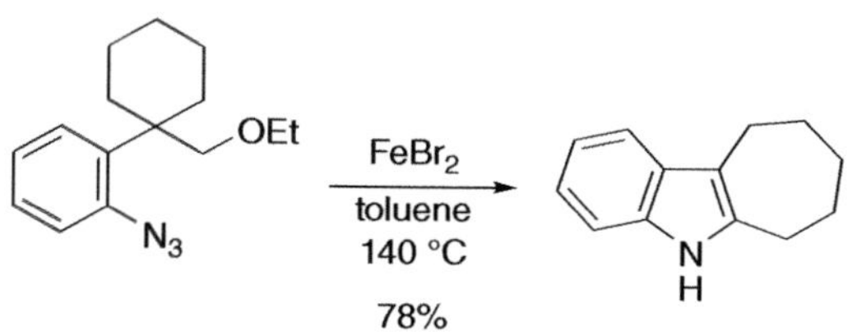

[eqn. 2] [6]

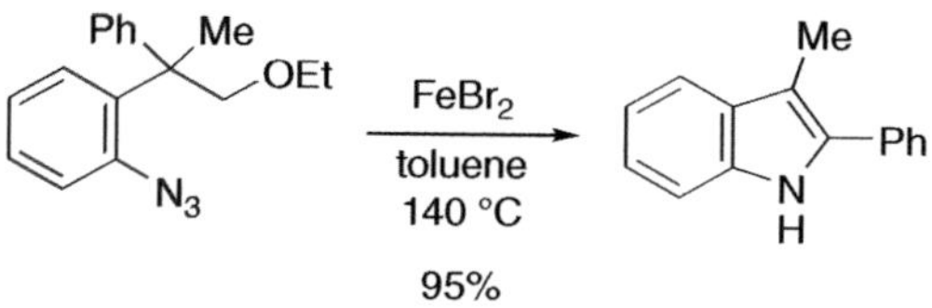

[eqn. 3] [6]

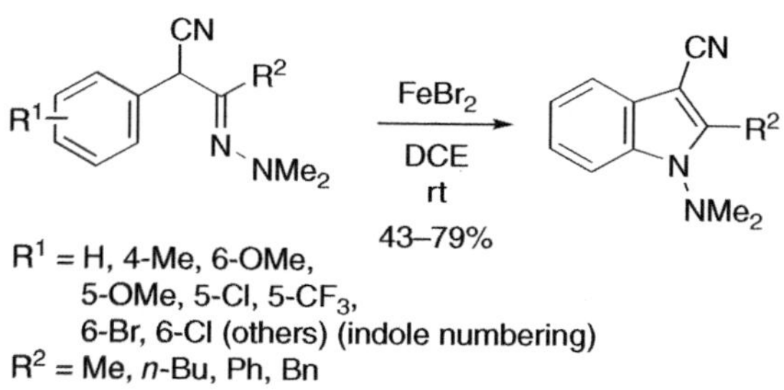

[eqn. 4] [7]

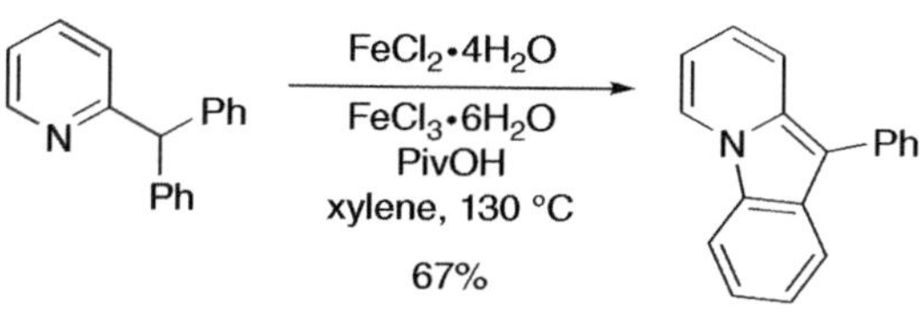

[eqn. 5] [8]

Scheme 2 *Iron-Catalyzed Indole Syntheses – 2*

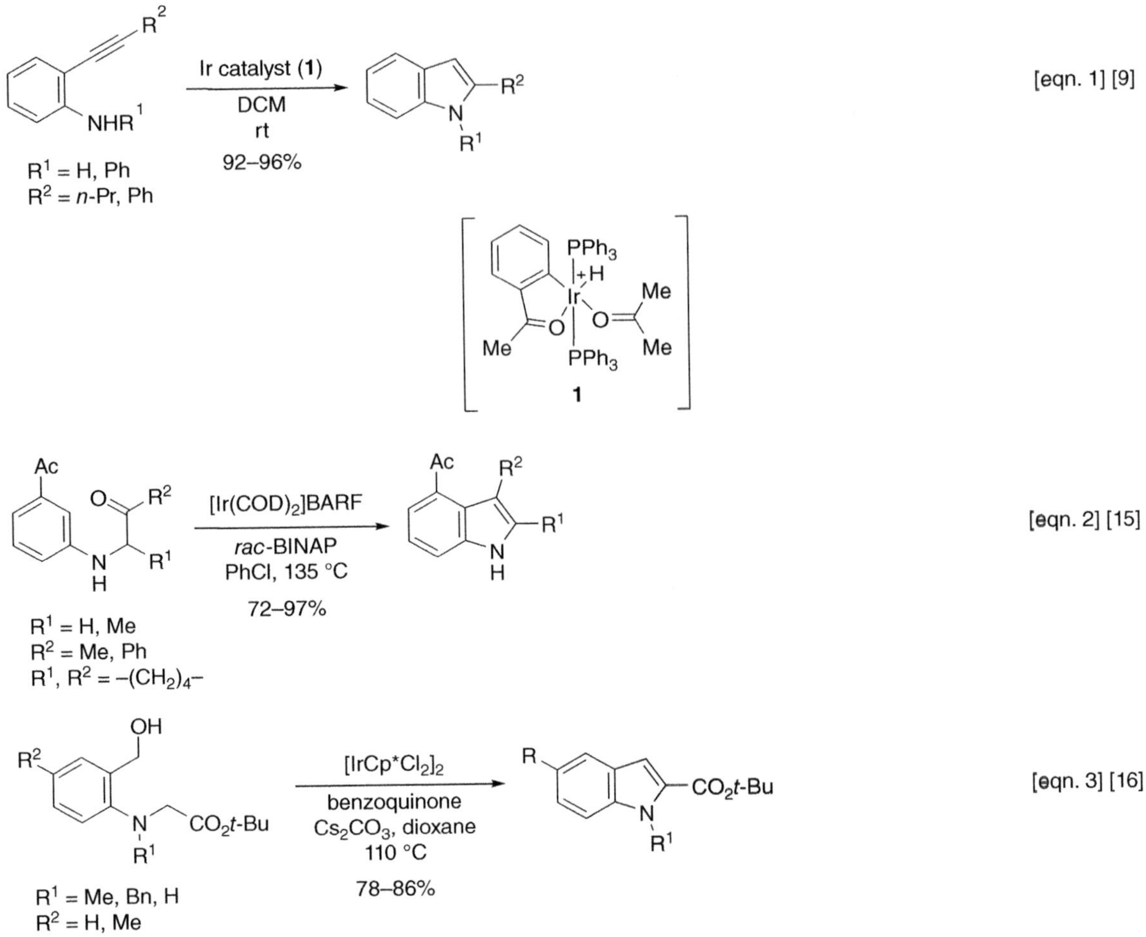

Scheme 3 *Iridium-Catalyzed Indole Syntheses*

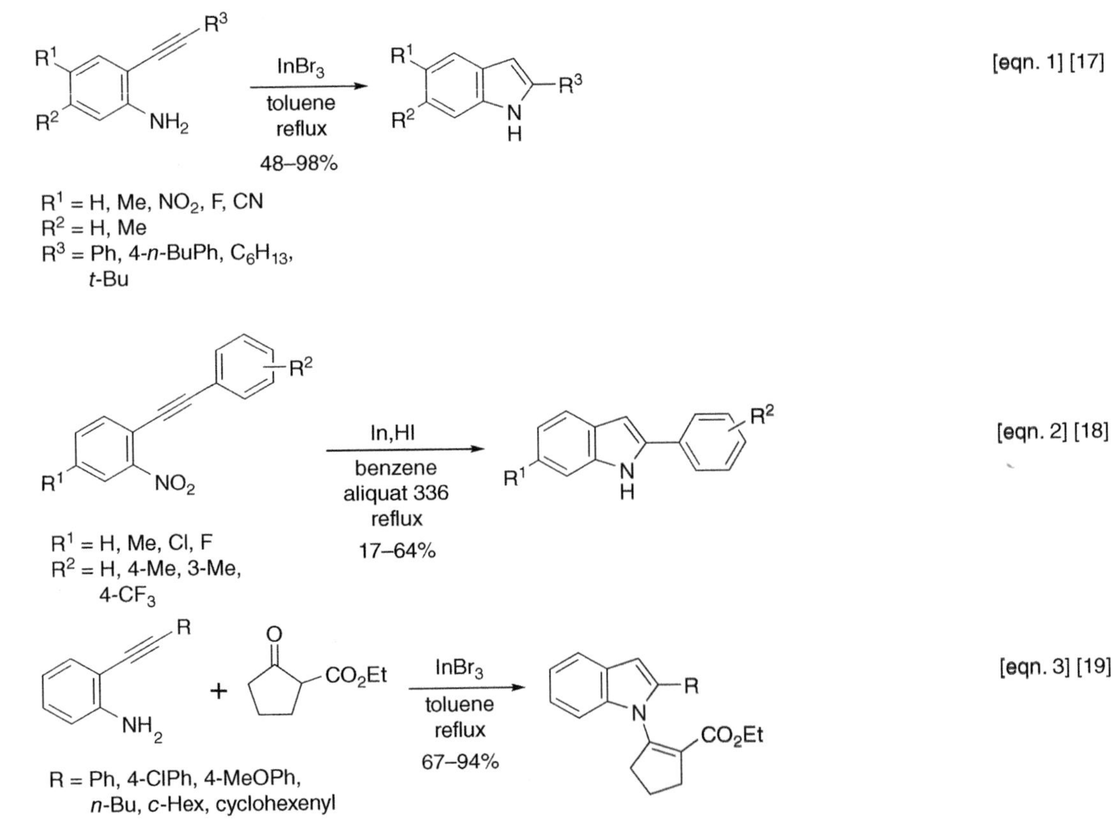

Scheme 4 *Indium-Catalyzed Indole Syntheses*

References

[1] A.A. Lamar and K.M. Nicholas, *Tetrahedron*, 2009, **65**, 3829–3833.
[2] Z.-H. Guan, Z.-Y. Yan, Z.-H. Ren, *et al.*, *Chem. Commun.*, 2010, **46**, 2823–2825.
[3] S. Jana, M.D. Clements, B.K. Sharp, and N. Zheng, *Org. Lett.*, 2010, **12**, 3736–3739.
[4] Y. Liu, J. Wei, and C.-M. Che, *Chem. Commun.*, 2010, **46**, 6926–6928.
[5] J. Bonnamour and C. Bolm, *Org. Lett.*, 2011, **13**, 2012–2014.
[6] Q. Nguyen, T. Nguyen, and T.G. Driver, *J. Am. Chem. Soc.*, 2013, **135**, 620–623.
[7] Z. Zheng, L. Tang, Y. Fan, *et al.*, *Org. Biomol. Chem.*, 2011, **9**, 3714–3725.
[8] I. Karthikeyan and G. Sekar, *Eur. J. Org. Chem.*, 2014, 8055–8063.
[9] X. Li, A.R. Chianese, T. Vogel, and R.H. Crabtree, *Org. Lett.*, 2005, **7**, 5437–5440.
[10] R.-Y. Lai, K. Surekha, A. Hayashi, *et al.*, *Organometallics*, 2007, **26**, 1062–1068.
[11] T. Shibata, H. Hirashima, M. Kasagawa, *et al.*, *Synlett*, 2011, 2171–2176.
[12] E. Kumaran and W.K. Leong, *Tetrahedron Lett.*, 2014, **55**, 5495–5498.
[13] E. Kumaran, W.Y. Fan, and W.K. Leong, *Org. Lett.*, 2014, **16**, 1342–1345.
[14] K. Sun, R. Sachwani, K.J. Richert, and T.G. Driver, *Org. Lett.*, 2009, **11**, 3598–3601.
[15] K. Tsuchikama, Y. Hashimoto, K. Endo, and T. Shibata, *Adv. Synth. Catal.*, 2009, **351**, 2850–2854.
[16] B. Anxionnat, D.G. Pardo, G. Ricci, *et al.*, *Org. Lett.*, 2013, **15**, 3876–3879.
[17] N. Sakai, K. Annaka, A. Fujita, *et al.*, *J. Org. Chem.*, 2008, **73**, 4160–4165.
[18] J.S. Kim, J.H. Han, J.J. Lee, *et al.*, *Tetrahedron Lett.*, 2008, **49**, 3733–3738.
[19] K. Murai, S. Hayashi, N. Takaichi, *et al.*, *J. Org. Chem.*, 2009, **74**, 1418–1421.

86

Nickel-, Cobalt-, and Molybdenum-Catalyzed Indole Ring Syntheses

The first example of nickel-catalyzed indole ring formation was due to Mori and Ban (Scheme 1, equation 1) [1]. The $Ni(PPh_3)_4$ was prepared *in situ* from $Ni(acac)_2$, Et_3Al and PPh_3. The role of oxygen was to convert PPh_3 to Ph_3PO catalyst for "easy handling." Kurahashi and Matsubara reported the first Ni-catalyzed heteroannulation of *o*-haloanilines with alkynes (equation 2) [2]. The regiochemistry is as shown, except for $R^1 = i$-Pr, $R^2 = Me$ where the two regioisomers are equally formed. IPr is 1,3-bis(2,6-diisopropylphenyl)imidazol-2-ylidene. These investigators also found that the Ni-catalyzed heteroannulation of anthranilic acid derivatives with internal alkynes gave 2,3-disubstituted indoles of the *opposite* regiochemistry (equation 3) to that in equation 2 [3]. With $R^1 = n$-Bu, $R^2 = Et$; $R^1 = n$-Pr, $R^2 = (CH_2)_2OMe$; $R^1 = n$-Pr, $R^2 = (CH_2)_2OH$; $R^1 = i$-Pr, $R^2 = Me$, the ratio of regioisomers is 1:1. Garcia and colleagues observed the Ni-catalyzed cyclization of *o*-ethynylaniline to give indole $[Ni(dippe)\mu\text{-}H]_2$, albeit in low yield (15%–25%) [4].

The Pauson–Khand reaction is a cobalt-assisted [2+2+2] cycloaddition involving an alkyne, an alkene, and carbon monoxide to give a cyclopentenone [5, 6]. By suitable substrate design, several investigators have applied the Pauson–Khand reaction to the synthesis of indoles. Pérez-Castells and colleagues used $Co_2(CO)_8$ to transform *N*-(2-ethynylphenyl)-*N*-(prop-1-enyl) acetamide to a cyclopenta[*b*]indol-2-one (Scheme 2, equation 1) [7]. Saito and colleagues adapted a Pauson–Khand reaction, using either $Co_2(CO)_8$ or $Mo(CO)_6$, to the syntheses of 1*H*-pyrrolo[2,3-*b*]indol-2-ones and 2*H*-thieno[2,3-*b*]indol-2-ones (equation 2) [8, 9]. Molybdenum is actually superior to cobalt in these Pauson–Khand reactions (*vide infra*). Mukai and coworkers used a cobalt Pauson–Khand reaction to synthesize pyrrolo[2,3-*b*]indoles including the alkaloid physostigmine (equation 3) [10, 11].

Sanz, Arnáiz, and colleagues discovered that a dichlorodioxomolybdenum(VI) complex can reductively cyclize nitrobiphenyls to carbazoles and nitrostyrenes to indoles (Scheme 3, equations 1 and 2) [12]. The Mo-catalyzed cyclization was applied to the synthesis of 2-carbonyl indoles in one pot from *o*-nitrobenzaldehydes (equation 3) [12]. Waldvogel and colleagues used $MoCl_5/TiCl_4$ to synthesize carbazoles via an oxidative cyclization of diaryl amines (equation 4) [13]. These conditions were superior to $TiCl_4$, $FeCl_3$, and $Pd(OAc)_2/HOAc$. A Pauson–Khand reaction using $Co(CO)_6$ was employed by Saito's team to prepare 1*H*-pyrrolo[2,3-*b*]indol-2-ones (equation 5) [8].

Indole Ring Synthesis: From Natural Products to Drug Discovery, First Edition. Gordon W. Gribble.

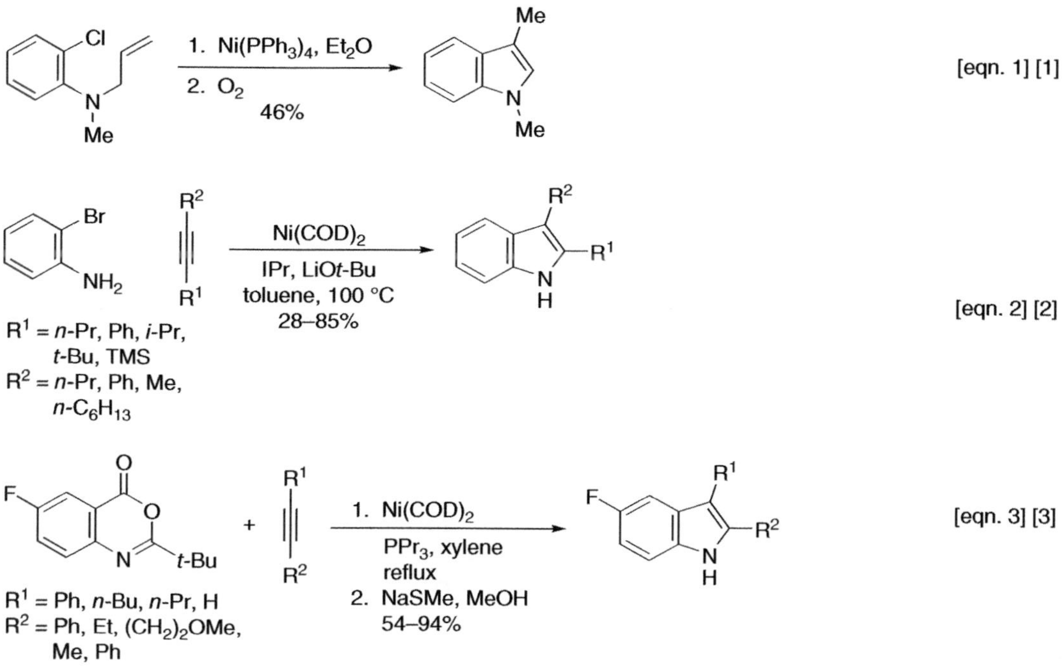

Scheme 1 *Nickel-Catalyzed Indole Syntheses*

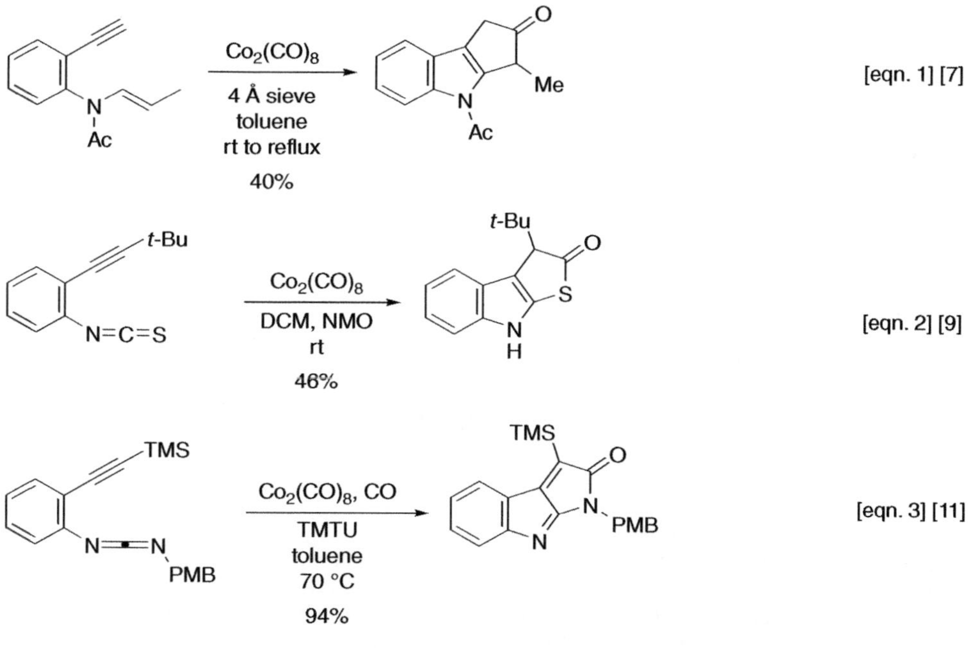

Scheme 2 *Cobalt-Catalyzed Indole Syntheses*

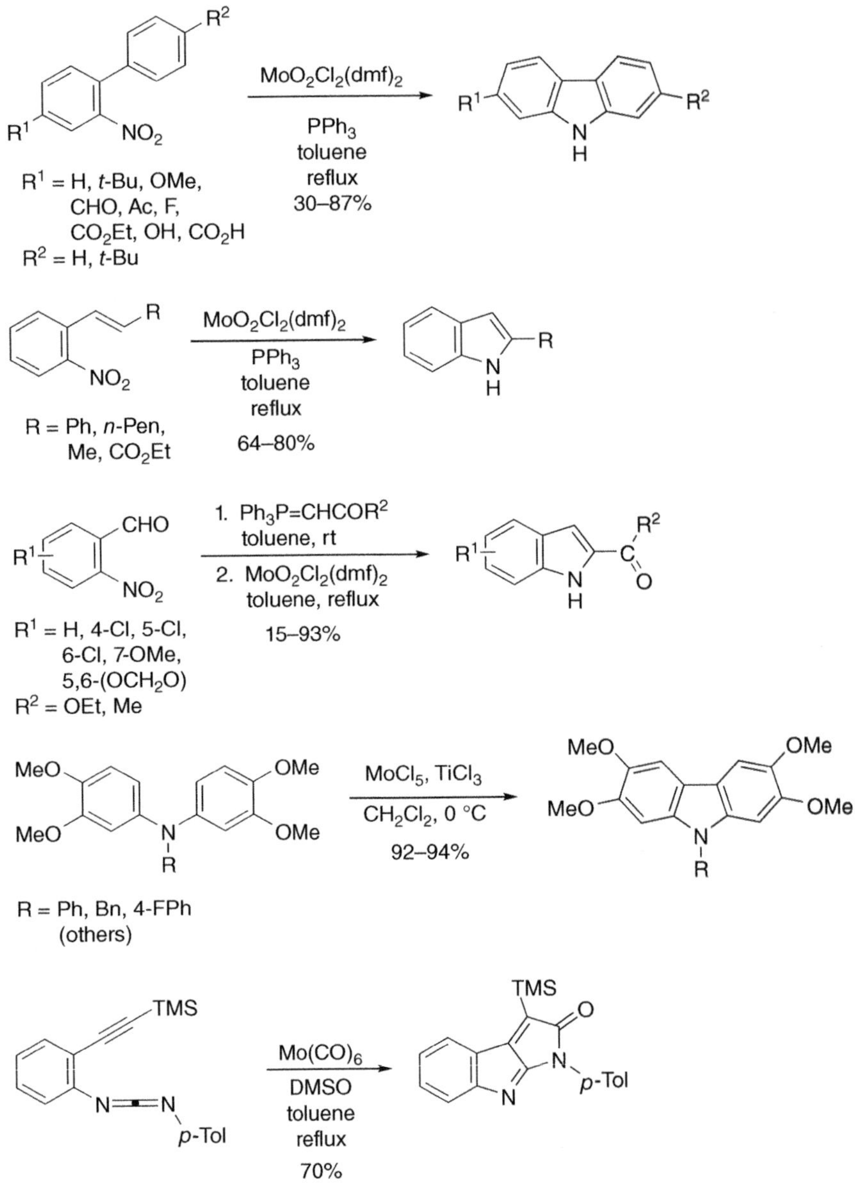

[eqn. 1] [12]

[eqn. 2] [12]

[eqn. 3] [12]

[eqn. 4] [13]

[eqn. 5] [8]

Scheme 3 *Molybdenum-Catalyzed Indole Syntheses*

References

[1] M. Mori and Y. Ban, *Tetrahedron Lett.*, 1976, **21**, 1803–1806.
[2] Y. Yoshida, T. Kurahashi, and S. Matsubara, *Chem. Lett.*, 2011, **40**, 1067–1068.
[3] N. Maizuru, T. Inami, T. Kurahashi, and S. Matsubara, *Org. Lett.*, 2011, **13**, 1206–1209.
[4] A. Reyes-Sánchez, F. Cañavera-Buelvas, R. Barrios-Francisco, *et al.*, *Organometallics*, 2011, **30**, 3340–3345.
[5] I.U. Khand, G.R. Knox, P.L. Pauson, and W.E. Watts, *J. Chem. Soc. D*, 1971, 36.
[6] I.U. Khand, G.R. Knox, P.L. Pauson, *et al.*, *J. Chem. Soc., Perkin Trans 1*, 1973, 977–981.
[7] G. Domínguez, L. Casarrubios, J. Rodríguez-Noriega, and J. Pérez-Castells, *Helv. Chim. Acta*, 2002, **85**, 2856–2861.
[8] T. Saito, M. Shiotani, T. Otani, and S. Hasaba, *Heterocycles*, 2003, **60**, 1045–1048.
[9] T. Saito, H. Nihei, T. Otani, *et al.*, *Chem. Commun.*, 2008, 172–174.
[10] C. Mukai, T. Yoshida, M. Sorimachi, and A. Odani, *Org. Lett.*, 2006, **8**, 83–86.
[11] D. Aburano, T. Yoshida, N. Miyakoshi, and C. Mukai, *J. Org. Chem.*, 2007, **72**, 6878–6884.
[12] R. Sanz, J. Escribano, M.R. Pedrosa, *et al.*, *Adv. Synth. Catal.*, 2007, **349**, 713–718.
[13] S. Trosien, P. Böttger, and S.R. Waldvogel, *Org. Lett.*, 2014, **16**, 402–405.

87

Mercury- and Chromium-Catalyzed Indole Ring Syntheses

Despite its notoriety as a toxin, mercury has played a role in organic synthesis, including indole ring formation. Larock and Harrison found in 1984 that mercuric acetate induces the cyclization of a few *o*-alkynylanilines in modest yield (Scheme 1, equation 1) [1]. The reaction was not general. Russell and Yao observed indole formation from the photolysis of β-nitrostyrenes in the presence of *tert*-butylmercury iodide [2], but this reaction is not synthetically useful. In contrast, Majumdar and Das developed a good synthesis of 1-alkoxy-1,2,3,4-tetrahydrocarbazoles (equation 2) [3]. Nishizawa and colleagues developed $Hg(OTf)_2$ as a catalyst to cycloisomerize *o*-ethynylanilines to indoles in excellent yields, under mild conditions, and with high catalytic turnover (up to 100 times) (equations 3 and 4) [4–7].

Whereas chromium plays a large role in inorganic and biological chemistry, its activity in organic chemistry is limited. However, Fischer chromium carbenes have found a niche in indole synthesis. Auman and colleagues crafted 3-aminoindoles from isocyanides and Fischer carbene complexes [8], and Söderberg's team synthesized several simple indoles from amino carbene complexes (Scheme 2, equation 1) [9]. Barluenga and colleagues expanded the scope of indoles that can be prepared from Fischer carbene complexes (equations 2 and 3) [10, 11].

Indole Ring Synthesis: From Natural Products to Drug Discovery, First Edition. Gordon W. Gribble.

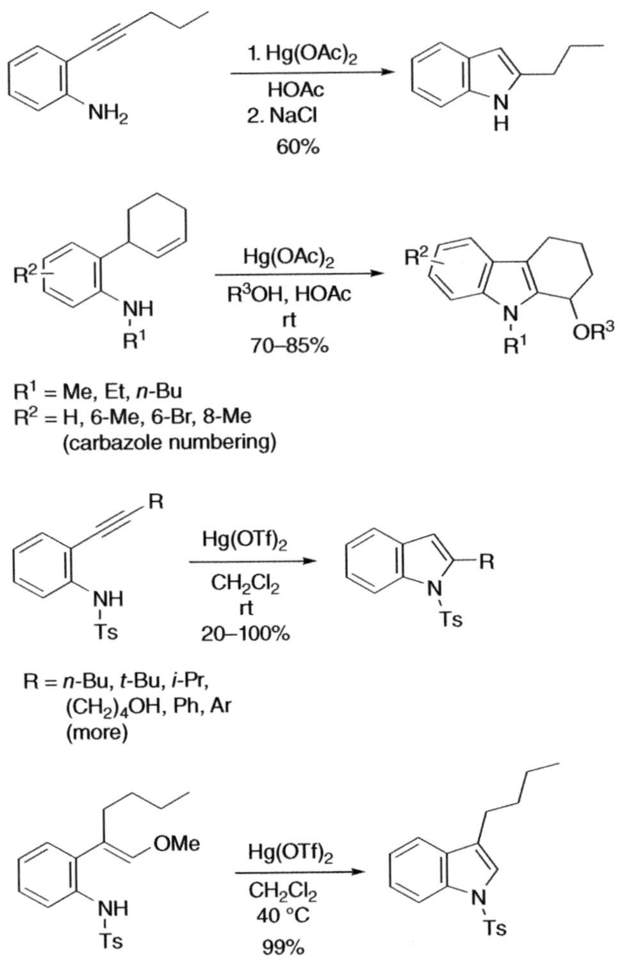

[eqn. 1] [1]

[eqn. 2] [3]

[eqn. 3] [4]

[eqn. 4] [5]

Scheme 1 *Mercury-Catalyzed Indole Syntheses*

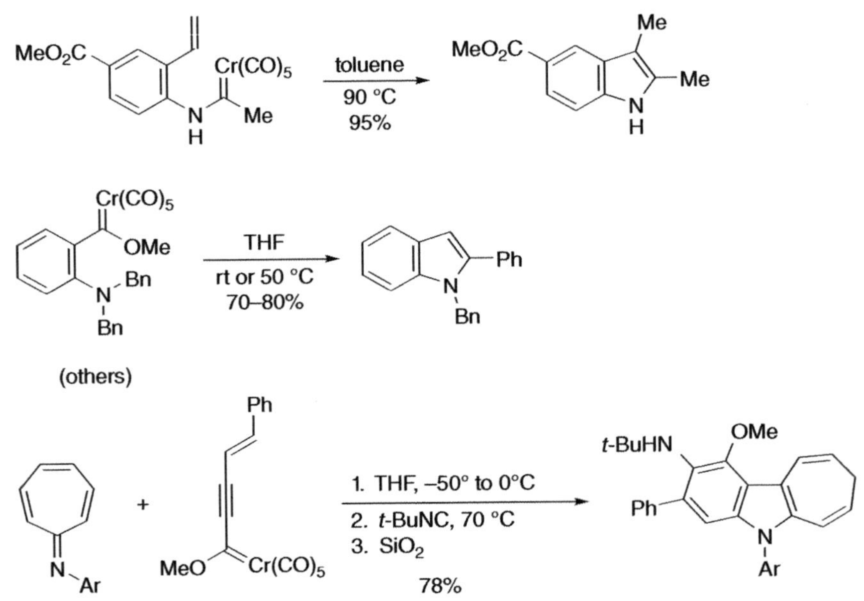

[eqn. 1] [9]

[eqn. 2] [10]

[eqn. 3] [11]

Scheme 2 *Chromium-Catalyzed Indole Syntheses*

References

[1] R.C. Larock and L.W. Harrison, *J. Am. Chem. Soc.*, 1984, **106**, 4218–4227.
[2] G.A. Russell and C.-F. Yao, *Heteroatom Chem.*, 1992, **3**, 209–218.
[3] K.C. Majumdar and U. Das, *Can. J. Chem.*, 1996, **74**, 1592–1596.
[4] T. Kurisaki, T. Naniwa, H. Yamamoto, *et al.*, *Tetrahedron Lett.*, 2007, **48**, 1871–1874.
[5] K. Namba, Y. Nakagawa, H. Yamamoto, *et al.*, *Synlett*, 2008, 1719–1723.
[6] H. Yamamoto, I. Sasaki, Y, Hirai, *et al.*, *Angew. Chem. Int. Ed.*, 2009, **48**, 1244–1247.
[7] M. Nishizawa, H. Imagawa, and H. Yamamoto, *Org. Biomol. Chem.*, 2010, **8**, 511–521.
[8] R. Aumann, E. Kuckert, and H. Heinen, *Angew. Chem. Int. Ed. Engl.*, 1985, **24**, 978–979.
[9] B.C.G. Söderberg, J.A. Shriver, S.H. Cooper, *et al.*, *Tetrahedron*, 2003, **59**, 8775–8791.
[10] J. Barluenga, M. Fañanás-Mastral, and F. Aznar, *Chem. Eur. J.*, 2008, **14**, 7508–7512.
[11] J. Barluenga, J. García-Rodríguez, Á.L. Suárez-Sobrino, and M. Tomás, *Chem. Eur. J.*, 2009, **15**, 8800–8806.

88

Miscellaneous Metal-Catalyzed Indole Ring Syntheses

This final chapter illustrates metals that catalyze or promote the formation of indoles but are not in the mainstream of indole synthetic methods and were not previously covered.

Iwasawa and coworkers discovered that tungsten pentacarbonyl facilitated indole ring formation from the appropriate aryl acetylenes (Scheme 1, equations 1 and 2) [1–3]. The $W(CO)_5$ can be generated *in situ* by the irradiation of $W(CO)_6$ in THF. The indolization in equation 2 is proposed to be a rare example of a 1,5-dipolar cycloaddition. Liang and colleagues observed that scandium triflate catalyzed the Friedel–Crafts cyclization of 5-(arylamino)pent-3-yn-2-ones to afford an array of indoles (equation 3) [4]. Interestingly, $Cu(OTf)_2$, AgOTf, $BF_3{\bullet}Et_2O$, $SnCl_4$, $Bi(OTf)_3$, $TiCl_4$, $AuCl_3$, $FeCl_3$, and $AlCl_3$ gave no indole product. Akiyama and colleagues reported that niobium(V) chloride, in conjunction with the $NaAlH_4$ reducing agent, converted *o*-(trifluoromethyl)anilines to indoles (equation 4) [5]. Some of the indoles synthesized in this study are **1–8**.

Knochel and coworkers discovered that phenylmagnesium chloride transforms 2-nitrostilbenes to 2-arylindoles (Scheme 2, equation 1) [6]. Hao's team reported a novel preparation of 2-fluoroalkyl indoles via the Grignard cyclization of *N*-(2-bromoalkyl)phenyl imidoyl chlorides, which were prepared in two steps from the appropriate *o*-alkylaniline (equation 2) [7].

Indole Ring Synthesis: From Natural Products to Drug Discovery, First Edition. Gordon W. Gribble.

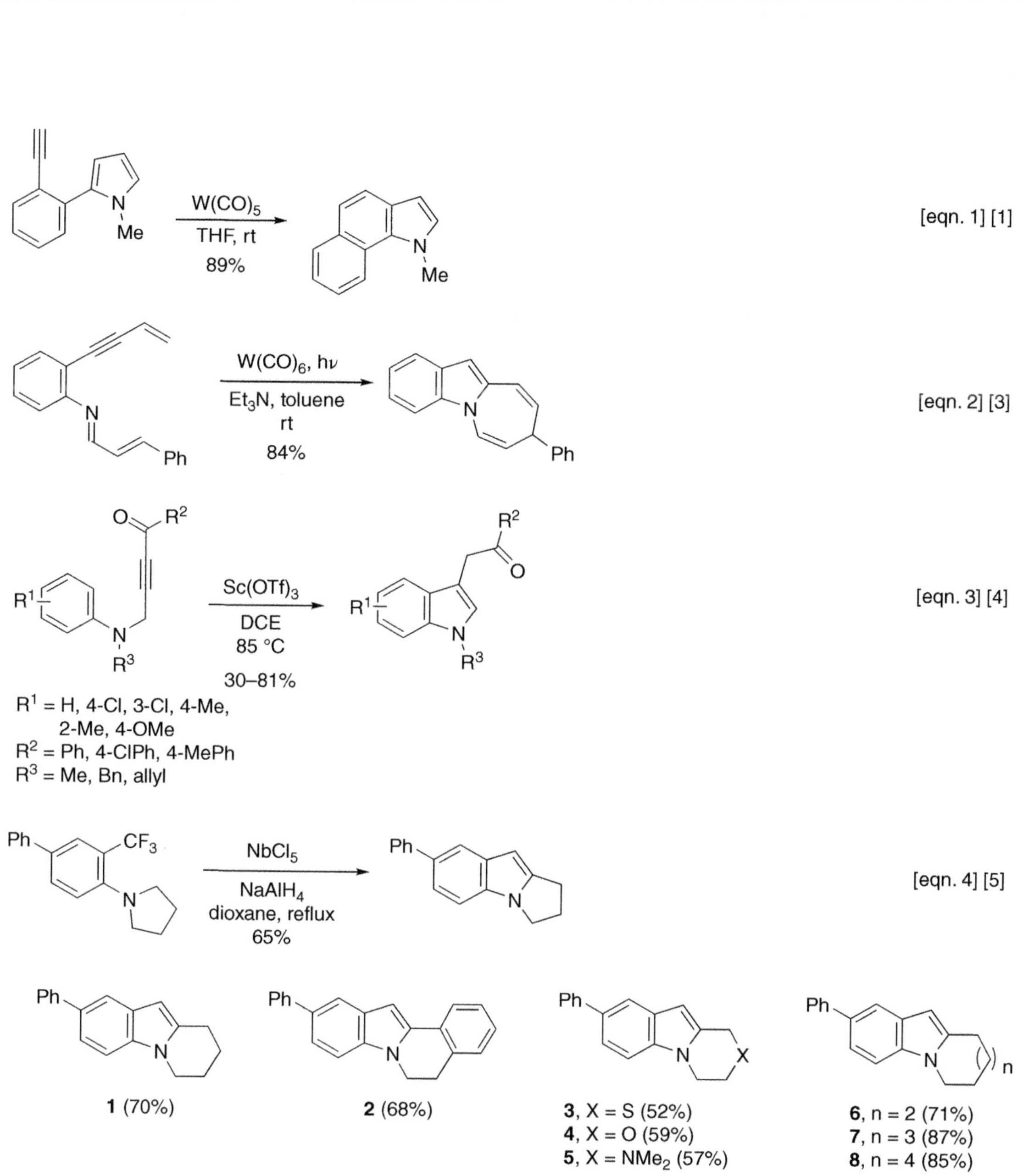

Scheme 1 *Miscellaneous Metal-Catalyzed Indole Syntheses – 1*

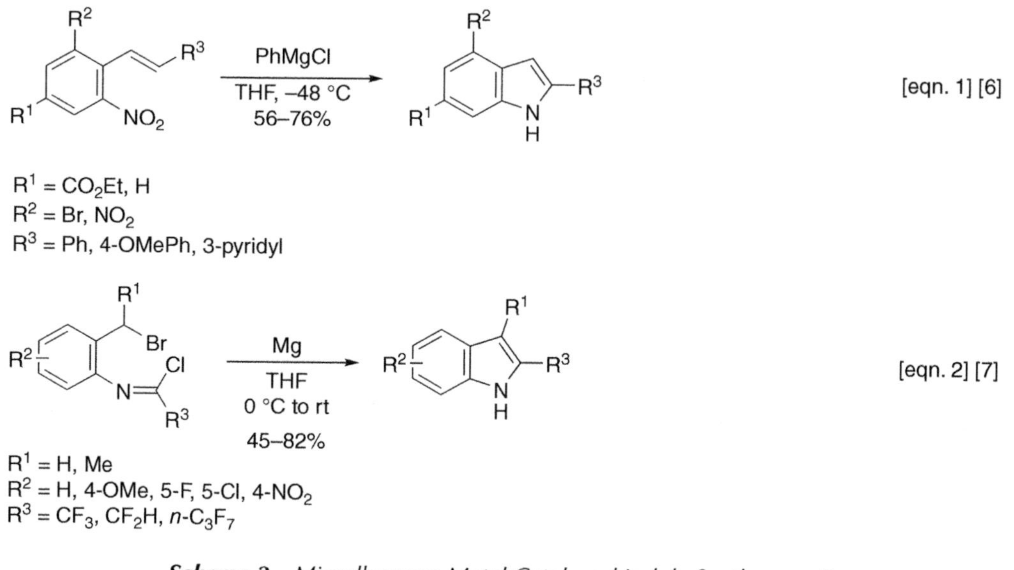

Scheme 2 *Miscellaneous Metal-Catalyzed Indole Syntheses – 2*

References

[1] K. Maeyama and N. Iwasawa, *J. Org. Chem.*, 1999, **64**, 1344–1346.
[2] H. Kusama, J. Takaya, and N. Iwasawa, *J. Am. Chem. Soc.*, 2002, **124**, 11592–11593.
[3] H. Kusama, Y. Suzuki, J. Takaya, and N. Iwasawa, *Org. Lett.*, 2006, **8**, 895–897.
[4] F. Yang, K.-G. Ji, S. Ali, and Y.-M. Liang, *J. Org. Chem.*, 2011, **76**, 8329–8335.
[5] K. Fuchibe, T. Kaneko, K. Mori, and T. Akiyama, *Angew. Chem. Int. Ed.*, 2009, **48**, 8070–8073.
[6] W. Dohle, A. Staubitz, and P. Knochel, *Chem. Eur. J.*, 2003, **9**, 5323–5331.
[7] F. Ge, Z. Wang, W. Wan, and J. Hao, *Synlett*, 2007, 447–450.

PART XI

Miscellaneous

A number of elegant indole ring syntheses are not logically categorized into the foregoing chapters and are therefore relegated to Chapter 89.

Indole Ring Synthesis: From Natural Products to Drug Discovery, First Edition. Gordon W. Gribble.

89

Miscellaneous Indole Ring Syntheses

This final chapter is a collage of indole ring syntheses that either do not come under any of the previous chapters or were inadvertently omitted from their appropriate chapter.

Although not considered to be a mainstream synthetic method, flash vacuum pyrolysis (FVP) often provides unique indole syntheses. Storr and Randles found that FVP of *o*-substituted amidines gave indoles in modest yield

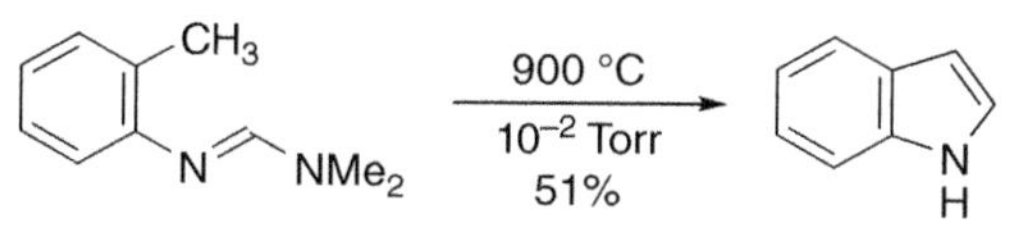

[eqn. 1] [1]

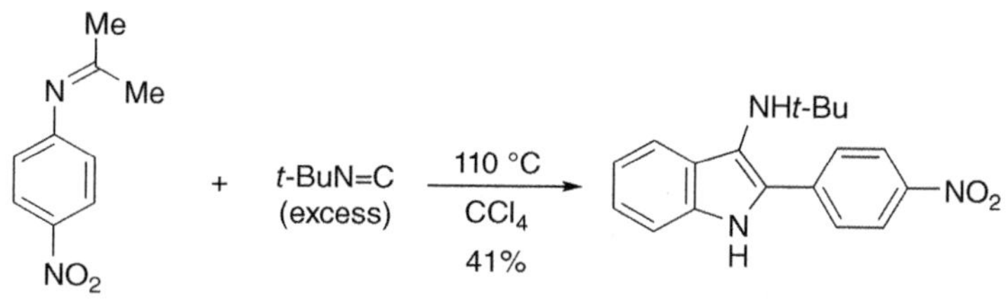

[eqn. 2] [3]

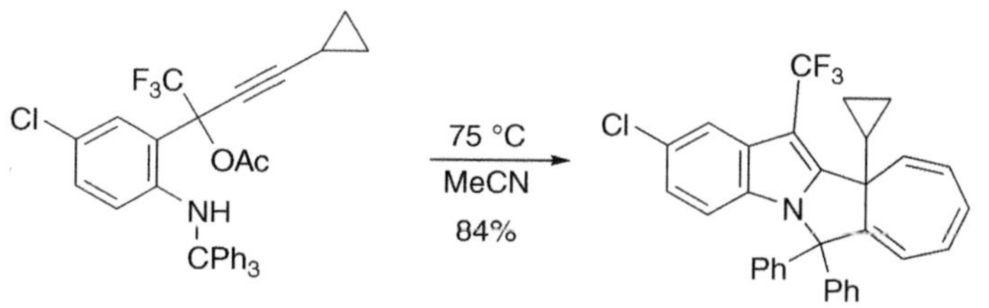

[eqn. 3] [4]

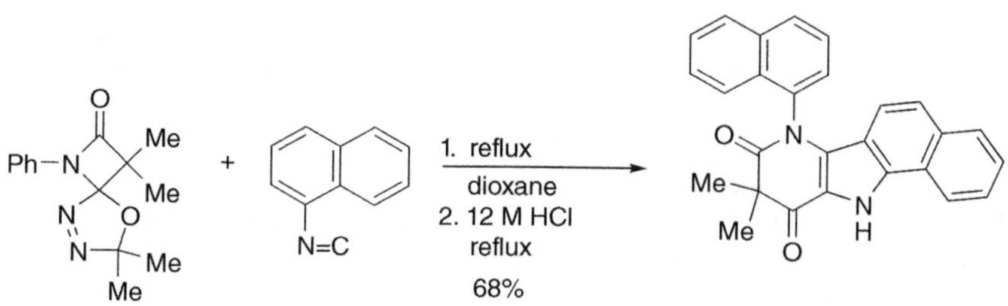

[eqn. 4] [5]

Scheme 1 *Miscellaneous Indole Syntheses by Pyrolysis*

Indole Ring Synthesis: From Natural Products to Drug Discovery, First Edition. Gordon W. Gribble.

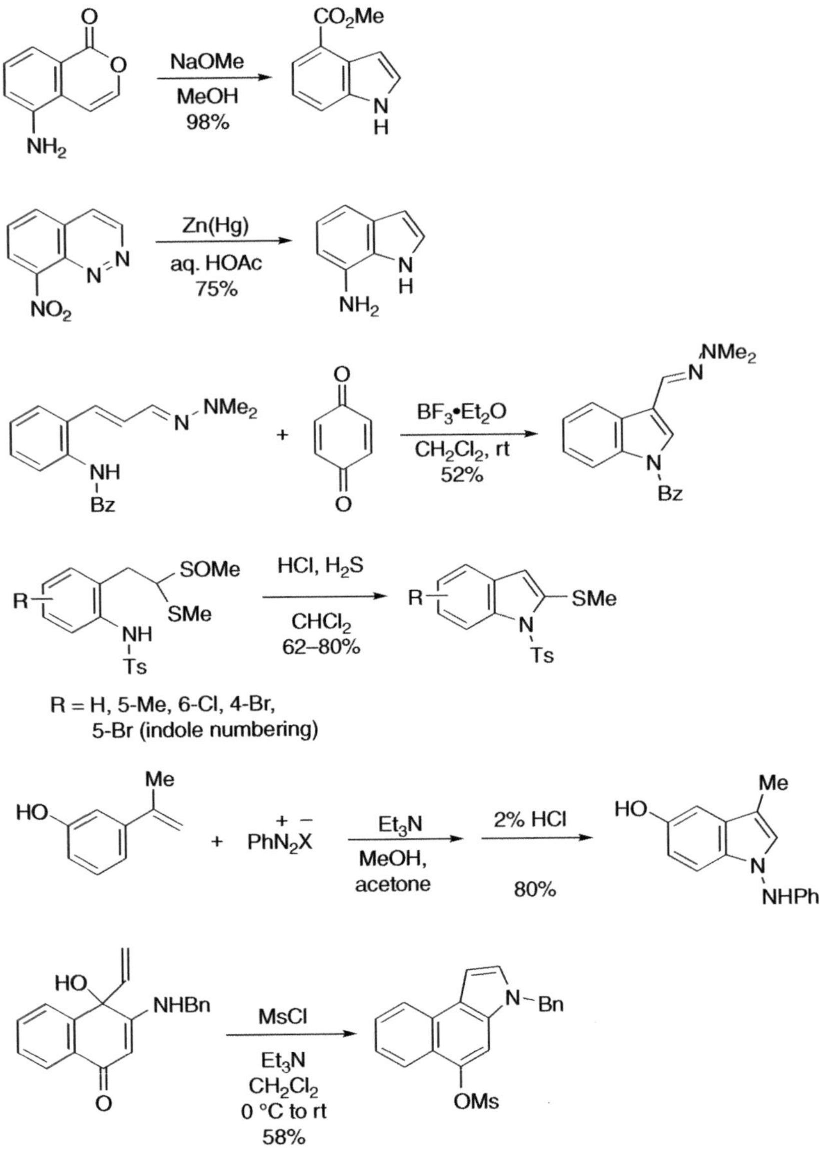

[eqn. 1] [6]

[eqn. 2] [7]

[eqn. 3] [8]

[eqn. 4] [9]

[eqn. 5] [10]

[eqn. 6] [12]

Scheme 2 *Miscellaneous Indole Syntheses*

(Scheme 1, equation 1) [1, 2]. Deyrup and colleagues discovered a novel indole synthesis from the reaction of *N*-arylimines with *t*-butyl isocyanide (equation 2) [3]. An interesting rearrangement of tertiary propargyl alcohols led to pyrrolo[1,2-*a*]indoles as discovered by Frey and coworkers (equation 3) [4]. Cheng's group generated β-lactam carbenes that reacted with naphthyl isonitriles to form naphthylbenzo[*h*]-δ-carbolin-2,4-diones (equation 4) [5].

Somei and colleagues reported that 5-aminoisocoumarin was converted to methyl indole-4-carboxylate in excellent yield (Scheme 2, equation 1) [6]. This group also discovered that both 4- and 7-aminoindoles were obtained from the corresponding nitrocinnolines (equation 2) [7]. Echavarren described a novel oxidative rearrangement to give indoles that was induced by 1,4-benzoquinone (equation 3) [8]. The acid-catalyzed cyclization of *o*-sulfonamido aryl ketene dithioacetal *S*-oxides in the presence of H_2S led to indoles, as reported by Hewson and colleagues (equation 4) [9]. The function of H_2S is to preclude formation of 3-chloroindoles. Satomura discovered a new synthesis of 5-hydroxyindoles from the coupling of 3-hydroxystyrenes with aryl diazonium salts (equation 5) [10, 11]. Raney nickel reduction cleaved the aniline group in quantitative yield. Kshirsagar and Hurley devised a simple route to the benz[*e*]indole ring system that may have involved a *p*-quinone methide intermediate (equation 6) [12].

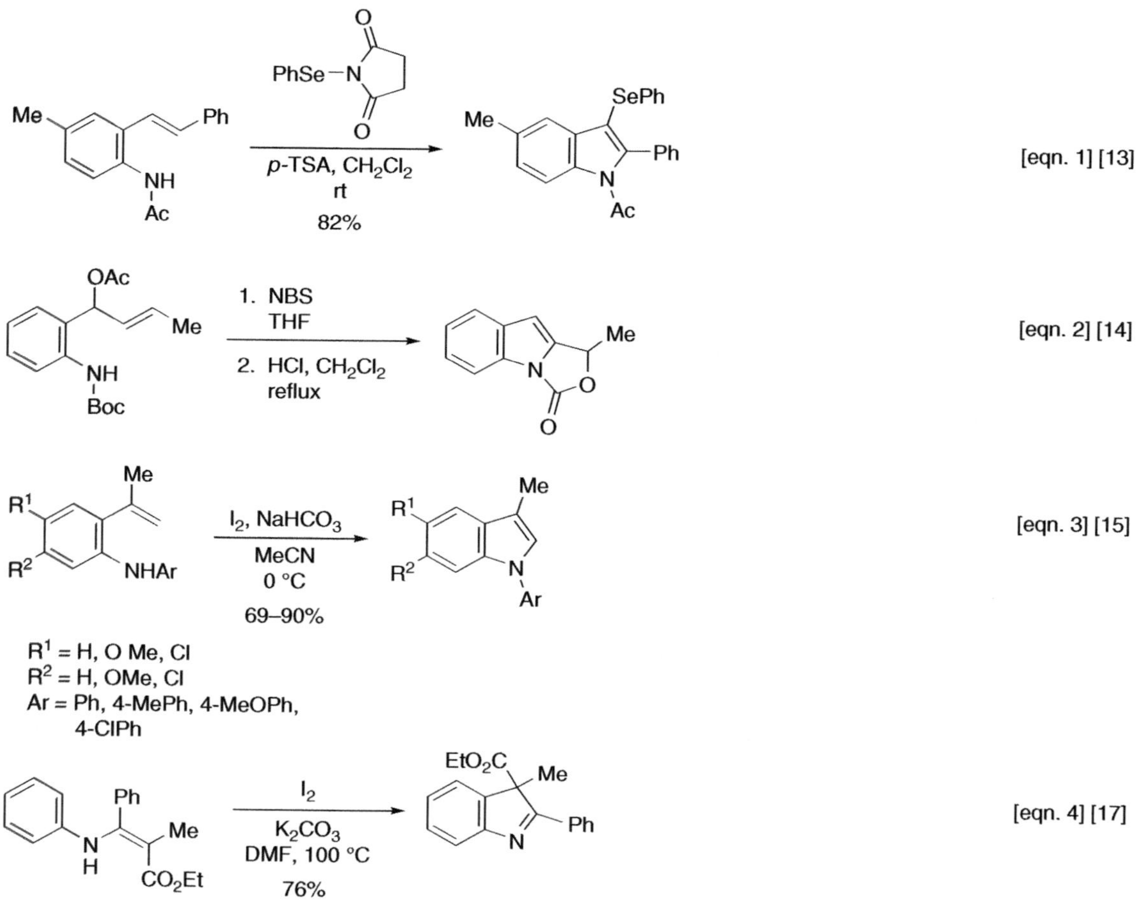

Scheme 3 *Electrophilic Cyclization of 2-(Arylamino)styrenes*

Several groups have accessed indoles via electrophilic activation of a styrene double bond, followed by intramolecular cyclization. Izumi and coworkers effected amidoselenation to afford indoles (Scheme 3, equation 1) [13]. In addition, treatment of 2-vinyltrifluoroacetanilide with PhSeBr gave indole (40%). Deselenation was accomplished with either n-Bu_3SnH/AIBN or $NiCl_2/NaBH_4$. Lamas and colleagues recorded an indole synthesis via an NBS bromocyclization (equation 2) [14]. Kobayashi and coworkers reported a simple indole synthesis from the iodocyclization of 2-(arylamino)styrenes (equation 3) [15]. Majumdar described a similar synthesis of hexahydrocarbazoles using *N*-iodosuccinimide, followed by Pd/C dehydrogenation to carbazoles or tetrahydrocarbazoles [16], Li's team employed an iodine-mediated cyclization of *N*-phenyl enamines to give indolenines (3*H*-indoles) (equation 4) [17].

The halocyclization of *o*-alkynylanilines to give indoles has attracted the interest of several research groups. Larock and Yue described this iodocyclization, which was accompanied by *N*-dealkylation (Scheme 4, equation 1) [18]. Several 3-iodoindoles were prepared in this study. Larock and colleagues also adapted their iodocyclization to a synthesis of coumestrol and coumestans [19]. Flynn and Hessian developed a 2-acylindole synthesis using an iodocyclization strategy (equation 2) [20]. Bräse and Balova developed a synthesis of 2-alkynyl-3-iodoindoles using iodocyclization of *o*-amino(buta-1,3-diynyl)arenes (equation 3) [21]. Wu and coworkers discovered an iodine-promoted domino reaction leading to benzo[*a*]carbazoles (equation 4) [22]. Several examples of this reaction were described. Wu's group employed iodocyclization to a synthesis of 3-iodo-*N*-amidines (equation 5) [23].

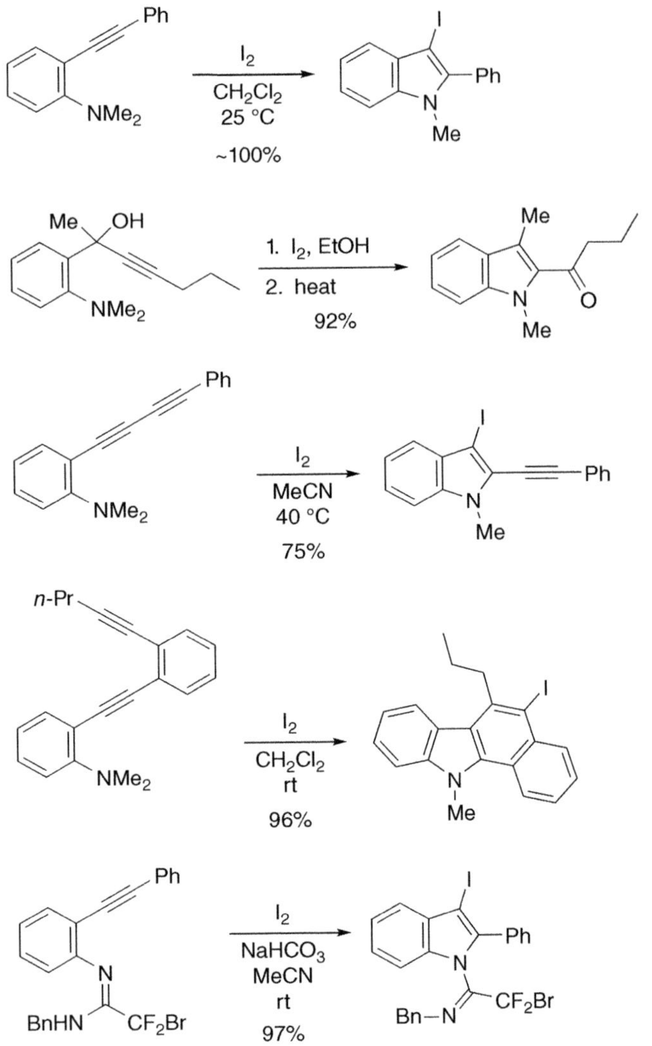

[eqn. 1] [18]

[eqn. 2] [20]

[eqn. 3] [21]

[eqn. 4] [22]

[eqn. 5] [23]

Scheme 4 *Halocyclization Indole Syntheses*

References

[1] K.R. Randles and R.C. Storr, *Tetrahedron Lett.*, 1987, **28**, 5555–5558.
[2] C.W.G. Fishwick, K.R. Randles, R.C. Storr, and P.W. Manley, *Tetrahedron Lett.*, 1985, **26**, 3053–3056.
[3] J.A. Deyrup, M.M. Vestling, W.V. Hagan, and H.Y. Yun, *Tetrahedron*, 1969, **25**, 1467–1478.
[4] L.F. Frey, R.D. Tillyer, S.G. Ouellet, *et al.*, *J. Am. Chem. Soc.*, 2000, **122**, 1215–1216.
[5] G.-F. Shi, Z.-M. Kang, X.-P. Jiang, and Y. Cheng, *Synthesis*, 2008, 2883–2890.
[6] M. Somei, Y. Karasawa, T. Shoda, and C. Kaneko, *Chem. Pharm. Bull.*, 1981, **29**, 249–253.
[7] M. Somei, S. Inoue, S. Tokutake, *et al.*, *Chem. Pharm. Bull.*, 1981, **29**, 726–738.
[8] A.M. Echavarren, *J. Org. Chem.*, 1990, **55**, 4255–4260.
[9] A.T. Hewson, K. Hughes, S.K. Richardson, *et al.*, *J. Chem. Soc., Perkin Trans. 1*, 1991, 1565–1569.
[10] M. Satomura, *J. Org. Chem.*, 1993, **58**, 3757–3760.
[11] M. Satomura, *J. Org. Chem.*, 1993, **58**, 6936–6938.
[12] T.A. Kshirsagar and L.H. Hurley, *J. Org. Chem.*, 1998, **63**, 5722–5724.
[13] T. Izumi, M. Sugano, and T. Konno, *J. Heterocycl. Chem.*, 1992, **29**, 899–904.
[14] J. Agejas, F. Delgado, J.J. Vaquero, *et al.*, *Tetrahedron Lett.*, 2002, **43**, 8025–8027.
[15] K. Kobayashi, K. Miyamoto, T. Yamase, *et al.*, *Bull. Chem. Soc. Jpn.*, 2006, **79**, 1580–1584.

[16] K.C. Majumdar, U.K. Kundu, U. Das, *et al.*, *Can. J. Chem.*, 2005, **83**, 63–67.
[17] Z. He, H. Li, and Z. Li, *J. Org. Chem.*, 2010, **75**, 4636–4639.
[18] D. Yue and R.C. Larock, *Org. Lett.*, 2004, **6**, 1037–1040.
[19] T. Yao, D. Yue, and R.C. Larock, *J. Org. Chem.*, 2005, **70**, 9985–9989.
[20] K.O. Hessian and B.L. Flynn, *Org. Lett.*, 2006, **8**, 243–246.
[21] N.A. Danilkina, S. Bräse, and I.A. Balova, *Synlett*, 2011, 517–520.
[22] C.-C. Chen, S.-C. Yang, and M.-J. Wu, *J. Org. Chem.*, 2011, **76**, 10269–10274.
[23] J. Zhu, H. Xie, Z. Chen, *et al.*, *Org. Biomol. Chem.*, 2012, **10**, 516–523.

Index

Page numbers in **bold** indicate tables

Indole Ring Synthesis: From Natural Products to Drug Discovery, First Edition. Gordon W. Gribble.

Printed and bound by CPI Group (UK) Ltd, Croydon, CR0 4YY
01/10/2026
14974210-0001